Concise Encyclopedia of Science and Technology of Wine

Concise Encyclopedia of Science and Technology of Wine

Edited by
V. K. Joshi

Associate Editors
Matteo Bordiga
Fernanda Cosme
Laura Fariña
Ronald S. Jackson
António Manuel Jordão
Aline Lonvaud-Funel
Creina Stockley

CRC Press
Taylor & Francis Group
Boca Raton London New York

CRC Press is an imprint of the
Taylor & Francis Group, an **informa** business

First edition published 2022
by CRC Press
6000 Broken Sound Parkway NW, Suite 300, Boca Raton, FL 33487-2742

and by CRC Press
2 Park Square, Milton Park, Abingdon, Oxon, OX14 4RN

Library of Congress Cataloging-in-Publication Data

Names: Joshi, V. K., 1955- editor.
Title: Concise encyclopedia of science and technology of wine / edited by V.K. Joshi ; associate editors : Matteo Bordiga, Fernanda Cosme, Laura Fariña, Ronald S. Jackson, António Manuel Jordão, Aline Lonvaud-Funel, Creina Stockley.
Description: First edition. | Boca Raton : CRC Press, 2021. | Includes bibliographical references and index.
Identifiers: LCCN 2020036566 (print) | LCCN 2020036567 (ebook) | ISBN 9781138092754 (hardback) | ISBN 9781315107295 (ebook)
Subjects: LCSH: Wine and wine making--Encyclopedias.
Classification: LCC TP548 .C653 2021 (print) | LCC TP548 (ebook) | DDC 641.2/203--dc23
LC record available at https://lccn.loc.gov/2020036566
LC ebook record available at https://lccn.loc.gov/2020036567

ISBN: 978-1-138-09275-4 (hbk)
ISBN: 978-0-367-56190-1 (pbk)
ISBN: 978-1-315-10729-5 (ebk)

Typeset in Times
by Deanta Global Publishing Services, Chennai, India

Dedicated respectfully to

the late Dr M.R. Thakur, the first Vice-Chancellor of the Dr Y.S. Parmar University of Horticulture and Forestry, Nauni, Solan, India: well-known plant breeder, able teacher, dedicated researcher, highly disciplined and unbiased administrator with an eye for picking up the best talent based on merit but without any consideration to religion, caste, creed, region, gender or political allegiance. He was tough from the outside but very soft from the inside. He laid strong foundations in the university for research, teaching, and extension education, which will be cherished for years to come. His contribution to the development of the technology of winemaking from its beginnings to its eminence is immense. He remains a source of inspiration to dedicated and meritorious scientists and was a model Vice-Chancellor. He was an ocean of knowledge with an immeasurably profound depth, and his contribution to the academic world will always be remembered.

Contents

UNIT 5 Process of Winemaking

UNIT 6 Technology for Production of Wine and Brandy

Preface

Wine is extolled as a food, a social lubricant, an antimicrobial and antioxidant, and a product of immense economic significance. It has been an integral component of the human diet since its discovery by a caveman around 6000 B.C. No other beverage except milk could earn such esteem from consumers as wine has. It has travelled a long distance with human civilization and has witnessed many ups and downs. It has given inspiration to numerous poets and artists, and consumers of diverse origins, regardless of different opinions, movements and cultures. From the ancient times, many healthful properties have been associated with wine consumption, and recently therapeutic properties have also been added to this association. What is a wine? When did its production start and what are its different types? Is wine nutritious, does it have any therapeutic values or have any role to play in health, or is it simply an intoxicating beverage? How are the qualities of wine determined or marketed, and how does this relate to tourism? The present volume attempts to answer some of these questions.

There are several books dealing with wine that focus on different aspects, ranging from the scientific angle such as the microbiology of winemaking or principles and techniques of wine production to those books of an exclusively literary facet. A manuscript giving the holistic view of wine seems to be lacking. A need was, therefore, felt to have a book where readers would have ample opportunities to peep into the holistic view of wine, having the basic principles and practices of winemaking, grouping relevant topics together, thus allowing readers to gain a better appreciation of the subject matter along with knowledge of various innovations and future trends made in this direction. This book is a broad introduction to the technology of wine production where the production processes are explained and illustrated, but it is certainly not an operating manual. Rather, it is slanted to guide the reader through the principles and practices underlying the wine production. At the same time, I would confess that it is impossible to have all and everything of the subject in a book of a modest size like this. Nevertheless, readers have not been deprived of experiencing the enthusiasm of this fascinating and unique product which encompasses both science and art of wine. The other driving force behind the publication of this manuscript is that books, unlike wine, do not always mature well with time, and so there is a dire need of a book with updated knowledge.

Intentionally, the book has been styled as an "encyclopedia," which could be used as a textbook as well as a reference book but is unlike those encyclopedias intended to be used exclusively as a reference book. This book has also deviated from the general encyclopedia-style by providing chapters in logical sequence rather than in alphabetical order. Attempts have been made to cover subject matter to the highest possible extent, 45 chapters with 8 units. The various chapters at the beginning of the book provide an introduction to the scientific basis of types of wines, composition of wine, nutritive value, therapeutic value, viticulture, microbiology, biochemistry and finally, fermentation technology. This is followed by those chapters that describe the factors influencing the production of different wines and brandies. After this, the chapters focus on quality determination, marketing, tourism, innovations in wine production and the status of the wine industry.

The text has been divided into eight units, and each unit is headed by an associate editor who introduces the respective unit and also gives some ideas for future research and direction on the subject matter. Each chapter starts with a preliminary, then moves to advances in the respective area, thus making it useful for students of various subjects and levels. The entire text is based on the basics of the subject matter, and wherever applicable, the latest trends or the innovations in the field are dotted throughout. Every chapter is illustrated with tables, figures and plates. The book concentrates essential technological information in one handy volume and provides references for further reading at the end of each chapter to enable its readers to search for more information on various aspects. As subject areas frequently cross section or chapter boundaries, every effort has been made to cross-reference the related topics as a means to reduce difficulties in finding information and to remove repetition in the text. Hopefully, the reader will find these cross-references interesting and a pleasant distraction from the normal text, especially given that the dual purpose of the book is for its use as a textbook and also as a general reference book for academia, industrialists and government agencies.

The process of winemaking is as old as human civilization, and its preparation has often been more of an art than a science. It has already been established that every technology is based on art, science, or a blend of art and science – commonly called "know-how." The science and technology of winemaking involves a wide spectrum of physical, biological, chemical, psychological and engineering sciences. It has, therefore, progressed from merely an art to the present level of scientific eminence.

Our understanding of the technology of wine production is based on the knowledge of fruit cultivation, ripening of fruits, diseases of fruits, biochemistry and microbiology of fermentation, methods of wine production, coupled with advances in plant and machinery design, and the analytical techniques employed in quality control to produce this fascinating beverage of the best quality. Developments could have happened organically, through slow but gradual improvements, keen observations and, of course, sheer luck and possible failure. Today, winemaking involves a series of technical steps right from harvesting the grapes, fermenting the must and finally, providing the bottled wine to the consumers. The fruit growers, wine manufacturers and the consumers are, thus, intimately connected with wine, rather are the main pillars. But the large number of cultivars, a number of methods and various styles of wine along with presentation and consumption patterns makes the process more complex and comprehensive. The anxiety of the fruit grower is to produce fruit of the specific quality needed for wine preparation. Here, the microbiologist, biochemist and biotechnologist come into the picture, where their findings are applied in production, for example in relation to the suitable microorganisms, their growth, fermentation behaviour and control, and various enzymes production and their applications. Quite often, corrective measures are applied if the wine so demands. The most important component of this scenario is the consumer, who judges the quality and finally accepts the product. A source of scientific information naturally comes as a helping hand. Moreover the consumer, equipped with greater knowledge, continues to demand a better product, and the winemaker endeavours to satisfy this by providing a diverse and quality product by imbibing the latest technological and scientific innovations, and marketing strategies.

During the last decade, a number of developments have taken place in Oenology viz., wine microbiology, genetically modified fruit crops and microorganisms, the therapeutic and medicinal role of wine in human health, bioreactors etc., that demand collective documentation. The microorganisms found in grape must, juice and wines are yeasts, lactic and acetic acid bacteria and moulds. Out of these, the yeasts, especially *Saccharomyces cerevisiae*, are very important for wine fermentation, lactic acid bacteria for malo-lactic acid fermentation while the growth of acetic acid bacteria is associated with preventing wine spoilage. The role played by non-*Saccharomyces* yeasts is increasingly being acknowledged. As is the case with killer yeasts, recombinant DNA technology, application of enzyme technology and the new analytical methods of wine evaluation have been developed that call for a comprehensive documented knowledge. Optimization of wine fermentation involves the growth of the microorganisms and the production of ethanol – the major compound of significance. Optimization is achieved by performing fermentation under rigorously controlled conditions in large fermenters with suitable cultural capacities. Thus, the design and operation of fermenters is also discussed in the text. The book also focuses on the waste management of wineries, especially wastewater and by-products of winemaking, and their income-generating potential as the winemaking process generates a significant amount of waste, such as prunings, stems and pomace, that is a source of bioactive compounds. These compounds could be used for various purposes in the pharmaceutical, cosmetic and food industries. Other by-products of wineries are seeds, skin (a source of anthocyanins), yeast lees, tartrates and carbon dioxide.

The book draws upon the expertise of leading researchers in winemaking all over the world and should be viewed as a continuance of texts by the great peers who have set the stage for the scientific and technological advancement of winemaking. In the preparation of this manuscript, the greatest contribution has been made by the contributory authors who have written the precise and well-documented chapters. To further improve each chapter, the associate editors spent a lot of time in reviewing and offering suggestions, for which I am extremely grateful. I must, therefore, acknowledge the input of the associate editors as well as the contributors in this regard. I am very grateful to each of them for listening to my comments and responding to my requests, time and again. The associate editors themselves have contributed to those chapters of their specialization, for which my special thanks are due to them. The assistance received from Prof R.S. Jackson, Prof Aline Lonvaud-Funel, Dr Creina Stockley, Dr Matteo Bordiga, Dr Laura Fariña, Dr Fernanda Cosme and Dr António Manuel Jordão is gratefully acknowledged. I really appreciate the advice of Prof Jackson. Being a well-established oenologist and a writer of high scientific repute, his criticism is equally useful as is his advice. I have always regarded him as my *de facto* mentor, and though I did not have any opportunity to work under him, I am sure it would have been a very worthwhile experience. Prof Aline Lonvaud-Funel has been a great source of inspiration due to her research contributions in wine microbiology. Besides her scholarly papers that I have read, she has always been a forthcoming and dedicated contributor to my books and handled two units in this text as an associate editor very successfully. High appreciation and thanks are due to Dr Laura Fariña for her assistance in various issues beyond that of editorship. The same appreciation and thanks are due to Dr Matteo Bordiga, Dr Fernanda Cosme and Dr António Manuel Jordão for their sincere and dedicated help in writing the chapters as co-authors with me besides, the editing work of their units. The contribution of Dr Creina Stockley is unforgettable, as she contributed when she was critically ill and even in the midst of this, offered to contribute a chapter and carry out the work of an associate editor. I salute her bravery and dedication.

My special thanks are due to Mr Stephen Zollo, Acquisitions Editor at Taylor & Francis Group, and his team, for their tireless job in the successful execution of this project and for always extending a helping hand, whenever a problem arose and providing a solution that made all of us comfortable. Thanks are also due to Ms Laura Piedrahita at Taylor & Francis Group in the publication of this book. She was always available to help me in every possible manner, whenever I requested her assistance. Thanks are due to Ms Iris Fahrer, Project Editor, CRC Press/Taylor & Francis Group and her team for the smooth and professional printing of the book. Sincere thanks would be inadequate to Ms Rachel Cook, Senior Project Manager at Deanta Global, for putting production back on track of this book. Thanks are due to Prof P.K. Khosla, the Vice-chancellor of Shoolini University of Biotechnology and Management, Solan, who gave me the first opportunity to publish a book on fruit wines as Director of Extension Education at Dr YSPUHF, Nauni, Solan, HP, India. In preparation of this manuscript, I received a lot of assistance from my family members and friends. I remember here my respected father, the late Sh M.L. Joshi, and my mother, the late Smt Bimla Joshi, who supported, encouraged and inspired me like a beacon of light during their lifetimes and now, from their heavenly abode. Their inspiration will always steer me in my future life endeavours. Special thanks are to my wife Mrs Sushma Joshi for her constant encouragement and support, to my sons, Dr Bharat Joshi and Er Sidharath

Joshi, and to my daughters-in-law, Dr Shruti and Ms Bharti. I owe them special thanks for their contribution in facilitating Internet operations, with emailing, formatting, printing, corrections etc., and the biggest task of the collection and supply of literature from various sources at their disposal. Finally, with great pleasure, I am placing this book before the readers, who will be the final judge. Their support and criticism is welcome for further improvements in the book.

V.K. Joshi

Editors

EDITOR

Dr V.K. Joshi

Professor V.K. Joshi is an eminent scientist and teacher with more than 40 years varied research experience in fruit fermentation technology and fermented foods, food toxicology, bio-colour, quality assurance and waste utilization. He has made a very significant contribution to the development of technology of non-grape fruit wines, apple pomace utilization, lactic acid fermented foods and indigenous fermented alcoholic beverages. He is a former Head of Department of Postharvest Technology (PHT) and former Head of Department of Food Science and Technology in Dr Y.S.P. University of Horticulture and Forestry, Nauni, Solan (HP), India. He has earned a BSc (med) from Guru Nanak Dev University, Amritsar, India, an MSc (microbiology) from Punjab Agriculture University, Ludhiana, India and a PhD (microbiology) specializing in food fermentation from Guru Nanak Dev University, Amritsar, India. He has received training related to several aspects of food, such as the sensory evaluation of food, processing and nutrition, and the postharvest technology of fruits and vegetables, from prestigious institutes such as the Central Food Technological Research Institute (CFTRI), Mysore and the Punjab Agricultural University (PAU), Ludhiana.

He became a fellow of the Biotechnological Research Society of India (BRSI) in 2005 and a fellow of the Indian Society of Hill Agriculture (ISHA) in 2010. He has authored/edited more than 12 books, 5 practical manuals, 150 research papers in international and national journals, more than 60 book chapters, 40 review/technical articles, and several popular articles, besides presenting more than 50 papers and talks at many seminars and conferences. He has chaired technical sessions in several seminars and has been a guest of honour in scientific seminars and workshops.

Professor Joshi has edited special issues of the *Journal of Scientific and Industrial Research* (JSIR) and the *Natural Product Radiant* as a guest editor. He has been a reviewer for several prestigious journals such as the *Journal of the Institute of Brewing, Journal of Food, Science and Technology, International Journal of Food Microbiology, Food Bioscience, Food Chemistry*, the *Journal of Food Processing and Preservation* and the *Journal of Catalysis*.

The first book he authored was about fruit wines and was published by the Directorate of Extension Education, Nauni, Solan (HP), India, which made his entry into the journey of writing and editing of scientific books. He has authored/edited several books, including *Indigenous Fermented Foods of South Asia* (CRC) and *Science and Technology of Fruit Wine Production* (Elsevier). A complete treatise on winemaking was the *Handbook of Enology* in three volumes and *Wine Making: Basic and Applied* (CRC–Food Biology series), while the others are two volume sets on *Food Processing and Preservation* and *Technology of Harvesting, Storage, Preservation and Processing of Fruits and Vegetables* (NIPA). Two of his books, *Postharvest Technology of Fruits and Vegetables*, and *Biotechnology: Food Fermentation* are prescribed for MSc/PhD courses for the subjects of Food and Fermentation Technology, Food Science and Technology, Microbiology and Biotechnology at the Dr Y.S.P. University of Horticulture and Forestry and other universities in India by the Indian Council of Agricultural Research (ICAR). Two other books which cover the complete course of ICAR are *Food Industrial Waste Management Technology* (NIPA) and *Sensory Science: Principles and Applications in Food* (Agrotech), the latter being a book of immense utility for food science and technology students as a textbook for the course and as a research aid.

Professor Joshi has guided several postgraduate students for their dissertations in postharvest technology, and food science and technology. He is a regular reviewer of research projects and research papers, and a paper-setter and thesis-evaluator in the subject. Professor Joshi has successfully conducted research and handled the research projects of ICAR, the National Horticultural Board (NHB), the Department of Biotechnology (DBT), the Ministry of Food Processing Industry (MOFPI) and the Department of Science and Technology (DST). He remained a member of the Board of Studies of several universities and institutes, a member of the Food Safety and Standards Authority of India, an advisor to the Horticultural Produce Processing and Marketing Corporation (HPMC) and a wine consultant (and has been involved in the establishment of an apple wine factory).

He has been honoured with several awards for his research contributions, such as the N.N. Mohan, Kejeriwal, Pruthi, N.A. Pandit awards for the best research and review papers, and the Himachal Shree for teaching contribution, among others. Professor Joshi successfully conducted the ICAR-sponsored Summer School, 2012 as a Course Director, and a Khadi and Village Industry Corporation (KVIC)-sponsored workshop on Food Processing in March, 2013.

As a chairman of the equal opportunities cell of the university, he initiated measures for the welfare of physically disabled persons, including the construction of disabled-friendly buildings, the grant of transport allowances and employment creation, and he helped organize several training programs as well as mock tests for the students of the university for JRF (Junior Research Fellowship),

SRF (Senior Research Fellowship) and ARS (Agricultural Research Services).

He established a fruit processing unit at the Department of FST (now, FSSAI) for the manufacture of fruit and vegetable products with FPO license, which generated income for the university. As a principal investigator of the experimental learning program of ICAR, he organized training for undergraduate students of the university and established a processing plant for fruit and vegetable products.

After his retirement in 2015, he joined the faculty of Shoolini University of Bioengineering and Management as Adjunct Professor. Besides this, he remained a consultant in CSIR, Palampur (HP), India between 2018–2019. At present, he is an editor-in-chief of the *International Journal of Food Fermentation Technology*, former editor of *Indian Food Packer* presently, the editor of the *Journal of Food and Nutrition*, *World Journal of Biotechnology*, *Acta Microbiology* and *Novel Techniques in Food Science and Nutrition*. He is also chairman of the Panel on Alcoholic Beverages and a Member of Scientific committee, of FSSAI. Professor Joshi is actively engaged in editing two books to be published by CRC and a volume to be published by NIPA. Additionally, he is involved in wine consultancy.

ASSOCIATE EDITORS

Dr Matteo Bordiga

Dr Matteo Bordiga is currently an Assistant Professor of Food Chemistry at Università del Piemonte Orientale (UPO), Novara, Italy. He earned his PhD in Food Science and MS in Chemistry and Pharmaceutical Technologies from the same university. His main research activity concerned food chemistry, investigating the different classes of polyphenols from an analytical, technological and nutritional point of view. More recently, he moved his research interests to wine chemistry, focusing his attention on the entire production process. Dr Bordiga has published more than 40 research papers in peer-reviewed international and national journals. Since 2013, he has been an editorial board member of the *International Journal of Food Science & Technology*. He is editor of the books *Valorization of Wine Making By-Products* (CRC Press, Taylor & Francis Group, 2015) and *Post-Fermentation and Distillation Technology: Stabilization, Aging, and Spoilage* (CRC Press, Taylor & Francis Group, 2018). All of these research activities have also been developed through important collaborations with foreign institutions (Foods Science & Technology Department and Foods for Health Institute, University of California, Davis, United States; Fundación Parque Científico y Tecnológico de Albacete; and Instituto Regional de Investigación Científica Aplicada, Universidad de Castilla-La Mancha, Ciudad Real, Spain).

Dr Fernanda Cosme

Fernanda Cosme is currently an Assistant Professor of Food Science/Oenology at University of Trás-os-Montes and Alto Douro (UTAD), School of Life Sciences and Environment (Department of Biology and Environment), Vila Real, Portugal. She graduated in Agro-Industrial Engineering from the Technical University of Lisbon, Portugal. In 2008, she earned a PhD in Food Science at the University of Trás-os-Montes and Alto Douro of Vila Real. Dr Cosme is currently a member of the Chemistry Research Centre, Vila Real (CQ-VR), research group of Food Chemistry and Biochemistry. Her main research interests include grapes and wines in respect to grape composition, wine technology, wine stability and quality, and aging, fining, specifically phenolic compounds, polysaccharides and proteins. Recently, she edited three books on oenology as a scientific editor. She is author/co-author of 42 articles in peer-reviewed scientific journals, author/co-author of 22 book chapters and 175 papers in conference proceedings. She has participated in 12 funded research projects and supervised PhD and MSc student theses in the fields of food science and oenology.

Dr Laura Fariña

Dr Laura Fariña is at present an Associate Professor in the Enología y Biotecnología de las Fermentaciones group at Facultad de Química of Universidad de la República (UdelaR), Uruguay. She earned her PhD from the Facultad de Química-UdelaR, her research work being focused on the characterization of the aroma profile of the *Vitis vinifera* cv *Tannat* wines and the evolution phenomena related to aging. Currently, she teaches at a university in the areas of enology, food analysis, chromatographic techniques and mass spectrometry.

Dr Fariña's current research interests are focused on the study of volatile metabolites and precursors in *Vitis viniferas* and other complex matrices such as honey, aromatic plants and native fruits from Uruguay. She is also involved in the development of new applications for the fast techniques in the analysis of volatile organic compounds through the use of spectroscopic tools such as NIR and NMR.

She has supervised three post-graduate theses. To date, she has more than 40 scientific articles in peer-reviewed journals, has presented more than 120 papers in meetings, regional, national and international congresses, and is author of a book chapter on the multidisciplinary approaches to food science and nutrition for the 21st century. She is level 1 in the National Research System from Uruguay, was member of the journal editorial board of *Food Research International* and reviewer for publications linked to the study of volatile matrices and chromatographic applications.

Dr Ronald S. Jackson

Dr Jackson is a world authority on Oenology. He earned his BA and MA degrees from Queen's University, Canada, and doctorate from the University of Toronto, Canada. His time in Vineland, Ontario, and subsequent sabbatical at Cornell University, United States, redirected his interest in botrytis toward viticulture and enology. As part of his teaching duties at Brandon University, Canada, he developed the first wine technology course in Canada. For many years, he was a technical advisor to the Manitoba Liquor Control Commission, developing sensory tests to assess the candidates of its sensory panel, and was a member of its external tasting panel. Although officially retired, he is a Fellow of the Cool Climate Oenology and Viticulture Institute, Brock University, Canada, and continues working on his texts: *Wine Science: Principles & Applications* and *Wine Tasting: A Professional Handbook*, as well as preparing various chapters and technical reviews in other works. He is author of *Wine Science: Principles & Applications, 3e* (2008), *Conserve Water Drink Wine* (1997), several technical reviews, and annual articles in Tom Stevenson's *The Wine Report*. Dr Jackson has retired from teaching to devote his time to writing, but remains allied with the Cool Climate Oenology and Viticulture Institute, Brock University. Besides others, he has edited a special issue of *Advances in Food and Nutritional Research*, 2011(63), Elsevier, London. He has done pioneering work on cool climate oenology. His book on wine science is a fundamental textbook for teaching and research, and is considered as the most authentic treatise. Recently, his book on wine science earned him an international award from OIV. In his spare time, he loves cycling, hiking, swimming and yoga.

Dr António Manuel Jordão

António Manuel Jordão is currently an Associate Professor of Oenology at the Polytechnic Institute of Viseu, Agrarian Higher School, Portugal, and a member of the Chemistry Research Centre, Vila Real, Portugal. He graduated in Agro-Industrial Engineering and earned an MA in Food Science and Technology from the Technical University of Lisbon, Portugal. In 2006, he earned a PhD in Agro-Industrial Engineering at the Agronomy Higher Institute of the Technical University in Lisbon. Dr Jordão's research interests are phenolic compounds from grapes and wines, wine technology and in particular the topic of the wine aging process. He is a scientific editor of four books on oenology and author/co-author of 45 articles in peer-reviewed scientific journals, 15 book chapters and 70 papers in scientific congresses. Dr Jordão is also a member of the editorial board of several scientific journals, supervisor of several master theses in oenology and has developed consultancy services to wine companies.

Dr Aline Lonvaud-Funel

Aline Lonvaud-Funel is a Professor Emeritus of the University of Bordeaux, France, at the Institute of Vine and Wine Sciences (ISVV) in the Oenology Research Unit. After earning her National Diploma in Oenology and a University Diploma in Ampelology in 1970, she continued her studies in Biochemistry at the University of Bordeaux. Holder of an MA in Biochemistry, she was admitted to the laboratory of Professor Pascal Ribéreau-Gayon to carry out research under her supervision and a first thesis entitled "Carbon dioxide dissolved in wines" was completed in 1975. Subsequently, her research was exclusively devoted to wine microbiology, which first materialized in the PhD thesis entitled "Research on lactic acid bacteria in wine: metabolic functions, growth, plasmid genetics" in 1986.

Prof Lonvaud-Funel then continued her research on lactic acid bacteria by addressing various physiological and metabolic aspects, by introducing in the laboratory the methods of molecular biology and the new approaches made possible by genomics. While directing the microbiology laboratory of the ISVV Research Unit, she supervised several doctoral theses. In addition to the fundamental aspects of the knowledge of wine microorganisms, the results led to the development of practical tools for the detection of wine spoilage microorganisms. In 2003, Prof Lonvaud-Funel created a spin-off to exploit the results of the research laboratory and offer them to wine professionals. In 2020, this activity still operates at the ISVV in conjunction with the research laboratory. During the last years of her activity as professor, she directed laboratory studies towards the study of oenological microbiota to address the important questions of the interactions between yeasts and bacteria in wine.

Throughout her career as a researcher, Prof Lonvaud-Funel was a teacher for the National Diploma of Oenologists at the University of Bordeaux and was called upon to lecture students at other universities in France and abroad. She has published numerous articles in scientific journals and has participated in many collaborative books on wine microbiology. Since 2007, she is part of the group of experts in microbiology of the International Organization of Vine and Wine (OIV). Today, she is mainly involved in scientific publishing, as editor and reviewer for international scientific journals and books specializing in oenology.

Dr Creina Stockley

Dr Creina Stockley, PhD, MBA, has 28 years of experience in the alcohol and health arena and was based at the Australian Wine Research Institute (AWRI) from 1991 to 2018. Her academic background is clinical pharmacology and physiology, and she has also been associated with public health projects via the Faculty of Medicine, Nursing and Health Sciences at Flinders University, Australia. She is currently a consultant to the alcohol beverage industry and an Adjunct Senior Lecturer in the School of Agriculture, Food and Wine at the University of Adelaide, Australia. In 1997, Dr Stockley was appointed Australian Government Representative on the Health and Safety Commission of the Organisation *International de la Vigne et du Vin* (OIV) and served as President of the Commission IV Safety and Health, being awarded the Knight of the Order of Agricultural Merit (France) in 2015 and, more recently, the OIV Merit Award. She has been actively involved in the preparation of alcohol policy, such as reviews of the National Alcohol Strategy, the NHMRC Australian Alcohol Guidelines and warning labelling, as well as actively being involved in wine research projects on a range of health, nutrition and safety-related issues. These have included the potential allergenicity of wine and the effects of wine and wine-derived phenolic compounds on cardiovascular diseases, cognitive function and cancers. She has presented papers at in excess of 115 conferences and workshops and published more than 70 peer-reviewed papers, 85 non-peer-reviewed papers and 12 book chapters.

Contributors

Ghanshyam Abrol
Department of Horticulture and Forestry
Rani Lakshmi Bai Central Agricultural University
Jhansi, Utter Pradesh, India

César Elosúa Aguado
Electric, Electronic and Communications Engineering Department
Universidad Pública de Navarra
Pamplona, Spain

Arminda Alves
LEPABE – Laboratory for Process Engineering, Environment, Biotechnology and Energy
University of Porto
Porto, Portugal

Arina O. Antoce
Department of Bioengineering of Horti-Viticultural Systems
University of Agronomic Sciences and Veterinary Medicine
Bucharest, Romania

M. Remedios Marín-Arroyo
Department of Agricultural Engineering, Biotechnology and Food
Universidad Pública de Navarra
Pamplona, Spain

Surekha Attri
Department of Food Science and Technology
Dr Y.S. Parmar University of Horticulture and Forestry
Nauni, Solan, Himachal Pradesh, India

Wamik Azmi
Department of Biotechnology
Himachal Pradesh University
Shimla, Himachal Pradesh, India

Vandana Bali
Food Biotechnology Research Laboratory
Department of Food Engineering & Technology
Sant Longowal Institute of Engineering and Technology
Longowal, Punjab, India

Eveline Bartowsky
Lallemand Australia
Edwardstown

and

Department of Agriculture, Food and Wine
University of Adelaide
Adelaide, Australia

Fábio Bernardo
LEPABE – Laboratory for Process Engineering, Environment, Biotechnology and Energy
Universidade do Porto
Porto, Portugal

Tek Chand Bhalla
Department of Biotechnology
Himachal Pradesh University
Shimla, Himachal Pradesh, India

Shashi Bhushan
Dietetics and Nutrition Technology Division
CSIR - Institute of Himalayan Bioresource Technology
Palampur, Himachal Pradesh, India

Eduardo Boido
Departamento de Ciencia y Tecnología de los Alimentos
Universidad de la República
Montevideo, Uruguay

Matteo Bordiga
Department of Pharmaceutical Sciences
Università del Piemonte Orientale (UPO)
Novara, Italy

Carlos A.M. Callegaro
Centro Interdisciplinar de Pesquisasem Agronegócios
Universidade Federal do Rio Grande do Sul
Porto Alegre, Brazil

Daniela Callegaro-de-Menezes
Centro Interdisciplinar de Pesquisasem Agronegócios
Universidade Federal do Rio Grande do Sul
Porto Alegre, Brazil

J.M. Cantoral
Laboratorio de Microbiología y Genética
Universidad de Cádiz
Cádiz, Spain

Francisco Carrau
Departamento de Ciencia y Tecnología de los Alimentos
Universidad de la República
Montevideo, Uruguay

Nilakshi Chauhan
Department of Food Science and Technology
Dr Y.S. Parmar University of Horticulture and Forestry
Nauni, Solan, Himachal Pradesh, India

Luca Cocolin
Department of Agricultural Forest and Food Sciences
University of Torino
Turin, Italy

Giuseppe Comi
Department of Agricultural, Food, Environmental and Animal Science
University of Udine
Udine, Italy

G. Cordero-Bueso
Laboratorio de Microbiología y Genética
Universidad de Cádiz
Spain

Fernanda Cosme
Department of Biology and Environment
University of Trás-os-Montes and Alto Douro
Vila Real, Portugal

Eduardo Dellacassa
Departamento de Química Orgánica
Universidad de la República
Montevideo, Uruguay

Anju K. Dhiman
Department of Food Science and Technology
Dr Y.S. Parmar University of Horticulture and Forestry
Nauni, Solan, Himachal Pradesh, India

Bhawna Dipta
Department of Fruit Science
Dr Y.S. Parmar University of Horticulture and Forestry
Nauni, Solan, Himachal Pradesh, India

Arturo Esnoz
Food Engineering Department
Technical University of Cartagena
Cartagena, Spain

Laura Fariña
Departamento de Ciencia y Tecnología de los Alimentos
Universidad de la Republica
Montevideo, Uruguay

Vivienne Gepp
Departamento de Protección Vegetal
Universidad de la República
Montevideo, Uruguay

B.H. Gump
Professor of Beverage Management
Florida International University
Miami, Florida

Paulo Herbert
LEPABE – Laboratory for Process Engineering, Environment, Biotechnology and Energy
University of Porto
Porto, Portugal

R.S. Jackson
Cool Climate Oenology and Viticulture Institute
Brock University
St Catharines, Canada

Philippe Jeandet
Research Unit "Induced Resistance and Plant Bioprotection"
University of Reims
Reims, France

António Manuel Jordão
Department of Food Industries
Polytechnic Institute of Viseu
Portugal

Chetan Joshi
Formerly Himachal Pradesh Pollution Control Board
Shimla, Himachal Pradesh, India

V.K. Joshi
Department of Food Science and Technology
Dr Y.S. Parmar University of Horticulture and Forestry
Nauni, India

Kriti Kanwar
Department of Biotechnology
Himachal Pradesh University
Shimla, Himachal Pradesh, India

Savitri Kapoor
Department of Biotechnology
Himachal Pradesh University
Shimla, Himachal Pradesh, India

Stylianos D. Karagiannis
General Chemical State Laboratory of Greece
Directorate of Alcohol and Food stuffs, Department of Alcohol and Alcoholic Beverages
Athens, Greece

R. Kaur
Department of Biotechnology
Dr Y.S. Parmar University of Horticulture and Forestry
Nauni, Solan, Himachal Pradesh, India

Julie Kellershohn
Hospitality and Tourism Management
Ryerson University
Toronto, Canada

Rena I. Kosti
Department of Dietetics and Nutrition
School of Physical Education, Sport Science and Dietetics
University of Thessaly
Trikala, Greece

K. Kumar
School of Agriculture
Shoolini University of Management and Biotechnology
Bajhol, Solan, Himachal Pradesh, India

Naveen Kumar
Department of Food Science and Technology
Dr Y.S. Parmar University of Horticulture and Forestry
Nauni, Solan, Himachal Pradesh, India

Gérard Liger-Belair
Team "Effervescence, Champagne and Applications"
Group of Molecular and Atmospheric Spectroscopy
University of Reims
Reims, France

Rocco Longo
Horticulture Centre
University of Tasmania
Launceston, Australia

Aline Lonvaud-Funel
Unité de recherche Oenologie
Université de Bordeaux
Villenave d'Ornon, France

Antonio López-Gómez
Food Engineering Department
Technical University of Cartagena
Cartagena, Spain

Valentina Martín
Departamento de Ciencia y Tecnología de los Alimentos
Universidad de la República
Montevideo, Uruguay

S.S. Marwaha
Punjab Pollution Control Board
Patiala, India

U. Marwaha
Punjab Pollution Control Board
Patiala, India

Karina Medina
Departamento de Ciencia y Tecnología de los Alimentos
Universidad de la Republica
Montevideo, Uruguay

Sergio Moser
Fondazione Edmund Mach
Istituto Agrario San Michele All'Adige
Trento, Italy

Ginés Navarro
Department of Agricultural Chemistry
University of Murcia
Murcia, Spain

Simón Navarro
Department of Agricultural Chemistry
University of Murcia
Murcia, Spain

Bhanu Neopaney
Department of Environment, Science and Technology
Govt of Himachal Pradesh
Shimla, Himachal Pradesh, India

Antonio D. Padula
Centro Interdisciplinar de Pesquisasem Agronegócios
Universidade Federal do Rio Grande do Sul
Porto Alegre, Brazil

Parmjit S. Panesar
Food Biotechnology Research Laboratory
Department of Food Engineering & Technology
Sant Longowal Institute of Engineering and Science
Longowal, Punjab, India

Preethi Ramachandran
Department of Food Science and Technology
Dr Y.S. Parmar University of Horticulture and Forestry
Nauni, Solan, Himachal Pradesh, India

Cintia Lacerda Ramos
Department of Basic Science
Federal University of Vales
Diamantina, Brazil

Ajay Rana
Dietetics and Nutrition Technology Division
CSIR-Institute of Himalayan Bioresource Technology
Palampur, Himachal Pradesh, India

Neerja S. Rana
Department of Basic Sciences
Dr Y.S. Parmar University of Horticulture and Forestry
Nauni, Solan, Himachal Pradesh, India

Vishal S. Rana
Department of Fruit Science
Dr Y.S. Parmar University of Horticulture and Forestry
Nauni, Solan, Himachal Pradesh, India

Alka Rani
Department of Biotechnology
Himachal Pradesh University
Shimla, Himachal Pradesh, India

L. Rebordinos
Laboratorio de Microbiología y Genética
Universidad de Cádiz
Cádiz, Spain

L. Veeranjaneya Reddy
Department of Microbiology
Yogi Vemana University
Kadapa, India

M.E. Rodríguez
Laboratorio de Microbiología y Genética
Universidad de Cádiz
Cádiz, Spain

Inge Russell
Heriot-Watt University
International Centre for Brewing and Distilling
Edinburgh, Scotland

Michikatsu Sato
Lifelong Wine Education Program
University of Yamanashi
Kofu, Japan

Leigh M. Schmidtke
National Wine and Grape Industry Centre
Charles Sturt University
Wagga Wagga, Australia

Rosane Freitas Schwan
Department of Biology
Federal University of Lavras
Minas Gerais, Brazil

D.R. Sharma
Department of Biotechnology
Dr Y.S. Parmar University of Horticulture and Forestry
Nauni, India

Nivedita Sharma
Department of Basic Science
Dr Y.S. Parmar University of Horticulture and Forestry
Nauni, India

Poonam Sharma
Department of Basic Science
Dr Y.S. Parmar University of Horticulture and Forestry
Nauni, Himachal Pradesh, India

Satish K. Sharma
Directorate of Research
Dr Y.S. Parmar University of Horticulture and Forestry
Nauni, India

Somesh Sharma
Department of Bioengineering and Food Science
Shoolini University of Management and Biotechnology
Bajhol, India

Sunny Sharma
Department of Fruit Science
Dr Y.S. Parmar University of Horticulture and Forestry
Nauni, India

Swati Sharma
Department of Food Science and Technology
Dr Y.S. Parmar University of Horticulture and Forestry
Nauni, Solan, Himachal Pradesh, India

Shilpa
Department of Biotechnology
Dr Y.S. Parmar University of Horticulture and Forestry
Nauni, Punjab, Himachal Pradesh, India

R.S. Singh
Department of Biotechnology
Punjabi University
Patiala, Punjab, India

R. Soni
Department of Biotechnology
D.A.V. College
Chandigarh, Punjab, India

S.K. Soni
Department of Microbiology
Panjab University
Chandigarh, Punjab, India

Creina S. Stockley
Stockley Health & Regulatory Solutions
Malvern, Australia

Urvashi Swami
Department of Microbiology
Panjab University
Chandigarh, Punjab, India

Zoltán Szakál
Department of Marketing and Tourism
University of Miskolc
Miskolc, Hungary

Abhimanyu Thakur
Department of Food Science and Technology
Dr Y.S. Parmar University of Horticulture and Forestry
Nauni, Solan, Himachal Pradesh, India

Navdeep Thakur
Department of Biotechnology
St. Bede's College
Shimla, Solan, Himachal Pradesh, India

N. S. Thakur
Department of Food Science and Technology
Dr. Y.S. Parmar University of Horticulture and Forestry
Nauni, India

Devina Vaidya
Department of Food Science and Technology
Dr Y.S. Parmar University of Horticulture and Forestry
Nauni, Solan, Himachal Pradesh, India

Manoj Vaidya
Directorate of Research
Dr Y.S. Parmar University of Horticulture and Forestry
Nauni, Solan, Himachal Pradesh, India

Yann Vasserot
Laboratory of Enology
University of Reims
Reims, France

B.W. Zoecklein
Enology-Grape Chemistry Group
Department of Food Science and Technology
Virginia Tech
Blacksburg, Virginia

Unit 1

Introduction, Role, Composition and Therapeutic Values

INTRODUCTION

Given that the products of the grape vine are amongst the most diverse and oldest of any agricultural crop, this unit commences with an overview of the history as well as science of wine and brandy production internationally (Chapter 1). It touches on topics such as different wine regions, types and styles, as well as the world wine market, trade and tourism, as wine consumption has become an integral part of many cultures in many countries.

This second chapter segways into a catalogue and discussion of the main characteristics of wines and wine products such as aromatised cocktails and drinks. Wines are divided into three basic categories, for example, still wines, sparkling wines and liqueur-wines. Two other types of wine-products are also discussed: the beverages obtained by de-alcoholisation of wine, which are not considered as wines; and the alcoholic beverages produced by fermentation of fruits other than grapes, for example, fruit wines.

The detailed composition of wine, that is, various components such as sugars, acids, alcohols, colours, phenols minerals, esters, minerals, other by products of wine fermentation etc. have been described in Chapter 3. There are hundreds of volatile components, both pleasant and unpleasant, that are able to be analysed in wine, each of which has a sensory descriptor and can be present in quantities ranging from ng/L to mg/L. The influence of various viticultural practices on the composition of wine has also been documented here.

A detailed description of the chemical components of wine aroma is provided in Chapter 4, along with the primary viticultural and winemaking practices that impact on them. Given that aroma is strongly linked to taste, a lot of what consumers experience when drinking wine actually occurs through their sense of smell, not taste. Details are provided on the chemical changes that occur to the aroma compounds during fermentation and other winemaking practices, as well as specific factors that alter aromas. The case study grape variety in this chapter is Tannat, which is the most widely cultivated variety in Uruguay. Ageing wine in wood such as oak also influences wine chemistry, such as that of the phenolic compound-associated aroma and flavour, as well as ability to 'age'.

The perception of wine consumption has changed, however, throughout the centuries. Although considered a therapeutic agent in ancient times, this became less accepted when the relationship between alcohol and human health was shown to also have harmful effects to both the person and their society. An extreme reaction to societal harms occurred in the United States with prohibition, a nationwide constitutional ban on the production, importation, transportation and sale of alcoholic beverages from 1920 to 1933. During the mid and late 20th century, a J-shaped relationship between the consumption of alcoholic beverages, such as wine, and human health was seen in scientific studies, and these are elaborated in Chapter 5. It has been proved the Mediterranean-style diet is an example of a health-promoting diet, thus representing a reference diet for the prevention of cardiovascular and metabolic diseases. The diet includes the consumption of relatively high intakes of olive oil, legumes, fruits and vegetables, unrefined cereals, fish and wine. Moderate consumption of wine is an essential component of such a diet as bioactive components present in it are able to exert various protective functions like free radical scavenging, decreasing oxidative stress and reducing inflammatory atherosclerotic lesions. Thus, the potentially therapeutic effects of wine consumption in moderation include a reduction in risk of cardiovascular diseases, diabetes, certain cancers and cognitive dysfunction such as dementia.

The final chapter in Unit 1 outlines the food safety aspects of wine production, as wine is one of the most rigorously regulated beverages internationally with an exclusive list of allowable additives and processing aids, and limits on potential contaminants and toxicants. This chapter also discusses the harmful effects of wine consumption on human health, which generally occur when wine is consumed in amounts above moderation.

Dr Creina Stockley, PhD, MBA
Principal, Stockley Health and Regulatory Solutions
Adjunct Senior Lecturer, The University of Adelaide, Australia

1 Wine and Brandy
An Overview

V.K. Joshi and Matteo Bordiga

CONTENTS

1.1 INTRODUCTION

Wine is an alcoholic agricultural product that has been made and consumed all over the world for over 6,000 years. Louis Pasteur called wine 'the most hygienic and healthiest of drinks'. Its production has grown both in popularity and in sophistication. Wine has a rich history dating back thousands of years, since the dawn of civilization, and has followed humans and agriculture along diverse migration paths (Figure 1.1). Wine has been closely associated with Jewish and Christian religious services. Ancient scriptures suggest that the origin of wine might have been accidental, when the juice of some fruit transformed itself into a beverage having exhilarating or stimulating properties. The consumption of wine induced euphoria and pleasing relaxation from the strains of life, and it eventually gained social importance and acceptance. Today, wine is an integral component of the cultures of many countries.

Wine is a completely or partially fermented juice of grapes, but other fruits are also utilized for the production of wines. Another grape product is brandy, made through the distillation of wine. The word 'brandy' comes from the Dutch word '*brandewejn*', which means burnt wine, while '*al-koh'l*' is an Arabic word. Brandy is made by distilling wine or fruit and then, aging it in oak barrels. Differences between brandies vary from country-to-country. Soil, climate, grapes,

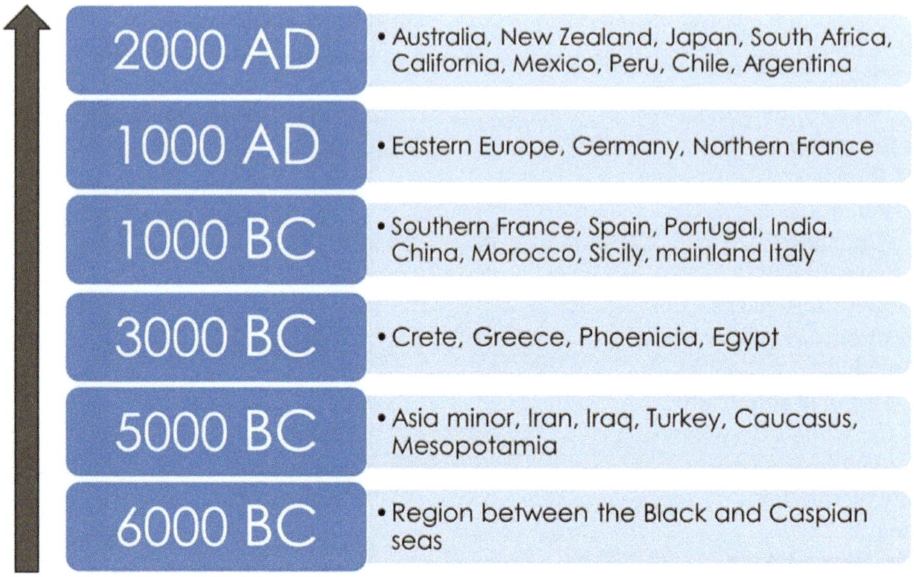

FIGURE 1.1 A generalized scheme of the spread of *Vitis vinifera* noble varieties of grapevine and winemaking from their center of origin in Asia Minor to other parts of the world

production methods, and blending give each brandy its own unique flavor and style.

The perception that the wine is good for the body is now scientifically supported and there is no dispute that moderate wine consumption is associated with lower mortality from coronary heart disease. The antimicrobial activities of wine are associated with ethanol, while antioxidant properties with phenolic components and flavonoids. Wines are nutritious, safe, provide psychotropic effects, and are preferable to distilled liquors (see Chapter 3 for details). Wine is an important adjunct to the human diet and increases satisfaction thus, contributing to the relaxation necessary for proper digestion and absorption of food. It is an integral part of the Mediterranean diet. Wine has also been associated with therapeutic benefits (see Chapter 5 for details). Excessive alcohol consumption, both acute and chronic, can have devastating effects on the physical and mental well-being of the consumer (see Chapter 6 for details).

Wines may be classified broadly based on geographic origin, grape variety, or fermentation and maturation process, or may bear generic names. Based on color, wines could be red, rosé (pink) or white, or could be differentiated as table, sparkling, or fortified wines. Such wines with additional flavoring are called aromatic wines, like vermouth. Sweet wines are high in sugar content, while dry wines have a negligible amount of sugar (see Chapter 2). *Saccharomyces cerevisiae* is the microorganism on which the entire gamut of wine and brandy preparation rests. The leading wine-producing countries include France, Italy, Spain, Argentina, Portugal, Germany, South Africa, and the United States. This chapter introduces the origin of vine and wine, the history of wine making, wine regions, the science and technology of wine making, the process of wine making, quality evaluation, and finally wine consumption and marketing.

1.2 SCIENCE AND TECHNOLOGY OF WINEMAKING

The process of wine making is unique in the sense that nearly all the physical, biological (especially microbiological), and chemical sciences are involved, along with engineering processes. Because of the botanical nature of the raw materials and its microbial transformation into wine, knowledge of the physiology and genetics of the vine, yeasts, and bacteria is crucial to understand the origins of wine quality. Microclimatology and soil physico-chemical can clearly reveal the vineyard origins of the grape quality. Wine science, or oenology, has three pillars – grape culture, wine production, and sensory analysis evaluation. It is also concerned with chemistry, biochemistry, microbiology, and large scale production of wine or engineering. Oenology is defined as the science of wine making, comprising the horticultural, biological, and food sciences, technological and engineering aspects of the process, as well as economics and marketing angles.

Thus, it is a combination of several aspects of knowledge and is an interdisciplinary subject. Principles of chemistry, microbiology, food science, biochemistry, genetic engineering, chemical engineering, nutrition, economics and marketing, and sensory science are the pillars on which oenology rests. Finally, a knowledge of human sensory psychophysiology is essential for interpreting wine quality data, especially the sensory quality and consumer acceptance. So, covering all these aspects, wine science or enology (or precisely oenology) is a multidisciplinary science, and a successful enologist will possess in-depth knowledge of all these aspects, thus making wine of superior quality.

1.3 ORIGIN AND HISTORY OF VINE AND WINE

1.3.1 Viticulture

Plant taxonomists consider the region between the Black and Caspian seas to be the original home of the Old World grape. Grape culture first began in Asia Minor, from where it spread to both the west and east. It was underway in the Near East as early as the 4th millennium BC. Before 600 BC, the Phoenicians probably carried wine varieties to Greece, then to Rome and to southern France. Fossil vines which are 60 million years old are the earliest scientific evidence of the existence of grapes. The earliest documentation of viticulture is in the Old Testament of the Bible. *Vitis vinifera* was being cultivated in the Middle East by 4000 BC, and probably earlier. The vine was evident in 6000 BC in Egypt, 3000 BC in Phoenicia, and 2000 BC in Greece and in China. The Greeks introduced viticulture in France, North Africa, and Egypt, while the Romans exported the vines to Bordeaux, the valleys of the Rhone, Marne, Seine, etc., and to Hungary, Germany, England, Italy, and Spain. The Greeks also planted grapes in their colonies from the Black Sea to Spain. Following the voyages of Columbus, grape culture and wine making were transported from the Old to the New World. Spanish missionaries took viticulture to Chile and Argentina in the mid-16th century, and to lower California in the 18th century. The prime wine-growing regions of South America, however, were established in the foothills of the Andes Mountains. The centre of viticulture, however, shifted from the southern missions to the Central Valley and Sonoma, Napa, and Mendocino in California. British settlers planted European vines in Australia and New Zealand in the early 19th century, while Dutch settlers took grapes from the Rhine region to South Africa as early as 1654. The ancient Aryans possessed knowledge of grape culture as well as beverage preparation from it. Famous Indian scholars Sasruta and Charaka mentioned the medicinal properties of grapes in their medical treatises, entitled *Sassuta Samhita* and *Charaka Samhita*, respectively, and written between 1356–1220 BC. Kautilya, in his *Arthashastra* written in the 4th century BC, also mentioned the type of land suitable for grape cultivation.

1.3.2 Wine Mission for California

Jean-Louis Vignes planted California's first documented imported European vines in Los Angeles in 1833. During the 1850s and 1860s, Agoston Harazsthy, a Hungarian

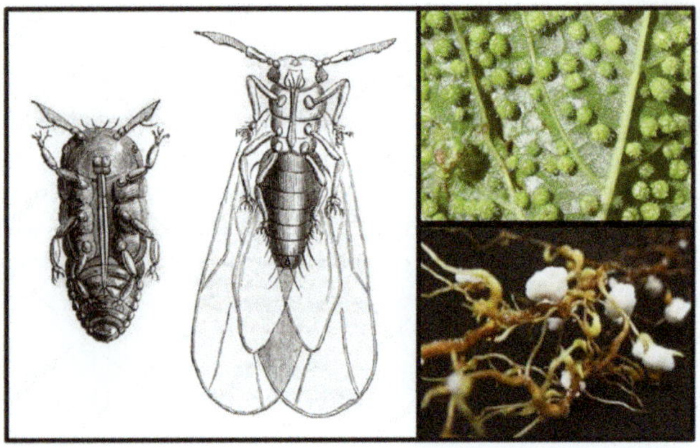

FIGURE 1.2 A sketch of *Phylloxera vastartrix*. Leaves and roots showing the typical damages made by phylloxera

soldier, merchant, and promoter, imported grape cuttings from the greatest European vineyards to California, and is rightly known as the founder of the California wine industry. Unfortunately, Haraszthy's success was almost negated by the vine root louse *Phylloxera vastatrix* (Figure 1.2) which was discovered in England in 1863. By 1865, *Phylloxera* had spread to vines, and over the next 20 years, it inhabited and decimated nearly all the vineyards of Europe. Grafting every vine in Europe to the phylloxera-resistant American rootstock was undertaken and thus, the European wine industry could be retrieved from extinction. During the period when Europeans were contending with *Phylloxera*, the American wine industry was flourishing.

1.3.3 Wine: Historical Aspects

Indirect evidence about the existence of wine has come from different sources (paintings, engravings on buildings, writings of poets, mythological evidence, religious commands, etc.), including the occurrence of numerous grapes in the crypts of pyramids. Examples of the development of wooden cooperage for wine storage by the Romans and wine amphorae (sealed and containing wine remnants) have been found on several occasions. Ancient relics of wine growing can be observed at the wine museum in Beaune (Burgundy, France). The process of wine making existed long before the chronicles found in Egyptian hieroglyphics. The vine was evident in 6000 BC in Egypt, 3000 BC in Phoenicia, and 2000 BC in Greece. It is not known who might be the first human who might have tasted wine. Some early Cro-Magnon cave paintings show the use of libation cups some tens of thousands of years ago. Wine drinking had started by about 4000 BC and possibly as early as 6000 BC. The texts from tombs in ancient Egypt amply prove that wine was in use around 2700 to 2500 BC, when priests and royalty were using it. The Phoenicians from Lebanon introduced wine to the Romans and Greeks, who subsequently propagated wine making.

1.3.4 New Alcoholic Beverages

By the 15th century, based on the property of alcohol to act as a solvent for a number of essences of fruit, spices, and herbs, a number of drinks such as Benedictino and Chartreuse were prepared. During the 17th century, alcoholic beverages like aperitifs or vermouth emerged. The sparkling nature of wines had certainly been established by 1676, as evidenced when Sir George Etheridge mentioned 'sparkling Champagne' in one of his plays (see Chapter 34). In Spain, the addition of cooked wines to wines meant for export made the wine sweet, due to evaporation by cooking, and prevented their spoilage. Such wines were named *Sherris-sack*, or presently known as sherry. By 1775, elongated bottles with short necks were developed, which could be kept on their sides, enabling the wine to stay in contact with the cork (see Chapter 33). Port, a type of fortified wine, was a new wine created at the end of the 17th century and was named after the town Porto in Portugal. Its diffusion, in some respects, was related to the use of bottles whereas its shelf-life to the high alcohol content. Cider was established in the Basque Country well before the 12th century, and by the 11th and 12th centuries it was being produced in Cotentin and in Pays d'Auge (see Chapter 35).

Traditional fortified wines are generally associated with areas where the climate and soil conditions do not favor the production of high-quality grapes. Each wine type has its own history and winemaking characteristics and accordingly, the quality characteristics. In fact, the warm and hotter climates of the sherry wine region produce bland, white table wines, while the red wines from the Douro region are astringent, and the cool, damp summers of the island of Madeira produces wines with acidic characters. However, the application of fortification, maturation, and blending techniques produces wines of great type and style. One of these wines is Madeira wine, which is produced in a variety of styles, ranging from dry (consumed as an *aperitif*) to sweet (usually consumed with desserts). Historically speaking, only 25 years after colonization had begun, the island was already exporting this wine, and by the middle of the 17th century, most producers had already started fortification of their wines.

Port is known to have a long history dating back to 1756, when the DDR (Douro Demarcated Region) was declared the third protected wine region in the world, after Chianti (1716) and Tokaj-Hegyalja (1730). Its name comes from the name of Oporto (Porto), the second largest city in Portugal. Port is produced in the Douro valley of northern Portugal, in wineries called *quintas*. Sherry is the most famous fortified Spanish wine which comes from the Southern regions of *Jerez de la Frontera* and *Montilla-Moriles*. In Spain, sherry is referred to as *vino de Jerez* and its earliest records date back to the 8th century, when Moorish Muslims controlled Spain.

1.3.5 Yeast and Alcoholic Fermentation – Origin and History

Historical developments in the microbiology and biochemistry of alcoholic fermentation paralleled the progress in wine fermentation. Scientific research rose at an exponential rate, and nowhere is this more evident than in the historical milestones of chemistry/biology that have improved the knowledge of the biology of the microorganisms that drive fermentation

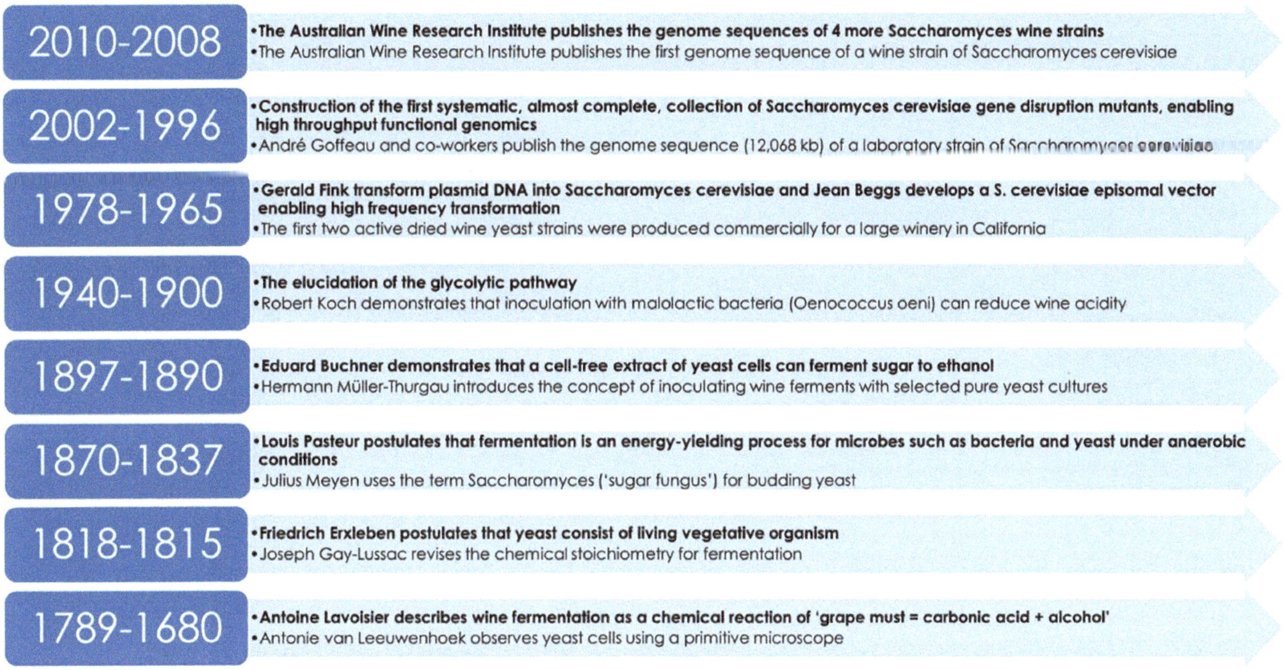

FIGURE 1.3 Selected milestones that mark the path of research in microbiology and yeast biology that have affected wine science and winemaking

(Figure 1.3). This progress has been possible thanks to some of the most significant names in the chemical and biological sciences, including van Leeuwenhoek, Lavoisier, Gay-Lussac, Pasteur, Buchner, and Koch. It is well established that grapes readily ferment, due to the prevalence of fermentable sugars, but the wine yeast (*Saccharomyces cerevisiae*) is not a major indigenous member of the grape flora. Other yeasts indigenous to grapes, such as *Kloeckera apiculata* and various *Candida spp.*, can readily initiate fermentation. However, they seldom complete fermentation, being sensitive to the accumulating alcohol content and limited fermentative metabolism that curtails their activity. Further, it was revealed that the natural habitat of the ancestral strains of *S. cerevisiae* appears to be the bark and sap exudates of oak trees. If so, the tendency of grapevines to climb trees, such as oak, and the joint harvesting of grapes and acorns, could have encouraged the inoculation of grapes and grape juice with *S. cerevisiae*.

Specific words referring to yeast action (fermentation) begin to appear about 2000 BC. The earliest evidence of the connection between wine and *Saccharomyces cerevisiae* came from an amphora found in the tomb of Narmer, the Scorpion King (c. 3150 BC). *Saccharomyces cerevisiae* was confirmed by the extraction of DNA from one of the amphoras. The DNA showed more similarity with modern strains of *S. cerevisiae* than closely related species, *S. bayanus* and *S. paradoxus* single-strain inoculations.

In today's winemaking, pure cultures of *S. cerevisiae* (selected strains) are added (See chapter 26) to grape must as soon as possible after crushing, thus ensuring greater control of vinification (more predictable outcomes and less risk of spoilage by other microorganisms). In modern oenology, there are many different yeast strains available, and the winemaker's choice can substantially affect the wine quality. *S. cerevisiae* produces many secondary metabolites (aroma-active) and releases aroma compounds from inactive precursors in grape juice, thus affecting the sensory properties of the wine. For this reason, any genetic variation in wine yeast that affects the production or release of important molecules will certainly affect the wine quality (yeast-induced variation). It has been shown that the different commercial yeast strains generate wines with very different profiles of volatile thiols. *S. cerevisiae* has been one of the most important model organisms in molecular biology and emerging fields (see Chapters 16 and 19). It must be emphasized that innovations in molecular systems, and now synthetic biology rarely happen without *S. cerevisiae* featuring somewhere prominently in the story (Figure 1.4). *S. cerevisiae* was the first eukaryote to have its genome sequenced, a feat that was achieved through an international effort (large collaborative projects that involved 600 scientists), thus paving the way for the first chip-based gene array experiments.

1.3.6 Barrels and Silicone Bung

Oak is the material used for the manufacture of barrels – containers where the wine is preserved during the aging process. The predominance of oak in the manufacture of barrels is explained by its physical properties: resistance, flexibility, watertightness, natural durability, and by its ability to crack by splitting. Initially, the choice of this material was made, in addition to its physical properties, because it was considered inert, that is, it does not add undesirable tastes to wine. Nowadays, it is known that oak barrels are not a simple container but contribute favorably to the sensory evolution of the great wines. Oak from France is considered to be the most superior for wineries.

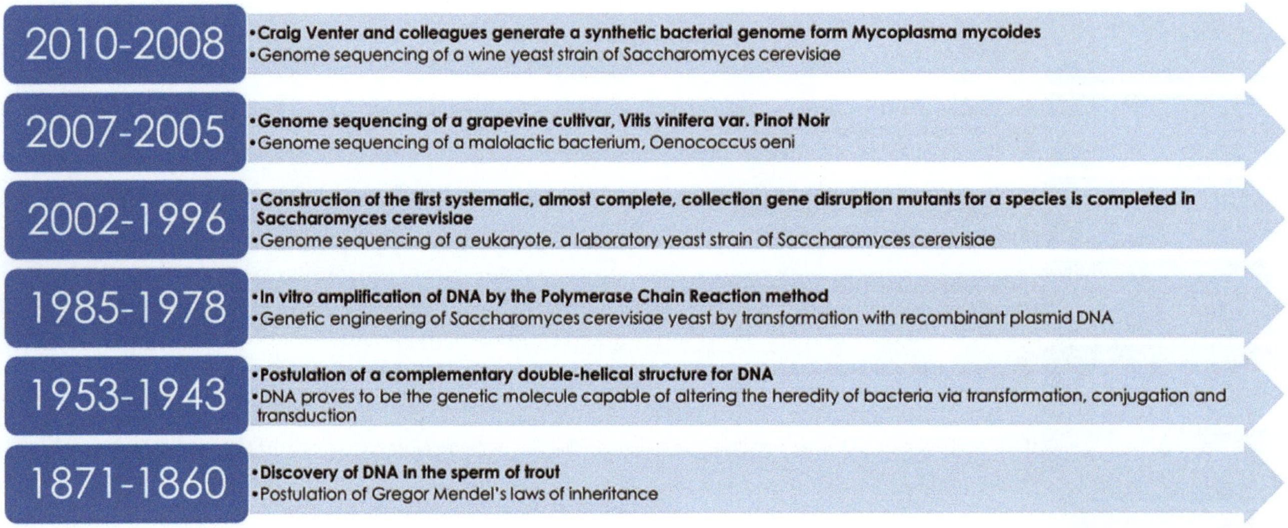

FIGURE 1.4 Selected milestones that mark the path of research in genetics and molecular biology that have affected wine science and winemaking

Oak from forests like Nevers, Vosges, Allier, Limousin, and Troncaisare is known for producing unique flavors in wine. The leading sources of white oak in the United States are Kentucky, Minnesota, Missouri, Ohio, Tennessee, and Wisconsin. In Europe, the primary sources of oak are France and the former Yugoslavia. Oak barrels impart flavors and tannins, both of which are desirable for most red wine as well as some white wines. Oak is slightly porous, which creates an environment ideal for aging wines. Redwood and chestnut are distant second choices to oak, but neither work as well as oak. Sometimes these woods are used for larger casks because use of oak is expensive. Oak flavors and tannins can overpower a wine's varietal character, which results in a poorly balanced wine.

Choosing the right barrel requires some knowledge and experience with various types of oak. The favorite wood for wine barrels is white oak (red oak is too porous), with the U.S. species differing slightly from the European. Oak barrels lose their ability to impart flavor in 4 to 5 years, and the most high-quality wine estates and châteaus replace all or part of their oak barrels with new ones each year.

In 1980, Vincent Bouchard created the first silicone bung. Prior to this, all barrels in Europe were plugged with a burlap cloth wrapped underneath the wooden bung, while a wooden bung with paraffin or tar was used in the New World. The disadvantages of these two methods for sealing the bunghole were the occurrence of leaks, cracked/broken wooden bungs, and oxidation. The silicone bung invented by Vincent Bouchard was called 'Bouchard Versilique'. Unfortunately, Vincent Bouchard never patented this invention and it has been widely copied. The versilique bung is reputed to be 'indestructible', due to its high percentage of silicone, and is still widely available and used in Europe.

Around 1981, Vincent Bouchard was the first to recognize the importance of the charring process. Prior to this, Burgundy and Bordeaux barrels were banded with heat regardless of the source of heat. Heat was simply used to make the wood pliable enough to bend and shape the barrel. One of the sources of heat was an oak fire, while others included steam, warm water, and hot air. The coopers fired the barrels (Figure 1.5) according to tradition and the shape of the barrels (Figure 1.6), but not with the intention of adding a flavor contribution. Vincent Bouchard observed a difference between the ways that various Burgundy and Bordeaux

FIGURE 1.5 Barrel toasting process. *Source:* created using information from Bordiga (2018)

FIGURE 1.6 Charring process and silicone bungs

coopers were firing their barrels with different appearances. The different toasting appearances between each of the coopers he studied and worked with produced very different flavor results. Consequently, Vincent Bouchard proposed different toast levels or colorations (the concept of 'light', 'medium' *vs* 'medium plus' toast, etc.), and for the first time this concept of specifically charring barrels to different levels was carried out by any cooper. Experiments showed the more you toast or char the barrel, the less astringency is obtained, together with a softer, creamy character. However, this depends on the penetration of the heat in the barrel stave and whether it is well-seasoned wood. A light char would still give a tannic or astringent character; a medium, deep toast normally offers a creamy, smooth flavor throughout the life of the barrel (three to four passes of white wine); a heavy toast starts to offer some smoky characteristics (see Chapter 28 for more details).

1.4 WINE REGIONS

Production of wine in different regions of the world is well established and regulated by various acts and regulations. Besides varietal differentiation, the regions where grapes are grown also influence the quality of the wine. Naturally, consumers expect wine from a particular region to possess unique qualities that differentiate it from other regions. Indeed, quality wines are currently being produced in all six arable continents, and emerging nations are active in the international wine trade (see Chapter 3).

The unique characteristics of grape influence the style and taste of wine within each wine region. Today, the finest wines originate anywhere from the famous, established vineyards of Italy and France, to the newer wineries of Australia and the United States. The vineyards are limited by appellation laws to a certain few designated varieties, thus maintaining the characteristics that made their wines famous.

Old World Wine
Old World wines (Figure 1.7) are from countries or regions where winemaking (with *Vitis vinifera* grapes) first originated. Old World wine countries include: Italy, England, France, Spain, Portugal, Greece, Austria, Hungary, and Germany. Based on the definition, countries like Turkey, Georgia, Armenia, and Moldova are also considered Old World wine regions.

New World Wine
New World wines (Figure 1.7) are from countries or regions where winemaking (and *Vitis vinifera* grapes) were imported during (and after) the age of exploration. New World wine countries include: the United States, Australia, South Africa, Chile, Argentina, and New Zealand. Based on the definition, China, India, South Africa, and Japan are also considered New World wine regions. Different regions of the world where wine is produced, along with their typical wines, are described in the subsequent sections.

1.4.1 Old World Wine

1.4.1.1 Italian Wines

With nearly 400 grape varieties – representing over one-quarter of the world's 1400 grape varieties that produce wine in commercial quantities – Italy offers something new and engaging for even the most passionate wine drinker. The current international wine market contains a large amount of Italian wine. Despite being the world's biggest producer of wine, Italy only exports about a quarter of her wine. Much of this product is used to fortify the wines of other nations. However, these wines have very high quality standards, placing Italian wines among the best in the world. Several Italian wines are recognized as among the finest of their type: mainly Piedmontese and Tuscan reds from the Nebbiolo and

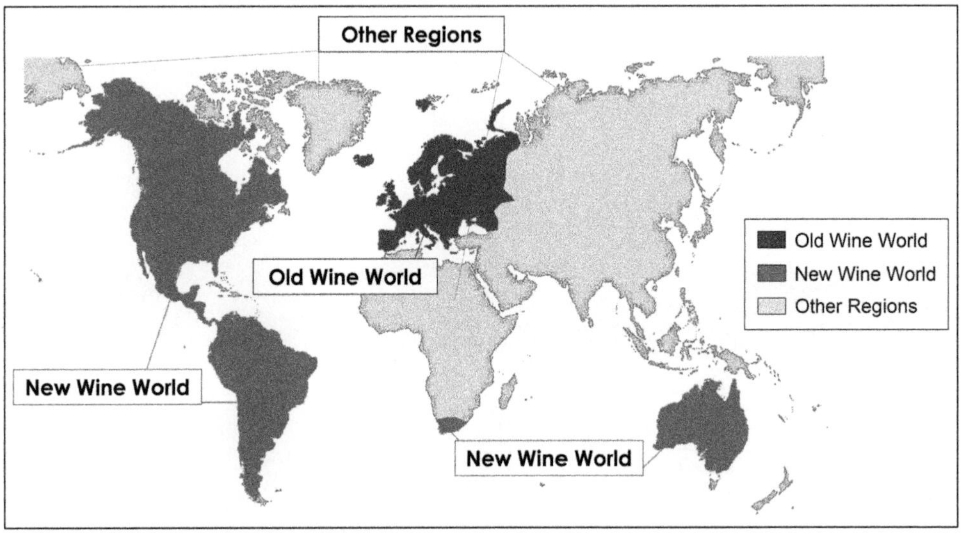

FIGURE 1.7 A simplified map showing Old and New World wine regions. *Source:* Created using informations from the OIV (2018)

Sangiovese varieties, but also white wines, still and sparkling, dry or sweet, have merited international respect. Growers have complemented their local varieties with foreign vines, such as Cabernet, Merlot and Pinots. There has been evidence, in the past as well as today, that Italy's multifarious climates and terrains favor vines of many different types and styles, and consumers elsewhere in Europe and in North America have come to appreciate these new examples of class.

1.4.1.2 French Wine Regions

Each wine-producing region in France specializes in the production of specific types of wines and flavors. But the industry is facing a great challenge from the New World regions such as California. France is still considered the best region in the world to find excellent wines with great finesse and elegance. The records kept by Christian monks have enabled the French to continue to make the quality wine they have perfected over centuries of work. The major wine areas in France are: Burgundy, Bordeaux, Alsace, the Rhone Valley, Champagne, and the Loire Valley (Table 1.1).

1.4.1.3 Spanish Wines

The wine industry in Spain is as old and established as that of France. Despite this similarity, the wines produced by these two countries are vastly different. Spain produces a large amount of red wine, sparkling wines, and sherry. Spain's products are very much a part of the international wine market.

***The Rioja*:** This is one of the best-known wine regions of Spain. When the phylloxera epidemic struck France, many of its wine makers moved to northern Spain in order to continue their trade. The French taught the local Spaniards how to make wine from their local red Tempranillo grapes. The result was production of a series of red wines that are very flavorful and strong. Enjoyed globally today, many of the Rioja reds are aged for ten years in large wooden barrels.

***Catalonia*:** Even though this region does not share the same world recognition as Rioja, it is where most Spanish wines are produced. Recently, more standard wines like merlot and cabernet are made in Catalonia, along with its own traditional wines. Catalonia also specializes in the production of Cavas or sparkling wines. It is a good alternative to French

TABLE 1.1
Major Regions of France and the Types of Wines Each Produces

Regions	Grapes	Wines	Qualities
Burgundy	Pinot Noir, Gamay and Chardonnay	Red, white	Dry whites, richly textured reds
Bordeaux	Cabernet Sauvignon, Cabernet Franc and Merlot	Red, white and sweet wines	Deeply flavored reds, sweet whites
Rhone	Syrah, Cabernet Sauvignon and Muscat	Red, white and sparkling	Earthy, big wines, some with a slight fizz
Loire	Sauvignon Blanc, Pinot Noir and Pinot Gris	Rose, red, white and sparkling	Rich reds and dry whites that go well with shellfish
Alsace	Sauvignon Blanc, Pinot Noir and Pinot Gris	Riesling and some rose	Sweet wines that are Germanic in flavors
Champagne	Pinot Noir, Pinot Meunier and Chardonnay	White, red and champagne	Thin and tart, excellent sparkling wines
Beaujolais	Chardonnay and local Gamay	Reds, whites and slightly fizzy wines	Fresh tasting, slight fruity flavour

Source: WineScience.com, 1620 I Street NW, Suite 615, Washington, DC 20006

champagne and is of very high quality. Currently, consumers around the world are starting to drink more and more Cava.

***Jerez*:** This city is one of the most famous of the Spanish wine regions. Its sherry is perhaps what has made Spain's wine industry famous. In fact, the name 'sherry' is itself an English form of Jerez. Sherry is still shipped to different parts of the globe from this city and is found in markets everywhere.

1.4.1.4 German Wines

Generally, production focuses on white wines, although there are some quality reds, but these are rarely exported. Germany's global recognition for production of the sweet riesling wine is gaining momentum. These wines have a reputation for being too sweet; however, some very dry versions are also manufactured in Germany, specifically for export purposes. Since German vineyards are in the far north, it is difficult to get the grapes to ripen to a point where they contain a large quantity of sugar, consequently the sweet wines produced from these grapes are highly sought after. The German government has a regulated system for ranking their wines which is printed on all wine labels. This makes choosing a German wine easier for international consumers.

1.4.1.5 Portuguese Wines

Portuguese wines are produced in various regions and do very well in markets around the world, except the United States. However, Portugal has avoided producing varieties like chardonnay and merlot in order not to Americanize its wines. The wine makers of Portugal use local grape varieties to add flavor and color to their products. At least 18 different types of grapes are used to make certain varieties of distinct port worth trying.

1.4.1.6 English Wines

Over the last twenty years, the production of wine in England has undergone a quiet revolution. Winemaking in England has never been more professional or focused on achieving the highest levels of quality. Though the English climate is not the ideal one to grow vines, it has successfully grown grapes in the vineyards in the southern regions (mainly Surrey, Sussex, Hampshire, and the South West). In fact, the climatic conditions and chalky soils in these southern regions are remarkably similar to the famed Champagne region of France. It is not surprising that many successes of English vineyards have been in the production of sparkling wines. Still wines are also produced in abundance, and the whites are typically more successful than the reds. Indeed, the cooler climate and grape varieties used in England are responsible for imparting delicate floral and grapefruit character to the white wines.

1.4.1.7 Wines of the Balkans

The wines of the Balkans differ greatly, and only few of them have share in the international wine market.

***Former Yugoslavia*:** The wines of this region seem to have an acquired taste to many. Many wine drinkers believe that this region's most famous wine, Zilvaka, tastes like bruised apples.

***Bulgaria*:** Bulgaria exports about 80% of its wines due to the high Muslim population there. Bulgarian red wines are usually pale and high in quality. Two of the more famous Bulgarian wines are Pamid and Gamza.

***Hungary*:** While many other nations attempt to follow the French method of wine production, Hungary holds onto its traditional methods. It produces fiery red wines and strong white wines that can accompany spicy foods. Hungary also produces Tokaji, the famous sweet wine produced from grapes with noble rot.

***Greece*:** It is believed that the Ancient Greeks were the first to study the process of wine production. The prototype for vine cultivation that the Greeks provided us with has been modified and improved over the centuries. Today, the Greeks make excellent table wines of both the red and white varieties. Although Greece is not a major exporter of its wines, its large tourism industry helps promote them.

***Romania and Moldova*:** Although Moldavian wines were greatly sought after in 19th century Paris, they do not have the same level of international fame today. The best exported wines of Romania are the rich honeyed wines Traminer and Pinot Gris.

***Turkey*:** Even though Turkey's vineyard land is the fifth largest in the world, it produces little wine. When phylloxera struck Europe, Turkey began to make wine for export. Today however, these wines are not well known globally, since many do not appeal to the western tastes.

1.4.2 New World Wine

1.4.2.1 Wines of the United States

The west coast of the United States is a major producer of wine in today's market. The United States interest in wine has helped its industry to grow during the past a few decades, turning the United States into one of the largest producers of quality wines.

***California*:** California is perhaps the biggest producer of US wine. Many of the new technologies used in wine making were pioneered in this state, and Californian wines now rank among the highest quality wines in the world. The wines from the Napa Valley of California are very popular and are even competitive with French wines. Napa Valley, with its rich soil and balanced weather, uses about 80% of its cultivable land for the cultivation of grapes. The wines are extremely flavorful. California wine makers have begun to explore the Dry Creek and Russian River lands for the cultivation of grapes. Russian settlers were the first to plant grapevines in this area, and the wines from these regions have recently begun to rival those of Napa. The abundant fog of these areas produces grapes that make soft, fruity wines, which the American public tends to favor. Californians also make wine in the Central Valley, and even as far south as Temecula. Generally, these wines are not as flavorful as those of Sonoma and Napa, yet the market for them is growing.

***The Pacific Northwest*:** Although much of the focus of American wine making is on California, the Pacific Northwest region also produces some very fine wines. The cool climate allows its wine makers to copy the methods of their European counterparts with greater ease than that of Californians.

Washington: The wine industry is currently expanding in Washington State. White wines are most successful, but some quality merlots and cabernets are also produced here. The grapes are high in quality, and vineyards from Oregon use them to boost the flavor of their own wines.

Idaho: The vineyards of Idaho are at a very high altitude so the grapes tend to have a high natural sugar content, which gives the wine a good body and taste. Although the reds of this region are very light, they have delicate fruity flavors that many Americans find appealing.

1.4.2.2 Chilean Wines

South America's Chile produces some very high quality wines that are exported to markets throughout the world. Although Chilean wine does not taste quite like conventional wines, it is earning a reputation as a high quality product. The Chilean climate is very beneficial to wine production. In fact, this climate acted as a deterrent to the phylloxera epidemic, and Chile is the only wine-producing region in the New World that does not have to graft its vines onto American roots. The chief wine region in Chile is the Maipo Valley that produces good reds and quality whites. The cabernets made in the Maipo Valley are world famous and have a smoky flavor unlike other cabernets. Wineries such as Linderos, Conchay Toro, and Undurraga produce some of the Maipo Valley's highest quality wines. The major grape varieties of Chilean wines are Periquita, Tinta Roriz, Touriga Francesa, Esgana Cao, Verdelho, and Arinto.

1.4.2.3 Australian Wines

Australia has enjoyed many years of producing world famous wines, with the highest level of wine consumption of any English-speaking nation. Even though Australia did not escape the phylloxera outbreak, its wine industry eventually recovered to become one of the best in the world. Like California, Australia benefits from experimenting with new technologies and unconventional methods of producing wine. New South Wales is one of the most successful wine regions in Australia. The Hunter Valley in this area grows about 60 different varieties of grapes. Unfortunately, this region is very hot and many grapes rot before they are harvested. The wines of this valley are also said to have a distinct taste that some individuals refer to as the 'sweaty saddle'. Despite this strange description, the wines of the Hunter Valley are of high quality and sought after throughout the world. During the 19th century, Victoria was the largest producer of wine in Australia. The phylloxera outbreak devastated the vineyards of Victoria. It recovered 15 years ago, and now has a very modern collection of grapes that are mixed with bordeaux varieties, producing wines that are fragrant, full, and minty. About 60% of Australia's wine is produced in South Australia. The red soil in the Coonawara region is rich in minerals and produces wines very rich in flavor and texture. A large number of Australia's most famous table wines come from this area, and its wines are known throughout the world. Recently, Tasmania has begun to develop vineyards that are expected to produce quality grapes and wines in the near future.

1.4.2.4 Canadian Wines

The cool climate of Canada is not very suitable for the cultivation of grapes, but with the current expansion of American wineries across the border, this country is beginning to produce some very nice wines. Recently, the creation of rootstocks that can cope with the harsh Canadian winters has helped Canada to become better known in the world's wine industry. Vineyards in British Columbia and Ontario are some of the best in today's Canadian wine industry. Although vine has been cultivated in the Ontario region since 1811, it was mainly based on the strong Labrusca grapes and dull hybrids that produced some very good ports and sherries but did not make very good table wines. Canada has also started using a more varied array of grapes in its wine making procedures, using grapes such as Riesling, Chardonnay and Gamay to boost the local flavors. The red wines of Canada possess a very pleasant flavor.

1.4.2.5 South African Wines

The fine mixture of soil and climate in South Africa makes it an ideal place to cultivate vine. That is distributed throughout the country. Although South Africa has been producing fine wines since the 17th century, there has been little interest in its industry until recently. The white wines of South Africa have a very good reputation in the international wine industry, being light and crisp, and even fruity in flavor. The *Chenin blanc* and *Chardonnay* grapes especially seem to thrive in the South African climate. These grapes produce aromatic and spicy wines that complement many different foods and tastes. The white wines of South Africa have a better image than the reds. One of the most flavorful South African reds is Pinotage, made from grape developed from a cross between the *Pinot Noir* and the *Cinsault*, a grape from southern Rhone. It is exported to most parts of the world with great success. Shiraz (a light and fruity wine) is another wine enjoyed globally.

1.4.2.6 New Zealand Wines

The vineyards of New Zealand are grouped into ten main regions, mainly in the country's coastal areas. The maritime climate in these areas results in a long and slow ripening period yielding the best possible grapes. The North Island has about 240 wineries within the main regions, which include the Auckland, Gisborne, Hawkes Bay, and Martin borough areas. The South Island has about 160 wineries. However, it includes the heavily grape-planted Marlborough region, as well as Nelson and Canterbury, two smaller growing regions. Recently, Otago, a large region in the cool, southern part of the South Island, has been a booming area for new vineyards, with many among them planting the *Pinot Noir* grape variety. New Zealand is famous for world-class wine from white grape varieties, particularly *Sauvignon Blanc, Chardonnay, Muller-Thurgau*, and *Riesling*. Of these, the *Sauvignon Blanc* is the one most associated with the New Zealand wine industry. Though this famous grape is the main reason for New

Zealand's rise in standing, it would be wrong to ignore some of the other varieties. In fact, *Chardonnay*, not Sauvignon Blanc, is the most planted white grape variety in New Zealand at the moment, while more quality red grape types such as *Pinot Noir, Merlot*, and *Cabernet Sauvignon* grape varieties have also made great strides.

1.4.2.7 India

India is historically known for wines making, though very few wines are made at present. Most of them are made by western missionaries but recently both wine making and wine consumption is picking up. Consequently, production of sparkling wine, cider, and apple wine has been started in states like Mahrashtra and Himachal Pradesh.

1.4.2.8 China

China has been producing wines since 2140 BC when wine production was completely alien to the western world. However, these were originally made for medicinal purposes and not for social enjoyment. Currently, China is producing wines Chaosing Chefoo, Tsingtao, Great Wall, and Meikuishanputaochu, using western grape varieties such as *Pinot Noir* and *Merlot*.

1.4.2.9 Japan

The first Japanese wine was produced only for the medicinal purposes. Modern wine making in Japan began in the mid-19th century. At present, the major wines come from vineyards spread throughout Japan, but are found mostly on the main island of Honshu. Japan's best wines come from the slopes of Mt. Fujiyama on Honshu. Most of the Japanese wines are not well known in western markets, though Japan does produce a large amount of Sake for export, which is popular in the West.

1.5 BRANDY: ORIGIN, HISTORY, AND PRODUCTION

The stories of how brandy was first invented are many, but the true origin of brandy may never be known. Preparation of brandy is an ancient process that has therefore lost its origin. Not only this, but whether it was invented first by the Greeks, the Arabs, or by the Chinese is debatable. The derivations of the words 'alcohol' and 'alembic' from Arabic, indicate that it was from the Arabic word that the practice of distillation first entered Europe. Until the end of the 15th century, distilled wine (*aqua vitae*) seems to have been used largely as a medicine. The production of brandy within France, only emerged from the control of doctors and apothecaries when Louis XII granted the privilege of distilling to the guild of vinegar makers in 1514. Brandy production got further encouragement when Francis I enabled victuallers to distill it in 1537. Brandy is made in most of the countries that produce wine. 'Brandy' is derived from the Dutch term *brandewijn*, meaning 'burnt wine'. The term was known as *branntwein* in Germany, *brandevin* in France, and *brandywine* in England. Today, the word has been shortened to *brandy*. When brandy is produced, it undergoes four basic processes: fermentation of the grape, distillation to brandy, aging in oak barrels, and blending by the master blender.

1.5.1 French Brandies

France produces two world famous type of brandy namely Cognac and Armagnac. Cognac is the best-known type of French brandy in the world. The Cognac region is located on the west-central Atlantic coast of France and is further subdivided into six growing zones: Grande Champagne, Petite Champagne, Bois Ordinaries, Borderies, Fins Bois, and Bons Bois. The first two of these regions produce the best Cognac. The primary grapes used in making Cognac are Ugni Blanc, Folle Blanche, and Colombard. Cognacs are often a blend of brandies from different vintages, and even different growing zones. Armagnac is the oldest type of brandy in France, with documented references to distillation dating back to the early 15th century. The Armagnac region is located in the heart of the ancient province of Gascony in the southwest corner of France, that include Bas-Armagnac, Haut Armagnac, and Tenareze. The primary grapes used in making Armagnac are likewise the Ugni Blanc, Folle Blanche, and Colombard. Blended Armagnacs frequently have a greater percentage of older vintages in their mix than comparable Cognacs. Marc (Pomace brandy) from France is produced in all the nation's wine-producing regions, but is mostly consumed locally. *Marc de gewürztraminer* from Alsace is particularly noteworthy because it retains some of the distinctive perfumed nose and spicy character of the grape.

1.5.2 Spanish Brandies

Brandy de Jerez is made by the Sherry houses centered around the city of *Jerez de la Frontera* in the southwest corner of Spain. Virtually, all brandy de Jerez, however, is made from wines produced elsewhere in Spain, primarily from the Airen grape in *La Mancha* and *Extremadura*, as the local sherry grapes are too valuable to divert into brandy production. Nowadays, most of the distilling is done elsewhere in Spain using column stills which are then shipped to Jerez for aging in used sherry casks in a solera system similar to that used for sherry wine (see Chapters 2 and 38 for details on the *Solera system*). Basic Brandy de Jerez Solera must age for a minimum of six months, Reserva for a year, and Gran Reserva for a minimum of three years (sometimes for up to 12 to 15 years). Penedès brandy comes from the Penedès region of Catalonia in the northeast corner of Spain, near Barcelona. Modeled after the Cognacs of France and made from a mix of regional grapes and locally grown Ugni Blanc of Cognac, it is distilled in pot stills. One of the two local producers (Torres) ages in Soleras consisting of butts made from French Limousin oak, whereas the other (Mascaro) ages in the standard non-Solera manner, but also in Limousin oak. The resulting Brandy is heartier than Cognac, but leaner and drier than *Brandy de Jerez*.

1.5.3 Italian Brandies

Italy has a long history of brandy production dating back to at least the 16th century, but unlike Spain or France, there are no specific brandy-producing regions. Italian brandies are made from regional wine grapes, and most are produced in column stills and are aged in oak for a minimum of 1 to 2 years, with 6 to 8 years being the industry average. Italian brandies tend to be on the light and delicate side with a touch of residual sweetness. Italy also produces a substantial amount of grappa (grape pomace distillate) that can be un-aged, or aged for a few years in old casks.

1.5.4 German Brandies

German monks were distilling brandy by the 14th century and the German distillers had organized their own guild as early as 1588. Yet almost from the start, German brandy has been made from imported wine rather than the more valuable local varieties. Most German brandies are produced in pot stills and must be aged in oak for a minimum of six months. The best German brandies are smooth, somewhat lighter than Cognac, and finish with a touch of sweetness.

1.5.5 United States Brandies

Brandy production started in California in the late 18th and early 19th centuries by the Spanish missionaries. In the years following the Civil War, brandy became a major industry. At one time, Leland Stanford, founder of Stanford University, was the world's largest brandy producer. Soon after the end of the World War II, the industry commissioned the Department of Viticulture and Oenology at the University of California in Davis, to develop a prototype 'California-style' brandy. It had a clean palate, was lighter in style than most European brandies, and had a flavor profile that made it a good mixer. Starting in the late 1940s, the California brandy producers began to change over to this new style. Contemporary California brandies are made primarily in column stills from table grape varieties such as the Thompson Seedless and Flame Tokay, although a handful of small new-generation Cognac-inspired pot distillers, such as Jepson and RMS, are using the classic *Ugni Blanc, Colombard*, and *Folle Blanche* grapes. California Brandies are aged for 2 to 12 years in used American oak (both brandy and bourbon casks) to limit woodiness in the palate.

1.5.6 Latin American Brandies

South American brandies are generally confined to their domestic markets. The best known type is Pisco which is a clear raw brandy from Peru and Chile, made from Muscat grapes and double-distilled in pot stills, has a perfumed fragrance, and serves as the base for a variety of mixed drinks, including the famous Pisco Sour. Mexico produces a large amount of wine, most of which is used for brandy production. Mexican brandies are made from a mix of grapes, including *Thompson Seedless, Palomino*, and *Ugni Blanc*.

1.5.7 Brandies from around the World

Greece produces pot-distilled brandies, many of which are flavored with Muscat wine, anise, or other spices. In Israel, brandy production dates back only to the 1880s. In the Caucasus region, along the eastern shore of the Black Sea, the ancient nations of Georgia and Armenia draw on monastic traditions to produce rich, intensely flavored pot still brandies both from local and from imported grape varieties such as *Muscadine* (from France), *Sercial*, and *Verdelho* (most famously from Madeira). South Africa has produced brandies since the arrival of the first Dutch settlers in the 17th century, but these early spirits from the Cape Colony earned a reputation for being harsh firewater ('witblits' or 'white lightning' were typical nicknames).

1.5.8 Fruit Brandies

Normandy traditionally produces a substantial amount of hard and sweet cider that in turn is distilled into an apple brandy called Calvados. The local cider apples (small and tart, closer in type to crab apples) are used for this purpose. This spirit has its own appellations, with the best brands coming from Appellation Controlee Pays d'Auge near the Atlantic seaport of Deauville, and the rest in 10 adjacent regions that are designated Appellation Reglementee. All varieties of Calvados are aged in oak casks for a minimum of two years. In the United States, Applejack, as apple brandy is called locally, is thought by many to be the first spirit produced in the British colonies. Apple brandies that are more like *eau-de-vie* are produced in California and Oregon. The fruit-growing regions of the upper Rhine River are the prime *eau-de-vie* production areas of Europe. The Black Forest region of Bavaria in Germany, and Alsace in France, are known for their cherry brandies (Kir in France, Kirschwasser in Germany), raspberry brandies (Framboise and Himbeergeist), and Pear brandies (Poire). Similar *eaux-de-vie* are now being produced in the United States in California and Oregon. Some plum brandy is also made in these regions (Mirabelle from France is an example), but the best known type of plum brandy is Slivovitz, which is made throughout Eastern Europe and the Balkans from the small blue Sljiva plum. In the Goa region of India, a type of brandy called 'Fenni' is also made from Cashew apple.

1.6 WINE AS FOOD, MEDICINE AND HEALTH

Different epidemiological and experimental evidence has clearly supported the idea that that moderate consumption of wine is not only beneficial in decreasing atherosclerosis and prolonging the average life span, but is also associated with making life more pleasant and making one socially successful (Figure 1.8). These observations and research-based findings led to in-depth investigations of the cause of the beneficial effects of wine consumption. Accordingly, plant foods have attracted great interest as sources of antimicrobial substances such as polyphenols, terpenoids, alkaloids, lectins, polypeptides, and polyacetylenes. These secondary metabolites are known to be antimicrobial agents. Most of these metabolites are safe, showing negligible toxicity and side effects. Green tea, cranberry, and cocoa are considered as the most promising plant sources. Furthermore, many studies have shown that

FIGURE 1.8 Health benefits of wine. *Source:* Created using informations from Bordiga (2018)

grapes and related products (grape juice and wine) exerted antibacterial activity against a wide range of bacteria, due to the high content of polyphenols responsible for wine quality and its positive effects on human health.

The phenolic composition of red wine is strictly influenced by different factors, such as variety, location, climatic condition, winemaking process, and aging). Red wines generally contain up to 4,000 mg/L of phenolic substances, including non-flavonoids and flavonoids – the latter representing an important percentage of phenolic compounds, with amounts ranging from 1,000 to 2,000 mg/L. Flavonols, dihydroflavonols, anthocyanins, proanthocyanidins, and flavan-3-ols are the main classes of flavonoids in wine. Anthocyanins and proanthocyanidins (mainly stored in berry skin) are responsible for sensory properties such as color, astringency, bitterness, and aroma. These compounds are also important for the chemical stability of wine against oxidation. Tannins are able to form complexes with saliva proteins, thus conferring astringency and structure to the final beverage. Non-flavonoid compounds such as hydroxycinnamic and hydroxybenzoic acids together with phenolic acids commonly reach concentrations of up to 200 mg/L in red wines. Wine contains small quantities of the B-vitamins such as B_1 (thiamin), B_2 (riboflavin), and B_{12} (cyanocobalamin), but it is lacking in vitamins A, D, and K.

Consumption of wine with food has the advantage of slowing the rate of alcohol absorption into the blood and stimulates the production of gastric juices. Most of the positive effects of wine have been attributed to phenolic compounds, thus delaying the development of some forms of diabetes. The benefits of moderate wine consumption on increasing the plasma levels of high-density lipoprotein (HDL), which is helpful in lowering the blood cholesterol concentration and the incidence of arteriosclerosis, is well established. Several studies suggest that moderate intake of red wine exerts positive and protective effects on human health, strongly related to its polyphenolic composition. Wine is a particularly rich dietary source of flavonoid phenolics to inhibit platelet aggregation, eicosanoid synthesis, and oxidation of human low-density lipoproteins (LDL). The antioxidants in wine have recently been linked to the prevention of cancer and cardiovascular diseases. Polyphenols can inhibit some transcriptional factors and can modulate enzyme activity and metabolic pathways thus, exerting health benefits with particular regard to chronic degenerative diseases such as cardiovascular pathologies, cancer, and diabetes. This subject has been elaborated further in Chapter 5.

1.7 DISTRIBUTION OF THE WORLD'S GRAPEVINES

1.7.1 Distribution

Over the last 15 years, the world's vine stock has undergone considerable variations owing to the grubbing up of vineyards and restructuring activities. Indeed, some traditionally high-production varieties have seen a significant decline in their surface area, no longer corresponding to the tastes of consumers or the market. Recent changes in the world's vine surface area are primarily due to the grubbing up and restructuring of the European Union's vineyards, combined with the overall net growth of vineyards on other continents. The world's vineyard area has experienced a slight decline over the last decade (-3% since 2010). However, this downward trend is not uniform across the world. Some European countries, such as Spain, have recorded significant declines, reducing their vineyard area by more than 10% in 10 years. In other countries, such as China, the opposite has occurred, reporting an increase of 177% in its area under vines since 2000.

1.7.2 Diversity of Varieties

The Vitis genus, including 80 species identified, is composed of two sub-genera: *Muscadinia* and *Euvitis*. The Muscadinia sub-genus comprises three species, including *M. rotundifolia*. Grown in the southeast of North America, *M. rotundifolia* is considerably resistant to the main cryptogamic diseases to which most *Vitis vinifera* varieties are prone. Most cultivated grapevines belong to the Euvitis sub-genus. These fall under three groups: American, East Asia, and Eurasian. The American group is composed of more than 20 species, including V. *berlandieri*, V. *riparia*, and V. *rupestris*, and was commonly used as rootstock to address the phylloxera crisis. The East Asia group consists of around 55 species, and is currently considered to be of limited importance to viticulture. The Eurasian group is composed of one single species, *Vitis vinifera* L., which accounts for most of the world's *Vitis* varieties. There are 2 sub-species: *sylvestris*, the wild form of the vine, and *vinifera*, referring to the cultivated form. Over the years, the cultivation of grapevines has led to a significant increase in genetic diversity, thus contributing to the creation of new varieties. The hybridization of *Vitis* that took place from the end of 19th century until the mid-20th century to address the phylloxera issue resulted in the creation of direct producer hybrids and rootstock, ultimately contributing to an increase in the diversity of plant material. To date, with its several thousand varieties, the *Vitis* genus is characterized by high levels of genetic diversity. The Vitis international variety catalogue identifies more than 20,000 names of varieties (including about 12,000 for V. *vinifera*) but includes a considerable number of synonyms and homonyms. The number of vine varieties for the V. *vinifera* species in the world is estimated at about 6,000.

Table 1.2 reports the classification of most planted varieties.

1.8 WINE PRODUCTION

1.8.1 Fruits for Wine Making

Grape is the principal crop across the world to be processed into wine, and, unlike other crops, can be grown in diverse climates and soils. Similar to other fruit crops, however, grapes are subject to environmental stress. To make each type of wine, the specific variety of grape (or any other fruit), together with which wine yeast culture to be used, are some of the aspects which must be decided very carefully. For making wine, grape is the perfect fruit for fermentation. The color of wine is almost always derived from the skins and seeds. Wines of different styles and types are produced and consumed the world over and have been classified accordingly (see Chapter 2).

Development and sustainable growth of grapes and other fruits is the most essential component for growth of the wine industry across the world. Different cultivars of grapes, other fruit crops, and some of the cultivation conditions are described in Chapter 7. Selection of new varieties of grape and other fruits for wine making has been done in the past and will continue in the future also. Genetic engineering is expected to play a very significant role in developing various varieties, by incorporating into grapes those characteristics that could improve the quality of wine (see Chapter 8). To protect the environment, the industry would have to mobilize against pests and diseases of grape and other fruits used in wine production (See chapter 11). The industry has to adopt eco-friendly techniques, including biological control methods, to control the pests and diseases of fruit crops. Efforts to protect future crops have heightened the need for new disease-resistant grape cultivars. Because *V. vinifera* does not possess any particular disease resistance, new cultivars must incorporate resistance genes for a wide range of diseases in *V. vinifera*, as the production of wine is almost exclusively made from these cultivars.

1.8.2 Winemaking – the Process

In general, the nature and sequence of winemaking operations are the same for the various types and styles of wine. Preparation of red and white wines (Figure 1.9) includes the procurement of fruits, crushing or pulping the fruits, extraction of juice or on skin fermentation, racking, maturation, clarification, filtration, and bottling. Some operations may not be necessary for all the wines (sterile filtration of microbial stable dry wines, for example), and others are applicable only where a specific style is sought (carbonic maceration, *sur lies*, etc.), but there are general steps that are required for all the wines. The number of times a process such as centrifugation, fining, or filtration is applied affect costs and ultimately quality, owing to the potential for loss of wine volume, loss of volatiles, exposure to air, etc. The young wine needs to be clarified, processed, and distributed to the consumer. These have been described in detail in Chapter 24. Nevertheless, each time a new wine type is produced, the relevant regulation(s) should be consulted.

TABLE 1.2
The Most Planted Varieties. *Varieties With a Total Surface Area above 100,000 ha

Variety*	Color	Destination	Area (ha)	Main zones
Kyoho	Black	Table	365,000	Japan, South Korea
Cabernet Sauvignon	Black	Wine	341,000	widely distributed across the world
Sultanina	White	Table, drying, wine	273,000	Middle East and Central Asia
Merlot	Black	Wine	266,000	Present in 37 countries
Tempranillo	Black	Wine	231,000	Spain
Airen	White	Wine, Brandy	218,000	Spain
Chardonnay	White	Wine	210,000	Present in 41 countries
Syrah	Black	Wine	190,000	Present in 31 countries
Red Globe	Black	Table	159,000	China
Garnacha Tinta	Black	Wine	153,000	France and Spain
Sauvignon Blanc	White	Wine	123,000	Major wine-producing countries
Pinot Noir	Black	Wine	112,000	Major wine-producing countries
Trebbiano Toscano	White	Wine, Brandy	111,000	Italy, France and Portugal

Source: OIV, International Organisation of Vine and Wine. (2018). OIV Statistical Report on World Vitiviniculture

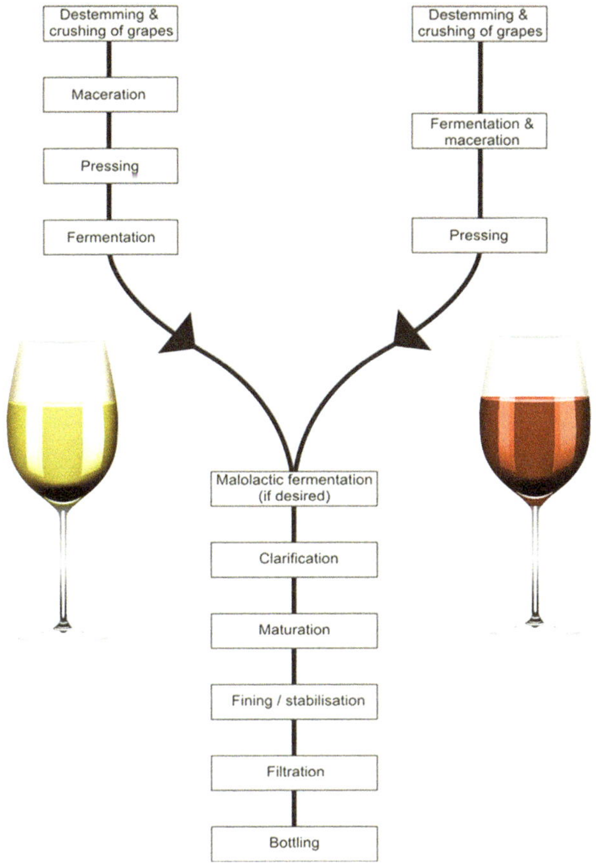

FIGURE 1.9 Diagrammatic representation of white and red wine production. *Source:* Created using informations from Bordiga (2018)

There are several other wines, produced from grape, that are consumed. These include dessert, fortified, specialty products such as sparkling wines, flavored wines, and coolers, and are known to be preferred by particular groups of consumers (See chapters 32, 33, 34, and 35). Using a wide range of grape varieties, fortified wines derive their distinctive attributes from the specific and unique winemaking and aging processes employed in their production. Low or non-alcoholic wines are also catching the fascination of consumers (see Chapter 36). Different types of wines such as vermouth, sparkling wine, sherry-type wines, or brandies from non-grape fruits (see Chapter 37) are also prepared and consumed. The detailed process of the making of different styles of wine has been covered in a separate unit about wine making technology. More details can be found in various chapters of this book and in the literature in the bibliography. Another important grape product is brandy (see Chapter 38) which is made by distillation of wine, while other fruits are also converted into corresponding fruit brandies. See chapter 44 for more innovations in wine.

1.8.2.1 Extraction of Juice and Must Preparation

Grapes of specific varieties and of proper maturity (See chapter 9 and 10 of this text) are used to make must and the fermented product, wine, as described earlier. How these are crushed or how the juice is extracted has been described, emphasizing the type of juice extractor, filter press, centrifuging units, etc. In preparation of must, what additives are added, and in which concentration these should be used is also decided. In some cases, amelioration is done to make the juice or pulp suitable for wine making. In later chapters of the book, specific operations involved in making specific wines are discussed (see Chapter 37 for details). Although focusing on wines made from grapes, the production of wines from non-grape fruits has not been ignored. These fruits require amelioration (for example acid dilution, sugar addition), and all these aspects have been described in depth (see Chapter 37).The only additive needed by the winemaker in wine making is sulphur dioxide (SO_2), which helps to control wild yeast and bacteria during the fermentation process. These sulfites have been used by vintners for centuries. In a well-made wine, there may be as little as 100–150 parts per million (ppm) (see Chapter 14 for details of this aspect).

1.8.3 Microorganisms and Wine Fermentation

Significant microorganisms in the process of wine making are yeasts, lactic bacteria and acetic acid bacteria, which are involved in one way or another in normal transformation of juice into wine or can cause spoilage, so their growth must be controlled. A separate chapter (See chapter 41) that focuses on the spoilage of wine by microorganisms can be found in this book. In the future, these techniques are likely to be used more frequently, such as indirect profiling of yeast dynamics in wine fermentation. As stated earlier, the microorganisms in the production of wine play a pivotal role in alcoholic fermentation (see Chapters 12, 13, 14, 15, and 16) *i.e.* they start alcoholic fermentation and produce alcohol, and some may inhibit fermentation as killer yeast (see Chapter 20). Others transform malolactic acid into lactic acid, in order to lower the acidity and make the wine palatable, otherwise the wine could become spoiled. In the preparation of wine and brandy, wine microbiology (see Chapter 14) plays a central role. Different microorganisms involved in the process are yeasts, lactic bacteria, and acetic acid bacteria, which cause normal transformation of juice into wine or cause spoilage (Chapter 14).

The entire gamut of wine fermentation is dependent upon the wine yeast known as *Saccharomyces cerevisiae.* Yeasts are natural in the skin of grapes. But very often a wine maker adds a special yeast to ensure successful fermentation. There are some very famous yeast strains used for wine fermentation, such as UCD 522, UCD 505, and UCD 519, for example, developed by the University of California, Davis. Other institutes, similarly, have developed yeast strains for wine fermentation. A separate chapter (Chapter 26) is devoted to the culture of yeast and bacteria for wine making, while another chapter (Chapter 18) describes the application of malolactic acid bacteria in wine fermentation. Further information as to the microbiology of wine making has been described in a separate section of this book (Chapter 14).

*Saccharomyces cerevisiae var ellipsoideus*is – the well-known wine yeast is involved in the production of all alcoholic beverages. Yeasts are known to have an impact on the formation and the liberation of aroma compounds during alcoholic fermentation, thus, various yeast strains could play a role in the quantitative and qualitative production of these aromas (See chapter 4). Presently, various strains of yeast are commercially available and promise aroma enhancement.

It is generally accepted that the wealth of yeast biodiversity has a lot of hidden potential for oenology, which unfortunately has yet remained largely untapped. The yeasts indigenous to grapes, such as *Kloeckera apiculata* and various *Candida* spp., can readily initiate fermentation, but only a few strains of non-*Saccharomyces* can complete alcoholic fermentation as these are sensitive to high levels of ethyl alcohol. These include *Candida, Kloeckera, Hanseniaspora, Zygosaccharomyces, Schizosaccharomyces, Torulaspora, Brettanomyces, Saccharomycodes, Pichia*, and *Williopsis* genera (see Chapter 15). Not only this, these yeasts often produce undesirable secondary metabolites (acetic acid, ethyl acetate, ethyl phenols, aldehydes, and acetoin). Nowadays, however their presence in wines is marginal since they are sensitive to SO_2 (universally used in current oenological procedures). Recently, the positive influence of non-*Saccharomyces* yeast(s) on wine quality has been documented. In the use of non-*Saccharomyces*, a better option is to use mixed starter cultures. Using modern technology, novel wine-based beverages can be produced using mixed yeast starter cultures which are tailored to reflect the characteristics of a given wine region, along with the use of indigenous yeast species.

This aspect is likely to provide a greater possibility to make wine with low alcohol, to produce with wine improved flavor, or similar characteristics. To conduct fermentation, the choice of a fermentation container (or precisely the fermenter) is essential. Conventionally, the wine fermenter could be as simple as a wooden or stainless steel container, or a temperature-controlled vessel. It can be conical, cylindrical, or similar. Construction and operation of bioreactors for fermentation of wine (see Chapter 27) is also likely to receive priority.

The future could see the use of biological de-acidification using malolactic bacteria or de-acidifying yeasts like *Schizosaccharomycespombe*, and could also provide solutions to several other problems like stuck and sluggish fermentation, and killer yeast phenomenon. Using selective yeast, immobilized yeast, or malolactic acid bacteria, and modeling wine fermentation (see Chapters 14, 15, 18, and 23) are comparatively new aspects and are expected to be a promising area of future research.

1.8.4 Wine Clarification

After the completion of fermentation, the sediment settles down at the bottom of a fermenter. It consists of lees (the yeasts) and the insoluble material of the wine (mainly the components of the must and wine). The separation of the sediment from the fermented liquid is the most essential operation and is normally carried out by siphoning or racking. The sediment is removed. The operation is repeated several times (4–5 times at least) over a period of 2–3 months, depending upon the clarity of the wine. If needed, the wine is clarified using a centrifuge or filter press. This simple operation of siphoning usually makes the wine clear with acceptable flavor.

1.8.5 Wine Maturation and Aging

Wine maturation and aging are very important operations. As one of the most important practices, the wine is matured with cooperage wherein the sensory quality of wine is immensely improved. Wine aging usually refers to maturation in a bottle, where many chemical changes occur. Most wines can safely be stored for six months or more at home. It should be borne in mind that wine is a living (or rather dying) thing and has some formidable enemies such as air, direct sunlight, vibration, and temperature variation. For long-term storage of wines and spirits, the perfect place to keep them in is a cool, dark location, free from drafts and vibrations. All these aspects have been discussed in Chapter 28.

1.8.6 Waste from a Winery

Like other food industries, the waste from wine and brandy is highly polluting, but rich in several useful and recoverable constituents. Besides minimizing the waste, the strategy should be for maximum use of the waste (see Chapter 31) for recovery of value added compounds still present in these by-products. If is treated only to meet the requirements of pollution control authorities, it becomes a colossal loss of precious natural resources. As mentioned earlier, various products from the waste of wineries can be made, which, besides reducing the pollution load, can also improve the economy of a winery. How waste products from the wine industry can be utilized has been well illustrated in Chapter 31 of this MSS. It can be seen that the waste from a winery is a good source of tartaric acid, anthocyanin, polyphenolic compounds, and fibers, etc. The winery waste can also be fermented into ethanol, acetic acid, etc.

1.9 EVALUATION OF WINE COMPOSITION AND QUALITY

1.9.1 Wine Composition

Wine is a complex symphony of natural chemicals, minerals, and vitamins revealed by analysis of its composition. It has a plethora of components like ethyl alcohol, sugars, esters, tannins, glycerin, and several other congeners (See Chapters 3, 17). A few come from the grapes (or from other fruits), others from the fermentation or aging process. Sugar is natural to the grape and is the component which is converted into alcohol in the fermentation process. The sweetness, however, must be balanced by good acidity. Wine contains many alcohols, the most common is ethyl alcohol, which adds hotness or even sweetness to the body of the wine. If alcohol is not balanced it gives a burning taste to the wine. Glycerin, another by-product of alcoholic fermentation (See Chapter 17), is a tri-alcoholic product that adds richness and texture to the wine. Acid is the skeleton of a wine. Wine is comprised of several other components which are volatile, and which impart a specific smell or aroma (See Chapter 4), while other components like tartaric acid or malic acid are not volatile but have specific taste. An acid with volatile acidity (V.A.) is acetic acid, which can be found in vinegar. The pungent fragrance and prickly taste of wine are clear signs that the wine has been exposed to oxygen, either during the fermentation or aging process, or due to defective corking, resulting in the production of acetic acid. Methyl Anthranilate is the constituent that adds the 'foxy' character to native American *Labrusca* varietals, like *Concord*. Detailed information about the composition of wine has been given separately in Chapter 3. For further details of the various components which impart aroma to the wine based on grape varieties, see Chapter 10. A good wine has a balanced composition that makes it acceptable to consumers.

1.9.2 Quality of Wine

Similar to other food products, the quality of wine is of utmost importance. There are five factors on which it is dependent, viz., the vine, the soil, climate, geographic location, and the human factor. When these factors are in perfect harmony, each supporting and complementing each other, wines of great quality and popularity can be created. The quality of the grapes ultimately determines the wine quality, including stability and sensorial characteristics, especially flavor. In addition, the technology of wine making has a tremendous effect on the quality of wine. New and better performing techniques make the production of wine and brandy of superior quality possible (see Chapters 24 to 38) and this aspect would continue to receive top priority in future too. The wine maker should therefore, consider these factors.

1.9.3 Evaluation of Wine

Wines are evaluated for physical, chemical, microbiological, and sensory characteristics. Chemical analysis of wine constitutes a significant component of quality measures, especially those prescribed by the specific laws of nutritional significance, in particular that they are free from toxic compounds (see Chapter 6 for details). New techniques have emerged to monitor and ensure the quality of wine, including the use of molecular biology techniques like PCR (Chapters 39 and 41). Before recently, the available methods used for analysis only allowed the determination of metabolites that were present in high concentration, like sugars, amino acids, and some important carboxylic acids. Because of this, the roles of many significant but less concentrated metabolites during the winemaking process remained unknown. A large number of metabolites can now be quantified easily, due to the development of analytical instruments with high resolution and exceptional sensitivity. Metabolomics, combined with other 'omics' approaches, is becoming extremely useful, and thus, can be applied in all areas of enology and viticulture, mainly due to its capability of analyzing more than 1000 metabolites in a single run with its high resolution and sensitive analytical instruments.

It is known that the sensory characteristics depend on the content and composition of several different groups of compounds from grapes. One of these groups of compounds are sugars, and consequently, the alcohol content quantified in wines after alcoholic fermentation. Over the past few decades, advancement in the field of sensory science has enabled us to understand the variables that influence and contribute to the sensory perception of alcoholic beverages like wines. Recent efforts have been directed to linking chemical and sensory measurements of flavor, *i.e.* a trained human subject sniffs the effluent from a gas chromatogram (see Chapter 40). Gas chromatography–olfactometry can link the detection and quantification of odorants to their sensory impact in wine. The subset of compounds upon recombination that closely mimic the properties of the original aroma of the mixture can be chosen from such chromatograms. To characterize wine flavor quantitatively, the application of descriptive analysis (DA) has led to the improvement of wine quality.

A consumer's preference for certain foods and beverages is determined by mouth-feel and texture. Viscosity, density, and surface tension are the essential rheological properties which affect the mouth-feel of liquid food products, such as wine. Not only this, they also modify other sensory properties like saltiness, sweetness, bitterness, flavor, and astringency. It is very important to understand how and where the interactions are generated, as these have impacts on the flavor perception and the key sensory profile of food products. There are physical interactions between the components in the food or beverage matrix, which influence the volatiles release and/or viscosity, and multi-modal interactions resulting from the cognitive or psychological integration of the anatomically independent sensory systems.

Various scientific advances have been made recently, such as identification of receptors and other important molecules involved in the transduction mechanisms of flavor. Alcohol is the most abundant volatile compound in wine that can modify both the sensory perception of aromatic attributes and the detection of volatile compounds. Therefore, alcohol is not only important for wine sensory sensations but also for their interaction with other wine components, such as aromas and tannins, which also influence the wine viscosity and body and our perceptions of astringency, sourness, sweetness, aroma, and flavor. Alcohol activates olfactory, taste, and chemesthetic receptors, and each modality is carried centrally by different nerves; thus, these inputs affect the perception evoked by alcohol.

The oral consumption of alcohol by humans is accompanied by chemosensory perception of flavor, which plays an important role in its acceptance or rejection. Three independent sensory systems, taste, olfaction, and chemosensory irritation, are involved in the perception of flavor in food. In wine, humans perceive alcohol as a combination of sweet and bitter tastes and odors, and oral irritation (burning sensation).

Many factors underline the role that alcohol plays in the development of flavor preference and consumption pattern, such as: the activation of peripheral chemo-receptors by central mechanisms that mediate the hedonic responses to alcohol flavor; learned associations of alcohol's sensory attributes and its post-digestive effects; early postnatal exposure to alcoholic flavor; and genetically determined individual variation in chemo-sensation. The role of chemosensory factors in alcohol intake and preferences is therefore of great interest, and has been witnessed in research during the past decade. Consequently, significant scientific advances have been made, such as the identification of receptors and other key molecules involved in the transduction mechanisms of olfaction, chemosensory irritation, and taste. The enologist must strive to achieve this excellence in the wine.

1.9.4 Sensory Quality of Wine and Consumerisms

The wine industry caters to the specific needs and financial means of consumers by making wines with different prices for which consumers will have certain expectations of product quality. Several factors influence the adoption and diffusion of new products like wine into the marketplace; thus, placing wine on the food menu would have a significant impact on wine sales. Out of several factors, the sensory quality of wine is of great significance, as discussed earlier. The actual sensory characteristics desired in wine may differ, but the pleasing sensory experience is always expected to be the same. It has been established that consumers are always willing to pay more for a product with a satisfying sensory experience. Consumers' preferences are now shifting towards wines with low alcohol.

Consumers are also aware that wines with a high alcohol content can cause a gustatory disequilibrium affecting wine sensory perceptions, resulting in an unbalanced wine. But at the same time, consumption of alcoholic beverages is accompanied by chemosensory perception of flavor – an important factor for their acceptance or rejection. Thus, the main factors for selection are maintaining quality, in terms of flavor, in the final wine, and with the lowest cost. The preference for wines with lower alcohol levels has led to technological innovations in low alcohol content without changing the wine's sensorial profile. The alcohol content of wine and consequently, the study of the role of chemosensory factors in alcohol intake and consumer preferences, continues to be a challenge in oenology.

Wines not only have to be healthy, but produced in an environmentally sustainable manner and likely to become a stimulator of profitability. The 'French Paradox' has enhanced consumer interest in the role of diet in health and health promotion, and it would be wise to exploit this phenomenon with the aim of designing a better, yet equally attractive, human diet. Wine is likely to be a component of such a diet. The consumption of wine, therefore, would be influenced by other variables too, such as the therapeutic properties associated with the consumption of wine in moderation (see Chapter 5).

At the same time, it must be borne in mind that the excessive consumption of alcohol is also associated with several problems including cirrhosis of the liver, road accidents, some cancers, social problems, etc. (Chapter 6). Besides this, some toxic components (methanol, higher alcohols, carbamates, aflatoxins, etc.) have been detected in wine, and a few of these are associated with production of grapes as well as some wine production practices. The levels of these in wine and spirits, as well as how these can be eliminated, has also been discussed (Chapter 6).

1.10 SETTING UP A WINERY

1.10.1 Requirements and Regulations

Production of wine at a commercial level requires the setup of a factory, or more precisely, a winery. For this purpose, if any entrepreneur wanted to set up a winery, they would have to meet certain guidelines and regulations set by the respective government. But it is outside the scope of this chapter to discuss these aspects in detail. Prior to creating a winery, it needs to be decided who the proprietor will be and what type of company it will be. The amount of money to be spent should also be decided and the risks involved anticipated. Similarly, the projected start-up or schedule time, operating costs involved, and when the sales would begin, how much profit there might be and when a return on investment is likely to occur, should be considered.

Planning a winery must take into account all the relevant factors. Sparkling wine facilities have a unique position in a winery setup, and its production by *methode champenoise* requires that the fermentation and clarification of the base wine occur rapidly, so that the bottling and *tirage* step may proceed as quickly as possible. The Charmat process (bulk sparkling wine production) is very similar to table wine production in terms of process steps, except retention of CO_2. In the bottle fermentation process, the long (2-year) bottle fermentations are practised, as is the disgorging step, which in *methode champenoise* sparkling wine facilities can be a significant source of alcohol, yeast lees, and metabolites in the waste stream. The required utility services include a number of elements such as electrical power, refrigeration, ventilation and air conditioning, telecommunications, sanitation, steam and hot water, potable water supply, irrigation, fire protection and solid waste management systems, etc.

The wine maker has to focus on the availability of raw materials at comparative and affordable cost to make a wine. Acquiring grapes of suitable variety is a prerequisite for making wine of acceptable quality. Further considerations would be how much functional and efficient the winery will be, with regard to its layout, site, maintenance, expansion options, waste disposal, etc., and its architecture. Successful winemakers need to undertake initially modest-sized wine production, preferably with some of their own vineyards for initial ease of operation. The winemaker must also consider the right type of the machinery, and the necessary staff needed to operate the winery (see Chapter 24).

The regulations to set up a winery vary from country to country. This is however, beyond the scope of this text, except for a few general considerations. Nevertheless, it is essential to know such regulations with respect to the country or state, and the local licenses that have to be obtained. Prior to setting up a winery, rules regarding environmental impact have to be complied with. The wine maker also ought to have sound knowledge of various general and special taxes that apply to wine, especially if any change in wine type and style could be possible in the future.

Regarding marketing, consideration should be given as to what kinds of wines be created for selling (generic, varietal, white, red, etc.), and what price category should be targeted (super premium, house blend, standard), especially when good wines are to be made. The most important aspect of marketing is the marketplace where the wines will be offered (local, state, national, export, specialty, varieties, catalog, etc.), and the distribution and method of the sale of the wines. Finally, knowledge of how public relations should be handled and the most effective publicity required is important for the marketing of the wines. A bonded cellar is another winery sub-category, that by federal permit is prohibited from crushing or fermenting grapes but is allowed to receive bulk wines in bond for blending, aging, and bottling for eventual sale. Bonded cellars often fall into the urban location category because they have no marketing or other pressure to be 'estate', or to even be near the vineyards.

1.11 SOME HAZARDS SPECIFIC TO WINEMAKING

It is not the intention to describe in detail all hazards that threaten winemakers or their employees. But a few important risks will be discussed here, as many of the risks common to any production industry (rather than everyday life risks) are the same for the wine industry. At present, with proper care and attention, wineries are relatively safe and pleasant places to work. However, one risk is carbon dioxide poisoning, probably the greatest risk in a winery. To prevent this, the proper procedure is to use fans to actively ventilate containers and rooms having, or potentially having, appreciable CO_2 levels, until they are shown to be safe before entering. The use of CO_2 sensors and constant attention have also drastically reduced such accidents. Another hazard is that of fire, if alcohol vapor or other combustible vapor is a diluent in air. With the introduction of electronic meters reading oxygen content in the air, asphyxiation deaths have nearly disappeared from the wine industry (for more details, see the literature cited).

1.12 CLIMATIC CHANGE, ENVIRONMENTAL ISSUES, AND WINE PRODUCTION

Change in climate is threatening our ability to ensure global food security and achieve sustainable development. The change in climate has both direct and indirect effects on agricultural productivity and the factors effecting it. The effect that climate change has on oenology has been reviewed, and using food microbiology to design climate-smart food systems has been proposed to mitigate its effect. It has been established that the climate effects enology in several ways, such as: increased undesired microbial proliferation; increased sugars, and consequently, increased ethanol content; reduced acidity and increased pH; imbalanced perceived sensory properties (*e.g.* color, flavor); and intensified safety issues (*e.g.* mycotoxins, biogenic amines). The potential microbial-based strategies are proposed to be suitable to cope with the five challenges listed above. In designing a smart system, the focus proposed was on a recognized positive role in oenological processes of microorganisms, namely *Saccharomyces* spp. (*e.g.*, *Saccharomyces cerevisiae*), non-*Saccharomyces* yeasts (*e.g. Metschnikowia pulcherrima, Torulaspora delbrueckii, Lachancea thermotolerans*, and *Starmerella bacillaris*), and malolactic bacteria (*e.g.*, *Oenococcus oeni, Lactobacillus plantarum*). The potential of microbial activities as mitigating strategies in the wine sector has enhanced our interest in the continuous exploration of solutions associated with microbial diversity, both in the vine and wine (see the literature cited).

1.13 WINE PRODUCTION, CONSUMPTION, MARKETING, AND INTERNATIONAL TRADE

1.13.1 Area under Vineyards and Wine Production

According to OIV (the International Organization of Vine and Wine), the total world area under vines in 2017 (corresponding to the total surface area planted with vines, including that not yet in production or not yet harvested) is estimated to be almost equivalent to that of 2016 (–22,000 ha), reaching 7.6 mha. With 7.6 mha in 2017, the size of the global area under vines appears to have stabilized (OIV, 2018) (Figure 1.10).

The areas under vines in European vineyards. Since the end of the European Union program (2011/2012 harvest) to regulate viticultural production potential in the EU, the rate of decline of EU vineyards has significantly slowed. It is estimated that EU vineyards cover 3.3 mha, a reduction of 5.6 kha compared with 2016. The implementation of the new system regarding the management of viticultural production potential, including the possibility of annual growth being limited to 1% of the planted vineyards per member state, and the management methods for old rights held in portfolios, have led – within this context of regulatory transition – to contrasting developments for the area under vines in different EU countries. The latest available data show a trend towards the stabilization of the overall area under vines in France (787 kha), Romania (191 kha), and Germany (102 kha), and more recently in Greece. In contrast, vineyards in Spain (967 kha) are estimated to have decreased by about 8 kha between 2016 and 2017, while those in Italy (695 kha) are estimated to have grown by 5 kha (OIV, 2018) (Figure 1.11).

Outside Europe. Vineyards outside Europe appear to have remained stable between 2016 and 2017, reaching an estimated 3.6 mha. This apparent stability is the result of contrasting developments. In Asia, the expansion of Chinese vineyards (870 kha) slowed after 10 years of strong growth, while Turkey (448 kha) saw its vineyards decline at a sustained pace, falling by 19.7 kha between 2016 and 2017 (Figure 1.12). North and South America did not record any significant variations in the size of their vineyard surface area between 2016 and 2017. The same was almost true in Oceania: the recent downturn in Australian vineyards (145 kha) seems to have slowed, while New Zealand's vineyards have remained more or less stable

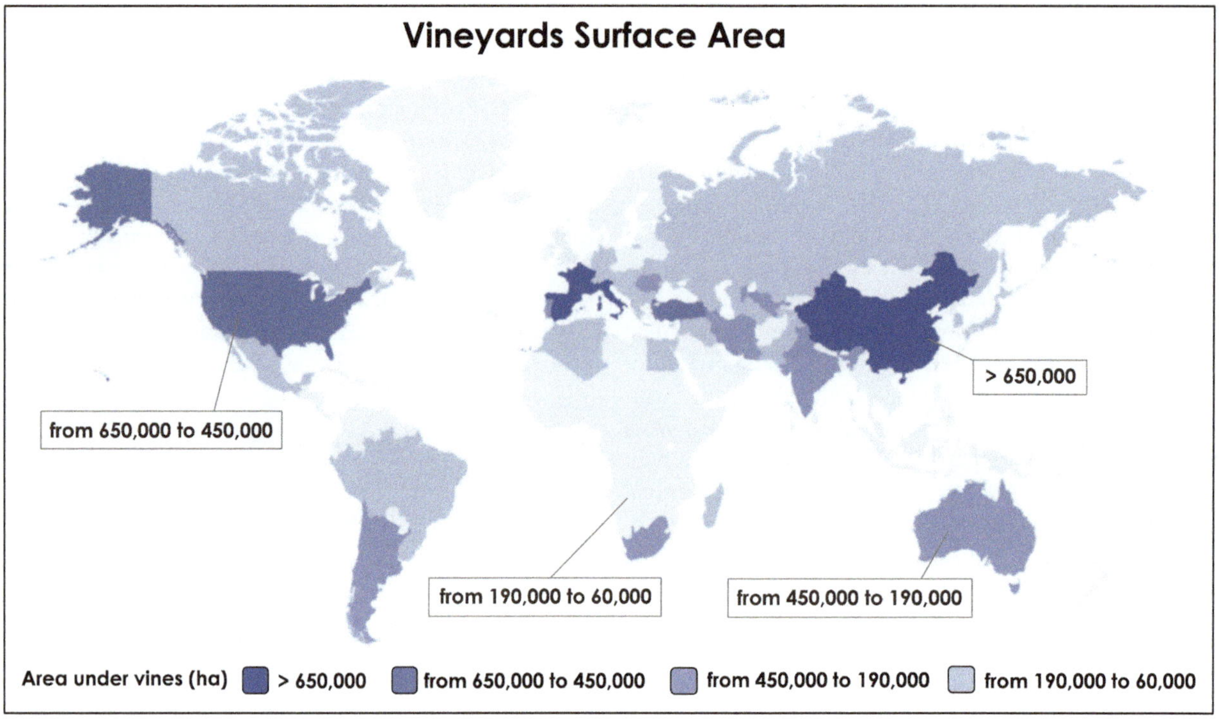

FIGURE 1.10 Vineyards surface area. *Source:* Created using informations from the OIV (2018)

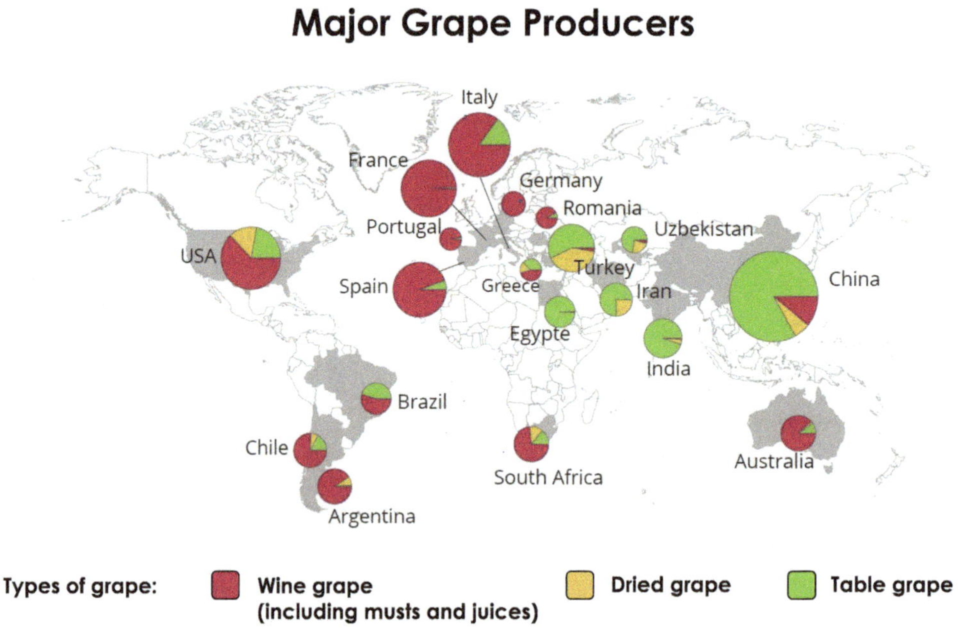

FIGURE 1.11 Major grape producers. *Source:* Created using informations from the OIV (2018)

at around 40 kha. Finally, South African vineyards (125 kha) have been slowly declining since 2012.

Wine production (resulting from grapes harvested in autumn 2017 in the northern hemisphere and in spring of the same year in the southern hemisphere). In 2017, global wine production (excluding juice and musts) fell to 250 mhl, a decline of 23.6 mhl compared with 2016 production. This production volume can be described as historically low. However, 2018 world production of wine (preliminary estimate) could reach the value of 279 mhl (Figure 1.10). With a 13% increase compared to 2017, wine production (excluding juice & musts) in 2018 is estimated to be one of the highest since 2000 (OIV, 2018) (Figure 1.11).

Within the European Union. EU vinified production (2017) is estimated at 141 mhl, a 14.6% drop compared with 2016. This situation is the result of adverse weather conditions in the main wine-producing countries in Europe. This production figure is 4.5% lower still than the very low

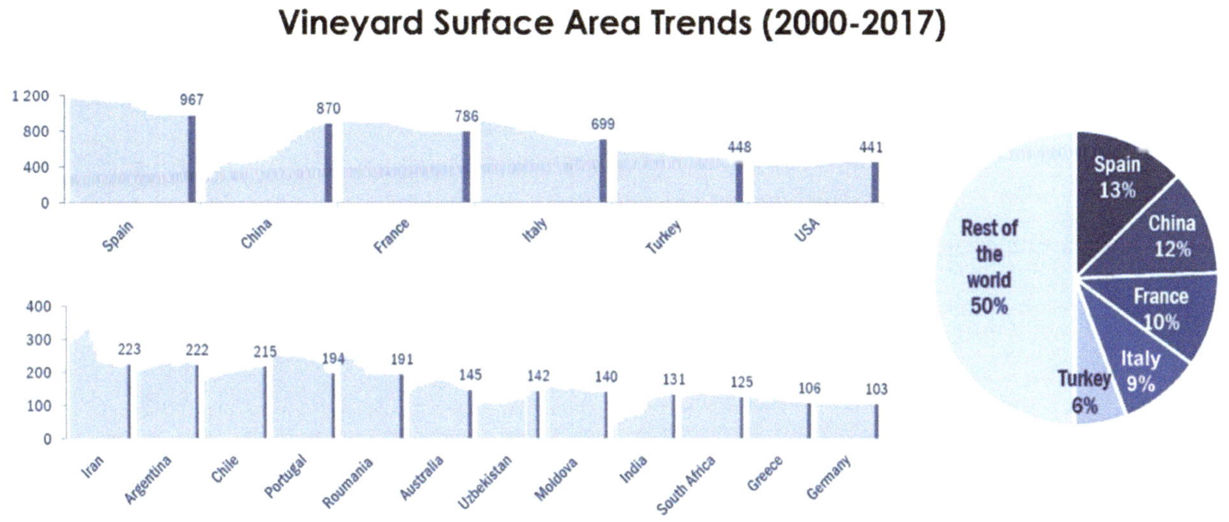

FIGURE 1.12 Vineyard surface area trends (2000–2017) within Europe. *Source:* Created using informations from the OIV (2018)

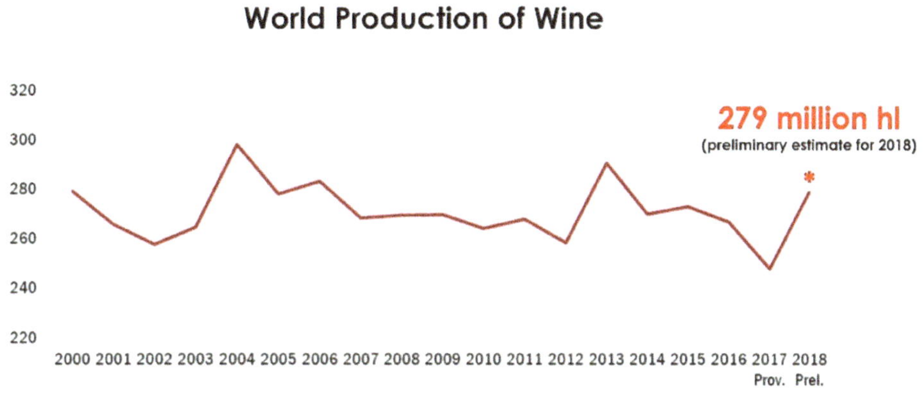

FIGURE 1.13 World production of wine. *Source:* Created using informations from the OIV (2018)

volumes produced in 2012 (147 mhl). When compared with 2016, production in Italy (42.5 mhl), France (36.7 mhl), Spain (32.1 mhl), and Germany (7.7 mhl) declined by 17%, 19%, 20%, and 15% respectively. Moderate production in 2017 in Portugal, Romania, and Austria was still an increase relative to the modest levels of 2016. However, looking at the 2018 forecast, the data show a marked increase in production (high production levels in major producing countries in Europe and the Americas) (Figure 1.13). Conversely, Oceania and South Africa show average production levels.

Outside the European Union. Estimated at 23.3 mhl excluding juice and musts, 2017 wine production in the United States remained very high (almost as high as that of 2016, estimated at 23.6 mhl and of 2013 at 24.4 mhl). After the significant impact of El Niño on 2016 production, wine production in South America has evolved differently in different countries. Wine production in Argentina, at 11.8 mhl, grew in relation to the low levels of 2016, but failed to return to production levels generally achieved at the start of the decade. Following the catastrophic 2016 production, in 2017 it was more than a return to normal for Brazilian production (3.4 mhl), which rose to the same level as for the productive harvest of 2011. However, 2017 production in Chile declined once more after the low levels of 2016, reaching only 9.5 mhl. South African production excluding juice and musts reached 10.8 mhl in 2017. At 13.7 mhl, Australian wine production continued to grow in volume to return to levels from around 2005, in the context of a vineyard area that remains virtually unchanged. New Zealand's production reached 2.9 mhl, close to the 2012–2016 five-year average (2.6 mhl).

1.13.2 Wine Consumption

World wine consumption in 2017 is estimated at 2,436 mhl, an increase of 1.8 mhl compared with 2016. The United States, with consumption estimated at 32.6 mhl, confirmed its position as the top global consumer since 2011, and saw domestic demand grow compared with the previous year (+2.9%) (OIV, 2018). There was a break in the declining consumption of traditional producer and consumer countries in Europe (Figure 1.14). There was a very moderate decrease in France to 27 mhl, and increases in Italy to 22.6 mhl, in Spain to 10.3 mhl, and in Germany to 20.2 mhl. In 2017, the United Kingdom returned to its 2015 level of 12.7 mhl, after a slight

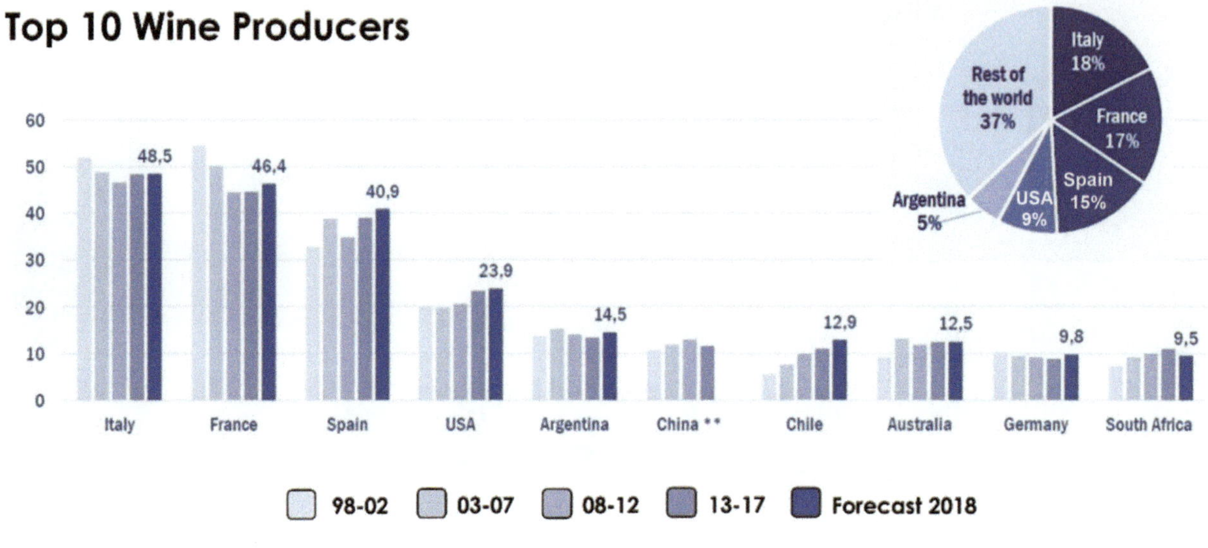

FIGURE 1.14 Top 10 wine producers. *Source:* Created using informations from the OIV (2018)

increase in consumption in 2016. With regard to China, consumption shows a positive variation of 3.5% compared with 2016. In Oceania, overall consumption in the Australian and New Zealand markets stabilized in 2017. South African consumption rose again between 2016 and 2017 to reach 4.5 mhl. In South America, domestic consumption was lower than in 2016, especially in Argentina (–5%) and in Chile (–10%).

1.13.3 International Trade

In 2017, the global market – considered here as the total exports of all countries – is estimated at 107.9 mhl in terms of volume (+3.4%), and 30.4 billion EUR in terms of value (+4.8%). In terms of volume, Spain remained the biggest exporter with 22.1 mhl and a global market share of 20.5%. Exports from New Zealand, Chile, Portugal, France, Italy, and South Africa all increased by more than 3% in relation to 2016. In terms of value, France was the biggest world exporter in terms of value, with 9.0 billion EUR of exports in 2017 (OIV, 2018). It has been recorded a rise of +4.8%, with substantial increases in Australia, France, Spain, Italy, Portugal, and New Zealand. The most significant decreases relate to the United States, Argentina and South Africa. In terms of volume, the share of bottled still wine sales is estimated to have increased from 54% to 57% between 2016 and 2017. In 2017, the bottled (< 2L) export share by volume was very high in Germany, Portugal, Argentina, and France. In terms of export value, bottled wine represented 72% of the total value of wine exported in 2017. Sparkling wines (8.6 mhl exported in 2017) once again saw the biggest growth, both in terms of volume and total value (+11.2% and +8.9%, respectively). In terms of volume, a significant share of Italy and France's wine exports relate to sparkling wine (18% and 13% respectively). Sparkling wine exports are also on the increase in Spain and South Africa. By value, sparkling wines account for 19% of the global market (although they only represent 8% of the total volume exported). The volume of bulk wine exports significantly decreased in 2017 compared with 2016. The 2017 bulk export share in terms of volume continues to be significant in Spain, South Africa, Chile, Australia, and the United States. By volume, bulk exports fell sharply in Germany, Argentina, and Portugal, but rose in New Zealand. In 2017, bulk wines accounted for 8% of the total value of wine exports but represented 35% of the global market in terms of volume.

1.13.4 Main Exporters

An analysis by country shows that the wine trade was largely dominated by Spain, Italy, and France, which together accounted for 54.6% (58.9 mhl) of global market volumes in 2017, and 58.2% (17.7 billion EUR) of exports in terms of value (OIV, 2018) (Figure 1.15). Chile and New Zealand recorded relatively significant increases, as did Australia between 2016 and 2017. Argentina and the United States recorded the most significant decreases in relative terms (respectively -14.0% and -13.5%), followed by Spain (–9.7%). In terms of value, Italy and France continued to dominate the market with shares of 29.6% and 19.3%, respectively. Despite the high volume of Spanish exports, the significant share of bulk wines (55% in 2017 in terms of volume but 20% in terms of overall value) resulted in a lower overall weighted average price for its total exports.

1.13.5 Main Importers

The five main importing countries – Germany, the United Kingdom, the United States, France, and China – which typically represent more than half of all imports, imported a total of 55.3 mhl at a value of 14.4 billion EUR in 2017 (OIV, 2018) (Figure 1.16). The top importer by volume in 2017 is still Germany, which recorded a slight decrease in its imports (–0.1%). The United Kingdom remains the second biggest global importer by volume, with 13.2 mhl, and by value, with 3.5 billion EUR (–1.3%). In 2017, the United States reported

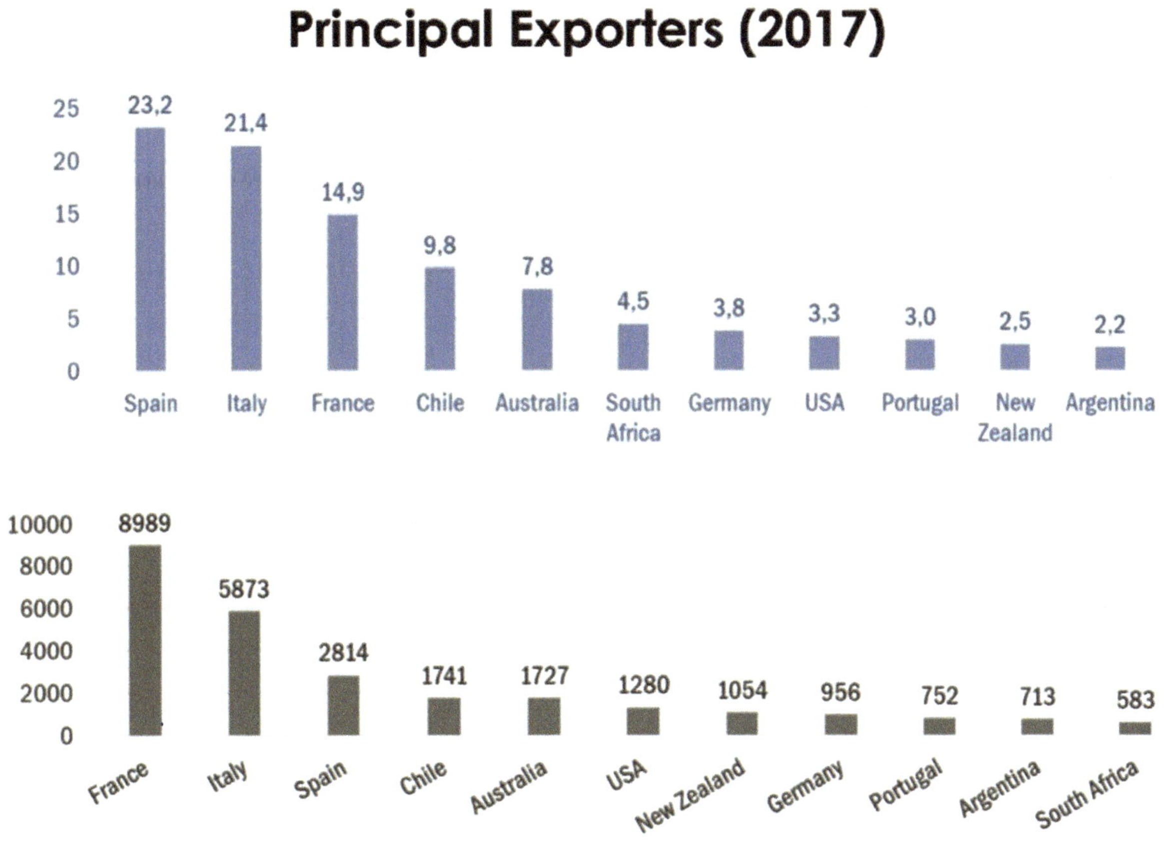

FIGURE 1.15 Million hectoliters (blue) and Million Euros (gray). *Source:* Created using information from the OIV (2018)

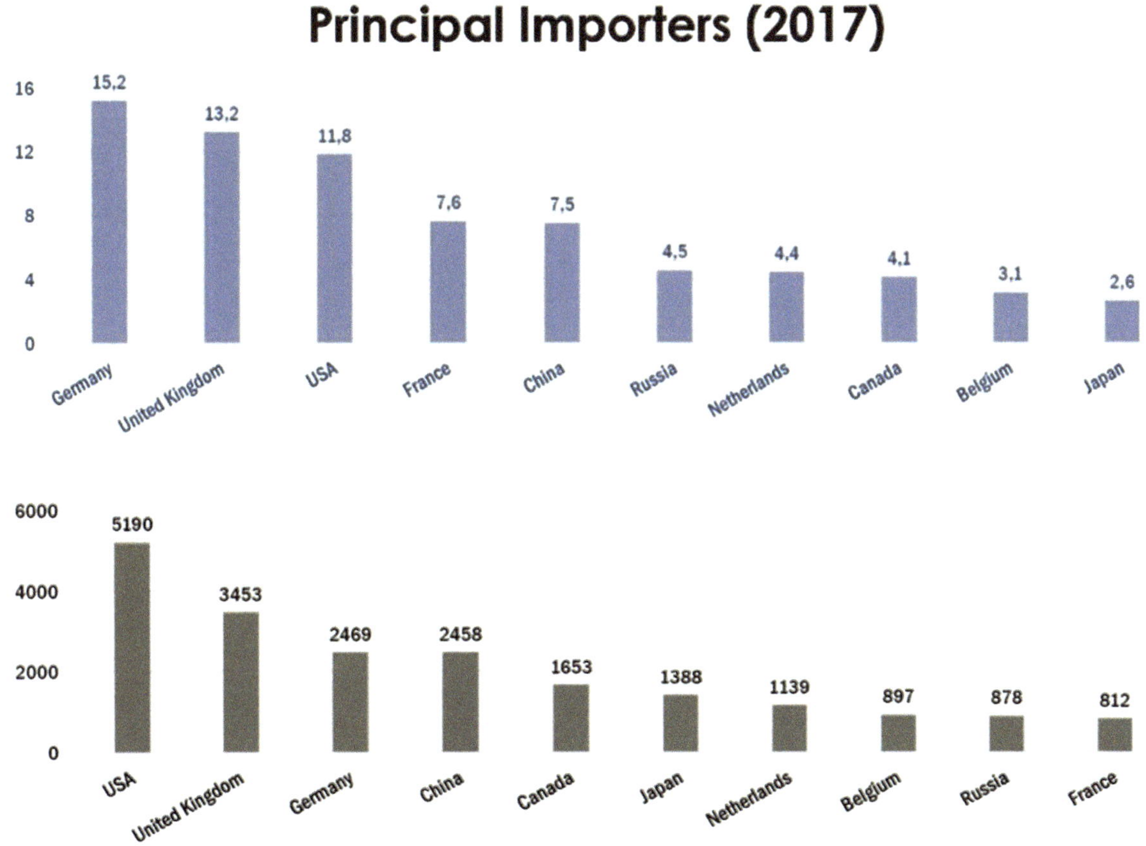

FIGURE 1.16 Million hectoliters (blue) and million Euros (gray). *Source:* Created using informations from the OIV (2018)

a significant increase in wine imports, in terms of volume and value (+5.7% in volume, +3.6% in value). As a consequence, it consolidates its position as top importer by value (5.2 billion EUR in 2017). China saw another significant rise in its imports in terms of volume (+17%; 7.5 mhl in 2017). The share of bottled wine imports (+15%) helped China reach fourth position by value (+14.7%; 2.46 billion EUR in 2017). Domestic demand in China was still the biggest contributory factor, in terms of volume, to global trade growth in 2017. In Russia, following the embargo and the economic difficulties it caused, there was an increase in terms of volume (+10.4%) and 32.6% (878 million EUR) growth in terms of value in 2017. It is also worth noting the significant increase in imports to the Netherlands. If compared with 2016, they went up by 10.9% in terms of volume and 16.2% in terms of value. (See chapter 42 for international market of organic wine and chapter 45 for an overview of wine Industry).

1.14 WINE TOURISM

Winery is a source of attraction and a properly designed and constructed winery becomes a source of tourism in any country. Tourism plays a very significant role in promoting the consumption of wine and the development of wine tourism in conjunction with the spread of the industry (See chapter 43). Many wineries world over organize specific wine fairs to attract tourists, families, and friends, organized wine tours, and larger tour groups on general sightseeing trips. Those with restaurants and other facilities can tap into a variety of other market segments, including private functions and corporate meetings. For most visitors, buying wine is a major motivation. Information regarding the product, the environment and the behavior of the visitors to the wineries is important. The wine industry, increasingly involved in experimenting with greening processes, influences its related businesses. One of these is represented by wine tourism. To preserve resources represents one of the main points in the wine industry, with particular attention to vineyards and landscapes (two key resources for wineries). Many wineries have improved their approach to vineyard exploitation, moving from a performance-oriented productive approach to a quality and sustainability one. In recent years, more and more wineries manage their vines in order to improve the overall quality of the grapes and to extend the average life of high quality vines. Therefore, with the boom in wine tourism, wineries have to ponder on how to balance the economic profits arising from tourism business with the carrying capacity of their surrounding area. Many countries are including sustainability when planning the expansion of their wine tourism offers; wine-producing countries are not only seeking a differentiation strategy, but they are also trying to optimize tourism flows and natural resources. The development of this tourism itself requires deep attention to resource preservation and exploitation in order to support a strategic progress.

BIBLIOGRAPHY

Acree, T.E. (1997). GC/Olfactometry: GC with a sense of smell. *Anal. Chem.*, 69(5), 170A.

Amerine, M.A., Berg, H.W., Kunkee, R.E., Ough, C.S., Singleton, V.L. and Webb, A.D. (1980). *The technology of wine making*, 4th edn., AVI Publishing, Westport, CT.

Anon. (2018). *OIV, International Organisation of Vine and Wine. OIV Statistical: Report on World Viti Viniculture.*

Attri, Devender and Joshi V. (2005). A panorama of research and development of wines in India. *J. Sci. Indust. Res.*, 64(1) , 9–18.

Barber, N., Taylor, D.C. and Deale, C.S. (2010). Wine tourism, environmental concerns, and purchase intention. *J. Travel Tour. Mark.*, 27(2), 146–165.

Berbegal, C., Fragasso, M., Russo, P., Bimbo, F., Grieco, F., Spano, G. and Capozzi, V. (2019). Climate changes and food quality: The potential of microbial activities as mitigating strategies in thewine sector. *Fermentation*, 5, 85. doi:10.3390/ fermentation 5040085.

Bission, L.F., Waterhouse, A.L., Ebeler, S.E., Walker, A.M. and Lapsley, J.T. (2002). The present and future of the international wine industry. *Nature*, 418(6898), 696.

Bordiga, M. (2019). *Post-Fermentation and -Distillation Technology: Stabilization, Aging, and Spoilage*, CRC Press, Boca Raton, FL.

Bordiga, M., Locatelli, M., Travaglia, F., Coïsson, J.D., Mazza, G. and Arlorio, M. (2015). Evaluation of the effect of processing on cocoa polyphenols: Antiradical activity, anthocyanins and procyanidins profiling from raw beans to chocolate. *Int. J. Food Sci. Technol.*, 50(3), 840–848.

Bordiga, M., Montella, R., Travaglia, F., Arlorio, M. and Coïsson, J.D. (2019). Characterization of polyphenolic and oligosaccharidic fractions extracted from grape seeds followed by the evaluation of prebiotic activity related to oligosaccharides. *Int. J. Food Sci. Technol.* doi:10.1111 /ijfs.14109.

Bordiga, M., Piana, G., Coïsson, J.D., Travaglia, F. and Arlorio, M. (2014). Headspace solid-phase micro extraction coupled to comprehensive two-dimensional with time-of-flight mass spectrometry applied to the evaluation of Nebbiolo-based wine volatile aroma during ageing. *Int. J. Food Sci. Technol.*, 49(3), 787–796.

Bordiga, M., Travaglia, F., Locatelli, M., Coisson, J.D. and Arlorio, M. (2011). Characterization of polymeric skin and seed proanthocyanidins during ripening in six Vitis vinifera L. cv. *Food Chem.*, 127(1), 180–187.

Boulton, R.B., Singleton, V.L., Bision, L.F. and Kunkee, R.F. (1995). *Principles and Practices of Wine Making*, Chapman and Hall, New York, p. 13.

Cavalieri, D., McGovern, P.E., Hartl, D.L., Mortimer, R. and Polsinelli, M. (2003). Evidence for *S. cerevisiae* fermentation in ancient wine. *J. Mol. Evol.*, 57, S226–S232.

Cocolin, L., Bission, L.F. and Mills, D.A. (2000). Direct profiling of the yeast dynamics in wine fermentations. *FEMS Microbiol. Lett.*, 189(1), 81.

Deng, Q. and Zhao, Y. (2011). Physicochemical, nutritional, and antimicrobial properties of wine grape (cv. Merlot) pomace extract-based films. *J. Food Sci.*, 76(3), 309–317.

Dennison, C. (1999). Historical note: A version of the invention of barrels and barrel alternatives. International symposium on oak in wine making. *Am. J. Enol. Vitic.* 50(4), 539.

Duarte, S., Gregoire, S., Singh, A.P., Vorsa, N., Schaich, K., Bowen, W. and Koo, H. (2006). Inhibitory effects of cranberry polyphenols on formation and acidogenicity of Streptococcus mutans biofilms. *FEMS Microbiol. Lett.*, 257(1), 50–56.

Dutraive, O., Benito, S., Fritsch, S., Beisert, B., Patz, C.D. and Rauhut,D. (2019). Effect of Sequential Inoculation with Non-*Saccharomyces* and *Saccharomyces* Yeasts on Riesling Wine chemical composition. *Fermentation*, 5(3), 79. doi:10.3390/ fermentation5030079 www.mdpi.com/journal/fermentation.

Ebeler, S.E. (1999). *Flavor Chemistry –30 Years of Progress*, R. Teranishi, E.L. Wick and I. Horstein (Eds.), Kluwer Academic/Plenum, New York, p. 409.

Forbes, R.J. (1965). *Studies in Ancient Technology*, 2nd edn.. Vol. III, E. J. Brill, Leiden, p. 83, n. 17.

Fugelsang, K.C. and Muller, I.J. (1996). The *in vitro* effect of wine of red wine on *Helicobacterium pylori*. In: A.L. Waterhouse and J.M. Rantz (Eds.), *Wine in Context: Nutrition, Physiology, Policy*. Proceedings of the Symposium on Wine and Health. American Society of Enology and Viticulture, Davis, CA, p. 43.

Gasteineau, F.C., Darby, J.W. and Turner, T.B. (1979). *Fermented Food Beverages in Nutrition*, Academic Pres, New York.

Getz, D. and Jamal, T.B. (1994). The environment-community symbiosis: A case for collaborative tourism planning. *J. Sustain. Tour.*, 2(3), 152–173.

Giovinazzo, G. and Grieco, F. (2015). Functional properties of grape and wine polyphenols. *Plant Foods Hum. Nutr.*, 70(4), 454–462.

Hyams, E. (1987). *Dionysus: A Social History of the Wine and Vine*, 2nd edn., Sidgwick and Jackson, London.

Ignatow, G. (2006). Cultural models of nature and society reconsidering environmental attitudes and concern. *Environ. Behav.*, 38(4), 441–461.

Jackson, R.S. (1999). Grape based fermentation products. In: V.K.Joshi and Ashok Pandey (Eds.), *Biotechnology: Food Fermentation (Microbiology, Biochemistry and Technology)*, Vol. II, Educational Publishers and Distributors, New Delhi, pp. 583–647.

Jackson, R.S. (2000). *Wine science – Principles, practices, perception*, 2nd edn., Academic Press, San Diego, CA.

Jolly, N.P., Augustyn, O.P.H. and Pretorius, I.S. (2006).The role and use of non-*saccharomyces* yeasts in wine production. *S. Afr. J. Enol. Vitic.*, 27(1), 15–39.

Jordão, António M., Vilela, Alice and Cosme, Fernanda. (2015). From sugar of grape to alcohol of wine: Sensorial impact of alcohol in wine. *Beverages*, 1(4), 292–310. doi:10.3390 /beverages 1040292.

Joshi, V.K. (1997). *Fruit Wines*, 2nd edn., Directorate of Extension Education, Dr, Y.S. Parmar University of Horticulture and Forestry, Nauni, Solan, India, p. 255.

Joshi,V.K. (Ed.) (2011). *Handbook of Enology: Principles, Practices and Recent Innovations*, 3 Volumes set. Asia -Tech Publisher and Distributors. New Delhi, p. 1450.+ Plates, Fig.

Joshi, V.K. and John, Siby (2002). Antimicrobial activity of apple wine against some pathogenic and microbes of public health significance. *Alimentaria*, 67–72.

Joshi, V.K. and Pandey, A. (Eds.) (1999). *Biotechnology: Food Fermentation: (Microbiology, Biochemistry and Technology)*, Vol. I & II. Educational Publisher & Distributors, Ernakulum, New Delhi, pp. 1450.

Joshi, V.K. and Parmar, Mukesh (2004). Present status , scope and Future strategies of fruit wines production in India. *Indian Food Ind.*, 23(4), 48–52.

Joshi, V.K. and Singh, R.S. (Eds.) (2012). *Food Biotechnology: Principles and Practices*, IK International Publishing House, New Delhi, p. 920.

Joshi, V.K. and Attri, Devender (2005). A Panorama of research and development of wines in India. *J. Sci. Indust. Res.*, 64(1), 9–18.

Kosseva, M.R., Joshi, V.K. and Panesar, P.S. (Eds.) (2017). *Science and Technology of Fruit Wine Production*. Academic Press is an imprint of Elsevier, London, UK, p. 705.

Lachman, J., Sulc, M., Faitová, K. and Pivec, V. (2009). Major factors influencing antioxidant contents and antioxidant activity in grapes and wines. *Int. J. Wine Res.*, 101, 101–121.

Lambrechts, M.G. and Pretorius, I.S. (2000). Yeast and its importance to wine aroma - A review. *S Afr. J. Enol. Vitic.*, 21(1), 97–129.

Leifert, W.R. and Abeywardena, M.Y. (2008). Cardioprotective actions of grape polyphenols. *Nutr. Res.*, 28(11), 729–737.

Liu, S.Q. (2002).Malolactic fermentation in wine – beyond deacidification-A review.Journal of Applied *Microbiology*, 92, 589–601.

Marchese, A., Coppo, E., Sobolev, A.P., Rossi, D., Mannina, L. and Daglia, M. (2014). Influence of in vitro simulated gastroduodenal digestion on the antibacterial activity, metabolic profiling and polyphenols content of green tea (*Camellia sinensis*). *Food Res. Int.*, 63, 182–191.

Nabavi, S.F., Sureda, A., Daglia, M., Rezaei, P. and Nabavi, S.M. (2015). Anti-oxidative polyphenolic compounds of Cocoa. *Curr. Pharm. Biotech.*, 16(10), 891–901.

Nijveldt, R.J., van Nood, E., van Hoorn, D.E., Boelens, P.G., van Norren, K. and van Leeuwen, P.A. (2001). Flavonoids: A review of probable mechanisms of action and potential applications. *Am. J. Clin. Nutr.*, 74(4), 418.

Percival, R.S., Devine, D.A., Duggal, M.S., Chartron, S. and Marsh, P.D. (2006). The effect of cocoa polyphenols on the growth, metabolism, and biofilm formation by *Streptococcus mutans* and *Streptococcus sanguinis*. *Eur. J. Oral Sci.*, 114(4), 343–348.

Perestrelo, R. Silva, C. and Pereira, J. Caˆmara. (2016). *Wines: Madeira, Port and Sherry Fortified Wines – The Sui Generis and Notable Peculiarities*, Major Differences and Chemical Patterns Encyclopedia of Food and Health, Elsevier, London, 534, 554. doi:10.1016/B978-0-12-384947-2.00758-3.

Phaff, H.J. (1986).Ecology of yeasts with actual and potential value in biotechnology. *Microb. Ecol.*, 12(1), 31.

Pinu, R.F. (2018).Grape and wine metabolomics to develop new insights using untargeted and targeted approaches. *Fermentation*, 4(4), 92. doi:10.3390/fermentation4040092.

Pretorius, I.S. (2000). Tailoring wine yeast for the new millennium: Novel approaches to the ancient art of winemaking. *Yeast*, 16(8), 675–729.

Rathore,A.S. (2006). *The Complete Indian Wine Guide*, The Lotus Collect. Imprint RoliBooks N, Delhi.

Robinson, J. (1986). *Vines, Grape and Wines*, Mitchell, Beazley, London.

Romano, P., Fiore, C., Paraggio, M., Caruso, M. and Capece, A. (2003). Function of yeast species and strains in wine flavour. *Int. J. Food Microbiol.*, 86, 169–180.

Romano, P., Marchese, R., Laurita, C., Saleano, G. and Turbanti, L. (1999). Biotechnological suitability of *Saccharomycodes ludwigii* for fermented beverages. *World J. Microbiol. Biotech.*, 15(4), 451–454.

Romano, P. and Suzzi, G. (1993a). Potential use for *Zygosaccharomyces* species in winemaking. *J. Wine Res.*, 4(2), 87–94.

Sandhu, D.K. and Joshi, V.K. (1995). Technology, quality and scope of fruit wines with special reference to apple. *India Food Industry*, 14(1), 24–32.

Santini, C., Cavicchi, A. and Casini, L. (2013). Sustainability in the wine industry: Key questions and research trends. *Agric. Food Econ.*, 1, 1.

Schena, M., Shalon, D., Davis, R. and Brown, P. (1995). Quantitative monitoring of gene expression patterns with a complementary DNA microarray. *Science*, 270(5235), 467–470.

Sickles, N., Verdoni, A., Wolkomir, J. and Stolarz, R. (1991). *Vintage Wine Book History-A Practical Guide to the History of Winemaking, Classification, and Selection*, 2nd edn., Food Products Press, NY.

Soobrattee, M.A., Bahorun, T. and Aruoma, O.I. (2006). Chemopreventive actions of polyphenolic compounds in cancer. *BioFactors*, 27(1–4), 19–35.

Storm, R.D. (1997). *Winery Utilities Planning, Design and Operation*, Springer Science+ Business Media Dordrecht, New York, Chapman and Hall, pp. 1–17.

Swiegers, J.H., Bartowsky, E.J., Henschke, P.A. and Pretorius, I.S. (2005). Yeast and bacterial modulation of wine aroma and flavour. *Aust. J. Grape Wine Res.*, 11(2), 139–173.

Swiegers, J., Kievit, R.L., Siebert, T., Lattey, K.A., Bramley, B.R., Francis, I.L., King, E.S. and Pretorius, I.S. (2009). The influence of yeast on the aroma of Sauvignon blanc wine. *Food Microbiol.*, 26(2), 204–211.

Swiegers, J. and Pretorius, I.S. (2007). Modulation of volatile sulfur compounds by wine yeast. *Appl. Microbiol. Biotechnol.*, 74(5), 954–960.

Travaglia, F., Bordiga, M., Locatelli, M., Coisson, J.D. and Arlorio, M. (2011). Polymeric proanthocyanidins in skins and seeds of 37 *vitis vinifera* L. cultivars: A methodological comparative study. *J. Food Sci.*, 76(5), C742–C749.

Unwin, T. (1991). *Wine and the Vine: An Historical Geography of Viticulture and the Wine Trading*, Routledge, London.

Verma, L.R. and Joshi, V.K. (2000). An overview of post-harvest Technology. In: *Postharvest Technology of Fruits and Vegetables*, Vol. I, Verma and Joshi (Eds.), Indus Publishing Co., New Delhi, p. 1.

Vine, R.P. (1981). Wine and the history of western civilization. In: *Commercial Wine Making, Processing and Controls*, The AVI Publishing Co., Inc., Westport, CT.

Watkins, T.R. (Ed.) (1997). *Wine: Nutritional and Therapeutic Benefits*. ACS Symp. S. 661, American Chemical Society, Washington, DC.

2 Categories and Main Characteristics of Wines, Wine-Products and Wine Spirits

Stylianos D. Karagiannis

CONTENTS

2.1 INTRODUCTION

According to the basic definition given by OIV (International Organisation of Vine and Wine), wine is the beverage resulting exclusively from the partial or complete alcoholic fermentation of fresh grapes, whether crushed or not, or of grape must. Its actual alcohol content should not be less than 8.5% vol. Nevertheless, taking into account climate, soil, vine variety, special qualitative factors, or traditions specific to certain vineyards, the minimum total alcohol content may be able to be reduced to 7% vol. by legislation particular to the region considered.

There is no generally accepted way of classifying wines except in the broadest sense but they can be grouped according to their basic characteristics (such as colour, sugar content, alcoholic strength, carbon dioxide content, age), the vinification process involved, the geographic origin, or the different taxation (excise duties) imposed on these products.

Wines are often divided into three basic categories, namely still wines, sparkling wines, and liqueur-wines; the latter two typically are taxed (excise duties) at a higher rate. Still wines are, in most cases, intended to accompany a meal and can be further classified into white, red, and rosé groups or into dry, medium-dry, semi-sweet, and sweet wines. Sparkling wines can be consumed before, during, or after meals, depending on their sugar content and are often classified by the method used to achieve the high carbon dioxide content. Liqueur-wines can be consumed as an aperitif before meals, especially those which do not contain residual sugars, such as Fino-style Sherries, but more commonly they contain high amounts of sugars and are consumed after meals as a dessert, such as Ports, Madeiras and Samos.

An important category of wine-products, although not considered as wines, is the group of aromatised wine-products, which comprises the aromatised wines (such as Vermouth), the aromatised wine-based drinks (such as Sangria and Glühwein), and the aromatised wine-product cocktails.

The classification of wines can be further extended by including beverages obtained by the alcoholic fermentation of fruits other than grapes, although the term "wine" should normally apply only to the naturally fermented juice of grapes.

Concerning spirit drinks of wine origin, wine spirits and brandies are very popular spirit drinks produced in most of the winemaking countries worldwide. In this chapter, their classification is mainly done according to their geographical indication, especially in EU countries where the terms "wine spirit" and "brandy" refer to spirits produced only by distillation of wine and where these drinks are mostly (and traditionally) produced.

Here an attempt has been made to present all known types of wines, wine-products, and wine spirits, grouped in their classification systems along with their general characteristics.

2.2 WINES

2.2.1 Still Wines

The "still" category comprises most wines and therefore, the most complex system is required for their classification. They may be grouped into white, rosé, and red wines, according to the grapes used for their production and the length of time that skins have been left to ferment with the juice. They can also be classified into dry, medium-dry, semi-sweet, and sweet wines, depending upon whether all the grape sugar has been allowed to ferment into alcohol or some residual sugar has been left. However, the most acceptable division is based on wine colour, and reflects distinct differences in flavour, use, and production throughout the world. The main wine regions of the world are included in Table 2.1.

2.2.1.1 White Wines

Still white wines are a large category of wines, which can be subdivided by sugar content or age. Classification by sugar content gives two major groups: dry and sweet white wines. Semi-dry wines and semi-sweet wines correspond to much lower production worldwide compared to dry and sweet wines. For this reason, medium-dry wines may be considered as part of dry wines, whereas semi-sweet wines may be classified as sweet wines in most cases, although in the relevant OIV resolutions, as well as in the European legislation [Commission Delegated Regulation (EU) 2019/33], these groups are clearly distinguished.

2.2.1.2 Dry White Wines

According to the OIV definition, the wine is said to be dry when it contains a maximum of either 4 g/l sugar or 9 g/l, when the level of total acidity (expressed in grams of tartaric acid per liter) is no more than 2 g/l less than the sugar content.

Dry white wines are a large group of wines intended to be consumed with meals and designed to have an acidic finish.

TABLE 2.1
Main Wine Regions of the World

Country	Main wine regions
France	Burgundy, Bordeaux, Rhone, Loire, Alsace, Champagne, Midi, Jura
Italy	Tuscany, Piedmont, Veneto, Abruzzo, Emilia-Romana, Apulia, Sicily
Spain	La Rioja, Catalonia, La Mancha, Valencia, Jerez, Malaga, Navara
Portugal	Minho, Douro, Dão, Bairrada, Colares, Bucelas, Setúbal
Greece	Nemea, Naoussa, Samos, Macedonia, Achaia, Santorini
Argentina	Mendoza, San Juan, Salta, Catamarka, La Rioja
Australia	New South Wales, Victoria, Riverland, Swan Valley
Chile	Maipo, Aconcagua, Santiago, Talca, Kuriko
United States	California, Pacific Northwest
South Africa	Western Cape
Germany	Rhine, Moselle, Franconia
Austria	Niederosterreich, Burgenland, Steiermark
Croatia	Dalmatia, North-western Croatia, Slavonia, Istria
Switzerland	Vaud, Lavaux, Valais, Zürich
Hungary	Alföld, Northeastern Hungary, Balaton,
Romania	Cotnari, Murfaltar, Tîrnave, Dealul-Mare, Vrancea
Brazil	Serra Gaúcha, Campanha
New Zealand	Auckland, Gisborne, Marlborough, Hawkes Bay, Martinborough
China	Shandong, Xinjiang, Beijing, Ningxia, Tianjin,
India	Maharashtra, Punjab
Japan	Hokkaido, Yamanashi
Canada	Niagara, British Columbia, Ontario
Peru	Ica, Moquega
Mexico	Baja California
Uruguay	Montevideo, Canelones, San José, Maldonado

Combined with food proteins, the acidic aspect of the wine becomes balanced and can both accentuate and harmonize with food flavours. White wine making demands immediate pressing of grapes, juice clarification, cold fermentation, wine racking, and cold stabilization before bottling. Dry white wines are produced in all winemaking countries of the world, and a great number of such wines produced in Europe bear protected designations of origin or protected geographical indications.

Dry white wines can be divided into three main types:

a) Plain dry or medium-dry wines of no intense aromatic character, fully fermented, and not intended to be aged. They are usually made with non-aromatic grapes, but during fermentation they obtain the characteristic aromatic profile of the volatile compounds formed during this process. Winemaking follows the standard procedure with emphasis on techniques that enhance fresh character of the resulting wines (oxygen exclusion, low temperature fermentation). Usually, these wines have low-moderate alcohol content (9% vol.–12% vol.), high titratable acidity (greater than 7 g/l), and low pH values (less than 3.3).

PLATE 2.1 Vineyard in Mount Olympos slopes (Greece). *Courtesy:* "Pieria Eratini Estate"

b) Fresh, fruity, dry to semi-sweet wines, for drinking young, made from aromatic grape varieties: Muscat Blanc, Muscat of Alexandria, Muscadelle, Riesling, Traminer, Sauvignon Blanc, Gewürztraminer, Müller-Thurgau, Moschofilero, and others. Extreme emphasis is given to picking at the right moment (Plate 2.1), clarified juice, cold fermentation, and early bottling. Their alcoholic strength ranges between 10.5% vol. and 12.5% vol., but their total acidity and pH are similar to those of the previous category.

c) Dry but full-bodied and smooth white wines, usually made with a degree of "skin-contact" and fermentation at a relatively higher temperature. These wines have higher alcoholic strength (13% vol.–14% vol.), lower titratable acidity (5.5–7 g/l), and higher pH values than those of the previous categories. They are bottled after a minimum of 2–3 months of conservation in barrels and are intended for further aging in bottles. Characteristic examples are wines made from cultivars Chardonnay, Semillon, and Sauvignon Blanc, and the classic wines of origin for this style are the Grand Cru whites of Burgundy.

2.2.1.3 Sweet, Not Fortified White Wines

Wines of this group are intended to be consumed alone as "sipping" wines, to accompany dessert, or to replace dessert. Their alcoholic strength may vary from 6% vol. to 14% vol., sugar content ranges from 45 g/l to 250 g/l, and may have from aromatically simple to complex fragrance. Sweet, not fortified, white wines can be classified into two main groups, botrytised and non-botrytised wines.

a) **Botrytised wines:** Botrytised wines are derived from grapes infected by *Botrytis cinerea*. Typically, its effect on the quality of the resultant wine is negative, but under unique climatic conditions, the infected grapes develop the so-called noble rot. One of the most notable changes during noble rotting is an increase in sugar concentration. Infection by *B. cinerea* also affects the aroma and taste of the resultant wine. The formation of glycerol during noble rotting contributes to the smooth mouth-feel of these wines. The main cultivars used for the production of botrytised wines are Riesling and Sémillon. The Hungarian cultivars Furmint and Hárslevelü are used for vinification of Tokaji Aszú botrytized wines. Other cultivars, such as Sauvignon, Gewürztraminer, Picolit and Chenin Blanc are also used, depending on tradition and adaptation to local conditions. Botrytised wines are mainly produced in France, Hungary, Germany, and Austria. Subdivision of botrytised wines would give three types:

i) **French botrytised wines** that contain high alcohol (over 11% vol.) and high sugar concentration. Botrytised wines of France are produced in the regions of Bordeaux (Sauternes, Barsac, Cérons, Loupiac, Ste Croix du Mont), Val de Loire (Côteaux du Layon), and Alsace (Vallée du Rhin).

ii) **German style botrytised wines** that contain low alcohol (commonly 6 to 8% vol.) and very high sugar content. A typical example of this group is Auslesen wines, which are derived from specially selected clusters of late-harvested fruit. Subcategories of this type are Beerenauslese and Trockenbeerenauslese wines. German style botrytised wines are produced in Germany and Austria.

iii) **Tokaji Aszú wines** are produced in the northeast region of Hungary. Most *aszú* (botrytised) wines are made from mixing young white wines, or juice derived from healthy grapes, with *paste* made from pulverized *aszú* berries. The alcohol content of the resulting wine varies from 12 to 14% vol., and maturation in barrels (136 to 240 l) takes place for a variable period. The most famous version of Tokaji wines is, however, Aszú Eszencia, derived from juice that seeps out of highly botrytised berries placed in small tubs. Normally, it has a dark brown color and odors reminiscent of honey and dried figs with a greasy, very sweet, and viscous taste, with a long after-taste.

b) **Non-botrytized sweet wines:** Production of non-botrytised sweet white wines takes place in many winemaking regions of the world. Several vinification techniques have been developed in these regions, depending upon the climatic conditions and consumer demand.

i) **Sweet reserve addition** is the standard procedure for producing the non-botrytised sweet and semi-sweet wines of Germany and is increasingly used in many countries. The method comprises sterilisation of a portion of the juice (sweet reserve) instead of fermenting it. The majority of grape juice is fermented to wine in the normal way, until no sugar is left. The "sweet reserve" (*süssreserve*) is then added and the blend is bottled under sterile conditions.

ii) **Sweet wines from partially dehydrated grapes**, in this case, the grapes are left to dehydrate for a variable period in shade or sunlight. Then, the dried grapes are crushed and the concentrated juice is fermented. The sugar concentration is typically so high that fermentation ends leaving the wine with high sugar content. Variations on the procedure are mainly found in southern Europe.

A classic example of this group is the famous "Vinsanto" from the indigenous grape cultivar Assyrtiko, cultivated in some of the world's oldest vineyards and dating back 3.500 years on the volcanic island of Santorini (Greece) (Plates 2.2, 2.3, and 2.4). This terroir-driven sweet wine of

PLATE 2.2 Vine pruning known as **"kouloura"** or "wreath" or "basket", developed on the volcanic island of Santorini (Greece). *Courtesy:* "Gaia Wines"

PLATE 2.3 Vine pruning known as **"kouloura"** or "wreath" or "basket", developed on the volcanic island of Santorini (Greece). *Courtesy:* "Gaia Wines"

PLATE 2.4 Winery in island of Santorini (Greece). *Courtesy:* "Gaia Wines"

distinct character is structured on exciting minerality and density, which both reflect the unique volcanic and anhydrous soil of Santorini, an internationally acclaimed unique ecosystem.

iii) **Other types of (nonbotrytised) sweet wines** several other techniques such as juice heating, grape freezing, cryo-extraction, and reverse osmosis are used in some winemaking regions for the production of sweet wines.

Grape freezing leads to the production of Icewine (Vin de glace or Eiswein), well known in Alsace (France), Germany, Austria, and Canada. Grapes are left on the vine until winter temperature falls to –7°C or lower.

Then, harvest takes place, and by pressing the frozen grapes, juice from the most mature berries is extracted. The potential alcohol strength by volume for musts cannot be increased and should be 15% vol. as a minimum. The minimal acquired alcoholic strength of the finished wine should be 5.5% vol.

Cryo-extraction is the technological approach of the aforementioned natural procedure. Some winemakers prefer to maintain a small degree of residual sugar in certain wines (California Rieslings and Gewürztraminers for example) by cooling the vat to stop fermentation at the appropriate moment, then using a centrifuge and fine filtering to remove the yeasts and sterilise the wine.

2.2.2 Rosé Wines

Still rosé wines are generally classified by their colour, regardless of their vinification technique. Normally, they are produced by similar techniques of white winemaking and this is the reason why they are considered as more similar to white wines than to red ones. To achieve the desired rosé colour, the grape skins are removed from the juice shortly after fermentation has begun. Thus, the uptake of substances that give red wines their chromatic characteristics is limited. Very few rosé wines age well and most are made to have a semi-sweet finish.

Rosé wines are mainly produced in France followed by Spain, Portugal, the United States (Californian "blush" wines), Italy, South Africa, Chile, Argentina, Australia, Germany, Morocco, Greece (Plate 2.5), Croatia (Plate 2.6), and many other countries.

Three types of still rosé wines can be distinguished according to the vinification technique:

a) Pale rosé wines from red grapes pressed immediately to extract juice with very little colour, sometimes called "grey" wines (Vins Gris) or Blancs de Noirs.
b) Rosé wines made by *saignée* technique. These wines are produced from red grape juice left in contact with skins from 2 up to 20 hours in controlled conditions, and then pressed and fermented like white wines. These rosé wines are the most aromatic, and depending on the cultivars as well as the techniques

PLATE 2.5 Winery in Attiki (Greece). *Courtesy:* "Papagiannakos winery"

PLATE 2.6 Stainless steel tanks in traditional winery in Croatia

and practices used for their production, may give exceptional wines.

c) Rosé wines derived by blending red and white wines.

2.2.3 Red Wines

2.2.3.1 Dry Red Wines

Unlike white wines, still red wines are almost exclusively dry. The absence of a detectable sweet taste is consistent with their intended use as a food beverage. Rich in phenolic compounds, derived from grape skins, red wines possess a bitter and astringent sensory character, which perfectly combines with food proteins. However, some well-aged red wines are better appreciated after a meal.

Generally, red wine making comprises the following basic procedure: a) Grape crushing and stem removal, b) extraction of anthocyanins and other substances with organoleptic properties from skins to juice, c) fermentation at higher temperature than those of white wine making, d) pressing by the time fermentation is finished or nearly finished, e) malolactic fermentation, and f) aging (oxidative-reductive) in most cases.

The diverse range of flavours that exist within the red grape cultivars in commercial production has enabled a wide spectrum of distinctive red wines styles to exist. As with white wines, differences in red wines may be adjusted along a range based on the degree to which wines reflect varietal character as well as on production procedures. For example, most varietally designated wines attempt to accentuate the varietal aroma, such as Tempranillo, Sagniovese, and Xinomavro. On the other hand, several other red wines express primarily processing features, such as Amarone and Beaujolais wines. Alternately, many red wines attempt to blend both varietal flavour and fermentation aroma along with maturation/aging characteristics, such as premium Bordeaux wines.

As with dry white wines, dry red wines are produced in all winemaking countries of the world and a great number of such wines bear protected designations of origin or protected geographical indications.

Dry red wines can be further classified into two groups:

a) **Short-aging wines**, made with low tannin content and processed for early consumption, have a light and fruity character with little varietal aroma (Gamay, Grenache, Garnacha, and Carignan wines), or with more intense varietal aroma (wines made from Freisa, Bonarda, and Dolcetto cultivars).
b) **Long-aging wines**, initially rich in tannins and phenolic compounds, are then subjected to maturation in a barrel or tank followed by additional in-bottle aging, and have a complex subtle bouquet derived from the association of varietal aroma with oak flavours. Typical examples are wines made from the cultivars Cabernet Sauvignon, Syrah, Merlot, Tempranillo, Cabernet Franc, Malbec, Nebbiolo, and Barbera.

2.2.3.2 Sweet, Not Fortified Red Wines

As mentioned before, unlike white wines, few still red wines are made to be sweet. The most famous wines of this category are the result of grapes drying before fermentation.

However, in some winemaking regions, especially in Italy, botrytized or non-botrytised sweet red wines are produced, such as Recioto della Valpolicella produced in the Italian region Veneto. The name of the wine is due to the selected *recie* (ears) or wings of the grape clusters, which are dried before winemaking, leading to a relatively dark garnet-red wine in colour, with a characteristic and accentuated odor and sweet flavour. Vin Santo di Carmignano Occhio di Pernice is another sweet red wine made from dried Sangiovese (50% minimum) grapes, in Tuscany. Piedirosso semi-sweet wine is produced from grapes subjected to drying either on the vine or after the harvest, in the region of Campania (Italy).

Famous wine of this category is Commandaria (protected designation of origin) which has been traditionally produced in Cyprus for centuries from the indigenous vine cultivars Mavron and Xynisteri. Although Commandaria is made from overripened and sun-dried grapes, it may also be fortified. Commandaria, by law, is aged for at least 4 years in oak barrels.

2.2.4 Sparkling Wines

Sparkling wines, according to the common European legislation [Regulation (EU) 1308/2013], are the wines which when their container is opened, release carbon dioxide derived exclusively from fermentation, and which have an excess pressure, due to carbon dioxide in solution, of not less than 3 bar when kept at a temperature of 20°C in closed containers. Semi-sparkling wines have an excess pressure, due to endogenous carbon dioxide in solution, of not less than 1 bar and not more than 2.5 bar when kept at a temperature of 20°C in closed containers. Additional restrictions are provided for the production of these wines, according to the relevant European legislation.

Most sparkling wines are white, fewer are rosé, and very few are red, and range from dry to sweet, and from subtly to highly aromatic or fruit-flavoured. These wines can be consumed before and during meals if they are dry or with fruits and dessert, after meals if they are sweet.

According to European legislation [Commission Delegated Regulation (EU) 2019/33], a sparkling wine is named as: "Brut nature" if its sugar content is less than 3 g/l (this term may be used only for products to which no sugar has been added after the secondary fermentation), "Extra Brut" if its sugar content is between 0 and 6 g/l, "Brut" when it contains at the most 12 g/l of sugar, "Extra-dry" when it contains at least 12 g/l and at most 17 g/l, "Dry" when it contains at least 17 g/l and at most 32 g/l, "Demi-sec" when it contains 32 to 50 g/l, and "Sweet" when it contains sugars more than 50 g/l.

The classification of sparkling wines described here is made by the method of production, which is the most widely used method of classification of sparkling wines, according to the relevant sources.

2.2.4.1 Traditional Method

This is the method required for sparkling wine manufacture in the province of Champagne in France. Many other sparkling wines produced outside the Champagne region and not entitled to the appellation "Champagne" are produced by the same technique in France *i.e.* the "Crêmants" of Alsace, Burgundy, the Loire and Die, and the sparkling wines of Gaillac and Seyssel. The term "traditional method" replaced the term "méthode champenoise" which is not allowed to be used for sparkling wines produced by this technique in regions other than Champagne. In traditionally produced sparkling wines, the grapes are pressed whole, without prior stemming or crushing, and fermentation of juice is carried out in the presence of selected yeast strains, at about 15–18°C. The final steps of the procedure are removal and ejection of yeast sediment (riddling, disgorging), *tirage liqueur* addition, corking, and maturation.

2.2.4.2 Transfer Method

Currently used in France and Australia, although the method does not have the prestige and pricing advantage of the traditional method and its continued existence is in doubt. Similarly to the traditional method, it leads to the production of dry and semi-dry wines with limited varietal aroma and possessing a "toasty" flavour.

2.2.4.3 Cuvée Close Method

The "cuvée close" or "Charmat" process is mainly used for the production of sparkling wines derived from aromatic cultivars and especially Muscats. It avoids the first bottling by inducing the secondary fermentation in a tank, then filtering and bottling. It is the most common method for production of sparkling wines cheaper than those of the traditional method, although in many cases not of low quality. Although wines of cuvée close technique tend to be sweet (such as "Asti Spumante" sugars: 60–100 g/l) and aromatic, some are made dry with subtle fragrances.

2.2.4.4 Other Types of Sparkling Wines

A method of sparkling wine elaboration was initiated after 1950 in the Soviet Union (Russia), based on a continuous fermentation process. Though extensively used in Russia, it has limited application in other countries such as Portugal. Other sparkling wines have derived their sparkle from malolactic fermentation ("Vinho verde" from northern Portugal, although currently the wine may be carbonated to reintroduce the characteristic *pétillance).* Fermentation of grapes with the skins leads to the production of rosé or light red base wines that can be made into rosé sparkling wines, but are usually made by blending small amounts of red wine into the white *cuvée.* Commonly, rosé and red sparkling wines are semi-sweet or sweet.

2.2.4.5 Aerated Sparkling Wines

Aerated sparkling wines (carbonated wines) are the wine-products which release carbon dioxide when their container is opened, derived wholly or partially from an addition of that gas. They are much less costly to prepare and least prestigious compared to sparkling wines but, in some cases, are pleasing in character if made of table wine of good quality. Carbonation has the advantage of leaving the aroma and taste character of the base wine almost unmodified. Many carbonated wines are also fruit flavoured (coolers).

2.2.5 Liqueur Wines

Two main properties characterise the large group of liqueur wines (otherwise known as fortified wines). They are the high alcohol content and, at some stage of production, the addition of ethyl alcohol of vine origin (or distillates of vine origin). They are produced in many parts of the world particularly in Spain, France, the United States, Italy, Portugal, Greece, South Africa, and Australia. Liqueur-wines can be made in a wide range of styles such as white or red, dry or sweet, and with or without added flavours. Because of their elevated alcohol content and their flavour intensity and complexity, they are not intended to accompany a meal. Instead, they are served as aperitifs, dessert wines or cocktail beverages. Dry liqueur-wines, such as Fino-style Sherry, are generally considered to be the perfect aperitifs, especially before a meal at which fine wines are to be served. More commonly, liqueur-wines possess a sweet character, such as Ports, Oloroso-style Sherries, Samos and Madeiras, and for this reason are consumed after meals or as a dessert. The distinctive types of liqueur-wines developed during the last centuries, especially in Southern Europe, are the base for the most appropriate classification.

2.2.5.1 Sherry and Similar Wines

The base wine for Sherry (and similar wine) production is commonly a white wine fermented to dryness or to low sugar content. Then, a small quantity of ethyl alcohol of vine origin or wine spirit is added to stabilise it while it matures in contact with air. The various types of Sherries and similar type wines are:

2.2.5.1.1 Sherry (Jerez)

The term "Sherry" (Jerez) is a protected designation of origin restricted to wines produced in and around Jerez de la Frontera of Spain. Similar wines produced elsewhere are not permitted to use the Sherry appellation. Other areas of Spain producing similar wines are Montilla-Moriles and Malaga of Andalucia. The grape varieties preferred for Sherry production in Spain are Palomino and Pedro Ximénez. Production of the base wine generally follows standard procedures. The wines are carefully tasted and classified into potential *Flor* and *Oloroso* wines. *Flor* is a film of yeast cells that grows on the surface of the wine. Several yeast strains may be part in the *flor* of Sherries. Three main groups may be produced as:

a) **Fino**, is very pale, dry, light in body, and characteristically has a slightly pungent bouquet and flavour derived from the action of *flor*. After the first racking, the wine is fortified to bring the alcohol content up to 15–15.5% vol. At 15% vol., alcohol favours *flor* development and prevents the development of acetic acid bacteria and other contaminating microorganisms. The wine is then stored in barrels (*butts* 500 l capacity) that are arrayed in rows in above ground buildings called *bodegas*. During the preliminary storage, *flor* film develops on wine in many barrels. Wines in this preliminary storage stage are known as *añada* wines, which are then matured through the *solera* technique. Fino Sherries may be sweetened slightly by the addition of a small amount of sweet Pedro Ximénez wine. The characteristic properties of Fino Sherries are undoubtedly due mainly to the aforementioned biological aging procedure under the *flor*.
b) **Amontillado**, is a Fino Sherry that has been aged for a long period in wood and lost much of its original Fino character. Its production is similar to that of Fino at the first stages, but subsequently, the frequency of transfers is reduced, leading to progressive elimination of the *flor*, due to a decrease of nutrients and increase of alcohol content. Without *flor* protection, the wine becomes dark-coloured and develops a richer oxidized flavour. Its alcohol content averages 18% vol., but in some cases it may reach 22% vol.
c) **Oloroso**, is a wine of deeper color than Fino, usually fairly sweet. The first step in production of an Oloroso Sherry involves fortification of the *añada* wine to about 18% vol. alcohol. This inhibits yeast and bacterial growth, and makes Oloroso maturation less sensitive to temperature fluctuation compared to other Sherry types. There are typically few *criaderas* stages in an Oloroso *solera* system with slow transfer rates. Because Oloroso Sherries are not produced in the presence of *flor*, they do not have the sensory properties of Fino Sherries. On the other hand, Olorosos contain more phenolic compounds and are more full-bodied than Fino and Amontillado Sherries because of their longer conservation in oak. The average alcoholic strength is 18% to 20% vol. although they may be brought up to 21% vol. or even to 22% vol. Olorosos are commonly blended with sweetening and colour wines. Special Oloroso types are Palo Cortado (with characteristics of Fino group), Raya (rougher version), Amoroso (sweetened Oloroso of a dark colour), and Cream Sherry (heavily sweetened Oloroso).

2.2.5.1.2 Flor-Matured Liqueur-Wines

a) **Vin Jaune** is exclusively produced from the cultivar Savagnin in four regions of appellation: Château Chalon, Arbois, l' Étoile, and Côtes du Jura in the Jura district of France. The wine is vinified under the classic white winemaking, followed by malolactic fermentation. *Flor* development occurs spontaneously and the wine is aged for approximately 6 years.
b) **Sardinian liqueur-wines** are Vernaccia di Oristano and Malvasia di Bosa wines. Vernaccia di Oristano has been produced in Sardinia since the 11th century (alcohol content 15.5%–16.5% vol. for sweet type or 18% vol. for dry type).
c) **South Africa liqueur-wines** are produced from grapes of the cultivars Palomino and Chenin Blanc (Steen). Wines designed to become *flor*-matured are fortified to 15.5% vol. whereas wines intended for Oloroso production are fortified to 17% vol. A system of maturation similar to that used for Sherry is used but with fewer *criaderas* stages and little fractional blending.
d) **Australian liqueur-wines** are produced from Pedro Ximénez grapes. The grapes are seldom fractionally blended, and when the desired *flor* character has been obtained, the wine is fortified up to 19% vol. prior to maturation.
e) **Submerged culture liqueur-wines** is practiced in Australia, California, and Canada, and involves a submerged-culture technique where the respiratory growth of the *flor* yeasts is maintained with agitation and aeration throughout the whole volume of the wine. After *flor* treatment, the wines are fortified up to 19% vol. but they lack the complexity and finesse of *solera*-aged wines and are commonly used to improve the characteristics of baked sherry-like wines.

2.2.5.1.3 Baked-Type Liqueur-Wines

They are produced in the United States and Canada from the grape cultivars Thompson Seedless (Sultanina), Palomino, Tokay, Niagara, Malaga, and Emperor. The basic procedure includes fermentation to dryness or to a low sugar content but employs the usual baking practice (heating the wine at 49°C for 4 weeks), leading to slight caramelisation of the sugar and a certain degree of oxidation. The organoleptic characteristics of baked type fortified wines resemble the wines of the island of Madeira and are always finished sweet.

2.2.5.2 Port

Port (or Porto), protected designation of origin, is a liqueur-wine produced in the demarcated area of the Douro valley, in northern Portugal. The Douro wine region is a delimited area mainly bordering the Douro River from Régua some distance eastward. Vines are grown on the terraced slopes of the rocky sides of the river or its tributaries. The unfavorable locale of the vineyards combined with the hard climatic conditions makes this area a unique vine growing ecosystem. Final alcohol content of wine, after fortification, must be around 19% vol. in order to prevent any further fermentation. Most Port wines are not vintage dated. Producers blend samples from several vintages to produce the formula for brand name wines of consistent character. However, the best wines produced during an excellent year mature separately as Vintage ports in oak casks for 2–6 years and in bottles for a long period of up to 20 years. After long bottle aging, vintage Port develops a distinctive and highly complex fragrance. Wine from a single vintage, aged in cooperage for about 4–6 years before bottling, may be designated Late-bottling vintage (LBV) Port, that matures more rapidly than vintage port. After 2 to 3 years oak cask maturation, most Port wine is bottled and sold as Ruby Port. Blending small quantities of white Port into a Ruby Port may be used to produce brands of tawny Ports. High quality tawny Ports are derived from long aging in oak and this aging process is applied to blends sold as "10 years", "20 years", "30 years", and "More than 40 years".

Similar wines are also produced in various other countries, such as Australia and South Africa. Nevertheless, these Port-like wines cannot achieve the quality of Port wine. The free amino acid content may be a good indicator in discriminating between Port wine and Port imitations.

2.2.5.3 Madeira

Madeira (or Madère), protected designation of origin, is a liqueur-wine produced in the Portuguese island of the same name. It is a white wine of various styles with naturally high acidity and fortified with alcohol (up to 20% vol.) before fermentation has stopped. Then, it is baked in cement tanks or wooden casks before aging. Wines from different vintages and varieties usually are kept separate for many months during wood conservation, before final blending. Madeira wines are produced in a wide range of styles. Malmsey Madeiras are very sweet wines, Verdelho and especially Sercial styles are fermented to, or near, dryness, and Buals are fortified when about half the sugars have been fermented.

2.2.5.4 Other European Sweet Liqueur-Wines

In France, "vins doux naturels" are made in the areas of Pyrénées Orientales, Aude, Hérault, Vaucluse, and the island of Corse. Banyuls, Muscat de Rivesaltes, Grand Rousillon, Muscat de Frontignan, and Rasteau are some examples of these wines.

In Italy, the most typical examples of sweet liqueur-wines are Marsala and Moscato di Pantelleria.

In Greece, famous liqueur-wines are produced in the islands of Samos, Lemnos, Rhodes, as well as in the Achaia region (the Mavrodaphne of Patras) all of them deserving protected designations of origin. The rich terpenic profile of Samos liqueur-wines aroma, deriving from the Muscat lefko grapes used for their production, is well indicated in the relevant bibliography. Lemnos liqueur-wines are made from grapes of the aromatic cultivar Muscat of Alexandria.

2.2.6 Mistelles

Mistelles are fortified wines produced from unfermented fresh grapes or grape musts (1% vol. actual alcohol is tolerated) and rendered non-fermentable by addition of wine spirit or neutral alcohol of vine origin.

2.3 AROMATISED WINE-PRODUCTS

2.3.1 Aromatised Wines

Aromatised wines may contain alcohol from 14.5% to 22% vol. and a total alcoholic strength by volume of not less than 17.5% vol., according to the relevant European legislation [Regulation (EU) 251/2014]. These aromatised wine-products are obtained from grapevine products (representing at least 75% of the total volume of the final product) to which ethyl alcohol of agricultural origin may have been added and which have been flavoured with the aid of natural flavouring substances and/or aromatic herbs and/or spices and/or flavouring foodstuffs.

The main categories of aromatised wines are the following:

2.3.1.1 Vermouth

A large number of herbs and spices are used for vermouth production. The different parts of these herbs and spices, such as seeds, wood, leaves, bark, or roots, in dry form, are used. The most important herbs and spices are classified as bitter, aromatic, or bitter-aromatic.

Two classes of vermouth are recognised in the trade, the sweet or Italian-type vermouth and the dry or French type.

a) **Sweet vermouths** are produced in Italy, Spain, Argentina, and the United States. Typical Italian vermouth (at least 15.5% vol.) is dark amber in colour, with a light muscat, sweet nutty flavour, and a well-developed and pleasing fragrance. It has a generous and warming taste, with a slightly bitter but agreeable aftertaste.
b) **Dry vermouths** are mainly produced in France, the United States, and Hungary. They are usually not only much lower in sugar content and lighter in colour than the sweet, but also are often higher in alcohol content. The French methods of vermouth production are different from those of California.

2.3.1.2 Bitter Aromatised Wines

These are aromatised wines with a characteristic bitter flavour (Quinquina wine, Bitter vino, Americano).

2.3.1.3 Egg-Based Aromatised Wines

These are aromatised wines to which good quality egg yolk or extracts thereof have been added.

2.3.1.4 Väkevä viiniglögi/Starvinsglögg

These are aromatised wines, the characteristic taste of which is obtained by the use of cloves and/or cinnamon.

2.3.2 Aromatised Wine-Based Drinks

According to the relevant European legislation [Regulation (EU) 251/2014], aromatised wine-based drink is a drink obtained from one or more of the grapevine products (representing at least 50% of the total volume) to which no alcohol has been added (except in specific cases), and which has an actual alcoholic strength by volume of not less than 4.5% vol. and less than 14.5% vol.

The main categories of aromatised wine-based drinks, according to Regulation (EU) 251/2014, are the following:

2.3.2.1 Sangría/Sangria

This is an aromatised wine-based drink which is obtained from wine and is aromatised with the addition of natural citrus-fruit extracts or essences, with or without the juice of such fruit to which spices may have been added, which have an actual alcoholic strength by volume of not less than 4.5% vol., and less than 12% vol. It may contain solid particles of citrus-fruit pulp or peel, and its colour must come exclusively from the raw materials used.

The term "Sangría" or "Sangria" may be used as a sales denomination only when the relevant product is produced in Spain or Portugal.

2.3.2.2 Clarea

This is an aromatised wine-based drink, which is obtained from white wine under the same conditions as Sangría/Sangria. The term "Clarea" may be used as a sales denomination only when the relevant drink is produced in Spain.

2.3.2.3 Zurra

This is an aromatised wine-based drink obtained by adding brandy or wine spirit to Sangría/Sangria and Clarea, possibly with the addition of pieces of fruit. The actual alcoholic strength by volume must be not less than 9% vol. and less than 14% vol.

2.3.2.4 Bitter Soda

This is an aromatised wine-based drink which is obtained from 'bitter vino' (bitter aromatized wine), the content of which in the finished product must not be less than 50% by volume, to which carbon dioxide or carbonated water has been added, and which has an actual alcoholic strength by volume of not less than 8% vol., and less than 10.5% vol.

2.3.2.5 Kalte Ente

Aromatised wine-based drink which is obtained by mixing wine, semi-sparkling wine, or aerated semi-sparkling wine with sparkling wine or aerated sparkling wine, to which natural lemon substances or extracts thereof have been added, and which has an actual alcoholic strength by volume of not less than 7% vol.

2.3.2.6 Glühwein

This is an aromatised wine-based drink which is obtained exclusively from red or white wine, which is flavoured mainly with cinnamon and/or cloves, and which has an actual alcoholic strength by volume of not less than 7% vol.

2.3.2.7 Viiniglögi/Vinglögg/Karštas vynas

This is an aromatised wine-based drink which is obtained exclusively from red or white wine, which is flavoured mainly with cinnamon and/or cloves, and which has an actual alcoholic strength by volume of not less than 7% vol.

2.3.2.8 Maiwein

This is aromatised wine-based drink which is obtained from wine in which Galium odoratum (L.) Scop. (*Asperula odorata* L.) plants or extracts thereof have been added, so as to ensure a predominant taste of *Galium odoratum* (L.) Scop. (*Asperula odorata L.*), and which has an actual alcoholic strength by volume of not less than 7% vol.

2.3.2.9 Maitrank

Aromatised wine-based drink which is obtained from white wine in which Galium odoratum (L.) Scop. (Asperula odorata L.) plants have been macerated, or to which extracts thereof have been added with the addition of oranges and/or other fruits, possibly in the form of juice, concentrated or extracts, and with maximum 5% sugar sweetening, and which has an actual alcoholic strength by volume of not less than 7% vol.

2.3.2.10 Pelin

This is an aromatised wine-based drink which is obtained from red or white wine and specific mixture of herbs, which has an actual alcoholic strength by volume of not less than 8.5% vol., and which has a sugar content expressed as invert sugar of 45–50 grams per litre, and a total acidity of not less than 3 grams per litre expressed as tartaric acid.

2.3.2.11 Aromatizovaný Dezert

This is an aromatised wine-based drink which is obtained from white or red wine, and a sugar and dessert spices mixture, which has an actual alcoholic strength by volume of not less than 9% vol. and less than 12% vol., and which has a sugar content expressed as invert sugar of 90–130 grams per litre and a total acidity of at least 2.5 grams per litre expressed as tartaric acid.

"Aromatizovaný dezert" may be used as a sales denomination only when the product is produced in the Czech Republic.

2.3.3 Aromatised Wine-Product Cocktails

According to the common European legislation [Regulation (EU) 251/2014], aromatised wine-product cocktails are considered the drinks obtained from one or more of the grapevine products (representing at least 50% of the total volume), to which no alcohol has been added, and which have an actual alcoholic strength by volume of more than 1.2% vol. and less than 10% vol.

An aromatised semi-sparkling grape-based cocktail is considered the aromatised wine-product cocktail obtained exclusively from grape must, which has an actual alcoholic strength by volume of less than 4% vol., and which contains carbon dioxide obtained exclusively from fermentation of the products used.

Sparkling wine cocktail is the aromatised wine-product cocktail, which is mixed with sparkling wine.

2.4 BEVERAGES OBTAINED BY DEALCOHOLISATION OF WINE

Several factors, such as the perceived social effects of excessive alcohol consumption, severe drunk-driving penalties, and the demand for beverages destined for "easy" consumption, have led some winemakers to produce wine-products with lower than normal alcoholic strength. However, the production and especially the labelling of such beverages can lead to legal complexities in many countries worldwide.

Two categories of products obtained by dealcoholisation of wine are foreseen in the OIV International Code of oenological practices:

a) **Beverage obtained by dealcoholisation of wine** is a beverage obtained exclusively from wine or special wine as described in the International Code of oenological practices of the OIV, which has undergone exclusively specific treatments, for this type of product, in accordance with the OIV International Code of oenological practices, in particular a dealcoholisation, and that has an alcoholic strength by volume below 0.5% vol.
b) **Beverage obtained by partial dealcoholisation of wine** is a beverage obtained exclusively from wine or special wine as described in the International Code of oenological practices of the OIV, which has undergone exclusively specific treatment, for this type of product, in accordance with the OIV International Code of oenological practices, in particular a dealcoholisation, and that has an alcoholic strength by volume equal or above 0.5% vol. and less than the applicable minimum alcoholic strength of wine or special wine.

2.5 FRUIT WINES

Apart from grapes, several other fruits are used in many parts of the world for the production of alcoholic beverages by fermentation, called in many countries fruit "wines". Nevertheless, in some European countries, the term "wine" is not allowed to be used, as part of a compound term, for the description of alcoholic beverages obtained by fermentation of fruits other than grapes.

Apples, berries, cherries, wild apricots, kiwifruits and many other fruits are converted into the corresponding "wines", in some cases in important quantities.

2.5.1 Apple Wines

Undoubtedly, apple wine (cider) is the most known beverage of this category. Ciders are traditional products in some European countries such as the United Kingdom, France, Germany, and Ireland, but are also made in significant quantities in the United States, Argentina, Australia, India, Canada, and other countries around the world. Apple wines are produced dry or sweet, low alcoholic or high alcoholic, non-sparkling or sparkling. While ciders can be made from any apple variety, they are typically produced from specific cultivars that are high in sugar and phenolic content. The technology of apple wine making is relatively different in the producing countries, but in most cases it comprises fermentation of the juice of fresh apples, possibly in combination with concentrated apple juice, as well as the addition of sugar before fermentation for the increase of the alcoholic strength of the final product.

2.5.2 Other Fruit Wines

Red currant, blackberry, raspberry, elderberry, and even strawberry juices are fermented in order to produce berry wines. Berry wines should not be aged too long as they may lose their colour and flavour.

Cherry wine is made under similar techniques for berry wine making in Europe and America. Sour cherries are preferred to the sweet dessert varieties because they contain higher acidity and richer flavour than the latter.

Wild apricots are found in abundance in dry and temperate regions of India and the fruit can be utilised for making wine of acceptable quality.

A significant proportion of the kiwifruit production of some regions (in New Zealand, California, Australia, Israel, etc.) is converted into kiwifruit wine, which is different from most fruit wines in that its flavour bears no relationship to this of the fruit it was made from. Young kiwifruit wines compare favorably with young Müller-Thurgau wines.

In France, Switzerland, and Germany, special varieties of pears of high tannin content are grown and used for making perry, a cider-like fermented pear beverage, or the juice is blended with apple juice before fermentation. The fermentation product is dry and fairly palatable but may be sweetened and pasteurised before bottling.

Other fruits that are used for the production of fruit wines are plums (especially in Japan, the United States, Korea, and China), pomegranates, pineapples, peaches, mango, oranges, grapefruits, bananas, plantains, guava, kinnows, palms, muskmelon, and dried figs. Many of them are produced commercially either as "table" fruit wines or as "dessert" fruit wines after being fortified and/or sweetened.

2.6 WINE SPIRITS AND BRANDIES

2.6.1 Wine Spirits

According to the definition given by OIV, wine spirit is the spirit beverage obtained exclusively by the distillation of wine, fortified wine, wine possibly with the addition of wine distillate, or by re-distillation of a wine distillate, with the result that the product retains the taste and aroma of the above-mentioned raw materials and the final product must not be less than 37.5% vol.

Based on the relevant European legislation [Regulation (EU) 2019/787], wine spirit is the spirit drink that has the following characteristics:

(a) It is produced exclusively by distillation at less than 86% vol. of wine, wine fortified for distillation or wine distillate, distilled at less than 86% vol.
(b) It has a volatile substances content equal to or exceeding 125 grams per hectolitre of 100% vol. alcohol.
(c) It has a maximum methanol content of 200 grams per hectolitre of 100% vol. alcohol.
(d) It has minimum alcoholic strength by volume 37.5% vol.
(e) It has no added alcohol.
(f) It is not flavoured, but this does not exclude traditional production methods.
(g) It may only contain added caramel as a means of adjusting the colour.
(h) It may be sweetened in order to round off the final taste. However, the final product may not contain more than 20 grams of sweetening products per litre, expressed as invert sugar.
(i) Where wine spirit has been matured, it may continue to be placed on the market as 'wine spirit', provided that it has been matured for as long as, or longer than, the maturation period provided for "Brandy", as this spirit drink is determined in the same Regulation.

2.6.1.1 Wine Spirits of France

The most famous wine spirits are those of France, and some of them that deserve geographical indications according to the relevant European legislation constitute a major market for French economy, such as:

a) **Cognac:** The denomination "*Cognac*" may be supplemented by the following terms: *Fine, Grande Fine Champagne, Grande Champagne, Petite Fine Champagne, Petite Champagne, Fine Champagne, Borderies, Fins Bois, Bons Bois,*
b) **Armagnac:** The denomination "*Armagnac*" may be supplemented by the following terms: *Bas-Armagnac, Haut-Armagnac, Armagnac-Ténarèze, Blanche Armagnac,*
c) **Other:** "*Eaux-de-vie de vin*" *Eaux-de-vie de vin de la Marne, Eau-de-vie de vin des Côtes-du-Rhône,* etc.

The wines that are used for Cognac production derive mainly from the grape varieties Ugni Blanc, Colombard, and Folle Blanche. Some other varieties such as Sémillon, Blanc Ramé, and Jurançon Blanc may be also used in maximum percentage of 10%. The principal grape varieties for Armagnac production are the same in the case of Cognac, as well as the variety Baco 22A. The secondary cultivars are Blanquette, Clairette, Jurançon, Meslier, and Mauzac.

In both cases, by the end of alcoholic fermentation the new wine must be quickly distilled so that no oxidation or development of microorganisms in the resultant wine appears. It has been shown that wine spirits made by traditional process (Cognac, Armagnac) have exactly the same composition as wine; only the heaviest and most polar products are rectified (acetic acid, phenyl ethanol, polyols, etc.) during distillation. Major defects can only be eliminated with the distillation column but at the cost of eliminating components contributing to quality. Good quality wine spirits can be made only with good quality wines and therefore expensive aging in oak casks should be reserved only for noble products.

Cognac usually acquires its best quality after an aging of 15 to 20 years in oak casks (Limousin or Tronçais), but in some cases this improvement may continue to 40 or even 50 years. Armagnac is aged in native Gascon oak casks. While some Armagnac is aged for 20 or more years, much is sold as blend after 2 to 10 years of aging.

2.6.1.2 Wine Spirits of Other Countries

Wine spirits are also produced in many other winemaking parts of Europe (Spain, Italy, Germany, Portugal, Greece, Romania, Bulgaria, etc.) and other countries such as the United States, Russia, and South Africa.

Well known wine spirits from Portugal are Aguardente de Vinho Douro, Aguardente de Vinho Ribatejo, Aguardente de Vinho Alentejo, Aguardente de Vinho da Região dos Vinhos Verdes, and Aguardente de Vinho Lourinhã, all deserving geographical indications according to the relevant European legislation.

A famous wine spirit of South America is Pisco which is produced by the distillation of wine of Muscat cultivars, Quebranta Moscatel and Albilla. The grapes are crushed mechanically or by the traditional method of stepping on the grapes. The must is kept in big deposits during fermentation or until ready for distillation. The distillation is done either in batch distillers or in falcas (very old distillers), and the final spirit is generally consumed without having being aged, aiming at preserving its terpenic aromatic profile. The origin of Pisco is the area of Ica (Peru), but nowadays is produced, apart from Peru, in Chile, Argentina, and Bolivia.

2.6.2 Brandies

Following the definition given by OIV, brandy is a spirit beverage obtained exclusively by the distillation of wine, fortified wine, and wine possibly with the addition of wine distillate or by re-distillation of a wine distillate, with the result that the product retains the taste and aroma of the above-mentioned raw materials. Also, a certain period of aging in oak wood containers is obligatory before marketing. The final product must not be less than 36% vol.

According to the relevant European legislation [Regulation (EU) 2019/787], brandy (or weinbrand) is the spirit drink that has the following characteristics:

(a) It is produced from wine spirit to which "wine distillate" may be added, provided that that wine distillate has been distilled at less than 94.8% vol. and does not exceed a maximum of 50% of the alcoholic content of the finished product. [The term "wine distillate", according to Regulation (EU) 2019/787, refers to an alcoholic liquid which is the result of the distillation of wine which does not have the properties of ethyl alcohol and which retains the aroma and taste of the raw material used].
(b) It is matured for at least one year in oak receptacles with a capacity of at least 1000 litres each, or for at least six months in oak casks with a capacity of less than 1000 litres.
(c) It has a volatile substances content equal to or exceeding 125 grams per hectolitre of 100% vol. alcohol, and derived exclusively from the distillation of the raw materials used.
(d) It has a maximum methanol content of 200 grams per hectolitre of 100% vol. alcohol.
(e) It has a minimum alcoholic strength by volume of 36% vol.
(f) It contains no added alcohol.
(g) It is not flavoured, but this does not exclude traditional production methods.
(h) It may only contain added caramel as a means of adjusting the colour.
(i) It may be sweetened in order to round off the final taste. However, the final product may not contain more than 35 grams of sweetening products per litre, expressed as invert sugar.

Brandies are produced in many European countries including Spain, Germany, Italy, Austria, Portugal, Romania, Moldova, Greece, Bulgaria, Cyprus, Georgia, and Armenia. Some of them deserve geographical designations such as Brandy de Jerez (Spain), Brandy de Penedés (Spain), Brandy Italiano (Italy), Deutscher Weinbrand (Germany), Pfälzer Weinbrand (Germany), and Wachauer Weinbrand (Austria).

In the Caucasus region, along the eastern shore of the Black Sea, in the ancient nations of Georgia and Armenia, intensely flavoured brandy-style spirit drinks are produced from wines deriving from grapes of local cultivars as well as imported cultivars such as Muscadine from France, and Sercial and Verdelho from Madeira. Most of the Armenian brandies, especially those aged for 8, 12, or 18 years in oak casks and considered as very high quality products, are exported to Russia and a little quantity to western countries.

Brandies and similar spirits are also produced in many other non-European countries around the globe such as South Africa, the United States (especially in California), Mexico, Australia, Argentina, and Israel.

In South Africa three types of brandies are found: Pot still brandies (which have to be aged for a minimum of three years), vintage brandies (which have to be aged for a minimum of eight years), and blended brandies. The wines that are mainly used for South Africa brandies derive from grapes of Ugni blanc, Colombard, Chenin blanc, and Palomino cultivars. Pot still and vintage South African brandies have a very pleasant character and have recently gained a remarkable reputation.

Although, as mentioned before, the term "brandy" refers only to the distilled spirit originating from fermented grapes and aged in oak casks, sometimes it is incorrectly used to refer to "grape marc spirit" (such as the Grappa of Italy) under the name "pomace brandy" or to other fruit spirits under the names "apple brandy" (such as Calvados of France and Applejack of the United States), "plum brandy", "apricot brandy", "peach brandy", and "cherry brandy", etc. Nevertheless, throughout this paragraph, the term "brandy" has been used only to refer to its proper meaning, which is a spirit originating from wine.

2.6.3 Raisin Brandy

A different category of brandy is "raisin brandy" or "raisin spirit" which, according to the relevant European legislation [Regulation (EU) 2019/787], is the spirit drink produced exclusively by the distillation of the product obtained by the alcoholic fermentation of extract of dried grapes of the "Corinth Black" or "Moscatel of the Alexandria" varieties, distilled at less than 94.5% vol., so that the distillate has an aroma and taste derived from the raw material used.

Raisin brandy has the following characteristics:

(a) The minimum alcoholic strength by volume is 37.5% vol.
(b) No addition of alcohol takes place.
(c) It is not flavoured.
(d) It may only contain added caramel as a means for adjusting the colour.
(e) It may be sweetened in order to round off the final taste. However, the final product may not contain more than 20 grams of sweetening products per litre, expressed as invert sugar.

2.6.4 Grape Marc Spirit

According to the relevant European legislation [Regulation (EU) 2019/787], "grape marc spirit" or "grape marc" is a spirit drink which meets the following requirements:

(a) It is produced exclusively from grape marc fermented and distilled, either directly by water vapour, or after water has been added and both of the following conditions are fulfilled:

i. each and every distillation is carried out at less than 86% vol.,
ii. the first distillation is carried out in the presence of the marc itself .

(b) A quantity of lees may be added to the grape marc that does not exceed 25 kg of lees per 100 kg of grape marc used.
(c) The quantity of alcohol derived from the lees shall not exceed 35% of the total quantity of alcohol in the finished product.
(d) It has a volatile substances content equal to or exceeding 140 grams per hectolitre of 100% vol. alcohol and has a maximum methanol content of 1000 grams per hectolitre of 100% vol. alcohol.
(e) The minimum alcoholic strength by volume of grape marc spirit or grape marc is 37.5% vol.
(f) No addition of alcohol takes place.
(g) It is not flavoured, but this does not exclude traditional production methods.
(h) It may only contain added caramel as a means for adjusting the colour.
(i) It may be sweetened in order to round off the final taste. However, the final product may not contain more than 20 grams of sweetening products per litre, expressed as invert sugar.

Grape marc spirits are mainly produced in the winemaking countries of Europe (France, Italy, Portugal, Greece, Spain, Cyprus, Hungary, Georgia, Serbia, Bulgaria, etc.) but also in other parts of the world (such as South Africa, Israel and South America), sometimes sold under the name "pomace brandy".

Well known European grape marc spirits are Marc de Champagne/Eau-de-vie de marc de Champagne (France), Marc d'Aquitaine/Eau-de-vie de marc originaire d'Aquitaine (France), Marc de Bourgogne/Eau-de-vie de marc de Bourgogne (France), Marc d'Alsace (France), Grappa (Italy), Aguardente Bagaceira (Portugal), Tsipouro (Greece) (Plate 2.7), Tsikoudia (Greece), and Zivania (Cyprus), all deserving geographical indications according to the relevant European legislation.

2.7 LEGISLATION

2.7.1 OIV

All vitivinicultural products (grapes, musts, wines, special wines, mistelles, other vitivinicultural products), as well as spirits, alcohols, and spirituous beverages of vitivinicultural origin, are well defined and regulated in the relevant Resolutions of the International Organisation of Vine and Wine (OIV). As known, the OIV is an intergovernmental organisation of a scientific and technical nature, of recognised competence for its works concerning vine, table grapes, raisins, and other vine-based products, wines, and wine-based beverages. As of 6 November 2018, the International

PLATE 2.7 Distillery for the production of grape marc spirit (GI: Tsipouro) in Greece

Organisation of Vine and Wine was made up of 47 Member States.

OIV publications comprise the definitions of vitivinicultural products and spirituous beverages of vitivinicultural origin (International Code of Oenological Practices, Part I), as well as their labelling (International Standard for the Labelling of Wines, International Standard for the Labelling of Spirituous Beverages of Vitivinicultural Origin), the names of vine varieties (International List of Vine Varieties and their synonyms), the oenological treatments and practices (International Code of Oenological Practices, Part II), the maximum acceptable limits for several analytical parameters of wines (International Code of Oenological Practices, Annex), the specifications of oenological products (International Oenological Codex), the methods of analysis (Compendium of International Methods of Analysis of Wines and Musts, vol. 1, 2, and Compendium of International Methods of Analysis of Spirituous Beverages of Vitivinicultural Origin), and the good practices guidelines.

2.7.2 European Legislation

Regarding European legislation of vitivinicultural products, the main provisions for this sector are comprised, among other agricultural sectors, in Regulation (EU) 1308/2013, establishing a common organisation of the markets in agricultural products. According to Article 80 par. 3(a) of this Regulation, when the EU authorises oenological practices for wine, the EU Commission must take into account the oenological practices and methods of analyses recommended and published by the OIV, as well as the results of experimental use of as-yet unauthorised oenological practices. The issues of oenological treatments and practices are covered in detail by Regulation (EU) 2019/934, as regards (among other topics) the authorised oenological practices and restrictions applicable to the production and conservation of grapevine products, according to the OIV publications.

Detailed provisions concerning labelling of wines are included in Regulation (EU) 2019/33 as regards the protection of designations of origin, geographical indications, and traditional terms in the wine sector and the labelling and presentation of wines.

The issues of aromatized vitivinicultural products are regulated in Regulation (EU) 251/2014 on the definition, description, presentation, labelling, and the protection of geographical indications of aromatized wine products.

The legislation that concerns spirits of vitivinicultural origin (wine spirit, brandy, grape marc spirit, and raisin brandy) is comprised in Regulation (EU) 2019/787 on the definition, description, presentation, and labelling of spirit drinks, the use of the names of spirit drinks in the presentation and labelling of other foodstuffs, the protection of geographical indications for spirit drinks, and the use of ethyl alcohol and distillates of agricultural origin in alcoholic beverages. This Regulation also includes the legislation for the production, geographical indications, and labelling of any other kind of spirit drink (of non-vitivinicultural origin) that is placed in the EU market, such as rum, whisky, vodka, liqueur, etc.

BIBLIOGRAPHY

Amerine M.A., Berg H.W. and Cruess W.V. (1972). *The Technology of Wine Making*. 3rd ed. AVI Publ. Co. Inc., Westport, p. 419, 455, 525.

Amerine M.A., Kunkee R.E., Ough C.S., Singleton V.L. and Webb A.D. (1980). *Technology of Wine Making*. AVI Publ. Co. Inc., Connecticut, CT.

Baron R., Mayen M., Merida J. and Medina M. (2000). Comparative study of browning and flavan-3-ols during the storage of white sherry wines treated with different fining agents *Journal of the Science of Food and Agriculture*, 80(2): 226.

Bertrand A. (2003). Armagnac and wine-spirits. In: *Fermented Beverage Production*. Eds. A.G.H. Lea and J.R. Piggott. Springer Science and Business Media, New York, p. 213–238.

Blouin J., Boulet J.C., Escudier J.L., Feuillat M., Flanzy C., Peyron D. et Razungles A. (1998). Vinifications en rouge. In: *Enologie, fondements scientifiques et technologiques*. C. Flanzy, Coord. TEC & DOC Lavoisier, Paris, p. 759.

Boulton R. (2003). Red wines. In: *Fermented Beverage Production*. Eds. A.G.H. Lea and J.R. Piggott. Springer Science and Business Media, New York, p. 107–137.

Castro R. and Barroso C.G. (2000). Behavior of a hyperoxidized must during biological aging of Fino Sherry wine *American Journal of Enology and Viticulture*, 51(2): 98.

Charpentier C., Etiévant P. and Guichard E. (1998). Vinifications des vins de voile: Vin Jaune Xérès et autres. In: *Enologie, fondements scientifiques et technologiques*. C. Flanzy, coord. TEC & DOC Lavoisier, Paris, p. 875–879.

Commission delegated regulation (EU) 2019/33 (Official Journal of the EU, L 9, 11/01/2019).

Commission delegated Regulation (EU) 2019/934 (Official Journal of the EU, L 149, 07/06/2019).

Dhiman A.K. and Attri S. (2011). Production of brandy. In: *Handbook of Enology, Principles, Practices and Recent Innovations. Volume III, Technology of Production and Quality Control*. Eds. V.K. Joshi. Asiatech Publishers, New Delhi, India, p. 1222–1283.

European Parliament and Council Regulation (EU) 1308/2013 (Official Journal of the EU, L 347, 20/12/2013).

European Parliament and Council Regulation (EU) 251/2014 (Official Journal of the EU, L 84, 20/03/2014).

European Parliament and Council Regulation (EU) 2019/787 (Official Journal of the EU, L 130, 17/05/2019).

Enoteca Italiana (1999). The list of Italian DOC and DOCG wines. In *Ente Mostra Vini, Enoteca Italiana,* XIV ed., Siena, Italy. p. 71, 145, 212.

Ewart A. (2003). White wines. In: *Fermented Beverage Production*. Eds. A.G.H. Lea and J.R. Piggott. Springer Science and Business Media, New York, p. 89–106.

Herbert P., Barros P. and Alves A. (2000). Detection of Port wine imitations by discriminant analysis using free amino acids profiles *American Journal of Enology and Viticulture*, 51(3): 262.

Jackson R.S. (2000). *Wine Science: Principles, Practice, Perception*. Academic Press, San Diego.

Jackson R.S. (2003). Wines: Types of table wine. In: *Encyclopedia of Food Sciences and Nutrition*. 2nd ed. Eds. B. Caballero, L. Trugo and P.N. Figlas. Elsevier Science Ltd., UK.

Jackson R.S. (2011). Red and white wines. In: *Handbook of Enology, Principles, Practices and Recent Innovations. Volume III, Technology of Production and Quality Control*. Ed. V.K. Joshi. Asiatech Publishers, New Delhi, India p. 981–1020.

Joshi V.K., Bhutani V.P. and Sharma R.C. (1990). The effect of dilution and addition of nitrogen source on chemical, mineral and sensory qualities of wild apricot wine *American Journal of Enology and Viticulture*, 41(3): 229.

Joshi V.K., Attri D., Singh T.K. and Abrol G.S. (2011). Fruit wines: Production Technology. In: *Handbook of Enology, Principles, Practices and Recent Innovations. Volume III, Technology of Production and Quality Control*. Ed. V.K. Joshi. Asiatech Publishers, New Delhi, India, p. 1177–1221.

Joshi V.K., Panesar P.S., Rana V.S. and Kaur S. (2017). Science and technology of fruit wines: An overview. In: *Science and Technology of Fruit Wine Production*. Eds. M.R. Kosseva, V.K. Joshi and P.S. Panesar. Elsevier Inc., AP, USA, pp. 1–72.

Karagiannis S. (2011). Classification and characteristics of wines and brandies. In: *Handbook of Enology, Principles, Practices and Recent Innovations. Volume I, Introduction to Vine and Wine*. Ed. V.K. Joshi. Asiatech Publishers, New Delhi, India, pp. 46–65.

Karagiannis S., Economou A. and Lanaridis P. (2000). Phenolic and Volatile composition of wines made from Vitis vinifera Cv Muscat Lefko grapes from the islands of Samos . *Journal of Agriculture and Food Chemistry*, 48(11): 5369–5375.

Lichine A. (1988). *Encyclopédie des vins & des alcohols*. Ed. Robert Laffont, Paris, France.

Louw L. and Lambrechts M.G. (2012). Grape-based brandies: Production, sensory properties and sensory evaluation. In: *Alcoholic Beverages. Sensory Evaluation and Consumer Research*. Ed. J. Piggott. Woodhead Publishing, Cambridge, UK.

Navarre C. (1998). *L'oenologie*. 4e ed. TEC & DOC Lavoisier, Paris, pp. 149–176.

Organisation International de la Vigne et du Vin (OIV). *Norme Internationale pour l'étiquetage des boissons spiritueuses d' origine vitivinicole*. Edition (2013). OIV: Paris, 75008.

Organisation International de la Vigne et du Vin (OIV). *Norme Internationale pour l'étiquetage des vins*. Edition (2015). OIV: Paris, 75008.

Organisation International de la Vigne et du Vin (OIV). *Code International des pratiques oenologiques*. Edition (2019). OIV, Paris, 75008.

Organisation International de la Vigne et du Vin (OIV). *Codex Oenologique Intrenational*. Edition (2019). OIV, Paris, 75008.

Organisation International de la Vigne et du Vin (OIV). *Receuil des Methodes Internationales d' analyse des vins et des moûts*. Vol. 1, 2. Edition (2019). OIV: Paris, 75008.

Organisation International de la Vigne et du Vin (OIV). *Receuil des methodes internationales des boissons spiritueuses d' origine vitivinicole.* Edition (2019). OIV: Paris, 75008.

Panesar P.S., Marwaha S.S., Sharma S. and Kumar H. (2011). Preparation of fortified wines. In: *Handbook of Enology, Principles, Practices and Recent Innovations. Volume III, Technology of Production and Quality Control.* Ed. V.K. Joshi. Asiatech Publishers, New Delhi, India p. 1021–1063.

Ribéreau, Gayon P., Dubourdieu D., Doneche B. and Lonvaud A. (2004). Traité d' oenologie. 1. *Microbiologie du vin. Vinifications.* 5ᵉ ed. Dunod.

Ribéreau, Gayon P., Dubourdieu D., Doneche B. and Lonvaud A. (2004). Traité d' oenologie. 2. *Chimie du vin. Stabilisation et traitements.* 5ᵉ éd. Dunod.

Robillard B., Delpuech E., Viaux L., Malvy J., Vignes, Adler M. and Duteurtre B. (1993). Improvements of methods for sparkling base wine foam measurements and effect of wine filtration on foam behavior *American Journal of Enology and Viticulture*, 44(4): 387.

Robinson J. (1999). *Oxford Companion to Wine.* 2nd ed. Oxford University Press, Oxford UK.

Schmidtke L.M., Delves T. and Agboola S. (2011). Technology of production of reduced alcoholic wines. In: *Handbook of Enology, Principles, Practices and Recent Innovations. Volume III, Technology of Production and Quality Control.* Ed. V.K. Joshi. Asiatech Publishers, New Delhi, India p. 1152–1176.

Schreier P. (1979). Flavor composition of wines: A review *CRC Critical Reviews in Food Science and Nutrition*, 12(1): 59.

Somers T.C. and Evans M.E. (1986). Evolution of red wines. I. Ambient influences on colour composition during early maturation *Vitis*, 25: 31.

Vaidya M.K., Vaidya D. and Joshi V.K. (2011). Wine regions and status of world wine production. In: *Handbook of Enology, Principles, Practices and Recent Innovations. Volume I, Introduction to Vine and Wine.* Ed. V.K. Joshi. Asiatech Publishers, New Delhi, India p. 66–82.

Vlassov V.N. and Maruzhenkov D.S. (1999). Application of GC/MS method for the identification of Brandies and Cognacs *Analusis*, 27(7): 663.

3 Wine
Composition and Nutritive Value

S.K. Soni, S.S. Marwaha, R. Soni, Urvashi Swami and U. Marwaha

CONTENTS

3.1 INTRODUCTION

Wine making as a technology involves several unit operations and they are likely to impact the levels of different components of wine. There are several types of wines available around the world based on the type of fruit used, source of yeast inocula, color, alcohol content, presence or absence of carbon dioxide, and processing of the fruit prior to fermentation, etc. The composition and nutritive value are the characteristics of every wine which are affected by the fruit and the fermentation process. This has been discussed in this chapter.

3.2 COMPOSITION OF GRAPE MUST AND WINE

Grapes are the principal fruit used in the preparation of various varieties of wine and the composition of berry at harvest is very important. Grape berry contains a number of components like water, other inorganic substances, carbohydrates, acids, phenolics, nitrogenous components, terpenoids, fats, volatile compounds, odorants, flavor compounds, and vitamins which are passed into the resulting wine after the fermentation of grapes (Table 3.1).

TABLE 3.1
Approximate Composition of Grapes and Wine

Component/compound	% in grapes	% in wine
Water	75.0	86.0
Sugars (fructose, glucose with minor levels of sucrose)	22.0	0.3
Alcohols (ethanol with trace levels of terpenes glycerol, higher alcohols)	0.1	11.2
Organic acids (tartaric, malic, with minor levels of lactic, succinic, oxalic acids, etc.)	0.9	0.6
Minerals (potassium, calcium, with minor levels of sodium, magnesium, iron, etc.)	0.5	0.5
Phenols (flavonoids such as color pigments along with nonflavonoids such as cinnamic acid and vanillin)	0.3	0.3
Nitrogenous compounds (protein, amino acids, humin, amides, ammonia, etc.)	0.2	0.1
Flavor compounds (esters such as ethyl caproate, ethyl butyrate, etc.)	Trace	Trace

Source: Amerine (1956)

Chemically, wine is a complex beverage. In addition to water and ethanol there are a variety of minerals, vitamins, esters, aldehydes, and phenolic compounds present in it. As the fermentation process includes crushing and pressing of the fruits, it enables the extraction of the fruit components into must, which thereby get imparted into wine. Fruits and herbs contain various phyto-nutrients, whose extraction and bio-availability in the wines depend upon the vinification process, ageing of wine, and many other factors. The major constituents of grape, must and wine, have been described here.

3.2.1 Water

Water is a major part of the wine and can be present up to 90 % v/v. While being a major ingredient, it is not given much importance. However, it is noteworthy that water plays a major role in extraction of bio-actives into wine and is necessary for the action of many enzymes. There are hardly any compounds in wine which are water insoluble. Therefore, it is an integral solvent in wine. It is also an essential component for the fermentation process and ageing.

3.2.2 Alcohols

3.2.2.1 Ethanol

Ethanol is the main alcohol present in wine and is produced during fermentation of sugars by yeasts whose growth and metabolism are affected by temperature, which determines the final alcohol content. Besides adding their own characteristic flavors and odors, alcohols are the main carriers of aroma or bouquet. Ethyl alcohol has high miscibility with water and acts as a solvent for a number of minor components, especially those resposible for the flavor of wines. Quantitatively, ethyl alcohol is the most important component present in all alcoholic beverages, and its content in wines generally varies with the type of the product (Table 3.2). Ethanol concentration can be expressed as % w/v or % v/v, specifically, gravity or proof. The generalized method of expression is % v/v. Ethanol, being the second most abundant component of wine, is an important co-solvent along with water. It aids the extraction of grape constituents into wine. The most important health beneficial constituents of wines, polyphenolics, are non-polar. Therefore, ethanol is a crucial factor for maintaining the phenolic content of wines. Not only this, ethanol effects the volatile aromatic compounds and adds to the beautiful aroma of wines. At low ethanolic concentrations, ethanol takes the form of a mono-dispersed, aqueous solution. Concentrations below 7% v/v favor the discharge of volatile compounds. This influences the sensorial attributes and the overall structure of a wine. However, with the increase in ethanol levels, the hydrophobic hydration is reduced. This leads to an increase in the solubility of nonpolar compounds. High ethanol concentration also reduces tannin-protein interactions and leads to decreased astringency of wines and therefore, a moderate quantity of ethanol is desirable in wines as inadequate quantities hamper the overall quality of wine.

TABLE 3.2
Ethyl Alcohol Content of Different Alcoholic Beverages

Beverages	Ethyl alcohol content (% v/v)
Wines	12.2
Table wines	11–14
Dry red wines	12.6
Sweet white wines	19.3
Sweet red wines	19.3
Sparkling wines	13.2
Champagne	11.5–13.0
Fortified wines	> 15
Dry white wines	12.4

Source: Amerine (1953); Amerine *et al.* (1972)

3.2.2.2 Higher Alcohols

In addition to ethyl alcohol, wines also contain several other alcohols including polyalcohols and cyclic alcohols, which add to the degree of sweetness and aroma of wines and whose quantities vary with the type of beverage (Tables 3.3). Higher alcohols are among the most vital volatile compounds produced during wine fermentation and are also known as fusel

TABLE 3.3
Higher Alcohols and Polyols of Wine

	Concentration (mg/l)			
	White wines		Red wines	
Alcohols	Minimum	Maximum	Minimum	Maximum
2-Methyl-propanol	28	170	45	140
1-Butanol	0.5	8.5	0.5	2.3
2-Methyl-butanol	17	82	48	150
3-Methyl-butanol	70	320	117	490
1-Hexanol	1	10	12	10
2-Phenyl-ethanol	15	250	42	129
Glycerol	5600	9460	7900	9200
2,3 Butanediol	300	600	486	576

Source: Bertrand, (1975)

oils. Deamination and decarboxylation of amino acids result in the formation of higher alcohols which are dependent upon yeast species. After initiation of fermentation, the levels of amino acids like isoleucine, leucine, and valine diminish rapidly within 18–38h, though the formation of higher alcohols continues throughout the fermentation period.

Most of the higher alcohols, excluding ethanol, are formed during maceration process. They may generate from grape-derived aldehydes, by the reductive denitrification of amino acids, or via synthesis from sugars. Higher alcohols contribute to aged wine bouquet, as they react with organic acids and lead to the production of esters. Other alcohols include diols like 2,3-butanediol, polyols like glycerol, and sugar alcohols. In dry wine, after water and ethanol, glycerol is the most abundant compound. It is present in higher quantities in red wine (~10 g/l) as compared to the white ones (~7 g/l). Quantities of higher alcohols up to 400 mg/l improve the aroma, while a concentration higher than this level deteriorates the quality of wine and results in a hangover after consumption of such wines.

3.2.2.3 Methanol

Of all the alcohols, methanol is most undesirable and is toxic at 4% concentration, causing blindness or even death. Hydrolysis of pectin in the fruit results in the formation of methanol and trace levels of it are in wine (Table 3.4). The concentration of methanol is influenced by the type of fermenting organism used, the raw materials, and the fermentation temperature.

3.2.3 Acids

Organic acids represent the second largest group of compounds in grapes next to carbohydrates. Small amounts of different acids ranging from 0.4 to 1.0% occur in grapes and their products. Of the three organic acids that originate in grapes, tartaric is the major acid, followed by malic and citric. In addition, small amounts of isocitric, aconitic, glutaric, fumaric, pyrrolidone carboxylic, 2-ketoglutaric; and shikimic

TABLE 3.4
Methanol Content of Some Alcoholic Beverages

Alcoholic beverage	Quantity of methanol
Plum brandy (g/hac l)	1200
Cognac (g/hac l)	59
Apple (g/hac l)	100
Pear (g/hac l)	700
Spanish wines (g/l)	Traces–0.635
Brandy (%)	Traces–0.188
Pomace brandy (%)	0.039–0.86

Source: Bertrand (1975); Sandhu and Joshi (1995)

acids may also be found in musts. Tartaric acid is a relatively strong acid and contributes to the biological stability of wine, as it is buffered to a relatively low pH. The total acid ranges from 0.3 to 1.5% and the amount varies with the season and variety of grape (Table 3.5). In warmer climates, the predominant acid found in the grape berries is tartaric acid (2.0–8.0 g/l), while in cooler climates where the grapes are harvested at early maturity, the malic acid content is higher than tartaric acid. Small amounts of citric, isocitric, cis-aconitic, glutaric, fumaric, pyrrolidone carboxylic, keto glutaric acid, and shkimic acids are present in musts. The musts from moldy grapes contain glucuronic and gluconic acids. Ascorbic acid is also present (5–15 mg/l) in the grapes. Most fruits other than grapes have citric acid as the principal acid, while some have malic acid as the main acid. Acids give wine the sour or sharp aspect that enhances flavor when in balance with other components. The sum total of acids determine the amount of tartness a wine will deliver on the palate, and the resultant low pH ranging from 3.1–3.5 may help to keep the microbiological and chemical reactions properly controlled. The volatile acids content of different wines is variable, and a value more than 0.04% as acetic acids is considered undesirable.

3.2.4 Carbohydrates

Grape contains various carbohydrates in the form of reducing and non-reducing sugars (Table 3.5). The most important and

TABLE 3.5
Total Acids, Volatile Acids and Reducing Sugars Contents of Various Wines

Wine	Total acids (g/100 ml)	Volatile acids (g/100 ml)	Reducing sugars (g/100 ml)
Dry white	0.586	0.101	0.134
Dry red	0.649	0.128	0.146
Sweet white	0.412	0.092	11.30
Sweet red	0.502	0.122	10.26
Sparkling	0.658	0.082	3.409

Source: Amerine *et al.* (1972)

abundant sugars in grapes are hexoses; glucose and fructose. The sugar content in mature berries may range from 20 to 24° Brix. Fully ripened grapes may contain up to 1% of sucrose, which is inverted later by yeast-synthesized invertase during fermentation. The sugars are utilized by wine yeast and fermented to produce ethanol. Apart from hexoses, pentoses and other sugars are also present but in insignificant amounts. The wine yeast, *Saccharomyces cerevisiae* has a restricted ability to utilize hexoses. Therefore, other sugars are of least importance in wine production. Some other carbohydrates can be problematic if present in abnormal quantities. For example, the presence of galacturonic acid and pectin, if not broken, create an undesirable haziness in wine. Therefore, their quantity should be monitored in the wines. Concentration of glucose and fructose ranges from 15–25% in grapes, and are essential substrates for winemaking as these are the main source of ethanol. About one-third of the sugars can be attributed to various polysaccharides, which are complex forms of sugar unfermentable by yeasts, but may be the substrates for malolactic fermentation. Several other sugars like arabinose, rhamnose, ribose, xylose, maltose, mannose, melibiose, raffinose, and stachyose have also been identified in the grapes. Grape berries also contain in their vascular and other other tissues some cellulosic polymers, which are insoluble and do not appear in juice or wine. Pectins are also present in grape berries which amount to about 0.75% of the fresh weight of ripe berries. The whole of the sugar of fruits is not fermented into alcohol and a part of it remains as unfermented in the wines (Table 3.5). Many wines are designed to contain some specific amount of sugar to give a particular degree of sweetness, which traditionally results from natural grape hexoses or artificially added sucrose.

Many wines contain mannoproteins, a category of polymeric compounds, produced by wine yeast during alcoholic fermentation and represent 32.2% of the total wine polysaccharides. Their removal in wine filtration can affect sensory properties of wine and tartaric stability. Mannoproteins act as natural inhibitors of potassium hydrogen tartrate crystallization, preventing the occurrence of precipitates in wine and the formation of protein haze in the wines, besides a probable role in the volatility of aroma compounds.

The carbohydrate content of the non-grape wines depends upon the substrate used and extragenous supplementation of the sugar. Certain fruits like apple, mango, melon, jamun, etc., have high sugar content. However, the herbal substrates of wines like aloe-vera, mentha, and amla have insignificant sugar content and have to be supplemented with sugar.

3.2.5 Phenolic Compounds

Phenolic substances include a major group of compounds, accounting for 0.01–0.5% in grapes, and are responsible for color, taste, mouth feel, oxidation, and other chemical reactions, giving a characteristic color and flavor to the wines, besides giving UV protection, disease resistance, haze formation, hue, and health benefits including protection against cardiovascular diseases and cancer. These are formed from primary phenols during the ripening of grapes and maturation of wines and are also called polyphenols. The principal types of phenolics in grapes and wines include phenolic acid (hydroxy benzoic acid (O,P), salicylic, cinnamic, cumaric and ferulic derivatives, gallic esters), flavonols (kaempferol and quercetine), flavan-3-ols (catechin, epicatechin and its derivatives), flavanonols (Dihydroquercetin, dihydrokaempferol, ramnoside), and anthocyanins (cyanidin, pernonidin, petunidin, malvidin, coumarin, caffeine glycosides). Phenolic acids in grape berries are located primarily in the skin and in pulp, where they are present at much lower concentrations than anthocyanins. Polyphenols are categorized into two major groups: flavonoids and non-flavonoids. Flavonoids are large polymer molecules accounting for color. Also known as diphenylpropanoids, flavonoids possess two phenyl groups and are derived from a reaction between a phenylpropanoid and three malonylCoA moieties. They contain 15 carbon atoms. Non-flavonoids are generally smaller molecules which are primarily associated with oak flavor. The non-flavonoids include hydroxybenzoic acids, hydroxycinnamic acids, and stilbenes. Hydroxybenzoic acids occurring in grapes and wine include gallic acid, protocatechuic acid, ellagic acid, syringic acid, and vanillic acid. The occurrence of ellagic acid and its derivatives is mostly due to their extraction from wood during maturation in wooden barrels. The hydroxycinnamic acids, namely ferulic, p-coumaric, sinapic, and caffeicacids, infrequently occur in free form in fruits, but may be present in wine due to the vinification process. The soluble derivatives of these compounds have one of the alcoholic groups esterified with tartaric acid and these can also be glycosylated and acylated in different positions. Stilbenes have two ethane bonded benzene rings. Among these trans-isomer compounds, resveratrol is the major stilbene, along with infrequent piceids, viniferins, and pterostilbenes, respectively, dimethylated resveratrol derivatives, resveratrol glucosides, and resveratrol oligomers.

The phenolic compounds in the wine generally increase during the ageing process, and their levels are also affected by the type of the ageing process (Table 3.6). Aged red wines possess significantly different polyphenolic composition compared with young ones, due to formation of polymeric compounds, oxidation, hydrolysis, and other transformations that may occur in native grape phenolics during aging. The major phenolics present in white wines are the hydroxycinnamic acids, especially caftaric acid. The total ranges from 50 to 350 mg/L and in red wines, varies from 800 mg/l to 4 g/l.

In non-conventional wines like aloe-vera, jamun, and fruit wines, the phenolic content is mainly dependent upon the phenolic composition of the substrate. The phenolic content of the wines prepared from various parts of jamun ranges from 1000 to 5000 mg/l. A slight decrease in the phenolic content has also been found in the case of jamun wine, owing to the polymerization, condensation, and degradation of phenolic compounds. *Aloe vera* and mentha wines have been reported to have a total phenolic content of 1785 mg/l, including alantoin, myricetin, luteolin, and quercetin, etc.

TABLE 3.6
Concentration of Phenolics in New and Aged Wines

	Concentration (mg/l gallic acid equivalents)	
Compounds	**New Wine**	**Aged wine**
Non-flavonoids (Total)	235	> 250
Cinnamates (derivatives)	165	150
Benzene (derivatives)	50	60
Phenols (volatile)	5	15
Tannins	Traces	Traces–260
Flavonoids (Total)	1000	700
Catechins	200	150
Flavanols	50	10
Anthocyanins	200	20
Tannins (soluble)	550	450
Phenols (Total)	1300	900–1200

Source: Singleton (1982)

3.2.5.1 Tannins

Tannins are condensed giant phenolic polymers having a strong affinity for proteins, some of which they render insoluble, accounting for the "puckery" astringent effect. The tannins occur in the skin, stem, and seeds of the grape of grape berries. The free run juice of white grapes usually contains less than 0.02% tannins. The tannins of grapes are classified as hydrolyzable and condensed. The tannins which have been isolated include catechin, l-epicatechin, l-epigallocatechin, and dl-gallolectin. Catechin constitutes about 73% of the total tannins. It has also been found to be present in jamun, aloe, and seabuckthorn wines. Presence of certain tannins in moderate quantities is beneficial, as they also have medicinal properties apart from their immense contribution to the color of a wine.

3.2.5.2 Pigments

One of the phenolic flavonoids responsible for the pigmentation in grape skins and which determine the wine color are called anthocyanins. Traces of other pigments like chlorophyll, carotene, and xanthophyll are also found in grapes. Anthocyanin compounds generally occur in combination with sugar residues and are called glucosides, which are of five types. The main pigment of *Vitis vinifera* varieties is the monoglucoside of malvidin, which is more stable than the diglucoside form and generally found in red varieties belonging to *V. rotundifolia* types. Some varieties, such as Concord, also possess a significant level of monoglucoside malvidin, but in combination with the other four monoglucoside anthocyanins including cyanidin, delphinidin, peonidin, and petunidin. Interspecific hybrid grapes, such as the French-American varieties, have a unique phenolic color profile inherited from their breeding lines. Varying density of purple hues are typical to the predominance of the monoglucoside anthocyanin malvidin in Cabernet Sauvignon, Merlot, and other viniferas. The total amount of pigments in berries of a dark red wine variety such as Cabernet and Sauvignon is about 1000 mg/kg fresh weight. Lighter varieties may have a value of 100 mg/kg for pink variety. The anthocyanidins composition of grapes depends mainly on the type of cultivar, the location and soi, the growing season, and the ripening of grapes. Their concentration changes significantly over the early stage of development up to harvest, and the astringency is reported to decrease to tasteless compounds as the polyphenol polymerizes. This period is characterised by the color change for red varieties, as a result of the accumulation of anthocyans in skins. Virtually, all of the anthocyanin pigments are extracted from grape skins by the time a properly conducted on-skin fermentation has reached 5.0° Balling. The most abundant red wine pigments are the 3-*O*-glucosides of malvidin and peonidin, as well as their 6-acetylated and coumaroylated derivatives. The 3-*O*-glucoconjugates of petunidin, delphinidin, and cyanidin are also widespread. During wine aging, these pigments polymerize to a large extent and form a heterogeneous and not well characterized group of compounds, having major importance in the color of aged wines.

Anthocyanins coming from grapes also interact with pyruvic acid produced by wine yeast during fermentation and affect the color in the wines. Vitisin A, an anthocyanin-derived pigment, belongs to a group of minor pigments that is detected in red wine but not in fresh grapes. This pigment is formed by reaction between malvidin 3-glucoside and pyruvic acid, which is present in relatively high amounts during fermentation. The color expressed by anthocyanins is strongly dependent on the pH of the solution. In highly acidic media, the anthocyanins occur mainly as their red to bluish-red flavylium cations. An increase in pH leads to the formation of colorless carbinol bases, which in turn are in an equilibrium with the open chalcone form. In neutral and alkaline medium, the blue quinonoidal base is responsible for the perceived color.

3.2.6 Flavoring Compounds

Traces of diverse chemical compounds of a volatile nature account for certain sensory characteristics in wine, such higher alcohols, aldehydes, esters, and acids ketones, and constitute the aroma and flavor of the wines. These compounds originate from the native fruit, transformations by yeast during fermentation, and the aging process. A wide range of flavor compounds have been described in grapes and wines. Concentrations of volatile compounds are generally higher in wines produced at lower temperatures and depend on the yeast strain.

3.2.6.1 Terpenes

Terpenes are the metabolites of mevalonic acid, characterized by multiples of branched 5-carbon units resembling isoprene. Monoterpenes are 10-carbon compounds, many of which are volatile and odorous. Some like pinene in terpentine have a solvent-like, resinous odor, which is not always pleasant. Many of the primary floral and fruit aromas exist in the form of higher terpene alcohols, such as citronellol, linalool, and geraniol. These are commonly found in Johannisberg Riesling, Gewurztraminer, Vidal Blane, Vignoles, and most of the Muscat varieties. A considerable portion of terpenoid odorants of grapes occurs in bound form, particulary glycosides, which are too large and water soluble to have odors, but do not seem to contribute appreciably to bitterness, though, upon hydrolysis, the odor is released. The distribution and amount of monoterpenes fluctuate in response to temperature. The bound forms increase even in over-ripening, but the free forms decrease by volatilization.

3.2.6.2 Esters

Esters are essential components of wine quality, as they impart the fruity flavor to the wines and are formed by esterification of acids. Their concentrations are low in wines and may include the esters of acetic, lactic with methyl, ethyl, propyl, and amyl alcohol, but ethyl acetate predominates quantitatively. Esters are synthesized by the action of enzyme alcohol acyl-coA transferases on acyl-coA and alcohols. In wine, normally two types of esters are present: higher alcohols' acetates, which impart diverse odors, such as rose (phenylethanol acetate), banana (isoamyl acetate), and glue (ethyl acetate); the second ones being the esters of ethanol and fatty acids, which produce a fruity aroma. Apart from ethyl acetate, all esters contribute positively to the wine aroma.

Esters are produced in grapes, but rarely in significant quantities. Most wine esters are the secondary products of yeast metabolism. Over 160 wine esters have been identified up to now. Some esters, including isoamyl acetate associated with bananas, and ethyl propionate associated with apples, are found in the native American *Vitis labrusca* species, used in the production of various wine varieties. Primary fruit aromas in red grapes, which are often identified as cherry, black currant or cassis, strawberry, raspberry, and plum, are also found in ester forms and used as a flavor descriptor in the evaluation of wines. Strawberry flavor is attributed principally to several butyrate compounds, while raspberry is identified with ethyl caproate. Many of the esters responsible for the distinctive flavor of wine are also produced during alcoholic fermentation. The levels of various volatile and non-volatile esters detected in wines are shown in Table 3.7. The characteristic fruity odors of the fermentation bouquet are primarily due to a mixture of hexyl acetate, ethyl caproate (apple-like aroma), iso-amyl acetate (banana-like aroma), ethyl caprylate (apple-like aroma), and 2-phenylethyl acetate (fruity, flowery flavor with a honey note). The synthesis of acetate esters by the wine yeast *Saccharomyces cerevisiae* during fermentation is ascribed to at least three acetyltransferase activities, namely, alcohol acetyltransferase (AAT), ethanol acetyltransferase, and iso-amyl AAT. Over-expression of acetyltransferase genes such as ATF1 profoundly affects the flavor profiles of wines where the levels of ethyl acetate, iso-amyl acetate, and 2-phenylethyl acetate increase 3- to 10-fold, 3.8- to 12-fold, and 2- to 10-fold, respectively, depending on the fermentation temperature, cultivar, and yeast strain used. The concentrations of ethyl caprate, ethyl caprylate, and hexyl acetate only show minor changes, whereas the acetic acid concentration decreases by more than half. The amino acid profile of the grape variety also influences the aromatic composition in terms of the levels of ethanol, ethyl acetate, acetic acid, higher alcohols and some of their acetates, methionol, isobutyric acid, ethyl butyrate, and hexanoic and octanoic acids, in the resulting wine. The levels of some of the volatile compounds are well correlated with the aromatic composition of wines made with grapes of the same varieties.

TABLE 3.7
The Range of Various Esters Identified in Wines

Ester	Range (mg/l)
Ethyl acetate	11–232
n-Propyl acetate	0.04–0.8
Isobutyl acetate	0–0.5
Isoamyl acetate	Traces–9.3
n-Hexyl acetate	0–1.0
Phenylethyl acetate	0–1.14
Ethyl butyrate	0.2–0.44
Ethyl caproate	Traces–1.8
Ethyl caprylate	Traces–2.1
Ethyl caprate	Traces–0.9
Ethyl succinate	0.2–6.3
Mono-caffecyl tartarate	70.9–233.8
Mono-p-coumaroyl tartarate	8.3–33.3
Mono-feruloyl tartarate	1.6–15.9
Methyl anthranilate	0.14–3.50

Source: Ough (1991)

3.2.6.3 Carbonyl Compounds

Another flavor group called primary vegetal aromas, including anise, green pepper, tobacco leaf, and mint, is often structured in the form of carbonyl compounds, such as methoxy-isobutyl pyrazine, which is the principal aroma component associated with bell peppers. These compounds can be very complex such as the aroma associated with Cabernet Sauvignon, 2-methoxy-3-isobutylpyrazine. Wood aromas can also be carbonyl compounds, such as benzaldehyde, the aroma associated with bitter almond. The carbonyl compounds, except for acetaldehyde (3–49 mg/l), acetoin (0.7–3.8 mg/l), and diacetyl (0.1–7.5 mg/l), are found in traces in table wines, while in dessert wines and sherries, hydroxymethyl furfural is found at fairly significant levels. In most of the table wines, the average values for hydroxymethyl furfural are less than 3 mg/l.

3.2.6.4 Phenols

Vegetal aromas can also exist as phenols, such as cinnamic acid found in grass and tobacco, which is often desirable in very heavy red wines such as those traditionally made from Cabernet Sauvignon, Chancellor, and Merlot. Primary wood aromas are exemplified by briar, cedar, hazelnut, resin, oak, and eucalyptus, which generally exist in the form of phenolic compounds such as vanillin, an aroma component associated with oak. Some of the traces of the votatile compounds including whiskey-lactones, eugenol, furfural, and vanillin, also come from the oak wood. The chemical composition including the volatile composition of oak wood is highly variable, depending on the tree species, its geographic location, origin of oak, volume, and age of the barrel, and the single-tree effect.

3.2.6.5 Acetic Acid

A vinegary flavor in wine is generally associated with the oxidation of ethyl alcohol to acetic acid by various species of spoilage bacteria. Ethyl acetate, described as a "paint thinner" aroma, is often associated with acetic acid formation.

3.2.6.6 Mercaptan Compounds

Another set of spoilage flavors found in wine is due to one of the forms of sulfur degradation, which usually arises from elemental sulfur dusted on vines to control various types of mildew and molds in vineyards. Sulfur is also an important natural constituent in the synthesis of essential proteins in grapevines. Irrespective of source, it can be transformed by both cultured and wild yeasts into hydrogen sulfide or a "rotten eggs" character in wine. Further degradation of sulphur compounds by bacterial action results in one of the mercaptan complex compounds. At best, it is manifested in an "asparagus," "cabbage," or "green bean" flavor. Mercaptan spoilage is expressed as a "rubbery" aroma, or a foul "wet dog" smell, or worst of all, a "skunky" character.

3.2.6.7 Miscellaneous Compounds

Moldy/musty flavors arise from either moldy grapes, moldy barrels, or some other exposure of the wine to a source of mold, though molds can not grow directly in wine due to its alcohol content. An earthy flavor, sometimes described as "barnyard", is attributed to fermentation products resulting from the action of *Brettanomyces* yeast. Flavors that seem "cooked' or "brown apple' in character are often due to oxidation of wine resulting from extended exposure to air and/or higher temperatures (over 65°F) in processing and storage. Another flavor, "prune", comes from grapes that are over-ripened and/or have decayed. Wines such as Madeira are purposely heavily oxidized by exposure to relatively high temperature during maturation to produce high levels of aldehyde compounds that result in a pronounced "nutty-caramel" flavor. The formation of heterocyclic acetals (1,3-dioxanes and 1,3-dioxolanes) is generally taken as an indicator of wine age.

3.2.7 Nitrogenous Components

The total nitrogen content of grape musts vary between 100 and 2000 mg/l, the usual amount being about 600 mg/l, depending on the soil conditions. Ammonium salts, amino acids, peptides, proteins, and nucleic acid derivatives are the major nitrogenous components of grapes, and are important as nutrients for yeast fermentation, as enzymes such as phenolase, and as a factor involved in haze formation, especially in white wines. Minor constituents include pyrazines in aromas of Cabernet or Sauvingnon blanc varieties. Amino nitrogen in the musts is usually in the range of 100–400 mg/l, polypeptide nitrogen seldom exceeds 100 mg/l, while protein nitrogen amounts to less than 50 mg/l in ripe grapes. Urea is not present in grapes but is produced during fermentation, depending partly on the arginine content of the must, and combines with ethanol-yielding urethane under warm conditions. The nitrogen composition of grape musts affects fermentation kinetics and production of aroma and spoilage compounds in wine. It is common practice in wineries to supplement grape musts with diammonium phosphate to prevent nitrogen-related fermentation problems. The DAP affects the expression of 350 genes in the commercial wine yeast strains, in which 185 genes are down-regulated and 165 genes are up-regulated. The low levels of yeast assimilable nitrogenous compounds in grape juice are associated with the sluggish or struck fermentations (See Chapter 21 on this aspect). In contrast, excessive levels may lead to the increased formation of ethyl carbomate.

Amino acid composition and its level depend on grape variety, vineyard locale, and climate, among other factors as indicated in Table 3.8. Amino acids are amino derivatives and contain a carboxyl group attached to the amine-linked carbon. They may originate from grapes and are also released into the wine by yeast autolysis. They are critically significant components and are involved in biosynthesis of enzymes and other proteins. In addition, they are a major source of nitrogen and energy for metabolism of yeast. They constitute ≥90% of the nitrogen content of musts. In addition to this, they also add to the flavor of wine after being metabolized to aldehydes, phenols organic acids, lactones, and higher alcohols. Few amino acids have undesirable odors and therefore, above-threshold concentrations may lead to a bitter taste. Amino acids can be condensed into peptides and proteins, which creates the common problem of protein instability in white table wines. Proline accounts for half of the total juice nitrogen and about 10% of it is found in peptides. Another amino acid, glutamic acid, constitutes a larger portion in the peptides, and is the second most abundant amino acid. Yeasts have no known direct requirement for amino acids, but fermentation requires amino acids as a catalyst in synthesizing nitrogen into the free ammonium state needed by yeasts. Most of these amino acids are readily utilised by the yeast, and their contents decrease with the fermentation, except that of proline. Amino acids have a profound influence on wine quality, and their composition can be used to differentiate wines from different varieties, geographical origin, and year of production. Yeast can synthesize amino acids from intermediates of glycolytic pathways, such as glutamate, arginine, proline from α-ketoglutarate; lysine, aspartate, methionine, asparagine, threonine from oxaloacetate; cysteine, alanine from pyruvate; and tryptophan, tyrosine, and phenylalanine from enolpyruvate, etc.

The protein content in most wines is negligible, and fermentation by yeast does not increase the content except where a nitrogen source is added. In wine, the amount of nitrogenous compounds, including amino acids, is greatly reduced because of their utilization by yeast, or precipitation of some

TABLE 3.8
Amino Acids (mg/l) Composition of Musts of Chardonnay and Pinot Noir

	Chardonnay[2]		Pinot noir[3]		
Amino acid	**1983**	**1984**	**1983**	**1984**	**1987[4]**
Aspartic acid	25.1	24.7	36.8	32.8	134.2
Threonine	90.3	147.4	131.1	136.7	178.4
Serine	157.2	211.4	88.8	134.6	125.8
Glutamic acid	66.3	86.9	74.6	73.2	406.7
Glutamine	410.9	973.2	218.8	626.0	-
Proline	611.2	329.4	119.8	109.9	232.8
Alanine	366.2	527.2	264.2	424.0	89.9
Citrulline	18.7	37.4	36.3	94.9	-
Valine	43.0	119.0	38.5	73.0	77.4
Cystine	10.8	2.4	1.1	0.0	0.0
Methionine	5.2	44.1	5.4	24.3	9.2
Isoleucine	24.2	124.5	35.0	90.2	64.3
Leucine	23.7	139.3	36.3	106.8	103.9
Tyrosine	12.1	27.9	10.1	18.5	44.9
Phenylalanine	31.8	126.9	35.5	84.7	78.7
Ornithine	0.0	3.3	6.5	14.6	9.7
Lysine	10.3	7.9	13.7	9.9	6.5
Histidine	26.4	42.7	23.1	35.5	42.5
Arginine	220.5	374.8	574.4	695.3	737.0
Total	2258.0[1]	3414.0[1]	1859.0[1]	2873.0[1]	2629.6[1]

Source: Huang and Ough (1989). [1]Includes ethanolamine, γ-aminobutyric acid, β-alanine but not ammonia, [2]For the Chardonnay 10 samples, in 1983 and 65 in 1984,[3]For the Pinet noir 9 samples in 1983 and 27 in 1984, [4]Pinot noir grape must from Bourgogne, 15 samples. Includes α-amino butyric acid, 23.9 mg/l and glycine, 71.2 mg/l

proteins followed by their removal by filtration, decantation, centrifugation, or filtration. The nature of the nitrogen compounds present in wines also determines fermentation time and the ageing conditions. The yeast autolysis which occurs during prolonged fermentation and/or secondary fermentation of champagne also makes available more amino acids in wines, probably due to an increased protease activity after the exhaustion of sugar. The amount of total nitrogen present in the autolysates and the concentration of most free amino acids is significantly affected by the yeast strain used in the manufacture of sparkling wines and the aging process. Grape plants also activate several different defense mechanisms against both biotic and abiotic stresses, by synthesising heat shock proteins for thermo tolerance and pathogenesis-related (PR) proteins, which are resistant to proteolysis. The PR proteins present in wines cause haze formation, lowering the commercial value of these beverages.

3.2.8 Minerals

There are various mineral elements present in both grapes and wine. In adequate range, they act as important cofactors for vitamins and enzymes. However, some heavy metals like cadmium, lead, selenium, and mercury are actually toxic, and they usually precipitate during fermentation. The elemental composition of wine adds to the nutritional quality of the wine.

Wines contain several inorganic constituents in the form of many anions and cations (Table 3.9), imparting the nutritive value of the wine and impart freshness to its flavor. Ash content of some dry and sweet wines varies from 0.196 to 0.311 g/100 ml. The salts amount to 0.2 to 0.4%, and are generally derived from mineral acids or organic acids. Table 3.9 summarizes the range of various cations found in different wines. Amongst the minerals, K, Ca, Na, and Mg are the major elements, while Fe, Cu, Mn, and Zn are the trace elements found in wines. Some Spanish red wines contain significant levels of Fe (12 mg/l), Se (0.0005 ppm), Cr (047 ppm), and Cu (0.23 ppm). High concentration of some minerals leads to the reduction in yeast cell growth and fermentation rate. Bentonites have the capacity to remove the unwanted metals like Cu, Fe, and Zn. The Zn content of wines causes an astringent taste when present at a level above 5mg/l, which normally occurs through contact with utensils of galvanized iron.

Several factors, such as environmental contamination, agricultural practices, climatic changes, and vinification processes, markedly affect the elemental composition of grapes and the resulting wine. During fermentation and fining of wines, the concentration of several elements (Al, Cd, Co, Cr, Cu, Fe, Mn, Pb, V, and Zn) decreases. In contrast, during the maturation of the wine, there is a slight contamination of Cd, Cr, and Pb, which are released from wine cellar equipment made of brass and stainless steel.

Analysis of mineral content is important for assessing the safety of the wines, as some of the heavy metals when present in higher quantities may cause damage to the body. Therefore, there are regulations in different countries to limit the heavy metal concentration in wines.

3.2.9 Vitamins

In grapes, vitamins are primarily important as accessory growth factors for microorganisms involved in wine fermentation. Vitamins are involved in the regulation of cellular activity. They are found in minor quantities in wine and decrease during fermentation and aging due to oxidation by light, SO_2, or chemical reactions. Fat soluble vitamins and their precursors found in grape seed oil, with the exception of vitamin E, are low or absent in grapes or wines, but the water soluble vitamin B group generally occurs, although not at levels sufficient to make these sources nutritionally attractive for the human diet. Thiamine is a normal constituent of grapes and fresh grapes usually contain about 600 μg/kg, but sulphiting, pasteurization, or filtering grape juice through bentonite markedly reduces its content in grape juice. Riboflavin occurs in musts at up to 1.45 mg/kg but is easily destroyed by light. Pyridoxin is present up to 1.81 mg/kg in must. Pantothenic acid is also found in musts in an amount up to 15 mg/kg, and it is retained during storage. Nicotinic acid has been reported to be 2.8 mg/kg in Thompson seedless variety. Among other vitamins of the B-group, B_6 and biotin are also present in

TABLE 3.9
Mineral Contents of Different Fruit Wines

	Minerals (mg/l)							
Type of wine	Na	K	Ca	Mg	Cu	Fe	Mn	Zn
Grape wine	51	803	106	88	3.0	0.13	0.66	0.70
Apricot wine (New Castle)	11	1481	18	71	2.72	0.96	1.92	0.88
Wild apricot wine (*Chulli*)	43	2602	25	94	5.97	0.50	2.69	0.99
Apple wine (Golden Delicious)	18	1044	11	144	3.68	0.21	0.76	0.84
Cider (hpmc apple juice concentrate)	61	1900	23	137	4.31	0.32	1.54	0.10
Hard cider (Golden Delicious)	19	1069	17	97	3.03	0.19	0.91	0.82
Pear wine (Sand Pear)	87	1906	37	122	8.91	0.16	0.80	1.10
Plum wine (Santa Rosa)	20	1008	18	82	12.73	0.20	1.04	0.95
Grape Vermouth	111.64	735.64	89.25	62.18	0.53	6.95	0.58	-
Aperitivos vinicos	45.56	297.62	57.06	53.03	0.46	5.13	0.47	-
Vermuts balanceos	58.70	225.00	59.50	37.17	0.48	4.31	0.38	-
Vermuts rosas	73.65	524.72	54.96	57.57	0.42	7.13	0.34	-
Plum Vermouth	41.00	973	101	17.0	1.07	1.30	1.07	0.82
Sand Pear Vermouth	45.0	967	43	15.0	1.23	7.11	1.23	2.39

Source: Amerine *et al.* (1980); Bhutani and Joshi (1996); Bhutani *et al.* (1989); Joshi *et al.* (1999)

trace amounts. Ascorbic acid is typically present at about 100 mg/kg in fresh grapes, but more may be added into the wine for preventing oxidative browning. The levels of vitamin A, vitamin B, and vitamin C are influenced both by the fermentation and maturation process. Wine yeasts are good source of many B-vitamins and thus, account for an increase in the levels of these in wine, whereas vitamins A and C content in wine is negligible. The presence of thiamine, riboflavin, pyridoxine, and nicotinic acid varies from 0 to 50 µg/100ml, 5 to 120 µg/100ml, 70 µg/100ml, and 65 to 120 µg/100ml, respectively.

3.2.10 Sulfites

Hydrogen sulfide (H_2S), having the odor of "rotten eggs," often results in wines made from grapes having been dusted with sulfur in the vineyard. Excessively high inoculations with yeast cells or from sluggish, nitrogen-deficient, and/or extended low-temperature fermentations, or any other condition which may inhibit healthy yeast growth, results in the production of H_2S. Accordingly, the wine yeast strains fall into three major groups, (i) non H_2S producers, (ii) must-composition-dependent H_2S producers, and (iii) invariable H_2S producers.

3.2.11 Enzymes

A variety of enzymes including tannase, invertase, pectinase, ascorbase, catalase, dehydratase, esterase, proteases, and polyphenol oxidase are found in musts (See chapter 22 for more details). Enzymatic oxidation by polyphenol oxidase is important for many wines, particularly when made from moldy grapes but the activity is completely inhibited by SO_2.

3.3 VINIFICATION PRACTICES AND COMPOSITION OF WINE

It is well-known that the quality and composition of wine is influenced by a broad spectrum of factors, such as the grape's varietal factors, vintage, and production factors, including cultural and enological practices.

3.3.1 Vintage Factors

Grape is the raw material of wine and its proper quality is a prerequisite to the quality of the final product. Various factors such as climatic and soil conditions, including temperature, rainfall, sunshine, humidity and wind, cultural practices, yield, grade of ripening, variety, and sanitary conditions. Grape composition affects the organoleptic characteristics of wine, including flavor, color, and foam capacity. The must quality, in terms of maturation, determines the foam capacity of base wine. The yield and maturation index (ratio between soluble solid content and titratable acidity) is independent of enological practices, but it affects grape juice composition and varies according to the climatic and cultural conditions, the ground, and the grape variety. There is a correlation between maturation index, yield, and grape composition. The grapes with yield >10,500 kg/ha have a significantly higher content of soluble proteins and lower content of total polysaccharides than the grapes of the low-yield group (<10500 kg/ha). Several factors may modify the polyphenol content in grapes. The evolution of polyphenols during maturation is not uniform, and there is a negative correlation between polyphenols and maturation index.

The variables which affect the fermentation also influence the quality of fruit-based beverages. The volatile compounds in wines which are biochemically related to the yeast amino acid metabolism depend on the variety of grape from which the wine is made. The levels of isobutyric and isovaleric acids, their ethyl esters, isobutanol, isoamyl alcohol, β-phenylethanol, methionol, and isoamyl and phenylethyl acetates, differ according to the variety of grape in Spanish red wines. The levels of isobutyric and isovaleric acids depend on the grape variety in the case of French red wines. Isoamyl acetate is identified as a key odorant of red wines made with the Pinotage variety. There exists a more or less specific amino acid profile for each grape variety, and the variations are related to area, vintage, or the maturity level. Musts may have similar amino acid profile but may influence the order in which the different amino acids are taken up by the yeast, which in turn would influence the ratio of secondary metabolites produced, and finally the aroma profile of the wine. There is a correlation between the aromatic composition of wines made with grapes of the same varieties and the quantities of some of the volatile compounds, as some byproducts of fatty acid synthesis are related to threonine and serine, the concentration of β-phenyletanol is closely related to the level of phenylalanine, and methionol is stongly correlated to the must methionine contents.

3.3.2 Enological Practices

The process in which grapes are converted into wine is called as vinification. Vinification systems differ depending on the region of grape cultivation, type of grape variety, and financial state of the winery. The same is also applicable to the non-grape wines. The substrate and production processes influence the changes in the chemical composition of the final product, and any modification in the process is associated with the changes in the properties of wine. The application of pectolytic enzymes induces changes in "the classical enological parameters, methanol content, phenolic composition, technological variables including filterability, turbidity, chromatic parameters of wines, and the composition of phenols" depending on the variety of grapes employed for the wine, the vintage, and the type of enzyme preparation used. The modifications of these characteristics induced by the use of pectolytic enzymes are advantageous to the wine.

3.3.3 Yeast Strains

The influence of wine yeast on wine composition and quality is well-known. Concentration of the substrate utilized and which product is being developed depend on the yeast species and its growth extent. Along with this, wine flavors are also dependent on yeast strain and are formed on the basis of the particular metabolic activities of different strains. Besides ethanol and CO_2, the metabolism of yeasts yields a large number of byproducts including glycerol, acetic acid, succinic acid, and lactic acid (see Chapter 13 and 19 of this text). Moreover, the aromatic wine properties can be deeply affected by production of higher alcohols and other volatile substances. In addition, yeast species and different strains within each species have great differences in volatile compound production. Therefore, the yeast-induced fermentative aroma is responsible for great differences in composition, as well as in the taste of wine.

The cell wall of *Saccharomyces cerevisiae* is made of mannoproteins bound to oligopolysaccharides, glucanose and chitin. The different polarities and the hydrophilic or hydrophobic nature of these wall polymers define the capacity of yeast to retain or adsorb different wine molecules such as volatile compounds, fatty acids, and pigments. The porosity of the wall also influences adsorption, and an increased surface area provided by interstitial spaces favors adsorption of anthocyanins during fermentation by hydrophobic interactions. The acyl derivatives of anthocyanins (acetyl and *p*-coumaryl compounds) are more strongly adsorbed than nonacyl derivatives. Anthocyanins with a greater degree of methoxylation (malvidin and peonidin) are more adsorbed than those most hydroxylated (delphinidin and petunidin). Peonidin-3G and its acyl derivatives are also strongly adsorbed.

3.3.4 Biological Aging

Phenolic compounds are directly related to the quality of wines and influenced by several factors including grape variety, especially the reactions that take place during aging in wood. The phenolic composition is influenced during the time the yeast remains in contact with the wine during aging. Young wines are harsh and yeasty in flavor. Maturation mellows the taste of wine. Aging is a two-phase process, first is maturation, which occurs between the transfer of wine from fermenter to bottle. This time period may vary from 6 months up to years. During this, malolactic fermentation takes place. In this secondary fermentation, malic acid is converted into lactic acid with the loss of CO_2. Acidity of the wine decreases after this. Many other compounds are also produced during malolactic fermentation, which adds to the body of the wine. Aging also leads to increase in mannoprotein concentration in wine, which improves the overall structure of the wine.

Wine aging with yeasts takes place in three types of great wines: the *crianza* wines from Jerez, the *sur lie* from Burgundy, and the sparkling wines produced by the *champenoise* method. Biological aging of the wines from Jerez and the *sur lie* wines from Burgundy takes place under a layer of yeasts in wooden barrels, and in the same barrel a population of viable yeasts coexists with dead yeasts. After approximately the first month, there are no viable yeasts in the wine. Changes also occur in low molecular weight phenolic compounds during the manufacture of sparkling wines of white and red grape varieties. The yeast cells also undergo autolysis during biological ageing and during the preparation of most of the wines, and influence the wine composition. Breakdown of cell membranes, release of hydrolytic enzymes, liberation of intracellular constituents, and hydrolysis of intracellular biopolymers into products of a low molecular weight, are the main events that occur during this process. Yeast autolysis also occurs in the case of sparkling wines or in wines elaborated by the traditional practice of the "sur lies" method, by contact between wine and lees during aging. This results in an enrichment of different components, such as nitrogen compounds, volatile substances, lipids, and carbohydrates, and enhancement of the wine's body and flavor. In addition, yeast metabolites are made available, which greatly enhance the growth of *Oenococcus oeni*, a bacterial species frequently responsible for malolactic fermentation.

3.3.5 Gluconic Acid Consumption

The presence of mycoses caused by *Botrytis cinerea* in the form of "noble rot" in grapes gives rise to special wines such as the French Sauternes or the Hungarian Tokay; however, it is also the origin of severe problems in the winemaking process. Mycosis occurs during the grape ripening stage, and its development is influenced by meteorological factors such as moisture and rainfall, and physiological factors such as vine variety and bunch shape. Rot increases the activity of oxidase enzymes such as tyrosinase and laccase, which cause substantial color changes in wine, decrease titratable acidity, and increase volatile acidity as a result of *B. cinerea* infections being accompanied by the presence of acetic bacteria of the genera *Acetobacter* and *Gluconobacter.* The type of rot affecting grapes and its extent are usually determined from the gluconic acid content in the resulting must, and a content above 1 g/l reflects a substantial proportion of rotten grapes. Further, low levels of gluconic acid (up to 1–2 g/l) indicate an initial stage of grape infection mainly by fungi (*B. cinerea*), whereas higher levels (up to 2–3 g/l) are attributed to the activity of acetic acid bacteria.

The sensory properties of wine are altered by the presence of gluconic acid, which additionally renders it microbiologically unstable and results in long-term storage problems that can be solved only by reducing the concentration of the acid. *Saccharomyces bulderi* also converts a glucono-δ-lactone into ethanol. Assimilation of gluconic acid during the aerobic biological aging process of sherry wines using Flor yeast strains decreases volatile acidity and butanoic, isobutanoic, 2-methylbutanoic, and 3-methylbutanoic acids. A decrease in the content of gluconic acid and an increase in volatile acidity during biological aging processes have been ascribed to the presence of lactic bacteria, which metabolize gluconic acid and produce lactic acid, glycerol, ethanol, acetic acid, and carbon dioxide.

3.3.6 Aging in Wood

Aging in wooden barrels is a traditional practice in the production of quality wines, and this involves many physicochemical processes in wine. Certain components are also extracted from the wood into the wines, giving the wines their characteristic aromas and tastes. The phenolic components are involved in oxidation processes during aging, which are directly related to color and organoleptic properties of wine, including astringency and bitterness, depending on the barrel wood quality. Oak is the most commonly used wood for barrels used in aging, which leads to the reduction of certain undesirable components including n-propanol, n-butanol, iso-butanol, iso- and active amyl alcohols, and an increase in desirable components including ethyl acetate, and the extraction of phenols and volatile compounds. This also enhances color stability, brings about spontaneous clarification and a more complex aroma. Only some of the voltiles are originally present in wood in significant amounts, depending on the oak species and geographical origin, but many of them are intensified during the processing of wood in cooperage, through seasoning and the toasting process. During this phase, wine may be stored in oak barrels or steel tanks, depending on the requirements. Wine absorbs as much as 40 mL O2/year in an oak barrel during this period. The second phase of aging starts with bottling. During barrel aging, physical, chemical, and physical-chemical processes take place in wine, in which the phenolic composition of the oak wood plays an important role. As no oxygen comes into contact with wine during this phase, it is also known as reductive aging. This involves the settling of flavors of the wine. Oak wood aging leads to an increase in titratable acidity and a decrease in the pH of the wines. This is due to the conversion of the acids into their respective esters. The decrease in the acidity of wines may vary, depending on the type and time of storage. During this period, the wood permits atmospheric oxygen to pass slowly through its pores, leading to the gentle oxidation of certain components of the wines, a reduction in astringency, and changes in color and taste. At the same time, certain components are extracted from the wood into the wines, giving the wines their characteristic aromas and tastes. It is evident that the phenolic components are involved in oxidation processes during aging, which are directly related to color and organoleptic properties of wine, including astringency and bitterness, and depend on the barrel wood quality.

During the process of barrel toasting, the polyosides degradation produces furanic aldehydes, mainly furfural and 5-hydroxymethylfurfural. The hexoses produce 5-hydroxymethyl-furfural and 5-methylfurfural, whereas the pentoses, principal constituent of hemicelluloses, produce furfural through reactions catalyzed by acetic acid. These compounds contribute to the wine aroma with a burnt almond odor. Other volatile odorous compounds derived from sugar degradation, such as pentacyclic and hexacyclic ketones, were also identified in toasted wood, but with a limited aromatic role. The degradation of wood lipids by heat leads to lactones that contribute to the flavor of barrel-aged alcoholic beverages. The oak wood has two isomers, cis and trans, of β-methyl-γ-octalactone, responsible for the characteristic odor of oak, which is present in nontoasted wood in substantial quantities depending on the species and origin, and in toasted wood in quantities depending on the toasting conditions. American or French oak is preferred, but American oak contributes more cis and trans isomers of whiskylactone than French oak. Both forms of whiskylactone have low thresholds and are usually identified as having "woody" or "oak-like" aromas. These compounds, together with other lactones, volatile phenols, and related compounds such as γ-butyrolactone, 4-ethylphenol, 4-ethylguaiacol, guaiacol, eugenol, furfural, and vanillin, form the most important group of aromas in wines obtained by aging in wood.

3.3.7 SO_2 Concentration in the Must

Sulfur dioxide is widely used in wine fermentation. The total SO_2 added to wine consists of bound and free forms in equilibrium, with different antimicrobial activity, the degree and speed of this equilibrium being dependent on pH and temperature. The sulfur dioxide-binding compounds in wines are derived from three sources: (i) fruit components (such as glucose and arabinose), (ii) metabolites produced by bacteria (such as by *Gluconobacter* and *Acetobacter*, growing in rotting fruit), and (iii) during fermentation, by *Saccharomyces* spp. that can produce acetaldehyde, pyruvic, and 2-oxoglutaric acids. The binding of the bisulfite ion and acetaldehyde reduces the amount of free SO_2 available, but this bound form may also be inhibitory to lactic acid bacteria conducting malolactic fermentation, probably due to SO_2 release coupled to bacterial metabolism of the acetaldehyde moiety. Microorganisms vary greatly in their sensitivity to SO_2: bacteria, particularly Gram-negative rods, are markedly sensitive; aerobic microorganisms are more sensitive than fermenting microorganisms. It is essential that only a minimum effective quantity of SO_2 should be used for a product, depending on the pH level of the product and the concentrations of the sulfite-binding compounds present.

BIBLIOGRAPHY

Aiken, J.W. and Noble, A.C. (1984). Composition and sensory properties of Cabernet Sauvignon wine aged in French versus American oak barrels. *J. Grapevine Res.* 23: 27–36.

Alexandre, H., Heintz, D., Chassagne, D., Guilloux-Benatier, M., Charpentier, M. and Feuillat, M. (2001). Protease A activity and nitrogen fractions released during alcoholic fermentation and autolysis in enological conditions. *J. Ind. Microbiol. Biotechnol.* 26(4): 235.

Amerine, M.A. (1953). The composition of wines. *Sci. Mon.* 77: 254.

Amerine, M.A. (1956). The maturation of grape wines. *Wine Vines* 37: 27.

Amerine, M.A., Berg, H.W. and Cruess, W.V. (1972). *The Technology of Wine Making.* AVI Publsihing Co., Inc, Westport, CT.

Amerine, M.A., Berg, H.W., Kunkee, R.E., Ough, C.S., Singleton, V.L. and Webb, A.D. (1980). *The Technology of Wine Making.* 4th edn. AVI, Westport, CT.

Antonelli, A., Castellari, L., Zambonelli, C. and Carnacini, A. (1999). Yeast influence on volatile composition of wines. *J. Agric. Food Chem.* 47(3): 1139.

Baron, R., Mayen, M., Merida, J. and Medina, M. (1997). Changes in phenolic compounds and browning during biological aging of Sherry-type wine. *J. Agric. Food Chem.* 45(5): 1682.

Başkan, K.S., Tütem, E., Akyüz, E., Özen, S. and Apak, R. (2016). Spectrophotometric total reducing sugars assay based on cupric reduction. *Talanta* 147: 162–168.

Belda, I., Navascués, E., Marquina, D., Santos, A., Calderón, F. and Santiago, Benit (2016). Outlining the influence of non-conventional yeasts in wine ageing over lees. *Yeast* 33(7): 329–338.

Bertrand, A. (1975). *These Doctorate d'Etat*, Universite de Bordeaux, II. Bordeaux, France

Bhutani, V.P. and Joshi, V.K. (1996). Mineral composition of experiment sand pear and plum vermouth. *Alimentaria* 272: 99.

Bhutani, V.P., Joshi, V.K. and Chopra, S.K. (1989). Mineral composition of fruit wines produced experimentally. *J. Food Sci. Technol.* 26: 332.

Boulton, R.B., Singleton, V.L., Bission, L.F. and Kunkee, R.E. (1997). *Principles and Practices of Winemaking.* CBS Publishers and Distributors, New Delhi, India.

Cadahía, E., Fernández de Simón, B. and Jalocha, J. (2003). Volatile compounds in Spanish, French, and American Oak woods after natural seasoning and toasting. *J. Agric. Food Chem.* 51(20): 5923.

Carrau, F., Gaggero, C. and Aguilar, P.S. (2015). Yeast diversity and native vigor for flavor phenotypes. *Trends Biotechnol.* 33(3): 148–154.

Chamkha, M., Cathala, B., Cheynier, V.and Douillard, R. (2003). Phenolic composition of champagnes from Chardonnay and Pinot noir vintages. *J. Agric. Food Chem.* 51(10): 3179.

Chatonnet, P. and Dubordieu, D. (1998). Comparative study of the characteristics of American white oak (*Quercus alba*) and European oak (*Quercus petraea* and *Quercus robur*) for production of barrels used in barrel aging of wines. *Am. J. Enol. Vitic.* 49: 79.

Chauhan, A. (2015). *Development of a Non-Traditional Probiotic Aloe-Mentha Herbal Wine and Evaluation of Its In-Vitro Efficacy against Common Food Borne Pathogens* (Master's thesis). Panjab University, Chandigarh, India.

Chen, E.C.H. (1978). The relative contribution of Ehrlich and biosynthetic pathways to the formation of fusel alcohols. *J. Am. Soc. Brew. Chem.* 35(1): 39–43.

Crippen, D.D. and Morrison, J.C. (1986). The effects of sun exposure on the compositional development of Cabernet Sauvignon berries. *Am. J. Enol. Vitic.* 37: 235–242.

Cristino, R., Costa, E., Cosme, F. and Jordão, A.M. (2013). General phenolic characterisation, individual anthocyanin and antioxidant capacity of matured red wines from two Portuguese appellations of origins. *J. Sci. Food Agric.* 93(10): 2486–2493.

D'Angelo, M., Onori, G. and Santucci, A. (1994). Self-association of monohydric alcohols in water: Compressibility and infrared absorption measurements. *J. Chem. Phys.* 100(4): 3107–3113. doi:10.1063/1.466452.

Desportes, C., Charpentier, M., Duteurtre, B., Maujean, A. and Duchiron, F. (2001). Isolation, identification and organoleptic characterization of low-molecular-weight peptides from white wines. *Am. J. Enol. Vitic.* 52: 376–380.

Díaz-Plaza, E.M., Reyero, J.R., Pardo, F., Alonso, G.L. and Salinas, M.R. (2002). Influence of oak wood on the aromatic composition and quality of wines with different tannin contents. *J. Agric. Food Chem.* 50(9): 2622.

Doussot, F., De Jeso, B., Quideau, S. and Pardon, P. (2002). Extractives content in cooperage oak wood during natural seasoning and toasting; Influence of tree species, geographic location, and single-tree effects. *J. Agric. Food Chem.* 50(21): 5955.

Fabios, M., Lopez-Toledano, A., Mayen, M., Merida, J. and Medina, M. (2000). Phenolic compounds and browing in Sherry wines subjected to oxidative and biological aging. *J. Agric. Food Chem.* 48(6): 2155.

Fernández de Simón, B., Hernández, T., Cadahía, E., Dueñas, M. and Estrella, I. (2003). Phenolic compounds in a Spanish red wine aged in barrels made of Spanish, French and American oak wood. *Eur. Food Res. Technol.* 216(2): 150.

Fiket, Z., Mikac, N. and Kniewald, G. (2011). Arsenic and other trace elements in wines of eastern Croatia. *Food Chem.* 126(3): 941–947.

Fulcrand, H., Benabdeljalil, C., Rigaud, J., Cheynier, V. and Moutounet, M. (1998). A new class of wine pigments generated by reaction between pyruvic acid and grape anthocyanins. *Phytochemistry* 47(7): 1401.

Geana, E.I., Popescu, R., Costinel, D., Dinca, O.R., Ionete, R.E., Stefanescu, I., Artem, V. and Bala, C. (2016). Classification of red wines using suitable markers coupled with multivariate statistic analysis. *Food Chem.* 192: 1015–1024.

Gomez, E., Martinez, A. and Laencina, J. (1995). Changes in volatile compounds during maturation of some grape varieties. *J. Sci. Food Agric.* 67(2): 229.

Goncalves, F., Heyraud, A., Norberta De Pinho, M.. and Rinaudo, M. (2002). Characterization of white wine mannoproteins. *J. Agric. Food Chem.* 50(21): 6097.

Guth, H. (1998). Comparison of different white wine varieties by instrumental and analyses and sensory studies. In: *Chemistry of Wine Flavor,* Waterhouse L.A. and S.E. Ebeler Eds. American Chemical Society, Washington, DC. ACS Symposium Series #714.

Hernandez Orte, P., Cacho, J.F.. and Ferreira, V. (2002). Relationship between varietal amino acid profile of grapes and wine aromatic composition. Experiments with model solutions and chemometric study. *J. Agric. Food Chem.* 50(10): 2891.

Hossain, K.M., Abdal Dayem, A., Han, J., Yin, Y., Kim, K., Kumar Saha, S., Yang, G.M., Choi, H.Y. and Cho, S. (2016). Molecular mechanisms of the anti-obesity and anti-diabetic properties of flavonoids. *Int. J. Mol. Sci.* 17(4): 569. doi:10.3390/ijms17040569.

Huang, Z. and Ough, C.S. (1989). Effect of vineyard location, varieties and woodstocks on the juice amino acid composition of several cultivars. *Am. J. Enol. Vitic.* 40: 135.

Jackson, D.I. and Lombard, P.B. (1993). Environmental and Management practices affecting grape composition and wine quality – A review. *Am. J. Enol. Vitic.* 44: 409.

Jackson, R. (2014). *Wine Science: Principles and Applications-IV.* Academic Press, New York, NY.

Jeandet, P., Douillet-Breuil, A., Bessis, R., Debord, S., Sbaghi, M. and Adrian, M. (2002). Phytoalexins from the Vitaceae: Biosynthesis, phytoalexin gene expression in transgenic plants, antifungal activity and metabolism. *J. Agric. Food Chem.* 50(10): 2731–2741.

Joshi, V.K. (1977). *Fruit Wines.* 2nd edn. Directorate of extension education, Dr. Y.S. Parmar University of Horticulture and Forestry, Nauni-Solan, India, p. 256.

Joshi, V.K., Attri, V., Singh, T.K. and Abrol, G.S. (2011). Fruit wines: Production Technology. In: *Handbook of Enology: Principles, Practices and Recent Innovations-II,* Joshi V.K. Ed. Asiatech Publishers, Inc, New Delhi, India, pp. 1117–1221.

Joshi, V.K., Bhutani, V.P. and Thakur, N.K. (1999). Composition and nutrition of fermented products. In: *Biotechnology: Food Fermentation Microbiology, Biochemistry and Technology.* Vol. I., V. K. Joshi and A. Pandey Eds. Educational Publishers & Distributors, New Delhi, Ernakulum and Calcutta, India, pp. 259–320.

Joshi, V.K., Sandhu, D.K. and Thakur, N.K. (1999). Fruit based alcoholic beverages. In: *Biotechnology: Food Fermentation (Microbiology, Biochemistry and Technology),* Vol. II. Joshi V.K. and A. Pandey Eds. Educational Publishers & Distributors, New Delhi, Ernakulum and Calcutta, India, pp. 647–744.

Karatas, D.D., Aydin, F., Aydin, I. and Karatas, H. (2015). Elemental composition of red wines in Southeast Turkey. *Czech J. Food Sci.* 33(3): 228–236.

Lee, S.J., Rathbone, D., Asinont, S., Adden, R. and Ebeler, S.E. (2004). Dynamic changes in ester formation during chardonnay juice fermentations with different yeast inoculation and initial Brix conditions. *Am. J. Enol. Vitic.* 55: 346–354.

Macheix, J.J. and Fleuriet, A. (1998). Phenolic acids in fruits. In: *Flavonoids in Health and Disease,* Rice-Evans C.A. and L. Packer Eds. Marcel Dekker, Inc., New York, NY, pp. 35–59.

Machiex, J.J., Sapis, J. and Fleuriet, A. (1991). Phenolic compound and polyphenol oxidase in relation to browning in grapes and wine. *CRC Rev.* 30: 441.

Marais, J. and Pool, H.J. (1980). Effect of storage time and temperature on the volatile composition and quality of dry white table wines. *Vitis* 19: 151–164.

Marais, J., van Rooyen, P.C., and du Plessis, C.A. (1979). Objective quality rating of pinotage wine. *Vitis* 18: 31–39.

Marisa, C., Almeida, R., Teresa, M. and Vasconcelos, S.D. (2003). Multielement composition of wines and their precursors including provenance soil and their potentialities as fingerprints of wine origin. *J. Agric. Food Chem.* 51(16): 4788.

Martinez-Rodriguez, A.J., Gonza´lez, R. and Carrascosa, A.V. (2004). Morphological changes in autolytic wine yeast during aging in two model systems. *J. Food Sci.* 69(8): 233–239.

McRae, J.M., Ziora, Z.M., Kassara, S., Cooper, M.A. and Smith, P.A. (2015). Ethanol concentration influences the mechanisms of wine tannin interactions with poly(l-proline) in model Wine. *J. Agric. Food Chem.* 63(17): 4345–4352.

Monteiro, S., Piçarra-Pereira, M.A., Teixeira, A.R., Loureiro, V.B. and Ferreira, R.B. (2003). Environmental conditions during vegetative growth determine the major proteins that accumulate in mature grapes. *J. Agric. Food Chem.* 51(14): 4046.

Noble, A.C. and Bursick, G.F. (1984). The contribution of glycerol to perceived viscosity and sweetness in white wine. *Am. J. Enol. Vitic.* 35: 110–112.

Ough, C.S. (1991). *Winemaking Basics.* Food Products Press. An Imprint of the Haworth Press Inc, New York, NY.

Peinado, R.A., Moreno, J.J., Ortega, J.M. and Mauricio, J.C. (2003). Effect of gluconic acid consumption during simulation of biological aging of sherry wines by a flor yeast strain on the final volatile compounds. *J. Agric. Food Chem.* 51(21): 6198.

Pérez, L., Valcarcel, M.J., González, P. and Domecq, B. (1991). Influence of *Botrytis* infection of the grapes on the biological aging process of Fino sherry. *Am. J. Enol. Vitic.* 42: 58.

Perez-Prieto, L.J., Lopez-Roca, J.M., Martianez-Cutillas, A., Pardo Mianguez, F. and Gomz-Plaza, E. (2002). Maturing wines in oak barrels. Effects of origin, volume, and age of the barrel on the wine volatile composition. *J. Agric. Food Chem.* 50(11): 3272.

Pozo-Bayon, M.A., Hernandez, M.T., Martin-Alvarez, P.J. and Polo, M.C. (2003). Study of low molecular weight phenolic compounds during the aging of sparkling wines manufactured with red and white grape varieties. *J. Agric. Food Chem.* 51(7): 2089.

Rapp, A. and Güntert, M. (1986). Changes in aroma substances during the storage of white wines in bottles. In: *The Shelf Life of Foods and Beverages,* Charalambous G. Ed. Elsevier, Amsterdam, Netherland, pp. 141–167.

Revilla, I. and González-SanJosé, M.L. (2001). Effect of different oak woods on aged wine color and anthocyanin composition. *Eur. Food Res. Technol.* 213(4–5): 281.

Riu-Aumatell, M., Loapez-Barajas, M., Loapez-Tamames, E. and Buxaderas, S. (2002). Influence of yield and maturation index on polysaccharides and other compounds of grape juice. *J. Agric. Food Chem.* 50(16): 4604.

Robinson, A.L., Ebeler, S.E., Heymann, H., Boss, P.K., Solomon, P.S. and Trengove, R.D. (2009). Interactions between wine volatile compounds and grape and wine matrix components influence aroma compound headspace partitioning. *J. Agric. Food Chem.* 57(21): 10313–10322. doi:10.1021/jf902586n.

Romero, C. and Bakker, J. (2000). Effect of acetaldehyde and several acids on the formation of vitisin A in model wine anthocyanin and colour evolution. *Int. J. Food Sci. Technol.* 35(1): 129.

Sandhu, D.K. and Joshi, V.K. (1995). Technology, quality and scope of fruit wines especially apple beverages. *Indian Food Ind.* 11: 24–34.

Schwarz, M., Quast, P., von Baer, D. and Winterhalter, P. (2003). Vitisin A content in Chilean wines from *Vitis vinifera* Cv. Cabernet Sauvignon and contribution to the color of aged red wines. *J. Agric. Food Chem.* 51(21): 6261.

Sharma, R., Soni, S.K. and Gupta, L.K. (2001). Comparison of carrot wine stored in glass bottles and matured in oakwood barrels for home consumption. In: *Plant Diversity of Himalaya,* Pande P.C. and S.S. Samant Eds. Gyanodaya Prakashan, Nainital, India, pp. 593–598.

Singleton, V.L. (1995). Maturation of wines and spirits: Comparisons, facts and hypotheses. *Am. J. Enol. Vitic.* 46: 98.

Singleton, V.L. (1982). Grape and wine phenolics. In: *Symposium Proceedings: Grape and Wine Centennial*, 1982, 215. Univ. of Calf., Davis, CA.

Soleas, G.J., Diamandis, E.P. and Goldberg, D.M. (1997). Wine as a biological fluid: History, production, and role in disease prevention. *J. Clin. Lab. Anal.* 11(5): 287–313.

Sumby, K.M., Grbin, P.R. and Jiranek, V. (2010). Microbial modulation of aromatic esters in wine: Current knowledge and future prospects. *Food Chem.* 1(1): 1–16.

Swami, U. (2016). *Production of Non-Traditional Wines fromSyzygium cumini and Assesement of Their Medicinal Efficacies in Animal Model* (Doctoral thesis). Panjab University, Chandigarh, India.

Van Dijken, J.P., van Tuijl, A., Luttik, M.A.H., Middelhoven, W.J. and Pronk, J.T. (2002). Novel pathway for alcoholic fermentation of -gluconolactone in the yeast *Saccharomyces bulderi. J. Bacteriol.* 184(3): 672.

Varela, F., Calderon, F., Gonzalez, M.C., Colomo, B. and Suarez, J.A. (1998). Effect of clarification on fermentation kinetics and production of volatile compounds in white wines. *Alimentaria* 291: 109.

Vine, R.P., Harkners, E.M. and Wagner, C. (1999). *Winemaking.* Aspen Publishers Inc., Gailthersburg, MD.

Williams, A.A. and Rosser, P.R. (1981). Aroma enhancing effects of ethanol. *Chem. Senses* 6(2): 149–153.

Zoecklein, B.W., Fugelsang, K.C., Gump, B.H. and Nury, F.S. (1995).*Wine Analysis and Production.* Chapman and Hall, New York.

4 Aromatic Composition of Wine

Laura Fariña, Francisco Carrau, Sergio Moser, Eduardo Dellacassa and Eduardo Boido

CONTENTS

4.1 INTRODUCTION

Wine preference is related to the individual sensory perception of consumers, involving aspects such as colour, aroma and taste in the mouth (*e.g.* astringency). However, there is no doubt about the dominant role that flavour has, influencing what type of wine we select for consumption.

Knowledge of the importance of smell to the perception of flavour and the formation of emotional responses to wines could help to improve viticultural and oenological practices in order to offer typicity and particular attributes to consumers. As an important trait of wine quality, aroma has gained increasing attention in recent years, but the knowledge behind the sensorial attributes remains almost a precious biochemical mystery for most people involved in the winemaking process.

Thanks to the development of science and technology, particularly the application of fast and precise techniques such as GC-MS and other analytical techniques, progress in aroma research has been made in the oenological field. Using combinations of gas chromatographic methods linked to odour testing, it is possible to create a great deal of information providing essential links between the physical and chemical properties of wines and the flavour perceptions they produce.

Wines produce a range of volatile compounds that make up their characteristic aromas and contribute to their flavour. The volatile profiles of wines are complex and vary depending on the *Vitis vinifera* variety, vineyard management, environmental conditions, ripeness and the winemaking and analytical methods utilized.

Wine volatile compounds are mainly comprised of chemicals closely related to the different steps involved in winemaking, from grapes to fermentation and ageing (both in barrels and bottles).

Grape volatile compounds proceed from several biosynthetic pathways involved, starting from lipids, amino acids, terpenoids and carotenoids, while the diversity of volatiles is achieved *via* additional modification reactions. Some grape aroma compounds are also bound to sugars such as glycosides, mainly *O*-β-D-glucosides and *O*-diglycosides. The proportion of glycosidically-bound volatiles is usually greater than that of free volatiles, making them an important potential source of flavour compounds. Odorous aglycones may be released from the sugar moiety during maturation, wine processing and storage, or by the action of enzymes, acids or heat, constituting an important aroma reserve in wines (aromatic potential).

In this chapter, the composition of wine aroma, the characteristic aroma compounds present, the factors affecting aroma volatiles, and the effect of the different steps involved in winemaking have been briefly reviewed. Many of the points are discussed with in relation to the previous work performed by our research group on the Tannat variety, the red grape variety most widely cultivated in Uruguay.

In brief, this chapter intends to provide a convenient reference for those involved in winemaking, both researchers and producers, and those interested in the study of wine aroma and recognition of the chemical structures of aroma compounds.

4.2 ROLE OF AROMA COMPOUNDS

Aroma components are represented by terpenes, phenols, norisoprenoids and ethyl esters among other volatile compounds, while taste is associated mainly with acids, glycerol and the astringency of the polyphenolic compounds and their polymeric forms (tannins). The equilibrium and balance of the different components characterizes the top wines, and no one component exceeds the others. Irrespective of their origin and process of elaboration, in wine there are hundreds of volatile components. However, not all are sensorially active.

Volatile sensorially active compounds are those that, when detected by the nose or taste bud receptors, are able to trigger a nervous impulse and finally a response at brain level.

Each volatile sensorially active compound can be associated to one descriptor able to describe the aroma, appealing to the memories and previous experiences of a person smelling this compound. In addition to the descriptor, each volatile sensorially active compound has an associated aroma threshold. This value represents the minimum concentration at which the odour of the volatile compound can be perceived. This value can be determined using a hydroalcoholic solution (similar to wine) or a wine for which the absence of the compound of interest has been previously demonstrated.

Usually, the aroma threshold is determined by a triangular test with a panel of sensory judges by the addition of increasing concentrations of the volatile sensorially active compound. The minimum concentration at which 50% of the sensory judges perceive the aroma is considered as the aroma threshold value, which in wines ranges from a few ng/L to mg/L. The lower the aroma threshold value, the greater the possibility of the component being detected in a wine.

Finally, another useful term in evaluating the incidence of an aromatic compound in wine is the so-called odour active value (OAV), which can be defined as "the ratio of the compound concentration to its perception threshold". Even when the OAV is just an approximation, it gives an idea of the final impact of a volatile sensorially active compound on the wine aroma.

4.3 CHEMISTRY OF AROMA

Wine is one of the most complex alcoholic beverages, and its aroma components are responsible for much of this complexity. Describing the aroma of wines is not a simple task for researchers, because more than 800 aroma compounds, with a wide concentration range varying between hundreds of mg/L to the μg/L or ng/L level, and their combinations, form the character of wine and differentiate one wine from another.

In order to understand the molecular role of aroma compounds in wines, it is necessary to consider that odourants are volatile compounds that are carried by inhaled air to the olfactory epithelium located in the nasal cavities of the nose. The odourant must possess certain molecular properties in order to produce a sensory impression. It must have a certain degree of lipophilicity and sufficiently high vapour pressure so it can be transported to the olfactory system, some water solubility to permeate the thin layer of mucus and must occur at a sufficiently high concentration to be able to interact with one or more of the olfactory receptors.

The simultaneous presence of many different odour chemicals causes the final perception to be the result of complex brain processing in which some odours are integrated into a single perception; some act in a competitive or even destructive way. Beyond any particular cases or exceptions, families of structurally related compounds have enhanced aromaticity descriptors; this is the case for ethyl esters (fruity aromas), medium-chain fatty acids (milk flavours) and higher alcohols (fusel aromas).

From a chemical point of view, the high degree of diversity in physicochemical properties of the different wine impact compounds represents another step in wine aroma complexity. Some of them are quite hydrophobic and are released very easily from the wine matrix, being main constituents of the headspaces of wine. In contrast, others are quite hydrophilic and are released with difficulty, reaching the olfactory receptors only when the level of liquid in the glass is very small or when the wine is swallowed.

Finally, some of the most powerful wine aroma compounds take part in reversible interactions, which evolve during wine ageing and can be reversed, at least in part, when the wine makes contact with air, as in the case of reversible associations of carbonylic compounds with sulfur dioxide.

4.4 ORIGIN OF VARIOUS COMPONENTS

The compounds responsible for the aroma of a wine can be classified into four categories according to their formation, also corresponding to each stage in the biotechnological winemaking process.

1. Varietal aromas come from the grapes, constitute their aromatic potential and are responsible, in large part, for the typical characteristics of the wine. They can be divided into:
   1a. Free aromatic compounds (can be directly detected by the smell) consist essentially of terpenols and pyrazines.
   1b. Aroma precursor compounds are not odouriferous in grapes, but under certain circumstances they may become volatile and participate in the aroma.

2. Pre-fermentative aromas are those that are formed in the period from the time the grape is harvested until the moment of fermentation. Essentially, they are produced by enzymatic or chemical reactions with lipid precursors present in the grape.
3. Fermentative aromas are formed in the fermentation process of vinification, from the secondary metabolism of yeasts or possibly of lactic bacteria in the case of malolactic fermentation.
4. Bouquet results from the transformation of the aroma during the ageing period of wine. It occurs because of the formation of new compounds or rearrangement of components present through chemical or biochemical reactions. The results can be different if the process is carried out in a reducing environment, such as a bottle, or in an oxidative situation as in a barrel.

4.4.1 Varietal Aromas

Distinct classes of varietal aromas can be defined, comprising monoterpenes, C_{13}-norisoprenoids, substituted methoxypyrazines, sulfur compounds with a thiol function and others.

Even when grape juice has very little flavour, a small number of impact compounds, such as the monoterpene linalool or methoxypyrazines, present in their free form in the grape and in the must, impart the varietal typicity. However, most varietal aroma compounds are present as glycosylated forms, making them non-volatile and hence they have no odour.

Monoterpenes, norisoprenoids, aliphatic alcohols and phenols can be bound to monosaccharides, disaccharides or thiols such as cysteine or glutathione conjugates, which can be enzymatically split into important odourants.

It is well known that monoterpenes play a key role in the typical floral and sweet notes of the so-called aromatic grape varieties like Muscat, Riesling and Gewürztraminer. However, lately sesquiterpenes have attracted the interest of wine researchers. This interest is a direct consequence of the discovery of the sesquiterpene ketone rotundone, a key aroma compound for the peppery character of Syrah wines.

Whereas numerous studies on monoterpenes and C_{13}-norisoprenoids in grapevine have been published, works focused on the role of sesquiterpenes are rare.

In the case of C_{13}-norisoprenoids, this is a very diverse group of natural compounds generated by oxidative cleavage of carotenoid molecules between the C_9 and C_{10} positions, yielding norisoprenoids with 13 carbon atoms. Although other norisoprenoids are present in nature, for grapes only the C_{13}-norisoprenoids are of importance. Most of the C_{13}-norisoprenoids are present as glycosides.

Methoxypyrazines are extremely potent odourants with very low odour thresholds in wine. Some 3-alkyl-2-methoxypyrazines (MPs), such as 3-isopropyl-2-methoxypyrazine (IPMP) and 3-isobutyl-2-methoxypyrazine (IBMP), are particularly relevant in the aroma of wine due to their low olfactory thresholds (1–10 ng/L) and their characteristic green or vegetative aroma. Depending on their concentration, these compounds are considered to give wine an off-flavour.

In grape juice, research regarding thiol compounds has focused mainly on 3-mercaptohexan-1-ol (3MH), its acetate (3MHA) and 4-methyl-4-mercaptopentanone (4MMP). These compounds possess an extremely low organoleptic threshold and a strong tropical fruity aroma. Although research on thiols has been focused on Sauvignon Blanc, specifically the *V. vinifera* variety most characterized by thiols which significantly contribute to the wine quality, occurrence of thiols has also been observed in a number of other cultivars, both in juices as non-volatile precursors (as their odourless cysteine or glutathione conjugates) and in wines in their free forms.

4.4.2 Pre-Fermentative Aromas

The main grape lipids are esters of unsaturated fatty acids, the most important being linoleic and linolenic acid esters, which are also wine aroma precursors.

These compounds, mainly located in the solid parts of the berry, are degraded by grape enzymes to C_6-compounds when grapes are crushed in air in the pre-fermentative stage. The powerful odourants hexanal and 2-hexenal are the major C_6-compounds produced in grape must, while hexanol and hex-2- and 3-en-1-ols are minor products. The enzymes responsible for this degradation are lipoxygenases, which oxidize only unsaturated fatty acids containing a *cis*, *cis*-1, 4-pentadiene system to hydroperoxides with a *cis-trans*-diene system (linoleic and linolenic lipids), and hydroperoxidelyases which cleave hydroperoxides to C_6-aldehydes.

These C_6-compounds are odourants possessing green and grassy olfactive notes. However, C_6-aldehydes are reduced during alcoholic fermentation by yeast to hexanol, which hardly exceeds its olfactive perception threshold in wine. By contrast, the hex-2- and 3-en-1-ols found in wine usually occur at much lower levels and contribute to unpleasant herbaceous wine off-odours.

4.4.3 Fermentative Aromas

Alcoholic and malolactic fermentations are important stages in winemaking, being conducted by yeasts and by lactic bacteria, respectively.

The chemical identification of aroma compounds in wine derived from the metabolic activity of yeasts has been the objective of research during the last decades. Even though *Saccharomyces cerevisiae* is responsible for most of the ethanol in wine, it also has a significant effect on the production of aroma compounds including ethyl and acetate esters, higher alcohols, fatty acids, lactones, sulfur compounds, monoterpenes and benzenoids.

The formation of these by-products, within the genus *Saccharomyces*, is determined by the strain, with remarkable variability depending on the conditions of aeration, turbidity, fermentation temperature and pH. The main compounds produced within higher alcohols are 2- and 3-methylbutanol, propanol, 2-methylpropanol, butanol, pentanol, 2-phenylethanol and 3-methylthio-1-propanol.

Yeasts also produce fatty acids of different chain length, those whose chain length is less than 12 carbons being volatile. Acetic acid is the one present in the highest concentration and, having a relatively low threshold of perception, it can be perceived in some cases. In yeasts, the production of acetic acid has important metabolic roles, and its biosynthesis is from the enzyme acetaldehyde decarboxylase.

Long-chain fatty acids, between 4 and 12 carbons, are produced as by-products of the synthesis of saturated fatty acids by yeasts.

Esters are the volatile compounds of fermentative origin that have the greatest impact on young white and red wines. Esters can be divided into ethyl esters of fatty acids and acetates of higher alcohols. In the case of the ethyl esters, their formation occurs generally through the esterification of activated fatty acids in the form of acyl-Co-A. The limiting factor during their biosynthesis is the concentration of the fatty acid. Acetate esters are formed through the condensation of alcohol with acetyl-CoA catalysed in the cell by alcohol acetyltransferase enzymes. The expression of alcohol acetyl transferases is the most important factor in determining acetate ester levels during fermentation.

Malolactic fermentation is required for most of the red wines and some white wines; it reduces the tart taste associated with malic acid and provides additional advantages like microbial stability and improved aroma complexity.

Oenococcus oeni is the typical bacterial species responsible for malolactic fermentation processes and is used as a widespread starter culture for this purpose.

4.4.4 Post-Fermentative or Ageing Aromas

Maturation, whether in a vat or barrel, is essential to promote oxidative reactions, also allowing the extraction of wood compounds. Ageing wine in wooden barrels is an industrial process widely used in oenology. It is carried out to stabilize the colour, improve limpidity and to enrich the sensorial characteristics of the product.

In-bottle ageing does not involve oxygen or wood extractives, although many important chemical changes are still undertaken. During the ageing process, wine aroma undergoes modifications leading to the so-called 'bouquet' or post-fermentative aroma. At the end of this stage the wine has reached a final balance, characterized by delicate and penetrating aromas. This balance is achieved through the transformation of volatile compounds during conservation, in which there is a decrease in the fruity aromas characteristic of young wines, evolving towards more complex aromas.

4.5 SENSORY IMPACT OF DIFFERENT AROMA COMPOUNDS

Volatile wine compounds according to their impact on the perception of aroma have been classified. These aroma compounds have been classified as basic, subtle and impact. These groups were defined as follows: basic aromas are represented by about 20 compounds present in all wines, with an odour activity value (OAV) ranging between 5 and 20; their origin is fermentative except for the case of beta-damascenone, which is the only varietal component considered. These components act as a buffer, and ethanol is found among them. The second group is represented by the subtle aromas: components present in almost all the wines, with an OAV below 1, but which acting individually or in a group can break the aromatic buffer. In this group, volatile phenols, ethyl esters and fatty acids can be found, among others. Aromas of impact: this third group characterizes some special wines, giving their essence to them. The compounds involved are characterized by their very low thresholds of perception. Examples are pyrazines (Cabernet), monoterpenes such as linalool, *cis*-rose oxide (Muscat) and mercaptans (4MMP, 3-mercaptohexyl acetate) in Sauvignon Blanc. Even when those compounds can be individually active, resulting in aroma descriptors which identify the wine, it must be considered that the aroma components present in the wine can show additive, synergic and also antagonistic effects. Although these effects have been investigated, much deeper research remains to be done in this regard.

4.6 FACTORS INFLUENCING VARIOUS AROMA COMPONENTS

4.6.1 Factors Influencing the Biosynthesis of Varietal Aroma Compounds

The grape quality, and therefore its composition, is influenced by many factors of which the climate, soil, water status, variety and practices at the vineyard level, among others, are of fundamental importance. The varietal aromas, being secondary metabolites generated by the grape, are influenced by the factors mentioned above.

4.6.1.1 Luminosity Effect on the Profile of Carotenoid Precursors

The Tannat variety is characterized by its glycosylated aroma compounds, particularly norisoprenoids derived from carotenoid breakdown. In fact, phenols and norisoprenoids represent almost 80% of the total glycosidic components. In other *V. vinifera* grape varieties, terpenes have been reported as possessing an important aromatic role, but in Tannat they represent only 12% of the aroma fraction. Figure 4.1 shows the importance of the different groups of varietal glycoside compounds.

In order to show the norisoprenoid dynamics, the carotenoid content and profile evolution in two differently managed vineyards, exposed to different sunlight and temperature conditions, were studied over two successive ripening periods. The sunlight and temperature effects were managed through the addition of stone mulch soil (SMS) and compared to a conventional vineyard with bare soil (BS). In these conditions, seven carotenoids were identified: neoxanthin, violaxanthin, luteoxanthin, lutein 5,6-epoxide, flavoxanthin, lutein and β-carotene. Irrespective of species, the highest carotenoid concentrations were found at initial sampling, immediately decreasing sharply to a minimum

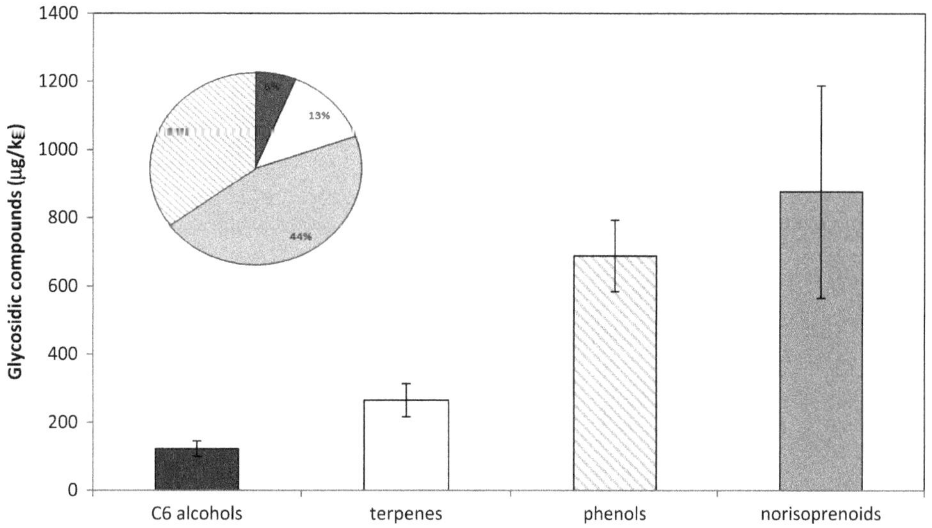

FIGURE 4.1 Glycosylated compounds in *Vitis vinifera* grapes of the Tannat variety

at harvest. However, a different variation was found for each carotenoid, and while significant carotenoid degradation was observed in both vineyards, the effect was greater in grapes of vines growing in SMS. The evolution of carotenoid breakdown products was also studied and, in fruit from SMS, the norisoprenoid content increased noticeably (by 85% of the initial concentration) over the ripening season, whereas no such increase was found in fruit from BS (Figure 4.2).

4.6.1.2 Effect of Pruning System on Glycosidic Tannat Grape Compounds

Among the practices of vineyard management, pruning is a key stage and different systems can be used, with different results on the quality of the grapes and the wines obtained. We have studied the effect of two pruning systems very widespread among Uruguayan winemakers and their effect on classical grape quality parameters (sugar, pH, acidity, berry weight, total anthocyanins, total polyphenols) as well as the potential aroma (glycosidic compounds). Using trellis conduction, cane and spur pruning systems were evaluated during maturity in three consecutive harvests. When assessing the effect of different factors on metabolite biosynthesis in plants, the load (kg/plant) obtained at the time of harvest should be considered either by previous adjustment or in the order of expressing the results. Spur pruning presented a lower fruit load per plant for the Tannat variety, and significant differences in total anthocyanin content were obtained with cane pruning. No significant differences were observed for the total polyphenol index, but for some groups of glycosidic precursor analysed, the concentrations of terpenes and norisoprenoids in cane pruning grapes were higher and did not correlate with the load per plant. The influence of these two vineyard management strategies on the quality of the fruit for the Tannat variety was demonstrated. However, the maximum concentration of aromatic precursors did not match the maxima of the classical physicochemical parameters that were analysed. Consequently, the aromatic maturity in grapes can only be evaluated by monitoring this parameter, since it does not present correlation with other physicochemical parameters like the content of total anthocyanins, as shown in Figure 4.3.

4.6.1.3 Effect of Over-Ripening on the Appearance of Eucalyptus Notes

Waiting for the optimum moment of maturation in grapes, the fruit can reach a state of over-ripening, causing harmful effects on the aromatic profile of the wine. In Tannat variety grapes, the transformation of limonene to 1,8-cineole when grapes are subjected to a period of over-ripening has been verified. The precursor compounds postulated are 1,8-terpin and limonene. The concentration in wines of this variety reached or exceeded the threshold of perception of this component, established as 1.8 µg/L (Figure 4.4).

4.6.2 Factors Influencing the Biosynthesis of Aromatic Components by Yeasts

It is well known that *S. cerevisiae* produces different concentrations of aroma compounds as a function of fermentation conditions and must treatments, including temperature, grape variety, micronutrients, vitamins and the nitrogen composition of the must. However, there are other relevant factors influencing the profile of volatile compounds produced by yeasts; some of them will be discussed in this section.

4.6.2.1 Effect of Yeast Assimilable Nitrogen in Fermentation Media

Yeast assimilable nitrogen (YAN) level and amino acid profile measure key nutrients in producing fermentative compounds,

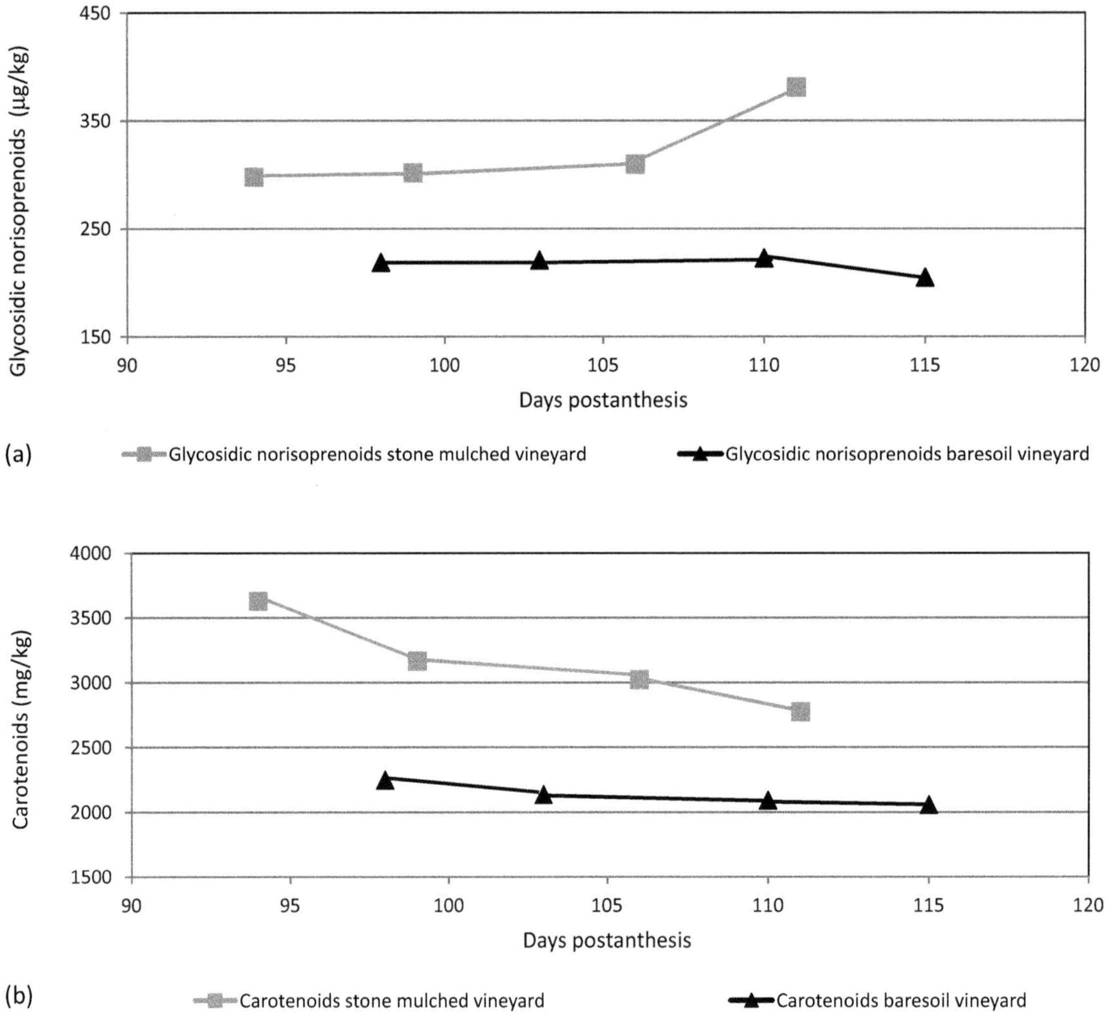

FIGURE 4.2 (a and b). Evolution of norisoprenoids and carotenoids in Tannat grapes in two different sunlight and luminosity conditions

meaning that different yeast strains growing in a standardized culture medium can generate different volatile profiles. Based on this, a yeast classification system was postulated based on nitrogen utilization capacity. To verify this approach, two yeasts were selected, fermented in different culture media with nitrogen concentrations in a range of 75 to 400 mg/L and the generated volatile compounds analysed (Figure 4.5). The results obtained allowed the identification of yeasts able to make efficient use of nitrogen (KU1); in contrast, the strain M522 had a less effective use of the nitrogen. The KU1 group, which produced a lower concentration of higher alcohols and isoacids at all nitrogen concentrations tested when compared to the M522 group, could more effectively regulate the carbon flux at any given nitrogen level (isoacids and higher alcohol).

4.6.2.2 Effect of the Inoculum Size

The wine industry uses commercial preparations of *S. cerevisiae* to establish a population of selected yeasts and thus, avoid problems during the fermentation process, assuring the complete consumption of sugars, but what quantity of initial cells need to be added to be sure of this? There are only a few antecedent investigations focusing on this problem, so in line with the work mentioned in Section 4.6.2.1, was performed experiments using three initial inoculum sizes (104, 105 and 106 cells/mL), using yeasts M522 and KU1 at two nitrogen levels (148 and 211 mg N/L). A significant effect on the final concentrations of higher alcohols, esters, fatty acids, free monoterpenes and lactones was attributed to the size of inoculum in both strains but not in an easily predictable way. However, a consistent increase of desired aroma compounds (esters, lactones and free monoterpenes) and a decrease of compounds less desired for white wine (higher alcohols and medium-chain fatty acids) was shown at inoculum sizes of 10^5 cells/mL for both strains in real winemaking conditions (Figure 4.6).

4.6.2.3 Effect of Redox State on Volatile Compounds Produced During the Fermentation Process

Redox state is a well-known key process parameter for microbial activity. In order to examine the effect of this factor in

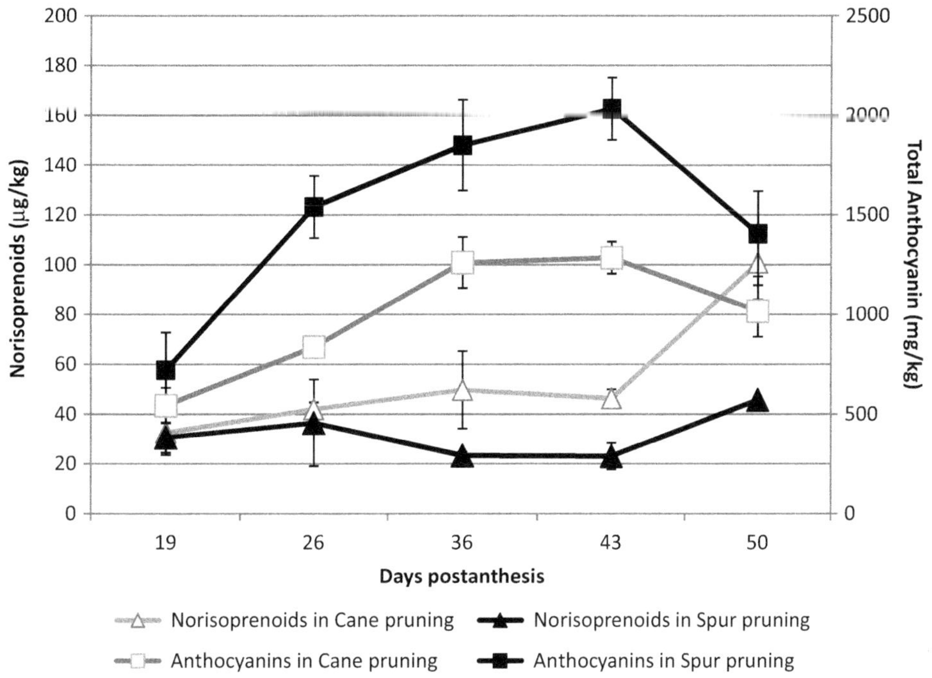

FIGURE 4.3 Evolution of norisoprenoid and total anthocyanin content during the grape maturity period

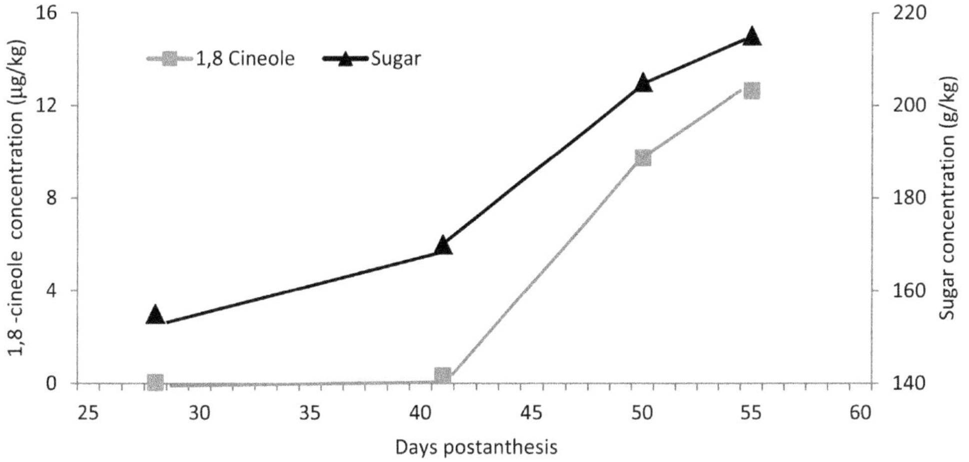

FIGURE 4.4 Evolution of 1,8 cineole and sugar content during the Tannat grape maturity period

the biosynthesis of fermentative compounds in more depth, was realized an experiment in two conditions: reductive and micro-aerobic, using a simil grape must medium and two *S. cerevisiae* strains commonly used in winemaking. The different redox state conditions were obtained using flasks plugged with Müller valves filled with sulfuric acid (reductive) and defined cotton plugs (micro-aerobic conditions). The difference in the potential between the two conditions studied was over 100 mV during the main fermentation period. Significant differences in the final concentration of higher alcohols, esters and fatty acids were attributed to differences in the redox state in both strains. A consistent increase in esters and medium-chain fatty acids, as well as a decrease of higher alcohols and isoacids, was seen under reductive fermentation conditions. Interestingly, 1-propanol, gamma-butyrolactone and ethyl lactate concentrations showed no significant variation under the different redox conditions (Table 4.1). A better understanding of the influence of the redox state of the fermentation medium on the composition of volatile compounds in wine could enable improvement of vinification management.

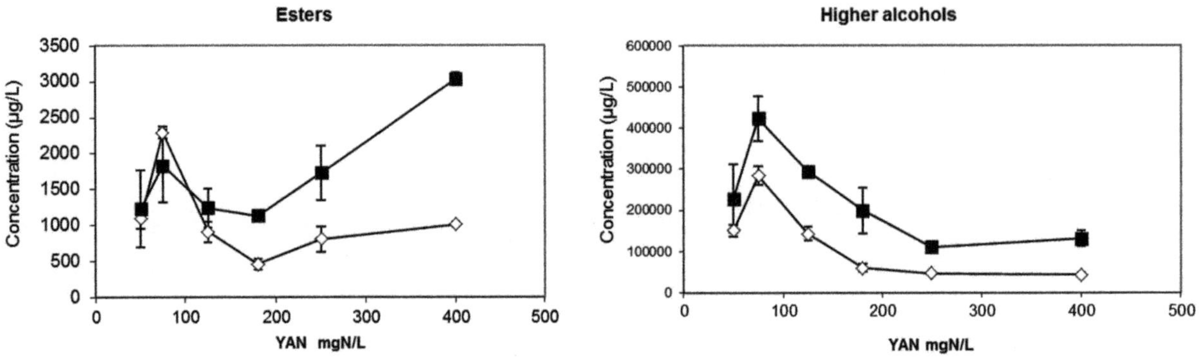

FIGURE 4.5 Concentration of higher alcohols and esters in artificial grape must medium fermented with two *Saccharomyces cerevisiae* yeast strains in response to the initial nitrogen concentration. These compounds were produced by yeast M522 () and KU1 ( ) and measured using GC-MS. M522 is a black square; KU1: empty rhombus

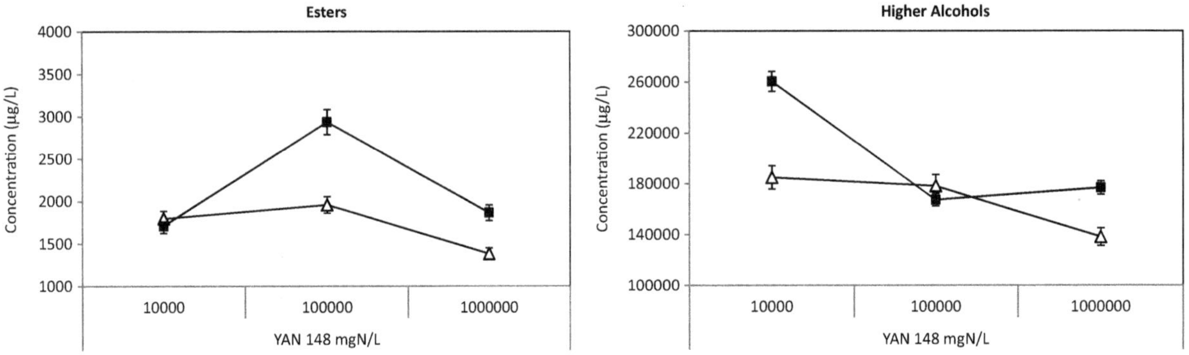

FIGURE 4.6 Inoculum size effect on aroma compound formation for model strains M522 (■) and KU1 (△) at 148 mg N/L. Inoculum sizes are indicated (1 × 10^4, 10^5 and 10^6 cells/mL) on Muscat Alexandria grape must

TABLE 4.1
Compounds With an OAV Greater Than 1 for Strain M522 at Two YAN Levels and Two Redox Conditions

	Treatments at two YAN levels					
	180 mg N/L		400 mg N/L		ANOVA	
Compounds	Cotton plug	Müller valve	Cotton plug	Müller Valve	YAN	Redox
2-Phenylethyl alcohol	1.1 ± 0.1	1.0 ± 0.1	0.7 ± 0.1	0.4 ± 0.1	**	**
3-Methyl thio-1-propanol	3.2 ± 0.1	2.5 ± 0.1	2.4 ± 0.6	0.6 ± 0.2	*	*
Isoamyl acetate	5.1 ± 0.5	11.0 ± 0.4	5.4 ± 1.2	14.8 ± 2.5	NS	**
Ethyl hexanoate	3.9 ± 0.1	4.4 ± 0.3	3.9 ± 0.2	7.7 ± 1.3	NS	**
Butanoic acid	1.5 ± 0.1	1.9 ± 0.1	2.2 ± 0.1	2.6 ± 0.2	*	*
Isobutanoic acid	1.7 ± 0.1	1.2 ± 0.1	1.7 ± 0.1	1.0 ± 0.1	NS	**
Isovaleric acid	2.2 ± 0.2	1.5 ± 0.1	1.4 ± 0.2	0.7 ± 0.1	**	**
Hexanoic acid	1.8 ± 0.1	2.8 ± 0.1	2.7 ± 0.1	4.1 ± 0.7	NS	NS
Octanoic acid	2.9 ± 0.2	4.9 ± 0.2	4.1 ± 0.1	6.8 ± 0.8	NS	**
gamma-Butyrolactone	0.7 ± 0.1	0.8 ± 0.1	1.1 ± 0.1	1.1 ± 0.1	**	NS
3-Methyl-1-butanol	2.9 ± 0.1	2.9 ± 0.1	2.3 ± 0.4	1.8 ± 0.2	*	NS
Acetaldehyde	250 ± 8.3	125.0 ± 25.0	166.7 ± 8.3	133.3 ± 16.7	NS	**
Ethyl acetate	0.5 ± 0.1	1.0 ± 0.1	6.0 ± 0.1	1.9 ± 0.1	**	***

*$p < 0.05$; **$p < 0.01$; ***$p < 0.001$
OAV= Odour activity value

TABLE 4.2
Compounds With an OAV Greater Than 1 for Volatile Compound Content in Co-Fermentation of *H. vineae* With Commercial Yeast and Conventional Fermentation With Commercial Yeast

Aroma compounds with OAV > 1	Odour active value (OAV)			
	Co-fermentation with *H. vineae*	Commercial	ANOVA results	Odour descriptor
	Acetates			
Ethyl acetate	1.2 ± 0.1	0.7 ± 0.1	**	Fruity, solvent
Isoamyl acetate	31.6 ± 1.4	29.0 ± 8.4		Banana, pear
Hexyl acetate	15.4 ± 0.6	37.4 ± 0.6	***	Fruity
2-Phenylethyl acetate	17.4 ± 2.4	1.7 ± 0.6	**	Rose, honey, tobacco
	Ethyl esters			
Ethyl hexanoate	19.4 ± 3.7	36.5 ± 1.6	*	Apple, fruit
Ethyl octanoate	1.4 ± 0.4	2.2 ± 0.5		Sweet, banana
Ethyl decanoate	3.4 ± 0.4	2.8 ± 0.4		Sweet, hazelnut oil
	Alcohols			
3-Methyl-1-butanol	1.4 ± 0.3	1.6 ± 0.1		Like wine, nail polish
2-Phenylethanol	3.0 ± 0.4	2.4 ± 0.6		Honey, rose, spicy
	Acids			
2-Methylpropanoic acid	22 ± 0.4	1.3 ± 0.1		Acid, fatty
Butanoic acid	2.1 ± 0.3	2.5 ± 0.6		Cheese, rancid
3-Methylbutanoic acid	0.7 ± 0.5	3.0 ± 0.5	*	Blue cheese
Hexanoic acid	4.5 ± 1.0	9.1 ± 2.2		Fatty, cheese
Octanoic acid	8.4 ± 2.7	17.0 ± 6.0		Fatty
Decanoic acid	4.6 ± 0.5	3.8 ± 1.6		Rancid, fat
	Other compounds			
3-Methyl-thio-1-propanol	0.7 ± 0.1	1.3 ± 0.1	*	Sweet, potato
4-Vinyl guaiacol	3.0 ± 0.7	2.4 ± 0.3		Clove, curry

4.6.2.4 Effect of Yeast Diversity on Flavour

The use of native apiculate strains, selected from grapes, is a current research field in oenological microbiology. The main objective is to offer winemakers the option of variation between wines and subtle characteristic differences in fine wines. One experience with *Hanseniaspora vineae* isolated from Uruguayan vineyards, fermented in triplicate in 225 L oak barrels using a Chardonnay grape must, had successful results, and this experience has been transferred to the industrial sector. In this research, the following fermentation strategies were compared: conventional inoculation (commercial *S. cerevisiae*) and sequential inoculation (*H. vineae* and then *S. cerevisiae* strain). Basic winemaking parameters and some key chemical analysis, such as the concentration of glycerol, biogenic amines, organic acids and aroma compounds and sensory analysis, were analysed. Sequential inoculation with *H. vineae* followed by *S. cerevisiae* resulted in dry wines with increased aroma and flavour diversity, compared with wines resulting from inoculation with *S. cerevisiae* alone. Besides that, a winemaker's panel considered wines produced from sequential inoculations as having an increased palate length and body. Characteristics of wines derived from sequential inoculation could be explained by the significant increases in glycerol and acetyl and ethyl ester flavour compounds, and relative decreases in alcohols and fatty acids. Aroma sensory analysis of wine character and flavour, attributed to winemaking using *H. vineae*, indicated a significant increase in fruit intensity described as banana, pear, apple, citric fruits and guava (Table 4.2).

4.7 DIFFERENT TECHNIQUES OF WINEMAKING AND AROMA

The vinification techniques in use are very diverse but, in general, for white wines, vinification is realized through minimum contact of must with the solid parts during the fermentation. In contrast, red wine vinification implies prolonged contact with the solid parts during the fermentation and sometimes, post-fermentative maceration is performed. During the different options of vinification, grape-derived aroma compounds are extracted, including monoterpenes, norisoprenoids, aliphatics, phenylpropanoids, methoxypyrazines, volatile sulfur compounds and cinnamic acids. In addition, compounds derived from microbial metabolism such as sugars, fatty acids and organic nitrogen compounds (pyrimidines, proteins and nucleic acids) are incorporated into the must. In a second winemaking stage, wines can be aged in barrels, where oak-derived aroma compounds are incorporated into wine whose composition will depend on the origin and heat treatment of the wood.

The volatile profile of red wines bottled for different periods is characterized by a decrease in the concentration of esters (fermentative aroma), where only ethyl lactate and diethyl succinate show increasing values. This behaviour does not occur at the same rate during ageing, but seems to be faster at the beginning.

Most importantly, as mentioned above, many compounds (terpenes, norisoprenoids, phenols) also occur as bound forms. During ageing, the glycosylated compounds are slowly converted to free volatile ones over time. These chemical changes are associated with the presence of beta-glycosidases and acid media; when these reactions occur, the aroma complexity of a wine is enhanced during maturation.

Analytical characterization of the complex flavour profile of red Tannat wine determined the presence of high concentrations of aromatic glycosylated compounds derived from aromatic aglycones, such as C_{13}-norisoprenoids, benzene derivatives and aliphatic alcohols. From a sensory point of view, Tannat wine aromas are described as raspberry, plum, quince and small berry-like scents. In aged wines, the aroma descriptors change to dry fruits, smoked and liquorice characters. Possible peculiarities of the free forms and also the heteroside fraction may explain some of these aroma descriptors.

Under certain circumstances, during the ageing period, off-flavours or defects may appear. Aromatic defects are compounds associated with unpleasant descriptors, whose aroma usually covers or masks the characteristic aroma profile of the wine. In white wines, the chemical modifications associated with oxidative processes can produce off-flavour compounds such as 2-aminoacetophenone (AAP), known by its characteristic impact responsible for the so-called 'untypical ageing off-flavour' (UTA). The formation of AAP is caused by oxidative degradation of the phytohormone indole-3-acetic acid (IAA), produced as a consequence of the addition of SO_2 after fermentation. However, UTA is not correlated with the IAA content of the must or wine. Differences in the release of IAA during fermentation of musts derived from both early- and late-harvested grapes indicate a correlation between the ripeness of the processed grapes and UTA formation. Late-harvested grapes usually lead to lower free IAA levels throughout the course of fermentation than early-harvested grapes from the same vineyard.

Moreover, sotolon (3-hydroxy-4,5-dimethyl-2(5H)-furanone), a chiral furanone, is known to be responsible for premature ageing flavour in dry white wines. Aldol condensation between 2-ketobutyric acid and acetaldehyde was found to be the chemical mechanism responsible for the low concentrations of sotolon found in prematurely aged white wines.

Premature ageing aroma phenomena may also reflect a defective ageing process in red wines, developing several aromatic nuances reminiscent of prunes and figs. The compound suggested as being responsible, 3-methyl-2,4-nonanedione, is one of the most odourant non-thiol compounds in red wines and is specifically associated with the prune aroma of old red wines.

4.8 AROMA PROFILE OF DIFFERENT WINE STYLES

Young wines are characterized by their dominant aromatic profile of the fruit series, predominantly tropical fruits, mainly due to their important content of ethyl esters and acetates. The concentration of these compounds decreases during ageing, an important decrease in content being reported during the first year, especially for acetates, producing changes in the aromatic profile of aged wines. The profile of aged wines is also characterized by an increase in the content of norisoprenoids due to hydrolysis of their glycosylated forms produced in acidic media. The aroma profile is differentiated according to the *V. vinifera* variety considered, with descriptors that are grouped into ripe fruit, tea and tobacco.

Winemaking can also be conducted following different vinification styles. Typical examples are those relating to the production of wines where the fermentation is incomplete, resulting in high sugar content. This style of wine is particularly appreciated by consumers for its sweetness and characteristic fruity aromas.

The vinification options for elaboration of sweet wines involve two distinct processes. One option is fortified wines (Port wine or French ones such as Rivesaltes, Maury and Muscat appellations), obtained from the partial fermentation of fresh grapes or grape juice, prematurely stopped by adding grape brandy. The other typical wines, dessert wines, are produced from overripe grapes. The juice is concentrated by drying (Passito in Italy, Vin de Paille in France, Pedro Ximenez, Alexandria Muscat, etc.), freezing (ice wine, eiswein) or noble rot (Sauternes-Barsac, TokajiAszu, Auslese).

The characteristic aroma profile of these wines is different according to the technique of vinification used. Porto wines are produced by adding aguardente (about 20%), whose contribution to the characteristic aromatic profile of this wine (fruity, balsamic and spicy) is mainly due to ethyl hexanoate, ethyl octanoate, ethyl decanoate, benzaldehyde, alpha-terpineol, linalool and ethyl hydroxycinnamate.

On the other hand, wines made from botrytized grapes, with an exceptional range of aromas evoking citrus, orange peel and dried fruit, have molecules positively linked to noble-rotted grapes, such as phenylacetaldehyde, sotolon, furaneol, gamma-lactone and 4-carboethoxy-gamma-butyrolactone.

Another vinification technique, a prolonged microbiological ageing process, is used to produce the Fino type of sherry wines in Spain. The microorganisms growing spontaneously on the wine surface, flor yeasts, develop an aerobic metabolism responsible for the typical aromatic profile of the finished wines. The 'criaderas and solera' method of winemaking is based on mixing young wine with an old one by periodical blending, the production of acetaldehyde and glycerine by flor yeast being the characteristic process in the production of these wines.

4.9 BRANDY

The word 'brandy' originates from the Dutch word 'brandewijn', meaning 'burnt wine', an appropriated terminology, as most brandies are made by heating wine (without burning it), distilling over alcohol (ethanol) and volatiles, and ageing the distillate in small oak barrels.

Probably, wine was initially distilled to decrease its volume before transport on ships. Distillate may also have been added to wine to preserve it during transit. The containers in which the fortified wines and distillates are stored and transported are made from wood, and extraction of various aromatics and phenolics intensifies flavours further during lengthy transport. The back-addition of water to the distilled liquor before consumption results in a product with greater complexity.

Brandy is usually made from the fermented juice of *V. vinifera* grapes, but distillates can be made from any fermented beverage that originally contained carbohydrates (for example, whisky is the distillate of fermented grains).

Brandies are classified by region of production, even though the soil conditions, climate and grapes used to produce the base wine have less influence on the final product than the distillation process and ageing practices used.

Brandy aromas are described as herbaceous (grass, mint or eucalyptus) and like fresh or dried fruits such as gooseberry, apple, citrus and even flowers. The impact of oak wood will be detected as vanilla, cedar wood, nuttiness, toast, cloves or cigar box aromas, as well as chocolate and mocha. Generally, brandy that has spent more time in oak casks tends to have a smoother, more harmonious mouth-feel.

BIBLIOGRAPHY

Amerine, M. and Singleton, V. (1977). Distillation and brandy. In: *Wine: An Introduction*, 2nd ed., pp. 173–188, University of California Press, London.

Atanasova, B., Thomas-Danguin, T., Langlois, D., Nicklaus, S. and Etievant, P. (2004). Perceptual interactions between fruity and woody notes of wine. *Flav. Frag. J.* 19(6):476.

Bartowsky, E.J. and Pretorius, I.S. (2008). Microbial formation and modification of flavour and off-flavour compounds in wine. In: *Biology of Microorganisms on Grapes, in Must and Wine*, H. König, G. Unden and J. Fröhlich, eds., pp. 211–33, Springer, Heidelberg.

Baumes, R. (2009). Wine aroma precursors. In: *Wine Chemistry and Biochemistry*. M.V. Moreno-Arribas and M.C. Polo, eds., pp. 251–273, Springer, NYC.

Bayonove, C., Baumes, R., Crouzet, J. and Günata, Z. (1998). *Arômes. Oenologie, fondementsscientifiques et technologiques. C. Flanzy*, pp. 1311, Tec & Doc, Paris.

Bartowsky, E.J., and I.S. Pretorius. 2009. Microbial formation and modification of flavor and off-flavor compounds in wine. *In* Biology of Microorganisms on Grapes, in Must and in Wine. H. König *et al.* (eds.), pp. 209–231. Springer-Verlag, Berlin.

Boido, E., Lloret, A., Medina, K., Carrau, F. andDellacassa, E. (2002). Effect of β-glycosidase Activity of *Oenococcus oeni* on the Glycosylated Flavor Precursors of Tannat Wine during Malolactic Fermentation. *J. Agric. Food Chem.* 50(8):2344.

Boido, E., Lloret, A., Medina, K., Fariña, L., Carrau, F., Versini, G. and Dellacassa, E. (2003). Aroma composition of *Vitis vinifera* cv. Tannat: The typical red wine from Uruguay. *J. Agric. Food Chem.* 51(18):5408.

Boido, E., Medina, K., Fariña, L., Carrau, F., Versini, G. and Dellacassa, E. (2009). The effect of bacterial strain and aging on the secondary volatile metabolites produced during malolactic fermentation of Tannat red wine. *J. Agric. Food Chem.* 57(14):6271

Buglass, A., McKay, M. and Lee, C. (2011). Destilled spirits. In: *Handbook of Alcoholic Beverages*. Technical, Analytical and Nutritional Aspects, A.J. Buglass, ed., pp. 329–350, John Wiley & Sons, UK.

Cacho, J. and Ferreira, V. (2010). The aroma of wine. In: *Handbook of Fruit and Vegetable Flavors*, Y.H. Hui, ed., pp. 303–317, John Wiley & Sons, Inc., Hoboken, NJ.

Carrau, F., Medina, K., Fariña, L., Boido, E. and Dellacassa, E. (2010). Effect of *Saccharomyces cerevisiae* inoculum size on wine fermentation aroma compounds and its relation with assimilable nitrogen content. *Int. J. Food Microbiol.* 143(1–2):81.

Carrau, F., Medina, K., Fariña, L., Boido, E., Henschke, P. and Dellacassa, E. (2008). Production of fermentation aroma compounds by Saccharomyces cerevisiae wine yeasts: Effects of yeast assimilable nitrogen on two model strains. *FEMS Yeast Res.* 8(7):1196.

Chatonnet, P., D. Dubourdieu, J.N. Boidron, and M. Pons. 1992. The origin of ethylphenols in wines. *J. Sci. FoodAgric.* 60:165–178.

Cordonnier, C. and Bayanove, C. (1978). Les composantesvariétales et préfermentaires de l'arôme des vins. *Perfums, Cosmét. Arômes* 24:67.

Drawert, F. (1974), *The Chemistry of Winemaking as a Biological-Technological Sequence.* A.D. Web, (Ed), Washington, DC.

Ebeler, S. and Thorngate, J. (2009). Wine chemistry and flavor: Looking into the crystal glass. *J. Agric. Food Chem.* 57(18):8098.

Etiévant, P.X. (1991). *Wine. Volatile Compounds in Foods and Beverages*, pp. 483–546. H. Maarse, Marcel Dekker, New York.

Fariña, L. (2008), *Caracterización del perfil aromáticos de vinos Tannat y su evolución durante la crianza.* PhD thesis, Universidad de la República, Uruguay.

Fariña, L., Boido, E., Carrau, F., Versini, G. and Dellacassa, E. (2005). Terpene compounds as possible precursors of 1,8-cineole in red grapes and wines. *J. Agric. Food Chem.* 53(5):1633.

Fariña, L., Carrau, F., Boido, E., Disegna, E. and Dellacassa, E. (2010). Carotenoid profile evolution in *Vitis vinifera* cv. Tannat grapes during ripening. *Am. J. Enol. Vitic.* 61(4):451.

Fariña, L., Medina, K., Urruty, M., Boido, E., Dellacassa, E. and Carrau, F. (2012).Redox effect on volatile compound formation in wine during fermentation by *Saccharomyces cerevisiae. Food Chem.* 134(2):933.

Ferreira, V. and Cacho, J. (2009). Identificaction of impact odorants of wine. In: *Wine Chemistry and Biochemistry*, M.V. Moreno-Arribas and M.C. Polo, eds., pp. 393–415, Springer, NYC.

Flanzy, C., López Gomez, A., Macho Quevedo, J. and Madrid Vicente, A. (2002), *Enologia: Fundamentos científicos y tecnológicos.* Madrid Vicente, A, Madrid.

Garde-Cerdán, T. and Ancín-Azpilicueta, C. (2006). Review of quality factors on wine ageing in oak barrels. *Trends Food Sci. Technol.* 17(8):438.

González-Barreiro, C., Rial-Otero, R., Cancho-Grande, B. and Simal-Gándara, J. (2015). Wine aroma compounds in grapes: A critical review. *Crit. Rev. Food Sci. Nutr.* 55(2):202.

Gonzalez-Viñas, M., Perez-Coello, M. and Cabezudo, M. (1998). Sensory analysis of aroma attributes of young airen white wines during storage in the bottle. *J. Food Qual.* 21(4):285.

Guitart, A., P. Hernández Orte, V. Ferreira, C. Peña, and J. Cacho. 1999. Some observations about the correlation between the amino acid content of musts and wines of the Chardonnay variety and their fermentation aromas. *Am. J. Enol. Vitic.* 50:253–258.

Guymon, J. (1974). Chemical aspects of distilling wines into brandy. In: *Chemistry of winemaking*, A. Dinsmoor Webb, ed., pp. 232–253, American Chemical Society, Washington, DC.

Hernández-Orte, P., Cacho, J. and Ferreira, V. (2002). Relationship between varietal amino acid profile of grapes and wine aromatic composition. Experiments with model solutions and chemometric study. *J. Agric. Food Chem.* 50(10):2891.

Hoenicke, K., Simat, T., Steinhart, H., Christoph, N., Geßner, M. and Köhler, H. (2002). 'Untypical aging off-flavor' in wine: formation of 2-aminoacetophenone and evaluation of its influencing factors. *Anal. Chim. Acta* 458(1):29.

Herraiz, T., and C.S. Ough. 1993. Formation of ethyl esters of amino acids by yeasts during the alcoholic fermentation of grape juice. *Am. J. Enol. Vitic.* 44:41–48.

Mosedale, J.R. and Puech, J.L. (1998). Wood maturation of distilled beverages. *Trends Food Sci. Tech.* 9(3):95.

Malcorps, P., Cheval, J.M., Jamil, S. and Dufour, J. (1991). A new model for the regulation of ester synthesis by alcohol acetyltransferase in *Saccharomyces cerevisiae*. *J. Am. Soc. Brew. Chem.* 49(X):47–53.

Mason, A.B. and Dufour, J.P. (2000). Alcohol acetyltransferases and the significance of ester synthesis in yeast. *Yeast* 16(14):1287.

Medina, K., Boido, E., Fariña, L., Gioia, O., Gomez, M., Barquet, M., Gaggero, C., Dellacassa, E. and Carrau, F. (2013). Increased flavour diversity of Chardonnay wines by spontaneous fermentation and co-fermentation with *Hanseniaspora vineae*. *Food Chem.* 141(3):2513.

Monterio, F.F. and L.F. Bisson (1991). Biological assay of nitrogen content of grape juice and prediction of sluggish fermentations. *Am. J. Enol. Vitic.* 42: 47–57.

Nakamura, S., Crowell, E., Ough, C. and Totsuka, A. (1988). Quantitative analysis of γ-nonalactone in wines and its threshold determination. *J. Food Sci.* 53(4):1243.

Ough, C., Huang, D. and Stevens, D. (1991). Amino acid uptake by four commercial yeasts at two different temperatures of growth and fermentation: Effect on urea excretion and readsorption. *Am. J. Enol. Vitic.* 42(1):26.

Pérez-Coello, M., Martín-Álvarez, P. and Cabezudo, M. (1999). Prediction of the storage time in bottles of Spanish white wines using multivariate statistical analysis. *Z. Lebensm. Unters. Forsch.* 208(1):408.

Pons, A., Lavigne, V., Landais, Y., Darriet, P. and Dubourdieu, D. (2008). Distribution and organoleptic impact of sotolon enantiomers in dry white wines. *J. Agric. Food Chem.* 56(5):1606.

Pons, A., Lavigne, V., Landais, Y., Darriet, P. and Dubourdieu, D. (2010). Identification of a sotolon pathway in dry white wines. *J. Agric. Food Chem.* 58(12):7273.

Rogerson, F. and de Freitas, V. (2002). Fortification spirit, a contributor to the aroma complexity of port. *J. Food Sci.* 67(4):1564.

Schwab, W. and Wüst, M. (2015). Understanding the constitutive and induced biosynthesis of Mono- and sesquiterpenes in grapes (*Vitis vinifera*): A key to unlocking the biochemical secrets of unique grape aroma profiles. *J. Agric. Food Chem.* 63(49):10591.

Swiegers, J.H., E.J. Bartowsky, P.A. Henschke, and I.S. Pretorius. 2005a. Yeast and bacterial modulation of wine aroma and flavour. *Aust. J. Grape Wine Res.* 11:139–173.

Swiegers, J.H., P.J. Chambers, and I.S. Pretorius. 2005b. Olfaction and taste: Human perception, physiology and genetics. *Aust. J. Grape Wine Res.* 11:109–113.

Tamang, J. (2010). Diversity of fermented beverages and alcoholic drinks. In: *Fermented Foods and Beverages of the World*, J. Tamang and K. Kailasapathy, eds., pp. 86–117, Taylor & Francis Group, Boca Raton, FL.

Thibon, C., Dubourdieu, D., Darriet, P. and Tominaga, T. (2009). Impact of noble rot on the aroma precursor of 3-sulfanylhexanol content in *Vitis vinifera* L. cv Sauvignon blanc and Semillon grape juice. *Food Chem.* 114(4):1359.

5 Wine
Therapeutic Potential

Creina Stockley, Arina O. Antoce and Rena I. Kosti

CONTENTS

5.1 INTRODUCTION

Historically, wine has been used as a therapeutic agent since ancient times. The physician Hippocrates of Kos (c. 460–370 BC) prescribed it as a wound dressing, a nourishing dietary beverage, a cooling agent for fevers, a purgative and a diuretic, as did Claudius Galenus (Galen) (c. AD 130–201). Wines then fell from favor as a medicine, which is attributed to the Puritan religious movement led by Oliver Cromwell (c. AD 1599–658) that spread to the New World. These puritan movements in England and the United States eventually led to the temperance movement of the 19th century, which condemned alcohol in all its forms. In the 21st century, however, the potential therapeutic effects of moderate wine consumption were rediscovered, such as a reduction in the risk of cardiovascular diseases (CVD), a slowed rate of aging and a reduction in the risk of degenerative diseases such as certain cancers, diabetes mellitus, Alzheimer's disease and dementia. The chemical components of wine purported to be primarily responsible for these therapeutic effects are ethanol and the phenolic compounds. This chapter mainly, but not exclusively, focuses on the potential therapeutic effects of wine on cardiovascular diseases.

5.2 HEALTH EFFECTS

5.2.1 Potential Cardiovascular Effects of Wine

Early in the 20th century, the first population-based study of wine observed a linear relationship between the prevalence of hypertension or high blood pressure and the amount of wine consumed by French troops on the Western front during World War I. A J-shaped relationship between amount of wine consumed and risk of CVD (*i.e.* hypertension) was first observed in 1974 and independently in 1979. The J-shaped relationship only came into public prominence, however, in June 1992, which has been popularized as the 'French Paradox'. The Monitoring of Trends and Determinants in Cardiovascular Disease (MONICA) project found that the French, whose traditional cardiovascular risk factors are at least as great as those of other Western countries, had a lower cardiovascular risk than the latter societies, 36% compared to 75%. The gist was that the moderate consumption of wine (1–2 glasses of wine per day), in particular with meals, could reduce the risk of CVD (cardiac and vascular diseases such as atherosclerosis, coronary heart disease, myocardial infarction

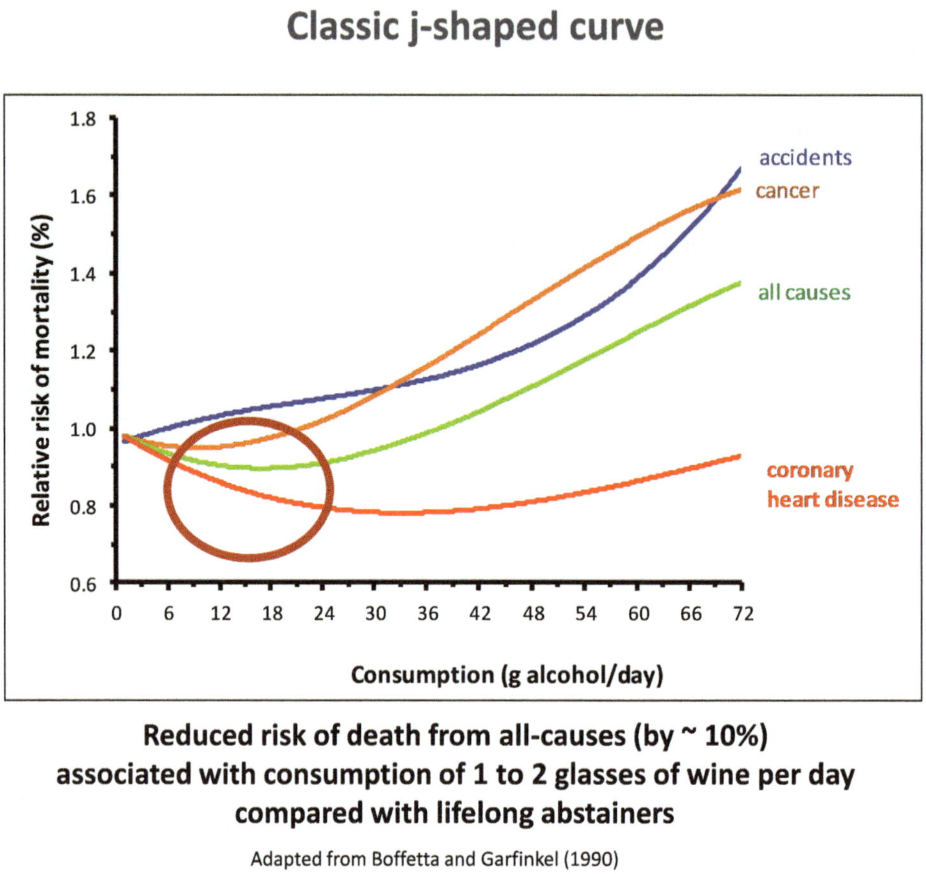

FIGURE 5.1 Reduced risk of death from all-causes (by ~ 10%) associated with consumption of 1 to 2 glasses of wine per day compared with lifelong abstainers. *Source:* Adapted from Boffetta and Garfinkel (1990)

and ischaemic stroke, as well as hypertension) by up to 40% of that of abstainers.

This finding is of crucial importance, given that CVD is the leading cause of death globally, where more people die annually from CVD than from any other cause. For example, an estimated 17 million people died from CVD in 2018, representing 31% of all global deaths. Of these deaths, an estimated 7.4 million were due to coronary heart disease and 6.7 million were due to stroke, where 87% are ischemic, caused by thrombosis or embolisms.

The protective effect is first observed when risk factors for cardiovascular disease begin to influence medium and long-term health, at approximately the age of 40 years for men and 50 years for women, depending on the age of onset of menopause and use of hormone replacement therapy. This effect generally continues in men until about 75 years of age but continues in women beyond 75 years. The consumption of alcohol at a younger age also positively influences the risk of cardiovascular disease at a later age. At a younger age, however, other mortality risks from alcohol consumption are more likely, such as motor vehicle accidents and drownings. Nowadays, most epidemiological studies reveal that low-to-moderate alcohol consumption is associated with an increase in HDL cholesterol, a reduction of platelet aggregation and promotion of fibrinolysis. Wine consumption, in particular, has an inverse association with ischemic heart disease, where in most studies the association follows a J-shaped curve. A recent population-based prospective study in current drinkers of alcohol, however, suggested that consumption of 100 g/week was the threshold for lowest risk of all-cause mortality. In particular, for the other subtypes of CVD, apart from myocardial infarction, no clear risk thresholds were determined below which lower alcohol consumption was associated with lower disease risk. Conversely, the findings of a recent randomized nutrigenomic trial suggested that low to moderate alcohol consumption as part of a Mediterranean-style diet could reduce the risk of atherosclerosis by modulating antioxidant gene expression, thereby preventing inflammatory and oxidative damage. Therefore, the existing scientific literature supports alcohol consumption up to 1-2 glasses per day, in particular with meals, as part of a healthy diet and lifestyle (Figure 5.1).

5.2.2 Other Potential Inter-Related Health Effects of Wine

Given the relationship and similarities between risk factors for CVDs and those for diabetes and cognitive decline/dementia, it is not unreasonable to expect to see a similar J-shaped relationship between risk of these non-communicable diseases and alcohol consumption. For example, individuals with diabetes have coexistent lipid disorders, as well as blood clotting and blood flow abnormalities similar to individuals with, or at risk of, CVD, while damage to DNA, lipids and proteins by oxidative free radicals have been implicated in accelerated ageing. Most, but not all, published papers suggest that such a relationship exists. For example, the relationship between alcohol consumption and risk of developing both Type 2 diabetes and the inter-related Metabolic Syndrome is characterized by a J-shaped curve. Accordingly, moderate alcohol consumers have a lower risk of developing both Type 2 diabetes and Metabolic Syndrome than abstainers and heavier consumers, which is independent of gender, age, ethnicity, body mass index, diet quality and exercise. Furthermore, among individuals with Type 2 diabetes, moderate alcohol consumption may reduce the risk of developing CVD, and among individuals with Metabolic Syndrome, it may reduce the risk of developing Type 2 diabetes and CVDs. A similar J-shaped relationship is observed between alcohol consumption, cognitive dysfunction or impairment and risk of developing dementias. Long-term heavier consumption, however, amplifies cognitive impairment and dementias associated with aging, as the brain of people aged over 65 years is the most sensitive to the toxic effects of the alcohol, which acts both directly and indirectly on the central nervous system. The suggested optimal amount of alcohol to lower risk is approximately two glasses of wine daily.

Furthermore, moderate wine consumption may also protect against tumor cell proliferation associated with cancer. A relatively recent study attempting to elucidate the effect of wine and its chemical components on tumor cells showed that even though ethanol can promote cell proliferation, red wine significantly repressed cell proliferation of different human cancer lines, in a dose-dependent manner, by reducing the RNA Pol III gene transcription. A similar J-shaped relationship is also seen here, as the rate of cell proliferation induced by ethanol is higher at lower concentrations of ethanol.

5.3 POTENTIAL BIOLOGICAL WINE COMPONENTS

More than 500 compounds have been identified in *Vitis vinifera* grapes and wine to date, but only those of therapeutic significance are discussed here, with particular emphasis on alcohol/ethanol and phenolic compounds. Aside from phenol derivatives, other classes of compounds derived from grapes, such as monoterpenoids, sesquiterpenoids and norisoprenoids, as well as indoleamines, melatonin and serotonin, also show potentially beneficial human health effects.

5.3.1 Ethanol Component

Wine typically contains an assortment of alcohols in measurable amounts other than ethanol; the concentration of ethanol in 'table' wine generally ranges between 8 and 15% (v/v), which is determined by a number of factors.

5.3.2 Polyphenolic Component

The polyphenolic compounds (polyphenols) found in wine originate primarily from the grapes, but may also be imparted from yeast and oak during winemaking and maturation. Chemically, they are cyclic benzene compounds, possessing one or more hydroxyl groups associated directly with an aromatic ring structure. All polyphenols are synthesized from the amino acid phenylalanine through the phenylpropanoid pathway. Phenylalanine is in turn a product of the shikimate pathway, which links carbohydrate metabolism with the biosynthesis of aromatic amino acids and secondary metabolites. Two main classes of compounds can be produced: flavonoids (by chalcone synthase) and stilbenes (by stilbene synthase), which are generally considered with the other non-flavonoids. The flavonoids comprise greater than 85% of the total polyphenol content of red wine but less than 20% of white wine and are distributed differently within different parts of the grape berry, being mainly located in the seeds and skin.

5.3.2.1 Flavonoids

Wine flavonoids are all polyphenols, having multiple aromatic rings possessing hydroxyl groups (Figure 5.2). They have a very specific three-ring structure – two phenols (rings A and B) joined by a pyran or oxygen-containing hexagonal (ring C) structure – and have hydroxyl substitution groups on ring A at positions 5 and 7. The different classes of flavonoids are characterized by differences in the oxidation state of the C in position 4 and the substitution of the C in position 3 on ring C. They exist free and also conjugated to other flavonoids, sugars, non-flavonoids or a combination of these compounds. Those flavonoids conjugated to sugars or non-flavonoids are called glycosides and acyl derivatives, respectively. There are six main classes of flavonoids: flavanols (flavan-3-ols), flavonols, flavones, flavanones, anthocyanidins and isoflavones. The major classes of flavonoids in grapes and wines are flavanols, flavonols and anthocyanins, but some flavones such as apigenin or luteolin have also been reported.

Levels of certain of these flavonoid classes are reported in the USDA database for the flavonoid content of selected foods, and are expressed in mg/100 g. The levels range as follows: flavones – 1.3 in red grapes and 0.04–0.17 in red wine; flavonols – 1.05–2.39 in grapes and 0.01–1.94 in wine; flavanones – 0.78–2.40 in wine; flavanols – 2.03–21.63 in grapes and 1.32–18.4 in wine; and anthocyanidins – 47.7–120.1 in red grapes and 0.06–153.0 in wine.

Flavanols, which are found in both the seed and skin of the grape as well as in the leaves and stems, are the major class of flavonoids found in wine. The alcohol group is on position 3 of ring C and hence they may be named flavan-3-ols. They comprise monomeric catechins (catechin and its stereoisomer epicatechin) and their hydroxylated forms (gallocatechin and epigallocatechin) that are not glycosylated but may be free or in the form of gallate esters (catechin-3-gallate, epicatechin-3-gallate, gallocatechin-3-gallate and epigallocatechin-3-gallate). They are, however, mainly present in wine as oligomeric and polymeric forms, proanthocyanidins and condensed tannins, respectively, and are responsible for the astringency of grape skins, seeds and wines. Unlike the monomeric forms (concentration in wine 40–120 mg/L), these oligomeric and polymeric forms (concentration in red wine 500–1500 mg/L and white wine 10–450 mg/L) have a high molecular weight,

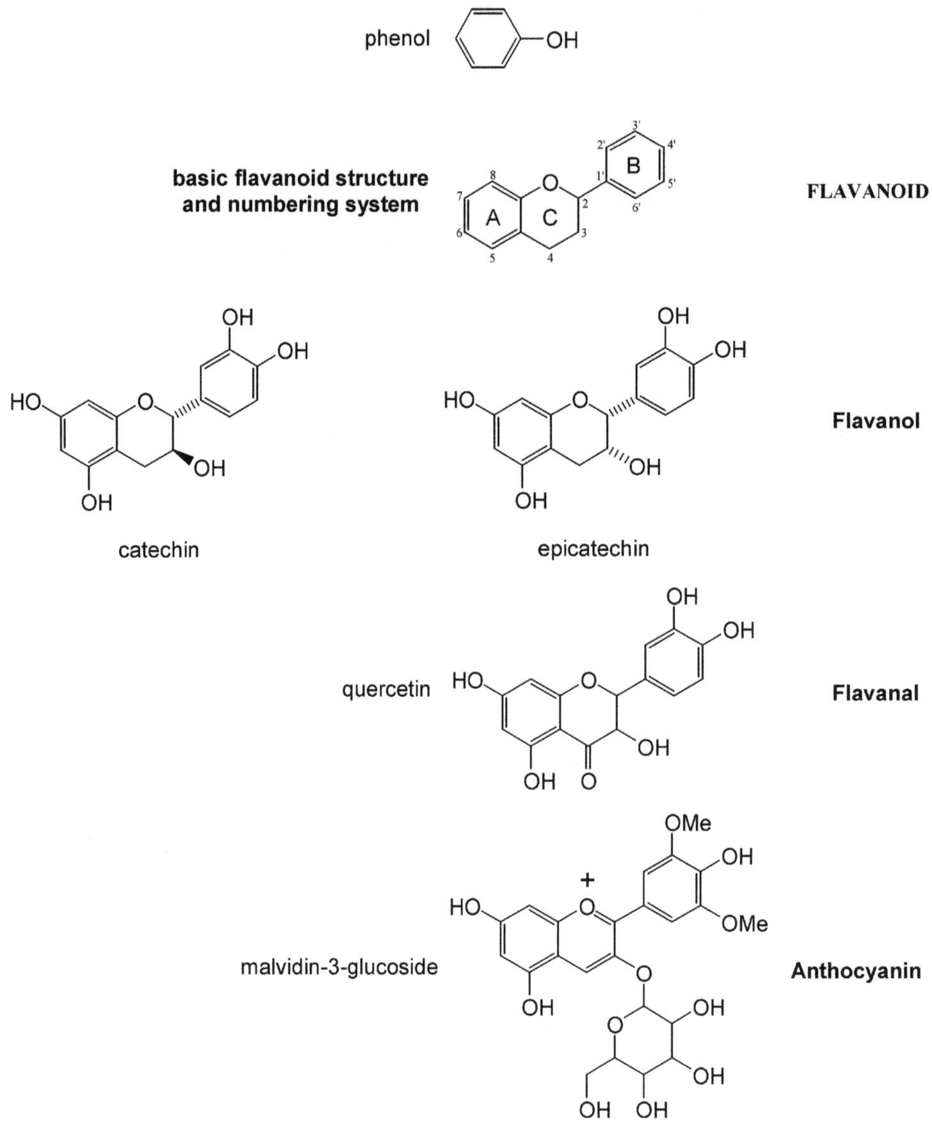

FIGURE 5.2 Wine flavonoids

appear not to be absorbed into the bloodstream and also have lower antioxidant activity than the monomeric forms.

Flavonols are produced and found in the skin of the grape, rachis and leaves and have a similar chemical structure to flavones but are differentiated by the addition of hydroxyl or methoxy groups on ring B. Unlike flavanols, they are primarily glycosylated. During winemaking and maturation, the sugar can be cleaved off to leave the aglycone form. The major flavonols found in wine (concentration in red wine 50–200 mg/L) are quercetin, myricetin and kaempferol and their glycosidic forms, mainly 3-glucosides (quercetin-3-O-glucoside, myricetin-3-O-glucoside and kaempferol-3-O-glucoside), as well as 3-glucuronides and 3,5-diglycosides. Less common flavonols (isorhamnetin) are also found. Compared to red grapes, however, white grapes do not contain myricetin, due to lack of expression of the flavonoid 3',5'- hydroxylase enzyme. The colorless flavonols are reactive molecules and participate in copolymerization reactions with anthocyanins, leading to copigmentation.

Anthocyanins are the major pigment of red grapes, found in the skin of the grape and also in the pulp of teinturier grapes. Being water soluble, they are dissolved in their free forms in the vacuolar sap of the skin cells. They are glycosides (of glucose or *p*-coumaroyl-glucose) and acylglycosides of insoluble anthocyanidins. The glucosides may be mono- or diglucosides, 3-O-glucoside or 3,5-O-diglucoside, respectively. Major anthocyanins found in red wine are malvidin-3,5-O-diglucoside and 3-O-glucoside, cyanidin-3-O-galactoside, cyanidin-3-O-glucoside and 3,5-O-diglucoside, peonidin-3-O-glucoside, delphinidin-3-O-glucoside, and petunidin-3-O-glucoside. Substitutions determine the type of anthocyanin, and the five basic anthocyanidins are cyanidin, peonidin, delphinidin, petunidin and malvidin, which predominate in grapes and wines. Concord grapes are a notable exception in which delphinidin predominates, followed by cyanidin. Another exception is the Pinot Noir grape, which does not synthesize acetylated or *p*-coumaroylated anthocyanins. Being unstable, only traces of anthocyanidins are found in grapes and wine (90–400 mg/L). During fermentation, the monomeric anthocyanins also interact with tannins to form complex polymeric pigments which are more stable to oxidation, bleaching and changes in the pH value of wine and which generally persist during aging, although they can also be hydrolysed during fermentation.

Recently anthocyanins have received more media attention, as higher consumption appears to be associated with anti-inflammatory effects linked to the reduction of risk of certain chronic diseases, including CVD. Malvidin-3-O-β glucoside reduces the inflammatory mediators TNFα, IL1, IL-6 and iNOS-derived nitric oxide (NO) secretion from activated macrophages, confirming the inhibition of the transcription of their encoding genes. In addition, recent research into the potential cardioprotection of a red grape extract containing a high concentration of 63.93±12.50 mg malvidin/g of fresh grape skin, indicated that malvidin was able to protect the perfused rat heart against ischemia/reperfusion damage.

5.3.2.2 Non-Flavonoids

The main non-flavonoid polyphenols found in wine include derivatives of hydroxycinnamic and hydroxybenzoic acids, stilbenes and so-called hydrolysable tannins (Figure 5.3). Hydroxycinnamic acids are found in the flesh of grapes (mesocarp) and are often esterified to sugars, organic acids or alcohols. The most common are based on coumaric acid, caffeic acid, caftaric acid and ferulic acid, where caftaric acid is the most common. Caffeic acid, along with tyrosol and hydroxytyrosol, are linked to the cardioprotective effects associated with white wine consumption. The concentration of hydroxycinnamic acids is higher in white wines (130 mg/L) than in red wines (60 mg/L). Hydroxybenzoic acids are structurally simpler than the other polyphenols and include benzoic acid, gallic acid, ellagic acid, vanillic acid, protocatechuic acid and their derivatives, which together with those of hydroxycinnamic acid, form the primary non-flavonoid polyphenols in wines not aged in oak. These are stored primarily in the vacuoles of grape cells and are easily extracted on crushing. Wines aged in oak, however, have a higher concentration of hydroxybenzoic acid derivatives from the degradation of hydrolysable tannins. Tannins are organic molecules that form stable combinations with proteins and have a high molecular weight (500–3000 Da). Hydrolysable tannins in wine are imparted by oak wood during maturation and include ester-linked oligomers of gallic acid or ellagic acid with glucose or related sugars (gallotannins and ellagitannins). Correspondingly, the average concentration of hydrolysable tannins varies, but typically ranges from 100 mg/L for white wines aged in wood for six months up to 250 mg/L for red wines ages in wood for two years. Oak bark extracts appeared cardioprotective in rats fed high-fat or high-carbohydrate diets, improving both the structure and function of the heart and reducing the signs of metabolic syndrome.

Phytoalexins are low molecular weight compounds, such as the stilbene monomer, resveratrol, that are anti-microbial and produced by the skin cells of grapes and by the leaves in

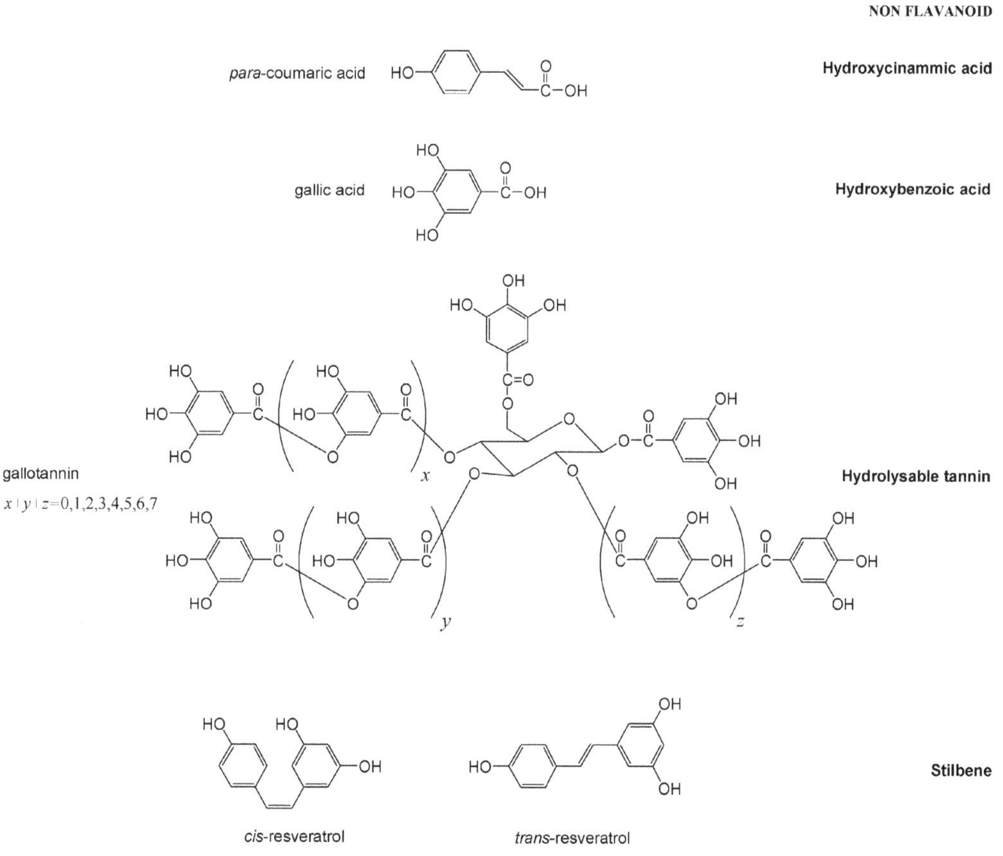

FIGURE 5.3 Non-flavonoid polyphenols

response to *Botrytis cinerea* and other fungal infections on grapevines. Resveratrol exists in both *cis* and *trans* forms in wine, but *trans*-resveratrol predominates in grapes and may be methylated and polymerized to produce piceid, pterostilbene and the viniferins, respectively, or glycosylated during fermentation. Other hydroxyl-derivatives of resveratrol such as piceatannol are also present in grapes and wine. The average concentration of total resveratrol ranges from 7 mg/L in red wine to 0.5 mg/L in white wine. Fermentation with recombinant yeast strains expressing particular genes from *Aspergillus niger* or *Candida molischiana* led to an increase in the resveratrol content of white wine, but these transgenic yeasts are not approved for winemaking.

5.4 POTENTIAL BIOLOGICAL ACTIVITIES OF WINE COMPONENTS ON THE CARDIOVASCULAR SYSTEM

There is continuing controversy regarding the relative health effects of the major types of alcoholic beverages: spirits, beer, and wine. Most, but not all, studies suggest that red wine uniquely reduces morbidity/mortality compared with other alcoholic beverages. The ethanol component, common to beer, wine and spirits, improves haemostasis and serum lipoprotein concentrations. Consumers of wine, however, may have a greater reduction in the risk of CVD than consumers of beer or spirits, or consumers of certain fruits, grains and vegetables, based on concentration of polyphenols. Presumably, therefore, both ethanol and the monomeric and polymeric phenolic compounds act synergistically to provide wine with cardioprotective and other healthy properties. For example, in conferring immunity, the polyphenol content of wine is found to modulate leukocyte function and adhesion molecules, while both the ethanol and polyphenol contents modulate inflammatory mediators.

5.4.1 Effects of the Ethanol Component of Wine

5.4.1.1 Peroxidative Effects

The effects of the ethanol component in wine on lipids are dose dependent, and ethanol may have both positive and negative effects on atherosclerosis and CVD. Indeed, CVD is a pathological condition in which predominantly non-specific lipid peroxidation occurs *in vivo*, such that the accumulation of oxidized low-density lipoprotein (LDL) in the intima of the arteries and in atheroslerotic lesions leads to the pathogenesis of CVD. While ethanol can indirectly prevent the oxidative modification of LDL when consumed in moderation, excessive ethanol can also induce the oxidative modification (peroxidation) of LDL, by inducing the generation of excess oxidative free radicals (also known as reactive oxygen species or ROS) which peroxidise lipids. For example, the hepatic metabolism of ethanol results in the formation of molecules such as acetaldehyde, the further metabolism of which in the cell leads to ROS production. Higher hepatic concentrations of ethanol also stimulate the activity of the hepatic enzyme cytochrome P_{450} that metabolizes ethanol and other molecules generating ROS in the process and can also increase the level of free iron in the body's cells, which promotes ROS generation, as well as decreasing the amount and altering the activity of anti-oxidant enzymes, increasing the susceptibility of lipids to peroxidation. The resulting state, defined as a disturbance in the balance between the production of reactive oxygen species and antioxidant defenses, is known as oxidative stress.

Ethanol-induced peroxidation is implicated in the etiology of alcoholic liver disease and in the pathogenesis of diseases including alcoholic cardiomyopathy. Indeed, ingestion of alcoholic beverages other than wine, particularly in excess, has dose-dependent pro-oxidant effects in both animals and humans. Beer has been shown to increase the susceptibility of LDL to *in vitro* oxidation, but brandy had no effect on LDL oxidation although the plasma and lipid concentration of β-carotene decreased significantly, indicating a predominant pro-oxidant effect.

Conversely, moderate alcohol consumption, and therefore ethanol, can also increase the plasma concentration of paraoxonase 1 (PON1), an antioxidant enzyme associated with HDL, that prevents the oxidation of LDL by ROS, hydrolyses oxidised LDL and detoxifies the homocysteine metabolite, homocysteine thiolactone. Homocysteine thiolactone can cause protein damage by homocysteinylation of the lysine residues, leading to atherosclerosis. Hyperhomocysteinaemia, which occurs in 5–7% of the general population, is a risk factor for CVD.

5.4.1.2 Effect on High Density Lipoproteins

Ethanol protects against the initiation and progression of atherosclerosis, however, by increasing the production of plasma concentration of high density lipoprotein (HDL) and its HDL_2 and HDL_3 sub-fractions. As HDL transports excess cholesterol from the arteries and extra-hepatic or peripheral tissues to the liver for secretion (reverse cholesterol transfer), the plasma concentration of HDL is inversely related to atherosclerosis. Unlike LDL, HDL is assembled from constituents within plasma and is positively correlated with the plasma concentration of apolipoprotein A-I, but negatively associated with the plasma concentration of triglycerides. Ethanol increases the concentration of HDL by stimulating the hepatic synthesis and secretion of its sub-components, apolipoproteins A-I and A-II. Triglycerides reduce the cholesterol content of HDL by exchanging cholesterol ester for triglyceride from VLDL. Ethanol, however, also decreases the activity of the cholesteryl ester transfer protein and hence, this exchange to the more atherogenic LDL particles. Ethanol also increases the rate of HDL cholesterol esterification. HDL can also inhibit numerous changes associated with the oxidative modification of LDL by endothelial and smooth muscle cells. The enzymatic mechanism is dependent on the concentration of HDL in the artery wall intima.

A four-week randomized dietary intervention of red wine, red grape extract dissolved in water and placebo showed that wine consumption significantly increased fasting HDL by 11–16% as compared to the other two groups, while fasting fibrinogen decreased by 8–15% in the wine consumption group. Thus, the effect of wine on these parameters seems to be mainly due to the presence of ethanol.

5.4.1.3 Haemostatic Effects

Normal haemostasis involves a delicate balance between coagulation and fibrinolysis (lysis of the fibrin clot), which is regulated through the synthesis of the fibrinolytic proteins, tissue type plasminogen activator (t-PA), urokinase type plasminogen activator (u-PA) and plasminogen activator inhibitor-1 (PAI-1). Ethanol decreases coagulation and increases fibrinolysis. Specifically, it affects the following haemostatic variables: fibrinogen; factor VII and factor VIII; platelet aggregability; PAI-1; t-PA; and u-PA. (Figure 5.4).

Ethanol *in vitro* and *in vivo* reduces the concentration of fibrinogen (soluble plasma glycoprotein converted to an insoluble fibrin polymer clot) and thus reduces clot formation. Consumption of 30g of an alcoholic beverage per day is associated with an approximate 17–33% decrease in the concentration of fibrinogen. Ethanol may also inhibit the transcription of fibrinogen genes. Furthermore, factor VII is a vitamin K-dependent plasma clotting factor, that complexes with a tissue factor to form a procoagulant enzyme which initiates clot formation, and Factor VIII is a glycoprotein that acts as a cofactor in blood coagulation. From epidemiological and *in vivo* studies, consumption of an alcoholic beverage appears to decrease their activity.

Platelets adhere to collagen fibers in the blood vessel wall when they are damaged or when atherosclerotic plaques rupture, leading to a chain of events. Briefly, the adhering platelets secrete granules containing active clotting factors and adenosine diphosphate (ADP) into the blood. The platelet enzyme phospholipase A_2 then releases arachidonate from membrane phospholipids, which degrades to thromboxane A_2, which is also secreted into the blood. The secreted ADP and thromboxane A_2 induce further activation, adhesion and aggregation of platelets in a positive feedback loop. Eventually fibrinogen polymerizes into fibrin threads which bind to activated adjacent platelets, forming a fibrin clot that continues to grow until its growth is arrested. In animal models and *in vitro*, ethanol significantly inhibits collagen and ADP-induced platelet aggregability and secondary aggregation induced by ADP, associated with the release of thromboxane A_2. This inhibition is mediated by inhibition of phospholipase A_2 activation, that reduces the release of arachidonate from platelet membranes and the subsequent formation of thromboxane A_2, thus interrupting the positive feedback loop. Ethanol may also directly inhibit thrombin-stimulated platelet aggregation and secretion of granule contents by platelets, and the collagen and thrombin-induced hydrolysis and mobilization of arachidonate.

In fibrinolysis, the pro-enzyme plasminogen (Pmg) is converted to the active serine protease plasmin by two Pmg activators, tissue type PA (t-PA) and urokinase type-PA (u-PA). Plasmin initiates lysis of the fibrin clot. The PAs are regulated by the primary inhibitor of fibrinolysis, PA inhibitor type-1 (PAI-1), in endothelial cells. The complexing of PAI-1 with t-PA serves to inactivate t-PA and to transport it to the liver, such that the activity of t-PA is inversely related to that of PAI-1. Ethanol increases the secretion of t-PA and u-PA by stimulating t-PA and u-PA gene expression. It also independently down-regulates PAI gene expression, which decreases the circulating concentration and activity of PAI and hence,

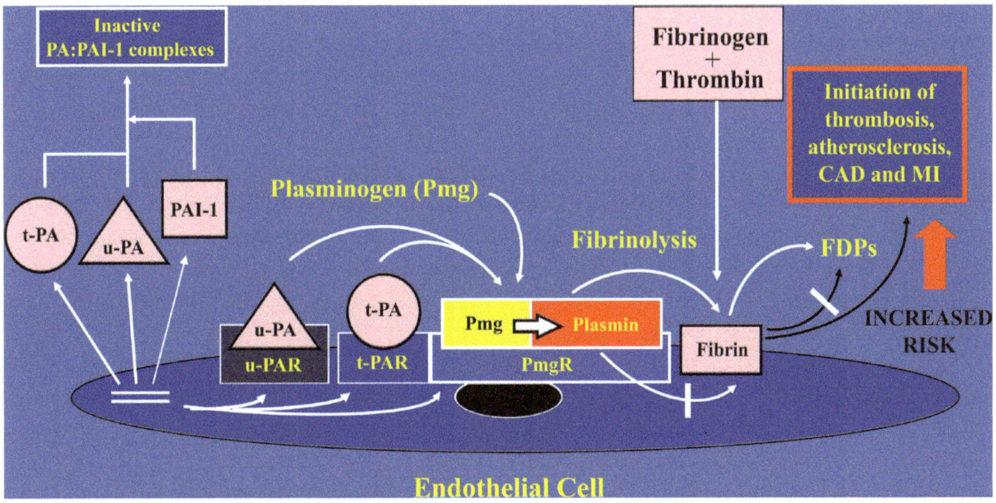

FIGURE 5.4 (a) Proposed cardioprotective mechanisms of ethanol and the wine-derived phenolic compounds following moderate wine consumption. Reproduced with permission from Dr F.M. Booyse, Division of Cardiovascular Disease, University of Alabama at Birmingham, Alabama. (b) Sites of action on the endothelial cell for ethanol and the wine-derived phenolic compounds relating to fibrinolysis and haemostasis, where moderate wine consumption is proposed to increase endothelial cell fibrinolysis and hence reduce the risk of cardiovascular disease. Reproduced with permission from Dr F.M. Booyse, Division of Cardiovascular Disease, University of Alabama at Birmingham, Alabama. (c) Proposed cardioprotective mechanisms of ethanol and the wine-derived phenolic compounds following moderate wine consumption. Reproduced with permission from Dr F.M. Booyse, Division of Cardiovascular Disease, University of Alabama at Birmingham, Alabama

increases the activity of t-PA, leading to increased fibrinolysis. The effects of ethanol on t-PA, u-PA, PAI and fibrinolysis are relatively rapid and appear to be both short-term (acute) and long-term (sustainable) for up to approximately 24 hours post ethanol addition *in vitro*. The short-term effect of ethanol on fibrinolysis is by directly effecting one or more surface localized fibrinolytic components. The long-term effect of ethanol on fibrinolysis is by stimulating the gene expression of PAs while down-regulating that of PAI-1 to increase fibrinolysis. Indeed, *in vivo*, PAI-1 activity initially increases, while that of t-PA decreases, but t-PA activity increases significantly by approximately 9 hours post consumption of beer or wine.

5.4.1.4 Other Effects

The ethanol component of wine also exhibits other cardioprotective mechanisms such as: reducing coronary artery spasm in response to stress; increasing coronary blood flow by vasodilatation; decreasing blood pressure by vasodilatation; and decreasing the plasma concentration of insulin. Insulin-dependent diabetics are observed to have a higher concentration of oxidatively-modified LDL.

Moderate alcohol consumption is also related to decreased insulin resistance in skeletal muscle, and such insulin sensitizing activity may be related to improved production of AMP-activated protein kinase, generated by the metabolism of acetate in peripheral tissues and involved in glucose uptake (among other functions).

Other dose-dependent effects of ethanol on the cardiovascular system include possible protection of the body's cells and tissues against injury from cerebral and myocardial ischaemia, where moderate but not excessive ethanol ingestion 24 hours prior to ischaemia/reperfusion, known as ethanol preconditioning, completely prevented ischaemia/reperfusion-induced leukocyte-endothelial cell adhesive interactions by a mechanism that involves the ability for ethanol to activate an oxidant-dependent signaling pathway. More than a light amount of an ethanol-containing beverage, approximately 10-20g ethanol/day, is also observed to incrementally increase blood pressure, in particular systolic. This effect is more pronounced in males than females and in older rather than younger individuals, leading to hypertension, which is a risk factor for CVD. Although the link between ethanol consumption and hypertension is well established, the mechanism through which ethanol increases blood pressure remains elusive. Possible mechanisms underlying ethanol-induced hypertension based on clinical and experimental observations include an increase in sympathetic nervous system activity, stimulation of the renin-angiotensin-aldosterone system, an increase of intracellular Ca^{2+} in vascular smooth muscle, increased oxidative stress and endothelial dysfunction.

In addition, ethanol has a specific dose-dependent effect on heart rate and heart rate variability. When compared with water, approximately 12g ethanol, either as red wine or alcohol per se, lowered time- and frequency-domain markers of vagal heart rate, while 24g raised heart rate by diminishing vagal and increasing sympathetic heart rate modulation. Heart rate variability associated with ethanol consumption was also similar irrespective of the type of alcoholic beverage.

In contrast, moderate drinking of red wine and ethanol showed opposite effects on myocardial recovery after ischemia-reperfusion injury in rat hearts.

5.4.2 Effects of the Polyphenolic Components of Wine

Wine typically contains a higher concentration of polyphenols than other alcoholic beverages. These are non-flavonoid classes of compounds (*i.e.* hydroxycinnamates, hydroxybenzoates and the stilbenes) and flavonoid classes of compounds (*i.e.* flavones, flavanones, flavan-3-ols, flavonols and anthocyanins). As the grape and wine phenolic compositions differ, even when ethanol is not present, the effect of wine and grape extract differ; only the black-grape wine extract and not the grape extract proved to be effective in reduction of blood pressure in mildly hypertensive people.

While polymeric condensed tannins and pigmented tannins constitute the majority of red wine phenolic compounds, their large size and combination with salivary proteins precludes absorption as such and thus, they are unlikely to contribute to any beneficial biological mechanisms for human health. The total amount of polyphenols in a 100 ml glass of red wine is approximately 200 mg versus 40 mg in a glass of white wine. Approximately 90% of the ethanol in wine is readily absorbed across the small intestine into the blood stream, and *in vivo* data has shown that the wine-derived polyphenols are also absorbed into the blood stream in measurable quantity. After passing across the intestinal wall, red wine polyphenols are adsorbed at the LDL surface, transported and later easily dissociated. Indeed, to have any direct effect on lipids or on haemostasis, the ingested polyphenols and/or their active metabolites have to be absorbed into the bloodstream. And the presence of ethanol may also contribute to the bioavailability of polyphenols.

While most of the studies based their conclusions on the potent *in vitro* antioxidant activity and free radical scavenging of various polyphenols, recently the focus changed towards other action mechanisms. These include metal chelation, modulation of enzymic activity and nuclear receptors, subcellular signaling pathways modulation and gene expression effects, in addition to protection of DNA from damage by antioxidant actions. *In vivo,* polyphenols may act as potential "multi-targeting agents", with complementary and overlapping mechanisms of action. Polyphenols extracted from grapes tend to attenuate postprandial hypertriglyceridemia, potentially by inhibition of intestinal diglyceride acyltransferase activity.

5.4.2.1 Bioavailability of Wine-Derived Phenolic Compounds

Irrespective of *in vitro* research showing the antioxidant and other activities of wine-derived phenolic compounds, if these compounds or their active metabolites are not absorbed in sufficient amounts and in a readily available form for cells, they are less likely to have any *in vivo* activity. Therefore, contradictory or inconsistent *ex vivo* results on the antioxidant activity of polyphenols may reflect poor bioavailability in some studies, and any biological activity observed in these studies results primarily from the alcohol component of the wine. It is hypothesized that the polyphenols are present as soluble forms in wine and should be more bioavailable than those forms in fruits and vegetables where they are present as polymeric, insoluble or tightly bound and in compartmentalized forms. Comparison of the bioavailability of the flavonol, quercetin, in humans shows that although quercetin and other flavonols are absorbed from the red wine, one 100 ml glass of red wine provides less available flavonols than one 125 ml cup of tea or one 15 g portion of onions. A 4-week treatment with quercetin lead to a relevant increase in plasma levels to 2.47 μmol/L. The absorption of flavonols from foods depends on the form in which the compound is present in the food, *e.g.*, glycosylation enhances absorption of quercetin, or on the sugar moiety, *e.g.*, the bioavailability of quercetin glucoside is 20% greater than that of quercetin rutinoside and may depend upon the site of absorption. Hydroxycinnamate and caffeic acid appear to be rapidly absorbed from wine and reach a maximum concentration in the plasma at approximately 60 minutes post consumption. The time taken for wine-derived polyphenols (free, conjugated or total) to reach the maximum plasma concentration ranges from approximately 30–50 minutes. Anthocyanins and their metabolites were detected in plasma about 30 minutes after ingestion, their glucosides and glucuronides reaching their maximum values after 1.6 h and 2.5 h, respectively. Approximately 94% of the anthocyanins were excreted in urine within 6 hours. Another study showed that after both 10 and 30 minutes, anthocyanins concentrations in the liver of a rat were similar with those in plasma, while in the kidneys they were 3- and 2.3-fold higher, after 10 and 30 minutes, respectively. In a comparison of the bioavailability of three wine derived polyphenols (catechin, quercetin and *trans*-resveratrol) in three different matrices (white wine, grape juice and vegetable juice) in humans, *trans*-resveratrol was the most bioavailable phenolic compound followed by quercetin and then catechin. The maximum serum concentration of 10–40 nmol/L was found, however, to be 100-fold less than the 5–100 μmol/L required in *in vitro* studies to demonstrate antioxidant and other biological activities. In addition, the absorption of the wine-derived polyphenols does not require the presence of ethanol, as aqueous and vegetable matrices are as effective as wine in facilitating absorption, as evidenced by the equal absorption of catechin from dealcoholized red wine and red wine containing alcohol. Some polyphenols are also found primarily conjugated in serum and urine (*e.g.*, glucuronidation, glycosylation, methylation and/or sulfation) facilitating their elimination from the body, such that it may be their metabolites and not the free forms found in grapes and wine that have antioxidant and other activities. The conjugated metabolites of the polyphenols may also have antioxidant activity. For example, no unmetabolized quercetin or epicatechin was found in the blood of rats following oral consumption of the flavonoids, although plasma antioxidant capacity was significantly increased post-consumption. The *in vivo* or *ex vivo* activity of the metabolized forms of the polyphenols has not been analyzed, however, although *in vitro* conjugated quercetin metabolites inhibit copper-mediated lipid oxidation in human LDL. Recent studies show that the majority of grape and wine polyphenols are metabolized by the gut microflora in the gastrointestinal tract, being first broken down in their aglycones or further into phenolic acids and aldehydes and only then absorbed.

5.4.2.2 Antiatherogenic Effects of the Wine-Derived Phenolic Compounds

From *in vitro* data, it is purported that the primary cardioprotective effect of wine and the wine-derived polyphenols is prevention and reversal of atherosclerosis. The primary biological mechanism initially proposed was the prevention of oxidation of plasma lipids such as LDL or antioxidation. A high plasma concentration of LDL is another risk factor for cardiovascular disease, where oxidized-LDL aids all stages of the atherosclerotic process. Another biological mechanism of these compounds may be the restoration of endothelial function. Dysfunctional endothelium (induced by high oxidative stress and a concomitant high concentration of lipid peroxides and oxidized lipids) is an early marker of atherosclerosis. It actively contributes to the development of atherosclerotic plaques by, for example, inducing the proliferation of monocyte-derived macrophages that ingest oxidized-LDL to form foam cells, and also contributes to the development of blood clots or thrombi (Figure 5.5).

Antioxidation mechanisms include the binding of polyphenols to the LDL particle by forming a glycoside bond to protect the lipoprotein against lipid peroxidation. Another antioxidation mechanism is the scavenging of free radicals, such as $LO_2^{\cdot}$ (lipid peroxyl radical), $HO^{\cdot}$ (hydroxyl radical), $O_2^{-\cdot}$ (oxygen radical) and $NO_2^{-\cdot}$ (nitric oxide radical), by phenolic compounds, *e.g.*, free radical scavenging occurs by hydrogen donation to lipid peroxyl radicals competing with the chain reaction propagation as follows:

$$\mathrm{LOO}\cdot + \mathrm{AH} \rightarrow \mathrm{LOOH} + \mathrm{A}\cdot$$
$$\mathrm{LO}\cdot + \mathrm{AH} \rightarrow \mathrm{LOH} + \mathrm{A}\cdot$$
$$\mathrm{LOO}\cdot + \mathrm{LH} \rightarrow \mathrm{LOOH} + \mathrm{L}\cdot$$
$$\mathrm{A}\cdot + \mathrm{LH} \rightarrow \mathrm{AH} + \mathrm{L}\cdot,$$

where
- $LOO^{\cdot}$ = fatty acid peroxyl radical,
- AH = phenolic antioxidant,
- LOOH = fatty acid hydroperoxide,
- $A^{\cdot}$ = phenoxyl radical,
- $LO^{\cdot}$ = alkoxyl radical,
- LOH = alcohol and LH = polyunsaturated fatty acid and $L^{\cdot}$ = alkyl radical.

Metal chelation and protein binding are also antioxidant-related mechanisms. *In vitro* studies have demonstrated that under certain conditions, the wine-derived polyphenols increase the antioxidant capacity of blood plasma and serum.

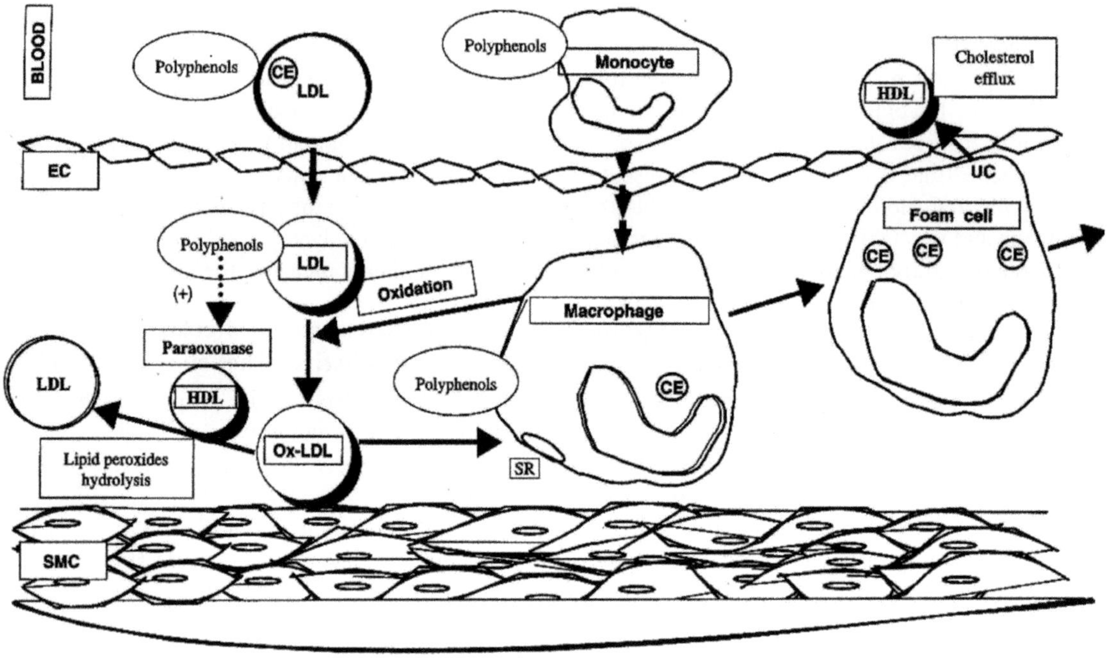

FIGURE 5.5 Summary of proposed effects of wine-derived phenolic compounds on atherosclerosis. Phenolic compounds may associate directly with LDL, which inhibits the oxidation of LDL. They may also be associated with arterial cells such as monocytes and macrophages, which inhibits the macrophagemediated oxidation of LDL. Phenolic compounds may also preserve the cativity of the LDL-associated enzyme, paraoxonase (PON1), which further protects LDL from oxidation. Altogether, these effects maylead to a reduced formation of macrophage-foam cells and attenuate the development of atherosclerotic lesions. Abbreviations: CE=cholesteryl ester; EC=endothelial cells; SMC=smooth muscle cells;SR=scavenger receptor; and UC=unesterified cholesterol. Reproduced with permission from Aviram,M., and Fuhrman, B. (2002) Wine flavonoids protect against LDL oxidation and atherosclerosis. Ann. N.Y. Acad. Sci. 957:146-61

Specifically, wine does confer protection against the oxidation of LDL by free radicals in blood plasma, in particular when the plasma concentration of αtocopherol was depleted in a dose-dependent manner (red wine being 10–20-fold more protective than white wine). Such compounds are also, collectively and individually, 10–20-fold more anti-oxidative than αtocopherol, an endogenous antioxidant. Whether the wine-derived polyphenols are absorbed into the blood stream in sufficient amount to be active antioxidants has not yet been demonstrated. More than 50 *in vivo* studies have been undertaken to date to assess the bioavailability and antioxidant activity of the wine-derived phenolic compounds. In the majority of the studies it is demonstrated that the wine-derived polyphenols are absorbed into the blood stream in sufficient amount to be active antioxidants and hence, increase the total antioxidant capacity of plasma, although some inconsistencies were also evident. Significant antioxidant activity, however, may only be observed following the medium- to long-term consumption of wine, although these compounds are absorbed in significant amount after the acute or short-term consumption of wine. An increase in the total antioxidant status of plasma may only occur with red wine and not white wine, such that the anti-oxidative effects of white wine may not completely counter any pro-oxidant effects of ethanol on lipids. *In vivo* studies have also been undertaken in a cigarette smoking population group, which found that smoking significantly increases oxidative damage and stress in otherwise healthy male subjects. Phenolic acid compounds in plasma act as biomarkers of the absorption of wine-derived polyphenols *per se*, while free and esterified 8-isoprostanes are oxidation products of arachidonic acid that accumulate in plasma and are excreted in urine, and thus provide a measure of oxidative damage and stress. The concentration of these oxidation products increases in cigarette smokers and in the presence of ethanol but decreases in the presence of antioxidants *in vitro* and *in vivo.* Phenolic acids have been detected in plasma and urine following the short-term consumption of red, dealcoholized red and white wine, confirming that polyphenols are absorbed. Following the consumption of dealcoholized red wine, the plasma and urine concentration of free and esterified 8-isoprostane oxidation products decreased, suggesting that the polyphenols act as antioxidants in the plasma. The concentration of these oxidation products did not decrease, however, following consumption of either the red or white wine, but remained at the baseline concentration and was similar for the two wines. The white wine contained an equivalent concentration of ethanol to the red wine but had approximately two thirds less phenolic compounds. Thus, the antioxidative phenolic acids absorbed from the ethanol-containing red and white wine may have merely countered the pro-oxidative ethanol concomitantly absorbed, as there was also no increase in the concentration of plasma oxidation products. In context with *in vitro* studies that have demonstrated that the wine-derived polyphenols are antioxidative in plasma, the wine-derived polyphenols may independently protect against LDL and plasma oxidation, but when consumed with ethanol, they may merely act to counter the pro-oxidant effects of ethanol. The antioxidant activity of the polyphenols may thus be dose dependent. The net effect may be dependent on the relative concentration of ethanol and polyphenols in the environment of LDL and accordingly, in wine. This was observed in a similar study in non-smoking healthy male subjects, which were also on a low plant phenolic compound diet to reduce the background level of phenolic compounds. In this study, an increase in the plasma concentration of HDL from baseline was observed in 74% of subjects only after the consumption of red wine for four weeks; no increased was observed for the other alcoholic beverages. In addition, a concomitant improvement in the ratio of HDL to LDL was observed. There was, however, no change in the plasma concentration of LDL or in the antioxidant capacity of the plasma, which may reflect the short-term duration of the study as polyphenols may need more than 4 weeks to accumulate in LDL particles. At the same time, the plasma concentration of the phenolic compound, catechin, decreased continuously over the 12-week study period from baseline, with a concomitant increase in the plasma concentration of F_2-isoprostane oxidation products. Thus, when combined with a low plant phenolic compound diet, the regular consumption of red wine in the short-term is unable to improve vascular function or increase the antioxidant capacity of the plasma, or prevent the oxidation of lipids such as LDL. The antioxidative and/or other cardioprotective effects of red wine may, therefore, best be observed against a 'normal' or 'higher' background diet of plant polyphenols, like a Mediterranean-style diet, against a background diet high in fat, when oxidative stress is increased. This suggests that anti-oxidation is not the primary cardioprotective activity of the wine-derived polyphenols as previously proposed.

In humans, the most biologically active fat-soluble antioxidant is αtocopherol (vitamin E) which is transported in the blood by LDL. The oxidation of LDL may occur when the concentration of αtocopherol in the LDL particle is depleted. In addition, αtocopherol requires co-antioxidants to protect LDL from oxidation under all conditions. Endogenous compounds such as ubiquinol-10 have co-oxidant activity, but are depleted in LDL particles in atherosclerotic plaque, suggesting that LDL oxidation associated with atherosclerosis may result from a lack of co-oxidants rather than depletion of αtocopherol, which is not depleted in atherosclerotic plaques. *In vitro* studies have shown that wine-derived polyphenols can react with, and reduce, αtocopherol radicals, the one electron oxidation product of αtocopherol produced in micelles; this reduction is a prerequisite for compounds to synergize with LDL-associated αtocopherol.

An animal study was designed to determine whether wine-derived phenolic compounds, in addition or separate to free radical scavenging, interact synergistically with αtocopherol to protect LDL and other plasma lipids from oxidation, even when absorbed ineffectively and present at a low micromolar level in the circulation and sub-endothelial space. Apolipoprotein E gene knockout (apoE –/– mice are an animal model commonly used for atherosclerosis, as the mice when fed on a high fat or western diet, develop atherosclerotic lesions, similar to humans. In apoE –/– mice, the short-term consumption of dealcoholized red wine increased the resistance of LDL and other plasma lipids to oxidation when compared to the control and other non-dealcoholized wine, such that these compounds may only be countering the pro-oxidant effect of ethanol, consistent with that seen in human studies. Also in apoE –/– mice, the long-term consumption of dealcoholized red wine significantly decreased (5%) the plasma concentration of triglycerides without affecting the concentration of LDL, other plasma lipids and αtocopherol, and there were no detectable primary products of lipid oxidation. The total antioxidant status of the plasma was increased, however, which implies that the wine-derived polyphenols were supplementing the endogenous plasma antioxidants, such as αtocopherol.

Stable compounds resulting from the oxidation of LDL and F_2-isoprostanes have also been identified in the advanced development of atherosclerosis, and an increased concentration is found in human atherosclerotic plaques. An increase in the antioxidant capacity of the aorta wall of these apoE –/– mice was also observed, as well as a non-significant increase in the tissue concentration of αtocopherol, but without a concomitant decrease in the aortic tissue concentration of oxidized LDL and other lipid oxidation products. There was a 20% decrease in the progression of atherosclerosis, however, in the aortic arch of mice given dealcoholized red wine. In a similar study, an approximate 40% decrease in aortic arch atherosclerotic lesions and a non-significant increase in the tissue concentration of αtocopherol was observed in apoE –/– mice that had ingested ethanol-containing red wine or specific wine-derived polyphenols for six weeks. In another study with control and apoE –/– mice placed on a high fat diet and given dealcoholized red wine *ad libitum* as their sole fluid for 20 weeks, no change was observed in the concentration of plasma oxidation products for any group of mice, suggesting that the wine-derived polyphenols did not act as antioxidants. Lipid deposition, however, was significantly increased only in the apoE –/– mice, which suggests that deposition or uptake of lipids in the aorta was independent of lipid oxidation. Furthermore, the consumption of dealcoholized red wine by the apoE –/– mice attenuated this deposition, which then suggests that the wine-derived polyphenols may protect against the aortic deposition of lipids independent of any antioxidant activity. Overall, these studies add to the evidence dissociating the oxidation of LDL in blood vessel walls from atherogenesis, although some studies still support the LDL oxidation hypothesis of atherogenesis, such that decreased atherosclerosis might result from a decreased uptake of oxidized LDL by arterial macrophages and thus, attenuation of foam cell formation and lesion development.

The endothelium overlies the developing atherosclerotic lesion and a dysfunctional endothelium is a risk factor for the initiation and development of atherosclerosis and hypertension, often observed before any anatomical signs of atherosclerosis appear. Another potential mechanism of action for the wine-derived polyphenols is related to endothelial function. Endothelial dysfunction may result from a decreased production of endothelial nitric oxide (NO) or its inactivation by oxidative free radicals generated by the metabolism of dietary fat, and hence may be induced by a high-fat diet. Plant-derived phenolic compounds, *in vitro* and *in vivo*, modify or reverse the endothelial dysfunction by restoring or increasing the production of NO or by protecting against its degradation by superoxide anions. This restores the ability of the endothelium to relax with concomitant dilation of the blood vessels having beneficial effect on blood pressure. NO-mediated vasorelaxation is a short-term response to wine polyphenols, apparently as a result of increasing the influx of extracellular Ca^{2+} and the mobilization of intracellular Ca^{2+} in endothelial cells. This occurs, however, only after medium- to long-term moderate wine consumption. Conversely, the ethanol component of wine *in vitro* may initially improve, but then impair, endothelial function with increasing doses.

A 4-week treatment with quercetin reduced *in vivo* the levels of asymmetric dimethylarginine (ADMA), a novel risk marker of cardiovascular disease, which, in elevated concentrations, inhibits NO synthesis, thus impairing endothelial function and promoting atherosclerosis.

In an *in vivo* model of coronary occlusion, quercetin and also rutin showed cardioprotective properties on both normal and on streptozotocin-induced Type I diabetic rats. The conclusion was based on evaluation of infarct size and the oxidative stress markers. Quercetin and rutin significantly limited the infarct size, rutin offering complete cardioprotection at a dose of 10 mg/kg.

As opposed to the cardiomodulation effect elicited by quercetin, myricetin induced selective vasodilation only, without affecting relaxation or contractility.

Besides the cardioprotection supposedly conferred by the antioxidant properties of quercetin, its anti-ischemic effects may also be induced by the observed reduction of expression of inflammatory mediators, such as TNF-α, IL-10 and reduction of inflammatory cytokine levels.

Supplementation of a high-fat diet with red wine has been found to reverse or prevent endothelial dysfunction. Indeed, red grape juice, wine and dealcoholized red wine have been found in animals, *in vitro* and *in vivo*, to relax vascular endothelial smooth muscle, by inducing or up-regulating NO synthase gene expression (eNOS), thereby inducing the synthesis and release of NO from endothelial cells. In addition, the induced synthesis of NO by the wine-derived polyphenols has other protective effects on both the early and late phases of atherogenesis. The proposed cascade effect of wine-derived polyphenols on endothelial function can be summarized as follows: inhibition of transcription of prepro-ET-1 gene that in turn inhibits synthesis of ET-1 and restores endothelial function; induction of eNOS, which increases the synthesis and release of endothelial cell NO, that in turn inhibits platelet aggregation and adhesion to endothelium; decreases expression of monocyte chemoattractant protein-1 (MCP-1); decreases expression of leukocyte adhesion glycoproteins CD11/CD18; decreases expression of adhesion molecule P-selectin; decreases expression of vascular cell adhesion molecule-1; and decreases expression of intercellular adhesion molecule-1 (ICAM-1), the consequence of which is the prevention of leukocyte adhesion to endothelium and migration into the vascular wall, thereby restoring endothelial function.

Other mechanisms can also be involved. At very low doses, similar to those measured after moderate white wine consumption, caffeic acid seems to protect endothelial cell function by modulating NO release independently from eNOS expression and phosphorylation.

Furthermore, endothelium dysfunction is prevented by endothelin-1 (ET-1) antagonists. ET-1 is a potent vasoconstrictor peptide and its overproduction is another risk factor for atherosclerosis. Resveratrol, for example, has been observed to inhibit *in vitro* the synthesis of ET-1 by suppressing transcription of the prepro-ET-1 gene and by changing the morphology of the endothelial cell to modify tyrosine-kinase signaling and hence, tyrosine phosphorylation. Inhibition is, however, dose dependent.

Atherosclerosis is actually an inflammatory disease involving numerous cell types, such as macrophages, T-lymphocytes, platelets and endothelial cells that produce and secrete numerous cytokines and growth factors, which act locally to promote atherosclerosis. Ethanol and the phenolic compound components of wine could also affect inflammatory effects or responses associated with atherosclerosis. As atherosclerosis develops, vascular smooth muscle cells are released by platelets and endothelial cells, which proliferate and accumulate within the intima of the blood vessel wall, to develop the lesion or plaque. The primary mitogenic and chemotactic compound for the release of the vascular smooth muscle cells is platelet-derived growth factor, which exerts its effects via activation of two subtypes of trans-membrane receptor tyrosine kinases, α and β platelet-derived growth factor receptors. Wine-derived phenolic compounds, in particular flavonoids such as catechin, may inhibit the activation of the β platelet-derived growth factor receptors and hence, platelet-derived growth factors and the subsequent proliferation and migration of vascular smooth muscle cells. The postprandial suppression of smooth muscle cell proliferation by ethanol has also been observed but the mechanism of action is independent to that of the wine-derived phenolic compounds. Three principal cytokines are involved in the acute phase of inflammation: interleukin 1-β; interleukin-6; and tumor necrosis factor-α (TNF-α). The hydroxycinnamate, caffeic acid, modulated *in vitro* the expression and release of the cytokine TNF-α from human monocytes, even at low doses, while the flavonoid apigenin inhibits the TNF-α-stimulated release of the cytokine interleukin-8 from human intestinal cells, such that wine-derived polyphenols may be protective against other inflammatory diseases, *i.e.* Crohn's Disease. In addition, C-reactive protein is also a biomarker for the acute phase of inflammation as well as a risk factor for cardiovascular disease. The moderate consumption of alcoholic beverages lowers the concentration of C-reactive protein independent of effects on lipoproteins and on fibrinogen.

5.4.2.3 Haemostatic Effects of the Wine-Derived Phenolic Compounds

As discussed previously, the primary cardioprotective activity of the polyphenols may not be anti-oxidation as initially proposed. The polyphenols may also affect coagulation and fibrinolysis as does ethanol, although the effects of red wine and alcohol *per se* are difficult to distinguish. Also, the effects of red wine on haemostasis are confounded and cannot be generalized. For example, there are gender differences in fibrinolytic responses after the consumption of red wine. The fibrinolytic response of post-menopausal women, who consume red wine moderately, is similar to that of middle-aged men who consume red wine moderately. Fibrinolytic activity is initially decreased in these two groups and then, subsequently increased post-prandially, related to changes in plasma t-PA and PAI activity. Moderate red wine consumption has minimal effects on fibrinolysis, however, in pre-menopausal women using oral contraceptives. Diet is also a confounder where wine consumers on a diet high in saturated fat may experience more significant effects on coagulation and fibrinolysis than those on a different diet. Furthermore, the fibrinolytic responses to the acute consumption of red wine appear different to that of short-term and longer-term consumption.

Platelets interact with each other and adhere to form a haemostatic plug or thrombus. Induced (and increased) platelet aggregation is consequently a risk factor for cardiovascular disease. Epidemiological studies showed an inverse relationship between the consumption of alcohol, particularly wine, and induced platelet aggregation, similar to the relationship between alcohol consumption and risk of mortality from cardiovascular disease. There is *in vitro* evidence that wine-derived polyphenols may have an independent and additive effect on the reduction of platelet aggregation to ethanol, but the different flavonoid classes may exhibit different effects, particularly on arachidonic acid metabolism. Specifically, wine-derived polyphenols down-regulate cellular adhesion processes responsible for the recruitment and activation of platelets and their aggregation at the site of vascular damage, hence reducing platelet aggregation. Red and white wine appear equally effective *in vivo* in reducing the plasma concentration of thromboxane B_2 and platelet aggregation, while a red wine inhibited platelet aggregation at a significantly lower blood alcohol concentration than an ethanol solution did.

Platelet aggregability is dependent, however, on the amount and pattern of the alcoholic beverage consumed, such that 'binge' and/or excessive alcohol consumption is associated with platelet rebound effects or hyperaggregability, implicated in sudden deaths after episodes of excessive consumption and in alcoholics. The type of beverage consumed may also affect platelet aggregability. For example, only the consumption of red wine, irrespective of quantity consumed, is not associated with platelet rebound effects. It is attributed to the inhibition of ethanol-induced lipid peroxidation by the wine-derived phenolic compounds, as ethanol in excessive amounts is a pro- rather than an anti-oxidant, such that the initiation of platelet aggregation may be induced by a high concentration of lipid peroxides.

The consumption of both red and white wine increases fibrinolytic activity, and the wine-derived polyphenols increase fibrinolytic activity independent of ethanol. For example, catechin, epicatechin, quercetin and resveratrol individually upregulate both t-PA and u-PA gene transcription, resulting in the sustained increased expression of surface-localized fibrinolytic activity in cultured human umbilical vein endothelial cells. Also, wine-derived resveratrol and quercetin act as suppressors of tissue factor (TF induction), whose expression is triggered by chronic inflammation and atherosclerosis, promoting blood coagulation and subsequent thrombosis.

Grape seed extract in concentrations of 1.25–50 μg/ml has also shown dose-dependent reduction of platelet aggregation. Grape seed extract inhibits P-selectin expression on endothelial cells and platelet-derived microparticles formation, showing a greater effect than the solution of pure resveratrol. However, not all fractions of a grape seed extract may have the desirable inhibitory effects on platelet aggregation. When the grape seed extract was separated into six fractions, only the fractions 4–6, rich in polygalloyl polyflavan-3-ols, significantly decreased platelet aggregation and inhibited LDL oxidation. The fractions 1–3, containing mostly hydroxycinnamic acids, anthocyanins, flavanols and oligosaccharides were found to actually increase platelet aggregation, this detrimental effect reducing the overall efficacy of the grape products.

In a comparison of moderate red and white wine consumption in conjunction with a Mediterranean-style diet on coagulation and fibrinolytic factors and on the oxidizability of the inter-related plasma membrane lipids, both red and white wine had similarly positive effects. The consumption of red wine during a Mediterranean-style diet as compared with a high fat diet appears synergistic in decreasing platelet aggregation by increasing the concentration of platelet membrane unsaturated fatty acids and thereby, platelet membrane microviscosity. While the diet alone decreases the concentration of plasma fibrinogen and factors VII and VIII compared to baseline, red wine consumption further reduces the plasma concentration of fibrinogen and factor VII but increases the plasma concentration of t-PA and, paradoxically, that of PAI-1, inconsistent with other observations.

Therefore, considering all the potential cardioprotective mechanisms postulated for the wine-derived phenolic compounds, the effect of an alcoholic beverage on lipid oxidation and coagulation and fibrinolysis *in vivo* may depend on the balance between the pro-oxidant effects of the ethanol component and the concentration of phenolic antioxidant compounds in the beverage, as well as on the diet of the consumer. Thus, the overall haemostatic effects purportedly associated with moderate wine consumption are attributable to the combined, additive or perhaps synergistic effects of the ethanol component and the wine-derived polyphenols (Figure 5.6).

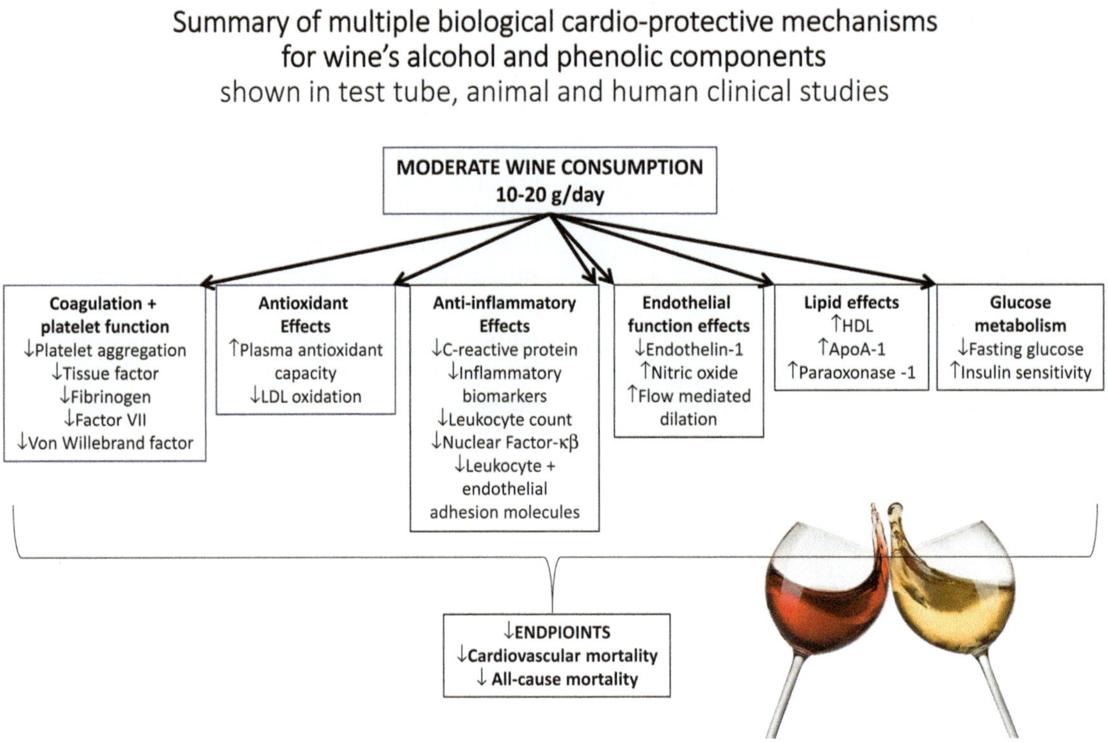

FIGURE 5.6 Summary of multiple biological cardio-protective mechanisms for wine's alcohol and phenolic components shown in test tube, animal and human clinical studies

5.4.2.4 Other Potential Cardioprotective Effects of the Phenolic Compounds

In addition to the antiatherogenic and haemostatic effects, wine polyphenols may also exhibit a cardioprotective effect by decreasing the plasma concentration of homocysteine and inhibiting the induced growth of vascular smooth muscle cells by angiotensin II. High plasma concentration of the amino acid, homocysteine, is a recently recognized risk factor for atherosclerosis, endothelial dysfunction and cardiovascular disease. The American Heart Association recommends that the concentration of homocysteine in plasma should be maintained below 10 μmol/L. Homocysteine is closely linked to the metabolism of the essential amino acid, methionine and has a direct toxic effect on the endothelium of blood vessels altering their function, leading to key early steps in the atherogenic process. In human umbilical vein endothelial cells, the addition of homocysteine resulted in a dose-dependent increase in the generation of reactive oxygen species and a correlated decrease in intracellular NO. Homocysteine also increased tyrosine kinase activity, phosphokinase stimulation of ERK5, Scr and p38, activated NF-κB, increased the concentration of nitrotyrosine and induced VCAM-1, which are markers of oxidative stress or endothelial dysfunction. All these effects were inhibited *in vitro* by the wine-derived phenolic compounds.

Three micronutrients are important cofactors in homocysteine metabolism. Folate and vitamins B_{12} are cofactors for the methylation of homocysteine to methionine while vitamin B_6 is a cofactor for the trans-sulphuration of homocysteine to cysteine. Deficiency of any of these micronutrients leads to a higher concentration of homocysteine and increased risk of atherosclerosis and endothelial dysfunction.

Although excessive alcohol consumption is associated with an increased plasma concentration of homocysteine, the effect of low to moderate consumption on the concentration of homocysteine has been variable. A J-shaped relationship is observed in severely obese patients. Light to moderate alcohol consumption is associated with a lower and more favorable plasma concentration of homocysteine while patients consuming up to 100 g/week of alcohol have a significantly lower homocysteine concentration (a higher concentration of folate) compared with non-consumers. Furthermore, red wine consumers have a significantly lower mean fasting concentration of homocysteine than non-consumers, and beer, spirit and white wine consumers. The mechanisms for the beneficial effect of red wine consumption on the plasma concentration of homocysteine, however, are unclear. The concentration of micronutrients is unlikely to provide the answer, as the effect of red wine is independent of the plasma concentration of folate and vitamin B_{12}, and red wine contains negligible quantities of vitamin B_6. An alteration in the relationship between the homocysteine concentration and that of folate and vitamin B_{12} has been observed as people lose weight. A higher plasma concentration of folate and vitamin B_{12} is needed to maintain the concentration of homocysteine as weight is lost. Thus, the wine-derived polyphenols may shift the dose-response curve such that a lower plasma concentration of homocysteine is achieved with an equivalent micronutrient concentration.

Angiotensin II is the main peptide hormone in the rennin-angiotensin system. In addition to regulating blood pressure and circulating volume, angiotensin II also induces the growth of vascular smooth muscle cells (VSMC), *i.e.* hypertrophy, by stimulating the G protein-coupled angiotensin type 1 (AT_1) receptor in VSMC that activates multiple protein kinase pathways leading to VSMC protein synthesis. On vascular injury, the hypertrophied VSMC migrate, which may result in rupturing the plaque and a myocardial infarct (heart attack). An increased plasma concentration of angiotensin II is thus considered to contribute to the development of diseases characterized by VSMC growth, such as hypertension, atherosclerosis and restenosis after vascular injury. *In vitro*, the polyphenols quercetin and resveratrol appear to independently inhibit or suppress the angiotensin II-induced growth of VSMC and interfere with the multiple protein kinase pathways and VSMC protein synthesis activated by angiotensin II. In addition, both the polyphenols inhibit the enzyme angiotensin-converting enzyme, which converts angiotensin I to angiotensin II, and thus, they may also decrease the plasma concentration of angiotensin II, thereby inhibiting the VSMC hypertrophy.

In grape-fed rats, the activity of peroxisome proliferator activating receptor (PPAR) is enhanced, while the levels of nuclear factor κB (NF-κB) are reduced, as the formation of cytosolic NF-κB inhibitor is upregulated and the formation of tumor necrosis factor-α and transforming growth factor-$\beta 1$ are downregulated. Reduced PPAR activity and high activity of NF-κB is correlated with cardiac inflammation and fibrosis, thus a grape-based diet is linked to the reduction in fibrosis.

Resveratrol also inhibits or antagonizes the nuclear factor-κB activity, modulating the mechanisms involved in the inflammatory-oxidative stress cycle. Another beneficial effect of resveratrol is the ability to promote autophagy by increasing the expression level of sirtuin-1 (SIRT-1). Piceatannol, a hydroxylated resveratrol derivative also found in wine, acts synergistically with the later in inducing autophagy. Piceatannol itself is also under investigation for other possible cardioprotective effects.

5.4.2.5 Structure Activity Relationships

The chemical structures of the polyphenols are largely predictive of their potential anti-oxidant activity regarding radical scavenging, hydrogen or electron donating and/or metal-chelating capacities. Their unique structure facilitates their ability to scavenge ROS due to resonance stabilization of the captured electron. Anti-oxidant activity is also influenced by the stability of the resulting phenoxyl radical. A potential order of cardioprotectiveness for the various polyphenols has been proposed, where flavonoids appear more cardioprotective than non-flavonoids. Flavanols and anthocyanins are structurally related, being biogenetically derived from a common C-15 tetrahydroxychalcone precursor, naringenin, formed by a condensation reaction between 4-coumaroyl-coenzyme A and malonyl coenzyme A in the pivotal step of flavonoid biosynthesis. The antioxidant activity of flavonoids is determined by the position and number of hydroxyl groups on the B-ring. There are three primary chemical structure characteristics inferring antioxidant activity. An ortho 3'4', hydroxy moiety in the B-ring for electron delocalization and stability of the phenoxyl radical is the most critical, and others are: a 2,3-double bond in combination with the 4-keto group for electron delocalization in the C-ring; and 3- and 5-hydroxyl groups in the C- and A-rings, respectively, in combination with the 4-keto group in the C-ring for maximum radical scavenging activity. Indeed, it has been recently observed for anthocyanins, that antioxidant activity increases with the number of hydroxyl groups on the B-ring, where the flavylium ion of ring-C contributes to the overall stabilization of the resulting phenoxyl radical. A 3'4'-dihydroxyphenol B-ring, however, is critical for the antioxidant activity of flavanols such as catechin, where *cis-trans* isomerization, epimerization and racemization do not influence antioxidant activity. In contrast to other flavonoid classes, for anthocyanins substitution of the B-ring hydroxyls with methoxyl groups significantly decreases antioxidant activity, while the conjugation of certain substituents at position 3 of ring-C decreases the antioxidant activity for all compounds, and antioxidant activity also decreases as the number of glycosyl or sugar residues at position 3 increased.

The antioxidant capacity of non-flavonoids such as hydroxycinnamic acids and hydroxybenzoic acids is related to the acid moiety and the number and position of hydroxyl groups on the aromatic ring structure. Hydroxycinnamic acids are more effective antioxidants than hydroxybenzoic acids due to increased potential for delocalization of the phenoxyl radical. The total antioxidant activity capacity of wine, however, decreases during aging as the total concentration of polyphenols in wine decreases, although the relative antioxidant capacity of polyphenols such as flavonoids is increased gradually during storage.

5.5 DIETARY PATTERN AND LIFESTYLE OF WINE CONSUMERS FOR HEALTH

A series of large longitudinal studies looking at nutrition and healthy ageing in Australia (Dubbo Study of the Elderly), Italy (Moli-Sani) and Spain (PrediMed) have all shown that the inclusion of the moderate consumption of alcoholic beverages such as wine, as part of a healthy diet and lifestyle, reduces the risk of CVD, diabetes and dementias.

Subjects in large population studies who are non-smokers, are not obese, eat a Mediterranean-style diet, and get regular exercise have much lower risk of non-communicable diseases and total mortality, whether or not they consume alcoholic beverages. One such study observed, however, that regardless of other 'healthy' lifestyle factors, moderate consumption of wine provided greater reduction in risk of these diseases than was seen for the other factors in isolation.

A similar, prospective Australian study of 7989 individuals aged between 65–83 years and followed for five years showed consistent results with these international studies. Eight selected low-risk behaviors included having no more than two drinks daily (20 g alcohol). Individuals with five or more of the selected low-risk behaviors had a lower risk of death from any cause within five years, compared with those having less than five. More importantly, the study showed that while most individuals already have some healthy habits, almost all could make changes to their diet and lifestyle to improve their health. The study did not suggest abstinence, but avoidance of heavier consumption is also inferred.

Further, it has been shown that in addition to lower mortality, women who were moderate wine consumers surviving to age 70 years and older generally had less disability and disease and more signs of 'successful ageing.' For 'regular' light-to-moderate wine consumers (on 5–7 days/week), there was an approximately 50% greater chance of such successful ageing compared with non-drinkers.

These studies suggest that the alcohol and polyphenolic components of wine, for example, may act in synergy with each other and with other dietary components, meaning that overall diet may be more important than a single dietary component. For example, the moderate consumption of wine may either supplement the effect of an already phenolic-rich diet and/or counter the harmful or oxidative stressing effects of a high fat diet or ethanol ingestion.

Consumers of wine also purportedly have different consumption patterns, dietary and lifestyle characteristics and hence have fewer risk factors for cardiovascular disease than beer and spirit consumers. This is reflected in a lower risk of CVD for wine consumers compared to consumers of beer and spirits, respectively. Wine, in comparison to beer and spirits, is usually consumed with food, slowly or over a longer period of time, which would attenuate a high blood alcohol concentration, prolong any acute and short-term plasma anti-oxidative and haemostatic effects and prevent any rebound effects of the ethanol and phenolic components of the beverage. Regular wine consumption, determined as daily, promotes long-term effects on the plasma anti-oxidative capacity and on systolic blood pressure associated with maximal cardioprotection. For example, the acute local effects of moderate wine consumption on various haemostatic variables are short-term or temporary and return to normal within 24 hours. Furthermore, if the anti-oxidative effects of the wine-derived polyphenols are dose-dependent, the daily consumption of wine in moderation would be needed to maintain an appropriate concentration (10 μmol/L) in blood plasma and tissues for significant anti-oxidative activity. In addition, any effect of wine on systolic blood pressure is readily reversible, within 7–4 days, such that regular consumption may also be necessary to maintain a lowering effect on blood pressure. Conversely, binge drinking (consumption of approximately more than six drinks per drinking session), is seen to significantly increase systolic blood pressure, which significantly increases the risk of a heart attack or stroke.

BIBLIOGRAPHY

Antoce, A.O. and Stockley, C. (2019). An overview of the implications of wine on human health, with special consideration of the wine-derived phenolic compounds. *AgroLife Sci. J.*, 8(1):21–34.

Abu-Amsha Caccetta, R., Burke, V., Mori, T.A., Beilin, L.J., Puddey, I.B. and Croft, K.D. (2001). Red wine polyphenols, in the absence of alcohol, reduce lipid peroxidative stress in smoking subjects. *Free. Radic. Biol. Med.*, 30(6):636–642.

Andriambeloson, E., Stoclet, J.C. and Andriantsitohaina, R. (1999). Mechanism of endothelial nitric oxide-dependent vasorelaxation induced by wine polyphenols in rat thoracic aorta. *J. Cardiovasc. Pharmacol.*, 33(2):248–254.

Angelone, T., Pasqua, T., Di Majo, D., Quintieri, A.M., Filice, E., Amodio, N., Tota, B., Giammanco, M. and Cerra, M.C. (2011). Distinct signalling mechanisms are involved in the dissimilar myocardial and coronary effects elicited by quercetin and myricetin, two red wine flavonols. *Nutr. Metab. Cardiovasc. Dis.*, 21(5):362–371.

Annapurna, A., Reddy, C.S., Akondi, R.B. and Rao, S.R. (2009). Cardioprotective actions of two bioflavonoids, quercetin and rutin, in experimental myocardial infarction in both normal and streptozotocin-induced type I diabetic rats. *J. Pharm. Pharmacol.*, 61(10):1365–1374.

Aviram, M. and Fuhrman, B. (2002). Wine flavonoids protect against LDL oxidation and atherosclerosis. *Ann. N.Y. Acad. Sci.*, 957:146.

Bell, J.R., Donovan, J.L., Wong, R., Waterhouse, A.L., German, J.B., Walzem, R.L. and Kasim-Karakas, S.E. (2000). (+)-Catechin in human plasma after ingestion of a single serving of reconstituted red wine. *Am. J. Clin. Nutr.*, 71(1):103–108.

Bertelli, A.A. (2007). Wine, research and cardiovascular disease: Instructions for use. *Atherosclerosis*, 195(2):242–247.

Cassidy, A., Rogers, G., Peterson, J.J., Dwyer, ,J.T., Lin, H. and Jacques, P.F. (2015). Higher dietary anthocyanin and flavonol intakes are associated with anti-inflammatory effects in a population of US adults. *Am. J. Clin. Nutr.*, 102(1):172–181.

Chen, S., Yi, Y., Xia, T., Hong, Z., Zhang, Y., Shi, G., He, Z. and Zhong, S. (2019). The influences of red wine in phenotypes of human cancer cells. *Gene*, 702:194–204.

Chiva-Blanch, G., Urpi-Sarda, M., Llorach, R., Rotches-Ribalta, M., Guillén, M., Casas, R., Arranz, S., Valderas-Martinez, P., Portoles, O., Corella, D., Tinahones, F., Lamuela-Raventos, R., Andres-Lacueva, M.C. and Estruch, R. (2012). Differential effects of polyphenols and alcohol of red wine on the expression of adhesionmolecules and inflammatory cytokines related to atherosclerosis: A randomized clinical trial. *Am. J. Clin. Nutr.*, 95(2):326–334.

Corder, R., Douthwaite, J.A., Lees, D.M., Khan, N.Q., Viseu Dos Santos, A.C., Wood, E.G. and Carrier, M.J. (2001). Endothelin-1 synthesis reduced by red wine. *Nature*, 414(6866):863–864.

Decendit, A., Mamani-Matsuda, M., Aumont, V., Waffo-Teguo, P., Moynet, D., Boniface, K., Richard, E., Kris, A.S., Rambert, J., Mérillon, J.M. and Mossalayi, M.D. (2013). Malvidin-3-O-β glucoside, major grape anthocyanin, inhibits human macrophage-derived inflammatory mediators and decreases clinical scores in arthritic rats. *Biochem. Pharmacol.*, 86(10):1461–1467.

Dell'Agli, M., Busciala, A. and Bosisio, E. (2004). Vascular effects of wine polyphenols. *Cardiovasc. Res.*, 63(4):593–602.

Di Renzo, L., Cioccoloni, G., Sinibaldi Salimei, P., Ceravolo, I., De Lorenzo, A. and Gratteri, S. (2018). Alcoholic beverage and meal choices for the prevention of noncommunicable diseases:Arandomized nutrigenomic trial. *Oxid. Med. Cell. Longev.* doi:10.1155/2018/5461436.

Dixon, J.B., Dixon, M.E. and O'Brien, P.E. (2002). Reduced plasma homocysteine in obese red wine consumers: A potential contributor to reduced cardiovascular risk status. *Eur. J. Clin. Nutr.*, 56(7):608–614.

Draijer, R., de Graaf, Y., Slettenaar, M., de Groot, E. and Wright, C.I. (2015). Consumption of a polyphenol-rich grape-wine extract lowers ambulatory blood pressure in mildly hypertensive subjects. *Nutrients*, 7(5):3138–3153.

Faggio, C., Sureda, A., Morabito, S., Sanches-Silva, A., Mocan, A., Nabavi, S.F. and Nabavi, S.M. (2017). Flavonoids and platelet aggregation: A brief review. *Eur. J. Pharmacol.*, 807:91–101.

Fernandes, I., Pérez-Gregorio, R., Soares Nuno Mateus, S. and de Freitas, V. (2017). Wine flavonoids in health and disease prevention. *Molecules*, 22(2):292. doi:10.3390/molecules22020292.

Flamini, R., Mattivi, F., De Rosso, M., Arapitsas, P. and Bavaresco, L. (2013). Advanced knowledge of three important classes of grape phenolics: Anthocyanins, stilbenes and flavonols. *Int. J. Mol. Sci.*, 14(10):19651–19669.

Forester, S.C. and Waterhouse, A.L. (2009). Metabolites are key to understanding health effects of wine Polyphenolics. *J. Nutr.*, 139(9):1824S–1831S.

Garcia-Alonso, M., Minihane, A.M., Rimbach, G., Rivas-Gonzalo, J.C. and de Pascual-Teresa, S. (2009). Red wine anthocyanins are rapidly absorbed in humans and affect monocyte chemoattractant protein 1 levels and antioxidant capacity of plasma. *J. Nutr. Biochem.*, 20(7):521–529.

Goldberg, D.M., Yan, J. and Soleas, G.J. (2003). Absorption of three wine-related polyphenols in three different matrices by healthy subjects. *Clin. Biochem.*, 36(1):79–87.

González-Candelas, L., Gil, J.V., Lamuela-Raventós, R.M. and Ramón, D. (2000). The use of transgenic yeasts expressing a gene encoding a glycosyl-hydrolase as a tool to increase resveratrol content in wine. *Int. J. Food Microbiol.*, 59(3):179–183.

Grønbæk, M., Becker, U., Johansen, D., Gottschau, A., Schnohr, P., Hein, H.O., Jensen, G. and Sorensen, T.I. (2000). Type of alcohol consumed and mortality from all causes, coronary heart disease and cancer. *Ann. Intern. Med.*, 133(6):411.

Hansen, A.S., Marckmann, P., Dragsted, L.O., Finné , Nielsen, I.L., Nielsen, S.E. and Grønbaek, M. (2005). Effect of red wine and red grape extract on blood lipids, haemostatic factors, and other riskf actors for cardiovascular disease. *Eur. J. Clin. Nutr.*, 59(3):449–455.

Haseeb, S., Alexander, B., Santi, R.L., Liprandi, A.S. and Baranchuk, A. (2019). What's in wine? A clinician's perspective. *Trends Cardiovasc. Med.*, 29(2):97–106.

Hollman, P.C., Bijsman, M.N., van Gameren, Y., Cnossen, E.P., de Vries, J.H. and Katan, M.B. (1999). The sugar moiety is a major determinant of the absorption of dietary flavonoid glycosides in man. *Free Radic. Res.*, 31(6):569–573.

He, F., Mu, L., Yan, G.L, Liang, N.N., Pan, Q.H, Wang, J., Reeves, M.J. and Duan, M.J. (2010). Biosynthesis of anthocyanins and their regulation in colored grapes. *Molecules*, 15(12):9057–9091.

Huang, Y., Li., Y., Zheng, S., Yang, X., Wang, T. and Zeng, J. (2017). Moderate alcohol consumption and atherosclerosis: Meta-analysis of effects on lipids and inflammation. *Wien. Klin. Wochenschr.*, 129(21–22):835–843.

Iriti, M. and Faoro, F. (2006). Grape phytochemicals a bouquet of old and new nutraceuticals for human health. *Med. Hypotheses.*, 67(4):833–838.

Jin, H.B., Yang, Y.B., Song, Y.L., Zhang, Y.C. and Li, Y.R. (2012). Protective roles of quercetin in acute myocardial ischemia and reperfusion injury in rats. *Mol. Biol. Rep.*, 39(12):11005–11009.

Kaur, G., Roberti, M., Raul, F. and Pendurthi, U.R. (2007). Suppression of human monocyte tissue factor induction by red wine phenolics and syntheticderivatives of resveratrol. *Thromb. Res.*, 119(2):247–256.

Kerry, N.L. and Abbey, M. (1997). Red wine and fractionated polyphenols prepared from red wine inhibit low density lipoprotein oxidation *in vitro. Atherosclerosis*, 135:193–102.

Klatsky, A.L. (2003). Drink to your health? *Sci. Am.*, 288(2):74–81.

Leger, C.L., Carbonneau, M.A., Carton, E., Cristol, J.P. and Descomps, B. (2000). Wine polyphenols and cardiovascular protection in humans: Evaluation in vitro of protection secondary indicators must take into account the real polyphenol concentration in vivo. *Bulletin de l'O.I.V.*, 73:833–834, 481–488.

McCarty, M.F. (2001). Does regular ethanol consumption promote insulin sensitivity and leanness by stimulating AMP-activated protein kinase? *Med. Hypotheses.*, 57(3):405–407.

Mansvelt, E.P., van Velden, D.P., Fourie, E., Rossouw, M., van Rensburg, S.J. and Smuts, C.M. (2002). The *in vivo* antithrombotic effect of wine consumption on human blood platelets and hemostatic factors. *Ann. N.Y. Acad. Sci.*, 957:329–332.

Mezzano, D., Leighton, F., Martinez, C., Marshall, G., Cuevas, A., Castillo, O., Panes, O., Munoz, B., Perez, D.D., Mizon, C., Rozowski, J., San Martín, A. and Pereira, J. (2001). Complementary effects of Mediterranean diet and moderate red wine intake on haemostatic cardiovascular risk factors. *Eur. J. Clin. Nutr.*, 55(6):444–451.

Migliori, M., Cantaluppi, V., Mannari, C., Bertelli, A.A., Medica, D., Quercia, A.D., Navarro, V., Scatena, A., Giovannini, L., Biancone, L. and Panichi, V. (2015). Caffeic acid, a phenol found in white wine, modulates endothelial nitric oxide production and protects from oxidative stress-associated endothelial cell injury. *PLoS One*, 10(4):e0117530.

Millwood, I.Y., Walters, R.G., Mei, X.W., Guo,Y., Yang, L., Bian, Z., Bennett, D.A, Chen, Y., Dong, C., Hu, R., Zhou, G., Yu, B., Jia, W., Parish, S., Clarke, R., Smith, G.D., Collins, R., Holmes, M.V., Li, L., Peto, R. and Chen, Z., for the China Kadoorie Biobank Collaborative Group. (2019). Conventional and genetic evidence on alcohol and vascular disease aetiology: A prospective study of 500000 men and women in China. Conventional and genetic evidence on alcohol and vascular disease aetiology: A prospective study of 500 000 men and women in China. *Lancet*, 393(10183):1831–1842.

Modun, D., Katalinić, V., Salamunić, I., Kozina, B. and Boban, M. (2005). Effects of four-weeks moderate drinking of red wine and ethanol on the rat isolated heart and aortic rings reactivity during ischemia and hypoxia. *Period. Biol.*, 107(2):165–173.

Nickel, T., Hanssen, H., Sisic, Z., Pfeiler, S., Summo, C., Schmauss, D., Hoster, E. and Weis, M. (2011). Immunoregulatory effects of the flavonol quercetin *in vitro* and *in vivo. Eur. J. Nutr.*, 50(3):163–172.

Olas, B., Wachowicz, B., Stochmal, A. and Oleszek, W. (2012). The polyphenol-rich extract from grape seeds inhibits platelet signaling pathways triggered by both proteolytic and non-proteolytic agonists. *Platelets*, 23(4):282–289.

Panchal,S.K. and Brown, L. (2013). Cardioprotective and hepatoprotective effects of ellagitannins from European oak bark (*Quercus petraea* L.) extract in rats. *Eur. J. Nutr.*, 52(1):397–408.

Pietrocola, F., Mariño, G., Lissa, D., Vacchelli, E., Malik, S.A., Niso-Santano, M., Zamzami, N., Galluzzi, L., Maiuri, M.C. and Kroemer, G. (2012). Pro-autophagic polyphenols reduce the acetylation of cytoplasmic proteins. *Cell Cycle*, 11(20):3851–3860.

Quintieri, A.M., Baldino, N., Filice, E., Seta, L., Vitetti, A., Tota, B., De Cindio, B., Cerra, M.C. and Angelone, T. (2013). Malvidin, a red wine polyphenol, modulates mammalian myocardial and coronary performanceand protects the heart against ischemia/reperfusion injury. *J. Nutr. Biochem.*, 24(7):1221–1231.

Renaud, S.C., Gueguen, R., Schenker, J. and d'Houtaud, A. (1998). Alcohol and mortality in middle-aged men from Eastern France. *Epidemiology*, 9(2):184–188.

Rigacci, S. Olive (2015). Oil phenols as Promising multi-targeting agents against Alzheimer's disease. *Adv. Exp. Med. Biol.*, 863:1–20.

Rimm, E.B., Williams, P., Fosher, K., Criqui, M. and Stampfer, M.J. (1999). Moderate alcohol intake and lower risk of coronary heart disease: Meta-analysis of effects on lipids and haemostatic factors. *Br. Med. J.*, 319(7224):1523–1528.

Rodrigo, R., Miranda, A. and Vergara, L. (2011). Modulation of endogenous antioxidant system by wine polyphenols in human disease. *Clin. Chim. Acta*, 412(5–6):410–424.

Saldanha, J.F., Leal, Vde O., Stenvinkel, P., Carraro-Eduardo, J.C. and Mafra, D. (2013). Resveratrol: Why is it a promising therapy for chronic kidney disease patients? *Oxid. Med. Cell. Longev.*, 2013:963217.

Seeram, N.P. and Nair, M.G. (2002). Inhibition of lipid peroxidation and structure-activity- related studies of the dietary constituents: Anthocyanins, anthocyanidins, and catechins. *J. Agric. Food Chem.*, 50(19):5308–5312.

Seymour, E.M., Bennink, M.R., Watts, S.W. and Bolling, S.F. (2010). Whole grape intake impacts cardiac peroxisome proliferator-activated receptor and nuclear factor kappaB activity and cytokine expression in rats with diastolic dysfunction. *Hypertension*, 55(5):1179–1185.

Shanmuganayagam, D., Beahm, M.R., Kuhns, M.A., Krueger, C.G., Reed, J.D. and Folts, J.D. (2012). Differential effects of grape (*Vitis vinifera*) skin Polyphenolics on human platelet aggregationand low-density lipoprotein oxidation. *J. Agric. Food Chem.*, 60(23):5787–5794.

Sierksma, A., van der Gaag, M.S., Kluft, C. and Hendriks, H.F. (2002). Effect of moderate alcohol consumption on fibrinogen levels in healthy volunteers is discordant with effects on C-reactive protein. *Ann. N.Y. Acad.Sci.*, 936:630–633.

Soares, S., Brandão, E., Mateusa, E.N. and de Freitas, V. (2015). Interaction between red wine procyanidins and salivary proteins: Effect of stomach digestion on the resulting complexes. *R.S.C. Adv.*, 5:12664–12670.

Spaak, J., Tomlinson, G., McGowan, C.L., Soleas, G.J., Morris, B.L., Picton, P., Notarius, C.F. and Floras, J.S. (2010). Dose-related effects of red wine and alcohol on heart rate variability. *Am. J. Physiol. Heart Circ. Physiol.*, 298(6): H2226–H2231.

Stivala, L.A., Savio, M., Carafoli, F., Perucca, P., Bianchi, L., Maga, G., Forti, L., Pagnoni, U.M., Albini, A., Prosperi, E. and Vannini, V. (2001). Specific structural determinants are responsible for the antioxidant and the cell cycle effects of resveratrol. *J. Biol. Chem.*, 276(25):22586–22594.

Stocker, R. and O'Halloran, R.A. (2004). Dealcoholized red wine decreases atherosclerosis in apolipoprotein E gene-deficient mice independently of inhibition of lipid peroxidation in the artery wall. *Am. J. Clin. Nutr.*, 79(1):123–130.

Tabengwa, E.M., Wheeler, C.G., Yancey, D.A., Grenett, H.E. and Booyse, F.M. (2002). Alcohol-induced up-regulation of fibrinolytic activity and plasminogen activators in human monocytes. *Alcohol. Clin. Exp. Res.*, 26(8):1121–1127.

Tang, Y.L. and Chan, S.W. (2014). A review of the pharmacological effects of piceatannol on cardiovascular diseases. *Phytother. Res.*, 28(11):1581–1588.

USDA Database for the Flavonoid Content of Selected Foods, Release 3.2 (2015).

Vanzo, A., Terdoslavich, M., Brandoni, A., Torres, A.M., Vrhovsek, U. and Passamonti, S. (2008). Uptake of grape anthocyanins into the rat kidney and the involvement of bilitranslocase. *Mol. Nutr. Food Res.*, 52(10):1106–1116.

Velliquette, R.A., Grann, K., Missler, S.R., Patterson, J., Hu, C., Gellenbeck, K.W., Scholten, J.D. and Randolph, R.K. (2015). Identification of a botanical inhibitor of intestinal diacylglyceride acyltransferase 1 activity via in vitro screening and a parallel, randomized, blinded, placebo-controlled clinical trial. *Nutr. Metab. (Lond.)*, 12:27.

Wallerath, T., Poleo, D., Li, H.G. and Forstermann, U. (2003). Red wine increases the expression of human endothelial nitric oxide synthase—A mechanism that may contribute to its beneficial cardiovascular effects. *J. Am. Coll. Cardiol.*, 41(3):471–478.

6 Wine Consumption
Toxicological Aspects

Fábio Bernardo, Paulo Herbert and Arminda Alves

CONTENTS

6.1 INTRODUCTION

Wine has been appreciated throughout the centuries for its aroma and taste, but new agricultural practices, technological innovations and the increasing quality control demands are currently placing food safety as one of the main factors that affect the consumer's opinion. For this reason, food safety is being used for marketing strategies. Two different types of wine constituents contribute to the so-called toxicological aspects of wine consumption: (i) the ethyl alcohol, traditional and predominant compound, whose consequences have been studied throughout the years, namely the detrimental effects of the excessive consumption of wine and alcohol in general on individuals and society, and (ii) some trace compounds, most of them residues of phytosanitary products applied to vine, and others also present in wine, such as ethyl carbamate, biogenic amines, nitrosamines, lead and mycotoxins (Figure 6.1).

The detection of contaminants responsible for several human diseases or minor undesirable reactions is nowadays the main goal of quality control of food products. Contamination may arise directly from agricultural practices (pesticides) and environmental pollution (atmospheric deposition and aqueous hazardous pollutants) or during and after industrial processing (added chemicals, unitary operations involved and packaging materials) (Figure 6.2).

The appearance of an increasing number of hazardous trace contaminants within food products in general, and particularly in wines, is demanding sophisticated and highly sensitive analytical techniques, able to detect the minor quantities in which they exist. However, the inverse also applies: the availability of today's powerful instrumentation allows the detection of a successively higher number of contaminants, undetected a few years ago. So, when is food safety really a problem?

The multidisciplinary efforts of scientists are necessary to obtain sustained conclusions from epidemiologists, toxicologists, analytical chemists and oenologists. These conclusions require time and are often difficult to gather in a useful period of time, because the gap between the detection of the contaminant and the product consumption is generally short, and decisions have to be taken rapidly. That is the main reason why concerns associated with the detection of some toxicological compounds recently found in wines have led to the imposition of, or the discussion to impose, maximum legal levels in wine without, however, consistent epidemiological research and interlaboratorial collaborative studies to uniform the methods of analysis. An example from the past is ethyl carbamate, nowadays it is ochratoxin A. Fortunately, for the majority of these compounds, it has been concluded that there is no reason for concern, although even the detection of a sporadic case of contamination urges the problem to be studied and strategies of control and prevention to be adopted. In addition to summarizing the deleterious effects of ethanol on excessive consumers, this chapter will highlight the most recent research on the main trace compounds with toxicological interference for wine consumers: methanol, ethyl carbamate, biogenic amines, nitrosamines and ochratoxin A. The residues of phytosanitary products will be dealt with separately in this book.

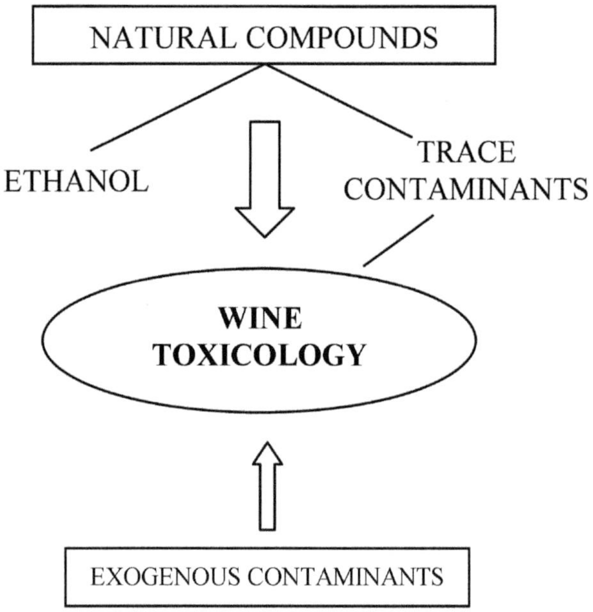

FIGURE 6.1 Chemical factors involved in wine toxicology

FIGURE 6.2 Interactions concerning food safety

6.2 THE ROLE OF ETHANOL

6.2.1 Ethanol Nutrient

Wine contains more than 1000 compounds, some of them arising from the grapes, others derived from the metabolic products of the yeast and, sometimes, bacterial activity, or even from chemical reactions during fermentation. Among the compounds present in wine at concentrations normally above 100 mg/l (such as water, alcohols, organic acids, sugars and glycerol), ethanol is the most important, with concentrations varying from 10 to 14% (v/v) for table wines, up to 20% for dessert wines and 60% for spirits.

Ethanol, being a small molecule soluble both in water and lipids, easily travels through all the tissues of the body, damaging them and affecting vital functions when consumption is not moderated. Although considered a non-essential nutrient, ethanol is a strong source of energy (29.7 kJ/g or 7.1 kcal/g), which exceeds the equivalent content of carbohydrates and proteins, thus is capable of providing energy for all the essential biological activities of the human organism: replication, function and maintenance of individual cells, energy for physical work and thermogenesis.

Ethanol metabolism and toxicity have been extensively studied. Ethanol's oxidation occurs *via* conversion to acetaldehyde, mainly by means of the enzyme alcohol dehydrogenase. Acetaldehyde is highly toxic and rapidly metabolized to acetate, mainly by mitochondrial aldehyde dehydrogenase. Acetate is finally converted to carbon dioxide and water. Almost all the ethanol is metabolized in the gastrointestinal tract, with 2% being lost through urine and respiration.

6.2.2 Moderate Consumption and Cardioprotective Effects

Alcohol consumption has many different dimensions, and strong arguments can be gathered either for its advantages or disadvantages. In a moderate dose, combined with a balanced diet and a healthy lifestyle, alcohol and wine seem to be beneficial to health. The relationship between the global mortality rate and alcohol consumption is expressed in a U-shaped curve. Hence, the global mortality rate of those who drink moderately is smaller than of those who either abstain or drink excessively. Moderation is defined as consumption up to 24 g/day (two glasses) for men and 16 g/day (1.5 glasses) for women.

A 2015 study assessed European consumers' perception of moderate wine consumption on health, specifically in Italy, France and Spain. These countries collectively produce 48% of the world's wine volume. The results showed that consumers were not aware of what the rates of moderate wine consumption are, with the majority of respondents tending to underestimate it. Consumers from France and Spain with high environmental awareness perceived a wine with an eco-label to be healthier than a conventional one.

According to the World Health Organization (WHO) in their 2014 Global status report on alcohol and health, with data from 2010, individuals above 15 years of age drank, on average, 6.2 litres of pure alcohol per year, which translates into 13.5 grams of pure alcohol per day. Spirits accounted for 50.1% of total consumption of recorded alcoholic beverages, followed by beer (34.8%) and wine (8.0%).

Ethanol and its first metabolite acetaldehyde were both evaluated by the International Agency for Research on Cancer (IARC) as group 1 carcinogens. For low or moderated ethanol intake, the putative cardioprotective effects of alcohol and other substances in alcoholic beverages were evidenced by the French Paradox. Many epidemiological studies have established a reduced incidence of death from coronary artery diseases and atherosclerosis among moderate consumers when comparing with abstainers. This cardioprotective effect is due to the increased plasma levels of high-density lipoproteins (HDL), an inhibition of platelet aggregation and improved endothelium function. Controversy about the primary factor responsible for these effects arises between those that defend the non-alcoholic fraction of wine, mainly the phenolic compounds, and others that suggest ethanol *per se*. For more details, please see the exhaustive chapter 'Therapeutic value of wine' in this text.

6.2.3 Alcoholism – Individual and Social Consequences

When we are dealing with excessive alcohol intake (acute or chronic), damage of digestive organs, the liver in particular, is the worst consequence, which leads most of the time to cirrhosis (Figure 6.3). Hepatotoxicity of ethanol results from three main pathways for ethanol metabolism: (i) alcohol dehydrogenase; (ii) the microsomal ethanol oxidizing system (MEOS), mainly the cytochrome P450 system and (iii) catalase. Each of these pathways produces a specific metabolic and toxic disturbance and all result in excessive hepatic generation of acetaldehyde. It seems that about 90-95% of ethanol metabolism is due to alcohol dehydrogenase, being 5% attributed to MEOS, but this value may increase up to 25% in chronic drinkers. Other complications such as chronic pancreatitis and malabsorption of nutrients (proteins and vitamins), may also result in decreased insulin production and diabetes.

The prevalence of alcoholic cardiomyopathy (heart muscle disease) is not well known, but heart damage appears to increase with a lifetime abuse of alcohol. There is evidence of a direct relationship between heavy alcohol consumption and increased blood pressure, known as hypertension. Ethanol also depresses brain function. Low activity of the nervous system, loss of self-control, mental confusion, inability to walk properly, delirium tremens and memory gaps are some of the symptoms experienced by alcohol abusers. The neurochemical basis for the rewarding effects of alcohol may involve the potentiation of GABA and GABA receptors (causing relaxation) and release of dopamine from mesolimbic neurons (causing euphoria). Neurological disturbance (intoxication, withdrawal seizures, delirium tremens and Wernick-Korsakoff syndrome) may appear by disruption of glutamatergic neurotransmission. Long-term heavy alcohol consumption may also be toxic for brain serotonergic neurons, diminishing the serotonergic neurotransmission. In some cases, it can result from the interaction of ethanol toxicity and nutrient shortage, such as thiamine deficiency, that occurs in long-term alcohol users as a consequence of both inadequate ingestion and malabsorption of the vitamin.

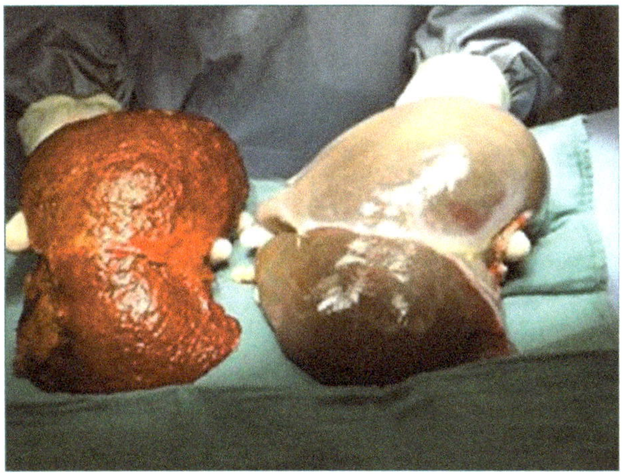

FIGURE 6.3 Photograph of a healthy (right) and a damaged liver (left) caused by heavy alcohol consumption. *Source:* Courtesy of Dr. Pedro Correia da Silva, F.C. Porto physician and Head of Surgery I Department, Hospital S. João – Porto

Other effects include: electrolytic depletion (rapid loss of water due to the decreased secretion of antidiuretic hormone, a pituitary peptide); magnesium shortage, due to renal wasting, enhanced by dietary Mg-privation and gastrointestinal losses through diarrhoea and vomiting; deficiency of several vitamins, mainly retinol (vitamin A), thiamin (vitamin B1), riboflavin (vitamin B2), pyridoxine (vitamin B6) and cobalamin (vitamin B12), caused by impaired absorption.

Failure of reproductive function and cancers of mouth, larynx and esophagus, although not clearly proven as consequences of alcohol abuse, are linked to it by statistical evidence. The U-shape curve of the global mortality rate in relation to alcohol consumption equally applies to cancer. Patients with alcohol-use disorders (AUDs) have a high prevalence of anxiety disorders (AnxDs), and women suffer from higher levels of stress and AnxDs than men. Social Anxiety Disorder symptoms are common among help-seeking alcohol-dependent individuals. Alcohol-dependent patients with co-morbid anxiety disorders are particularly prone to relapse during the first months of treatment. Finally, the social dimension of alcoholism may affect established market economies, especially in the areas of morbidity and disability, and may carry substantial social costs. Besides the detrimental effects of alcoholism on chronic diseases as mentioned above, social problems, both chronic (family disruption and unemployment) and acute (accidents, violence, public disorder), play an important role in the worldwide acceptance of the need to establish strong preventive measures to avoid ethanol abuse. When alcohol consumption is used to predict either medical or social consequences, usually the volume, the frequency and the variability of drinking are the main factors used in these correlations. It has been suggested that future research in alcohol epidemiology should distinguish different dimensions of consumption.

In particular, occasions of heavy drinking within a specific time period, the impact of drinking with or without meals and the beverage type, etc. should be properly investigated. A 2016 review assessed different aspects of drinking patterns and effects on human health, addressing the effects of food on blood alcohol concentration, as well as wine's antimicrobial properties and its potential to act as a digestive aid.

Wine has been regarded throughout the centuries as a stimulant, an appetizer, an analgesic or just as a pleasant and tasteful partner of a good meal or a drink to share with friends. Still, the health aspects have been and continue to be discussed.

6.3 THE ROLE OF METHANOL

Methanol is a natural fermentation product and its concentration may be up to 300 mg/l in table wines, or even higher in spirits. It is a toxic substance to humans at doses higher than 340 mg/kg of body weight, being well absorbed by the gastrointestinal tract mucosa, and through the skin and lungs. It is

TABLE 6.1
Methanol Limits in Wine

Country	Methanol limit based on volume of wine
Argentina	350 mg/L
Chile, China, Vietnam, International Office of Vine and Wine (OIV)	400 mg/L (Red Wine) 250 mg/L (White and Rosé Wine)
Japan, Korea, USA	1000 mg/L
South Africa	300 mg/L
Switzerland	300 mg/L (Red Wine) 150 mg/L (White and Rosé Wine)
Taiwan	2000 mg/L
Turkey	10 mg/L

Source: Adapted from Dyer (1994)

mainly oxidized in the liver, producing lactic acidosis. Once absorbed, methanol is rapidly distributed in the body, with a peak level in blood occurring about 30 to 90 minutes after exposure.

A 2017 study, conducted by FIVS Scientific and Technical Committee, reviewed the toxicology of methanol and the associated regulatory limits established by competent authorities in various parts of the world. Table 6.1 summarizes the methanol limits in wine in different global markets.

The metabolic pathway of methanol includes its transformation to formaldehyde by alcohol dehydrogenase and subsequently, to formic acid. The former causes the profound metabolic acidosis that is typical of methanol poisoning. The symptoms are weakness, vomiting and early visual disturbances. Gastritis may be severe and is occasionally hemorrhagic in acute intoxications. Other complications arising from severe intoxication include coma, blindness, cardiac failure, pulmonary edema and ultimately death.

Curiously, ethanol decreases the metabolism of methanol. In fact, alcohol dehydrogenase, which breaks down both ethanol and methanol, has a greater affinity for the former. So, in the presence of ethanol, the metabolic conversion of methanol to its toxic metabolites is greatly slowed.

Some authors found that pectolytic enzymes added in the fermentation stage may induce the release of methanol, although other factors such as grape variety, some enological practices and the yeast strain used may also play an important role. Commercial preparations of pectolytic enzymes may contain two distinct groups of enzymes: depolymerases and pectin-methyl-esterases. The responsibility of methanol being released to musts and wines rests upon the latter. In the production of red wine, grape skins are retained and in white wine they are usually removed. The skin is the major source of pectin, thus on average, red wine contains about 200 mg/l of methanol and white wine contains about 50 mg/l. Remedy of this problem is hot pressing at 85°C for 15 minutes, which reduces the level of pectin esterase and, consequently, the level of methanol, but it is expensive in terms of energy and has significant effects on taste. Vacuum distillation is another technological process used to reduce methanol, but it also lowers ethanol levels.

6.4 THE ROLE OF ETHYL CARBAMATE

In the beginning of the 1960s, a very promising compound with sterilizing properties was used by the food industry – diethyl pyrocarbonate (DEP). In the wine industry, DEP kills yeasts and bacteria in a short period of time (in minutes for yeasts), and it could allow the cold pasteurization of wine. Upon addition to alcoholic solutions, DEP breaks down to ethanol and carbon dioxide (main end products).

In 1971, Löfroth e Gejvall described the formation of ethyl carbamate (EC, also called urethane) in concentrations up to 2.6 mg/l in white wine, 1.3 mg/l in beer and 0.46 mg/l in orange juice, as a consequence of the treatment of these beverages with doses of DEP ranging from 250 g/l to 1000 g/l. The EC formed was analyzed by isotopic dilution techniques and was seen to be dependent on the quantity of DEP added and on the ammonium concentration of the medium. The proposed reaction is:

$$\mathbf{(EtO)_2\,CO + NH_3 \rightarrow EtOCONH_2 + EtOH}$$
$$(\text{diethyl pyrocarbonate})\ (\text{ammonium})\ (\text{ethyl carbamate})\ (\text{ethanol}) \tag{6.1}$$

Because ethyl carbamate is a possible human carcinogen (it was upgraded to Group 2A by the IARC in 2007), and because it exhibits carcinogenic properties in laboratory animals when administrated in high doses (mg/l), its presence in food has called for strict control. Consequently, Löfroth and Gejvall's work led to the prohibition of DEP utilization in the US food industry, based on the fact that it was impossible to unequivocally conclude that ethyl carbamate was present or absent, due to analytical difficulties. Further work was carried out to elucidate the origin and levels of EC in alcoholic beverages, being EC quantified by different techniques. It was shown that with great probability, Löfroth and Gejvall's results were affected by an error that led to concentrations of EC about 100 times higher than the real levels.

In one of those works, Ough (1976) reported that EC exists naturally in certain fermented foods and beverages, suggesting that its formation is due to the reaction between ethanol and carbamyl phosphate naturally occurring in those aliments. Yeasts produce carbamyl phosphate in the arginine synthesis catalysed by carbamyl synthase, involving ATP, CO_2 and ammonia. In addition, DEP addition in doses ranging from 50 mg/l to 100 mg/l resulted in an EC formation of less than 1 g/l in red, rosé and white wines, proposing the re-evaluation of DEP utilization in the food industry.

Equation 6.2 shows the reaction between carbamyl phosphate and ethanol to produce EC:

$$\mathbf{O_2P(O)OC(O)NH^-{}_2 + EtOH \rightarrow EtOCONH_2 + HPO_4{}^{2-}}$$
$$(\text{carbamyl phosphate})\ (\text{ethanol})\ (\text{ethyl carbamate})(\text{hydrogen phosphate}) \tag{6.2}$$

The EC question finally became public when, in the mid-1980s, the Canadian government ordered the withdrawal of several alcoholic beverages that were found to contain EC in concentrations up to 13400 μg/l and set limits for EC concentrations in such products. Research concerning the development of new analytical methods and the elucidation of ethyl carbamate formation in alcoholic beverages was also conducted. Table 6.2 presents the maximum allowed levels of ethyl carbamate in different types of beverages. In Canada, ethyl carbamate level is now restricted to 30 μg/l in table wines, 100 μg/l in fortified wines, 150 μg/l in distilled spirits, and 400 μg/l in fruit brandies and liqueurs.

Nowadays, much is known on the processes that lead to ethyl carbamate formation in alcoholic beverages, where it exists naturally in vestigial concentrations (ppb range). Furthermore, adequate analytical techniques have also been developed towards EC quantification. However, not much information was reported on the levels of EC in alcoholic beverages, at least coming from representative samples on the international market. However, one of the few exceptions has been reported, where an analysis of "about 16 000 wines [...] from every viticultural region of the world" was collected in 1994 and published in a technical report in 1995. In this study, 1.3% of the samples presented EC levels between 20 and 30 μg/l, and only 0.3% above 30 μg/l. Ethyl carbamate formation in alcoholic beverages is believed to be a non-enzymatic process, where ethanol reacts with compounds bearing a carbamyl group. Besides its formation from DEP and carbamyl phosphate, EC can also be formed from urea (EC is industrially produced by direct reaction of urea with ethanol) (Equation 6.3).

$$\mathbf{NH_2CONH_2 + EtOH \rightarrow EtOCONH_2 + NH_3}$$
$$(\text{urea}) \quad (\text{ethanol}) \quad (\text{ethyl carbamate}) \quad (\text{ammonium}) \tag{6.3}$$

Moreover, it is suggested that EC can also be formed from azodicarbonamide, used as a blowing agent in beer bottle cap liner and as a bread improver, and from cyanate, where copper (II) can act as a catalyst in the production of spirits.

Alcoholic beverages are therefore prone to produce EC and methyl carbamate, especially those with higher ethanol content. However, the presence of ethanol may itself be a protector from the deleterious effects of EC in the human body, as suggested by Stoewsand *et al.* in 1996. In their work, the administration of ethanol to laboratory animals, together with EC, reduced the number of tumours induced by EC. It was thus suggested that the normal ethyl carbamate levels in wine might have little or no effect on consumers, given the protective role of ethanol. In juices or musts, EC is not detected due to the absence of ethanol, although juices and musts have several carbamyl compounds capable to produce ethanol. Urea and citrulline are considered the main ethyl carbamate precursors in wines. These compounds are usually formed in arginine metabolism by yeasts, arginine being one of the most important free amino acids in musts and wines.

With regard to ethyl carbamate formation in distillates, other precursors such as cyanate may be involved, because neither urea, citrulline nor ethyl carbamate distillate during spirit production, and spirits usually contain higher levels of EC than wines.

A two-way preventive approach has been made to avoid a significant amount of EC formation in wines keeping EC precursors at low levels, namely urea and citrulline, or to promote its degradation when high levels have already been found. High urea levels can occur in wines produced from musts with high arginine contents. Such musts tend to be produced with grapes that have come from heavily fertilized vineyards. The evaluation of arginine concentration in musts prior to fermentation is therefore very important, in order to prevent musts with increasing arginine levels. In normal situations, it is desirable that arginine content in musts does not exceed 1000 mg/l, thus avoiding urea concentrations in wines above 5 mg/l. Recently, a simple and fast spectrophotometric method (derivatization with OPA and N-acetyl-L-cystein) that allows arginine determination in musts has been developed.

Besides arginine levels, however, there are other factors that influence arginine metabolism during fermentation and ultimately urea levels. High levels of glutamic acid, glutamine or ammonia reduce arginine catabolism by yeasts. However, it also seems that, like high assimilable nitrogen levels, they reduce urea re-absorption at the end of fermentation, leading to wines with high urea levels. Yeast strain also affects arginine metabolism during fermentation. In addition, technological factors such as degree of aeration, temperature, storage temperature and time can also influence the EC levels of wines.

Although citrulline reacts with ethanol much slower than urea to produce EC, its presence in wines at high concentrations during long storage times can contribute to increasing EC levels significantly. Citrulline can be produced during malolactic fermentation if bacteria incompletely metabolize arginine via arginine deiminase. It is now possible to use malolactic starters with low citrulline production.

In wines presenting high urea levels, acid urease was used to lower the levels of urea. However, the utilization of this enzyme still needs further research, because the presence of ethanol and other substances such as fluoride, malic acid or gallic acid can interfere with enzyme activity.

Like other vestigial compounds present in wines, the determination of EC in alcoholic beverages poses problems due to its low concentration and to matrix interference. Most of the reported methods for EC analysis employ liquid-liquid extraction and a pre-concentration step prior to gas chromatographic separation. Some methods still include an additional extraction clean-up step. EC detection is usually performed by mass spectrometry, either on electron or chemical ionization mode. Other detection techniques have also been applied, such as use of the Hall electrolytic conductivity detector and a thermal energy analyzer in nitrogen mode.

The reference method adopted by both the Office International de la Vigne et du Vin (O.I.V.) and the European Union, employs the extraction with methylene chloride, evaporation to dryness and subsequent recovery with ethyl acetate solution, prior to gas chromatographic injection and electron ionization mass spectrometric detection.

A previous work focusing on avoiding some of the sample preparation steps reported the use of direct capillary GC/MS/MS introduction of the extracts and a stable isotope-labeled internal standard. Although it has been demonstrated that the MS/MS approach (or other similar techniques presented in the literature) is capable of providing routine quantification, the high cost of the instrumentation and the complexity of the techniques, which require highly qualified personnel, constitute negative factors in the acceptance of its implementation in routine analysis. Recently, a new methodology for EC determination in alcoholic beverages based on the reaction with 9-xantydrol to form a fluorescent derivative was developed, employing HPLC and fluorescence detection. This methodology is suitable for implementation as a screening method for ethyl carbamate in beverages, as it only requires sample dilution prior to HPLC analysis.

6.5 THE ROLE OF BIOGENIC AMINES

Recent trends on food safety are promoting the search for compounds that can affect human health. Biogenic amines, the so-called natural amines with physiological significance, belong to this group of substances. These compounds can be naturally present in animals, plants and microorganisms, playing different roles. They can serve as a nitrogen pool, growth regulators in plants and microorganisms or as neurotransmissors in complex organisms, amongst other functions.

The presence of amines in food, even at low concentrations, is sometimes associated with the occurrence of undesirable effects in sensitive individuals, such as hypertensive crises, headaches, intestinal symptoms or histamine intoxication. The amines that are able to produce or enhance those effects comprise the aromatic, the heterocyclic and the polyamines. Depending upon the amine, medium levels (low ppm range) are usually observed in wine (Table 6.3), beer or other fermented beverages. In nonfermented foods, the biogenic amines appear as a result of undesirable microbial activity. High levels of histamine are found in spoiled fish, especially from the scombroid group. Although histamine and tyramine are the major biogenic amines investigated in wines and other beverages, several different ones (*ca* 25) have also been detected, with total amounts varying from below one to several dozen mg/l.

Taking into account that this is not an extensive list of all biogenic amines, the chemical characteristics and also the different consequences that they may induce can be used to divide the most common biogenic amines into three groups:

- the aromatic and the heterocyclic amines – histamine, tyramine, β-phenethylamine, dopamine, noradrenaline, adrenaline, pyrrolidine, serotonin and tryptamine – to which pronounced toxicological effects (vomiting, swelling, shock, etc.) are seldom attributed, rather being simply responsible for slight alterations of human well-being (nausea, headache, etc.);
- the aliphatic di-, tri- and poli-amines – putrescine, cadaverine, agmatine, spermidine and spermine – closely related by the same metabolic pathway, and traditionally associated with deficient sanitary conditions of musts and/or vinification;
- the aliphatic volatile amines – ethylamine, methylamine, isoamylamine and ethanolamine – with no described significant expression of adverse effects on humans but need to be recognised and controlled in wines, in order to prevent the possible alteration of sensorial properties.

Although some amines are usually absent or present in low quantities in wines and other alcoholic beverages, they exhibit interactions with the normal human metabolism (*e.g.* vasoactive or psychoactive properties) that justify research based on their presence and their origin in wines and the possibly related toxicological effects that they may cause. Moreover, the consumption of alcoholic beverages is usually concomitant with other food, that can itself contain high levels of biogenic amines or enhance its effect, such as tuna fish. Some canned fish was found to contain histamine, tyramine and cadaverine up to 278 mg/kg, 12 mg/kg and 78.2 mg/kg, respectively. A 2015 study determined biogenic amines in fish products in 63 samples (from different cities of Turkey), using HPLC. Maximum levels of histamine, putrescine, cadaverine, tyramine and tryptamine were 110.33, 116.53, 122.18, 48.63 and 190.61 mg/kg, respectively. The presence of histamine, tyramine and β-phenylethylamine in cheese up to 140 mg/kg, 625 mg/kg and 46 mg/kg, respectively, and putrescine and cadaverine in "choucroute" up to 550 mg/kg and 311 mg/kg, respectively, has been documented.

The main defense mechanisms of humans against biogenic amines are the physical barrier formed by intestinal mucosa, the mucines (glicoproteins) produced by the latter that bind biogenic amines and at least two enzymatic systems based on the activity of monoamine oxidase and diamine oxidase. These enzymatic systems are inhibited by polyamines such as putrescine and cadaverine, as well as by tryptamine, tyramine or b-phenethylamine. Moreover, they are inhibited by ethanol and acetaldehyde. Adverse reactions described upon the oral absorption of similar quantities of the main biogenic amines

TABLE 6.2
Maximum Allowed Level of Ethyl Carbamate in Different Types of Beverages

	Ethyl carbamate concentration (μg/L)		
Country	Wine	Fortified wine	Distilled spirits
Canada	30	100	150
USA	15	60	–
Czech Republic	30	100	150
France	–	–	150

Source: Adapted from Fujinawa *et al.* (1990)

(histamine, tyramine and β -phenethylamine), revealed that while 100-200 mg of oral intake of the first two were not enough to promote alterations, only 5 mg of the latter was sufficient to advance adverse reactions in sensitive individuals. Hypertensive crisis may also occur in patients undergoing treatment with monoamine oxidase inhibitor drugs (MAOI) after ingesting foods with high contents of tyramine.

Thus, aside from assessing a maximum overall quantitative intake of biogenic amines in wines or other alcoholic beverages, it is important to discriminate between them.

Until now, the difficulty of detection and reliable quantification of amines in wines has been responsible for the scarce information about their occurrence in the different types of wines worldwide. These problems relate to matrix interferences (*e.g.* presence of free amino acids) and the low levels at which they are found (usually in the order of few ppm).

The biological formation of biogenic amines in food proceeds by the decarboxylation of each precursor amino acid by the respective decarboxylase enzyme. This process occurs normally in spoiled meat or fish, or processed food where microbial growth takes place.

In wine, prerequisites for a considerable formation of biogenic amine are the availability of precursor free amino acids, the presence of decarboxylase-positive microorganisms, and conditions that allow bacterial growth, decarboxylase synthesis and decarboxylase activity. The presence of amines in musts and wines is well documented in the literature. However, the processes that generate these amines, together with the factors that influence their quantitative and qualitative presence are not yet well defined. Amines in wine may have two different sources: grapes and fermentation processes. Some amines have already been found in grapes, namely histamine and tyramine, although in small quantities, as well as several volatile amines and polyamines.

Concerning the levels of biogenic amines in grapes and musts, the general opinion points towards the existence of low concentrations of histamine and tyramine, usually not exceeding 1 mg/l. However, polyamines, as natural grapevine constituents, can be found in grapes, in different parts of the plant, with concentrations that may not be negligible or influenced by physiological stress. This natural pool also contributes to the final content in biogenic amines of the wines.

Other studies have shown that both alcoholic and malolactic fermentations may originate amines in wines. Several workers have stated that at least part (and sometimes all) of the final content of tyramine in the tested wines was due to alcoholic fermentation, although not confirming an increase in the histamine content during this process, nor finding any increase in histamine or tyramine levels during malolactic fermentation in 50% of the wines studied. Production of histamine during alcoholic fermentation has been reported, although a highly significant correlation between levels of histamine, tyramine and lactic acid/malic acid ratio for red wines, and a significant correlation for white and rosé wines, was observed.

Work concerning the factors that influence type and quantity of amines in wines, particularly biogenic, indicates lactic acid bacteria mainly responsible for the significant generation of these substances, mainly some strains of *Lactobacillus* and *Pediococcus*. Saccharomyces species were shown to be weak producers of histamine. However, some *Oenococcus oeni* strains possess decarboxylase activity capable of producing amines as well as some *Saccharomyces*, yeast extracts may contain noticeable amounts of histamine and tyramine, and some "non-*Saccharomyces*" yeasts (which may proliferate in the early stages of alcoholic fermentation) are capable of producing histamine at concentrations as high as 8.3 mg/l.

It also seems that even bacteria showing decarboxylase activity may not produce biogenic amines. Amino acid decarboxylation seems to have an energetical function in bacteria. It has been suggested that biogenic amines production would be more intense in conditions where nutrients, namely glucose or malic acid, were scarce. In such cases, histidine concentration in the culture medium would influence histamine production.

Biogenic amines may be produced during (or after) malolactic fermentation, as reported. Nevertheless, it is now believed that the formation of histamine and other biogenic amines in wine is not only due to spoilage pausing bacteria, mainly *Pediococcus* damnosus strains, but also due to the metabolism of lactic acid bacteria responsible for malolactic fermentation: *Oenococcus oeni*. A direct PCR detection for the presence of histamine-producing bacteria (which are not rare) that might be done during alcoholic fermentation when the bacterial population is still low, has been reported. In the case of positive results, suitable malolactic starters could be used in order to supplant and to eliminate the undesired indigenous flora.

The presence of *Botrytis cinerea* in grapes seems to influence both quantity and quality of amines in musts. Among others, vintage and technology employed, variety of grape, levels of precursor amino acids in musts, assimilable nitrogen content, sulfur dioxide content, malic acid and/ or lactic acid contents, malolactic fermentation intensity and pH seem to be relevant factors to account for. Nevertheless, biogenic amines concentration in wines is normally low when compared with other fermented products.

Even besides the complexity of factor interactions in biogenic amines production and the great range of concentrations sometimes found amongst different wines, some countries have already recommended maximum limits for the histamine content of wines, namely: Germany (2 mg/l), Switzerland (10 mg/l), Belgium (5–6 mg/l), France (8 mg/l), Finland (5 mg/l), Austria (10 mg/l) and Netherlands (3,5 mg/l).

6.6 THE ROLE OF NITROSAMINES

Nitrosamines did not receive much attention from the toxicological point of view until 1956, when it was found that dimethylnitrosamine produced liver tumours in rats. Approximately 300 of these compounds have been tested, and 90% of them have been found to be carcinogenic in a wide variety of test animals. Most nitrosamines are mutagens and a number are transplacental carcinogens. Most are also organ specific. Some N-nitrosamines with carcinogenic properties may form in foodstuffs mainly through the interaction of secondary amines with nitrite or nitrogen oxides, or both. In most of the cases, the concentrations found have been in the parts per billion (ppb) range. It is pertinent to add that the amounts of volatile N-nitrosamines found so far in wines do not pose the same problems as its presence in beers and malt beverages did twenty years ago. Red wines studied along with foodstuff samples collected in the French market were found to contain trace levels of volatile N-nitrosamines (NDMA essentially), similar to those found in non-alcoholic beverages (fruit juices, lemonades, etc.).

It has been reported that some phenolic compounds and biogenic amines present in wine, became directly genotoxic after *in vitro* nitrosation. Moreover, this study suggests synergistic effects may occur between various nitrosatable compounds in the wine.

Tetrahydro-β-carbolines are another group of nitrosatable substances present in wines. These are heterocyclic compounds formed by the condensation of indole ethylamine-type compounds with various aldehydes. Both 1,2,3,4-tetrahydro-c arboline-3-carboxilic acid (THβCCA) and 1-methyl-1,2,3,4-tetrahydro-β-carboline-3-carboxilic acid (MTHβCCA) can be formed from L-tryptophan upon reaction with formaldehyde and acetaldehyde, respectively. Their reported properties as neuromodulators and as inducers of alcoholism in laboratory animals stresses the importance of the knowledge of their presence in foods and beverages.

Both THβCCA and MTHβCCA have been reported to occur at variable but normally low ppm levels (ppm) in wines and beer, although the reported data on the occurrence of THβCCAs in alcoholic beverages is very limited. Besides their properties, some THβCCAs such as the referred THβCCA and MTHβCCA have been shown to produce, upon nitrosation, highly mutagenic N-nitroso compounds not yet characterized. Furthermore, nitrosation of THβCCAs can also occur *in vivo* in the stomach due to the interaction between ingested THβCCAs and salivary nitrite. Nevertheless, phenolic acids, catechins, monomeric anthocyanidins and other reducing agents present in red wine aid in restricting the formation of nitrosamines and other nitroso derivatives (at least partially) in model systems. Further research, however, is needed to fully estimate the effective contribution of nitrosable wine compounds to wine toxicity.

6.7 THE ROLE OF A MYCOTOXIN – OCHRATOXIN A (OTA)

Mycotoxins are toxic metabolites produced by various fungi growing in several foodstuffs. The most commonly found are aflatoxins, ochratoxin A, patulin, trichothecenes and zearalenone. Since the 1970s, the occurrence of OTA has been systematically investigated in coffee, beer, cocoa and cereals. In the 1990s, the sporadic detection of ochratoxin A in wines made this a food safety problem.

Ochratoxin A is a mycotoxin produced by some mould fungi, that is a derived isocoumarin, linked through the carboxy-group to a L-β-phenylalanine by an amide bond. This metabolite is a potent nephrotoxic substance, with immunosuppressive and teratogenic properties and its carcinogenicity has been detected in different rodent species. Therefore, the International Agency for Research on Cancer (IARC) has classified it in group 2B, as a possible human carcinogen. A kidney disease (Balkan endemic nephropathy) was also associated with the presence of OTA in food. The increase in renal disease is accompanied by a high risk of tumours of the urinary tract.

Because OTA has been found in a large variety of foods – cereals, dried fruits, nuts and meats – discussion around maximum dietary intake still exists, pointing to 1–8 ng/kg body weight/day, with proposals of different countries varying from 0–2 to 16 ng/kg/day. In wines, while the European Union is currently discussing the adoption of a maximum level, a recommended value of 2 μg/l in alcoholic beverages has been reached.

Ochratoxin A was first isolated from *Aspergillus* and *Penicillium* cultures, specifically from *Aspergillus ochraceus* and *Penicillium verrucosum* species. The former is abundant in tropical climates and the latter in temperate ones. Other Aspergilli fungi, from the *Aspergillus niger* group, the *Aspergillus* section *Nigri* and the *Aspergillus carbonarius*, may be also capable of producing OTA.

The origin of OTA in agricultural products has not been established yet, but it is accepted that humidity, high temperature (or heat) during growth, aeration, substrate and time of infection are favorable factors. While data from countries where *Penicillium verrucosum* is the major ochratoxin producer show that the contamination is associated to post-harvest conditions, it is believed that the toxin production from *Aspergilli* may already exist in the growing plants.

In the 1990s, OTA was reported in musts and wines at levels from 3 to 338 ng/l. Several studies were conducted in different countries at that time, using several analytical methodologies (Table 6.4). Differences of OTA content in white and red wines were evidenced, referring to average levels of 7 ng/l for white and 200 ng/l for red. An overall mean level of OTA of 116 ng/l and 235 ng/l in white and red juices, respectively, was found. It can be explained by different winemaking methods: while white grapes are directly pressed out, red ones are macerated. Grape juices seem to contain more OTA than wines.

Although it is expected that OTA is formed prior to fermentation, and thus the best eliminating step is prevention, some attempts have been made to reduce the mycotoxin in wines. Some clarifying agents, such as activated carbon or silica gel+gelatin can reduce OTA levels to about 50% and 90%, respectively. However, it seems that the quality of wines is severely decreased. Therefore, some microbiological methods may constitute a good alternative in mycotoxin detoxification, but more research is still needed on this topic. Major conclusions drawn from the studies point to:

(i) analytical methods able to detect OTA in wines are available, validated by international collaborative trials;
(ii) the contamination of OTA in wine is sufficiently described by more than a thousand results;

TABLE 6.3
Average Amine Levels in Wines from Different Origins (mg/l)

Origin	Wine	n	Hista		Tyra		Ethyla		Put		Cad		Methyla		Phen		Iso	
			Min	Max	Min	Max	Min	Max	Min	Max	Min	Max	Min	Max	Min	Max	Min	Max
Several	R	102	**5.73** ± 0.59		**5.18** ±0.43				**5.13** ± 0.50		**0.66** ± 0.11							
	W	99	**3.35** ± 0.32		**4.41** ± 0.48				**1.94** ± 0.18		**0.92** ± 0.17							
Portugal[a]	R	2	**1.24** ± 0.50		nd				**0.91** ± 0.17		nd							
	W	4	**1.12** ± 0.75		**4.43**±2.92				**2.38** ± 1.22		**1.45** ± 0.99							
France[a]	R	50	**8.14** ± 1.25		**7.27**±0.72				**7.63** ± 0.92		**0.99** ± 0.23							
	W	52	**4.35** ± 0.52		**6.54**±0.76				**2.26** ± 0.29		**1.43** ± 0.28							
France	R+W	117	**4.78**		**5.21**		**1.03**		**12.54**		**0.77**		**0.26**		**1.31**		**1.53**	
			0.03	31.63	0.13	22.10	0.28	2.15	0.25	137.86	0	9.91	0	2.26	0	9.83	0	10.49
USA[c]	R	158	**1.85**															
			0.2	15.5														
	W	79	**1.78**															
			0.3	11.4														
Spain	R	127	**4.07**		**3.03**													
			0	34.25[f]	0	7.80[f]												
	W	65	**0.81**		**1.49**													
	RO	34	**0.86**		**1.66**													
Portugal[e]	R	6	**1.21**		**0.82**		**9.15**		**3.61**		**0.64**		**0.68**		**4.87**		**2.89**	
			0.39	1.66	0	2.01	5.38	11.18	1.76	6.01	0.56	0.73	0.30	1.54	0.85	12.73	0.44	6.17
	W	6	**1.20**		**0.54**		**23.04**		**2.14**		**0.63**		**0.96**		**0.75**		**1.17**	
			0.84	1.70	0	0.78	18.94	28.55	1.31	3.63	0.40	0.71	0.53	1.56	0.29	2.05	0.22	4.21
	F	18	**0.52**		**1.24**		**10.46**		**1.28**		**0.33**		**0.58**		**0.39**		**0.74**	
			0	0.99	0.06	3.10	5.73	24.55	0.25	4.42	0	0.51	0.09	0.91	0.10	1.65	0.18	3.63
Total		**919**																
Highest			**34.25**		**22.10**		**28.55**		**137.86**		**9.91**		**2.26**		**12.73**		**10.49**	
Lowest			**0**		**0**		**0.28**		**0.25**		**0**		**0**		**0**		**0**	

Legend: a – [296]; b - [255]; c – [206]; d - [287]; e – [169]; f – maximum and minimum relative to all samples; n – sample number; R – red; W – white; RO – rosé; F – fortified wine; nd – not detected; Hista – histamine; Tyra – tyramine; Ethyla – ethylamine; Put – putrescine; Cad – cadaverine; Methyla – methylamine; Phen – phenylethylamine; Iso – isoamylamine; Try – not detected in any sample, Eta – detected in Portugal only

Note: reference a – mean ± standard deviation.

TABLE 6.4
Reported Ochratoxin A Concentrations in Wines from Different Regions

Average OTA concentration μg/l	Sample number (n)	Origin (observations)	Reference
White table wines – < 3 to 0.178 μg/l	118	Switzerland	Zimmerli e Dick, 1996
Red table wines – < 3 to 0.338 μg/l	15		
Dessert – 0,003 – 0.017 μg/l			
White table wines – 0.007 μg/l	114 wines	Europe (one from Algeria – 1.85 μg/l)	Majerus *et al.*, 1996
Red table wines – 0.200 μg/l			
Red wine – < 10 to 7.63 μg/l	56	Italy	Visconti *et al.*, 1999
Red wines – 0.054 μg/l av.	91	Wines from Spain, France, Italy,	Burdarspal *et al.*, 1999
Rosé – 0.031 μg/l av.	32	Portugal and Hungary	Otteneder *et al.*, 2000
White wines – 0.02 μg/l av.	69		Festas *et al.*, 2000
Dessert – 1.048 μg/l av.	47		
Red and white wines <1 to 3.856 mg/l	96 red 15 white	Italy (Northern – 100 % < 0.2 μg/l and South – 80% > 1 μg/l)	Pietri *et al.*, 2001
Red wines – 0.68 μg/l av.	14	Greek wines (Maximum – 3.2 μg/l)	Soufleros *et al.*, 2003
White wines – 0.27 μg/l av.	13		
Rosé – not detected	1		
Dessert – 0.94 μg/l av.	7		
Red dry wines – 0.09 μg/l av.	104	Greek commercial wines	Stefanaki *et al.*, 2003
White dry wines – 0.06 μg/l av.	118		
Rosé – 0.08 μg/l av.	20		
Dessert wines – 0.33 μg/l av.	18		
Red wines – 0.03 to 0.62 μg/l	16 red	Brazil	Terra *et al.*, 2012
White wines – 0.03 μg/l av.	7 white		
Red wines – 0.36 μg/l av.	183 red	China	Zhong *et al.*, 2014
White wines – 0.13 μg/l av.	40 white		
Red wines – 0.02 to 0.98 μg/l	136	Argentine	Mariño-Repizo *et al.*, 2016

(iii) the OTA intake by wine is very small, but the maximum residue level should be below 2 g/l;
(iv) systematic research on the reasons for OTA contamination in wine is urgent in order to avoid extreme values.

6.8 SCOPE FOR FUTURE RESEARCH

Apart from getting more and more insight on some of the problems closely related to the specific toxic compounds as described above, in order to prevent health damage, new scopes for investigators have come to light.

The consequences of climate change for enological productivity, land use and related policy are now being discussed worldwide. Climate change may produce positive effects on agriculture through the introduction of new crop species and varieties, but an increase in the need for plant protection, a risk of nutrient leaching and the turnover of organic matter in the soils may also happen. The effect of secondary factors of agricultural production and the interactions with the surrounding natural ecosystems would be the issues to be dealt with by researchers in the future, to assure better quality of crops and food safety.

ACKNOWLEDGMENTS

Contributors are financially supported by: Base Funding – UIDB/00511/2020 of the Laboratory for Process Engineering, Environment, Biotechnology and Energy – LEPABE – funded by national funds through the FCT/MCTES (PIDDAC).

BIBLIOGRAPHY

Ahmedullah, M. and Himelrick, D.G. (1990). Grape management. In: *Small Fruit Crop Management*, G.J. Galletta and D.G. Himelrick (eds.). Prentice Hall, Englewood Cliffs, NJ, p. 383.

Beach, F.W. and Carr, J.G. (1977). Cider and Perry. In: *Economic Microbiology. Vol. 6. Alcoholic Beverages*, A.H. Rose (ed.). Academic Press, London, p. 139.

Bisson, L.F., Waterhouse, A.L., Ebeler, S.E., Walker, M.A. and Lapsley, J.T. (2002). The present and future of the international wine industry. *Nature* 418(6898): 696.

Boulton, R.B., Singleton, V.L., Bisson, L.F. and Kunkee, R.E. (1995). *Principles and Practices of Winemaking*. Chapman & Hall, New York, p. 13.

Campos, I., Neale, C.M.U. and Calera, A. (2017). Is row orientation a determinant factor for radiation interception in row vineyards? *Australian Journal of Grape and Wine Research* 23(1): 77–86.

Degaris, K.A., Walker, R.R., Loveys, B.R. and Tyerman, S.D. (2017). Exogenous application of abscisic acid to root systems of grapevines with or without salinity influences water relations and ion allocation. *Australian Journal of Grape and Wine Research* 23(1): 66–76.

Fidelibus, M.W., Christensen, L.P., Katayama, D.G. and Verdenal, P.T. (2005). Performance of Zinfandel and Primitivo Grape wine Selections in the Central San Joaquin Valley, California. *American Journal of Enology and Viticulture* 56(3): 294.

Food and Agricultural Organization. (2016). FAOSTAT (food and agricultural organization of the United Nations statistic division: Rome, Italy). http://faostat3.fao.org/browse/Q/QC/E.

Fuentes, S. and De Bei, R. (2016). Advances of the vineyard of the future initiative in viticultural, sensory science and technology development. *Wine and Viticulture Journal*: 53–57.

Jackson, R.S. (2003). Wines: Types of table wine. In: *Encyclopedia of Food Sciences and Nutrition*, 2nd edn., B. Caballero, L.Trugo and P.N. Finglas (eds.). Elsevier Science, UK.

Joshi, V.K., Sandhu, D.K. and Thakur, N.S. (1999). Fruit based alcoholic beverages. In: *Biotechnology: Food Fermentation*, Vol. I., I.I.V.K. Joshi and Ashok Pandey (eds.). Educational Publishers and Distributors, New Delhi, pp. 647–744.

Kicherer, A., Klodt, M., Sharifzadeh, S., Cremers, D., Topfer, R. and Herzog, K. (2017). Automatic image-based determination of pruning mass as a determinant for yield potential in grapevine management and breeding. *Australian Journal of Grape and Wine Research* 23(1): 120–124.

Rathore, D.S. (1998). All India coordinated research project on subtropical fruits. In: *50 Years of Horticultural Research*, S.P. Ghosh, P.S. Bhatnagar and N.P. Sukumaran (eds.). ICAR, New Delhi, India, p. 141.

Reynolds, A.G., Wardle, D.A., Cliff, M.A. and King, M. (2004a). Impact of training system and vine spacing on vine performance, berry composition and wine sensory attributes of Seyval and Chancellor. *American Journal of Enology and Viticulture* 55(1): 84.

Reynolds, A.G., Wardle, D.A., Cliff, M.A. and King, M. (2004b). Impact of training system and vine spacing on vine performance, berry composition and wine sensory attributes of Riesling. *American Journal of Enology and Viticulture* 55(1): 96.

Smart, R.E., Dick, J.K., Gravett, I.M. and Fisher, B.M. (1990). Canopy management to improve grape yield and wine quality–principles and practices. *South African Journal of Enology and Viticulture* 11(1): 3.

Tinello, F. and Lante, A. (2017). Evaluation of antibrowning and antioxidant activities in unripe grapes recovered during bunch thinning. *Australian Journal of Grape and Wine Research* 23(1): 33–41.

Weaver, R.J. (1976). *Grape Growing*. Wiley-Interscience, New York.

Winkler, A.J. (1970). *General Viticulture*. University of California Press, CA.

Unit 2

Viticulture for the Winemaker

INTRODUCTION

Humans have been drinking wine for the last 9,000 years and using it as a medicine for the past 5,000 years, making it the oldest medicine. Thus, considering the parallelism between the history of civilization and that of wine, its impact on our society (past, present and future) is clear. For this reason, to continue investigating new effects of well-known compounds and, at the same time, searching novel bioactive molecules, appears fundamental. However, while the importance of wine as a final product is undeniable, it is also significant to focus our attention on the entire production process: the developing scientific and technological practices which are now affecting how grapes are grown and how wine is produced.

Unit 2 covers different aspects of vine growing and partially wine making. This section is characterized by seven chapters focused on topics including grape cultivars, genetic engineering, grape diseases and varietal influence on the quality of wine up to vineyard mechanization.

Chapter 7 reports how grape varieties and climate represent the indispensable keys to achieve success in making distinctive and high-quality wine. Worldwide, wines are produced from several different grape varieties, each one imparting specific characteristics. Different characteristics of typical fruits are described, also including various cultural practices needed for growing. Moreover, the discussion covers the concept of terroir, stating that the local environment influences the composition of grapes produced in a specific growing region. Thus, in contrast to other agricultural commodities, wine is marketed by the geographical location of production, and its quality is associated with minimal vineyard input or manipulation.

Several aspects of the alteration produced in the fruits, what techniques are used, how these are applied, what characteristics are produced and also includes future projections on genetic engineering of fruit crops (Chapter 8). More than two-thirds of grapes produced worldwide are used in making wine while the remaining part is either consumed as fresh berries or as raisins. Breeding programs have generated improved varieties/rootstocks characterized by a better fruit quality, disease resistance and tolerance to abiotic stresses. Molecular technology and genetic modification of grapevines either applied to new/common varieties offer the opportunity to improve specific traits of cultivars of known performance and acceptance.

The bouquet of wine is associated with the winemaker's knowledge in winemaking, stabilization and aging process, but mainly with the grape varietal character and its particular expression in a given terroir. Chapter 9 describes the evolution of grape berry and consequently, its chemical composition during the maturation process which is mainly dependent on the biosynthesis process of the diverse compounds and influences of the different environmental and viticultural practices. Grape berries present a complex chemical composition, and a great number of chemical compounds have a considerable impact on wine composition and quality (namely sugars, acids, phenolic compounds and aromatic compounds). Furthermore, main aspects of different methodologies used for grape maturation evaluation and quantification are also reported.

Chapter 10 describes how wine quality and its aroma depend upon several pre-harvest factors affecting growth and development of grape varieties as well as the subsequent enological processes. There is a limited number of varieties suitable for noble wines. How much these varieties and/or the enological practices may influence the quality of wine are reported. Beside this, there are a large number of grape berry components, including several organic substances, carbohydrates, phenolics, nitrogenous compounds, terpenoids, fats and lipids and other aroma compounds. Each variety has certain aromatic characteristics that, through a rigorous oenological process, can be maintained or enhanced in the finished product.

Chapter 11 covers the main diseases which infect grapevines worldwide, with special reference to wine grapes, and the consequences they can have on the enological process. Diseases represent a crucial factor to be considered when grapes are grown for enological purposes because they can cause substantial losses in production, also affecting wine quality. The symptoms of disease and the main effects of downy mildew, powdery mildew or bunch rots on wines are reported. Bunch rot management in wine grape production depends on a combination of cultural, biological and chemical control practices, which are thoroughly discussed throughout the text.

Chapter 12 describes one of the most important fungi in viticulture and enology: *Botrytis cinerea Pers.* (*B. cinerea*). Botrytized wines are produced all over Europe (mainly Germany, Hungary and France), South Africa and Australia. However, this fungus is responsible for causing bunch rot disease, known as noble rot, which is of great usefulness in the winemaking process. Various enzymes of this fungus are responsible for the degradation of pectin compounds and further alteration of cell walls. During rotting, aromatic compounds are highly modified and lead to the formation of characteristic compounds like 3-hydroxy-4,5-dimethyl-2-furanone, furfural, benzaldehyde, phenylacetaldehyde and benzyl cyanide. Sweet wines are prepared from botrytized grapes containing high concentrations of volatile thiols and are famous for their distinctive aroma.

Labor accounts for a considerable fraction of the production costs of most fruits, including grapes. Manual harvesting is particularly laborious, and was thus, among the first tasks to be mechanized. Starting from the 1960s, when commercial grape harvesters first came into use, vine training and trellis systems have been developed to facilitate mechanical harvesting. Chapter 13 is focused on vineyard mechanization and how its importance and diffusion is constantly growing among grape growers worldwide. Since mechanization affects many of the key factors identified with terroir, an analysis of the interaction between vineyard mechanization and terroir may help in this direction.

Matteo Bordiga, PhD
Associate Editor
Department of Pharmaceutical Sciences
Università del Piemonte Orientale (UPO)
Novara, Italy

7 Winemaking
Fruit Cultivars

K. Kumar, R. Kaur and Shilpa

CONTENTS

7.1 INTRODUCTION

Wine production is a fine blend of art, skill and science emanating from individual creativity coupled with innovative technology evolved over thousands of years. Most of the vine is produced between 30° and 50° N and 30° and 50° S of the Equator, with major countries France, Italy, Spain, the United States, Chile, Argentina, South Africa, Australia and New Zealand, etc. having warm summers and relatively mild winters. Grapes represent the world's largest fruit crop. More than 77 million tonnes were produced in 2013, and after wine was made, three to six million tonnes of grape marc per year in the 2000–2013 period remained. All of these wine industry by-products, including skins, seeds, stems and lees, are rich in phenolic antioxidants as reported by Food and Agricultural Organization. Suitable grape varieties and climate are the indispensable requirements to achieve success in making distinctive and high-quality table wine. The French concept of terroir states that the composition of grapes produced in a specific growing region will be influenced by the local environment, which will carry through to the wines of the area. Thus, in contrast to other agricultural commodities, wine is marketed by the geographical location of production, and quality is associated with minimal vineyard inputs or manipulation. Optimal yield combined with high disease resistance and a good climatic adaption are the major objectives in wine grape breeding.

Globally, wines are made from hundreds of different grape varieties, each one imparting specific characteristics. In some regions, one grape variety is used to make a wine; in others, winemakers blend several varieties in a single wine. Differences between grape varieties are the largest and most easily recognizable factors which affect wine composition and quality. Essentially all the great table and dessert wines of the world, whether varietal labeled or not, owe their important characteristics in large part to a specific grape variety or a small group of grape varieties. Nuances that make the wines vary by vineyard, climate and vintage are important, but they tend to be small compared to the differences possible among grape varieties. Clearly the choice of varieties is a crucial one. The wine industry of the world is based primarily on one European grape species *i.e. Vitis vinifera*. A few may disagree, but all of the world's greatest and most popular wines come from the fruit of *Vitis vinifera* vines. Other grape species, such as *V.labrusca, V.rotundifolia, V.rupestris* and *V.riparia*, have been used to make wine or to produce hybrids by crossbreeding with *V.vinifera*. Nevertheless, vinifera wines are famous worldwide along with local interest as the native species are well adapted and resistant to pests and diseases. Here in this chapter, fruit cultivars used in wine making and associated aspects are discussed within the scope of this text.

7.2 WINEGRAPE VARIETIES

Aromas and tastes of wines can be affected by several factors: grape variety, vineyard location, soil type, method of growing, weather conditions, wine making techniques, winemaking skill, etc. Of these, perhaps the single most important factor is the variety or varieties of grape. Varietal origin sets the outer limits of wine's sensory attributes, followed by vintage conditions and production style. There are some 10,000 documented *vinifera* varieties, out of which 200 varieties are described to be suitable for wine making. However, there may not be more than 50 such varieties which are extensively used in wine making on a commercial scale in the major wine-producing regions of the world. Thus, wines of high reputation are often regarded as associated with a single grape variety, often referred to as a "varietal" especially in the English-speaking world. An attempt has been made to highlight distinguishing characteristics of globally famous wine grape varieties (Plates 7.1 and 7.2), with emphasis on their origin, distribution, berry characters and style of wine produced from each one of them.

7.3 WHITE WINE VARIETIES

7.3.1 Chardonnay

The queen of international white varieties is employed to make white Burgundy, Champagne and many New World wines, which makes it the most famous white member of the Noirien family of cultivars. It not only produces wines with an appealing fruit fragrance, but also tends to do well worldwide. Under optimal conditions, the wine develops aspects reminiscent of various fruits, including apple, peach and melon. But it is a less vigorous vine than Sauvignon Blanc, with leaves being a little larger and paler gold in colour. The clusters are smaller and are made up of smaller, amber-yellow berries in very tight bunches, having a sugary flavour. It buds early (March–April), thus running the risk of being affected by spring frosts, but has comparatively more cold hardiness than other vinifera cultivars in cooler climates.

PLATE 7.1 White Winegrape Varieties. *Courtesy:* http://www.bbr.com

7.3.2 Chenin Blanc

This is the main white grape of the Loire region, where it is used to make both dry and sweet wines, known as *Pineau de la Loire*. A widely grown variety called Steen in South Africa and White Pinot (Pinot Blanco) elsewhere in the world, it produces very fruity berries if harvested very ripe. The berries are golden, with fairly thick skin, firm and aromatic (grapefruit, cinnamon, hazelnut, honey).

7.3.3 Gewurztraminer

This is a very old vine variety, and it is thought that in ancient Rome it was cultivated under the name of *V.aminea*. A clone of the parent cv. Traminer, grown widely and a famous white wine variety of Alsace, France, it does well in the cooler coastal regions of western United States, Australia, New Zealand and British Columbia, Canada. It has a characteristic primary aroma before vinification, but after fermentation it achieves its aromatic splendor (roses, litchi, acacia, jasmine). It is a vigorous grower, with thick and large leaves which resist frost. The clusters are long and straggly, with small to medium sized berries. Its berries are pinkish and small with thick skin. As a fruit, it is delightful to eat, with a sweet, pungent and spicy character.

7.3.4 Muller Thurgau

This is possibly the most well known modern *V. vinifera* cultivar, constituting about 30% of the German hectarage. It is considered to have evolved from a cross between ‘Riesling’

PLATE 7.2 Red Winegrape Varieties. *Courtesy:* http://www.bbr.com

and 'Chasselas de Courtillier'. Its mild acidity, subtle fruity fragrance and early maturation are ideal for producing light wines in cooler climates. It is widely planted in Europe (known as Rivaner), New Zealand and some cooler regions of North America.

7.3.5 Muscat Blanc

This is one of the many members of another 'cepage' family of Muscat cultivars and its clones. Their aroma is so distinctive that it is described in terms of the cultivar name Mubeaty. Because of the intense flavour, slight bitterness and tendency to iodize, clones of Muscat grapes have been used in making medium-sweet and dessert-style table or fortified wines. An example of these is 'Constatia', a century-old wine blend, still made in South Africa from the Orange Muscat grape. A local variation of this wine from the same clone is also made in California and Australia. The reduced bitterness and lower oxidation susceptibility of the new Muscat cultivars 'Symphony' permit the production of dry wines with a better aging potential. 'Moscato Bianco' is the primary variety used in the flourishing sparkling wine industry in Asti, Italy.

7.3.6 Parellada

This is a variety distinctive to the Catalonian region of Spain and produces an aroma that is apple to citrus-like in character, occasionally showing hints of licorice or cinnamon. Its bunches are compact, with average sized berries. It makes aromatic dry white wines of medium alcohol content, fresh aroma and delicate fruity acidity.

7.3.7 Pinot Blanc

Also a mutation of Pinot Noir grown in Alsace, parts of Burgundy and Austria (called Weissburgunder), this variety is used to make dry, crisp, rather intense white wines. In California, a similarly named grape is used to make a fruity, rather subtle wine similar to the simpler versions of Chardonnay. It is used in many of the better champagne-style sparkling wines of California because of its acid content and clean flavour.

7.3.8 Pinot Gris

A variety from Burgundy, developed by mutation of Pinot Noir and cultivated throughout the cooler, climatic regions of Europe for the production of dry, botrytized and sparkling wines. Pinot Gris typically yields subtly fragrant wines with aspects of passion fruit. Gris meaning grey, the berries vary from almost golden through pink to almost blue. The clusters are small and tight, with the berries also being small. The fruit ripens earlier and has lower acid than almost all of the white wine varieties, but with careful handling, the unique flavours of stone fruit, such as apricots and peaches, and pear juice in older wines, are retained.

7.3.9 Riesling

A famous variety from Germany producing high quality dry and sweet white wines. Its complex floral aroma, commonly reminiscent of roses and pine has made it acclaimed throughout central Europe and much of the world. Outside Germany, its largest plantings are in California and Australia. It is one of the few varieties to display a distinctive primary aroma apparent before vinification. Vines are vigorous and moderately productive with cane pruning. The growing tip of the Riesling vine is bronzy, with quite short spaces between the leaves. The clusters are small, cylindrical and well filled. Its greenish or golden berries have thick skin according to maturity. Berries are juicy and aromatic in flavour and ripen in early midseason.

7.3.10 Sauvignon Blanc

This is a major French grape variety planted in the Bordeaux and upper Loire Valley but is being grown more and more all over the world. It retains its acid well into the ripening period, and this, with the characteristic flavours of gooseberry, passion fruit and capsicum, gives the wine its freshness and clean finish. The vines are vigorous with small, dark green leaves having five distinct indentations. Berries are small, beautifully golden, with thick skin and a slightly musky flavour, and are borne in medium sized clusters.

7.3.11 Semillion

A semi-classic grape widely grown in Bordeaux, France, that has a distinct fig-like character. In France, Australia and increasingly in California, it is often blended with Sauvignon Blanc to cut some of the strong "gooseberry" flavour of the latter grape and create better balance. It is also used to create dry single-varietal white wines. When infected by the 'noble rot' fungi (*Botrytis cinerea*), it is used to produce first class sweet white wines such as those of French Sauternes.

7.3.12 Traminer

This is a distinctively aromatic cultivar grown throughout the cooler regions of Europe and much of the world. Although possessing a rose blush in the skin, it produces a white wine and is fermented to produce both dry and sweet styles.

7.3.13 Trebhiano

This is the most widely planted white wine grape cultivar in Italy. High yields and disease resistance obtained throughout Italy makes it a popular variety. This variety has three clones: Tuscan, Romagnolan and Giallo. Like cv. Malvasia, it is also used in making Chianti, the finest wine of Italy.

7.3.14 Verdiccio

This is a variety grown in the Marche region in Italy since the 14th century. It makes a clean, crisp and virtually colourless wine. This grape also makes a good sparkling wine because of the natural acidity.

7.3.15 Viognier

This is a popular white wine cultivar in the United States and Australia but vanished from its home of the Rhone valley in France due to *Phylloxera* epidemic in the late 1800s. The wine matures quickly and is characterized by the development of a fragrant peach to apricot aroma.

7.3.16 Viura

This is a predominant white wine variety in Rioja, Spain and the Languedoc region of France. It is also known as Macabeo. In cooler areas, it produces a fresh wine possessing a subtle floral aroma with aspects of citron. After extended aging in large wooden cooperage, it develops a golden colour and a rich butterscotch to banana fragrance that characterizes the traditional white wines of Rioja.

7.3.17 Welschriesling

A white wine variety grown in former Yugoslavia and Hungary, it produces a pale, greenish straw-coloured wine and a neutral but vinous nose with flowery aroma. Table wines have a higher content of acidity; good quality wines are racy with crisp acidity, medium body and for easy drinking.

7.3.18 Xarel-Lo

One of the traditional white varieties in Spain, its spherical, golden berries have fleshy pulp with little flavour. If well made, it produces wines of an acceptable level of alcohol, which are distinguished by their fresh, fruity aroma (apple, lemon and grapefruit), typical of the 'terroir'.

7.4 RED WINE VARIETIES

7.4.1 Barbera

A semi-classic red grape commonly grown in the Piedmont region and most of northern Italy. It is now thought by some to resemble 'Perricone' or 'Pignatello', a variety from Sardinia. It was introduced in the United States in the 19th century and its plantings are mostly confined to the warm western coastal regions in North America. It usually produces an intense red wine with deep colour, low tannins and high acid, and is used in California to provide backbone for so-called "jug" wines. Century-old vines still exist in many regional vineyards and allow production of long-aging, robust red wines with intense fruit and enhanced tannin content.

7.4.2 Cabernet Sauvignon

Widely regarded as the emperor of red wines, is a well known member of the Carmenet family of grape cultivars which include 'Merlot' and 'Malbec'. It probably resulted from a chance cross with 'Sauvignon Blanc'. Under optimal conditions, it yields a fragrant wine possessing a blackcurrant aroma, but under less favourable conditions, it generates a bell pepper aroma. Cabernet Sauvignon produces long clusters of loosely-spaced, small, spherical berries with thick and tough skin. Their flesh is firm, crisp and astringent on the palate, with characteristic flavours of mint, blackberry and green plum.

7.4.3 Carignane

This is also known as Carinena and Mazuelo in Spain, and Gragnano in Italy. It is a semi-classic grape commonly used for making red wines in Southern France and Spain. It is also grown in California's Central Valley, often ending up in generic blends and 'jug' wines, although some old plantings allow small lots of premium extract wine to be made. Blended with other varieties such as Cinsaut, Grenache, Mourvedre and Syrah, it has been used to create French Rhone-style red wines in California similar to the famous Chateauneuf-du-Pape blend.

7.4.4 Gamay

This is a popular variety in the granite hills of Beaujolais, France and makes a light red wine with few distinctive characteristics. It is also grown in the Loire Valley in France, Switzerland and California. At its best, Gamay produces wine that is incomparably light, fruity and gulpable, and is pale red in colour. However, when processed by carbonic maceration, it yields a distinctively fruity wine.

7.4.5 Grenache

This is the second most planted red grape in the world, with major plantings in France, Spain and the United States. California mostly uses this red grape for jug wine blends, rose and the white Grenache. It is known as Garnacha in the Rhone Villages and Chateauneuf-du-Pape in northern Spain, and in Australia it represents the G in the Rhone-inspired GSM blends. It thrives well in dry, hot conditions, and its strong stalks make it well suited for windy conditions as well. When ripe, Grenache has a very high sugar level and can produce wines with 15–16 percent alcohol. The wines are sweet, fruity and contain very little tannins. Grenache Blanc is the white version of the variety and is not as popular as is its red version in California but widely planted in Spain and France. The white wines produced are high in alcohol and low in acid content.

7.4.6 Malbec

A semi-classic red grape grown in the Bordeaux region of France, known as 'Cot' in other areas and 'Auxerrois' in Alsace. It is also grown in California and Argentina where it is called 'Fer' and thought to be a clone of Malbec. It usually produces an inky red, intense, velvety wine. It is a concentrated wine of high alcohol content, fruit and extract content, along with soaring taste. But the vine planting is declining in France due to its susceptibility to frost, mildew and rot. The vine also requires tender loving care. That's why many producers have declined its propagation, when merlot, for instance, offers an easier proposition. The best wine is produced on infertile, high, rugged terrain, ideally with limestone soil. Cahors, in southwest France, provides such conditions, and here France's so-called 'black wines' are produced, which are full of rustic, gamey flavours and leathery nuances.

7.4.7 Merlot

A classic international grape variety widely grown in Bordeaux, France, and other countries like the United States, Canada, Chile, Argentina and New Zealand. It is an early budding and flowering variety and is a comparatively easy grape to grow. It produces excellent varietal wine and is distinguished by its fine nature, elegance and oily tannins of a velvety and rich nature. The red wine bears a resemblance to Cabernet Sauvignon wine with which it is sometimes blended, but this is usually not so intense, with softer tannins. Its tendency to mature more quickly has made it a popular substitute for 'Cabernet Sauvignon'. Merlot vines give long, straggly bunches of large, loose berries and very large leaves. Its berries are small, blackish-blue with medium thick skin and sweet flesh.

7.4.8 Monastrell

This is a legendary red grape variety adapted to Mediterranean climates and is widely grown in Spain. It produces heavy wines with deep colour and noble tannin. It is distinguished by its small black berries with a heavy bloom, thick skin and prickly-tasting flesh. Its ripeness requires perfect maturing, making it necessary to carefully choose the lay of the land and the position of the vineyard.

7.4.9 Nebbiolo

The Nebbiolo grape has proven to be one of the most difficult grapes to be grown in the world. Its name is derived from 'nebbia', an Italian word which means fog. It ripens late in the season (October), when Langhe hills in Piedmont are covered with fog, due to which it evolved into a thick-skinned grape. With traditional vinification, it produces a wine high in acid and tannin content that requires years to mellow. The colour has a tendency to oxidize rapidly. Common aroma descriptors include tar, violets, and truffles. Apart from producing one of the most sought-after wines in the world with its own grapes, Nebbiolo is also used as a blending grape with varieties like Bonarda, Arneis and Vespolina (Chiavennasca) grown in northern Italy and other areas.

7.4.10 Pinot Noir

This is one of the finest and most elegant red grape members of the Noirien family, but it is extremely difficult to grow and is particularly sensitive to environment, preferring cooler temperatures. It produces its 'typical' fragrance (beets, peppermint or cherries) only occasionally. Almost one third of all Pinot Noir grown is used for sparkling wines, especially in French and Californian champagnes. It exists as a varied collection of distinctive clones, and usually, the more prostrate, lower-yielding clones produce the more flavoured wines. The upright, higher yielding clones are more suited to the production of rose and sparkling wines. It sets in small, tight clusters of very small and violet black berries.

7.4.11 Sangiovese

This is an ancient cultivar consisting of many distinctive clones grown extensively throughout central Italy and is well known for the light to full bodied wines from Chianti but also produces many fine red wines elsewhere in Italy. Under optimal conditions, it yields a wine possessing an aroma reminiscent of cherries, violets and licorice. 'Sangiovese' is also grown under local synonyms, such as 'Brunello' and 'Prugnolo', used in producing Brunello di Montalcino and Vino Nobile di Montepulciano wines, respectively.

7.4.12 Shiraz/Syrah

This is a noble red wine grape of the Rhone valley in France. It is famous for yielding a deep red tannic wine with long aging potential in Australia, better known as Shiraz. Its wines are peppery with aspects reminiscent of violets, raspberries and currants. It sets in straggly, open clusters of large round berries. The vine is very productive and disease resistant. In order to retain quality, yield and productivity must be restrained, as deep, dark-fruit qualities fall as yield increases, along with aroma and acidity. In the southern Rhone and southern France, it is mostly found in blends with grenache, cinsault and mourvedre.

7.4.13 Temperanillo

This is perhaps one of best known Spanish red grape varieties used in the best Rioja wines, also grown under the names 'Cencibel' and 'Valdepenas'. It is also grown in Portugal (under the name Roriz), Argentina and California. In Spanish, temprano means 'tendency to ripen early'. Due to having a short cycle, it is not adversely affected by harsh climates. Under favourable conditions, it yields a delicate, subtle wine that ages well. It generates an aroma distinguished by a complex, berry jam fragrance, with nuances of citrus and incense.

7.4.14 Zinfandel

An important red grape variety, earlier known as Black St. Peter and grown extensively in California. Its origin is not clear however, and it is thought to be related to 'Primitivo' an Italian variety or 'Plavac' from Hungary. It is used to produce robust red wine as well as the very popular 'blush wines' called 'White Zinfandel'. Its wine is noted for distinct aroma and prickly taste characteristics in its red version and pleasant strawberry flavours when made into a 'blush' wine.

7.4.15 Local Wine Grape Varieties

There are few local grape varieties grown in Kinnaur, India which are used for wine making through traditional methods (Plates 7.3). However in recent years, some of the grape varieties developed through inter-varietal hybridization have been found to be suitable for making good quality wine (Table 7.1).

7.5 CULTIVATION PRACTICES

7.5.1 Effects of Vineyard Location

Commercial wine growing is largely confined to the north and south temperate regions of the earth. Although special techniques can extend these limits, lack of dormancy and problems with disease restrict viticulture on the equatorial sites, as does cold on the polar sides of these belts. Eliminating desert marshes, higher altitudes, overly rocky areas and prevailing conditions unsuitable for viticulture, large areas that are possibly available for viticulture are left. Generally, the best sites for grapes are those with full sun exposure, mild winter temperatures, freedom from frost and good soil drainage.

7.5.1.1 Climate

In general, wines produced from grapes grown in a hot climate can lack the fruitiness and complexity which is characteristic of wines from the same variety grown in a cooler climate.

Chholtu Red

Chholtu White

Roghi Black

PLATE 7.3 Local Winegrape Varieties from Kinnaur, India. *Courtesy:* Ms. Raj Kumari Negi

Fluctuating temperatures increase the degree of cold injury, especially when relatively warm autumns and early winters are followed by rapid or extreme temperature drops in midwinter. Spring frosts that occur after grapevines have broken bud and commenced shoot growth are not uncommon in wine grape growing regions. Severe frosts can sometimes kill the shoots and significantly reduce the ultimate fruit yield. A hot, humid growing season promotes the incidence of disease. Excessive moisture in the fruit maturation period often causes berry splitting and fruit decay. The impact of winter cold, spring frosts, high humidity and diseases can be reduced by appropriate variety selection, timely pesticide application, conscientious vineyard management and sprinkler irrigation, etc. Grapevines also respond to salinity through systemic internal disturbances, leading to reduction in vegetative growth and yield. Exposure to salinity or water deficit can lead to an increase in abscisic acid (ABA) of the grapevine, which can further reduce the flux of chloride from roots to shoots.

TABLE 7.1
Grape Varieties Suitable for Wine Developed in India

Hybrid	Parentage
G-68/2	Black Champa × Ruchi Red
Arka Sona (E-9/3)	Anab-e-Shahi × Queen of Vineyards
Arka Trishna (E-21/128)	Bangalore Blue × Convent Large Black
Pusa Navarang (76-2)	Madelein Angevine × Rubi Red
(76-3)	Maldelein Angevine × Rubi Red
(75-32)	Bangui Abayad × Perlette

Source: Rathore (1998)

7.5.1.2 Topography

Areas with gentle slopes which are elevated above surrounding areas are usually frost free. Cold air drains from elevated sites to lower areas, which reduces the risks of damaging frosts at elevated sites. Elevated sites also have better air drainage throughout the growing season, which promotes the rapid drying of foliage following dew or rain, greatly reducing the incidence of disease. However, steep slopes can create problems, particularly soil erosion and rendering the vineyard unfit for mechanized harvesting. In general, eastern, northern and northeastern facing slopes are considered superior to southern and western facing slopes. Being hotter, the latter advances bud break, thus making these wine vulnerable to frost damage. Bark splitting and trunk injury also occur more on southwest facing sites.

7.5.1.3 Soils

Grapes are adapted to a wide range of soil types from gravelly sands to clay loams, from shallow to very deep soils and those with low to high fertility. Optimum performance, however, is achieved if the vines have healthy, well-developed root systems. Soil conditions favourable to root growth include good aeration, loose texture, moderate fertility and good internal and surface drainage. Soils that are consistently wet during the growing season due to an impervious subsoil, high water table or other drainage problems should be avoided. Root growth in poorly drained soils is usually limited to the top two feet or less, whereas in deep, well-drained soils, roots may penetrate six feet or more. When root growth is restricted because of poor drainage, plant growth and fruit yields are generally low, and vine survival is limited to a few years. It is generally believed that famous high quality wines often come from vineyards with relatively low fertility. In fact, owing to such factors as excessively vigorous vine growth as opposed to fruit production, highly fertile soils are considered undesirable for wine vineyards.

7.5.2 Varietal Selection

Selection of variety is also important in determining the suitability of a given site. It primarily depends upon the genetics and biology of the variety, its ability to adapt under prevailing climatic conditions, wine characteristics and market value. No single variety can be recommended universally, nor should it be, considering the value of diversity among wines to maintain consumer interest. The number of varieties that are world famous for distinctive noble wines are relatively few, and even fewer are suited to any given vineyard or winery situation as described earlier (see Section 7.2).

7.5.3 Canopy Management

Grapevine produces fruits on fresh growth. It is a prolific producer of clusters and there is generally an over-abundant crop potential once the vine is large enough to produce fruits. High vegetative growth is counterproductive to desirable fruit production. Excessive vigour results in decreased yield and decreased wine quality. This is one reason to be skeptical that less vegetative growth gives the best wine as crop load is reduced. Heavy and dense grapevine canopies create a highly localized climate, distinctly different from that immediately outside the canopy. Uniformly positioned vine parts greatly facilitate the mechanization of vineyard operation and even simplify hand labour for certain practices.

Canopy management also includes spatial and three-dimensional characteristics including illuminated and shadowed backgrounds. In such cases, well-defined biophysical parameters, such as ground cover, leaf area index and fraction of photosynthetically active radiation absorbed by the plants (fPAR) play crucial roles.

7.5.3.1 Over-Cropping

If vines are over-cropped, the berries fail to develop normal sugar content by the time they are matured and picked. This can be overcome if harvesting is delayed, but in hot weather late harvesting results in flat wines with inadequate acidity. In cool weather, retained acid may be adequate but the wine remains unbalanced. Grape aroma and other constituents appearing late in ripening are likely to be deficient. On the whole, wine quality does get impaired due to excessive vegetative growth. In relation to both over-cropping and vigour management, the arrangement of the leaf canopy of the vines should be considered. Leaves should be directly exposed to the sun and not shaded by each other or other vine or trellis parts. It is important for the fruit to be exposed to light and air. Direct overhead sunlight puts the fruit at risk of becoming burned, but fruit buried within the canopy often fails to develop properly, particularly with regard to colour. Full fruit exposure by the positioning of clusters and the removal of basal leaf increases the sugar by nearly 1° Brix close to the time of maturity. The acidity may also be reduced proportionally to the additional ripening.

7.5.4 Training and Pruning

Fortunately, the grapevine makes itself amenable for training to a wide variety of shapes and forms. The training of the vine depends on two fundamental factors, namely the growth characteristics of a variety and the influence of the local climate on the growth of the variety. The purpose of training is to produce vines of a shape that facilitates cultivation, plant protection operations, pruning and harvesting, etc., that are economical to maintain and capable of producing fruit of the desired type and of good quality. The grapevine is a vigorous climber. If it is not properly trained and pruned, it does not bear fruit properly. Pruning is one of the most important operations in grape culture. The objectives of pruning are: to reduce the amount of old wood in order to keep the vine within manageable limits; to secure fruit-bearing branches in predetermined places; and to expose the fruiting branches to sufficient sunshine and to reduce the excessive vegetative growth.

7.5.4.1 Initial Training

It is important to properly train vines during the first few years of growth to establish a vine form that will be easy to manage. Regardless of the intended training system, the initial training of grapevines has different goals such as: to develop a large, healthy root system in the first year; to establish the initial components of the intended training system including at least one semi-permanent trunk in the second year; and to develop or complete the training system, harvest a partial crop and establish a second trunk in the third year. These goals can be achieved in the manner summarized here (also see Figure 7.1). After planting, but before growth begins, the top of the dormant plant should be pruned back to a single cane with two to five buds. After growth starts, all shoots except two to four of the best shoots should be removed. One or more of these retained shoots will become the trunks. Support should be provided for new shoots to keep them off the ground. This will greatly reduce disease problems and provide full sun exposure for maximum growth. The trellis should be established soon after planting to provide this support. String can be tied from a side shoot of the vine to the wires, and the new shoots should be wrapped around the string. Never tie around the main trunk of the plant, because the trunk will expand during the first growing season and can be girdled by the string. A stake can be driven next to each plant and the shoots tied to the stake instead of using string, especially if the trellis is not established in the first year.

7.5.4.2 Trellising

Grapevines require some kind of support for ease of management. This can vary from a simple wire trellis to an elaborate arbor. Vine support serves to hold the vine up where it can be managed and cared for efficiently and to expose a greater portion of the foliage to full sunlight which promotes the production of highly fruitful buds. Trellis design features

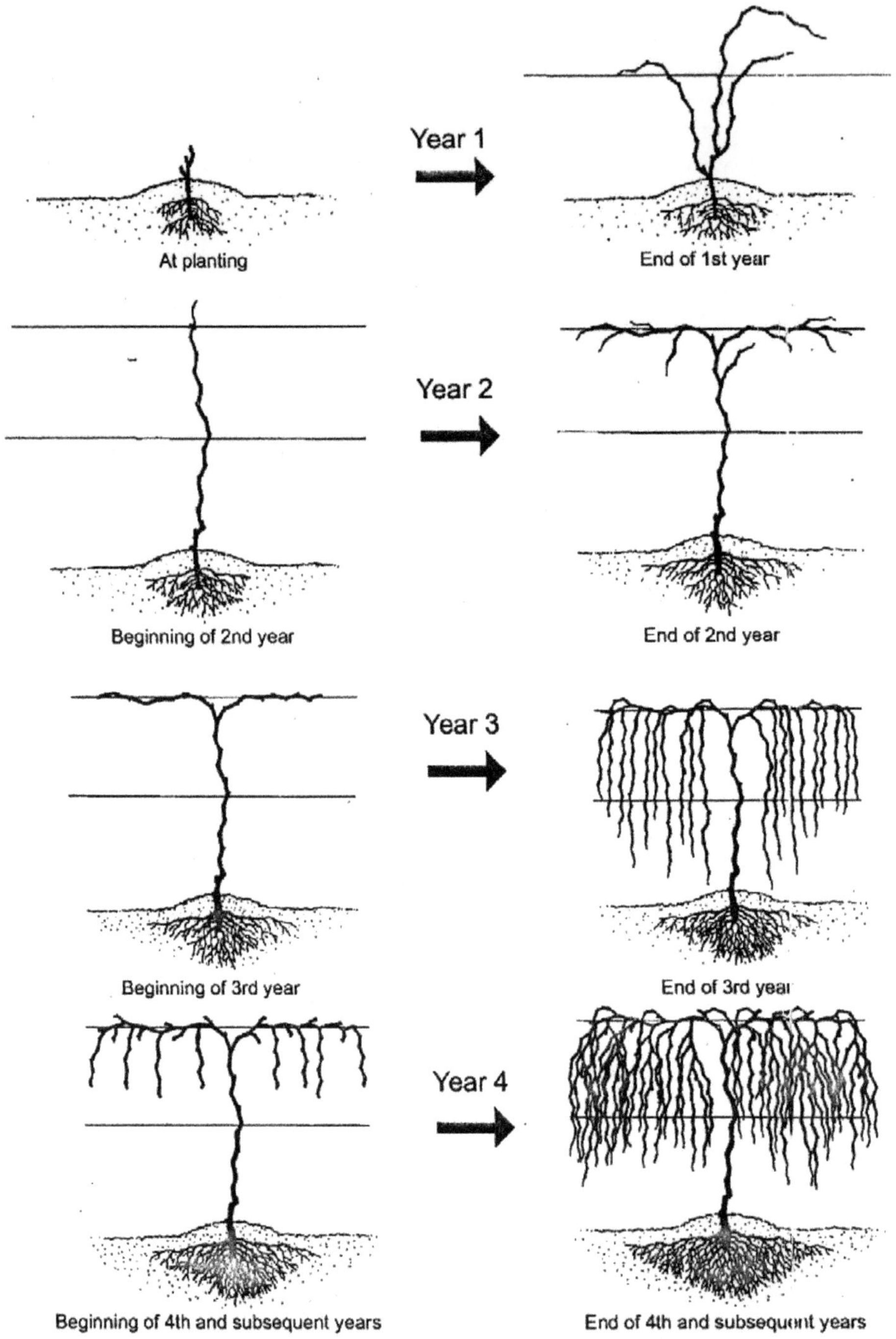

FIGURE 7.1 Pruning and training of a grapevine in initial years

such as number and placement of wires is dictated by the type of training system used. All training systems require a strong trellis assembly to support the weight of fruits and vines.

7.5.4.3 Training Systems

Like dormant pruning, grapevine training is essential for high quality grape production. The training system used depends on variety, costs involved, degree of mechanized harvesting, etc., besides location-specific problems. The most efficient methods provide well-spaced, even distribution of fruiting wood along the trellis and promote full sun exposure for clusters and basal nodes. The distinct features, merits and demerits of commonly used training methods can be seen in some of the well known literature on this subject. Appropriate grapevine canopy management is often hindered by excess vine vigour, an effect exacerbated by close vine spacing and inappropriate training. Recent studies on wine grape varieties Seyval, Chancellor and Riesling strongly suggest that high wine quality may be obtained from divided canopies.

7.5.4.4 Pruning

Grapevines require annual pruning to remain productive and manageable. There are two basic types of pruning: cane pruning and spur pruning differ only in the length of the fruiting wood that is retained and the method of training used to effectively display the fruiting wood. Although dormant pruning is the primary means of controlling the crop, it will not provide adequate control in all situations. Additional control through thinning of flower or fruit clusters is generally required with young vines (not more than two years old) with very fruitful varieties and in any case where the vigour and size of the vine are likely to fill the available trellis space.

To balance prune a grapevine, estimate the vine size, then prune the vine, leaving enough extra buds to provide a margin of error, usually 70 to 100 buds in total. Next, weigh the one-year-old cane prunings and use the pruning formula to determine the number of buds to retain per vine. After determining the appropriate number of buds to retain, prune the extra buds off, taking care of space the fruiting buds evenly along the trellis. Pruning formulas for many varieties have been developed to calculate the number of nodes to be retained for a given pruning weight based on its productivity. A pruning formula of 20+20 for example, would require leaving 20 nodes for the first pound of canes removed, plus an additional 20 nodes for each additional pound above the first. A 3.2-pound vine would therefore retain 64 nodes if the 20+20 schedule were used at pruning. In highly fruitful cultivars such as French hybrids, cluster thinning and shoot removal, as well as balanced pruning, may be required. Cluster thinning is the removal of all but one cluster per shoot. The basal cluster is usually the largest and should be retained. Select only canes or nodes that show good wood maturation. This criterion is far more important than selecting wood strictly on the basis of its location in relation to a desired training system. If other factors do not limit productivity, vines pruned correctly are likely to produce large crops of high quality fruit, but when pruned incorrectly, vines and crop will ultimately suffer.

7.5.4.5 Radiation Inception

Photosynthetically active radiation (PAR) absorption (APAR) and the fraction of PAR absorbed (fPAR) are affected by row direction (north-south, northeast-southwest, northwest-southeast and east-west) in vertical shoot-positioned trellises. The determination of the potential PAR absorption for any vineyard training system enables understanding of the effect of canopy management on vineyard productivity. This will also help in plot designing and vineyard development.

7.6 VINE BALANCE

The relation between vegetative growth (mass of dormant pruning wood) and generative growth (yield growth) is defined as vine balance. This is an important aspect of grapevine breeding in which mass of dormant pruning wood using an automated image-based method for estimating the pixel area of dormant pruning wood is explored on the basis of individual seedling evaluation. This represents a new, inexpensive, time saving and non-invasive tool for objective data acquisition, by evaluating digital images along with depth calculation and image segmentation.

7.7 ANTIBROWNING AND ANTIOXIDANT AACTIVITIES IN UNRIPE GRAPES

The use of unripe grapes with enhanced antioxidant properties to control anti-browning has created eco-friendly alternatives to various conventional treatments and additives. Previously used methods tended to alter the sensory, nutritional and health properties of the product. In one of the experiments, unripe berries were assessed during the bunch thinning stage. It was observed that the berries having the highest antioxidant activity in 2,2-diphenyl-1-picrylhydrazyl assays and ferric reducing ability during plasma assays were the most effective in preventing enzyme browning. Thus, unripe grapes which are otherwise discarded during bunch thinning are actually a significant source of bioactive compounds that are easy to produce and safe for human health, thereby converting these agricultural waste products to a beneficial purpose.

7.8 NEW TECHNIQUES FROM VINEYARD OF THE FUTURE (VOF)

VoF is a multinational project which has developed many new techniques like the VitiCanopy app, biological sensors in the vineyard and winery, modelling strategies using big data, robotic pourers and the use of unmanned aerial systems (UAS) and remote sensing in order to apply research directly into production strategies to obtain benefits and competitive advantages. VitiCanopy is a newly-developed smartphone and tablet/PC app; which estimates leaf area index crown cover, canopy cover, porosity and clumping index of grapevines on the basis of digital photography.

Modelling techniques and monitoring are used to assess the effects of climate change on various quality characters like effect of elevated temperature on berry cell death and shrivel and sensorial characters. The data obtained from weather, chemometry and wine quality traits have also been implemented to assess changes in the rotundone content, which produces a peppery aroma and is distributed within bunches and berries.

Novel robotic pourers and computer vision techniques to assess foamability and bubble dynamics of sparkling wines and beer have immerged as important new advanced technologies, as they can indicate direct and significant relationships between the foamability and the total protein content of wines. The robots are made using open hardware, smartphone cameras and affordable electronic sensors to obtain parameters, within controlled conditions, for conducting various farming operations such as pruning, crop monitoring, yield estimation and harverseting.

Computer vision analysis assesses the lifetime of foam, maximum volume and total life of foam, bubble count within foam, distribution of bubbles between small, medium and big, and colour of brewage. On the other hand, remote sensing performs an assessment of alcohol content, CO_2 content and the infrared temperature of brewage. From the output of these parameters gathered from different wines and beers, artificial neural network (ANN) models can be generated to automatically determine the specific types of fermentation or processes involved in the making of sparkling wines and beers.

Unmanned aerial systems have led to the development of automation processes, production of a pipeline of analysis from data uptake to automated processing algorithms and presentation of processed data in an app for smartphones and tablet PCs.

The development of all these techniques is highly promising for growers to have readily available, objective, affordable and reproducible methods to assess the quality and other characters of vines.

7.9 OTHER FRUITS SUITABLE FOR MAKING FRUIT WINES

It would be prudent to mention other fruit species, apart from grapes, which are used to make wine or other alcoholic beverages. These are usually small scale production in various parts of the world (Table 7.2).

TABLE 7.2
Fruits Other Than Grapes Suitable for Making Wine or Other Alcoholic Beverages

Fruit	Species
Mango	*Mangifera indica*
Banana	*Musa* spp.
Citrus fruits (orange, grapefruit, kinnow)	*Citrus* spp.
Plum	*Prunus salicina*
Ber	*Zizyphus mauritiana*
Apple	*Malus x domestica*
Jamun/Jambal	*Syzgium cumini*
Muskmelon	*C.melo*
Pomegranate	*Punica granatum*
Peach	*Prunus persica*
Kiwifruit	*Actinidia deliciosa*
Strawberry	*Fragaria x ananassa*
Raspberry	*Rubus ellipticus*
Sour cherry	*Prunus cerasus*
Date palm	*Phoenix dactylifera*
Apricot	*Prunus armeniaca*
Litchi	*Litchi chinensis*
Cider	
Apple	*Malus x domestica*
Perry	
Pear	*Pyrus communis*
Vermouth	
Mango	*Mangifera indica*
Plum	*Prunus palicina*
Sand pear	*Pyrus pyrifolia*
Apple	*Malus x domestica*
Brandy	
Apple	*Malus x domestica*
Plum	*Prunus salicina/ Prunus domestica*
Peach	*Prunus persica*
Apricot (wild)	*Prunus armeniaca*
Cashew	*Anacardium occidentale*

Source: Joshi *et al.* (1999)

7.9.1 Cider

Cider quality inevitably depends on the type of apple used. Cider is traditionally made with one third each of sweet, bittersweet and sharp apples. The principal characteristics of cider apples which contribute to this classification are the content of phenolic compounds (tannins) and the acidity. There are a great many varieties of apple (Table 7.3) which one can use for cider making, most of them now very rare. There are probably only ten or so varieties of apples widely grown for cider making. (For more details, see chapter 32 of this text).

7.9.2 Perry

Like apple cider, perry quality inevitably depends on the type of pear used. The classification of pears into different categories is more ambiguous than for apples. The best classification is probably that of Pollard and Beech, who defined the following categories: Sweet, Medium Sharp, Bittersweet and Bittersharp, although they state that the latter category would probably be better named as Astringent-sharp. The citric acid content of perry pears is also of importance but is not used for classification. Bittersharp (Astringent-sharp) pears have an acidity of greater than 0.45% (w/v) and a tannin content of greater than 0.2% (w/v). These pears have a penetrating flavour which is very striking since the tannin is astringent rather than bitter. This category is unsuitable for eating (due to the harsh flavour) but makes the best perries. A considerable number of varieties of pear suitable for perry making, many of which are now very rare, are Arlingham Squash, Barnet, Blakeney Red, Brandy, Brown Bess, Butt, Flakey Bark, Gin, Green Horse, Hellens Early, Hendre Huffcap, Judge Amphlett, Moorcroft, Newbridge, Oldfield, Parsonage, Red Longdon, Red Pear, Rock, Taynton Squash, Thorn, Turner's Barn, Winnal's Longdon and Yellow Huffcap.

TABLE 7.3
Common Cider Apple Varieties

SWEETS	BITTERSWEETS	SHARPS	BITTERSHARPS
Neutral	*Tannic, astringent*	*Acidic, tart*	*Tannic, acidic*
T<0.2, A<0.45	T>0.2, A<0.45	T<0.2, A>0.45	T>0.2, A>0.45
Cider apples	**Cider apples**	**Cider apples**	**Cider apples**
Berkeley Pippin	Ashton Brown Jersey	Breakwell	Cap of Liberty
Court Royal	Ball's Bittersweet	Brown's apple	Dufflin
Eggleston Styre	Bedan	Coleman's Seedling	Foxwhelp
Geeveston Fanny	Broadleaf Norman	Dymock Red	Improved Foxwhelp
Peau de Vache	Brown Snout	Fair Maid of Devon	Kingston Black
Pomme Gris	Bulmer's Norman	Frederick	Stoke Red
Sweet Alford	Cimitiere	Hereford Redstreak	Worcester Pearmain
Sweet Coppin	Chisel Jersey	Ponsford	
Vagon Flocher	Cow Jersey	Tom Putt	
Wayne	Dabinett	Winter Stubbard	
Woodbine	Gilpin	Yellow Styre	
	Harry Masters' Jersey, Knotted kernel, Medaille D'Or, Michelin, Nehou, Porter's Perfection, Reine des Hatives, Reine des Pommes, Royal Wilding, Sherrington Norman, Somerset Redstreak, Stembridge Jersey, Taylor's, Tremlett's Bitter, Vilberie, Yarlington Mill	York Imperial	
Standard apples	**Standard apples**	**Standard apples**	**Standard apples**
Baldwin, Ben Davis, English Golden, Russet, Fameuse, Grimes Golden, Hubbardston, McIntosh, Rambo, Rome Beauty, Roxbury Russet, Sops of wine, Stark, Westfield Seek No, Further, Winter Banana	Lindel,Newtown Pippin, Red Astrakhan	Bramley's Seedling, Cox's Orange, Pippin, Crimson King, Esopus Spitzenberg, Gravenstein, Jonathan, Northern Spy, Rhode Island, Greening, Ribston Pippin, Stayman, Wealthy, Winesap	Crabapples, Dolgo, Hagloe, Joeby, Martha, Red Siberian, Transcendant

BIBLIOGRAPHY

Ahmedullah, M. and Himelrick, D.G. (1990). Grape management. In: *Small Fruit Crop Management*, G.J. Galletta and D.G. Himelrick (eds.). Prentice Hall, Englewood Cliffs, NJ, p. 383.

Beach, F.W. and Carr, J.G. (1977). Cider and Perry. In: *Economic Microbiology. Vol.6. Alcoholic Beverages*, A.H. Rose (ed.). Academic Press, London, p. 139.

Bisson, L.F., Waterhouse, A.L., Ebeler, S.E., Walker, M.A. and Lapsley, J.T. (2002). The present and future of the international wine industry. *Nature* 418(6898): 696.

Boulton, R.B., Singleton, V.L., Bisson, L.F. and Kunkee, R.E. (1995). *Principles and Practices of Winemaking*. Chapman & Hall, New York p. 13.

Campos, I., Neale, C.M.U. and Calera, A. (2017). Is row orientation a determinant factor for radiation interception in row vineyards? *Australian Journal of Grape and Wine Research* 23(1): 77–86.

Degaris, K.A., Walker, R.R., Loveys, B.R. and Tyerman, S.D. (2017). Exogenous application of abscisic acid to root systems of grapevines with or without salinity influences water relations and ion allocation. *Australian Journal of Grape and Wine Research* 23(1): 66–76.

Fidelibus, M.W., Christensen, L.P., Katayama, D.G. and Verdenal, P.T. (2005). Performance of Zinfandel and Primitivo Grape wine Selections in the Central San Joaquin Valley, California. *American Journal of Enology and Viticulture* 56(3): 294.

Food and Agricultural Organization. (2016). FAOSTAT (food and agricultural organization of the United Nations statistic division: Rome, Italy). http://faostat3.fao.org/browse/Q/QC/E.

Fuentes, S. and De Bei, R. (2016). Advances of the vineyard of the future initiative in viticultural, sensory science and technology development. *Wine and Viticulture Journal* 31: 53–57.

Jackson, R.S. (2003). Wines: Types of table wine. In: *Encyclopedia of Food Sciences and Nutrition*, 2nd edn., B.Caballero, L.Trugo and P.N.Finglas (eds.). Elsevier Science, Norwich, UK.

Joshi, V.K., Sandhu, D.K. and Thakur, N.S. (1999). Fruit based alcoholic beverages. In: *Biotechnology: Food Fermentation*, Vol. I., I.I.V.K. Joshi and Ashok Pandey (eds.). Educational Publishers and Distributors, New Delhi, pp. 647–744.

Kicherer, A., Klodt, M., Sharifzadeh, S., Cremers, D., Topfer, R. and Herzog, K. (2017). Automatic image-based determination of pruning mass as a determinant for yield potential in grapevine management and breeding. *Australian Journal of Grape and Wine Research* 23(1): 120–124.

Rathore, D.S. (1998). All India coordinated research project on subtropical fruits. In: *50 Years of Horticultural Research*, S.P.Ghosh, P.S.Bhatnagar and N.P.Sukumaran (eds.). ICAR, New Delhi, India, p. 141.

Reynolds, A.G., Wardle, D.A., Cliff, M.A. and King, M. (2004a). Impact of training system and vine spacing on vine performance, berry composition and wine sensory attributes of Seyval and Chancellor. *American Journal of Enology and Viticulture* 55(1): 84.

Reynolds, A.G., Wardle, D.A., Cliff, M.A. and King, M. (2004b). Impact of training system and vine spacing on vine performance, berry composition and wine sensory attributes of Riesling. *American Journal of Enology and Viticulture* 55(1): 96.

Smart, R.E., Dick, J.K., Gravett, I.M. and Fisher, B.M. (1990). Canopy management to improve grape yield and wine quality-principles and practices. *South African Journal of Enology and Viticulture* 11(1): 3.

Tinello, F. and Lante, A. (2017). Evaluation of antibrowning and antioxidant activities in unripe grapes recovered during bunch thinning. *Australian Journal of Grape and Wine Research* 23(1): 33–41.

Weaver, R.J. (1976). *Grape Growing.* Wiley-Interscience, New York.

Winkler, A.J. (1970).*General Viticulture.* University of California Press, California.

8 Gendtic Engineering in Grapes

R. Kaur, K. Kumar, D.R. Sharma and Shilpa

CONTENTS

8.1 INTRODUCTION

Grape belongs to genus *Vitis,* a member of family Vitaceae. More than two-thirds of over 66 million metric tonnes of grapes produced from an area of 7.3 million hectares are used in making wine, while the remaining one-fourth is either eaten as fresh berries or as raisins after drying. Breeding programs have yielded improved varieties/rootstocks with better fruit quality, disease resistance and tolerance to abiotic stresses. Unlike in table and raisin industries, adoption by growers and wineries of newly released wine grape varieties has been extremely slow, owing to the distinctness and labeling of wine grapes which cannot undergo alteration in genetic makeup of genotypes through conventional breeding. Molecular technology and genetic modification of grapevines applied either to new varieties or to existing cultivars offer the opportunity to selectively improve specific traits of cultivars of known performance and acceptance. Directed genetic modification of existing cultivars by the introduction of single genes may result in improved genotypes that will be accepted as variants of the original cultivars. Thus, genetic transformation offers significant opportunities for the genetic improvement of grapes.

8.2 METHODS OF GENETIC TRANSFORMATION

Successful generation of transgenic plants of grapevine, or for that matter of any other plant, requires the combination of a cloned gene along with suitable elements such as promoters, enhancing and targeting sequences for its regulatory expression, a reliable method for delivery and stable integration of DNA into cells and a tissue culture method to recover intact plants, either from fully dedifferentiated tissues or more importantly from organs or tissues that are easy to regenerate from transformed methods. A variety of strategies have been developed over the past two to three decades to identify, isolate and clone desirable gene sequences from almost any organism. It is even possible to design and synthesize genes or gene sequences using automated machines. Also, the insight gained from molecular biological studies permits us to efficiently tailor genes from other organisms such as viruses, fungi, bacteria, animals etc., so that they express efficiently in transgenic plants. Some of the methods (for more details see selected references) to deliver DNA into cells are briefly explained here:

8.2.1 *AGROBACTERIUM*-MEDIATED GENE TRANSFER

The first major clue of the ability of *Agrobacterium* to form tumors or hairy roots is linked to the presence of tumor-inducing (Ti) or root-inducing (Ri) plasmids. Virulent strains of *A.tumefaciens* carry a large (200–250 kb) plasmid called Ti (tumor inducing) plasmid. Upon infection, a precise piece of DNA, the T-DNA or transferred DNA from the Ti-plasmid, is transferred to the plant cell and is targeted to the nucleus where it gets stably incorporated into the host chromosome. In wild type strains, T-DNA (~20 kb) carries genes for the biosynthesis of plant growth regulators – auxin and cytokinin. Also T-DNA contains genes coding for specific amino acid or sugar derivatives called opines. Using host cell machinery, the genes carried on the T-DNA are transcribed and translated. The overproduction of plant growth regulators by the infected/transformed cells leads to uncontrolled cell proliferation that causes tumors.

The key elements essential for T-DNA transfer have been described here. T-DNA is flanked by direct repeats of 24 to 25 bp. Usually all DNA between these border sequences are transferred to the host cell. Borders are the only sequences needed in *cis* for DNA transfer. By eliminating the undesirable DNA (*e.g.* genes for auxin, cytokinin or opine biosynthesis) present in the T-DNA of wild type strains and instead by inserting desirable genes, novel 'disarmed' strains of *Agrobacterium* have been developed for plant transformation. These strains do not produce tumors and allow regeneration of complete plants from transformed cells. These are present on Ti-plasmid of wild type strains and are necessary for cell infection, for excision, transfer and integration of T-DNA into the host chromosome. The *vir* genes are located outside the T-DNA borders and can function in *trans*. A number of *vir* genes have been characterized, and their exact role in transformation process has been identified. Many genes present on the bacterial chromosome play important roles in early stages of host recognition (chemotaxis) and attachment of bacteria to host cells of the plant. Chromosomal genes also have a role in transformation. These genes are necessary but can function in *trans* to bring about transformation.

Besides *A.tumefaciens, A.rhizogenes* is also capable of plant transformation by a similar process. Instead of tumors, *A.rhizogenes* causes "hairy-roots" upon infection due to production of auxins in transformed cells. *A.rhizogenes* strains have also been engineered for transformation but are not widely used. *Agrobacterium* mediated transformation, being simple and widely applicable, remains the method of choice for transformation of plants. Provided with a suitable protocol for regeneration of complete plants from transformed cells, a large number of transformants can be generated.

8.2.2 Biolistic – Mediated Transformation

Transformation of plant cells through micro-projectiles coated with DNA was developed almost at the same time that *Agrobacterium*-mediated transformation was being perfected. In this method, heavy DNA coated particles are mechanically "shot" directly into the plant tissue through the physical barriers of the plant cell wall and membrane. The most commonly used particles are tungsten or gold with a diameter between one and four μm. The particles are surface coated with DNA by precipitation with spermidine or polyethylene glycol and are accelerated at a high velocity from a custom-made device using either an electrical discharge, high gas pressure (such as compressed air, nitrogen or helium) or gun powder. The particles penetrate the cells, and the adsorbed DNA is delivered into the cells. Use of biolistic gun is relatively simple and rapid. Its main advantage is its non-specificity resulting from the physical properties of the method. For example, the particle gun method can be used to directly transform plant species like cereals which are recalcitrant to *Agrobacterium* mediated DNA delivery. Moreover, any type of cells can be targeted including embryos, pollen grains, microspores, leaf, stem, apical meristem, etc.

In this respect, biolistic-mediated gene transformation outdoes the *Agrobacterium* method because the latter specifically brings about nuclear transformation only and predominantly in the dicots. Though particle bombardment has yielded success with a range of plant species, the high cost of equipment and consumables make it expensive for individual laboratories with limited resources. Although DNA can be efficiently delivered, recovery of transgenics is rather low. Nevertheless, successful genetic transformation has been achieved by biolistic gun, mainly when host range restrictions limit the use of *Agrobacterium*.

8.2.3 Electroporation

Electroporation is perforation caused in the cell membranes by the use of electric current. Application of a short, high voltage electric pulse to cells leads to reversible membrane breakdown and pore formation, making the cells permeable to macromolecules. These pores are transient and reseal by natural processes. Thus, if cells are suspended in a solution containing DNA, electroporation will aid movement of the DNA into the cells. With the development of protoplast technology, electroporation was used to transform protoplasts. Later, the technique was applied to even intact tissues. This technique is basically simple. The tissue to be electroporated is held in a buffer containing the particular DNA, and a short pulse (micro to milliseconds duration) of high voltage (300–10,000 v/cm) direct current is applied to facilitate pore formation for DNA uptake.

8.2.4 Polyethylene Glycol (PEG) Mediated Uptake of DNA

PEG is widely used for fusion of animal cells and plant protoplasts. In a general procedure, protoplasts are isolated, and a particular concentration of protoplast suspension is taken in a tube followed by the addition of plasmid DNA (donor and carrier). To this, 40% PEG 4000 (w/v) dissolved in mannitol and calcium nitrate solution is slowly added because of its high viscosity, and this mixture is incubated for a few minutes. The application of PEG, which has a high affinity for water, to protoplast suspension leads to a contraction of their volume, resulting in endocytic vesiculation. DNA-cation complexes adsorbed to the membrane thus get introduced into the cells. PEG-mediated transformation is perhaps the least expensive of all the methods of transformation available at present.

8.2.5 Silicon Carbide Fibers

A simple alternative to microinjection, silicon carbide fiber-mediated transformation involves vortexing of cells and fibers in a buffer that contains DNA. The fibers puncture holes into the cells, and perhaps also through the force of vortexing, the DNA enters into the cell. It is a relatively simple and recent method and hence, has not been widely tested. However, the hazardous nature of silicon fibers may hamper its routine use.

8.2.6 Laser-Mediated Transformation

Use of focused, high-energy laser in surgery is well established. On a similar principle, laser is used to bore micropores into cells which are thus induced to take up DNA from the surrounding medium. The physiological condition of the external medium is manipulated to favor movement of macro molecules into the cells. However, the method is not well worked out or demonstrated for wider applicability, though the precision achievable with laser may prove useful in delivering DNA into cell organelles.

8.2.7 CRISPR/Cas Immune System Engineering

Clustered Regularly Interspaced Short Palindromic Repeats (CRISPR) is a prokaryotic defence system targeting the DNA of invading viruses and plasmids. This is a genome editing technology. In this system, CRISPR associated protein 9 (Cas9), which is an endonuclease, is directed to cut the invading DNA at a particular target, where the DNA sequence matches the sequence of RNA guide strand (gRNA) associated with Cas9. It has commonly been used for targeted mutagenesis and targeted modifications making very modest changes in existing genes in live cells. As genetic changes can be as limited as a single nucleotide change with no trace of introduced DNA, it is therefore possible to create a non-transgenic gene edit that cannot be distinguished from a mutation that has naturally occurred or that was introgressed through conventional breeding. Because of these reasons, many experts consider such gene edits as excludable from GMO regulation, arguing that they should be clearly distinguished from GMOs, referring to them as genetically edited crops (GECs).

8.3 GENETIC TRANSFORMATION SYSTEM IN GRAPES

Early transformation studies in grapes were unsuccessful. Grape is a host for *Agrobacterium* and adventitious regeneration can be induced from a range of explant tissues. The 1990s marked the era of development of transgenic grapevine rootstocks and scion cultivars. Success in grapevine transformation came only when researchers started using what are termed embryogenic cultures. To insert genes into embryogenic cultures, most researchers working on grape transformation rely upon modified strains of *Agrobacterium*. The efficiency of transformation in grapes is highly influenced by the strain, the genotype and culture conditions.

Some workers rely upon the biolistic process, whereby DNA-coated particles of an extremely minute size are used to carry foreign genes into grapevine cells. DNA coding for the genes of interest is coated onto the minute tungsten-microprojectiles. These are accelerated at extremely high speeds into the cultured cells using a biolistic device also known as "gene gun". There are usually several genes transferred into each cell penetrated by a microprojectile. One of these genes might be the gene of interest coding for a desired trait. Biolistic gene transfer is a more efficient method than *Agrobacterium*-mediated transformation, though it is less exploited in grapevines perhaps owing to high costs involved.

On the whole, genetic transformation of *Vinifera* grapes has been more difficult than of non-*Vinifera* grapes using embryogenic cultures, but the number of genotypes transformed is increasing steadily. Figure 8.1 depicts a summary of the processes involved in grapevine transformation, as

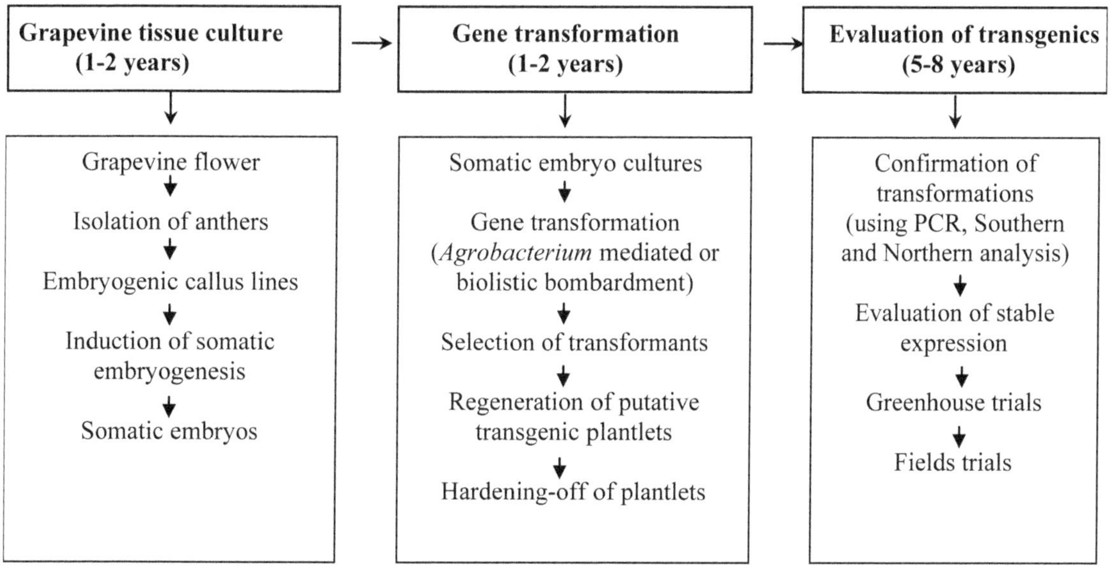

FIGURE 8.1 Summary of processes involved in grapevine transformation and the generalized time scale involved. *Source:* Vivier, M.A. and Pretorius, I.S. (2002). *Trends in Biotechnology*, 20: 472.

well as a generalized time scale of the individual components through to field-testing of the grapes.

8.3.1 Gene Introduction and Expression

Selection of transformed cells from non-transformed cells is a critical step in the recovery of transgenic plants successfully. A reporter gene such as uidA (GUS) is usually the first gene introduced when a transformation system is being developed for a particular genotype. Transformed cells can be selected by using genes like antibiotic resistant genes including npt II for kanamycin and paromomycin resistance, hpt for hygromycin resistance, bar gene for phosphinothricin resistance.

8.3.1.1 Gene Transfer for Specific Traits

The current approach of single gene transfers into plant genomes is perhaps best suited to the aim of enhanced disease resistance because single genes can confer disease resistance to plants. Most transformation strategies involve a gene product with known anti-pathogenic activity, that is introduced at high copies or in an inducible manner into the host of choice in an attempt to optimize parts of the plant's innate defense response. The other major approach of manipulated disease tolerance in grapevine and other plants relies on pathogen derived resistance and various applications thereof. In this approach, a pathogen derived gene and its encoding product is expressed at an inappropriate time or in an inappropriate form or amount during the infection cycle, thus preventing the pathogen from maintaining infection. Most of the antiviral strategies rely on some aspect of pathogen derived resistance and constitute a major portion of the activity in the genetic transformation of grapevine varieties. Researchers at the University of Florida introduced a silkworm gene into grapevines to make them resistant to Pierce's disease. Sugar in the ripe grape berry is an important quality factor for the dried fruit, table and wine grape industries. The basic processes of berry ripening and, more importantly, the elusive ripening signals are being researched. The primary aim is to change the metabolic pathways and processes that are active in the ripening berry to increase the formation of desirable or novel products linked to the quality parameters of grapes. During grape berry ripening, sucrose is transported from the leaves to the berry, where invertase converts it to the monosaccharides such as glucose and fructose that get accumulated in the vacuoles of the berry cells. Yeast invertase gene (suc II) under the control of a synthetic gibberellin promoter has been successfully introduced in grapevine, and transgenic plants have been obtained. Anthocyanin content is also a quality parameter for red and black grapes. Attempts have been made to introduce gene constructs to silence the polyphenol oxidase (PPO) gene. The PPO enzyme has been shown to be associated with browning of grape berries. Juvenility of transgenic grapevines requires three or more years before fruit appearance on field planted vines, which is a limitation when investigating genes that modify berry quality.

Tight control of introduced genes in transgenic plants, using gene promoters, will be necessary for optimum effectiveness of the introduced trait to prevent any side effects on overall plant performance or unwanted characteristics in grape products. Many of the genes introduced into grapevine are controlled either by the cauliflower mosaic virus (CaMV) 35S promoter or the nopaline synthase (NOS) promoter. Both of these promoters can be described as giving constitutive gene expression patterns. A number of gene promoters have been isolated from grapevine and some of them may have application as gene expression regulators in transgenic grapevines. The grapevine resveratrol synthase promoter has been isolated and analyzed in other species to understand its response to pathogens and environmental stimuli. A promoter controlling the expression of an alcohol dehydrogenase gene in grape berries has also been isolated. While promoters which control gene expression in response to external stimuli can be tested in other species, grapevine promoters which determine tissue specific expression in organs such as the berry are more difficult to characterize and test, due to the time involved in producing transgenic grapevines and waiting three or more years for fruiting. One way of overcoming this limitation may be the use of transient expression, where micro-projectile bombardment is used to introduce promoter reporter gene constructs into tissue such as leaves or berries. For confirmation of transgene insertion and expression, molecular techniques such as PCR, Southern blot and Western blot analysis have been routinely employed. A large number of putative transformants can be screened using simple assays for expression, such as the fluorescent assay for chitinolytic enzymes used in 'Merlot' and 'Chardonnay' plants, produced by biolistic transformation. Only a few studies provide information on stability and expression of transgenes in grapevines.

8.3.2 Mining R Genes

Cisgenics is a way of genetic engineering which involves the transfer of genes obtained from a sexually compatible gene pool. This technique has immense importance in resistance breeding especially in crops like grape, where breeding through conventional methods is very difficult and time consuming. R genes can also be introduced through transgenesis even from crops that are not a part of normal breeding pool.

8.3.3 Production of Antimicrobial Compounds

In order to restrict pathogen activity, expression of genes for antimicrobial compounds leads to increased disease resistance. In grapes, transformation with the chitinase gene of *Trichoderma* sp. resulted in production of chitin degrading enzymes and provided resistance to diverse fungal diseases.

8.3.4 Genome Editing

CRISPR technique of genome editing was used successfully to target a susceptible gene *'MLO-7'* in grape cultivar Chardonnay to increase powdery mildew resistance. To achieve efficient protoplast transformation, the molar ratio of Cas9 and sgRNAs were optimized along with analysis of targeted mutagenesis insertion and deletion rate using targeted deep sequencing, leading to the conclusion that direct delivery of CRISPR/Cas9 RNPs to the protoplast system enables targeted gene editing together with generation of DNA-free genome edited grapevine plants. The CRISPR/Cas9 system has also been used for efficient knockout of the *L-idonate dehydrogenase* gene (*IdnDH*) involved in the tartaric acid pathway. Successful targeted mutagenesis in grape cv. Neo Muscat for phytoene desaturase (VvPDS) gene has also been achieved. DNA sequencing confirmed that the VvPDS gene was mutated at the target site in regenerated grape plants.

8.4 DEVELOPMENT AND RELEASE OF TRANSGENICS IN GRAPES

Significant progress has been made in developing molecular technologies for grapevine improvement, however, the research is still in its infancy and it is expected that it will be quite some time before an improved transgenic grapevine is released to commercial growers. Although few identified genes have been isolated that may genetically improve grapevines, attempts are already in progress to modify berry quality and improve plant health by providing tolerance or resistance to viruses, bacteria and fungal pathogens. Tables 8.1 and 8.2 summarize the development of transgenic grapevine cultivars and rootstocks. Despite the strong and persuasive scientific case for the use of gene technology in the improvement of grapes and wine, the wine industry has entered the 21st century without a single transgenic grapevine variety being used on a commercial scale. The development of genetic transformation methods for *V.vinifera* and the advent of technology with which entire genomes, transcriptomes, proteomes and metabolomes can be analyzed has undoubtedly opened new horizons for the wine industry. The accessibility of the grapevine genome in terms of applications in molecular biology is currently relatively restricted owing to the size of genome (483 mb distributed along 38–40 chromosomes) and specifically its complexity (only 4% of the genome is transcribed). However, the grapevine genome is currently targeted for intense study, with multinational consortia collaborating in several initiatives to render molecular markers as well as the complete sequencing of the *Vitis* genome. Here, it is important to note that the information and technology that currently exist for model plant systems have yet to be expanded to the much more complex genome of the grapevine, before all of the requirements and concerns of the producers, consumers and regulatory authorities can be addressed satisfactorily. Transgenic plants of grape cultivars 'Thompson Seedless', 'Riesling' and 'Chardonnay' are currently being field tested for disease resistance in the United States, Germany and France, respectively.

TABLE 8.1
Development of Transgenic Plants of *Vitis* Species and Rootstocks

Cultivar	Selectable marker	Gene product (trait of interest)
V.rupestris	NPTII	(GUS reporter gene)
	NPTII	(GUS reporter gene)
	NPTII	Coat protein (GFLV resistance)
	NPTII	Anti-sense movement protein (virus resistance)
V.rupestris, 110 Richter	NPTII	Coat protein, antifreeze protein (virus resistance, freeze tolerance)
110 Richter	NPTII	Coat protein (GCMV resistance)
	NPTII	Coat protein (GFLV resistance)
	NPTII	(GUS reporter gene)
	NPTII	Coat protein (virus resistance)
	NPTII	Replicase (virus resistance)
	na	Eutypine-reducing enzyme (Eutypa toxin resistance)
41B	NPTII	Coat protein, replicase, proteinase (GFLV resistance)
SO4	NPTII	Coat protein (GFLV resistance)
Georgikon 28	NPTII	(GUS reporter gene)
3309C	NPTII	Translatable, anti-sense, non-translatable coat protein (virus resistance)
V.riparia	NPTII	Translatable, anti-sense, non-translatable coat protein (virus resistance)
MGT 101-14	NPTII	Translatable, anti-sense, non-translatable coat protein (virus resistance)
5C Teleki	NPTII	Translatable, anti-sense, non-translatable coat protein (virus resistance)
Freedom MGT 101-15 5C Teleki	NPTII	GNA (Homopteran insect resistance)
V.x labrusca	NPTII	GUS reporter gene
Interspecific hybrids, 110 Richter	NPTII	Coat protein (GFLV resistance)
V. vinifera cv. Chardonnay	CRISPR/Cas9 system	'*MLO-7*' for powdery mildew resistance
V. vinifera cv. Neo Muscat	CRISPR/Cas9 system	Phytoene desaturase (VvPDS)

Source: Adapted from Kikkert *et al.* (2001); Malnoy *et al.* (2016)

TABLE 8.2
Development of Transgenic Plants of *Vitis* Scion Cultivars

Cultivar	Selectable marker	Gene product (trait of interest)
Koshusanjaku	NPTII	(GUS reporter gene)
Chardonnay	NPTII	Coat protein (grapevine fanleaf virus resistance)
Chardonnay	HPT	(GUS reporter gene)
Chardonnay	NPTII	Chitinase (disease resistance)
Sultana	NPTII HPT	Shiva-I (disease resistance), TomRSV (virus resistance), GUS reporter gene, silencing of polyphenol oxidase to reduce browning
Superior Seedless	Bar	(Basta herbicide resistance)
Sugraone	NPTII	Anti-sense movement proteins (virus resistance)
Chancellor	NPTII	(GUS reporter gene)
Merlot	NPTII	Chitinase (fungal disease resistance)
Seyval blanc	Na	na
Riesling, Dornfelder	NPTII	Glucanase, chitinase (disease resistance)
Emperor	Na	na
Almeria		
Ugni blanc	Na	na
Red Globe	Na	Barnase gene (seedlessness)
Red Globe Velika	NPTII, HPT	(seedlessness)
Neo Muscat	NPTII	Class I chitinase (fungal disease resistance)
Cabernet Sauvignon, Podarok, Magaracha, Rubinovyi, Magaracha, Krona 42	NPTII,bar	(Basta herbicide resistance)
Niagara	NPTII	Glucanase, 206 disease resistance response gene (fungal disease resistance)
Cabernet Franc	NPTII	Fe-superoxide dismutase (freezing tolerance)
Shiraz, Chardonnay, Cabernet Sauvignon, Sauvignon Blanc, Chenin Blanc, Riesling, Muscat, Gordo, Blanco	NPTII	(GFP reporter gene)
Pusa Seedless, Beauty Seedless, Perlette, Nashik	NPTII	GUS reporter gene
Nebbiolo	NPTII	Coat protein (grapevine fanleaf virus resistance)
Beauty Seedless	BADH	GUS reporter gene
Chardonnay	NPTII, Th En42	Chitinase (disease resistance), peptide gene (antimicrobial)

Source: Adapted from Kikkert *et al.* (2001)

It is expected that time frames, methodologies and replicated field trials in different environments similar to those used for evaluation of genotypes from conventional grapevine breeding programs will be adopted. Intellectual property issues, such as patents or other forms of protection on genes, promoters and technologies, 'freedom to operate' capability through formal agreements where commercially improved plants may be difficult to release, should be addressed. Some critics and activists are whipping up public alarm and fuelling political agendas with protests against globalization and a universal borderless economy. Certain lobby groups also claim that patents on genetically engineered organisms confer an unfair advantage to certain producers, thereby concentrating economic power in the hands of a few large multinational producers. The initial problems with statutory approval for the use of genetically engineered plants and organisms in the agro-industry are now slowly being dissolved by a growing consensus that risk is primarily a function of the characteristics of a product rather than the use of genetic modification *per se*. The public should be informed about research and products and the consumer reassured of first-class, transparent regulatory systems and the meticulous implementation of bio-safety legislation with clear technical standards and definitions with regard to genetically modified products. Assurance must be given that GM wine and other grape-derived products will not be forced onto consumers for profit when there is no clear advantage for the consumer. The implementation and successful commercialization of genetically improved grapevine varieties will only be realized if an array of hurdles, both scientific and otherwise, are overcome.

BIBLIOGRAPHY

Bansal, K.C. and Koundal, K.R. (2003). *Recent Techniques in Plant Genetic Engineering and Molecular Breeding – A Laboratory Manual*. NRC on Plant Biotechnology. IARI, New Delhi, India.

Bisson, L.F., Waterhouse, A.L., Ebeler, S.E., Walker, A.M. and Lapsley, J.T. (2002). The present and future of the international wine industry. *Nature*, 418(6898): 696.

Colova-Tsolova, V., Perl, A., Krastanova, S., Tsvetkov, I. and Atanassov, A. (2001). Genetically engineered grape for disease and stress tolerance. In: *Molecular Biology and Biotechnology of the Grapevine*, p. 411, K.A. Roubelakis-Angelakis (ed.). Kluwer Academic Publishers, The Netherlands.

FAO website. http://www.fao.org/faostat.

Gheysen, G., Angenon, G. and Van Montagu, M. (1998). *Agrobacterium* – Mediated transformation: A scientifically intriguing story with significant applications. In: *Transgenic Plant Research*, p. 1, K. Lindsey (ed.). Harwood Academic Publishers, Chur, Switzerland.

Gray, D.J. and Meredith, C.P. (1992). Grape. In: *Biotechnology of Perennial Fruit Crops*, p. 229, F.A. Hammerschlag and R.E. Litz (eds.). CAB International, Wallingford, UK.

Gribaudo, I. and Schubert, A. (1990). Grapevine root transformation with *Agrobacterium rhizogenes*. In: *Proc. V Intl. Symp. on Grape Breeding*, p. 412. Vitis Special Issue, Pflaz, Germany.

Gribaudo, I., Scarlot, V., Gambino, G., Schubert, A., Golles, R., Laimer, M., Hajdu, E. and Borbas, E. (2003). Transformation of *Vitis vinifera* L. cv. Nebbiolo with the coat protein gene of grapevine fanleaf virus (GFLV). *Acta Horticulturae*, 603(603): 309.

Iocco, P., Franks, T. and Thomas, M.R. (2001). Genetic transformation of major wine grape cultivars of *Vitis vinifera* L. *Transgenic Research*, 10(2): 105.

Kikkert, J.R., Thomas, M.R. and Reisch, B.I. (2001). Grapevine genetic engineering. In: *Molecular Biology and Biotechnology of the Grapevine*, p. 393, K.A. Roubelakis-Angelakis (ed.). Kluwer Academic Publishers, The Netherlands.

Malnoy, M., Viola, R., Jung, M.H., Koo, O.J., Kim, S., Kim, J.S., Velasco, R. and Kanchiswamy, C.N. (2016). DNA-free genetically edited grapevine and apple protoplast using CRISPR/Cas9 ribonucleoproteins. *Frontiers in Plant Science*, 7: 1.

Nakajima, I., Ban, Y., Azuma, A., Onoue, N., Moriguchi, T., Yamamoto, T., Toki, S.. and Endo, M. (2017). CRISPR/Cas9-mediated targeted mutagenesis in grape. *PLoS One*, 12(5): e0177966.

Perl, A. and Eshdat, Y. (1998). DNA transfer and gene expression in transgenic grapes. In: *Biotechnology & Genetic Engineering Reviews*, p. 365, M.P. Tomes (ed.). Intercept, Andover.

Reisch, B.I., Kikkert, J., Vidal, J., Ali, G.S., Gadoury, D., Seem, R., Wallace, P., Hajdu, E. and Borbas, E. (2003). Genetic transformation of *Vitis vinifera* to improve disease resistance. *Acta Horticulturae*, 603(603): 303.

Ren, C., Liu, X., Zhang, Z., Wang, Y., Duan, W., Li, S. and Liang, Z. (2016). CRISPR/Cas9-mediated efficient targeted mutagenesis in Chardonnay (*Vitis vinifera* L.). *Science Reports*, 6: 32289.

Torregrosa, L., Iocco, P. and Thomas, M.R. (2002). Influence of *Agrobacterium* strain, culture medium and cultivar on the transformation efficiency of *Vitis vinifera* L. *American Journal of Enology & Viticulture*, 53: 183.

Vincelli, P. (2016). Genetic engineering and sustainable crop disease management: Opportunities for case-by-case decision-making. *Sustainability*, 8(495): 1–22.

Vivier, M.A. and Pretorius, I.S. (2002). Genetically tailored grapevine for the wine industry. *Trends in Biotechnology*, 20(11): 472.

9 Grape Maturation

António Manuel Jordão and Fernanda Cosme

CONTENTS

9.1 INTRODUCTION

Grapes are considered the world's most predominant fruit crop and have been consumed for a long time. According to the last data published by the International Organization of Vine and Wine, approximately 77 million tons of grapes are produced annually in the world and are destined mostly for wine production (36.6 million tons for wine production and 3.0 million tons for grape must and juice production) or for *in natura* consumption (27.7 million tons for table grapes and 6.5 million tons for dried grapes). In addition, in 2016, the area under vines in the world destined for the production or awaiting production of wine grapes, table grapes or dried grapes was about 7.5 million hectares. The main wine producing countries are Italy, France, Spain, United States, Australia, China, Chile, Argentina, Germany, and South Africa. In 2016, these countries produced about 221 million hectoliters of wine.

Grape berry chemical composition is complex, containing hundreds of different chemical compounds. Water (75–85%) is the main compound followed by sugars (glucose, fructose, and sucrose) and organic acids (malic, tartaric, and citric acid). Other important compounds include amino acids, proteins and phenolic compounds (phenolic acids, tannin, anthocyanins, and flavonols). All of these compounds are present in different percentages in each grape berry stage, and all of them have an important role in wine quality. Thus, for example, berry sugar composition has a key role in wine alcohol content, while organic acids and phenolic compounds have an important role in the sensory properties of the different wine types and styles. Also, in the last decades, the richness in phenolic compounds of grapes, especially for red grape varieties, has gained the attention of extensive studies about the grape phenolic composition and health-promoting properties of these compounds quantified in grapes. Considering the main grape compounds and respective impact on the wine's final composition and quality, to decide grape harvest, grape sugars, acidity, and phenolic compounds will be determined. *Bouquet* and flavor are related to the winemaker's expertise, stabilization of the wine and storage processes, but primarily they are related to grape varietal character and its particular expression in a given *terroir.* Besides this, the content and the evolution of the different grape compounds are dependent upon the individual biosynthesis process of the compounds that occur during the grape maturation process and also as a result of different environmental factors. Thus, the main purpose of this chapter is to summarize recent developments of grape composition and the dynamic of the evolution of the most important compounds during this process, as well as the different factors that could determine how this evolution occurs. In addition, several key aspects of different methodologies used for the evaluation and quantification of the most important parameters for the monitoring of grape maturation will also be considered.

9.2 GENERAL GRAPE BERRY COMPOSITION

Botanically, grapes can be classified as a fruit of the vine, belonging to the family of *Vitaceae* and the genus of *Vitis*, with the following main species: *Vitis vinifera*, *Vitis labrusca*, *Vitis rupestris*, *Vitis aestivalis*, *Vitis riparia*, *Vitis cinerea*, and *Vitis rotundifolia*. Grape is a non-climacteric fruit that grows in the temperate zones of the northern and southern

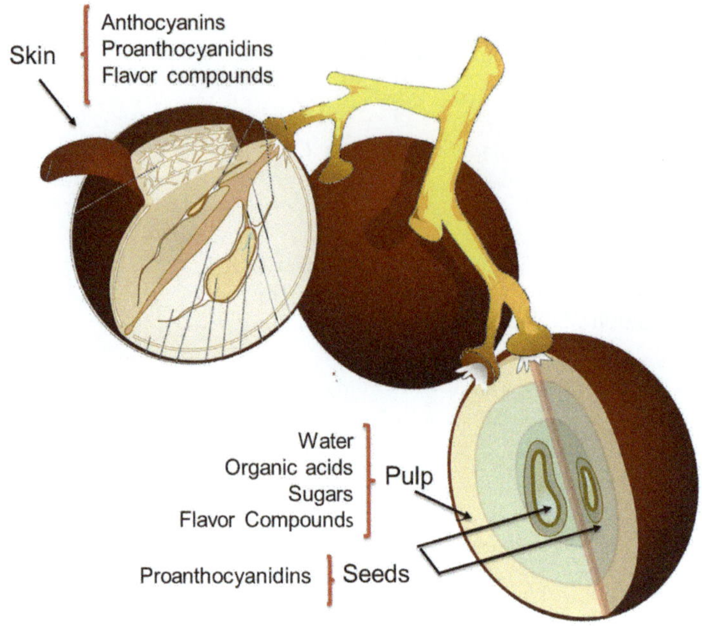

FIGURE 9.1 Grape berry fractions and main chemical compounds

hemispheres and distributed between America, Asia, Oceania, and partially in specific zones of Africa. Among the 60 vining species grown, it is the grapevine *Vitis vinifera* L. which is the species largely used in the global wine industry. The existence of numerous grape varieties means a large difference in their chemical composition, which allows the selection of the most suitable cultivars for wine production and for natural consumption (table and dried grapes). Grape berry 'quality' is one of the key variables that determine final wine quality and also the consumer quality of table grapes. Berry 'quality', however, is a generic term that refers to levels of a diverse range of berry chemical compounds including organic acids, sugars, phenolics, flavor compounds, etc. All of these compounds are detected and quantified in the different fractions of the grape bunch: skin, pulp, seeds, and stems. Figure 9.1 illustrates the different grape berry fractions and the main chemical compounds that are possible to detect.

In general, skins are characterized by high concentrations of phenolic compounds, namely anthocyanins for red grape varieties and tannins, and also a high diversity of phenolic acids and flavonols. Phenolics are important compounds of grape berry because they play a key role in determining the color of red wines and also wine astringency and bitterness. During grape maturation, these compounds are synthesized in the berry and concentrated in the berry skin, seeds, and pulp. In the case of seeds, this grape berry fraction is the most important source of proanthocyanidins. In skins, it is also important to consider the presence of aroma and flavor compounds. These compounds are decisive directly or indirectly for the aroma characteristics of red and white wines and are considered as varietal aromas. One of the odoriferous compounds detected in *Vitis vinifera* grapes belongs to the terpene family. These terpenoid compounds form the basis of the sensory expression of the wine bouquet and are decisive in differentiating grape varieties. As well as this, it is also possible to detect several aroma and flavor compounds in the pulp.

In pulp, besides aroma and flavor compounds, it is possible to detect other main chemical compounds, namely water, organic acids, and sugars. Grape berries require a significant amount of water for growth and development, and typically water contributes to 75 to 85% of berry fresh weight at harvest. Organic acids are a key factor to determine if the must has the potential to produce a balanced and stable wine. The major organic acids in grape berries are tartaric and malic acid. Other minor acids are also present in ripened grapes: citric, and acetic acids. Finally, sugars, in particular glucose and fructose, are important grape berry compounds, specifically in the pulp. A massive accumulation of these compounds is found in the vacuoles of mesocarp cells after *véraison*. Sugar content is an indicator often used to assess ripeness and to decide grape harvest date. However, in current grape ripening controls, "sugar ripeness" is not, by itself, the only and best index of optimal maturity to determine the grape harvest date.

9.2.1 Water

Water is by far the most abundant compound in grape must, acting as a solvent of volatile and non-volatile chemical compounds. Up to 99% of the water content in the grape must is absorbed by the vine roots from the soil. Vine growth and grape berry development are closely related to water availability in the soil. Berry volume per vine depends on berry number and water volume per berry. According to several authors, a great part of the fleshy fruits, including grapes, water is essential for volumetric growth and the accumulation of primary and secondary compounds, which determine the final fruit composition and quality. For Grenache Noir grape variety, the importance of plant water status, irrespective of the leaf: fruit ratio in berry compound accumulation, has been documented. Thus, according to several studies, vine water status is a major factor impacting source-sink relationships.

Water deficit generally leads to smaller berries since it inhibits both cell division and especially cell expansion. Water influx into fruits occurs via both the xylem and phloem, and most of the berry volume gain before *véraison* is due to water import from the xylem, whereas most of the water coming into the berry after *véraison* is imported by the phloem.

The effect of vine water status on the grape berry composition is not only related to berry size; we know that the effect of vine water status on the concentration of different grape compounds is decisive for grape and wine quality. The effect of vine water status on the concentration of skin tannins and anthocyanins is greater than the effect of fruit size *per se* on those same variables.

In the last thirty years, different research results have concluded that regulated deficit irrigation induces higher phenolic content, titratable acidity, malic acid, potassium contents and lower pH but has no significant effect on berry sugar accumulation or tartaric acid content. Also, just before harvest, excess water should be avoided since it can increase berry size and can cause a "dilution" of solutes (sugars, acids, anthocyanins, tannins, etc.) or cracking of the berries.

9.2.2 Sugars

Sugar composition is important for grape berry quality, contributing consequently to both wine quality and also to fresh table grapes quality. In general, sugar is often used to assess ripeness and to assess readiness for harvest. However, despite sugar content indicating the grape maturity level, "sugar ripeness" is not, by itself, the only and best indicator of optimal grape maturity. Most of the sugars are fermented to ethanol during the vinification process.

Sugar accumulation in the grape berry is regulated by complex mechanisms involving several steps. The transport and allocation of sugars between the photosynthetic "source tissues" and the heterotrophic "sink tissues" is known as assimilate partitioning and is a major determinant of plant growth and productivity. Grape sugar accumulation is regulated by complex mechanisms and is affected by several parameters including light, water, and ion status, wounding, fungal and bacterial attacks, and hormones.

The predominant sugars that are quantified in grapes are glucose and fructose, with only trace content of sucrose in the grape berries of most cultivars. According to several authors, only in a few cultivars, from *Vitis rotundifolia* and hybrids between *Vitis labrusca* and *Vitis vinifera*, was it possible to detect a high sucrose content. Also, the majority of grape species accumulate more glucose than fructose. In 98 different grape cultivars, it was found that glucose and fructose were the predominant sugars in grape berries, ranging from 45.86 to 122.89 mg/mL and from 47.64 to 131.04 mg/mL, respectively. For nine ancient grape cultivars (*Vitis vinifera* L.) from the Igdir province of Eastern Turkey, this sugar content variability is strongly influenced by cultivars, wherein the higher values were obtained for glucose (16.47 g/100 g) followed by fructose (15.55 g/100 g). A similar trend was also previously observed in five grapevine cultivars at final maturation which had been cultivated in a typical Mediterranean climate. According to the authors of the studies, glucose, and fructose content varied from 86.4 (Italia) to 107.0 g/L (Muscat of Hamburg), and from 73.1 (Italia) to 94.1 g/L (Alphonse Lavallée). In general, higher average temperatures and lower average precipitation during the last months of grape maturation contribute to higher sugars accumulation and consequently to higher soluble solids. This tendency was revealed in different Portuguese and French red grape varieties, where for general physicochemical composition, including sugar content, the higher average temperature, and lower average precipitation during August and September in one region induced a higher sugars accumulation in the grapes at technological maturity, compared to that observed for the same grape varieties collected in the vineyard located in the other region, where a lower average temperature and higher average precipitation were observed.

9.2.3 Organic Acids

In general, organic acids do not exceed more than 1% of the total grape juice, the most important being tartaric acid, followed by malic acid. Tartaric and malic acid accounts for 70–90% of the total acids present in the grape berry, existing at roughly a 1:1 to 1:3 ratio of tartaric to malic acid. However, the ratio of tartaric to malic acid is cultivar-specific and depends on the genetic background of the grape. Considering the organic acid composition for several table grapes, the Thompson Seedless grape variety showed the lowest tartaric/malic acid ratio (1:19). Both acids are produced in the grape berry along with small amounts of citric acid and several other non-nitrogenous organic acids. After tartaric acid, malic acid is the second most important organic acid. Malic acid confers a green taste to the grape berry and wine. This organic acid decreases during alcoholic fermentation as a consequence of yeast consumption. Further decrease in malic acid content are achieved during malolactic fermentation which occurs in wines (especially in red wines) when lactic acid bacteria convert malic acid into lactic acid.

The actual acid composition and concentration within the grape must or wine is influenced by many factors, such as grape variety, climatic conditions, soil composition, and cultural practices, thereby contributing to the wine's flavor, stability, color, and pH. These factors also have an important effect on individual malic and tartaric acid content. Cool regions typically produce grapes with a higher concentration of malic acid, while grapes grown in warmer regions tend to have lower acidity. This negative temperature correlation with malic acid levels are due to the effect of temperature on the balance between malic acid synthesis and catabolism.

Grape berry variety as well as the sensory properties and aging potential of wines are closely linked to the levels of tartaric acid present in the grape and those added during wine production. For example, 1.90 g/kg tartaric acid and 0.97 g/kg malic acid have been quantified in grapes from the Cardinal grape variety. For the Alphonse Lavellee, Muscat of Hamburg, Isabella, Italia, and Muscat of Alexandria grape cultivars reported values of tartaric acid from 3.8 to 5.0 g/L, malic acid from 3.6 to 3.4 g/L, and citric acid from 0.2 to 0.4 g/L have been documented. A similar trend was also quantified by other authors and for other grape varieties.

9.2.4 Phenolic Compounds

Phenolic compounds, also called polyphenols, constitute a diverse group of secondary metabolites which exist in grapes, mainly in the grape berry skins and seeds. Figure 9.2 shows the localization of the main phenolic compounds found in the different grape berry fractions. The total extractable phenolic compounds in grapes are only present at about 10% or less in the pulp, 60–70% in the seeds, and 28–35% in the skins. Red grape varieties contain higher concentrations of phenolic compounds compared to white grape varieties, since, for example, the latter contains no anthocyanins or, in the case of some specific grape varieties such as in grape skins of Gewürztraminer, a very low content. It has recently been reported for the first time that in the skin and pulp of Siria grape variety (a Portuguese variety) and in the skin of international white grape varieties (Chardonnay, Sauvignon Blanc, and Riesling) also contains measurable traces of anthocyanins.

This important class of compounds is classified in general into two groups: flavonoids and non-flavonoids. The non-flavonoid compounds include phenolic acids divided into hydroxybenzoic acids and hydroxycinnamic acids, but also other phenol derivatives such as stilbenes. Flavonoids cover a large number of subclasses, such as flavonols, flavanols, and anthocyanins. Flavonols are the most abundant phenolic compounds in grape skins, while grape seeds are rich in flavan-3-ols.

It is important to note that phenolic compounds play a key role in determining fruit color and astringency. Also, these compounds are considered to be one of the most important wine components, due to their direct relationship with the wine color, astringency, bitterness, and susceptibility to oxidation reactions. In the last 20 years, several published research works also reported that grape and wine phenolic compounds potentially play an important role in human health, namely oxidation inhibition of low-density lipoproteins (LDLs), a decrease in inflammatory and carcinogenic processes and inhibition of platelet aggregation. More recently, the effects of grape seed extract supplementation on exercise performance and oxidative stress in acutely and chronically exercised rats were investigated, and the results concluded that grape seed extracts supplementation prevents exercise-induced oxidative stress by preventing lipid peroxidation and increasing antioxidant enzyme activities.

Generally speaking, in each grape fraction, the average concentration of total phenolic compounds is around 2178.8 mg/g gallic acid equivalents in seeds, 374.6 mg/g gallic acid equivalents in skins, and 23.8 mg/g gallic acid equivalents in pulps. However, it is important to note that the grape phenolic composition is influenced by different factors, namely the grape variety, sunlight exposure, solar radiation, altitude, soil composition, climate, cultivation practices, exposure to

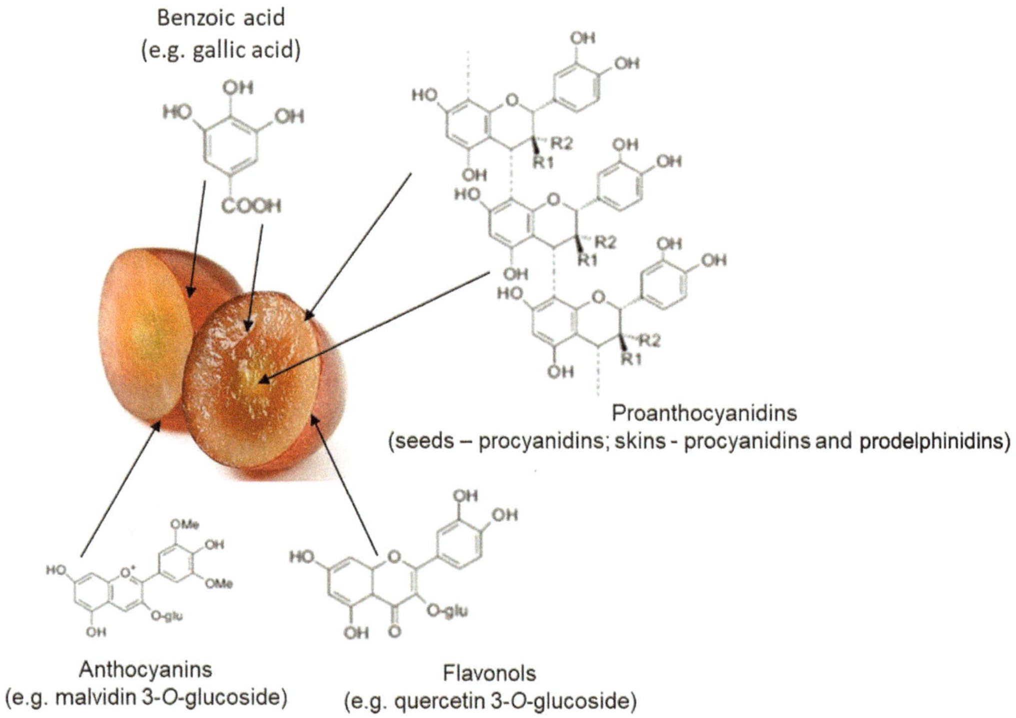

FIGURE 9.2 Localization of the main phenolic compounds found in *Vitis vinifera* red grape berry

diseases, and degree of grape ripeness. In addition, the influence of genetic factors and vintage in the flavan-3-ol composition of grape seeds of a segregation of *Vitis vinifera* population were analyzed. Several authors concluded that some phenolic compounds, namely (–)-epicatechin and (+)-catechin followed by proanthocyanidins, A2 and B2, showed high correlations between the vintage and the compound content, proving that the harvest year also has an important effect on phenolic content besides genetic factors.

9.2.5 Aroma and Flavor Compounds

In wines, the aroma composition is due to a great number and diversity of mechanisms, namely: grape metabolism (depending on the grape variety, climate, soil, and cultural practices); biochemical reactions (oxidation and hydrolysis) that occur before alcoholic fermentation and triggered during grape must extraction and maceration processes; the fermentation metabolisms of the microorganisms involved in alcoholic and malolactic fermentation (namely yeasts and lactic acid bacteria); and chemical or enzymatic reactions that occur during the aging process in the vat, bottle, and wood barrel. In general, the type and concentration of aroma and flavor compounds found in grapes vary extensively among grape varieties. There are a great number of factors that determine the grape aroma and flavor composition, such as grape variety, cultural practices, environmental conditions such as light environment and exposure, vineyard altitude and location, temperature and rainfall and harvest date.

The volatile compounds from grapes include monoterpenes, C_{13}-norisoprenoids, methoxypyrazines, and sulfur compounds. One of the most studied, volatile compounds from grapes, belongs to the terpene family and is responsible for the floral aroma normally associated with Muscat varieties. These compounds are located in the grape berry skin and about 50 monoterpene compounds are known. The main monoterpenols are linalool, α-terpineol, citronellol, nerol, geraniol, and hoterienol. The olfactory description is, respectively, rose, lily of the valley, citronella, rose, and linden. Geraniol, linalool, and α-terpineol are in general the dominant terpenes in Muscat grape varieties.

Including in the monoterpenes, free odorless polyols (diols and tiols) are also potent and reactive volatile compounds in which some can break easily giving rise to other compounds such as diendiol (3,7-dimethylocta-1,5-diene-3,7-diol) which can give hotrienol and neral oxide, these monoterpene polyols are present in grapes at concentrations up to 1 mg/L.

Norisoprenoids derivatives with 13 carbons atoms (C_{13}-norisoprenoids) also have an important role in grape and wine aroma. These derivatives include compounds like β-damascenone, which is usually associated with a complex smell of flowers, tropical fruit and stewed apple, and β-ionone with a characteristic aroma of violets. Several Portuguese red grape varieties (such as Tincadeira Preta, Moreto, Tinta Caiada, Aragonez, and Castelão) and white (Roupeiro, Antão Vaz, Rabo de Ovelha, Perrum, and Arinto) show different contents of norisoprenoids (3-hydroxy-β-damascone, 3-oxo-α-ionol, 3,9-dihydroxy-mega-5-ene, and vomifolio) in grape pulp and skins. Skins are always richer in these compounds than pulps. In addition, the values of the norisoprenoids derivatives varied from 71.8 (Trincadeira Preta grape variety) to 400 (Castelão grape variety) μg/kg for red grape varieties and from 52.6 (Roupeiro grape variety) to 322.1 (Arinto grape variety) μg/kg for white grape varieties.

Another important aromatic compound group from grapes are pyrazines. Amongst the methoxypyrazines, the most important ones are 2-isopropyl-3-methoxypyrazine, 2-*sec*-butyl-3-methoxypyrazine and 2-isobutyl-3-methoxypyrazine. In particular, the last one is responsible for the green pepper and asparagus aromas that are characteristic of several grape varieties, such as Cabernet Sauvignon, Sauvignon Blanc, Cabernet Franc and sometimes in Merlot.

9.2.6 Nitrogen and Mineral Compounds

In grapes, nitrogen is present in the form of ammonium cations (NH_4^+) and organic nitrogen such as amino acids, proteins, and other nitrogenated organic compounds. At harvest, arginine and proline are the main amino acids in the majority of grape varieties. The total nitrogen in the must can vary from 100 to 1200 mg/L, and generally, red wines possess a higher nitrogen content than white wines. About 20 different amino acids have been found in grapes, accounting for between 28 and 39% of the total nitrogen, depending on whether they are in a white or red grape variety. However, only around 7% are present in quantities above 100 mg/L (proline, arginine, glutamine, alanine, glutamate, serine, and threonine). Amino acids present in must, play an important role as nutrients required for the growth and development of yeasts and bacteria, during alcoholic and malolactic fermentation, respectively.

Thus, nitrogen is very important as a yeast nutrient, so that normal alcoholic fermentation of the musts can be carried out. Low nitrogen values in the must (150-200 mg/L) could induce yeasts to produce hydrogen sulfide and other sulfur odors that have a negative impact on final wine quality. Thus, nitrogen levels in grape must have an important impact on yeast growth, fermentation kinetics, and flavor. The nitrogen compounds, namely amino acids have an important role in wine flavor because they are precursors to an important number of compounds that contribute to the wine aroma, such as higher alcohols, aldehydes, ketones, and esters.

The natural assimilable nitrogen useful for yeast nutrition in the must is highly dependent on the vine nitrogen nutrition and can vary with the grape and rootstock, climate, soil and nitrogen fertilization. According to several authors, the grape amino acid composition is very much influenced by the grape variety used, vintage, vineyard fertilization, irrigation and the grape berry maturity degree.

The most consistent effect of vineyard nitrogen application on grape berry quality components is an increase in the nitrogenous compounds, such as total free amino acids, arginine, proline, ammonium, and total nitrogen concentration. It has also been reported that an excessive nitrogen supply combined with shoot trimming could induce a decrease in potential grape and wine quality. Also, several studies have shown that the best period to increase the nitrogen content in vineyards, and consequently the level in grapes and must through fertilizing, is at *véraison*.

With respect to proteins, the total content in the grape must is increased by nitrogen fertilization of the vine, and they are present in concentrations ranging from 15 to 230 mg/L.

Finally, for mineral compounds, the principal minerals found in grape berries include the cations potassium, calcium, and sodium. For anions, the most common minerals are phosphate and chloride. However, the literature describes that grape berry as an important source of potassium and is the main cation in must and wine (around 900 mg/L). The concentrations of different minerals in grapes principally derive from their absorption by the vines from the soil, and thus they provide information regarding the wine origin and authenticity. Different factors affect the potassium level in grape berries, including the potassium level in the soil, grape variety, viticultural practices, and soil fertilization. The grape juice pH and total soluble solids are positively correlated with potassium, whereas titratable acidity correlates negatively. In addition, the vines with no added nitrogen show more accumulation of polyphenols in the skin than those with other treatments, especially during the last few weeks of sampling.

Other minerals, such as calcium, magnesium, copper, sodium, and iron, are also present in grape berries. In general, all of these mineral compounds, in particular sodium and iron, are present in minimal amounts. However, an unusual increase in these compounds and others in grapevines and musts can occur as a result of viticulture, vinification practices, and also as a result of a pollutive environment. The presence of lead, copper, zinc, and cadmium in the separate tissues and organs of grapevines which were grown in an industrially polluted region indicated that their amounts were mainly due to the heavy-metal-containing aerosols falling from the atmosphere.

9.3 DYNAMIC OF MAIN INDIVIDUAL COMPOUND EVOLUTION DURING GRAPE MATURATION

Grape berry development is a dynamic process that involves a complex series of molecular, genetic, and biochemical changes generally divided into three major phases. During initial berry growth, berry size increases along a sigmoidal growth curve due to cell division and subsequent cell expansion, and organic acids (mainly malic and tartaric acids), tannins and hydroxycinnamates accumulate to peak levels. The second major phase is defined as a lag phase, in which cell expansion ceases, and sugars begin to accumulate. *Véraison* (the onset of ripening) marks the beginning of the third major phase, in which berries undergo the second period of sigmoidal growth due to additional mesocarp cell expansion, accumulation of anthocyanin pigments (for red grape varieties), accumulation of volatile compounds for aroma, softening, peak accumulation of sugars (mainly glucose and fructose) and a decline in organic acid accumulation. Figure 9.3 illustrates the major developmental events that occur during grape berry development.

In the following, the individual evolution of the main grape berry compounds during the maturation process are described and illustrated, with several practical examples which have been recently published.

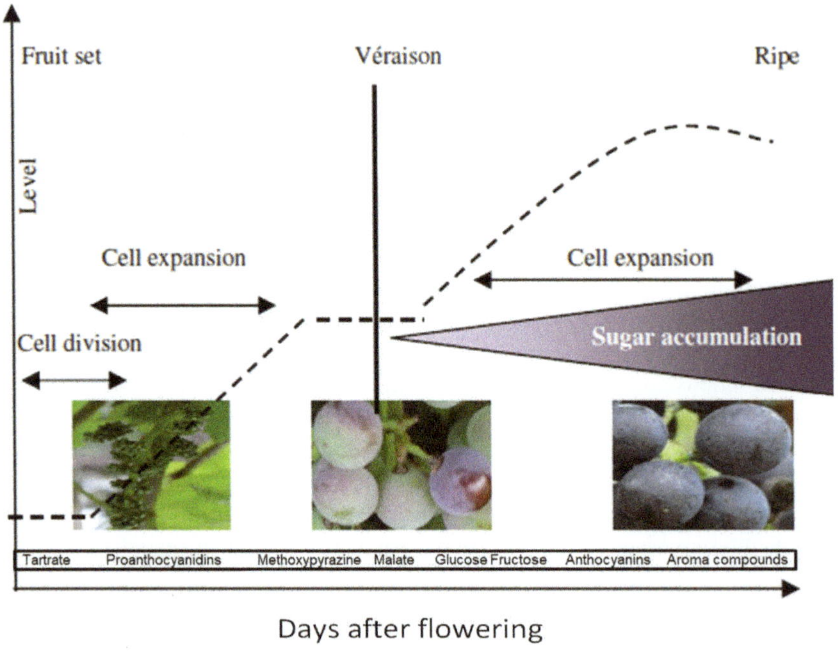

FIGURE 9.3 Schematic representation showing the major grape berry development during ripening. Legend: (-----) Berry changes in size

9.3.1 Sugars and Organic Acids

Sugar concentration is strongly influenced by grape berry transpiration, while grape berry acidity is linked to temperature. The defoliation in grapes (that contributes to higher light exposure and consequently higher temperature) could decrease titratable acidity. Generally, a higher temperature during grape ripening leads to an increase in the rates of sugar accumulation and organic acid degradation.

Grape berry sugar composition and concentration change during grape ripening and can be influenced by many factors, such as environmental conditions and viticultural practices. However, sugar composition is mainly determined by genotype, and sugar concentration is more dependent on berry development, environment, and cultural management practices. Grape berry sugar accumulation is regulated by complex mechanisms and diverse factors, such as light, water, ion status, fungal and bacterial attacks, and hormones.

In the first phase of grape berry development, an abundant quantity of sugar is imported into the berry which is metabolized. Grape berries have relatively small quantities of sugars. At *véraison*, sugar accumulation starts, and the imported sucrose is transformed into hexoses, which are deposited in the vacuoles. Glucose and fructose accumulate in the grape berries in similar quantities at a relatively constant rate throughout ripening.

Temperature is an important environmental factor and, monitored by several researchers, has been shown to influence the grape berry sugar concentration. It is shown that for a temperature above 25°C, net photosynthesis decreases even at constant sun exposure. Also, other authors have shown that a temperature higher than 30°C induces a reduction in the grape berry size and weight, and so the metabolic activity and consequently the sugar concentration may stop. Conversely, it has also been observed that a high temperature could accelerate grape berry maturation.

Irrigation is one of the viticultural practices that may greatly influence vine and grape metabolism, and different irrigation regimes may lead to physiological modifications that will disturb vine yield and grape berry composition. However, irrigation has a variable effect on grape berry sugar concentration. The influence of water regimes on grape must composition of *Vitis vinifera* L. cv. Tempranillo grapes was studied during a period of three years and showed that the sugar concentration was significantly higher in the irrigated vines compared to the non-irrigated vines, principally near the end of the grape ripening. Also, it was shown that irrigated grapes reached higher weights, but this did not affect sugar accumulation. This result is in line with other researchers. However, other authors have found a suspension in sugar accumulation related to irrigation. The dissimilar results stated in the research works may be explained by the different quantities of water provided in the trials and the environmental conditions in which the vine grew. Excess water raises the water flow into the grape berries which may result in excess berry growth and dilution of the sugar concentration. Excess irrigation can lead to a very high vegetative growth and crop load and, since berry size is increased, a dilution of certain important compounds may occur. Therefore, the sugar concentration may decrease due to competition between vegetative growth and grape berry development. On the other hand, severe water stress tends to reduce vigor, and sugar concentration as a photosynthetic activity may also be compromised. The effect of water deficit stress on grape berry sugar concentration is also yield-dependent; for low yields, vine water deficit improves grape berry sugar concentration, and for high yields, it reduces grape berry sugar concentration. Various authors also proposed that the water deficit effects on grape berries, sugar concentration is variety dependent. For example, no significant alterations were observed in the sugar concentration of Merlot grapes under a water deficit when compared to the irrigated control, whereas a significant increase in sugar concentration of Cabernet Sauvignon was observed. An increase in grape berry sugar concentration under a water deficit has been shown in Cabernet Sauvignon but not in Chardonnay, and a reduction in the sugar concentration in Grenache, but not observed in Syrah with the same amount of water stress, has been shown.

As seen in the studies, acidity decreased more slowly during ripening in non-irrigated grapes. Small differences in pH reflect large changes in titratable acidity. It has been shown that pH values increased linearly with grape berry ripening, while titratable acidity decreased exponentially. At the beginning of grape ripening, when titratable acidity was high, a decrease in titratable acidity did not bring about a substantial change in pH; as ripening advanced, the change in pH became larger. In addition, pH changes in grapes are affected by the metabolism of the main acids and the increase of cations, which convert the free acids into their corresponding salts.

Organic acids (tartaric and malic) are synthesized in the grape berry and accumulate during the first growth period. During the ripening period, the decrease observed for tartaric acid concentration was not so intense as that observed for malic acid concentration. The decrease in titratable acidity during grape ripening is usually due to the loss of malic acid during respiration, since tartaric acid concentration is considered to be unaffected, due to its difficulty to be metabolized, which is recognized to both its resistance to combustion at high temperatures and its tendency to form salts which are not easily degraded by any identified enzyme. Tartaric acid is in the acid form at *véraison*; during ripening, the bitartrate and tartrate salts are formed, and these salts are not very soluble. Thus, the reduction during grape ripening is principally caused by the tartaric acid changing from acid to salt forms and by a dilution effect as the grape berry weight increases. The decrease of malic acid concentration occurs essentially through respiration, and, as shown by the tartaric acid/malic acid ratio, this process is emphasized in non-irrigated grapes. In addition, the decrease of malic acid concentration is greater with water deficits and with higher temperatures, since the amount of malic acid respiration increases with temperature, therefore grapes grown in cooler climates have higher concentrations of tartaric and malic acid. The temperature has a high influence on the concentration of malic acid in ripe grapes. Therefore, all viticultural practices that led to the intensification of sunlight exposure in the vine canopy, such as vine training systems, leaf removal, shoot thinning, and cluster thinning induced a decrease in the tartaric and/or malic acid concentrations in the grape berries.

No titratable acidity modifications have been shown in grape must from moderately water stressed vines. Nevertheless, some studies report a reduction of titratable acidity under deficit irrigation, observing that titratable acidity and pH values were only slightly influenced by irrigation. Several authors observed that if grape berries are harvested on the same date, the grape berries from irrigated vines present higher titratable acid compared to the grape berries from non-irrigated vines. This seems to be due to the ripening delay induced in irrigated vines. The higher titratable acidity presented in grape berries from irrigated vines compared to non-irrigated vines are may be related to a decrease in the malic acid respiration rate, linked to higher grape berry shading and vegetative growth and yield.

9.3.2 Phenolic Compounds

The biochemical pathways that plants have evolved to produce phenolic compounds from primary photosynthetic products have been described over the past 70 years. During the last decades, the development of new advanced equipment and techniques have not only facilitated the identification and quantification of individual phenolic compounds in grapes but has also supported the identification and characterization of the key individual compounds (such as enzymes) in the individual phenolic compound's biosynthesis.

One of the most important phenolic compounds quantified in grapes and with an important role in red grape and wine quality are anthocyanins. They are localized in the grape skins and sometimes in the pulp of "teinturier" varieties that have colored flesh. Their structure is characterized by a flavylium cation, with two benzene rings, linked by an unsaturated cationic oxygenated heterocycle, derived from the 2-phenyl-benzopyrylium nucleus, glycosylated at position C_3. The variation of the degree of hydroxylation, methylation and/or glucosylation leads to five aglycones or anthocyanidins found in grape *Vitis vinifera*: delphinidin, cyanidin, petunidin, peonidin, and malvidin. The anthocyanin content in grapes is influenced mostly by differences in geographical conditions, climatic characteristics of individual years, the intensity of agronomical measures practices and, according to some authors, between clones that belong to the same grape cultivar.

In general, the evolution of anthocyanins during grape ripening is characterized by an increase which begins at *véraison*, even two or three weeks before the color of grapes becomes visible. From *véraison* to complete maturity, according to several authors, the development of anthocyanins is generally characterized by three phases: the first one presents a slow increase, followed in the second phase by a rapid increase (20 to 35 days after *véraison*), ending in the third phase in a stabilization phase before a decrease at the end of ripening and/or during over-ripening. However, a noticeable fall in the last two weeks of the ripening process was followed by a sharp increase in the last days of maturation, while other authors reported a continual increase in individual anthocyanins from Tannat and Shiraz grapes during ripening. Figure 9.4, shows an example that summarizes and illustrates the various stages of the evolution of skin anthocyanin glucoside derivatives during grape maturation of two Portuguese red grape varieties.

Another important phenolic compound group quantified in grape berry (in particular in seeds and skins) with a great impact on wine characteristics, in particular red wines, are flavanols. This phenolic compound group includes monomers and condensed tannins (proanthocyanidins mainly oligomers and polymers of flavan-3-ols). The monomers are (+)-catechin, (–)-epicatechin, (+)-gallocatechin, (–)-epigallochatechin and epicatechin-3-*O*-gallate. For this group of phenolic compounds, several research works have established that grape variety, vintage, environmental factors, geographical conditions and also viticultural practices are decisive for the grape flavanols content.

The tendency for higher concentrations of flavanols, in particular proanthocyanidins, in the early stages of grape berry development may be related to their metabolization throughout the ripening process. The pattern of decline after *véraison* of proanthocyanidins in seeds and skins could be explained by oxidation reactions and also by a reduction of the extractability which results from the conjugation of proanthocyanidins with other cellular components. Such a decrease in grape berry post *véraison* has been reported by various authors. Figure 9.5 shows an example of the evolution of the different seed and skin proanthocyanidins fractions during grape ripening (from *véraison* until grape harvest).

In grape skins, it is also possible to detect other phenolic compounds, namely hydroxycinnamic acids, which are characterized by a C_6–C_3 structure, the most common in grapes being caffeic, ferulic, *p*-coumaric, sinapic, caftaric and coutaric acids. Several factors induce a variation in the content of grape hydroxycinnamic acids. Grape variety, maturity degree of grape berry and geographical origin are the main factors that affect their concentrations in grape berry. These phenolic compounds are present inside the vacuole cells in grape skins and pulp in the form of tartaric esters or associated with a glucose molecule or polyamines. During grape maturation, there is an accumulation of hydroxycinnamic acids during the initial berry growth, followed by an increase of the values until *véraison*, and finally a stabilization of the values until grape harvest.

Finally, concerning stilbenes, several studies point out that the number of stilbenes present in the grapes is dependent on the amount of resveratrol, considered the precursor of these phenolic compounds. The largest amounts of *trans*-resveratrol

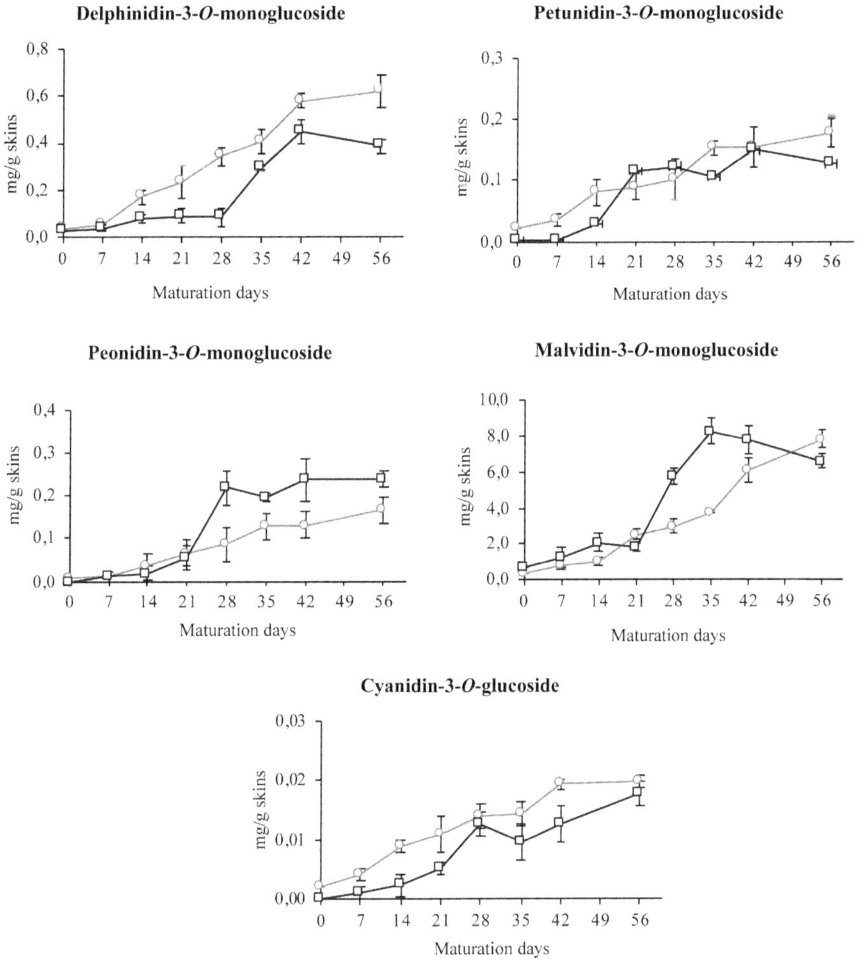

FIGURE 9.4 Evolution of skin anthocyanin glucoside derivatives during the ripening of two Portuguese red grape varieties (-o-Tinta Roriz, -□- Touriga Nacional)

are found in skins compared with seeds, whereas no *trans*-resveratrol was detected in the grape pulp. In general, these compounds are synthesized in response to abiotic stresses such as UV radiation, with a rapid response in grape berries throughout developmental stages. In grape berry, stilbenes are mainly located in the skins and, to a much lesser extent, in the pulp after *véraison*.

9.3.3 Aroma and Aroma Precursors

The grape aroma is composed of more than 100 different chemical compounds. In this complex combination, hydrocarbons, alcohols, esters, aldehydes, and other carbon-based compounds are present.

The grape varietal aroma is composed of free aroma compounds and bound aromas or precursors. The different types and quantities of aromatic precursors in grape berries are the main source of aromas, which differentiate grape variety. It is known that these aroma precursors are synthesized during grape ripening.

Terpenes are the varietal compounds that have been most extensively studied in *Vitis vinifera* grapes. In grape berries, around 40 terpenes have been identified; they can occur as hydrocarbons, alcohols, aldehydes, ketones, or esters, and their olfactory impression is synergistic. These compounds are mainly responsible for attributes like fruity (citric) and floral aromas, however, others have resin-like odors (α-terpinene, *p*-cimene, myrcene, and farnesol). Monoterpene alcohols, particularly linalool, α-terpineol, nerol, geraniol, citronellol, and hotrienol are considered the most odoriferous. Free and bonded forms of terpenols occur during grape ripening, and their concentration increases during berry development. The terpenicheterosides are abundant in green grape berries (250–500 µg/kg in fresh weight), however, the free terpenols are only present in lower quantities (30–90 µg/kg in fresh weight). Temperature and water supply may influence aroma development during ripening.

The presence of methoxypyrazine compounds in grape varieties such as Cabernet Sauvignon, Merlot, and Sauvignon Blanc has been established as a characteristic of the varietal aroma of these grapes. These aromatic compounds are produced during the first ripening stage and decrease during grape berry ripening. Their concentration in grape berries is also determined by different climatological and agronomic

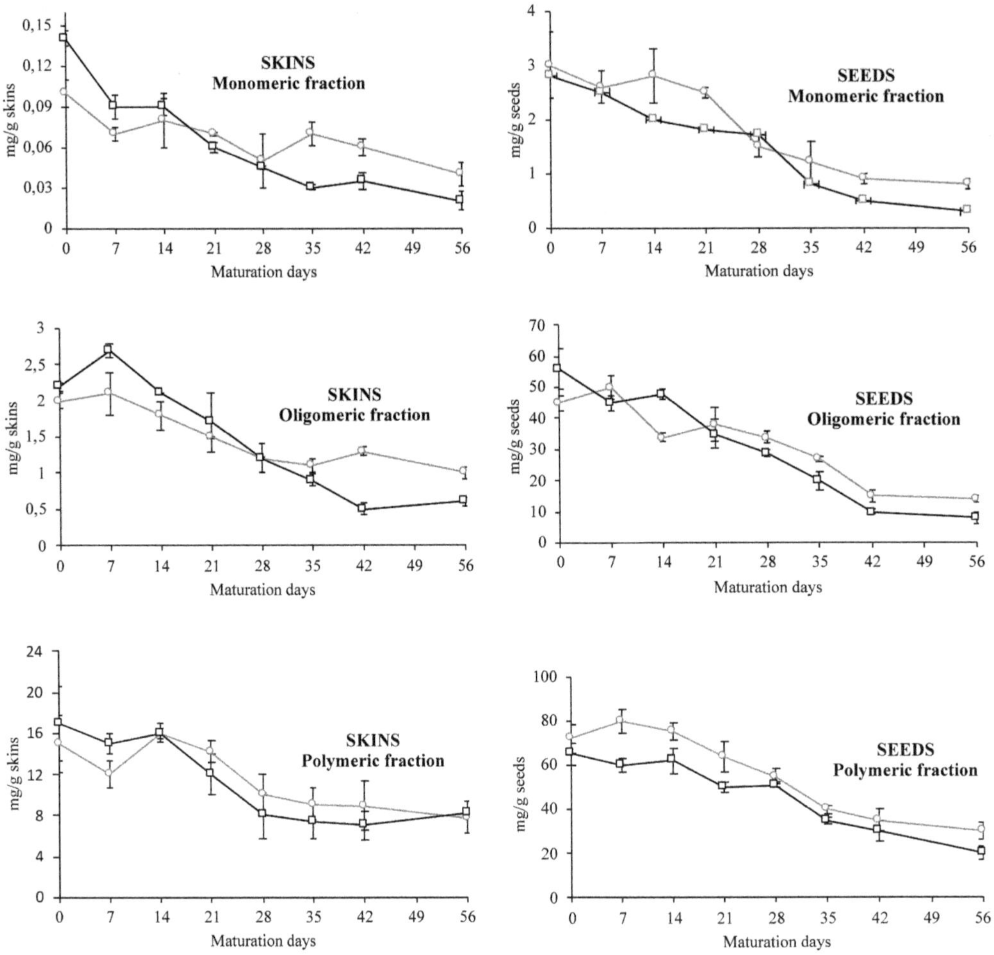

FIGURE 9.5 Evolution of the different skin and seed proanthocyanidin fractions during the ripening of two Portuguese red grape varieties (-o-Tinta Roriz, -□-TourigaNacional)

factors, such as the degree of grape berry ripening. For example, in grape berries from Cabernet Sauvignon, the concentrations of methoxypyrazine were around 100 ng/L at the beginning of *véraison* and decreased to around 8–15 ng/L in ripe grape berries. Correspondence between the decrease in methoxypyrazine concentration and malic acid degradation with temperature or water influence on the vine has also been established.

Grape-derived C_{13}-norisoprenoids are important odorants and are thought to originate from carotenoid precursors in grapes. The influence of a grape berry irrigation strategy on grape aromas compounds has been studied. The research showed that deficit irrigation changes the wine sensory profile if compared with standard irrigated vines. Water deficits in the Cabernet Sauvignon grape variety led to wine with fruitier and less vegetal aromas, compared to wine produced from the same grape variety with no water deficit. This observation may be explained by the fact that water deficits led to an increase in amino acids (precursors of esters in wines) and carotenoids, resulting in the fruitier aroma. Also, it has been observed that deficit irrigation led to an increase in the concentration of hydrolytically released C_{13}-norisoprenoids (β-damascenone, β-ionone and 1,1,6-trimethyl-1,2-dihydronaphthalene) in Cabernet Sauvignon grape berries at harvest and an increase in the berry-derived carotenoids (lutein, β-carotene, neoxanthin, violaxanthin and luteoxanthin) of up to 60%, compared to an irrigation treatment.

9.3.4 Mineral, Nitrogen and other Minor Compounds

The minerals present in grape berry can be divided into two categories: those minerals that accumulate continually throughout grape berry growth (potassium, phosphorus, sulfur and magnesium) which are generally considered to be phloem-mobile minerals, and those minerals that accumulate frequently before *véraison*. Calcium and manganese are considered minerals that have low phloem mobility. Calcium accumulation stops at the start of grape maturation, the same happens for manganese but to a minor extension. Ripe grape berries presented a potassium level of 15 times more than the green berries, but their calcium and manganese contents increased only two folds.

The principal minerals found in grape berries include the cations potassium, calcium, and sodium, and the most common mineral anions are sulfates, phosphate, and chloride. From this, potassium was the most abundant mineral with higher rates of accumulation after *véraison* compared to the other minerals. The cation mineral concentration in the fruit increases two- to threefold during ripening, with potassium as the dominant mineral.

In the grape berry, nitrogen can be found in mineral and organic forms. The nitrogen mineral in the form of NH_4^+ can represent up to 80% of the total nitrogen before *véraison*, but it decreases by 5 to 10% after grape berry maturation. During grape development, it is possible to observe two intense nitrogen incorporation phases: the first following grape berry set, and the second initiated at *véraison* and ending mid-maturation. In the last stages of grape maturity, the total nitrogen content may increase again. During the initial stages of grape berry growth (unripened grape berry), ammonium ions account for more than half of the total nitrogen in the grape berry, and amino acids are relatively low. After *véraison*, amino acid synthesis in the berry increases sharply, and the ammonium ion concentration declines as the ions are incorporated into amino acids. The amino acid concentration of the grape berry may increase two- to fivefold during ripening, with wide variations depending on variety and fruit maturity.

Soluble protein content in grape berries increases with maturity, reaching its maximum before complete maturity and then diminishing towards the end of grape maturation. The content of grape berry soluble protein changed with the grape variety and is depending on the geographical region ranging from 1.5 to 260 mg/L.

9.4 METHODOLOGIES FOR GRAPE MATURATION CONTROL – KEY CONCEPTS

Among the various grape constituents, an estimation of the alcohol level of the final wine (sugar content), pH, total acidity and, most recently, total phenol and anthocyanin content to determine the phenolic maturity, are the most frequent cluster parameters used to assess ripeness (technical maturity). Winemakers commonly use these parameters to determine the grape harvest date according to the wine they want to produce. For example, for sparkling wines and light white wine the harvest occurs earlier when it is possible to collect grapes with low sugar and a higher total acidity (high levels of organic acids, in particular, tartaric and malic acids), while for red wines with higher aging potential, it is important to harvest grapes with considerable sugar content and an adequate phenolic maturation (in particular anthocyanin and proanthocyanidin content). In addition, it is important to note that other non-compositional factors influence the decision to harvest the grape, in particular, labor availability, tank space limitations in the winery, seasonal changes, and climatic conditions (namely rainfall and hot temperatures) and other factors beyond the control of the winemakers.

During the last 20 years, efforts have been invested in accurately estimating grape maturation kinetics and the halfway point to the *vérasion* stage, in order to predict the best harvest date, to define homogeneous maturation zones in the vineyard and to select grapes at the weighbridge stage. Using a representative grape berry sample, different classical chemistry analytical methodologies are used: hydrometry, potentiometry, refractometry, titration, and spectrophotometry. However, in the last few years, new analytical techniques have been introduced, such as HPLC for phenolic composition analysis, NIR´S and FTIR to decrease the time needed to analyze a great number of parameters and also, more recently, the techniques based on intrinsic fruit fluorescence. Figure 9.6 shows several examples of different classical analytical methodologies (A to C) and also the more recently developed analytical techniques used to estimate grape maturation (D to E).

The classical analytical criteria, such as sugar content, total acidity and pH, have prevailed for a long time as the favored indicators of the grape maturation state at harvest. All of these analytical methodologies are simple, but a preparative grape sample is requested. However, independently of the method used and parameters to be determined, several aspects must be considered before grape analysis, namely the grape sampling method and the possibility to store grape samples, etc. There are several ways to collect a grape sample which are representative of a given vine plot: whole bunches, portions of bunches or single grape berries. In addition, before grape analysis, there are different sample treatments, such as grinding, centrifugation, samples treated when fresh or after lyophilization.

When the grape phenolic analysis is an option, two important points should be considered: the extraction of phenolic compounds (no extraction other than grinding treatment, centrifugation or maceration, under the conditions of temperature, length, agitation, solvents, etc.) and the methodologies of measurement of the different phenolic fractions. New methodologies, for example pressurized liquid extraction, electrically assisted extraction and microwave-assistance, have been developed in the last few years to increase the extraction of phenolic compounds from grape samples.

For methodologies of measurement of the phenolic potential of red grapes, there are numerous methods for the qualitative and quantitative determination; most of them remain experimental and are not usually used by winemakers. Presently, the methods most used are based on obtaining extracts from grapes by maceration in different solvents. In this case, the Glories method is the most employed, although this method is slow and laborious. The cellular maturity index also called anthocyanin extractability (EA%), is currently one of the most used indices to assess the extractability of the pigment compounds from red grapes. Another methodology was developed more recently and is known as the Cromoenos method. This method uses two commercial reagents and specific equipment to extract phenolics and give results in just 10 minutes. Other new techniques for phenolic maturity determination, such as spectrophotometric visible

FIGURE 9.6 Examples of grape maturation control procedures in the laboratory using current methodologies (A, B and C) and using recently developed technologies by the use of a FTIR equipment (D and E) and by the use of non-destructive optical monitoring Multiplex® in the vineyard (F)

TABLE 9.1
Main Characteristics of Several Methods for Grape Phenolic Potential Analysis

Main characteristics (analytical conditions)	Reference
Phenolic extration from entire grape berry with successive extractions with acid solution at 70ºC	Puissant and Léon (1967)
Phenolic extraction from skins with a hydroalcoholic solution at pH 3.2	Di Stefano and Cravero (1991)
Successive macerations of crushed grapes by the use of buffer solutions at pH 3.2 and pH 1.0	Glories and Augustin (1993)
Maceration of crushed grapes and seeds using hydroalcoholic solutions at pH 3.2 for 24 hours	Carbonneau and Champagnol (1993)
Maceration of skins and seeds using hydroalcoholic solutions at pH 3.2	Amrani and Glories (1994)
Macerations of crushed grapes with ethanol and HCl for two hours	Lamadon (1995)
Extractions with hydroalcoholic solution at pH 3.5 with 100 mg/L of SO_2 at 70ºC	Riou and Asselin (1996)
Extraction of entire grape berry for 30 minutes with a hydroalcoholic solution with HCl	Mattivi (2006)
Extraction of entire grape berry for 20 minutes at 40ºC with hydroalcoholic solution acidified with tartaric acid (pH 3.5)	Fragoso-Garcia *et al.* (2009)

and near-infrared (Vis-NIR) measurements, mid-infrared spectroscopy, fluorescence, and methods that indicate a relationship between phenolic maturity and appearance (color and morphology) of grape seeds, have been developed to allow for image analysis.

Methods based on tasting grape berries and seeds have also been introduced in recent years and are often used by wine professionals directly in the vineyard, being good tools for decision making. At the moment, several procedures are available to allow the sensory analysis of the grape ripening process. (For more information see Chapters 39, 40 and 44 of this text.)

Table 9.1 shows some methods used for phenolic compositional analysis in red grapes to help winemakers to determine the phenolic maturation.

BIBLIOGRAPHY

Alessandrini, M., Gaiotti, F., Belfiore, N., Matarese, F., D'Onofrio, C. and Tomasi, D. (2017). Influence of vineyard altitude on Glera grape ripening (*Vitis vinifera* L.): Effects on aroma evolution and wine sensory profile. *J. Sci. Food Agric.*, 97(9):2695–2705.

Allegro, G., Pastore, C., Valentini, G., Muzzi, E. and Filippetti, I. (2016). Influence of berry ripeness on accumulation, composition and extractability of skin and seed flavonoids in cv. *Sangiovese (Vitis vinifera* L.). *J. Sci. Food Agric.*, 96(13):4553–4559.

Almeida, C.M. and Vasconcelos, M.T. (2003). Multielement composition of wines and their precursors including provenance soil and their potentialities as fingerprints of wine origin. *J. Agric. Food Chem.*, 51(16):4788–4798.

Amerine, M.A. (1956). The maturation of wine grapes. *Wines Vines*, 37:53–55.

Amrani, K. and Glories, Y. (1994). Étude en conditions modéles de l'extractibilité des composés phénoliques des pellicules et des pépins de raisins rouges. *J. Int. Sci. Vigne Vin*, 28:303–317.

Andrea-Silva, J., Cosme, F., Filipe-Ribeiro, L., Moreira, A.S.P., Malheiro, A.C., Coimbra, M.A., Domingues, M.R.M. and Nunes, F.M. (2014). Origin of the pinking phenomenon of white wines. *J. Agric. Food Chem.*, 62(24):5651–5659.

Arapitsas, P., Oliveira, J. and Mattivi, F. (2015). Do white grapes really exist? *Food Res. Int.*, 69:21–25.

Azevedo, H., Conde, C., Gerós, H. and Tavares, R.M. (2006). The non-host pathogen *Botrytis cinerea* enhances glucose transport in *Pinus pinaster* suspension-cultured cells. *Plant Cell Physiol.*, 47(2):290–298.

Azuma, A., Yakushiji, H., Koshita, Y. and Kobayashi, S. (2012). Flavonoid biosynthesis-related genes in grape skin are differentially regulated by temperature and light conditions. *Planta*, 236(4):1067–1080.

Bais, A.J., Murphy, P.J. and Dry, I. (2000). The molecular regulation of stilbene phytoalexin biosynthesis in *Vitis vinifera* during grape berry development. *Plant Physiol.*, 27:425–433.

Bautista-Ortín, A.B., Jiménez-Pascual, E., Busse-Valverde, N., López-Roca, J.M., Ros-García, J.M. and Gómez-Plaza, E. (2013). Effect of wine maceration enzymes on the extraction of grape seed proanthocyanidins. *Food Bioprocess Technol.*, 6(8):2207–2212.

Bell, S.-J. and Henschke, P.A. (2005). Implications of nitrogen nutrition for grapes, fermentation and wine. *Aust. J. Grape Wine Res.*, 11(3):242–295.

Belviranli, M., Gökbel, H., Okudan, N. and Başaral, K. (2012). Effects of grape seed extract supplementation on exercise-induced oxidative stress in rats. *Brit. J. Nutr.*, 108(2):249–256.

Bergqvist, J., Dokoozlian, N. and Ebisuda, N. (2001). Sunlight exposure and temperature effects on berry growth and composition of Cabernet Sauvignon and Grenache in the Central San Joaquin Valley of California. *Am. J. Enol. Vitic.*, 52:1–7.

Bindon, K.A., Dry, P.R. and Loveys, B.R. (2007). Influence of plant water status on the production of C13-norisoprenoid precursors in *Vitis vinifera* L. cv. Cabernet Sauvignon grape berries. *J. Agric. Food Chem.*, 55(11):4493–4500.

Blouin, J. and Cruège, J. (2003). *Analyse et composition des vins: Comprendre le vin*. Editions La Vigne, Dunod, Paris, France, 304.

Blouin, J. and Guimberteau, J. (2000). *Maturation et maturité des raisins*, Editions Féret, Bordeaux, France, 151.

Bogicevic, M., Maras, V., Mugoša, M., Kodžulović, V., Raičević, J., Šućur, S. and Failla, O. (2015). The effects of early leaf removal and cluster thinning treatments on berry growth and grape composition in cultivars Vranac and Cabernet Sauvignon. *Chem. Biol. Technol. Agric.*, 2(1):13.

Boido, E., García-Marino, M., Dellacassa, E., Carrau, F., Rivas-Gonzalo, J.C. and Escribano-Bailón, M.T. (2011). Characterisation and evolution of grape polyphenol profiles of *Vitis vinifera* L. cv. Tannat during ripening and vinification. *Aust. J. Grape Wine Res.*, 17(3):383–393.

Bordiga, M., Travaglia, F., Locatelli, M., Coisson, J.D. and Arlorio, M. (2011). Characterisation of polymeric skin and seed proanthocyanidins during ripening in six *Vitis vinifera* L. cv. *Food Chem.*, 127(1):180–187.

Boussetta, N., Lebovka, N., Vorobiev, E.N., Adenier, H., Bedel-Cloutour, C. and Lanoisellé, J.-L. (2009). Electrically assisted extraction of soluble matter from Chardonnay grape skins for polyphenols recovery. *J. Agric. Food Chem.*, 57(4):1491–1497.

Brar, H.S., Singh, Z. and Swinny, E. (2008). Dynamics of anthocyanin and flavonol profiles in the 'Crimson Seedless' grape berry skin during development and ripening. *Sci. Hortic.*, 117:349–356.

Bravdo, B. and Hepner, Y. (1986). Water management and the effect on fruit quality in grapevines. In: *Proceedings of the 6th Australian Wine Industry Technical Conference*, Adelaide, pp. 150–158.

Bravdo, B.A., Hepner, Y., Loigner, C., Cohen, S. and Tabacman, H. (1985). Effect of irrigation and crop level on growth, yield, and wine quality of Cabernet Sauvignon. *Am. J. Enol. Vitic.*, 36:132–139.

Brunetto, G., de Melo, G., Toselli, M., Quartieri, M. and Tagliavini, M. (2015). The role of mineral nutrition on yields and fruit quality in grapevine, pear and apple. *Rev. Bras. Frutic.*, 37(4):1089–1104.

Bureau, S.M., Razungles, A.J. and Baumes, R.L. (2000). The aroma of Muscat of Frontignan grapes: Effect of the light environment of vine or bunch on volatiles and glycoconjugates. *J. Sci. Food Agric.*, 80(14):2012–2020.

Butkhup, L., Chowtivannakul, S., Gaensakoo, R., Prathepha, P. and Samappito, S. (2010). Study of the phenolic composition of Shiraz red grape cultivar (*Vitis vinfera* L.) cultivated in north-eastern Thailand and its antioxidant and antimicrobial activity. *S. Afr. J. Enol. Vitic.*, 31:89–98.

Cabrita, M.J., Freitas, A.M.C., Laureano, O. and Stefano, R.D. (2006). Glycosidic aroma compounds of some Portuguese grape cultivars. *J. Sci. Food Agric.*, 86(6):922–931.

Cadot, Y., Minana-Castello, M.T. and Chevalier, M. (2006). Anatomical, histological, and histochemical changes in grape seeds from *Vitis vinifera* L. cv Cabernet Franc during fruit development. *J. Agric. Food Chem.*, 54(24):9206–9215.

Carbonneau, A. and Champagnol, F. (1993). Nouveaux systèmes de culture integer du vignoble. *Programme AIR-3-CT*, 93.

Casazza, A.A., Aliakbarian, B., Sannita, E. and Perego, P. (2012). High-pressure-temperature extraction of phenolic compounds from grapes skins. *Int. J. Food Sci. Technol.*, 47(2):399–405.

Castellarin, S., Matthews, M.A., Gaspero, G.D. and Gambetta, G.A. (2007). Water deficits accelerate ripening and induce changes in gene expression regulating flavonoid biosynthesis in grape berries. *Planta*, 227(1):101–112.

Castillo-Muñoz, N., Gómez-Alonso, S., García-Romero, E. and Hermosín-Gutiérrez, I. (2007). Flavonol profiles of *Vitis vinifera* red grapes and their single-cultivar wines. *J. Agric. Food Chem.*, 55(3):992–1002.

Chapman, D.M., Roby, G., Ebeler, S.E., Guinard, J.X. and Matthews, M.A. (2005). Sensory attributes of Cabernet Sauvignon wines made from vines with different water status. *Aust. J. Grape Wine Res.*, 11(3):339–347.

Cheng, J. and Liang, C. (2012). The variation of mineral profiles from grape juice to monovarietal Cabernet Sauvignon wine in the vinification process. *J. Food Process. Pres.*, 36(3):262–266.

Cheynier, V. and Rigaud, J. (1986). HPLC separation and characterization of flavonols in the skins of *VitisVinifera* var. Cinsault. *Am. J. Enol. Vitic.*, 37:248–252.

Cheynier, V., Prieur, C., Guyot, S., Rigaud, J. and Moutounet, M. (1997). The structures of tannins in grapes and wines and their interaction with proteins. In: *Wine: Nutritional and Therapeutic Benefits*, ed. T.R. Watkins, Am. Chem. Soc., Washington, DC, 81–93.

Chorti, E., Guidoni, S., Ferrandino, A. and Novello, V. (2010). Effect of different cluster sunlight exposure levels on ripening and anthocyanin accumulation in Nebbiolo grapes. *Am. J. Enol. Vitic.*, 61:23–30.

Conde, C., Silva, P., Natacham, F., Dias, A.C.P., Tavares, R.M., Sousa, M.J., Agasse, A., Delrot, S. and Gerós, H. (2007). Biochemical changes throughout grape berry development and fruit and wine quality. *Food*, 1:1–22.

Conradie, W.J. (2001). Timing of nitrogen fertilisation and the effect of poultry manure on the performance of grapevines on sandy soil. II. Leaf analysis, juice analysis and wine quality. *S. Afr. J. Enol. Vitic.*, 22:60–68.

Coombe, B. (1986). Influence of temperature on composition and quality of grapes. In: *ISHS Acta Horticulturae. Proceedings of the International Symposium on Grapevine Canopy and Vigor Management*, Davis, CA, USA, 14 August, Volume XXII IHC, 23–35.

Coombe, B.G. (1992). Research on development and ripening of the grape berry. *Am. J. Enol. Vitic.*, 43:101–110.

Coombe, B.G. and Clancy, T. (2001). Ripening berries – A critical issue. *Australian Viticulture Editor*, March/April.

Coombe, B.G. and McCarthy, M.G. (2000). Dynamics of grape berry growth and physiology of ripening. *Aust. J. Grape Wine Res.*, 6(2):1–135.

Correia, A.C. and Jordão, A.M. (2015). Antioxidant capacity, radical scavenger activity, lipid oxidation protection analysis and antimicrobial activity of red grape extracts from different varieties cultivated in Portugal. *Nat. Prod. Res.*, 29(5):438–440.

Cosme, F., Ricardo-Da-Silva, J.M. and Laureano, O. (2009). Tannin profiles of Vitis vinifera L. cv. red grapes growing in Lisbon and from their monovarietal wines. *Food Chem.*, 112(1):197–204.

Costa, E., Cosme, F., Jordão, A.M. and Mendes-Faia, A. (2014). Anthocyanin profile and antioxidant activity from 24 grape varieties cultivated in two Portuguese wine regions. *J. Int. Sci. Vigne Vin*, 48(1):51–62.

Costa, E., Cosme, F., Rivero-Pérez, M.D., Jordão, A.M. and González-SanJosé, M.L. (2015a). Influence of wine region provenance on phenolic composition, antioxidant capacity and radical scavenger activity of traditional Portuguese red grape varieties. *Eur. Food Res. Technol.*, 241(1):61–73.

Costa, E., Da Silva, J.F., Cosme, F. and Jordão, A.M. (2015b). Adaptability of some French red grape varieties cultivated at two different Portuguese terroirs: Comparative analysis with two Portuguese red grape varieties using physicochemical and phenolic parameters. *Food Res. Int.*, 78:302–312.

Crippen, D.D. and Morrison, J.C. (1986). The effects of sun exposure on the compositional development of Cabernet Sauvignon berries. *Am. J. Enol. Vitic.*, 37:235–242.

De Bolt, S., Cook, D.R. and Ford, C.M. (2006). L-tartaric acid synthesis from vitamin C in higher plants. *Proc. Natl Acad. Sci. U. S. A.*, 103(14):5608–5613.

De Bolt, S., Ristic, R., Iland, P.G. and Ford, C.M. (2008). Altered light interception reduces grape berry weight and modulates organic acid biosynthesis during development. *Hortscience*, 43(3):957–961.

De Coninck, G., Jordão, A.M., Ricardo-da-Silva, J.M. and Laureano, O. (2006). Evolution of phenolic composition and sensory proprieties in red wine aged in contact with Portuguese and French oak wood chips. *J. Int. Sci. Vigne Vin*, 40:23–34.

De La Hera Orts, M.L., Martínez-Cutillas, A., López-Roca, J.M. and Gómez-Plaza, E. (2004). Effects of moderate irrigation on vegetative growth and productive parameters of Monastrell vines grown in semiarid conditions. *Spanish J. Agric. Res.*, 2(2):273–281.

De Orduña, R. (2010). Climate change associated effects on grape and wine quality and production. *Food Res. Int.*, 43(7):1844–1855.

Delcambre, A. and Saucier, C. (2012). Identification of new flavan-3-ol monoglycosides by UHPLC-ESI-Q-TOF in grapes and wine. J. Mass Spectrom., 47(6):727–736.

Delgado, R., Martín, P., del Álamo, M. and González, M.-R. (2004). Changes in the phenolic composition of grape berries during ripening in relation to vineyard nitrogen and potassium fertilisation rates. *J. Sci. Food Agric.*, 84(7):623–630.

Delrot, S., Atanassova, R. and Maurousset, L. (2000). Regulation of sugar, amino acid and peptide plant membrane transporters. *Biochim. Biophys. Acta*, 1465(1–2):281–230.

Delrot, S., Picaud, S. and Gaudillère, J.P. (2001). Water transport and aquaporins in grapevine. In: *Molecular Biology and Biotechnology of the Grapevine*, eds. K.A. Roubelakis-Angelakis, Kluwer Academic Publishers, Dordrecht, The Netherlands, 241–262.

Deluc, L.G., Decendit, A., Papastamoulis, Y., Mérillon, J.-M., Cushman, J.C. and Cramer, G.R. (2011). Water deficit increases stilbene metabolism in Cabernet Sauvignon berries. *J. Agric. Food Chem.*, 59(1):289–297.

Deluc, L.G., Quilici, D.R., Decendit, A., Grimplet, J., Wheatley, M.D., Schlauch, K.A., Merillon, J.M., Cushman, J.C. and Cramer, G.R. (2009). Water deficit alters differentially metabolic pathways affecting important flavour and quality traits in grape berries of Cabernet Sauvignon and Chardonnay. *BMC Genomics*, 10:212.

Di Stefano, R. and Cravero, M. (1991). Metodi per lo studio dei polifenoli dell'uva. *Rivista di Viticultura ed Enologia*, 2:37–45.

Dopico-Garcia, M.S., Fique, A., Guerra, L., Afonso, J.M., Pereira, O., Valentao, P., Andrade, P.B. and Seabra, R.M. (2008). Principal components of phenolics to characterize red Vinho Verde grapes: Anthocyanins or non-coloured compounds? *Talanta*, 75(5):1190–1202.

Downey, M.O., Dokoozlian, N.K. and Krstic, M.P. (2006). Cultural practice and environmental impacts on the flavonoid composition of grapes and wine: A review of recent research. *Am. J. Enol. Vitic.*, 57:257–268.

Downey, M.O., Harvey, J.S. and Robinson, S.P. (2003). Analysis of tannins in seeds and skins of Shiraz grapes throughout berry development. *Aust. J. Grape Wine Res.*, 9(1):15–27.

Downey, M.O., Harvey, J.S. and Robison, S.P. (2004). The effect of bunch shading on berry development on avonoid accumulation in Shiraz grapes. *Am. J. Enol. Vitic.*, 10:55–73.

Escribano-Bailón, M.T., Guerra, M.T., Rivas-Gonzalo, J.C. and Santos-Buelga, C. (1995). Proanthocyanidins in skins from different grape varieties. *Z. Lebensm. Unters.-Forsch.*, 200(3):221–224.

Esteban, M.A., Villanueva, M.J. and Lissarrague, J.R. (1999). Effect of irrigation on changes in berry composition of Tempranillo during maturation. Sugars, organic acids and mineral elements. *Am. J. Enol. Vitic.*, 50:418–434.

Esteban, M.A., Villanueva, M.J. and Lissarrague, J.R. (2001). Effect of irrigation on changes in the anthocyanin composition of the skin of cv. Tempranillo (*Vitis vinifera* L.) grape berries during ripening. *J. Sci. Food Agric.*, 81(4):409–420.

Etchebarne, F., Ojeda, H. and Hunter, J.J. (2010). Leaf:fruit ratio and vine water status effects on Grenache Noir (*Vitis vinfera* L.) berry composition: Water, sugar, organic acids and cations. *S. Afr. J. Enol. Vitic.*, 31:106–115.

Eyduran, S.P., Akin, M., Ercisli, S., Eyduran, E. and Maghradze, D. (2015). Sugars, organic acids, and phenolic compounds of ancient grape cultivars (*Vitis vinifera* L.) from Igdir Province of Eastern Turkey. *Biol. Res.*, 48:1–8.

Fernández-López, J.A., Hidalgo, V., Almela, L. and López-Roca, J.M. (1992). Quantitative changes in anthocyanin pigments of *Vitis vinifera* cv monastrell during maturation. *J. Sci. Food Agric.*, 58(1):153–155.

Ferreira, R.B., Piçarra-Pereira, M.A., Monteiro, S., Loureiro, V.B. and Teixeira, A.R. (2002). The wine proteins. *Trends Food Sci. Technol.*, 12(7):230–239.

Fragoso-García, S., Mestres-Sole, M., Busto-Busto, O. and Guasch-Torres, J. (2009). Estudio y optimización de un método analítico para la estimación de los parámetros de madurez fenólica. *Rev. Enólogos*, 58:38–43.

Galgano, F., Favati, F., Caruso, M., Scarpa, T. and Palma, A. (2008). Analysis of trace elements in southern Italian wines and their classification according to provenance. *LWT Food Sci. Technol.*, 41(10):1808–1815.

García-Escudero, E., López-Martín, R., Santamaría-Aquilue, P. and Zaballa-Ogueta, O. (1997). Ensayos de riego localizado en viñedos productivos de cv. Tempranillo. *Vitic. Enología Prof.*, 50:35–47.

Garde-Cerdan, T., Lorenzo, C., Lara, J.F., Pardo, F., Ancin-Azpilicueta, C. and Salinas, M.R. (2009). Study of the evolution of nitrogen compounds during grape ripening. Application to differentiate grape varieties and cultivated systems. *J. Agric. Food Chem.*, 57(6):2410–2419.

Gaudillère, J.P., Van Leeuwen, C. and Ollat, N. (2002). Carbon isotope composition of sugars in grapevine, an integrated indicator of vineyard water status. *J. Exp. Bot.*, 53(369):757–763.

Genebra, T., Santos, R.R., Francisco, R., Pinto-Marijuan, M., Brossa, R., Serra, A.T., Duarte, C.M.M., Chaves, M.M. and Zarrouk, O. (2014). Proanthocyanidin accumulation and biosynthesis are modulated by the irrigation regime in Tempranillo seeds. *Int. J. Mol. Sci.*, 15(7):11862–11877.

Ghozlen, B., Moise, N., Latouche, G., Martinon, V., Mercier, L., Besançon, E. and Cerovic, Z.G. (2010). Assessment of grapevine maturity using new portable sensor: Non-destructive quantification of anthocyanins. *J. Int. Sci. Vigne Vin*, 44:1–8.

Ghozlen, N.B., Cerovic, Z., Germain, C., Toutain, S. and Latouche, G. (2010). Non-destructive optical monitoring of grape maturation by proximal sensing. *Sensors*, 10(11):10040–10068.

Glories, Y. and Augsutin, M. (1993). Maturité phénolique du raisin, conséquaneces technologiques: Application aux millésimes 1991 et 1992. In: *Compte rendu colloque. Actes du colloque journée Technique CIVB.* Boredeaux, France, 21 janvier 1993, 56–61.

Hale, C.R. (1977). Relation between potassium and the malate and tartrate contents of grapes berries. *Vitis*, 16:9–19.

Harbertson, J.F., Picciotto, E.A. and Adams, D.O. (2003). Measurement of polymeric pigments in grape berry extracts and wines using a protein precipitation assay combined with bisulfite bleaching. *Am. J. Enol. Vitic.*, 54:301–306.

Hashizume, K. and Samuta, T. (1999). Grape maturity and light exposure affect berry methoxypyrazine concentration. *Am. J. Enol. Vitic.*, 50:194–198.

Hepner, Y. and Bravdo, B.A. (1985). Effect of crop level and drip irrigation scheduling on the potassium status of Cabernet Sauvingnon and Carignane vines and its influence on must and wine composition and quality. *Am. J. Enol. Vitic.*, 36:140–147.

Hernández, M.M., Song, S. and Menéndez, C.M. (2017). Influence of genetic and vintage factors inflavan-3-ol composition of grape seeds of a segregating *Vitis vinifera* population. *J. Sci. Food Agric.*, 97(1):236–243.

Hernandez-Jimenez, A., Gomez-Plaza, E., Martinez-Cutillas, A. and Kennedy, J.A. (2009). Grape skin and seed proanthocyanidins from Monastrell x Syrah grapes. *J. Agric. Food Chem.*, 57(22):10798–10803.

Hernandez-Orte, P., Concejero, B., Astrain, J., Lacau, B., Cacho, J. and Ferreira, V. (2015). Influence of viticulture practices on grape aroma precursors and their relation with wine aroma. *J. Sci. Food Agric.*, 95(4):688–701.

Heymann, H., Noble, A.C. and Boulton, R.B. (1986). Analysis of methoxypyrazines in wines. 1. Development of a quantitative procedure. *J. Agric. Food Chem.*, 34(2):268–271.

Hrazdina, G., Parsons, G.F. and Mattick, L.R. (1984). Physiological and biochemical events during development and maturation of grape berries. *Am. J. Enol. Vitic.*, 35:220–227.

Huglin, P. and Schneider, C. (1998). *Biologie et Ecologie de la Vigne.* 2 edn. Lavoisier, Technique et documentation: Commune, France.

Ignat, I., Volf, I. and Popa, V.I. (2011). A critical review of methods for characterization of polyphenolic compounds in fruits and vegetables. *Food Chem.*, 126(4):1821–1835.

Iland, P. (1987). Interpretation of acidity parameters in grapes and wine. *Aust. Grapegrower Winemaker*, 298:81–85.

Iland, P. (1989). Grape berry composition-the influence of enviromental and viticultural factors. *Aust. Grapegrower Winemaker*, 302:13–15.

Iland, P., Bruer, N., Edwards, G., Weeks, S. and Wilkes, E. (2004). *Chemical Analysis of Grapes and Wine: Techniques and Concepts.* Patrick Iland Wine Promotions, Campbelltwon, Australia, 110.

Jackson, D.L. and Lombard, P.B. (1993). Environmental and management practices affecting grape composition and wine quality review. *Am. J. Enol. Vitic.*, 4:409–430.

Jiranek, V., Langridge, P. and Henschke, P.A. (1995). Regulation of hydrogen sulfide liberation in wine producing Saccharomyces cerevisiae strains by assimilable nitrogen. *Appl. Environ. Microbiol.*, 61(2):461–467.

Jordão, A.M. and Correia, A.C. (2012). Relationship between antioxidant capacity, proanthocyanidin and anthocyanin content during grape maturation of Touriga Nacional and TintaRoriz grape varieties. *S. Afr. J. Enol. Vitic.*, 33:214–224.

Jordão, A.M., Correia, A.C. and Gonçalves, F.J. (2012). Evolution of antioxidant capacity in seeds and skins during grape maturation and their association with proanthocyanidin and anthocyanin content. *Vitis*, 51:137–139.

Jordão, A.M., Ricardo-da-Silva, J.M. and Laureano, O. (1998). Evolution of anthocyanins during grape maturation of two varieties (*Vitis vinífera* L.), Castelão Francês and Touriga Francesa. *Vitis*, 37:93–94.

Jordão, A.M., Ricardo-da-Silva, J.M. and Laureano, O. (1998). Influência da rega na composição fenólica das uvas tintas da casta Touriga Francesa (*Vitis vinifera* L.). *J. Food*, 2:60–73.

Jordão, A.M., Ricardo-da-Silva, J.M. and Laureano, O. (2001). Evolution of catechins and oligomeric procyanidins during grape maturation of Castelão Francês and Touriga Francesa. *Am. J. Enol. Vitic.*, 52:230–234.

Jordão, A.M., Vilela, A. and Cosme, F. (2015). From sugar of grape to alcohol of wine: Sensorial impact of alcohol in wine. *Beverages*, 1(4):292–310.

Ju, Z.Y. and Howard, L.R. (2003). Effects of solvent and temperature on pressurized liquid extraction of anthocyanins and total phenolics from dired red grape skin. *J. Agric. Food Chem.*, 51(18):5207–5213.

Keller, M., Pool, R.M. and Henick-Kling, T. (1999). Excessive nitrogen supply and shoot trimming can impair colour development in Pinot Noir grapes and wine. *Aust. J. Grape Wine Res.*, 5(2):45–55.

Keller, M., Smith, J.P. and Bondada, B.R. (2006). Ripening grape berries remain hydraulically connected to the shoot. *J. Exp. Bot.*, 57(11):2577–2587.

Kennedy, J.A., Mattheus, M.A. and Waterhouse, A.L. (2000). Changes in grape seed polyphenols during fruit ripening. *Phytochemistry*, 55(1):77–85.

Kennedy, J.A. (2008). Grape and wine phenolics: Observations and recent findings. *Cien. Cienc. Inv. Agr.*, 35(2):107–120.

Khairallah, R., Reynolds, A.G. and Bowen, A.J. (2016). Harvest date effects on aroma compounds in aged Riesling icewines. *J. Sci. Food Agric.*, 96(13):4398–4409.

Kingston-Smith, A.H. (2001). Resource allocation. *Trends Plant. Sci.*, 6(2):48–49.

Kliewer, W.M. (1964). Influence of environment on metabolism of organic acids and carbohydrates in *Vitis vinifera*. I. Temperature. *Plant Physiol.*, 39(6):869–880.

Kliewer, W.M. (1966). Sugars and organic acids of *Vitis vinifera*. *Plant Physiol.*, 41(6):923–931.

Kliewer, W.M. and Dokoozlian, N.K. (2005). Leaf area/crop weight ratios of grapevines: Influence on fruit composition and wine quality. *Am. J. Enol. Vitic.*, 56:170–181.

Kliewer, W.M. and Schultz, H.B. (1964). Influence of environment on metabolism of organic acids and carbohydrates in *Vitis vinifera*. II. Light. *Am. J. Enol. Vitic.*, 15:119–129.

Kontoudakis, N., Esteruelas, M., Fort, F., Canals, J.M. and Zamora, F. (2010). Comparison of methods for estimating phenolic maturity in grapes: Correlation between predicted and obtained parameters. *Anal. Chim. Acta*, 660(1–2):127–133.

Kontoudakis, N., Esteruelas, M., Fort, F., Canals, J.M., De Freitas, V. and Zamora, F. (2011). Influence of the heterogeneity of grape phenolic maturity on wine composition and quality. *Food Chem.*, 124(3):767–774.

Koundouras, S., Marinos, V., Gkoulioti, A., Kotseridis, Y. and Van Leeuwen, C. (2006). Influence of vineyard location and vine water status on fruit maturation of nonirrigated cv. Agiorgitiko (*Vitis vinifera* L.). Effects on wine phenolic and aroma components. *J. Agric. Food Chem.*, 54(14):5077–5086.

Kriedemann, P. and Smart, R. (1971). Effect of irradiance, temperature and leaf water potential on photosynthesis of vine leaves. *Photosynthetica*, 5:6–15.

Kristl, J., Veber, M. and Slekovec, M. (2002). The application of ETAAS to the determination of Cr, Pb and Cd in samples taken during different stages of the winemaking process. *Anal. Bioanal. Chem.*, 373(3):200–204.

Kyraleou, M., Kotseridis, Y., Koundouras, S., Chira, K., Teissedre, P.L. and Kallithraka, S. (2016). Effect of irrigation regime on perceived astringency and proanthocyanidin composition of skins and seeds of *Vitis vinifera* L. cv. Syrah grapes under semiarid conditions. *Food Chem.*, 203:292–300.

Lamadon, F. (1995). Protocole pour l'évolution de la richesse polyphénolique des raisins. *Revue. Françaised'Oenologie*, 76:37–38.

Lamikanra, O., Inyang, I. and Leong, S. (1995). Distribution and effect of grape maturity on organic acid content of red muscadine grapes. *J. Agric. Food Chem.*, 43(12):3026–3028.

Lang, A. and Thorpe, M.R. (1989). Xylem, phloem and transpiration flows in a grape: Application of a technique for measuring the volume of attached fruits to high resolution using Archimedes' principle. *J. Exp. Bot.*, 40(10):1069–1078.

Lee, J. and Schreiner, R.P. (2010). Free amino acid profiles from "Pinot Noir" grapes are influenced by vine N-status and sample preparation method. *Food Chem.*, 119(2):484–489.

Lee, S., Seo, M., Riu, M., Cotta, J., Block, D., Dokoozilian, N. and Ebeler, S. (2007). Vine microclimate and norisoprenoid concentration in Cabernet Sauvignon grapes and wine. *Am. J. Enol. Vitic.*, 58:291–301.

Liu, H.F., Wu, B.H., Fan, P.G., Li, S.H. and Li, L.S. (2006). Sugar and acid concentrations in 98 grape cultivars analyzed by principal component analysis. *J. Sci. Food Agric.*, 86(10):1526–1536.

Liu, H.F., Wu, B.H., Fan, P.G., Xu, H.Y. and Li, S.H. (2007). Inheritance of sugars and acids in berries of grape (*Vitis vinifera* L.). *Euphytica*, 153(1–2):99–107.

López, M.I., Sánchez, M.T., Díaz, A., Ramírez, P. and Morales, J. (2007). Influence of a deficit irrigation regime during ripening on berry composition in grapevines (*Vitis vinifera* L.) grown in semi-arid areas. *Int. J. Food Sci. Nutr.*, 58(7):491–507.

Lorrain, B., Kleopatra, C. and Teissedre, P.L. (2011). Phenolic composition of Merlot and Cabernet-Sauvignon grapes from Bordeaux vineyard for the 2009-vintage: Comparison to 2006, 2007 and 2008 vintages. *Food Chem.*, 126(4):1991–1999.

Mateo, J.J. and Jiménez, M. (2000). Monoterpenes in grape juice and wines. *J. Chromatogr. A*, 881(1–2):557–567.

Mateus, N., Machado, J.M. and De Freitas, V. (2002). Development changes of anthocyanins in Vitis vinifera grapes grown in the Douro Valley and concentration in respective wines. *J. Sci. Food Agric.*, 82(14):1689–1695.

Mateus, N., Marques, S., Gonçalves, A.C., Machado, J.M. and De Freitas, V. (2001). Proanthocyanidin composition of red *Vitis vinifera* varieties from the Douro Valley during ripening: Influence of cultivation altitude. *Am. J. Enol. Vitic.*, 52:115–121.

Matthews, M.A. and Anderson, M.M. (1988). Fruit ripening in *Vitis vinifera* L.: Responses to seasonal water deficits. *Am. J. Enol. Vitic.*, 39:313–320.

Matthews, M.A. and Anderson, M.M. (1989). Reproductive development in grape (*Vitis vinifera* L.): Responses to seasonal water deficits. *Am. J. Enol. Vitic.*, 40:52–60.

Mattivi, F. (2006). Gliindici di maturazione dele uve e la loroimportanza. *Quaderni di viticoltura ed Enologia della Università di Torino*, 28:27–40.

McCarthy, M.G. and Coombe, B.G. (1999). Is weight loss in ripening grape berries cv. Shiraz caused by impeded phloem transport? *Aust. J. Grape Wine Res.*, 5(1):17–21.

Melendez, E., Ortiz, M.C., Sarabia, L.A., Iniguez, M. and Puras, P. (2013). Modelling phenolic and technological maturities of grapes by means of the multivariate relation between organoleptic and physicochemical properties. *Anal. Chim. Acta*, 761:53–61.

Mendez-Costabel, M.P., Wilkinson, K.L., Bastian, S.E.P., Jordans, C., McCarthy, M., Ford, C.M. and Dokoozlian, N. (2014). Effect of winter rainfall on yield components and fruit green aromas of *Vitis vinifera* L. cv. Merlot in California. *Aust. J. Grape Wine Res.*, 20(1):100–110.

Mirás-Avalos, J.M. and Intrigliolo, D.S. (2017). Grape composition under abiotic constrains: Water stress and salinity. *Front. Plant Sci.*, 8:851.

Monagas, M., Gómez-Cordovés, C., Bartolomé, B., Laureano, O. and Ricardo-da-Silva, J.M. (2003). Monomeric, oligomeric, and polymeric flavan-3-ol composition of wines and grapes from *Vitis vinifera* L. Cv. Graciano, Tempranillo, and Cabernet Sauvignon. *J. Agric. Food Chem.*, 51(22):6475–6481.

Moreno-Arribas, M.V. and Polo, M.C. (2009). Amino acids and biogenic amines. In: *Wine Chemistry and Biochemistry*, Editors: Moreno-Arribas, M.V. and Polo, M.C. Springer, New York.

Mpelasoka, B.S., Schachtman, D.P., Treeby, M.T. and Thomas, M.R. (2003). A review of potassium nutrition in grapevines with special emphasis on berry accumulation. *Aust. J. Grape Wine Res.*, 9(3):154–168.

Muñoz-Robredo, P., Robledo, P., Manríquez, D., Molina, R. and Defilippi, B.G. (2011). Characterization of sugars and organic acids in commercial varieties of table grapes. *Chil. J. Agr. Res.*, 71(3):452–458.

Oberholster, A., Elmendorf, B.L., Lerno, L.A., King, E.S., Heymann, H., Brenneman, C.E. and Boulton, R.B. (2015). Barrel maturation, oak alternatives and micro-oxygenation: Influence on red wine aging and quality. *Food Chem.*, 173:1250–1258.

Obreque-Slier, E., López-Solís, R., Castro-Ulloa, L., Romero-Díaz, C. and Peña-Neira, A. (2012). Phenolic composition and physicochemical parameters of Carménère, Cabernet Sauvignon, Merlot and Cabernet Franc grape seeds (*Vitis vinifera* L.) during ripening. *Food Sci. Technol.-Leb*, 48(1):134–141.

Obreque-Slier, E., Pena-Neira, A., Lopez-Solis, R., Zamora-Martin, F., Ricardo-da-Silva, J.M. and Laureano, O. (2010). Comparative study of the phenolic composition of seeds and skins from Carménère and Cabernet-Sauvignon grape varieties (*Vitis vinifera* L.) during ripening. *J. Agric. Food Chem.*, 58(6):3591–3599.

Ojeda, H., Andary, C., Kraeva, E., Carbonneau, A. and Deloire, A. (2002). Influence of pre- and postveraison water deficit on synthesis and concentration of skin phenolic compounds during berry growth of *Vitis vinifera* cv. Shiraz. *Am. J. Enol. Vitic.*, 53:261–267.

Oliveira, J.M., Araújo, I.M., Pereira, O.M., Maia, J.S., Amaral, A.J. and Maia, M.O. (2004). Characterization of differentiation of five "Vinhos Verdes" grape varieties on the basis of monoterpenic compounds. *Anal. Chim. Acta*, 513:269–275.

Oliveira, J.M., Faria, M., Sá, F., Barros, F. and Araújo, I.M. (2006). C_6-alcohols as varietal markers for assessment of wine origin. *Anal. Chim. Acta*, 563(1–2):300–309.

Ollat, N., Diakou-Verdin, P., Carde, J.P., Barrieu, F., Gaudillère, J.P. and Moing, A. (2002). Grape berry development: A review. *J. Int. Sci. Vigne Vin*, 36:109–131.

Ollé, D., Guiraud, J.L., Souquet, J.M., Terrier, N., Ageorges, A., Cheynier, V. and Verries, C. (2011). Effect of pre- and post-veraison water deficit on proanthocyanidin and anthocyanin accumulation during Shiraz berry development. *Aust. J. Grape Wine Res.*, 17(1):90–100.

Ó-Marques, J., Reguinga, R., Laureano, O. and Ricardo-da-Silva, J.M. (2005). Changes in grape seed, skin and pulp condensed tannins during berry ripening: effect of fruit pruning. *Cienc. Tec. Vitivinic.*, 20:35–52.

Ortega-Heras, M., Pérez-Magariño, S., Del-Villar-Garrachón, V., González-Huerta, C., Moro Gonzalez, L.C., Guadarrama Rodríguez, A., Villanueva Sanchez, S., Gallo González, R. and Martín de la Helguera, S. (2014). Study of the effect of vintage, maturity degree, and irrigation on the amino acid and biogenic amine content of a white wine from the *Verdejo* variety. *J. Sci. Food Agric.*, 94(10):2073–2082.

Paneque, P., Álvarez-Sotomayor, M.T., Clavijo, A. and Gómez, I.A. (2010). Metal content in southern Spain wines and their classification according to origin and ageing. *Microchem. J.*, 94(2):175–179.

Pastore, C. (2010). Researches on berry composition in red grape: agronomical, biochemical and molecular approaches. Tesi di Dottorato di ricerca in Colture Arboree ed Agrosistemi Forestali e Paesaggistici, XXII ciclo, Faculta di Agraria. Universita di Bologna.

Pastrana-Bonilla, E., Akoh, C.C., Sellappan, S. and Krewer, G. (2003). Phenolic content and antioxidant capacity of muscadine grapes. *J. Agric. Food Chem.*, 51(18):5497–5503.

Piñeiro, Z., Palma, M. and Barroso, C.G. (2004). Determination of catechins by means of extraction with pressurized liquids. *J. Chromatogr. A*, 1016(1–2):19–23.

Pinelo, M., Arnous, A. and Meyer, A.S. (2006). Understanding of grape skins: Significance of plant cell-wall structural components and extraction techniques for phenol release. *Trends Food Sci. Tech.*, 17(11):579–590.

Prieur, C., Rigaud, J., Cheynier, V. and Moutounet, M. (1994) Oligomeric and polymeric procyanidins from grape seeds. *Phytochemistry*, 36(3):781–784.

Proffitt, T., Bramley, R., Lamb, D. and Winter, E. (2006). *Percision Viticulture. A New Era in Vineyard Management and Wine Production; Winetitles*, Ashfort, Winetitles Pty Ltd Australia, 90.

Puissant, A. and Lén, H. (1967). La matière colorante des grains de raisins de certains cépages cultivés en Anjou en 1965. *Annals de Tecnologie Agricole*, 16:217–225.

Ramos, M.C. and Romero, M.P. (2017). Potassium uptake and redistribution in Cabernet Sauvignon and Syrah grape tissues and its relationship with grape quality parameters. *J. Sci. Food Agric.*, 97(10):3268–3277.

Reynard, J.-S., Zufferey, V., Nicole, G. and Murisier, F. (2011). Soil parameters impact the vine-fruit-wine continuum by altering vine nitrogen status. *J. Int. Sci. Vigne Vin*, 45(4):211–221.

Reynolds, A.G. and Vanden Heuvel, J.E. (2009). Influence of grapevine training systems on vine growth and fruit composition: A review. *Am. J. Enol. Vitic.*, 60:251–268.

Ribéreau- Gayon, P. (1965). *Les composes Phenoliques du raisin et du vin.* Institut National de la Recherche Agronomique, Paris.

Ribéreau-Gayon, P., Glories, Y., Maujean, A. and Dubourdieu, D. (2006). *Handbook of Enology: Volume 2. The Chemistry of Wine Stabilization and Treatments.* John Wiley & Sons, LTD, Chichester, England.

Riou, V. and Asselin, C. (1996). Potential polyphénoloque disponible du rasin. Estimation rapide par extraction partielle à chaud. *Porgès Agricole et Viticole*, 113:382–384.

Robinson, S.P. and Davies, C. (2000). Molecular biology of grape berry ripening. *Aust. J. Grape Wine Res.*, 6(2):175–188.

Roby, G., Harbertson, J.F., Adams, D.A. and Matthews, M.A. (2004). Berry size and vine water deficits as factors in winegrape composition: Anthocyanins and tannins. *Aust. J. Grape Wine Res.*, 10(2):100–107.

Rodriguez, M.R., Romero-Peces, R., Chacon-Vozmediano, J.L., Martinez-Gascuena, J. and Garcia Romero, E. (2006). Phenolic compounds in skins and seeds of ten grape *Vitis vinifera* varieties grown in a warm climate. *J. Food Comp. Anal.*, 19:687–693.

Rodriguez-Lovelle, B. and Gaudillère, J.P. (2002). Carbon and nitrogen partitioning in either fruiting or non-fruiting grapevines: Effects of nitrogen limitation before and after veraison. *Aust. J. Grape Wine Res.*, 8(2):86–94.

Romero, R., Chacón, J., García, E. and Martínez, J. (2006). Pyrazine contents in four red grape varieties cultivated in a warm climate. *J. Int. Sci. Vigne Vin*, 40:203–207.

Roubelakis-Angelakis, K.A. and Kliewer, W.M. (1992). Nitrogen metabolism in grapevine. *Hort. Rev.*, 14:407–452.

Roujou De Boubée, D. and Dubourdieu, D. (1999). Incidence des conditions de maturation et des practiques viticoles sur la teneur en 2-methoxy-3-isobutylpyrazine des raisins de Cabernet Sauvignon et de Merlot à Bordeaux. In: *Oenologie 99, 6e Symp. Int. Œnol. Bordeaux 10 to 12 June 1999 Lonvaud and Funel Ed. Tec&Doc*, 126–130.

Rousseau, J. (2001). Quantified description of sensory analysis of berries. Relationships with wine profiles and consumer tastes. *Bulletin de l'OIV*, 74(849–850):719–728.

Sabir, A., Kafkas, E. and Tangolar, S. (2010). Distribution of major sugars, acids and total phenols in juice of five grapevine (*Vitis* spp.) cultivars at different stages of berry development. *Spanish, J. Agric. Res.*, 8(2):425–433.

Sala, C., Busto, O., Guasch, J. and Zamora, F. (2005). Contents of 3-alkyl-2-methoxypyrazines in musts and wines from *Vitis vinifera* variety Cabernet Sauvignon: Influence of irrigation and plantation density. *J. Sci. Food Agric.*, 85(7):1131–1136.

Sala, C., Zamora, F., Busto, O. and Guasch, J.S. (2004). Influence of vine training and sunlight exposure on the 3-alkyl-2-methoxypirazines content in musts and wines from the Vitis vinifera variety Cabernet-Sauvignon. *J. Agric. Food Chem.*, 52(11):3492–3497.

Samoticha, J., Wojdylo, A. and Golis, T. (2017). Phenolic composition, physicochemical properties and antioxidant activity of interspecific hybrids of grapes growing in Poland. *Food Chem.*, 215:263–273.

Sánchez-Palomo, E., Díaz-Maroto, M.C., GonzálezViñas, M.A., Soriano-Pérez, A. and Pérez-Coello, M.S. (2007). Aroma profile of wines from Albillo and Muscat grape varieties at different stages of ripening. *Food Control*, 18(5):398–403.

Santos, T.P., Lopes, C.M., Rodrigues, M.L., Souza, C.R., Ricardo-Da- Silva, J.M., Maroco, J.P., Pereira, J.S. and Chaves, M.M. (2007). Effect of deficit irrigation strategies on cluster microclimate for improving fruit composition of Moscatel field-grown grapevines. *Sci. Hort.*, 112:321–330.

Schultz, H.R. (1996). Water relations and photosynthetic responses of two grapevine cultivars of different geographical origin during water stress. *Acta Hort.*, 427(427):251–266.

Sefton, M.A., Francis, I.L. and Williams, P.J. (1993). The volatile composition of Chardonnay juices: A study of flavour precursor analysis. *Am. J. Enol. Vitic.*, 44:359–370.

Sefton, M.A., Francis, I.L. and Williams, P.J. (1996). The free and bound volatile secondary metabolites of VitisVinifera grape cv. Semillon. *Aust. J. Grape Wine Res.*, 2:179–183.

Serafini, M., Laranjinha, J.A.N., Almeida, L.M. and Maiani, G. (2000). Inhibition of human LDL lipid peroxidation by phenol-rich beverages and their impact on plasma total antioxidant in humans. *J. Nutr. Biochem.*, 11(11–12):585–590.

Sério, S., Rivero-Pérez, M.D., Correia, A.C., Jordão, A.M. and González-SanJosé, M.L. (2014). Analysis of commercial grape raisins: Phenolic content, antioxidant capacity and radical scavenger activity. *Ciência Téc. Vitiv*, 29:1–8.

Shi, J., Yu, J., Pohorly, J.E. and Kakuda, Y. (2003). Polyphenolics in grape seeds – biochemistry and functionality. *J. Med. Food*, 6(4):291–299.

Shiraishi, M. (2000). Comparison in changes in sugars, organic acids and amino acids during berry ripening of sucrose- and hexose-accumulating grape cultivars. *J. Japan. Soc. Hortic. Sci.*, 69(2):141–148.

Shiraishi, M., Fujishima, H. and Chijiwa, H. (2010). Evaluation of table grape genetic resources for sugar, organic acid, and amino acid composition of berries. *Euphytica*, 174(1):1–13.

Sidhu, D., Lund, J., Kotseridis, Y. and Saucier, C. (2015). Methoxypyrazine analysis and influence of viticultural and enological procedures on their levels in grapes, musts, and wines. *Crit. Rev. Food Sci. Nutr.*, 55(4):485–502.

Singleton, V.L., Timberlake, C.F. and Lea, A.G.H. (1978). The phenolic cinnamates of white grapes and wine. *J. Sci. Food Agric.*, 29(4):403–410.

Singleton, V.L., Zaya, J. and Trousdale, E.K. (1986). Caftaric and coutaric acids in fruits of vitis. *Phytochemistry*, 25(9):2127–2133.

Soufleros, E.H., Bouloumpasi, E., Zotou, A. and Loukou, Z. (2007). Determination of biogenic amines in Greek wines by HPLC and ultraviolet detection after dansylation and examination of factors affecting their presence and concentration. *Food Chem.*, 101(2):704–716.

Souquet, J.M., Cheynier, V., Brossaud, F. and Moutounet, M. (1996). Polymeric proanthocyanidins from grape skins. *Phytochemistry*, 43(2):509–512.

Souza, C.R., Maroco, J.P., Santos, T.P., Rodrigues, M.L., Lopes, C.M., Pereira, J.S.. and Chaves, M.M. (2005). Grape berry metabolism in field-grown grapevines exposed to different irrigation strategies. *Vitis*, 44:103–109.

Spayd, S.E., Tarara, J.M., Mee, D.L. and Ferguson, J.C. (2002). Separation of sunlight and temperature effects on the composition of *Vitis vinifera* cv. Merlot berries. *Am. J. Enol. Vitic.*, 53:171–182.

Spayd, S.E., Wample, R.L., Evans, R.G., Stevens, R.G., Seymour, B.J. and Nagel, C.W. (1994). Nitrogen fertilisation of white Riesling grapes in Washington. Must and wine composition. *Am. J. Enol. Vitic.*, 45:34–42.

Spring, J.L. and Lorenzini, F. (2006). Effet de la pulvérisation foliaire d'urée sur l'alimentation azotée et la qualité du Chasselas en vigne enherbée. *Rev. Suisse Vitic. Arboric. Hortic.*, 38:105–113.

Spring, J.L., Verdenal, T., Zufferey, V., Gindro, K. and Viret, O. (2012). Influence du porte-greffe sur le comportement du cépage Cornalin du Valais central. *Rev. Suisse Vitic. Arboric. Hortic.*, 44:298–307.

Sun, B.S., Leandro, C., Ricardo-da-Silva, J.M. and Spranger, I. (1998). Separation of grape and wine proanthocyanidins according to their degree of polymerization. *J. Agric. Food Chem.*, 46(4):1390–1396.

Sun, B.S., Ricardo-da-Silva, J.M. and Spranger, I. (2001). Quantification of catechins and proanthocyanidins in several Portuguese grapevine varieties and red wines. *Ciência Téc Vitiv*, 16:23–34.

Tapiero, H., Tew, K.D., Nguyen, B.A.G. and Mathe, G. (2002). Polyphenols: Do they play a role in prevention of human pathologies? *Biomed. Pharmacother.*, 56:200–207.

Teissedre, P.L., Frankel, E.N., Waterhouse, A.L., Peleg, H. and German, J.B. (1996). Inhibition of in vitro human LDL oxidation by phenolic antioxidants from grapes and wines. *J. Sci. Food Agric.*, 70(1):55–61.

Topalovic, A. and Milukovic-Petkovsek, M. (2010). Changes in sugars, organic acids and phenolics of grape berries of cultivar cardinal during ripening. *J. Food Agric. Environ.*, 2:223–227.

Torchio, F., Cagnasso, E., Gerbi, V. and Rolle, L. (2010). Mechanical properties, phenolic composition and extractability indices of Barbera grapes of different soluble solids contents from several growing areas. *Anal. Chim. Acta*, 660(1–2):183–189.

Tounsi, M.S., Ouerghemmi, I., Wannes, W.A., Ksouri, R., Zemni, H., Marzouk, B. and Kchouk, M.E. (2009). Valorization of three varieties of grape. *Ind. Crops Prod.*, 30(2):292–296.

Treeby, M.T., Holzapfel, B.P., Pickering, G.J. and Friedrich, C.J. (2000). Vineyard nitrogen supply and Shiraz grape and wine quality. *Acta Hort.*, 512(512):77–92.

Trégoat, O., Van Leeuwen, C., Choné, X. and Gaudillère, J.-P. (2002). Étude du régime hydrique et de la nutrition azotée de la vigne par des indicateurs physiologiques. Influence sur le comportement de la vigne et la maturation du raisin. *J. Int. Sci. Vigne Vin*, 36:133–142.

Ubalde, J.M., Sort, X., Zayas, A.A. and Poch, R.M. (2010). Effects of soil and climatic conditions on grape ripening and wine quality of Cabernet Sauvignon. *J. Wine Res.*, 21(1):1–17.

Versari, A., Parpinello, G.P., Tornielli, G.B., Ferrarini, R. and Giulive, C. (2001). Stilbene compounds and stilbene synthase expression during ripening, wilting, UV treatment in grape cv. Corina. *J. Agric. Food Chem.*, 49(11):5531–5536.

Vilela, A., Jordão, A.M. and Cosme, F. (2016). Wine phenolics: Looking for a smooth mouthfeel. *J. Food Sci. Technol.*, 1:1–8.
Welch, R.M. (1986). Effects of nutrient deficiencies on seed production and quality. *Ad Plant Nutri.*, 2:205–247
Yamane, T., Jeong, S.T., Goto-Yamamoto, N., Koshita, Y. and Kobayashi, S. (2006). Effects of temperature on anthocyanins biosynthesis in grape berry skins. *Am. J. Enol. Vitic.*, 57:54–59.
Yokotsuta, K., Makino, S. and Singleton, V.L. (1988). Polyphenol oxidase from grapes: Precipitation, re-solubilization and characterization. *Am. J. Enol. Vitic.*, 39:293–302.
Zarrouk, O., Brunetti, C., Egipto, R., Pinheiro, C., Genebra, T., Gori, A., Lopes, C.M., Tattini, M. and Chaves, M.M. (2016). Grape ripening is regulated by deficit irrigation/elevated temperatures according to cluster position in the canopy. *Front Plant Sci.*, 7:1640.
Zerihun, A., McClymont, L., Lanyon, D., Goodwin, I. and Gibberd, M. (2015). Deconvoluting effects of vine and soil properties on grape berry composition. *J. Sci. Food Agric.*, 95(1):193–203.

10 Wine Quality
Varietal Influence

Vishal S. Rana, Neerja S. Rana and Sunny Sharma

CONTENTS

10.1 INTRODUCTION

Generally, it is said that "great wines are made in the vineyards", which means that for commercial winemaking as a whole, high levels of cooperation are required between the grape growers and the wine makers. The wine industry of the world is built mainly upon one species, *Vitis vinifera* L., which is native to the area of Asia Minor near the Black and Caspian seas. It is generally termed as a 'European grape'. However, specific location choices and general considerations of vineyard site and associated climatic influences may also affect the choice of the suitable grape variety and wine type. Wine aroma and flavour are determined by a complex balance of several volatiles. The wide varieties of such volatile compounds have been identified in wine, which impart primary floral and fruit aromas.

A clear and distinct aroma and flavour difference between cultivars can be attributed to relatively minor variations in the ratios of the compounds that constitute the aroma profile of a grape. Only a few aroma compounds have been directly linked to specific varietal flavours and aromas. Although most of these compounds are present at low concentrations in both grapes and the fermented wine, they have a huge impact on the overall aroma profile. Varietal wine aroma from Muscat-related grapes is mainly due to the presence of various isoprenoid monoterpenes in the grapes, the most important being linalool, geraniol, nerol, and citronellol. These compounds are formed from the precursor mevalonate and which is found in free and odourless glycosidically-bound forms in grape berries. During fermentation, yeast can release glucosidases, and these enzymes can hydrolyse the glycosidic bonds of the odourless bound forms of monoterpenes, releasing more odour-contributing compounds to the wine. It has been found, however, that the formation of some aromas associated with varietal character can be an integral part of yeast metabolism and not a simple hydrolytic process. Another set of varietal aroma compounds released from odourless bound precursors are volatile thiols, *i.e.* 4-methyl-4-mercaptopentan-2-one and 3-mercapto-1-hexanol, that give Sauvignon blanc wines their characteristic bouquet. These compounds are not present in grape juice in their active form, but occur in grape must as odourless, non-volatile, cysteine-bound conjugates. The wine yeast is responsible for the cleaving of thiol from the precursor during alcoholic fermentation. It is interesting to note that some varietal aromas occur completely independently of each other. It is thought that the 'green' characters in Sauvignon blanc wines imparted by 3-isobutyl-2-methoxypyrazines can be manipulated through vineyard management. However, the 'tropical fruity' characters imparted by 4-methyl-4-mercaptopentan-2-one and 3-mercapto-1-hexanol appear to be largely dependent on the wine yeast strain used during fermentation. The carotenoids also play a role in varietal aroma. They originate from the precursor compound mevalonate. Oxidation of these carotenoids produces volatile and strong odour-contributing fragments known as C_{13}-norisoprenoids, including b-ionone (viola aroma), b-damascenone (exotic fruits), b-damascone (rose), and b-ionol (fruits and flowers). However, many of the aroma and flavour compounds found in the finished wine don't come from the grape, but rather from compounds formed during primary (essential) or secondary metabolism of the wine yeast during alcoholic fermentation. Besides this, esters also contribute to aroma by imparting fruity flavours to the wine and are the indispensable components of wine quality. Primary fruit aromas in red grapes are often identified as cherry, blackcurrant, strawberry, raspberry, and plum, which are often used as flavour descriptors in the evaluation of wines made from several grape varieties.

Certain wine making techniques allow a better extraction of volatile components, especially the phenols. Quantification of various aroma compounds in wine at very minute levels has been achieved by solvent extraction and concentration, followed by several sophisticated analytical techniques. The extent of varietal impact on quality and aroma of wine has been examined and discussed at length in this chapter.

10.2 FACTORS AFFECTING WINE AROMA AND QUALITY

10.2.1 Varieties

The varieties noted for making wines with desirably distinctive flavours include Grenache, Cabernet Sauvignon, Muscats as a group, Merlot, Semillon, and Riesling. A small number could be further added to this list, including famous but less cultivated varieties such as Chardonnay, Gewurztraminer, and Pinot noir. Distinctive and desirably flavoured varieties are sought by wine makers for new flavourful wines to distinguish their listing from others in the marketplace. For determining the suitability of the fruit variety in winemaking and the typical characteristics of grape varieties, see Chapter 7 of this text. From the vintner's point of view, the flavour characteristics of the wines are probably the most crucial factors in varietal selection.

Muscat varieties are the most extensively planted grape types for the production of Pisco under Chilean legislation. A distinctly muscat, fruity aroma distinguishes Pisco from other young distillates. Most significant terpenes like linalool, nerol, geraniol, and other aromatic components of the main grape varieties grown in the Pisco-producing region of Chile have been quantified. Moscatel de Alejandria and Moscatel Rosada have proven to be highly aromatic, while the terpene profiles of two little grown varieties, *i.e.* Early Muscat and Moscatel Amarilla, indicated that these could contribute much to the aroma of Pisco. Significant differences have also been detected in the concentrations of total free terpenols among different varieties. A large number of varieties have been studied in several areas of California, and wines made out of them were evaluated, both by chemical composition and sensory characteristics. Both established varieties and new hybrid grapes primarily used for small scale winemaking have been evaluated. The results from 20 years of wine variety evaluation and breeding at CSIRO, Division of Horticultural Research, Adelaide, South Australia, indicate that there is considerable potential to improve wine flavour and quality in hot irrigated vineyards which produce about 80 percent of Australian wine. An arbitrary interest ranking for promoting varieties within each wine style was developed from combined analysis of sugar (°Brix), wine pH, and titratable acid (g/l), also including the panel's comments regarding flavour and aroma, the panel's score, and the vine yield. An assessment of performance characters of the most promising varieties, the country of their origin, and ripening time are presented in Table 10.1.

10.2.2 Volatile Constituents Contributing to Wine Aroma and Quality

10.2.2.1 Glycosidic Precursors

Grape, being a major horticultural crop of the world, is among the earliest of many fruits to have been studied for flavour precursors. The presence of precursors in fruits that could act as a source of latent or potential flavour was recognized many years ago. These precursors are the non-volatile conjugates of mevalonic acid and shikimic acid derived from secondary metabolites. The conjugation found most commonly in fruit flavour precursors is glycosidic. Some grapes have a readily perceived and highly distinctive flavour. These are the fruits of floral varieties like various Muscats, Riesling, and Gewurztraminer. Free monoterpene compounds are responsible for the sensory properties of these grapes. Monoterpene glycosides were first isolated and identified as flavour precursors in grape berries and wines made from these varieties. Monoterpene concentration in different grape varieties has also been determined. Such varieties have been screened and classified into three categories: i) intensely flavoured Muscats, in which total monoterpene concentrations can be as high as 6 mg/l; ii) non-Muscat but aromatic varieties with total monoterpene concentration of 1-4 mg/l; and iii) neutral varieties which do not depend upon monoterpenes for their flavour (Table 10.2). The Chardonnay grapes have been found to contain volatiles that are particularly rich in norisoprenoid compounds. As the floral grape varieties are monoterpene, dependent for their flavour, Chardonnay can be analogously categorized as norisoprenoid dependent. The volatile norisoprenoids are glycosidase – released aglycons. Plant carotenoid pigments have been considered as progenitors of these norisoprenoid flavour and aroma compounds in a number of products like tobacco and roses. The great majority of norisoprenoid compounds in Chardonnay grapes could be derived from four major xanthophylls reported in grapes, *viz.*, lutein, antheraxanthin, violaxanthin, and neoxanthin.

10.2.2.2 Phenolic Substances

Phenols are very important to grape and wine quality characteristics for imparting the astringent flavours and the known bitter substances. Major groups of phenols include phenolic acids, flavonoids, and tannin polymers. Phenolic acids include two major sub-groups, viz. hydroxycinamates and hydroxybenzoates. Hydroxycinamates are derivatives of caffeic, *p*-coumaric, and ferulic acids and are present in the easily expressed juice and are the same in red and white wines. Wine sensory properties, especially colour and taste,

TABLE 10.1
Detailed Ratings of the Assessment Parameters of Promising Varieties

Variety	Country of origin	Harvest period	Brix	pH	Acid	Flavour	Score	Yield
		A. Light red wine varieties						
Carignan	Spain	Mid March	****	**	**	***	***	****
Gamay	France	Mid February	****	*****	*****	***	**	*
Grenache	Spain	Late February	****	***	**	***	**	***
Pinot noir	France	Mid February	****	**	**	***	**	***
Tarrango	Australia	Late March	****	***	***	****	***	****
Valdiguié	France	Early March	****	*****	****	****	****	**
Zinfandel	United States	Mid March	***	***	*****	***	*	**
		B. Full bodied red wine varieties						
Cabernet Franc	France	Late February	****	**	***	****	***	***
Cabernet Sauvignon	France	Late February	****	*	***	****	***	**
Chambourcin	France	Mid February	****	***	*****	****	***	***
Malbec	France	Mid February	****	**	***	****	****	**
Roboso Piave	Italy	Late March	****	****	*****	****	***	*
Ruby Cabernet	United States	Late February	****	**	***	****	***	***
Shiraz	France	Late February	****	*	**	***	**	****
Tannat	France	Early March	****	***	****	****	***	**
		C. Delicate white wine varieties						
Emerald Riesling	United States	Late February	****	*****	****	***	****	**
GF 31-17-115	Germany	Mid January	****	*****	****	***	**	***
Rieslina (CG38.049)	Argentina	Late February	****	*****	***	***	***	***
Riesling	Germany	Late February	****	*****	****	****	***	**
		D. Full Bodied white wine varieties						
Chenin Blanc	France	Late February	***	****	****	***	***	***
Colombard	France	Early March	****	***	***	***	***	***
Ehrenfelser	Germany	Early February	****	****	***	***	***	**
Goyura	Australia	Early March	****	****	****	***	**	***
Semillon	France	Late February	****	**	**	***	***	***
		E. Aromatic white wine varieties						
Bacchus	Germany	Late January	****	***	***	****	****	***
Sauvignon blanc	France	Early February	****	***	***	***	**	***
Schönburger	Germany	Mid January	****	***	****	****	***	**
Taminga	Australia	Early March	****	****	***	****	****	*****
Verdelet	France	Early February	**	****	****	****	***	***
		F. Muscat wine varieties						
Gordo Blanco	Egypt	Mid March	****	*	*	****	****	****
Irsay Oliver	Hungary	Mid January	***	*****	*****	****	*****	**
Muscat a' petits grains rouge	France	Mid February	****	***	*	****	****	***

are largely related to the phenolic compounds extracted from the grape. These include flavonoids, including anthocyanins, flavonols, and flavanols. Each of these phenolic classes comprises various structures, differing by the number and position of hydroxyl groups, which can also be diversely substituted. Sensory analysis of six-month-old red wines made from Cabernet franc grapes grown in different locations, pointed out site-related characteristics which were attributed to flavanoids. The flavonoid composition of grapes harvested from characteristic vineyards chosen from various places in the Loire Valley, France indicated that the major differences in phenolic composition of Cabernet franc grapes within the Loire Valley concerned the anthocyanin levels and anthocyanin to tannin ratios. Qualitative anthocyanin composition was same in all the samples and, therefore, was regarded as a varietal character. Tannins also depended on both growing site and vintage.

Maturation in oak barrels leads to wines with much more complex sensory properties and is largely attributed to the phenols extracted from oak wood (see Chapter 28 of this

TABLE 10.2
Classification of Some Grape Varieties Based on Monoterpene Content

Muscat varieties	Non-muscat aromatic varieties	Varieties independent of monoterpenes for flavour	
Canada Muscat	Traminer	Bacchus	Merlot
Muscat of Alexandria	Huxel	Cabernet Sauvignon	Nobling
Muscat a petits	Kerner	Chardonnay	Rkaziteli
grains blancs	Morio Muskat	Carignan	Ruländer
Moscato Bianco	Müller Thurgau	Chasselas	Sauvignon blanc
del Piemonte	Riesling	Chenin blanc	Semillon
Muscat Hamburg	Schemebe	Cinsault	Shiraz
Muscat Ottonel	Schönburger	Clairette	Sultana
	Siegerebe	Dattier de Beyrouth	Terret
		Doradillo	
	Sylvaner	Forta	Trebbiano
		Grenache	Verdelho
			Viognier

Source: Adapted from William *et al.* (1981)

book). The wines acquire wood components such as volatile and nonvolatile phenols that are responsible for the so-called 'woody' character of wine. The variability in the phenolic components depends on the species, its geographical origin, aging and growth rate, and seasoning and toasting of the barrel. Both qualitative and quantitative changes occur in wine during barrel aging. The main phenolic species that give the typical character to barrique wines are substituted benzoic and cinnamic acids, and aldehydes. These are well known products of lignin and tannin degradation. Recently, more attention has been paid to these substances, since they have been proven to be chemopreventive agents for carcinogenesis. Two white wines prepared from Chardonnay and Picapoll free run juices were fermented on an industrial scale in wooden barrels, while controls were fermented in stainless steel. There was a characteristic oak wood phenol, conifer aldehyde, in white wines fermented in barrels, that was not detected in stainless steel vats fermented wine. There are several reports on the use of barrels in the aging of table wine but only scanty information is available on the use of wood chips in wines. The addition of wood chips to wine when maturing it in glass containers is one of the reasons to impart barrel-aged characteristics to wine. Treatment of apricot, plum, and mixed fruit wines with wood chips has also been done. Further, the chemical composition and sensory qualities of peach wines made from eight cultivars and aged with three different species of wood chips, viz. *Quercus*, *Bombax* and *Albizia*, were determined and compared with those aged without chip. Significant changes in the biochemical characteristics of the wines were observed. Wines aged with *Quercus* wood had higher total phenols, aldehydes, and more ester contents.

10.2.2.3 Methoxypyrazines

Wines made from Cabernet Sauvignon and Sauvignon blanc grapes often have a characteristic aroma described as vegetative, herbaceous, grassy, and green. These aromas have been attributed to methoxypyrazine (MP) components (Figure 10.1). The occurrence of 2-methoxy-3- (2- methyl propyl) pyrazine has been reported in Cabernet Sauvignon grapes. The odour of isobutyl (2 methoxy-3-isobutyl pyrazine or 2-isobutyl-3-methoxy pyrazine) MP has been described as like bell pepper. It contributes to the characteristic aroma of

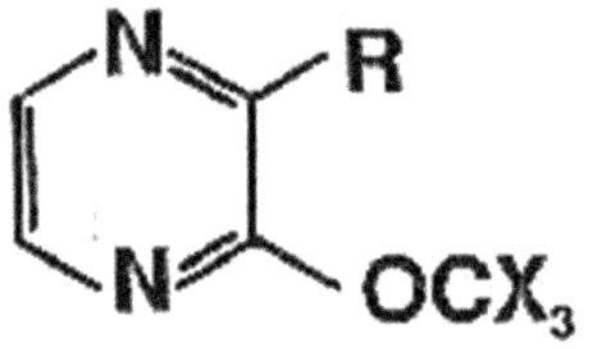

	R	Name of the methoxy pyrazine
a.	$CH_2CH(CH_3)_2$	2-Methoxy-3-(1-methylethyl) pyrazine OR Isobutyl MP
b.	$CH(CH_3)_2$	2-methoxy-3-isopropyl pyrazine OR Isopropyl MP
c.	$CH(CH_3)CH_2CH_3$	2-Methoxy-3 Sec. Butyl pyrazine OR Sec. Butyl MP

FIGURE 10.1 Methoxypyrazine identified in Sauvignon blanc grapes and wines. *Source:* Ough and Groat (1978)

some vegetables and has an extremely low sensory detection threshold in water of about 2 ng/l (2 parts per trillion). Three methoxypyrazines were identified and quantified by gas chromatography/mass spectrometry in Sauvignon blanc wine and juice samples. 22 wines of Australian, New Zealand, and French origin were analysed together with 16 juice samples from four Australian regions. 2-methoxy-3-(2-methyl propyl) pyrazine was present in all the wines (0.6-38.1 ng/l), whereas 2-methoxy-3-(1-methyl ethyl) pryazine was identified in few samples. Methoxypyrazine levels in New Zealand wines were significantly higher than the Australian wines. Fruits grown under cool conditions gave higher grape methoxypyrazine levels than the hot conditions, as was the case of fruits at véraison, but decreased markedly with ripening.

10.2.2.4 Lactones

Among many volatile constituents of wine, the lactones, particularly gamma lactones, occupy a place of prominence not only in terms of their contribution to the total aroma and bouquet picture but also because of their physiological properties. Delta lactones are also often associated with impact flavour compounds. Table 10.3 lists some of the important lactones found in wines.

10.2.2.5 Volatile Sulphur Compounds and Thiols

The mechanisms of formation of volatile sulphur compounds (VSC) during wine fermentation are only partially understood due to the complex nature of factors involved in winemaking. The potential precursors of VSCs during fermentation were studied in eight white grape musts, and the prominent VSC (H_2S) was continuously produced throughout the fermentation and was the highest during the rapid growth phase of the yeast. All juices produced H_2S most rapidly during the rapid growth phase of the yeast, but musts with low assimilable amino acids (EAA) generally produced higher levels of H_2S throughout fermentation. Total H_2S (ranging from 112–516 mg/l) inversely correlated with the concentration of assimilable amino acids and with total nitrogen content. The formation of some volatile sulphur compounds in Greek white wines of the cultivar Batiki and Muscat Hamburg showed the formation of hydrogen sulfide, methionol, 3-methyl thiopropionic acid, ethyl 3-methylthiopropionate, 2-methyl thioethanol, 2-methyltetrahydro-thiophenone-3, cis- and trans-2-methylthioptrano-3-ol during and after fermentation. This is directly influenced by the vinification parameters like bisulfite addition to the must, must turbidity, fermentation temperature, inoculation with different yeast strains, and period of contact of sulfated wines with their yeast sediment and pressings during must preparation.

Several volatile thiols have been identified in Sauvignon blanc wines. 4-mercapto-4-methylpentan-2-one (4 MMP) and 3-mercaptohexyl acetate (A3MH) have a strong box tree odour, whereas 3-mercaptohexane-1-ol (3 MH), 4-mercapto-3-methyl pentane-2-ol (4 MMPOH), and 3-mercapto-3-methyl butan-1-ol (3 MMB) have aromas reminiscent

TABLE 10.3
Lactones Isolated From Wines

Structure	Name	Occurrence
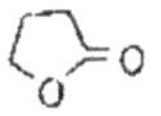	Gamma Butyrolactone	All wines
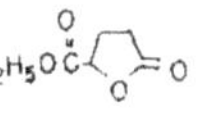	5-Carboethoxy-Dihydro-2(3H)-Furanone	Flor sherries, Cabernet Sauvignon, Ruby Cabernet
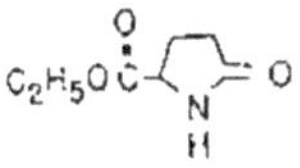	Ethyl Pyroglutamate	Flor sherries
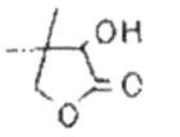	Pantolactone	Flor sherries
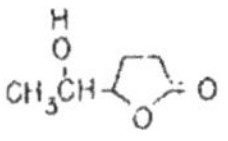	(+) 4R:5S or 4S:5R & (-) 4R:5R or 4S:5S 4,5-Dihydroxyhexanoic Acid Gamma-Lactones	Flor sherries
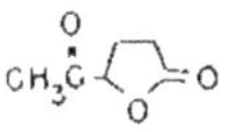	5-Acetyldihydro-2(3H)-Furanone	All wines
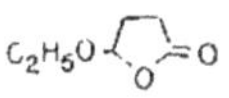	5-Ethoxydihydro-2(3H)- Furanone	Ruby cabernet
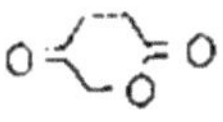	6-Methyldihydro-2,5(3H)-Pyrandione	Ruby Cabernet
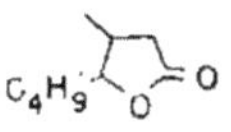	Trans-5-Butyl-4-Methyl-Dihydro-2(3H)-furanone	Oakwood-aged Cabernet

Source: Muller *et al.* (1973). Reprinted by permission of American Society for Enology and Viticulture of grapefruit, passion fruit, citrus zest, and cooked leeks, respectively. In fact, 4 MMP, 4 MMPOH, and 3 MH are present in must as odourless precursors, in the form of S-cysteine conjugates. The major role of these volatile thiols in the aroma of Sauvignon blanc wines has now been clearly demonstrated, but for the wines made from other grape varieties it is not well known. The impact of these volatile thiols on the aromas of wines made from different grape varieties, namely Gewurztraminer, Riesling, Colombard, Petit Manseng and botrytized Semillon, has been recorded. Interestingly, the five volatile thiols identified in Sauvignon blanc wines are also present in wines made from various other white grape varieties.

10.2.3 Pre-Harvest Factors Affecting Wine Quality and Aroma

Grape vines are grown in distinct climatic regimes worldwide that provide the ideal situation to produce high quality grapes. Interactions between the local climate, soil, and site locations play an important role in the ontogeny and yield of the grape vines. Mild to cool and wet winters followed by warm springs, then warm to hot summers with little precipitation provide adequate growth potential and increase the likelihood of higher wine quality. The influence of vineyard site and grape maturity on juice and wine quality of mature Riesling vines at three sites in Southern Australia, viz. Virginia (altitude 20m), Williamstown (altitude 280m), and High Eden (altitude 520m), harvested over a period of two months, showed that juice monoterpenes increased with the increasing maturity. Thus, vineyard altitude and grape maturity appear to be the main factors contributing to the final wine quality. A long term (1952–1997) climatology using reference vineyard observations was developed in Bordeaux, France. The procedure partitioned the season into growth intervals from one phenological event to the next, viz. bud burst, floraison, véraison, and harvest. Over last two decades, the phenology of grape vines in Bordeaux tended towards earlier phenological events, a shortening of phenological intervals, and lengthening of a growing season. Merlot and Cabernet Sauvignon varieties tended to produce higher sugar to total acid ratios, greater berry weights, and greater potential wine quality.

Terpene concentration was measured during the ripening of Müller-Thurgau grapes to provide objective information on grape and wine quality. Reduction in yield and change of the canopy microclimate by shoot and cluster thinning resulted in a higher terpene concentration in the first year which remained unchanged during the second year. Viticulturists and oenologists are now exploring different techniques to affect grape composition in the vineyard. The wine growers believe that much can be done in the vineyard to affect grape flavour. However, many new experiments and trials on clones, rootstock, spacing, canopy management, and water management are still in progress.

10.2.4 Processing Factors Affecting Wine Flavour, Aroma, and Quality

10.2.4.1 Influence of Vinification Treatments on the Aroma Constituents

Wine colour, aroma, and flavour depend primarily on the initial condition and composition of the grapes, as well as subsequent enological processes. Fermentation temperature can impact significantly in red wines. Colour extraction in Pinot noir was a problem, and raising temperatures from 10°C to 21°C improved both colour intensity and flavour in the resultant wines. When a range of fermentation temperatures, *i.e.* 12°C, 15.5°C, 21°C, and 27°C, were examined for Pinot noir wines, colour and flavour were adjudged the best with 21°C and 27°C fermation temperatures. Increasing fermentation temperature from 12°C to either 20°C or 30°C in Cabernet Sauvignon, Grenache, or Pinot noir wines increased both colour and tannin extraction. Subsequent work showed a significant linear relationship between temperature and wine quality for these three cultivars.

The quality of wine is also affected by yeast strain, grape cultivar, time of skin contact, oxygen level, and type of suspended solids, while insoluble materials, *i.e.* grape solids, bentonite, diatomaceous earth, etc., can also influence the fermentation environment and rate of sugar conversion. However, studies on the effects of suspended solids on wine quality have been mainly restricted to white cultivars. Three vinification methods in Pinot noir wines were compared to assess their effects on chemical composition, sensory descriptors, and headspace volatile constituents. Two methods involved standard vinification at fermentation temperatures of 20°C (VM_1) and 30°C (VM_2), respectively, while the third method (VM_3) included a two stage pre-fermentation treatment involving heat extraction, followed by fermentation at 15°C in contact with bentonite. VM_2 produced wines with the highest colour intensity, and other sensory characteristics. However, VM_3 wines contained twice the concentration of anthocyanins than VM_1 and VM_2 wines and possessed the most intense fruity aroma and flavour. Total ester concentration was fourfold higher in VM_3 compared to VM_1 and VM_2. Several esters were responsible for this difference, but isoamyl acetate was the predominant one.

10.2.4.2 Effect of Yeast Strains on Wine Quality

Although the influence of yeast strains on wine fermentation aromas consisting of esters, higher alcohols, and fatty acids has long been known, the effect of winemaking yeasts on grape aromas and their precursors has not been studied to any great extent. It is known that the monoterpenol composition and muscat aroma of wines vary little during fermentation, as the glycosidases of *Saccharomyces cerevisiae* have little effect on terpenic glycosides in normal pH of the must. It is not so in some of the varieties known as simple flavoured. The amplification of grape aroma by yeast has been clearly demonstrated for Sauvignon blanc. The volatile thiols responsible for the box tree, grapefruit and passion fruit nuances of

TABLE 10.4
Effect of Yeast Strain on 4 MMP, 4 MMPOH, and 3 MH Amounts in Sauvignon Blanc Wines after Alcohol Fermentation

Must	Wine a	Wine b	Wine c	Wine d	Average
4 MMP (ng/l): 4-Mercapto-4-methyl pentan-2-one					
VL3C	12	12	12	10	12a
EG8	8	9	16	8	10a
VL1	7	2	7	6	6b
522d	0	0	0	0	0c
4 MMPOH (ng/l): 4-Mercapto-4-methyl pentan-2-ol					
VL3C	28	12	27	41	27a
EG8	25	9	10.6	39	21ab
VL1	25	7	9	38	20ab
522d	25	6	2	32	16b
3 MH (ng/l): 3-Mercaptohexan-1-ol					
VL3C	2161	3261	413	991	1706
EG8	2994	4581	460	1135	2267
VL1	2077	2227	305	1457	1516
522d	2128	2890	235	1184	160

Source: Murat *et al.* (2001). Reprinted by permission of American Society for Enology and Viticulture

*Values denoted by different letters are statistically different (p = 0.01)

**Wine a, wine b, wine c, and wine d represent musts from four different vineyards in Bordeaux, France.

wines made from this variety are principally formed by yeast from cysteinylated precursors in the must. The ability of four different industrial strains of *S. cerevisiae* to release certain Sauvignon blanc aromas from their cysteinylated precursors showed statistically significant differences among the various yeast strains (Table 10.4). The VL3C and EG8 strains were found to be quite effective in enhancing the aroma components. During alcoholic fermentation, 4 MMP, 4 MMPOH, and 3 MH are released from S-4-(4-methylpentan-2-one)-L-cysteine, S-4-(4-methylpentan-2-ol)-L-cysteine and S-3-(hexan-1-ol)-L-cysteine, respectively. The mechanism by which these odourless precursors are converted into aromas by yeast during alcoholic fermentation has not been fully elucidated. Previous work has shown that β-lyase-type enzyme activity is found in several microorganisms, including Baker's and Brewer's yeasts, which is capable of hydrolysing S-conjugate cysteins and, thus, releasing the corresponding volatile thiols.

10.2.4.3 Qualitative Changes in the Volatiles During Aging

Aging brings about perfect maturity in wines which bear their own distinctive sensory makeup. Wines made from Riesling and Vidal blanc grapes are often consumed when they are one to two years old, since they quickly acquire their characteristic fruity aroma. Riesling is usually aged in glass, and examination of changes in the chemical composition of the aroma compounds upon storage has led to the description of a varietal bouquet for these wines. Off-odours do not contribute significantly to the aroma of aged Riesling. But the concentration of dimethyl sulfide in some white wines increases with time and temperature, and makes a significant contribution to the bouquet of aged wines. 1,1,6-trimethyl-1,2-dihydronapthalene (TDN) has been blamed for a characteristic kerosene aroma in Riesling wines, particularly those made from grapes grown in warmer climates such as South Africa and Australia. The level of TDN can be used to estimate the aging potential of Riesling wine. The role of monoterpenes during aging is critical to the quality of aged wine. Riesling wines are not affected negatively by aging, whereas Muscat-type wines, which have a higher level of monoterpenes, do not improve with bottle storage.

The aroma characteristics of aged Vidal Blanc wine revealed that the wine lost much of its characteristic fruity aroma. In some cases, the wine acquired a strong odour of asparagus after two to three years, while in other cases the dominant aroma of three-year-old wine was found to resemble straw. The occurrence of the asparagus aroma was not predictable and may depend upon the yeast used. Thus, the aging of Vidal blanc wine is most affected by changes in the terpene composition, and therefore, it does not maintain its varietal character with age, unlike Riesling wines. The amount of monoterpenes in young Vidal blanc wine is higher than that of Riesling, but lower than that of Muscat wines. Analysis of periodic changes in the volatile composition of Zinfandel wine during aging, starting with a newly fermented must and monitored every three months thereafter, showed that the most significant changes occurred at six months, which coincided with the completion of malo-lactic fermentation. Approximately 20 new compounds could be identified after six months of storage.

BIBLIOGRAPHY

Agosin, E., Belancic, A., Ibacache, A., Baumes, R., Bordeu, E., Crawford, A. and Bayonove, C. (2000). Aromatic potential of certain Muscat Varieties important for Pisco Production in Chile. *Am. J. Enol. Vitic.*, 51(4): 409.

Allen, M.S. (1997). Stable Isotope dilution gas chromatography – mass spectrometry for determination of methoxypyrazine ('green' aromas) in wine. In: *Modern Methods of Plant Analysis*, Vol. 19. H.F. Linskens and J.F. Jackson (Eds.). Springer Verlag, Berlin. p. 193.

Boulton, R.B., Singleton, V.L., Bisson, L.F. and Kunkee, R.F. (1996). *Principles and Practices of Winemaking.* Chapman & Hall, NY, p. 42.

Brossaud, F., Cheynier, V., Asselin, C. and Moutounet (1999). Flavonoid compositional differences of grapes among site test plantings of Cabernet Franc. *Am. J. Enol. Vitic.*, 50(3): 277.

Cassava, L.F., Beaver, C.W., Mireles, M., Larsen, R.C., Hopfer, H., Heymann, H. and Harbertson, J.F. (2013). Influence of fruit maturity, maceration length and ethanol amount on chemical and sensory properties on Merlot wines. *Am. J. Enol. Vitic.*, 64(4): 437–449.

Clingeleffer, P.R., Kerridge, G.H. and Possingham, J.V. (1986). Effect of variety on wine quality. In: *Proceedings of the Sixth Australian Wine Industry Technical Conference* held Adelaide, South Australia, 14–17 July, p. 78.

Esteban, M.A., Villanueva, M.J. and Lissarrague, J.R. (2002). Relationships between different berry components in Tempranillo (*Vitisvinifera* L.) grapes from irrigated and non-irrigated vines during ripening. *J. Sci. Foodagric.*, 82(10): 1136–1146.

Fontoin, H., Saucier, C.,Teissedre, P.L. and Glories, Y. (2008). Effect of pH, ethanol and acidity on astringency and bitterness of grape seed tannin oligomers in model wine solution. *Food Qual. Preference*, 19(3): 286–291.

Francis, I.L., Sefton, M.A. and Williams, P.J. (1994). The sensory effects of pre- or post-fermentation thermal processing on Chardonnay and Semillon wines. *Am. J. Enol. Vitic.*, 45: 243.

Fretz, C.B., Luisier, J.L., Tominaga, T. and Amado, R. (2005). 3-Marcaptohexanol: An aroma Impact Compound of Petite Arvinewine. *Am. J. Enol. Vitic.*, 56(4): 407.

Gawel, R. (1998). Red wine astringency: A review. *Aust. J. Grape Wine Res.*, 4(2): 74.

Gerbaux, V., Vincent, B. and Bertrand, A. (2002). Influence of maceration temperature and enzymes on the content of volatile phenols in Pinot noir wines. *Am. J. Enol. Vitic.*, 53(2): 131.

Gil, M., Estevez, S., Kontoudakis, N., Fort, F., Canals, J.M. and Zamora, F. (2013). Influence of partial dealcoholisation by reverse osmosis on red wine composition and sensory characteristics. *Eur. Food Res. Technol.*, 237(4): 481–488.

Godden, P. and Gishen, M. (2005). Trends in the Composition of Australian Wine. *Aust. N. Z. Wine Ind. J.*, 20: 21–46.

Hernandez-Orte, P., Cersosimo, M., Loscos, N., Cacho, J., Garcia-Moruno, E. and Ferreira, V. (2008). The development of varietal aroma from non-floral grapes by yeasts of different genera. *Food Chem.*, 107(3): 1064–1077.

Iriti, M. and Faoro, F. (2006). Grape phytochemicals: A bouquet of old and new nutraceuticals for human health. *Med. Hypoth.*, 67(4): 833–838.

Jones, G.V. and Davis, R.E. (2000). Using a Synoptic climatological approach to understand climatic viticulture relationship. *Inter. J. Cli. M.*, 20: 813.

Jordao, A.M., Vilela, A. and Cosme, F. (2015). From sugar of grape to alcohol of wine: sensorial impact of alcohol in wine. *Beverages*, 1(4): 292–310.

Joshi, V.K. and Shah, P.K. (1998). Effect of wood treatment on chemical and sensory quality of peach wine during ageing. *Acta Aliment.*, 27(4): 307.

King, E.S., Dunn, R.L. and Heymann, H. (2013). The influence of alcohol on the sensory perception of red wines. *Food Qual. Preference*, 28(1): 235–243.

King, E.S. and Heymann, H. (2014). The effect of reduced alcohol on the sensory profile and consumer preferences of white wine. *J. Sens. Stud.*, 29(1): 33–42.

Kotseridis, Y. and Baumes, R. (2000). Identification of impact odorants in Bordeaux red grape juice, in the commercial yeast used for its fermentation, and in the produced wine. *J. Agric. Food Chem.*, 48(2): 400–406.

Ligouri, L., Russo, P., Albanese, D. and di Matteo, M. (2013). Evolution of quality parameters during red wine dealcoholisation by osmotic distillation. *Food Chem.*, 140(1–2): 68–75.

Liu, H.F., Wu, B.H., Fan, P.G., Li, L.S. and Li, L. (2006). Sugar and acid concentrations in 98 grape cultivars analysed by principal component analysis. *J. Sci. Food Agric.*, 86(10): 1526–1536.

Loscos, N., Hernandez-Orte, P., Cacho, J. and Ferreira, V. (2007). Release and formation of varietal aroma compounds during alcoholic fermentation from non-floral grape odorless flavour pre-cursors fractions. *J. Agric. Food Chem.*, 55(16): 6674–6684.

Maga, J.A. and Sizer, C.E. (1973). Pyrazines in foods. A review. *J. Agric. Food Chem.*, 21(1): 22.

Meillon, S., Urbano, C. and Schlich, P. (2009). Contribution of the temporal dominance of sensations (TDS) method to the sensory description of subtle differences in partially dealcoholized red wines. *Food Qual. Preference*, 20(7): 490–499.

Muller, C.J., Richard, K.E. and Webb, A.D. (1973). Lactones in wines – A review. *Am. J. Enol. Vitic.*, 24(1): 4.

Murat, M.L., Masneuf, I., Darriet, P., Lavigne, V., Tominaga, T. and Dubourdieu, D. (2001). Effect of *Saccharomyces cerevisiae* yeast strains on the liberation of volatile thiols in Sauvignon blanc wine. *Am. J. Enol. Vitic.*, 52(2): 136.

Noble, A.C. (1994). Wine flavour. In: *Understanding Natural Flavours*, J.R. Piggott and A. Paterson (Eds.). Blackie Academic and Professional. London, p. 228.

Ough, C.S. and Groat, M. (1978). Partide nature, yeast strain, and temperature interactions on the fermentation rates of grape juice. *Appl. Environ. Microbiol.*, 35(5): 881.

Pankiewicz, U. and Jamroz, J. (2013). Evaluation of physicochemical and sensory properties of ethanol blended with pear nector. *Czech J. Foodsci.*, 31(1): 66–71.

Park, S.K., Boulton, R.B. and Noble, A.C. (2000). Formation of hydrogen sulfide and glutathione during fermentation of white grape musts. *Am. J. Enol. Vitic.*, 51(2): 91.

Polaskova, P., Herszage, J. and Ebeler, S. (2008). Wine flavor: Chemistry in a glass. *Chem. Soc. Rev.*, 37(11): 2478–2489.

Quiros, M., Rojas, V., Gonzalez, R. and Morales, P. (2014). Selection of non-*Saccharomyces* yeast strains for reducing alcohol levels in wine by sugar respiration. *Int. J. Food Microbiol.*, 181: 85–91.

Redondo, N., Gomez- Martineza, S. and Marcos, A. (2014). Review Article – sensory attributes of soft drinks and their influence on consumer's preferences. *Food Funct.*, 38: 2550–2560.

Schreier, P., Drawert, F. and Abraham, K.O. (1980). Identification and determination of volatile constituents in Burgundy Pinot noir wines. *Lebensm. Wiss. U. Technol.*, 13: 318.

Shiraishi, M., Fujishima, H. and Chijiwa, H. (2010). Evaluation of table grape genetic resources for sugar, organic acid and amino acid composition of berries. *Euphytica*, 174(1): 1–13.

Stern, D.J., Guadagni, D. and Stevens, K.L. (1975). Aging of wine: Qualitative changes in the volatiles of Zinfandel wine during two years. *Am. J. Enol. Vitic.*, 26(4): 208.

Styger, G., Prior, B. and Bauer, F.F. (2011). Wine flavour and aroma. *J. Ind. Microbiol. Biotechnol.*, 38(9): 1145–1159.

Swiegers, J. and Pretorius, I.S. (2007). Modulation of volatile sulfur compounds by wine yeast. *Appl. Environ. Microbiol.*, 74(5): 954–960.

Tominaga, T., Guyot, R.B., Gachons, C.P.D. and Dubourdieu, D. (2000). Contribution of volatile thiols to the aromas of white wines made from several *Vitis vinifera* grape varieties. *Am. J. Enol. Vitic.*, 51(2): 178.

Wattenberg, L.W. (1992). Inhibition of carcinogenesis by minor dietary compounds. *Cancer Res.*, 52: 208.

Williams, P.J., Strauss, C.R. and Wilson, B. (1981). Classification of the monoterpenoid composition of Muscat grapes. *Am. J. Enol. Vitic.*, 32: 230.

Williams, P.J., Strauss, C.R. Aryan, A.P. and Wilson, B. (1986). Grape flavour—a review of some pre and post harvest influences. In *Proc. 6th Aust. Wine Ind. Tech. Conf., Adelaide*, (ed.) Lee T.H. pp. 111–116.

11 Diseases of Grapes

Vivienne Gepp

CONTENTS

11.1 INTRODUCTION

The European grapevine, *Vitis vinifera*, is planted in a variety of temperate climates around the world and produces most of the grapes used for making wine. This species and others used for winemaking are prone to more than fifty diseases caused by fungi, viruses, bacteria and other agents which affect wine production both in quantity and quality. The incidence of each disease varies from one vine-growing area to another, according to the local climate, varieties and management, but a few are important in many areas.

Mankind has come to realize that plant pests and diseases are an integral part of the ecosystem and that it is unrealistic to expect to eliminate them. Knowledge of the different pathogens which cause grape diseases, their symptoms, epidemiology, conditions which favour them and possible management strategies and tactics is essential for successful wine production.

Diseases affect the wine grapes mainly in two ways: reducing the production of grapes per acre, and/or affecting the quality of the grapes for making wine. The latter effect is very important as consumer preferences depend on organoleptic characteristics of the wine, such as aroma, taste, colour, etc. Another negative consequence of disease is the effect on wine quality due to fungicide residue on bunches before harvest.

Diseases of worldwide relevance to grapevines which reduce yield through reducing foliage, weakening and/or killing vines will be briefly reviewed, but the main part of this chapter will refer to the diseases which affect the quality of grapes for winemaking. These diseases result in the rotting of grapes around harvest and/or the production of mycotoxins and other substances which are detrimental to the wine. This is especially significant, as little can be done to remove undesirable compounds, such as those that confer taints or unpleasant aromas to the wine, mycotoxins or pesticide residues, in the wine making process. A few possibilities exist, and will be mentioned in this chapter, but it is always preferable to avoid the accumulation of these compounds before harvest.

The fungi residing on the surface of grapes, even if they do not infect and rot the berries, influence wine quality and can produce or induce the production of undesirable compounds in must and wine. Although this little-studied aspect is important to the wine industry, it is not within the scope of the present chapter to discuss it in depth.

11.2 MAIN DISEASES OF GRAPES

11.2.1 Diseases Which Mainly Affect the Vine

Several of the following diseases are capable of infecting inflorescences or clusters, and some of producing berry rots, but only before véraison.

11.2.1.1 Downy Mildew

Downy mildew, caused by the Oomycete *Plasmopara viticola*, is the most destructive disease in humid climates. If left uncontrolled, it can kill most of the leaves, preventing the grapes from ripening properly and can also dry up the flowers and young grapes. Berries are susceptible until about pea-size, but berry stems remain susceptible until harvest. White down appears on recently infected stems, these turn brown and die. Berries on the infected stem dry up.

11.2.1.2 Powdery Mildew

Powdery mildew, on the other hand, is especially destructive in drier climates with little rain. It is caused by *Erysiphe* (*Uncinula*) *necator*, an Ascomycete which grows as an ectoparasite on young leaves and any other green organs, including young grapes. The fungus produces necrosis of the epidermis, from minute spots to large areas of the berry where the skin splits as the grape increases in size. The dead spots or areas and splits facilitate the infection of rotting microorganisms, such as *Botrytis cinerea*.

11.2.1.3 Black Rot

Black rot is a disease caused by *Guignardia bidwellii*, which can infect young leaves and other green parts of the vine. Grapes are susceptible until véraison. Spots on berries enlarge until the whole grape goes dark and rots with numerous pycnidia on the surface. *G. bidwellii* overwinters in mummies, and in spring, ascospores are ejected from perithecia by rain and carried by the wind to susceptible organs. Rain is also necessary for the secondary dispersal of conidia formed in pycnidia on infected leaves and grapes. Black rot can cause complete loss of the harvest in some areas with high summer rainfall.

11.2.1.4 Dead Arm or Excoriose

Dead arm, or excoriose, is caused by *Phomopsis viticola*, which can infect young shoots, leaves and clusters, weakening vines and reducing yield. It occasionally infects individual grapes, causing a rot with numerous pycnidia, similar to black rot (Figure 11.1). When *P. viticola* blights cluster stems, the whole cluster dies and falls.

FIGURE 11.1 Rot caused by *Phomopsis viticola*

11.2.1.5 Anthracnose

Anthracnose, caused by *Elsinoe ampelina*, is another common disease in warm humid areas. It infects green organs, producing characteristic lesions with pale centres on immature grapes. This disease is more common in table than wine grapes.

11.2.1.6 Bacterial Blight and Pierce's Disease

Three bacteria infect grapevines, two of which (*Xanthomonas ampelina* causing bacterial blight and *Xylella fastidiosa*, Pierce's disease) can cause stunting and death of infected vines.

11.2.1.7 Grapevine Leaf Roll

Over a dozen different viruses infect grapevine, including Grapevine Fanleaf Virus and the viral complex known as the Grapevine Leaf Roll-associated viruses. Of the latter, at least some strains decrease the amount of sugars in grapes as well as reducing yield, which means that infected vines produce less wine and lower alcohol content.

11.2.1.8 Grapevine Trunk Diseases

A group of emerging diseases are the grapevine trunk diseases, such as Eutypa dieback, Esca, Petri disease and others, which diminish yields by killing vines.

11.2.2 Diseases Which Mainly Affect the Grapes: Bunch Rots

Bunch rots are mainly caused by opportunistic fungi, most of which can only develop in high sugar tissues, so symptoms normally first appear after véraison, beginning in one or two berries, but often spreading through the bunch as grapes ripen. They are aerobic and incapable of tolerating conditions during alcoholic fermentation, so the damage to wine production is generated during the ripening period and post-harvest until fermentation gets underway.

They affect the wine industry in three ways:

- loss of bunches which must be discarded;
- production of undesirable aromas, taints and off-flavours, oxidation and loss of red colour and in some cases, the formation of mycotoxins;
- presence of residues of fungicides used to control them.

Frequently, more than one pathogen infects a bunch, and symptoms are often similar, so it may be difficult to determine which predominates. In spite of this, the characteristics of each type of bunch rot will be described below.

11.2.2.1 Grey Mould

Grey mould is caused by the fungus *Botrytis cinerea* (teleomorph *Botryotinia fuckeliana*) and is the most common and destructive of the bunch rots. It can cause organoleptic defects in wine, especially when associated with other fungi, as is often the case. When it infects grapes, *B. cinerea* secretes laccase enzymes which oxidize various compounds, causing detrimental effects on wine, including the loss of red colour.

B. cinerea can, on the other hand, aid in the production of special wines, producing what is known as "noble rot" when it grows on grapes in certain conditions (see Chapter 12).

Although *B. cinerea* can penetrate green berries, the infection remains quiescent until the grape ripens and sugars accumulate. Then, the grape takes on a brownish hue as it rots. The fungus grows out through pores or cracks in the skin and produces enormous quantities of multinucleate conidia (Figure 11.2 and 11.3). The rotted berries eventually dry out and remain as mummies over winter.

B. cinerea as a species has an extremely wide host range of dicotyledons, but there is a tendency for different strains to colonise certain plant species. So, although there may be many potential sources of inoculum in and around a vineyard, probably most of the spores which infect the grapes come from remains of infected bunches and from sclerotia on shoots (Figure 11.4). The abundant conidia are dispersed by air and need water or exudates from damaged tissues to germinate. *B. cinerea* penetrates through wounds caused by birds, insects, cultural practices, wind, hail, etc. or cracks due to fast enlargement after a dry period or even through minute necrotic spots resulting from powdery mildew infection. It also colonises senescent floral tissues which remain within the cluster (Figure 11.5) and are a source of inoculum for the grapes around them as they ripen. Infection at flowering may or may not be correlated with berry rot, depending on the local climate (timing of rainfall and temperatures).

Although *B. cinerea* can infect at temperatures between 34 and 86°F (1–30°C), it is most aggressive in the 60 to 68°F (15–20°C) range and is especially destructive in vineyards exposed to cool wet conditions as the grapes ripen.

FIGURE 11.2 Grey mould on a cluster of Tannat grapes

FIGURE 11.3 Grey mould sporulating over the whole surface or through cracks in the skin of rotten grapes

11.2.2.2 Sour Rot

Sour rot is a complex disease which infects grape bunches as they change colour and begin to ripen. It is caused by several yeasts and bacteria which, in association with pathogenic fungi, such as *Aspergillus* sp. or *Rhizopus* sp., degrade the pulp and produce acetic acid. The resulting wine can be two to three times more acidic and only fit for making vinegar. Wine spoilage may also be caused by some of the yeast species surviving the fermentation conditions.

The first symptoms are a change of colour to brick red in white varieties and to purplish brown in red ones. The pulp

FIGURE 11.4 Sclerotia of *Botrytis cinerea* on surface of previous years' shoot

FIGURE 11.5 Remains of flowers in a cluster at véraison, a probable source of *B. cinerea* inoculum for neighbouring grapes

FIGURE 11.6 Sour rot (above) and grey mould (below) on the same bunch

FIGURE 11.7 Fruit flies attracted to sour rot

liquefies and oozes out (Figure 11.6), giving off a smell of vinegar, which attracts fruit flies (*Drosophila* sp.) (Figure 11.7). The fly larvae feed on the rotting berries. The disease is favoured by warm, humid weather and thin skinned, tight clustered varieties.

11.2.2.3 Black Mould

Black mould is a rot caused by several species of *Aspergillus*, notably *A. carbonarius* and *A. niger*, both of which are capable of secreting the mycotoxin Ochratoxin A, which remains in the wine. Health authorities establish low tolerance levels in wine due to its nephrotoxic and carcinogenic properties.

The typical symptoms consist of a watery, tan to brown rot, which often starts near the pedicel. The skin turns light grey, and cracks appear. Masses of brown or black powdery spores soon appear on the surface, first along cracks in the skin but later all over the grape (Figure 11.8). If conditions are dry enough, the berry shrivels up. Species of *Aspergillus* which attack grapes prefer temperatures around 68–86°F (20–30°C), moisture and injured grapes.

11.2.2.4 Blue Mould

Blue mould is caused by *Penicillium* spp., in particular *P. expansum* normally associated with apples, commonly acting as secondary invaders and mainly in cool climates and on overripe grapes. *Penicillium* spp. can produce the mycotoxins patulin, which inhibits *Saccharomyces cerevisiae* fermentation, and citrinin, but both are destroyed during wine making. They can also secrete geosmin, which confers an earthy aroma to wines. They can cause loss of colour and decreased sugar concentration.

11.2.2.5 Alternaria Rot

Alternaria alternata infects grapes, producing fairly firm lesions, generally near the pedicel, at first tan and later dark brown to black, which are then covered by dark fluffy masses of conidia. *A. alternata* can be isolated from surface sterilized

FIGURE 11.8 Dark brown sporulation of *Aspergillus* sp

FIGURE 11.9 Ripe rot with characteristic damp-looking orange masses of spores

berries from end of full bloom onwards, but symptoms are seen on ripe, damaged grapes post-harvest.

Alternaria spp. are capable of producing at least one of three mycotoxins (alternariol, alternariol monomethyl ether or tenuazonic acid).

11.2.2.6 Brown Spot or *Cladosporium* Rot

Different species of *Cladosporium* such as *C. herbarum* and *C. cladosporioides* species complexes can infect ripe, injured grapes and especially overripe ones. *Cladosporium* spp. produce dark necrotic lesions, dehydration and a firm decay of part of the berry covered with an olive-green mould. They can develop at temperatures from 32 to 86°F (0–30°C) but grow best at 65 to 82°F (18–28°C).

11.2.2.7 Ripe Rot

Ripe rot is a disease caused by certain species of *Colletotrichum*, such as *C. acutatum* and *C. gloesporoides*, which infect grapes in subtropical and tropical areas. Infected berries, even those which have not yet developed symptoms, taste bitter. The bitterness and off-flavour are retained in wine made with as little as 1.5% of affected grapes. The colour of this wine is browner and less red, possibly due to laccase enzymes.

Colletotrichum sp. can be detected within flowers and young grapes, but remains quiescent until berries mature, when rot sets in. Ripe rot appears as soft, sunken, light brown areas. There may be cracking of the skin which releases juice, and a few days later, black spots (acervuli) appear which are quickly covered by distinctive salmon-pink or orange spore masses (Figure 11.9). The grapes later shrivel and may drop.

11.2.2.8 Bitter Rot

Greeneria uvicola, the causal agent of bitter rot, is also more common in subtropical areas. This rot confers a bitter flavour to the berries which is retained in the wine made from them. Infection may occur at flowering or more commonly around véraison but remains latent until berry maturity. White grapes turn brown with concentric rings of acervuli which appear as black spots. Red varieties hardly change colour, but their surface becomes rough when the acervuli break the cuticle.

Optimal conditions for *G. uvicola* are around 77–86°F (25–30°C) and 6 to 12 hours of wetness. Wounds aid infection although they are not strictly necessary.

11.2.2.9 White Rot

White rot is caused by species of *Coniella* (*Pilidiella*) such as *C. diplodiella*. The fungus infects grapes through wounds, especially those produced by hailstones or other mechanical injury, and in a few days the infected berries turn yellowish and then pinkish blue and begin to dry out. The fungus produces greyish-white pycnidia which push up the cuticle, leaving a layer of air below, which makes the grapes look whitish. The fungus also infects pedicels and rachis and sometimes green shoots. It remains viable in debris for many years.

11.2.2.10 Rhizopus Rot

Rhizopus spp. are typically postharvest fungi which infect damaged or overripe fruit and vegetables. They can infect injured grapes, often as secondary invaders, but in humid conditions, *Rhizopus* spp. can outgrow other fungi and colonise whole bunches and even spread from one bunch to another after harvest.

Rhizopus spp. produce a soft, watery rot which can decay the whole grape. The skin of the berry turns light grey and cracks, permitting the development of white cobweb-like hyphae which end in black sporangia containing numerous spores.

11.2.2.11 Other Bunch Rots

Fungi such as *Monilinia fructicola* (Figure 11.10), species of *Botryosphaeria*, such as *B. dothidea* and *B. ribis*, and *Fusarium* spp. can also infect ripening grapes. Symptoms of black rot and dead arm can also appear around véraison, so are included in Table 11.1.

11.2.3 Bunch Rot Management

Bunch rot management aims to prevent infections and requires a combination of different practices, starting even before the

FIGURE 11.10 Brown rot with clumps of buff-coloured spores

vineyard is planted. As most of the rots are caused by opportunistic microorganisms, the first goal is to avoid conditions which allow them to infect berries, such as injuries to the grapes and high humidity. The ubiquitous nature and easy dispersal of most of the causal agents means that control cannot depend only on removal of inoculum sources, though this can help. Other approaches which reduce the infection rate and/or the susceptible period must normally be employed, including the application of fungicides or biological control in order to prevent infections. The combination of different tactics in order to reduce losses to a tolerable level, the emblematic concept of integrated pest management (IPM), clearly applies here. But several difficulties are encountered in strict application of IPM strategies originally developed for insect pests to diseases, and in particular to bunch rots. IPM programs require:

- Valid economic thresholds which must be established locally, taking into consideration the type and quality of the wine to be produced. They may differ according to the bunch rotting organisms present, as their effects on wine vary.
- Simple, fast and accurate monitoring methods. Disease monitoring relies on symptoms, but once these appear, no curative options exist. As a result, instead of monitoring bunch rot symptoms, it is necessary to forecast conditions which are conducive to bunch rot development, mainly through weather forecasts.
- IPM also depends on availability of control tactics which can be applied when an action threshold is reached. In the case of bunch rots, these are limited, due to the proximity to harvest, and may consist of options such as selective harvesting, which may be uneconomical, or early harvesting, which may decrease wine quality, or rejection by wineries.

So, rather than using thresholds of disease intensity or pathogen numbers, those in charge of bunch rot management must rely on predicting the probability of bunch rot development in the pre-harvest period. There are methods for detecting the colonization by pathogens of unripe berries, such as incubation after freezing overnight or serological testing, but the results will only be useful in areas where a correlation between pre-harvest infection and bunch rots at harvest has been clearly established.

Other techniques used for estimating grape quality at harvest include measuring laccase activity, which is often taken as a measure of *B. cinerea* infection, in spite of the fact that other fungi, such as *Aspergillus* spp. and *C. gloesporoides*, are

TABLE 11.1
List of Bunch Rots, Causal Agents and Distinguishing Characteristics

Disease	Pathogen	Characteristic features	References
Sour rot	*Acetobacter* spp., *Gluconacetobacter* spp., etc.	Odour of vinegar, fruit flies	Barata *et al.* (2012a)
Grey mould	*Botrytis cinerea*	Grey fluffy spores	Elad *et al.* (2007), Pearson & Goheen (1988)
Black mould	*Aspergillus* spp.	Dark brown powdery spores	Latorre *et al.* (2002), Pearson & Goheen (1988)
Alternaria rot	*Alternaria alternata*	Black fluffy spores	Prendes *et al.* (2018), Swart & Holz (1994)
Blue mould	*Penicillium* spp.	Blue-green fluffy sporulation	La Guerche *et al.* (2005)
Cladosporium rot	*Cladosporium* spp.	Olive-green fluffy sporulation	Briceño & Latorre (2008), Swett *et al.* (2016)
Brown rot	*Monilinia fructicola*	Clumps of buff-coloured sporodochia	Sholberg *et al.* (2003)
Fusarium rot	*Fusarium* spp.	White or pink spore masses	Wang *et al.* (2015)
Rhizopus rot	*Rhizopus* spp.	Black sporangia on hyphae	Latorre *et al.* (2002)
Ripe rot	*Colletotrichum* spp.	Orange or salmon droplets of spores coming from acervuli	Meunier & Steel (2009), Pearson & Goheen (1988)
Bitter rot	*Greeneria uvicola*	Concentric rings of black acervuli	Longland & Sutton (2008), Pearson & Goheen (1988)
Black rot	*Guignardia bidwellii*	Black pycnidia, also on leaves	Pearson & Goheen (1988)
Botryosphaeria rot	*Botryosphaeria* spp.	Black pycnidia	Pearson & Goheen (1988)
Dead arm or excoriose	*Phompsis viticola*	Black pycnidia, also on previous years' shoots	Pearson & Goheen (1988)
White rot	*Coniella (Pilidiella)* sp.	Greyish-white pycnidia which push up the cuticle	Van Niekerk *et al.* (2004), Chethana *et al.* (2017), Pearson & Goheen (1988)

capable of producing laccase. Commercial serological tests can be used to specifically quantify *B. cinerea* in grapes and must. PCR methods have not proved to be suitable for analysing field samples.

11.2.3.1 Vineyard Site Selection, Preparation and Planting

When deciding where and how to plant vines, it is important to consider that once the plants are established, any conditions which favour rots will affect grape health for many years. The site selection and soil preparation must be considered to avoid high humidity in the canopy. Soils which allow the roots to grow to greater depths reduce the probability of grapes splitting due to variations in moisture supply, and so tend to reduce berry rots.

11.2.3.2 Variety Selection

Some varieties produce tight clusters in which the internal berries have a thin cuticle and little epicuticular wax and remain humid for longer. As the grapes increase in size, some are often squeezed off their pedicels, opening wounds and releasing juice, which promotes growth of microorganisms inside the bunch and infection of nearby berries. These varieties are more prone to rots, whereas those with loose, open clusters are less affected. Varieties which ripen before climatic conditions become favourable for bunch rots can make a great difference in bunch rot incidence.

Different varieties have been created by crossing *Vitis vinifera* with related species which are more resistant to several pathogenic fungi, including *B. cinerea*. But if the resistance is due to a hypersensitive reaction, the pathogens may overcome it after a few years. These hybrids may also need different treatment in the wine production process in order to obtain high quality wines.

Of course, variety selection in order to reduce bunch rot incidence is not an option when certain varieties must be grown in order to produce the desired wine. If very susceptible varieties must be used, extra care should be taken to apply other measures which tend to reduce the probability of rot development.

11.2.3.3 Vineyard Management Practices

These are essential and can be grouped according to the objectives pursued as detailed below:

- Practices which promote air circulation and drying of clusters and their exposure to light: training systems which separate clusters, adequate pruning, hedging and tying up of shoots, leaf plucking around clusters, limiting nitrogen supply, controlling irrigation, keeping vegetation between vines low, etc. (Figures 11.11 and 11.12).
- Avoiding injury to grapes: control of insects and other pests, care in manipulation of bunches around harvest.
- Reducing the susceptible period: early harvest, as long as this does not affect wine quality.

FIGURE 11.11 Vineyard ground cover cut short to aid air circulation

FIGURE 11.12 Leaves removed around clusters to increase ventilation and expose them to the light

11.2.3.4 Fungicides

In areas prone to bunch rots, fungicides should only be used in conjunction with several of the practices mentioned above, not as a stand-alone treatment, because the application of fungicides does not confer sufficient control and because of the negative effects of fungicide use.

Copper-based fungicides, which have been used since 1882 on vines, mostly for downy mildew, can also prevent black rot infection. Captan is another broad-spectrum fungicide which restricts infection by several of the bunch-rotting fungi, though *C. acutatum* may be resistant. Fungicides will not control the sour rot bacteria, but some of the associated fungi may be affected. There are many newer fungicides which prevent infection by *B. cinerea* and other pathogens. The choice of fungicide will depend on which pathogen(s) predominate in the vineyard and on what is locally available, the pre-harvest interval and possible effects on fermentation.

One of the difficulties in obtaining sufficient control through fungicide use is that the fungus may not be reached by the toxic substance. It may be inside a grape, having penetrated days or even months before, or between berries in the centre of a tightly closed bunch. Even systemic fungicides are unlikely to reach the fungus, because they are only partially absorbed by fruit and, even if they can travel in the xylem vessels, they cannot leave one berry and reach another. So, fungicide coverage of all grapes in a bunch is critical, and fungicides should be applied before the bunch closure stage in order to reach the inside of the cluster.

The fact that bunch rots develop rapidly close to harvest limits the options for fungicidal control for two reasons: the residues which can remain in the wine and the effects of these compounds on the winemaking process.

11.2.3.4.1 Fungicide Residues

Fungicides applied to grapes may persist for some time and show up in the wine. For example, some of the active ingredients often used to control grey mould, Benomyl, Iprodione, Procymidone and Vinclozolin, have been detected in wines.

11.2.3.4.2 Effects on Winemaking

All fungicides are unspecific to some degree, *i.e.* they affect fungi belonging to the same genera or family and even other microorganisms. Yeasts are often sensitive to fungicides, for instance Benomyl present in the must has been shown to delay the beginning of the fermentation process. Fungicide residues present at harvest can also cause variation in the colour of red wines and can affect the phenolic compounds.

11.2.3.4.3 Fungicide Resistance

An additional problem is that fungi can develop resistance to fungicides. The probability of resistance is great in species which produce enormous amounts of spores such as *B. cinerea* and *P. expansum*. The resistance also depends on the fungicide mode of action; in some cases (Methyl Benzimidazole Carbamates) the resistance is qualitative, whereas in others (Dicarboxamides) there is a gradient from the most sensitive to the most resistant (Figure 11.13).

11.2.3.4.4 Alternatives to Fungicides

Chitosan is a natural substance which can be used to control several diseases caused by fungi, increasing chitinase activity and other defence responses of the grape. It may induce an increase in resveratrol and other desired components of wine. It has been tried for the management of grey mould of table grapes. Potassium can also increase resistance in grapes, and potassium sorbate has been shown to reduce grey mould.

Ozone is used for surface decontamination of fruit and vegetables in order to avoid post-harvest rots, but when it is used on grapes prior to fermentation, the wine had higher acetic acid content, unless *Saccharomyces cerevisiae* was added to the must. Extracts of certain plants can inhibit infection, but their effect on wine making has not been studied.

11.2.3.4.5 Biological Control

Due to the negative effects of fungicides and public concern over their use, biological control of bunch rots has been intensively studied since the end of last century. The ideal place to look for antagonists of cluster rot organisms is on the surface of the grapes themselves, where the isolates can be expected to adapt to the vineyard conditions and are able to compete successfully with pathogens. If they are normally present on the grapes around harvest, they will remain in the must when winemaking begins and will not be expected to interfere with fermentation. Indeed some, such as *S. cerevisiae*, are even necessary for the desired winemaking process.

Several microorganisms have been studied for management of grey mould, black mould, Alternaria rot and blue mould (Table 11.2), and some have been registered as biocontrol agents for use on grapes and other fruit crops. Most prevent infection, but *Ulocladium oudemansii* colonises necrotic flower tissues more rapidly than *B. cinerea*, and diminishes the pathogenic inoculum within clusters. Some of the yeasts that diminish infection by toxigenic fungi also degrade the mycotoxins they secrete.

Apart from avoiding the use of synthetic fungicides, biocontrol agents may have other advantages; some are capable of increasing resveratrol synthesis, thus increasing resistance to plant pathogens and conferring health benefits to consumers of the wine made with the grapes.

There is a great deal of information about *B. cinerea* and the epidemiology and management of grey mould, but little on

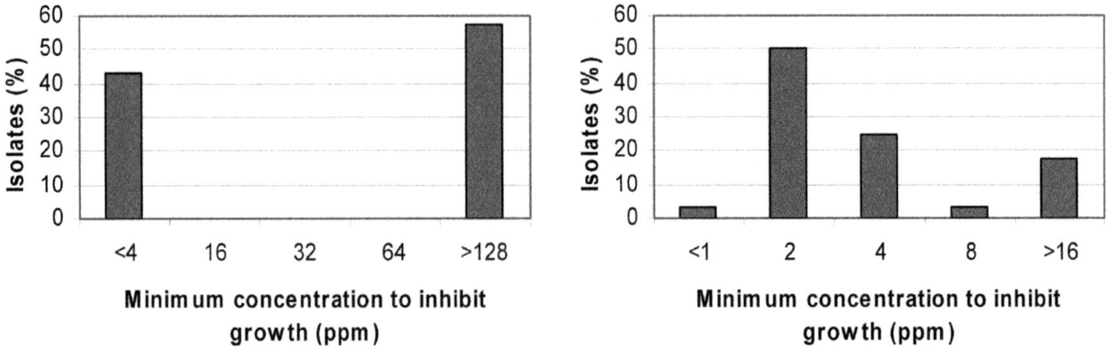

FIGURE 11.13 Distribution of *B. cinerea* isolates obtained from grapes according to their ability to grow in medium amended with Carbendazim (left) or Iprodione (right)

TABLE 11.2
Some Microorganisms Antagonistic Towards Bunch Rot Pathogens

	Antagonistic towards:	References
Acremonium cephalosporium	*Botrytis cinerea, Aspergillus niger, Rhizopus stolonifer*	Zahavi *et al.* (2000)
Aureobasidium pullulans	*B. cinerea, Aspergillus carbonarius*	Dimakopoulou *et al.* (2008), Parafati *et al.* (2015), Raspor *et al.* (2010)
Bacillus subtilis, Bacillus sp.	*B. cinerea, Aspergillus carbonarius*	Elad *et al.* (2007), Jiang *et al.* (2014), Otoguro & Suzuki (2018)
Candida guilliermondii	*B.cinerea, A. niger, R. stolonifer*	Zahavi *et al.* (2000)
Candida incommunis	*A. carbonarius, A. niger*	Bleve *et al.* (2006)
Candida intermedia	*A. carbonarius*	Fiori *et al.* (2014)
Candida oleophila	*B. cinerea*	Elad *et al.* (2007)
Candida sake	*B. cinerea*	Calvo-Garrido *et al.* (2012)
Candida zemplinina	*Alternaria alternata*	Prendes *et al.* (2018)
Hanseniaspora uvarum	*B. cinerea, A. alternata*	Liu *et al.* (2010), Prendes *et al.* (2018)
Issatchenkia orientalis, I. terricola	*A. carbonarius, A. niger*	Bleve *et al.* (2006)
Lanchancea thermotolerans	*Aspergillus sp.*	Fiori *et al.* (2014), Ponsone *et al.* (2016)
Metschnikowia fructicola	*B. cinerea, Alternaria spp., A. niger*	Karabulut *et al.* (2003)
Metschnikowia pulcherrima	*B. cinerea, A. carbonarius, A. niger, A. alternata*	Bleve *et al.* (2006), Parafati *et al.* (2015), Prendes *et al.* (2018), Raspor *et al.* (2010)
Metschnikowia reukaufii	*B. cinerea*	Raspor *et al.* (2010)
Pichia guilliermondii	*B. cinerea*	Raspor *et al.* (2010)
Pseudomonas syringae	*B. cinerea*	Elad *et al.* (2007)
Saccharomyces cerevisiae	*B. cinerea*	Nally *et al.* (2012), Parafati *et al.* (2015)
Schizosaccharomyces pombe	*B. cinerea*	Nally *et al.* (2012)
Trichoderma harzianum, Trichoderma spp.	*B. cinerea*	Elad *et al.* (2007), Harman *et al.* (1996), Elad (1994) , Otoguro & Suzuki (2018)
Ulocladium oudemansii	*B. cinerea*	Elad *et al.* (2007)
Wickerhamomyces anomalus	*B. cinerea*	Parafati *et al.* (2015)

the other bunch rots. Though grey mould is the most common disease of grape bunches worldwide, some of the others cause important losses in certain conditions, and some of these, such as brown rot, appear to be increasing in severity, so more research is necessary for their management.

11.3 APPROACHES IN DISEASE MANAGMENT

Fungi and other microorganisms present on the surface of grapes, even if they do not infect the berries, may produce compounds, such as geosmin, which cause off-flavours in wine. As the composition of the microflora on grape berries may be very different according to the variety, grape development stage, environmental factors, etc., research is needed to characterise the biodiversity on grapes in different situations in order to enable prediction and management of the negative effects on the wine. For this research, techniques for rapidly identifying and quantifying fungal species and strains on berries which do not require culture need to be developed. The effects on wine quality of species individually and in different combinations must also be studied in depth.

In a competitive market, the highest-valued wines will be those with excellent organoleptic qualities, made from healthy grapes and free from mycotoxins and pesticide residues. Therefore, more research should be carried out on alternatives to pesticides which are more acceptable, both ecologically and by consumers. Three areas of research should be emphasised. 1) Varieties which combine resistance to bunch rots and other diseases with the characteristics necessary for good wines require many years to obtain but are needed for consumers who are increasingly demanding wines produced organically or at least free of pesticide residues. 2) More research should be carried out in order to obtain consistent results from biological control agents in each grape producing area. As yeasts growing naturally on the surface of grapes can aid in protecting them from rots but are susceptible to several fungicides, alternatives should be looked for, which would preserve the natural yeast populations on grapes. 3) Several plant extracts and other natural substances have been studied for rot management in table grapes but cannot be applied to grapes for wine until they have been shown not to affect yeasts and the fermentation process.

BIBLIOGRAPHY

Barata, A., Malfeito-Ferreira, M. and Loureiro, V. (2012a). Changes in sour rotten grape berry microbiota during ripening and wine fermentation. *Int. J. Food Microbiol.* 154(3):152–161.

Barata, A., Malfeito-Ferreira, M. and Loureiro, V. (2012b). Review: The microbial ecology of wine grape berries. *Int. J. Food Microbiol.* 153(3):243–259.

Bentley, W. (2009). The integrated control concept and its relevance to current integrated pest management in California fresh market grapes. *Pest Manag. Sci.* 65(12):1298–1304.

Bleve, G., Grieco, F., Cozzi, G., Logrieco, A. and Visconti, A. (2006). Isolation of epiphytic yeasts with potential for biocontrol of *Aspergillus carbonarius* and *A. niger* on grape. *Int. J. Food Microbiol.* 108(2):204–209.

Boiteux, J., Hapon, M., Fernandez, M., Lucero, G.and Pizzuolo, P. (2015). Effect of Aqueous Extract of Chañar (*Geoffroea decorticans* Burkart) on *Botrytis cinerea*, as Possible Alternative for Control in Post-Harvest of Table Grape. *Rev. FCA, UNCUYO* 47(1):241–250.

Briceño, E.X. and Latorre, B.A. (2008). Characterization of *Cladosporium* rot in grapevines, a problem of growing importance in Chile. *Plant Dis.* 92(12):1635–1642.

Briz-Cid, N., Figueiredo-González, M., Rial-Otero, R., Cancho-Grande, B.and Simal-Gándara, J. (2015). The measure and control of effects of botryticides on phenolic profile and color quality of red wines. *Food Control* 50:942–948.

Calhelha, R.C., Andrade, V.J., Ferreira, I.C. and Estevinho, L.M. (2006). Toxicity effects of fungicide residues on the wine-producing process. *Food Microbiol.* 23(4):393–398.

Calvo-Garrido, C., Elmer, P.A.G., Viñas, I., Usall, J., Bartra, E. and Teixodó, N. (2012). Biological control of botrytis bunch rot in organic wine grapes with the yeast antagonist *Candida sake* CPA-1. *Plant Pathol.* 62(3):510–519.

Calvo-Garrido, C., Viñas, I., Elmer, P., Usall, J., Bartra, E. and Teixodó, N. (2013). Candida sake CPA-1 and other biologically based products as potential control strategies to reduce sour rot of grapes. *Lett. Appl. Microbiol.* 57(4):356–361.

Chethana, K.W.T., Zhou, Y., Zhang, W., Liu, M., Xing, Q.K., Li, X.H. and Yan, J.Y. (2017). *Coniella vitis* sp. nov. is the common pathogen of white rot in Chinese vineyards. *Plant Dis.* 101(12):2123–2136.

Cravero, F., Englezos, V., Rantsiou, K., Torchio, F., Giacosa, S., Río Segade, S., Gerbi, V., Rolle, L. and Cocolin, L. (2016). Ozone treatments of post harvested wine grapes: Impact on fermentative yeasts and wine chemical properties. *Food Res. Int.* 87:134–141.

Dantas Guerra, I.C., Lima de Oliveira, P.D., Fernandes Santos, M.M., Carneiro Lúcio, A.S., Tavares, J.F. and Barbosa-Filho, J.M. (2016). The effects of composite coatings containing chitosan and *Mentha* (*piperita* L. or x *villosa* Huds) essential oil on postharvest mold occurrence and quality of table grape cv. Isabella. *Innov. Food Sci. Emerg. Technol.* 34:112–121.

Dimakopoulou, M., Tjamos, S.E., Antoniou, P.P. and Pietri, A. (2008). Phyllosphere grapevine yeast *Aureobasidium pullulans* reduces *Aspergillus carbonarius* (sour rot) incidence in wine-producing vineyards in Greece. *Biol. Control* 46(2):158–165.

Elad, Y. (1994). Biological control of grape grey mould using *Trichoderma harzianum. Crop Prot.* 13(1):35–38.

Elad, Y., Williamson, B., Tudzynski, P. and Delen, N. (2007). *Botrytis: Biology, Pathology and Control.* Springer, Dordrecht, The Netherlands, 412 p.

Feliziani, E., Smilanick, J.L., Margosan, D.A., Mansour, M.F., Romanazzi, G., Gu, S., Gohil, H.L. and Rubio Ames, Z. (2013). Preharvest fungicide, potassium sorbate, or chitosan use on quality and storage decay of table grapes. *Plant Dis.* 97(3):307–314.

Ferrer, M., Gonzalez-Neves, G., Camussi, G., Echeverria, G. and Carbonneau, A. (2011). Variety, plant architecture and pruning methods: Influence on grey mould of grapevine. *Prog. Agric. Viticole* 128(18):367–371.

Fiori, S., Urgeghe, P.P., Hammami, W., Razzu, S., Jaoua, S. and Migheli, Q. (2014). Biocontrol activity of four non- and low-fermenting yeast strains against *Aspergillus carbonarius* and their ability to remove ochratoxin A from grape juice. *Int. J. Food Microbiol.* 189:45–50.

Guillamón, J.M. and Mas, A. (2017). Acetic acid bacteria. In: König, H., Unden, G., Fröhlich, J., eds. *Biology of Microorganisms on Grapes, in Must and in Wine.* 2nd ed. Springer International Publishing AG, Cham, Switzerland, pp. 43–64.

Harman, G.E., Latorre, B., Agosin, E., San Martin, R., Riegel, D.G., Nielsen, P.A., Tronsmo, A. and Pearson, R.C. (1996). Biological and integrated control of Botrytis bunch rot of grape using *Trichoderma* spp. *Biol. Control* 7(3):259–266.

Hed, B., Ngugi, H.K. and Travis, J.W. (2009). Relationship between cluster compactness and bunch rot in Vignoles grapes. *Plant Dis.* 93(11):1195–1201.

Jiang, C., Shi, J., Liu, Y.and Zhu, C. (2014). Inhibition of Aspergillus carbonarius and fungal contamination in table grapes using Bacillus subtilis. *Food Control* 35(1):41–48.

Karabulut, O.A., Smilanick, J.L., Mlikota Gabler, F., Mansour, M. and Droby, S. (2003). Nearharvest applications of *Metschnikowia fructicola*, ethanol, and sodium bicarbonate to control postharvest diseases of grape in central California. *Plant Dis.* 87(11):1384–1389.

La Guerche, S., Chamont, S., Blancard, D., Dubourdieu, D. and Darriet, P. (2005). Origin of (-)-geosmin on grapes: On the complementary action of two fungi, *Botrytis cinerea* and *Penicillium expansum. Antonie Leewenhoek* 88(2):131–139.

Latorre, B.A., Viertel, S.C. and Spadaro, I. (2002). Severe outbreaks of bunch rots caused by *Rhizopus stolonifer* and *Aspergillus niger* on table grapes in Chile. *Plant Dis.* 86(7):815–815.

Liu, H.M., Guo, J.H., Cheng, Y.J., Luo, L., Liu, P., Wang, B.Q., Deng, B.X. and Lo, C.A. (2010). Control of gray mold of grape by *Hanseniaspora uvarum* and its effects on postharvest quality parameters. *Ann. Microbiol.* 60(1):31–35.

Longland, J.M. and Sutton, T.B. (2008). Factors affecting the infection of fruit of *Vitis vinifera* by the bitter rot pathogen *Greeneria uvicola. Phytopathology* 98(5):580–584.

Machota, R., Bortoli, L., Botton, M. and Grützmacher, A. (2013). Fungi that cause rot in bunches of grape identified in adult fruit flies (*Anastrepha fraterculus*) (Diptera: Tephritidae). *Chil. J. Agric. Res.* 73(2):196–201.

Marois, J.J., Bledsoe, A.M. and Bettinga, L.J. (1992). Bunch rots. In: Flaherty, D.L., ed. *Grape Pest Management*, University of California, Davis, pp. 63–70.

Meunier, M. and Steel, C.C. (2009). Effect of *Colletotrichum acutatum* ripe rot on the composition and sensory attributes of Cabernet Sauvignon grapes and wine. *Aust. J. Grape Wine Res.* 15(3):223–227.

Moreno, D., Valdé, E., Uriarte, D., Gamero, E., Talaverano, I. and Vilanova, M. (2017). Early leaf removal applied in warm climatic conditions: Impact on Tempranillo wine volatiles. *Food Res. Int.* 98:50–58.

Nally, M.C., Pescea, V.M., Maturano, Y.P., Muñoz, C.J., Combina, M., Toro, M.E., Castellanos de Figueroa, L.I. and Vazquez, F. (2012). Biocontrol of *Botrytis cinerea* in table grapes by non-pathogenic indigenous *Saccharomyces cerevisiae* yeasts isolated from viticultural environments in Argentina. *Postharvest Biol. Technol.* 64(1):40–48.

Otoguro, M. and Suzuki, S. (2018). Status and future disease protection and grape berry quality alteration by micro-organisms in viticulture. *Lett. Appl. Microbiol.* 67(2):106–112.

Parafati, L., Vitale, A., Restuccia, C. and Cirvilleri, G. (2015). Biocontrol ability and action mechanism of food-isolated yeast strains against *Botrytis cinerea* causing post-harvest bunch rot of table grape. *Food Microbiol.* 47:85–92.

Pearson, R.C. and Goheen, A.C. (1988). *Compendium of Grapevine Diseases.* APS Press, St. Paul, 93 p.

Pedneault, K. and Provost, C. (2016). Fungus resistant grape varieties as a suitable alternative for organic wine production: Benefits, limits, and challenges. *Sci. Hortic.* 208:57–77.

Ponsone, M.L., Nally, M.C., Chiotta, M.L., Combina, M., Köhl, J. and Chulze, S.N. (2016). Evaluation of the effectiveness of potential biocontrol yeasts against black sur rot and ochratoxin A occurring under greenhouse and field grape production conditions. *Biol. Control* 103:78–85.

Prendes, L., Merín, M., Fontana, A.R., Bottini, R.A., Ramirez, M.L. and Morata de Ambrosini, V. (2018). Isolation, identification and selection of antagonistic yeast against *Alternaria alternata* infection and tenuazonic acid production in wine grapes from Argentina. *Int. J. Food Microbiol.* 266:14–20.

Raspor, P., Miklič-Milek, D., Avbelj, M. and Čadež, N. (2010). Biocontrol of grey mould disease on grape caused by *Botrytis cinerea* with autochthonous wine yeasts. *Food Technol. Biotechnol.* 48(3):336–343.

Rousseaux, S., Diguta, C.F., Radoï-Matei, F., Hervé Alexandre, H.and Guilloux-Bénatier, M. (2014). Non-Botrytis grape-rotting fungi responsible for earthy and moldy off-flavors and mycotoxins. *Food Microbiol.* 38:104–121.

Samuelian, S.K., Greer, L.A., Savocchia, S. and Steel, C.C. (2011). Detection and monitoring of *Greeneria uvicola* and *Colletotrichum acutatum* development on grapevines by real-time PCR. *Plant Dis.* 95(3):298–303.

Sanzani, S.M., Miazzi, M.M., di Rienzo, V., Fanelli, V., Gambacorta, G., Taurino, M.R. and Montemurro, C. (2016). A rapid assay to detect toxigenic *Penicillium* spp. contamination in wine and Musts. *Toxins* 8(8):235–247.

Serra, R., Braga, A. and Venancio, A. (2005). Mycotoxin-producing and other fungi isolated from grapes for wine production, with particular emphasis on ochratoxin A. *Res. Microbiol.* 156(4):515–521.

Sholberg, P.L., Haag, P.D., Hambleton, S. and Boulay, H. (2003). First report of brown rot in wine grapes caused by *Monilinia fructicola* in Canada. *Plant Dis.* 87(10):1268.

Sonker, N., Pandey, A.K. and Singh, P. (2016). Strategies to control post-harvest diseases of table grape: A review. *J. Wine Res.* 27(2):2.

Steel, C.C., Blackman, J.W. and Schmidtke, L.M. (2013). Review: Grapevine bunch rots: Impacts on wine composition, quality, and potential procedures for the removal of wine faults. *J. Agric. Food Chem.* 61(22):5189–5206.

Swart, A. and Holz, G. (1994). Colonization of table grape bunches by *Alternaria alternata* and rot of cold-stored grapes. *S. Afr. J. Enol. Vitic.* 15(2):19–25.

Swett, C.L., Bourret, T. and Gubler, W.D. (2016). Characterizing the brown spot pathosystem in late-harvest table grapes (*Vitis vinifera* L.) in the California central valley. *Plant Dis.* 100(11):2204–2210.

Van Niekerk, J.M., Groenewald, J.Z., Verkle, G., Paul, H., Fourie, P.H., Wingfield, M.J. and Crous, P.W. (2004). Systematic reappraisal of *Coniella* and *Pilidiella*, with specific reference to species occurring on *Eucalyptus* and *Vitis* in South Africa. *Mycol. Res.* 108(3):283–303.

Wang, Y., Wang, C.W. and Gao, J. (2015). First report of *Fusarium proliferatum* causing fruit rot on grape (*Vitis vinifera*) in China. *Plant Dis.* 99(8):1180.

Zahavi,T., Cohen, L., Weiss, B., Schena, L., Daus, A., Kaplunov, T., Zutkhi, J., Ben-Arie, R. and Droby, S. (2000). Biological control of *Botrytis*, *Aspergillus* and *Rhizopus* rots on table and wine grapes in Israel. *Postharvest Biol. Technol.* 20(2):115–124.

12 Wine Production and *Botrytis*

N.S. Thakur, Satish K. Sharma and Abhimanyu Thakur

CONTENTS

12.1 INTRODUCTION

Grapes are grown in both tropical and temperate climates, although the majority of grapevines are located in temperate regions, mostly in Europe. The main producers are Spain, France, Italy, Russia, Turkey, and Portugal, followed by the United States and countries from the southern hemisphere, namely Australia, South Africa, Chile, and Argentina. There are a number of pathogens which are responsible for infecting the grapes and further affecting the quality of wine produced. Grape microflora is dependent upon various factors like rainfall, humidity, altitude, nitrogen fertilization, insect vector, pesticide sprays, and winery waste disposal practices. Various fungi like *Aspergillus*, *Penicillium*, *Rhizopus*, *Botrytis*, and *Mucor* attack the grapes. *Botrytis cinerea* is a ubiquitous, filamentous, and necrotrophic fungus, which is one of the principal causes of quantitative and qualitative degradation of grapes. This pathogen is responsible for causing *Botrytis* bunch rot and/or grey mould in many vineyards around the world. *B. cinerea* is unique in parasitology; its development results in poor grape quality referred as pourriture grise, gray rot, bunch rot, or graufaule by French and German wine makers (Figure 12.1). Only under certain conditions does *B. cinerea* produce an over mature condition where it is known as noble rot or *edelfaule*, indispensable in the production of sweet white wine, known as great sweet Sauternes in France, Trockenbeerenauslese in Germany, Tokay Aszu in Hungary, and with other such names throughout the world.

Prevailing warm, sunny, and windy weather conditions during the infection of *B. cinerea* leads to the shriveling of berries due to loss of moisture, which ultimately leads to the increase in sugar concentration and is termed as pourriture noble or noble rot (Figure 12.2).

The fungus is responsible for the consumption of some portion of the grape sugar, but it is countered by an increase in sugar due to berry dehydration during the infection process. The dehydration of the grape berries leads to an increase in the concentration of sugars, acids, glycerol, minerals, and certain aroma components. To exploit the higher concentration of various chemical constituents including sugars, sweet white table wines can be produced commercially from these infected

FIGURE 12.1 Gray mould caused by *B. cinerea* on a cluster of grapes. *Source:* https://learn.winecoolerdirect.com/Botrytis/

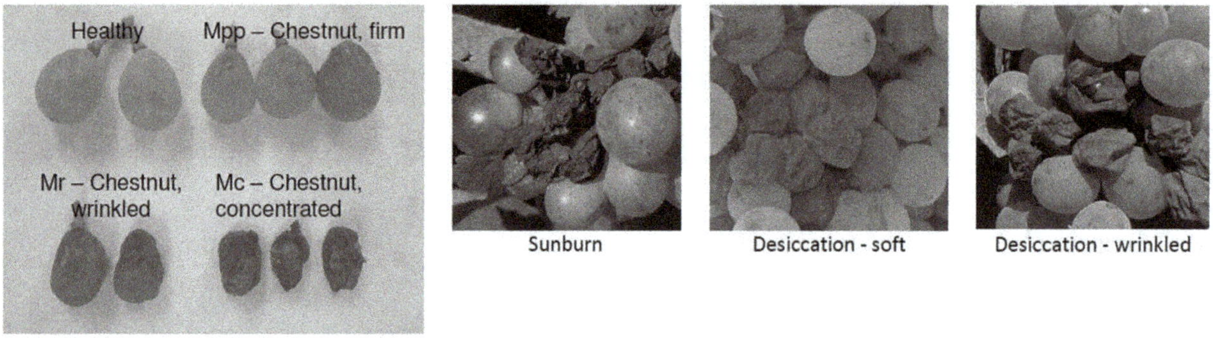

FIGURE 12.2 Classification scheme for symptoms of *B. cinerea* and natural desiccation of grape berries. *Source:* Carey *et al.* (2004)

TABLE 12.1
Comparison of Juice from Healthy and *B. Cinerea* Infected Grapes

	Sauvignon berries		Semillon berries		Germany		Tokaj	
Component	**Healthy**	**Infected**	**Healthy**	**Infected**	**Healthy**	**Infected**	**Healthy**	**Infected**
Fresh weight/100 berries (g)	225	112	202	98	85	36	–	–
Sugar content (g/litre)	281	326	247	317	295	500	685	708
Acidity (g/litre)	5.4	5.5	6.0	5.5	15.20	20.80	16.55	14.70
Tartaric acid (g/litre)	5.2	1.9	5.3	2.5	2.60	2.40	4.81	4.44
Malic acid (g/litre)	4.9	7.4	5.4	7.8	8.00	10.10	5.82	7.42
Citric acid (g/litre)	0.3	0.5	0.26	0.34	0.20	0.24	0.11	0.99
Gluconic acid (g/litre)	0	1.2	0	2.1	1.50	2.17	3.20	3.88
Ammonia (g/litre)	49	7	165	25	–	–	–	–
pH	3.4	3.5	3.3	3.6	–	–	–	–

Source: Charpentie (1954); Ribereau-Gayonet *et al.* (1980); Linssen (1986); Dittrich (1989); Magyar (2006)

berries. A comparison of juice from healthy and *B. cinerea*-infected grapes has been made in Table 12.1. In this chapter, the historical developments in botrytised wine making, types of botrytised wines, pathogenesis of *B. cinerea*, and prerequisites for noble rot development, methods for alcoholic fermentation, and wine making have been discussed.

12.2 HISTORICAL DEVELOPMENTS

The history of the production of botrytised wine is very old, and various explanations about how and when botrytised grapes were first used for wine making have been given. The first intentional use of infected grapes for production of wine is however unknown. The production of botrytised wines

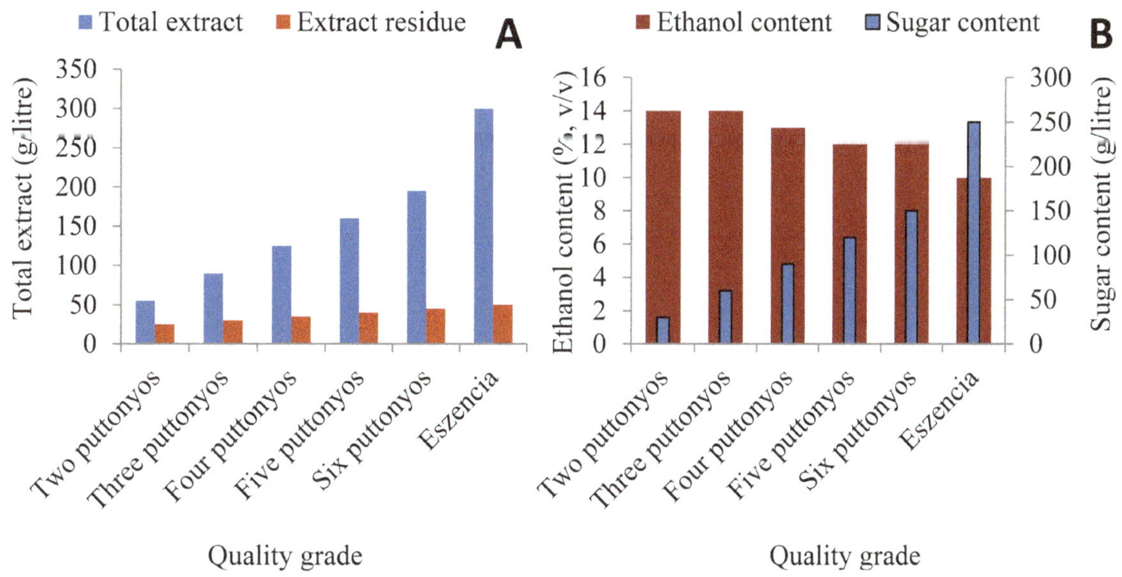

FIGURE 12.3 (a and b): Chemical composition of Tokaji wines. *Source:* Based on Farkas (1988)

for the first time has been reported in the Tokaj region of Hungary (ca. 1560) and the Schloss Johannisberg region of Germany (ca. 1750). In France, the production of botrytised wines started in the Sauternes region between the years 1830 and 1850. The French botrytised wines produced in the Sauternes region became so popular that Sauternes became the brand name of sweet white table wine. In 1775, the owner of Schloss Johannisberg forgot to send permission to start the harvesting, and so the advantages of harvesting grapes very late in the Rheingau region in Germany was originated accidentally. The botrytised fruit is harvested at the desired stage of shriveling and involves picking off portions of the cluster in successive harvests, since not all the berries are infected equally or shriveled sufficiently at the same time. The number of pickings depends upon conditions prevailing from vineyard to vineyard and the degree of sugar concentration desired. The sugar content may go up to 50% in botrytised *musts*. The shriveled grapes which had nearly become raisins were used for the production of botrytised wines. For the production of botrytised wines, various picking processes of botrytised fruits in France were explained during 1863, and the procedure of wine making from botrytised grapes was described as early as 1872. *Botrytis* was prevalent in Napa Valley in California in 1893, which caused many grapes to be rejected. At least one wine from botrytised grapes from St. Helena was produced, which remained sweet for about a year. In 1899, a *Botrytis* infection of Semillon grapes of Crimea in Russia was reported and led to the production of a sweet table wine in the same year. The sweet white wines of Gramdjo in northern Portugal are also produced from *B. cinerea*-infected grapes. In the 1950s, studies were carried out for commercial implementation of *B. cinerea* inoculation. Palletized trays of grapes were inoculated with *B. cinerea*, produced in one-ft^2 custom-made petri dishes containing a two percent agar-grape juice medium. In later periods, several researchers conducted investigations to produce botrytised wines by adding *Botrytis*, or enzymes from *Botrytis*, directly to juice or *must* and succeeded in introducing the sensory characteristics of botrytised wines.

12.3 TYPES OF BOTRYTISED WINES

12.3.1 Tokaji Aszu

Tokaji Aszu was the first botrytised wine prepared from the noble rotted grapes. It was initially prepared in Hungary, and Tokaj is the name of a town which is a wine district in Hungary. A very small amount (1 to 1.5 litres) of Aszu Eszencia is collected from 30 litres of juice extracted from botrytised berries. The collected Eszencia is kept in small wooden barrels for further fermentation and maturation. The categories of Aszu are based on the amount of Aszu added to the mixture and termed as *puttony* (a *puttony* equals 28–30 litres). Aszu wines are matured for a time period of three years, and out of this time interval the use of small oak barrels for two years is compulsory. The chemical composition of Tokaji wines had been mentioned in Figure 12.3 (a and b). Fordias, Malslas, and Szamorodni are internationally well known Tokaj specialty wine, as well as Eszencia and Aszu.

12.3.2 German Botrytised Wines

The production of botrytised wines from noble rot grapes in Germany started in 1775 at Schloss Johannisberg vineyards. Riesling, Gewurztraminer, Rulander, Scheurebe, Silvaner, and Huxelrebe are the predominant cultivars grown, which are prone to noble rot. These wines are divided into three categories, namely *Auslesen* wines (may or may not be *Botrytis*-concentrated), Beerenauslesen wines, and Trockenbeerenauslesen wines. Out of these, Beerenauslesen and Trockenbeerenauslesen wines are made from *B. cinerea*-infected berries which possess a higher sugar content. The

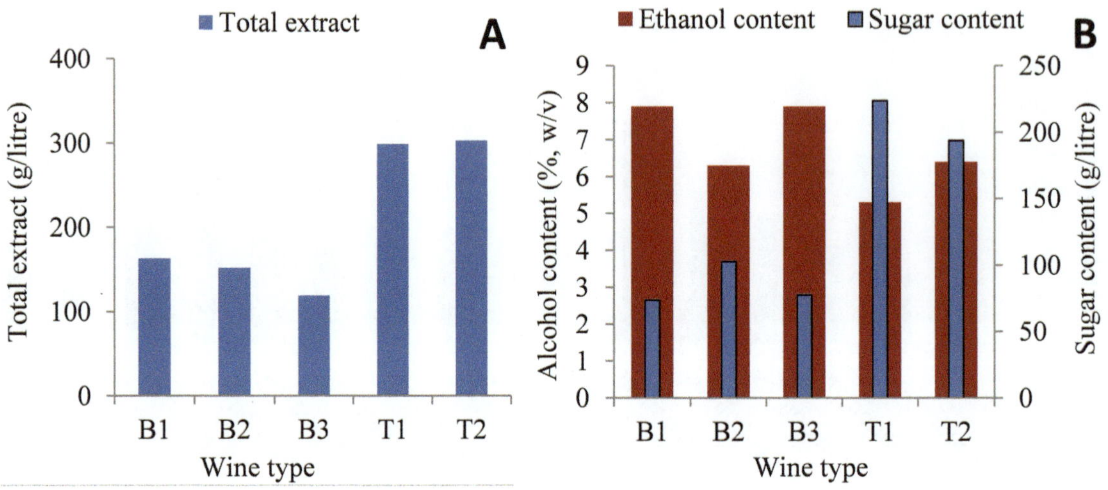

FIGURE 12.4 (a and b): Chemical composition of Beerenauslesen (B1, B2, B3) and Trockenbeerenauslesen (T1, T2) wines. *Source:* Based on Watanabe and Shimazu (1976)

German botrytised wines have relatively higher sugars and a lower alcohol content in comparison to Tokaji wines (Figure 12.4).

12.3.3 French Botrytised Wines

French botrytised wines are white and sweet and are produced in Sauternes. Sauternes is located near two rivers, namely the Garonne and the Ciron, where favourable conditions prevail for the development of noble rot. The intentional production of botrytised wines from infected grapes started in this region two centuries later than in Tokaj. French botrytised wines contain a higher alcohol content (≥11–13%) when compared to German wines.

12.4 PATHOGENESIS

Although grey mould is one of the most serious diseases in the vineyards but for wine makers it is highly desirable for the production of high value, sweet dessert wine. The main factors that lead to noble rot are: 1) appropriate climatic conditions, 2) grape variety (ripening time, anatomy of berry and bunch, phytoalexins production, etc.), and 3) full or over-ripeness of the grape at the time of *Botrytis* invasion (when the vascular connection between the vine and berry ceases). However, it is of foremost importance to understand the pathogenesis process of the fungus for the production of botrytised wines.

12.4.1 Dissemination of Conidia, Germination, and Penetration of Berries

B. cinerea is characterized by abundant conidia (asexual spores), which are borne on branching trees like conidiophores. *B. cinerea* is an Ascomycete fungus, belonging to order Halotiales and family Sclerotiniaceae. In old cultures, sclerotia are also produced, which overwinters and germinates to produce conidiophores in the spring season. The dissemination of fungus takes place through conidia which are dispersed by wind and rainwater and leads to infection. Two types of infection usually occur on grape berries. First is the grey rot which occurs due to consistently wet or humid conditions, and the other is the noble rot which occurs when wet conditions are followed by dry conditions (Figure 12.5).

The fungus attacks weaker or damaged parts of grapevines before spreading to whole bunch. Grape berry moth larvae (*Lobesia botrana*) is one of the important vectors in dissemination of fungus by carrying the conidia both on and in their bodies. Due to easy disruption of the grape cuticle wax, the tightly clustered varieties are more prone to *Botrytis* degradation. Under moist conditions, the conidia germinate on the host surface, producing a germ tube which further develops into appressorium and penetrates the host surface. Due to the weak enzymatic profile of the fungus, it can't penetrate directly and requires physical injuries/wounds or natural openings for inserting a penetration peg into the host tissue. After penetration, the underlying cells are killed, and the fungus establishes a primary lesion in which necrosis and a host defence response may occur. The infection cycle is completed in three to four days depending on the type of host tissue attacked. The attachment of conidia to the host tissue is a two-step process. The first step involves weak adhesive forces resulting from hydrophobic interactions between the host and the conidial surface, and after several hours of inoculation the conidia gets germinated, leading to stronger binding. High humid conditions (≥93%) are essential for conidia germination and penetration of the host epidermis. After attachment and germination of conidia, the host tissue invasion takes place by means of active penetration or passive ingress. After physical penetration, if the fungus is not capable of growing hyphal branches, it further attacks the berries by releasing various chemicals or enzymes.

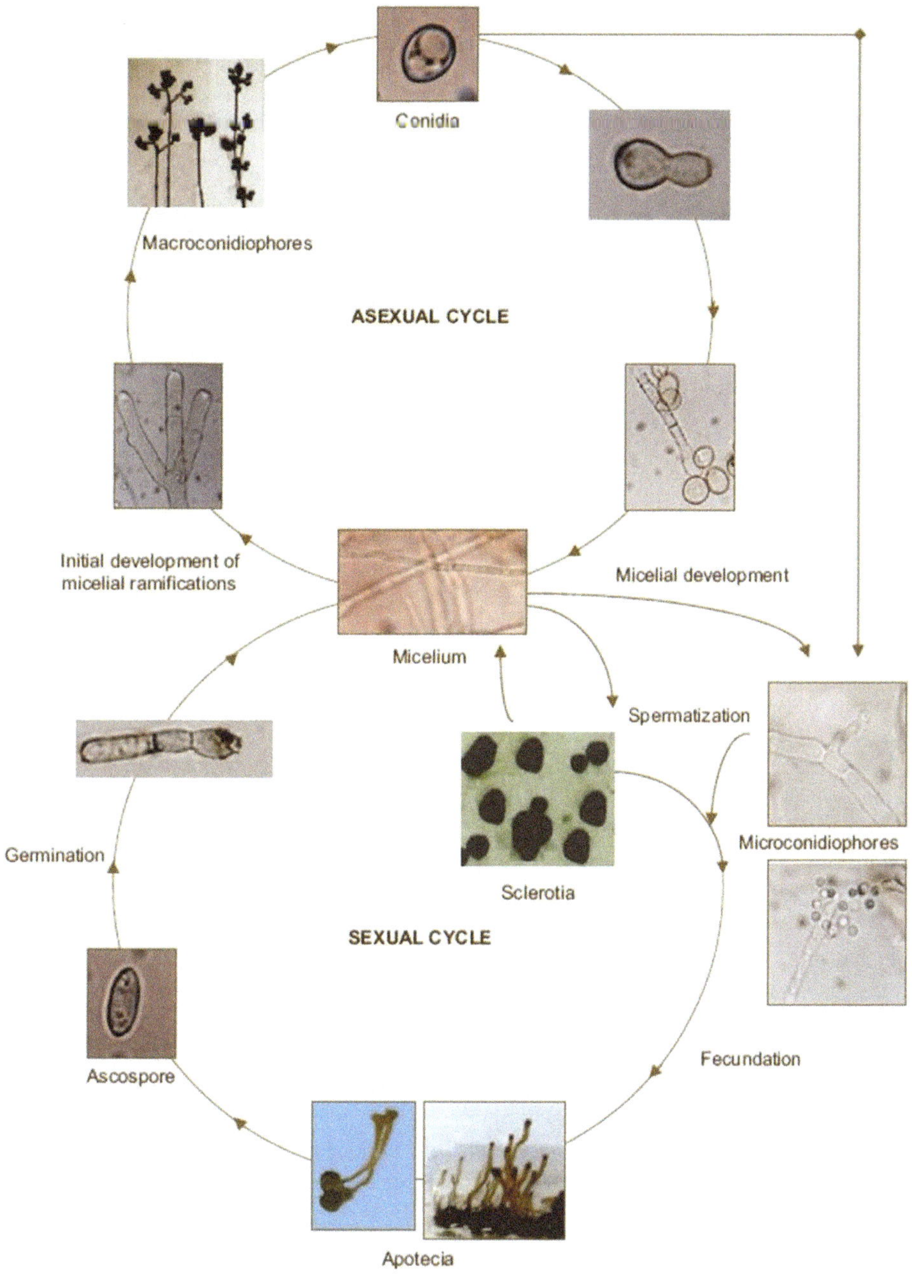

FIGURE 12.5 Life cycle of *B. cinerea*. *Source:* Cantoral and Collado (2011)

12.4.2 Role of Fungal Enzymes in Invasion of Tissues

The initial attack of the fungus is near the cell wall, and it triggers a biochemical attack on the plant tissue and cells to aid the spread of infection. A large number of complex interactions take place during the cell invasion process after the fungal attack (Figure 12.6). The fungus uses two chemical weapons at this point: high-molecular-weight enzymes that break down the cell wall and membrane and lead to tissue maceration, and low-molecular-weight toxins that kill the plant cells as the hyphae advance through the host tissue. As the conidial germination starts, the endopolygalacturonase enzyme is synthesized. The pectin compounds are degraded completely, and the cell wall composition is greatly altered by other enzymes, such as exopolygalacturonase, pectinmethylestrase, transeliminase, phospholipase, and protease. After the complete mycelium invasion into the pectocellulosic cell walls, the white grape berry converts into a characteristic chocolate hue known as "*pourri plein*". The filaments produced on the mycelium pierce through the cuticles or take a route through the various fissures already used for penetration. After the invasion process, small circular partly clear spots are the first symptoms visible on berries. The berries become dark coloured with typical greyish hairy mycelium all over their surface, but under highly humid conditions the mycelial growth may be cottony and white. Often the fungus can be seen growing along the cracks or splits on the berries.

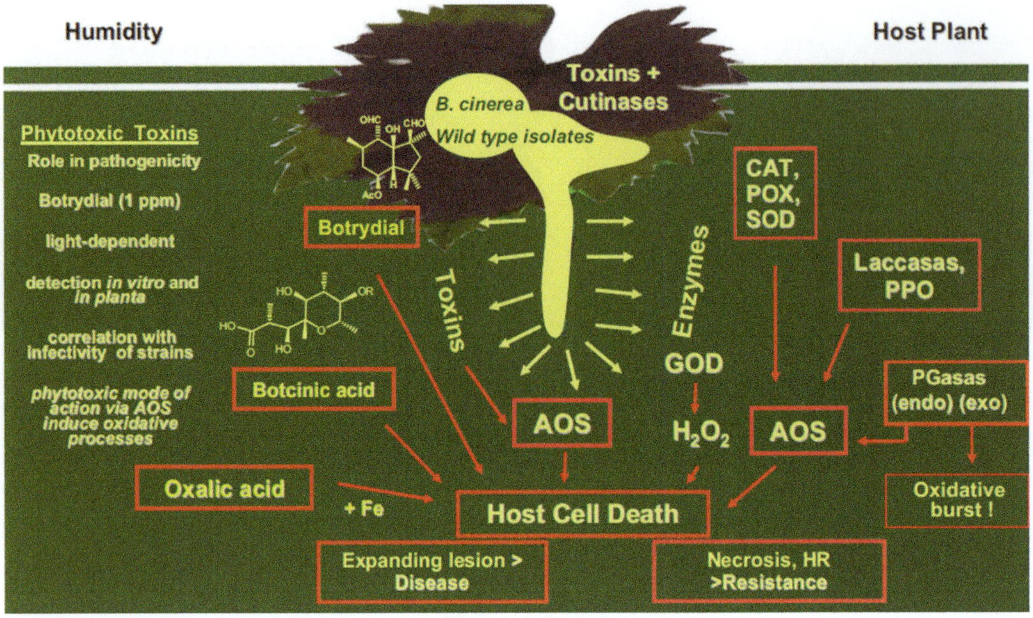

FIGURE 12.6 Enzymes and toxins involved in the infective processes of *B. cinerea*; CAT: catalase; POX: peroxidase; SOD: superoxide dismutase; GOD: glucose oxidase; PPO: polyphenoloxidase; AOS: active oxygen species. *Source:* Cantoral and Collado (2011)

With the development and differentiation process of fungus, conidiophores develop on these filaments, which are erect, hyaline, and unbranched or seldomly branched. The fungus-infected cell walls of berries are so greatly modified that they are no longer functional. The hydration of the berry cells varies according to climatic conditions and usually leads to a characteristic withering which ultimately causes the cytoplasmic death of the epidermal cells. This stage is known as *confit* or "*pourri roti*". The infected berries are brown in colour and dry into a sort of moist raisin with low acid and low nitrogen and are expected to reach extremely high levels of sugars (30–40%). In Sauternes and Tokaj, the sugar level of these intensively infected berries commonly reaches up to 350 and 700 g/L. These berries are then picked and utilized for the preparation of highly aromatic sweet wines.

12.4.3 Defence Response of Infected Tissues

At the initiation of a fungal attack, there is a development of necrotic lesions and rot, along with secretion of different compounds like phytoalexins, which are stilbenic derivatives (*trans*-resveratrol, ε-viniferin-dimer, ε-viniferin-trimer, ε-viniferin-tetramer). Phytoalexins have fungicidal properties and lead to the restriction of mycelium into the epidermal cells. The phytoalexins consist of various substances which are of isoflavonoid or terpenoid nature that are normally specific to the plant host. α-Viniferina (resveratrol trimer), which is a stilbene synthesized by the phenylalanine-polymalonate pathway, has been identified in the *B. cinerea*-infected leaves of *Vitis vinifera*. Ripe berries with low stilbenes are more prone to fungal infection, which is favoured through the exposure of berries to UV light. When the sugars level is 100–150 g/L in the berries, there is maximum production of stilbenes which then decrease rapidly. Lower production of phytoalexins in grape berries greatly favours the development of noble rot. In addition to phytoalexins, other fungal inhibitors, mainly phenolic compounds, are produced. The main classes of phenolic compounds in grape berries, other than stilbenes, are anthocyanins and condensed tannins (proanthocyanidins). The plant proanthocyanidins keep the *B.cinerea* in a quiescent stage and delay the development of rotting symptoms.

12.4.4 Predisposing Factors for Berry Infection

Development of noble rot is dependent on various factors and conditions which exist in certain regions of the world. Hot and dry regions are less prone to the incidence of this rot, and berries with open fruit clusters and tough skin are also less prone to the fungus. The favourable conditions for noble rot development include: excess water supply, a temperature range from 15–20°C, humidity of 90% or above, dew or morning mist, hot sunny days with high wind velocity, evaporation rate, fully ripened berries, level of skin thickness, number of stomata, antifungal substance formation, etc. Conidia can germinate within a wide range of temperatures (10–25°C), the optimum temperature for its germination being 18°C. The berries are prone to bursting at high temperatures, and high evaporation rates are responsible for the production of concentrated juice. The dissemination of the fungus by conidial dispersion is favoured by alternate dry and humid situations. As *B. cinerea* attack is less frequent in hot and dry conditions, artificial inoculation of berries with fungus is practiced. For artificial inoculation of mature grapes with *B. cinerea,* the berries are kept at a temperature range of 20–25°C along with >90% relative humidity for 24 hours. The berries are exposed to cold and dry conditions (20°C temperature and 50% relative humidity) to inhibit the fungal growth and facilitate the dehydration of berries. The berries treated with artificial inoculation of fungus are high in sugar and low in acid content, with a similar aromatic profile of naturally botrytised grapes. The artificial botrytisation of grapes is carried out by spraying with berries with ground hyphae powder (1 × 10^6 CFU/ml) of *B. cinerea* dissolved in 20 g/L glucose. Botrytised wines prepared from these botrytised grapes were found to be richer in aroma compounds than dry white wines and delayed harvest sweet wines. During inoculation of these berries, the favourable conditions of temperature (20–25°C) and humidity (>90% for 24 hours) were maintained for fungal growth. The commercial production of botrytised sweet wines is limited due to the high cost of production, along with other technological problems.

12.4.5 *In Vitro* Establishment of *B. cinerea* in the *Must*

Many attempts or investigations have been made to produce botrytised wines with the addition of inoculum into the juice or *must* rather than in the grape berries. *B. cinerea* can be produced in liquid as well as solid-phase cultures in different media. The solid-fermentation process can be used as a large-scale production method for controlled introduction of *B. cinerea* into vineyards, as the solid-state fermentation products can be stored for longer than liquid fermentation products. Fermenters have been developed for the treatment of grape *must* with this fungus, and addition of compound-like cyclic adenosine mono-phosphate into the fermenters can help with the controlled production of gluconic acid and glycerol during the fermentation process. It is difficult to produce a special aroma and to control the levels of acetic acid and acetaldehyde in botrytised wines. The commercial production of these wines is a costly affair, due to various complexities. However, an automatic system for grape treatment in drying rooms in various drying cycles has been internationally patented. The first cycle involves dehydration of berries at 76% relative humidity and 11°C for 30 days, which is further followed by exposure at 100% RH for 7 days and finally kept at 76% and 8–10°C in the last cycle. This leads to an induction of *B. cinerea*, and the wine produced in this way typically has characteristics similar to botrytised wines.

12.5 GRAPE, *B. CINEREA*, AND WINE CHEMISTRY

Changes in the composition of fruit caused by *B. cinerea* are of prime importance in botrytised wine production. Dehydration of berries leads to an increase in grape sugar levels and a decrease in acids like tartaric and malic acid. Various physico-chemical changes in the berries are caused by noble rot, and the effect of these changes on the quality of botrytised wine has been studied by various researchers.

12.5.1 Changes in Grape Sugars

The attack of *B. cinerea* leads to the concentration of the sugar content of berries, which is then further utilized for the production of botrytised wines. *Botrytis* infection followed by a dehydration of berries results in the development of berries which are suitable for making high quality, sweet botrytised wines. The initial sugar content of berries may rise as high as 30–40%, but the cost of production of the fruit is high, as there is reduction of grape volume. The fruit sugar enables various metabolic activities. The conidia and young mycelium of fungus contain enzymes similar to the TCA cycle, the Embden-Meyerhof pathway, the monophosphate shunt, and the glyoxylate cycle. These enzymes are responsible for glucose catabolism, and this oxidation of glucose is similar to the Entner-Doudoroff pathway where cellular material is synthesized during the exponential growth phase. The glucose is oxidized directly by glucose oxidase, leading to accumulation of gluconic acid during the stationary phase. The mycelial growth beneath the grape skin is poor in oxygen concentration and reduces the glucose catabolism process and growth of hyphae. Regeneration of NAD is due to the enzyme glycerol dehydrogenase under semi-anaerobic conditions, leading to glycerol accumulation. The accumulation of glycerol and gluconic acid are the chief indicators of *B. cinerea* infection.

The degradation of sugar is also limited on account of insufficient supply of oxygen during the initial development of the fungus under the grape skin. True noble rot is characterized by a higher ratio of glycerol to gluconic acid, whereas sour rot is indicated by a lower ratio. The glycerol-3 phosphate dehydrogenase enzyme is responsible for the accumulation of glycerol in the infected berries. The decrease in osmotic potential may explain the inability of *B. cinerea* to metabolize a higher proportion of sugars during infection. When the fungus emerges out of the grape, it adopts an oxidative metabolism because of the predominance of oxidase enzyme in the presence of enough oxygen responsible for the accumulation of gluconic acid and the assimilation *de novo* of some of the glycerol formed previously. So, as a result of an increase in internal osmotic pressure due to dehydration of the berries, the development of fungus consequently stops, as it is unable to draw nutritive substances for its growth. The authentication of botrytised wines is determined by the measure of a minimum amount of gluconic acid. Gluconic acid remains as such in the wine, as it is not fermented by yeasts. However, some wild yeasts may contribute towards the elevated levels of glycerol and gluconic acid content of juice from botrytised berries. The gluconic acid level in wine prepared from botrytised fruits ranges from 1 to 5 g/L, whereas it is 0.5 g/L in wine prepared from clean fruits. The glycerol concentration is highest during the *pourri plein* stage and is about 5–60 micromoles per berry, whereas it is 10–40 micromoles per berry at the *pourri roti* stage. The oxidation of glucose leads to production of 5–7 g/L of glycerol, and it may exceed 30 g/L after further berry dehydration. The glycerol (0.03 to 2.61 g/L) and gluconic acid content (0.02 to 0.13 g/L) is higher in juice obtained from grapes inoculated with *B. cinerea* compared to juice from non-inoculated berries. The glycerol production may be as high as 20 g/L, which can be metabolized by bacteria before harvest. Other microbial species like *Gluconobacter oxydans* and *Acetobacter aceti* are responsible for the oxidation of glycerol to dihydroxyacetone, which further affects the concentration of glycerol in *must*.

In addition to glucose, other polysaccharides are decomposed by the fungal enzymes, leading to the accumulation of polyols (mannitol, erythritol, and meso-inositol), arabinose, rhamnose, mannose, galactose, xylose, and galacturonic acid. Besides these polysaccharides, the presence of trehalose has also been reported. These compounds are accumulated around the first cell layer of the pulp close to the skin and have been divided into two groups. The compounds low in molecular weight (20,000 to 50,000 Da) such as polymers of mannose and galactose are included in first group, whereas the second group consists of β-glucans and polymers of glucose having high molecular weight (100,000 to 10,00,000 Da). The compounds, having low molecular weight, enhance the production of acetic acid and glycerol. Depending upon the severity of *B. cinerea* infection, wines made from botrytised grapes can have 100 times more manitol than non-botrytised wines, and which may impart a sweet and sour character to the wines. These polysaccharides, especially b-D-glucan, may cause clarification problems in wine, and a little concentration (2–3 mg/litre) can retard the filtration process. The other enzymes which are pectolytic (pectin methylestrase, polygalacturonase), cellulolytic, and proteolytic in nature are responsible for partial liquification of the berry cell wall. As these glucans are located just under the skin, the berries are harvested gently along with slow pressing so as to minimize their release into the juice. Enzymes like beta glucanase hydrolyse the beta-glucan and can improve the filterability of wines prepared from botrytised grapes.

12.5.2 Changes in Grape Acids

Tartaric and malic acid are the predominating acids present in grape berries, whereas other acids like citric, isocitric, aconitic, glutaric, fumaric, shikimic, lactic, and succinic acids are found in minute concentrations. *B. cinerea* affects the composition of fruit acids during the noble rotting process. Malic acid degradation is as high as 70 to 90%, whereas tartaric acid is degraded up to the extent of 50 to 70%. Juice obtained from grapes inoculated with *B. cinerea* was reported to have lower tartaric acid (1.99 g/L) compared to juice from non-inoculated berries (3.48 g/L). The total acidity of grapes decreases and pH increases, although the synthesis of citric acid, gluconic acid, galactaric acid, oxo-gluconic acid, and glucuronic acid occurs. *B. cinerea* synthesizes gluconic acid in *must* and wine, while *Saccharomyces cerevisiae* cannot utilize gluconic acid in reductive conditions during alcoholic fermentation. Tartaric acid is metabolized more completely than malic acid, as the latter is assimilated during the external phase of fungal development. The various exocellular enzymes (pectinolytic, cellulosic complex, protease, and phospholipase enzymes) lead to the degradation of galacturonic acid. In botrytised *must*, galacturonic acid is oxidized into galactaric acid (mucic acid), which forms salts with calcium ions and gets precipitated in wine. Berries with high potassium ions have high pH values, and acid degradation is higher when the initial acid level is high in healthy grapes. Lactic and succinic acids are formed during alcoholic and malolactic fermentation, and the average concentration of succinic acid in wine reaches 1 g/L.

12.5.3 Evolution of Nitrogenous Substances

During the infection process, there is a detrimental effect on assimilable nitrogen and various nitrogen-containing components like ammonium salts, amino acids and peptides, and nucleic acid derivatives. There is a 30–80% degradation of total amino acid content in infected grapes. The exocellular proteolytic enzymes like proteases and amino oxidases act upon the proteins, and the nitrogen component of amino acids is liberated. The fungus utilizes this liberated nitrogen for various metabolic processes during infection, and ammonia is secreted at the end of fungal growth. The *must* from *pourri roti* grapes has been found to be rich in complex nitrogenous forms, since it is prepared mainly from the fungus-infested pulp and contains less ammonia compared to *must* from healthy grapes. *B. cinerea* is responsible for the degradation of ammonium compounds, and vitamins like thiamine and pyridoxine. So, during the preparation of botrytised wines, there is a need for supplementation with nitrogen and vitamins to avoid the chances of stuck fermentation and possible H_2S formation. The addition of thiamine (0.5 mg/L) to the juice before fermentation can compensate with the thiamine deficiency, and production of sulphur-binding compounds can be reduced during the fermentation process. The effect of *B. cinerea* on nitrogenous compounds has been shown in Figure 12.7 and 12.8.

12.5.4 Evolution of Enzymes

Various enzymes which evolve during *B. cinerea* infection are responsible for the oxidation process which plays a vital role in the preparation of botrytised wines. These enzymes can convert the phenolic compounds into brown-coloured quinones. Laccase (phenoloxidase) is one of the chief enzymes produced by fungus infection, it utilizes oxygen and causes browning of the *musts*. These laccases might be responsible for the inactivation of antifungal phenolic compounds like pterostilbene and resveratrol in grape berries. The enzyme production is based on the induction of phenolic compounds (gallic acid, hydroxycinnamic acid) that are toxic to the fungus and are degradation products of pectic substances of cell walls. The caffeic and p-coumaric acids present in white grapes are converted into quinines with the activity of laccase enzyme and further polymerized into brown compounds. The typical colour of grape skin at the *pourri plein* stage is brown as a result of these browning reactions. At the end of development, since the fungus no longer renews the enzyme, the *pourri roti* grapes have a lower laccase activity than at the *pourri plein* stage. At high grape sugar levels, the laccase activity is partly inhibited in the *must*, as it limits the dissolution of O_2. Laccase activity in wine has been found even after 12 months in storage. Some of the factors optimum for the activity of laccase enzymes are: temperature (50°C); pH

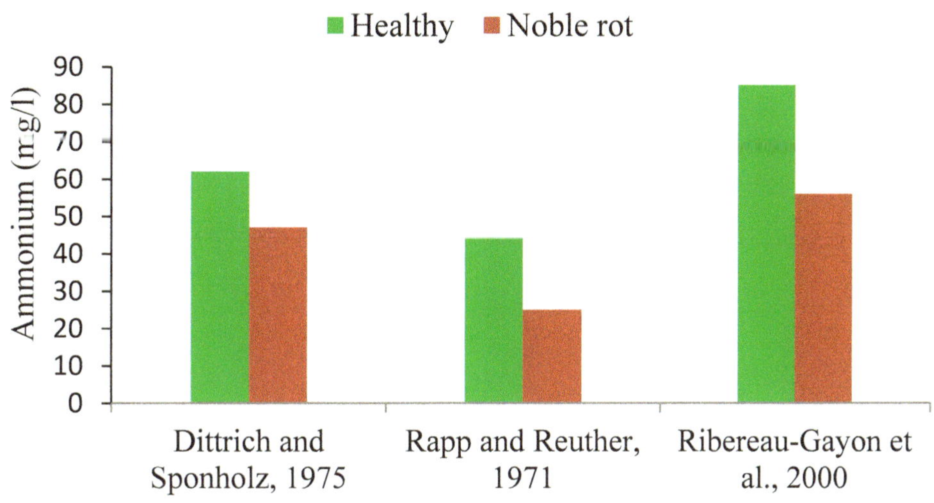

FIGURE 12.7 Effect of *Botrytis cinerea* on ammonium compound of grape juice. *Source:* Based on Magyar (2011)

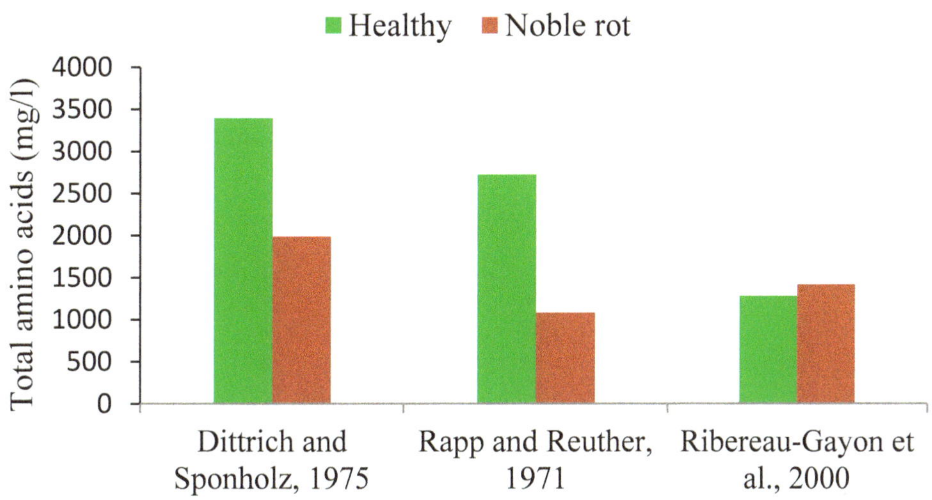

FIGURE 12.8 Effect of *Botrytis cinerea* on total amino acids of grape juice. *Source:* Based on Magyar (2011)

(2.5–7.0); and a low level of grape sugars and high concentration of oxygen. Juice settling and the use of SO_2 to minimize the browning effect have been suggested by various people working in this area. The SO_2 concentration of 50 mg/L at 3.40 pH has been reported to inhibit the action of laccase in wine.

12.5.5 Changes in Phenolic Compounds

Grapevines are generally able to mount a spectrum of defence responses as the pathogen infection begins. In addition to mechanical barriers in the skin and epidermis cell layers, grape skin also contains preformed and/or induced fungal inhibitors, mainly phenolic compounds. The phenolic composition of grape skin has been found to be significantly affected by *B. cinerea* infection. A significant decrease in TPC (total phenolic content) was observed in seeds from healthy to infected grapes (51.1 *vs* 42.6 mg GAE/g dw, respectively), whereas total tannins decreased from 137.60 to 120.80 mg GAE/g dw with the *Botrytis* infection process. Major content variations between healthy and rotten grapes were detected for all phenolic compounds as shown by TPC, total tannin, and anthocyanin contents, which all decreased markedly by 84, 70, and 82%, respectively (Figure 12.9).

Among various polyphenols of wine, anthocyanins have shown significant and consistent variation with the addition of rotten berries to the *must*. The concentration of phenolic compounds in botrytised wines prepared from 15% rotten berries decreased in the range of 15–17% compared to non-botrytised wines.

12.5.6 Evolution of Aromatic and Other Compounds

Botrytised wines produce complex aroma compounds during the botrytization process. During the vinification process, these compounds can easily be transferred from skin to *must*. The characteristic aroma of wines made from *Botrytis*-infested fruit is due to the formation of 3-hydroxy-4,5-dimethyl-2-furanone. These wines also possess a higher level of other

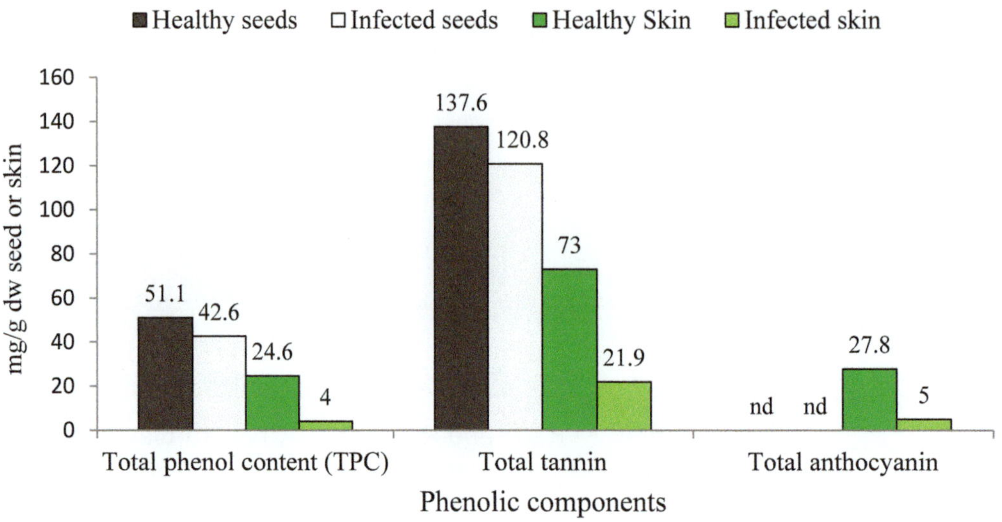

FIGURE 12.9 Concentration of phenolic components in seeds and skins of healthy grapes and botrytised grapes; nd: not detected. *Source:* Based on Ky *et al.* (2012)

TABLE 12.2
Quantitative Assays (μg/l) of Furaneol, Homofuraneol, Norfuraneol, and Phenylacetaldehyde Depending on the Stage of Botrytization

***Botrytis* stage**	**Mean grape volume**	**Homo-furaneol**	**Furaneol**	**Norfuraneol**	**Phenyl-acetaldehyde**
		Must			
Healthy[a]	0.85	nd[b]	nd	nd	1 ± 0.10
Pourri plein	0.68	nd	nd	nd	22 ± 2.0
Pourri roti	0.37	nd	nd	nd	40 ± 4.0
Late pourri roti	0.38	nd	nd	nd	27 ± 3.0
		Wine			
Healthy	0.85	87 ± 5.0	27 ± 1.0	1918 ± 2.0	3 ± 0.2
Pourri plein	0.68	145 ± 6.0	53 ± 6.0	3524 ± 423	281 ± 82.0
Pourri roti	0.37	390 ± 43	73 ± 21	5609 ± 1570	187 ± 21.0
Late pourri roti	0.38	300 ± 30	76 ± 8.0	3593 ± 359	214 ± 21.0

Comparison with decrease in mean grape volume (ml/grape).

[a] healthy (grapes not infected by *B. cinerea*), pourri plein (grapes entirely botrytised but not desiccated, picked two weeks after healthy grapes), pourri roti (grapes botrytised and desiccated, picked two weeks after full-rotten grapes) and late pourri roti (shrivelled grapes left a further 10 days before picking).

[b] nd: not detected.

aroma compounds like homofuraneol, furaneol, norfuraneol, phenyl-acetaldehyde, and methional. Monoterpene glycosides are degraded by the fungal metabolism, and ester hydrolysis is due to the fungal esterase enzyme activity. The primary monoterpenes, such as nerol, geraniol, and linalool, are degraded up to 90% with the infection process. Various aroma compounds like homofuraneol, furaneol, norfuraneol were not detected in grapes, whereas these compounds were found to be present in the prepared wine which might have been formed during alcoholic fermentation. The concentration of phenylacetaldehyde was found to have increased many times during the botrytization process up to the *pourri roti* stage compared to healthy grapes (Table 12.2).

A comparative analysis of wines obtained from manually selected healthy grapes and botrytised grapes was carried out, and aroma analysis revealed that most compounds varied significantly according to the percentage of botrytised berries utilized (Table 12.3 and Figure 12.10). Botrytised wines contain fewer ethyl esters (C6–C10) and related fatty acids than healthy wines, as well as more fruity acetates, such as isoamyl and 2-phenethyl acetate, than healthy wines.

Grape varieties like Sauvignon blanc and Semillon contain cysteine-S conjugate molecules, which are precursors of various strongly odiferous thiols released by yeasts during fermentation. The quantities of these precursors increase dramatically due to noble rot. In addition, specific volatile

TABLE 12.3
Aroma Compounds in Wines Prepared from *Must* from Healthy Grapes (W-100) and Botrytised Grapes in Different Proportions (W-80 and W-60)

	γ/(μg/L)		
Aroma compound	**W-100**	**W-80**	**W-60**
C6 alcohols	1656 ± 76	1855 ± 33	1691± 114
Other alcohols	13978 ± 1845	14233 ± 522	13593 ±366
Fermentative esters	7099 ± 309	8370 ± 913	8569 ± 929
Fatty acids	4422 ± 144	3985 ± 67	3663 ± 456
Aldehydes and ketones	79 ± 6	110 ± 10	114 ± 8
Terpenes	47 ± 1	58 ± 4	61 ± 2
C130-norisoprenoides	34 ± 1	42 ± 2	43 ± 2
Lactones	4403 ±146	4880 ± 276	5066 ± 461
Benzenoids	480 ± 38	501 ± 37	599 ± 41
Phenols	7 ± 0.3	7.2 ± 0.6	8.2 ± 0.6
Others	65 ± 3	915 ± 142	1878 ± 138
N-(3-methylbutyl)-acetamide	1995 ± 179	2598 ± 369	3287 ± 80
Not identified			
Total	34265 ± 1865	37554 ± 1629	38563 ± 550

Source: Fedrizzi *et al.* (2011)

compounds like furfural, benzaldehyde, phenylacetaldelyde, and benzyl cyanide are synthesized by the fungus. Sotalone (3-hydroxy-4,5-dimethyl-2(5H)-furanone) is one of the main compounds which has been found to cause the "*roti*" characteristic in botrytised grapes. After infection, the grape berries also produce some polysaccharides, which have phytotoxic activity. This fungus also produces various antibiotic substances in grapes like botrytidial, norbotryal acetate, and botrylactone. A large number of lactones have also been identified in botrytised grapes as well as botrytised wines. These aromatic compounds induce various aroma characteristics in the wine including citrus fruit, orange peel or grapefruit, honey, caramel, crystallized fruit, walnut, and spice-curry overtones. The higher alcohols, aldehydes, and esters like ethyl-hexanoate (pineapple and banana), phenylethanol (rose), phenylacetaldehyde (honey), beta-damascenone (fruity, quince, and canned apple) and different furanones (caramel), have important roles in the distinct aroma of botrytised wines.

12.6 DETERMINATION OF *B. CINEREA* INFECTION

The *B. cinerea* infection on grape berries can be determined by various visual, physiological, biochemical, and molecular methods. Visual determination is time consuming and may be unreliable. To overcome this, RotBot software has been developed using NIR (near-infrared) and MIR (mid-infrared) spectroscopic techniques, in which each bunch was photographed on a blue background. RotBot differentiates each pixel in the RGB image based on hue, and designates the pixel as either healthy, rotten or background (Figure 12.11). The level of gluconic acid in berries or juice is also an important indicator of the *Botrytis* infection. The Organization Internationale du Vin (OIV) has recommended levels of gluconic acid lower than 0.2–0.3 g/L, whereas levels up to 1.0 g/L indicate an initial stage of fungus infection. A screen-printed amperometric enzymatic biosensor has also been developed. This is used for the determination of gluconic acid based on gluconate kinase (GK) and 6-phospho-D-gluconate dehydrogenase (6PGDH), co-immobilized onto polyaniline/poly (2-acrylamido-2-methyl-1-propanesulfonic acid; PANI-PAAMPSA). The linear range, low detection limit, high sensitivity, operational and storage stability have shown the potential of this biosensor as a highly capable analytical device for a fast gluconic acid measurement of *B. cinerea* in berries and *must*. Test kits are available for the estimation of laccase content, and 3 units/ml is

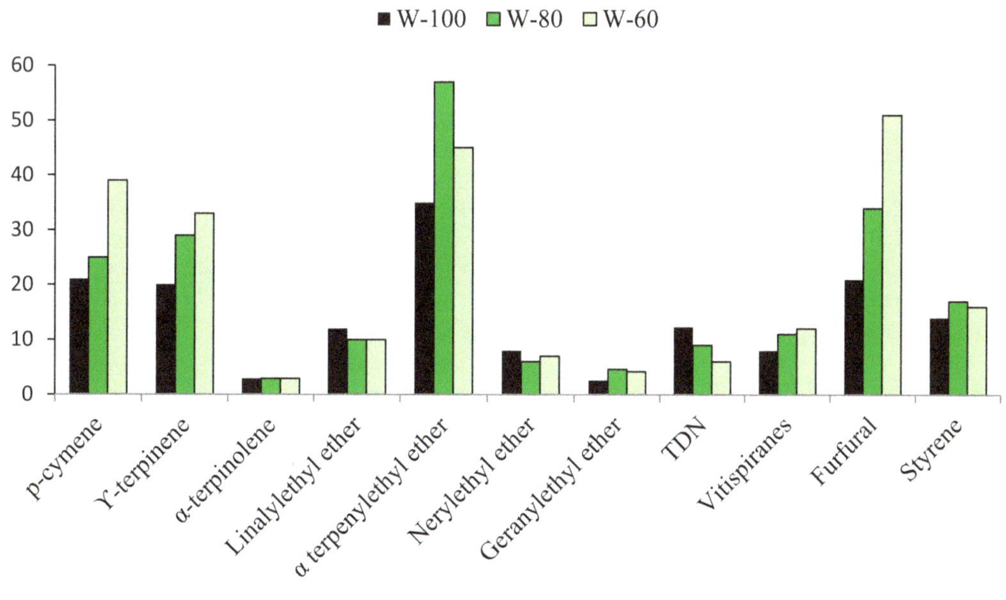

FIGURE 12.10 Aroma compounds in wines prepared from *must* from healthy grapes (W-100) and botrytised grapes in different proportions (W-80 and W-60) TDN: 1,1,6-Trimethyl-1,2-dihydronaphthalene. *Source:* Based on Fedrizzi *et al.* (2011)

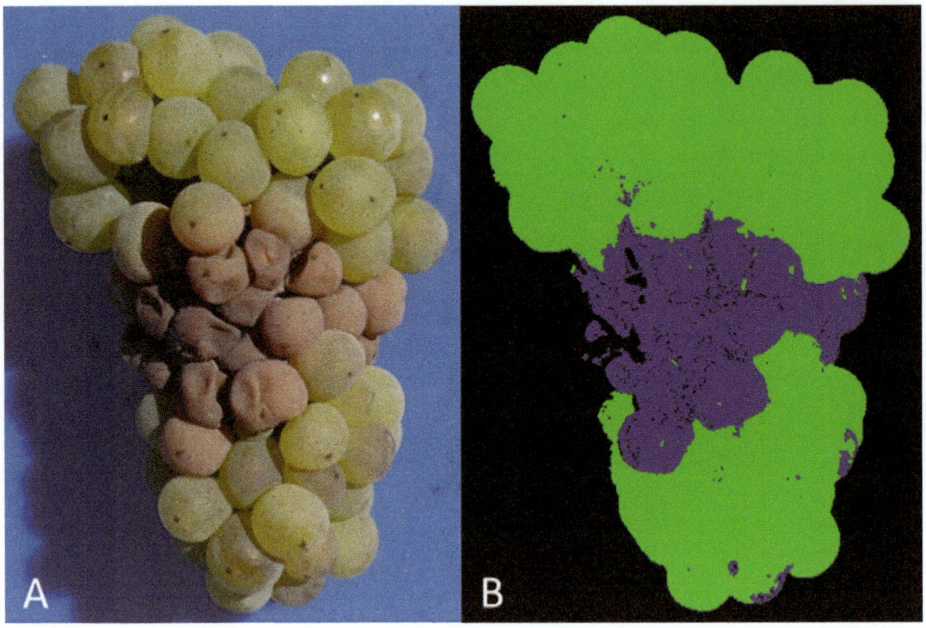

FIGURE 12.11 Symptoms of *Botrytis* bunch rot, caused by *Botrytis cinerea*, on *Vitis vinifera* Riesling before (A) and after (B) analysis by RotBot software: pixels are designated as either healthy (green), rotted (purple), or background (black). RotBot estimated 25% BBR severity for this image. *Source:* Hill *et al.* (2013)

considered significant for fungal activity. After exposure of a small quantity of juice or a wine sample to air at 20–30°C for several hours, a loss of red colour or formation of brown hues and brown deposits can indicate the presence of laccase. Recently, real-time quantitative PCR (qPCR) applications have been developed, and this new technique has provided the ability to simultaneously detect and quantify DNA based on nucleic acid sequences and concentrations. A dual system for the detection, isolation, and quantification of *B. cinerea* has been developed. This real-time PCR assay is highly expensive, but it does appear to be more rapid and sensitive than the conventional selective medium assay, allowing both detection and quantification of *B. cinerea* within three hours.

12.7 WINEMAKING FROM BOTRYISED GRAPES

The wine making process from healthy berries is already well established. Due to various technological problems like juice extraction, uncertainty of noble rot development, low juice yield, control of alcoholic fermentation, and final stability, the production of botrytised wines are challenging tasks for winemakers. The process of winemaking from botrytised grapes is described in this chapter.

12.7.1 Harvesting of Grapes

The harvesting period of berries is prolonged if all the berries are not at perfect noble rot state. Individual berries which are perfect for the preparation of botrytised wines are harvested. Most berries are harvested at the *pourri roti* stage which is termed as *triage.* Depending on the climatic conditions, there may be three to four selective pickings in a year. The wine prepared from the selective harvest of individual berries is of superior quality. As multiple harvesting is a costly procedure, mostly a single picking followed by a separation of the botrytised berries is practiced.

12.7.2 Preparation of *Must*

The various steps followed for the preparation of *must* include the removal of deteriorated fruits, thermo-vinification, whole cluster pressing, separation of press fractions, cryoextraction, and post-fermentation heat treatment. It is difficult to extract juice from botrytised berries by traditional pressing techniques, and so high-pressure presses are used to extract the juice. Slow pressing in two to three cycles with high pressure is practised for better juice yield with lesser browning. Cryoextraction technology has been developed for better extraction of the juice, especially when the level of rot is not homogeneous and the extracted juice has a low sugar content. In this technique, the berries are kept at freezing temperature and the berries with a low sugar content get frozen. The frozen berries are immediately pressed and juice is only released from the ripe berries. The extensive modifications caused by freezing to the pectocellulosic components in the cell wall of the grapes facilitate easy release of cell contents and the resultant juice has a higher sugar content. During juice extraction, various polysaccharides like β-glucan may get released from the berry skin and can cause clarification problems in the *must.* This β-glucan forms a localized network between the epidermal cells and the pulp of the infected berries. In presence of ethyl alcohol, the glucan chains get aggregated and block the filters during the winemaking process. In the 1980s, it was suggested that *Trichoderma* β-glucanase could be successfully used for clarification of juice from grapes infected with *B. cinerea*, where pectolytic enzymes were ineffective on β-glucans. As the cost of *Trichoderma* β-glucanases is high, settling and decanting processes for the juice clarification are generally practiced.

12.7.3 Alcoholic Fermentation

Botrytised grapes exhibit a wide range of yeasts and other microbial species in comparison to the fresh grapes. *B. cinerea* infection encourages the proliferation of acetic acid bacteria and yeasts like *Kloeckera apiculata* and *Torulopsis stellata* on grapes, which in turn can affect the chemical composition and microbial makeup of the *must*/juice. The yeasts, like the *Saccharomyces* species, contribute significantly to the alcoholic fermentation process, whereas the total population of bacteria and fermentative yeasts decreases. The build-up of glycerol and acetic acid in the musts which have high sugar and low nitrogen was reported by Hermann Müller-Thurgau in 1888. To improve the fermentation process, the addition of nitrogenous substances and ethanol-tolerant strains of yeasts is generally practised.

Botryticine, which is released by *B.cinerea* and is a heteropolysaccharide rich in rhamnose and mannose, affects fermentation due to its inhibitory nature. It initiates the glycero-pyruvic pathway in the yeast, leading to the production of acetic acid and glycerol. Alcohol-tolerant strains of *Saccharomyces cerevisiae*, in practice, are preferred to as *Saccharomyces bayanus*, because these strains are less affected by the botryticine of *B. cinerea*. *Saccharomyces uvarum* has also been isolated from botrytised wines by various researchers. As soon as the sufficient alcohol content has been achieved, fermentation should be stopped without delay so as to control the volatile acidity in wine. The fermentation of sweet white wines is stopped either by pasteurization or by the use of SO_2, and the latter practice is most commonly used. The SO_2 also protects against oxidation of phenolic compounds by laccase. Botrytised grapes produce wines rich in substances likely to combine with SO_2, thereby decreasing the proportion of its active free form. A concentration of 50 mg/L of free sulphite is necessary to terminate the fermentation process. Among certain substances, pyruvic acid and α-ketoglutaric acids are mainly produced at the beginning of fermentation, and their levels decrease at the end of fermentation. Dry white wines have lower levels of these acids than sweet white wines. The degradation products like galacturonic and glucuronic acids formed during *B. cinerea* infection which are found in sweet white wines can also combine with SO_2.

The concentration of various carbonyl compounds in botrytised *must* is very high compared to normal *musts*. When compared to the normal *musts*, acetaldehyde, pyruvic acid, and 2-ketoglutaric acid content can be 60, 350, and 500% higher, respectively, in botrytised *musts*. Organic acids (acetic and gluconic etc.), sugars, and some volatile compounds like benzaldehyde and 4-nonanolide are present in higher concentrations in botrytised wines, whereas other volatile compounds such as isoamyl alcohol and ethyl octanoate are found in low concentrations. The high levels of gluconic acid and glycerol produced affects the biological aging process of the wine. During maturation, gluconic acid content is more than 600 mg/L, heterolactic fermentations appear with certain intensity, producing high concentrations of lactic acid and volatile acidity (up to 2 g/L), and the latter is known to affect the quality of the wine. To control this increase in volatile acidity, *Torulaspora delbrueckii* can be used during fermentation as a mixed culture with other strains. The production of other undesirable by-products like acetaldehyde, ethyl acetate, and acetoin may also be minimized with the use of this strain during fermentation. At lower concentrations of gluconic acid, no significant increase in volatile acidity has been observed, as the acetic acid produced during the infection process is consumed in the metabolism of the floor-forming yeasts. Yeasts also metabolize the lactic acid present in the wines during the last phase of the biological aging process. The yeast metabolizes the high glycerol content in the *must* to very low level during the final phase. Premium Sauternes wines are aged in barrels for 12–18 months and sometimes up to a period of two years or more, whereas barrel-aging of German wines is avoided so as to lower down the risk of re-fermentation as they have a lower alcohol content.

BIBLIOGRAPHY

Akau, H.L., Miller, K.M., Sabeh, N.C., Allen, R.G., Block, D.E. and Vander Gheynst, J.S. (2004). Production of *Botrytis cinerea* for potential introduction into a vineyard. *Bioresour. Technol.*, 92(1): 41–48.

Albanese, D., Malvano, F., Sannini, A., Pilloton, R. and Marisa, D.M. (2014). A doped polyaniline modified electrode amperometric biosensor for gluconic acid determination in grapes. *Sensors*, 14(6): 11097–11109.

Bailly, S., Jerkovic, V., Meurée, A., Timmermans, A. and Collin, S. (2009). Fate of key odorants in sauternes wines through aging. *J. Agric. Food Chem.*, 57(18): 8557–8563.

Bely, M., Stoeckle, P., Masneuf-Pomare`de, I. and Dubourdieu, D. (2008). Impact of mixed *Torulasporadelbrueckii-Saccharomyces cerevisiae* culture on high-sugar fermentation. *Int. J. Food Microbiol.*, 122(3): 312–320.

Boillot, J.M. (1986). *Botrytis cinerea*. In: *Chardonnay Sym. Proc. Focus on Chardonnay*, Sonoma-Cutrer Vineyards, Inc, Sonoma, California, p. 157.

Cantoral, Jesus M. and Collado, I.G. (2011). Filamentous fungi (*Botrytis cinerea*). In: Alfonso V. Carrascosa, Rosario Mun˜ Oz and Ramo´ N. Gonza´ Lez, eds. *Molecular Wine Microbiology*. Academic Press, Amsterdam, p. 287–277.

Carey, V., Barbeau, G., Cadot, Y., Conradie, W.J. and Vermeulen, A. (2004). The implication of terroir for development of *Botrytis cinerea* into grey rot or noble rot on grapes of *Vitis vinifera* L. In: *CV.Chenin Blanc.*, OIV Congress, p. 1–10

Charpentié, Y. (1954). Contribution à l'étude biochemique des facteurs de l'acidité des vins. Doctoral Thesis, University of Bordeaux, France.

Debernet, M. and Ribereau-Gayon, P. (1973). Presence of signification dans les mouts et les vins de la tyrosinases du raisin. *Connaiss Vigne Vin*, 7: 283.

Dittrich, H.H. (1989). Die Veränderungen der Beereninhaltsstoffe und der Weinqualität durch *Botrytis cinerea*—Übersichtsreferat. *Vitic Enol. Sci.*, 44: 105–131.

Dittrich, H.H. and Sponholz, W.R. (1975). Die Aminosäure Abnahme in *Botrytis*-infizierten Traubenbeeren und die Bildunghöher Alkohole in diesem Mosten bei ihrerVergärung. *Weinwiss.*, 30: 188–210.

Dittrich, H.H., Sponholz, W.R. and Gobel, H.G. (1975). Vergleichende Untersuchungen von Mosten und Weinen aus gesunden und *Botrytis*-infizierten Traubenbeeren. II. Modellversuche zur Veränderung des Mostes durch *Botrytis*-Infektionung ihre Konsequenzen für die Nebenproduktbildung bei Gärung. *Vitis*, 13: 336–347.

Doneche, B. (1989). Carbohydrate metabolism and gluconic acid synthesis by *Botrytis cinerea*. *Can. J. Bot.*, 67(10): 2888.

Doneche, B. (1990). Metabolisme de l'acide tartrique du raisin par *Botrytis cinerea*. Premiers resultats. *Science des Aliments*, 10: 589.

Doss, R.P., Potter, S.W., Chastagner, G.A. and Christian, J.K. (1993). Adhesion of non-germinated *Botrytis cinerea* conidia to several substrata. *Appl. Environm. Microbiol.*, 59(6): 1786–1791.

Doss, R.P., Potter, S.W., Soeldner, A.H., Christian, J.K. and Fukunaga, L.E. (1995). Adhesion of germlings of *Botrytis cinerea*. *Appl. Environm. Microbiol.*, 61(1): 260–265.

Dubourdieu, D. (1999). La vinification des vins liquoreux de pourriture noble. *Rev. Fr. Oenol.*, 176: 32–35.

Dubourdieu, D., Grassin, C., Deruche, C. and Ribereau-Gayon, P. (1984). Mise au Point d'une mesur rapide de l'activite laccase, dans les mouts et dans les vins par methode a la syringaldazine. Application a l'appreciation de l'estat sanitaire des vendages. *Connaissance Vigne Vin*, 18: 237–252.

Everett, W. (1954). The wines of Portugal. In: *House and Garden Wine Book*. House and Garden, London, p. 90.

Farkas, J. (1988). *Technology and Biochemistry of Wine*, Vols 1 and 2, Gordon & Breach, New York.

Fedrizzi, B., Tosi, E., Simonato, B., Finato, F., Cipriani, M., Caramia, G. and Zapparoli, G. (2011). Changes in wine aroma composition according to botrytized berry percentage: A preliminary study on Amarone wine. *Food Technol. Biotechnol.*, 49(4): 529–535.

Fermaud, M. and Le Menn, R. (1989). Association of *Botrytis cinerea* with grape berry moth larvae. *Phytopathology*, 79(6): 651.

Ferrarini, R., Casarotti, E.M. and Zanella, G. (2009). *Botrytis cinerea* noble form induction on grapes during withering. *Am. J. Enol. Vitic.*, 60: 300.

Fregoni, M., Lacono, F. and Zamboni, M. (1986). Influence du *Botrytis cinerea* sur les caracteristiques physicochimiques du raisin. *Bulletin de l'office International du Vin*, 667: 995.

Genovese, A., Gambuti, A., Piombino, P. and Moio, L. (2007). Sensory properties and aroma compounds of sweet Fiano wine. *Food Chem.*, 103(4): 1228–1236.

Hill, G., Evans, K., Beresford, R. and Dambergs, B. (2013). Quantification of *Botrytis* bunch rot in white wine grape varieties. *Plant Food Res.* http://www.utas.edu.au/__data/assets/pdf_file/0003/410772/Hill_AWITC_Poster_July2013.pdf.

Jackson, R.S. (2008). Specific and distinctive wine styles. In: *Wine Science: Principles and Applications*, Academic Press, Amsterdam, p. 520–576.

Jan, A.L. van Kan (2005). Infection strategies of *Botrytis cinerea*. *Acta Hort.*, 669(669): 77–89.

Jarvis, W.R. (1980). Epidemiology. In: J.R. Coley-Smith, K. Verhoef and W.R. Javis, eds. *The Biology of Botrytis*. Academic Pres Inc., London, p. 219.

Jonson, H. and Robinson, J. (2001). *The World Atlas of Wine*. 5th edn., Mitchel Beazley, London, p. 102–103, 250–251.

Karmona, Z., Barnun, N. and Mayer, A.M. (1990). Utilization du *Botrytis cinerea* pers. Comme source d'enzymes pour la clarification des vins. *J. Int. Sci. Vigne Vin*, 24: 103.

Kovac, V. (1979). Étude de l'inactivation des oxidases du raisin par des moyens chimiques. *Bull. O.I.V.*, 53: 809–815.

Ky, I., Lorrain, B., Jourdes, M., Pasquier, G., Fermaud, M., Geny, L., Rey, P., Doneche, B. and Teissedre, P.L. (2012). Assessment of grey mould (*Botrytis cinerea*) impact on phenolic and sensory quality of Bordeaux grapes, *musts* and wines for two consecutive vintages. *Australian Soc. Viticulture Oenol.*, 18(2): 215–226.

Lafon-Lafourcade, S., Lucmaret, V., Joyeux, A. and Ribereau-Gayon, P. (1981). Utilization de levains mixtes dans l'elaboration des vins de pourriture noble en vue de reduire l'acidite volatile. *Comptes- Rendus de l'Academie d'Agriculture*, 67: 616.

Landrault, N., Larronde, F., Delaunay, J.C., Castagnino, C., Vercauteren, J., Merillon, J.M., Gasc, F., Cros, G. and Teissedre, P.L. (2002). Levels of stilbene oligomers and astilbin in French varietal wines and in grapes during noble rot development. *J. Agric. Food Chem.*, 50(7): 2046–2052.

Langecake, P. and Pryce, R.J. (1977). A new class of phytoalexins from grapevines. *Experientia*, 33(2): 151.

Le-Roux, G., Eschenbruch, R. and De Bruin, S.I. (1973). The microbiology of South African wine making. Part VIII- The microflora of healthy and *Botrytis cinerea* infected grapes. *Phytophylactica*, 5: 51.

Linssen, U. (1986). Diploma Arbeit.cit. Dittrich, H.H., and Grossmann, 2011. *Mikrobiologie des*. 4th edn., Verlag Eugen Ulmer, Stuttgart, p. 236.

Lorenzini, M., Azzolini, M., Tosi, E. and Zapparol, G. (2013). Postharvest grape infection of *Botrytis cinerea* and its interactions with other moulds under withering conditions to produce noble-rotten grapes. *J. Appl. Microbiol.*, 114(3): 762–770.

Magyar, I. (2006). *Microbiological Aspects of Winemaking*, Corvinus University of Budapest, Faculty of Food Science, Budapest.

Magyar, I. (2011). Botrytized wines. *Adv. Food Nutr. Res.*, 63: 147–206.

Magyar, I. and Soos, J. (2016). Botrytized wines – Current perspectives. *Int. J. Wine Res.*, 8: 29–39.

Magyar, I., Toth, T. and Pomazi, A. (2008). Oenological characterization of indigenous yeasts involved in fermentation of Tokaji aszu. *Bull.*, OIV 81: 35–43.

Marbach, I., Harel, E. and Mayer, A.M. (1985). Pectin, a second inducer for laccase production by *Botrytis cinerea*. *Botrytiscinerea Phytochem.*, 24(11): 2559.

Million, L., Reboux, G., Bellanger, P., Roussel, S., Sornin, S., Martin, C., Deconinck, E., Dalphin, J.C. and Piarroux, R. (2006). Quantification de Stachybotryschartarum par PCR en temps reel dans l'environnement domestique, hospitalier, et agricole. *J. Mycol. Med.*, 16(4): 183–188.

Müller-Thurgau, H. (1888). Die edelfaule der Trauben. *Landwirt Jahrbucher*, 17: 83.

Nelson, K.E. and Nightingale, M.S. (1959). Studies in the commercial production of natural sweet wines from botrytized grapes. *Am. J. Enol. Vitic.*, 10: 135–141.

Nelson, K.E., Kosuge, T. and Nightingale, A. (1963). Large-scale production of spores to botrytised grapes for commercial natural sweet wine production. *Am. J. Enol. Vitic.*, 14: 118.

Pezet, R., Pont, V. and Hoang-Van, K. (1991). Evidence for oxidative detoxification of pterostilbene and resveratrol by laccase-like stilbene oxidase produced by *Botrytis cinerea*. *Physiol. Molec. Plant Pathol.*, 39(6): 441–450.

Pryce, R.J. and Langcake, P. (1977). Alpha viniferin: An antifungal resveratrol trimer from grapevines. *Phytochemistry*, 16(9): 1452.

Rapp, A. and Reuther, K.H. (1971). Der Gehalt an freien Aminosäuren in Traubenmosten von gesunden und edelfaulen Beeren verschiedener Rebsorten. *Vitis*, 10: 51.

Reboredo-Rodríguez, P., González-Barreiro, C., Rial-Otero, R., Cancho-Grande, B. and Simal-Gándara, J. (2015). Effects of sugar concentration processes in grapes and wine aging on aroma compounds of sweet wines – a review. *Crit. Rev. Food Sci. Nutr.*, 55(8): 1053–1073.

Renault, P., Miot-Sertier, C., Marullo, P., Herna'ndez-Orte, P., Lagarrigue, L., Lonvaud- Funel, A. and Bely, M. (2009). Genetic characterization and phenotypic variability in *Torulasporadelbrueckii* species: Potential applications in the wine industry. *Int. J. Food Microbiol.*, 134(3): 201–210.

Ribereau-Gayon, P. (1988). *Botrytis*: Advantages of producing quality wines. In: R. Smart, R. Thornton, Rodriquez and J. Young, eds. *Proceedings of the. Second International Symposium for Cool Climate Viticulture*. New Zealand Soc. Vitic.Oenol, Auckland, p. 319.

Ribereau-Gayon, P., Dubourdieau, D., Done`che, B. and Lonvaud, A. (2000). *Handbook of oenology, Vol. 1.The Microbiology of Wine and Vinifications*. John Wiley and Sons, Ltd, Chichester, p. 255–268, 410–419.

Ribereau-Gayon, J.R., Bereau-Gayon, P. and Seguin, G. (1980). *Botrytis cinerea* in Enology. In: J.R. Coley, K. Smith, K. Verhoeff and W.R. Jarvis, eds. *The Biology of Botrytis*. Academic Press Inc., London, p. 251.

Ribereau-Gayon, J., Peynaud, P., Ribereau-Gayon, P. and Sudraud, P. (1976). Sciences et technique du vin, Tome 3. *Vinifications Transformations du Vin*, Dunod, Paris.

Robinson, J. (2006). *Oxford Companion to Wine*. 3rd edn., Oxford University Press, Oxford, p. 840.

Sarrazin, E., Dubourdieu, D. and Darriet, P. (2007). Characterization of key-aroma compounds of botrytized wines, influence of grape botrytization. *Food Chem.*, 103(2): 536–545.

Shaw, T.G. (1863). *Wine, the Vine and the Cellar*. Longman, Green, Longman, Robertsand Green, London, 505 p.

Somers, T.C. (1984). Interpretation of colour composition in young red wines. *Vitis*, 17: 161.

Sponholz, W.R., Brendel, M. and Periadnadi (2004). Bildung von Zuckersäuren durch Essigsäurebakterien auf Trauben und im Wein. *Mitt. Klosterneuburg*, 54: 77–85.

Sponholz, W.R. and Dittrich, H.H. (1985). Über die Herkunft von gluconsaure, 2- und 5-oxoglyconsaure sowie glucuron und galacturonsäure in Mosten und Weinen. *Vitis*, 24: 51.

Tosi, E., Azzolini, M., Guzzo, F. and Zapparoli, G. (2009). Evidence of different fermentation behaviors of two indigenous strains of *Saccharomyces cerevisiae* and *Saccharomyces bayanus var. uvarum* isolated from Amarone wine. *J. Appl. Microbiol.*, 107: 201–218.

Touzani, A., Muna, J.P. and Doneche, B. (1994). Effect du poeenzymatiqueexocellulaire de *Botrytis cinerea*sur les cellules de *Vitis vinifera*. Application a la pellicule du raisin. *J. Int. Sci. Vignevin*, 28: 190.

Van Baarlen, P., Legendre, L. and Van Kan, J.A.L. (2007). Plant defence compounds against *Botrytis* infection. In: Y. Elad, B. Williamson, P. Tudzynski and N. Delen, eds. *Botrytis: Biology, Pathology and Control*. Springer, Dordrecht, p. 143–161.

Vannini, A. and Chilosi, G. (2013). *Botrytis* infection: Grey mould and noble rot. In: *Sweet, Reinforced and Fortified Wines: Grape Biochemistry, Technology and Vinification*. John Wiley and Sons, Oxford, p. 159–169.

Villettaz, J.C., Steiner, D. and Trogus, H. (1984). The use of beta glucanase as enzyme in wine clarification and filtration. *Am. J. Enol. Vitic.*, 35: 253.

Wahab, H.A. and Younis, R.A.A. (2012). Early detection of gray mold in grape using conventional and molecular methods. *Afr. J. Biotechnol.*, 11(86): 15251–15257.

Wang, X.J., Tao, Y.S., Wu, Y., An, R.Y. and Yue, Z.Y. (2017). Aroma compounds and characteristics of noble-rot wines of Chardonnay grapes artificially botrytized in the vineyard. *Food Chem.*, 226: 41–50.

Watanabe, M. and Shimazu, Y. (1976). Application of *Botrytis cinerea* for wine making. *J. Ferment. Technol.*, 54: 471–478.

Zeng, Q.Y., Westermark, S.O., Rasmuson-Lestander, Å. and Wang, X.R. (2006). Detection and quantification of *Cladosporium* in aerosols by real-time PCR. *J. Environ. Monit.*, 8(1): 153–160.

13 Vineyard Mechanization

Matteo Bordiga

CONTENTS

13.1 INTRODUCTION

By the early 1950s, in California, harvest mechanization trials had started to use cutter bars on clusters hanging from specific Pergola vines (roof-like trellis). Using this approach, vines have to be shaped to a horizontal plane at a given height, allowing cutter bars to properly operate. However, because it proved impossible to place the clusters under the roof-like trellis and away from the wires in a homogeneous manner, these efforts were abandoned. In fact, because of the vagaries of climate and other environmental factors, every cultivar showed variable growth and cropping responses. Moreover, the effectiveness of the cutters was only partial, and so forced a manual intervention for the remaining part. Towards the end of the same decade, new trials focused on harvest mechanization were made by researchers at New York's Cornell University on the Concord grape (*Vitis labrusca* L.). The aim was focused on integrating or matching training systems to machines. This system, known as the Geneva Double Curtain (GDC), consists of a horizontally divided canopy. The goal of the system is to manage a dense canopy by dividing it in two, allowing more sunlight to reach the fruit renewal zone. Using this approach, grapevines are trained from the trunk to bilateral cordons (spaced 1.2–1.4 m) and pruned to retain short canes able to grow downward freely. Flexible cross arms allowed mechanical harvesting, performed by an over-row unit equipped by two spiked wheel shakers. Thanks to their vertical action on the lower part of the cordons, these indirectly transmit energy towards the cropping area and detach the grapes. This represented an integrated approach to vineyard design. In fact, using this system, there was a need to plant in a different way instead of the usual single high hedge wall of limited depth. In Europe in 1968, harvest mechanization trials had started testing over-row horizontal units, first in France and immediately thereafter in Italy. European viticulture was based on *Vitis vinifera* L. cultivars and mostly managed under single hedgewall system. However, discordant voices emerged. In fact, French researchers found good results applying harvest mechanization trials in flat lowland regions characterized by low-trained, single hedgerows of weak vigour vineyards. Conversely, their Italian counterparts observed several problems related to hillside plantings with vines of stronger vigour (long cane pruned). For this vineyard structure, harvester machines proved to be too heavy, damaging the cultivated soil. These machines needed lower trellising for over-row operations, which led to notable crop loss (both berries left on vines and must) because of the relatively dense foliage. Given these disappointing results in Italy, a group of farm equipment manufacturers (Montanari, Tanesini and Bubani), designed the MTB, a motorized over-row vertical-shaker featuring two star-shaped heads, following the model of the vertical shaking harvester in the United States. While the MTB, because of its weight and handling problems, proved to be largely unsuited to small Italian holdings, the short-pruned two permanent cordons of the Geneva Double Curtain (GDC) obviated the defects related to the non-uniform ripening of long-pruned traditional systems. Efforts to modify the original GDC to the Double Curtain (DC), a system more adapted to these conditions, followed. Its use spread quickly and reached the vineyards located in the country's northern and central areas. When compared to traditional hedgewall system, the DC delivered better berry quality, good yield and quicker pruning/harvesting, even when manually performed. After 1970, the issues of structural and management factors of the DC itself and the need to integrate it with a more streamlined and economical vertical slapper were addressed and resolved, leading to simpler single-row units working one side of the DC per passage. The University of Bologna, Italy developed the prototype and subsequently had it industrially manufactured. The studies that took place in parallel regarding the modified DC system led to a prototype of an original cutter bar, followed in recent years by a more upgraded model.

Vineyard operators faced with financial pressures and an inconsistent, costly labor supply are considering mechanization to reduce operating costs, execute timely cultural practices and increase flexibility within their operations. In this chapter, an overview of the vineyard mechanization and how its importance and diffusion is constantly growing among grape growers worldwide have been described. A description of the principles of new vineyard planning, different fully mechanizable training systems and how mechanization can be extended to other managment practices has been discussed in the Sections 13.3, 13.4 and 13.5. The list of tasks that can be performed by machine has grown extensively over the past decade and now includes pruning, suckering, shoot thinning, shoot positioning, fruit thinning, and leaf removal, in addition to harvesting. Because tasks typically can be performed more efficiently by machine than by hand, using machines not only increases an operator's ability to complete tasks at the appropriate times but also often lowers production costs. Moreover, an exhaustive description of the different machines used has been reported from Section 13.6.1 to Section 13.6.16.

13.2 FULL MECHANIZATION OF THE DOUBLE CURTAIN

These developments in machinery and training systems stimulated efforts to begin further trials to determine the efficiency of the horizontal and vertical harvesters and the amount of crop loss they were responsible for. To design an integrated vineyard system represented the final goal of these studies. Data from several comparative field tests (SPC *vs.* DC) did not however, highlight any significant differences between the two units (*e.g.* crop lost on vines or ground loss). However, differences emerged towards a total must loss, a direct consequence of berry rupture during harvest operations. The horizontal unit showed around a twofold loss compared to the vertical one (12.2% *vs.* 6.5%) and an amount of free must three times higher (18.1% *vs.* 5.9%). The results showed that vertical harvesting of the DC affected indirect crop detachment by inertia before the clusters were actually hit by the slapper. This happened as the pivoting star-shaped head caused the main support wire to vibrate. Conversely, the horizontal unit directly hit clusters on the hedgerow fruiting belt, causing berries to rupture and spill their must. The next step was to test the performance of the new cutter-bar units in winter pruning trials, representing the first attempt in Italy to combine machine units and training systems in an integrally mechanized vineyard design. Table 13.1 summarizes the data obtained, comparing hand management and full mechanization during a lengthy period from 1981 to 1987. It emerged that the data related to cropping and berry quality of DC vines pruned and picked by hand were identical to those of DC vines integrally pruned and harvested by machine. Labour costs appeared significantly different between the two approaches. While hand pruning and picking amounted to 320 man-hours per hectare per year, mechanization of the two operations resulted in just over 33 hours. Since mechanical pruning is not selective, leaving a bigger bud load than routine hand pruning, data showed that the machine-pruned vines reacted in three physiological ways: reduced bud sprouting, flower-bud differentiation and average cluster weight. The combined effect of these three reactions generated a self-regulation of plant growth, enabling the vines to preserve a performance response close to that of hand-pruned control. Since 1980, most of the DC vines planted in Italy are to be found in the Veneto and Emilia-Romagna regions. All these have been mechanically pruned, often with manual retouching, and mechanically harvested.

TABLE 13.1
Long-Term Comparison of Hand Management and Full Mechanization of *V. vinifera* (cv. Montuni; Double Curtain Trained; Period 1981–1988)

Treatment	Grape yield/quality		Labour demand (person-hours per ha per year)				
	Yield (t/ha)	°Brix (%)	Pruning and shoot positioning		Harvesting		Total
			Manual	Mechanical	Manual	Mechanical	
Hand harvest and pruning	17.0	19.7	80.0 h	1.0 h	239.0 h	–	320.0 h
Mechanical harvest and pruning	18.7	19.6	17.5 h	4.5 h	–	12.0 h	34.0 h

Source: Created using information from Intieri *et al.* (1988)

Today the country has nearly 10,000 hectares of DC vineyards and about 150 vertical harvester units. This number is still well below the nearly 350,000 hectares of new or converted hedgerow plantings, although the nearly 2,000 horizontal harvesters working these plantings are not suited to several plantings of small holders, especially hillside vineyards. Moreover, the rest of the wine-growing areas are still characterized by vertical trellises which are too high or open-canopy systems like Pergola, Raggi and Tendone that are unsuited to mechanical operations.

13.3 PRINCIPLES OF NEW VINEYARD PLANNING

Since 1980, several trials have been conducted for developing innovative and balanced training systems as alternatives to traditional ones employing long-cane pruning. The starting point was the spur-pruned permanent cordon. In fact, under equal vigour conditions, this system usually delivers better berry quality even though it is characterized by a lower yield. The main advantage of the spur-pruned permanent cordon is to apportion bud load to a given number of short fruiting spurs (1-3 count nodes). For this reason, and also because of the effects of acrotony, the growth of cropping shoots is uniform; hence, competition among young shoots appears reduced. Short pruning results in a uniform leaf-area distribution and grape ripening, as clusters may exploit the nutrients produced by the leaves on their shoots. Since these parameters are essential to achieve the uniformity required for mechanical pruning and harvesting, new plantings have also had to incorporate management practices into these systems. Quality grapes, optimum cropping, low environmental impact and reduced overheads represent four further key areas that a vineyard, properly integrated for full mechanical management, should also comply with. These concepts have been included in design and management practices employed in traditional SPC vineyards planted after 1980. These vines were characterized by trellises no higher than two metres to accommodate over-row units and a single cordon 80–90 cm from the ground, leaving at least 1.1–1.2 m of trellis space above the cordon for the leaf wall. In addition to this, "mobile" secondary wires were included to position the thrust bearings upwards from the initial growth, thereby limiting mechanized summer pruning (top and sides of the hedge walls). Generally, the usual SPC spacing appeared narrowed to about 0.8–1.0 m in rows and 2.3–2.5 m between rows. This layout is able to produce 4,000–5,500 vines per ha, with 4–6 spurs of 2–3 buds per metre of row. This set-up is well suited to winter pre-pruning using rotating heads or cutter bars and manual touch-up for accurate bud-load reduction. With this configuration, the SPC lends itself to traditional horizontal harvesters whose slappers can easily work directly on the cropping belt above the cordon.

13.4 NEW FULLY MECHANIZABLE TRAINING SYSTEMS

13.4.1 Free Cordon

The Free Cordon (FC) represents the first example of the new training systems integrating these principles. Modelled on the DC in order to allow shoots to grow freely, it features poles of 1.4–1.6 m out of the ground and a permanent cordon secured to a single horizontal support wire on top of the poles. Planting is conducted in accordance with strict regulatory requirements, designed to satisfy the dual demands of the vine and its growing environment (1.0 m intra-row and 2.5 m inter-row spacing for a planting of 4,000 vines per ha). Since its introduction over 20 years ago, the FC has undergone significant modifications. Currently it features tautly coiled support wire to keep the permanent cordon straight. It is characterized by a permanent cordon of two entwined canes in order to increase bud number and spurs (vertically positioned), ensuring an "open" canopy of upward-growing shoots. The advantage of this system (compared to the DC) is that it does not need combing, but it naturally balances light and shade around clusters. Interestingly, this effect may be achieved by choosing upper spurs of cultivars of upright or semi-upright tendencies like 'Cabernet Sauvignon', 'Cabernet Franc', 'Sauvignon', 'Chardonnay' and 'Sangiovese'. In fact, shoots of these cultivars tend to grow vertically to form a well-defined leaf area above the row-long cordon, and the resulting canopy appears "open" with clusters well exposed to light. Usually, an early shoot trimming is applied to increase and maintain the shoot's verticality.

13.4.2 Moveable Free Cordon

Even though initially the FC was harvesting using horizontal over-row units, fine-tuning led to a bending of the upper trunk by forming an arc, thereby making the top wire moveable. Thanks to this, the vineyard will not suffer any damage. This development led to the creation of the Moveable Free Cordon (MFC). Its main characteristic is that it can harvest even with vertical over-row pivoting head units. This system was set up in many vineyards together with a developed harvester unit known as the TRINOVA-Harvester presenting two pivoting heads. Related trials showed a low level of berry damage caused by the indirect action of the harvester on the cropping zone. There are no obstacles for both FC and MFC above the one-cordon support wire, and each system is suited to cutter-bar pruning both in winter and in summer. Following this, a custom-built over-row prototype with multiple cutter bars known as TRINOVA-Pruner was realized, containing two rear-end platforms with extra pneumatic shears for manual touch-up. Trials showed good results, reporting that the bars worked close enough to the cordon, leaving very short spurs and that the need for manual touch-up was minimal. In summary, FC and MFC proved easy to set up and suited to integral pruning and harvesting processes. Thanks to these factors, their spread has been quite fast in major vine growing areas. Interestingly, since 1996, the FC has been adopted in France, first in the Rhone Valley and more recently in the regions of Gard, Vaucluse and Haute Garonne.

13.4.3 Moveable Spur-Pruned Cordon

Based on advantages like intact berry retention and consequently low must loss compared to horizontal slappers, vertical harvesting was applied to SPC vines by stringing the cordon support wire through an eyelet, placed on the side of the trellising, thereby bending the upper unit's head. The new configuration, known as the Moveable Spur-pruned Cordon (MSPC), utilizes the same structural concept behind the MFC. Since MSPC also has the advantage of harvesting with a horizontal component, like any SPC, further trials were set up to test the comparative performance of both horizontal and vertical units. The results showed that the vertical harvesting of the MSPC improved grape quality by picking more clusters with intact berries and reduced free must by a third compared to horizontal harvesting.

13.4.4 COMBI System

In the early 2000s, a novel training system (COMBI) was developed, combining the advantages of the other two systems. Combining elements of the DC with those of the California U trellis, this system is characterized by vines with an intra-row spacing of 0.7–1.0 m, split at 1.1–1.2 m from the ground, thus creating two permanent cordons (1.2–1.4 m horizontally spaced). The spurred cordons form two vertical walls as the seasonal shoots are trained upwards by two pairs of foliage wires secured to the vertical arms of the U. Usually, the U trellises are located with a pole ratio of one to three and are only located every 10–15 m along the row. The intermediate poles are equipped with mobile arms like those of the DC. The COMBI's cordon-support wires pass through the eyelet on each U arm. For this reason, the cordons and the canopies of each hedgewall are able to move under the action of the pivoting star-heads of the vertical harvester. Therefore, the same inter-row-single-row mechanical unit used for the DC can also be utilized in the COMBI system. Moreover, considering the particular structure of the vineyard, the inter-row spacing can be reduced to about 3.2 m (in the DC it is 4.0 m) without causing any shoot damage by the harvester. Summarizing, while the DC has a total length of 5 km/ha of cropping cordons, the COMBI has a length that can extend up to 6.2 km/ha, giving the COMBI a higher cropping capacity per hectare. The preliminary COMBI trials using a routine, single-row vertical unit showed positive results, as many intact clusters were picked, free must was very low (an average of 2.3%) and no canes were damaged. The vines positively responded to the treatment by producing a good quality grape and, depending on the cultivar, were characterized by a yield from 2.6 to 3.6 kg per linear metre of cordon. The lower yield (per linear m) of cordon that results from keeping spurs short and hence reduces the related bud load, goes hand in hand with an optimum 1:1 ratio of inter-wall spacing and height. This is able to prevent internal shade cones, representing a good balance between leaf area and hanging clusters (1.5–2.0 m^2 leaf area/kg of crop), for a planting density of about 4,500 vines per hectare.

13.4.5 Semi-Minimal Pruned Hedge

The Semi-Minimal Pruned Hedge (SMPH) represents a variation on the Minimum Pruning (MP) system devised and tested over the last three decades in Australia. Using this system, good results arose in dry hot areas and in irrigated vineyards of Australia, especially using such premium cultivars such as 'Riesling', 'Chardonnay', 'Shiraz' and 'Cabernet Sauvignon'. However, trials in cooler climate growing areas of Australia, the United States and Europe have often highlighted contrasting results. This showed that the MP is susceptible to weather and management practices and that its best performance can be achieved in warm regions with early- and mid-season cultivars. Trials performed in Italy have confirmed this aspect, indicating the MP's suitability for early-ripening cultivars like Chardonnay. Later trials on the MP with nearly 600 buds/m, in comparison to a routine SPC system used as a control (18 buds/m), actually showed that this system cropped more than double than control, and the buds had a notably slower sugar accumulation and hence, were incompletely ripened. At the same time, however, more clusters were present; their berries were smaller and less tightly packed compared to what is typical of 'Sangiovese', thereby evincing less susceptibility to botrytis blight. This suggested that a well-adjusted reduction of bud load using the MP system may maintain vine performance and berry quality in 'Sangiovese', without losing the capability to produce fewer compact clusters (less susceptible

TABLE 13.2
Comparison of Total Costs (Pruning, Shoot Thinning and Fruit Thinning) by Hand or Machine of *V. vinifera* Grapes (3 Trellis Systems at French Camp Vineyards, Santa Margarita (CA) 2005 Season)

	Total cost ($/ha)		
	Vertical shoot-positioned trellis	(0.6 m) Lyre trellis	(0.9 m) Quadrilateral trellis
	Machine-farmed		
Prune	297.5	597.5	392.5
Follow-up	167.5	180.0	117.5
Shoot thin	195.0	390.0	292.5
Fruit thin	196.0	195.0	145.0
Total	856.0	1,362.5	947.5
	Hand-farmed		
Pre-prune	67.5	132.5	92.5
Prune	627.5	965.0	792.5
Shoot thin	580.0	1,157.5	1,157.5
Fruit thin	272.5	435.5	437.5
Total	1547.5	2,690.5	2,480.0
Difference (Hand-Machine)	691.5	1,328.0	1,532.5

Source: Created using information from Morris (2007)

to bunch rot). The first SMPH was an evolution, derived from vines traditionally SPC trained, and was developed in order to design a model suited to maintaining pruning efficiency, coupled with the severe mechanical pruning needed to reduce bud load and balance cropping. In this system, canes of several years old were maintained during winter pruning and secured to the horizontal foliage wires of the trellis, forming a hedge. A multi-blade unit characterized by an upside-down, U-shaped cutting profile was used in the following years to prune the hedge's top and sides short in winter, to eliminate some of the buds. Trials comparing SMPH and SPC (control) showed that the SMPH reduced the bud load by about 50% when compared to the original MP trials. Moreover, SMPH maintained the physiological characteristics in the vines of reduced bud sprouting, flower-bud differentiation and average cluster weight. Clusters resulted smaller and less compact, thereby hardly affected by rot. When compared to the SPC and MP, the SMPH allows a more streamlined mechanical pruning, earlier leaf-area development (marked photosynthesis capacity and sugar accumulation), good quality grapes, uniform ripening and greater ease for mechanical harvesting.

13.5 EXTENDING MECHANIZATION TO OTHER MANAGEMENT PRACTICES

As previously reported, labour represents the largest share of overheads, reaching a peak of about 400 hours per hectare per year in non-mechanized traditional vineyards, with 80–90% of hours dedicated to pruning and harvesting. When considering modern systems which are manually managed, the number of hours still remains significantly high. While pruning and harvesting mechanization reduces this demand to about 50 hours, manual operations (*e.g.* shoot thinning, leaf removal, etc.) require an average of an additional 100 hours, even in vineyards with modern systems. Thus, the aim is to mechanize at least part of these operations in order to further reduce labour costs and to improve vine performance and crop quality.Table 13.2 shows the economic analysis of balanced cropping operations. The data shows that the costs saved through mechanization are economically significant. For the operations considered, machine farming resulted in almost a 50% saving compared to hand farming for grapes produced using the lyre trellis, a 49% saving on the vertical shoot-positioned system and around a 60% saving on the quadrilateral trellis. The major cost savings were gathered from shoot-thinning and fruit-thinning operations.

13.5.1 Post-Sprouting Shoot Thinning

After mechanical pruning, the spur-pruned systems (*e.g.* SPC, DC, FC, MFC and COMBI) are often integrated by manual retouching in order to reduce node count and balance cropping. Using a tractor-mounted cutter bar unit integrated with a tractor-towed cart equipped with power shears (2–3 workers), the operating speed that can be achieved is about 0.8–1.0 km/h. As the pruning unit moves along the row, hand removal of leftover spurs can be done, requiring about 20–25 person-hours per hectare as additional labour. An alternative way is by pruning in winter without manual touch-up, leaving extra bud thinning until spring when the new shoots are about 15–20 cm long. Using just a driver, the unit can reach a speed of 2 km/h, also allowing the use of mechanical rotating whisks, instead of cutter bars, to remove excess bearing shoots on the sides of the cordons. Related trials, performed during 2010 and 2011 on FC-trained vines, showed regular cropping levels and improved crop quality.

13.5.2 Pre-Bloom Leaf Removal

The carbohydrates synthetized by leaves represent a key factor in determining fruit set. As previously reported, these compounds in leaves near clusters migrate to flower buttons at bloom. These insights suggest that basal-shoot defoliation at bloom may trigger a more severe than usual shelling of the flowers, inducing a physiological regulation of cropping. Trials have shown that early shoot leaf removal, as soon as the first cluster flowers open up, leads to a better performance, of the vine, inducing clusters with fewer, less compact berries that are less susceptible to a botrytis attack. Berry quality also improves, with grapes having a higher sugar content and anthocyanin count than the control. Manual defoliation appears to be unrealistic because of labour costs and time required, and mechanical removal of about 50% of leaves adjacent to flower clusters showed results comparable to the manual approach. A defoliator unit working on one side of the row per run can cover a hectare in 5 hours at a drive speed of about 2 km/h. Using a bilateral unit, working on both sides of the row at the same time, time may be further reduced. Interestingly, defoliated clusters showed a greater tolerance to sunburn, an effect presumably imputable to earlier exposure of berries to direct sunlight and a consequent adaptive response of their tissues.

13.5.3 Mechanical Cluster Thinning after Berry Development

During 2009, some trials were performed in several COMBI-trained vineyards, where the canopy grows above the cordon, in order to thin clusters (pea-sized berry stage) using a vertical harvester unit, whose pivoting head works below the cordon so as not to damage vines. The aim was to eliminate the time-consuming manual cluster thinning normally performed at berry véraison stage. With the harvester moving at 1.5 km/h and the harvester head spinning at about 300 rpm, many clusters and parts thereof, as well as individual berries, were easily removed, thereby halving the vine's crop. However, after several days, all the remaining clusters gradually withered as the shoot oscillation had damaged the young tissue of the cluster stems. During 2010, another trial was arranged on SPC vines using a horizontal shaker device instead of the harvester unit. The machine was a simple prototype-head, mounted on the side of the tractor in order to shake the trunks on one side when cropping was at berry pea-size phase. The shaker, driven at 1.5 km/h with a shake frequency of 500 rpm and able to ensure a good transmission of trunk oscillation to the canopy, removed some clusters and individual green berries, with some clusters on the vines withering within a few days. However, unlike the previous trial, most of the clusters left on the vines were undamaged, ripening as usual. In 2011, a new vertical shaking head, connected to a tractor, was tested in the COMBI system. The running speed was set at 1.5 km/h as it was in previous trials, however, shake frequency was set to reach only 250 rpm, in order to reduce vibrations on the clusters. Again, while a few clusters and berries were removed and others withered wholly or in part on the vine, most of the crop appeared intact. However, during harvest, the mechanical machine collected the ripened grapes along with many of the withered clusters; the effect of the impurities of the withered clusters on a real commercially harvested crop intended for winery delivery remains to be determined.

13.5.4 Post-Véraison Shoot Trimming, Mechanical Defoliation or Antitranspirant Canopy Spray

Under the influence of climate change, testing post-véraison trimming or defoliation began only during the last decade. In many vineyard areas of Europe and worldwide, climate change has brought more hours of sunlight and raised temperatures, thereby shortening the period from new shoot growth to ripening. This problem is of particular relevance in areas characterized by hot summers where yields are low for natural reasons (*e.g.* poor soils, lack of rainfall) or for constraints imposed by production standards. The rise in hot days, for example, can increase must sugar content to optimum levels in red grapes ahead of the usual date but induce an excess loss of acidity. Moreover, in many cases, the loss of positive traits emerge, such as typical aromas, skin colour, seed browning and full synthesis of phenol compounds, while astringency still appears overly high. These traits represent the result of incomplete senescence that affects both skin and seeds ("cell maturity") independently from sugar accumulation in berry flesh. Delaying picking, for instance, is a common approach utilized by growers in order to guarantee a good maturity of skin and seeds. In that case, however, must sugar content can be too high, thereby raising the potential wine's alcohol level in excess of 14%. Thus, to delay the harvest date may increase negative factors relevant to the marketplace, which today demands well-balanced wines of moderate alcohol content. In addition, a proper balance of acidity, taste and flavour does not always match with an alcohol content of wines of above 14%. Yet delaying harvest has its advantages, mainly the completion of the senescence of skin and seed tissues. Two opposite but beneficial processes affect berry quality. The first one includes the softening and collapse of skin cells, thereby making extraction of polyphenols and anthocyanins easier. The second one is related to the progressive lignification of seed coats so that their compounds become less astringent and harder to extract. These issues have led several growers to ask themselves what steps could be taken to slow down berry sugar accumulation so as to keep it in line with skin and seed senescence. Trials have addressed these concerns of normalizing the seasonal cycle, through post-véraison trimming or leaf removal in the

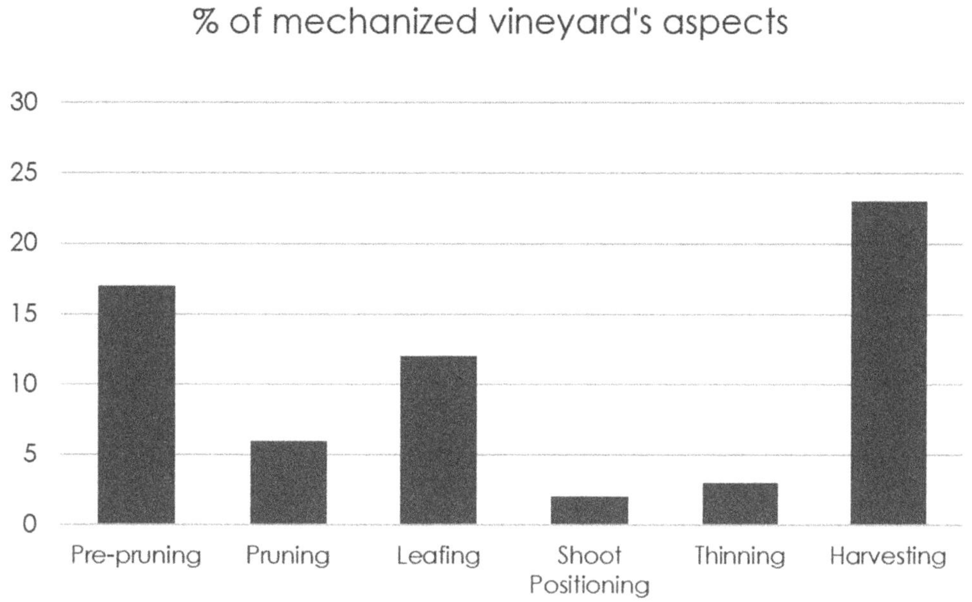

FIGURE 13.1 Percentage distribution of aspects most often mechanized in the vineyard operation. *Source:* Created using information from Greenspan (2009)

mid to upper canopy or post-véraison full canopy spray with antitranspirant. The aim was focused on both reducing total assimilation and limiting the sugar synthesis and uptake by berry flesh. Thus, the vine clusters have more time to complete skin and seed senescence without excess sugar accumulation. The results suggest that source limitation after véraison may slow down sugar accumulation, without significant differences in total anthocyanin accumulation. During 2010, trials on "late" defoliation were carried out by comparing relatively light mechanical leaf removal to routinely managed vines. Encouraging results highlighted how grapes, harvested eight days later than the control vines, showed a similar sugar content (13.5% potential alcohol) to the control vines. While the berry samples showed slight differences in overall skin phenol counts, the difference was negligible compared to those extracted from skin of the defoliated vines. Preliminary data indicate that these techniques are promising but need to be improved and utilized at a better time of year and intensity in order to delay harvest date, thus allowing proper sugar accumulation at a higher level of cell maturity. The prevalent opinion is that pruning and harvesting by machine cannot produce the grape quality achieved by hand labour. However, no one would argue that mechanization is more efficient than hand labour and that production costs are consequently much lower. Currently, thanks to gradual improvements, the majority of winegrapes, for example in California, are now machine-harvested. Harvesting is the task most often mechanized in the vineyard (Figure 13.1). After harvest, the next most laborious viticultural practice is pruning. Mechanical pruning equipment has also been available for decades, and many winegrape growers use this equipment for pre-pruning. Along with this operation, a crew finishes the job manually, because without additional manual pruning, too many nodes may be left behind, resulting in overcropped vines. Machine pruning shows undisputed advantages beyond labour savings. For this reason, more and more, the largest growers appear to be the most mechanized (Figure 13.2).

13.6 MACHINES

13.6.1 Trimming Machines

The trimming machine, usually applied to the front part of the tractor, operates with hydraulic feeding and is endowed with one or more cutting units. This working machine can be used in viticulture both for lopping vineyards which are still green and for stripping them of their shoots, potentially damaging the tree's natural growth and its consequent fructification. Its use substitutes the manual work done at the side of the row or over it.

13.6.2 Pre-Trimming Machines

This machine may be used for winter pruning of the "espaliers", "Guyot" and "spur pruned cordon" vineyard systems. Its utilization is useful in eliminating all vegetal parts of the over season (by now wooden part) that have settled upwards the fruitful line, among a row of supporting cables. This approach allows a final saving of 50–60% on labour. The machine is positioned in front of the tractor, and it operates by hydraulic power. The machine has been developed in order to fit mostly on tracked mini tractors, considering its lightness and simplicity. The ability to rise, a sideways movement and discs which open the nearby row's poles represent the three hydraulic movements of the machine, meaning it is better adapted to working through the rows. Many discs, depending on the required working height, can be assembled in both the two mechanical appendages of the machinery.

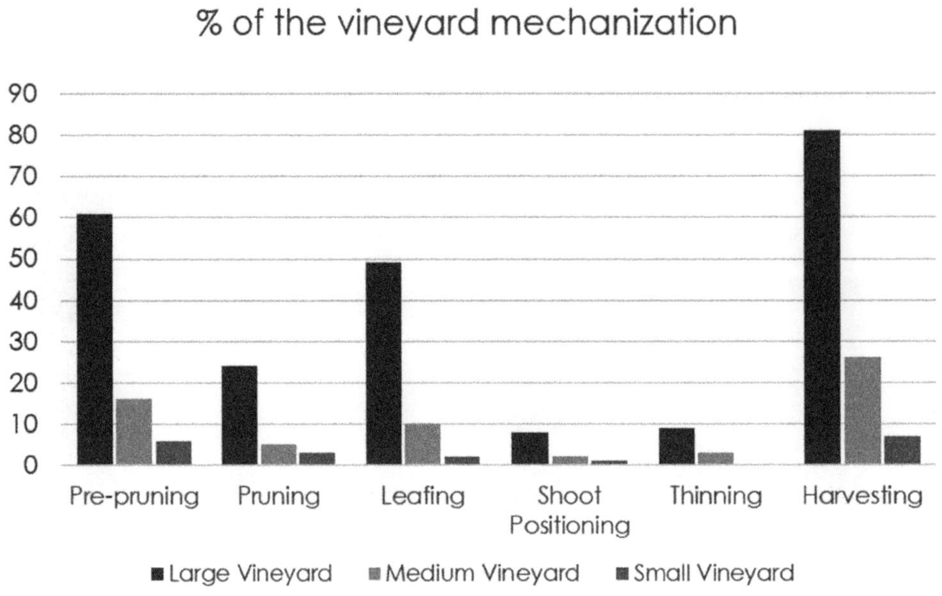

FIGURE 13.2 Comparison between small, medium and large vineyard with regard to mechanized tasks. *Source:* Created using information from Greenspan (2009)

13.6.3 Vine-Shoot Removers

The vine-shoot remover is able to eliminate the vegetation present on vine trunks, using special whips in special rubber (reinforced with canvas), especially developed for vine-shoot removal. It is specifically designed for working in vineyards characterized by high stems. The exclusive system of automatic roller retraction allows for a simple operation, making the machine fit for any kind of plantation and ground. The vine-shoot remover is equipped with two hydraulic movements in height and width, with a floating roller adjustable with a spring that allows smooth working around the trunks. The rotation is maintained by a hydraulic system, and the hydraulic flux regulator can control the rotor speed.

13.6.4 Electronic Pruner

These cutters have been designed to have characteristic qualities such as speed, power and precision, thereby representing effective and precise electronic pruning equipment. There are different typologies on the market, consisting of ergonomic and symmetric designs. These cutters are suitable for pruning vineyards, orchards and olive orchards. Thanks to their blade position control, the tool's cutting speed outperforms the first models of past years. They are characterized by very low vibration and are able to perform a huge number of cuts using only one battery charge. There is the possibility to select between cutting modes, impulse or proportional, and, depending on requirements and intermittent aperture based on the diameter of the branches, the cutters are programmable, thus ensuring clean and precise operations.

13.6.5 Hedge Bush Cutter

Hedge bush cutters have been designed and manufactured with a particular geometrical shape and are able to overcome obstacles such as bushes, nets or rows. The arm on which the working head is fixed is telescopic in order to better fit to the asperity of the ground.

13.6.6 Pruning Machines

The double barrel leaf remover represents a comprehensive piece of equipment for removing the leaves in front of the grape-producing part of the plant, thus promoting the growth potential and reducing the phytosanitary treatments. The machine consists of two opposite rollers, one with holes and the other with rubber, thereby allowing a perfect and safe removal of the leaf without damaging the grapes. A fan equipped with a hydraulic motor provides aspiration. The single head can be fitted in the frame with a rotating function that allows defoliating in a single step only to the less exposed side, having the ability to go to a 180° tilt. Moreover, there is the possibility to apply a second head in the same frame in order to defoliate both sides of the same row simultaneously. Furthermore, a double blade cutter bar can be applied to the remover head which, when placed in front of the head, cuts branches outside the wires, thus preventing a block of the rollers.

13.6.7 Shoot Thinning and Binding

Thinning. A flexible, rubber flapper that removes shoots for shoot thinning characterizes this dedicated thinner system. The machine is most effective on shoots that are shorter than 15 cm. However, it is important that growers do shoot counts prior to and immediately after use of the machine to make sure things are adjusted at the proper speed and the right number of shoots are removed. Trials involving Cabernet Sauvignon grapes showed good results, in fact at harvest, grapes easily reached 24° Brix with a yield of about 14 tons per hectare.

FIGURE 13.3 Shoot-binder. *Source:* Courtesy of Clemens GmbH & Co. KG

Binding. Since this development, patented more than forty years ago, it has been possible to successfully mechanize the time-consuming task of binding vine shoots. Today, the most recent shoot binders still work on the same basic principle of gently straightening the shoots, but they have been continually improved in important ways. According to the size of their vineyards, customers can choose between different models. Figure 13.3 shows a typical shoot binder. Two rubber belts with hydraulic gear raise the vegetation. The speed is proportionally adjusted to the driving speed. The belt's flexibility allows a gentle raising of the vegetation, so that nothing damages it.

13.6.8 Leaf Removal

This intervention increases air flow around the fruit and reduces humidity in the cluster zone, which in turn minimizes fungus and reduces botrytis bunch rot. The main advantage of leaf removal is improved ventilation and exposure, leading to faster drying of clusters, which reduces mildew and botrytis. With less risk of rot, it may be possible to postpone the harvest date, which may have a positive effect on grape maturity. Spray coverage and its efficacy improved. It is possible to increase exposure of the grape to sunlight which thickens the skin's wax cuticle, thereby improving its quality and protecting against infection. This process guarantees more energy for the grape, resulting in an increased concentration of phenol and polyphenols, thus affecting the taste, colour, and flavour of wine. In fact, good exposure of the grapes promotes the production of phenolic substances in the berry skins (*e.g.* anthocyanins and tannins) responsible for wine character and colour development of red varietals. A berry's aroma development is positively influenced by good sun exposure to the grapes. However, it is crucial to calibrate and optimize the use of this practice. For example, to thin the canopy too soon might negatively impact fruit quality by reducing the fruit set or berry size. Conversely, if removal is done too late during the season, there will be no effect on the amount of carbohydrates stored in the grape. Removing too many leaves at once can hinder photosynthesis and decrease the overall yield. Growers increasingly recognize that grapes with greater sun exposure develop more aroma and colour and are less susceptible to powdery mildew and botrytis. As a result, leaf removal has become a crucial step to optimize grape quality, and growers' increasing interest has brought about substantial technical development in leaf removal equipment. Leaf remover machines operate either by blowing compressed air with enough force to sever the leaves or by sucking the leaves into the path of a cutting tool. The device is connected to a rotary power transmitter on the tractor and is available for one-side, bilateral, or over-the-vine row operation (Figures 13.4 and 13.5). A large compressor delivers compressed air via flexible hoses to leaf-removing heads. Air is in the form of high-velocity streams through rotary nozzles (outward curved air tubes). The nozzles rotate under a cover plate pierced with narrow semicircular slots, which direct the air toward the leaves that are to be removed. The slots interrupt the airflow at short intervals, generating a sort of pulsation, so that leaves met by the air jet are not simply pressed to one side. Compressed-air leaf removers are most suitable for early leaf removal,

FIGURE 13.4 Leaf remover in action. *Source:* Courtesy of Clemens GmbH & Co. KG

FIGURE 13.5 After the passage of the remover. *Source:* Courtesy of Clemens GmbH & Co. KG

between bloom and pea-size berry development. During this period, the impact on remaining berries is very low with a negligible degree of damage. A further advantage is the possibility of cleaning the grape clusters, because the strong airflow is able to remove debris. However, in order to obtain the desired results, the airflow must be accurately adjusted and optimized. Leaf removal is done on the side of the canopy receiving morning sun, so as to expose clusters to the air and sunlight and allow better pesticide application. Leaves remain on the other side of the canopy to protect clusters from afternoon sun and the potential risk of sunburn. Another machine used for the same purpose is designed to blow hot air (about 150 °C) into the canopy from propane-powered jets in order to control pests and diseases without harming the grapevine. Additional benefits include assuring fruit set and improving wine quality.

13.6.8.1 Leaf Removal Machines

Trimmer. The leaf trimmer is an operating machine equipped with a fan that sucks the leaves and removes the sucked leaves with a series of rollers. Usually, it is positioned on the front side of the tractor. In order to obtain a better reduction of vegetation, there is the possibility of mounting an anterior pre-cutting blade. The machine allows the elimination of basal leaves, thereby improving the microclimate in the bunch zone and increasing exposure to the light. The air circulation is therefore optimized. Working times are 70% reduced, making vintage easier.

Cutter. Thanks to special manufacturing techniques, the most recent leaf cutters offer an interesting technical solution (Figures 13.6 and 13.7). For example, the aluminum blade holder of the leaf cutter characterized by a strong and very light frame. This increases its stability and makes it possible to work on steep slopes and gradients. Moreover, the sickle-shaped cutting tools made from tempered knife steel produce a clean cut.

FIGURE 13.6 Double-sided leaf cutter with unobstructed view. *Source:* Courtesy of Clemens GmbH & Co. KG

FIGURE 13.7 Double-sided leaf cutter prepared for a single row. *Source:* Courtesy of Clemens GmbH & Co. KG

13.6.9 Grape Harvesters (Harvesting Process and Machine Functionality)

A self-driven grape harvester harvests the grapes by shaking the grapevines. To do this, the grape harvester drives over the rows using the double shaker drive to shake the grapes off the grapevine. The separated grapes fall on the shingle conveyor and slide sideways onto the big conveyor belt. In the course of this movement, the residue leaves are removed by three fan drive systems. The de-stemmer, which is available as an option, separates the grapes from their stems in the next stage of the harvesting process. After passing the final transverse conveyor, the grapes will end up in the grape bin. They can be unloaded sequentially out of the bin onto a truck or continuously by using an additional conveyor belt.

13.6.10 Wind Machines

Active frost protection procedures allow for a simple modification of the vineyard climate immediately before and during frost occurrences. The climate of the vineyard may be altered by utilization of atmospheric heat such as wind machines or the addition of heat with heaters. Wind machines represent one of the best protections against radiant frosts. They work by directing warmer air from above the inversion layer downward around the vines, thereby displacing the colder air on the ground away from the vines. An effective procedure requires that a sufficiently strong temperature inversion exists within reach of the machine. Ideally, a temperature inversion of 3 to 5°C must exist between the warm air and the cold air situated at ground level. Usually, growers turn on wind machines at about 0 to 1°C, which is appropriate for many radiative frost situations. A critical temperature is one in which vine tissues will be damaged by cold weather. Critical temperatures will vary depending on the variety and stage of

FIGURE 13.8 Self-operating fan sprayer. *Source:* Courtesy of Clemens GmbH & Co. KG

FIGURE 13.9 Tunnel sprayer. *Source:* Courtesy of Clemens GmbH & Co. KG

development. Generally, a wind machine consists of a tower of about 10 m in height, with a propeller mounted at the top and drive units located at the bottom of the tower for ease of servicing. An efficient machine must have the rotor installed high enough to draw the warm air from the inversion down into the crop zone. The coverage of each machine is approximately 4 hectares, with the actual range depending on the site contours, the power of the machine, and the proximity of other machines. Multiple wind machines increase the range of effectiveness compared to individual ones. Generally, a single wind machine can cost up to $30,000 and can be powered by electric motors, gasoline/liquefied gas (LP) powered engines or diesel engines. Usually several temperature monitoring stations are distributed across a vineyard. These represent an essential part of frost risk management by providing warnings and data for tracking long-term trends.

13.6.11 Heaters

Heat machines provide frost protection by heating the air as uniformly as possible up to the inversion layer. As is well-known, warm air rises and cools until it reaches the height where the ambient temperature is the same as the inversion layer. Then, the air spreads out and creates a circular pattern as it descends again. The heaters are more effective when there is a strong inversion (smaller heated volume), and they are less efficient in weak inversion conditions because there is a bigger volume of air to be heated.

13.6.12 Over-Vine Sprinkler Systems

These systems remain among the most reliable methods of frost protection. They do not rely on access to warm air above the vineyard but are able to protect against severe frosts if sufficient water is applied. The system involves spraying the vines with a fine mist of water as the temperature falls to freezing. This water then freezes, encasing the canes and buds in ice. As the water changes to ice on the surface of the vine, it releases a small amount of heat which is able to protect the vine from any damage. This small amount of released heat, defined as latent heat, prevents the surface temperature of the vine tissue from falling below 0 °C. The standard practice is to start the sprinklers when the temperature drops to 1°C, thus providing a margin of safety. Frost protection with overhead sprinklers is dependent upon uniformly supplying water to the ice-encased surface on the vines. Intervals must be both regular and rapid. Different critical issues must be considered when planning a frost protection system for a vineyard. The first is determining the available water supply in terms of litres/minute. The second is determining how many liters of water are needed in case of a frost event.

13.6.13 Sprayers

Modern machines are characterized by a high volume, low pressure system of air movement that controls the spray droplets and targets them precisely where they need to go, thereby ensuring an excellent coverage from plant-to-plant and row-to-row. Recycling sprayers recapture any spray not deposited onto the foliage, thereby returning it into the tank to be quickly recycled; this system significantly reduces spray waste and eliminates spray drift, which can cause a problem in some vineyards (Figures 13.8 and 13.9).

13.6.14 Drones

Drones represent one of the most promising technological solutions that have emerged for vineyards nowadays. These devices already play a role in military operations and retail package delivery. Drones are unmanned aerial vehicles that fly autonomously or by remote control. Keeping an eye on thousands of hectares of crops each day is often difficult. This issue is what makes drone technology so attractive for vineyards. Using drones can be useful in determining the quality of grapes being grown and providing estimates of how many grapes will be produced in the next harvest. Overall, growers can obtain a clearer picture of what is happening in the vineyard. For example, they may be able to pinpoint zones where vigor is higher than desired and so can remove those zones from a fertilizer program. With drone technology, growers can also monitor water saturation and pest damage, thereby determining where to spray insecticides. Some vineyards have used tanks locked onto the drones to spray their vineyards for diseases. Drones have sensors able to measure temperature and humidity, and then they communicate that information through a transceiver. Based on this data, soil moisture monitors can measure water volume at different soil depths as well. Multi spectral imagery is able to detect anomalies that cannot be seen by the human eye. Drones can produce Normalized Difference Vegetation Index (NDVI) imagery that goes far beyond the manual capacity of any other monitoring techniques. Drones, specialized for vineyard use, have a field-uniformity algorithm that helps quantify the relative density and health of the grapes. These machines can be designed with 2D and 3D map processing, along with possessing crop analytics tools, early disease detection abilities and professional-grade drone flight servicing. There are many benefits of using drones in vineyards. In fact, they represent the fastest and most precise way to obtain high quality photos, which can both identify distinct soil types and monitor the soil's effects on the vines. Modern machines provide high image resolution that is very valuable, especially in determining the ripeness of grapes. They are easier to obtain and more affordable than aeroplanes. Moreover, since they can fly closer to the ground compared to planes, their images can be more accurate. Generally, drones promise a smarter and more sustainable form of agriculture. By some evaluations, vineyards may be able to reduce spraying and watering by about 30%, which matches with reduced costs and a lower environmental impact. Drones may help vineyards save money on labour costs, and so the labour force can be better used in other operations. Thanks to modern advances in technology, drones are no longer a niche solution for common vineyard issues. However, it is not enough to simply collect data, and vineyards must fully understand the potential, the risks and the applications of drones.

13.6.15 Vineyard Mowers

For many growers, the option of spraying weeds in the growth of the under-vine zone is not viable. Organic vineyards especially have a need for an under-vine mower machine, which can keep weed growth under control right up to the base of the plants. Mower machines have pivoting mowing heads that move into the rows between plants, retract when the base of a vine is approaching and pull back into the row. The operating speed of the mower ranges from 6 to 8 km/h.

13.6.16 Vineyard Cane Sweepers

Sweeping vineyard debris into the rows allows a more effective mulching of waste products like pruned canes and unwanted weeds. These can be mulched by inter-row mowing operations, returning the organic material to the soil, thus enriching it and improving the overall health of vineyard. Cane sweepers can have special features including a safety breakaway, preventing damage after an impact. Brushes are able to prevent stones and rocks being swept into the rows, and many controls for adjusting the height and contour of sweeping operations may be included.

13.7 FUTURE TRENDS

Additional studies in the area of total vineyard mechanization and its impact on quality are required on all species of grapes, with emphasis on the development of totally integrated systems. The timely performance of most of the mechanization procedures will be critical to the success of the total systems approach. The latest frontier in this direction seems to be robotic pruners – machines designed to replicate human skills but to greater efficiency. Generally, a standard machine features a set of cameras that take photos of dormant grapevines, and so building a 3D image to which it applies various pruning protocols. The machine, thanks to two robotic arms, makes the cuts and, once it has finished, moves forward taking pictures (feedback) along the way before the next series of cuts. Mechanized/automated winter pruning will ultimately result in labour saving. This is a necessary factor, but not the only one, to take into account. By effectively mechanizing cultural and harvesting operations, the production of grapes destined to become premium wines will become more competitive with regions characterized by available and inexpensive hand labour.

BIBLIOGRAPHY

Baldini, E., Intrieri, C. and Marangoni, B. (1976). P.I. Emme-V: Potatrice integrale monofilareper vigneti a doppia spalliera. *L'Informatore Agrario* 38:24075–24080.

Baldini, E., Intrieri, C., Marangoni, B. and Zocca, A. (1973). V.I.em me: un prototipo sperimentale di vendemmiatrice integrale monofilare. *L'Italia Agricola* 7/8:3–9.

Caspari, H.V. and Lang, A. (1996). Carbohydrate supply limits fruit-set in commercial Sauvignon blanc grapevines. In: *Proc. of the 4th Int. Symp.on Cool Climate Enology and Viticulture,* Rochester, NY, T. Henick-Kling, T.E. Wolf and E.M. Harkness (eds.), New York State Agricultural Experiment Station, Geneva, NY, vol. 2, pp. 9–13.

Clingeleffer, P.R. (1983). Minimal pruning and its role in canopy management and implications of its use for the wine industry. In: *Advances in Viticulture and Oenology for Economic Gain. Proc. 5th Austr. Wine Industry Tech. Conf.*, Perth, pp. 133–145.

Clingeleffer, P.R. and Possingham, J.V. (1987). The role of minimal pruning of cordontrained vines (MPCT) in canopy management and its adoption in Australian viticulture. *Austr. Grape Grower Wine Maker* 280:7–11.

Di Stefano, R., Borsa, D., Bosso, A. and Garcia-Moruno, E. (2000). Sul significato e sui metodi di determinazione dello stato di maturità dei polifenoli. *L'Enologo* 36(12):73–76.

Flippetti, I., Allegro, G., Movahed, N., Pastore, C., Valentini, G. and Intrieri, C. (2011). Effects of late-season source limitations induced by trimming and anti-transpirant canopy spray on grape composition during ripening. In: *Progress Agr. etVitic. Horsserie-Special. 17th GiESCO Symp*, Asti-Alba, Italy, August 29–September 2, pp. 259–262.

Filippetti, I., Intrieri, C., Silvestroni, O. and Poni, S. (1991). Effetti della potatura corta elunga sulla sincronizzazione fenologica e sul comportamento vegetativo e produttivo della cv. Sangiovese (V. vinifera L.). *Vignevini* 12:41–46.

Greenspan, M. (2009). Vineyard survey report: Mechanization and technology. *Wine Business Monthly*, November.

Intrieri, C. (1988). TRINOVA: Nuovo mezzo polivalente per la meccanizzazione delvigneto. *L'Informatore Agrario* 43:91–106.

Intrieri, C. and Filippetti, I. (2000). Innovations and outlook in grapevine training systems and mechanization in North-Central Italy. In: *Proc. of the ASEV 50th Anniversary Meeting*, Seattle, Washington, DC, June 19–23, pp. 170–184.

Intrieri, C. and Poni, S. (2004). Integration of grapevine training systems and mechanization in north-central Italy: Innovations and outlook. In: *Proc. Symp. on Quality Management in Horticulture and Viticulture*, Stuttgart, pp. 78–92.

Intrieri, C. and Silvestroni, O. (1983). Evoluzione delle forme di allevamento della vite nella pianura emiliano-romagnola. *Vignevini* 10:23–38.

Intrieri, C. and Filippetti, I. (2007). The semi-minimal pruned hedge (SMPH), a novel grapevine training system tested on cv. *Sangiovese*. In: *Proc. of the XV GiESCO Symp*, Porec, Croatia, June 20–23, vol. 2, pp. 860–873.

Intrieri, C. and Filippetti, I. (2009). Cambiamenti climatici, maturazione accelerata delleuve ed eccessivo grado alcolico del vino. Cosa può fare la ricerca se cambia il clima? *Frutticoltura* 9:76–78.

Intrieri, C., Silvestroni, O. and Poni, S. (1988). Long-term trials on winter mechanical pruning of grapes. *Riv. Ing. Agraria, Quaderno* 9:168–117.

Intrieri, C., Silvestroni, O. and Poni, S. (1994). PLUKER: Defogliatrice polivalente pervigneti a controspalliera e a cortina semplice o doppia. *Vignevini* 1/2:5–31.

Intrieri, C., Poni, S., Lia, G. and Gomez Del Campo, M. (2001). Vine performance and leaf physiology of conventionally and minimally pruned Sangiovese grapevines. *Vitis* 40:123–130.

Intrieri, C., Pratella, F., Poni, S. and Filippetti, I. (1995). TRIMMER: Nuova potatrice polivalente per vigneti. *L'Informatore Agrario* 38:61–69.

Intrieri, C., Filippetti, I., Allegro, G., Centinari, M. and Poni, S. (2008). Early defoliation (hand vs mechanical) for improved crop control and grape composition in Sangiovese (V. vinifera L.). *Austr. J. Grape Wine Res.* 1:25–32.

Jones, G.V., White, M., Cooper, O. and Storchmann, K. (2005). Climate change and global wine quality. *Clim. Change* 73(3):319–343.

Ollat, N., Sommer, K.I., Pool, R.M. and Clingeleffer, P.R. (1993). Quelques résultats sur la taille minimale en Australie et France. In: *GiESCO*, Compte Rendu, Reims, France, pp. 181–183.

Morris, J.R. (2007). Development and commercialization of a complete vineyard mechanization system. *HortTechnology* 17(4):411–420.

Poni, S., Intrieri, C. and Magnanini, E. (2000). Seasonal growth and gas-exchange of conventionally and minimally pruned Chardonnay canopies. *Vitis* 39:13–18.

Poni, S., Bernizzoni, F., Civardi, S. and Libelli, N. (2008). Effects of pre-bloom leaf removal on growth of berry tissues and must composition in two red vitis vinifera cultivars. *Austr. J. Grape Wine Res.* 15:185–193.

Poni, S., Casalini, L., Bernizzoni, F., Civardi, S. and Intrieri, C. (2006). Effects of early defoliation on shoot photosynthesis, yield components and grape quality. *Am. J. Enol. Vitic.* 57(4):397–407.

Quinlan, J.D. and Weaver, J.R. (1970). Modification of pattern of the photosyntate movement within and between shoots of Vitis vinifera L. *Plant Physiol.* 46(4):527–530.

Rombolà, A.D., Covarrubias, J.I., Boliani, A.C., Marodin, G.A., Ingrosso, E. and Intrieri, C. (2011). Post- véraison trimming practices for slow down berry sugar accumulation and tuning technological and phenolic maturity. In: *ProgressAgr. etVitic. Hors serie-Special, Hors. 17th GiESCO Symp*, Asti-Alba, Italy, August 29–September 2, pp. 567–569.

Schultz, H.R., Kraml, S. and Werwitzke, U. (1999). Distribution of glycosides, including flavour precursors in berries from minimal pruned (MP) and vertical shoot positioned grapevines (VSP). In: *Proc. of the XI: GiESCO Symp*, Marsala, Italy, June 6–12, pp. 271–279.

Shepardson, E.S., Shaulis, N. and Moyer, J.C. (1969). Mechanical harvesting of grape varieties grown in New York State. In: B.F. Cargill and G.E. Rossmiller (eds.), *Fruit and Vegetable Harvest Mechanization: Technical Implications.* Michigan State University, East Lansing, pp. 571–579.

Studer, H.E. and Holmo, H.P. (1969). Mechanical harvesting of grapes in California: cultural practices and machines. In: B.F. Cargill and G.E. Rossmiller (eds.), *Fruit and Vegetable Harvest Mechanization: Technical Implications.* Michigan State University, East Lansing, pp. 611–621.

Unit 3

Microbiology and Biochemistry of Wine Production

INTRODUCTION

Isolation, culture, identification, physiology, and metabolic pathways of yeasts and bacteria that turn grape must into wine continue to lead the work of research laboratories around the world. Associated with the more and more precise results of fine chemical analysis, this research now allows the interpretation of the sensory differences observed both in the development of wine aromas and in that of microbiological alterations.

More recently, "omics" have appeared in these same laboratories. A new boom has resulted. Now molecular methods, more and more accessible, allow more detailed study of the diversity of genera, species and strains. The laboratories facilitate large-scale studies on microbiota from grapes to wine. But these studies can only assist the wine producer if they are based on his or her observations and inquiries in the winery.

The chapters in Part 3 of the Encyclopedia provide an update on the basic knowledge of wine microbiology and recent developments that are changing practices. The characteristics of wine micro-organisms, the conditions of their multiplication in wine, their physiology and their metabolisms are exposed through the different chapters. Yeasts and bacteria transform many substrates of grape must by innumerable complex metabolic pathways. Some of these transformations are similar for yeast or bacterial strains of the same species. Others, on the other hand, are specific even at the strain level. This diversity plays an important role in the quality and typicality of the wine. For a long time, the attention of the researcher and the wine producer focused on *Saccharomyces cerevisiae*, but today the observations of the first wine microbiologists on non-*Saccharomyces* yeasts have been brought up to date with new molecular methods. These yeasts form almost all the microflora of yeast in grape must. They take part in alcoholic fermentation, joining their metabolic pathways with those of *S. cerevisiae* with which they interact. Following this, lactic acid bacteria continue the biochemical transformations of substrates coming directly from the grapes or products of the metabolism of the yeasts. The extraordinary complexity of the composition of grape must and microbiota determines the quality of finished wine. The fermentation phenomenon is spontaneous. For a long time, the producers lacked effective tools to ensure quality. Soon after Pasteur's discoveries, they used more or less well purified and selected yeast suspensions.

Based on this, the microbial industry was founded for winemaking. First, for *Saccharomyces* yeasts, whose use has gradually become widespread. Advocated at the beginning (1960–1970) to ensure the best course of alcoholic fermentation, they are more and more often selected and indicated for their effect on the aromas. Genetics has progressed at the same time, and it is now possible to obtain new strains (non-GMO) adapted more precisely to the demands of oenology. The yeast industry is diversifying, and now it also concerns non-*Saccharomyces* whose preparations are added in association with those of *Saccharomyces*. Similarly, lactic acid bacteria preparations are marketed for malolactic fermentation.

But their number is much lower than that of yeasts because isolation, cultivation and preparation of marketable forms are more difficult.

Today, winemaking in its microbiological aspects has achieved an undeniable quality of result. Analytical methods, microbiological tools and knowledge exist. It is important to use them well. Each wine is a special case. The producer in his or her winery must make the right observations and deduce the most appropriate operations.

Dr. Aline Lonvaud-Funel
Professor Emeritus of the University of Bordeaux
Institute of Vine and Wine Sciences (ISVV), France

14 Wine Microbiology

V.K. Joshi, Karina Medina, Valentina Martín and Laura Fariña

CONTENTS

14.1 INTRODUCTION

Wine is regarded as a gift from God, a divine fluid (*Soma*) in Indian mythology. It has been prepared and consumed by humans since antiquity and has also been used as a medicine. Wine is principally made from grapes, but a variety of other fruits – apples, mangos, peaches, plums, pears, kiwi-fruits, strawberries, citrus-fruits etc. are used in the preparation of fruit wines. Winemaking essentially involves the activities of microorganisms, which could be spontaneous or inoculated in the juice of grapes or other fruits. The metabolites produced by the microorganisms or their enzymes are reflected in the physico-chemical and sensory qualities of wine.

Factually speaking, the basic and primary microbiology of wine fermentation has remained the same since the time of Pasteur. But later on, the developments in microbiology, biochemistry, food science, technology, and engineering put the process of winemaking on a sound scientific footing. Since it is well known now that microorganisms (moulds, yeast and bacteria) are involved at one or another stage in winemaking, a sound knowledge of the microbiology of winemaking is essential to make quality wine. Thus, knowledge of microbiology in winemaking is indispensable to understand the complete process of wine production.

At different stages of the biotechnological process of winemaking, different microorganisms (mould, yeast, bacteria) can grow in a beneficial form, or conversely, in a non-beneficial form. Grapes and other fruits are subject to mould spoilage at different stages (harvesting, transportation), and this growth can result in the production of toxins like ochratoxin, patulin, etc. In wine, ethanol and the anaerobic condition has an inhibitory effect on mould growing.

Traditionally during winemaking, the yeast *Saccharomyces cerevisiae* carries out alcoholic fermentation resulting in ethanol production, the major compound of alcoholic beverages. Other non-*Saccharomyces* yeasts have recently acquired a considerable significance. Because they are involved in the production of various chemical components, especially volatile compounds, with an impact on sensory quality. Some microorganisms like lactic acid bacteria are involved in malo-lactic fermentation that make highly acidic wine palatable; their activity contributes immensely to the development of wine of desirable quality.

Various microorganisms are equally involved in the spoilage of wine, including the non-*Saccharomyces* yeast/wild yeast, lactic acid bacteria and acetic acid bacteria. Various aspects related to the microbiology of winemaking are discussed in this chapter.

14.2 DIVERSITY AND ECOLOGY OF MICROORGANISMS IN GRAPE MUST AND WINES

Microorganisms flourish in an extremely wide range of habitats and occur nearly everywhere in nature. They occur most abundantly where they find food, moisture and a temperature suitable for their growth and reproduction. One such niche for microorganisms is the grapevine and berry. Grape berry microbiota can be grouped according to their ecological significance, technological and economic importance. This consortium includes the yeasts responsible for alcoholic fermentation of wine must and spoilage of wine, lactic acid bacteria responsible for metabolizing malic acid and for some wine disorders, and acetic acid bacteria, capable of converting wine into vinegar – a type of spoilage. Thus, a large diversity of microorganisms are associated with the winemaking process – a complex ecological and biochemical interaction between species of yeasts, bacteria and fungi that are naturally found on the surface of grapes. Diversity of microorganisms and their populations are determined by the amount of rainfall prior to grape harvest, degree of physical damage to the berry, use of fungicides and time between harvest, crushing and fermentation. Table 14.1 shows the principal microorganisms associated with the process of winemaking.

14.2.1 Moulds

Grapes, like other fruits, are also subject to mould spoilage at different stages of harvesting, transportation and storage. Infected fruit may become soft due to the activity of pectinases. Some species may produce off-flavours in juices, wines and other fruit products. These species may be present on various processing machinery and equipment (*e.g.* wooden barrels/tanks), and the metabolites of many moulds are also toxic to human beings, as these fungi produce mycotoxins present in grapes, raisins or wines .

Besides damaging the fruits, moulds can also affect the flavour of wine by their growth on the cooperage. However, moulds cannot grow in the wine itself, mainly due to the inhibitory effect of ethanol and the anaerobic conditions - most of the moulds being aerobic in nature. Different moulds of significance in winemaking include *Aspergillus, Fusarium, Penicillium, Botrytis, Trichosporon* and *Aureobasidium*. Finally, there is a wide diversity of microbial cells present on, the berry surface which are unable to grow and consequently, lack any significance. The microbial community of the grape berry is composed of an array of species exhibiting different physiological characteristics and relevance to vine growing and winemaking.

There are several phytopathogens responsible for grapevine diseases worldwide, as discussed in Chapter 11 of this book. One of these diseases is rot. The causal agent of grey rot is the saprophytic mould *Botrytis cinerea*. In the manufacture of some white wines, use is sometimes made of the serendipitous growth of the fungus *Botrytis cinerea*, which, if it occurs at the right time, can cause "noble rot". The growth of this fungus damages the grape skin, causing dehydration and resulting in the production of a very concentrated must with a high sugar content and consequently, more ethyl alcohol (for more details see Chapter 12).

14.2.2 Yeasts

A wide diversity of yeast species are common contaminants of berry surfaces and are associated with wine fermentation. However, *Saccharomyces cerevisiae* is rarely present in grapes but plays a vital role in the production of all alcoholic beverages. So, the selection of suitable yeast strains is essential not only to maximise alcohol yield, but also to maintain the sensory quality of the beverage especially the flavour and aroma characteristics of different beverages. Highly complex interactions and chemical signalling take place among grapevines themselves and with the intervening biota, which also include insects, birds, and mammals. The exact microbiota and the resulting interactions depend basically on the berry development stage, the condition of the grape skin and on the prevailing environmental conditions, exerting a profound effect on the fruit quality.

Four distinct groups of microorganisms – residents, adventitious, invaders and opportunists – are defined on the basis of grape biochemical evolution, nutrient availability and ability to proliferate on berry surface. The vine bark which is present during the entire life of the vine plant could be considered as a potential niche to host *S. cerevisiae* during the period when grape bunches are not present. However, *S. cerevisiae* when present on grapes, occurs at concentrations lower than 10–100

TABLE 14.1
Diversity of Microorganisms During Winemaking

Microbial group	Source	Significance
	Fungi	
Penicillium spp. *Botrytis spp.* *Aspergillus spp.*	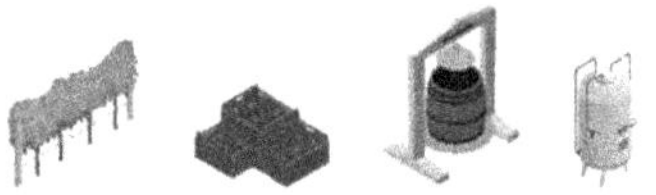	Spoilage of fruit (except in Botrytized wines which is considered useful), corky taints
	Yeast	
Saccharomyces cerevisiae *Schizosaccharomyces pombe* *Candida spp.* *Toruloppsis spp.* *Brettanomyces spp.* *Hansenula spp.* *Kloeckera spp.* *Pichia spp.* *Killer yeast*	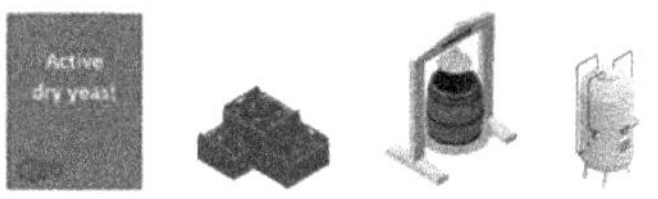	Alcoholic fermentation, spoilage, autolysis, deacidification
	Bacteria	
Lactic acid bacteria *Oenococcus oeni* *Pediococcus pentosaceus* *P. parvulus* *Lactobacillus plantarum* *L. fermentum*		Malolactic fermentation
Acetic acid bacteria *Acetobaceter* *Gluconobacter*		Vinegary, spoilage stuck fermentation
Other bacteria *Bacillus* spp. *Clostridium* spp.		Spoilage
	Actinomycetes	
Streptomyces spp.		Earthy, Corky taints
	Bacteriophages	
		Disrupt malolactic fermentation

CFU/g, and the cell number never exceeds 10 CFU/cm^2 of grape berries. Using meta-barcode DNA sequencing, *Saccharomyces* sp. is found to comprise less than 0.00005% of the fungal community on ripe grapes in vineyards, thus the vineyard cannot be thought of as the primary source. Therefore, it is a "transient" environment where *S. cerevisiae* presence is mainly associated with grape berry ripening, while habitats other than fruit may represent a refuge when the fruit is not available. Yeast populations on grapes increase from 10^2–10^3 CFU/g on immature berries to 10^3–10^6 CFU/g on mature ones.

Qualitative and quantitative data on the yeast ecology of grapes have revealed that the yeasts responsible for wine fermentations originate from one of three sources: (i) the surface of the grape; (ii) surfaces of the winery equipment; and (iii) inoculum cultures.

The total yeast population and the relative proportions of individual species on grapes are affected by several factors like temperature, rainfall, other climatic influences, physical damage due to mould, insect or bird attacks, use of fungicides, grape variety, degree of maturity at harvest, etc. Out of 500 species of yeasts recognised by current taxonomists, only 15–20 are relevant to winemaking. These are the genera belonging to *Saccharomyces* – the principal yeast in the process of winemaking. Morphologically, *Saccharomyces* is spherical to ellipsoidal in shape, with an approximate dimension of 8 x 7μm. Asexual reproduction takes place by multilateral budding. Most of the strains of *S. cerevisiae* are capable of producing alcohol levels up to 16% (v/v). During alcoholic fermentation of a grape must, *S. cerevisiae* remains the dominant species as the ethanol concentration increases. This yeast though, is of the fermentative type, yet may also grow oxidatively as a part of the surface film community. Different related species, based on the fermentation pattern of various carbohydrates, can be distinguished, *e.g.* *Saccharomyces bayanus* and *Saccharomyces fermentum* with an inability to ferment galactose, *Saccharomyces diastaticus* fermenting starch and *Saccharomyces uvarum* with the ability to assimilate melibiose, etc.

The efficient conversion of sugar into ethanol is the most important characteristic of the species of this genus. Out of four sub-divisions, *Saccharomyces cerevisiae*, the wine yeast (first isolated from beer by Hensen) and associated species are of immense value for the processing of wine, beer and other distilled alcoholic beverages. Though called wine yeast, it includes baker's yeast as well as beer yeast (*Saccharomyces carlbergenesis*). The growth gives haziness followed by intensive gas formation in the sugar rich juices. Later on, the cells of yeast fall to the bottom of the vessel. Figure 14.1 includes the shapes of typical yeast.

Hanseniaspora genera as a whole and particularly *Hanseniaspora uvarum* species are non-*Saccharomyces* yeasts commonly encountered at high concentrations on the grape surface and throughout the fermentation process.

Another non-*Saccharomyces*, *Metschnikowia pulcherrima* (Mp), is a globous/elliptical yeast that cannot be distinguished from *Saccharomyces cerevisiae* by microscopy. It can be observed sometimes in a single large, highly refractive oil droplet inside the cell. It is a ubiquitous yeast that frequently appears in spontaneous fermentations.

The term "wild" is normally applied to any organism other than the one being used or encouraged to grow and multiply in a medium. Thus, an organism employed in one process could be a wild organism in another. Examples of wild yeast in wine include *Candida, Kleockera, Pichia* etc. A typical example of the wild yeast that may occur in wine fermentation is the microorganism *Saccharomyces cerevisiae* var. *ellipsoideus,* which produces alcohol.

Most of the troublesome wild yeasts are asporogenous, also known as "false yeasts" or "wild yeasts", which do not display sexual reproduction and are all placed in the family cryptococcaceae. Some of these are yeast-like organisms which are sometimes placed in the fungi imperfecti group along with some of the more common moulds. The wild yeasts are, however, difficult to distinguish morphologically and culturally but may be identified serologically. While some are the strains of *Saccharomyces cerevisiae* or *S. carlbergensis*, others are representatives of various genera, including *Candida*, *Debaryomyces* and *Pichia*. Wine quality is significantly affected by the particular strain of *S. cerevisiae* conducting the fermentation. Thus, important quality parameters such as sensory characters, dryness of fermentation, volatile acidity and hydrogen sulphide production can be shown to be carried out by the strain of *S. cerevisiae*. In addition, there could be important strain influences within species such as *Kloeckera apiculata, Candida stellata* and other non-*Saccharomyces* yeasts that also contribute to some fermentations.

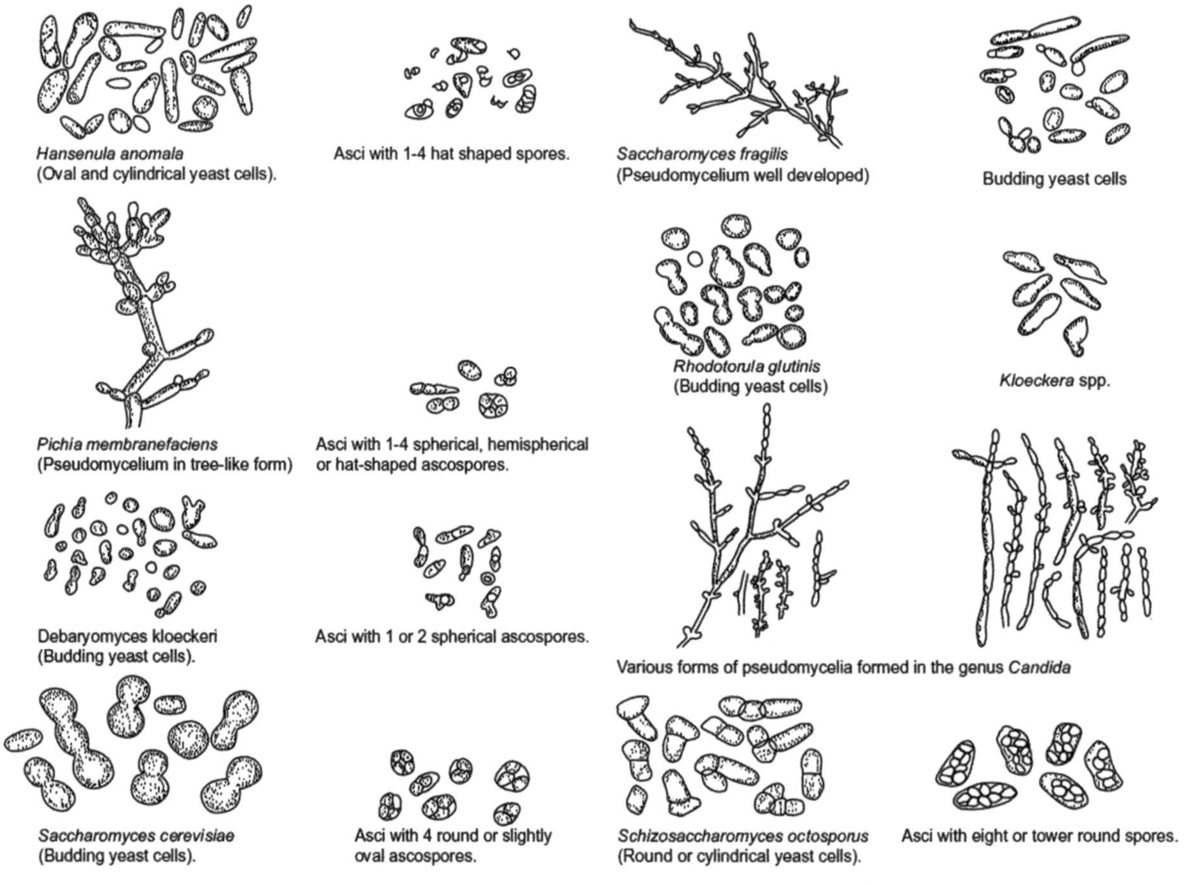

FIGURE 14.1 Types of yeast. *Source:* Adapted from Amerine *et al.* (1980); Joshi *et al.* (1999)

14.2.2.1 Non-Saccharomyces Wine Yeasts

Traditional wine fermentation depends on the presence of natural microflora, mainly the yeasts, occurring in the fruit must. At the initial stage of fermentation, the population of *Saccharomyces* is very insignificant, while several non-*Saccharomyces* yeasts such as *Aureobasidium pullulans, Candida stellata, Hanseniaspora uvarum, Issatchenkia orientalis, Kloeckera javanica, Metschnikowia pulcherrima* and *Pichia anomala* dominate in the early stages of fermentation. Other yeasts, such as the species of *Candida, Cryptococcus, Debaryomyces, Hansenula, Issatchenkia, Kluyveromyces, Metschnikowia, Pichia,* and *Rhodotorula* have also been found but in lower numbers.

It is generally accepted that in the case of undamaged grapes, the viable population of yeasts ranges from 10^3 to 10^5 CFU/mL. The most frequently isolated native yeasts are the apiculate yeasts of the genera *Kloeckera* (especially *K. apiculata*) and *Hanseniaspora*, which are the predominant species on the surface of grapes (counting for about 50–75% of the total yeast population). Table 14.2 shows the principal characteristics of non-*Saccharomyces* isolated from grapes. (For more details, see Chapter 15 of this text.)

Grape juice or concentrate is sometimes added to a grape must to increase its alcohol yield or as a sweetening agent in wine and can encourage the growth of yeasts like *Zygosaccharomyces* and *Saccharomyces*. These yeasts can be found in grape musts, but the populations are often less than 50 CFU/mL. It is applicable even to vineyards where fermented pomace is added as a soil amendment. However, failure to routinely isolate *Saccharomyces* from the vineyard could reveal the preference of this yeast for the high-sugar environments of grape juice and fermentation.

Many yeast species, particularly non-*Saccharomyces*, deserve special attention due to the great potential they provide to winemaking. These have different oenological characteristics than *S. cerevisiae*, so can be used to improve wine quality through the production of enhanced wine aroma and complexity and are thus, getting increased attention of oenologists all over the world. Some of the most studied non-*Saccharomyces* yeast species in winemaking are *Hanseniaspora uvarum, Kloeckera apiculata, Candida zemplinina, Torulospora delbruckii, Hanseniaspora vineae, Candida pulcherrima, Hansenula anomala, Schizosaccharomyces pombe* and *Lachancea thermotolerans*. Non-*Saccharomyces* yeast species are frequently found on grapes and in grape must, and are known to dominate the initial phases of spontaneous fermentations. Growth of non-*Saccharomyces* on the grapes or in the grape juice prior to fermentation alters the chemical composition of the juice and generates products which affect the subsequent growth of wine yeasts. *Hanseniaspora uvarum* demonstrates interesting enological characteristics

TABLE 14.2
Characteristics of Principal Species of Non-*Saccharomyces* Isolated from Grapes

	Enzyme production					
Yeast	**BGL**	**ARA**	**RHA**	**XYL**	**CSL**	**Metabolites**
Brettanomyces spp.	X					Volatile phenols
C. stellata	X		X	X		Glycerol, Volatile fatty acids, Aldehydes
C. zemplinina					X	Glycerol, Volatile fatty acids, Higher alcohols
Hanseniaspora sp.	X			X		Volatile fatty acids, Higher alcohols, Esters, Sulfur compounds
H. guilliermondii	X					2-phenylethyl acetate, Isoamyl acetate, Ethyl acetate
H. osmophila	X			X		Volatile phenols
H. vineae	X	X	X	X		2-phenylethyl acetate, 3-ethoxy-1-propanol, Esters, Volatile fatty acids, Volatile phenols
H. uvarum	X	X	X	X		Aldehydes
I. terricola	X					Esters
L. thermotolerans	X				X	Ethyl lactate, Lactic acid, Higher alcohols, Acetic acid
M. pulcherrima / C. pulcherrima	X			X	X	Isoamyl acetate, hexanol, 2-phenylethanol, Ethyl caprilate, Acetaldehyde, Higher alcohols
P. anomala	X	X	X	X		Volatilephenols
P. guilliermondii			X			Volatilephenols
P. kluyvery					X	Volatile sulfur compounds
P. membranifaciens	X			X		Volatile phenols
S. ludwigii	X					Acetoin, Ethyl acetate
S. pombe	X					Pyruvic acid
T. delbrueckii	X			X		3-ethoxy-1-propanol, Esters, Volatile fatty acids, Volatile phenols, Sulfur compounds

BGL, *β-D-glucosidase*; ARA, *α-L-arabinofuranosidase*; RHA, *α-L-rhamnosidase*; XYL, *β-D-xylosidase*; CSL, *carbon-sulfurlyase*.

Source: Adapted from Fugelsang and Edwards (2006); Padilla, Gil, and Manzanares (2016); Morata, Escott, Bañuelos, Loira, Del Fresno, González, and Suárez-Lepe (2019)

such as: capacity to grow at high sugar, ethanol and SO_2 contents; produces high concentrations of glycerol; low acetic acid and hydrogen sulfide levels; and the release of proteolytic enzymes. As well as this, *H. uvarum* produces less ethanol as it requires more than 19 g/L of consumed sugar to produce 1% v/v of ethanol.

Among the non-*Saccharomyces* yeasts is *Schizosaccharomyces pombe,* also known as fission yeast and discovered by Lindner in 1983. The cells of this species have a characteristic rod shape with sizes varying between 3–5 µm of diameter and 5–24 µm of length. The yeast has many oenological uses, such as the ability to reduce the acidity of high acid must (like those of grapes of low temperature climate, stone fruits like plum) by metabolising the malic acid into ethanol and carbon dioxide, and it helps in the formation of stable pigments in wine and releasing large quantities of polysaccharides during ageing on lees. The urease activity and its competition for malic acid with lactic acid bacteria limits the formation of ethyl carbamate and biogenic amines in wine. At the same time, it has certain disadvantages, such as its low fermentation speed and the development of undesirable flavours and aromas which hinders its application in winemaking commercially. In the latter section of this chapter, the application of *Schizosaccharomyces* yeast with *Saccharomyces* or with other non-*Saccharomyces* yeasts has been elaborated. For detailed information on the use of non-*Saccharomyces* in winemaking, readers are referred to a separate chapter in this text (Chapter 15).

14.2.3 Bacteria

Bacterial groups include acetic acid bacteria known for wine spoilage and lactic acid bacteria responsible for malolactic fermentation; the latter also resulting, *albeit* rarely, in wine spoilage. These microorganisms colonise grape surfaces from berry set to the ripening stage, following a repeatedly cyclic pattern year after year. The bacteria in winemaking are associated more with spoilage than the wine production. Acetic acid bacteria are the important groups of bacteria that spoil the wine. Acetic acid is the prime component of the volatile acidity of grape musts and wines. Acetic acid is formed as a by-product of mainly alcoholic fermentation or as a by-product of the metabolism of acetic bacteria, which can metabolize residual sugars to increase volatile acidity. This acid affects the quality of wines when it is present above a given concentration.

14.2.3.1 Lactic Acid Bacteria

Like yeasts, lactic acid bacteria are also present in vineyards, but due to their nutritional requirements, diversity and population density of species is limited. Sound, undamaged fruit contains 10^3 CFU/g, so populations in grape musts during the early stages of processing tend to be low. Species that have been isolated from grape musts include *Lactobacillus hilgardii, Lactobacillus plantarum, Lactobacillus casei, Oenococcus oeni, Leuconostoc mesenteroides* and *Pediococcus damnosus.* In the case of deteriorated fruit, however, substantial populations of lactic acid bacteria usually develop. Besides the grapes, an indigenous (native) population is frequently isolated from the cooperage, poorly sanitised pumps, valves and transfer lines as well as from fillers and drains in the bottling rooms.

The lactic acid bacteria cause spoilage of wine as well as reduce the acidity of high acid wines to an acceptable level and, therefore, are of high significance to the winemaker.

A group of lactic acid bacteria, called 'Malo-lactic acid bacteria', can degrade the malic acid into lactic acid and reduce the acidity of high acid wines to make them palatable. This aspect has been further developed under wine fermentation and further elaborated in the chapter about malolactic acid fermentation of this text (Chapter 18).

14.2.3.2 Acetic Acid Bacteria

Acetic acid bacteria are commonly associated with grapes and are normally present in musts. Sound, unspoiled grapes are reported to have 10^2 to 10^3 cells/g, whereas deteriorated fruit can have up to 10^6 cells/g. These bacteria belong to the genera *Acetobacter* and *Gluconobacter.* The *Acetobacter* genus includes three distinct species: *A. aceti, A. pasteurianus* and *A. peroxydans.* These bacteria are characterised by their ability to oxidise ethyl alcohol into acetic acid and subsequently, into carbon dioxide and water. These bacteria are of commercial importance for the production of vinegar but increases the volatile acidity of the wine, which affects its sensory quality of wine. Volatile acid is a good indicator of the growth and activity of acetic acid bacteria. At lower concentrations, only the sensory quality is affected, which can be rectified by blending. However, at higher concentrations, the wine remains suitable for vinegar production only. The acetic acid bacteria can spoil the must, fermenting wine and wine undergoing maturation, as they can tolerate the ethanol content in wine between 14–15%.

14.3 MICROBIOLOGY OF WINE PRODUCTION

14.3.1 Winemaking Process

Winemaking is a complex process involving the metabolism and interaction of different microbes. The complex microbial interactions start on the grape surface and continue throughout the fermentation. The two main groups of microorganisms involved in winemaking are yeasts and bacteria. The pioneering studies of Pasteur showed that yeasts are responsible for the alcoholic fermentation of grape juice into wine and that certain species of bacteria could grow in wine causing spoilage. Some yeasts generate metabolites that lead to wine faults that affect flavour, haze or CO_2 production in the final product (this will be discussed later in Chapter 41). Since that time, the microbiology of winemaking has been studied extensively, which has revealed the complexity of the ecology of the microorganisms involved.

The basic process of winemaking using grapes is fundamentally the same and involves similar practices for other fruits. However, a specific procedure is adopted both by traditional and modern wineries: the sugar-enriched juice is physically extracted from grapes or other fruits and is fermented to yield alcoholic beverages with the help of yeasts. Apart

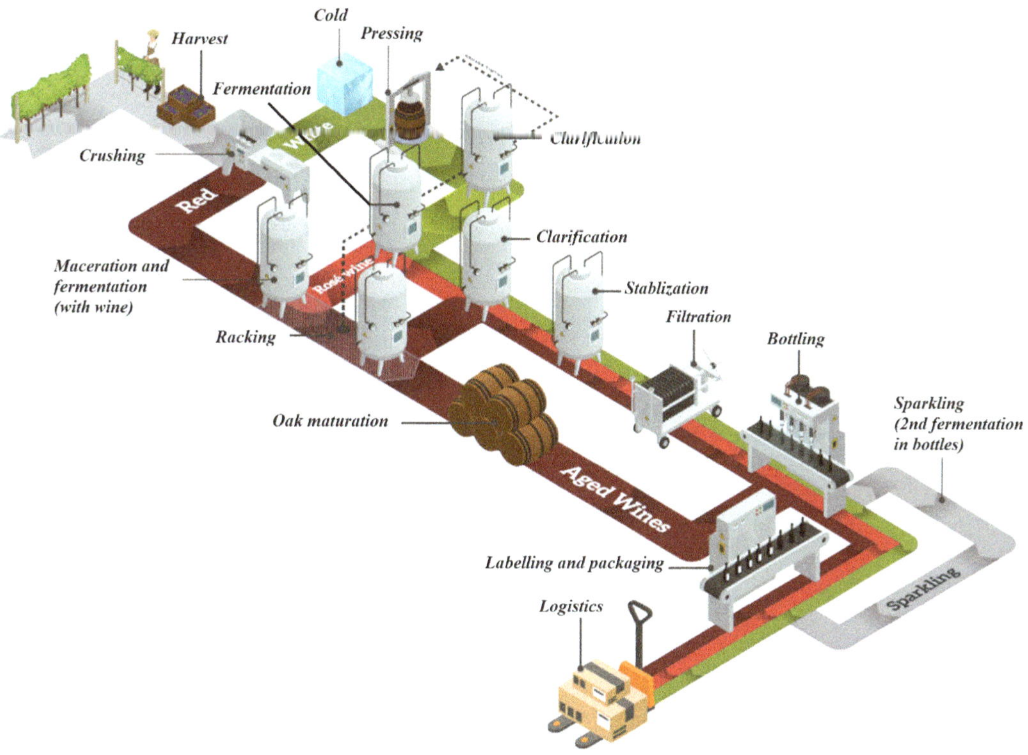

FIGURE 14.2 Flow sheet of wine production

from yeast, certain other microorganisms also play important roles in the winemaking process, such as the development of sensory qualities, *i.e.* colour, aroma and flavour of the wine following the maturation after primary fermentation; so that the wine end-product is of desirable quality. Wine production includes two important fermentation processes, *i.e.* alcoholic fermentation conducted by yeast, and malolactic fermentation (MLF) conducted by lactic acid bacteria (LAB). The yeasts drive the alcoholic fermentation by converting grape sugar to alcohol, carbon dioxide, and volatile compounds, which affect the aroma and taste of wine. At the onset of alcoholic fermentation, a large number of non-*Saccharomyces* species may be present, but the final stage is dominated by alcohol-tolerant *Saccharomyces cerevisiae* strains.

Traditionally, grape winemaking includes various steps, such as the physical-pressing of red and/or white grapes and other fruits to obtain the fruit-juice which is enriched in sugars, and these juices either ferment naturally or are inoculated with standard yeast to ferment the sugars to alcohol.

The wines from grapes are basically red and white wines, depending upon the variety of grape used and vinification methods employed. To prepare red wine, fermentation is essentially conducted on the grapes with skin and seeds, while in the case of white wine, the clear juice is fermented. A generalized flowsheet of wine production is shown in Figure 14.2.

During the process, there are key points for the production of quality wines. One of these is the vineyard, in which there are many parameters to control in order to avoid undesirable fermentations. The second step is the style of harvest and transport of the grapes. During these, it is very important to avoid breaking berries, which could impact the development of natural flora and is not desirable for sensory wine quality. At the winery, temperature, pH, aeration and gentle crushing when pumping the must should be controlled to obtain an optimal grape must to inoculate and start fermentation, stabilisation and barrel aging, depending on style and grape variety.

14.3.2 Fermentative Process

14.3.2.1 Microbiological Stages of Fermentation

*14.3.2.1.1 Non-*Saccharomyces *Yeasts*

After the crushing of grapes, non-*Saccharomyces* yeasts multiply and reach peak populations sometime during the early stages of alcoholic fermentation (Figure 14.3). *K. apiculata* and *C. stellata* tend to dominate in the early- to mid-stages of fermentation, but other genera of non-*Saccharomyces* may also be found with a peak population as high as 10^6 to 10^8 CFU/mL, depending on conditions. With a wide diversity of microorganisms, winemaking commonly involves a sequential development of microorganisms. In general, non-*Saccharomyces* yeasts are the first to grow and dominate during winemaking, followed by *Saccharomyces*, which normally completes alcoholic fermentation.

It is an established fact that many non-*Saccharomyces* yeasts possess lower ethanol tolerances compared to *Saccharomyces*, which contributes to the die-off phenomenon of these yeasts shortly after the start of alcoholic fermentation when the ethanol concentration reaches 5% to 6% (v/v), as confirmed experimentally. In a mixed culture fermentation at 10°C, however, *K. apiculata* or *C. stellata* could achieve populations as high as 10^7 CFU/mL and could complete the fermentation. Out of these yeasts, *K. apiculata* survived

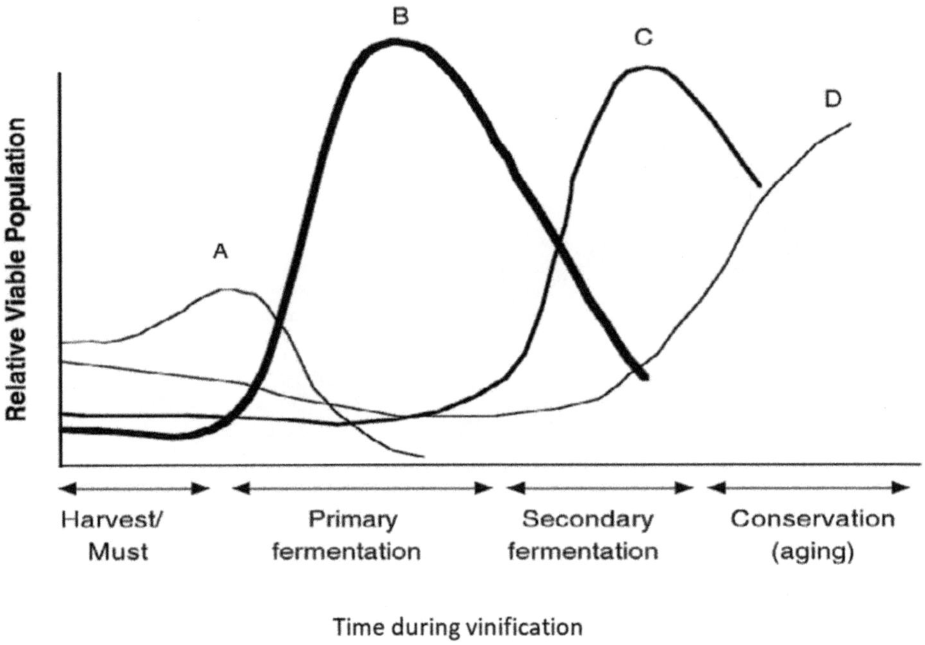

FIGURE 14.3 Generalised growth of (A) non-*Saccharomyces* yeasts, (B) *Saccharomyces*, (C) *Oenococcus oeni* and (D) spoilage yeasts and/or bacteria during the vinification of a wine. *Source:* Fugelsang and Edwards (2006)

longer during fermentations at 10°C and 15°C than at vinification conducted above 20°C. Therefore, the contributions of non-*Saccharomyces* yeasts to wine quality could increase with a reduction in fermentation temperature, and it has been found that the wines fermented at cooler temperatures and not inoculated with *Saccharomyces* were "more aroma intense" than those inoculated with *Saccharomyces.* Although ethanol tolerance is one of the major factors affecting the growth of non-*Saccharomyces* yeasts, there are clearly other factors that also affect the growth of this yeast.

Another variable effecting this phenomenon is a lack of oxygen and subsequently, a lack of sterol and phospholipid biosynthesis, resulting in the die-off of *Torulaspora delbruckii* during alcoholic fermentations. However, early deaths of *Kluyveromyces* and *Torulaspora* could not be solely explained by nutrient depletion or the presence of toxic compounds. Rather, it is affected by a cell-to-cell contact mechanism with high cell populations of *Saccharomyces.*

Different factors have been suggested that could affect the ethanol tolerance of non-*Saccharomyces* yeast species, which include the presence of oxygen and the survival factors (unsaturated fatty acids and sterols) which allow cells to maintain their viability and fermentative activity.

Dominance by a particular species or by a group of microorganisms at any given stage during vinification depends on many factors, including the microorganisms present and the conditions of grapes prior to harvest. Moreover, the chemical and physical environments of juice/must/wine, winery cleaning and sanitation programmes, as well as metabolic interactions between microorganisms, are also known to play key roles in wine fermentation.

The simple microbial ecology during vinification is complicated, however, by ever increasing evidence that microorganisms can also exist in a state known as "viable-but-non-culturable" (VBNC), which is induced in response to different stress conditions such as osmotic pressure, temperature, oxygen concentration and others. The cells may fail to grow on microbiological media (indicating the death of the cells) yet display low levels of metabolic activity and can be resuscitated under favourable environmental conditions. The net result is the potential failure to detect such microorganisms during vinification, leading to false conclusions regarding population dynamics. Microorganisms found in wine and believed to be able to enter a VBNC state include *Acetobacter aceti*, *Brettanomyces bruxellensis*, *Candida stellata*, *Lactobacillus plantarum*, *Saccharomyces cerevisiae* and *Zygosaccharomyces bailii.*

14.3.2.1.2 Yeast Saccharomyces

As populations of non-*Saccharomyces* yeasts decline, *Saccharomyces* dominates and completes the alcoholic fermentation. Recognising this, many winemakers inoculate musts with commercial cultures of *Saccharomyces* so as to control the microbial fermentation. When added to must/juice at recommended levels, the population of actively growing *Saccharomyces* should exceed $1x10^6$ to $3x10^6$ CFU/mL. During the peak of alcoholic fermentation, populations normally reach at least 10^7 CFU/mL or so. Generally, by the time peak populations are reached and the culture is in the stationary phase, at least half of the fermentable sugar may have been utilised. *Saccharomyces*, however, continues to utilise the remaining sugar, most notably glucose and fructose, until dryness is reached. *Saccharomyce* is known to be glucophilic and prefers glucose to fructose. Consequently, the glucose present in grape is exhausted before the complete utilisation of fructose.

Modern and large-scale wineries generally use specially selected starter cultures of this yeast rather than rely on the fermentative activities of naturally occurring yeasts. These special cultures are used as dried yeast – as powder, tablet, lyophilised in glass tubes (for more details, see Chapter 26 of this text)

The use of pure cultures of *S. cerevisiae* is generally recognised to minimise problems of controlling the fermentation and to produce a wine of a more consistent quality. The addition of *S. cerevisiae*, however, neither affects the presence of indigenous yeast nor, to a great extent, the pattern of fermentation. It is apparent that each stage of natural fermentation is characterised by the development of different strains of *S. cerevisiae* and that the main consequence of adding pure cultures is to influence the development of *S. cerevisiae* rather than to inhibit non-*Saccharomyces*.

14.3.2.1.3 Desirable Characteristics of Wine Yeast

As well as this, while conducting alcoholic fermentation, it is well known that the yeast *S. cerevisiae* plays an important role in the evolution of various flavour molecules. Not all wine yeast strains, however, are capable of imparting the characteristic flavours to the wines. Attempts have been made to find out the desirable characteristics of a strain of *Saccharomyces* for conducting alcoholic fermentation to prepare a wine of acceptable quality. However, the use of locally-selected *Saccharomyces cerevisiae* strains improves the sensory characteristics of some wines such as Malvar wines. The high concentrations of isoamyl acetate, hexyl acetate and high acidity are said to be responsible for the fruity and fresh character of these wines. A generalised list of desirable traits of wine yeasts can be found in Figure 14.4.

14.3.2.1.4 Impact of Pure Culture on Wine Fermentation

In 'pure culture' fermentation, selected commercial strains of *Saccharomyces cerevisiae* are inoculated into the juice at populations of 10^6–10^7 CFU/mL. Essentially, the pure culture approach gives a more rapid and predictable fermentation. *S. cerevisiae* eventually dominates most of the wine fermentations, as has been discussed earlier, so it has been commercialised as a starter culture.

In this type of fermentation, *S. cerevisiae* is usually added to achieve a population of about 10^5 to 10^6 cells/mL in the must. The addition of SO_2 as free molecular SO_2 has generally been thought to inhibit, if not kill, most of the indigenous yeast population of the grapes. However, the content of free SO_2 rapidly falls during maceration and fermentation, when indigenous strains may grow and multiply. Thus, indigenous strains of *S. cerevisiae* may play a significant sensory role during the early stages of fermentation.

In non-traditional winemaking areas, the inoculation of the grape or fruit must with *S. cerevisiae* is practiced, while commercial preparation of active dry yeast (ADY) is added to inoculate the fermentation medium. The same practice of using a commercial preparation of starter culture has also been started in traditional regions and is propagated in wine production. The yeast from master cultures held in-house is propagated in large scale production. However, commercially produced liquid or dried cultures are more convenient for use in many cases. Small wineries prefer the dried cultures, which can be added directly to the vat without any propagation.

It is now well established that for the purpose of producing wine of consistent quality, the use of commercial preparations of *S. cerevisiae* is essential. This practice also minimises the problems in producing wine with abnormal quality characteristics, as it controls the kinetics of fermentation in all batches. The other advantage of adding *S. cerevisiae* is that it does not affect the occurrence of the indigenous yeast without affecting the pattern of fermentation greatly. However, it appears that during natural fermentation or even the inoculated

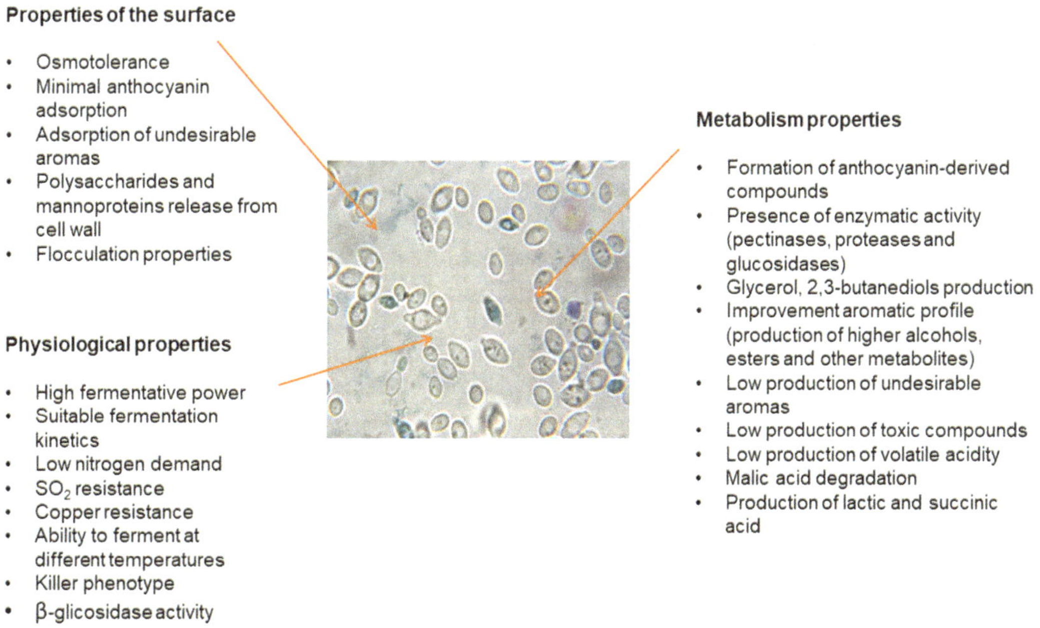
Properties of the surface
• Osmotolerance
• Minimal anthocyanin adsorption
• Adsorption of undesirable aromas
• Polysaccharides and mannoproteins release from cell wall
• Flocculation properties
Physiological properties
• High fermentative power
• Suitable fermentation kinetics
• Low nitrogen demand
• SO_2 resistance
• Copper resistance
• Ability to ferment at different temperatures
• Killer phenotype
• β-glicosidase activity
Metabolism properties
• Formation of anthocyanin-derived compounds
• Presence of enzymatic activity (pectinases, proteases and glucosidases)
• Glycerol, 2,3-butanediols production
• Improvement aromatic profile (production of higher alcohols, esters and other metabolites)
• Low production of undesirable aromas
• Low production of toxic compounds
• Low production of volatile acidity
• Malic acid degradation
• Production of lactic and succinic acid

FIGURE 14.4 Desirable characteristics of wine yeast. *Source:* Adapted from Suárez-Lepe *et al.* (2007)

fermentation, each stage is characterised by the development of different strains of *S. cerevisiae* and that the addition of an active dry yeast commercial preparation influences the development of *S. cerevisiae* rather than inhibiting other yeasts.

At the same time, it is now believed that some characteristics associated with particular natural micro-flora could be lost in their absence, as a few yeasts possess low fermentation efficiency, such as *K. apiculata*, but can produce significant quantities of volatile aroma compounds, particularly esters, as described earlier.

An alternative to either spontaneous or induced fermentations with a single yeast strain is inoculation with a mixed culture of local and commercial yeast strains. Different species appear to diminish the differences in the metabolic properties of one another, producing a more uniform and regionally distinctive character. In this case, intentional inoculation is needed to achieve a rapid initiation of fermentation

14.3.2.2 Malolactic Fermentation

A balanced wine in terms of flavour is always the goal of every winemaker. In some cases, grape must acidity may need to be corrected either by acid addition or acidity reduction. When the acidity needs reduction, the winemakers generally use physical methods (blending and amelioration), chemical methods (addition of bicarbonates), and biological methods (yeast and bacteria). Biological deacidification only affects the malic acid in the total acidity of the wine but does not reduce tartaric acid. Malolactic fermentation (MLF) is the most common method used for this purpose.

This other aspect of malolactic acid fermentation, is where L-malic acid is converted to L-lactic acid and CO_2. Although lactic acid bacteria can produce D-, L- or DL-lactic acid from glucose, only the L-isomer is produced during fermentation, resulting in a reduction in the acidity of wine with an increase in pH of about 0.2 units. The increased pH provides favourable conditions for bacterial propagation. The reduction in acidity by metabolising a dicarboxylic acid (malic acid) to a monocarboxylic acid (lactic acid) improves the sensory quality of highly acidic wine, and it becomes more palatable.

Once the fruit is crushed, the ecology of lactobacilli and other lactic acid bacteria become complex, with different species dominating at different times during vinification. Among the lactic acid bacteria, *L. jensenii*, *L. buchneri*, *L. hilgardii*, *L. brevis*, *L. cellobiosis*, *L. plantarum*, *Leuconostoc oenos* (*O. oeni*) and *Pediococcus* spp. have been isolated. Growth and the decline of any particular species is influenced by a number of conditions, including nutritional status, pH, alcohol and cellar temperature, as well as the interactive impact of yeast and other bacteria. For instance, wines from warmer regions typically have pH values in excess of 3.5, a condition favourable to the growth of lactobacilli and other bacteria.

Malolactic fermentation (MLF) can affects wine flavour by modifying the concentrations of aroma-impacting compounds such as diacetyl, esters, higher alcohols and volatile acids. Previously, *Oenococcus oeni* has been the lactic acid bacteria (LAB) of choice as a MLF starter, but recently more *Lactobacillus plantarum* starters have become available, which produce a broader range of extracellular enzymes than *O. oeni*, which enhance flavour development. Both yeast selection and MLF strategy effect the berry aroma, but MLF strategy also has a significant effect on acid balance and astringency of wines.

Malolactic acid fermentation is not free from possible collateral effects that on some occasions produce off-flavours, wine quality loss and human health problems. Winemaking decisions, such as yeast selection, fermentation and storage temperature, timing of SO_2 additions, racking, fining and filtration also play important roles in this context (for more details see Chapter 16 of this text).

14.3.3 Biological Deacidification by *Schizosaccharomyces pombe*

Another approach to solve the problem of higher acidity in wine is by biological deacidification by the use of non-*Saccharomyces* yeasts like *Schizosaccharomyces pombe*, which has the ability to degrade L-malic acid. In this case, the yeast has malic enzyme, which is responsible for the transformation of malic acid into pyruvic acid. The degradation of malic acid is complete in the yeasts of *Schizosaccharomyces* genus since these yeasts possess an active malate transport system. Unfortunately, growth of some strains of *S. pombe* yields undesirable sensory characteristics. The technique has successfully been applied by using immobilized or encapsulated cells of *Schiz. pombe*, which could be easily removed after consumption of malic acid but prior to synthesis of off-flavours (see Chapter 15 of this book for more details). Similarly, excessive volatile acidity in wine is also a problem in the wine industry, which can be solved by the use of *Schizosaccharomyces pombe* in wine fermentation.

14.3.4 Microbial Interactions in Wine Preparation

Microbial interactions that influence microbial ecology during vinification are the various interactions that occur between various microorganisms, and many of these result in the suppression and potential death of one or more species of the population. For example, the interactions between *Saccharomyces* and *O. oeni* during vinification may be stimulatory or inhibitory to the bacteria. Inhibitory interactions take place where the viability of *O. oeni* declines from 10^7 to 10^5 CFU/mL to undetectable populations soon after inoculation into wine, and rapid decline in bacterial viability takes place frequently, even when *Saccharomyces* and *O. oeni* are co-inoculated at similar population densities. It could be due to the possibility that faster growing *Saccharomyces* removes the nutrients from grape must that are important to the nutritionally fastidious malolactic bacteria. Secondly, wine yeasts are known to produce various compounds during alcoholic fermentation that are inhibitory to malolactic bacteria, including ethanol, SO_2 and medium-chain fatty acids.

A sluggish fermentation characterised by a rapid growth of *Lactobacillus* microorganisms, being very swift with abundant bacterial growth during the early stages of vinification, has been observed by a wine producer. The spoilage resulted in the premature arrest of alcoholic fermentation. In such cases, acetic acid levels were found to be extraordinarily high, generally ranging from 0.8 to 1.5 g/L and, on occasion, 2 to 3 g/L. However, a "cause and effect" relationship between the growth of these bacteria and cessation of alcoholic fermentation was not found, but it was experimentally found that *Lactobacillus* can inhibit *Saccharomyces*.

A number of mechanisms for the yeast inhibition by *Lactobacillus* have been proposed, including the production of acetic acid, which is well-known to be inhibitory to the yeasts, influencing both growth and fermentative abilities, but other mechanisms are probable. Processing strategies to control infections of *Lactobacillus* may include the use of SO_2, low-temperature storage and adjustment of must pH. Of these genus, the strain of *L. kunkeei* was found to be highly sensitive to SO_2. In addition, Gram-positive bacteria like *Lactobacillus* are sensitive to lysozyme.

Due to the increase in pH, MLF can be favourable to the growth of other lactic acid bacteria like *Pediococcus*. However, a definite antagonism by *O. oeni* against *Pediococcus* in Cabernet Sauvignon and Merlot wines that have undergone malolactic fermentation has also been seen. Here, the viability of all strains declined from an initial population of 10^4 to 10^5 CFU/mL shortly after inoculation, and the extent of this decline depended on the strain *Pediococci*. The wines that had undergone MLF were more resistant to ropiness caused by *Pediococci*. The effect was due to the accumulation of small (less than 1 kDa) compounds, quite possibly peptides or proteins. In other studies, synthesis of H_2O_2 by a strain of *L. hilgardii* was enough to inhibit *O. oeni* and *P. pentosaceus* in mixed cultures. Many species of lactic acid bacteria are known to produce antibacterial proteinaceous substances called bacteriocins, and a number of researchers have reported the production of bacteriocins by bacterial species present in wine. The use of bacteriocins as a preservative has generated interest among researchers as a means to reduce the use of SO_2, due to potential health concerns related to SO_2. Different bacteriocins as means to control malolactic fermentation and biofilm formation by *O. oeni* have also been applied.

Yeasts are well-known to produce and secrete the so-called killer factors that are inhibitory to other yeasts. These factors are either proteins or glycoproteins and are lethal to the sensitive yeasts. The ability to produce killer factors is widespread among yeasts including *Saccharomyces*, *Hansenula*, *Pichia*, *Kluyveromyces*, *Candida*, *Kloeckera*, *Hanseniaspora*, *Rhodotorula*, *Trichosporon*, *Debaryomyces* and *Cryptococcus*. Also, some authors identified strains of *H. uvarum* that produced killer toxins and that had activity toward sensitive strains of *S. cerevisiae*. Given their stability in wine, these toxins have the potential to be used as fungicidal agents during the aging. *Acetobacter* is capable of not only inhibiting *Saccharomyces* but also *Pichia*, *Schizosaccharomyces*, *Zygosaccharomyces*, *Candida* and even *Brettanomyces*. Although the mechanism of inhibition is not specifically known and a rise in acidity has not been documented, it was proposed that *Acetobacter* produces some type of antibiotic under aerobic conditions.

14.3.5 Sequential/Mixed Cultures in Improvement of Wine Quality

During alcoholic fermentation, a microbiological population evolves as a consequence of the chemical changes produced in the environment. The yeast succession of non-*Saccharomyces* to *Saccharomyces* during spontaneous fermentation of grape juice has been established. These yeasts are the predominant microbiota in grapes and are mainly responsible for starting the spontaneous alcoholic fermentation. The use of non-*Saccharomyces* yeasts has become a common trend in the main wine regions, particularly because of their effects on the composition, flavour and colour of the wine. *Torulaspora delbrueckii, Kloeckera apiculata, Hanseniaspora uvarum, Hanseniaspora vineae, Candida zemplinina, Candida pulcherrima, Schizosaccharomyces pombe, Hansenula anomala* and *Lachancea thermotolerans* are the non-*Saccharomyces* yeasts used for his purpose. The wines produced with non-*Saccharomyces* yeasts differed chemically and sensorially from wines produced with *S. cerevisiae*, only as these yeasts could present peculiar oenological characteristics that can influence the sensorial profile of the final wine. Indeed, these can be combined with *Saccharomyces cerevisiae* in mixed starter cultures to achieve objectives such as wine biological acidification or deacidification, reduction of ethanol and/or increase of glycerol concentrations, an increase of the final content of polysaccharides and enhancement of wine complexity and aroma.

14.3.5.1 Mixed Cultures of Yeasts in Winemaking

The use of non-*Saccharomyces* yeasts to improve, complexify and diversify wine style is increasing. The interactions of seven non-*Saccharomyces* yeast strains of the genera *Candida, Hanseniaspora, Lachancea, Metschnikowia* and *Torulaspora* in combination with *S. cerevisiae* and three malolactic fermentation (MLF) strategies in a Shiraz revealed that the wines produced with non-*Saccharomyces* yeasts had lower alcohol and glycerol levels than wines produced with *S. cerevisiae* only. Malolactic fermentation also completed faster in these wines.

Inoculation of mixed starter cultures made of *Saccharomyces cerevisiae* EC1118 and *Schizosaccharomyces japonicus* #13 in commercial grape wines produced a modulation in the concentration of malic and acetic acids, and of some of the most important volatile compounds, such as 2-phenyl ethanol, in an inoculum-ratio-dependent fashion. The wines obtained with *S. japonicus* #13 in mixed cultures reached concentrations of total polysaccharides significantly higher than those obtained with pure cultures of *S. cerevisiae* EC1118, and total polysaccharides increased with the increase of *S. japonicus* #13 cell concentration. So *S. japonicus* #13 could profitably be inoculated in combination with *S. cerevisiae* EC1118 to enhance the wine complexity and aroma and to improve wine stability by increasing the final concentration of polysaccharides.

Mixed inoculation of non-*Saccharomyces* yeasts and *S. cerevisiae* is of interest to oenologists. However, the interactions between these yeasts are not well understood, especially those regarding the availability of nutrients like nitrogen, sugar concentration and the effects of these on the evolution of mixed yeast populations in controlled laboratory-scale fermentations. The effect of the time of inoculation of *Saccharomyces cerevisiae* with respect to the initial co-inoculation of three non-*Saccharomyces* yeasts show that *S. cerevisiae* inoculation during the first 48 h conferred a stabilizing effect over the fermentations with non-*Saccharomyces* strains and generally, reduced the yeast diversity at the end of the fermentation, but the nitrogen limitation increased the time of fermentation and also the proportion of non-*Saccharomyces* yeasts at mid and final stage of fermentation. Further, high sugar concentration gave different proportions of the inoculated yeast depending on the time of *S. cerevisiae* inoculation.

The current interest in *Metschnikowia pulcherrima* is due to the expression of many extracellular activities, some of which enhances the release of varietal aromatic compounds. Its low fermentative power makes it necessary to use as a sequential culture or mixed with *Saccharomyces cerevisiae* to completely ferment grape musts. Its low respiratory metabolism helps to lower ethanol content when used under aerobic conditions. The fermentation shows good compatibility with Sc in producing a low to moderate global volatile acidity and, with suitable strains, a reduced level of H_2S. The fermentation process using mixed cultures, with sequential addition of non-*Saccharomyces* yeasts and *S. cerevisiae*, tends to mimic spontaneous fermentations, especially in terms of population dynamics, as it is one of the species most detected in the initial phase of alcoholic fermentation of grape musts. *M. pulcherrima* in mixed fermentation, although mainly recommended for white wine, was also tested for red wines, where higher glycerol, reducing sugars, total dry matter and lower alcohol content were detected, in line with the current market trend. Apparently, grape variety had more impact in studies with non-*Saccharomyces* yeasts.

Hanseniaspora vineae is an apiculate non-*Saccharomyces* yeast that generates interest due to its proven ability to produce aromas of interest such as phenylethyl acetate. The application of *H. vineae* as an inoculum starter in a co-fermentation was successfully used for the production of quality white barrel fermented Chardonnay wines. Furthermore, the wines produced had a uniquely fruity character, intense flavours as well as full body, and a relatively long palate length. In Macabeo must, there was a good fermentative rate which resulted in more flowery wines.

14.3.5.2 Sequential Inoculations

Sequential inoculations of a non-*Saccharomyces* and a *Saccharomyces cerevisiae* yeast improve the wine quality. *Lachancea thermotolerans* has been used in warm regions to produce more acidic wines with less volatile acidity and higher aroma complexity from low acidic musts. A red wine making technology that uses a combination of *Lachancea thermotolerans* and *Schizosaccharomyces pombe* as an alternative to the conventional malolactic fermentation has been proposed based on parameters such as aroma compounds, amino acids, ethanol index and sensory evaluation. In this fermentation, *Schizosaccharomyces pombe* totally consumes malic acid while *Lachancea thermotolerans* produces lactic acid, avoiding excessive deacidification of musts with low acidity in warm viticulture areas. This methodology can also reduce the problems associated with malolactic fermentation (low acidity, less acetic acid, less biogenic amines and ethyl carbamate precursors) than the traditional wines produced *via* conventional fermentation techniques.

Similarly, certain *H. uvarum* strains can also produce high levels of desirable compounds such as esters, higher alcohols and carbonyl compounds, although there is great variability among the strains. The mixed culture fermentation of *H. uvarum* and *S. cerevisiae* could possibly be used to enhance wine aroma and quality. Wines produced with *H. uvarum* in combination with *S. cerevisiae* completed MLF in a shorter period than wines produced with only *S. cerevisiae.* Wine produced by sequential MLF scored higher for fresh vegetative and spicy aroma than wines where MLF was induced as a simultaneous inoculation, while those produced with *H. uvarum* had better body than wines produced with only *S. cerevisiae.*

It is well known that high ethanol levels in wines adversely affect the taste perception of new wine consumers. High ethanol levels can also affect local laws, health of consumers and trade possibilities. The sensorial quality of wines with more alcohol content significantly affects an increase in the perception of bitterness, sweetness, astringency and hotness, and masking of volatile aromatic compounds. In addition to the aforementioned effects, this yeast group is also known to be less efficient in the production of ethanol from consumed sugars when compared with *S. cerevisiae* yeasts. The use of sequential or simultaneous inoculation of a non-*Saccharomyces* and a *Saccharomyces cerevisiae* strain is the best option to produce wine with low or no alcohol. Assessment of *Hanseniaspora uvarum* yeast strain in sequential inoculations with *S. cerevisiae* yeasts under optimized fermentation conditions shows a significant reduction in ethanol levels compared with the fermentations carried out with *S. cerevisiae* monocultures. But how ethanol reduction affects the sensorial perception of wine is not yet clear. Wines with lower alcohol content were connected to fruity aromas and more colour intensity. See the subsequent section for effect of sequential inoculations on sensory quality.

14.3.6 Contribution of Microorganisms to Sensory Properties of Wine

14.3.6.1 Impact on Colour – Formation of Derived Anthocyanin Compounds by Yeast

The release of secondary metabolic yeast products into the medium, such as pyruvic acid and acetaldehyde, was demonstrated for *Saccharomyces* strains reacting with anthocyanins

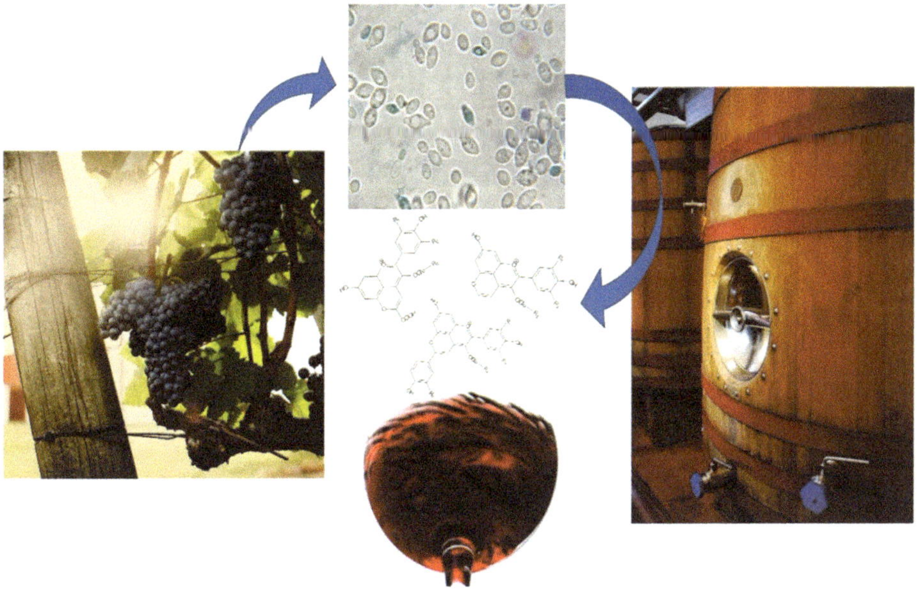

FIGURE 14.5 Yeast anthocyanin-derived compounds (vitisin A, vitisin B and vinylphenol adduct) result in increased wine-red colour stability

and producing derivatives like vitisin A, vitisin B and ethyl-linked anthocyanin-flavanol pigments (Figure 14.5). This suggests that yeast strain selection strongly effects colour intensity and the final concentration of anthocyanins and other phenolics. As mentioned, some of these reactions could be attributed to the variable levels of acetaldehyde synthesis by different yeast species. For example, *Pichia* species produced significantly higher levels of acetaldehyde compared to *Saccharomyces*, and it was shown that acetaldehyde increases linearly with increasing cell biomass concentration.

When fermented in pure cultures, *S. cerevisiae* produced higher concentrations of acetaldehyde and vitisin B (acetaldehyde reaction-dependent) compared with some non-*Saccharomyces*. However, co-fermentation of non-*Saccharomyces* with *S. cerevisiae* resulted in a significantly higher concentration of acetaldehyde compared with the pure *S. cerevisiae* control. In addition, the involvement of a mixed fermentation of *S. cerevisiae* and *Schizosaccharomyces pombe* produced large amounts of polymeric pigments (such as catechin), which predisposed the forming of such pigments. Moreover, vitisin A, vitisin B, malvidin-3-glucoside-4-vinylphenol and malvidin-3-glucoside-4-vinylguaicol were reported for the genera *Hanseniaspora*, *Metschnikowia* and *Schizosaccharomyces*.

14.3.6.2 Impact of Microorganisms on Wine Aroma

The secondary metabolism of yeasts generally forms higher alcohols, fatty acids, esters, aldehydes, sulphur compounds and phenolic compounds. In addition, yeasts are capable of producing terpenes and can modify varietal components (Figure 14.6).

Quantitatively, higher alcohols are the most important components. At high concentrations, they can have a negative impact and impart strong, pungent aromas, but in traditional fermentations, with the required concentration, they collaborate on the perception of fruit notes. They are produced anabolically from glucose and catabolically from amino acids (through the Erlich mechanism).

Organic acids are generally insufficiently volatile to modify the aromatic profile of the wine. There are some exceptions such as acetic acid, which is formed mainly by oxidative decarboxylation of pyruvic acid. The content of this acid that negatively impacts the quality of the wine varies considerably with the yeast strain used. The rest of the volatile fatty acids can be of two types: linear chain and even number of carbons (between 6 and 12) or iso acids of branched chain; all of these components are associated with dairy descriptors.

The esters formed by the yeasts are synthesised from the metabolism of lipids and acetyl CoA, and can be sub divided into ethyl esters and acetates. The first group, characterised by low perception thresholds and pleasant associated descriptors, is directly responsible for the quality of young wines. While the acetates are present in fermentations carried out by *Saccharomyces cerevisiae*, their production is markedly increased when it is fermented with non-*Saccharomyces* or when mixed cultures are used.

Volatile phenols are formed from hydroxycinnamic acid present in grapes. Vinyl phenols have a positive effect on the aroma, but if the growth of *Brettanomyces* bruxellensis occurs, which is capable of transforming these components into ethyl phenols, its compounds are associated with aromatic defects of the plastic type as well as curry and stable.

The yeast is able to produce sulphur compounds of various chemical structures, at low concentrations, most of which are associated with off-flavours (onion, rotten egg, cooked cabbage, etc.), with the exception of the varietal thiols that are transformed during alcoholic fermentation and are associated with notes of passion fruit and grapefruit.

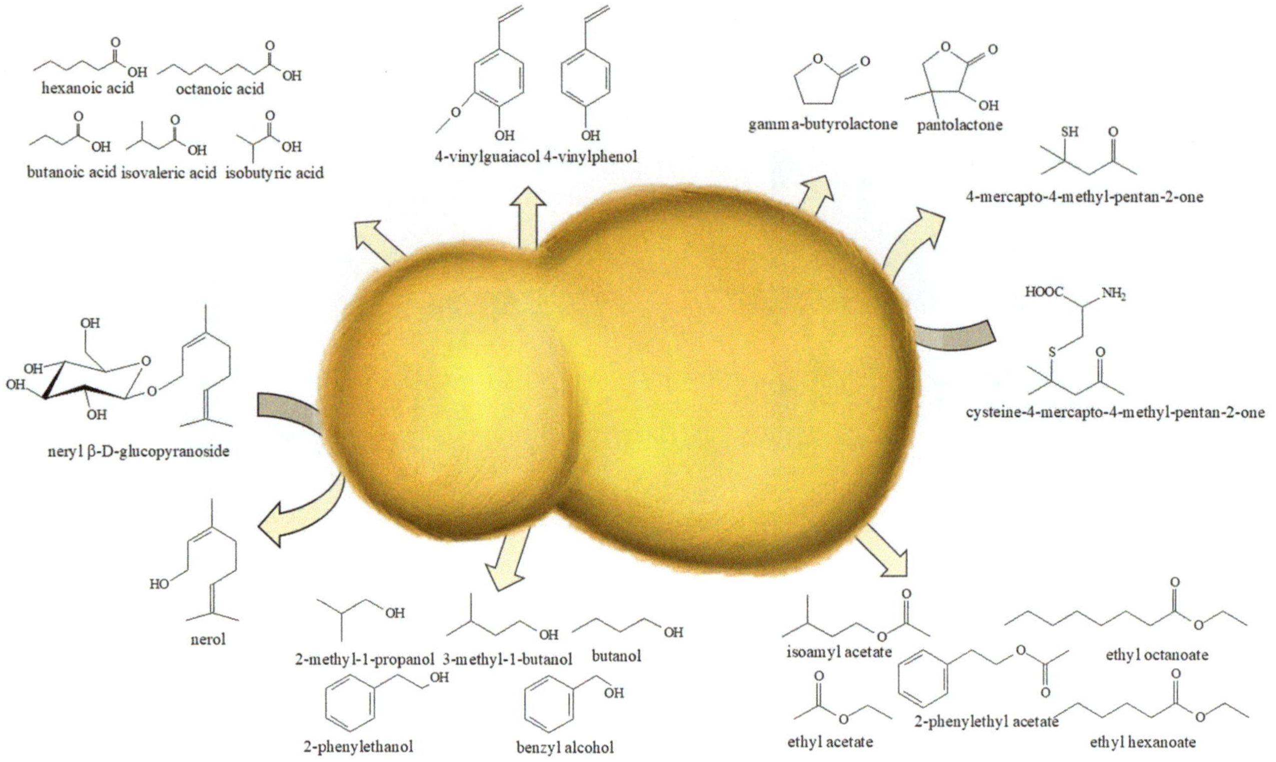

FIGURE 14.6 Various volatile compounds produced or modified by yeast during alcoholic fermentation

During MLF, the main product of the metabolism of lactic acid bacteria is lactic acid, which is produced in trace amounts during alcoholic fermentation as well as diacetyl – a carbonyl compound that drastically increases its concentration after this type of fermentation, providing notes of butter. At the same time, during MLF the level of some stresses that impact on the aroma of the final wine is increased: ethyl acetate, ethyl hexanoate, ethyl octanoate and ethyl lactate.

Some polysaccharides from the yeast (*Schizosaccharomyces pombe*) have many positive effects on wine quality, by reducing protein and tartrate instability, thus increasing the 'fullness' sensation. These interact with polyphenols aggregates and smoothens the perception of astringency and retains aromatic compounds. Thus, it can possibly be applied in to the improvement of wine quality.

14.3.7 Special Vinifications

14.3.7.1 Sherry Wines

Different types of sherry wines can be produced depending on winemaking conditions. Spanish Sherry can be classified as Fino, Oloroso, and Amontillado. Fino-type wines are produced by a biological aging process, in which the yeast employed are capable of forming a surface layer called velum. Oloroso-type wines are subjected to an oxidative aging process, while Amontillado wines are produced by combining both production systems.

During the preparation of sherry, a considerable microbial diversity occurs in the velum that develops on the surface of the wine. The microflora of velum consists mainly of the yeasts, but other fungi and bacteria could also be present. The restrictive conditions of biological aging (low pH, presence of sulphite, high ethanol and acetaldehyde concentrations, scarcity of sugars and low oxygen concentration) are compatible with only a few film-forming *S. cerevisiae*, so more than 95% of the film's microbiota consists of this species. The other 5% is explained by the presence of species of the genera *Debaryomyces, Zygosaccharomyces, Pichia, Hansenula* and *Candida*. For further information, please see the specific chapter of this text on this topic, as well as the literature, cited under Bibliography.

14.3.7.2 Sparkling Wine

Elaboration of sparkling wine consists of two steps. In the first step, the base wine is obtained after vinification similar to a table wine. The second step or second fermentation occurs after the addition of the "liqueur de tirage" – a mix of yeasts, sucrose, nutrients and adjuvants. This process could occur in the bottle (champenoise method) or in isobaric tanks (charmat method). The yeast used for second fermentation needs to be acclimatized for low temperature and alcohol concentration. The sparkling wines have a special biological aging, where the wine remains in contact with the lees, and they develop sensory notes such as toasty, lactic, sweet and yeasty, which can be attributed to proteolytic processes. During sparkling wine production, yeasts are subjected to particular stress conditions, and for this reason, they have to possess some additional characteristics such as autolytic ability and flocculation capacity. For more information, please see the chapter 34 in this text.

14.3.7.3 Non-Grape Wines

14.3.7.3.1 Honey Wine

Honey from honey bees is a natural source of abundant, readily fermentable sugar that is used for the production of a wine called mead. The microbial flora in honey may be classified as primary or secondary. The primary sources include pollen, the bee digestive tract, dust, air, soil, barrel and flowers.

A secondary source of microorganisms could be contamination from humans, packaging, and other equipment, or be wind borne, come from soil, insects, animals, or water.

When a diluted honey solution is fermented on its own without the addition of fruit juice or other additives, it creates band aid, and a traditionally strong (8–18% alcohol by volume, ABV) mead. Honey can be used for ameliorating a must to produce wines from fruits like grapes, apple and plum.

14.3.7.3.2 Fruit Wines

The microbiology of the production of fruit wines is similar to that of grape, where the natural microflora is contributed by the fruits used, machinery and equipment or during maturation in the cooperage, and the spoilage behaviour is also similar to wine production from grapes. However, in the production of fruit wine, more emphasis is placed on the preparation of a must of suitable composition that could undergo fermentation and produce wine of acceptable quality than on the microbiology of fermentation; this is because of the wide variation of composition of fruits, with factors like very high acidity, high tannins, low sugar or pulpy nature of fruits. The conventional yeast *Saccharomyces cerevisiae var. ellipsoideus* is employed, except under specific conditions, for specific fruit wine production (see the relevant chapter (37) on fruit wines). These days, the use of non-*Saccharomyces* species in winemaking can be found for several purposes especially the flavour development (chapter 15). An example can be cited of the use of *Schizosaccharmyces pombe*, which was employed to reduce the acidity of wine from acidic fruits like plum. However, the systematic research on the different aspects of fruit wines is lacking, and it is an exciting field for future research. The aspects mentioned here have been illustrated in some fruit wines like peach, citrus, pear, cider, plum wine, papaya wine and berry wines (see Chapter 37 on fruit wines). For further information about cider, see Chapter 35.

14.3.8 Wine Aging in Tanks, Barrels and Bottled Wine. Apparition of Undesirable Microorganisms

14.3.8.1 Yeast

Transitory exposure to oxygen may occur during pumping, transferring and/or fining. Even when wines are subsequently stored properly, such exposure may have been sufficient to stimulate rapid and continued growth of undesirable microorganisms.

Yeasts play an important role in the spoilage of beverages like wine. Two major spoilage yeasts in the wine industry, *Brettanomyces bruxellensis* and *Zygosaccharomyces rouxii*, produce off-flavours and gas, causing considerable economic losses. Film yeasts like *Saccharomycodes* and *Zygosaccharomyces* can cause serious spoilage in wines during storage.

Brettanomyces/Dekkera can be found in fermenting must and in wine. Wines contaminated with *Brettanomyces bruxellensis* yeast are characterized by the presence of off-flavours. Typically, it grows after alcoholic and malolactic fermentations during the storage of wines in tanks, barrels or bottles. Leveraging its ability to thrive in wine and cider conditions (low pH, high levels of ethanol, and low oxygenation levels), the yeast can proliferate inside beverage production tanks. It contributes to characteristic 'brett' flavours, which are described as smoky, plastic, burnt plastic, vinyl, band-aid and creosote. The compounds which are responsible for the 'bretty' flavour in a wine are mainly 4-ethylphenol, 4-ethylguaiacol and isovaleric acid. These flavours are often considered a defect. The wine's varietal and regional flavour characteristics might be completely masked by these flavours, and the wine can be unpleasantly bitter.

The metabolic activity of the spoilage yeast causes irreparable damage to many litres of final products every year. Specifically, their resultant by-products reduce the quality of the beverage. Therefore, winemakers and cider-house companies can suffer substantial economic impact. One alternative is biocontrol, which can be used either independently or in a complementary way to chemical control (SO_2).

Rapid and precise detection of spoilage yeasts is key to improve the quality of alcoholic fermentation beverages such as wine and cider. Thus, over the years, many detection techniques have been applied to control the occurrence of spoilage yeast. Traditional methods such as microscopy, cell plating, gas chromatography, mass spectrometry, etc. are often imprecise, expensive and/or complicated. New emerging spoilage yeast detection platforms, such as biosensors and microfluidic devices, have been developed to alleviate these constraints.

14.3.8.2 Acetic acid bacteria

Exposure to oxygen spoils the wine through the growth of acetic acid bacteria belonging mainly to the genera *Acetobacter* and *Gluconobacter.* These bacteria can spoil the must, fermenting wine and wine undergoing maturation. The maximum ethanol tolerance of these bacteria lies between 14–15%.

14.3.8.3 Lactic acid bacteria

Once alcoholic and malolactic fermentations are complete, lactic acid bacteria may still grow in the wine, resulting in spoilage. This happens because wines commonly contain small amounts of arabinose, glucose, fructose and trehalose – the sugars that can be metabolised by microorganisms, including lactic acid bacteria. The appearance and flavour of the wine is altered by the growth and activity of lactic acid bacteria, and the appearance of wine in a bottle or a test tube becomes silky/cloudy due to the growth. It gives off a disagreeable smell, known as a mousy smell, with the formation of flocculent or sediment. The lactic acid bacteria belonging to the genus *Lactobacillus, Leuconostoc* and *Pediococcus* are responsible for wine spoilage. These bacteria are acid tolerant, rod or cocci shaped, gram positive and could be homo- or heterofermentative. The homofermentative lactic acid bacteria chiefly convert glucose to lactic acid without the formation of appreciable amounts of carbon dioxide or acetic acid, while the latter forms lactic acid, carbon dioxide, acetic acid, ethanol and glycerol.

Besides that, they can also form mannitol from fructose and ferment malic and citric acid in wine. *Leuconostoc*, due to its ability to ferment sucrose into dextran, can cause ropiness in wine. Out of these, *Lactobacillus* and *Leuconostoc* are heterofermentative lactic acid bacteria, and *Pediococcus* is homofermentative.

14.4 FACTORS AFFECTING WINE MICROORGANISMS PREVENTION OF WINE SPOILAGE

14.4.1 Factors Affecting Wine Yeasts

The patterns of microbial succession and their dominance in wine are greatly influenced by certain factors including temperature, aeration, pH, key nutrients (see Chapter 15) and preservative used. Table 14.3 shows the principal factors affecting wine yeast performance. Controlling the growth of microorganisms at critical junctures during the process of winemaking is essential to produce a quality wine. The emphasis is not only on maximising the fermentative performance of *Saccharomyces* but also on managing the growth of undesirable yeasts and bacteria.

14.4.2 Preservative Used in Wine

14.4.2.1 Sulphur Dioxide

A number of additives, such as sulphur dioxide, lysozyme, dimethyl dicarbonate and sorbic acid, may be used to control the growth of microorganisms.

Preservatives such as SO_2 cannot be considered the only remedy for controlling undesirable microorganisms during vinification. Its use with other factors such as low pH and finning is very effective. Further, its combinations can often result in an enhanced effect. Sulphur dioxide is widely recognised in both the wine and food industries for its antioxidative and antimicrobial properties. It can be added to musts or wines in the form of compressed gas, potassium metabisulphite ($K_2S_2O_5$) or by burning candles containing sulphur in an enclosed container such as a barrel. Most commonly, sulphur dioxide is incorporated into grape must or wine as potassium metabisulphite.

TABLE 14.3
Factors affecting yeast growth during alcoholic fermentation

- Aeration level
- Composition of juice (key nutrients and pH)
- Clarification of juice
- Addition of sulphur dioxide or other preservative
- Temperature of fermentation
- Inoculation of juice or pulp
- Interaction with other microrganism

Once dissolved in water, sulphur dioxide exists in equilibrium between molecular SO_2, bisulphite $(HSO_3)^-$ and sulphite (SO_3^{2-}) species. However, the equilibrium is dependent on the pH of the medium, with the dominant species at a wine pH of 3 to 4 being bisulphite anion. As well as this, bisulphite also exists in "free" and "bound" forms. The molecule reacts with carbonyl compounds (*e.g.* acetaldehyde), forming additional products such as hydroxy sulphonic acids in addition to acetaldehyde. Molecules present in grape juice that react with bisulphite include pyruvic acid, keto-glutaric acid, dihydroxyacetone, diacetyl, anthocyanin pigments and others.

It is generally believed that the molecular sulphur species is the antimicrobial form of sulphur dioxide. Because SO_2 does not have a charge, the molecule enters the cell and undergoes rapid pH-driven dissociation at cytoplasmic pH (generally near 6.5) to yield bisulphite and sulphite. Due to a decrease of the intracellular concentration of molecular SO_2 and consequently the internal equilibrium, more molecular SO_2 enters the cell, further increasing intracellular concentrations. Mechanisms of SO_2 resistance differ but are related to variable rates of diffusion across cell membranes, biosynthesis of compounds that bind SO_2 and varying enzymes. Inhibition of microorganisms by using SO_2 takes place through different means, including the rupture of disulphide bridges in proteins and reaction with co-factors, including NAD and FAD. It also reacts with ATP and brings about deamination of cytosine to uracil, thus increasing the likelihood of lethal mutations as well as reducing the concentrations of crucial nutrients such as thiamine.

The addition of SO_2 to the must reduces the total number of yeasts present, makes qualitative changes in the yeast microflora and favours the resistant *S. cerevisiae* and *Saccharomycodes ludwigii* at the expense of more sensitive species such as *K. apiculata*. Sulphur dioxide delays yeast multiplication without blocking it and delays alcoholic fermentation. While the bacteria contributed by grapes (present as natural flora on grapes) at the same time as yeasts are introduced, are killed or deactivated to protect the fermentation medium from their development. The yeast *Saccharomyces cerevisiae* is introduced by inoculating and perform the alcoholic fermentation. Due to the domination of yeast over bacteria, the amount of total sugar is effectively transformed into alcohol, thus enhancing the fermentation efficiency of yeasts.

14.4.2.2 Dimethyl Dicarbonate (DMDC)

Dimethyl dicarbonate (DMDC) is approved for use in wines containing 5g/l or more of sugars at a maximum concentration of 200 mg/L. The biggest advantage of its use is that it does not have any residual activity because it undergoes hydrolysis to yield carbon dioxide and methanol. In a wine containing 10% (v/v) ethanol, a concentration of 25 mg/L was found effective against *Saccharomyces*, *Brettanomyces* and *Schizosaccharomyces*. It is also found to be inhibitory against acetic and lactic acid bacteria. A synergy between DMDC and SO_2 against *Saccharomyces* also exists. DMDC is more effective than a combination of SO_2 and/or sorbic acid and could suppress the fermentation of grape juice more effectively but at higher temperatures. The antimicrobial effect of DMDC results from the inactivation of microbial enzymes. The mechanism thus, involves denaturation of the fermentative pathway enzymes *viz.* glyceraldehyde-3-phosphate dehydrogenase and alcohol dehydrogenase.

DMDC, however, suffers from both formulation and dosing problems as it is only partially soluble in water and requires thorough mixing to ensure uniform distribution in wines to be bottled. Besides that, the compound has a melting point of 15.2°C, so it has to be slightly warmed prior to addition and is added to the wine using special equipment that delivers the optimal dose in each application.

14.4.2.3 Lysozyme

Lysozyme is a low molecular weight protein (14,500 Da) which is derived from egg whites and brings about lysis of the cell wall of Gram-positive bacteria (*Oenococcus, Lactobacillus* and *Pediococcus*). However, its activity toward Gram-negative bacteria (*Acetobacter* and *Gluconobacter*) is limited because of the protective outer layers in this group. The enzyme also has no effect on the yeasts or moulds. Because of its specificity, lysozyme is used by white, rosé and blush wine producers with a specific purpose to prevent malolactic fermentation as well as by the wineries to reduce the initial populations of lactic acid bacteria before fermentation. Pre-fermentation additions of 500 mg/L of lysozyme inhibits MLF. The enzyme concentrations required for microbial stability, if added post-alcoholic fermentation, could be reduced to 125–250 mg/L. At the same time, differential sensitivity to lysozyme among species of *Lactobacillus* or *Pediococcus* has also been documented. It is an approved additive for use in the United States at concentrations up to 500 mg/L. Since it is a protein in nature, the presence of phenolic compounds and degree of clarification can affect its activity because the plant phenolics are well-known to react with enzymes, thereby decreasing the activity. Probably due to this, the enzyme is more effective in white wines than red wines.

14.4.2.4 Sorbic Acid

Sorbic acid (2,4-hexandienoic acid) is a short-chain fatty acid. It is used in grape juices and in sweetened, bottled wines to prevent re-fermentation by *Saccharomyces*. The maximum concentration of sorbic acid allowed in the United States is 300 mg/L, whereas the *Organisation Internationale de la Vigne et du Vin* (OIV) has reduced the limit to 200 mg/L. Generally, concentrations of 100 to 200 mg/L are used in wine. At recommended levels, sorbic acid is generally effective in controlling *Saccharomyces*, but the sensitivity of other yeasts is variable. Yeasts like *Kloeckera apiculata* and *Pichia anomala* (formerly *Hansenula anomala*) are inhibited at concentrations of 156 and 168 mg/L, respectively, whereas *Schizosacccharomyces pombe* and *Zygosaccharomyces bailii* require at least 672 mg/L. Mechanisms underlying the inhibition by sorbic acid are not fully understood but could be attributed to the morphological differences in cell structure, changes in genetic material, alteration in cell membranes as well as inhibition of enzymes or transport functions.

In brief, the bacteria are not affected by sorbic acid, and rather several species can metabolise the sorbic acid to yield 2-ethoxyhexa-3,5-diene – a compound that imparts a distinctive "geranium" odour/tone to wines. Other odour/flavour-active compounds detected in spoiled wines treated with sorbic acid include 1-ethoxyhexa-2,4-diene and ethyl sorbate. The latter compound has also been associated with off-flavours in sparkling wines so should not be used in sparkling wine production. However, ethyl sorbate has been described to possess a "honey" or "apple" aroma but has also been found to impart a very unpleasant "pineapple-celery" odour after short-term (6 month) storage. Sorbic acid is relatively insoluble in water (1.5 g/L at room temperature). The additive is usually sold as potassium sorbate, which is readily soluble (58.2 g/L in water). Sulphur dioxide is thought to work synergistically with sorbic acid, lowering the concentration of the acid needed for the control of fermentative yeasts. SO_2 and sorbic acid added to sweetened table wine at 80 mg/L each had a greater inhibitory action than either SO_2 at 130 mg/L or sorbic acid alone at 480 mg/L. However, there was an antagonistic interaction between SO_2 and sorbate against *S. cerevisiae*, where the inhibition exerted by each compound individually was greater than the combination. It could be attributed to the fact that sorbic acid reacts with SO_2 to yield an adduct product. The amount of ethanol present, however, determines the appropriate amounts of sorbic acid to be added to the wine.

14.4.2.5 Fumaric Acid

Fumaric acid is used for controlling the growth of lactic acid bacteria. Being a relatively strong organic acid, it is used as an acidulant at a maximum concentration of 3.0 g/L and has received attention as an acidulating agent in wine instead of the more expensive tartaric acid, because its addition at the rate of 1 g/L is equivalent to the tartaric acid addition of 1.2 g/L. Fumaric acid has limited solubility in wine (6.3 g/L at 25°C). However, it is more soluble at higher temperatures (10.7 g/L at 40°C) or in 95% ethanol (57.6 g/L at 30°C). The efficacy of fumaric acid depends on pH where less activity is noted with increasing pH. However its use it not yet approved by the OIV.

The most important function of fumaric acid is its ability to inhibit MLF. Its effectiveness is indicated by the fact that none of the wines containing 1.5 g/L fumaric acid underwent MLF, even after 12 months of storage. However, it is degraded during alcoholic fermentation by *Saccharomyces*, forming L-malic acid. Nevertheless, fumaric acid might prove to be useful for reducing initial bacterial populations in musts, such as those of some species of *Lactobacillus*. Although it is not known whether these spoilage bacteria can metabolise fumaric acid, it is known that it could be degraded to L-lactic acid by wine *Leuconostocs* (*Oenococcus*), possibly by the same mechanism as that of yeast. With respect to sensory quality, it is known to have a "harsh" taste. For more details on microbiological spoilage of wine, see Chapter 41 in this book.

14.4.3 Biological Control of Wine Spoilage

Traditionally, SO_2 is used in winemaking to control microbial proliferation such as bacteria, yeasts and fungi, as described in an earlier section. There are however, strict regulations in relation to its use due to its toxic and allergenic effects on human health. International organizations, such as the OIV, encourage a reduction in usage of SO_2 in winemaking. Moreover, modern consumers prefer more natural, healthy foods and beverages that are minimally processed and are free from preservatives.

Therefore, biological control is an alternative proposal that can be used either independently or in a complementary way to chemical control (SO_2) of microorganisms. It is known that some *Saccharomyces* and non-*Saccharomyces* yeasts have the ability to bio-suppress other yeasts through different mechanisms (such as the production of toxic compounds, competition for limiting substrates and/or cell to cell contact). Native non-*Saccharomyces* yeasts have been explored for their biocontrol activity and their ability to be employed under fermentation conditions, as well as certain enological traits. Based on certain criteria (like assay of bio-controller yeasts for their ability to prevail in the fermentation medium and their positive/negative contribution to the wine spoilage), two yeasts have been proved to be effective. These include *Wickerhamomyces anomalus* BWa156 and *Metschnikowia pulcherrima* BMp29. Before these are applied to winemaking, more in-depth research is needed.

BIBLIOGRAPHY

Agnolucci, M., Tirelli, A. Cocolin, L. and Toffanin, A. (2017). *Brettanomyces bruxellensis* yeasts: Impact on wine and winemaking *bruxellensis* yeasts: Impact on wine and winemaking. *World J. Microbiol. Biotechnol.* 33(10): 180.

Agouridis, N., Bekatorou, A., Nigam, P. and Kanellaki, M. (2005). Malolactic fermentation in wine with *Lactobacillus casei* cells immobilized on delignified cellulosic material. *J. Agric. Food Chem.* 53(7): 2546.

Alonso-del-Real, J., Lairón-Peris, M., Barrio, E. and Querol, A. (2017). Effect of temperature on the prevalence of *Saccharomyces* non-*cerevisiae* species against a *S. cerevisiae* wine strain in wine fermentation: competition, physiological fitness, and influence in final wine composition. *Front. Microbiol.* 8: 150.

Amerine, M.A., Berg, H.W., Kunkee, R.E., Qugh, C.S., Singleton, V.L. and Webb, A.D. (1980). *The Technology of Wine Making*, fourth ed. AVI, Westport, CT.

Asenstorfer, R.E., Markides, A.J., Iland, P.G. and Jones, G.P. (2003). Formation of vitisin A during red wine fermentation and maturation. *Aust. J. Grape Wine R.* 9(1): 40.

Bartowsky, E.J. and Henschke, P.A. (1999). Use of polymerase chain reaction for specific detection of the malolactic fermentation bacterium *Oenococcu oeni* (formerly *Leuconostoc oenos*) in grape juice and wine samples. *Aust. J. Grape Wineres.* 5(2): 39.

Benito, Á., Calderón, F. and Benito, S. (2016). Combined use of *S. pombe* and *L. thermotolerans* in winemaking. Beneficial effects determined through the study of wines' analytical characteristics. *Molecules* 21(12): 1744. doi:10.3390/ molecules21121744.

Benito, Á., Calderón, F. and Benito, S. (2019). The influence of non-*Saccharomyces* species on wine fermentation quality parameters. *Fermentation* 5(3): 54. doi:10.3390/ fermentation 5030054.

Benito, Á., Calderón, F., Palomero, F. and Benito, S. (2015). Combined use of selected *Schizosaccharomyces pombe* and *Lachancea thermotolerans* yeast strains as an alternative to the traditional malolactic fermentation in red wine production. *Molecules* 20(6): 9510–9523. doi:10.3390/molecules20069510.

Benito, S., Palomero, F., Morata, A., Uthurry, C. and Suárez-Lepe, J.A. (2009). Minimization of ethylphenol precursors in red wines via the formation of pyranoanthocyanins by selected yeasts. *Int. J. Food Microbiol.* 132(2–3): 145–152.

Bhardwaj, J.C. and Joshi, V.K. (2009). Effect of cultivar, addition of yeast type, extract and form of yeast culture on foaming characteristics, secondary fermentation and quality of sparkling plum wine. *Natl Prod. Rad.* 8(4): 452.

Bisson, L.F. and Kunkee, R.E. (1993). Microbial interactions during wine production. In: *Mixed Cultures in Biotechnology*, J.G. Zeikus and E.A. Johnson (Eds.), McGraw-Hill, New York, p. 37.

Boido, E., Medina, K., Fariña, L., Carrau, F., Versini, G. and Dellacassa, E. (2009). The effect of bacterial strain and aging on the secondary volatile metabolites produced during malolactic fermentation of Tannat red wine. *J. Agric. Food Chem.* 57(14): 6271.

Boulton, R.B., Singleton, V.L., Bisson, L.F. and Kunkee, R.E. (1996). *Principles and Practices of Winemaking.* Chapman and Hall, New York.

Carrau, F., Medina, K., Fariña, L., Boido, E. and Dellacassa, E. (2010). Effect of *Saccharomyces cerevisiae* inoculum size on wine fermentation aroma compounds and its relation with assimilable nitrogen content. *Int. J. Food Microbiol.* 143(1–2): 81.

Ciani, M. (2008). Continuous deacidification of wine by immobilized *Schizosaccharomyces pombe* cells: Evaluation of malic acid degradation rate and analytical profiles. *J. Appl. Microbiol.* 79(6): 631.

Ciani, M. and Maccarelli, F. (1998). Oenological properties of non-*Saccharomyces* yeasts associated with wine-making. *World J. Microbiol. Biotechnol.* 14(2): 199.

Cordero-Bueso, G ., Esteve-Zarzoso, B., Gil-Díaz, M., García, M., Cabellos, J. and Arroyo, T. (2016). Improvement of Malvar wine quality by use of locally-selected *Saccharomyces cerevisiae* strains. *Fermentation* 2(4), 7. doi:10.3390/ fermentation2010007.

Del Fresno, J.M., Antonio Mora, A., Loira, I., Bañuelos, M.A., Escott, C., Benito, S., González, C., Chamorro, J. and Suárez-Lepe, A. (2017). Use of non-*Saccharomyces* in single-culture, mixed and sequential fermentation to improve red wine quality. *Eur. Food Res. Technol.* 243(12): 2175.

Di Gianvito, P., Arfelli, P., Suzzi, G. and Tofalo, R. (2019). New trends in sparkling wine production: Yeast rational selection. In: *Alcoholic Beverages*, A.M. Grumezescu and A.M. Holban (Eds.), Woodhead Publishing, pp. 347–386.

Drysdale, G.S. and Fleet, G.H. (1989a). The effect of acetic acid bacteria upon the growth and metabolism of yeasts during the fermentation of grape juice. *J. Appl. Bacteriol.* 67(5): 471.

Drysdale, G.S. and Fleet, G.H. (1988b). Acetic acid bacteria in winemaking: A review. *Am. J. Enol. Vitic.* 39(2): 143.

Duarte, F.L., Egipto, R. and Baleiras-Couto, M.M. (2019). Mixed fermentation with *Metschnikowia pulcherrima* using different grape varieties. *Fermentation* 5: 59. doi:10.3390/ fermentation 5030059.

du Plessis, H., du Toit, M., Nieuwoudt, H., van der Rijst, M., Martin, K.I.D. and Jolly, N. (2017). Effect of *Saccharomyces*, non-*Saccharomyces* yeasts and malolactic fermentation strategies on fermentation kinetics and flavor of Shiraz wines. *Fermentation* 3(4): 61. doi:10.3390/fermentation3040064

du Plessis, H., du Toit, M., Nieuwoudt, H., van der Rijst, M., Martin, K.I.D. and Du Plessis Toit, H., Du, M., Nieuwoudt, H., Van der Rijst, M., Ho, J. and Jolly, N. (2019). Modulation of wine flavor using *Hanseniaspora uvarum* in combination with different *Saccharomyces cerevisiae*, lactic acid bacteria strains and malolactic fermentation strategies. *Fermentation* 5(3): 64. doi:10.3390/fermentation5030064.

Du Toit, M. and Pretorius, I.S. (2000). Microbial spoilage and preservation of wine: Using weapons from nature's own arsenal – A review. *S. Afr. J. Enol. Vitic.* 21(1): 74.

Fariña, L., Villar, V., Ares, G., Carrau, F., Dellacassa, E. and Boido, E. (2015). Volatile composition and aroma profile of Uruguayan Tannat wines. *Food Res. Int.* 69: 244–255.

Fleet, G.H. (1998). The microbiology of alcoholic beverages. In: *Microbiology of Fermented Foods*, Vol. 1., 2nd edn. B.J.B. Wood (Ed.), Blackie Academic and Professional, London, p. 217.

Fleet, G.H. (2003). Yeast interactions and wine flavour. *Int. J. Food Microbiol.* 86(1–2): 11–22

Fleet, G.H. and Heard, G.M. (1993). Yeasts – Growth during fermentation. In: *Wine Microbiology and Biotechnology*, Chapter 2, G.H. Fleet (Ed.), Harwood Academic Publishers, Chur, Switzerland, pp. 27–55.

Fleet, G.H. and Heard, G.M. (1997). Wine. In: *Food Microbiology: Fundamentals and Frontiers*, M.P. Doyle, L. Beuchat and T. Montville (Eds.), American Society of Microbiologists, Washington DC, p. 671.

Fugelsang, K.C. and Edwards, C.G. (2006). *Wine Microbiology Practical Applications and Procedures*, 2nd edn. Springer, New York.

González, A., Hierro, N., Poblet, M., Mas, A. and Guillamón, J.M. (2005). Application of molecular methods to demonstrate species and strain evolution of acetic acid bacteria population during wine production. *Int. J. Food Microbiol.* 102(3): 295.

Joshi, V.K. (1997). *Fruit Wines*, 2nd edn. Directorate of Extension Education. Dr. YS Parmar University of Horticulture and Forestry, Nauni, Solan (HP), p. 255.

Joshi, V.K., Attri, B.L., Gupta, J.K. and Chopra, S.K. (1990). Comparative fermentation behaviour, physico- chemical characteristics of fruit honey wines. *Indian J. Hort.* 47(1): 49–54.

Joshi, V.K., Attri, D., Singh, T.K. and Abrol, G. (2011). Fruit wines: Production technology. In: *Handbook of Enology*, Vol. III, V.K. Joshi (Ed.), Asia Tech Publishers, Inc., New Delhi, pp. 1177–1221.

Joshi, V.K. and Bhardwaj, J.C. (2011). Effect of different cultivars yeasts (free and immobilized cultures) of *S. Cerevisiae and Schizosaccharomyces pombe on* Physico-Chemical and Sensory Quality of Plum Based Wine for Sparkling wine Production. *Int. J. Food Ferment. Technol.* 1(1): 69–74.

Joshi, V.K. and Sandhu, D.K. (2000). Quality evaluation of naturally fermented alcoholic beverages, microbiological examination of source of fermentation and ethanolic productivity of the Isolates. *Acta Aliment.* 29(4): 323–334.

Joshi, V.K., Sandhu, D.K. and Thakur, N.S. (1999). Fruit based alcoholic beverages. In: *Biotechnology: Food Fermentation*, Vol. II, V.K. Joshi and Ashok Pandey (Eds.), Educational Publishers and Distributors, New Delhi, pp. 647–744.

Joshi, V.K., Sandhu, D.K., Thakur, N.S. and Walia, R.K. (2002). Effect of different sources of fermentation on flavour profile of apple wine by descriptive analysis technique. *Acta Aliment.* 31(3): 211–225

Joshi, V.K., Sharma, R. and Abrol, G. (2011). Stone fruit: Wine and brandy. In: *Handbook of Food and Beverage Fermentation Technology*, Y.H. Hui and E.O. Evranuz (Eds.), CRC Press, Boca Raton, pp. 273–304.

Joshi, V.K., Sharma, P.C. and Attri, B.L. (1991). A note on the deacidification activity of *Schizosaccharomyces pombe* in plum musts of variable composition. *J. Appl. Bacteriol.* 70: 386–390.

Joshi, V.K., Sharma, S. and Sharma, A. (2016). Wines – White, red, sparkling, fortified, and Cider (Ch 12). In: *Current Developments in Biotechnology and Bioengineering (Elsevier Book Series)*, Ashok Pandey, Du, Guocheng, Maria Angeles Sanroman, Carlos Ricardo Soccol and Claude-Gilles Dussap (Eds.), Elsevier, London UK, pp. 353–407.

König, H. and Claus, H. (2018). A future place for *Saccharomyces* mixtures and hybrids in wine making. *Fermentation* 4(3): 67. doi:10.3390/fermentation4030067.

Kosseva, M.R., Joshi, V.K. and Panesar, P.S., (Eds.), (2017). *Science and Technology of Fruit Wine Production.* Academic Press is an imprint of Elsevier, London, UK, p. 705.

Kuchen, B., Maturano, Y.P, Victoria, M., María Combina, M., Eugenia, T.M. and Vazquez, F. (2011). Selection of native non-*Saccharomyces* yeasts with biocontrol activity against spoilage yeasts in order to produce healthy regional wines. *Fermentation* 5: 60. doi:10.3390/ fermentation5030060.

Liger-Belair, Gérard (2016). Wines: Champagne and sparkling wines – Production and effervescence. In: *Encyclopedia of Food and Health*, B. Caballero, P.M. Finglas and F. Toldrá (Eds.), Academic Press, London, pp. 526–533.

Loira, I., Morata, A., Palomero, F., González, C. and Suárez-Lepe, J.A. (2018). *Schizosaccharomyces pombe*: A promising biotechnology for modulating wine composition. *Fermentation* 4(3): 70. doi:10.3390/fermentation4030070.

Lonvaud-Funel, A. and Joyeux, A. (1993). Antagonism between lactic acid bacteria of wines: Inhibition of *Leuconostoc oenos* by *Lactobacillus plantarum* and *Pediococcus pentosaceus. Food Microbiol.* 10(5): 411.

Loureiro, V. (2000). Spoilage yeasts in foods and beverages: Characterisation andecology for improved diagnosis and control. *Food Res. Int.* 33(3–4): 247.

Loureiro, V. and Querol, A. (1999). The prevalence and control of spoilage yeasts in foods and beverages. *Trends Food Sci. Tech.* 10(11): 356.

Lu, Y., Huang, D., Lee, P.R. and Liu, S.Q. (2016). Assessment of volatile and non-volatile compounds in durian wines fermented with four commercial non-*Saccharomyces* yeasts. *J. Sci. Food Agr.* 96(5): 1511.

Martín, V. (2016). *Hanseniaspora vineae*: Caracterización y su uso en la vinificación. Unpublished Ph. D. thesis in chemistry. Universidad de la Republica, Uruguay.

Matei, F. and Kosseva, M.R. (2016). Microbiology of fruit wine production chapter 2. In: M.R. Kosseva, V.K. Joshi, and P.S. Panesar (2017) (Eds.), *Science and Technology of Fruit Wine Production*, Academic Press is an imprint of Elsevier, London, UK, pp. 73–99.

Medina, K., Boido, E., Dellacassa, E. and Carrau, F. (2005). Yeast interactions with anthocyanins during red wine fermentation. *Am. J. Enol. Vitic.* 56(2): 104.

Medina, K., Boido, E., Fariña, L., Dellacassa, E. and Carrau, F. (2016). Non-*Saccharomyces* and *Saccharomyces* strains co-fermentation increases acetaldehyde accumulation: Effect on anthocyanin-derived pigments in Tannat red wines. *Yeast* 33(7): 339.

Mestre, M.V., Maturano, Y., Combina, M., Candelaria, G., Mercado, L., Toro, M.E., Carrau, F., Vazquez, F. and Dellacassa, E. (2019). Impact on sensory and aromatic profile of low ethanol Malbec wines fermented by sequential culture of

Hanseniaspora uvarum and *Saccharomyces cerevisiae* Native Yeasts. *Fermentation* 5:65. doi:10.3390/ fermentation 5030065.

Millet, V. and Lonvaud-Funel, A. (2000). The viable but non-culturable state of wine micro-organisms during storage. *Lett. Appl. Microbiol.* 30(2): 136.

Mills, D.A., Phister, T., Neeley, E. and Johannsen, E.L. (2008). Wine Fermentation. In: *Molecular Techniques in the Microbial Ecology of Fermented Foods*, Vol. 6, Cocolin and D. Ercolini (Eds.), Springer, New York, pp. 162–192.

Monaga, S.M, Gómez-Cordovés, C. and Bartolomé, B. (2007). Evaluation of different *Saccharomyces cerevisiae* strains for red winemaking. Influence on the anthocyanin, pyranoanthocyanin and non-anthocyanin phenolic content and colour characteristics of wines. *Food Chem.* 104(2): 814.

Morata, A., González, C. and Suárez-Lepe, J.A. (2007). Formation of vinylphenolic pyranoanthocyanins by selected yeasts fermenting red grape musts supplemented with hydroxycinnamic acids. *Int. J. Food Microb.* 116(1): 144.

Morata, A., Escott, C., Bañuelos, M.A., Loira, I., Del Fresno, J.M., González, C. and Suárez-Lepe, J. A. (2019). Contribution of non-*Saccharomyces* yeasts to wine freshness. A review. *Biomolecules*, 10(1):34.

Morata, A., Gómez-Cordoves, M.C., Calderon, F. and Suárez, J.A. (2006). Effects of pH, temperature and SO_2 on the formation of pyranoanthocyanins during red wine fermentation with two species of *Saccharomyces. Int. J. Food Microb.* 106(2): 123.

Morata, A., Gómez-Cordovés, M.C., Colomo, B. and Suárez-Lepe, J.A. (2003). Pyruvic acid and acetaldehyde production by different strains of *Saccharomyces cerevisiae*: Relationship with vitisin A and B formation in red wines. *J. Agric. Food Chem.* 51(25): 7402.

Nadai, C., Vendramini, C., Carlot, M., Andrighetto, C., Giacomini, A. and Corich, V. (2019). Dynamics of *Saccharomyces cerevisiae* strains isolated from vine bark in Vineyard: Influence of plant age and strain presence during grape must spontaneous fermentations. *Fermentation* 5(3): 62. doi:10.3390/ fermentation 5030062.

Nigam, P. (2011). Microbiology and wine making. In: *Handbook of Enology*, Vol. III, V.K. Joshi (Ed.), Asia Tech Publishers, New Delhi, pp. 383–405.

Nigam, P. (2014). Wines: Production of special wines. In: *Encyclopedia of Food Microbiology*, C.A. Batt and M.L. Tortorello (Eds.), Elsevier Ltd, Academic Press, London, pp. 793–799.

Ough, C.S. and Amerine, M.A. (1966). *Effects of Temperature on Wine Making. California Agricultural Experiment Station Bulletin 827.* University of California, Davis, CA.

Padilla, B., Gil, J.V. and Manzanares, P. (2016). Past and future of non-*Saccharomyces* yeasts: From spoilage microorganisms to biotechnological tools for improving wine aroma complexity. *Front. Microbiol.* 7: 411.

Pinu, F.R . (2018). Grape and wine metabolomics to develop new insights using untargeted and targeted approaches. *Fermentation* 4(4): 92. doi:10.3390/fermentation4040092.

Pozo-Bayón, P. and Moreno-Arribas, M.V. (2011). Sherry wines. In: *Speciality Wines, Advances in Food and Nutrition Research*, R.S. Jackson (Ed.), Academic Press, London, pp. 17–40.

Pretorius, I.S. (2000). Tailoring wine yeast for the new millennium: novel approaches to the ancient art of winemaking. *Yeast* 16(8): 675.

Pretorius, I.S. (2016). Conducting wine symphonics with the aid of yeast genomics. *Beverages* 2:36. doi:10.3390/ beverages2040036.

Querol, A., Barrio, E. and Ramón, D. (1994). Population dynamics of natural *Saccharomyces* strains during wine fermentation. *Int. J. Food Microbiol.*, 21(4):315.

Querol, A. and Ramón, D. (1996). The application of molecular techniques in wine microbiology. *Trends Food Sci. Technol.*, 7(3):73.

Ramalhosa, E., Gomes, T., Pereira, A.P., Dias, T. and Estevinho, L.M. (2011). Mead Production: Tradition Versus Modernity. In: *Advances in Food and Nutritional Research*, R.S. Jackson (Ed.), Elsevier, Inc., London, U.K. pp. 1–19.

Raspor, P., Cus, F., Povhe, J.K., Zagorc, T., Cadez, N. and Nemanic, J. (2002). Yeast population dynamics in spontaneous and inoculated alcoholic fermentations of Zametovka must. *Food Technol. Biotechnol.*, 40(2):95.

Remize, F. and Montet, D. (2019). Safety and microbiological quality. *Fermentation*, 5(2):50. doi:10.3390/fermentation5020050.

Renouf, V., Strehaiano, P. and. Lonvaud-Funel, A. (2007). Yeast and bacteria analysis of grape, wine and cellar equipments by PCR-DGGE. *J. Int. Vigne Vin*, 41(1):51–61.

Ribéreau-Gayon, P., Dubourdieu, D., Donéche, B. and Lonvaud, A. (2000). *Handbook of Enology. Volume 1. The Microbiology of Wine and Vinifications.* John Wiley & Sons, New York.

Romano, P. (1997). Metabolic characteristics of wine strains during spontaneous and inoculated fermentation. *Food Technol. Biotechnol.*, 35(4):255.

Romano, P. and Suzzi, G. (1993). Sulphur dioxide and wine microorganisms. In: *Wine Microbiology and Biotechnology.* Chapter 13, G.H. Fleet (Ed.), Harwood Academic Publishers, Chur, Switzerland, pp. 373–393.

Santiago, B., Palomero, F., Morata, A., Fernando, C. and Suárez-Lepe, J.A. (2012). New applications for *Schizosaccharomyces pombe* in the alcoholic fermentation of red wines. *Int. J. Food Sci. Tech.*, 47(10):2101.

Sharma, S.K. and Joshi, V.K. (1996). Optimization of some factors for secondary bottle fermentation for production of sparkling plum (*Prunus salicina*) wine. *Indian J. Exp. Biol.*, 34(3): 235–238.

Soni, S.K. and Sandhu, D.K. (1999). Microbiology of food fermentation. In: *Biotechnology: Food Fermentation*, V.K. Joshi and Pandey, A. (Eds.), Educational Publisher and Distributors, New Delhi.

Soni, S.K., Sharma, S.C. and Soni, R. (2011b). Yeast genetics and genetic engineering in wine making. In: *Handbook of Enology*, Vol. III, V.K. Joshi, (Ed.), Asia Tech Publishers, New Delhi, pp. 441–501.

Splittstoesser, D.F. and Churney, J.J. (1992). The incidence of sorbic acid-resistant gluconobacters and yeasts on grapes grown in New York State. *Am. J. Enol. Vitic.*, 43(3):290.

Suárez Valles, B., Pando Bedriñana, R., Fernández Tascón, N., Querol Simón, A. and Rodríguez Madrera, R. (2007). Yeast species associated with the spontaneous fermentation of cider. *Food Microbiol.*, 24(1):25.

Swiegers, J.H., Bartowsky, E.J., Henschke, P.A. and Pretorius, I.S. (2005). Yeast and bacterial modulation of wine aroma and flavour. *Aust. J. Grape Wine R.*, 11(2):139.

Tubia, I., Prasad, K., Pérez-Lorenzo, Eva, Abadín, C., Zumárraga, M., Iñigo, O., Barbero, F., Jacobo, P. and Arana, S. (2018). Beverage spoilage yeast detection methods and control technologies: A review of *Brettanomyces. Int. J. Food Microbiol.*, 283:65–76.

Vilela, A. (2017). Biological Demalication and Deacetification of musts and wines: Can wine yeasts make the wine taste better? *Fermentation*, 3(51). doi:10.3390/fermentation3040051.

Vyas, K.K. and Joshi, V.K. (1988). Deacidification activity of *Schizosaccharomyces pombe* in Plum musts. *J. Food Sci Technol.*, 25(5): 306–307. , M.A., , H.W., , R.E.,

15 Wine Production
Role of Non-Saccharomyces Yeast

V.K. Joshi, Cintia Lacerda Ramos and Rosane Freitas Schwan

CONTENTS

15.1 INTRODUCTION

The fermentation of grape juice into wine is a complex microbiological process in which yeast plays an important role. Grape musts naturally contain a host of microorganisms including mixture of yeast species, and wine fermentation is not a 'single-species' fermentation. Historically, studies on enological microbiology have centered on yeasts belonging to the genus *Saccharomyces*, which are responsible for alcoholic fermentation. The dominance of *S. cerevisiae* (inoculated or indigenous) in fermentation is expected and desired. However, different yeast, especially non-*Saccharomyces* yeasts, present in the initial stages of the fermentation process may have an influence on initial sensory properties of wine. The indigenous non-*Saccharomyces* yeasts, already present in the must and often in greater numbers than *S. cerevisiae*, are adapted to the specific environment and are in an active growth state, which gives them a competitive edge. The main genera of these are *Kloeckera*, *Cryptococcus*, *Torulospora*, *Hanseniaspora*, *Candida*, *Pichia*, *Hansenula*, *Zygosaccharomyces*, *Metschnikowia*, *Debaromyces*, *Issatchenkia* and *Rhodotorula*. All the species of these genera grow in succession throughout the fermentation process. Non-*Saccharomyces* yeast are an ecologically and biochemically diverse group, which can alter fermentation dynamics, composition and flavor of wine. The non-*Saccharomyces* yeasts usually dominate in mature grape population. The persistence of non-*Saccharomyces* species during fermentation depends upon many factors, *viz.* temperature of fermentation, nutrient availability, inoculum, population of *Saccharomyces*, uses and levels of SO_2, number and kinds of organism present on the grape and, finally, vinification technology. In general, the natural fermentation of grape must is usually started by low-alcohol-tolerant apiculate yeasts (non-*Saccharomyces*) that predominate the first stages of fermentation. Then, after 3–4 days, they are replaced by elliptical yeasts (*Saccharomyces cerevisiae*) that continue and finish the fermentation process. The wine quality is a result of the grape and yeasts' enzymatic activities. Various enzymes produced by non-*Saccharomyces* yeasts can influence the process of winemaking as well as wine quality.

In this chapter, the presence and importance of non-*Saccharomyces* yeasts in grape and wine has been discussed along with their characteristation. The non-*Saccharomyces* yeasts that are responsible for improving the aroma and quality of wines are also described here.

15.2 WINE YEAST, ECOLOGY AND FERMENTATION

15.2.1 Wine Yeasts

Wines yeasts are group of fungi that are predominantly unicellular organisms which are widely distributed in nature. These yeasts are fermentative organisms that can metabolize a variety of sugars. Many types of yeasts naturally occur on the surface of grapes. For a long time, the methods used to identify yeasts have been especially based on morphology, biochemical characteristics and sexual reproduction. However, with the development of

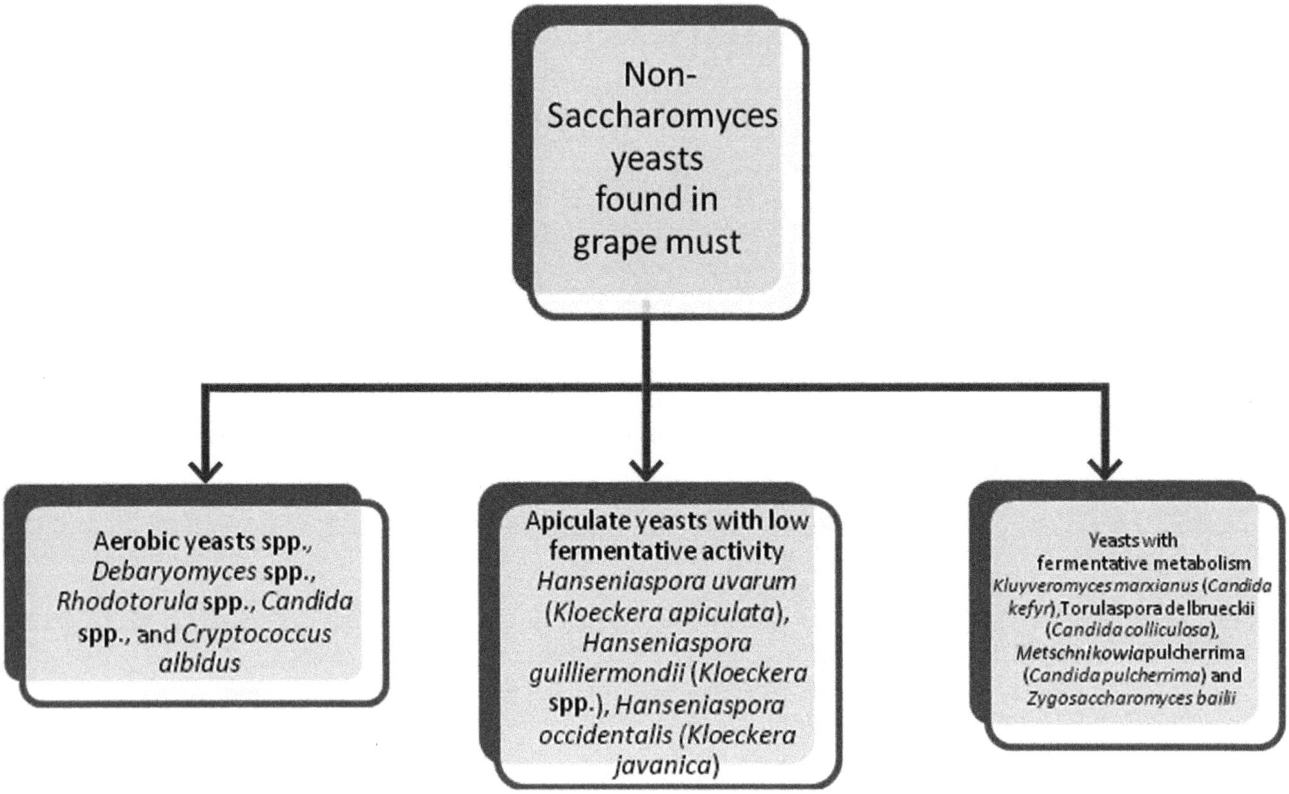

FIGURE 15.1 Hierarchical diagram of the group of non-*Saccharomyces* yeasts found on grapes

biological molecular techniques, methods have been developed that allow identification of micro-organisms with great accuracy. The commonly found yeasts on grapes and in wines belong to following the genera: *Saccharomyces*, *Kloeckera*, *Torulaspora*, *Kluyveromyces*, *Hanseniaspora*, *Candida*, *Hansenula*, *Pichia*, *Zygosaccharomyces* and *Brettanomyces*. The microbial diversity of grapes is affected by a number of factors including vineyard altitude and aspect, climatic conditions, grape variety, viticulture practices, developmental stage of grapes, health of grapes and winery waste-disposal practices.

15.2.2 Non-*Saccharomyces* Yeasts and Their Ecology

The *Saccharomyces* is the predominant yeast, and the yeasts other than these are known as the non-*Saccharomyces* yeasts, though the term non-*Saccharomyces* yeast is a loose term used among wine microbiologists for different yeast species. These yeasts are either ascomycetous or basidiomycetous and have vegetative states which predominantly reproduce by budding or fission and do not form their sexual states within or on a fruiting body. These yeasts are found throughout the winemaking environment (habitat); grape berry surfaces, cellar equipment surfaces and the grape can be considered specialized niches where wine-related yeasts form communities. Due to constant contact with grape must, cellar surfaces can harbor yeasts, depending on the cellar-hygiene procedures practiced. The grape must is a rich nutritive environment, but low pH, high osmotic pressure and the presence of SO_2 detract from this otherwise ideal yeast niche. Many external factors affect populations both on grapes and in must. During crushing, the non-*Saccharomyces* yeasts on the grapes, on cellar equipment and in the cellar environment (air- and insect-borne) are carried over to the must. However, cellar surfaces play a smaller role than grapes as a source of non-*Saccharomyces* yeasts, as *S. cerevisiae* is the predominant yeast inhabiting such surfaces, and hygienic procedures used in most modern cellars minimize the contamination of must by resident cellar flora. Non-*Saccharomyces* yeasts found in grape must and during fermentation can be divided into three groups as illustrated in Figure 15.1.

15.2.3 Behaviour During Wine Fermentation

The behavior of non-*Saccharomyces* during fermentation can be summarized as follows:

- During fermentation, especially in spontaneous fermentations, which lack the initial high-density inocula of *S. cerevisiae*, there is a sequential succession of yeasts.
- It is commonly believed that in a natural or spontaneous fermentation, non-*Saccharomyces* yeasts participate in early stages of fermentation.
- Initially, species of *Hanseniaspora (Kloeckera)*, *Rhodotorula*, *Pichia*, *Candida*, *Metschnikowia* and *Cryptococcus* are found at low levels in fresh grape must.
- Of these, *H. uvarum* is usually present in the highest numbers, followed by various *Candida* spp.
- Despite the sustained presence of certain non-*Saccharomyces* yeasts, the majority disappear during

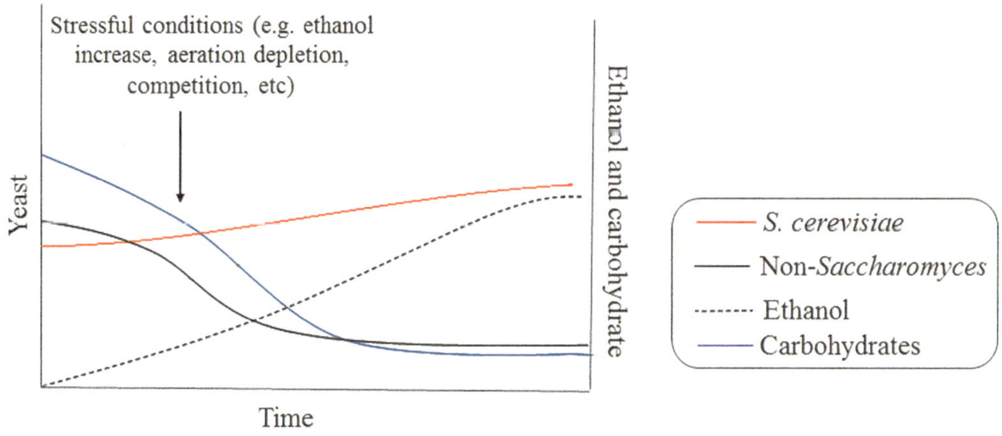

FIGURE 15.2 Scheme of non-*Saccharomyces* and *Saccharomyces* evolution, carbohydrate consumption and ethanol production during wine fermentation

the early stages of a vigorous fermentation, possibly due to their slow growth and inhibition by the combined effects of SO_2, low pH, high ethanol and oxygen deficiency, in line with their oxidative or weak fermentative metabolism.

- Besides this, nutrient limitation and size or dominance of *S. cerevisiae* inoculum can also have a suppressive effect, sometimes separate from temperature or ethanol concentration.
- *T. delbrueckii* and *Kluyveromyces thermotolerans* (now classified as *Lachancea thermotolerans*) have been found to be less tolerant to low oxygen levels, and this, rather than ethanol toxicity, affects their growth and leads to their death during fermentation.
- The non-*Saccharomyces* spp. that do survive and are present until the end of fermentation may also have a higher tolerance to ethanol, which would account for their sustained presence. Other species reported throughout fermentation are *Saccharomyces acidifaciens* (now classified as *Z. bailii*) and *Pichia* sp. of different growth kinetics. The standard practice of di-ammonium hydrogen phosphate (DAP) addition to grape must as a nitrogen source results in higher pH values.
- In the initial phases of fermentation, non-*Saccharomyces* yeast can deplete essential nutrients that, combined with toxic metabolites formed, can inhibit the growth of lactic acid bacteria essential for the secondary malolactic fermentation in wine.

The high ethanol in addition to other stressful condition (*e.g.* oxygen depletion, competition for nutrients and others) limits the growth of non-*Saccharomyces* and explains the domination of native *Saccharomyces* yeasts which then, conduct fermentation.

Figure 15.2 shows a general representation of the evolution of yeasts, carbohydrate consumption and ethanol production during spontaneous wine fermentation.

Defenders of natural fermentation believes that a mixed culture fermentation produces a more complex flavor in wine. On the other hand, winemakers find natural fermentations unpredictable, because these fermentations have been shown to develop off-flavors/odors. For these and other reasons, winemakers prefer to use yeast strains in the form of a starter culture or as an active dry wine yeast in order to conduct alcoholic fermentation.

Inoculation of a must by a pure yeast strain assumes that fermentation will be carried out by inoculated yeasts. However, research has revealed that in many cases, non-*Saccharomyces* and native *Saccharomyces* strains also contribute in various degrees to the fermentation, even though a pure strain is added to the must. The degree to which native strains contribute to fermentation varies greatly based on certain vinification variables such as amount of inoculum and fermentation temperature. However, to what extent these factors are effective in reducing the contribution of indigenous yeasts in fermentation is not clear. It is reasonable to assume that, in many cases, native yeast strains participate and contribute to the alcoholic fermentation, even if the must has been inoculated with a pure starter culture.

15.3 INFLUENCE OF NON-*SACCHAROMYCES* YEASTS OVER UNIQUE ENOLOGICAL CHARACTERISTICS

In the past, non-*Saccharomyces* yeasts were considered as undesirable, as most of them were involved in spoilage of wine, and *S. cerevisiae* in its pure form was preferred to conduct the fermentation. With advancements in the detection and cultural methods of yeasts and developments in wine microbiology, the role played by the non-*Saccharomyces* yeasts is now being elaborated. These yeasts have been found to improve the flavor of wine and make it more complex. They have the ability to create several metabolites, having significance in oenology. The metabolites include glycerol, esters, polysaccharides, succinic acid, terpenes and several flavoring compounds which improve the flavor/aroma of the treated wines. Some of these have been illustrated in Figure 15.3. As a result, the sensory quality of the treated wines is improved considerably. The yeasts involved in the improvement of sensory qualities have been listed in Table 15.1.

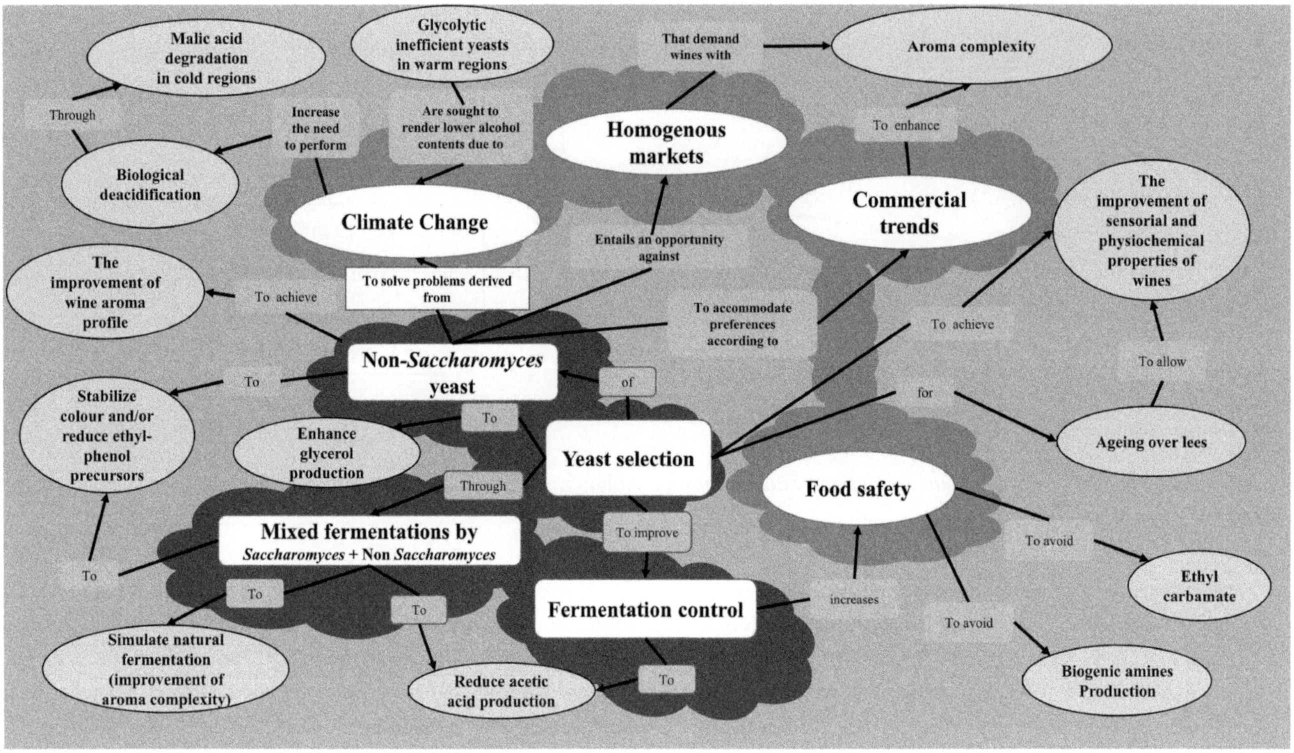

FIGURE 15.3 Role of non-*Saccharomyces* yeasts in winemaking and its evaluation. Light grey area encompasses novel problems in viticulture, oenology and wine marketing; darker grey area encompasses new tools and new ways of solving different issues. *Source*: Based on the work of Suárez-Lepe (2012); Ciani *et al.* (2010): Abalos *et al.* (2011); Pretorius and Bauer (2002); Clemente-Jiménez *et al.* (2005); Benito *et al.* (2009); Toro and Vázquez (2002); Palomero *et al.* (2009); Moreira *et al.* (2008); Moreira *et al.* (2008); De Fátima *et al.* (2007); Lodder J (1970); Kapsopoulou *et al.* (2007)

TABLE 15.1
Non-*Saccharomyces* Yeasts for Enhancing the Sensorial Profile of Wines

Yeast	Improvements	Drawbacks
Williopsis saturnus (Hansenula saturnus)	Acetic, ester producer (Isoamyl acetate, 2-phenylethyl acetate).	Low fermentative power.
Torulaspora delbrueckii	Reduction of volatile acidity and acetaldehyde, low production of volatile phenols.	Medium fermentative power.
Schizosaccharomyces pombe	Enhancement of aromatic profile, maloalcoholic fermentation. Improvement of acidity. Stability.	Production of hydrogen sulphide and mercaptans.
Pichia guillermondii	HCDC activity, pyranoanthocyanin formation.	Low fermentative power, high production of ethyl acetate and acetic acid in pure culture.
Pichia anomala	Isoamyl acetate production.	Low fermentative power, high production of ethyl acetate and acetic acid in pure culture.
Hanseniaspora osmophila	2-phenylethyl acetate production.	Low fermentative power.
Hanseniaspora guillermondii	2-phenylethyl acetate production.	Low fermentative power.
Candida stellata	Enhanced glycerol production.	Low fermentative power.

Note: Non-*Saccharomyces* yeasts with low or medium fermentative power are used in mixed or sequential cultures with *S. cerevisiae*. Normally, these techniques also control the production of compounds such as acetic acid and ethyl acetate.

Source: Lee, Ong, Yu, Curran, and Liu (2010); Erten and Tanguler (2010); Renault *et al.* (2009); Lallemand (2010); Volschenk *et al.* (2003); Benito *et al.* (2011); Rojas *et al.* (2001); Soden, Francis, Oakey, and Henschke (2000)

Some of the non-*Saccharomyces* yeasts can degrade malic acid, leading to the reduction of acidity and thus, making the wine palatable. Non-*Saccharomyces* yeasts also have antagonistic effect and therefore, can be used in the biological control. Since these yeasts are a low ethanol producer as well as tolerant, they are used mostly as a sequential culture or are co-cultured for getting the desired results. However, the contribution of non-*Saccharomyces* yeasts to wine flavor is species and strain specific, and related to their particular metabolic features, including the ability to secrete enzymes and metabolites with impact on the primary and secondary aroma of wines, glycerol production, release of mannoproteins, low volatile acidity or contributions to wine color stability. At the same time, an uncontrolled fermentation mainly driven by non-*Saccharomyces* yeasts and bacteria would normally result in must and wine spoilage. Thus, there is a need for extreme care in conducting such fermentations. Some of the positive aspects of non-*Saccharomyces* yeasts have been described in the subsequent sections of the chapter.

15.3.1 Enzymatic Activity of Non-*Saccharomyces* Yeasts

Yeast have enzymes that may be responsible for giving the wine unique characteristics from the region of its production. Different enzymes, such as β-glucosidase, polygalacturonase, amylase, hemicellulase, lipase, protease, pectinase and esterase are synthetized by different species of yeasts. The enzymatic activity is not characteristic of a particular genus or species but depends on the yeast strain analyzed. It is, therefore, worthwhile to characterize every isolate so as to determine its potential as an enzyme producer. Furthermore, it is known that grape and *Saccharomyces* enzymes are insufficient to carry on the transformation of odorless precursors from the grape and the synthesis of all the other aroma compounds. Thus, the non-*Saccharomyces* strains have been often mentioned as enzyme producers which effect the wine quality. Table 15.2 shows some non-*Saccharomyces* genera and the enzymes produced by them.

Protease: The action of non-*Saccharomyces* strain proteases on the hydrolysis of wine proteins has been investigated by several authors. Protease activity has been observed in strains of *Candida pulcherrima*, *Kloeckera apiculata* and *Pichia anomala*. Although grape proteins influence the clarification and stabilization of must and wine, the yeast proteases hydrolyze the peptide linkages between amino-acid units of protein, improving the clarification process. These enzymes also play a major role during the autolysis process in wines as is practiced in sparkling wine production production (See chapter 35). However, due to the particular conditions found in wine, only few proteases are active. Another important aspect of yeast proteolytic activity is its potential for use in protein haze reduction. The proteic instability induced by high alcohol concentrations and low pH could be the source of haze formation in wines. This instability is due to aggregation of macromolecules, the mechanism of which is unknown. However, it has been determined that this instability is due to group of proteins with a molecular weight of 16 and 25 KD. Some non-*Saccharomyces* yeasts secrete proteases into beer, which reduces the potential for haze. Studies have been carried out with the purpose of finding proteases that could reduce proteic instability in wines. *Kloeckera apiculata*, *M. pulcherrima*, *P. anomala*, *C. stellata* and *D. hansenii* isolates have been screened for this activity. Proteolytic activity was best observed at neutral and basic pH, but at acid pH it was absent.

Pectinase: With regard to pectinase action in the winemaking process, some applications can be mentioned here, such as mash treatment for juice extraction, juice clarification, wine filtration and also color extraction. The use of pectinolytic enzymes for maceration may also increase the terpenol content of juice, improving the aroma of the wine. The pectinase enzymes include pectin lyase, pectin methylesterase and polygalacturonase, and are responsible for breaking down the pectin present in the cell walls of grapes. Although pectin lyase and polygalacturonase activities increase during grape ripening and are produced by non-*Saccharomyces* yeasts present in must, the addition of a fungal pectinase preparation is a common industrial practice. Non-*Saccharomyces* yeast pectinolytic activity has been found in various species of *Candida*, *Cryptococcus*, *Kluyveromyces* and *Rhodotorula*.

Glycosidase: The glycosidase enzymes such as β-glucosidase, β-xylosidase, β-apiosidase, L-rhamnosidase and L-arabino-furanosidase have been described as being involved in the flavor-releasing process. Terpenes are considered important compounds in the varietal aroma of the wine. However, in grapes, there is a considerable portion in bound forms, particularly glycosides, which might not contribute to the aroma of wine. In general, the glycosylated terpenes are

TABLE 15.2
Enzymatic Activities of Non-*Saccharomyces* Wine Yeasts

Enzymatic activity	Genera
Protease	*Candida, Kloeckera, Pichia, Debaryomyces, Hanseniaspora, Metschnikowia*
β-glucosidase	*Candida, Debaryomyces, Hanseniaspora, Hansenula, Kloeckera, Kluyveromyces, Metschnikowia, Pichia, Saccharomycodes, Schizosaccharomyces, Zygosaccharomyces*
Esterase	*Brettanomyces, Debaryomyces, Rhodotorula*
Pectinase	*Candida, Cryptococcus, Kluyveromyces, Rhodotorula*
Lipase	*Candida, Issatchenkia, Torulaspora, Metschnikowia*

found in β-D-Glucopyranosides. Although many studies have been carried out to determine the presence of β-D-glycosidase in *Saccharomyces* yeasts, it is quite rare and generally, at lower levels than the non-*Saccharomyces* yeasts investigated. The non-*Saccharomyces C. pulcherrima, K. apiculata* and *P. anomala* have shown a significantly higher activity of β-glucosidase in the absence of glucose than *S. cerevisiae* (on a dry mass basis). This indicates that non-*Saccharomyces* yeasts may potentially have a greater role in flavor development through β-glucosidase activity than *S. cerevisiae.* The presence of glucose in the growth medium has an important effect on production of β-glucosidase by wine yeast. *Hansenula* sp. has been found in grape must, which is responsible for the production of β-glucosidase. Studies have shown that other non-*Saccharomyces* genera, *Debaromyces, Kloeckera, Kluyveromyces, Metschnikowia, Dekkera, Hansenula, Hanseniaspora* and *Zygosacchromyces*, are also important for this activity. Table 15.3 shows the β-Glucosidase activity of some wine yeasts strains.

The activity of glycosidases including β-D-glycosidase, L-rhamnosidase and β-D-xylosidase, has been described as being involved in flavor-releasing processes. However, many studies have only focused on β-D-glycosidase activities because of their widespread occurrence in plants, fungi and yeast. An extensive review of 317 strains from 20 wine yeast species indicated that yeasts belonging to *Schizosaccharomyces, Candida, Debaromyces, Hanseniaspora, Pichia, Kloeckera, Kluyveromyces, Metschnikowia, Saccharomyces* and *Zygosaccharomyces* possessed β-glycosidase activity.

Cellulase: It has been proposed that the cell wall degradation of grapes permits many aromatic or potential aromatic compounds to form as part of the must, resulting in a better aromatic profile and better sensory properties. This enzymatic activity has been described in numerous non-*Saccharomyces* isolates of *Candida. K. apiculata K. termotolerans, Z. cidri, Z. fermentans, M. pulcherrima, Candida* sp., *Dekkera* spp., *I. delbrueckii* and *Z. microellipsoideus* have also been found to produce cellulase activity.

Amylase: Amylase activity is of little importance in winemaking but could be of significance for large scale production of wine yeasts from the starchy substrates of waste products. The production of amylases by yeast is of interest in the bioconversion of starchy substrates to biomass or ethanol. *Saccharomyces cerevisiae* lack starch decomposing activity and lack L-amylase and isoamylase. However, amylase activity has been studied in non-*Saccharomyces* yeast strain such as *C. stellata, M. pulcherrima*, and *K. apiculata*. None of the wine yeasts were found to be amylolytic. Amylase secretion by yeasts is highly dependent on medium composition, with starch and dextrin being the best carbon source for inducing this activity. Some amylases are produced constitutively and require the presence of glucose as well as starch substrates. Secretion of amylase influences the pH during the fermentation process.

Other yeast enzymes such as esterases and lipases are also involved in the formation of aroma compounds. Yeast esterase-producers include *Brettanomyces* and *Rhodotorula mucilaginosa.* Lipases degrade the lipids originating from the grape or the autolytic reactions of yeasts, releasing free fatty acids into juice or wine, which may affect wine quality. Several wine yeasts, *C. stellata, C. pulcherrima, C. kruesi* and *C. colliculosa,* have shown the potential to produce extracellular lipases.

TABLE 15.3
β-Glucosidase Activity of Wine Yeasts

	β-Glucosidase activity⁺ glucose in medium	
Yeast strain	**0 g/l**	**5g/l**
C. pulcherrima 1000	3024	25
K. apiculata 610	3467	66
P. anomala 703300	3846	253
P. anomala 209	4376	269
S. cerevisiae V-1118	1590	22
S. cerevisiae HB350	1902	16

⁺Activity is expressed as nmole of hydrolysed β-glucoside analogue per ml of assay medium per gram dry mass of cells.

15.3.2 Non-*Saccharomyces* Strains for the Production of Esters

The use of selected non-*Saccharomyces* strains with enhanced ester production offers a potential to obtain aromatic profiles in fermented beverages, especially in wines. This strategy has been explored by many researchers and is used by the wine producers over the world. The yeast *Torulaspora delbrueckii* has been used in sequential fermentations or mixed cultures to reduce the volatile acidity of wine and enhance its aromatic profile. Other non-*Saccharomyces* yeasts, such as *Hanseniaspora* spp., *Pichia* spp., *Kloeckera* spp., *Metschnikowia* spp., *C. stellata, Zygosaccharomyces fermentati, Saccharomycodes ludwigii* and others, have been studied and are suggested to improve aromatic characteristics of wine, specially by the production of esters. In general, non-*Saccharomyces* yeasts are present at the beginning of spontaneous fermentations and are associated with a more complex aromatic profile of wines. However, non-*Saccharomyces* yeasts are unable to complete fermentation due to the increasing concentration of ethanol. For this reason, these yeasts are normally inoculated in mixed cultures or in sequential fermentations. A mixed fermentation with selected *S. cerevisiae* increases the formation of esters, with a positive sensorial impact, while at the same time the production of volatile acidity and ethyl acetate production is controlled. Some esters formed by yeasts during fermentation, especially those associated with floral and fruity aromas, are of great interest to winemakers. *Pichia anomala* is one of the stronger producers of isoamyl acetate (associated with banana notes). The volatile compound 2-phenylethyl acetate is associated with a floral aroma similar to sweet roses, and its production by yeasts has been explored for wine fermentation. Strains of *Hanseniaspora guillermondii, H. osmophila* and *H. vineae* have been selected as producers of high levels of this ester.

15.3.3 Ethanol Tolerance of Non-*Saccharomyces* Wine Yeast

The accumulation of ethanol in the microbial environment represents a form of chemical stress on the organisms living there, and several studies have shown that the plasma membrane is the prime target of ethanol action in such situations. Furthermore, yeasts' ethanol tolerance has been correlated to the capacity of a cell to modify its lipid composition in response to the disrupting action of ethanol. This is part of a regulatory system which ensures the adjustment of the physico-chemical properties of the membrane lipid matrix in a physiologically optimal range. The high ethanol tolerance of *S. cerevisiae* has received widespread attention, although the mechanism underlying it and the way membrane lipids modulate membrane fluidity are not yet fully understood.

However, it is well known that yeast supplemented with various sterols and unsaturated fatty acids (UFA) exhibits an increased ethanol tolerance as a result of the incorporation of some of these compounds, as has been demonstrated by several studies on *S. cerevisiae*. These substances can act as 'survival factors' when added to grape must. However, similar studies concerning non-*Saccharomyces* yeasts are very scarce, and this has renewed interest in the role of these organisms in the early stages of fermentation. The necessity to obtain more information concerning the physiological and metabolic features of these yeasts has also been highlighted.

Some of the most frequently employed methods to measure ethanol tolerance include yeast cell growth, cellular viability in the presence of ethanol, the maximal amount of ethanol produced during fermentation and the fermentation rate of glucose as measured by CO_2 production. A study showed that the kinetics of inactivation in the presence of high concentrations of ethanol (22.5% and 25% v/v) was used as an inference of the ethanol resistance status of five non-*Saccharomyces* strains and one strain of *S. cerevisiae.* Cultivation conditions prior to ethanol challenge were of particular interest, *e.g.* anaerobiosis, semi-aerobiosis, active-oxygenation, addition of ergosterol and Tween 80. Changes in certain fatty acids and sterols of cells were examined in order to derive a potential relationship between ethanol tolerance and lipid composition of different strains.

The role of non-*Saccharomyces* yeasts in wine must fermentation is usually limited by their inability to tolerate a fermenting must environment; this is most often considered to be due to their intolerance to ethanol. Studies have shown that by manipulating the cultivation conditions, the ethanol tolerance of some non-*Saccharomyces* strains can be increased. Thus, the capacity of these strains to participate in alcoholic fermentation as part of more complex starter culture systems, compared to *S. cerevisiae* alone, can be enhanced. Another important finding is that the high ethanol resistance exhibited by a *H. guilliermondii* strain is not in accordance with previously reported research, which has reported and concluded that apiculate yeasts (*Kloeckera* and *Hanseniaspora*) are low ethanol tolerant species. However, some quantitative studies on winemaking ecology have shown that non-*Saccharomyces* species survived at significant levels for longer periods during fermentation than previously thought. Cell lipid changes induced by certain growth factors play an important role in the adaptation of yeasts to ethanol-containing environments. However, a better understanding of the intrinsic resistance of yeast to ethanol is still required.

The oenological properties of non-*Saccharomyces* yeasts associated with winemaking have been tested. Five species were selected, *T. delbrueckii*, *C. stellata*, *S. ludwigii*, *H. uvarum* and *K. apiculata*, and the oenological properties of these selected yeasts were evaluated. Table 15.4 shows different fermentation by-products formed by the non-*Saccharomyces* yeast species tested.

Non-*Saccharomyces* yeast co-inoculated with the standard wine yeast at the start of fermentation leads to more efficient utilization of sugars. During fermentation, the glucophilic wine yeast *S. cerevisiae* can preferentially utilize glucose above fructose. Should the glucose/fructose ratio decrease to 0.1, it can result in a stuck fermentation with a higher residual fructose than glucose concentration. Inoculation at this stage with fructophilic non-*Saccharomyces* yeast can rectify the imbalance, whereas *S. cerevisiae* can start fermenting again. Changes in the glucose-fructose ratio during fermentation of Chenin blanc must by different combinations of active dried non-*Saccharomyces* and *S. cerevisiae* yeast species has been revealed.

15.3.4 Reduction in Ethanol Yield in Wines Using Non-*Saccharomyces* Yeasts

Global climate change, with warm climates and therefore, a lengthy maturation period in vineyards, can lead to grapes with high sugar concentrations and, in turn, wines with high concentrations of ethanol, *i.e.* above 15% v/v. In addition to negative consequences for consumer health, high alcohol content in wine may compromise wine quality. High alcohol content has a negative effect on the sensory attributes of wine since it may mask some flavor-related volatile compounds, increase the perception of hotness, body and viscosity of a wine and increase the perception of the bitterness of tannins. For these reasons, interest has substantially increased in the development of technologies for producing wines with reduced ethanol concentrations without affecting the quality of wines. There are different techniques that can be used to reduce ethanol content in wine which generally focus on vineyard management and winemaking practices, particularly in the de-alcoholization of wine.

Microbiological approaches for decreasing ethanol concentrations use strategies, including the use of genetically modified yeasts, for the production low-alcohol wine (see Chapters 19 and 36). Using evolution-based strategies together with breeding strategies, it has been shown that the evolved or hybrid strains produced an ethanol reduction of 0.6–1.3% (v/v) in comparison with indigenous parent strains. Another important, easy and cheap approach to reduce the production of ethanol might be the use of non-*Saccharomyces* wine yeasts. As already mentioned, the use of non-*Saccharomyces* yeasts in combination with *S. cerevisiae* may improve the quality and enhance the aroma complexity of wines. In

TABLE 15.4
Fermentation By-Products Formed by Non-*Saccharomyces* Yeasts Species Tested

Range of values	Acetic acid (g/l)	2,3 Butanediol (g/l)	Succinic acid (g/l)	Glycerol (g/l)	Acetaldehyde (mg/l)	Acetoin (mg/l)	Ethyl acetate (mg/l)
		***Saccharomyces cerevisiae*: average value of ethanol formed = 14.90% (v/v)**					
Min. value	0.53	0.42	0.21	6.53	25.9	26.3	18.4
Max. value	1.75	1.03	0.51	7.95	106.6	76.2	48.6
Average value	0.97	0.77	0.42	7.38	63.3	56.4	35.3
SE	0.12	0.07	0.03	0.16	9.7	5.19	3.3
		***Torulaspora delbrueckii*: average value of ethanol formed = 9.35% (v/v)**					
Min. value	0.02	0.21	0.33	4.35	6.02	0.0	18.9
Max. value	0.72	1.55	0.83	6.20	33.4	22.0	51.7
Average value	0.32	0.86	0.53	5.07	22.6	5.6	31.5
SE	0.07	0.15	0.05	0.19	2.66	2.9	3.19
		***Candida stellata*: average value of ethanol formed = 5.83% (v/v)**					
Min. value	0.24	0.39	0.68	9.06	34.9	34.9	3.4
Max. value	0.52	1.67	1.35	13.62	84.6	254.1	41.5
Average value	0.44	0.78	1.05	11.76	60.9	94.9	23.2
SE	0.03	0.13	0.10	0.47	5.7	21.2	4.7
		***Saccharomycodes ludwigii*: average value of ethanol formed = 12.63% (v/v)**					
Min. value	0.57	0.37	0.54	7.65	46.7	211.7	141.4
Max. value	1.10	1.48	1.40	11.53	124.3	478.3	540.8
Average value	0.89	0.64	0.98	9.91	87.6	310.2	289.0
SE	0.05	0.11	0.11	0.51	7.8	1.2	36.0
		***Hanseniaspora uvarum*: average value of ethanol formed = 6.07% (v/v)**					
Min. value	0.62	0.20	0.22	3.47	16.8	22.9	139.5
Max. value	0.98	1.29	0.29	5.22	67.9	408.7	453.4
Average value	0.73	0.80	0.26	4.26	32.1	144.9	311.4
SE	0.04	0.11	0.01	0.19	5.1	42.0	38.4
		***Kloeckera apiculata*: average value of ethanol formed = 3.63% (v/v)**					
Min. value	0.50	0.33	0.22	3.66	19.3	68.6	166.2
Max. value	1.51	3.80	0.40	4.31	63.5	225.0	763.0
Average value	0.81	1.10	0.31	3.98	39.6	138.1	421.2
SE	0.11	0.36	0.02	0.06	4.6	22.0	63.3

Source: Ciani and Maccarelli (1998)

addition, the co-inoculation of grape juice with selected cultures of non-*Saccharomyces* coupled with a *S. cerevisiae* starter strain could be an interesting way to reduce the ethanol content in wine. This fermentation strategy may affect ethanol yield, alcoholic fermentation efficiency, biomass and by-product formation with a diversion of carbon away from ethanol production. Furthermore, different respiration-fermentation regulatory mechanisms of some non-*Saccharomyces* yeasts, rather than *S. cerevisiae*, could be a way to reduce the ethanol production through partial and controlled aeration of the grape juice. Indeed, in this way, sugar is consumed *via* respiration rather than fermentation. Therefore, the use of non-*Saccharomyces* wine yeasts may indicate a definitive approach to limiting ethanol production. The non-*Saccharomyces* may be used in sequential fermentation with *S. cerevisiae*, as well as in a multi-starter culture. Sequential inoculation allows the exertion of the metabolism of the first inoculated yeast without the influence of the following yeast strain. In this way, the reduction of ethanol content could be obtained, depending on the metabolic characteristics of the non-*Saccharomyces* strain and the interval between its inoculation and *S. cerevisiae*. Table 15.5 shows some reports on the reduction of ethanol using sequential fermentations or mixed cultures with non-*Saccharomyces* species compared to *S. cerevisiae* pure fermentation. Ethanol reduction ranged from 0.64 to 1.60 (% v/v) with a percentage of reduction from 5 to 11.6% using strains belonging to *C. stellata*, *M. pulcherrima* and *Lachancea thermotolerans*.

15.3.5 Role of Yeast in Deacidification of Wine

In grapes grown in cool climates, the malic acid content is higher, which makes the wine highly acidic, hence unpalatable. Similarly, many stone fruits like plum and apricot have

TABLE 15.5
Reduction in Ethanol Production Observed in Sequential Fermentations or Multi-Starter Culture Using Non-*Saccharomyces* Yeast Compared to *S. cerevisiae* Pure Culture

Non-*Saccharomyces* yeast	Wine	Ethanol reduction (%)	Starter type	Reference
M. pulcherrima	White wine	6.0	Sequential	Contreras *et al.*, 2014
M. pulcherrima	Red wine	11.6	Sequential	Contreras *et al.*, 2014
C. stellata	Synthetic wine	10.4	Sequential	Ciani *et al.*, 1998
C. stellata	White wine	5.7	Sequential	Ferraro *et al.*, 2000
L. thermotolerans	Red wine	5.0	Sequential	Gobbi *et al.*, 2013
C. stellata	Synthetic wine	22.4	Sequential	Ciani and Ferraro, 1996
C. stellata	Synthetic wine	12.1	Mixed	Ciani and Ferraro, 1996
H. uvarum		2.2	Mixed	Ciani *et al.*, 2006
H. uvarum	Pinot Grigio	2.5	Sequential	Ciani *et al.*, 2006
C. membranifaciens	Chardonnay	19.2	Mixed	García *et al.*, 2010

higher acidity that makes the wine unacceptable. These fruits also contain malic acid. One non-*Saccharomyces* yeasts *Schizosaccharomyces pombe* has the ability to degrade the malic acid into ethanol and carbon dioxide. The yeast has successfully been employed to degrade the malic acid of wines produced from grape, apple and plum. The yeast no doubt reduced the acidity of wines, but controlling the activity of the yeast to the desired level of degradation has remained a problem. To overcome this, the yeast has also been used in an immobilized form as well as in the co-culture, so as to improve the physico-chemical characteristics as well as to improve the sensory quality of the wine.

15.3.6 Non-*Saccharomyces* Yeast in Sparkling Wine Production

Another role for the non-*Saccharomyces* has been investigated in sparkling wine production. For this purpose, there is a need to determine the nutrient supplementation products specifically for non-*Saccharomyces* yeasts and focus on the long-term aging ability of sparkling wines made from non-*Saccharomyces* yeast. Mixed inoculations, including testing foam ability and persistence, organic acid levels and mouth-feel properties, need to be evaluated before the utility of non-*Saccharomyces* yeast can be proposed. Another aspect in this regard is that non-*Saccharomyces* yeasts can influence the aromas of sparkling wines through the production of enzymes and metabolites during aging when in contact with yeast lees, as the non-*Saccharomyces* yeasts have shown significant differences in numerous VOCs (Volatile organic compounds) between species and strains. A sensory evaluation (Table 15.6) of the influence of application of non-*Saccharomyces* yeasts on sparkling wines reflected the use of yeasts as sole inoculations. It is possible to obtain specific sensory attributes and distinctive characters in mixed fermentations, however the relatively short aging times for lees (4, 6 and 12 months) needs to be extended, as sparkling wine is normally aged for a period of 3 years. Similarly, yeasts effect on the practical production stages of sparkling wine (*i.e.* riddling, disgorging), foam stability and flavor and aroma in wines that have been aged in cellars for long periods is not known.

Another interesting feature of *Schizosaccharomyces* spp. is the structure and composition of the cell walls that makes the yeast more suitable than lees production for aging wines like sparkling wine. The polysaccharide fraction released from the walls of the yeast through the action of the cells' own β-glucanases and wall mannosidases has an important influence on the sensorial and physico-chemical properties of wines aged by this technique. Aging over lees using *Schizo. pombe* reveals that the yeast has a complex polysaccharide profile in the cell wall, and that high molecular weight biopolymers are rapidly released from the walls during cellular autolysis. These cell wall fragments have good properties to maintain wine pigments in colloidal suspension, the anthocyanins adsorbed onto the walls of the living yeasts being released with these post-autolysis wall fragments. But there is a need to select the suitable *Schizo. pombe* strains for this purpose.

15.3.7 Killer Toxin Production

Non-*Saccharomyces* yeasts which are able to produce killer toxins that kill certain other yeasts can be used as biological antimicrobial agents in starter cultures (see Chapter 18 for more details.) How it can be applied is a future topic of research.

15.4 CONCLUSION

Non-*Saccharomyces* species have not been recommended for winemaking because most of them are simultaneously very weak producers of ethanol and strong producers of undesirable by-products, such as acetate and ethyl acetate. However, some species present in mature grape berries and at the beginning of alcoholic fermentation seem to play an important positive role in the aromatic profile of the wines, due to their high β-glucosidase activity, lipase, cellulase and protease production. In addition, non-*Saccharomyces* strains have been reported as important ester producers, important for the aromatic characteristics of wine. Furthermore, non-*Saccharomyces* inoculation, in sequential fermentation or multi-starter

TABLE 15.6
Summary of the Impact of Non-*Saccharomyces* Yeasts on Sensory Profiles of Base Wines and Sparkling Wines

Yeast	Production stage	Sensory evaluation	Effect on the sensory profile
T. delbrueckii + *S. cerevisiae*	First fermentation for base wine production	Sensory triangle test, panel with 9 tasters	It was distinguishable by 6 of the 9 tasters and 5 of them preferred them over control wine.
M. pulčherrima + *S cerevisiae*			It was distinguishable by 8 of the 9 tasters and 4 of them preferred them over control wine. Smoky and flowery aromas.
T. delbrueckii (sequential inoculation with *S cerevisiae)*	First fermentation followed by a second fermentation	Sensory triangle test, panel with 12 tasters	It was distinguishable by 9 of the 12 tasters and 8 of them preferred them over control wine. Better integrated effervescence and less aggressiveness in the mouth.
S. *ludwigii 979*	Second fermentation in bottle + 4 months of aging on lees	Prepared evaluation sheet panel with 11 tasters	In the red sparkling wines, higher limpidity and effervescence, in white sparkling wines higher limpidity but lower aroma intensity and quality in comparison to control.
S. pombe 7VA			In red sparkling wines, higher aroma intensity and higher scores for herbal, buttery, yeasty, acetic acid and oxidation aromas, in white sparkling wines higher limpidity; lower aroma quality, higher buttery, yeasty and reduction; lower flowery and fruity aromas in comparison to control.
T. delbrueckii 130 *T. delbrueckii 313*	Second fermentation in bottle +12 months of aging on lees	Prepared evaluation sheet panel with 11 tasters	It was characterized for the sensorial attributes of white flowers, bread crust, sapidity and acidity, with significant differences from other sparkling wines, except the attribute of sapidity.
S. cerevisiae + *T. delbrueckii 130* *S. cerevisiae* + *T. delbrueckii 313*			Significant differences were detected in the main sensory attributes in comparison to control wine. Higher scores for the aromatic descriptors (white flowers, citrus, honey, odor intensity, softness). Control wine showed significantly higher astringency in comparison to all other studied fermentations.

Source: Ivit and Kemp (2018); González-Royo *et al.* (2015); Medina-Trujillo *et al.* (2017); Canonico *et al.* (2018)

culture, may be used to produce wines with reduced ethanol content. So, it may be concluded that the selection of non-*Saccharomyces* yeasts could be useful for their potential application in the wine industry.

BIBLIOGRAPHY

Abalos, D., Vejarano, R., Morata, A., González, C. and Suárez-Lepe, J.A. (2011). The use of furfural as a metabolic inhibitor for reducing the alcohol content of model wines. *European Food Research and Technology* 232, 663-669. doi:10.1007/s00217-011-1433-9.

Benito, S., Morata, A., Palomero, F., González, M.C. and Suárez-Lepe, J.A. (2011). Formation of vinyl phenolic pyranoanthocyanins by *Saccharomyces cerevisiae* and *Pichia guillermondii* in red wines produced following different fermentation strategies. *Food Chemistry* 124(1), 15–23.

Benito, S., Palomero, F., Morata, A., Calderón F. and Suárez-Lepe, J.A. (2009). A method for estimating *Dekkera/Brettanomyces* populations in wines. *Journal of Applied Microbiology* 106(5), 1743–1751.

Benito, S., Palomero, F., Morata, A., Uthurry, C. and Suárez-Lepe, J. A. (2009). Minimization of ethylphenol precursors in red wines via the formation of pyranoanthocyanins by selected yeasts. *International Journal of Food Microbiology* 132(2–3), 145-152.

Bilinski, C., Russel, I. and Stewart, G. (1987). Applicability of yeast extracellular proteinases in brewing; physiological and biochemical aspect. *Applied and Environmental Microbiology* 53(3), 494–499.

Canonico, L., Comitini, F. and Ciani M. (2018). *Torulaspora delbrueckii* for secondary fermentation in sparkling wine production. *Food and Microbiology* 74, 100–106.

Ciani, M., Beco L. and Comitini, F. (2006). Fermentation behaviour and metabolic interactions of multistarter wine yeast fermentations. *International Journal of Food Microbiology* 108(2), 239–245.

Ciani, M., Canonico, L., Oro, L. and Comitini, F. (2014). Sequential fermentation using non-*Saccharomyces* yeasts for the reduction of alcohol content in wine. *BIO Web of Conferences* 3, 02015.

Ciani, M., Comitini, F., Mannazzu, I. and Domizio, P. (2010). Controlled mixed culture fermentation: A new perspective on the use of non-*Saccharomyces* yeasts in winemaking. *FEMS Yeast Research* 10(2), 123–133.

Ciani, M. and Ferraro, L. (1996). Enhanced glycerol content in wines made with immobilized *Candida stellata* cells. *Applied and Environmental Microbiology* 62(1), 128–132.

Ciani, M. and Maccarelli, F. (1998). Oenological properties of non-*Saccharomyces* yeasts associated with wine-making. *World of Microbiology and Biotechnology* 14(2), 199–203.

Clemente-Jimenez, J.F., Mingorance-Cazorla, L., Martínez-Rodríguez, S., Las Heras-Vázquez, F.J. and Rodríguez-Vico, F. (2005). Influence of sequential yeast mixtures on wine fermentation. *International Journal of Food Microbiology* 98(3), 301–308.

Contreras, A., Hidalgo, C., Henschke, P.A., Chambers, P.J., Curtin, C. and Varela, C. (2014). Evaluation of non-*Saccharomyces* yeasts for the reduction of alcohol content in wine. *Applied and Environment Microbiology* 80(5), 1670–1678.

Daniela, D., Machado, I.M.P., Chociai, M.B. and Bonfim, T.M.B. (2005). Influence of the use of selected and non-selected yeasts in Red Wine Production. *Brazilian Archives of Biology and Technology* 48(5), 747–775.

De Barros Lopes, M., Eglinton, J., Henschke, P., Høj, P. and Pretorius, I. (2003). The connection between yeast and alcohol: Managing the double-edged sword of bottled sunshine. *Australian and New Zealand Wine Industry Journal* 18, 17–22.

De Fátima, M., Centeno, F. and Palacios, A. (2007). *International Symposium of Microbiology and Food Safety in Wine "Microsafety wine" Villafranca del Penedes Spain,* 20–21 November 2007.

De Mot, R. and Verachtert, H. (1987). Some microbiological and biochemical aspects of starch bioconversion by amylolytic yeast. *Critical Review in Biotechnology* 5, 259–272.

Dizy, M. and Bisson, L. (2000). Proteolytic activity of yeasts strains during grape juice fermentation. *American Journal of Enology and Viticulture* 51, 155–167.

Erten, H. and Tanguler, H. (2010). Influence of *Williopsis Saturnus* yeasts in combination with *Saccharomyces cerevisiae* on wine fermentation. *Letters in Applied Microbiology* 50(5), 474–479.

Ferraro, L., Fatichenti, F. and Ciani, M. (2000). Pilot scale vinification process using immobilized *Candida stellata* cells and *Saccharomyces cerevisiae. Process Biochemistry* 35(10), 1125–1129.

Ferreira, A.M., Climaco, M.C. and Faia, A.M. (2001). The role of non-*Saccharomyces* sp. in releasing glycosidic bound fraction of grape aroma components – A preliminary study. *Journal of Applied Microbiology* 91, 67–71.

Fleet, G.H. (1990). Which yeast species really conducts the fermentation? In: *Proceedings 7th Aust. Wine Ind. Technical Conference, 13–17 August 1989*, eds Williams, P.J., Davidson, D.M. and T.H. Lee. Australian Wine Research Institute, Winetitles, Adelaide, Australia, pp. 153–156.

Fleet, G.H. (1992). Spoilage yeasts. *Critical Reviews in Biotechnology* 12(1–2), I-44.

Fleet, G.H. and Heard, G.M. (1993). Yeast growth during fermentation. In: *Wine Microbiology and Biotechnology*, ed Fleet, G.H. Harwood Academic Publishers, Lausane, pp. 27–54.

Gafner, J., Hoffmann-Boller, Porret N.A. and Pulver, D. (2000). Restarting sluggish and stuck fermentation. *Paper: 2nd International Viticulture and Enology Congress*, 8–10 November, Cape Town, South Africa.

Ganga, M.A. and Martinez, C. (2004). Effect of wine yeast monoculture practice on the biodiversity of non-*Saccharomyces* yeasts. *Journal of Applied Microbiology* 96(1), 76–83.

García, V., Vasquez, H., Fonseca, F., Manzanares, P., Viana, F., Martinez, C. and Ganga, M.A. (2010). Effects of using mixed wine yeast cultures in the production of Chardonnay wines. *Revista Argentina de Microbiologia* 42(3), 226–229.

Gawel, R., Francis, L. and Waters, E. (2007a). Statistical correlations between the in-mouth textural characteristics and the chemical composition of Shiraz wines. *Journal of Agricultural and Food Chemistry* 55(7), 2683–2687.

Gawel, R., van Sluyter, S. and Waters, E. (2007b). The effects of ethanol and glycerol on the body and other sensory characteristics of Riesling wines. *Australian Journal of Grape and Wine Research* 13(1), 38–45.

Gobbi, M., Comitini, F., Domizio, P., Romani, C., Lencioni, L., Mannazzu I and Ciani, M. and Ciani, M. (2013). *Lachancea thermotolerans* and *Saccharomyces cerevisiae* in simultaneous and sequential co-fermentation: A strategy to enhance acidity and improve the overall quality of wine. *Food Microbiology* 33(2), 271–281.

Gobbi, M., De Vero, L., Solieri, L., Comitini, F., Oro, L., Giudici, P. and Ciani, M. (2014). Fermentative aptitude of non-*Saccharomyces* wine yeast for reduction in the ethanol content in wine. *European Food Research and Technology* 239, 41–48.

Godden, P. (2000). Persistent wine instability issues. *Australian Grape Grower and Winemaker* 443, 10–14.

González, J.A., Gallardo, C.S., Pombar, A., Regu, P. and Rodríguez, L.A. (2004). Determination of Enzymatic activities in Ecotypic *Saccharomyces* and Non-*Saccharomyces* yeasts. *Electronic Journal of Environmental, Agriculture and Food Chemistry* 3(5), 743-750.

González-Royo, E., Pascual, O., Kontoudakis, N., Esteruelas, M., Esteve-Zarzoso, B., Mas, A., Canals, J.M. and Zamora, F. (2015). Oenological consequences of sequential inoculation with non-*Saccharomyces* yeasts (*Torulaspora delbrueckii* or *Metschnikowia pulcherrima*) and *Saccharomyces cerevisiae* in base wine for sparkling wine production. *European Food Research and Technology* 240(5), 999–1012.

Grassin, C. (1987). Recherches sur les enzymes extra cellulaires secretes par *Botrytis cinerea* dans la baie de raisin. Applications oenologiques et phytopathologiques. Ph.D. Thesis. Universite de Bordeaux II, Bordeaux, France.

Gunata, Y.Z. (1994). Etude et exploitation par voie enzymatique des précurseurs d'aromes du raisin de nature glycosidique. *Revue des Oenologues* 74, 22–27.

Guth, H. and Sies, A. (2002). Flavour of wines: Towards an understanding by reconstitution experiments and analysis of ethanol's effect on odour activity of key compounds. In: *Proceedings of the Eleventh Australian Wine Industry Technical Conference*, Adelaide, October 2001, pp. 4-7.

Hansen, H.E., Nissen, P., Sommer, P., Nielsen, J.C. and Arneborg, N. (2001). The effect of oxygen on the survival of non-*Saccharomyces* yeast during mixed culture fermentation of grape juice with *Saccharomyces cerevisiae. Journal of Applied Microbiology* 94, 541–547.

Ivit, N.N. and Kemp, B. (2018). The impact of non-*Saccharomyces* yeast on traditional method sparkling wine fermentation. *Fermentation* 4, 73. doi:10.3390/fermentation4030073.

Ivit, N.N., Loira, I., Morata, A., Benito, S., Palomero, F. and Suaréz-Lepe, J.A. (2018). Making natural sparkling wines with non-*Saccharomyces* yeasts. *European Food Research and Technology* 244(5), 925–935.

Jolly, N. (2008). *The Use of non-Saccharomyces Fructophilic Yeast for Efficient Fermentation of Grape Juice*. A Technical Guide for Wine Producers, Wynboer, pp. 1–8.

Jolly, N.P., Augustyn, O.P.H. and Pretorius, I.S. (2003). The effect of non-*Saccharomyces* yeasts on fermentation and wine quality. *South African Journal for Enology and Viticulture* 24(2), 55–62.

Jolly, N.P., Augustyn, O.P.H. and Pretorius, I.S. (2006). The role and use of non-*Saccharomyces* yeasts in wine production. *South African Journal for Enology and Viticulture* 27(1), 15–39.

Jolly, N.P., Varela, C. and Pretorius, I.S. (2014). Not your ordinary yeast: non-*Saccharomyces* yeasts in wine production uncovered. *FEMS Yeast Research* 14(2), 215–237.

Joshi, V.K. and Attri, B.L. (2017a). Pome fruit wines: Production technology 7.1. Specific features of table wine production technology Chapter. In: *Science and Technology of Fruit Wines*, eds Kossovea, Maria, V.K. Joshi and P.S. Panesar. Elsevier, London, pp. 295–347.

Joshi, V.K. and Attri, B.L. (2017b). Stone fruit wines: Production technology 7.2. Specific features of table wine production technology Chapter 7. In: *Science and Technology of Fruit Wines*, eds Kossovea, Maria, V.K. Joshi and P.S. Panesar. Elsevier, London, pp. 347–381.

Joshi, V.K., Rakesh and Ghanshyam, Abrol (2011). Stone fruit: Wine and brandy. In: *Handbook of Food and Beverage Fermentation Technology*, eds Hui, Y.H. and Evranuz E. Ozgul, CRC Press, Boca Raton, pp. 273–304.

Joshi, V.K., Sharma, P.C. and Attri, B.L. (1991). A note on the deacidification activity of *Schizosaccharomyces pombe* in plum musts of variable composition. *Journal of Applied Bacteriology* 70, 386–390.

Kapsopoulou, K., Mourtzini, A., Anthoulas, M. and Nerantzis, E. (2007). Biological acidification during grape must fermentation using mixed cultures of *Kluyveromyces thermotolerans* and *Saccharomyces cerevisiae. World Journal of Microbiology and Biotechnology* 23(5), 735–739.

Kaur, I., Misra, B.N. and Kohli, A. (2001). Synthesis of Teflon-FEP grafted membranes for use in water desalination. *Desalination* 135(1–3), 357–365.

Lallemand (2010). Winemaking update. http://www.lallemandwine.com/IMG/pdf_WUP_1_-_2010_-_TD_-_ESP.pdf.

Lamiknara, C. and Inyang, I. (1988). Temperature influence on muscadine wine protein characteristic. *American Journal of Enology and Viticulture* 41, 147–155.

Lea, A. and Arnold, G. (1978). The phenolics of ciders: Bitterness and astringency. *Journal of the Science of Food and Agriculture* 29(5), 478–483.

Lee, H., To, R.J.B., Latta, R.K., Bieley, P. and Schenidu, H. (1987). Some properties of extracellular acetylxylan esterase produced by yeast *Rhodotorula mucilaginosa. Applied and Environment Microbiology* 53, 2831–2834.

Lee, P.R., Ong, Y.L., Yu, B., Curran, P. and Liu, S.Q. (2010). Profile of volatile compounds during papaya juice fermentation by a mixed culture of *Saccharomyces cerevisiae* and *Williopsis saturnus. Food Microbiology* 27(7), 853–861.

Lodder, J. (1970). *The Yeast*, North Holland Pub Co., Amsterdam.

Lomolino, G., Zocca, F., Spettoli, P. and Lante, A. (2006). Detection of beta-glucosidase and esterase activities in wild yeast in a distillery environment. *Journal of Institute of Brewing* 112(2), 97–100.

Loureiro, V. and Malfeito-Ferreira, M. (2003). Spoilage yeasts in the wine industry (review). *International Journal of Food Microbiology* 86(1–2), 23–50.

Magyar, I. and Tóth, T. (2011). Comparative evaluation of some oenological properties in wine strains of *Candida stellata, Candida zemplinina, Saccharomyces uvarum* and *Saccharomyces cerevisiae. Food Microbiology* 28(1), 94–100.

Mateo, J.J. and Di Stefano, R. (1997). Description of β-glucosidase activity of wine yeasts. *Food Microbiology* 14(6), 583–591.

Mateo, J.J., Jimenez, M., Hureta, T. and Poastor, A. (1992). Comparison of Volatiles produced by four *S. cerevisiae* strain isolated from Monastrell must. *American Journal of Enology and Viticulture* 43, 206–209.

Medina-Trujillo, L., González-Royo, E., Sieczkowski, N., Heras, J., Fort, F., Canals, J.M. and Zamora, F. (2017). Effect of sequential inoculation (*Torulaspora delbrueckii/ Saccharomyces cerevisiae*) in the first fermentation on the foam properties of sparkling wine (Cava). *European Food Research and Technology* 243, 681–688.

Mendes, A., Climace, M. and Mendes, A. (2001). The role of non-*Saccharomyces* sp. in releasing glycosidic bound fraction of grape aroma components – A preliminary study. *Journal of Applied Microbiology* 91, 67–71.

Moreira, N., Mendes, F., Guedes de Pinho, P.P., Hogg, T. and Vasconcelos, I. (2008). Heavy sulphur compounds, higher alcohols and esters production profile of *Hanseniaspora uvarum* and *Hanseniaspora guilliermondii* grown as pure and mixed cultures in grape must. *International Journal of Food Microbiology* 124(3), 231–238.

Moreira, N., Mendes, F., Hogg, T. and Vasconcelos, I. (2005). Alcohols, esters and heavy sulphur compounds production by pure and mixes cultures of apiculate wine yeasts. *International Journal of Food Microbiology* 103(3), 285–294.

Nelson, G. and Young, T.W. (1986). Yeast extracellular proteolytic enzymes for chill proofing beer. *Journal of the Institute of Brewing* 92(6), 599–603.

Ollivier, C.H. (1987). *Recherches sur la Vinification des vins blancs secs. Diplome d'études et de recherches.* Universite de Bordeaux II, Bordeaux, France.

Palomero, F., Morata, A., Benito, S., Calderón, F. and Suárez-Lepe, J.A. (2009). New genera of yeasts for over-lees aging of red wine. *Food Chemistry* 112(2), 432–441.

Perez-Gonzalez, J., Gonzalex, R., Qverol, A., Sendra, J. and Ramon, D. (1993). Construction of a recombinant wine yeast strain expressing β-(1–4 endoglucanase and its use in micro vinification process. *Applied and Environmental Microbiology* 59, 2801–2806.

Pina, C., Santos, C., Couto, J.A. and Hogg, T. (2004). Ethanol tolerance of five non-*Saccharomyces* wine yeast in comparison with a strain of *S. cerevisiae* influence of different culture condition. *Food Microbiology* 21(4), 439–447.

Ponzzes-Gomes, C.M.P.B.S., Mélo, D.L.F.M., Santana, C.A., Pereira, G.E., Mendonça, M.O.C., Gomes, F.C.O., Oliveira, E.S., Barbosa Jr, A.M., Trindade, R.C. and Rosa, C.A. (2014). *Saccharomyces cerevisiae* and non-*Saccharomyces* yeasts in grape varieties of the São Francisco Valley. *Brazilian Journal of Microbiology* 45(2), 411–416.

Pretorius, I.P. (2000). Tailoring wine yeast for new millennium: Novel approaches to the ancient art of wine making. *Yeast* 16, 675–729.

Pretorius, I.S. and Bauer, F.F. (2002). Meeting the consumer challenge through genetically customized wine-yeast strains. *Trends in Biotechnology* 20(10), 426–432.

Quirós, M., Rojas, V., Gonzalez, R. and Morales, P. (2014). Selection of non-*Saccharomyces* yeast strains for reducing alcohol levels in wine by sugar respiration. *International Journal of Food Microbiology* 181, 85–91.

Renault, P., Miot-Sertier, C., Marullo, P., Hernandez-Orte, P., Lagarrigue, L., Lonvaud-Funel, A., *et al.* (2009). Genetic characterization and phenotypic variability in *Torulaspora delbrueckii* species: potential applications in the wine industry. *International Journal of Food Microbiology*, 134, 201–210.

Rojas, V., Gil, J.V., Piñaga, F. and Manzanares, P. (2001). Studies on acetate ester production by non-*Saccharomyces* wine yeasts. *International Journal of Food Microbiology* 70(3), 283–289.

Rojas, V., Gil, J.V., Piñaga, F. and Manzanares, P. (2003). Acetate ester formation in wine by mixed cultures in laboratory fermentations. *International Journal of Food Microbiology* 86(1–2), 181–188.

Romano, P., Fiore, C., Paraggio, M., Caruso, M. and Capece, A. (2003). Function of yeast species and strains in wine flavour. *International Journal of Food Microbiology* 86(1–2), 169–180.

Romano, P., Suzzi, G., Comi, G., Zironi, R. and Maifreni, M. (1997). Glycerol and other fermentation products of apiculate wine yeasts. *Journal of Applied Microbiology* 82(5), 615–618.

Rosi, I., Vinella, M. and Dormizio, P. (1994). Characterization of B-glucosidase activity in yeast of Oenological origin. *Journal of Applied Bacteriology*, 77, 519–527.

Sajbidor, J. (1997). Effect of some environmental factors on the content and composition of microbial membrane lipids. *Critical Reviews in Biotechnology* 17(2), 87–103.

Soden, A., Francis, I.L., Oakey, H. and Henschke, P. (1998). Effects of co-fermentation with *Candida stellata* and *S. cerevisiae* on aroma composition of chardonnay wine. *Australian Journal of Grape and Wine Research* 6, 121–130.

Soden, A., Francis, I.L., Oakey, H. and Henschke, P.A. (2000). Effects of co-fermentation with *Candida stellata* and *Saccharomyces cerevisiae* on the aroma and composition of Chardonnay wine. *Australian Journal of Grape and Wine Research* 6(1), 21–30.

Strauss, M.L.A., Jolly, N.P., Lambrechts, M.G. and van Rensburg, P. (2001). Screening for the production of extracellular hydrolytic enzymes by non-*Saccharomyces* wine yeasts. *Journal of Applied Microbiology* 91(1), 182–190.

Suárez-Lepe, J.A., Palomero, F.,Benito, S., Calderón, F. and Morata, A. (2012). Oenological versatility of *Schizosaccharomyces* spp. *European Food Research and Technology* 235(3), 375–383. doi10.1007/ s00217-012-1785-9.

Sutterlin, K.A., Hoffmann, Boller, P. and Gafner, J. (2004). Kurieren Von Garstoch Ungen mit der fructophilon weinhefe. *Zygosaccharomyces* Bailli. In: *Poster: 7th International Symposium on Innovations in Enology, Intervitis Interfrocta 2004*, 10–11 May, Stuttgart – Killesberg, Germany.

Tilloy, V., Ortiz-Julien, A. and Dequin, S. (2014). Reduction of ethanol yield and improvement of glycerol formation by adaptive evolution of the wine yeast *Saccharomyces cerevisiae* under hyperosmotic conditions. *Applied and Environment Microbiology* 80(8), 2623–2632.

Toro, M.E. and Vázquez, F. (2002). Fermentation behaviour of controlled mixed and sequential cultures of *Candida cantarelli* and *Saccharomyces cerevisiae* wine yeasts. *World Journal of Microbiology and Biotechnology* 18, 347–354.

Ubeda, J., Gil, M.M., Chiva, R., Guillamón, J.M. and Briones, A. (2014). Biodiversity of non-*Saccharomyces* yeasts in distilleries of the La Mancha region (Spain). *FEMS Yeast Research* 14(4), 663–673.

Viana, F., Belloch, C., Vallés, S. and Manzanares, P. (2011). Monitoring a mixed starter of *Hanseniaspora vineae-Saccharomyces cerevisiae* in natural must: Impact on 2-phenylethyl acetate production. *International Journal of Food Microbiology* 151(2), 235–240.

Viana, F., Gil, J.V., Genovés, S., Vallés, S. and Manzanares, P. (2008). Rational selection of non-*Saccharomyces* wine yeasts for mixed starters based on ester formation and enological traits. *Food Microbiology* 25(6), 778–785.

Viana, F., Gil, J.V., Vallés, S. and Manzanares, P. (2009). Increasing the levels of 2-phenylethyl acetate in wine through the use of a mixed culture of *Hanseniaspora osmophila* and *Saccharomyces cerevisiae. International Journal of Food Microbiology* 135(1), 68–74.

Volschenk, H., van Vuuren, H.J. and Viljoene Bloom, M. (2003). Malo-ethanolic fermentation in *Saccharomyces* and *Schizosaccharomyces. Current Genetics* 43(6), 379–391.

Zarzoso, B.E., Manzanares, P., Ramon, D. and Querol, A. (1998). The role of non-*Saccharomyces* yeast in industrial winemaking. *International Microbiology: The Official Journal of the Spanish Society for Microbiology* 1(2), 143–148.

Zohre, D.E. and Erten, H. (2002). The influence of *Kloeckera apiculata* and *Candida pulcherrima* yeasts on wine fermentation. *Process Biochemistry* 38(3), 319–324.

16 Understanding Wine Yeasts

L. Rebordinos, M.E. Rodríguez, G. Cordero-Bueso and J.M. Cantoral

CONTENTS

16.1 INTRODUCTION

Wine yeasts are the microorganisms responsible for the biochemical transformation of the natural media where they grow that lead to the elaboration of the wine. Such media can be fruit juice, normally grape juice (must) or a wine fortified with natural alcohol up to 15.5% (v/v) when required, where the yeast responsible for the biological aging of Sherry-type wines develops. Historically, winemaking and brewing technology, in general, without the knowledge of the actual agent of fermentation, was well developed in the earliest civilizations. However, Pasteur did not discover the direct relationship between yeast and fermentation until the middle of the nineteenth century. After that, investigations on the biochemical properties of yeasts started, and yeast taxonomy emerged. The early methods of taxonomy involved morphological studies and biochemical analysis of differences in fermentation tests. The chemical characterization of yeast DNA and the development of molecular technologies have made possible the development of a new yeast taxonomy named "molecular taxonomy", which has emerged as a powerful tool for the characterization of yeasts and for the differentiation of very similar strains. The chapter will focus on wine yeast growth and factors affecting it.

16.2 ORIGIN AND EVOLUTION OF WINE YEAST STRAINS

In contrast with other industrial processes carried out by yeast, an inoculum of yeast cells into the media is not necessary to start the winemaking process. Once the mature grapes are mixed and crushed, the must contains a natural population of yeasts, which are able to produce a spontaneous fermentation of sugars yielding ethanol, CO_2 and other compounds as final products. Such an important characteristic has allowed the elaboration of alcoholic beverages since the Late

Chalcolithic period (around 4000 BCE). Yeasts are present on the grapes and later in the fermentation, for only a few weeks in the year. Grapes and soil contain the highest number of yeasts in the vegetative period; the bark and the soil are the main natural reservoirs during the resting stages. In addition, there are several hypotheses on wild yeast reserves in nature and the mechanisms by which the yeast are transported to vineyards. Recent evidence suggests that the yeast is brought to the vineyard when the grapes are nearly ripe by insects, small mammals and human activities. Therefore, fermentative yeasts could be found as endophytic microorganisms in grape tissues, thus damaged berries would act as inoculums. Another model proposes that the yeast cells are present in the cellar environment and must becomes contaminated once it enters the cellar. This hypothesis could be true for some special situations, but it seems unlikely to occur in a modern winery with stainless steel tanks and management that includes a proper disinfection of the equipment.

In any case, the origin and mechanisms of the propagation of wine yeasts are very important because of their close relationship with human activities. In addition, the natural evolution of these yeasts is influenced by unique factors, such as the use of starter cultures in wine fermentations.

16.3 TECHNIQUES FOR IDENTIFICATION AND CHARACTERIZATION OF WINE YEAST STRAINS

Reliable and rapid identification of the yeast strains, constituting the populations responsible for the fermentation and biological aging of wines, is desirable to either control the process by monitoring the evolution of the different strains or to detect yeasts causing wine spoilage or generally to improve the wine making process.

16.3.1 Classical Techniques

Wine yeasts have traditionally been identified based on their morphological attributes at macro- and microscopic levels, as well as on their physiological behavior. In the case of *flor* yeasts, the most commonly analyzed substrates for their differential fermentation capabilities are galactose, dextrose, lactose, maltose, melibiose, raffinose and sucrose. These methods have shown than the *yeast velum* in southern Spain is composed of four *S. cerevisiae* species: *beticus, montuliensis, cheresiensis* and *rouxii* (also known as *Zygosaccharomyces rouxii).* However, both morphological and physiological characteristics may be influenced by culture conditions and can provide ambiguous results. Thus, classical techniques, in some cases, can lead to an incorrect classification of a species or a misidentification of strains. Moreover, such methodology often requires the evaluation of different parameters by complex, laborious and time-consuming processes that do not include the most important characteristics of the yeasts from the industrial point of view. Although in recent years new methods have been developed, such as rapid kits for yeast identification, analysis of total cell proteins and long-chain fatty acids using gas chromatography, determination of chemical compounds formed by the yeasts or the genetic marking of strains in order to monitor them, their reproducibility and/or suitability to wine fermentation yeasts is questionable.

16.3.2 Molecular Techniques

Recent progress in molecular biology has led to the development of several techniques for yeast identification based on similarity or dissimilarity of DNA, RNA or proteins. These include electrophoretic karyotyping, RFLP (Restriction Fragment Length Polymorphism) of mitochondrial DNA, random amplified polymorphic DNA analysis (RAPD), ribosomal internal transcribed spacers (ITS-PCR) or surveys of simple sequences repeats (SSR-Multiplex PCR). Some of these methods have been found to be of great interest in enology for the high level of resolution in yeast strain characterization and for making it possible to establish a correlation between the genetic variability and the most important industrial properties of the strains.

16.3.2.1 Study of Chromosomal Polymorphism by Electrophoretic Karyotyping

The group of electrophoretic techniques known as *Pulsed Field Gel Electrophoresis* (PFGE) allows the separation of the chromosomes of some yeasts and filamentous fungi. Samples of intact yeast cells embedded in agarose, lysed and deproteinized *in situ*, are loaded directly in an agarose gel. One of the most used PFGE systems are the Contour-Clamped Homogeneous Electric Fields (*CHEF*) that permit the separation of large DNA molecules ranging from 100 kb to 6 Mb in size. Variations in the number, intensities and electrophoretic mobility of the DNA bands displayed by the separation of the yeast chromosomes using PFGE allow the differentiation between wine yeast strains and other industrial yeast strains, and the characterization and monitoring of individual strains. This technique has been used with different applications for the control and improvement of different enological processes.

The utilization of starter cultures for must fermentation is a widespread practice in wineries, which assures a rapid fermentation and a total depletion of sugars by using a selected yeast strain with given properties, such as a high fermentation rate and ethanol resistance. However, a commercial yeast strain is not always able to adapt to the ecological conditions of a given enological region, and sometimes is even displaced by the wild yeasts, which are associated with grape berries. The application of the electrophoretic karyotyping to characterize the yeast strains, which compose the populations throughout the wine fermentation process, can monitor the evolution of the yeast, allowing the determination of the presence of either inoculated or wild yeast strains and their relative proportions within the population. In addition, in a wild fermentation, strains with a better capacity of adaptation to the process by using *PFGE* techniques can be identified (Figure 16.1). Once the strains are identified, they can be selected to study their main industrial properties. The wild selected strains could be

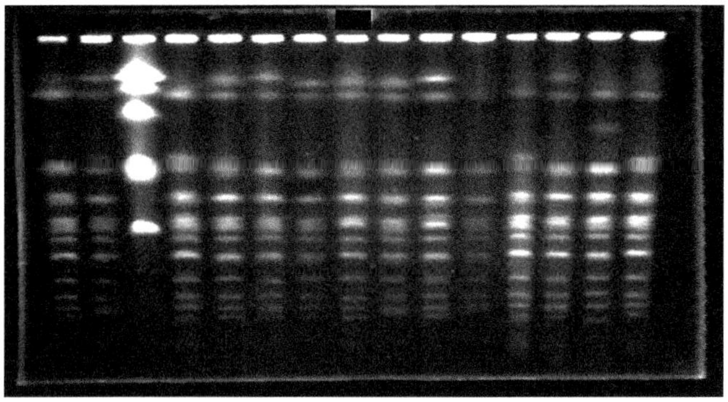

FIGURE 16.1 Electrophoretic karyotypes of different yeast colonies isolated from a fermentation tank at 24h from the beginning of the fermentation process. Up to 7 different *S. cerevisiae* strains can be distinguished based on differences in the number and electrophoretic mobility of the bands: I (lane 1); II (lanes 2, 9, 11); III (lanes 4, 12, 15); IV (lane 5), V (lanes 6, 7, 10, 13), VI (lane 8) and VII (lane 14). The strain with pattern VI was used as starter culture, the rest of the patterns correspond to wild *S. cerevisiae* yeast strains. The electrophoretic pattern in lane 3 corresponds to a non-*Saccharomyces* wild yeast strain

used as starter cultures in future fermentations and the microbial evolution of the process could be monitored by studying the electrophoretic karyotypes of colonies sampled from the fermentation tanks during the process.

16.3.2.2 RFLPs of Mitochondrial DNA

This is a rapid and simple method of yeast strain characterization that could also be applied to the control of wine fermentations conducted by either inoculated or wild yeast strains. In this method, the genomic DNA purified from a yeast strain and after a RNA-ase treatment is digested with one or several restriction enzymes, and the restriction fragments are separated by a conventional agarose electrophoresis. Mutations, which affect the position of the restriction sites in the mitochondrial DNA, lead to the generation of polymorphisms in the fragment sizes that could be used for the differentiation between two yeast strains. Endonucleases such as *Alu*I, *Hinf*I or *Rsa*I, which recognize very frequent restriction sites in the chromosomal DNA but not in the mitochondrial DNA, lead to a total cleavage of the chromosomal DNA in small pieces, so that the purification of mitochondria organelles is not required to get the *mt*DNA restriction pattern. *RFLP* of *mt*DNA has been shown to be as consistent as the electrophoretic karyotyping in order to demonstrate the dominance of an inoculated strain in wine fermentations as well as to study the evolution of the whole *S. cerevisiae* population in the fermentation process. But the feasibility of either one or another method mainly depends on the possibility to establish a correlation between the genetic variability and the more interesting phenotypic properties of the strains from an industrial point of view.

16.3.2.3 Methods Based on PCR Amplification

The wine industry needs a simple method for rapid diagnosis of the dominance of inoculated strains that could be performed routinely during the fermentation process. Molecular methods of characterization of industrial yeasts based on the amplification of certain genome fragments by Polymerase Chain Reaction (PCR) could present several advantages with respect to the techniques described above. First, a small amount of DNA template (at the level of 10 20 ng) is enough for the detection of a PCR amplification product. Second, PCR is highly selective due to the use of specific oligonucleotides for thc amplification. Third, previous extraction or purification of the DNA from the different strains is not always required. Nowadays, there are reliable, low-cost, time-saving and high-output protocols consisting of a pretreatment with activated carbon and the adaptation of yeast DNA extraction and multiplex PCR-SSR protocols, enabling discrimination on the strain level. The methods have been applied in a real commercial winery to detect yeast spoilages and to control inoculums purity in white and red wines, allowing winemakers to intervene rapidly during fermentation. Moreover, PCR-based techniques have provided very interesting results in the fields of wine yeast taxonomy at the species level and the identification of spoilage wine yeasts (see Chapter 38). Studies on Spanish wine fermentations using three different restriction patterns generated from the PCR-amplified region spanning the internal transcribed spacers (ITS 1 and 2) and the 5.8S rRNA gene showed a high length variation for the different species present during spontaneous fermentation and those *S. cerevisiae* flor yeasts responsible for the biological aging of sherry-type wines. Those identified as flor yeasts present a 24-bp deletion located in the ITS1 region. These results have been extended the analysis of this DNA region to flor yeast isolated in the Montilla-Moriles D.O. region and corroborate that this deletion is fixed in *flor yeast*. However, in view of the latest yeast taxonomy it is important to underline that there are no more species. *S. beticus* and *S. cheresiensis* are no longer considered species or subspecies of *S. cerevisiae* according to the latest taxonomic study. Indeed, all these previously named species or subspecies are now considered as *S. cerevisiae* synonyms based on nuclear DNA relatedness. *S. montuliensis* is now considered *Torulaspora delbrueckii*.

In addition, this technique can be applied to determine the quality of the musts and to quantify the presence of non-*Saccharomyces* species. PCR methods have been developed in order to detect yeasts contaminating Sherry wines. *Brettanomyces* sp. and its sexual state, *Dekkera sp.*, have been well documented as spoilage microorganisms, usually associated with barrel-aged red wines. By using a specific genomic DNA fragment of *Dekkera sp.*, a direct PCR amplification from intact cells is able to detect the presence of fewer than 10 yeast cells of *Dekkera-Brettanomyces* strains in contaminated samples of Sherry, which may help to prevent dissemination of these contaminant microorganisms in Sherry and other beverages.

16.4 WINE YEAST GROWTH AND NUTRITION

16.4.1 Alcoholic Fermentation

Alcoholic fermentation is a microbiological process involving more than 12 different reactions producing many intermediate metabolites. Gay-Lussac simplified this process in the following equation:

$$\underset{\text{Glucose}}{C_6H_12O_6} + 2ADP + 2H_3PO_4 \rightarrow \underset{\text{Ethanol}}{2C_2H_3OH} + 2CO_2 + 2AP + 2H_2O$$

The energetic balance of the reaction is the production of 2,54 Kcal, which is liberated to the medium.

The beginning of the fermentation is known as tumultuous fermentation, during which temperature must be controlled to avoid reaching a dangerous level for yeast activities. Very high temperatures inactivate the enzymes responsible for the conversion of hexoses into alcohol, generating an evaporation of alcohol and volatile aromatic compounds, making the development of nasty bacteria easier, and favor the dissolution of colored compounds coming from the solid parts of the grapes. On the other hand, very low temperatures would diminish speed of fermentation and some unwanted fermentative aromas are generated.

16.4.2 Yeast Nutrition

There are two different kinds of nutrients necessary for the development of yeasts: (*i*) essential nutrients for making the cell structure, and (*ii*) non-essential nutrients. The basic compounds necessary for growing and multiplication of yeasts have been discussed in the subsequent sections.

16.4.2.1 Source of Carbon

Sugars are metabolized, as carbon source, either aerobically (oxidation) or anaerobically (fermentation). Hexoses as D-glucose, D-fructose and D-mannose become metabolized in an oxidative way, although a fermentative one can follow. Disaccharides like sucrose and maltose are assimilated through fermentation, and lactose is assimilated in other ways. Several metabolic pathways assimilate polysaccharides as raffinose. Non-sugar compounds are used when sugars are lacking.

16.4.2.2 Nitrogen Source

Yeast cells have a nitrogen content of around 10% of their dry weight. Usually they utilize organic acids as nitrogen sources. Ammonium salts, sulfates, phosphates or nitrates an also be used. Ammonium sulfate is a commonly used nitrogen source in yeast growth media, since it also provides a source of assimilable sulfur. Amino acids are a simultaneous source of both carbon and nitrogen that can be used by yeasts.

16.4.2.3 Biogenic Compounds

Compounds such as oxygen, hydrogen, phosphorous, sulfur and magnesium are necessary for making new cell components. Yeasts are unable to grow well in the complete absence of oxygen. Thus, oxygen provides a substrate for respiratory enzymes during aerobic growth, and it is also required for certain growth-maintaining hydroxylations. Elemental hydrogen is present in yeast cellular macromolecules and is available from carbohydrates and other sources. Phosphorous is present in nucleic acids and in phospholipids and, therefore, is essential for yeast cells. Orthophosphate ($H_2PO_4^-$) and condensed inorganic phosphate are common sources of phosphorous in yeast growth media. Magnesium plays essential structural and metabolic functions and is an essential element for yeast growth. Yeasts require sulfur principally for the biosynthesis of sulfur-containing amino acids which can include sulfate, sulfite, thiosulfate, methionine and glutathione. Inorganic sulfate and the sulfur amino acid methionine are the two compounds central to the sulfur metabolism of yeasts. Methionine is the most effectively used amino acid in yeast nutrition.

16.4.2.4 Mineral Elements

Yeasts require minerals that are necessary for growth. Potassium and magnesium are macro elements needed to establish the main metallic cationic environment in the yeast cell. Potassium is essential as a cofactor for a wide variety of enzymes involved in oxidative phosphorylation, protein biosynthesis and carbohydrate metabolism. Other minerals, referred to as trace elements, include Mn, Ca, Fe, Zn, Cu, Ni, Co and Mo and are necessary as activators of a wide range of enzymes.

16.4.2.5 Growth Factors

These are the organic compounds required in very low concentrations for specific catalytic or structural roles in yeast but are not utilized as energy sources. Yeast growth factors include vitamins (which serve vital metabolic functions as components of coenzymes), purines and pyrimidines, nucleosides and nucleotides, amino acids, fatty acids, sterols and other compounds such as polyamines or inositol. Vitamins required include biotin, pantothenic acid, nicotinic acid and thiamine.

16.5 KINETICS OF GROWTH

During the growing of yeasts on a discontinuous system, seven different stages can be distinguished (Figure 16.2) : A) Latent phase: this constitutes the period of adaptation to the new medium, during which the number of inoculated cells remains constant, sometimes becoming even lower. B) Phase of acceleration of the growing: during this phase, the metabolic activity and the multiplication of microorganisms become faster. C) Phase exponential: in this phase, cell division is increased exponentially. D) Phase of slower growing: this phase is reached based on levels of limiting substrate, where this nutrient that has finished its process. E) Stationary phase of maximum population: in this phase the number of living cells is kept constant because the level of both cell divisions and dead cells has become equal. Nutrients start disappearing and some new nasty compounds for the yeasts arise on the medium. These compounds are named secondary metabolites and are excreted into the medium by the yeasts. F) Phase of increased death: the number of dead yeasts increases, and the equilibrium of the previous phase ends, so that the number of living cells becomes highly reduced. G) Exponential phase of death: this is produced by a constant decrease in cells until all of them become lysed and disappear.

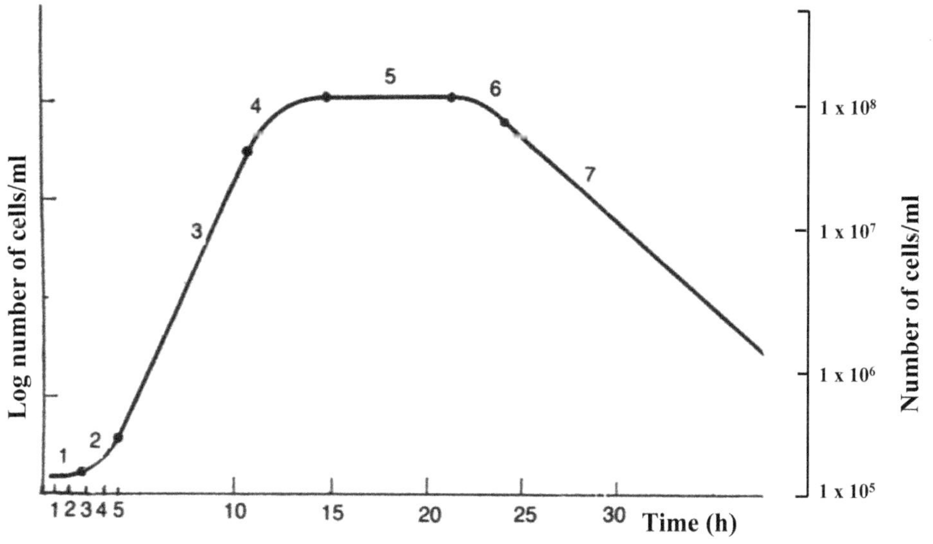

FIGURE 16.2 Yeast growth curve. Figure shows the phases of yeast in batch culture. 1: Latent phase; 2: Phase of accelerating growth; 3: Phase of exponential growth; 4: Phase of slower growth; 5: Stationary phase of growth; 6: Phase of increased dead cells, 7: Exponential phase of dead cells

16.6 FACTORS AFFECTING WINE YEAST GROWTH

The most important factors affecting wine yeast growth as well as the fermentative process performance can be classified into three groups: chemical, physical and biological.

16.6.1 Chemical Composition of Grape Must

a) *Sugar:* The sugar content in grapes ranges from 15 to 25%, and they are composed of the hexoses, *i.e.* glucose and fructose. In well-matured grapes, the concentration is quite similar, being a ratio of glucose-fructose of about 0.95. During the fermentation process, this relation decreases due to a faster transformation of glucose by the yeasts. The sugar content in the grape juice notably influences the final product of the fermentation process. Two functionally different hexose-uptake mechanisms have been proposed for yeast: A) the high affinity uptake, which is expected to play a role near to the end of must fermentation when the concentration of glucose is decreased. B) The low-affinity uptake, which is exhibited by yeast, cells in grape musts, a highly concentrated sugar media. In addition, sugar uptake changes depending on several factors such as the sugar content or the yeast species present in the media.

b) *Mineral salts and vitamins:* the concentration of these substances in grape musts is generally high enough for the yeast to carry out the transformation of the sugar during the fermentation process. Mineral salts contained in the must are sulfate, phosphate, calcium chloride, potassium, sodium, iron and copper. The vitamins or growth factors mostly needed for yeast development are: biotine, pyridoxine, panthothenic acid, myo-inositol, nicotinamide and thiamine.

c) *Nitrogen compounds:* a wide variety of assimilable nitrogen compounds present in the grape musts (particularly ammonium ion, amino acids, peptides, and small polypeptides) can be used as nitrogen sources by yeasts. Nitrogen-containing compounds are important not only for yeast growth and metabolism (percentage of nitrogen in cell yeasts is 25–60%) but also because low levels of nitrogen are known to produce sluggish and stuck fermentation. Kinetics of utilization of individual amino acids has been shown to vary between different *S. cerevisiae* strains. The two major amino acids in must are L-arginine and L-proline, in addition to ammonium ions. Yeasts use much of the ammonium and L-arginine along with other amino acids during alcoholic fermentation. However, under enological conditions, L-proline is used only to a limited extent due to both the inhibition of proline permease by nitrogen catabolites and the restricted supply of molecular oxygen, which is essential for the activity of proline oxidase.

d) *Acidity:* acids present in the grapes will be present in the musts after grape crushing and mainly include tartaric acid, malic acid and citric acid. Acidic pHs are not necessary for yeast cells' growth; indeed, the sugar fermentation is more efficient when taking place in low acidity media. The acidification of the must carried out before the onset of the fermentation process is important for damaging bacteria that may arise in the case of the fermentation process becoming stuck.

16.6.2 Antimicrobial Components

a) *Sulfur dioxide:* adding sulfur dioxide (SO_2) to either the crushed grapes or the must before starting the fermentation process is a common enological practice, which has been developed based on multiple properties of this compound. First, sulfur dioxide exerts an antiseptic action, thereby inhibiting bacterial growth and limiting the development of the non-*Saccharomyces* yeast cells. The dosage of SO_2 is important, however, so as to be controlled. A high dosage of SO_2 has been shown to restrict not only the growth of non-*Saccharomyces* species but also that of the *Saccharomyces* species (used as starter inocula). The correct choice of a starter culture may efficiently control both the yeast population and the sensory characteristics of the final product, thereby avoiding the use of a high dosage of SO_2 and its negative effects on the process.
b) *Ethanol:* during alcoholic fermentation, yeast cells are subjected to a number of stresses, including ethanol stresses. Despite a large number of studies on the ethanol tolerance of yeasts, the mechanisms by which this compound inhibits yeast growth remains unclear. Ethanol toxicity has been related to temperature (an increase in the latter variable reduces the ethanol tolerance in yeasts). Furthermore, the more tolerant the yeast is to ethanol, the more tolerant to temperature it would be. In addition, the presence of ethanol along with other compounds, such as glucose and organic acids yielded during fermentation, may increase the ethanol toxicity. Ethanol stress is related to the progressive production of this compound throughout vinification. However, the fermentative activity of the yeasts is not inhibited until high concentrations of ethanol are reached in the medium (15% ethanol or higher, depending on the strain). An accumulation of ethanol inside yeast cells occurs as fermentation proceeds, so that the intracellular and extracellular ethanol concentrations become similar. Increases in osmotic pressure are associated with increased intracellular accumulation of ethanol, although at higher osmotic pressure and temperature, and nutrient limitation has been shown to be responsible for the decreases in growth and fermentation activity of yeast cells.

16.6.3 Physical Factors Affecting Yeast Growth

16.6.3.1 Temperature

Temperature is a very important factor influencing yeast growth. The optimal temperature for development of yeast is about 20°C, being difficult for the fermentation to take place at temperatures below 13–14°C and above 35°C. As the temperature increases, the sugar transformation reactions speed up, and if this process finishes earlier, the yielded alcoholic degree may be low. Under these conditions, yeast cells behave in such a way that their tolerance to ethanol decreases, their nitrogen assimilation is poorer, their growth rate is lower, and, eventually, the fermentation process becomes stuck. Therefore, the correct procedures of fermentation are partly linked to the maintenance of a non-elevated temperature during the process. At temperatures above 30°C, yeasts become exhausted and stop fermenting. Also, the older the yeast culture, the more sensitive the cells are to the temperature of the medium. This correlation causes sluggish and sometimes stuck fermentation, so that sugars remain in the medium without being transformed into alcohol. Some additional problems derived from an elevated temperature, due to the development of undesirable bacteria whose optimal temperature is elevated, can alter the sensory features of the final product, such as the loss of primary aroma, and the evaporation of ethanol. Therefore, temperature is one of the most important factors to be controlled during the winemaking process, the optimal temperature for making red wines being between 25 and 30°C, whereas that for making white wines is about 20°C.

16.6.3.2 Aeration

Aeration promotes yeast growth and makes yeasts ferment sugars more rapidly and more efficiently, thus speeding up the transformation reactions in the beginning of the fermentation process. In a must without oxygen, cell reproduction is very low, and after a few generations will stop. In addition, to extend the timing of the fermentation process and thus yielding wines with an alcoholic degree about 11°, yeast reproduction must be kept active. Therefore, aeration of a healthy grape juice is important to obtain as much dissolved oxygen as possible. In general, the treatment to which grapes are subjected before the onset of fermentation provides a first aeration needed for starting the process. Aeration is also important to avoid stuck fermentation, so is the addition of ammonium into the fermenters. However, both practices are only effective in the early stages of the process, which infers that aeration is necessary for the assimilation of nitrogen (proline) by the yeast.

Aeration has also been found to influence many physiological aspects of the yeasts during development of the *flor velum* and biological aging of wines, by increasing adenylate energy and affecting the intracellular redox equilibrium and the consumption and production of compounds such as acetoin, acetaldehyde, higher alcohols, ethanol, glycerol and acetic acid. Growth of *S. cerevisiae* growing under agitation and aeration conditions have revealed that production of some enzymes, such as glucose 6-phosphatedehydrogenase and hexokinases, and cell growth are connected events, and depend on the oxygen concentration in the fermentation containers. Yeasts also need oxygen to assimilate long-chain fatty acids and synthesize sterols, which are essential constituents of the cell membranes to maintain membrane integrity, and play an active role in the movement of molecules across the cytoplasmic membrane.

16.6.4 Biological Factors Affecting Yeast Growth

16.6.4.1 Microbe-Microbe Interactions

In the natural environment, different yeast populations perform the transformation processes. Interactions between species of these populations might result in a synergistic or antagonistic effect on their metabolic activities. Microbial antagonism takes place in highly dynamic systems in which one of the microorganisms develops faster than the others do. This development of a given species can be caused by (*i*) competition for nutrients or substrate, (*ii*) changing the environmental conditions of the culture, or (*iii*) producing an array of microbial inhibitors, *i.e.* antibiosis, (*iv*) iron depletion, (*v*) cell wall degrading enzymes, (*vi*) diffusible and volatile antimicrobial compounds, and (*vi*) biofilm formation. In every case, the prevalent species can affect the remaining microorganisms of the system in several ways: (*i*) generating an undesirable environment for their development (antiseptic action), (*ii*) causing their death (antibiotic action), or (*iii*) causing their death and cell lysis (lytic action). In recent studies, microbial antagonism has been increasingly implicated as a cause of stuck fermentations. One such phenomenon is the production of killer toxins by killer yeasts. Killer yeasts kill sensitive strains by secreting a proteinaceous killer toxin into the medium to which they themselves are immune. Neutral yeast cells however, are immune to the killer toxin and do not kill sensitive yeast cells. Subsequently, killer-sensitive (KS) strains were discovered. These strains are immune to their own toxins but may be sensitive to toxins of other killer strains. The capability of killer yeast strains to eliminate sensitive strain depends on the initial proportion of killer yeasts, the susceptibility of sensitive strains and the treatment of the must. The distribution of the killer character, actually, has been linked to both the production area and the production technology used. For instance, the suspended solids that remains in the must after cold-settling decreased the killer toxin effect (More details can be seen in Chapter 20).

16.7 AUTOLYSIS

One of the most important processes taking place during the aging of sparkling wines in contact with yeast is yeast autolysis, which releases substances that exert a major impact on the sensory characteristics especially on the foam of the wine. Autolysis has been defined as the hydrolysis of intracellular biopolymers under the effect of hydrolytic enzymes associated with cell death, forming low molecular weight products. This process is accompanied by a loss of dry matter, a decrease in the percentage of protein and nucleic acids in this dry matter, and by an intracellular proteolytic activity. Autolysis is considered important in enology due to the products that are released from the yeast cells into the wine. For biochemical aspects of autolysis in relation to sparkling wine production see Chapters 17 and 34.

16.8 SPANISH "FINO" SHERRY WINES

Production of 'Fino' Sherry follows two successive processes: firstly, an alcoholic fermentation of must by yeasts in order to produce a "young" wine, and then, a biological aging of this wine by other yeasts. Aging of these wines involves two phases: a static one ('añadas') during which the wine is kept in a butt for a variable number of years, followed by the dynamic phase known as 'criaderas solera' (or 'soleraje'), consisting of a series of oak barrels of sherry in the process of maturation divided into a varying number of stages (Figure 16.3). The final stage (also called 'solera') contains the oldest wine; from it, the finished wine is withdrawn, but in doing so the butts with a capacity of 500 liters are only partially emptied, not more than one-third being removed in any one withdrawal. Withdrawal is usually made twice a year, and the complete process takes no less than three years and involves five or six 'criaderas' (with a variable number of barrels). Then present is the common characteristic of a film of yeasts (*flor velum*) of up to 1 cm thick growing on the free surface of the wine (barrels are only filled up to 5/6 parts of their total volume), which contains about 15 degrees of ethanol (Plate 16.1).

The yeast film through its oxidative metabolism is mainly responsible for the sensory characteristics of Sherry wines. The active consumption of oxygen and the isolation effect exerted by the yeast layer also prevent wine oxidation. Development of *flor velum* yeasts depends not only on a number of ambient factors but also on the chemical composition of the wine. Acetaldehyde displays an inhibitory effect on velum formation when its concentration is above the tolerance threshold. Different properties of industrial interest have been detected, such as the physiology, genetics and metabolic characteristics in the different identified strains, as well as clear differences using molecular analysis techniques. Therefore, for a better understanding, we will treat the different genetic profiles as different *S. cerevisiae* flor yeast strains. These strains are classically grouped as *beticus, montuliensis, cheresiensis* and *rouxii*, based on their differential fermentation capabilities for sucrose, maltose and raffinose.

In a previous work or research, 640 isolated flor yeasts were characterized and classified, falling into two main groups: *S. cerevisiae* derivatives and non-*Saccharomyces spp.*, of which most belonged to the *S. cerevisiae* species, following the metabolic pattern of strains *beticus* or *cheresiensis*. The absence of the *montuliensis* and *rouxii* race in the studied samples could be due to several reasons: first, they have always been described in very low proportions and furthermore have been found in some samples only from the same winery. On the other hand, the presence or absence of these races could be due to different winery cellar practices. The amount of oxygen acts in a different way on the races, and high concentrations favor the growth of *beticus* and *cheresiensis* compared to *montuliensis* and *rouxii*, hence, frequent withdrawals of wine that increase the presence of oxygen could be another possibility for the absence of those races. Moreover, the presence of some yeasts that produce killer toxins, although a minority, has been proposed as an important factor in driving yeast populations during wine fermentation, besides other practices relating to the preparation of the musts, the inoculation of strains during the biological aging, etc. On the other hand, *montuliensis* and *rouxii* races have been connected to

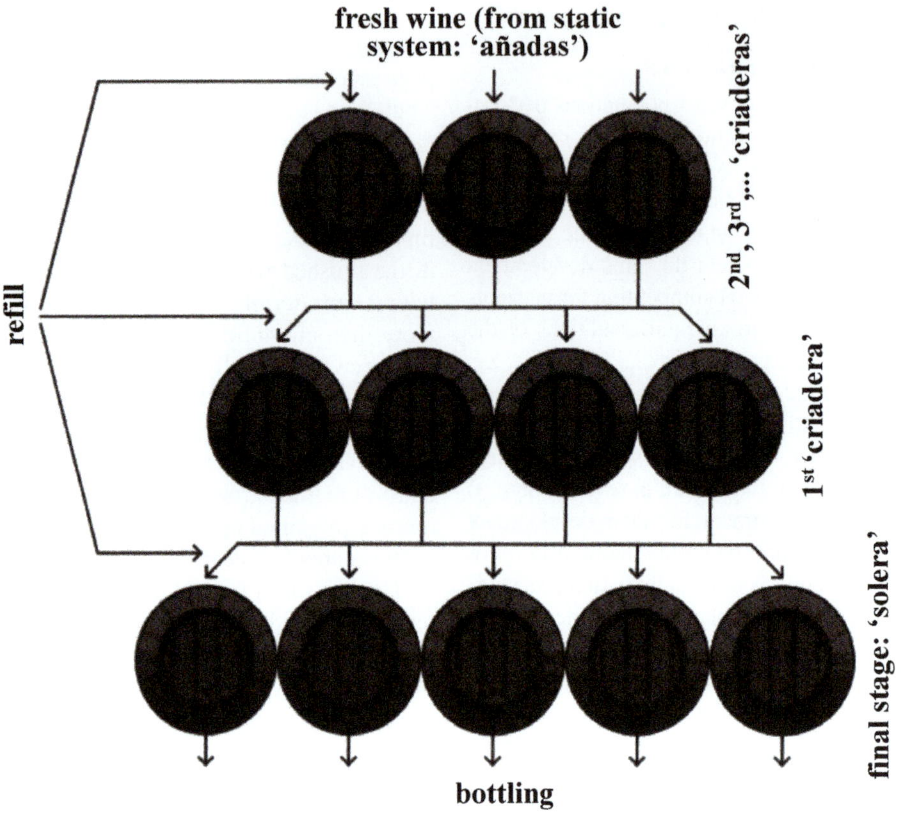

FIGURE 16.3 Biological aging takes place in oak barrels (500–600 L) through a dynamic process that involves a number of intermediate steps (or scales) and is called the "criaderas and solera" system. The system consists of a series of casks holding wine in the process of blending. The scale containing the oldest wine is called "solera" and is followed by the first, second and third "criadera" in a four-scale system. Sherry wine for marketing is collected from the solera and replaced with an identical amount of wine (one-fourth of the total volume) from the first criadera, which in turn is replenished with wine from the second criadera, and so on, young wine being added to the last scale closes the cycle

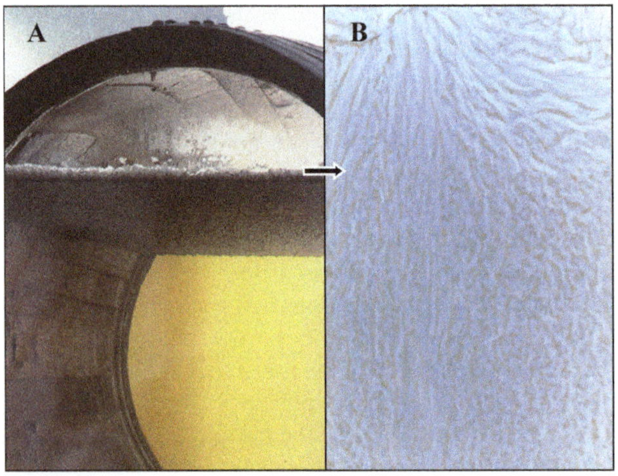

PLATE 16.1 A Yeast biofilm (*flor* velum) developed on the surface of wine kept in American oak barrels for aging. The light color of the wine is due to the barrier between wine and air formed by the biofilm, which prevents oxidation. B: Enlarged view of the yeast biofilm look

concentrations of acetaldehyde higher than 500 mg/L, which produces an inhibitory effect upon *beticus* and *cheresiensis*. Although acetaldehyde exerts positive effects over the aromatic compounds of the wine, and inoculation of high acetaldehyde-producing strains seems a straight forward way to the improvement of Sherries, the deleterious effect of the product on both chromosomal and isolated DNA still has to be considered. The determination of growth rate indicated a higher growing speed of *beticus* in relation to *cheresiensis*, whose duplication time being lower can also be the reason to first colonizing the velum. The growing rate of *beticus* was 0.526 ± 0.016 cell/h, and its time of generation 1.32 ± 0.04 h, while that of *cheresiensis* was of 0.298 ± 0.032 cell/h and the generation time 2.36 ± 0.14 h.

The electrophoretic karyotype of the strains was also determined. Polymorphism in *flor* yeasts is achieved by a balance of forces tending to induce chromosomal changes (high concentrations of ethanol or acetaldehyde) against a high selection pressure acting upon them (lack of fermentable sugars, absence of sexual reproduction, etc.) and low proportions of transposons of yeasts (Ty elements) in terms of selecting the best adapted karyotype to such conditions. More than just specific patterns of races, there were patterns related to the blending stages. In this sense, three different kinds of genotypes could be distinguished; one was present only in 'soleras' of both the races (pattern VIII), another was representative of the youngest stages ('5th criadera') (pattern XXII), and another in a transition pattern (III) which was shown both in the youngest and oldest stages. The use of highly specific techniques would facilitate the search for some relationship between sensory characteristics of Sherries and yeast properties, which is necessary in order to improve the production of Sherry wines.

To establish particular relationships between sensory properties and *flor* yeast strains of Sherry wines, samples were taken from two static systems ('añadas'), which were maintained without blending or adding of ethanol. Electrophoretic karyotype and mitochondrial DNA restriction analysis were carried out as a practical method to monitor *Saccharomyces* strains in the wine fermentation. Mitochondria are especially relevant in *velum flor* yeasts because they accommodate all the enzymatic systems necessary for the oxidative metabolism that differentiates these yeasts from the fermentative ones. In fact, there are some respiratory mutants which generate the phenotype called "petite" due to the small size of the colonies developed, which cannot develop themselves normally because they carry some deletions on their mtDNA. RFLPs of mitochondrial DNA of 60 strains with the enzymes *Alu*I, *Hinf*I and *Rsa*I yielded three, four and four different patterns, respectively, indicating a high degree of polymorphism (Table 16.1).

Combined patterns of chromosomal and mtDNA profiles suggested that the different strains identified are very closely related, since patterns were not randomly combined and some preferential associations could be distinguished, supported by the fact that only 24 out of 648 possible combined patterns appeared. Relationships between yeast populations and analytical parameters showed that 'añada' A, containing mainly *beticus*, displayed the best values in the three parameters studied: ethanol, acetaldehyde and volatile acidity. 'Añada' B, which also started with 100% *beticus*, showed a lower ethanol content, and, mainly due to the acetaldehyde concentration, the development of *cheresiensis* was allowed. Data on growth rates indicated a faster growing speed for *beticus* compared to *cheresiensis*. Its duplication time is lower and this could be the reason for being the first one in colonizing the *velum*. Only when conditions are not favorable for the development of *beticus* is the development of *cheresiensis* strains enabled. During summer, a decrease in the acetaldehyde content was observed in both 'añadas'. At the end of the summer, when the temperature decreased, a higher production of acetaldehyde was found that decreased once again at the beginning of the next spring. Volatile acidity was inversely related to acetaldehyde content. Ethanol concentration always decreased because the 'añadas' were not fortified. The ability of yeasts to resist the inhibitory effect of ethanol decreases as the temperature increases, and this explains the significant deterioration of the yeast film observed when the temperature of the wine rises during the summer. Moreover, elevated temperatures and inhibitory ethanol concentrations have a lethal effect on yeast cells, and it has been proposed that these factors act upon the mitochondria, suggesting that the production of respiratory deficient mutants in the *flor* yeast population may cause the seasonal deterioration of the yeast film. On the other hand, thermal tolerance and ethanol tolerance are genetically determined properties that may vary from one strain to another. These properties are susceptible to biological improvement, although the deterioration may be attributed to factors other than heat, such as a limited amount of essential nutrients or the opportunistic proliferation of contaminant yeasts. However, the large number of genes involved in the control of some of these characteristics has to be considered. They can impede genetic improvement, and changes in the thermal acclimatization of the winery, for example, could lead to better results in shorter times.

TABLE 16.1
Frequency and Distribution of Electrophoretic Chromosomal and mtDNA Restriction Patterns of *S. cerevisiae* Strains Isolated of Flor Velum from Different Blending Stages During the Biological Aging of Sherry Wines

		RFLP				
Race	Karyotype	*Alu*I	*Hinf*I	*Rsa*I	Origin[a]	Strains
beticus	III	A2	H2	R2	5th Cra[b]	4
beticus	III	A2	H2	R2	4th Cra	5
beticus	III	–	–	–	5th Cra	1
beticus	III	A2	H2	R2	Solera	1
beticus	VIII	A2	H4	R4	Solera	2
beticus	XXII	A2	H3	R5	5th Cra	2
cheresiensis	III	–	–	–	4th Cra	4
cheresiensis	III	A2	H2	R2	6th Cra	1
cheresiensis	III	A2	H2	R2	5tha Cra	3
cheresiensis	III	A2	H2	R2	Solera	2
cheresiensis	III	–	–	–	1st Cra	1
cheresiensis	III	A2	H2	R2	2nd Cra	1
cheresiensis	VIII	A2	H4	R4	Solera	1
cheresiensis	XXII	A2	H3	R3	5th Cra	1
cheresiensis	XXII	–	–	–	3rd Cra	1
cheresiensis	XXIV	A2	H3	R5	5th Cra	1

[a] Blending stage, [b] Cra: "criadera"

BIBLIOGRAPHY

Aguilera, F., Peinado, R.A., Millán, C., Ortega, J.M. and Mauricio, J.C. (2006). Relationship between ethanol tolerance, H^+ATPase activity and the lipid composition of the plasma membrane in different wine yeast strains. *Int. J. Food Microbiol.*, 110(1): 34–42.

Alexandre, H. (2013). Flor yeasts of *Saccharomyces cerevisiae* – Their ecology, genetics and metabolism. *Int. J. Food Microbiol.*, 167(2): 269–275.

Barnett, J.A., Payne, R.W. and Yarrow, D. (1990). *Yeasts: Characteristics and Identification.* 2nd edition. Cambridge University Press, Cambridge.

Cordero-Bueso, G., Arroyo, T., Serrano, A. and Valero, E. (2011). Remanence and Survival of commercial yeast in different ecological niches of the vineyard. *FEMS Microbiol. Ecol.*, 77(2): 429–437.

Cordero Bueso, G., Rodríguez, M.E., Garrido, C. and Cantoral, J.M. (2017). Rapid and not culture-dependent based on multiplex PCR-SSR analysis for monitoring inoculated yeast strain in industrial wine fermentations. *Arch.Microbiol.*, 199(1): 135–142.

Diderich, J.A., Schepper, M., van Hoek, P., Luttik, M.A., van Dijken, J.P., Pronk, J.T., Klaasseni, P., Boelens, H.F., deMattos, M.J., van Dam, K. and Kruckeberg, A.L. (1999). Glucose uptake kinetics and transcription of *HXT* genes in chemostat cultures of *Saccharomyces cerevisiae. J. Biol. Chem.*, 274(22): 15350–15359.

Esteve-Zarzoso, B., Peris-Torán, M.J., García-Maiquez, E., Uruburu, F. and Querol, A. (2001). Yeast population dynamics during the fermentation and biological ageing of sherry wines. *Appl. Environ. Microbiol.*, 67(5): 2056–2061.

Fernández-Espinar, M.T., Esteve-Zarzoso, B., Querol, A. and Barrio, E. (2000). RFLP analysis of the ribosomal Internal Transcribed Spacers and the 5.8S rRNA Gene Region of the Genus *Saccharomyces*: A fast method for species identification and the differentiation of *flor* yeasts. *Antonie Leeuwenhoek*, 78(1): 87–92.

Guillamón, J.M., Querol, A., Jimenez, M. and Huerta, T. (1993). Phylogenetic relationships among wine yeast strains based on electrophoretic whole-cell protein patterns. *Int. J. Food Microbiol.*, 8(2): 115–118.

Ibeas, J.I., Lozano, I., Perdigones, F. and Jiménez, J. (1996). Detection of *Dekkera-Brettanomyces* strains in sherry by a nested PCR method. *Appl. Environ. Microbiol.*, 62(3): 998–1003.

Infante, J.J., Dombek, K.M., Reborditos, L., Cantoral, J.M. and Young, E.T. (2003). Genome-wide amplifications caused by chromosomal rearrangements play a major role in the adaptive evolution of natural yeast. *Genetics*, 165(4): 1745–1759.

Jiranek, V., Langridge, P. and Henschke, P.A. (1995). Amino acid and ammonium utilization by *Saccharomyces cerevisiae* wine yeasts from a chemically defined medium. *Am. J. Enol. Vitic.*, 46(1): 75–83.

Kurtzman, C., Fell, J.W. and Boekhout, T. (2011). The yeasts; a taxonomic study. Elsevier Science, Amsterdam (The Netherlands), p. 2354.

Liu, S. and Pilone, G.J. (2000). An overview of formation and roles of acetaldehyde in winemaking with emphasis on microbiological implications. *Int. J. Food Sci. Technol.*, 35(1): 49–61.

Martínez, P., Pérez, L. and Benítez, T. (1997). Velum formation by flor yeast isolated from Sherry wine. *Am. J. Enol. Vitic.*, 48: 55–62.

Martínez, P., Valcárcel, M.J., Pérez, L. and Benítez, T. (1998). Metabolism of *Saccharomyces cerevisiae* flor yeasts during fermentation and biological ageing of fino sherry: By-products and aroma compounds. *Am. J. Enol. Vitic.*, 49(3): 240–250.

Mendes-Ferreira, A., Mendes-Faia, A. and Lea, C. (2004). Growth and fermentation patterns of *Saccharomyces cerevisiae* under different ammonium concentrations and its implications in winemaking industry. *J. Appl. Microbiol.*, 97(3): 540–545.

Menguiano, M., Romero-Sánchez, S.,Barrales, R. and Ibeas, J.I. (2016). Population analysis of biofilm yeasts during Fino Sherry wine ageing in the Montilla-Moriles D.O. region. *Int. J. Food Microbiol.*, 244: 67–73.

Mesa, J.J., Infante, J.J., Rebordinos, L. and Cantoral, J.M. (1999). Characterization of yeasts involved in the biological ageing of sherry wines. *Food Sci. Technol.*, 32(2): 114–120.

Mesa, J.J., Infante, J.J., Rebordinos, L., Sánchez, J.A. and Cantoral, J.M. (2000). Influence of the yeast genotypes on enological characteristics of sherry wines. *Am. J. Enol. Vitic.*, 51: 15–21.

Mortimer, R.K. (2000). Evolution and variation of the yeast (*Saccharomyces*) genome. *Gen. Res.*, 10(4): 403–409.

Quesada, M.P. and Cenís, J.L. (1995). Use of random amplified polymorphic DNA (RAPD-PCR) in the characterization of wine yeasts. *Am. J. Enol. Vitic.*, 46: 204–208.

Redzepovic, S., Orlic, S., Sikora, S., Majdak, A. and Pretorius, I.S. (2002). Identification and characterization of *Saccharomyces cerevisiae* and *Saccharomyces paradoxus* strains isolated from Croatian vineyards. *Lett. Appl. Microbiol.*, 35(4): 305–310.

Rodríguez, M.E., Infante, J.J., Espinazo, M.L., Rebordinos, L. and Cantoral, J.M. (2004). Selection and monitoring of wild yeast as starter cultures and study of yeast evolution in industrial fermentation. In: *Yeast Genetics and Molecular Biology Meeting,* Seattle (EEUU).

Silva, D.P., Pessoa, A. Jr., Roberto, I.C. and Vitolo, M. (2001). Effect of agitation and aeration on production of hexokinases in *Saccharomyces cerevisiae. Appl. Biochem. Biotechnol.*, 91: 605–611.

Török, T., Rockhold, D. and King, A.D. (1993). Use of electrophoretic karyotyping and DNA-DNA hybridization in yeast identification. *Int. J. Food Microbiol.*, 19(1): 63–80.

Versavaud, A. and Hallet, J.N. (1995). Pulsed-field gel electrophoresis combined with rare-cutting endonucleases for strain identification of *Candida farmata*, *Kloeckera apiculata* and *Schizosacharomyces pombe* with chromosome number and size estimation of the two former. *Syst. Appl. Microbiol.*, 18(2): 303–309.

17 Biochemical Facets of Winemaking

Neerja S. Rana, Vishal S. Rana and Bhawna Dipta

CONTENTS

17.1 INTRODUCTION

Among the alcoholic beverages, wine was the first to be made and has been used as a food adjunct by humans ever since their settlement in the Tigris-Euphrates Basin. Wine has been used as a therapeutic agent and has served as an important adjunct to the human diet. In addition, the compounds bonded to insoluble plant compounds are released into the aqueous ethanolic solution during the winemaking process, which makes them more biologically available for absorption during consumption. Wines, because they are not distilled, have more nutrients such as vitamins, minerals, and sugars than distilled beverages like brandy and whisky, especially the polyphenolic compounds that act as antioxidants and antimicrobials. Wine has also been found to be a tranquilizer, to be a diuretic, to reduce muscle spasms and stiffness associated with arthritis, to delay the development of some forms of diabetes and cardiovascular diseases, to have antioxidant effects, and to inhibit platelet aggregation. The conversion of grape juice into wine is a complex biochemical process and constitutes the fundamental basis of winemaking. Alcoholic fermentation includes some important reactions, which have been studied since the pioneer investigation of Louis Pasteur about 140 years ago. During the process of winemaking, yeast utilizes sugars and other constituents of grape juice for their growth, converting them into ethanol, carbon dioxide, and other metabolites formed during the process, that contribute to the chemical composition and sensory characteristics of the wine. The carbon metabolism is the prime pathway of winemaking, as the ultimate concentration of ethanol in a wine depends on the initial concentration of sugars in the must/juice and the conditions prevailing during fermentation. Nitrogen is the second most important element, required for yeast growth, and, therefore, the knowledge of nitrogen metabolism is equally important from a wine quality point of view. Nitrogen composition of the must sometimes also leads to problems like sluggish and stuck fermentation, formation of reduced sulfur compounds, and certain other end products like ethyl carbamate. Besides this, the yeast also utilizes sulfur for its biosynthesis. The sulfur-containing compounds, especially hydrogen sulfide, impart unpleasant aromas during yeast formation. Therefore, knowledge about the complete biochemistry of different carbon, nitrogen, and sulfur compounds is important. The current chapter deals specifically with the various metabolic events leading to the translocation and metabolism of major ingredients like sugars, nitrogen, organic acids, and sulfur compounds, and elucidates the formation of various end products, byproducts, and the factors influencing the rate of fermentation.

17.2 CARBON METABOLISM

Limited amount of carbon compounds are utilized by the *Saccharomyces* species. Among the monosaccharides, glucose, fructose, mannose, and galactose can be utilized. The disaccharides, namely sucrose, maltose, and melibiose can be utilized by most wine yeast strains, the trisaccharides (raffinose) can also serve as a substrate, but pentoses are not utilized by the *Saccharomyces* strains. The operation of these pathways depends upon the substrate availability and other conditions like oxygen. The major pathway for glucose (fructose, mannose) catabolism is glycolysis.

17.2.1 Glycolysis

The glycolytic pathway involves the conversion of glucose to pyruvate. It is virtually a universal pathway for glucose catabolism and is operational under fermentative and respiratory modes of metabolism. Winemaking involves fermentation of the soluble sugars, viz. glucose and fructose of grape juice into carbon dioxide and ethanol. The process of glucose metabolism by *Saccharomyces* for wine production includes three steps, viz. transport of sugars into the cell, their phosphorylation, and conversion of glucose-6-phosphate to pyruvate.

17.2.1.1 Transport of Sugars into the Cell

The first step of glycolysis is the transport of sugars into the cell. Glycolysis is operational under both fermentative and respiratory modes of metabolism. Basically, two types of mechanisms, *i.e.* passive transport and active transport, operate for the entry of sugars into the cell.Unlike active transport, the passive transport of molecules across a membrane does not require an input of metabolic energy. The rate of transport is proportional to the concentration gradient of the molecule across the membrane. There are further two types of passive transports, *i.e.* simple diffusion and facilitated diffusion.

17.2.1.2 Phosphorylation of Sugars

The complete pathway of glucose conversion to pyruvate, *i.e.* glycolysis, has been shown in Figure 17.1. In step 1 of glycolysis, one molecule of ATP is used for the transfer of the phosphoryl group to the 6-hydroxyl position of glucose. The reaction is catalyzed by a relative non-specific enzyme, hexokinase. Mg^{2+} ions are required for the catalysis of this reaction. The standard change in free energy (ΔGo) for this reaction is 16.7 KJ mole^{-1} that indicates the irreversible nature of the reaction in the existing cell conditions. The next step in the pathway is the conversion of glucose-6-phosphate to pyruvate. The glucose molecule is broken down into two molecules of pyruvate through a series of reactions catalyzed by different enzymes and co-enzymes. Glucose-6-phosphate is converted into its isomeric form, fructose-6-phosphate. This conversion is catalyzed by the enzyme phosphoglucoisomerase. This enzyme is specific for glucose and requires Mg^{2+} ions for its activity. Mutants deficient in phosphoglucoisomerase activity have been identified, *i.e.* unable to grow on glucose, suggesting that very little carbon metabolism occurs via pentose phosphate pathway in *Saccharomyces,* in contrast to other yeasts. Fructose-6-phosphate formed is phosphorylated by the enzyme phosphofructokinase (PFK). This is another step of the pathway where a second molecule of ATP is consumed, and phosphorylation occurs at carbon-1 position to produce fructose 1,6 bisphosphate. This enzyme also requires Mg^{2+}ions. This phosphofructokinase-catalyzed reaction is the second irreversible step of the pathway. It is a key regulatory enzyme based upon the number of allosteric effectors. Activity is modulated by AMP, ADP, Pi, fructose 2,6 bisphosphate, and other metabolites, and activity is inhibited by PEP, ATP, and citrate. The yeast PFK is composed of 8 subunits, *i.e.* α_4 and β_4. Fructose 1,6bisphosphate is cleaved into two triose phosphate sugars, viz. glyceraldehyde-3-phosphate and dihydroxy acetone phosphate, by the enzyme aldolase. Only glyceraldehyde-3-phosphate formed in the previous reaction out of two triose sugars is required for further metabolism, and therefore, to utilize the entire glucose molecule for the production of alcohol, the dihydroxyacetone phosphate is isomerized to glyceraldehyde-3-phosphate by the enzyme triose phosphate isomerase in a reversible reaction outlined here. Glyceraldehyde-3-phosphate in the presence of an enzyme, glyceraldehyde-3-phosphate dehydrogenase, is converted to 1,3 bisphosphoglycerate. The energy released in this oxidation process is conserved through phosphorylation of carboxylic group with inorganic phosphate and with formation of NADH. The catalytic activity of this enzyme requires the cysteinyl-SH group at the active site. Iodoacetamide, which has strong affinity for the SH group, is able to inhibit the reaction and the pathway through inactivation of the active site. The high energy substrate formed in the previous reaction undergoes transfer of the phosphoryl group from ADP to form ATP, and a low energy substrate, 3 phosphoglycerate, is generated by the enzyme, phosphoglycerate kinase. This phenomenon of generation of ATP from a high energy substrate is known as substrate level phosphorylation. Conversion of 3 phosphoglycerate to 2 phosphoglycerate is brought about by phosphoglycerate mutase. This enzyme catalyzes the transfer of the phosphoryl group from the 3-hydroxyl position to its 2-hydroxyl position to form 2 phosphoglycerate. The enzyme requires Mg^{2+} for its activity. This repositioning of the phosphoryl group brings the molecule to a high energy state and allows substrate level phosphorylation to yield another molecule of ATP. The enzyme enolase in this step catalyzes the dehydration of 2 phosphoglycerate and produces phosphoenolpyruvate (PEP) by reversible reaction and requires Mg^{2+}ions. This is the last step of the glycolytic pathway, catalyzed by pyruvate kinase. This is the second reaction which generates ATP by phosphorylating ADP in substrate level phosphorylation. $Mg^{2+}$and K^+ ions are essential for the enzyme activity. This enzyme is also under the control of allosteric effectors. The activity is inhibited by ATP, citrate, acetyl CoA, long chain fatty acids, and modulated by ADP and Pi. In glycolysis,

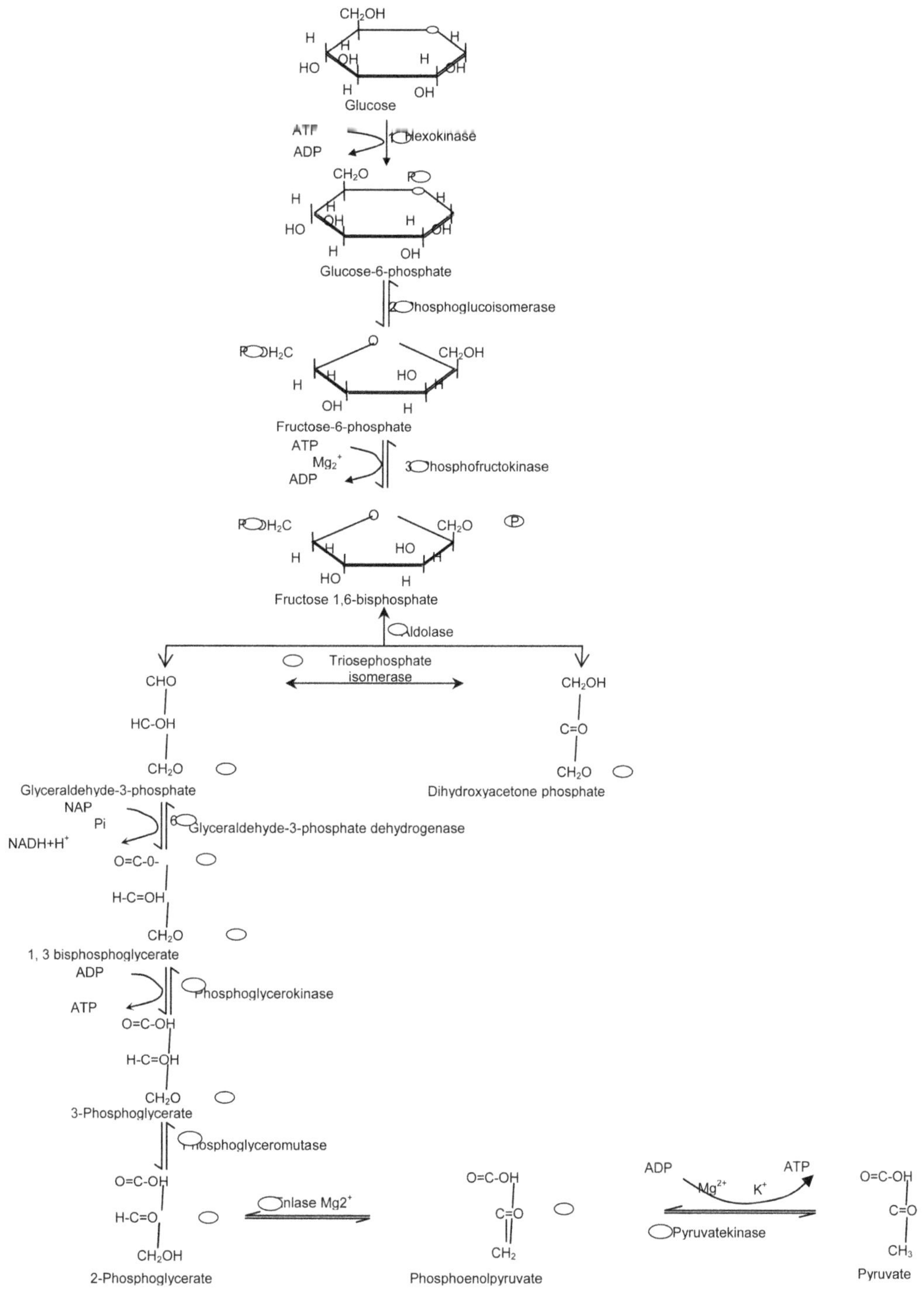

FIGURE 17.1 The reactions of glycolytic pathway

there is a net production of two molecules of ATP and NADH. The overall reaction of breakdown of one molecule of glucose to pyruvic acid is described below.

17.2.2 Metabolism of Pyruvate

The majority of industrially important microorganisms metabolize carbohydrates by glycolytic pathway and produce two molecules of pyruvate. NADH and ATP are also produced in the process by the oxidation of $NAD^+$and ADP, respectively. Both NAD^+ and ADP are present in small quantities in cells, and the basic necessity of the cell is to generate NAD^+ and ADP, if metabolism is to be continued. Therefore, each living organism has been provided with a mechanism to oxidize NADH for the continuity of substrate oxidation. Mainly, there are three types of mechanisms which metabolize pyruvate and regenerate NAD^+ during the process.These include fermentation, anaerobic respiration, and aerobic respiration. Fermentation, especially in relation to wine, has been described in this section.

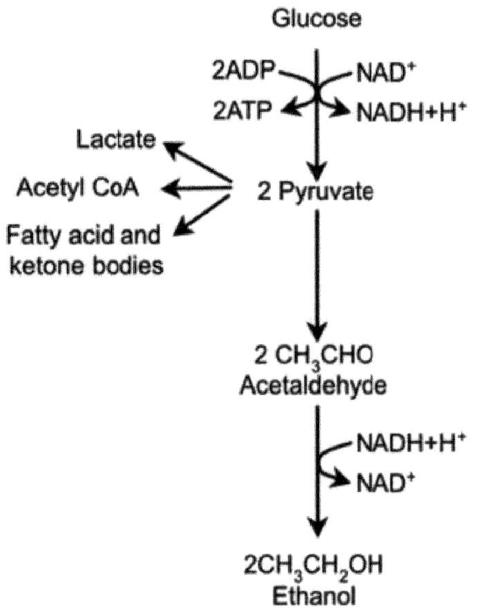

FIGURE 17.2 NAD^+ molecule regeneration and conversion of pyruvate to ethanol during alcoholic fermentation

17.2.2.1 Fermentation

This is the anaerobic breakdown of pyruvate. The metabolism of pyruvate varies considerably among different microorganisms and leads to the formation of characteristic fermentation products. Depending upon the end product, fermentation has been classified into various types.

17.2.2.1.1 Neuberg's First Fermentation (Alcoholic Fermentation)

The fermentation of pyruvate to ethanol is also called Neuberg's first fermentation and was named after the German scientist Carl Neuberg. Yeast cells, especially of the species of *Saccharomyces*, are the main alcohol producer under anaerobic conditions by converting pyruvate to ethanol and CO_2. This conversion of pyruvate to ethanol takes place in two steps. NAD^+ molecules reduced during glucose to pyruvate conversion are regenerated when pyruvate is metabolized to ethanol and CO_2. The process takes place in two steps (Figure 17.2).

17.2.2.1.2 Neuberg's Second Form of Fermentation

During the fermentation of glucose by yeast at pH 6 or below, only small amounts of glycerol and other products are formed. The addition of sulfite in the cultures of *S. cerevisiae* enhances glycerol production. Acetaldehyde, produced by the action of carboxylase during ethanol production, combines with sulfite to form an additional compound. Consequently, acetaldehyde is no longer able to act as a hydrogen acceptor for reduced NAD, which would accumulate and soon be exhausted unless some other hydrogen acceptor is provided.

17.2.2.1.3 Neuberg's Third Form of Fermentation

Under alkaline conditions, dihydroxyacetone, an intermediate in the glycolytic pathway, replaces acetaldehyde as a hydrogen acceptor, and is reduced to glycerol-3-phosphate which is dephosphorylated to glycerol with the regeneration of NAD^+(Figure 17.3). Many osmophilic yeasts carry out fermentations corresponding to Neuberg's third form in the absence of alkaline steering agents. The inhibition of alcohol dehydrogenase or pyruvate decarboxylase activities can produce glycerol in proportionally higher yields. The different species of *Saccharomyces cerevisiae* vary in their ability to produce glycerol.

17.2.2.1.4 Yield

The yield of ethanol in alcoholic fermentation with a surplus amount of fermenting sugars depends on the ethanol tolerance of the yeasts. A higher level of ethanol inhibits the enzymes and disintegrates the cell membrane through unfavorable interaction with cell wall components. Ethanol tolerance of yeast is linked with membrane fluidity. The amount of fatty acyl residues in membrane phospholipids play a significant role in membrane fluidity *vis-a-vis* ethanol tolerance. The amount of ethanol produced per unit of sugar during wine fermentation is of considerable commercial importance. As shown in the previous section, during alcoholic fermentation one molecule of glucose produces two molecules, each of ethanol and CO_2 under anaerobic conditions. In other words, 180 g of glucose (1 mole) should yield 92 g of ethanol (2 moles) and 88 g of CO_2 (2 moles). The metabolism, however, yields equimolar quantities of CO_2 and ethanol; the actual amount of CO_2 liberated is less than the calculated values. This is because of the practical utilization of CO_2 in anaerobic carboxylation reactions.

17.2.3 Regulation of Carbon Metabolism and Glycolysis

Control of carbon flux through the glycolytic pathway is a complex process. Many researchers have investigated

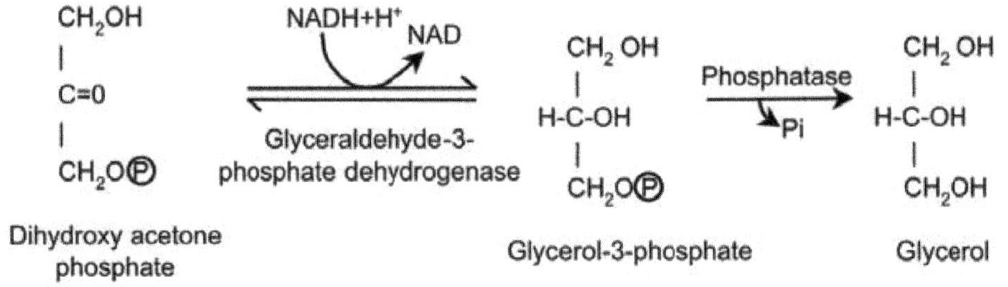

FIGURE 17.3 Neuberg's third form of fermentation

glycolysis in *S. cerevisiae* under aerobic and anaerobic conditions using ^{13}C and ^{31}P NMR spectroscopy combined with determination of the ratio of different end products using ^{14}C labelled glucose. These studies revealed that the major point of control during sugar utilization is at the step of transport or phosphorylation or both, but differentiation between uptake and subsequent phosphorylation of glucose in their analysis could not be made. Under some conditions of growth, control of flux may be mediated by modulation of phosphofructokinase activity or by a step later in glycolysis. Loss of enzyme activity at any of the three irreversible steps of glycolysis, viz. sugar phosphorylation, phosphofructokinase, and pyruvate kinase, immediately impacts the glucose transporters. In fact, they may not be rate-limiting steps in this pathway but there must be a coordinated balancing of uptake and flux through both higher (energy consumption) and lower (net energy production) glycolysis rate. There is mounting evidence that the glycolytic enzymes exist as multi-enzyme complexes in eukaryotic cells, and certain kinds of regulatory interactions may be possible which cannot be duplicated with purified enzymes in an *in vitro* system.

Control of glycolytic flux is also modulated by the availability of other nutrients, in particular by nitrogen. Nitrogen limitation causes a loss of transporter activity. The inactivation of glucose uptake resulting from nitrogen limitations causes a decrease in overall glycolytic rate. Ammoniacal nitrogen (NH_4+)is an allosteric effector of both phosphofructokinase (PFK) and pyruvate kinase (PK) activities, so its effect on transporter activity may be indirect.

17.2.4 Minor End-Products of Sugar Metabolism

When grape must or juice is inoculated with *Saccharomyces*, alcoholic fermentation is associated with the synthesis of some secondary metabolites. These compounds include different alcohols, esters, organic acids, fatty acids, aldehydes, ketones, lactones, and terpenes, which determine the characteristic aroma and flavor of alcoholic beverages, and their formation has been described below.

17.2.4.1 Higher (Fusel) Alcohol and Polyols and Other Related Compounds

Higher alcohol collectively referred to as fusel alcohol and constitute a major portion of secondary products of yeast metabolism. Those alcohols which contain more than 2 carbons are called as higher alcohols. These are n-propanol, isobutyl alcohol, 2-methyl butanol, isoamyl alcohol and 2-phenyl ethanol. The nitrogenous compound have major role in the formation of higher alcohols. They are mainly produced during amino acid metabolism as well, as from the intermediates of carbohydrate metabolism (Figure 17.4).Glycerol is the main by-product of alcoholic fermentation. No constant ratio of alcohol to glycerol has been established. Low temperature, high tartaric acid, and SO_2 favor the production of glycerol in alcoholic fermentation. Increase in sugar content decreases glycerol content relative to ethanol, and most of it is produced during the early stages of fermentation. Different species of yeasts show differences in their glycerol yield. Moldy grapes contain higher quantities of glycerol and mucic acids. Glycerol is of considerable sensory importance due to its sweet taste and oiliness. Glycerol is formed from dihydroxy acetone phosphate, which is one of the intermediate products of glycolysis. Dihydroxy acetone phosphate is reduced to glycerol phosphate by dihydroxyacetone phosphate reductase enzyme. This reaction requires $NADH^{+}H^{+}$ as a cofactor. Glycerol phosphate yields are illustrated in Figure 17.5. Wine makers believe that increased glycerol production will improve wine quality, leading to better mouth-feel and enhanced viscosity.

Methanol or methyl alcohol (CH_3OH) is a component of alcoholic beverages and is responsible for intoxication, blindness, or even death. Eyes are highly sensitive to formaldehyde, the immediate product of methanol, hence these are affected first. It is also formed as a by-product of alcoholic fermentation under certain conditions. The presence of high levels of pectin in a must leads to the formation of methanol. The methanol is liberated upon pectin esterase action (Figure 17.6).

17.2.4.2 Volatile and Non-Volatile Organic Acids

Grape must and wine contain a variety of organic acids. Their concentrations vary greatly and depend on the condition and maturity of grapes as well as the biochemistry of fermentation. During alcoholic fermentation, the limiting activities of both pyruvate decarboxylase and alcohol dehydrogenase coupled with high sugar concentration lead to the formation of many organic acids. In respiring cells, pyruvate decarboxylase and alcohol dehydrogenase activities are not highly expressed. Both of these enzymes are inducible by glucose. As a consequence, compounds other than ethanol are produced at the beginning of grape juice fermentation. Succinic acid, malic acid, and acetic acid account for the major organic acids formed by yeasts. Succinic acid is the main carboxylic acid produced by yeast during wine fermentation. The ratio of acids varies with different yeasts strains.

17.2.4.2.1 Synthesis of Succinic Acid

This is the main carboxylic acid produced by yeasts during wine fermentation (a concentration upto 2.0 g/L). Addition of sugars to the resting cells of yeasts results in rapid formation of succinic acid. Amino acid and glutamate favor the production of succinic acid. When the yeasts are grown aerobically on glutamate, the oxidative pathway involving succinyl CoA becomes operative and succinic acid is synthesized. The important enzyme, 2-oxoglutarate dehydrogenase of TCA cycle, is present in fermenting yeast, which is responsible for the production of succinate during fermentation by oxidative reactions. A small amount of succinate can also be synthesized by the reductive pathway from oxaloacetate or malate or aspartate. Malate formed from oxaloacetate may be converted to small amounts of succinate by enzymes of the reductive pathway.

17.2.4.2.2 Synthesis of Malic Acid

The synthesis of malic acid during alcoholic fermentation depends upon the yeast strain and cultural conditions. The

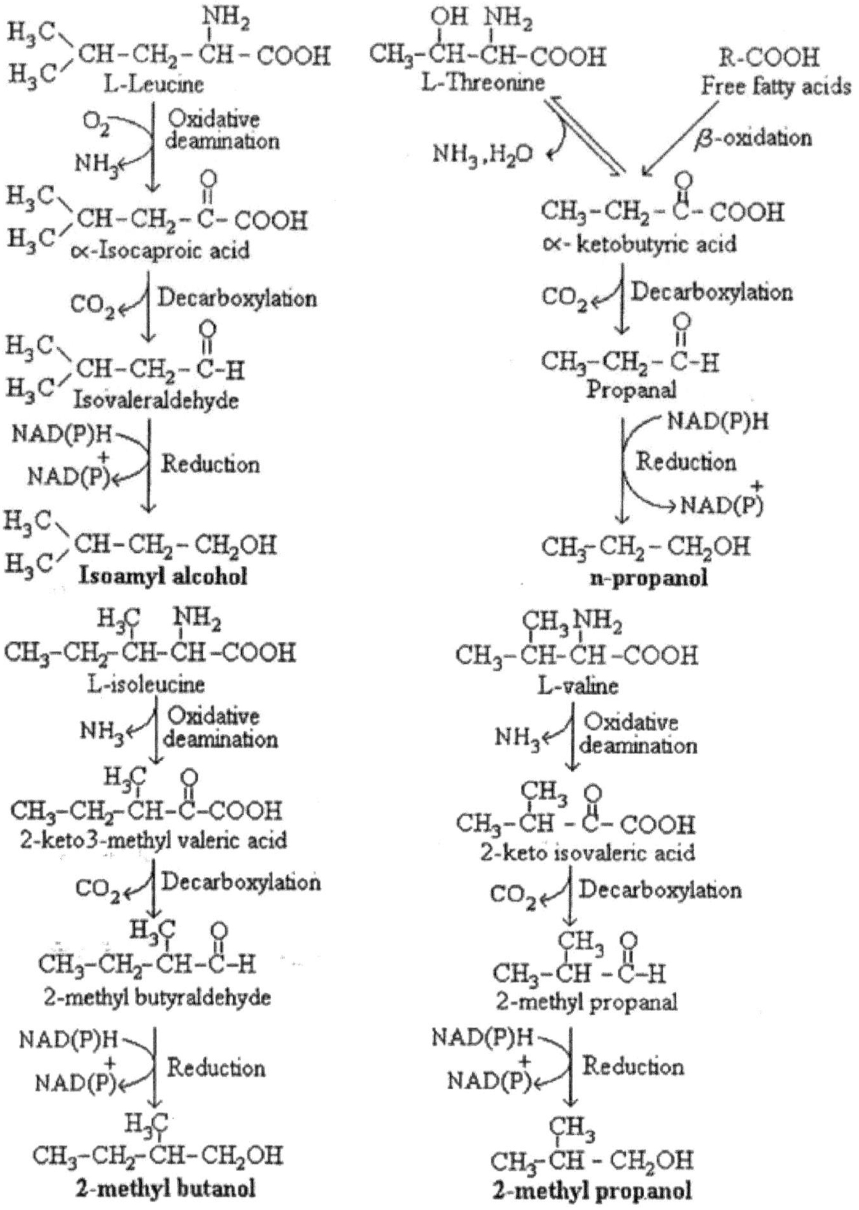

FIGURE 17.4 Biochemical reactions for the synthesis of isoamyl alcohol, n-propanol, 2-methyl butanol, and 2-methyl propanol. R: CH_3-$(CH_2)_n$-

synthesis of malic acid is favored by the concentration of sugars (20–30%), pH value of nearly 5, and limiting concentration of nitrogen compounds (100–200 mg N/L). The presence of CO_2 is also necessary for its formation. The concentration of malate formed is strictly dependent upon the yeast strain. *S. cerevisiae* results in the formation of approximately equal quantities of succinate and malate, whereas the strain *S. uvarum* produces about ten times more malate than succinate. Malate is formed by the reduction of oxaloacetate. The synthesis is catalyzed by pyruvate carboxylase. In wine, a high concentration of amino acid and a low pH may prevent the formation of malate. Malate can only accumulate if it is not further converted to fumarate by fumarase. The enzyme has a higher affinity for fumarate than malate, which shifts the equilibrium of a reversible reaction in favor of malate synthesis.

17.2.4.2.3 Synthesis of Acetic Acid

Acetic acid is produced by many strains of the yeast besides being main product of ethanol oxidation of acetic acid bacteria. Acetic acid is the main volatile acid in fermented beverages and constitutes more than 90% of the volatile acidity in wine. The presence of acetic acid beyond certain limits spoils the alcoholic beverage. Some yeast strains like *Hansenula* produce high amount of acetic acid and hence, are not suitable for brewing. Because of the negative sensory attributes of this acid at higher concentrations and its association with *Acetobacter* spoilage, its production during grape juice fermentation is highly undesirable. Acetic acid appears to be formed early in fermentation (range of 100 to 200 mg/L) by *Saccharomyces* and is also influenced by yeast strain, fermentation temperature, and juice composition especially sugars

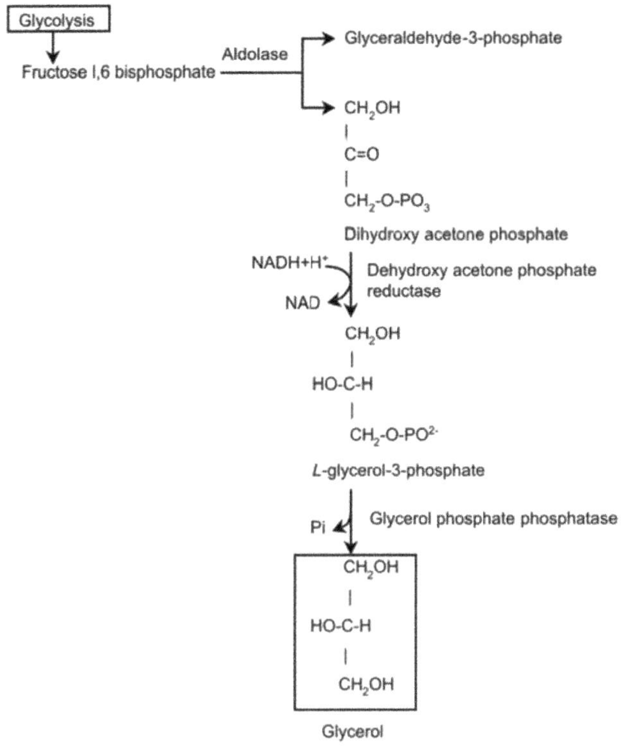

FIGURE 17.5 Pathway showing glycerol production

and nitrogen content of the juice. Under anaerobic conditions, carbohydrates form acetic acid by converting glucose to pyruvate, which is further converted to acetic acid instead of ethanol by the enzyme aldehyde dehydrogenase. Aeration of yeast increases the specific activity of pyruvate decarboxylase, resulting in a higher acetic acid concentration in comparison to anaerobic pitching yeast. Acetic acid is also formed from citrate present in the grape must by wine lactic acid bacteria. These bacteria possess an enzyme, citrate lyase, that can split citrate into oxaloacetate and acetate. Oxaloacetate is again decarboxylated to pyruvate and further to acetate and many other metabolites. Investigation of the production of acetic acid is complicated because during must fermentation, other organisms such as acetic acid bacteria, lactic acid bacteria, and the yeast *Brettanomyces* are also capable of producing acetic acid in high concentrations under favorable conditions. The presence of other micro organisms and the resulting competition with *Saccharomyces* may impact the metabolic activities of the yeast and thus, affect the end products produced.

17.2.4.2.4 Synthesis of Lactic Acid

Lactic acid is a constant by-product of alcoholic fermentation (Figure 17.7) and is a weak acid with slight odor. Usually, yeasts produce only a small amount (0.04 to 0.75 g/L) of free carboxylic acid during fermentation, and lactic acid constitutes a small fraction of these acids. Among several hundred yeast strains investigated, only a few strains have been observed to produce large amounts of lactate. These yeasts belong to the species *Torulopsis pretoriensis* (synonym *S. pretoriensis*) which resembles *S. cerevisiae*. Lactic acid is synthesized from sugar by reduction of pyruvate by α-lactate dehydrogenase.

17.2.4.2.5 Other Acids

Yeasts are able to produce several oxo acids like 2-oxoglutarate, 2-oxobutyrate, 2- oxoisovalerate, 2-oxo-3methyl valerate, and 2-oxoisocaproate. These acids are the products of amino acid metabolism. In addition, grape must also contains low concentrations of pyruvate and 2- oxoglutarate. During fermentation, their amount increases, and pyruvate is later partially metabolized by the yeast. Both acids may be formed from the corresponding amino acids, alanine and glutamate, but they are also excreted by yeast cells growing in the presence of low concentrations of nitrogen compounds. The oxo-acids are important in wine fermentation because they are able to bind to SO_2, thus lowering the content of free SO_2 in wine, which is necessary for the safe preservation of wine. The other fixed acids are of less importance. Glyoxylic acid is found in wine and in grapes. In diseased grapes, upto 0.13% glucuronic and 1.0% gluconic acids may be formed. Only

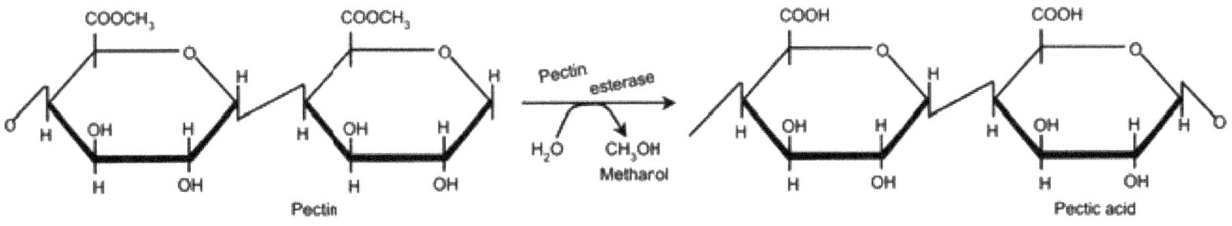

FIGURE 17.6 Methanol formation by de-sterification of pectin

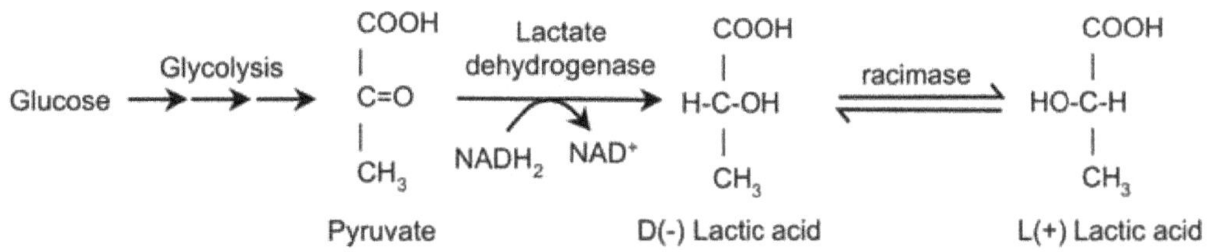

FIGURE 17.7 Mechanism of lactic acid synthesis

traces of fumaric acid are present in grape must, but in musts from moldy grapes it may be upto 130 mg/ L.

17.2.4.2.6 Fatty Acids

Although acetic acid is the main volatile acid comprising of about 90%, a variety of other fatty acids are also present. Except for formic acid, which originates from moldy grapes, other fatty acids like hexanoic acid, octanoic acid, decanoic acid, propionic acid, and butyric acid are also present in trace amounts in the wine. Some of them, in particular octanoic and decanoic formed by yeast, might also act as inhibitors of fermentation.

17.2.5 Decomposition of Organic Acids

Wine yeast (*S. cerevisiae*) is capable of metabolizing malate during fermentation but generally in small amounts, *i.e.* 3–5%, depending on the strain, as is the case of the wine species of *Kloeckera*, *Candida*, *Pichia*, and *Hansenula*. In contrast, however, strains of *Schizosaccharomyces pombe* (Figure 17.8) and *Schizosaccharomyces malidevorans* can completely degrade this acid. The deacidification activity of *Schizosaccharomyces pombe* during plum must fermentation was rapid at pH 3.0–4.5 but was adversely affected at pH 2.5 in the initial stages of fermentation. 150 ppm of SO_2 was effective in enhancing the activity, and allow the yeast to be quite susceptible to a higher concentration of ethanol (5–15%). Malate decomposition is important as the acidity of wine can be reduced by this process. In malo-lactic fermentation (MLF), malate is metabolized to lactic acid and CO_2. The phenomenon is of immense importance to the enologist. The malate is a dicarboxylic acid while the lactate has only one carboxylic acid, so this phenomenon reduces the acidity of wine. In certain wines, MLF fermentation brings about an improvement in flavor. MLF is carried out by the lactic acid bacteria.

The control of lactic acid bacteria during and after vinification is essential in order to obtain wines of consistently high quality. Lysozyme is an enzyme present in hen egg white, and its lytic activity against lactic acid bacteria has been documented. An addition of 500 mg Lysozyme per liter of grape must inhibited malo-lactic fermentation, while 250 mg/l to red wine malolactic fermentation promoted the microbiological stabilization. Among other acids, only traces of fumaric acid are present in grape must, but in musts from moldy grapes, upto 130 mg/ L have been observed. It is also not known whether these small amounts of fumaric acid are affected by yeasts. However, if the fumarate is added to must prior to fermentation, it is metabolized by fumarase to malate, which is then further metabolized by malic enzyme. Thus,fumarate, that is known to inhibit the bacterial malolactic fermentation, cannot be used as an acidulant for grape must before alcoholic fermentation is complete.

17.3 NITROGEN METABOLISM

Yeasts require an exogenous source of nitrogen mainly for the synthesis of proteins, nucleic acid, and also for the growth and metabolism of yeast. Nitrogen is the second important nutrient after carbon which is utilized by yeast during the fermentation of grape must. The major nitrogen components in grape juice are proline, arginine, alanine, glutamate, glutamine, serine, and threonine. Ammonium ion levels and γ-aminobutyrate may also be present in high concentrations in juice, depending upon the variety of grapes and time of harvest. Nitrogen has both positive and negative implications on wine production. It is responsible for the production of reduced sulfur compounds and the formation of ethyl carbamate. On the other hand, it is involved in the biosynthesis of esters, which are responsible for the fermentation bouquet. So, to enhance wine aroma and eliminate sluggish fermentation, it is necessary to improve knowledge about nitrogen metabolism. *S. cerevisiae* can grow on a diverse range of nitrogen compounds including ammonium, urea, amino acids, small peptides, purines, and pyrimidine-based compounds.

17.3.1 Nitrogen Source

In grape juice, nitrogen is present in a complex range of compounds, as their concentration varies according to the types of grape musts. A great variation has also been observed for amino acid concentration and composition in different grape juice varieties. A preferred yeast nitrogen source is one which is most readily converted into biosynthetically useful nitrogen compounds like ammonia, glutamate, or one that requires the least energy input or cofactors for mobilization of nitrogen moiety and should not lead to undesirable metabolites or residues in wine.

17.3.2 Nitrogen Supplements

In addition to the natural nitrogen present in grape juice, must is also supplemented with assimilable nitrogen to avoid

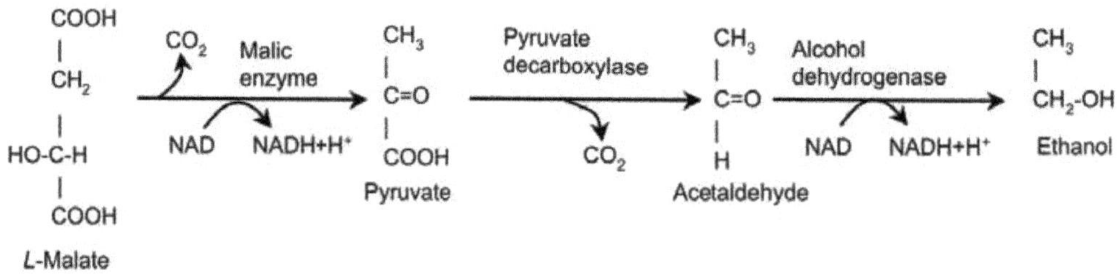

FIGURE 17.8 Pathway showing metabolism of malate to ethanol in *S. cerevisiae* and *Schizosaccharomyces pombe*

the problems associated with nitrogen deficiency during fermentation. Various salts of ammonium can be used for this purpose. Generally, di-ammonium phosphate (DAP) is added to the must prior to inoculation with yeast. Urea and other commercial yeast foods are alternate nitrogen supplements, but the addition of urea is prohibited in most wine-producing countries due to its involvement in the production of ethyl carbamate.

17.3.3 Amino Acid Utilization Profile

Amino acid composition is of great importance in wine production, since amino acids act as a source of nitrogen for yeast during fermentation. Amino acids also have a direct influence on the aromatic composition of wines. The amino acid in wines has a variety of origins. Those indigenous to grape can be partially or totally metabolized by yeasts during the growth phase; some are excreted by yeasts at the end of fermentation or released during the autolysis of dead yeasts, while others are produced by enzymatic degradation of the proteins. The pattern of amino acid utilization during fermentation reflects both the initial distribution of nitrogen compounds as well as strain-dependent preferences. When yeasts were presented with a mixture of amino acids in a model medium, the most important source of nitrogen was arginine, which provided 30–50% of the total nitrogen. Lysine, serine, threonine, leucine, aspartate, and glutamate were the next most accumulated compounds, while glycine, tyrosine, tryptophan, and alanine were the least utilized.

17.3.4 Uptake and Transport of Nitrogen Compounds

In grape juice fermentations, nitrogenous compounds having low concentration are taken up very quickly. Before the degradation of nitrogen compounds, amino acid are supplied for the yeast growth, and when the growth has started, various nitrogen compounds are taken up and degraded in specific order of preference (depending upon environmental, physiological, and strain-specific factors), such as ammonium ion, glutamate, and glutamines, which are utilized directly in biosynthesis. These three nitrogen sources are depleted first from the medium before utilization of other nitrogen sources. The next group of nitrogen compounds in terms of preference includes alanine, serine, threonine, aspartate, asparagine, urea, arginine, proline, glycine, lysine, histidine, and pyrimidine. Thiamine cannot be utilized as a source of nitrogen by most strains of *Saccharomyces* but can be readily utilized directly as a biosynthetic precursor. Metabolism of aromatic amino acids is complex, with some reactions requiring oxygen as a cofactor, that may be limiting during fermentation. The important amino acids, as evident by their uptake coinciding with cell growth, are arginine, glutamate, valine, isoleucine, leucine, histidine, aspartate tryptophan, phenylalanine, and methionine. About a 75–95% removal of these acids causes cessation of growth. The pattern of amino acid utilization during fermentation reflects both the initial distribution of nitrogen compounds as well as strain-dependent preferences. Amino acid transport in yeast is coupled with the movement of ions. In addition to this transport, both D and L isomers of neutral and basic amino acids, and proline to a limited extent, are transported by a group specific transport system, known as general amino acids permease (GAP).

17.3.5 Factors Affecting Nitrogen Accumulation

Many factors influence the accumulation of nitrogen compounds in grape must, and these include cultural conditions, medium composition, and yeast strain (Table 17.1).

17.3.6 Metabolism of Nitrogen

The general aspects of nitrogen metabolism in yeasts have been reviewed extensively. Grape must contains a varying proportion of nitrogenous compounds. Depending upon the type of amino acid accumulated, it may be utilized as follows: i) an amino acid is incorporated directly into the protein; ii) an amino acid can be degraded to liberate nitrogen which is used for the biosynthesis of other nitrogen cell constituents; and iii) the carbon component of an amino acid is released and used for the biosynthesis of other cell carbon constituents. Thus, amino acid/nitrogen compounds are catabolized or anabolized to synthesize various constituents during fermentation through various pathways. Essentially, all nitrogen compounds accumulated are degraded to either of the products ammonium or glutamate. The nitrogen catabolic pathways of yeasts are depicted in Figure 17.9. There have been some recent studies on the pattern of nitrogen compound utilization during grape juice fermentation. The two nitrogen compounds, *i.e.* NH_4^+ and glutamate are required by the cell to coordinate biosynthesis of all biologically-active, nitrogen-containing components. These end products of nitrogen metabolism are inter-converted. In general, two types of glutamate dehydrogenases (GDH) are expressed depending upon the available nitrogen source, and rarely are they expressed equally. $NADP^+$-GDH is expressed when ammonium ions are the sole source of nitrogen in the medium; in contrast, NAD-GDH is expressed when glutamate, aspartate, or alanine are the sole source of nitrogen. Different amino acids or nitrogen compounds catabolize to yield glutamate or NH_4^+ used for various biosynthetic processes. The amino acids lysine, histidine, cysteine, and glycine, are the good sources of nitrogen for many yeasts, but none of these compounds is utilized efficiently by *Saccharomyces* as a nitrogen source.

Glutamate and NH_4^+ are the two main degradation products of nitrogen metabolism, as these two are directly used for biosynthesis. Glutamine generates both glutamate and NH_4^+ and therefore, is also a preferred nitrogen source. In general, most yeast depletes these three compounds first. The next group of compounds includes alanine, serine, threonine, aspartate, asparagines, urea, and arginine. Proline is the nitrogen source under aerobic conditions. Glycine, lysine, histidine, and pyrimidine cannot be metabolized by most yeast strains as a

TABLE 17.1
Factors Affecting Nitrogen Accumulation

Sr. No.	Factor	Research findings	Reference(s)
1.	Ethanol	• Amino acid transport is strongly inhibited by ethanol by increasing the membrane permeability to proton thus inhibiting proton pumping membrane ATPases.	Ferreras *et al.* (1989)
		• Ethanol dissipates proton motive forces across the plasma membrane.	-do-
		• Ethanol may also affect active transport by reduction in number of carriers located in the plasma membrane.	-do-
		• Glucose transport is less sensitive to ethanol showing 50 per cent inhibition at 1.44 M ethanol concentration.	Van Uden (1989)
		• Progressive reduction in active transport activity of general amino acid permease occurs due to ethanol accumulation.	Iglesias *et al.* (1990)
2.	pH	• Active mechanism of amino acid transport through symport of hydrogen ion is affected at low pH and the pH optima of several permeases are generally above the pH of grape juice.	Kotyk and Dvorakova (1990)
3.	Temperature	• The rate of amino acid transport and accumulation decreases at low temperature. Different activation energies for maximum transport of amino acids has been studied and the influence of temperature on the sequence of amino acids is conflicting.	Horak and Kotyk (1977)
		• Rate and extent of uptake is low at 15°C than 20°C. However, ammonium accumulation is largely independent of temperature.	Ough *et al.* (1991)
4.	Carbon dioxide	• Carbon dioxide pressure reduces the rate and extent of amino acid absorption from the wort and grape juice.	Kumada *et al.* (1975)
		• The rates of valine, leucine, and isoleucine absorption are progressively reduced with an increasing CO_2 pressure. Exogenous CO_2 enters into carbon metabolism and causes the increase of α-Keto acid formation.	Knatchbull and Slaughter (1987)
5.	Degree of aeration and plasma membrane composition	• Anaerobic yeast growth condition induces an absolute requirement for the structural lipids, sterols, and unsaturated fatty acids.	Ratledge and Evans (1989)
		• Aeration of grape must sometimes adversely affects wineflavor by utilizing nitrogen sources like proline or urea which are often found in concentrated form in grape must.	Houtman and du Plessis (1986)
		• Small amounts of ergosterol and unsaturated fatty acid, *i.e.* Tween 80, favors yeast growth.	David and Kirsop (1973)
		• The change in membrane composition during fermentation reduces transport efficiency.	Henschke and Rose (1991)
		• Sterols are largely required for structural composition of yeast membrane.	Rattray *et al.* (1975)
		• Phospholipids play role in membrane structure and functions.	Rattray *et al.* (1975)
6.	Yeast strain	• Yeast strain show differences in amino acid utilization pattern, total nitrogen demand and fermentation under limited nitrogen.	Jiranek *et al.* (1990)
		• Epinary – 2 strain accumulate low alanine, especially at low temperature.	Ough *et al.* (1991)

source of nitrogen but can be readily and directly utilized for biosynthetic precursors. Metabolism of aromatic amino acid also does not occur during fermentation, as it is complex and requires oxygen and other cofactors. The ability of arginine to support high growth rates is due to the fact that it is rich in nitrogen, and for symport requires only one proton, making it economical to transport. The enological importance of arginine is evident, based on the observation which shows that this amino acid is the most predominantlyassimilable nitrogen source in Australian juice and alone satisfies 30 to 50% of total yeast nitrogen requirements. Some strains excrete urea during arginine metabolism. The urea is reabsorbed and degraded further when concentration of assimilable nitrogen becomes low. Arginine and proline are the two major amino acids which are found in grape juice.

Degradation of arginine is first catalyzed by arginase and yields ornithine and urea that are further degraded by the enzyme urea amidolyase. Urea amidolyase comprises of two activities, *i.e.* urea carboxylase and allophanate hydrolase, producing two molecules, each of ammonia and carbon dioxide. *Saccharomyces* does not degrade urea by the urease enzyme which is present in many other organisms. Ornithine, the other end product of arginase reaction, can be further degraded via glutamate semialdehyde to proline or can be converted to polyamines. Since proline cannot be metabolized under anaerobic fermentation conditions,

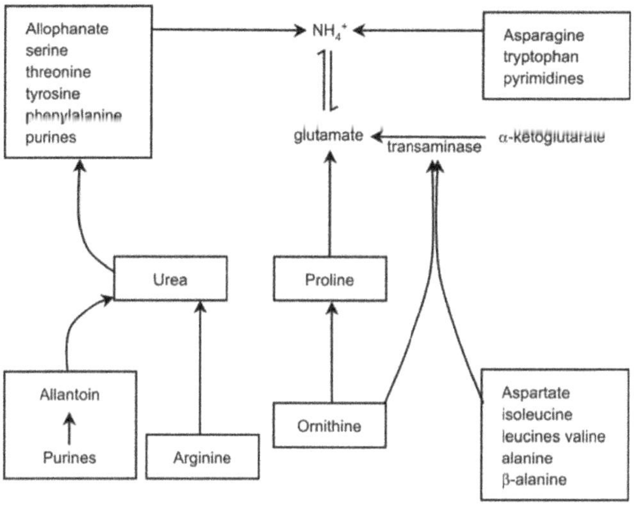

FIGURE 17.9 Major products of nitrogenous compound degradation

ornithine cannot be used as nitrogen source in the absence of oxygen. The high content of arginine thus leads to an increase in the proline content, and therefore, wine contains a higher concentration of proline than the musts. However, ornithine also acts as a precursor of polyamines, spermines, spermidine, and putrescine. These polyamines are needed at higher concentrations during growth, although the exact physiological function is not known. More than 20 amines have been identified in wines. Most reports on amines in fermented beverages and their substrates deal with the non-volatile amines, mainly histamine and tyramine, due to their biogenic effects. Histamine plays a special role as an indicator amine to assess the freshness and quality of wine. When there is no histamine, there are no other biogenic amines.

The metabolism of nitrogen-containing compounds and their end products are of sensory importance. Amino acids that are deaminated catabolically in order to release their nitrogen components leave behind a carbon skeleton which is regarded as a waste product. Deamination of amino acids can result in the formation of αketo acids or higher (fusel) alcohols via a metabolic pathway as shown in Figure 17.10. In addition to deamination and decarboxylation, higher alcohols are also produced during the biosynthesis of aminoacids from their corresponding keto acids. The principal higher alcohols produced by the yeast are the aliphatic alcohols like n-propanol, isobutanol, amyl alcohol, isoamyl alcohol, and aromatic alcohols. Although they exhibit harsh unpleasant aromas at a concentration below 300 mg/L in the wine, they are considered desirable. Various factors, viz. yeast strain, growth, temperature, ethanol production, must pH, level of solids, and degree of aeration, affect the production of alcohols.

17.4 SULFUR METABOLISM

Assimilable sulfur is essential for the growth of yeast, and the growth of yeasts is poor or negligible under the deficiency of sulfur. All yeasts utilize sulfate, sulfite, or thiosulfate as a source of sulfur for biosynthesis. Organic sulfur compounds such as cysteine, methionine, homocysteine, and S-adenosyl methionine can also serve as sulfur sources. Thiosulfate is first cleaved to sulfite and sulfide prior to utilization, and therefore, both sulfur atoms can be used. Utilization of sulfite is however limited by the toxicity of this compound. The deficiency of sulfur causes poor yeast growth and cells are largely depleted of glutathione. A sulfate compound, giving 10 mg/L of total sulfur is sufficient for the growth of *S. cerevisiae*. The maximum permitted total sulfur dioxide (SO_2) concentration in wine is 350 mg/L, but its concentration in wines produced in the United States was found to be 74.1mg/L. Yeasts differ in their requirement of sulfur which contains vitamins biotin and thiamine, so for optimum growth, supply of these vitamins is necessary. Biotin deficiency causes an inhibition of glucose and fructose during fermentation and reduces C_{18} fatty acids, DNA, RNA, and total protein contents. Sulfate is one of the most common sulfur sources, and its complete reduction pathway has been studied.The sulfate reduction proceeds by the formation of phosphosulfate intermediates. Sulfate anion in the first step is activated by using phosphate (ATP) and forms adenosine 5-phosphate (APS) followed by the formation of 3-phospho adenosine-5-phosphosulfate (PAPS). The initial reaction is catalyzed by ATP sulfurylase. The reduction of PAPS to sulfite and 3,5-diphosphoadenosine (PAP) involves PAPS reductase and $NADPH^+$ H^+ as an electron donor. In the initial step, sulfate is transferred from PAPS to the carrier proteins having one or more thiol group. This carrier protein is then reduced by ferredoxin to form sulfite. The bisulfate form of protein is then reduced by NADPH, and another molecule of PAPS can act on it. The intermediate, sulfite formed can be excreted to the medium either actively with sulfite permease or by diffusion.If sulfite is not excreted, it is reduced to sulfide by sulfite reductase. There appears two distinct sulfite reductases in *Saccharomyces,* one that utilizes carrier-bound sulfite and another that utilizes free sulfite. The

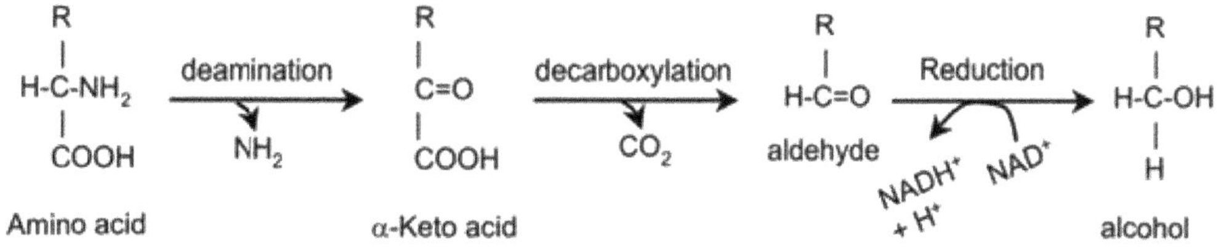

FIGURE 17.10 Pathway of higher alcohol formation from amino acid

reduced sulfur is utilized in various pathways and results in the formation of different sulfur compounds.

17.4.1 Assimilation of Reduced Sulfur

Reduced sulfur is incorporated into carbon compounds to synthesize sulfur-containing amino acids, like cysteine and methionine. Two pathways exist to form cysteine. The first pathway involves the acetylation of serine by serine acetyl transferase to form O acetyl serine, followed by sulfydration to form cysteine using free sulfide or bound sulfide. In the second pathway, the condensation of homocysteine and serine by βthione synthase, followed by cleavage by cystathionelyase, leads to the formation of cysteine. The formation of methionine begins with an acetyl CoA dependent acetylation of homoserine by the enzyme, homoserine acetyl transferase to form O-acetyl homoserine. These are the two possible ways to transform O-acetyl homoserine to homocysteine. One is direct sulfydration of O-acetyl homoserine using free or bound sulfide and to form homocysteine. The second possibility is the condensation of O-acetyl homoserine to cysteine by cystathionin synthase to form cystathionin. Then cystathionine splits by βcystathionase to form homocysteine pyruvate and ammonium. It is not clear which of the two pathways is predominant.

Methionine is then synthesized from homocysteine by methyl derivative of tetrahydrofolate, and the enzyme is known as homocysteinemethyltransferase. Methionine can also act as a precursor of cysteine. The biosynthesis of methionine and cysteine is regulated by feedback inhibition and control of enzyme formation. Cysteine and methionine are necessary for the synthesis of peptides (glutathionine) and proteins. The cysteine sulfhydryl group is important for the catalytic activities of numerous enzymes and is involved in binding of metal ions. Figure 17.11 illustrates the utilization and production of various metabolites by the yeast during wine production.

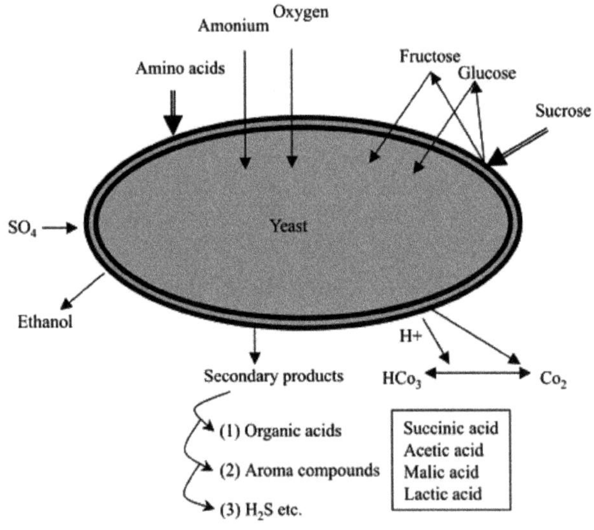

FIGURE 17.11 Diagram showing utilization and formation of various compounds by yeast during wine production

17.5 PRODUCTION OF OFF-FLAVOR

This is the main problem in the production of quality wine. The main components responsible for off-flavors are acetic acids, sulfur containing volatiles, free amino nitrogen, vitamins, and other factors. The mechanism of the formation of volatile sulfur compounds (VSC) during wine fermentation is only partially understood, due to the complex nature of the factors involved in winemaking. The production of VSCs varies widely in composition in model systems with nutrient deficiencies or with residual elemental sulfur. These are spoilage-causing compounds which are very volatile and have unpleasant odors. Although their production is very minute (10–100 μg/L), their sensory impact is huge during fermentation and poses a significant problem to wine makers. This group includes compounds which are very volatile and that have unpleasant odors, generally described in terms of rotten egg, skunk aroma, garlic, or onion etc. Among sulfur-containing volatiles, hydrogen sulfide (H_2S) creates the biggest problem with a rotten egg-like flavor. Several factors may contribute to H_2S production. It can beproduced by yeast during fermentation due to several factors: the presence of elemental sulfur ingrape skin; inadequate levels of free α- amino nitrogen (FAN) in the must; and a deficiency of pantothenic acid or pyridoxin or higher than usual level of cysteine in the juice and yeast strains. The odor threshold value of H_2S in wine varies from 50–80 μg/L, and above this concentration, it causes an off-flavor. Sulfate uptake and reduction is regulated by sulfur containing amino acids *i.e.* methionine. In general, the highest concentration of H_2S is produced during the rapid phase of fermentation. The deficiency of assimilable amino acids may be a major factor for the H_2S production at the later stages of fermentation.

Sulfur elements present in grapes and musts may undergo enzymatic and non-enzymatic reactions to produce H_2S during fermentation. The unavailability or deficiency of free amino nitrogen in the must also causes H_2S formation by yeast. It has been also suggested that a low concentration of the free amino nitrogen might stimulate proteases, which cause hydrolysis of juice proteins. A deficiency of the vitamins pantothenate or B_6 in the medium can cause an increased H_2S production by yeasts. A deficiency of pantothenate leads to the deficiency of CoA, which is essential for methionine biosynthesis. Pyridoxin is necessary for several reactions in the pathway of methionine synthesis, and a large amount of H_2S is producedif the concentration is below 2 mg/L.Normally, most of the H_2S formed during fermentation is carried away with carbon dioxide, so in normal faultless wines, H_2S is not detectable despite a detection limit of 1.5 μg/L. In most cases, H_2S disappears after fermentation slowly. Increased amounts of H_2S can be purged out or oxidized by aeration or oxidized by treatment with sulfite. Several other sulfur compounds are also detected, which are responsible for the off-flavor,such as dimethyl sulfide (concentration ranging 0–47 mg/L), synthesized from S-methylmethionine. Dimethylsulfide, with the flavor of cooked vegetables, onion, and garlic, and mercaptans, having a rotten egg-like flavor, are formed in synthetic media during fermentation containing cysteine, methionine or sulfate and thioester like S-methyl thioacetate. The factors leading to production of these compounds have not been thoroughly elucidated.

17.6 FERMENTATION BOUQUET AND YEAST FLAVOR COMPOUNDS

Yeast strains conducting a clean fermentation without any negative characteristics produce a special fruity aroma in the wine, known as a fermentation bouquet. It is very attractive odor in young wines and can be quite stable in wines kept cold. It is, however, very unstable and disappears rapidly at room temperature. This instability arises because the flavor is caused by volatile esters and the equilibria at the conditions in wine favors hydrolysis. The special fruity odor is primarily due to a mixture of hexyl acetate, ethyl caproate, and isoamyl acetate in the ratio of about 3:2:1. Odor-wise, the hexyl acetate appears to be the most important and isoamyl acetate the least important to the special fermentation bouquet. There is no apparent way, other than low temperature, to stabilize these esters in the wines. The other compounds of fermentation bouquet are the organic acids, higher alcohols, and to lesser extent aldehydes, which are influenced to various degrees by the nitrogen source. Lower fermentation temperature (about 15°C) encourages the production of volatile esters by yeasts. Two major groups of esters are formed during fermentation, the ethyl esters of straight chain fatty acids, and acetates of higher alcohol. Some of the fatty acid CoA takes part in ester formation by reacting with alcohol. Its formation is positively correlated with the must total nitrogen. The two important esters of the fermentation bouquet are also positively correlated with must nitrogen. The concentration of acetate ester is more dependent on the yeast strain and sugar concentration than the fatty acid ester. The acetate esters of ethanol and the higher alcohols often provide a major aroma impact in freshly prepared wines. The formation of the fermentation bouquet is little influenced by the cultivars involved or the yeast strain conducting fermentation. There are, however, some strains that can produce significantly higher concentrations of individual esters. The fermentation bouquet compounds such as acetaldehyde, acetic acid, ethyl acetate, higher alcohols, and diacetyl, if present in excess, are regarded as undesirable. Esters can also be produced from the alcoholysis of acyl CoA compounds which occur when fatty acid biosynthesis or degradation is interrupted in the cell and generate free coenzymes. Esters can also be formed from the carbon skeleton of amino acids. The pathway for bouquet formation in wine fermentation is depicted in Figure 17.12.

The higher alcohols which react with acetyl CoA are produced from sugar, amino acid, and sulfate during yeast metabolism. Small acyl chain esters are typically fruity or floral, while the longer acyl chain esters are sweeter or soap-like. If they have more than 12 carbons, they are not very volatile and hence, do not have much odor effect. The formation of esters is influenced primarily by temperature, the amino nitrogen content of the juice, and the yeast strain employed.

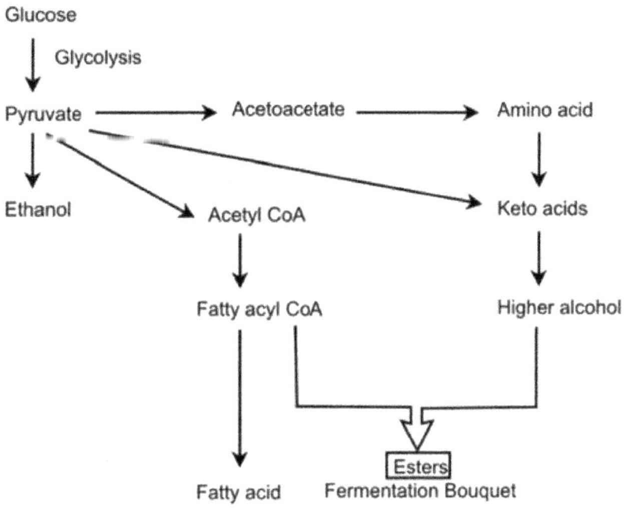

FIGURE 17.12 Pathway of synthesis of fermentation bouquet

The factors which affect the acyl CoA production also affect the ester formation.

In fact, a very large number of compounds play a role in determining the flavor of alcoholic beverages. Over 200 compounds have been identified in cider that influence the flavor. The number may be higher for other alcoholic drinks. The contribution of several volatile thiols to the aromas made from different grape varieties has been investigated more recently. Among these, 4-mercapto-4-methylpentan-2-one (4MMP), 3-mercaptohexan-1-ol (3MH), and 3-mercaptohexyl acetate (A3MH) were found considerably higher in certain grape varieties. The important flavoring compounds in alcoholic beverages include esters, alcohols, carbonyls, acids, sulfur derivatives, and phenolic compounds.

BIBLIOGRAPHY

David, M.H. and Kirsop, B.H. (1973). Yeast growth in relation to the dissolved oxygen and sterol content of wort. *J. Inst. Brew.*, 79(1):20–25.

Ferreras, J.M., Iglesias, R. and Gibbes, T. (1989). Effect of the chronic ethanol action on the activity of the general amino-acid permease from *Saccharomyces cerevisiae* var. *ellipsoideus*. *Biochim. Biophys. Acta*, 979(3):375–377.

Goni, D.T. and Azpilicueta, C.A. (2001). Influence of yeast strain on biogenic amines content in wines: Relationship with utilization of amino acid during fermentation. *Am. J. Enol. Vitic.*, 52(3):185–190.

Henschke, P.A. and Rose, A.H. (1991). Plasma membranes. In: *The Yeasts*, A.H. Rose and J.S. Harrison, eds., pp. 297–345, Academic Press, London.

Horak, J. and Kotyk, A. (1977). Temperature effects in amino acids transport by *Saccharomyces cerevisiae*. *Exp. Mycol.*, 1(1):63–68.

Houtman, A.C. and du Plessis, C.S. (1986). Nutritional deficiencies of clarified white grape juices and their correction in relation to fermentation. *S. Afr. J. Enol. Vitic.*, 7(11):39–46.

Iglesias, R., Ferreras, J.M., Arias, F.J., Munoz, R. and Girbes, T. (1990). Changes in the activity of the general amino acid permease from *Saccharomyces cerevisiae* var. *ellipsoideus* during fermentation. *Biotechnol. Bioengg*, 36(8):808–810.

Jiranek, V., Langridge, P. and Henschke, P.A. (1990). Nitrogen requirement of yeasts during wine fermentation. In: *Proceedings of the 7th Australian Wine Industry Technical Conference*, P.J.Williams, D.M.Davidson and T.H.Lee, eds., p. 181, Australian Industrial Publishers, Adelaide, Australia.

Joshi, V.K., Panesar, P.S., Rana, V.S. and Kaur, S. (2017). Science and technology of fruit wines: An overview. In: *Science and Technology of Fruit Wine Production*, M.R.Kosseva, V.K.Joshi and P.S.Panesar, eds., pp. 1–58, Elsevier Inc., the Netherlands.

Joshi, V.K., Thakur, N.S., Anju, B.and Chayanika, G. (2011). Wine and brandy: A perspective. In: *Handbook of Enology*, Vol. 1, V.K.Joshi, ed., pp. 3–45, Asia Tech Publishers, Inc., New Delhi.

Knatchbull, F.B. and Slaughter, J.C. (1987). The effect of low CO_2 pressures on theabsorption of amino acids and production of flavour-active volatiles by yeast. *J. Inst. Brew.*, 93(5):420–424.

Kotyk, A. and Dvorakova, M. (1990). Transport of L-tryptophan in *Saccharomyces cerevisiae. Folia. Microbiol.*, 35(3):209–217.

Moreno-Arribas, M.V. and Polo, M.C. (2005). Winemaking biochemistry and microbiology: Current knowledge and future trends. *Crit. Rev. Food Sci. Nutr.*, 45(4):265–286.

Ough, C.S., Huang, Z., An, D. and Stevens, D. (1991). Amino acid uptake by four commercial yeasts at two different temperatures of growth and fermentation: Effect on urea excretion and reabsorption. *Am. J. Enol. Vitic.*, 42:26–40.

Ratledge, C. and Evans, C.T. (1989). Lipids and their metabolism. In: *The Yeasts*, 2nd ed., Vol. 3, A.H.Rose and J.S.Harrison, eds., pp. 367–455, Academic Press, London.

Rattray, J.B.M., Schibeci, A. and Kidby, D.K. (1975). Lipids of yeasts. *Bacteriol. Rev.*, 39(3):197–231.

Van Uden, N. (1989). *Alcohol Toxicity in Yeasts and Bacteria*. CRC Press, Boca Raton, FL.

18 Malolactic Fermentation in Winemaking

Eveline Bartowsky

CONTENTS

18.1 INTRODUCTION

The term 'malolactic fermentation' (MLF) is derived from the bacterial conversion of L-malic acid to L-lactic acid and generally encompasses all the metabolic aspects of bacterial metabolism in wine during this process, not only the important deacidification reaction. The role of MLF is three fold; wine deacidification by the conversion of L-malic acid to the 'softer' L-lactic acid, microbial stability of wine and wine flavour modification. Historically, early microbiologist, Müller-Thurgau had noted the bacterial causes of MLF, and over the next few decades its significance was realised with the perception that these wines were of a higher quality. The majority of red wines, a number of white wines and sparkling wines can undergo MLF. It is particularly beneficial to the wines made from grapes grown in a cool climate, where the acid content is high.

The main bacterium responsible for MLF was originally classified as *Leuconostoc oenos* by Garvie, however it was renamed *Oenococcus oeni* by Dicks following an examination of the *Leuconostoc* genus using molecular techniques. Numerous species in the Lactic Acid Bacteria (LAB) family are able to perform MLF, however *Oenococcus oeni* is the LAB species most adapted to the harsh environment of wine. Most often MLF will occur after the alcoholic fermentation, but it is not limited to this stage of vinification. Its occurrence in the early stage of alcoholic fermentation gives a simultaneous MLF with alcoholic fermentation. The initiation of MLF by the indigenous bacterial population was unpredictable, however the introduction of commercially prepared bacterial cultures for inoculating wine for MLF in the 1990s provided better control over the MLF process.

18.2 DEACIDIFICATION BY MALOLACTIC CONVERSION

L-malic acid is one of the two principal organic acids present in grapes, the other being tartaric acid. The concentration of L-malic acid in grape juice from cool climate regions can often be 2–5 g/L, whereas in warm climate regions, the grape juice often contains less than 2 g/L. The bacterial conversion of L-malic acid to L-lactic acid is well understood and has been established as being a direct decarboxylation, with NAD and Mn^{++} as cofactors and without free intermediates. The single enzyme which catalyses this reaction is malate decarboxylase, often referred to as the malolactic enzyme. This enzyme is different from malic enzyme, malate dehydrogenase or other malate decarboxylases. The major end product with other malate decarboxylases is either pyruvic acid or oxaloacetic acid.

$$\text{L-malic acid} \xrightarrow{NAD^{+}Mn^{++}} \text{L-lactic acid} + CO_2 \quad (18.1)$$

Two consequences of the deacidification reaction are an increase in pH and a decrease in titratable acidity. The perception of sourness is a consequence of the titratable acidity. Wines produced from grapes grown in cool climate regions often have a high acid content, thus benefit from deacidification, whereas those produced in warmer regions often do not require the MLF, as the resultant wine would have too little acidity and too high a pH. For the latter wines, however, it is usual to add acid (typically tartaric acid) so that MLF can be undertaken so as to give the flavour benefits of MLF.

The rate of MLF in *O. oeni* is regulated by the rate of malate transport and mediated via the L-malate specific permease. When the external malate concentration is greater than 5 mM, malate-dependent ATP will be produced during the decarboxylation reaction. By the early 1990s, it had been established that the MLF reaction confers an energetic advantage to the cell from an increased intracellular pH and an increased proton motive force (Δp), and a model was proposed (Figure 18.1). In this model, one molecule of malate which enters the cell is decarboxylated and one molecule of lactate leaves the cell with one H^+, which is equivalent to the translocation of one H^+ to the external environment of the cell. The export of the lactate provides energy through the proton gradient for transport processes, and this can be converted by the membrane ATPase into ATP, energy for the cell. The ATP yield for *O. oeni* cells from low glucose concentration (2 g/L) is much higher (55% ATP yield) than high glucose concentration (10 g/L glucose and 10% ATP yield). Thus, when malate is metabolised under low rather than high carbohydrate conditions, the increase in ATP yield from the low glucose concentration metabolism is more significant for the bacterial cell. The rate of malate decarboxylation by malolactic bacteria is directly related to the cell density and to the specific malolactic activity of the bacterial cell. A significant rate of L-malate metabolism is not usually observed until the cell density exceeds 1×10^6 cells/mL.

The metabolic conversion of malic acid is usually also accompanied by the metabolism of citric acid by malolactic bacteria. Citric acid is usually present in low concentrations in wine (0.1–0.7 g/L), but its metabolism can have important consequences on the formation of flavour compounds, in particular diacetyl in wine (see Section 18.7). The relationship of the metabolism of various organic acids and sugars is also discussed further in Section 18.7.

18.3 BACTERIOLOGICAL STABILITY FOLLOWING MALOLACTIC FERMENTATION

An important outcome of MLF is the stabilisation of wine against further growth by other LAB. Some LAB species have a malate dehydrogenase and thus are able to use malic acid as a carbon source. The increase in wine pH, usually by 0.2–0.5 units, may allow for a second and even a third round of growth of other resident LAB species, and although suppressed by the acidic conditions prior to MLF, these bacteria may now proliferate. Nevertheless, with good wine management practices, wines are usually rendered microbiologically stable following MLF. Spoilage which does occur post-malolactic fermentation is more likely to be due to yeast, or acetic acid bacteria. For more details on wine spoilage, see Chapter 41.

18.4 INOCULATION AND CONTROLLING THE MALOLACTIC FERMENTATION

Malolactic fermentation can either occur naturally (spontaneously) by the indigenous bacterial population or can be

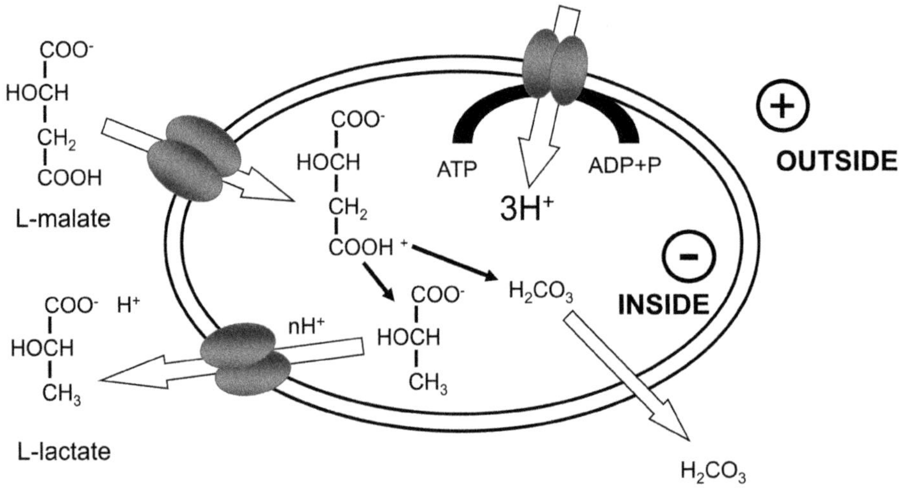

FIGURE 18.1 Model of the ATP-generating mechanism for the malolactic conversion. *Source:* Adapted from Henick-Kling in Fleet (1993)

induced by the winemaker with a selected bacterial culture. It can also be either deferred or inhibited in several ways. Wine conditions can be made less favourable for the bacteria, such that there is minimal growth or a low initial bacterial population. Alternatively, the bacteria can be eliminated from the wine.

18.4.1 Inoculation for Malolactic Fermentation

Spontaneous MLF in wine, conducted by indigenous LAB, originating from the vines and grape skins and surviving on winery equipment, can often be quite unpredictable, and may begin many months after alcoholic fermentation has completed. More often, MLF occurs in low acid wines, where a further reduction in acidity is not usually desired, and the extent of flavour modifications are unpredictable due to the different micro-flora which may have carried out the MLF. Spontaneous MLF becomes even more unpredictable as winery hygiene is improved, and typically, it occurs in wines above pH 3.4, as low pH strongly inhibits the growth of LAB. Different bacterial populations will dominate throughout the spontaneous MLF, especially *Lactobacillus* and *Pediococcus* species in wines with a higher pH. Undesirable flavour characters including mousy, bitter and acetic, as well as a ropy texture, are more likely to be caused by *Lactobacillus* and *Pediococcus* species. The development of further flavour defects after MLF in the presence of *Lactobacillus* and *Pediococcus* species, which can utilise small amounts of sugar, may also occur. The growth of various yeasts such as *Brettanomyces* can also further modify the wine post MLF.

Wine conditions that are conducive for MLF are summarised in Table 18.1. To have more control over the time of onset and rate of MLF, as well as to reduce the potential for wine spoilage by other bacteria, winemakers have moved to inoculate wines with a known bacterial culture. Inoculation for MLF with a starter culture may occur at any stage throughout the alcoholic fermentation, but, most commonly, it is induced either towards completion of or at the end of alcoholic fermentation. However, co-inoculation of bacteria, usually 24–48 hours after the yeast inoculation, is becoming more common. The development of commercial starter cultures of *O. oeni* strains over the last 15–20 years has greatly facilitated the induction and success of MLF. The bacterial starter culture needs to be pre-adapted to the harsh wine environment (low pH, high alcohol content, low nutrient status and SO_2 content). A loss of initial viability was very high with early commercial industry starter cultures, due to the lack of adaptation to the wine. Early starter cultures of *O. oeni* strains required one or more reactivation and adaptation steps to the wine before use, in order to enhance the survival of the bacteria. This reactivation and pre-adaptation of the starter culture can be time consuming and labour intensive. By the mid 1990s, freeze-dried concentrated bacterial starter cultures were developed which could be added directly into wine, without the reactivation and pre-adaptation steps.

TABLE 18.1
A General Guide of Wine Conditions for MLF

Conditions	Wine type	Limiting	Optimal
Temperature (°C)	White	< 16 > 24	18–22
	Red	< 15 > 25	18–22
pH	White	< 3.1	> 3.2
	Red	< 3.0	> 3.1
Alcohol (% vol.)	White	13.5	< 12.5
	Red	14	<13
Total SO_2 (mg/L)	White	> 30	< 15
	Red	> 30	< 15

Numerous commercial strains of *O. oeni* and a few specially selected strains of *Lactobacillus plantarum* are available, either as a direct inoculation product or as a standard culture which requires one or two reactivation steps. To use these starter cultures effectively, the winemaker must evaluate the wine's chemical properties, pH, alcohol content and free and total sulphur dioxide concentration, to ensure a successful bacterial establishment and MLF induction.

18.4.2 Cultivation of Wine Related-Lactic Acid Bacteria

Wine-associated LAB species tend to be nutritionally fastidious, often slow growing, and require complex cultivation media which includes pantothenic acid (4'-O-(β-D-glucopyranosyl)-D(R)-pantothenic acid; tomato juice factor). Artificial media are complex and often contain peptone, tryptone or yeast extract. A medium, originally developed for the cultivation of lactobacilli MRS medium, is usually used for the cultivation of LAB and often is enriched with a pantothenate source such as tomato juice or apple juice. Wine LAB tend to grow much better when the pH is adjusted to 4.5 or lower. When isolating bacteria from a wine sample, the inclusion of a yeast inhibitor, such as cycloheximide, is recommended.

18.4.3 Constraining Bacterial Growth

The wine conditions that would be antagonistic to bacterial growth can also be stimulatory. In general, the control of these conditions is not difficult. Sulphur dioxide (SO_2) has been used in the wine industry for centuries as an antioxidant and antimicrobial agent. It is the molecular form of SO_2 which exerts the antimicrobial effect, and only 5–10% sulphur dioxide is present in the molecular form in wines with pH 3 and close to zero in wines with pH 4. Usually, recommended concentrations of molecular SO_2 in wine (approximately 0.8 mg/L) and taking the temperature and wine pH into consideration with a low population of bacteria will only delay MLF rather than prevent it from occurring. Sulphur dioxide has been shown to induce a stress response in *O. oeni* and at concentrations of 20 to 40 mg/L; SO_2 will decrease the ATPase activity of *O. oeni*, up to 70%. Thus, free SO_2 diffuses into the cell, some of which is converted into the sulphite form which may react with proteins, nucleic acids and can inhibit enzymes, such as ATPase, resulting in reduced cell viability, even death, and decreased MLF activity. Wine acidity cannot be used alone as a mechanism to reduce the risk of MLF starting. Low pH and acid additions (such as is provided by tartaric acid) can inhibit bacterial growth, however, other wine parameters will come into play. Many spoilage bacteria, *Lactobacillus* and *Pediococcus*, do not readily proliferate in wine below pH 3.5. Wines with a pH below 3.3 are generally more microbiologically stable than those above 3.3–3.5. Wines with high pH should be acidified with tartaric acid, or treated with SO_2 and stored at low temperature. Concentrations of ethanol over 14% will hinder bacterial growth and the initiation of MLF. The wine ethanol concentration, like wine pH and fermentation temperature are not necessarily inhibitory but rather restrictive.

18.4.4 Elimination of Viable Bacteria

The risk of MLF occurring in bottled wine must be reduced and is usually most important in white wines which have not undergone MLF. White wines can often be a blend of wines which have or have not undergone MLF. The method which will be used to eliminate bacteria from the wine must be chosen and considered carefully so as to have minimal or no effect on the sensory properties of the wine. Methods which can be employed for the inhibition or elimination of bacteria from wine include sterile filtration, heating the wines and treating with different chemical inhibitors. The use of bacteriophage to eliminate bacteria should be approached with extreme caution. Alternative options for the elimination of bacteria from wine include the treatment of the wine with ultraviolet radiation, high pressure pulsing or high powered ultrasonics, but the long-term effects of these on wine flavour are not clear.

18.4.4.1 Treatment of Wine with Chemical Inhibitors

The interest in the use of chemical inhibitors for the elimination of bacteria from wine has increased over the last years, with pressure to reduce the concentrations of SO_2 in the wine. Dimethyl carbonate (DMDC, also known as dimethyl pyrocarbonate [DMPC]) has been shown to be an effective inhibitor of yeast but is also effective against LAB commonly found in wine. This compound is an inactivator of several glycolytic enzymes (including yeast alcohol dehydrogenase and glyceraldehyde-3-phosphate dehydrogenase) and hydrolyses quickly to methanol and carbon dioxide (within one hour at 30°C and within five hours at 10°C), which are natural constituents of grape juice or wine, not affecting the taste, flavour or colour of wine. The effectiveness of DMDC on yeast and bacteria can be enhanced when applied in conjunction with free SO_2 in wines with a pH of 3.6 or lower. DMDC is an approved additive in the United States, South Africa, Australia and New Zealand.

18.4.4.2 Treatment of Wine with Natural Products

The use of a 'natural product' such as a protein to inhibit bacterial growth in wine has been successful in numerous food industries. Lysozyme has been used successfully in both the pharmaceutical and food industries for almost 50 years. Bacteriocins, particularly nisin, have been successful in the dairy industry. Chitosan is a more recent natural product that inhibits bacteria.

18.4.4.2.1 Lysozyme

Lysozyme is a natural bacteriolytic enzyme that can have beneficial effects in winemaking by either inhibiting bacteria or delaying the onset of MLF. It is a low molecular weight single peptide (129 amino acids, 14.4 kDa) with muramidase (EC 3.2.1.17) activity, and was originally isolated by Alexander Fleming in 1922 from human mucosal secretions. This enzyme is present in various bodily fluids and is an important part of the eukaryotic immune system, playing a key role in the defence of bacterial infections. Lysozyme is effective against prokaryotic but not eukaryotic cells. It causes the lysis of Gram-positive bacteria cell walls, but not Gram-negative bacteria, by cleaving the β(1-4) glycosidic linkage between *N*-acetylmuramic acid and *N*-acetylglucosamine in the peptidoglycan layer. Thus, lysozyme will be effective in grape juice and wine against genera of LAB (*Lactobacillus*, *Pediococcus* and *Oenococcus*). However, it is not effective against spoilage acetic acid bacteria (*e.g. Acetobacter*) and yeast (*e.g. Brettanomyces*).

In recent years, there has been considerable interest in the use of lysozyme to control the proliferation of spoilage wine LAB genera during vinification, as an alternative to the antimicrobial activity of SO_2. Lysozyme only replaces the antibacterial action of SO_2 and not the antioxidative activity, thus it cannot eliminate the use of SO_2 in winemaking completely. Lysozyme is usually added at a rate of 250–500 mg/L to wine at the end of alcoholic fermentation as a method to prevent or delay MLF. Unlike SO_2, lysozyme is effective at higher wine pH. As with all treatments of wine, the addition of lysozyme must be considered carefully, as it is able to bind with tannins and polyphenols in red wines and typically results in a slight decrease of wine colour and a loss of enzymatic activity. The addition of lysozyme (protein) into a white wine might result in the formation of a wine haze but does not affect the flavour qualities of the wine. Table 18.2 summarises the factors that can affect lysozyme activity in wine.

Lysozyme is an approved wine treatment in Europe, Canada, the United States and Australia. However, as lysozyme is an egg product, several countries might require specific label statements when it is used.

18.4.4.2.2 Bacteriocins

Bacteriocins are small polypeptides produced by some LAB species and are inhibitory to other bacteria. Nisin, produced by *Lactococcus lactis* has been widely used in the dairy industry. Nisin acts on the cell wall, however lysis occurs via a different mechanism from lysozyme. The nisin molecule adheres to the cell wall and alters the cell membrane, resulting in leakage of low molecular weight cytoplasmic components and destruction of the proton motive force. As nisin and lysozyme act

TABLE 18.2
Factors Affecting Lysozyme Activity in Wine

Wine variable	Effect on lysozyme activity
Wine type	Effective in both red and white wines
Initial lactic acid bacteria concentration	Effective against bacteria at 10^6 cfu/mL and lower
Strain of lactic acid bacteria	Variation between genera, species and strains of bacteria
Wine pH	Activity increases as the wine pH increases
Wine alcohol content	No influence on lysozyme activity
Phenolics	Wine phenolics may decrease lysozyme activity Lysozyme may precipitate with phenolics
Protein stability (haze)	May induce haze formation in white wines
Bentonite	Decrease in antimicrobial effect if used in conjunction with bentonite
Other fining agents (Carbon, silica sol, oak chips and tannin)	These will bind lysozyme and result in reduced lysozyme activity
Sparkling wine foaming properties	No effect or may slightly increase the foamability
Other wine additives	Lysozyme should be used with caution in conjunction with metatartaric acid. A heavy haze may form.
Organoleptic properties of the wine	There is minimal sensory impact on the wine following the addition of lysozyme

differently on the bacterial cell wall, the combination of the two can cause more cell damage. Wine associated LAB species have been shown to produce different bacteriocins.

The concept of a yeast producing a bacteriocin to inhibit LAB during grape vinification is perhaps not so far away. For example, the gene for pediocin (*pedA*), produced by *Pediococcus acidilactici*, has been successfully expressed by *Saccharomyces cerevisiae*, and numerous wine LAB were found to be sensitive. This could then lead onto the development of bactericidal yeast strains, where the yeast strain not only conducts the alcoholic fermentation but also acts as a biological control agent to inhibit the growth of spoilage LAB.

The bacteriocin nisin has been shown to be effective against wine LAB, however, as with lysozyme, yeast is unaffected by nisin, and therefore, it does not affect the progression of alcoholic fermentation. Wine composition and flavour are not affected by the addition of nisin, whether at the grape must or fermentation stages. Yet, bacteriocins have only been used in a limited manner in controlling LAB in wine. Bacteriocins nisin, pediocin and plantaricin (produced by *Lactobacillus plantarum*) have been shown to successfully kill all viable cells of *O. oeni* in biofilms on stainless steel surfaces. This may well lead to more use of bacteriocins in winemaking.

18.4.4.2.3 Bacteriophage

Bacteriophage or phage, a virus that infects bacteria, have been isolated from wine *O. oeni* strains, such as in Australian and German wines. Bacteriophage require a host to replicate and produce progeny phage. Unlike a bacterial cell which produces two progeny in each cell cycle, a bacteriophage has the potential to produce 100–200 progeny. Bacteriophage may be responsible for problematic or stuck MLF, but this has not been conclusively established. The survival of bacteriophage in wine is questionable. Several factors which are not conducive to the survival of phage in wine include pH values below 3.5 and the presence of SO_2. Bentonite, a common fining agent used during vinification was also found to be antagonistic to the survival of phage in wine. As bentonite is a negatively charged clay material that has the capacity to bind proteins, it is likely to adsorb bacteriophage as well. Phenolic components of wine are also likely to act against wine bacteriophage, being inhibitory to animal viruses. The various bacteriophage which have been isolated from *O. oeni* have been characterised and may have the potential in transduction and genetic manipulations of *O. oeni*. It appears to be very unwise to use bacteriophage as a means to control or stop MLF. The dairy industry has experienced great economic problems with bacteriophage infection in starter culture preparations.

18.4.4.2.4 Chitosan

Chitosan is a deacetylated derivative of chitin which is found in shells of crustaceans and fungal cell walls (*Aspergillis* sp). Chitosan has been shown to have antimicrobial activity against a broad range of yeast, bacteria and moulds.

The OIV has approved the use of chitin and derivatives in winemaking, where the fungal derived source is used. The antimicrobial activity has been attributed to its positive charges that would interfere with the negatively charged macromolecules on the bacterial cell surface resulting in membrane disruption. In winemaking, chitosan has been shown to be effective as a fining and protein stabilisation, preventing wine oxidation and as an antimicrobial against LAB, acetic acid bacteria and the spoilage yeast *Brettanomyces*.

18.5 MONITORING OF MALOLACTIC FERMENTATION

The course of MLF is usually followed by monitoring the metabolism of L-malic acid, however the product, L-lactic acid, can also be measured. There are three commonly used monitoring techniques for malic acid; enzymatic, paper chromatography and liquid chromatography. The first two methods are most likely to be used by winery laboratories, with the last most often used in research laboratories or consulting winery laboratories. L-lactic acid can also be determined using these techniques. Enzymatic determination of malic acid utilises the conversion of L-malic acid to oxaloacetic acid (catalysed by L-malate dehydrogenase) which is then converted to L-aspartate in the presence of L-glutamate. The production of NADH is measured spectrophotometrically at 340 nm and is stoichiometrically related to the concentration of L-malic acid present in the sample. There are numerous commercial kits available that use this enzymatic procedure. The progress of MLF can be monitored by using paper chromatography or thin layer chromatography (TLC). Liquid chromatography, typically high-performance liquid chromatography (HPLC), can be used to determine the concentration of various organic acids in a wine sample, including L-malic acid.

18.6 IDENTIFICATION OF MALOLACTIC BACTERIA

Malolactic bacteria are firstly distinguished from other LAB by their ability to directly decarboxylate malic acid to lactic acid. These bacteria are encompassed in four genera of the lactic acid bacteria family: *Lactobacillus*, *Leuconostoc*, *Oenococcus* and *Pediococcus*. The species which are found in these genera can be characterised by their ability to tolerate low pH, high ethanol concentration and growth in wine.

The classical identification approach of wine bacteria was microscopic examination and biochemical reaction profile (substrates and products of metabolism). More recently, DNA composition and molecular based techniques are commonly employed to identify bacteria.

18.6.1 Biochemical Identification

The definitive biochemical test is the catalase test, which should be negative for wine-related lactic acid bacteria. The catalase enzyme converts hydrogen peroxide to oxygen and water, and all aerobic organisms have this enzyme. Those wine bacteria which are catalase positive are usually acetic acid bacteria. LAB are microaerophilic (they grow best in the presence of low concentrations of oxygen [facultative aerobes/anaerobes]).

The LAB can be distinguished by their utilisation of glucose to lactic acid either via the Embden-Meyerhoff Parnas pathway (homofermentative) or phosphoketolase pathway (heterofermentative). *Pediococcus* species are homofermentative, while *Leuconostoc* and *Oenococcus* are heterofermentative bacteria. The genus *Lactobacillus* comprises both homo- and heterofermentative species. As the name suggests, the homofermentative bacteria will primarily produce one product from glucose, namely lactic acid, while the heterofermentative bacteria will yield a mixture of products (about one-sixth carbon dioxide, one-third ethanol, acetic acid or acetaldehyde and the remainder as lactic acid).

A combination of microscopic examination, usually under 1000-fold magnification with oil, and biochemical tests will aid in the separation of the wine related LAB into the appropriate genus. Figure 18.2 gives a summary of the above described techniques to distinguish the wine related genera. The LAB which are likely to be isolated from wine are from four genera: *Lactobacillus* (*Lb. brevis*, *Lb. plantarum*, *Lb. hilgardii*), *Leuconostoc* (*Lc. mesenteroides*), *Oenococcus* (*O. oeni*) and *Pediococcus* (*P. damnosus*, *P. pentosaceus*). *O. oeni* is the most commonly associated LAB species with wine, as it is particularly well adapted to the harsh wine environment (low pH, high ethanol content, low nutrients).

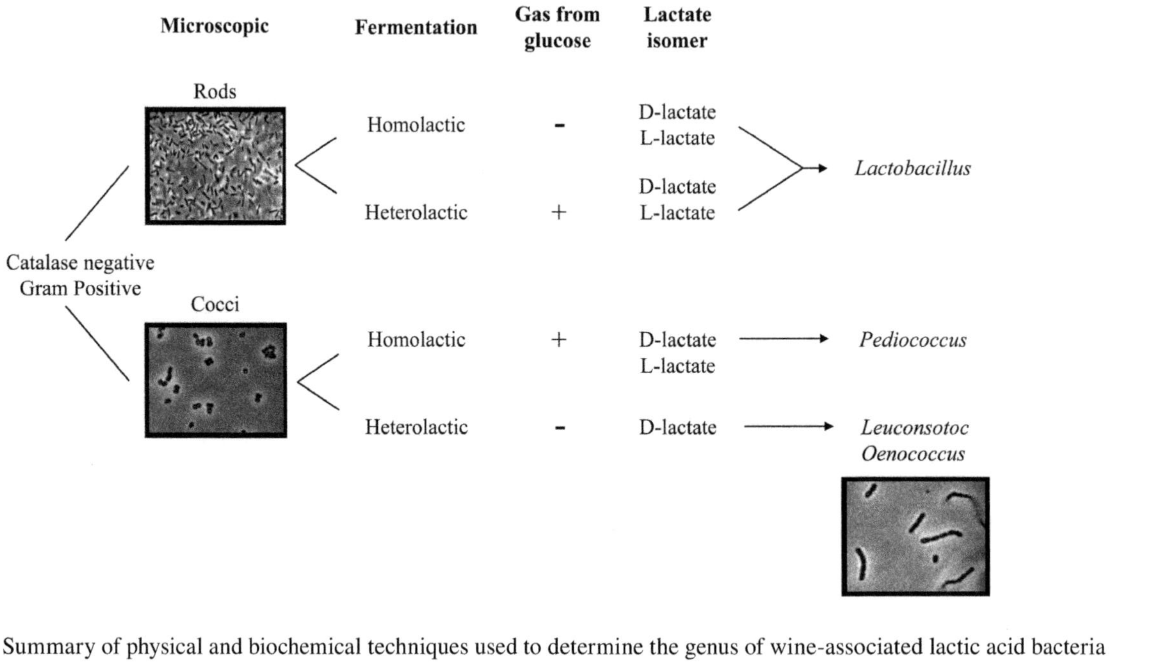

FIGURE 18.2 Summary of physical and biochemical techniques used to determine the genus of wine-associated lactic acid bacteria

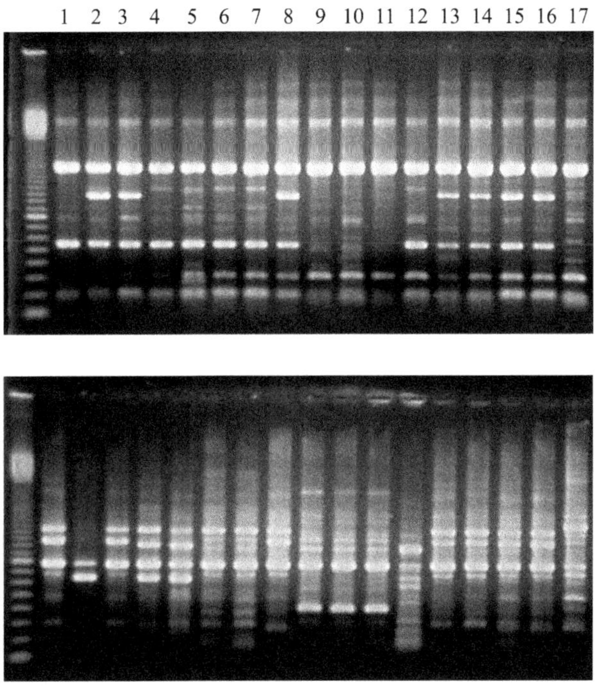

FIGURE 18.3 DNA fingerprinting of *Oenococcus oeni* winery isolates (1–17) using the RAPD technique with two RAPD primers. *Source:* Bartowsky *et al.* (2003)

18.6.2 Genetic Identification

The DNA composition of bacteria has been used extensively in phylogenic studies to determine their relatedness. The polymerase chain reaction (PCR) has opened the door for numerous molecular based techniques to be used to distinguish bacteria to the genus, species or strain level.

The reclassification of *Leuconostoc oenos* to *Oenococcus oeni* relied on both biochemical characteristics and the 16S rRNA sequence. Specific PCR tests are available to rapidly identify *O. oeni* isolates from other wine LAB species, based on the malolactic enzyme gene or the 16S rRNA gene. Similar PCR techniques have been applied to *Lactobacillus* and *Pediococcus* species. The differentiation of strains within a bacterial species requires the use of a variation of PCR which generates a DNA fingerprint, using methods such as restriction fragment length polymorphism (RFLP), amplified fragment length polymorphism (AFLP) and randomly amplified polymorphic DNA (RAPD). The RAPD technique has been used to distinguish strains of *O. oeni* from various wineries and showed that several strains of *O. oeni* can be present concurrently during MLF. An example of a RAPD-PCR DNA fingerprint of Australian *O. oeni* winery isolates is shown in Figure 18.3. Differentiation of species and strains using the RAPD-PCR technique has been successful both for *Lactobacillus* and *Pediococcus* species. The RAPD technique can also provide some information regarding the relatedness or similarity of strains. Various studies have demonstrated that no single strain of *O. oeni* dominates MLF, and the dynamics of *O. oeni* strains can change throughout MLF. Whole genome sequencing is now also used for species and strain identification.

18.7 FLAVOUR CHANGES AND MALOLACTIC FERMENTATION

Modification of wine organoleptic properties is one of the three main reasons why malolactic fermentation (MLF) is conducted in red and white wines. Flavour is usually associated with the aroma of the wine due to volatile compounds, however, it can also include the non-volatile components which can affect the palate or mouth-feel of the wine. The metabolism of malic acid induces an important change to the palate of the wine due to the increase in pH and decrease of titratable acidity. A wide range of terms have been used to describe wines that have undergone MLF, such as malolactic, buttery, butterscotch, lactic, nutty, sweaty, honey, vanilla, wet leather, fruity, vegetative, spicy, toasty, round, long after taste, body, sauerkraut, acetate, bitter and ropy. It is important to note that some compounds are desirable at low concentrations, but at higher concentrations many are considered to be undesirable. Various chemical changes that can occur in the wine during malolactic fermentation are due to the bacterial metabolism (Figure 18.4).

18.7.1 Organic Acids and Production of Diacetyl from Citric Acid

The major organic acids present in wine are generally tartaric, malic and citric acids. Malic acid metabolism has been discussed in Section 18.2. Tartaric acid can only be metabolised aerobically by certain *Lactobacillus* species and results in wine spoilage. Citric acid can be metabolised by lactic acid bacteria. A major intermediary compound of citric acid metabolism is diacetyl. For more details refer to Chapter 19; however, the wine flavour aspect of diacetyl is discussed here. The metabolism of organic acids during malolactic fermentation can have important consequences for the flavour of the wine.

The metabolism of sugars and organic acids during malolactic fermentation has been summarized in Figure 18.5. It is proposed that during the growth phase (Phase I), sugar catabolism occurs with little production of acetic acid and lactic acid and that little citric and malic acid is metabolised in this phase. As the bacterial cell number increase over $5x10^6$ cfu/mL, the catabolism of sugar ceases, and malic acid metabolism proceeds with a concomitant production of lactic acid (Phase II), however, the citric acid remains untouched at this stage. There is no acetic acid produced during malic acid degradation. Phase III is characterised by the metabolism of citric acid which is accompanied by an increase in acetic acid. This sequential metabolism of malic and citric acid is also illustrated in Figure 18.6. The increase of lactic acid content in the wine results in a softer mouth-feel and the acetic acid will contribute to the volatile acidity of the wine.

Diacetyl is a diketone (2,3-butandione), which is best known for being the compound responsible for the characteristic aroma and flavour of butter and dairy products. It is produced by several of the genera and species of the LAB family

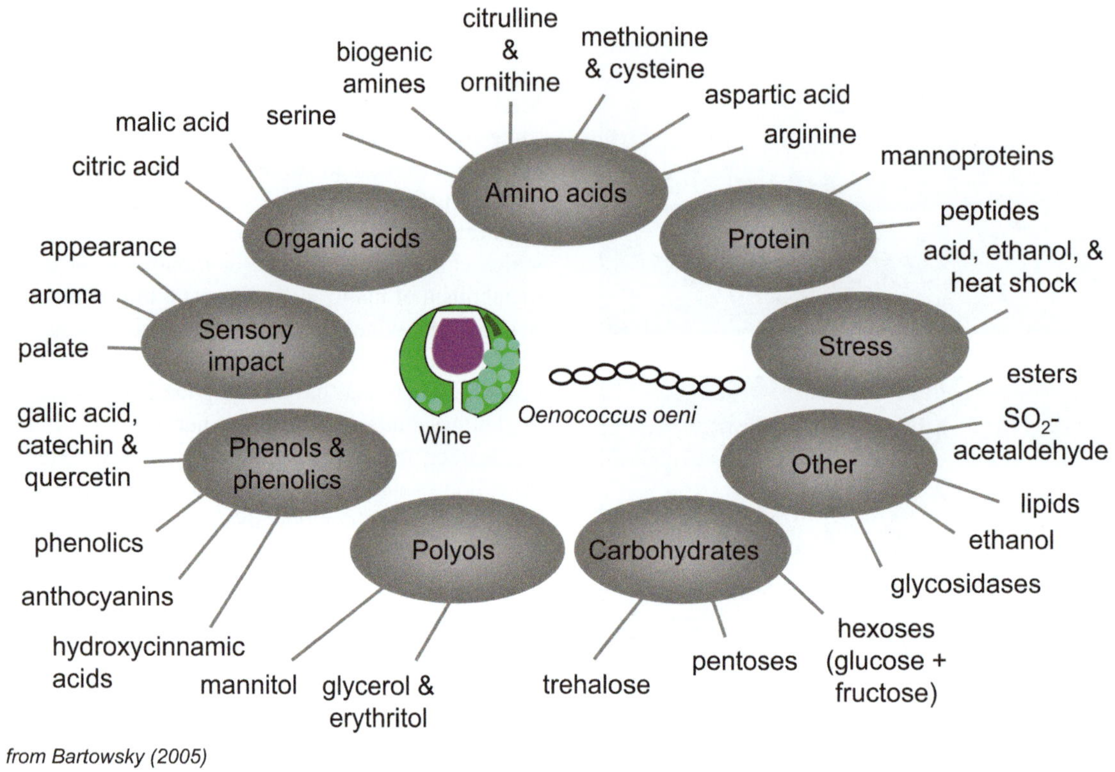

FIGURE 18.4 An overall summary of activities which can occur during malolactic fermentation by *Oenococcus oeni*. *Source:* From Bartowsky (2005)

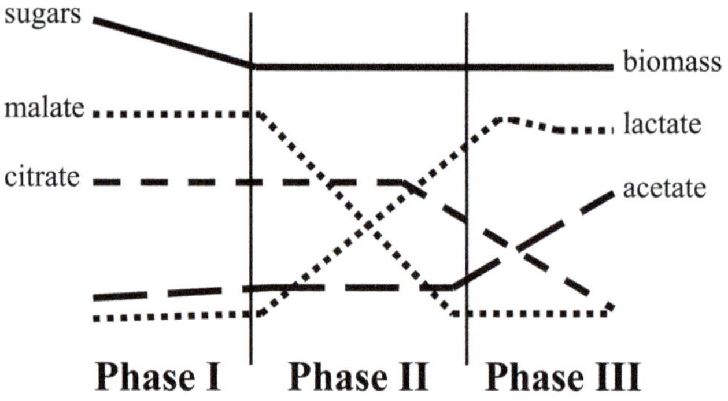

FIGURE 18.5 Metabolism of sugars and organic acids by bacteria during MLF. *Source:* Adapted from Krieger *et al.* (2000)

(*Leuconostoc*, *Lactobacillus*, *Oenococcus*, *Pediococcus* and *Streptococcus*). Diacetyl is an intermediary compound of citric acid metabolism and can be further metabolised to acetoin and 2,3-butanediol, both of which are considered to be flavourless in wine due to their low concentration and high taste threshold. As the 'buttery', 'nutty' or 'butterscotch' notes are an obvious consequence of MLF, considerable research work has focused on the manipulation of diacetyl, which is largely responsible for this character in wine. The sensory threshold of diacetyl in wine has been shown to strongly depend upon the style and the type of wine. In white wines, the taste threshold is approximately 0.2 mg/L, whereas in red wines it can be as high as 2.8 mg/L. The sensory perception of diacetyl in wine is also highly dependent upon the presence of other compounds present in the wine.

Using diacetyl as an example, it is possible to illustrate how numerous winemaking techniques can affect the final concentration of diacetyl (and other flavour compounds) in wine and as such be used to alter the sensory profile of wine (Table 18.3). The stage at which maximal diacetyl production occurs during MLF is an important aspect with respect to manipulating its desired concentration in wine. The diacetyl concentration tends to peak after the degradation of malic acid has been completed and when citric acid is being actively metabolised (Figure 18.6). The diacetyl concentration then declines due to re-metabolism by the bacteria, and yeast lees if present, subsequent to the completion of MLF. The ability of different bacterial strains to synthesise diacetyl can vary greatly, thus, in determining the final diacetyl content of a wine, the bacterial strain and wine type are important factors. The rate of MLF in a particular wine will depend on many factors, such as wine type, nutrient status, physical and chemical properties and the presence of growth inhibitory compounds. It is the combination of several winemaking factors that can be used to achieve the desired diacetyl concentration in the wine (Table 18.3). For example, to achieve a wine with a higher diacetyl content, select a bacterial strain which has high potential to synthesise diacetyl, use a higher wine redox potential to facilitate the formation of diacetyl from its precursor, α-acetolactate, minimise the time that the wine spends on yeast and bacterial lees, and stabilise the wine immediately after the citric and malic acids have been metabolised.

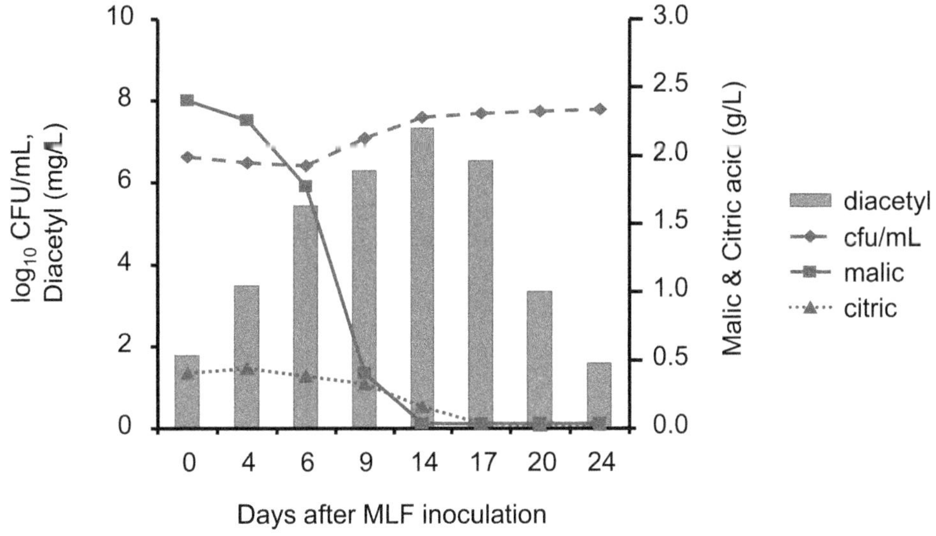

FIGURE 18.6 Dynamics of malic and citric acid metabolism, bacteria cell growth and diacetyl synthesis during malolactic fermentation in a Cabernet Sauvignon wine by *Oenococcus oeni*. *Source:* Bartowsky and Henschke (2004)

TABLE 18.3
Effects of Various Wine Compositional Factors and Winemaking Techniques on the Diacetyl Concentration in Wine

Winemaking factors	Effect on diacetyl concentration and/or sensory perception
Malolactic bacteria strain	*Oenococcus oeni* strains vary in diacetyl production
Wine type	Red wine versus white wine favours diacetyl
Inoculation rate	Lower inoculation rate (10^4 versus 10^6 cfu/mL) favours diacetyl
Contact with active yeast lees	Yeast contact decreases diacetyl content of wine
Contact of wine with air	Oxygen favours oxidation of α-acetolactate to diacetyl
SO_2 content	• SO_2 binds diacetyl, therefore sensorially inactive • SO_2 addition inhibits yeast/bacteria activity & stabilises diacetyl content at time of addition
Citric acid concentration	Favours diacetyl production, acetic acid also produced
Temperature of MLF	18°C versus 25°C may favour diacetyl production
pH of wine	Lower pH may favour diacetyl production
Fermentable sugar concentration	Conflicting information; residual sugar may reduce diacetyl production

18.7.2 Carbohydrates

Wine contains a range of carbohydrates, including monosaccharides (hexoses and pentoses), disaccharides (trehalose), trisaccharide sugars, glycosides, hexitols, glycerol and polysaccharides. Secondary metabolites produced by the metabolism of these carbohydrates during MLF can influence the wine flavour. Glucose and fructose can be fermented by either homolactic or heterolactic pathways (summarized in Section 18.6). The specific wine conditions, such as low pH and concentration of ethanol, will significantly influence the ability of LAB to utilise wine carbohydrates. Fructose can be reduced to mannitol, a six-carbon sugar alcohol or polyol by both homo- and heterofermentative LAB, which can serve as a sole carbon source for *Lactobacillus plantarum*, which in turn results in wine spoilage. Biosynthesis of other polyols, such as glycerol and erythritol from glucose has been demonstrated in *O. oeni*. The formation of these sugar alcohols may play a role in the sensory properties of the wine, in particular body and mouthfeel. The formation of a bitter taste, due to acrolein, which is formed from glycerol, can occur.

Polysaccharides might play a role in wine sensory attributes, such as mouth-feel and body. As *O. oeni* has been shown to have extracellular glucanase activity, there is potential for these bacteria to degrade polysaccharides such as glucan. Preliminary evidence suggests that in a model system, wine polysaccharide may bind to aroma compounds (flavour aglycones) and thus reduce their volatility.

18.7.3 Phenolic Compounds

The influence of bacterial metabolism on red wine colour is of considerable interest to winemakers. Phenolic compounds, such as gallic acid and anthocyanins (for example, malividin-3-glucoside), are able to activate early cell growth of *O. oeni* and also affect the rate of malolactic activity. The stimulation of *O.*

oeni growth in the presence of phenolic compounds is proposed to be due to these compounds acting as hydrogen acceptors, thus helping to create an anaerobic environment. Anecdotal evidence suggests that MLF reduces the colour intensity of wine. Free anthocyanins, such as malvidin-3-glucoside, an abundant coloured form in red wine, has a stimulatory effect on MLF. Viable cells of *O. oeni* appear to be able to metabolise anthocyanins via β-glucosidase activity, which releases the glucose moiety, thus providing an energy source for the *O. oeni*. The released malvidin may then precipitate with polysaccharides or peptidoglycan to give a reduction in wine colour. The inactivated bacterial cells might have the ability to adsorb free anthocyanins to the cell walls, providing another possible mechanism whereby the wine colour is reduced during MLF.

18.7.4 Glycosidase Activity

The complex array of flavour and aroma compounds found in wine largely originate from the grape, yeast metabolism during alcoholic fermentation and the oak wood when used. Bacterial metabolism during MLF may contribute to wine flavour by the formation of additional compounds and the modification of grape, yeast and oak derived compounds. Wine bacteria, in particular *O. oeni*, are able to cleave the glucose moiety from the major red wine anthocyanin, malvidin-3-glucoside, and use it as a carbon source. *O. oeni* strains possess various glycosidase activities, however these activities on synthetic glycosides were dependent on wine conditions, such as pH, ethanol and residual sugar content. Initial work in our laboratory, using an isolated Chardonnay wine glycosidic extract in a synthetic wine medium, demonstrated that there was some, albeit limited, release of glycosylated wine volatiles by *O. oeni* during MLF. These observations concur with that shown in Tannat wine, with the release of glycosylated compounds by the action of *O. oeni* β-glucosidase activity.

18.7.5 Mousy Off-Flavor

The metabolism of certain amino acids (lysine and ornithine) in wine, can lead to the formation of some extremely potent compounds which are sensorially undesirable. Strains of *O. oeni*, heterofermentative *Lactobacillus* species can synthesise; however, *Pediococcus* species generally cannot synthesise the three sensorially potent nitrogen-heterocycle compounds, 2-acetyltetrahydropyridine (ACTPY), 2-acetyl-1-pyrroline (ACPY) and 2-ethyltetrahydropyridine (ETPY), which contribute to a mousy off-flavour. The presence of ethanol, a fermentable carbohydrate (D-fructose) and iron are necessary for the formation of these N-heterocycle compounds by bacteria. In addition, L-ornithine is required for ACPY formation, whereas L-lysine is required for ACTPY formation.

18.7.6 Other LAB Metabolisms Involved in Wine Quality

Changes in the aromatic compounds that are associated with oak have been observed in wines that have undergone MLF in oak barrels. Oak lactone and vanilla increased in a Sauvignon blanc wine, and these changes could be perceived by a sensory panel. However, several oak compounds like eugenol and furfural were found to decrease, and a reduction in sensory impact was noted.

Some strains of *O. oeni* and *Lactobacillus*, but not that of *Pediococcus*, are able to metabolise acetaldehyde to acetic acid and ethanol. The metabolism of acetaldehyde has implications for sensory, colour stability and use of SO_2 in wine. Acetaldehyde mainly originates from yeast metabolism and is a highly volatile compound with apple-like and sherry aromas. The ability to metabolise acetaldehyde bound to SO_2 can be inhibitory to the growth of bacteria due to the release of SO_2, which in turn accumulates to form an inhibitory concentration for bacteria. Even though the impact of the ethanol and acetic acid formed by the metabolism of acetaldehyde by LAB is believed to have a limited chemical and sensory impact, the reduction in the acetaldehyde pool in wine is believed to influence final wine colour. Incomplete or prolonged MLF might be due to the release of SO_2 bound to acetaldehyde.

Ethyl esters, such as ethyl acetate, ethyl hexanoate, ethyl lactate and ethyl octanate can be formed during malolactic fermentation. Ethyl esters have a sensory impact on the fruity aroma of wine. Esterase activity has been shown in most LAB strains. Changes in ester concentration following MLF may either enhance or depreciate the wine quality, dependent upon the ester metabolised.

Biogenic amines have undesirable physiological effects when absorbed at too high a concentration. The major biogenic amines which are found in wine are histamine, phenylethylamine, putrescine and tyramine. Their concentration is lowest after alcoholic fermentation and increases in most wines during MLF to a variable extent. Wine-related LAB have been shown to be able to decarboxylate amino acids to their corresponding amines. Even though a few *O. oeni* strains have been shown to have the ability to produce biogenic amines, it is more prevalent in *Lactobacillus* and *Pediococcus* species. The decarboxylation reaction is purported to provide energy to the cell, favour growth and survival in acidic media, since it induces an increase in pH. Wines with extended lees contact may exhibit a higher concentration of amines, because the bacteria are able to find more peptides and free amino acids to hydrolyse and decarboxylate. The wine pH also plays a crucial role in amine production, as the microflora will be more complex at a higher wine pH, and so biogenic amines are produced in higher concentrations in higher pH wines. White wines, which are generally more acidic than red wines, contain lower biogenic amine concentrations. A PCR test based on the *O. oeni* histidine decarboxylase gene (*hdcA*) has been developed for the detection of amino acid decarboxylating genes in LAB. For toxicity of amines, see Chapter 6.

Ethyl carbamate, also referred to as urethane, an animal carcinogen, is formed through the chemical reaction of ethanol and citrulline, urea or carbamyl phosphate. Arginine, a quantitatively important amino acid of grape must and wine, is a precursor to citrulline, and LAB vary in their ability to degrade arginine. The metabolism of arginine, which provides energy for the bacteria, occurs via the arginine deaminase pathway with three enzymes, arginine deaminase (ADI), ornithine transcarbamylase (OTC) and carbamate kinase (CK). The three step reaction is as follows;

$$\text{L-arginine} + H_2O \overset{ADI}{\rightleftharpoons} \text{L-citrulline} + NH_4^+ \,(\text{Reaction 1}) \tag{18.2}$$

$$\text{L-citrulline} + P_i \xrightarrow{OTC} \text{L-ornithine} + \text{carbamyl-P} \,(\text{Reaction 2}) \tag{18.3}$$

$$\text{carbamyl-P} + ADP \xrightarrow{CK} ATP + CO_2 + NH_4^+ \,(\text{Reaction 3}) \tag{18.4}$$

O. oeni and *Lactobacillus buchneri* are able to metabolise arginine and citrulline. Arginine degradation is inhibited by ribose, fructose and glucose and is not pH dependent. However, there is an accompanying increase in pH with

TABLE 18.4
Changes in Aroma/Flavour Attributes of Wines Having Undergone MLF

	Change of attribute following MLF			
Attribute	**Increase**	**No change**	**Decrease**	**Wine variety**
Banana			+	Chardonnay
Bitter	+			Cabernet Sauvignon
Body	+			Chardonnay
	+			Cabernet Sauvignon
		+		Chardonnay
Buttery	+			Cabernet Sauvignon
	+		+	Chardonnay
Burnt-sweet	+			Chardonnay
			+	Riesling
Caramel				Pinot Noir
(honey, buttery, butterscotch)	+			
	+			Pinot Noir
Chemical (pungent, sulphur, ethanol)	+			Pinot Noir
Cheesy		+		Chardonnay
Citrus	+			Chardonnay
			+	Riesling
Cooked garlic			+	Chardonnay
Earthy	+			Pinot Noir
	+			Chardonnay
Finish	+			Cabernet Sauvignon
	+			Chardonnay
Floral		+	+	Chardonnay
Fruity	+			Pinot Noir
	+	+		Chardonnay
	+			Cabernet Sauvignon
		+	+	Chardonnay
		+		Chardonnay
Maple syrup	+			Riesling
Microbiological (lactic)	+			Pinot Noir
	+			Pinot Noir
Oaky	+			Chardonnay
Sauerkraut		+		Chardonnay
Spicy		+		Chardonnay
Spicy (black pepper, clove)	+			Pinot Noir
	+			Pinot Noir
Spicy/earthy	+			Cabernet Sauvignon
Sweaty	+			Chardonnay
Urine-like			+	Chardonnay
Vegetative	+			Pinot Noir
	+	+		Chardonnay
		+	+	Chardonnay
	+			Cabernet Sauvignon
Yeasty	+			Chardonnay

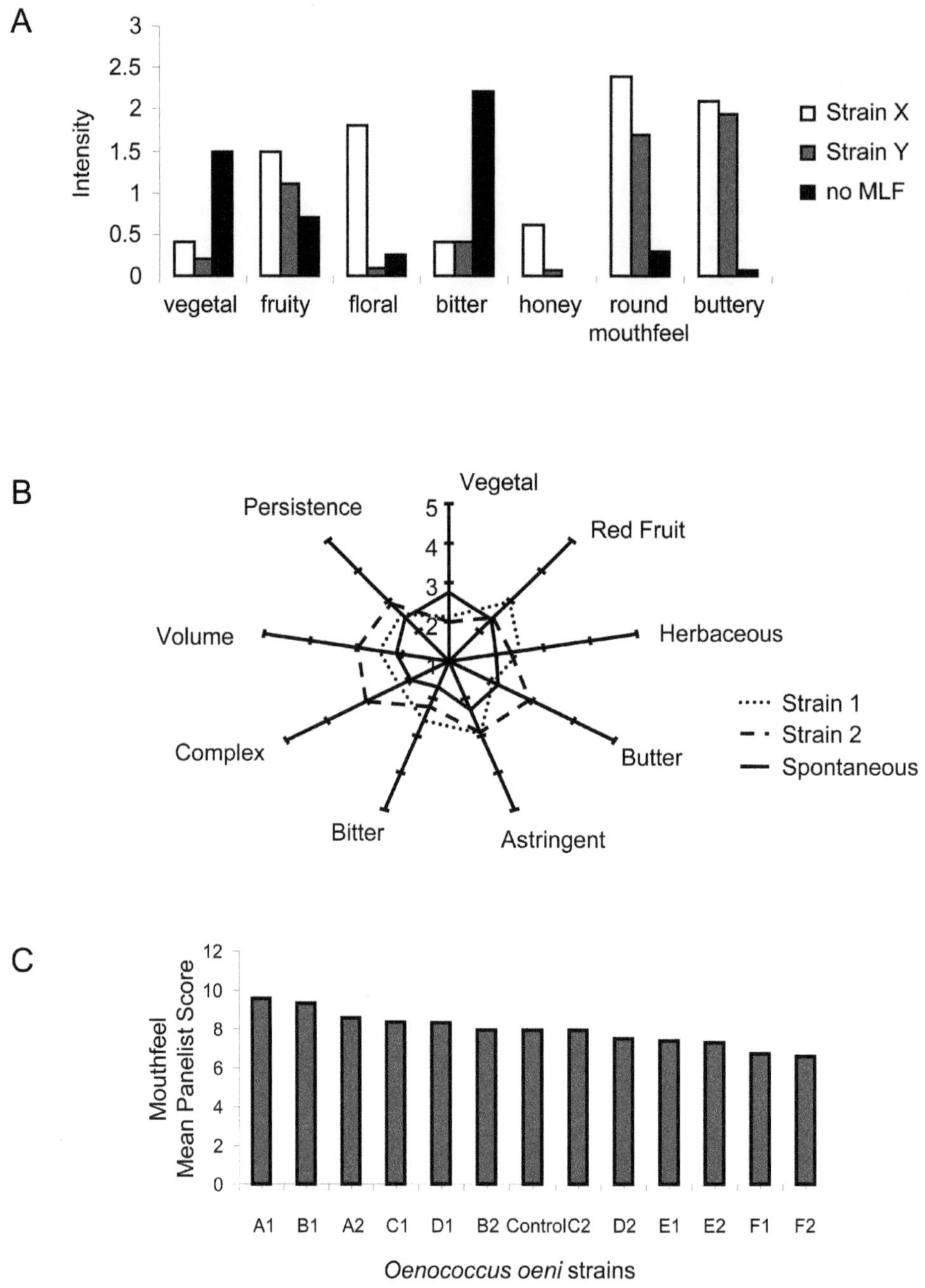

FIGURE 18.7 Sensory effect of MLF on different wines. A. The sensory effect of MLF conducted with different *O. oeni* strains in a Chardonnay wine (1997, Catania, Argentina) (data kindly provided by L. Dulau). B. Sensory evaluation of Merlot (2000, Buzet, France) in stainless steel tanks which have undergone MLF with different strains (data kindly provided by S. Krieger). C. Wine mouth-feel in Cabernet Sauvignon (1998, New York State, USA) which has undergone MLF with different *Oenococcus oeni* strains. *Source:* Data kindly provided by J. Richardson, T. Henick-Kling and S. Krieger (1999)

arginine metabolism, which may be physiologically important for tolerance and adaptation of wine LAB to the wine environment. The role that MLF plays in potential ethyl carbamate formation remains unclear, however experiments conducted in a synthetic wine and a laboratory vinified wine demonstrated a correlation between arginine degradation, citrulline production and ethyl carbamate formation during MLF. Even though the formation of this compound does not necessarily affect the wine sensory quality, it is an important aspect of wine bacteria metabolism.

18.7.7 Sensory Aspects of Malolactic Fermentation

Most red wines and to a lesser extent, sparkling base wines and several varieties of white wine, in particular Chardonnay, benefit from a reduction in wine acidity. There is also a growing interest in the desirable flavour changes associated with MLF. Given that MLF can add 1–5% to the overall cost of wine production, it is important to have a better understanding of the impact that MLF can have on the sensory attributes of wine. While the reduction in acidity represents the major organoleptic change associated with MLF, changes in the aroma profile, palate structure and wine colour are increasingly being recognised, but are still poorly understood. Descriptive terms for aromas which either increase, decrease or show minimal change during MLF in Chardonnay, Pinot Noir and Cabernet Sauvignon wines are summarised in Table 18.4. Generally, fruity aromas may be enhanced by MLF, whereas vegetative characters are generally reduced during MLF. Over the years, it has been continually noted that the aroma profile of wine which has undergone MLF is significantly modified, however, to date, no relationship has yet been established of the chemical composition, with the exception of diacetyl (discussed in Section 18.7.1.1). Changes in the texture and body of Chardonnay, Merlot and Cabernet Sauvignon wines have been observed: fuller, richer and a longer aftertaste (Figure 18.7). The chemical basis of the palate changes noted following MLF has not yet been established.

It is widely accepted that MLF will alter the aroma and palate of wine, beyond the pH change, however the biochemical nature of these changes are still poorly understood. It is well known that the physical and chemical environment of a microorganism can dramatically affect its metabolic responses. The inoculation regime for alcoholic fermentation and MLF induction can influence the metabolism of the bacteria because of competition with yeast for the limited available nutrients and grape secondary metabolites. Yeast too can metabolise bacterial metabolic products, thus both these scenarios can affect the final organoleptic profile of the wine.

BIBLIOGRAPHY

Axelsson, L.T. (1993). Lactic acid bacteria: Classification and physiology. In: *Lactic Acid Bacteria*, S. Salminen and A. von Wright (Eds.), Marcel Dekker Inc., New York.

Bartowsky, E.J. (2003). Lysozyme and winemaking. *Aust. Grapegrower Winemaker*, 473a: 101–104.

Bartowsky, E.J. (2005). *Oenococcus oeni* and malolactic fermentation – Moving into the molecular arena. *Aust. J. Grape Wine Res.*, 11(2): 174–187.

Bartowsky, E.J. (2017). *Oenococcus oeni* and the genomic era. *FEMS Microbiol. Rev.*, 41(Supp_1): S84–S94.

Bartowsky, E.J. and Henschke, P.A. (2004). The 'buttery' attribute of wine – Diacetyl – Desirability, spoilage and beyond. *Int. J. Food Microbiol.*, 96(3): 235–252.

Bartowsky, E.J., McCarthy, J.M. and Henschke, P.A. (2003). Differentiation of Australian wine isolates of *Oenococcus oeni* using random amplified polymorphic DNA (RAPD). *Aust. J. Grape Wine Res.*, 9(2): 122–126.

Bartowsky, E.J., Costello, P.J., Villa, A. and Henschke, P.A. (2004). The chemical and sensorial effects of lysozyme addition to red and white wines over six months' cellar storage. *Aust. J. Grape Wine Res.*, 10(2): 143–150.

Bartowsky, E.J., Francis, I.L., Bellon, J.R. and Henschke, P.A. (2002). Is buttery aroma perception in wines predictable from the diacetyl concentration? *Aust. J. Grape Wine Res.*, 8(3): 180–185.

Boido, E., Lloret, A., Medina, K., Carrau, F. and Dellacasa, E. (2002). Effect of β-glycosidase activity of *Oenococcus oeni* on the glycosylated flavor precursors of Tannat wine during malolactic fermentation. *J. Agric. Food Chem.*, 50(8): 2344–2349.

Cappello, M.S., Zapparoli, G., Logrieco, A. and Bartowsky, E.J. (2017). Linking wine lactic acid bacteria diversity with wine aroma and flavour. *Int. J. Food Microbiol.*, 243: 16–27.

Chung, W. and Hancock, R.E.W. (2000). Action of lysozyme and nisin mixtures against lactic acid bacteria. *Int. J. Food Microbiol.*, 60(1): 25–32.

Costello, P.J. and Henschke, P.A. (2002). Mousy off-flavor of wine: Precursors and biosynthesis of the causative N-heterocycles 2-ethyltetrahydropyridine, 2-acetyltetrahydropyridine, and 2-acetyl-1-pyrroline by *Lactobacillus hilgardii* DSM 20176. *J. Agric. Food Chem.*, 50(24): 7079–7087.

Coton, E., Rollan, G., Bertrand, A. and Lonvaud-Funel, A. (1998). Histamine-producing lactic acid bacteria in wines: Early detection, frequency, and distribution. *Amer. J. Enol. Vitic.*, 49: 199–204.

Cox, D.J. and Henick-Kling, T. (1989). Chemiosmotic energy from malolactic fermentation. *J. Bacteriol.*, 171(10): 5750–5752.

Cox, D.J. and Henick-Kling, T. (1990). A comparison of lactic acid bacteria for energy-yielding (ATP) malolactic enzyme systems. *Amer. J. Enol. Vitic.*, 41: 215–218.

Daeschel, M.A., Jung, D. and Watson, B.T. (1991). Controlling wine malolactic fermentation with nisin and nisin-resistant strains of *Leuconostoc oenos*. *Appl. Environ. Microbiol.*, 57(2): 601–603.

Davis, C.R., Wibowo, D., Fleet, G.H. and Lee, T.H. (1988). Properties of wine lactic acid bacteria: Their potential enological significance. *Amer. J. Enol. Vitic.*, 39: 137–142.

Davis, C.R., Wibowo, D., Eschenbruch, R., Lee, T.H. and Fleet, G.H. (1985). Practical implications of malolactic fermentation: A review. *Amer. J. Enol. Vitic.*, 36: 290–301.

de Revel, G., Martin, N., Prips-Nicolau, L., Lonvaud-Funel, A. and Bertrand, A. (1999). Contribution to the knowledge of malolactic fermentation influence on wine aroma. *J. Agric. Food Chem.*, 47(10): 4003–4008.

Delfini, C., Gaia, P., Schellino, R., Strano, M., Pagliara, A. and Ambrò, S. (2002). Fermentability of grape must after inhibition with dimethyl dicarbonate (DMDC). *J. Agric. Food Chem.*, 50(20): 5605–5611.

Gao, Y.C., Zhang, G., Krentz, S., Daruis, S., Power, J. and Lagarde, G. (2002). Inhibition of spoilage lactic acid bacteria by lysozyme during wine alcoholic fermentation. *Aust. J. Grape Wine Res.*, 8(1): 76–83.

Garvie, E.I. (1967). *Leuconostoc oenos* sp. nov. *J. Gen. Microbiol.*, 48(3): 431–438.

Guzzo, J., Jobin, M.-P., Delmas, F., Fortier, L.-C., Garmyn, D., Tourdot-Marechal, R., Lee, B. and Divies, C. (2000). Regulation of stress response in *Oenococcus oeni* as a function of environmental changes and growth phase. *Int. J. Food Microbiol.*, 55(1–3): 27–31.

Henick-Kling, T. (1993). Malolactic fermentation. In: *Wine Microbiology and Biotechnology*, G.H. Fleet (Ed.), Harwood Academic Publisher, Amsterdam, p. 289–326.

Krieger, S.A., Lemperle, E. and Ernst, M. (2000). Management of malolactic fermentation with regard to flavor modification in wine. In: *Proceedings of the 5th International Symposium on Cool Climate Viticulture and Oenology*, 16–20 January, 2000, Melbourne, Australia.

Kunkee, R.E. (1991). Some roles of malic acid in the malolactic fermentation in winemaking. *FEMS Microbiol. Rev.*, 88(1): 55–72.

Liu, S.-Q. (2002). Malolactic fermentation in wine - beyond deacidification. *J. Appl. Microbiol.*, 92(4): 589–601.

Lonvaud-Funel, A. (2001). Biogenic amines in wines: Role of lactic acid bacteria. *FEMS Microbiol. Lett.*, 199(1): 9–13.

Matthews, A., Grimaldi, A., Walker, M., Bartowsky, E., Grbin, P. and Jiranek, V. (2004). Lactic acid bacteria as a potential source of enzymes for use in vinification. *Appl. Environ. Microbiol.*, 70(10): 5715–5731.

Mira de Orduna, R., Liu, S.-Q., Patchett, M.L. and Pilone, G.J. (2000). Ethyl carbamate precursor citrulline formation from arginine degradation by malolactic wine lactic acid bacteria. *FEMS Microbiol. Lett.*, 183(1): 31–35.

Nielsen, J.C., Prahl, C. and Lonvaud-Funel, A. (1996). Malolactic fermentation in wine by direct inoculation with freeze-dried *Leuconostoc oenos* cultures. *Amer. J. Enol. Vitic.*, 47: 42–48.

Osborne, J.P., Mira de Orduna, R., Pilone, G.J. and Liu, S.-Q. (2000). Acetaldehyde metabolism by wine lactic acid bacteria. *FEMS Microbiol. Lett.*, 191(1): 51–55.

Schoeman, H., Vivier, M.A., Du Toit, M., Dicks, L.M.T. and Pretorius, I.S. (1999). The development of bactericidal yeast strains by expressing the *Pediococcus acidilactici* pediocin gene (*pedA*) in *Saccharomyces cerevisiae*. *Yeast*, 15(8): 647–656.

Sternes, P.R. and Borneman, A.R. (2016). Consensus pan-genome assembly of the specialised wine bacterium *Oenococcus oeni*. *BMC Genomics*, 17(308): 1–15.

Swiegers, J.H., Bartowsky, E.J., Hemschke, P.A. and Pretorius, I.S. (2005). Yeast and bacterial modulation of wine aroma and flavour. *Aust. J. Grape Wine Res.*, 11(2): 139–173.

Vivas, N., Augustin, M. and Lonvaud-Funel, A. (2000). Influence of oak wood and grape tannins on the lactic acid bacterium *Oenococcus oeni* (*Leuconostoc oenos*, 8413). *J. Sci. Food Agric.*, 80(11): 1675–1678.

Wibowo, D., Eschenbruch, R. Davis, C.R., Fleet, G.H. and Lee, T.H. (1985). Occurrence and growth of lactic acid bacteria in wine: Review. *Amer. J. Enol. Vitic.*, 36: 302–313.

19 Genetic Engineering of Microorganisms in Winemaking

S.K. Soni and R. Soni

CONTENTS

19.1 INTRODUCTION

The production of good quality wines involves a complex biochemical process with the sequential appearance of different microbial species including yeast and bacteria. Most wine production is based on the use of starter cultures consisting of selected strains of *Saccharomyces cerevisiae*, able to ensure quick and controlled fermentations. Though the eminent role of *S. cerevisiae* in wine making has been fully established, there is an ever-growing demand for innovations in the winemaking process, with research relying on the isolation and selection of new oenological strains of *S. cerevisiae* and non-*Saccharomyces* species showing interesting metabolic or technological features, or on the improvement of wine yeasts at a genetic level. In the case of the latter approach, examples of obtaining both non-genetically modified (GM) and GM organisms (GMO), which are capable of more efficient fermentation processing; controlling microbial spoilage; improving the sensory quality of wine; and enhancing health benefits, are available in literature. The genetic engineering of the microorganisms important in wine making is discussed in this chapter.

19.2 LIFE CYCLE OF WINE YEAST

Saccharomyces cerevisiae is an ascomycetous fungus, whose vegetative growth results predominantly from budding. When yeast is nutritionally stressed, diploid yeast sporulates by undergoing meiosis, producing 4 haploid stress-resistant ascospores, encapsulated in the ascus. Wine yeast can be maintained as either haploid cells or diploid cells, which continue to divide by budding. The cell division cycle begins with a single, unbudded cell. The transition from the haploid to the diploid phase results from mating between gametes that leads to the formation of a zygote. Diffusible molecules, the pheromones, mediate mating in yeast. When cells of opposite mating type are mixed on the surface of an agar growth medium in a petri plate, changes become apparent within 2–3 hours, and their effects on cells of the opposite mating haploid types (*MATα* and *MATa*) are easy to demonstrate. The cells stop dividing, elongate, and become pear-shaped, termed as 'shmoos'. Normal yeast can grow either aerobically or anaerobically. Under aerobic growth conditions they can support growth by oxidizing simple carbon sources, and under anaerobic conditions yeast can convert sugars only to carbon dioxide and ethanol, recovering less of the energy. Strains that can be maintained stably for many generations as haploids are termed as heterothallic, while those in which sex reversals, cell fusion, and diploid formation occur are termed as homothallic. The naturally occurring strains of *S. cerevisiae* are either haploid or diploid. However, industrial wine yeast strains are predominantly diploid or aneuploid and, occasionally, polyploid too.

19.3 GENETIC BACKGROUND OF WINE YEASTS

The genetic system of *S. cerevisiae* has been studied for decades to identify the molecular components that control the key biological pathways. There exist several biotechnological approaches, which can be applied for tailoring the wine yeasts according to the requirements of the type of wine to be prepared. Before going into the details of these approaches, the genetic background of the yeast is described.

19.3.1 Chromosomal DNA

The genome of *S. cerevisiae* is relatively a small, infrequently-repetitive DNA containing only 239 introns. Haploid strains contain approximately 14 mb of nuclear DNA, distributed along 16 linear chromosomes with sizes ranging from 250–2200 kb. The genome of *S. cerevisiae* has been completely sequenced and found to contain roughly 6300 protein-encoding genes which are relatively rich in guanine and cytosine content (% G+C of 39–41).

The development of a genetic map is initially required for many genetic investigations. The complete genome sequence of *S. cerevisiae,* with sixteen haploid chromosomes, was submitted to public databases in April 1996. The genome is apparently rich in information, with over 70% of data being devoted to coding sequence. In total, 6,215 potential protein-coding sequences have been identified. The average size of the part of the reading frame, a continuous stretch of codons that begins with a start codon and ends at a stop codon that has the ability to be translated (open reading frame), is 1450 bp, with an average intergenic distance of 472 bp. Introns are rare, and those that exist are short (the longest is 1 kb) and primarily found at the 5′ end of the gene. An outline of the physical and genetic map(s) of *S. cerevisiae* is depicted in Figure 19.1. The complete genetic maps of 16 chromosomes have also been obtained.

19.3.2 Extra-Chromosomal DNA

The genome of *S. cerevisiae* contains approximately 35–55 copies of transposable elements (Ty elements), which move from one genomic location to another via an RNA intermediate using reverse transcriptase. The 2 μm plasmid DNA is the only naturally occurring, stably maintained, circular nuclear plasmid, existing in two forms in *S. cerevisiae.* This 6.3 kb extrachromosomal element is also inherited in a non-Mendelian fashion, and although most strains of *S. cerevisiae* contain 50–100 copies of 2 μm DNA per cell, its biological function has not yet been discovered.

19.3.3 Mitochondrial DNA

The mitochondria of *S. cerevisiae* possess their own genetic system consisting of 75 kb circularly permuted DNA molecules, which are adenine-thymine (A-T) rich and carry the genetic information for only a few and essential mitochondria components. Unlike the replication of nuclear DNA, replication of mtDNA is not limited to the S phase but takes place throughout the cell cycle. The mtDNA polymerase also lacks

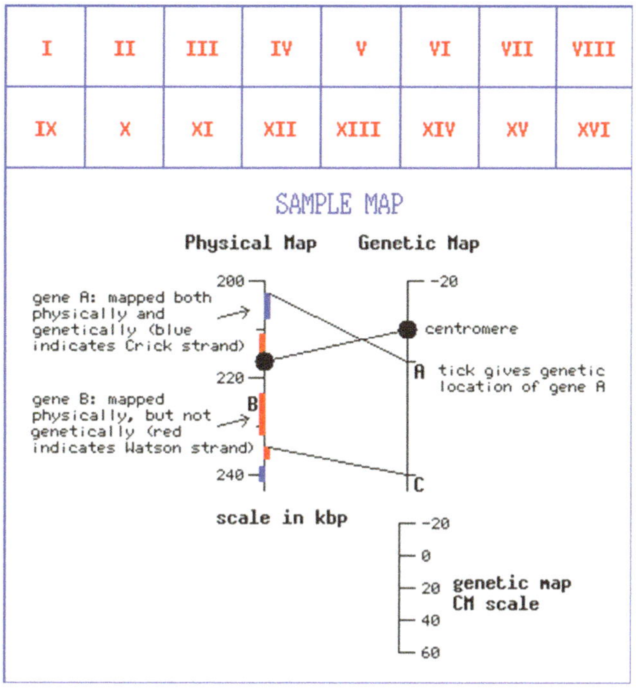

Chromosome	I	II	III	IV	V	VI	VII	VIII	IX	X	XI	XII	XIII	XIV	XV	XVI
cM/Kbp	0.45	0.30	0.48	0.31	0.35	0.48	0.37	0.30	0.45	0.30	0.38	0.36	0.34	0.37	0.33	0.29

FIGURE 19.1 The Combined Physical and Genetic Map for *Saccharomyces cerevisiae* chromosomes containing genetically and/or physically mapped open reading frames (ORFs). Each map is derived from the systematic genomic sequencing data and from data contained within *Saccharomyces* genome database. In the case where an ORF has been defined, its standard gene name is indicated. The roman numeral at the top of the Sample Map provides link to the Combined Physical and Genetic Maps for each chromosome. The chromosome numbers in the table of cM/Kbp values provide links to the Genetic Distance *vs.* Physical Distance Ratio graphs. *Source:* Adapted from Cherry *et al.* (1997)

proofreading (exo-nuclease) activity, resulting in a much higher mutation rate within the mtDNA than within nuclear genes, so mtDNA can evolve very rapidly. Mitochondrial mutants usually lack vital oxidative enzymes, enabling them to generate ATP oxidatively and grow slowly and form smaller *(petite)* colonies on solid agar surfaces than the wild-type *(grande)* cells.

19.3.4 Killer Factors

Some yeasts belonging to *Saccharomyces*, *Candida*, *Cryptococcus*, *Hansenula*, *Kluyveromyces*, *Kloeckera*, *Pichia* secrete toxins, which are proteins or glycoproteins in nature and are lethal to other yeasts. Based on the cross reactions between killer strains of various genera and species, these yeasts have been classified into 10 groups from K_1 to K_{11} (For more details, see Chapter 20).

19.4 GENETIC TECHNIQUES FOR DEVELOPMENT OF WINE YEAST STRAINS

In many ways, *S. cerevisiae* can be modified genetically using various techniques. The techniques having the greatest potential in genetic programming of wine yeast strains include selection of variants, mutation and selection, hybridization, rare-mating, protoplast fusion, gene cloning, and transformation. The combined use of tetrad analysis, replica plating, mutagenesis, hybridization, and recombinant DNA methods have dramatically increased the genetic diversity that can be introduced into yeast cells. Several attempts were made previously to generate recombination of desirable qualities by protoplast fusion with encouraging results. Recently, several important results have been obtained by transformation of yeast strains with recombinant plasmids bearing desirable genes.

Over time, advancements have been made in the DNA technologies which enable the chemical synthesis of long DNA stretches and their accurate editing. These breakthrough technologies have changed the researchers' approach from DNA sequencing to DNA synthesis. Thus, this became the turning point in every field of biotechnology along with wine yeast strain development. The purpose of the development of wine yeast strains is to, primarily, produce quality wine which has a sturdy fermentation with a brief phase of latency. The strain must be able to reproduce fermentation characteristics and must function in a foreseeable fashion.

Besides these, the strain must have high tolerance for ethanol, increased pressure and temperature, and sulfur dioxide, and it must yield glycerol and α-glycosidase without any off-flavour. Also, the strain should be able to flocculate easily to facilitate cell removal, especially for secondary fermentation to produce sparkling wine.

19.4.1 Selection of Natural Variants

It is very complex to maintain the genetic identity of strains in a 'pure culture', a term that denotes that it has been derived from a single cell but does not imply that the culture is genetically identical. Even under closely controlled conditions of growth, a yeast strain reveals slow but distinct changes after many generations, due to a number of different processes, including spontaneous mutation, Ty-promoted chromosomal translocations, and, more frequently, mitotic crossing-over or gene conversion. Over the years, the phenomena of heterogeneity in pure yeast cultures and natural genetic variations have been exploited to improve wine yeast strains. The successive single-cell cultures of commercial wine yeast strains could result in strains with considerably improved characteristics. Mating between *MATa* and *MATα* ascospores, generated by sporulation, can also cause genetic instability. Selection of variants, having molecular and physiological differences from the parent strains, is thus a direct means of strain development, and successful isolation of variants with desirable characteristics depends on the frequency at which mitotic recombination and spontaneous mutation occur, as well as the availability of selection procedures to identify those variants exhibiting improved enological traits.

19.4.2 Random Mutagenesis and Selection

Through mutations, the genetic constitution of wine yeasts can be altered. In nature, the appearance of mutation is a chance event with no regard to environmental conditions. Spontaneous mutations occur in the absence of any agent known to alter the genetic material. Induced mutations are produced mutagens such as base analogs, alkylating agents, deaminating agents, acridine derivatives, X-rays, gamma rays, UV rays, etc. The average spontaneous mutation frequency in *S. cerevisiae* at any particular locus is approximately 10^{-6} per generation. But use of mutagen greatly increases the frequency of mutations in a wine yeast population. The *in vivo* induction of random mutations by chemical or physical mutagens and subsequent selection of phenotypically improved cells is one of the most widely used techniques. Mutation and selection is a rational approach to strain development, where a large number of performance parameters are to be kept constant while only one is targeted for change, but it can improve certain traits with the simultaneous debilitation of other characteristics. Although mutations are probably induced with the same frequency in haploid, diploid, or polyploids, these are not easily detected in diploid and polyploid cells because of the presence of non-mutated alleles. Mutagenesis has the potential to eliminate undesirable characteristics, other favorable properties, and to isolate new variants of wine yeasts prior to further genetic manipulation.

19.4.3 Adaptive Evolution

A couple of mutations that arise as a result of a peculiar challenge that provides the organism with superiority over the unchanged cells under the same conditions are referred to as adaptive evolution. The organism here is subjected to a series of or repeated cultivation for a number of generations in an environment where the organism is not acclimatized, such as exposure to oxidative stress, freeze-thawing, high temperature, ethanol stress, etc. Because of natural selection, the organism which adapts to the change gets selected. Polyploid yeast strains are best-suited for such adaptations, thereby rendering it a resourceful non-recombinant method.

19.4.4 Hybridization

Hybridization is usually only feasible between two haploid cells of opposite mating types belonging to the same species and accordingly by recombination of existing variants between the commercial strains, a new yeast is produced. Conventional crossbreeding for yeast encounters problems relating to the life cycle of commercial strains, so repeated backcrossing of a derived strain to its original parent to remove undesirable characteristics is considered both tedious and inefficient. The hybrids obtained by crossing the mating derivatives of commercial yeasts displays intermediate properties when compared to the parental strains, a result which would be expected for characteristics under polygenic control, *e.g.* fermentation rate. Intra-species hybridization involves the mating of haploids of opposite mating-types to yield a heterozygous diploid. Recombinant progenies are recovered by sporulating the diploid, recovering individual haploid ascospores, and repeating the mating/sporulation cycle, as required. Haploid strains from different parental diploids, possessing different genotypes, can be mated to form a diploid strain with properties different from that of either parental strain. Thus, crossbreeding can permit the selection of desirable characteristics and the elimination of undesirable ones. Generally, hybridization provides a solution for precise objectives such as the development of non-H_2S-producing flocculant hybrids, by conjugating spores of two homothallic commercial strains as this combination is rarely found in indigenous populations. Conjugating spores introduced the flocculant property of a heterothallic haploid laboratory strain into a non-flocculant homothallic commercial strain in some reports. The original genome of the commercial strain was generated by five successive back crosses, yielding clones that did not differ from the original strain for fermentation kinetics and organoleptic properties and had the flocculant property that allowed better separation of yeast cells from the wine, thereby facilitating the process of clarification. Unfortunately, many wine yeasts are homothallic, so the use of hybridization techniques has proved difficult. The problem could however be circumvented by direct spore-cell mating using a micromanipulator, in which four homothallic ascospores from the same ascus are placed into direct contact with heterothallic haploid cells. Mating takes place between compatible ascospores/cells, and the resulting diploid is sporulated and screened.

19.4.5 Rare Mating and Cytoduction

Rare mating is based on the fact that in a population of diploid/polyploid cells, some individual cells become homozygous for the mating type and can, therefore, be crossed with a haploid cell of the opposite type without the need for sporulation that generally does not occur in commercial strains and runs the risk of genome alteration. This cross is a rare event that can only be detected with a highly effective screening process. The hybrids are selected on the basis of maximum DNA content. Rare mating is also used to introduce cytoplasmic genetic elements into wine yeasts without the transfer of nuclear genes from the non-wine yeast parent and is termed as cytoduction. Cytoductants/heteroplasmons, which receive cytoplasmic contributions from both the parents, retain the nuclear integrity of only one and require that a haploid mating strain carry the *kar1* mutation, *i.e.* a mutation that impedes karyogamy after mating.

It has also been applied to construct cryotolerant wine yeasts, dextrin-fermenting, and high ethanol-producing yeasts. Triploid interspecific hybrids of *S. cerevisiae* and other *Saccharomyces sensu stricto* species (like *Saccharomyces mikatae*) have been constructed using this rare mating technique to achieve improvements in flavour profile of wines.

19.4.6 Mass Mating and Genome Shuffling

Mass mating is a technique in which large numbers of haploid yeast cells, often from different parental strains are mixed and allowed to randomly mate. Mass mating is a particularly useful improvement technique for homothallic strains, for strains that show low mating efficiency, or for the creation of interspecific hybrids if strong selective markers for outbreeds are available.

19.4.7 Protoplast Fusion

Protoplast fusion is a direct, asexual technique employed in crossbreeding as a supplement to mating, whereby the cells can't have sexual reproduction. Like rare mating, protoplast fusion can be used to produce either hybrids or cytoductants. Both these procedures overcome the requirement for opposite mating types to be crossed, thereby extending the number of crosses along with the ploidy of the cells, thus enhancing the cell productivity. Lytic enzymes in the presence of an osmotic stabilizer can remove cell walls of yeasts. The protoplasts from the different parental strains are mixed together in the presence of polyethylene glycol and calcium ions and then allowed to regenerate their cell walls in an osmotically stabilized selective-agar medium (Figure 19.2). The method is useful for obtaining interspecific and intergeneric hybrids, and the final result is often a large predominance of the genome

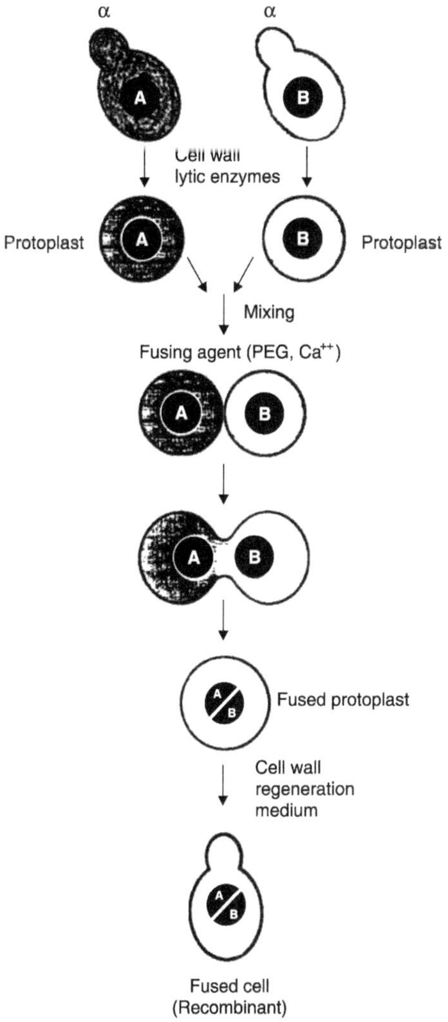

FIGURE 19.2 The method of yeast transformation. *Source:* Adapted from Srivastava and Raverkar (1999)

of one of the strains plus several genes of the other strain. It also greatly increases the size of the 'gene pool' that may be exploited further.

19.4.8 Recombinant DNA Technology

Recombinant DNA technology offers the possibility of altering the characteristics of wine yeasts with surgical precision: the modification of an existing property, the introduction of a new characteristic without adversely affecting other desirable properties, and by elimination of an unwanted trait. By these procedures, it is possible to transform yeast cells with DNA fragments or purified genes from the same or other organisms. The yeast transformation requires i) cell permeability to donor DNA, ii) an appropriate DNA vector, iii) an easily selectable gene marker, and iv) a suitable recipient strain. Gene cloning, being a straightforward procedure, can provide a pure sample of an individual gene, separated from all other genes that it normally shares with the cell. After cloning a gene, there is almost no limit to the information that can be obtained about the structure and expression of gene. Gene

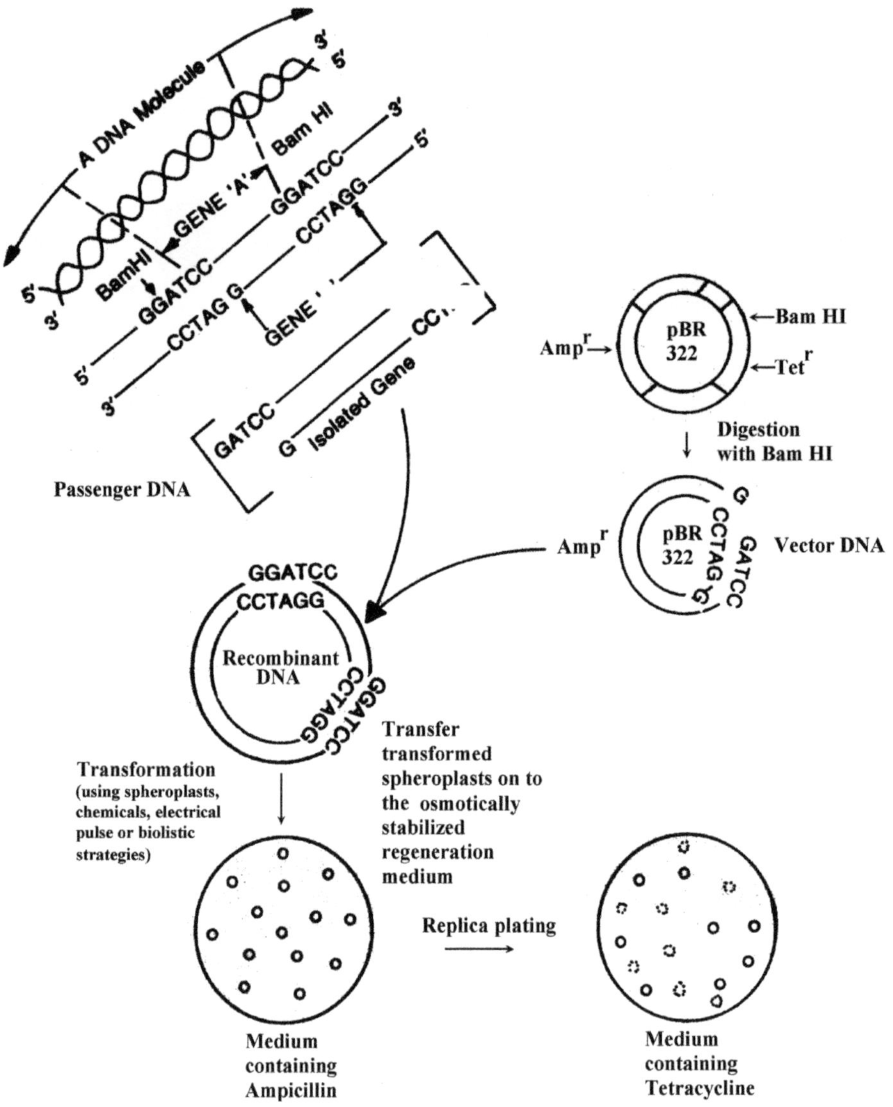

FIGURE 19.3 Outline of gene cloning and transformation indicating the introduction of recombinant DNA molecule into wine yeast. *Source:* Adapted from Srivastava and Raverkar (1999)

cloning requires DNA molecule to be cut in a very precise fashion (Figure 19.3). The gene of interest is cleaved from the DNA molecules by the specific restriction endonucleases (BamH1). The vector used in this case is pBR322, which contains a unique recognized sequence in which BamH1 cuts the DNA molecule, two selectable markers amp^r and tet^r. When the vector is digested with BamH1, the site of tet^r gets disrupted, and the gene of segment is introduced at the region of disrupted tet^r gene expression. This recombinant DNA is then transformed by using various techniques to the yeast cells. Replica plating of the cells is regenerated in a medium containing ampicillin and tetracycline. All colonies will consist of transformants (untransformed cells are amp^s, so do not produce colonies), but only a few contain recombinant PBR322 molecules; most contain normal plasmid. The recombinant DNA approaches offer wider applicability including: (a) amplification of gene expression by maintaining a gene on a multi-copy plasmid, integration of a gene at multiple sites within chromosomal DNA, or splicing a structural gene to a highly efficient promoter (region of DNA involved in binding of RNA polymerase to initiate binding) sequence, (b) releasing enzyme synthesis from a particular metabolic control or subjecting it to a new one, (c) in-frame splicing of a structural gene to a secretion signal to engineer secretion of a particular gene product into the culture medium, (d) developing gene products with modified characteristics by site-directed mutagenesis, (e) eliminating specific undesirable strain characteristics by gene disruption, and (f) incorporation of genetic information from diverse organisms such as fungi, bacteria, animals, and plants. Genetic techniques of mutation, hybridization, cytoduction, and transformation will most likely be used in combination for commercial wine yeast improvement. Strain modification has been revolutionized by recombinant DNA technologies; still it remains difficult to clone unidentified genes. Thus, mutation and selection will persist as an integral part of many breeding programmes. Furthermore, although recombinant DNA methods are the most precise way of introducing novel traits encoded by single genes into commercial wine yeast strains, hybridization remains the most effective method for improving and combining traits under polygenic control.

19.4.9 Vectors Used in Yeast Transformation

Yeast transformation vectors can be classified into two main categories: integrative and autonomously replicating, which bear extra-chromosomal plasmid vectors that are absent in the former.

19.4.9.1 Yeast Centromere Plasmid (YCp)

Here centromeric sequences are incorporated, thereby such plasmids exhibit three characteristics of chromosome in yeast cells: i) they are mitotically stable in the absence of selective pressure ii) they segregate during meiosis in a Mendelian manner, and iii) they are found at low copy number (one per cell) and are lost by 1% for each generation. A number of hybrid plasmids containing DNA segments from around the centromere-linked '*leu2*', '*cdcl0*', and '*pgk*' loci on chromosome III of yeast have been isolated which are unstable in yeast. However, one of them was maintained stably through mitosis and meiosis. The stability segment has been confined to a 1.6 Kb region lying between the '*leu2*' and '*cdcl0*' loci, and its presence on plasmids carrying either of two '*ars*' has confirmed the plasmids behaving like mini chromosomes. Centromeres have also been isolated from chromosomes IV and XI of yeast, and it has been found that they confer similar properties on plasmids to the centromere (*cen*) of chromosome III.

19.4.9.2 Yeast Replicating Plasmids (YRp)

The third type of yeast is replicating plasmids (YRp), which are able to multiply autonomously using yeast chromosomal sequences due to the presences of ARS acting as origins of replication. These plasmids transform yeast at high efficiencies (10^4–10^5 transformants/μg DNA) and are present in high copy numbers up to 100 per cell. The transformants are very unstable, and the loss of the transforming phenotype is much greater than 10–20% per generation in non-selective medium. Instability is attributed to a strong bias in segregation that favors the mother cell. A useful vector, which consists of a 1.4 Kb yeast DNA fragment containing the TRP1 gene inserted into the pBR322, was constructed. Replication origins are known to be located very close to several yeast genes, including one or two which can be used as selectable markers. YRp7 is an example of a replicating plasmid and is made up of pBR322 plus the yeast gene TRP1. This gene, which is involved in tryptophan biosynthesis, is located adjacent to a chromosomal origin of replication. No transformants were found in which the vector had integrated into the chromosomal DNA. Since pBR322 alone cannot replicate in yeast cells, yeast chromosomal sequence must permit the vector to replicate autonomously and to express yeast structural genes in the absence of recombination with host chromosomal sequences. A similar vector based on pBR313 was developed. In both the cases, the yeast gene was linked to a centromere, which, initially, was considered to be very important but later appeared non-essential; rather the vector carried on autonomously replicating sequences (*ars*) derived from the chromosomes. Although plasmids containing an '*ars*' transform yeast very efficiently, the resulting transformants are exceedingly unstable. Yeast replicating plasmids tend to remain associated with the mother cell and are not efficiently distributed to daughter cells. Occasional stable transformants are also found, and these appear to be cases in which the entire yeast replicating plasmid has integrated into a homologous region on a chromosome in a manner identical to that of yeast integrative plasmids.

19.4.9.3 Integrating Plasmid Vector (YIp)

These don't replicate efficiently in the yeast as they lack autonomous replication features and thereby integrate into the homologous sequences of the recipient cells for its stable maintenance. These are quite useful for generating mutation by homologous recombinations and are quite stable but can be lost during culturing. The efficiency of transformation is very low, *i.e.* 1–10 transformants/μg DNA/10^7 cells. The frequency of transformation can, however, be increased by about 1000-fold simply by cutting the vector within *Saccharomyces cerevisiae* DNA with a restriction enzyme due to recombinogenic free ends.

19.4.9.4 Yeast Episomal Plasmids (YEp)

These carry sequences of the 2-μm circle that is usually present in most strains of *S. cerevisiae*, where it seems to have no function. These provide an excellent module in cloning vector and have an ideal size of 6 kb and exist in the yeast cell at a copy number of between 70 and 200 depending upon selectable marker. Plasmid uses its origin of replication and makes use of several other enzymes provided by the host cell and the protein coded by the REP1 and REP2 genes carried by the plasmid.

19.4.10 Yeast Artificial Chromosomes (YACs)

In contrast to above-mentioned plasmids, these have a linear structure. The ends of all the yeast chromosomes, like those of all other linear eukaryotic chromosomes, have unique structures that are called telomeres, which preserve the integrity of the ends of DNA molecules which often cannot be finished by the conventional mechanisms of DNA replication. A linear yeast vector, known as artificial yeast chromosome, has been developed by combining *ars*, *cen*, and *tel* sequences. A method for using yeast (YACs) to clone large DNA fragments of up to 500 Kb has also been evolved. The basic vector yeast artificial chromosome (YAC2) is a plasmid that can be readily propagated in *E. coli*. It can be cleaved into two fragments that constitute chromosome arms. The left arm has a telomere, a CEN4, an *ARS5*, and the *TRP1* as a selectable marker. The right arm has only a telomere and a different selectable marker, URA3. The cleavage site that separates the two arms also divides the yeast SUP4, which encodes an ochre suppressor allele of a tyrosine tRNA. The arms are mixed with exogenous DNA that has been cleaved to produce a population of greater than 100 Kb fragments. YAC technology has also provided a very powerful new technique to study regulation and function of human genes, however, it has also been limited by its stability.

19.4.11 Gene Transfer Techniques

Among the various techniques known for the transfer of genes to the host cell, the most widely used is the process of transformation that includes uptake of any foreign DNA molecule by any type of cell, irrespective of whether the uptake results in a detectable change in the cell and whether the cell involved is bacterial, fungal, animal, or plant. The next step is to introduce the DNA into the yeast cells. Treatment with lithium acetate does enhance DNA uptake by yeast cells and is frequently used in the transformation of *S. cerevisiae*. The main barrier to DNA uptake by the yeast cells is a cell wall that is generally degraded by enzymes, and under the right conditions intact protoplasts can be obtained. Protoplasts generally take up DNA quite readily. Alternatively, transformation can be stimulated by a special technique called electroporation, where the cells are subjected to a short electrical pulse. It induces transient formation of pores in the cell membrane through which the DNA molecule can enter the cell. After the process of transformation, the protoplasts are washed to remove the degradative enzymes, and the cell wall spontaneously reforms. There are also two physical methods for introducing DNA into cells: microinjection, which makes use of a very fine pipette to inject DNA molecules directly into the nucleus of the cells to be transformed, and biolistics.

19.4.12 Stabilization of Transformants

Stabilization is an essential step in the construction of recombinant industrial strains. In addition, regulations require that all DNA sequences unnecessary for gene expression be removed, notably those from *E. coli*. The most frequently used stabilization system involves integrating the sequence of interest into the genome by homologous recombination. Different strategies are used, for example, during co-transformation the gene is introduced in linear form and transformants are selected using a replicating plasmid containing a dominant resistance marker that can easily be removed by segregation. Another possibility is two-step integration using an integrating vector-containing the selectable marker and the gene of interest, which enables the removal of the vector sequences after integration. The integration target can be the homologous sequence of a modified gene, in which it is often necessary to replace all the alleles of a polyploid strain by successive transformations. With a heterologous gene, the integration target must be a sequence whose inactivation in no way changes the industrial performance of the strain.

19.5 TARGETS FOR WINE YEAST STRAIN DEVELOPMENT

The principal targets for strain development fall into two broad categories: i) improvement of fermentation performance and simplification of the process, and ii) improvement of organoleptic and hygienic characteristics. The majority of commercial wine yeasts are strains of *S. cerevisiae*. The demand for more specialized wine yeasts has been growing, and the number of commercialized, selected wine yeast strains has increased from about 20 to over 100 during the past 10 years. Substantial amounts of work took place during the 1990s to develop new strains, which mainly involved recombinant DNA approaches. Some of the requirements of wine yeasts as discussed in earlier sections of this chapter are complex and difficult to define genetically without understanding of the biochemistry and physiology involved, and no wine yeast in commercial use to date has all the characteristics listed. While some degree of variation can be achieved by altering the fermentation conditions, a major source of variation is the genetic constitution of the wine yeasts. Nevertheless, many of the properties have a genetic basis and have been manipulated by genetic techniques. Various targets (Table 19.1) generally hit for the improvement of wine yeasts are discussed in following sections.

19.5.1 Improved Viability and Vitality of Active Dried Wine Yeast Starter Cultures

Both the genetic and physiological stability of stock cultures of seed yeast and wine yeast starter cultures are essential for optimal fermentation performance. The manufacturers of active dried wine yeast starter cultures can positively influence the degree of viability and vitality, as well as the subsequent fermentation performance of their cultures, by the way they cultivate their yeasts. Trehalose is associated with several functions of yeast cell and its accumulation on both sides of the plasma membrane confers stress protection by stabilizing the yeast's membrane structure. Glycogen accumulation by yeast propagated for drying has been linked to enhanced viability and vitality upon reactivation and provides a readily mobilizable carbon and energy source during the adaptation phase.

19.5.2 Efficient Sugar Utilization

The main sugars present in grape must (glucose and fructose) are metabolized to pyruvate via the glycolytic pathway in *S. cerevisiae* and finally converted to ethanol. Notwithstanding the theoretical yield in model fermentation, about 95% of the sugar is converted into ethanol and carbon dioxide, 1% into cellular material, and 4% into other products such as glycerol. The first step to ensure efficient utilization of grape sugar by wine yeasts is to replace any mutant alleles of genes encoding the key glycolytic enzymes, namely hexokinase (*HXK*), glucokinase (*GLK*), phosphoglucose isomerase (*PGI*), phosphofructokinase (*PFK*), aldolase (*FBA*), triosephosphate isomerase (*TPI*), glyceraldehyde-3-phosphate dehydrogenase (*TDH*), phosphoglycerate kinase (*PGK*), phosphoglycerate mutase (*PGM*), enolase (*ENO*), pyruvate kinase (*PYK*), pyruvate decarboxylase (*PDC*), and alcohol dehydrogenase (*ADH*). Because of the impermeability of the plasma membranes of yeast cells to highly polar sugar molecules, various

TABLE 19.1
Targets for the Genetic Improvement in Wine Yeasts

Desirable characters	Focus areas	Potential genetic targets	References
		A) IMPROVED FERMENTATION PERFORMANCE	
Improved general resistance and stress tolerance	Stress response, sterol, glycogen and trehalose accumulation	Modification of glycogen or trehalose metabolism [action on *GSY1* and *GSY2* (glycogen synthetase), *TPS1* (trehalose-6-phosphate synthase), *TPS2* (trehalose-6-phosphate phosphatase)]	Spencer *et al.* (1983) Thevelein (1996) Thevelein and De Winde (1999)
Improved efficiency of sugar utilization	Hexose transporters, hexose kinases	Overexpression and modification of *HXT1-HXT18, SNF3, FSY1*	Johannsen *et al.* (1984)
Improved efficiency of nitrogen assimilation	Improved utilization of less efficient N-sources	Proline catabolism [*PUT1* (proline oxidase) and *PUT2* (pyrroline-5-carboxylate dehydrogenase)], decreased expression of *URE2*	Salmon and Barre (1998)
Improved efficiency of fermenting high sugar juices	High-sugar grape juice fermentation resulting in wines with lower concentrations of acetic acid	Mating a robust wine strain of *Saccharomyces cerevisiae* with a wine isolate of *Saccharomyces bayanus*	Bellon *et al.* (2015)
Improved ethanol tolerance	Sterol formation, membrane ATPase activity	Modification of expression of *PMA1* and *PMA2* (ATPase)	Piper (1995) Takahashi *et al.* (2001)
Increased tolerance to antimicrobial compounds	Resistance to killer toxins, sulphur dioxide	Inclusion of *KIL2* (zymocin and immunity factor), overexpression of *CUP1* (copper chelatin)	Pretorius (2000)
Reduced foam formation	Cell surface proteins	Deletion of *FRO1* and *FRO2* (froth proteins)	Pretorius (2000)
		B) IMPROVED PROCESSING EFFICIENCY	
Improved protein clarification	Proteases	Overexpression of PEP4 (protease A)	Pretorius (2000)
Improved polysaccharide clarification	Glucanases, pectinases, xylanases, arabinfuranosidases	Overexpression of *END1* (endoglucanase), *EXG1* (exoglucanase), *CEL1* (cellodextrinase, *BGL1* (β-glucosidase), *PEL5* (pectate lyase), *PEH1* (polygalacturonase), *XYN1* (xylanases) and *ABF2* (arabinofuranosidase)	Pretorius (2000)
Controlled cell sedimentation and flocculation	Flocculins	Late expression of flocculation genes (*FLO1, FLO5,FLO8, MUC1*) under control of promoters (*HSP30*)	Querol and Ramon (1996)
Controlled cell flotation and flor formation	Cell wall hydrophobic proteins	Late expression of *MUC1* under control of promoters (*Hsp30*)	Pretorius (2000)
		C) IMPROVED BIOLOGICAL CONTROL OF WINE SPOILAGE MICROORGANISMS	
Wine yeasts producing antimicrobial enzymes	Lysozyme, glucanases, chitinases	Expression of *HEL1* (hen egg white lysozyme), *CTS1* (chitinase), *EXG1* (exoglucanase)	Pretorius (2000)
Wine yeasts producing antimicrobial peptides	Bacteriocins	Expression of *PED1* (pediocin), *LCA1* (leucocin)	Pretorius (2000)
Wine yeasts producing sulphur dioxide	Sulphur metaboilism and SO_2 formation	Overexpression of *MET14* (adenosylphosphosulphate kinase) and *MET16* (phospho adenosylphosphosulphate reductase) and deletion of *MET10* (sulphite reductase)	Pretorius (2000)
		D) IMPROVED WINE WHOLESOMENESS	
Increased production of resveratrol	Stilbene synthesis	Expression of *4CL9/216* (co-enzyme A ligase), *VST1* (stilbene synthase)	Pretorius (2000)
Reduced formation of ethyl carbamate	Amino acid metabolism, urea formation	Deletion of *CAR1* (arginase) or expression of *URE1* (urease)	An and Ough (1993)Kitamoto *et al.* (1991)
Reduced formation of biogenic amines	Bacteriolytic enzymes, bacteriocins	Expression of *HEL1* (hen egg white lysozyme), *PED1* (pediocin), *LCA1* (leucocin)	Pretorius (2000)
Decreased levels of alcohol	Carbon flux, glycerol metabolism and glucose oxidation.	Overexpression of *GPD1* and *GPD2* (glycerol-3-phosphate dehydrogenase), modification of *FPS1* (glycerol transport facilitator), expression of *GOX1* (glucose oxidase)	Michnick *et al.* (1997) Scanes *et al.* (1998)

(Continued)

TABLE 19.1 (CONTINUED)
Targets for the Genetic Improvement in Wine Yeasts

Desirable characters	Focus areas	Potential genetic targets	References
		E) IMPROVED WINE FLAVOUR AND OTHER SENSORY QUALITIES	
Enhanced liberation of grape terpenoids	Glycosidases, glucanases, arabinofuranosidases	Expression of *END1* (endoglucanase), *EXG1* (exoglucanase), *CEL1* (cellodextrinase), *BGL1* (β-glucosidase), *PEL5* (pectate lyase) and *PEH1* (polygalacturonase) and *ABF2* (arabinofuranosidase)	Van Rensburg *et al.* (1996) Van Rensburg *et al.* (1998a & b)
Enhaced production of desirable volatile esters, fatty acids	Esterases	Modified expression of *ATF1* (alcohol acetyl transferase), *IAH1* (esterase)	Lilly *et al.* (2000) Saison *et al.* (2009)
Optimised fusel oil production	Amino acid metabolism	Deletion of *ILE1, LEU* and *VAL* genes	Snow (1983)
Enhanced glycerol production	Glycerol metabolism	Overexpression of *GDP1* and *GDP2* (glycerol-3-phosphate dehydrogenase), *FPS1* (glycerol transport facilitator) and deletion of *ALD6*	Michnick *et al.* (1997) Scanes *et al.* (1998) Papapetridis *et al.* (2016)
Bioadjustment of wine acidity	Maloethanolic and malolactic fermentation, lactic acid production	Expression of *MAE1* (malate permease), *MAE2* (malic enzyme), *mleS* (malolactic enzyme), *LDH1* (lacticodehydrogenase)	Dequin *et al.* (1999) Volschenk *et al.* (1997) Eglinton *et al.* (2000)
Optimisation of phenolics	Phenolic acid metabolism	Modified expression of *PAD1* (phenyl acrylic acid decarboxylase) *PDC* (p-coumaric acid decarboxylase), *PADC* (phenolic acid decarboxylase)	Pretorius (2000)
Elimination of phenolic off-flavours	Phenolic acid metabolism	Disruption of *POF1* gene	Henschke (1997)
Reduced sulphite and sulphide production	Sulphur metabolism, H_2S formation	Deletion of *MET14* (adenosylphosphosulphate kinase) , *MRX1* (methionine sulphoxide reductase), *MET3*	Henschke (1997)

complex mechanisms are required for efficient translocation of sugars into the cell. The main point of control of glycolytic flux as one of the key targets for the improvement of wine yeasts could be considered, as some members of the *HXT* permease gene family are being over-expressed in an effort to enhance sugar uptake, thereby improving the fermentative performance of wine yeast strains. Protoplast fusion between industrial yeast strains (*e.g. S. cerevisiae* × *Candida utilis*; *S. cerevisiae* × *S. mellis*) using nutritional requirements and differential carbohydrates metabolism reveals that the fusion product produces more ethanol than its parental, presumably by utilizing both the starch and glucose, as was the case with a fusion of osmotolerant species, *S. mellis* with *S. cerevisiae*. Genes for D-xylose isomerase and D-xylulose kinase from *S. violaceoniger* have been cloned, and the resulting transformants showed higher isomerase and kinase activity than the wild types.

19.5.3 Fermenting High-Sugar Juices

In the attempt to produce sweet dessert wines, major challenges researchers faced was fermenting high-sugar juices that have the ability to elevate volatile acidity levels and extend fermentation times. *S. bayanus* contains FSY1 gene encoding active transporter, having a high affinity for fructose. By utilizing this sugar, it prevents the chances of suboptimal fermentation by accumulating fructose. In one of the studies, interspecific non-genetically modified hybrids of *Saccharomyces* spp. were constructed by mating a robust wine strain of *Saccharomyces cerevisiae* with a wine isolate of *Saccharomyces bayanus*, a species known to produce wines with low concentrations of acetic acid. These hybrids were successful in fermenting high-sugar grape juice resulting in botrytised Riesling wines with much lower concentrations of acetic acid relative to the industrial wine yeast parent. Additionally, this hybrid yeast produced wines with novel aroma and flavour profiles. In an environment of high sugar concentration, glycerol is produced as a reconcilable solute. NADH is utilized in this process, which gets regenerated to sustain redox balance and is carried out by oxidizing acetaldehyde that produces acetic acid.

19.5.4 Improved Nitrogen Assimilation

Of all the nutrients assimilated by yeast during wine fermentations, nitrogen is quantitatively second only to carbon. Carbon-nitrogen imbalances and, more specifically, deficiencies in the supply of assimilable nitrogenous compounds remain the most common causes of poor fermentative performance and sluggish fermentations. Such problematic and incomplete fermentations occur because nitrogen depletion irreversibly arrests hexose transport and the formation of reduced-sulphur compounds, especially hydrogen sulphide and the potential formation of ethyl carbamate from metabolically produced urea. Unlike grape sugars that are usually present in the large amounts needed for maximal yeast growth and alcohol production, the total nitrogen content of grape juices ranges from 60 to 2400 mg/L and can therefore be growth-limiting. Supplementation of juice with ammonium salts such as diammonium phosphate (DAP) is one solution. However, excessive addition of inorganic nitrogen often results in excessive levels of residual nitrogen, leading to microbial instability and ethyl carbamate accumulation in wine. *S. cerevisiae* is incapable of adequately hydrolyzing grape proteins to supplement nitrogen-deficient musts, and, therefore, relies on the ammonium and amino acids present in the juice. Wine yeasts differ from each other in terms of their ability to use various nitrogen sources, thus, ammonium is the preferred nitrogen source, and when a readily used nitrogen source, such as ammonium, glutamine, or asparagine is present, genes involved in the uptake and catabolism of poorly utilized nitrogen sources including proline are repressed.

19.5.5 Improved Ethanol Tolerance

The physiological basis of ethanol toxicity is complex and is not well understood, but it appears that ethanol mainly impacts upon membrane structural integrity and membrane permeability. The chief sites of action include the yeast cell's plasma membrane, hydrophobic proteins of mitochondrial membranes, nuclear membrane, vacuolar membrane, endoplasmic reticulum, and cytosolic hydrophilic proteins. Several intrinsic and extrinsic factors which synergistically aggravate the inhibitory effects of ethanol include high fermentation temperatures, nutrient limitation certain fatty acids and carbonyl and phenolic compounds. By manipulating the physico-chemical environment during the cultivation and manufacturing of active dried wine yeast starter cultures and during the actual vinification process, the yeast cells' protective adaptations which are acquired can be promoted, and the yeast cells containing high levels of the survival factors can be passed onto the progeny cells during the six or seven generations of growth in a typical wine fermentation. Wine yeast strains usually contain higher levels of survival factors than non-wine *Saccharomyces* strains, and their physiological response to ethanol challenge is also greater than non-wine strains. These defensive adaptations of wine yeasts, conferring enhanced ethanol tolerance, include several strategies including increased synthesis of cytochrome P450, alcohol dehydrogenase activity, and ethanol metabolism. Thus, the genetics of ethanol tolerance is multifarious and a very complex process with the involvement of more than 250 genes in the control of ethanol tolerance in yeast. Nevertheless, continuous culture of yeasts in a feedback system where ethanol is controlled by the rate of CO_2 release enables the selection of viable mutants with improved ethanol tolerance and fermentative ability.

19.5.6 Improvement of Wine Flavour and Other Sensory Qualities

The single most important factor in wine making is the sensory quality of the final product, which is due to the presence of desirable flavour compounds and metabolites in a well-balanced ratio and the absence of undesirable ones. One of the sensory properties that is evaluated is the colour of the wine. Brown colouring of wine can be a detrimental factor. Glutathione traps caftaric acid quinines in the form of 2-S-glutathionyl caftaric acid, also known as grape reaction product (GRP). This GRP limits the browning of the juice.

Many variables contribute to the distinctive flavours of wine, such as oenological practices including the yeast and fermentation conditions, the primary flavours of *V. vinifera* grapes, and the components formed during fermentation, aging, etc. The harmonious complexity of wine can subsequently be further increased by volatile extraction during oak barrel aging. Despite the extensive information published on flavour chemistry, odour thresholds, and aroma descriptions, the flavour of complex products such as wine cannot be predicted. With a few exceptions, perceived flavour is the result of specific ratios of many compounds rather than being attributable to a single impact compound. In wines, the major products of yeast fermentation, esters and alcohols, contribute to a generic background flavour, whereas subtle combinations of trace components derived from the grapes usually elicit the characteristic aroma notes of these complex beverages.

19.5.7 Enhanced Liberation of Grape Terpenoids

The varietal flavour of grapes is due to the accumulation of volatile secondary metabolites in *V. vinifera*, but a high percentage of these metabolites occur as their respective non-volatile o-glycosides. Increased enzymatic hydrolysis of aroma precursors present in grape juice can intensify the varietal character of wines, for instance terpenes such as geraniol and nerol can be released from terpenyl-glycosides by the β-D-glucosidases activities present in grape juice. However, these enzymatic activities are inhibited by glucose and exhibit poor stability at the low pH and high ethanol levels of wine. The less efficient grape glucosidases are replaced by aroma-liberating β-glucosidases from *Aspergillus* and other fungal species, added to fermented juice/young wine. To avoid the addition of expensive exogenous enzyme preparations to wine, interest has been renewed in producing the more active β-glucosidase from certain strains of *S. cerevisiae* and other wine-associated yeasts. Unlike the grape β-glucosidases, yeast β-glucosidases are not inhibited by glucose, so the liberation of terpenols during fermentation takes place ascribed to their action on the terpenyl-glycoside precursors. Since these β-glucosidases are absent in most of the *S. cerevisiae* starter culture strains, researchers have functionally expressed the β-glucosidase gene *(BGL1)* of the yeast *Saccharomycopsis fibuligera* in *S. ceresisiae*. When the β-1,4-glucanase gene from *Trichoderma longibrachiatum* was expressed in wine yeast, the aroma intensity of wine increased, presumably due to the hydrolysis of glycosylated flavour precursors. Likewise, the *S. cerevisiae* exo-β-1,3-glucanase gene *(EXG1)* has been over-expressed and introduced the endo-β-1,4-glucanase gene *(EVD1)* from *Butyrivibrio fibrisovrens*, the endo-β-1,3, 1,4-glucanases *(BEG1)* from *Bacillus subtilis*, and the α-arabinofuranosidase *(ABF2)* in *S. cerevisiae*.

19.5.8 Enhanced Production of Desirable Volatile Esters

During alcoholic fermentation, besides the primary product, a number of by-products including esters, in which are alcohol acetates and C_4–C_{10} fatty acid ethyl esters are formed and found in a high concentration in wine, Although these compounds are ubiquitous to all wines, but their level varies significantly. Apart from cultivation factors, the ester concentration produced during fermentation is dependent on the yeast strain, fermentation conditions, vinification practices, and composition of the must. The characteristic fruity odours of wine are primarily due to a mixture of hexyl acetate, ethyl caproate, caprylate, isoamyl acetate, and 2-phenyethyl acetate. The synthesis of acetate esters such as iso-amyl acetate and ethyl acetate in *S cerevisiae* are accomplished as a result of at least three acetyltransferase activities: alcohol acetyltransferase (*AAT*), ethanol acetyl transferase (*EAT*), and iso-amyl alcohol acetyl transferase (*IAT*). Isoamyl acetate (banana-like aroma) is an important determinant for sake and young wines, and is produced by yeast from isoamyl alcohol, which itself is a by-product of leucine synthesis. Over-expression of one of the genes responsible for leucine synthesis (*LEU4* encoding-isopropylmalate synthase) in a sake yeast strain results in a very slight increase in isoamyl alcohol concentrations and the corresponding ester. Another strategy in a brewing strain was to increase the formation of acetate esters by over-expression of the *ATF1* gene, which encodes the alcohol acetyltransferase, catalyzing the formation of esters from acetyl CoA and the relevant alcohols. The *ATF1*-encoded alcohol acetyltransferase activity (*AAT*) is the most studied acetyl transferase of *S. cerevisiae*. To investigate the role of *AAT* in wine, the *ATF1* gene from a widely used commercial wine yeast strain (*VIN13*) has been cloned and placed under the control of the constitutive yeast phosphoglycerate kinase gene (*PGK1*) promoter and terminator. Integration of this modified copy of *ATFI* into the genomes of three commercial wine yeast strains (VIN7, VIN13, and WE28) resulted in the over expression of AAT activity and increased levels of ethyl acetate, iso-amyl acetate, and 2-phenylethyl acetate in wine, while the concentration of ethyl caprate, ethyl caprylate, and hexyl acetate showed only minor changes, and the acetic acid concentration decreased by more than half. Thus, the over-expression of acetyl-transferase genes such as *ATF1* could profoundly affect the flavour profiles of wines that are deficient in aroma, thereby paving the way for the production of products maintaining a fruitier character for longer periods after bottling.

19.5.9 Optimized Fusel Oil Production

Fusel oils or higher alcohol are produced by wine yeasts during fermentation from intermediates in the branched chain amino acids biosynthetic pathways, for the production of Ile, Leu, and Val by decarboxylation, transamination, and reduction. Higher alcohols do not affect the wine flavour and aroma when their concentration is low; rather, they contribute to wine quality. Initial attempts to use Ile^-, Leu^-, and Val^- mutants succeeded in lowering the levels of isobutanol, active amyl alcohol, and isoamyl alcohol production in fermentations, but these mutants were of no commercial use as their growth and fermentation rates were compromised. A Leu^- mutant derived from the widely used Montrachet wine yeast has been reported to produce more than 50% less isoamyl alcohol during fermentation than the prototrophic parent. Further research on the capabilities of Ile^-, Leu^-, and Val^- to lower the levels of fusel oil in wine would be interesting.

19.5.10 Enhanced Glycerol Production

Glycerol is the most abundant by-product of alcoholic fermentation after ethanol and carbon dioxide. Due to its non-volatile nature, glycerol has no direct impact on the aromatic characteristics of wine but does impart certain other sensory qualities. Having a slightly sweet taste and a viscous nature, it contributes to the smoothness, the perceived sweetness consistency, and overall body of wine especially at higher levels. The amount of glycerol in wines depends on several factors including yeast strains. Wine yeast strains overproducing glycerol could, therefore, be of considerable value in improving the sensory quality of wine. Interest in rerouting the carbon flux towards the glycerol pathway is expected at the cost of ethanol production and could be an alternative approach to the current trend for removing ethanol from beverages. The overproduction of glycerol at the expense of ethanol also could be employed to produce table wine with lower levels of ethanol. About 4–10% of the carbon source is usually converted to glycerol, resulting in glycerol levels of 7–10% of that of ethanol. *FPS1*, which encodes a channel protein, was shown to act as a glycerol transport facilitator controlling both glycerol influx and efflux. Further, the over expression of *GPD1*, together with the constitutive expression of *FPS1*, successfully re-directed the carbon flux towards glycerol and the extracellular accumulation of glycerol varying from 1.5 to a four-fold increase was obtained, depending on the strains. Such yeasts offer new prospects to improve the quality of wines lacking in smoothness and body and to the production of low-alcohol wines. In the pentose-phosphate pathway, the cofactor specificity alteration of the oxidative branch of *S. cerevisiae* can be employed to increase glycerol production in wine fermentation and to enhance the generation of NADH or other precursors attained from pentose-phosphate pathway.

19.5.11 Bio-Adjustment of Wine Acidity

Wine contains many organic and inorganic acids, but the predominant organic acids are tartaric and malic acids, accounting for about 90% of the titratable acidity of grapes. Under certain climatic conditions, the grapes are produced with high acidity – consequently, the production of wines with high acidity. Unless the acidity of such wines is adjusted, these would be considered as unbalanced or spoilt. The most commonly used practice at present involves the bio-deacidification of wine by malolactic fermentation using lactic acid bacteria including *Oenococcus oeni*. Several alternatives have been explored, including the possible use of malate-degrading yeasts. During malo-ethanolic fermentations by *Schizosaccharomyces pombe*, malate was effectively converted to ethanol, but off-flavours were produced. Attempts to fuse wine yeasts with malate-assimilating yeast also failed. Genetic engineering of wine yeast to conduct alcoholic fermentations and malate degradation simultaneously has been explored by several groups. In order to engineer a malolactic pathway in *S. cerevisiae,* the malolactic genes (*mleS S*) from *Lactococcus lactis*, *Lactobacillus delbrueckii*, and the *mleA* from *O. oeni* were cloned and expressed in *S. cerevisiae*. However, due to the absence of an active malate transport system in *S. cerevisiae,* these engineered strains could still not metabolize malate efficiently. The same was achieved only when the *mleS* was co-expressed with the *Schiz. pombe MAE1*-encoding malate permease. Similarly, an efficient malo-ethanolic *S. cerevisiae* was constructed by co-expressing *MAEl* and *MAE2* from *Schiz. pombe* in *S. cerevisiae.* A functional malolactic wine yeast could replace the unreliable bacterial malolactic fermentation, whereas a malo-ethanolic strain of *S. cerevisiae* would be more useful for the production of fruity floral wines besides reducing the pH due to production of lactic acid. *S. bayanus* has been stated to produce lower levels of volatile acid in wine making.

19.5.12 Elimination of Phenolic Off-Flavour

The presence of excess amounts of volatile phenolic compounds is undesirable, due to their off-flavours. The *POF1* gene in some strains of *S. cerevisiae* encodes a substituted cinnamic acid carboxylase that is able to decarboxylate grape hydroxycinnamic acids in a non-oxidative fashion to vinyl alcohols. It thus appears that the disruption of *POF1* could provide a way to reduce the content of volatile phenols in wines. The production of acetic acid in large quantities during fermentation can cause esterification of acetic acid by ethanol, which can lead to the production of volatile compounds such as ethyl acetate having a peculiar solvent aroma similar to that of nail polish. At times, an increase in the concentration of a flavour or aroma has a tendency to not only enhance the sensory impact but also can aid in masking non-desirable flavours or aromas.

19.5.13 Reduced Sulphite and Sulphide Production

The sulphur-containing compounds have a pronounced effect on the flavour of wines. Unlike SO_2, which has some beneficial effects, H_2S is a highly undesirable compound in wine. Sulphite is only formed from sulphate, while sulphide is formed from sulphate, sulphite, elemental sulphur, and from cysteine. Hydrogen sulfide is produced during wine fermentation, mainly in response to the depletion of nitrogen and possibly certain vitamins, and its production is influenced by many environmental factors and by the yeast strain. In conditions of nitrogen starvation, hydrogen sulfide may accumulate and diffuse out of the cells. Wine yeast strains with low H_2S production have been obtained by hybridization. Alternatively, reduced sulfide production might be achieved by manipulating the sulfate metabolic pathway. Fortunately, H_2S is highly volatile and can usually be removed by the stripping action of CO_2 produced during fermentation. Yeast strains differ in their ability to produce sulphite and sulphide, and one way is to select or develop a wine yeast strain that will either produce less H_2S or that will retain most of the H_2S produced intracellularly; here, hybridization can be used. In addition, the deliberate introduction of mutations in certain enzymes of sulphur, sulphur amino acids, pantothenate, and pyridoxine pathways might well enable a stepwise elimination of these characteristics in wine yeasts. The *MET3* gene, encoding for ATP sulphurylase, the first enzyme involved in the conversion of sulphate to sulphite, has been cloned and shown to be regulated at the transcriptional level, which can provide a target for regulation of H_2S formation in wine yeasts.

19.5.14 Reduced Formation of Ethyl Carbamate

Ethyl carbamate, also known as ‘urethane’, is a suspected carcinogen that occurs in wine, sherry, and brandy. Due to potential health hazards, demand is growing to reduce the allowable limits of ethyl carbamate in wines. High-alcohol wines contain much higher levels of ethyl carbamate than low alcohol table wines. It is believed that ethyl carbamate is formed in ageing wines and fortified wines by a reaction between urea and ethanol. It is mainly formed by the spontaneous chemical reaction of ethanol and urea at elevated temperatures in acidic media. Urea is produced mainly from the cleavage of arginine by arginase. In *S. cerevisiae,* urea is formed during the breakdown of arginine by the *CAR1*-encoded arginase. Although all *S. cerevisiae* strains secrete urea, the extent to which they re-absorb the urea differs. Strain selection is the most important means of reducing the accumulation of urea in wine, but prevention of ethyl carbamate formation has been attempted in the end product by disrupting the *CAR1* in the wine yeasts, which proved to be successful in eliminating the urea accumulation in the wine. Yeast strains that produce urease extracellularly can also be developed, and a novel urease gene was constructed by fusing α, β, and γ subunits of *Lactobacillus fermentum* urease operon, and this gene construct was successfully expressed under the control of *S. cerevisiae PGK1* promoter and terminator signals in *S. cerevisiae* and *Schiz. pombe.* Although the level of transcription in *S. cerevisiae* was much higher than in *Schiz. pombe*, the secretion of urease was extremely low. However, *S. cerevisiae*-derived urease is unable to convert urea into ammonia and carbon dioxide. The absence of recombinant urease activity in transformed *S. cerevisiae* cells was probably due to the lack of the essential auxiliary proteins present only in urease-producing species such as *Schiz. pombe.* Without these proteins, *S. cerevisiae* is unable to assemble the various subunits into an active urease. Thus, accessory genes of *L. fermentum* need to be cloned and expressed, in addition to the structural urease genes to enable *S. cerevisiae* to express an active urease.

19.5.15 Improved Biological Control of Wine Spoilage Microorganisms

Various bacteria, wild yeasts, and moulds develop during wine production, which alter the chemical composition, appearance, aroma, and flavour of the end product. Of the various microorganisms, bacteria play a major role in spoiling wines, and the routine practice is to control the proliferation of such organisms by use of efficient *S. cerevisiae* and *O. oeni* starter cultures at appropriate inoculum levels, besides the addition of chemical preservatives. Because of growing consumer concern about preservatives, interest in the use of bio-preservatives, including lysozyme, zymocins, and bacteriocins, which protect the wines from spoilage by lactic acid bacteria, is growing. Purified forms of the bio-preservatives being expensive, attempts are being made to develop wine yeast starter cultures by cloning the genes for the antimicrobial enzymes and peptides. In wine making, there is always a risk that wild yeast becomes predominant and secretes zymocin, a toxin that kills sensitive wine yeast strains. The production and immunity to this toxin is determined by a cytosolic, double-stranded RNA. Out of three major types of killer toxins (K1-K3), most of the yeast strains found in grape musts produce the K2 zymocin, so the K1 dsRNA has been integrated into the genome of a K2 wine yeast. Attempts to develop bactericidal wine yeast strains have also been made, wherein two bacteriocin genes, encoding a pediocin and a leucocin gene from *Pediococcus acidilactici* and *Leuconostoc carnosum* respectively, have been expressed in *S. cerevisiae*. Thus, the use of bacteriocidal yeasts would be useful for the production of wine with reduced levels of sulphur dioxide and other chemical preservatives.

19.5.16 Flocculation

Yeast flocculation, the asexual aggregation of cells into flocs and their subsequent removal from the fermentation medium by sedimentation, is of major interest for some fermentation processes including the elaboration of sparkling wines such as in the Champagne method; a secondary fermentation is conducted in the bottle to develop specific sensory characteristics. This process can be advantageously simplified by the sedimentation of flocculent yeast cells. Several attempts have been made to construct flocculent industrial strains by classical genetic approaches. At a molecular level, several dominant flocculation genes, *FLO1*, *FLO5*, and *FLO8*, were found involved in flocculation, and the dominant *FLO1* gene has been isolated, characterized, and shown to encode a cell-wall protein containing a lectin domain. By transforming the strains with a multicopy vector containing *FLO1* isolated from a flocculent *S. cerevisiae* strain, this property can be transformed to a wine yeast strain. A major difficulty in the development of precise control of gene expression is the lack of regulated promoters that can be used under industrial conditions. The most well-known inducible promoters for yeasts cannot be used, due to the composition of the industrial medium and because of regulatory constraints preventing major modifications to it. An example of a promoter that is induced by environment signals is Hsp30. This gene is induced by factors that occur late or towards the end of fermentation. When the native *FLO1* promoter is replaced by the Hsp30 promoter, flocculation occurs towards the end of fermentation.

19.6 STATUTORY REGULATION AND CONSUMER DEMANDS

S. cerevisiae has been playing a major role in the fermented food and beverage industries for a long time and is mankind's oldest domesticated organism, generally regarded as safe. Its metabolism has made possible thanks to the development of various biotechnological processes, especially in wine and brandy making. It was also the first genetically modified organism (GMO) to be cleared for food, baking, and brewing use, but the genetically modified strains have not yet been commercially used for wine making in many countries, apparently due to public fear and a perception of risk with regard to genetically modified foods and the concern for environmental safety, although without scientific base. For that, a combination of both GM and non-GM methods needs to be implemented. It is assumed that such a combination might cater to the industrial needs and in this way benefit the consumer, thus addressing the concern of using GMO. But fear about food and environmental safety has evoked a plethora of strict legislations and regulatory guidelines which may differ in detail but are broadly similar in most countries. Guidelines for approval of GM products and the release of GMOs usually require a number of clear guarantees, such as a complete definition of the DNA sequence introduced, the elimination of any sequence that is not indispensable for expression of the desired property, the absence of any selective advantage conferred on the transgenic organism that could allow it to become dominant in natural habitats, no danger to human health and/or the environment from the transformed DNA, and a clear advantage to both the producer and the consumer. Consumer education is essential in overcoming doubts regarding the application of recombinant DNA technology in the wine industry.

BIBLIOGRAPHY

Adams, A., Gottschling, D.E., Kaiser, C.A. and Sterns, T. (1997). *Methods in Yeast Genetics*. Cold Spring Harbor Laboratory Press, Cold Spring Harbor, New York.

An, D. and Ough, C.S. (1993). Urea excretion and uptake by wine yeasts as affected by various factors. *Am. J. Enol. Vitic.*, 44: 35.

Astromoff, A. and Egerton, M. (1999). *Saccharomyces cerevisiae*: Genetics and genomics. In: *Manual of Industrial Microbiology and Biotechnology*, A.L. Demain and J.E. Davies (Eds). ASM Press, Washington, DC, p. 435.

Bellon, B.R., Yang, F., Day, M.P., Inglis, D.L. and Chambers, P.J. (2015). Designing and creating *Saccharomyces* interspecific hybrids for improved, industry relevant, phenotypes. *Appl. Microbiol. Biotechnol.*, 99(20): 8597.

Blomberg, A. and Adler, L. (1989). Roles of glycerol and glycerol-3-phosphate dehydrogenase (NAD^+) in acquired osmo-tolerance of *Saccharomyces cerevisiae*. *J. Bacteriol.*, 171(2): 1087.

Bonciani, T., Solieri, L., De Vero, L. and Giudici, P. (2016). Improved wine yeasts by direct mating and selection under stressful fermentative conditions. *Eur. Food Res. Technol.*, 242(6): 899.

Borneman, A.R., Schmidt, S.A. and Pretorius, I.S. (2013). At the cutting-edge of grape and wine biotechnology. *Trends Genet.*, 29(1): 63.

Cherry, J.M., Ball, C., Weng, S., Juvik, G., Schmidt, R., Adler, C., Dunn, B., Dwight, S., Riles, L., Mortimer, R.K. and Botstein, D. (1997). Genetic and physical maps of *S. cerevisiae*. *Nature*, 387: 67.

Chinery, S.A. and Hinchlifte, E. (1989). A novel class of vector for yeast transformation. *Curr. Genet.*, 16(1): 21.

Clarke, L. and Carbon, J. (1980). Isolation of a yeast centromere and construction of functional small circular chromosomes. *Nature*, 287(5782): 504.

Cliff, M.A. and Pickering, G.J. (2006). Determination of odour detection thresholds for acetic acid and ethyl acetate in ice wine. *J. Wine. Res.*, 17(1): 45.

Cole, V.C. and Noble, A.C. (1995). Flavour chemistry and assessment. In: *Fermented Beverage Production*, A.G.H. Lea and J.R. Piggott (Eds). Blackie Academic and Professional, London, p. 361.

Costa, V., Reis, E., Quintanilha, A. and Moradas-Ferreira, P. (1993). Acquisition of ethanol tolerance in *Saccharomyces cerevisiae*: The key role of the mitochondrial superoxide dismutase. *Arch. Biochem. Biophys.*, 300(2): 608.

Dequin, S. (2001). The potential of genetic engineering for improving brewing, wine making and baking yeasts. *Appl. Microbiol. Biotechnol.*, 56(5–6): 577.

Dequin, S., Baptista, E. and Berre, P. (1999). Acidification of grape musts by *Saccharomyces cerevisiae* wine yeast strains genetically engineered to produce lactic acid. *Am. J. Enol. Vitic.*, 50: 45.

Dujon, B. (1996). The yeast genome project: What did we learn? *Trends Genet.*, 12(7): 263.

Eglinton, J.M., McWilliam, S.J., Fogarty, M.W., Francis, I.L., Kwiatkowski, M.J., Høj, P.B. and Henschke, P.A. (2000). The effect of *Saccharomyces bayanus*-mediated fermentation on the chemical composition and aroma profile of Chardonnay wine. *Aust. J. Grape Wine Res.*, 6(3): 190.

Fleet, G.H. (Ed.) (1993). *Wine Microbiology and Biotechnology*. Harwood Academic Publishers, Switzerland.

Fleet, G.H. (1998). The microbiology of alcoholic beverages. In: *Microbiology of Fermented Foods*, Vol. I., B.J.B. Wood (Ed.). Blackie Academic and Professional, London, U.K. p. 217.

García-Ríos, E., Nuévalos, M., Barrio, E., Puig, S. and Guillamón, J.M. (2019). A new chromosomal rearrangement improves the adaptation of wine yeasts to sulfite. *Environ. Microbiol.*, 21(5): 1771.

Goffeau, A. *et al.* (1997). The yeast genome directory. *Nature*, 385(6611): 5.

Goffeau, A., Barrel, B.G., Bussey, H., Davis, R.W., Dujon, B., Feldmann, H., Galibert, F., Hoheisel, J.D., Jacq, C., Johnston, M., Louis, E.J., Mewes, H.W., Murakami, Y., Philippsen, P., Tettelin, S. and Oliver, S.G. (1996). Life with 6000 genes. *Science*, 274: 546, 563.

Gonzalez, R., Tronchoni, J., Quirós, M. and Morales, P. (2016). Genetic improvement and genetically modified microorganisms. In: *Wine Safety Consumer Preference, and Human Health*, Moreno-Arribas M., Bartolomé Suáldea B. (Eds.), Springer International Publishing, Switzerland. pp. 71–96.

Guerin, B. (1991). Mitochondria. In: *The Yeasts*, 2nd edn., Vol. 4, Yeast Organelles. A.H. Rose and J.S. Harrison (Eds). Academic Press, London. p. 541.

Guthrie, C. and Fink, G.R. (1991). Methods in enzymology, vol 194. *Guide to Yeast Genetics and Molecular Biology*. Academic Press Inc., San Diego.

Hart, R.S., Jolly, N.P. and Ndimba, B.K. (2019). Characterisation of hybrid yeasts for the production of varietal Sauvignon blanc wine – A review. *J. Microbiol. Method.*, 165. doi:10.1016/j.mimet.2019.105699.

Henschke, P.A. (1997). Wine yeast. In: *Yeast Sugar Metabolism*, F.K. Zimmermann and K.D. Entian (Eds). Technomic Publication, Lancaster, p. 527.

Jagtap, U.B., Jadhav, J.P., Bapat, V.A. and Pretorius, I.S. (2017). Synthetic biology stretching the realms of possibility in wine yeast research. *Int. J. Food Microbiol.*, 252: 24.

Jiranek, V., Langridge, P. and Henschke, P.A. (1995a). Amino acid and ammonium utilization by *Saccharomyces cerevisiae* wine yeasts from a chemically defined medium. *Am. J. Enol. Vitic.*, 46: 75.

Jiranek, V., Langridge, P. and Henschke, P.A. (1995b). Regulation of hydrogen sulphide liberation in wine-producing *Saccharomyces cerevisiae* strains by assimilable nitrogen. *Appl. Environ. Microbiol.*, 61(2): 461.

Johannsen, E., Halland, L. and Opperman, A. (1984). Protoplast fusion within the genus *Kluyveromyces*. *Can. J. Microbiol.*, 30: 540.

Kimura, A. (1986). Molecular breeding of yeasts for production of useful compounds: Novel methods of transformation and new vector systems. *Biotechnol. Genet. Eng. Rev.*, 4(1): 39.

Kitamoto, K., Oda, K., Gomi, K. and Takahashi, K. (1991). Genetic engineering of a sake yeast producing no urea by successive disruption of arginase gene. *Appl. Environ. Microbiol.*, 57(1): 301.

Kong, Q.X., Gu, J.G., Cao, L.M., Zhang, A.L., Chen, X. and Zhao, X.M. (2006). Improved production of ethanol by deleting *FPS1* and over expressing *GLT1* in *Saccharomyces cerevisiae*. *Biotechnol. Lett.*, 28(24): 2033.

Kurtman, C.P. and Fell, J.W. (1998). *The Yeasts: A Taxonomic Study*, 4th edn. Elsevier Science, Amsterdam.

Lilly, M., Lambrechts, M.G. and Pretorius, I.S. (2000). The effect of increased yeast alcohol acetyltransferase activity on the sensorial quality of wine and brandy. *Appl. Environ. Microbiol.*, 66: 774.

Martini, A. (1993). The origin and domestication of the wine yeast *Saccharomyces cerevisiae*. *J. Wine Res.*, 4(3): 165.

Masneuf-Pomarede, I., Bely, M., Marullo, P. and Albertin, W. (2016). The genetics of non-conventional wine yeasts: Current knowledge and future challenges. *Front. Microbiol.*, 6: 1563.

Michnick, S., Roustan, J.L., Remize, F., Barre, P. and Dequin, S. (1997). Modulation of glycerol and ethanol yields during alcoholic fermentation in *Saccharomyces cerevisiae* strains overexpressed or disrupted for *GDP1* encoding glycerol 3-phosphate dehydrogenase. *Yeast*, 13(9): 783.

Mortimer, R.K., Cherry, M.C., Dietrich, F.S., Riles, L., Olson, M.V. and Botstein, D. (1995). *Genetic and Physical Maps of Saccharomyces Cerevisiae*, 12th edn. http://genome-www.stanford.edu/sacchdb/edition.12html.

Noble, A.C. (1994). Wine flavour. In: *Understanding Natural Flavours*, J.R. Piggott and A. Patterson (Eds). Blackie Academic and Professional, Glasgow, p. 228.

Old, R.W. and Primrose, S.B. (1995). *Principles of Gene Manipulation: An Introduction to Genetic Engineering*. Blackwell Scientific Publications, London.

Oliver, S.G., van der Aart, Q.J., Agostoni-Carbone, M.L., Aigle, M., Alberghina, L., Alexandraki, D., Antoine, G., Anwar, R., Ballesta, J.P. and Benit, P. (1992). The complete DNA sequence of yeast chromosome III. *Nature*, 357(6373): 38.

Panchal, C.J., Bast, L.J., Dowhanick, T.M., Jones, R.M., Russell, I. and Stewart, G.G. (1987). Construction and characterization of an all yeast DNA vector. In: *Biological Research on Industrial Yeasts*, Vol. 2, G.G. Stewart, I. Russell, R.D. Klein and R.R. Hiebsch (Eds). CRC Press, Boca Raton, p. 49.

Papapetridis, I., van Dijk, M., Dobbe, A.P.A., Metz, B., Pronk, J.T. and van Maris, A.J.A. (2016). Improving ethanol yield in acetate-reducing *Saccharomyces cerevisiae* by cofactor engineering of 6-phosphogluconate dehydrogenase and deletion of ALD6. *Microb. Cell Fact.*, 15: 67.

Perez-Gonzalez, J.A., Gonzalez, R., Querol, A., Sendra, J. and Ramon, D. (1993). Construction of a recombinant wine yeast strain expressing β-(1,4)-endoglucanase and its use in microvinification processes. *Appl. Environ. Microbiol.*, 59(9): 2801.

Perez-Torrado, R., Querol, A. and Guillamon, J.M. (2015). Genetic improvement of non– GMO wine yeasts: Strategies, advantages and safety. *Trends Food Sci. Technol.*, 45(1): 1.

Pérez-Través, L., Lopes, C.A., Barrio, E. and Querol, A. (2012). Evaluation of different genetic procedures for the generation of artificial hybrids in *Saccharomyces* genus for wine making. *Int. J. Food Microbiol.*, 156(2): 102.

Piper, P.W. (1995). The heat shock and ethanol stress responses of yeast exhibit extensive similarity and functional overlap. *FEMS Microbiol. Lett.*, 134(2–3): 121.

Porro, D.L., Brambilla, L., Ranzi, B.M., Martegani, E. and Albershina, L. (1995). Development of metabolically engineered *Saccharomyces cerevisiae* cells for the production of lactic acid. *Biotechnol. Prog.*, 11(3): 294.

Pretorius, I.S. (2000). Tailoring wine yeasts for the new millennium: Novel approaches to the ancient art of winemaking. *Yeast*, 16(8): 675.

Pretorius, I.S. and Bauer, F.F. (2002). Meeting the consumer challenge through genetically customized wine yeast strains. *Trends Biotechnol.*, 20(10): 426.

Pringle, J.R. and Hartwell, L.H. (1981). The *Saccharomyces cerevisiae* cell cycle. In: *The Molecular Biology of the Yeast Saccharomyces*, J.N. Strathern, E.W. Jones and J.R. Broach (Eds). Cold Spring Harbor Laboratory, New York, p. 97.

Prior, B.A. and Hohmann, S. (1997). Glycerol production and osmoregulation. In: *Yeast Sugar Metabolism*, F.K. Zimmermann and K.D. Entian (Eds). Technomic Publication, Lancaster, p. 313.

Querol, A. and Ramon, D. (1996). The application of molecular techniques in wine microbiology. *Trends. Food Sci. Technol.*, 7(3): 73.

Remize, F., Roustan, J.L., Sablarrolles, J.M., Barre, P. and Dequin, S. (1999). Glycerol overproduction by engineered *Saccharomyces cerevisiae* wine yeast strains leads to substantial changes in by-product formation and to a stimulation of fermentation rate in stationary phase. *Appl. Environ. Microbiol.*, 65(1): 143.

Rodrigues de Sousa, H., Spencer-Martins, I. and Gonçalves, P. (2004). Differential regulation by glucose and fructose of a gene encoding a specific fructose/H+ symporter in *Saccharomyces sensu* stricto yeasts. *Yeast*, 21(6): 519.

Saison, D., De Schutter, D.P., Uyttenhove, B., Delvaux, F. and Delvaux, F.R. (2009). Contribution of staling compounds to the aged flavour of lager beer by studying their flavour thresholds. *Food Chem.*, 114(4): 1206.

Salgues, M., Cheynier, V., Gunata, Z. and Wylde, R. (1986). Oxidation of grape juice 2-S-glutathionyl caffeoyl tartaric acid by *Botrytis cinerea* laccase and characterization of a new substance: 2,5-di-Sglutathionyl caffeoyl tartaric acid. *J. Food Sci.*, 51(5): 1191.

Salmon, J.M. and Barre, P. (1998). Improvement of nitrogen assimilation and fermentation kinetics under ecological conditions by derepression of alternative nitrogen-assimilatory pathways in an industrial *Saccharomyces cerevisiae* strain. *Appl. Environ. Microbiol.*, 64(10): 3831.

Scanes, K.T., Hohmann, S. and Prior, B.A. (1998). Glycerol production by the yeast *Saccharomyces cerevisiae* and its relevance to wine: A review. *S. Afr. J. Enol. Vitic.*, 19(1): 17.

Snow, R. (1983). Genetic improvement of wine yeasts. In: *Yeast Genetics – Fundamental and Applied Aspects*, J.F.T. Spencer, D.M. Spencer and A.R.W. Smith (Eds). Springer-Verlag, New York, p. 439.

Solieri, L., Verspohl, A., Bonciani, T., Caggia, C. and Giudici, P. (2015). Fast method for identifying inter- and intra-species *Saccharomyces* hybrids in extensive genetic improvement programs based on yeast breeding. *J. Appl. Microbiol.*, 119(1): 149.

Spencer, J.F.T. and Spencer, D.M. (1983). Genetic improvement of industrial yeasts. *Ann. Rev. Microbiol.*, 37: 121.

Srivastava, D.K. and Raverkar, K.P. (1999). Genetic manipulation of industrially important microorganisms. In: *Biotechnology: Food Fermentation Microbiology, Biochemistry and Technology*, Vol. I, V.K. Joshi and A. Pandey (Eds). Educational Publishers and Distributors, New Delhi, p. 173.

Steensels, J., Snoek, T., Meersman, E., Nicolino, M.P., Voordeckers, K. and Verstrepen, K.J. (2014). Improving industrial yeast strains: Exploiting natural and artificial diversity. *FEMS Microbiol. Rev.*, 38(5): 947.

Szostak, J.W. and Blackburn, E.H. (1982). Cloning yeast telomeres on linear plasmid vectors. *Cell*, 29(1): 245.

Takahashi, T., Shimoi, H. and Ito, K. (2001). Identification of genes required for growth under ethanol stress using transposon mutagenesis in *Saccharomyces cerevisiae. Mol. Genet. Genomics*, 265(6): 1112.

Thevelein, J.M. (1996). Regulation of trehalose metabolism and its relevance to cell growth and function. In: *The Mycota III*, R. Brambl and G.A. Marzluf (Eds). Springer-Verlag, Berlin and Heidelberg, p. 395.

Thevelein, J.M. and De Winde, J.H. (1999). Novel sensing mechanisms and targets for the cAMP-protein kinase A pathway in the yeast *Saccharomyces cerevisiae. Mol. Microbiol.*, 33(5): 904.

Tofalo, R., Perpetuini, G., Di Gianvito, P., Arfelli, G., Schirone, M., Corsetti, A. and Suzzi, G. (2016). Characterization of specialized flocculent yeasts to improve sparkling wine fermentation. *J. Appl. Microbiol.*, 120(6): 1574.

Van Rensburg, P., Van Zyl, W.H. and Pretorius, I.S. (1996). Co-expression of a *Phanerochaete chrysosporium* cellobiohydrolase gene and a *Butyrivibrio fibrisolvens* endo-β-1,3-glucanase in *Saccharomyces cerevisiae. Curr. Genet.*, 30(3): 246.

Van Rensburg, P., Van Zyl, W.H. and Pretorius, I.S. (1998a). Engineering yeast for efficient cellulose degradation. *Yeast*, 14(1): 67.

Van Rensburg, P., Van Zyl, W.H. and Pretorius, I.S. (1998b). Over-expression of the *Saccharomyces cerevisiae* exo-β-1,3-glucanase gene together with the *Bacillus subtilis* endo-β-1,3–1,4-glucanase gene and the *Butyrivibrio fibrisolvens* endo-β-1,4-glucanase gene in yeast. *J. Bacteriol.*, 55: 43.

Vigentini, I., Gonzalez, R. and Tronchoni, J. (2019). Genetic improvement of wine yeasts. In: *Yeasts in the Production of Wine*, Romano P., Ciani M., Fleet G. (Eds.) Springer, New York, US. pp. 315–342.

Volschenk, H., Viljoen, M., Grobler, J., Petzold, B., Bauer, F.F., Subden, R., Young, R.A., Lonvaud, A., Denayrolles, M. and Vuuren, H.J.J. (1997). Engineering pathways for malate degradation in *Saccharomyces cerevisiae. Nat. Biotech.*, 15: 253.

Unit 4

Factors Affecting Winemaking, Control and Improvement

INTRODUCTION

This Unit comprises a series of four chapters of the *Concise Encyclopedia of Science and Technology of Wine*. The Unit concerns the factors affecting winemaking, a central theme in the current winemaking industry where consumers are increasingly demanding wine quality.

The first chapter is regarding winemaking and killer yeasts. Some yeast strains secrete proteins or glycoproteins known as killer toxins, which kill sensitive cells of the same or related yeast genera. The authors in this chapter highlighted several *Saccharomyces cerevisiae* and *S. bayanus* strains, isolated from grapes, grape musts, and wines, showing interesting killer activity, and consequently, they are suggested as starter cultures to inhibit spoilage yeasts like *Brettanomyces* spp. and other non-*Saccharomyces* strains in winemaking.

The second chapter focuses on the problematic issue of stuck and sluggish fermentation, where the winemaker needs an immediate solution. Naturally, factors affecting fermentation would definitely influence wine quality. The authors in this chapter emphasise an integrated approach concerning this problem, from recognising the concept to focusing on the causes and losses of a stuck fermentation, as well as the factors that influence the occurrence of stuck and sluggish fermentation.

The third chapter focuses on the use of enzymes in oenology that plays an important role in the winemaking process. Enzymes are considered key elements that catalyse a number of reactions from the pre-fermentation stage until wine ageing. The main sources of many of these enzymes are the grapes themselves and the microorganisms present during the winemaking process. In addition, exogenous enzymes, often commercial enzyme preparations, are also widely used as supplements to get the best wine quality. The authors provide a summary of the most important endogenous enzymes derived from grapes and microorganisms present in grape must and in wine, and the role of commercial enzymes applied during the winemaking process.

The last chapter is related to the control and modelling of winemaking. The winemaking process involves a large number of operations, but the fermentation process is a "key step", because of its important effect on wine quality. Winemaking technology has developed significantly in the last years, which has consequently led to wine quality improvement besides making it possible to produce quality wines with a wide range of characteristics. The use of winery equipment with adequate control of fermentation is strongly linked to the existence of an accurate fermentation model as described in this chapters by the authors capable of predicting the development of the fermentation process in advance with the use of suitable sensors which permit the key process variables to be measured. The main purpose of the application of models in the fermentation process is the prediction of fermentation kinetic behaviour based on the initial characteristics of the grape must.

Dr. Fernanda Cosme
Assistant Professor of Food Science/Oenology
Chemistry Research Centre – Vila Real,
School of Life Sciences and Environment,
University of Trás-os-Montes and Alto Douro, Portugal

20 Winemaking and Killer Yeasts

Giuseppe Comi and Luca Cocolin

CONTENTS

20.1 INTRODUCTION

In 1963, Bevan and Makover discovered a yeast-yeast interaction called the killer phenomenon, which consists of a particular action of certain yeast strains that produce and secrete toxins. These toxins are either proteins or glycoproteins and are able to kill sensitive yeast strains. The toxins are known as killer factors and are produced and secreted in the wine processing environment. Yeasts that produce toxins are called killer yeasts and are usually resistant to their own toxins, but a toxin coming from other killer yeasts can kill them. Among *Saccharomyces cerevisiae* strains, killer (K), sensitive (S), and neutral (N) yeasts are selected. The killer strains cause the inactivation of the sensitive ones, but they cannot kill the neutral ones. Sometimes other killer toxins can kill killer yeasts when they are called killer-sensitive yeasts. The neutral strains do not produce killer factors, but they are immune to those produced by killer strains. Lastly, the sensitive strains do not produce killer factors and they are killed by the yeasts that produce them. The killer phenomenon is widespread in nature, and many strains have been selected including those from wine fermentations. Because of their presence on grapes and in musts, efforts have been made to define their importance in the improvement of alcoholic beverages or in stuck and slow fermentations, as wild killer yeasts can stop or slow down wine fermentation. For wine yeast, the killer factor is a selected character, and in the last decades many yeasts have been selected from grapes and musts and are marketed and sold as dry active yeasts to improve wine production.

20.2 BIOLOGY AND GENETICS OF KILLER YEASTS

The killer toxins are associated with either chromosomal genes or with cytoplasmic virus-like particles (VLP) or mycoviruses. In particular, *S. cerevisiae* VLP contains an L (large) and M (medium) double-stranded RNA (dsRNA) genome. The L genome encodes the protein of the capsid of the VLPs, while the M genome encodes the killer toxins and the immunity factors. It is accepted that the neutral strains (K^-R^+) cannot encode the killer factor, but they possess the immunity determinant; the K^+R^+ can encode for both the factors and the K^-R^- does not have the killer and immunity ability. In some non-*Saccharomyces* strains, the killer factor can be encoded by the chromosomal genes, by linear DNA plasmids or by dsRNA plasmids. The genetic determinants of the killer property are not the same in all the yeast species, but in most of them, they are cytoplasmically inherited. Killer production in *Kl. lactis*, *P. acaciae* or *P. inositovora* is associated with linear double-stranded DNA (dsDNA) plasmids. In *S. cerevisiae*, as well as in *Hanseniaspora uvarum*,

Sporiodiobolus jahnsonii and *Cystofilobasidium bisporidii*, the killer property is associated with double-stranded RNA (dsRNA) plasmids. The dsRNA plasmids are encapsulated in virus-like particles that are non-infectious and, subsequently, are transmitted by vegetative cell division or hyphal sexual fusion, both very frequent in nature. Finally, the KHR and KSR systems of *S. cerevisiae* and some killer proteins of *C. glabrata* are encoded by chromosomal genes. However, in many other cases, the genetic basis of the killer activity is still unknown.

20.3 CLASSIFICATION OF KILLER YEASTS, THEIR PREVALENCE AND ISOLATION

20.3.1 Classification

The killer yeasts are classified into 11 groups (K_1–K_{11}) that differ due to the difference in the size of the dsRNA molecules and the properties of the killer toxins. *Saccharomyces* strains have been classified into three main groups, K_1, K_2, K_3, by determination of the spectrum of activity against sensitive strains, by assay activity against mutants resistant to certain toxins, by assay of the cross-reactivity of the toxin producing organism and the interaction between the yeasts. In particular, K_1 toxin is coded by M1, K_2 by M2, and K_3 by M3 VLPs. K_2 and K_3 types seem to be very similar, because it is possible that M3 VLP is a mutation of M2 VLP. Two additional *Saccharomyces* killer strains, K3GR1 and KT28, have been described which are different from each other and from the three killer strains, and therefore, it is suggested that they can be new killer phenotypes. Further, the K_1, K_2, and K_3 toxins of *Saccharomyces* strains are active between different pH ranges, have different heat stability and molecular weights, which are respectively 1.9 kb, 1.5 kb, and 1.3 kb. These characteristics influence the behaviour of the different toxins in must in other ecosystems. The K_1 toxin is inactive in grape musts because it is active and stable at a pH between 4.2 and 4.6; vice versa, the K_2 is active between pH 2.8 and 4.8, with a maximum activity between 4.2 and 4.4. Recently, a novel *S. cerevisiae* Kx strain able to produce an X factor involved in the inhibition of all known *S. cerevisiae* killer toxins strains (K1, K2, K28) was described. The Kx type yeast strains produced a small but clear lysis zones at a temperature between 20 and 30 °C. In addition, the killer/antikiller effect reached its maximum level at pH 5.2 and was associated with the presence of M-dsRNA, bigger than the typical M-dsRNA of K1, K2 and K28 *S. cerevisiae* killer strains. Consequently, the molecular weight of this new killer toxin, producing Kx type, might be 454 kDa. In addition, another *S. cerevisiae* strain (N1) showed killer activity against standard K1, K2, and K28 killer strains without displaying any antikiller activity.

20.3.2 Site of Action

The sites of the attack on the cell wall and the mode of action of the *Saccharomyces* killer toxins are quite similar. A primary receptor located in the cell wall β-1,6 glucan is the binding site of K_1 and K_2, and two chromosomal genes, called *KRE1* and *KRE2*, allow the linkage. If mutations occur in these genes, the strains become resistant. The mode of action includes attachment of the killer toxins to a glucan receptor, subsequent transfer to a membrane receptor site, by energy consumption, and the alterations of some functional activities, such as the interruption of the coupled transport of amino acids and protons, the acidification of the cellular contents and potassium and ATP leakage. Lastly, the formation of pores in the plasmic membrane causes the deaths of the cells within 2 or 3 hours after contact. Because of this, the cells in the log phase are more sensitive than the cells in the stationary phase. Mannoproteins, mannans and chitin are also primary receptors for killer toxins. Secondary receptors were identified. Indeed, after binding to the primary receptor in the cell wall of the sensitive cells, the killer toxins are translocated to the secondary receptors in the plasma membrane, even though little is known about these secondary receptors. In particular, it was discovered that some killer toxins of *S. cerevisiae* strains include a second step, involving their translocation to the cytoplasmic membrane and interaction with a secondary membrane receptor.

20.3.3 Prevalence of Killer Strains

The killer phenomenon has been recognised in many yeast strains isolated from the laboratory, industrial, clinical or natural, and environmental samples. Initially, it was discovered in wine, brewing, sake, or in fermented products. In particular, it can be present in *S. cerevisiae* strains or in other yeasts which contaminate fermentations. Subsequently, the killer phenomenon was also demonstrated in other genera like *Candida*, *Debaryomyces*, *Hansenula*, *Kluyveromyces*, *Pichia*, *Ustilago*, *Torulopsis*, and *Criptococcus*. In each case, the killer property is principally restricted to strains of species within a genus, but interactions between species of different genera are recognised. *Saccharomyces* strains, producing K_1, K_2, and K_3 toxins, are isolated from different ecosystems, including alcoholic beverages and musts. In winery areas, on grapes and in musts, they can be present in various percentages (12% to 90%). K_2 strains are more widespread than the K_1 and K_3 ones.

20.3.4 Method of Isolation

The killer strains can be isolated using different methods. The simplest method is by adding the sensitive strains (4×10^5 Colony Forming Unit – CFU/mL) in a nutrient agar medium (pH 4.2). After distribution in petri dishes, the killer strain is streaked on the surface and the dishes are incubated at 18 °C and 20 °C. The presence of a clear zone surrounding the streaked strain indicates the killer activity. It is possible to add methylene blue to the nutrient agar medium (MBNA), buffered at a pH of 4.2–4.7, and to spread the sensitive strain, before streaking the killer one on the surface of the MBNA agar. When the methylene blue is present, the killer strain is surrounded by a clear zone, indicating no growth of the sensitive one, bounded

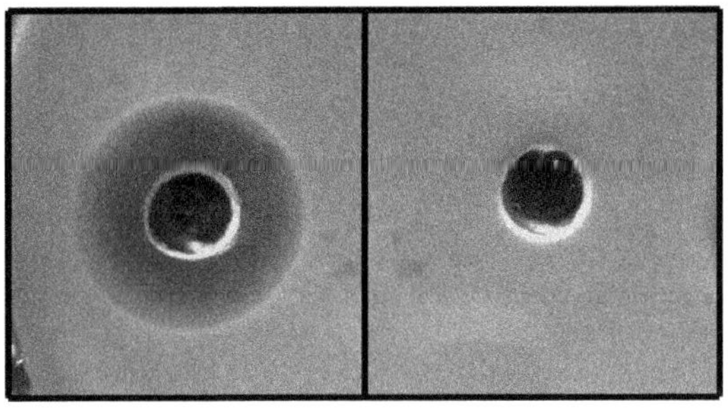

FIGURE 20.1 Agar-well assay to determine killer toxin production. The sensitive *S. cerevisiae* strain DBVPG 6500 (panel A) and the neutral *S. cerevisiae* strain DBVPG 1968 (panel B) were included in a medium containing 5 g/L yeast extract, 5 g/L meat extract, 20 g/L glucose and 15 g/L agar, pH 4.6. After solidification, a hole with a 100 μL was created and filled up with an 18–24 h culture of *S. cerevisiae* DBVPG6499. After 48 h of incubation at 25 °C, a visible inhibition in growth was evident in the case of the sensitive strain

by a dark blue zone of dead cells. In each case, it is important to remember that the strains in the exponential growth phase are more susceptible than those in the stationary phase, and the sensitive strain concentration can influence the final result of the assay. For this reason, it is advisable to add a maximum of 4.10^5 CFU/mL of the sensitive strain. To quantify the killer activity, a fluorescence microscopy method was proposed, but other studies included agar diffusion techniques, where different aliquots of killer suspensions were added to wells, placed in agar plates seeded with the sensitive ones. After keeping the plates for two hours at room temperature and overnight at 25 °C, the results were obtained by measuring zones of inhibition surrounding the wells (Figure 20.1), proportional to the activity of the killer toxins.

20.4 KILLER SYSTEMS

20.4.1 Killer Systems Associated with dsRNA Plasmids

S. cerevisiae, *Hansenula uvarum*, and its anamorph *K. apiculata* are characterised by the presence of killer systems controlled by two dsRNA plasmids. The two known dsRNA viruses of *S. cerevisiae* are called L-A and L-B. Each of them is a family of structurally and functionally distinct viruses, even though both have a genome size of 4.6 kb. Many strains containing L-A also carry a satellite dsRNA called M dsRNA, that codes for the killer toxin as well as for the immunity to that toxin, while the L genome codes for the polymerase and capsid proteins for both the genomes. In *S. cerevisiae*, different M dsRNAs have been described as each encoding a different killer protein, namely K_1, K_2, K_3, $K_{\text{"}3\text{"}}$ and K_{28}. The killer protein is produced as a pre-protoxin that consists of a signal peptide followed by a peptide (δ component) of unknown function, and then by two toxin subunits, α β and β, which are separated by the central glycosylated γ peptide. Once the pre-protoxin protein is secreted in its functional form, the β component would be mainly responsible for binding to the glucans of the cell wall, giving each killer protein a specific cell-wall binding site. The α component is involved in the interaction with the toxin cytoplasmic membrane receptor and subsequent ion channel formation or inhibition of DNA synthesis that constitute the effective toxin action. The N-terminus of γ and the central domain of α are essential for immunity, and it has been suggested that the pre-protoxin may bind to the toxin membrane receptors in such a way as to prevent the active toxin from binding, thus conferring immunity.

The toxin genes corresponding to killer proteins K_1, K_2 and K_{28} have been cloned and sequenced. However, no amino acid sequence similarity was found among them, in spite of the significant parallels between the toxins, such as the presence of δ, α, β, and γ components, processing of the toxins and binding to the cell wall and to cytoplasmic membrane receptors. The M1 dsRNA genome, carrying the K_1 toxin precursor gene, consists of two segments, M1-1 (1 kb) and M1-2 (0.6 kb), separated by a highly adenine-uracil (AU) rich sequence (0.2 kb), of which M1-1 encodes the K_1 toxin precursor, pre-protoxin. The expression of plasmid functions depends upon numerous chromosomal genes. The *MAK* and *SKI* genes affect the maintenance of dsRNA plasmids and their replication. The *KEX1* and *KEX2* genes encode for carboxypeptidases and endopeptidases respectively, which are involved in the processing of the killer toxins. Furthermore, several additional genes including *SEC* (toxin production), *END* (endocytosis), *KRE* (toxin binding), *VLP* (vacuolar protein localisation), and *REX* (resistance expression) have been reported. In addition, a novel killer system associated with L and M dsRNAs was found in *C. stellata*. Moreover, the so-called neutral strains of *S. cerevisiae* with M dsRNAs that do not encode for killer toxins but confer only immunity to either K_1 or K_2 toxin have been identified.

20.4.2 Killer Systems Associated with Linear DNA Plasmids

The killer systems associated with DNA plasmids have been described in *P. acaciae* or *P. inositovora* and in *Kl. Lactis.* The killer strains of *Kl. lactis* harbour two cytoplasmically inherited double stranded linear DNA plasmids, pGKL1 (8.9 kb) and pGKL2 (13.4 kb), of which pGKL1 codes for the production of the killer toxin (α, β and γ subunits) and the immunity determinant to the toxin, while pGKL2 is required for the maintenance of pGKL1. These pGKL plasmids which possess terminal proteins attached at their 5'-end can be transferred into *S. cerevisiae* by protoplast or transformation, and are stably maintained in *S. cerevisiae*, where the plasmids express the killer character. Unlike dsRNA killer systems, no viral nature is known in the pGKL plasmids. Cells harbouring these plasmids secrete a heterotrimetric killer toxin complex composed of 97, 31, and 28 kDa subunits into the culture medium, that kills yeast cells from a number of genera. Due to the exposure to the pGKL killer toxin, cells sensitive at the unbounded G1 phase of the cell cycle are finally killed, suggesting that the pGKL killer toxin may distort some essential process involved in cell cycle regulation or may interact with a target which is expressed only at G1 phase. But how it functions in the G1 phase arrest mechanism or the killing of sensitive cells remains unknown. The three sub-units of the killer toxin must have distinct functions in the killing process. The intracellular expression of the 28 kDa subunit in killer sensitive cells caused the death of the host cell, and that this killing activity by the 28 kDa subunit was prevented by the expression of the killer immunity indicated that the killing activity of the toxin complex was performed by the 28 kDa subunit. As found with dsRNA killer systems, chromosomal gene function is required for the expression of killer character in *Kl. lactis*. The *Kl. lactis KEX1* gene, which has been cloned and shown to be functionally homologous to *S. cerevisiae* KEX2 endopeptidase is thought to perform the processing of the αβ precursor into the mature α and β subunits of the killer toxin.

20.4.3 Other Killer Systems

In addition to the killer systems encoded by cytoplasmic plasmids, some killer phenotypes may be associated with determinants in chromosomal genes. Novel killer systems, designated *KHR* and *KHS* killers, have been identified in *S. cerevisiae* and showed that both killer characters were encoded by nuclear genes. Furthermore, it was shown that *KHR* and *KHS* genes were located on the left arm of chromosome IX and the right arm of chromosome V, respectively. Possible chromosomal location of killer determinants has been suggested in strains of *C. glabrata*, *P. kluyveri*, *P. farinose*, *H. anomala*, and *Hanseniaspora* spp.

20.5 CHARACTERISATION OF KILLER TOXINS

The main killer yeast species reported so far have been characterised (Table 20.1). They produce toxins that are either proteins or glycoproteins and some of which consist of subunits. Most of them are unstable at high temperatures and at high pH (Figures 20.2 and 20.3), but the toxins produced by the genus *Hansenula* seems to have relatively high stability. Halophilic killer yeast strains, *P. farinose*, *P. membranifaciens* and *H. anomala*, produce toxins whose killer activities increase with an elevation of NaCl concentration. The killer toxin from the K_3 strain of *S. cerevisiae* was found to be an 18 kDa glycoprotein, while the toxin of *C. glabrata* consisted of several molecular species, at least one of which was glycoprotein. The killer toxins of *S. cerevisiae*, except the *KHR* toxin, are active only against sensitive strains of a few yeast species, including *S. cerevisiae* and *C. glabrata*, while those of *H. mrakii* and *Kl. lactis* show much broader anti-yeast spectra.

TABLE 20.1
Killer Toxins from Different Yeast Strains and Their Characteristics

Species of Killer yeast	Nature	Reference
Candida sp. SW-55	Glycoprotein	Yokomori *et al.* (1988)
Hanseniaspora uvarum	Protein	Radler *et al.* (1990)
Hansenula anomala	Glycoprotein	Kagiyama *et al.* (1988)
Hansenula mrakii	Protein	Ashida *et al.* (1983)
Hansenula saturnis	Protein	Henschke (1979)
Kluyveromyces lactis	Glycoprotein	Stark *et al.* (1990)
Kluyveromyces phaffi	Protein	Ciani and Fatichenti (2001)
Pichia farinosa	Protein	Suzuki and Nikkuni (1989)
Pichia kluyveri	Glycoprotein	Middlebeek *et al.* (1979)
Pichia membranifaciens	Protein	Llorente *et al.* (1997)
Saccharomyces cerevisiae (K_1)	Protein	Bostian *et al.* (1983)
Saccharomyces cerevisiae (K_2)	Glycoprotein	Pfeiffer and Radler (1984)
Saccharomyces cerevisiae (KHR)	Protein	Goto *et al.* (1990)
Schwanniomyces occidentalis	Protein	Chen *et al.* (2000)
Zygosaccharomyces bailii	Protein	Radler *et al.* (1993)

20.6 BIOSYNTHESIS AND SECRETION OF KILLER TOXINS

S. cerevisiae K_1 and *Kl. lactis* toxins have been extensively investigated from the point of view of their biosynthesis and secretion. In *S. cerevisiae* K_1, the M1-1 segment of the M1 dsRNA plasmid encodes the pre-protoxin (32 kDa) (Figure 20.4). Genetic evidence clearly demonstrates that M-dsRNA is the determinant of both toxin production and immunity. This was first shown by dsRNA curing (heat-shock treatment with cycloheximide) studies and then by *in vitro* translation of denatured M1-dsRNAs, which program the synthesis of a 32 to 34 kb pre-protoxin polypeptide, designated M1-P1. As previously described, M1 is composed of δ, α and β subunits, where δ is the leader sequence, α one subunit of the toxin, γ the immunity determinant and finally β, the second subunit of the toxin. The γ component, consisting of 44 amino acid residues, contains the amino-terminal signal peptide of 26 amino acids that allow entry into the endoplasmic reticulum. This component has three sites that are glycosylated, whereas others are not. After this glycosylation in the endoplasmic reticulum, the 43 kDa K_1 protoxin is transported in the Golgi apparatus and converted to the mature K_1 toxin mostly in the secretion vesicles. The K_1 protoxin is attached to the cytoplasmic membrane by the hydrophobic amino-terminal leader peptide; in fact, it is not removed in the endoplasmic reticulum. The chromosomal genes of the K_1 killer yeasts *KEX1* and *KEX2* are responsible for maturation. *KEX2* produces proteolytic enzymes that cleave the membrane-associated K_1 protoxin generating the α (9.5 kDa) and β (9 kDa) toxin subunit, while the *KEX1* gene encodes for a carboxypeptidase that removes the COOH-terminal basic residues from the α-subunit. After this processing, the mature K_1 toxin is secreted as a disulphide-bonded α-β dimmer via the constitutive secretory pathway as observed for the K_2 protoxin, where the presence of a 43 kDa intracellular and a 21 kDa extracellular species may correspond to the glycosylated K_2 protoxin and a glycosylated subunit derived from the protoxin, respectively. Moreover, studies implicate the precursor protein as the immunity component. Immunity is conferred by the precursor, which can act as a competitive inhibitor of a mature toxin by saturating a cell membrane receptor that normally mediates

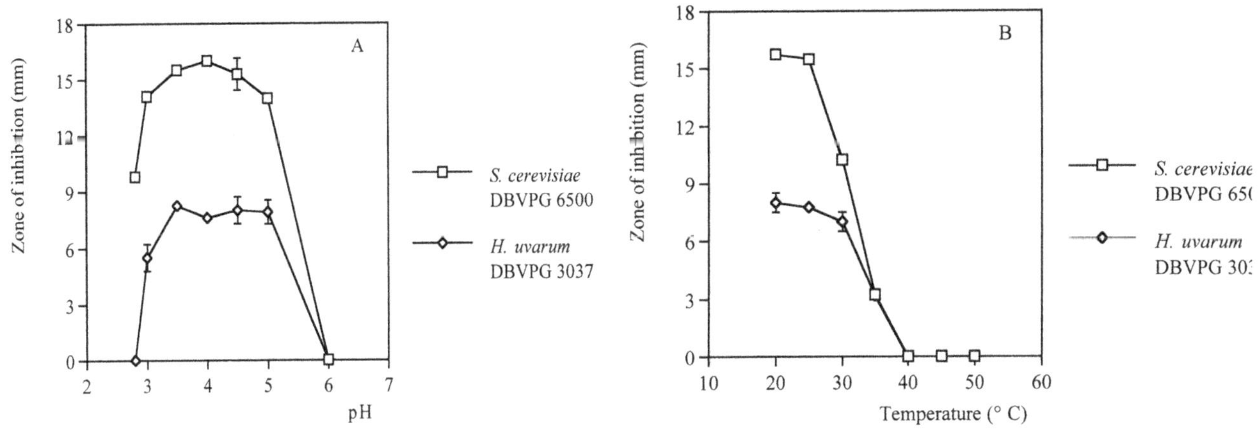

FIGURE 20.2 Effects of pH (A) and temperature (B) on the activity of *Kl. phaffii* killer toxin. pH effects were evaluated by agar-well assay. (A) Malt agar was buffered from 2.8 to 6 with 10 mM citrate-phosphate buffer. (B) After 2h incubation at various temperatures, the killer activity was evaluated by agar-well assay, pH 4.5. In both cases, toxin concentration was 14.3 AU/mL. Data are given as means ± standard deviations of at least duplicate experiments. *Source:* Ciani and Fatichenti (2001)

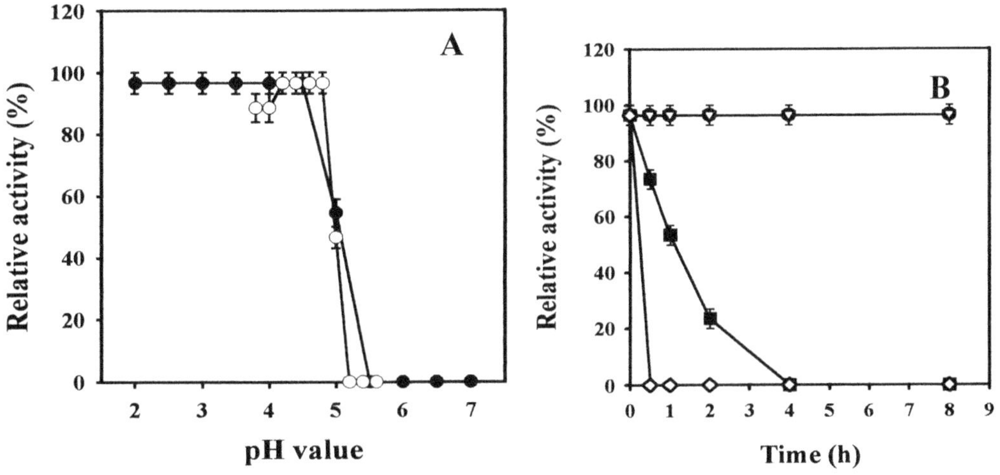

FIGURE 20.3 Effect of pH (A) and temperature (B) on the activity and stability of the purified killer protein from *Sch. occidentalis.* (A) Killer activity [O] and stability [□] of the toxin was tested against different pH values. The 100% killer activity is 500 AU, under conditions containing 0.4 μg of killer protein. To determine the optimal pH, the killer protein solution was adjusted to various pH values addition of 0.1 M citrate-phosphate buffer, and the killer activity of samples was then determined with the agar-well assay in media with identical pH values. For pH stability, after incubation in different pH values at 24 °C for 8h, the pH value of samples was adjusted to a final pH of 4.4, and the residual killer activity was determined. Error bars represented the mean ± the standard deviation of at least duplicate samples. (B) [□], 20 °C; [∇], 30 °C; [■], 40 °C; [◇], 50 °C. The 100% killer activity is 500 AU, under conditions containing 0.4 μg of killer protein. Error bars represented the mean ± the standard deviation of at least duplicate samples. *Source:* Chen *et al.* (2000)

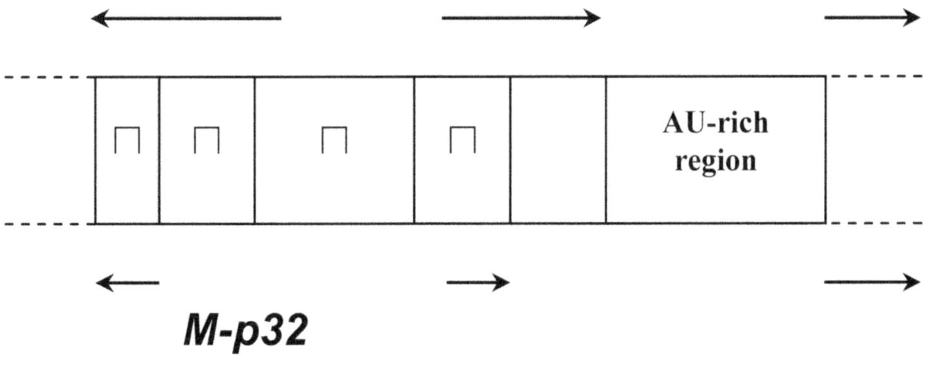

FIGURE 20.4 The structure of M1 dsRNA plasmid. *Source:* Wickner (1986)

toxin action. *Kl. lactis* secretes a killer toxin that consists of three sub-units, α (99 kDa), β (30 kDa), and γ (27.5 kDa). Only the subunit α was glycosylated. The α, β, and γ subunits are encoded by the pGKL1 plasmid, but the maturation of the toxin is performed by the KEX1 endopeptidase. Although the γ subunit has its own secretion signal, as is the case with the α β precursor, its secretion is blocked in the absence of the α β precursor. This suggests that specific assembly steps are required for the biogenesis of this toxin. Since all the proteins secreted by yeasts are either glycosylated or cleaved from glycosylated precursors, it would be necessary for non-glycosylated species, such as the γ subunit, to associate with a glycosylated species during secretion. pGKL1 in *Kl. lactis* also encodes for the immunity determinant. It seems to be a cytoplasmic protein because it has neither an amino-terminal signal peptide nor any other strongly hydrophobic regions.

Recently, a *Torulaspora delbrueckii* strain isolated and selected for winemaking was described to produce a new killer toxin (Kbarr-1), that killed either many *Saccharomyces cerevisiae* killer strains or non-*Saccharomyces* yeasts. This toxin derived from a medium-size 1.7 kb dsRNA, TdV-Mbarr-1 and depended on a large-size 4.6 kb dsRNA virus (TdV-LAbarr). Studies revealed there existed a relevant identity of TdV-Mbarr-1 RNA regions and the putative replication and packaging signals of most of the M-virus RNAs and suggested that they are all evolutionarily related.

A new toxin (Klus), able to kill different *S. cerevisiae* killer strains and many *Kluyveromyces lactis* and *Candida albicans* strains, was identified in the wine yeast *Saccharomyces cerevisiae*. The Klus phenotype is conferred by a medium-size double-stranded RNA virus (ScV-Mlus virus) of about 2.1–2.3 kb. Mlus dsRNA sequence was different from M1, M2, or M28 dsRNA ones, and consequently also Klus toxin was different from K1, K2, or K28 toxin.

20.7 MODE OF ACTION OF KILLER TOXINS

So far, most of the studies on the killer toxin action have been conducted with the K_1 killer protein which is a potent toxin, and only 6×10^3 molecules are needed to kill a cell of a sensitive strain. The primary event in the action is the binding of the killer toxin to a cell wall receptor, that is a 1,6-β-D-glucan for K_1 and K_2 killer yeasts. In contrast, the cell wall receptor for the K_{28} toxin contains mannan. Both the sensitive and killer strains contain large numbers of these receptors. Binding to the receptors on the cell wall of a susceptible yeast cell is an energy-dependent step that is complete within three minutes at 20 °C at pH 4.7. The existence of wall receptors is indicated by mutations of the chromosomal genes *KRE1* and *KRE2*, which substantially decreases the binding of the killer toxins to the cell wall and modifies the binding sites.

Although the cell wall receptor is necessary for toxin action, it does not appear to be the only component involved in the killing process. Several lines of evidence suggest that a second receptor is possibly on the plasma membrane. The *KRE* mutants are resistant to a number of killer toxins. Protoplasts of *KRE1* mutants and *KRE2* mutants are sensitive, favouring the existence of a membrane receptor. The events occurring after the binding of the toxin to the plasmid membrane remain unclear. An early observation was that the sensitive cells in the exponential growth phase were more sensitive to the killer toxin than cells in the stationary growth phase. This observation was extended to show that toxin-induced membrane damage was an energy, ATP-dependent event. It appears that shortly after its binding to the receptor in the cell wall, the killer toxin decreases the ion gradient across the membrane, interrupting the coupled transport of protons and amino acids. Consequently, it may be assumed that pores are formed through which ions such as potassium and low molecular weight metabolites can penetrate. This assumption is supported by the results of studies investigating ion conductivity changes across an artificial lipid membrane after the addition of the killer protein. Hence, the formation of pores is assumed to be associated with the lethal effect of the killer toxin. Similar physiological effects have been observed with the toxin of *P. kluyveri*. This toxin is thought to perturb membrane permeability by non-selective channel formation.

Maximum killing is attained after 2 or 3 hours, depending upon the strain. Measurement of metabolic and macromolecule biosynthetic events after toxin addition shows that there is a lag period of about 40 minutes during which no effects are seen, followed by a shut-off of macromolecule synthesis. The biosynthetic shut-off coincides with the plasma membrane damage, as indicated by the loss of potassium ions or ATP. The membrane-damaged cells shrink in volume owing to the loss of small metabolites through the large pores in the membrane, but no cell lysis occurs.

Both K_1 and K_2 killer toxins exhibit their lethal effect on sensitive cells by disrupting cytoplasmic membrane function. In contrast, K_{28} toxin has no such ionophoric effect, rather inhibits nuclear DNA synthesis. Killing of a sensitive yeast cell by killer toxin K_{28} is a two-stage process involving initial binding of K_{28} toxin to the outer 1,3-α-mannotriose side chains of a cell-wall mannoprotein (R_1) dependent on the gene products of *MNN1*, *MNN2*, and *MNN5* (Figure 20.5). The toxin is thus accessible to the postulated plasma membrane receptor (R_2), where it interacts with the cytoplasmic membrane. Interaction with this target is postulated to activate a signal transduction pathway leading to inhibition of DNA synthesis. It has been suggested that *Kl. lactis* toxin causes specific inhibition of cell division at the G1 stage of the cell cycle by affecting adenylate cyclase of sensitive cells. However, a strong argument against the involvement of adenylate cyclase in toxin action has been put forward. Among the three subunits of the toxin, the γ subunit is known to be responsible for toxin activity but the functions of the other two subunits are still uncertain. The killer toxin of *H. mrakii* is thought to act by selective inhibition of cell wall glucan synthesis in sensitive yeast cells.

20.8 ECOLOGY OF KILLER YEASTS

Killer yeasts are widely distributed in many environments, where they interact with other micro-organisms and

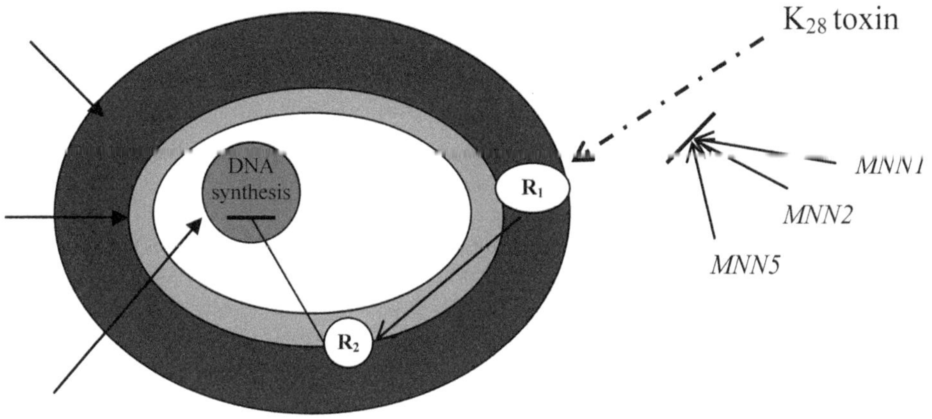

FIGURE 20.5 Receptor-mediated killing of a sensitive yeast cell by killer toxin K_{28}. The potential mechanisms of toxin resistance are indicated by bars. *Source:* Schmitt *et al.* (1996)

sometimes predominate. They have often been isolated from natural ecosystems including fermenting juice and in particular they seem closely related to vineyards, cellars, musts, and wine, as demonstrated by many studies. In Rioja, a district of Spain, the incidence of killer characteristics in 1,320 yeast strains isolated from various stages of spontaneous fermentation from red and white wines in 12 wineries were detected and different *S. cerevisiae* strains including killer, neutral and sensitive phenotypes were found. Non-*S. cerevisiae* strains isolated included both neutral and sensitive phenotypes, but not killer phenotypes. *S. cerevisiae* strains with different killer phenotypes co-existed throughout the fermentation. Despite the fact that dominant phenotypes varied between years, and sensitive strains were sometimes dominant, it was suggested that the killer phenotype is of less technological importance than was previously thought. Seven fermentations of red wines produced in south-western Slovenia for the presence of indigenous wine killer yeasts was investigated. To select starter cultures, 22 of the 613 yeast strains isolated from the red wine fermentations expressed killer activity that were isolated at various stages of wine fermentation and were identified as *S. cerevisiae*, *P. anomala*, *P. kluyveri*, *P. pijperi*, *Hans. uvarum* and *C. rugosa*. Some indigenous *S. cerevisiae* strains possessed both the stable killer activity and good fermentation properties. From a Brazilian Riesling Italico grape must, 85 yeasts were isolated consisting of killer, sensitive and neutral strains, and these properties were dependent on the temperature, the incubation medium, and the interaction among the yeasts. Strains with a temperature-dependent killer activity were isolated and those with resistance activity seemed to protect sensitive strains, so they suggested that sometimes killer strains could not predominate in a must with sensitive ones. From four regions of Argentina, killer strains were isolated at all stages of must fermentation and that the distribution of these strains differed from region to region. Wild-type killer yeast strains isolated from six South African wineries belonged to the K_2 phenotype and had different abilities to kill sensitive yeast. The killer strains were identified as *S. cerevisiae* and *S. bayanus*. In the last decade, yeasts with killer, sensitive and neutral characteristics have also been isolated from different types of grapes, musts and wines in wine regions of Turkey, Spain, and Italy. In Turkey, a total of 73 yeast strains were isolated and 60 of the strains were sensitive, while 13 were neutral. Flor yeasts on the surface of sherry wine were isolated, which produced important changes in the characteristics of the wine, and the killer properties were race-specific. In an area of Tuscany (central Italy), it was demonstrated that killer yeasts were present in many spontaneous wine fermentations and that the incidence of killers varied with respect to fermentation stage and vintage period, and could increase from the vintage to the subsequent periods and from the start to the end of fermentation.

Saccharomyces cerevisiae cells synthesised killer toxins, such as K1, K2, and K28, that can modulate the growth of other yeasts giving an advantage to the killer strains. Strains producing heterologous toxins K1 and K28 were less sensitive to K2 than the non-toxin producing one, suggesting partial cross-protection between the different killer systems.

Wickerhamomyces anomalus YF07b, coming from a marine environment, was described as able to produce a toxin with the killer and β-1,3-glucanase activity. A second toxin was identified producing only a killer activity against some yeast strains. Both the toxins acted in 3.5 % salt solution, at 16 °C and 4.0 % (w/v) against *Yarrowia lipolytica*, *Saccharomyces cerevisiae*, *Metschnikowia bicuspidata WCY*, *Candida tropicalis*, *Candida albicans* and *Kluyveromyces aestuartii*, involving the disruption of cellular integrity by permeabilizing cytoplasmic membrane function.

A *Torulaspora delbrueckii* NPCC 1033 (TdKT producer) strain with potential biocontrol activity of *Brettanomyces bruxellensis*, *Pichia guilliermondii*, *Pichia manshurica* and *Pichia membranifaciens* wine spoilage was recently selected. This TdKT toxin was biochemically characterized, aging on cell wall, showed glucanase and chitinase enzymatic activities, demonstrated a stable activity between pH 4.2 and 4.8 and was inactivated at a temperature above 40 °C. Considering it was active at oenological conditions, it was suggested as a biocontrol tool in winemaking.

Hanseniaspora uvarum strains were isolated, which release a killer toxin lethal to sensitive *S. cerevisiae* strains.

Because these yeasts are present in spontaneous fermentation, they could possibly be responsible for a stuck fermentation. The killer toxin of *Hans. uvarum*, like the *S. cerevisiae* one, was found to bound by beta-1,6-D-glucans.

In another study, *Kl. waltii* IFO 1666T strain was characterised for its property to kill *Schizosaccharomyces pombe*, aimed to control the extent of deacidification in winemaking. The killer toxin was found to have a spectrum different from those of the other known killer yeasts, and the activity found in the culture medium was lost by proteinase treatment. Further purification of the toxin was made by ultrafiltration and had a molecular weight larger than 10,000 Da. Lastly, non-*Saccharomyces* killer strains could contaminate industrial baking strains and can retard the fermentation of *S. cerevisiae*.

Mrakia frigida 2E00797, a psychrotolerant yeast, isolated from sea sediments in Antarctica, was found to be able to produce a killer toxin (55.6 kDa) that shared 35.1% sequence homology with a protein kinase and acted against *Metschnikowia bicuspidata*, *Candida tropicalis*, and *Candida albicans* preferable in 3 % NaCl at 16 °C and pH 4.5.

Two killer toxins, CpKT1 and CpKT2 produced by *Candida pyralidae* and isolated from wine, exhibited killer activity against several *Brettanomyces bruxellensis* strains especially in grape juice. Both the toxins were active and stable at pH 3.5–4.5 and temperatures between 15 and 25 °C. Consequently, they were suggested to kill spoilage yeast strains considering they were compatible with winemaking conditions, not affected by the ethanol and sugar concentrations, and did not inhibit *Saccharomyces cerevisiae* nor lactic acid bacteria strains.

20.9 INFLUENCE OF KILLER YEASTS IN WINE FERMENTATION

Killer yeast character has a great incidence in spontaneous fermentation and it has been discovered in many wine regions. Killer *S. cerevisiae* strains grow fast in musts, interacts with other *Saccharomyces* and non-*Saccharomyces* strains and influences fermentation and the final aroma of the wine. For this reason the inoculation of selected strains of *S. cerevisiae* that possess killer activity becomes very important during fermentation. Vice versa, it is also possible that selected yeasts could be killed by wild killer strains, and a decrease in the quality of the wine or a stuck fermentation can occur. Selected killer yeasts can improve fermentation by the suppression of other undesirable yeasts such as wild *S. cerevisiae* and non-*Saccharomyces* strains, but the implanted killer strains will rapidly predominate due to the toxin production, and their fast fermentation rate in must has to be considered.

The killer/sensitive ratio is an important factor in determining the killer yeast implantation and its prevalence over other undesirable yeasts, though it is difficult to calculate this ratio without any doubt, as the results of various authors, are in disagreement. The initial ratio of killer/sensitive strains, the growth phase and the environmental conditions of the sensitive cell, the presence of protein absorbing molecules, the presence of protective, neutral, or other killer strains, the specificity and the activity of the killer toxin, the killer yeast concentrations, the inoculum concentrations, and the vitamin and nitrogen availability are the main causes that can influence killer yeast activity and growth in musts. This evidence becomes important considering that the killer character is an important criterion for selecting yeasts, because, at a certain level, it guarantees the success of the fermentation and is also considered an indispensable way of ensuring the success of the inoculation and a tool for improving must fermentation in regions where native killer yeasts are present. The wild and native yeasts can be a problem for wine production because they can cause stuck fermentation, increase the level of acetaldehyde, lactic acid, acetic acid, and undesirable sensory qualities. The use of selected killer strains is necessary to prevent the growth of the autochthonous flora and its participation in vinifications, considering that the killer effect has been observed both in spontaneous and commercial fermentations and in experimental cultures. Killer character is only one of the several criteria used for the selection of yeasts. Other selection criteria are: high fermentation speed, the secretion of killer toxins, stability factors, survivor factors, aromas and flavour. In particular, it seems that K_2 toxins have the best performance and activity in must, considering the pH and the fermentation temperature usually utilised in wine production. Selected killer strains must, therefore, include *S. cerevisiae* K_2. The way to guarantee the success of the inoculum is either to add a high number of the selected killer strains or to add K_2 resistant strains to the musts.

20.10 ROLE AND APPLICATION OF KILLER YEASTS IN WINEMAKING

The principal application of killer yeasts in winemaking are in improvement of wine characteristics, as in the case of Malvasia wine, Franconia wine, Bobel wine and in red and white wines by strain S6u, and sparkling wine production and improvement of its characteristics.

20.10.1 Winemaking

In wine production, *S. cerevisiae* killer yeasts are utilised for different reasons: for starting wine fermentation and prevailing over native flora, for improving winemaking and quality, for increasing autolysis allowing the secondary fermentation and for running over stuck fermentations. An increase in wine quality is often obtained by *S. cerevisiae* killer yeasts. Indigenous wine killer yeasts are selected and, when tested, improve fermentation and produce favourable characteristics of Malvasia wine, as was the case with the production of the main red winemaking grape cultivars grown in the Franconia region of Germany, though there is also a report contrary to this. Two *S. cerevisiae* killer strains (D47, KIM) were inoculated in rosé wines obtained by fermentation of Garnacha must and similar production of volatile compounds total concentration of esters, alcohol, and carbonyl compounds were found in the wine. It was, therefore, concluded that the yeast strain employed exerted more influence on the evolution of volatile compounds during fermentation than on the final composition of the wine, so the use of a killer *S. cerevisiae* L-2226 to produce wines made from the grape cv. Bobal was suggested, as it produced a lower concentration of many minor volatile components which adversely affected sensory quality and its use can improve the quality of wines with high sugar concentrations, such as those produced in the Valencia region. Red and white wines have been produced by using different yeasts, including the *S. cerevisiae* physiological race *uvarum* (killer sensitive) strain S6u, the *S. cerevisiae* physiological race *cerevisiae* (killer) strain K1, the S. *cerevisiae* physiological race *bayanus* (killer) strain QU23 and the *S. cerevisiae* physiological race *cerevisiae* (killer neutral) strain D254. It was concluded that the sensory quality of wines made with S6u was better than the others, because it produced a lower concentration of SO_2, volatile acids, acetate, and ethyl esters but a higher concentration of fatty acids and phenylethanol than the other yeasts. Therefore, the use of this strain as a yeast starter was suggested. The application of pure yeast starters (a dry wine yeast preparation with killer activity) in the fermentation of Chardonnay and Grechetto wines in a commercial winery was not successful.

Recently, different strains of *Torulaspora delbrueckii* have been selected and used in order to confer specific characteristics to the wines. Despite being rapidly replaced by *Saccharomyces cerevisiae*-like yeasts during must fermentation, the researchers focused the studies on the killer toxin in order to favour the *T. delbrueckii* predominance. An isolated strain was able to kill *S. cerevisiae* yeasts and to dominate and complete the fermentation of sterile grape must. In particular, in sequential yeast inoculation, *T. delbrueckii* killer producer dominated and completed the must fermentation to reach over 11% (v/v) ethanol. However, it was concluded that the *S. cerevisiae*-dominated wines were preferred over the *T. delbrueckii*-dominated ones because of their fresh fruit aromas deriving from isoamyl acetate, ethyl hexanoate, and ethyl octanoate presence.

20.10.2 Sparkling Winemaking

Sparkling wine is characterised by a secondary fermentation during the prolonged ageing of wine, and the use of killer strains are proposed in its production. During several months of ageing, yeast cells release various cellular components such as polysaccharides, lipids, nucleic acids, proteins, and their breakdown products in wine and improve the quality of sparkling wines. All these products come from yeast autolysis. This phenomenon is natural in yeasts, but it is too slow in must. It can last from eighteen months to over four years. For this reason, the use of killer yeasts can produce a fast yeast autolysis and can consequently improve the aroma and quality of sparkling wines. The use of killer yeasts is suggested to destroy sensitive wine yeasts and to ensure the presence of a sufficient quantity of dead yeast cells to form autolysis products that enhance sensory properties and accelerate secondary fermentation of the wine. Use of killer strains to promote autolysis via their interactions with sensitive yeasts in sparkling wine production from two Australian Chardonnay/Pinot noir sparkling wines during ageing for over 18 months showed that mixed fermentation with killer and sensitive yeasts resulted in rapid death of the sensitive cells and a significantly high protein release in synthetic medium and, as concluded earlier, also in the sensory properties of sparkling wines, using *S. cerevisiae* subsp. *bayanus* (strain T18). This yeast was selected and used for its good characteristics including killer factor and other desirable physico-chemical and sensory quality characteristics.

20.10.3 Stuck and Sluggish Fermentation

It is well known that wild killer yeasts can be the cause of stuck fermentations. The toxins, secreted by the yeasts, can kill other flora and in particular the strains of native or added *S. cerevisiae*. So, the killer yeasts could be a problem for fermentation by stopping or slowing down the fermentations. The cause is recognised in the effect of killer/sensitive yeast interactions and in the K/S ratio. So if the stuck and sluggish fermentations are due to killer/sensitive strain interaction, the causes must be identified in the killer/sensitive yeast proportion at the start of fermentation, the concentration of nitrogen source in the grape must and the presence of bentonite during fermentation. However, as previously mentioned, killer strains are able to dominate a fermentation, even when present in a low K/S ratio. For this reason, the use of selected killer *S. cerevisiae* strains in commercial fermentations is needed to improve the quality of wines. Up to now, it seems that killer *S. cerevisiae* strains represent the only way of preventing stuck fermentations caused by native killer yeasts, considering their high frequency in grapes and musts.

Killer yeasts are sold by a large number of manufacturers. They have been isolated in many countries from grapes, musts, wines, and vineyards. Despite the fact that killer yeasts cannot solve all wine production problems, a large variety of selected dry killer yeasts are available. The most widely sold include *S. cerevisiae* and *S. bayanus*. Sales appear to have increased lately due to the number of wineries using them. As previously reported, the reason for their use is due to improvement of the wine quality and sometimes to eliminate the risks of a stuck fermentation. However, the use of killer yeast in preventing stuck fermentation has progressively declined and can even contribute to the problem.

BIBLIOGRAPHY

Ashida, S., Shimazaki, T., Kitano, K. and Hara, S. (1983). New killer toxin of *Hansenula mrakii*. *Agr. Biol. Chem.*, 47: 2953–2955.

Bevan, E.A. and Makover, M. (1963). The physiological basis of the killer character in yeast. In: *Proceedings of the Eleventh International Congress of Genetics (Hague, 1963)*, Ed. S.J. Goerts. p. 202.

Bostian, K.A., Jayachandran, S. and Tipper, D.J. (1983). A glycosylated protoxin in killer yeast: Models for its structure and maturation. *Cell*, 32(1): 741–751.

Chen, W.B., Han, Y.F., Jong, S.C. and Chang, S.C. (2000). Isolation, purification and characterization of a killer protein from *Schwanniomyces occidentalis*. *Appl. Environ. Microbiol.*, 66(12): 5348–5352.

Ciani, M. and Fatichenti, F. (2001). Killer toxin of *Kluyveromyces phaffii* DBVPG 6076 as a biopreservative agent to control apiculate wine yeasts. *Appl. Environ. Microbiol.*, 67(7): 3058–3063.

Goto, K., Iwase, T., Kichise, K., Kitano, K., Totsuka, A., Obata, T. and Hara, S. (1990). Isolation and properties of a chromosome-dependent *KHR* killer toxin in *Saccharomyces cerevisiae*. *Agr. Biol. Chem.*, 54(2): 505–509.

Guo, F.J., Ma, Y., Xu, H.M., Wang, X.H. and Chi, Z.C.. (2013). A novel killer toxin produced by the marine-derived yeast *Wickerhamomyces anomalus* YF07b. *Antonie Leeuwen*, 103: 737–746.

Henschke, P.A. (1979) Killer Factors of the Genus *Hansenula*, in Particular *Hansenula saturnis*. PhD thesis, The Univeristy of Adelaide, Australia.

Kagiyama, S., Aiba, T., Kadowaki, K. and Mogi, K. (1988). New killer toxins of halophilic *Hansenula anomala*. *Agr. Biol. Chem.*, 52: 1–8.

Liu, G.L., Chi, Z., Wang, G., Wang, Z.P., Li, Y. and Chi, Z.M. (2013). Yeast killer toxins, molecular mechanisms of their action and their applications. *Crit. Rev. Biotechn.*: 1–13. doi:10.3109/07388551.2013.833582.

Liu, G.L., Wang, K., Hua, M.X., Aslam Buzdar, M. and Chi, Z.M. (2012). Purification and characterization of the cold-active killer toxin from the psychrotolerant yeast *Mrakia frigida* isolated from sea sediments in Antarctica. *Process Biochem.*, 47(5): 822–827.

Llorente, P., Marquina, D., Santos, A., Peinado, J.M. and Spencer-Martins, I. (1997). Effect of salt on the killer phenotype of yeasts from olive brines. *Appl. Environ. Microbiol.*, 63(3): 1165–1167.

Mehlomakulu, N.N., Setati, M.E. and. Divol, B. (2014). Characterization of novel killer toxins secreted by wine-related non-*Saccharomyces* yeasts and their action on *Brettanomyces* spp. *Int. J. Food Microbiol.*, 188: 83–91.

Melvydas, V., Bružauskautė, I., Gedminienė, G.and Šiekštele, R. (2016). A novel *Saccharomyces cerevisiae* killer strain secreting the X factore related to killer activity and inhibition of *S. cerevisiae* K1, K2 and K28 killer toxins. *Indian J. Microbiol.*, 56(3): 335–343.

Middelbeek, E.J., Hermans, J.M.H. and Stumm, C. (1979). Production, purification and properties of a *Pichia kluyveri* killer toxin. *Ant. van Leeuwen*, 45(3): 437–450.

Orentaite, I., Poranen, M.M., Oksanen, H.M., Daugelavicius, R. and Bamford, D.H. (2016). K2 killer toxin-induced physiological changes in the yeast *Saccharomyces cerevisiae*. *FEMS Yeast Res.*, 16(2): 1–8.

Pfeiffer, P. and Radler, F. (1984). Comparison of the killer toxin of several yeasts and the purification of the toxin of type K_2. *Arch. Microbiol.*, 137(4): 357–361.

Radler, F., Herzberger, S. and Schwarz, P. (1993). Investigation of a killer strain of *Zygosaccharomyces bailii*. *J. Gen. Microbiol.*, 139(3): 495–500.

Radler, F., Schmitt, M.J. and Meyer, B. (1990). Killer toxin of *Hanseniaspora uvarum*. *Arch. Microbiol.*, 154(2): 175–178.

Ramírez, M., Velázque, R., Maqueda, M., López-Piñeiro, A. and Riba, J.C. (2015). A new wine *Torulaspora delbrueckii* killer strain with broad antifungal activity and its toxin-encoding double-stranded RNA virus. *Front. Microbiol.*, 6: 983. doi:10.3389/fmicb.2015.00983.

Rodríguez-Cousiño , Maqueda, M., Ambrona, J., Zamora, E., Esteban, R. and Ramírez, M. (2011). A new wine *Saccharomyces cerevisiae* Killer toxin (Klus), encoded by a double-stranded RNA virus, with broad antifungal activity is evolutionarily related to a chromosomal host gene. *Appl. Environm. Microbiol.*, 77(5): 1822–1832.

Schmitt, M.J., Klavehn, P., Wang, J., Schonig, I. and Tipper, D.J. (1996). Cell cycle studies on the mode of action of yeast K_{28} killer toxin. *Microbiol.*, 142(9): 2655–2662.

Stark, M.J.R., Boyd, A., Mileham, A.J. and Romanos, M.A. (1990). The plasmid-encoded killer system in *Kluyveromyces lactis*: A review. *Yeast*, 6(1): 1–29.

Suzuki, C. and Nikkuni, S. (1989). Purification and properties of the killer toxin produced by a halotolerant yeast *pichia farinose*. *Agr. Biol. Chem.*, 53: 2599–2604.

Velázquez, R., Zamora, E., Álvarez, M.L., Hernández, L.M. and Ramírez, M. (2015). Effects of new *Torulaspora delbrueckii* killer yeasts on the must fermentation kinetics and aroma compounds of white table wine. *Front. Microbiol.*, 6: 1222. doi:10.3389/fmicb.2015.01222.

Villalba, M.L., Sáez, J.S., del Monaco, S., Lopes, C.A. and Sangorrín, M.P. (2016). TdKT, a new killer toxin produced by *Torulaspora delbrueckii* effective against wine spoilage yeasts. *Int. J. Food Microb.*, 217: 94–100.

Wickner, R.B. (1986). Double-stranded RNA replication in yeast: The killer system. *Annu. Rev. Biochem.*, 55: 373–395.

Yokomori, Y., Akiyama, H. and Shimizu, K. (1988). Toxins of wild *Candida* killer yeast with a novel killer property. *Agr. Biol. Chem.*, 52: 2797–2801.

21 Winemaking Problem
Stuck and Sluggish Fermentation

Ginés Navarro and Simón Navarro

CONTENTS

21.1 INTRODUCTION

Winemaking can basically be divided into four phases. The first phase consists of finding a source of high quality fruit and making sure the grapes are harvested in an optimum condition. The second phase consists of fermenting the grapes into wine. Winemakers manage the fermentation by controlling several different fermentation parameters such as temperature, skin contact time, pressing technique, etc. During the third phase, the new wine is clarified and stabilized. Finally, in the fourth phase the winemaker ages the wine.

Nowadays, "sustainable" has been co-opted as a buzzword for various methods of culture, including "integrated", "biodynamic", and/or "organic". Sustainable production has both viticultural and economic dimensions and can be defined as a collective methodology that produces higher yields of ripe fruit per unit land area with no reduction in vine vegetative growth and does so over a period of years at a cost which returns a net profit.

It has often been said that wine quality is made in the vineyard, and few experienced winemakers disagree with this statement. The soil, the climate, the viticultural practices, and all other aspects of the vineyard environment contribute to the quality of a wine. Even if the winemaker does a perfect job, the quality of the starting grapes always determines the potential quality of the wine. Grape quality is extremely important. Many winemakers feel that when a grape-growing problem develops, the difficulty must be recognized and promptly dealt with to assure fruit quality.

Any factor (abiotic and/or biotic) that has a negative influence on the quantity and/or quality of the crop of the grapevine and its processed products is of direct economic importance. Natural pathogens, including fungi, bacteria, and viruses, play a crucial role in this regard, as they may have a negative influence on all stages of growth and berry development.

The disease control mechanisms currently used to prevent and control fungal infections mostly involve repeated spraying of chemicals combined with the labor-intensive practices of canopy management. The use of chemicals is undesirable on many levels since, inter alia, it is very expensive, must be regularly repeated and causes a build-up of resistant strains in the pathogen population. Moreover, the application of chemicals is increasingly being regarded as undesirable by a growing number of consumers who insist on healthier and more *naturally* produced products. Furthermore, the long-term adverse effects of agricultural chemicals on the environment are well known, and it is generally accepted now that the agricultural community should scale down its dependency on these products.

In the light of the above, the use of transgenic grapevine cultivars focuses mainly on increasing the resistance of grapevines to pathogen infections by using recombinant DNA technology. The strategy used relies on the strengthening of the plant's natural defense mechanisms against pathogen attacks. Plants have developed a variety of mechanisms to limit and/or resist such attacks. Of these, the first line of defense is formed by structural obstacles, such as waxy layers or strategically positioned hydrolyzing enzymes and/or antimicrobial compounds to stop the initial colonization by the pathogen. In addition to these pre-formed defense elements, the plant is also able to orchestrate an induced response (known as active defense) as soon as the structural obstacles have been breached.

21.2 MICROBIAL ACTIVITY AND FERMENTATION

Except for the first phase, the other three winemaking phases overlap. Each phase makes a specific contribution to a wine's characteristics, but the first and second phases are the most important for wine quality. Fermentation problems are often vineyard specific. Table 21.1 shows the different phases of winemaking and their effects on the wine.

Two different fermentations (alcoholic or primary and malolactic or secondary) occur in most of the wines. Yeasts, in the first case, and bacteria, in the second, are the microorganisms responsible.

Wine yeast consists of microscopically small, single-cell organisms that are found naturally on grapes and which turn the sugar in grape juice into alcohol and carbon dioxide. While some winemaking regions only use the natural yeasts that come with the grapes, others kill them with sulfites and then add a special yeast that is known to work well with their grapes.

Like every living organism, yeasts need energy to survive, and the necessary energy is obtained by metabolizing grape sugars. Ethyl alcohol is produced as an end product, but the yeasts do not use the alcohol. They are only concerned with the energy produced when the sugars are converted into alcohol. Besides sugar, yeasts must have access to many other materials to reproduce new cells, and yeasts are sensitive to their environments. The conversion of glucose into alcohol is a complicated, multi-step, biochemical process. Several different enzymes are needed to convert the sugars, while yeasts must have access to vitamins, minerals, oxygen, nitrogen, etc. to produce the required enzymes.

Yeast reproduces rapidly when sufficient oxygen is available, and in optimal conditions, yeast populations can double in less than an hour. This rapid period of yeast growth is called the "exponential" growth phase, and an enormous yeast population (10 million cells per mL of juice) may develop in less than 24 hours. Despite this rapid cell growth during the exponential growth phase, little alcohol is produced. However, the situation is different when oxygen is restricted. With little oxygen, yeast cell reproduction is much slower, but the yeast produces larger amounts of ethyl alcohol.

Furthermore, yeast must have protein to make new cells, and yeast must have nitrogen to produce the protein. Normally, grapes contain enough nitrogen to meet the yeast's needs. However, vineyards needing fertilization often produce fruit excessively low in nitrogen content, and then the yeast has problems in producing the large number of cells needed to complete fermentation.

Yeasts also need very small quantities of an assortment of vitamins, minerals, and other growth factors termed "micronutrients". Grapes normally contain adequate quantities of these micronutrients, but some vineyards consistently produce grapes deficient in some particular growth factors.

On the other hand, although dry yeast looks almost indestructible, it consists of live cells and must be handled with care. Yeast weakened by mishandling often requires an unusually long time to start fermenting, and damaged yeasts may have trouble fermenting the juice due to dryness. Dry yeast may be damaged by, for example: a) prolonged storage at temperatures above 35 °C; b) storage in a freezer; c) prolonged exposure to air after the package is opened; d) rehydrating yeast in water that is too hot or too cold; e) excessive amounts of sulfur dioxide in the juice and f) juice temperatures below 15 °C or above 30 °C.

The conversion of the two major grape sugars (glucose and fructose) into ethyl alcohol is called alcoholic fermentation. Yeasts in the wine produce enzymes, and the enzymes convert the sugars into alcohol. However, the conversion of grape sugars into alcohol is not a simple process. Many steps are

TABLE 21.1
Phases of Winemaking and Effects on the Wine

I	Stage	RW	WW	Results and considerations
I	*Vintage*	Yes	Yes	Obtaining the healthy fruit in an optimum state of maturity.
II	*Crushing*	Yes	Yes/No	Obtaining the must from selected grapes. Contribution of primary aromas to the must.
	Pressing	No	Yes	Separation skin/must to avoid the color increment starting from the skin.
	Maceration	Yes	Yes/No	Obtaining color starting from the skin.
	Fermentation	Yes	Yes	Biochemical transformations by microbiological activity. Appearance of fermentation aromas (secondary aromas).
III	*Pressing*	Yes	Yes/No	Separation of skin and must.
	Clarification	Yes	Yes	Removal of particles in suspension and obtaining of limpidity.
	Stabilization	Yes	Yes	Removal of bitartrates by cold to avoid later precipitations of crystals.
IV	*Aging*	Yes	Yes/No	Oxidations in wood with the appearance of tertiary aromas (Bouquet).
	Bottling	Yes	Yes	Reductions in the glass.

Source: Navarro and Navarro (2011)
RW: red wine; WW: white wine

involved in this transformation, and the yeast must produce several different enzymes. In addition, fermentation generates minor amounts of numerous incidental by-products that affect the aroma and taste of wine including, among others, acetaldehyde, acetic acid, ethyl acetate, glycerol, and/or alcohols other than ethanol.

Some important microbial infections in wines are especially related with the growth of acetic and lactic acid bacteria. The former constitute a heterogeneous group of microorganisms previously encapsulated under the name of *Mycodermaaceti*, which was given by Pasteur (1868) to the vegetations seen to develop in wine and whose consequences are well known. These bacteria have a common characteristic in that they are Gram negative and catalase positive. Their metabolism is strictly aerobic, and they are sensitive to SO_2. The most outstanding aspect of their physiology is their capacity to oxidize ethanol to acetic acid, which they may even partially or totally oxidize to CO_2 and H_2O. On the other hand, lactic acid bacteria, which are responsible for transforming sugars into lactic acid, are Gram positive. This non-taxonomic group includes species of *Lactobacillus, Leuconostoc*, *Pediococcus* and *Streptococcus.*

Since Pasteur's time, many studies have described two extremely important phenomena related to the activity of microbial populations in wines and other fermented products: a) the inhibition of yeast growth and fermentative activity by the presence of certain species of bacteria and b) the co-existence of certain populations of different bacterial and yeast species, on the one hand, and bacteria and molds on the other. It has been shown that bacteria are capable of living in the adverse conditions in the presence of other bacteria, yeasts or molds when they live longer than in their absence. Subsequently, the negative influence of certain bacterial species of the family *Lactobacillaceae*, on the growth of certain species of yeasts, both in static cultures and vinification processes, was described. It has also been shown that l-ornetine, which arises from the hydrolytic activity of lactic bacteria, acts on the arginine in must, where it inhibits the transport of amino acids and reduces the nitrogenated nutrition of certain species of yeast of the genera *Saccharomyces, Hansenula* and *Kloeckera*, leading occasionally to a substantial delay in alcoholic fermentation.

Malolactic fermentation (MLF) entails the bacterial conversion of L(-)-malic acid to L(+)-lactic acid and carbon dioxide. This secondary fermentation, which may occur during or after alcoholic fermentation, is usually conducted by *Oenococcus oeni*, previously *Leuconostoc oenos*, but also by other lactic acid bacteria (LAB). *O. oeni* is the preferred starter culture due to its tolerance of low pH and high alcohol levels. Depending on the strain(s) of LAB involved, several by-products are produced that may impact on the sensory properties of wine. Chemically, the most significant changes observed during the course of MLF are increases in pH and corresponding decreases in titratable acidity. MLF is important for three reasons: 1) deacidification of the wine, 2) flavor modification and 3) microbiological stability.

A variety of yeast and bacteria can grow in wine, and many of these microorganisms cause other fermentations that reduce the quality of the wine.

Depending upon the winemaking conditions, several other fermentations can and often do occur in wine. Some bacteria can ferment the glycerol in the wine into lactic and acetic acids. The natural grape sugars can be transformed into lactic and acetic acid by other types of bacteria. A few species of bacteria can ferment the tartaric acid in the wine into lactic acid, acetic acid, and carbon dioxide gas. Vinegar bacteria can convert the alcohol into acetic acid. Then the same bacteria convert the acetic acid into water and carbon dioxide gas. These other transformations can produce several materials that detract from wine quality. By-products of these undesirable fermentations can be devastating, and when these fermentations occur, wine is often called "diseased" or "sick".

Grape juice normally contains all the materials necessary for the development and growth of yeast, and this is why wine from grapes is much easier to produce than wine from other fruits. In general, if the yeast strain and yeast population are suitable for the particular grapes, the conditions for yeast fermentation are good. These conditions are detailed in Table 21.2.

Sometimes fermentation lacks one or more of the critical factors for yeast growth, in which the yeast cannot convert all of the grape sugars into alcohol, and sugar remains in the wine. This causes other problems, some of which are detailed in Table 21.3.

TABLE 21.2
Optimal Conditions for Yeast Fermentation

Parameter	Fermentation conditions
Temperature	18 °C–25 °C
Sugar	Not more than 1 kg per 4.5 liters or 2 lb per gallon
pH	3–4
Nutrients	N, vitamins and other ingredients needed by yeasts during their growth phase
Oxygen	The first part of fermentation needs oxygen for yeast replication
Toxic substances	Absence in the grapes and must (*e.g.* pesticide residues)

TABLE 21.3
Some Common Problems During Winemaking

Winemaking problem	Cause
Stuck fermentation	The yeast died before fermentation was completed
Cooked flavors	Fermentation temperatures too high; overripe grapes; too many raisins
Burnt match odor	Sulfur dioxide gas
Sherry or Madeira odors	Oxidized wine
Swampy odors	Bad corks
Rotten egg odors	Hydrogen sulfide gas
Vinegar and fingernail polish odors	Acetic acid and ethyl acetate produced by *Acetobacter bacterium*
Rancid butter odor	Excessive amount of diacetyl
Moldy or mildew odor	Moldy barrels; rotten fruit; mold in hoses and equipment
Barnyard, horsy, mousy or wet-dog odors	Fermentation products produced by *Brettanomyces* film yeast

Source: Navarro and Navarro (2011)

21.3 STUCK FERMENTATION: CONCEPT AND IMPORTANCE

Slow (sluggish) fermentation, stuck (incomplete, arrested) fermentation and off-character production such as hydrogen sulfide, sulfur volatiles, acetic acid, and undesired esters are the main problems that can arise during alcoholic fermentation.

By definition, a stuck fermentation is a fermentation that has stopped before all the available sugar in the wine has been converted to alcohol and CO_2. Were you to give up on the wine at this point, it would taste semi-sweet and undesirable.

The serious dangers arising from the premature arrest of alcoholic fermentation are well known. Generally, residual sugar in wine is a dangerous and undesirable condition. If sugar is still present, bacteria may multiply and increase volatile acidity. Residual sugar in wine represents major biological instability because fermentation may restart at any time. If this happens late in the winemaking cycle, much of the work carried out to clarify and stabilize the wine must be repeated. More processing is required, and the additional handling will not help wine quality. Sometimes, fermentation resumes after the wine is bottled, and the yeast produces unsightly sediment in the bottle. The wine becomes effervescent and the bottles may even explode.

Figure 21.1 shows in a schematic way the prevention systems, main causes and treatment for a stuck fermentation.

There are several reasons why better results are more common today: 1) higher levels of alcohol; 2) the growing use of closed vats for red wine where there is no contact with the air; 3) intensive clarification of white musts; 4) better management of problems related to difficult fermentation processes.

Apart from insufficient yeasting and nutritional deficiencies (nitrogen), other causes of premature arrest are as follows. *In red wines*: 1) inadequate sulfiting of the harvest, leading to early development of lactic acid bacteria; 2) poorly controlled or excessive temperature, especially during yeast growth; 3) insufficient aeration of yeasts during their growth. *In white wines*: 1) excessively fast must extraction or excessive clarification; 2) inadequate aeration; 3) excessive temperature variations (thermal shocks). If difficulties occur despite adequate precautions, the reasons for this may be found in the specific constitution of the must (*e.g.* presence of toxic substances). However, a more likely cause may be human error.

The first thing to do with a stuck fermentation is to gather the evidence and work out the underlying cause (Table 21.4).

Problems with slow and stuck fermentation in wines are not new, since these problems have been known since the late 1970s, and many research projects have tried to identify the possible causes. In the past, fermentation problems occurred mostly in white wines. In red wines, the main consequence of stuck fermentation is that the wines develop a high concentration of volatile acid and later spoil microbiologically. However, retarded, sluggish, or stuck fermentation of grape must in winemaking not only has an unfavorable effect on wine quality but also on the economy of wine production.

The loss of revenue caused by incomplete or "stuck" fermentation is a perennial problem in almost every American winery and also in some European countries. One out of every ten or fifteen fermentations demands special attention from the winemaker and may take as much time to manage as all of the others combined. When the problem batch gets stuck, the unwanted residual sugar which remains presents a special challenge for the winemaker. Blending may salvage the wine, but this may result in the undesirable loss of a special vineyard or district designation. To further complicate matters, an off-dry wine may not fit into a producer's product line. As a result, the smaller winery may decide to recover what remains of its wine's value on the bulk market. In either event, the outcome is similar. The winery has lost significant time and revenue to a problem that is now detectable and preventable.

The loss of potential earnings caused by stuck fermentations reveals just how serious this problem can be to a winery's profitability. For example, Napa Valley Chardonnay grape prices average $1500/ton. After crushing and pressing, a relatively small fermentation batch of 4000 gallons represents an investment of about $50,000. While the potential return (wholesale price $120/case) for this investment should

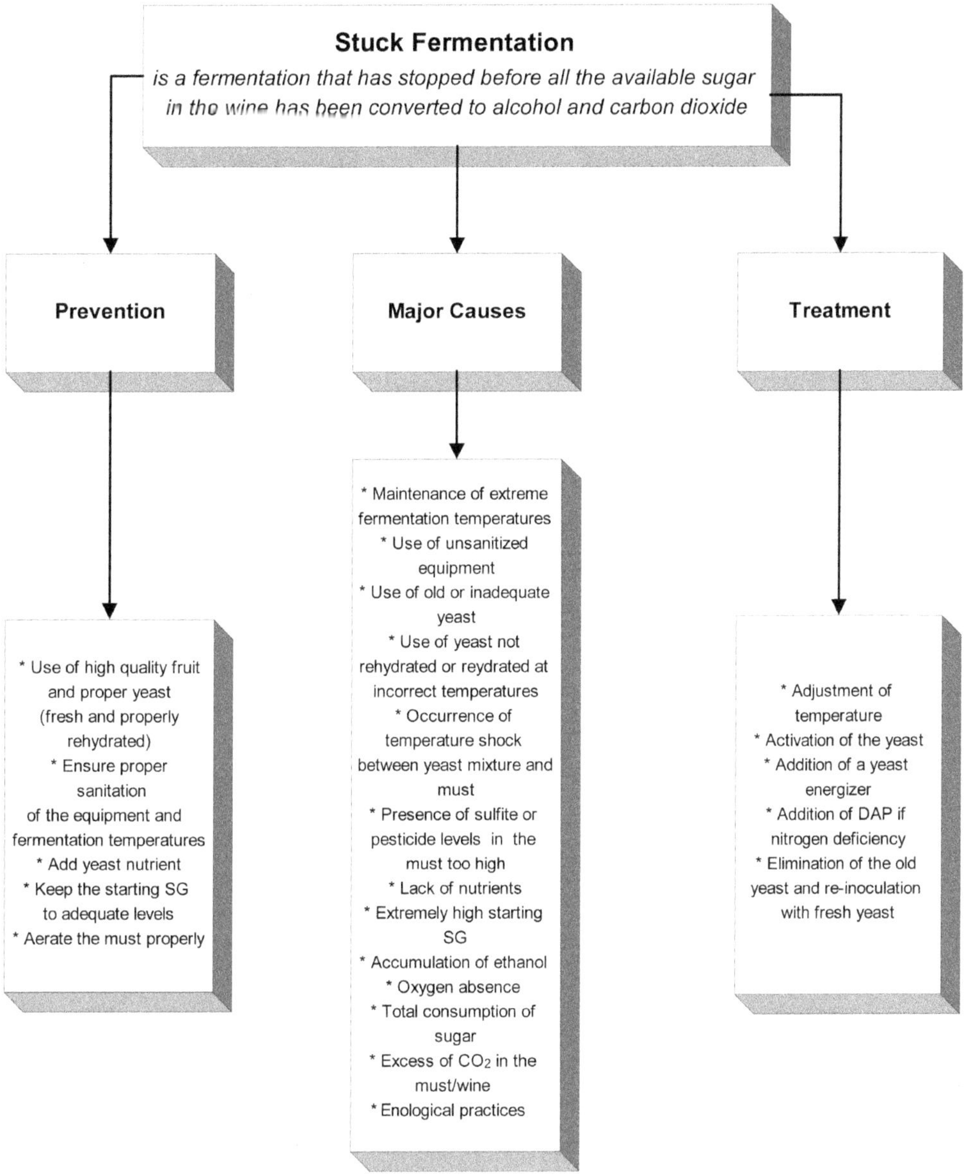

FIGURE 21.1 Stuck fermentation: Concept, prevention, causes and treatment. *Source:* Navarro and Navarro (2011)

be around \$181,000, the actual return from the bulk market (\$15/gallon) would be \$51,000. While barely breaking even on their return from the bulk market, the small premium winery has lost \$130,000 in potential earnings as the result of just one stuck fermentation.

21.4 INFLUENTIAL FACTORS FOR STUCK AND SLUGGISH FERMENTATION

To proceed properly, alcoholic fermentation needs certain environmental and chemical conditions, and these more or less coincide with those that yeasts need to grow. These conditions may change in the face of interference by other factors (Figure 21.2). Some of these conditions are inherent in the vegetal material itself and others arise from modifications in the optimal conditions necessary for the fermentative biochemistry to function adequately.

The factors that may delay or even stop fermentation are: reduced supply of nutrients in the medium, ion imbalances, substrate inhibition, ethanol toxicity, the presence of toxic substrates (pesticide residues), alterations in pH and temperature. As can be seen from the figure, some of these factors may intervene in several of the winemaking stages, as is the case of the nutrients, whose short supply in the original vegetal material, if uncorrected, may continue to influence the process until the very end, delaying or stopping fermentation.

21.4.1 Nutrient Restriction

The main cause of stuck and sluggish fermentation is the restriction of nutrients in the must, since they are essential for the development and metabolic activity of the yeasts. In general, mineral elements (N and P), amino acids, and vitamins are the main nutrients for the activity of the yeasts, although

TABLE 21.4
Trouble-Shooting for Sluggish and Stuck Fermentation

Gather evidence	What the evidence means
Could the temperature have exceeded 26 °C?	Temperature higher than 26 °C kills the yeast.
Is the temperature too low?	Temperatures of below 18 °C result in a very slow fermentation.
Fermentation never started.	Dead yeast.
Specific Gravity reading (1080–1090) exceeds the yeast's ability to wine yeast.	Sugar levels too high. Sauternes yeast is better able to resist high sugar levels.
Specific Gravity reading of 1010–1030.	Probably started with slightly too much sugar or the temperature fluctuated in the final stages, causing the yeast to stop.
Specific Gravity 1005–1010. Does not taste very sweet.	Normal termination for medium to sweet wines.
Specific Gravity less than 1005. Does not taste sweet.	Normal termination for dry to medium wines.
Think of the original ingredients; was there a source of yeast nutrients?	Without yeast nutrients the yeast does not replicate and fermentation is too slow.
Think of the ingredients; was there a source of acid? For those that can measure the pH this should be around 4.	Yeast needs a source of acid for fermentation, without which fermentation is too slow and has a clinical aroma.
Think of the original ingredients; was there a good source of tannin? Tannin is found in grapes, raisins and most fruits. Tannin is very low in honey and flowers.	Yeast needs the right level of tannin to proceed with fermentation correctly. Alas it is not easy to measure.
Think of the original ingredients; was there any source of preservatives?	Many preservatives kill yeast. Preservatives such as sodium benzoate and potassium sorbate are used abundantly. These two are particularly good at killing yeast.

Source: Navarro and Navarro (2011)

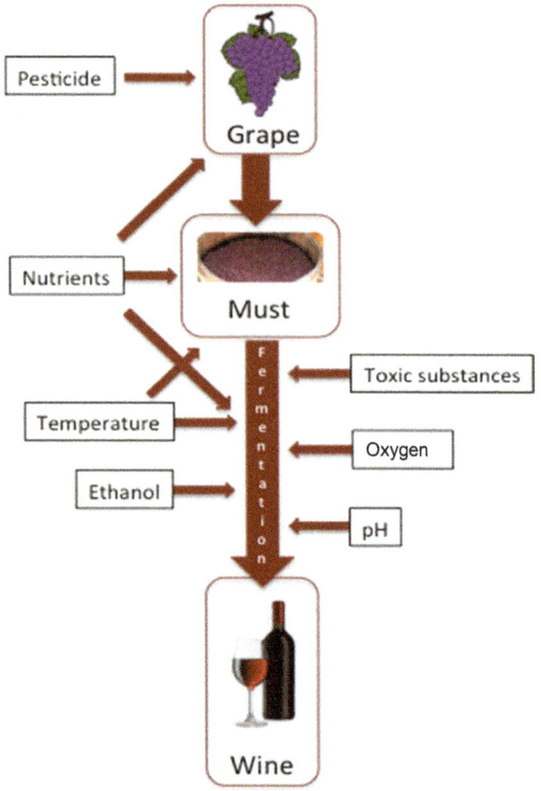

FIGURE 21.2 Factors intervening in the different stages of winemaking. *Source:* Navarro and Navarro (2011)

other compounds (survival factors) have a great influence on the development of the yeasts.

In many cases, stuck and sluggish fermentation is caused by insufficient concentrations of assimilable nitrogen in the juice; levels of nitrogen depend on a great proportion of agronomic factors. For example, nitrogen content in grape juice ranges from 60 to 2,400 mg/L of total nitrogen, from 19 to 240 mg/L of NH_4^+; various studies have shown that a minimum of 120–140 mg N/L values are considered sufficient for most fermentations. However, situations may arise where nitrogen is a limiting factor, as is the case with pre-fermentative clarifications that decrease the content of this element. A low initial level of nitrogen acts by limiting growth rate and biomass formation of yeast, resulting in a low rate of sugar catabolism. When nitrogen shortage exists in the medium, it has also been proven that a drastic decrease in the activity of the transport of sugars in the yeasts occurs.

Among other nitrogenated compounds, ammonium ion is the one metabolized most readily by yeasts, followed by different amino acids. These have been classified into different groups as a function of the speed with which they are metabolized (Table 21.5).

Amino acids can be directly used or can be degraded as N source via transamination:

$$\text{Glutamate} + \text{X} \rightarrow \alpha - \text{Ketoglutarate} + \text{N} - \text{X}$$

$$\text{Glutamine} + \text{X}' \rightarrow \text{Glutamate} + \text{N} - \text{X}'$$

$$\text{Alanine} + \text{X}'' \rightarrow \text{Pyruvate} + \text{N} - \text{X}''$$

Where X is an intermediate in the amino acid/nucleotide biosynthesis, and N-X is an amino acid or nucleotide base or can be interconverted with related amino acids:

$$\text{NH}_4^+ + \alpha - \text{Ketoglutarate} \rightarrow \text{Glutamate}$$

$$\text{NH}_4^+ + \text{Glutamate} \rightarrow \text{Glutamine}$$

TABLE 21.5
Assimilation Rate of Amino Acids by Yeasts in Enological Conditions

Assimilation rate	Compounds
Rapid	Arginine, aspartic ac., asparagine, glutamine, isoleucine, leucine, lysine, serine.
Slow	Glutamic acid, alanine, histidine, methionine, phenylalanine, valine.
Very slow	Glicine, tryptophan, tyrosine.
Partial or nil	Proline.

Source: Jiranek *et al.* (1990)

In the phase of cellular growth, more nitrogen than in the stationary phase is required. Only 36 hours after the start of fermentation, a period coinciding with the exponential phase of their growth, the yeasts have used up all the available nitrogen existing in the must and continue having nitrogenated necessities during the rest of the process. During this phase, the preferred source of inorganic nitrogen is ammoniacal. The addition of amino acids in the stationary phase prolongs the maximum fermentative activity, and some amino acids are more effective than others or a mix of them.

When the harvest provides little nitrogen, as happens when the grapes are very mature or have been attacked by *Botrytis cinerea*, ammoniacal nitrogen can be added to prevent the fermentation from slowing down or stopping.

The addition of ammoniacal nitrogen, generally in the form of ammonium phosphate, is indispensable in some cases, for example when the NH_4^+ concentration in the must is below 25 mg/L; it is also recommendable when the concentration of the ammonium ion is between 25 and 50 mg/L. It may be noted in passing that even if the must contains an acceptable concentration of NH_4^+, the addition of ammoniacal nitrogen is not harmful since it is not used by bacteria and therefore causes no problems.

Ammonium should always be added before fermentation begins in concentrations of 10–20 g of ammonium salts per hectoliter. As fermentation progresses, the yeasts use less ammoniacal nitrogen. When reactivating a slow or stuck fermentation, the ammonium salts are added in small doses of approximately 10 g/L.

However, for yeasts to act it is not only nitrogenated compounds (ammonium and amino acids) that are necessary as nutrients. For example, vitamins such as biotin, pyridoxine, pantothenic acid, inositol, B_1, B_2, and nicotinamide are also vital for the activity of *Saccharomyces cerevisiae*, because they stimulate cellular growth and the fermentative activity of the yeasts, since they act as co-enzymes or enzyme precursors. Although thiamine may be synthesized by the yeasts themselves, they act with greater intensity when the concentration of this vitamin is high. The normal concentration of thiamine in musts from healthy grapes is around 0.1–1 mg/L. When the grapes are rotten or the musts are stuck (sulfited) the content is zero, since its biological activity is annulled by SO_2, when this compound is used at high concentrations on the must. For this reason, it is convenient to add thiamine before fermentation starts in a concentration that does not exceed 60 mg/hL. The presence of acetic acid reduces the transport and retention of thiamine by *Saccharomyces cerevisiae*.

Finally, within this section on nutrients, mention should be made of the mineral elements as the third group of components essential for yeasts to function. Imbalances between minerals and cations can reduce the velocity of fermentation. These are essential components of many enzymes and contribute to cell metabolism by maintaining the pH and ionic balance. This way, it has been proven that a deficit of Ca increases the sensibility to ethanol, or that Mn and Mg have opposed activities. Mg is important for many metabolic and physiologic functions of the yeasts, being implied as important for cell-integrity, generally to stabilize nucleic acids, proteins, polysaccharides, and lipids. Mg also has a fundamental mission in the metabolic control, growth, and cellular proliferation of the yeasts. The concentrations of the different elements differ widely: from 0.5 µm/L in the case of Mo^{2+}, Co^{2+} and B^{2+} to $20x10^3$ µm/L in the case of K^+.

The most important of all these mineral elements is potassium; the K^+/H^+ratio must be at least 25/1 and must be adjusted at the outset of fermentation. High concentrations of potassium can inhibit the efficient assimilation of amino acids as glycine. In anaerobic conditions, the excretion of potassium may be necessary to maintain a correct ionic balance with the consequent incorporation of H^+. High potassium concentrations may also increase tartrate instability, if this is associated with a high pH in the must and wine. High pH values encourage microbial instability, increase the tendency to browning and induce color instability.

21.4.2 Ethanol

The accumulation of ethanol during the must fermentation by yeasts may inhibit the process and cause other unfavorable effects in the cells of the yeasts. The best situation takes place when the intracellular and extracellular concentrations of ethanol are similar. This case may also occur in other types of fermentation, for example in the beer industry.

Ethanol acts on the transporter system of glucose in the yeast. Below 8.5% (v/v) of ethanol, changes do not take place in the transformation rate of glucose or in the activity of glycolytic enzymes; above this value, the velocity can be reduced until 50%, which bears a decrease in the production

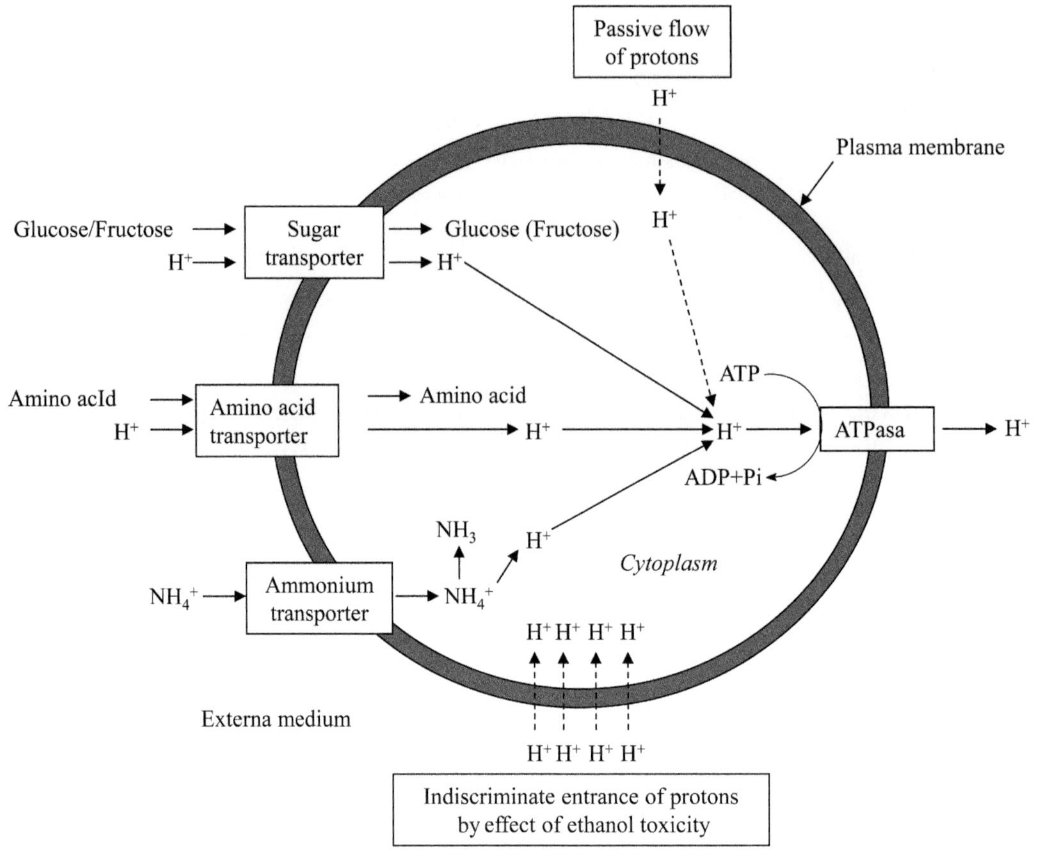

FIGURE 21.3 Schematic diagram of how the yeast cell works and its alteration due to the toxic effect of ethanol. *Source:* Navarro and Navarro (2011)

of CO_2 when decreasing the effectiveness of the sugar transport. The increase in the ethanol content as fermentation proceeds is directly proportional to the decrease in yeast activity, the main cause probably being that ethanol alters the cytoplasm membrane, modifying its permeability and lowering its selectivity.

The plasma membrane is made up of approximately 50% proteins, 40% lipids, and 10% of other compounds, and its main functions are to act as a permeable barrier, regulate the capture of nutrients, respond to alterations in the external environment and maintain the electrochemical gradients (Figure 21.3).

The impact of the ethanol content of the medium is based on the perturbation of the protein structure of the membrane, the increased flow of protons with the consequent acidification of the cytoplasm, the inhibition of protein activity and, lastly, on the alteration of membrane fluidity. The maintenance of cell pH, which is of fundamental importance for yeast activity, is based on the activity of membrane ATPase, whose job it is to displace protons to the external medium and so maintain internal pH, although it has also been shown that ATPase is sensitive to ethanol. Furthermore, the effect of ethanol in preventing sugars and amino acids from entering the cell is well known.

In order to decrease the sensitivity of the yeast cell membrane to ethanol, it is necessary to:

- Raise the content of sterols, since an increase in alcoholic degree is associated with a fall in their levels.
- Presence of protector compounds for ethanol toxicity: trehalose, proline, and glycine.
- Increase proteic synthesis to reinforce the membrane structure, as ATPase activity increases.
- Increase the level of fatty acid desaturation, since an increase in ethanol implies a lower degree of desaturation.
- Synthesize different phospholipids (Phosphatidylinositol, cephalin, ethanolamine, and choline), which alter their equilibrium in the cell in the presence of ethanol.

To correct these imbalances, oxygen, nitrogen, and some other compounds can be used to restore the cellular membranes damaged by the effect of the originated ethanol during alcoholic fermentation.

The need for oxygen is an indirect one. Sterol synthesis and the assimilation of long chain unsaturated fatty acids by yeasts can only occur in the presence of oxygen. When fermentation begins, the yeasts use the sterols of the cells and then those present in the medium. If there is no oxygen, all the sterols are used up and are not synthesized again. As a consequence, the plasma membrane is still more sensitive to the presence of ethanol in the medium and is altered to such an extent that the cell dies.

21.4.3 Oxygen

The fermentation process itself does not require oxygen, although the metabolism of the yeasts is dependent of the amount of dissolved oxygen to begin the fermentation. However, certain concentrations of this element favor fermentation since it encourages the direct oxidation of precursors in the biosynthesis of sterols and long chain unsaturated fatty acids. For this reason, a decrease in the oxygen availability has as consequence the inhibition of the biosynthesis of fatty acids and sterols. As already indicated, to function properly, cells and the plasmatic membrane need sterols (ergosterol and lanosterol) and the fatty acids C_{16} and C_{18}, and for these and other compounds, such as nicotinic acid to be synthesized, oxygen is necessary.

The effects of the oxygen absence can be decreased by addition of small quantities of sulfur dioxide to the must immediately after squeezing the grape, since this compound is an inhibitor of the oxidases of the grape, especially of the polyphenol oxidases. To reactivate sluggish or stuck fermentation, therefore, the must needs to be aerated whenever the mentioned enzymes have not been inhibited.

21.4.4 Toxic Substances

The presence of toxic substances in the must (which negatively affect alcoholic fermentation) may be due to several reasons. They may originate through microbial activity in the must (organic acids, mid-chain fatty acids, and *cis*-fatty acids, etc.) or they may be the consequence of *Saccharomyces* activity. They may also be substances present in the grape (pesticide residues). It should be noted that the effect of each toxic substance individually is less than when it is in the presence of substantial quantities of ethanol, which potentiates their activity.

Of the fatty acids produced during fermentation, the octanoic and decanoic are the most active in the case of a stuck fermentation. Organic acids inhibit yeast activity since they enter the cell and reduce the cytoplasm pH.

Some strains of *Saccharomyces* produce a small peptide known as *killer factor,* which when excreted to the medium kills other yeast strains, thus altering the way in which fermentation proceeds, although there are some strains that are resistant to this effect. There are three different killer factors, K1, K2 and K3, which differ in their degree of activity and resistance in the medium. The so-called K2 is the most active and widespread of those isolated in wine. The consequence for fermentation may be grave if a killer strain that is not very resistant to ethanol comes into contact with a sensitive population.

Some other compounds present in must and formed during alcoholic fermentation are also considered toxic. These include acetic acid, acetaldehyde, higher alcohols and aldehydes, carbon anhydride, and the bicarbonate ion.

21.4.4.1 Exogenous Toxic Substances: Pesticide Residues

The vinegrower has to defend his crop with all available means, which implies using phytosanitary products in most cases. The problem emerges when pesticides are used massively, with no respect for the most elementary specifications such as dose, pre-harvest time, etc. because the vinegrower thinks that his crop is at risk. In this case, residues on grapes can pass to the must during the first steps of the winemaking process (*i.e.* crushing, draining, and pressing) and later to the wine, causing an increased health risk for health; in certain cases, the fermentative process, and the final quality of the wine could be altered. In this sense, the elaboration system (with or without maceration, carbonic maceration, etc.) and the correct carrying out of the winemaking stages play a decisive role in the dissipation and/or elimination of the residues in grapes and must.

Although many fermentative pathways exist, *Saccharomyces cerevisiae* possesses the most common: alcoholic fermentation. In this, ethanol is a by-product while glucose is the preferred substrate. In spontaneous fermentation, there is an early and rapid succession of yeast species. Firstly, fermentation may involve the action of species such as *Kloeckera apiculata* and *Candida stellata*. The former commonly grows in the initial phase but quickly decreases its presence in the must. More significant may be the growth of *Candida stellata*. Both yeasts may ferment only up to about 6 and 10% (v/v) alcohol respectively, and they are responsible for producing compounds such as acetic acid, glycerol, and various esters. Other members of the indigenous grape flora such as *Pichia, Cryptococus*, and *Rhodotorula* grow content, or are inhibited by sulfur dioxide, low pH, high ethanol content or oxygen deficiency and consequently do not contribute significantly to fermentation. Most bacteria that could grow during fermentation are inhibited by *S. cerevisiae*, with the occasional exception of lactic acid bacteria, mainly *Leuconostoc oenos*.

The activity of the above mentioned microorganisms (yeasts and bacteria) can be affected by the presence of pesticide residues in grapes.

However, insecticides and acaricides, if their concentration in the must is not excessively high, do not usually have any effect on the development of the fermentation. For example, studies carried out with methomyl chlorpyrifos or parathion-methyl and fenitrothion did not show significant effects on the fermentative kinetic.

Among pesticides, fungicide residues show the greatest effect on the growth and fermentability of yeasts. Fungicides can have direct and indirect effects on fermentation. Delaying the start of fermentation is probably the most common, but since the depression primarily affect the lag phase (Figure 21.4), subsequent fermentation is unaffected.

Mineral products such as sulfur do not harm the yeasts although may confer the wrong taste to the wine at high concentrations. However, the presence of copper in a concentration of 10 ppm significantly inhibits the growth of *S. cerevisiae*.

Fungicides such as benomyl have little action on determined yeasts, although species like *Zigosaccharomyces fermentati, Metschnikovia purcherrima* and *Rhodorotula glutinis* are inhibited even by low concentrations. Other

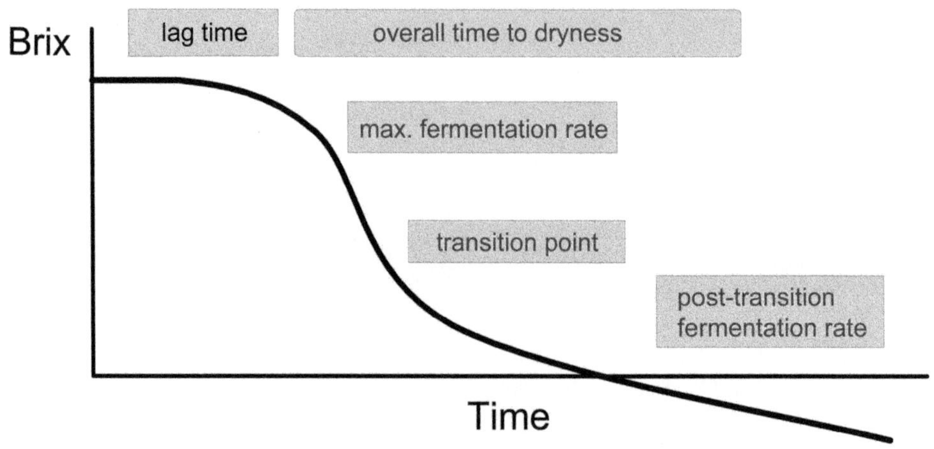

FIGURE 21.4 Fermentation profile. *Source:* Navarro and Navarro (2011)

fungicides of the same group, such as carbendazim or thiophanate-methyl, do not influence the yeasts or they affect the yeast but only at a very high dose. Dithiocarbamic fungicides such as mancozeb at concentrations higher than 50 ppm can inhibit to certain strains of *Saccharomyces sp*, although at their normal concentrations on grapes they have no effect on the fermentation. Other fungicides such as vinclozolin and iprodione have no influence on the development of the fermentative process, while only at concentrations higher than 50 ppm does procymidone influence the fermentative yeasts. More marked is the effect of other pesticides such as dichlofuanid and phthalimide fungicides including captan and folpet, which may delay fermentation even at small doses while high doses cause stuck fermentation. For thirty years it has been known that these fungicides are powerful fermentation inhibitors, retarding onset at concentrations of <1 ppm. Even at a concentration of 0.1 ppm, they may inhibit yeast cell development and reproduction. They affect both the quantity and quality of the spontaneous yeast microflora in the grape and must, reducing the beneficial fermentation yeasts (*Hanseniaspora uvarum, S. cerevisiae*, and *S. bayanus*) while increasing the weak alcoholigens (*Saccharomyces ludwicii, Torullopsis bacillaris, and Candida mycoderma*). Triazolic fungicides (inhibitors of the sterol synthesis), which include cyproconazol, triadimefon, and triadimenol, have demonstrated their ability to influence the fermentative kinetic although in a dose dependent way. Figure 21.5 shows the evolution of yeast number and density during the winemaking process in the presence of cyproconazole (azole) and benalaxyl (phenylamide) residues.

Other pesticides such as 2,4-D or simazine (herbicides) have been found to be non-toxic to wine yeasts. In some cases, the presence of certain pesticides, such as azoxystrobin, mepanipyrim or pyrimethanil may stimulate the yeasts, especially *Kloeckera apiculata*, to produce more alcohol.

Much less known is the effect of pesticide residues on malolactic fermentation. Vinclozolin and iprodione have been reported to depress this process while increasing the growth of acetic acid bacteria. Cymoxanil, dichlofuanid and copper have also been reported to inhibit malolactic fermentation. Except for some cases, pesticides have not shown any negative effects on malolactic fermentation.

21.4.5 Temperature

The exposition to extreme temperatures can paralyze the alcoholic fermentation. The optimum temperature for yeast activity is around 25–30 °C. However, between 5 and 40 °C, and if there are no toxic compounds in the medium, a variation in fermentation temperature affects mainly enzymatic activity, although cell death does not occur at significant levels.

An increase in cell death with increasing temperatures has been observed in the presence of certain components in the medium. For example, given that temperature affects cell membrane fluidity, the presence of a certain quantity of ethanol in the medium strongly increases the effect of temperature, making the plasma membrane of the microbial cell much more sensitive.

The same occurs when other toxic compounds are present, as has been seen with certain short chain organic acids and acetic acid. The negative effect of the temperature-toxic compound combination is more pronounced in the range 25–40 °C than at 5–25 °C.

During alcoholic fermentation, the temperature should not be allowed to reach a dangerous level in the hope that it can be controlled. Instead, the must should be cooled earlier, since once the danger limit has been reached it is difficult to return to the normal temperature because the yeast will have begun to be destroyed.

21.4.6 pH

When fermentation proceeds normally, the pH inside the cell is approximately neutral, even if the external medium is acidic. The yeasts are tolerant to low pHs, and they are developed correctly when the pH of the medium varies from 2.8–4.2. However, when the pH values are very low, the tolerance of the yeasts to the alcohol and organic and fatty acids is reduced.

The entrance of protons into the cell interior by means of a passive diffusion process contributes to the acidification of

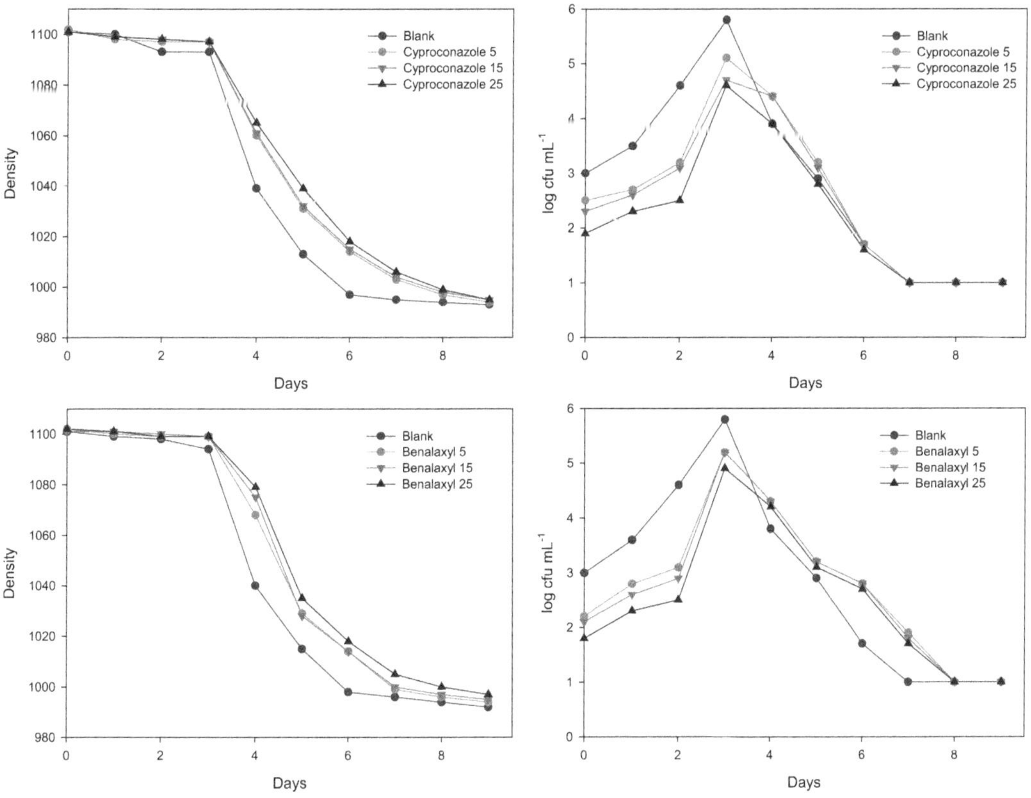

FIGURE 21.5 Number of viable yeasts and evolution of density during the winemaking process with added fungicides (cyproconazole and benalaxyl) at three doses, 5, 15 and 25 mg/L. *Source:* Zamorano *et al.* (1999)

the cytoplasm. However, other compounds, such as acetic acid and weak organic acids, can also cross the plasma membrane and, once inside the cell, dissociate and so contribute to lowering the intracellular pH. This forces the ATPase to step up its activity since it needs to expel protons from the cell to raise the pH to normal levels.

As already stated, the presence of ethanol potentiates the decrease in intracellular pH as plasma membrane stability diminishes and encourages the massive entrance of H^+ ions.

21.4.7 Enological Practices

Together with the previously mentioned factors, the enological practices have a fundamental effect on the rate of must fermentation. The treatments usually used are sedimentation, clarification, filtration, and centrifugation. When the clarification of the must is intense, it decreases the fermentation velocity and production of biomass. It has been proven that the must clarification originates a decrease (around 40–100%) in the content of fatty acids, sterols, and macromolecules (15–50%) related to sluggish fermentation. Furthermore, clarification increases the content of acetic acid and fatty acids of mid-chain that are inhibitors of the fermentative activity.

21.5 RESTORING STUCK FERMENTATION

When fermentation slows down or, in the worst of cases, stops, several different factors may be responsible, as we have seen. However, the most common reasons are the absence of nutrients or the scarcity of oxygen.

The first of these may have two possible causes: the grapes themselves or the technological process. In the first case, the grapes will probably have a low amino acid or available nitrogen content, while in the second case some particular treatment, such as must clarification, may result in there being insufficient quantity of nutrients for the yeasts to act.

Must clarification should be carried out in such a way that the lees responsible for organoleptic alterations are eliminated without the suspended solids that come from the grapes being eliminated at the same time, since these contribute nutrients.

The addition of yeast barks is also a common practice in such situations. Their way of acting is not entirely understood, although it seems that, besides eliminating acids from the medium, they may also provide lipids that strengthen the yeast plasmatic membrane structure.

To correct any nutritional deficiency in musts it is customary to use compounds that contain nitrogen, usually thiamine and ammonium phosphate alone or with vitamins or mineral elements.

Another alternative is to diminish the toxic effect of ethanol. Although it cannot be eliminated since it is the principal product of the process, its effects can be lessened by, for example, adding oxygen, whose beneficial influence has already been described. Such a step is reinforced if nitrogen, in an assimilable form, is added at the same time.

This addition of nitrogen can be carried out before or during fermentation, although the most suitable time is during the stationary phase of yeast development. However, this does not involve any increase in growth of the yeasts, but rather a reactivation of proteic synthesis, and a delay in their death is caused. Nitrogen is added in the form of phosphate or ammonium sulfate and in concentrations not exceeding 60 mg/L.

Oxygen should be added when ethanol begins to have a toxic effect at a concentration of 5–10 mg/L.

Fermentation may be slowed down by the effect of temperature, especially when it is too low. It is usually sufficient to raise the temperature to reactivate the process, although too high a temperature may induce cell death, which is an irreversible process. For this reason, a temperature of 25–30 °C is recommended to avoid sluggish or stuck fermentation.

The pH must also be controlled. If the pH of the must is very acidic, the indiscriminate entrance of protons into the cell increases with the increasing ethanol content of the medium, profoundly altering the plasma membrane. For this reason, the pH must be corrected to be neither excessively acidic (pH<3) nor excessively high (pH>4), which would favor bacterial growth.

Finally, it is important to remark that pesticide residues on grapes can pass to the must during the first steps of the winemaking process (*i.e.* crushing, draining, and pressing). Therefore, any enological practice that reduces the presence of residues in the must is very important. This points out a great affinity of some pesticides for the solid phase, cake and lees. If some pesticide residues are eliminated or their concentration is decreased in the must, certain problems (slow or arrested fermentation) caused by exogenous substances may be avoided in some cases. In this sense, the winemaking technique (with or without maceration, addition of pectolytic enzymes for white and rosé wines, addition of seeds and tannins for red wines, cryomaceration, cold pre-fermentation, fermentation in cask, etc.) also influences the decrease and/or elimination of pesticide residues in the must.

BIBLIOGRAPHY

Agembach, W.A. (1977). A study of must nitrogen content in relation to incomplete fermentations, yeast production and fermentation activity. In: *Proceedings South African Soc. Enol. Vitic*, Capr Town, pp. 66–88.

Alexandre, H. and Charpentier, C. (1998). Biochemical aspects of stuck and sluggish fermentation in grape must. *J. Ind. Microb. Biotech.*, 20(1):20–27.

Alexandre, H., Berlot, J.P. and Charpentier, C. (1994). Effect of ethanol on membrane fluidity of protoplasts of Saccharomyces cerevisiae and *Kloeckera apiculata* grown with or without ethanol, measured by fluorescence anisotropy. *Biotechnol. Techn.*, 8(5):295–300.

Andreasen, A.A. and Stier, T.J.B. (1953). Anaerobic nutrition of *Saccharomyces cerevisiae*. I. Ergosterol requirement for growth in a defined medium. *J. Cell. Comp. Physiol.*, 41(1):23–36.

Andreasen, A.A. and Stier, T.J.B. (1953). Anaerobic nutrition of *Saccharomyces cerevisiae*. II. Unsaturated fatty acid requirement for growth in a defined medium. *J. Cell. Comp. Physiol.*, 43(3):271–281.

Arias-Gil, M., Garde-Cerdán, T. and Ancín-Azpilicueta, C. (2007). Influence of addition of ammonium and different amino acid concentrations on nitrogen metabolism in spontaneous must fermentation. *Food Chem.*, 103(4):1312–1318.

Aries, V. and Kirsop, B.H. (1978). Sterols biosynthesis by strains of *Saccharomyces cerevisiae* in presence and absence of dissolved oxygen. *J. Inst. Brew.*, 84(2):118–122.

Barre, P., Blondin, B., Dequin, S., Feuillat, M., Sablayrolles, J.M. and Salmon, J.M. (2000). La levadura de fermentaciónalcohólica. In: *Enología: fundamentos científicos y tecnológicos*, C. Flanzy, ed. AMV-Mundi-Prensa Ediciones, Madrid, pp. 274–322.

Beavan, M.I., Charpentier, C. and Rose, A.H. (1982). Production and tolerance of ethanol in relation to phospholipid fatty-acyl composition of *Saccharomyces cerevisiae*. NCYC 431. *J. Gen. Microbiol.*, 128:1445–1447.

Bely, M., Salmon, J.M. and Barre, P. (1994). Assimilable nitrogen addition and hexose transport activity during enological fermentations. *J. Inst. Brew.*, 100(4):279–282.

Bisson, L.F. (1999). Stuck and sluggish fermentations. *Am. J. Enol. Vitic.*, 50:107–119.

Bisson, L.F. (2004). Stuck and sluggish fermentations. In: *Tópicos de actualización en Viticultura. y Enología (CEVIUC)*, 22 y 23 de julio de 2004, Santiago de Chile, Chile.

Bisson, L.F. and Butzke, C. (2000). Diagnosis and rectification of stuck and sluggish fermentations. *Am. J. Enol. Vitic.*, 51:168–177.

Boulton, R.B., Singleton, V.L., Bisson, L.F. and Kunkee, R.E. (1996). *Principles and Practices of Winemaking*. Ed. Chapman Hall, New York.

Cabras, P., Meloni, M. and Pirisi, F.M. (1987). Pesticide fate from vine to wine. *Rev. Environ. Contam. Toxicol.*, 99:83–117.

Cabras, P., Garau, V.L., Pirisi, F.M., Cubeddu, M., Cabitza, F. and Spanedda, L. (1995). Fate of some insecticides from vine to wine. *J. Agric. Food Chem.*, 43(10):2613–2615.

Cabras, P., Angioni, A., Garau, V.L., Pirisi, F.M., Cabitza, F., Pala, M. and Farris, G.A. (2000). Fate of quinoxyfen redidues in grapes, wine and their processing products. In: *Abstracts of First Mediterranean Workshop. Research and European Policy on Pesticide Residues in Mediterranean Countries*, Athens, p. 34.

Cabras, P., Angioni, A., Garau, V.L., Melis, M., Oirisi, F.M., Farris, G.A., Sotgiu, C. and Minelli, E.V. (1997). Persistence and metabolism of folpet in grapes and wine. *J. Agric. Food Chem.*, 45(2):476–479.

Carrau, F.M., Neirotti, E. and Gioia, O. (1993). Stuck wine fermentations: Effect of killer/sensitive yeast interactions. *J. Frem. Bioeng.*, 76(1):67–69.

Cartwright, C.P., Veazey, F.J. and Rose, A.H. (1987). Effect of ethanol on activity on the plasma membrane ATPase in and accumulation of glycine by *Saccharomyces cerevisiae*. *J. Gen. Microbiol.*, 133(4):857–865.

Casey, G.P., Magnus, C.A. and Ingledew, W.M. (1984). High-gravity brewing: Effects of nutrition on yeast composition fermentative ability and alcohol production. *Appl. Env. Microbiol.*, 48(3):639–646.

Cavazza, A., Poznanski, E. and Trioli, G. (2004). Restart of fermentation of simulated stuck wines by direct inoculation of active dry yeast. *Am. J. Enol. Vitic.*, 55:160–167.

Conner, A.J. (1983). The comparative toxicity of vineyard pesticides to wine yeasts. *Am. J. Enol. Vitic.*, 34:278–279.

David, M.H. and Kirsop, B.H. (1973). A correlation between oxygen requirements and the products of sterol synthesis in strain *Saccharomyces cerevisiae*. *J. Gen. Microbiol.*, 77:527–531.

Delfini, C. and Costa, A. (1993). Effects of the grape must lees and insoluble materials on the alcoholic fermentation rate and the production of acetic acid, pyruvic acid and acetaldehyde. *Am. J. Enol. Vitic.*, 44:86–92.

Delfini, D. and Parvex, C. (1989). Study on pH and total acidity variations during alcoholic fermentation. Importance of the ammoniacal salt added. *Riv. Vitic. Enol.*, 42:43–56.

Delfini, C., Cocito, C., Ravagkia, S. and Conterno, L. (1993). Influence of clarification and suspended grape solid materials on sterol content of free run and pressed grape must in the presence of growing yeast cells. *Am. J. Enol. Vitic.*, 44:86–92.

DeSante, D.M. (1996). *A New Idea for Detecting and Preventing*, Stuck Fermentatios and Butzke, C.E., ed. American Vineyard: UC Davis Viticulture & Enology Lab, 12/96, Davis, CA.

Dombeck, K.M. and Ingram, L.O. (1986). Magnesium limitation and its role in apparent toxicity of ethanol in yeast fermentation. *Appl. Environ. Microbiol.*, 52(5):975–981.

Domínguez, J., Hernáez, J.L., Otero, D., Pastrana, L. and Pazos, I. (1995). Los residuos antibotriticosen las fermentaciones vinicas. Incidenciaen la calidad del vino. *Nutri.-Fitos*, 95:111–123.

Drysdale, G.S. and Fleet, G.H. (1988). Acetic acid bacteria in winemaking: A review. *Am. J. Enol. Vitic.*, 39:143–154.

Edwards, C.G.R., Beelman, R.B., Bartley, C.E. and McConnell, L.A. (1990). Production of decanoic acid and other volatile compounds and the growth of yeasts and malolactic bacteria during vinification. *Am. J. Enol. Vitic.*, 41:48–56.

Fatichenti, F., Farris, A., Deiana, P., Cabras, P., Meloni, M. and Pirisi, F.M. (1984). The effect of *Saccharomyces cerevisiae* on concentration of dicarboxymide and acylamide fungicides and pyrethroid insecticides during fermentation. *Appl. Microbiol. Biotechnol.*, 29:419–421.

Fernandes, L.M., Corte-Real, V. and Loureiro, V. (1997). Glucose respiration and fermentation in *Zygosaccharomyces bailii* and *Saccharomyces cerevisiae* express different sensitivity patterns to etanol and acetic acid. *Letts. Appl. Microbiol.*, 25(4):249–253.

Fleet, G.H. and Heard, G.M. (1993). Yeasts-growth during fermentation. In: *Wine Microbiology and Technology*, G.H. Fleet, ed. Harwood Academic Pub., Chur, Switzerland, pp. 27–55.

Fleet, G.H., Lafon-Lafourcade, S. and Ribéreau-Gayon, P. (1984). Evolution of yeasts and acid lactic bacteria during fermentation and storage of Bourdeaux wines. *Appl. Environ. Microbiol.*, 48(5):1034–1038.

Flori, P. and Brunelli, A. (1995). Residues of EBI fungicides on grape, must and wine. *Med. Fac. Landbouww. Univ. Gent.*, 60:503–509.

Flori, P., Frabboni, B. and Cesari, A. (1999). Pesticide decay models in wine-making process and wine storage. In: *Proceedings International Symposium of Pesticides in Food in Mediterranean Countries*, Cagliary, pp. 167–173.

Fort, F., Arola, L. and Zamora, F. (1987). Efecto de la presencia de diferentespesticidassobre la cinéticafermentativa. *AEC Rev. D'enologia*, 4:5–14.

Gafner, J. and Schütz, M. (1996). Impact of glucose-fructose ratio on stuck fermentations: Practical experiences to restart stuck fermentations. *Vitic. Enol. Sci.*, 51:214–218.

Gao, C. and Fleet, G.H. (1988). The effects of temperature and pH on the ethanol tolerance of the wine yeasts, *Saccharomyces cerevisiae*, Candida stellata and Kloeckeraapiculata. *J. Appl. Bacteriol.*, 65(5):405–410.

García, J. and Xirau, M. (1994). Persistence of dicarboximidic fungicide residues in grapes, must and wine. *Am. J. Enol. Vitic.*, 45:338–340.

Garde-Cerdán, T. and Ancin-Azpilicueta, C. (2007). Effect of the addition of different quantities of amino acids to nitrogen-deficient must on the formation of esters, alcohols, and acids during wine alcoholic fermentation. *LWT*, 41(3):501–510.

Geneix, C. (1984). Recherches sur la simulation et l'inhibition de la fermentation alcoholique du moût de raisin. Thèse de Docteur Ingenieur en Oenologie-Ampélologie. Institutd'Oenologie, Université de Bordeaux.

Ghareib, M., Youssef, K.A. and Khalil, A.A. (1988). Ethanol tolerance of *Saccharomyces cerevisiae* and its relationship in lipid content and composition. *Folia Microbiol.*, 3:447–452.

Gnaegi, F. and Aerny, J. (1984). Influendes des fungicides inhibiteurs de la byosynthese des sterols sur la fermentation alcoolique et la qualite du vin. *Bull. O.I.V.*, 57:995–999.

Gutiérrez, A.R., Epifanio, S., Garijo, P., López, R. and Santamaría, P. (2001). Killer yeasts: Incidence in the ecology of spontaneous fermentation. *Am. J. Enol. Vitic.*, 52:352–356.

Hazzimitriou, E., Darriet, P., Bertrand, A. and Dubourdieu, D. (1997). Folpet hydrolisis incidence on the initiation of the alcoholic fermentation. *J. Int. Sci. Vigne Vin*, 31:51–55.

Heard, G.M. and Fleet, G.H. (1988). The effects of temperature and pH on the growth of yeast species during the fermentation of grape juice. *J. Appl. Bact.*, 65(1):23–28.

Henry, S.A. (1982). The membrane lipids of yeasts: Biochemical and genetic studies. In: *The Molecular Biology of the Yeast Saccharomyces: Metabolism and Gene Expression*, J.N. Stratern, Jones, E.N. and Broach, J.R., eds. Cold Spring Harbor, New York: Cold Spring Harbor Laboratory, pp. 101–158.

Henschke, P.A. (1997). Stuck fermentation: Causes, prevention and cure. *ASVO Seminar – Advances in Juice Clarification and Yeast Inoculation*, pp. 30–41.

Henschke, P.A. and Jiranek, V. (1991). H_2S formation during fermentation: Effect of nitrogen composition in model grape musts. In: *International Symposium on Nitrogen in Grapes and Wines, American Society of Enology and Viticulture*, Seattle.

Henschke, P.A. and Jiranek, V. (1993). Metabolism of nitrogen compounds. In: *Wine Microbiology and Technology*, G.H. Fleet, ed. Harwood Academic Pub., Australia, pp. 77–164.

Howell, G.S. (2001). Sustainable grape productivity and the growth-yield relationship: A review. *Am. J. Enol. Vitic.*, 52:165–174.

Ingledew, W.M. and Kunkee, R. (1985). Factors influencing sluggish fermentations of grape juice. *Am. J. Enol. Vitic.*, 36:65–76.

Ingram, L.O. and Buttke, T.M. (1984). Effects of alcohol on micro-organisms. *Adv. Microbiol. Physiol.*, 25:253–300.

Iwashima, A., Nishino, H. and Nose, Y. (1973). Carrier-mediated transport of thiamine in baker's yeast. *Biochem. Biophys. Acta*, 330(2):222–234.

Jackson, R.S. (1994). *Wine Science: Principles and Applications.* Academic Press, New York.

Jiménez, J. and Van Uden, N. (1985). Use of extracellular acidification for the rapid testing of alcohol tolerance in yeast. *Biotech. Bioeng.*, 27:196–1598.

Jiraneck, V., Langridge, P. and Henschke, P.A. (1990). Nitrogen requirements of yeast during wine fermentation. In: *Proceedings of the 7 Australian Wine Industry Technical Conference*, P.J. Williams, Davidson, D.M. and Lee, T.H., eds. Adelaide, 1989, Australian Publishers SA, Adelaide, pp. 166–171.

Jones, R.P. (1989). Biological principles for the effects of ethanol. *Enzyme Microbiol. Technol.*, 11(3):130–153.

Jones, R.P. (1990). Roles for replicative deactivation in yeast ethanol fermentations. *Crit. Rev. Biotechnol.*, 10(3):205–222.

Jones, R.P. and Greenfield, P.F. (1987). Ethanol and the fluidity of the yeast plasma membrane. *Yeast*, 3(4):223–232.

Jones, R.P., Pamment, N. and Greenfield, P.F. (1981). Alcohol fermentation by yeasts. The effect of environmental and other variables. *Process. Biochem.*, 39:42–49.

Juroszek, J.R., Feuillat, M. and Charpentier, C. (1987). Effect of ethanol on the glucose induced movements of protons across the plasma membrane of *Saccharomyces cerevisiae* NCYC 431. *Can. J. Microbiol.*, 66(2):93–97.

Kilian, S.G., Du Preez, J.C. and Ericke, M. (1989). The effects of ethanol on growth rate passive proton diffusion in yeast. *Appl. Microb. Biotechnol.*, 32:90–94.

Koshinsky, H.A., Cosby, R.H. and Khachatourians, G.G. (1992). Effects of T-2 toxin on ethanol production by *Saccharomyces cerevisiae*. *Biotech. App. Biochem.*, 16(3):275–286.

Kudo, M., Vagnoli, P. and Bisson, L.F. (1998). Imbalance of potassium and hydrogen ion concentrations as a cause of stuck enological fermentations. *Am. J. Enol. Vitic.*, 49:295–301.

Kunkee, R.E. and Bisson, L.F. (1993). Winemaking yeasts. In: *The Yeasts*, 2nd ed, Vol. 5, A. H. Rose and Harrison, J.S., eds. Academic Press, London, pp. 69–127.

Lafon-Lafourcade, S. and Ribéreau-Gayon, P. (1984). Developments in the microbiology of wine production. *Prog. Indust. Microbiol.*, 19:1–45.

Lafon-Lafourcade, S., Geneix, C. and Ribéreau-Gayon, P. (1984). Inhibition of alcoholic fermentation of grape musts by fatty acids produced by yeasts and their elimination by yeasts ghosts. *Appl. Environ. Microb.*, 47:1246–1249.

Lagunas, R., Bominguez, C., Busturia, A. and Saez, M.J. (1982). Mechanism of appearance of the pasteur effect in *Saccharomyces cerevisiae*: Inactivation of the sugar transport systems. *J. Bacteriol.*, 152(1):19–25.

Leao, C. and Van Uden, N. (1982). Effects of ethanol and other alkanols on the glucose transport system of *Saccharomyces cerevisiae*. *Biotechnol. Lett.*, 4:721–724.

Leao, C. and Van Uden, N. (1984). Effect of ethanol and other alkanols on the general amino acid permease of *Saccharomyces cerevisiae*. *Biotecnol. Bioeng.*, 26(4):403–405.

Maisonnave, P., Sánchez, I., Moine, V., Dequin, S. and Galeote, V. (2013). Stuck fermentation: development of a synthetic stuck wine and study of a restart procedure. *Int. J. Food Microbiol.*, 163(2–3):239–247.

Malfeito-Ferrera, M., Miller-Guerra, J.P. and Loureiro, V. (1990). Proton extrusion as an indicator of the adaptative state of yeast starters for the continuous production of sparkling wines. *Am. J. Enol. Vitic.*, 41:219–222.

Malherbe, S., Bauer, F.F. and Du Toit, M. (2007). Understanding Problem Fermentations. A review. *S. Afr. J. Enol. Vitic.*, 28:169–186.

Manginot, C., Roustan, J.L. and Sablayrolles, J.M. (1998). Nitrogen demand of different yeast strains during alcoholic fermentation. Importance of the stationary phase. *Enzyme Microb. Tech.*, 23(7–8):511–517.

Martí-Raga, M., Sancho, M., Guillamón, J.M., Mas, A. and Beltrán, G. (2015). The effect of nitrogen addition on the fermentative performance during sparkling wine production. *Food Res. Int.*, 67:126–135.

Mauricio, J.C. and Salmon, J.M. (1992). Apparent loss of sugar transport activity in *Saccharomyces cerevisiae* may mainly account for maximum ethanol production during alcoholic fermentation. *Biotechnol. Lett.*, 14(7):577–582.

Mezieres, R. and Carbonell, M. (1988). Les residus de produits de traitements dans la vendage et leur influence sur la vinification. *Viticulture*, 114:19–22.

Monk, P.R. (1982). Effect of nitrogen and vitamin supplements on yeast growth and rate of fermentation of Rhine Riesling grape juice. *Food Technol. Aust.*, 34:328–332.

Monteil, H., Blazy-Maugen, F. and Michel, G. (1986). Influence des pesticides sur la croissance des levures des raisins et des vins. *Sci. Aliments*, 6:349–360.

Monteiro, F.F. and Bisson, L.F. (1991). Biological assay of nitrogen content of grape juice and prediction of sluggish fermentation. *Am. J. Enol. Vitic.*, 42:47–57.

Muñoz, E. and Ingledew, W.M. (1989). Effect of yeast hulls on stuck and sluggish wine fermentations: Importance of the lipid component. *Appl. Environ. Microbiol.*, 55(6):1560–1564.

Muñoz, E. and Ingledew, W.M. (1990). Yeast hulls in wine fermentations. A review. *J. Wine Res.*, 1(3):197–209.

Navarro, G. and Navarro, S. (2011). Stuck and sluggish fermentation. In: *Handbook of Enology, Vol. II. Principles and Practices*, V.K. Joshi, ed. Asiatech Publishers, Inc., New Delhi, pp. 591–617.

Navarro, J.M. and Durand, G. (1978). Fermentation alcoolique; influence de la température sur l'accumulation de l'alcooldans les cellules de levures. *Ann. Microbiol. Inst. Pasteur*, 1219:215–221.

Navarro, S. (2000). Pesticide residues in enology. In: *Research Advances in Agricultural and Food Chemistry*, Vol. 1, R.M. Mohan, ed. Global Research Network, India, pp. 101–112.

Navarro, S., Barba, A., Oliva, J., Navarro, G. and Pardo, F. (1999). Evolution of residual levels of six pesticides during elaboration of red wines. Effect of winemaking procedures in their dissappearance. *J. Agric. Food Chem.*, 47(1):264–270.

Nilov, V.I. and Valuiko, G.G. (1958). Changes in nitrogen during fermentation. *Vinodel Vinograd SSSR*, 18:4–7.

Norton, J.S. and Krauss, R.W. (1972). The inhibition of cell division in *Saccharomices cerevisiae* (Meyen) by carbon dioxide. *Plant Cell Physiol.*, 13(1):139–149.

Oliva, J., Barba, A., Navarro, G., Alonso, G.L. and Navarro, S. (1998). Effect of pesticide residues in the content of organic acids in red wines elaborated in Jumilla wine-producing region. *Vitic. Enol. Prof.*, 59:35–43.

Oliva, J., Bernal, C., Barba, A., Navarro, S. and Pardo, F. (1999). Efecto de los residuos de diclofuanida y miclobutanil durante la elaboración de vinos rosados en la D.O. *Jumilla. Sevi*, 2763/4:2621–2627.

Oliva, J., Navarro, S., Barba, A., Navarro, G. and Salinas, M.R. (1999). Effect of pesticide residues on the aromatic composition of red wines. *J. Agric. Food Chem.*, 47(7):2830–2836.

Otero, D., Mañas, L. and Domínguez, J. (1993). Efecto de los residuos de antibotrit icoss obrep oblac iones levad urifo rmes. Suincidenciasobre la calidad del vino (I). *Vitivinicoltura*, 5/6:35–39.

Ough, C.S., Davenport, M. and Joseph, K. (1989). Effects of certain vitamins on growth and fermentation rate of several commercial active dry wine yeasts. *Am. J. Enol. Vitic.*, 40:208–213.

Pampulha, M.E. and Loureiro-Dias, M.C. (1989). Combined effects of acetic acid and ethanol on intracellular pH of fermenting yeasts. *Appl. Microbiol. Biotechnol.*, 31:547–550.

Pascual, C., Alonso, A., García, I., Romay, C. and Kotyk, A. (1988). Effect of etanol on glucosa transport, key glycolytic enzymes and proton extrusion in *Saccharomyces cerevisiae*. *Biotechnol. Bioeng.*, 32(3):374–378.

Petrov, W. and Okorokov, L.A. (1990). Increase of the anion and proton permeability of *Saccharomyces carlsbergensis* plasmalemma by n-alcohols as a possible cause of its deenergization. *Yeast*, 6(4):311–318.

Piper, P.W., Taljera, K., Panaretou, B., Moradas-Ferreira, P., Byrne, K., Praekelt, U.M., Meacok, P., Recnacq, M. and Boucherie, H. (1994). Induction of major heat-stock proteins of *Saccharomyces cerevisiae* including plasma membrane Hsp30, by ethanol levels above a critical threshold. *Microbiology*, 104:3031–3038.

Rosa, M.F. and Sa-Çorreia, I. (1991). In vivo activation by ethanol of plasma membrane ATPase of *Saccharomyces cerevisiae*. *Appl. Environ. Microbiol.*, 57(3):830–835.

Rosi, J. and Bertuccioli, M. (1984). Effect of lipids on yeasts growth and metabolism under simulated vinification conditions. In: *Annual Meeting, American Society for Enology and Viticulture.1984 Technical Abstracts*, June 21–23, 1984, San Diego, CA, p. 20.

Rowe, S.M., Simpson, W.J. and Fammond, J.R.M. (1994). Intracellular pH of yeasts during brewery fermentation. *Lett. Appl. Microbiol.*, 18(3):135–137.

Sablayrolles, J.M. and Barre, P. (1986). Evaluation du besoin en oxygène de fermentations alcooliques en conditions oenologiques simulées. *Sci. Alim*, 6:373–383.

Sablayrolles, J.M., Dubois, C., Manginot, C., Roustan, J.L. and Barre, P. (1996). Effectiveness of combined ammoniacal nitrogen and oxygen additions for completion of sluggish and stuck wine fermentations. *J. Ferm Bioeng.*, 82:361–365.

Sa-Correia, I. (1986). Synergistic effects of ethanol, octanoic and decanoic acids on kinetiks and the activation parameters of thermal death in *Saccharomyces bayanus*. *Biotechnol. Bioeng.*, 28(5):761–763.

Sa-Correia, I. and Van Uden, N. (1986). Etanol-induced death of *Saccharomyces cerevisiae* at low and intermediate growth temperatures. *Biotechnol. Bioeng.*, 28:301–303.

Sajbidor, J. and Grego, J. (1992). Fatty acids alterations in *Saccharomyces cerevisiae* exposed to etanol stress. *FEMS Microbiol. Lett.*, 93(1):13–16.

Sala, C., Fort, F., Busto, O., Zamora, F., Arola, L. and Guasch, J. (1996). Fate of some common pesticides during vinification Process. *J. Agric. Food Chem.*, 44(11):3668–3671.

Salmon, J.M. (1989). Effect of sugar transport inactivation on sluggish and stuck oenological fermentations. *Appl. Environ. Microbiol.*, 55(4):953–958.

Salmon, J.M. (1996). Sluggish and stuck fermentations: Some actual trends on their physiological basis. *Vitic. Enol. Sci.*, 51:137–140.

Salmon, J.M., Vincent, O., Mauricio, J.C., Bely, M. and Barre, P. (1993). Sugar transport inhibition and apparent loss of activity in *Saccharomyces cerevisiae* as a major limiting factor of enological fermentations. *Am. J. Enol. Vitic.*, 44:56–64.

San Romáo, M.V. and CosteBelchior, A.P. (1982). Study of the influence of some antifungal products on the microbial flora of grapes and musts. *Cienc. Tec. Vitivinic*, 1:101–112.

Santos, J., Sousa, M.J., Cardoso, H., Inácio, J., Silva, S., Spencer-Martins, I. and Leao, C. (2008). Ethanol tolerance of sugar atrnsport, and the rectification of stuck wine fermentations. *Microbiology (Reading Engl.)*, 154(2):422–430.

Sapis-Domerq, S., Bertrand, A., Mur, F. and Sarre, C. (1975). Influence des produit de traitement de la vigne sur la microflorelevurienne. *Conn. Vigne et Vin*, 10:369–389.

Shart, R. and Margalith, P. (1983). The effects of temperature on spontaneous wine fermentation. *Appl. Microbiol. Biotech.*, 17(5):311–313.

Slaughter, J.C., Flint, P.W.N. and Kular, K.S. (1987). The effect of CO_2 on the absorption of amino acids from a malt extract medium by *Saccharomyces cerevisiae*. *FEMS Microbiol. Lett.*, 40(2–3):239–243.

Specht, G. (2003). Overcoming stuck amd sluggish fermentations. Practical Winery and Vineyard, September 2003, 1–5.

Stanley, G.A. and Pamment, N. (1993). Transport and intracellular accumulation of acetaldehyde in *Saccharomyces cerevisiae*. *Biotechnol. Bioeng.*, 42(1):24–29.

Suutari, M., Liukkonen, K. and Laakso, S. (1990). Temperatura adaptation in yeast: The role of fatty acids. *J. Gen. Microbiol.*, 136(8):1469–1474.

Thomas, D.S. and Rose, A.H. (1979). Inhibitory effect of ethanol on growth and soluble accumulation by *Saccharomyces cerevisiae* as affected by plasma-membrane lipid composition. *Arch. Mikrobiol.*, 122:19–25.

Thomas, D.S., Hossack, A.J. and Rose, A.H. (1978). Plasma membrane lipid composition and ethanol tolerance. *Arch. Microbiol.*, 117(3):239–245.

Thomas, K.C., Hynes, S.H. and Ingledew, W.M. (1994). Effects of particulate materials and osmoprotectants on very-high-gravity ethanolic fermentation by *Saccharomyces cerevisiae*. *Appl. Environ. Microbiol.*, 60(5):1519–1524.

Tuduri, P., Nso, E., Amory, A. and Goffeau, A. (1985). Decrease of the plasma membrane H-ATPase during exponential growth of *Saccharomyces cerevisiae*. *Biochem. Biophys. Res. Comm.*, 133:917–922.

Ubeda, J., Briones, A.I. and Izquierdo, P.M. (1996). In vitro behaviour of winemaking strains of *Saccharomyces cerevisiae* in relation to fungicides used in viticulture. *Aliment. Equip. Tecnol.*, 15:117–120.

Van Uden, N. (1985). Ethanol toxicity and ethanol tolerance in yeasts. *Ann. Rep. Ferment. Proc.*, 8:11–58.

Van Vuuren, H.I.J. and Wingfield, B.D. (1986). Killer Yeasts. Cause of stuck fermentations in a wine cellar. *S Afr. Enol. Vitic.*, 7(2):113–118.

Vidal, M.T., Poblet, M., Constantí, M. and Bordons, A. (2001). Inhibitory effect of cooper and dichlofuanid on *Oenococcus oeni* and malolactic fermentation. *Am. J. Enol. Vitic.*, 52:223–229.

Viegas, C.A., Sa-Correia, I. and Novais, J.M. (1985). Synergic inhibition of the growth of *Saccharomyces bayanus* by etanol and octanoic acids. *Biotechnol. Lett.*, 7(8):611–614.

Walker, G.M. and Maynard, A.I. (1997). Accumulation of magnesium ions during fermentative metabolism in *Saccharomyces cerevisiae*. *J. Indus. Microbiol. Biotech.*, 18(1):1–3.

Walker-Caprioglio, H.M., Casey, W.M. and Parks, L.W. (1990). *Saccharomyces cerevisiae* membrane sterol modifications in response to growth in the presence of ethanol. *Appl. Environ. Microbiol.*, 56(9):2853–2857.

Wibowo, D., Eschenbruch, R., Davis, C.R., Fleet, G.H. and Lee, T.H. (1985). Occurrence and growth of lactic acid bacteria in wine: A review. *Am. J. Enol. Vitic.*, 36:302–313.

Winter, J. (1988). Fermentation alcoolique par *Saccharomyces cerevisiae*. Contribution a l'étude du controle de la dynamiquefermentaire par l'inhibition et les facteurs nutritionnels. These, INSA Toulouse.

Young, T.W. (1987). Killer yeasts. In: *The Yeasts*, A.H. Rose and Harrison, J.S., eds. Academic Press, New York, pp. 131–164.

Zamorano, M., García, M.A., Pardo, F., Oliva, J., Barba, A. and Navarro, S. (1999). Influence of benalaxyl and cyproconazole in the viability of yeasts during the fermentation of monastrell grapes. In: *Proceedings International Symposium of Pesticides in Food in Mediterranean Countries*, Cagliary, Italy, pp. 175–180.

22 Enzymes in Enology

Michikatsu Sato, L. Veeranjaneya Reddy and V.K. Joshi

CONTENTS

22.1 INTRODUCTION

The importance of enzymes involved in winemaking is clear because grapes and microorganisms contain and produce divergent kinds of enzymes related to the characteristics of wines. Although grape ripening is proceeded by biochemical processes, winemaking processes are primarily catalyzed by the enzymes of not only grapes and microorganisms (yeasts, fungi, bacteria) but also enzyme preparations added exogenously (Table 22.1; Figure 22.1). The processes such as juice and color extraction, alcohol and malolactic fermentation (MLF), extraction and formation of aromas, juice, and wine filtration, etc. are all influenced both by endogenous and exogenous enzymes. The control and use of enzymes are therefore crucial to produce quality wines. Here all these aspects have been described.

22.2 ENZYMES FROM GRAPES

Enzymes in grapes are related to grape sugar content, anthocyanin formation, juice browning, liberation of amino acids in wine, etc. Enzymes such as pectinases, polyphenoloxidases (PPO), proteases, glycosidases, and lipid or fatty acid-related enzymes are involved in the quality characteristics of juice and wine, while oxido-reductase (related to winemaking practices), pectinases, proteases, and glycosidases (associated with the wine aroma) are illustrated in Figure 22.2.

22.2.1 Oxido-Reductases

Grape juices naturally turn brown, and the enzymes involved in the phenomena are called polyphenoloxidases (PPOs). PPO is a ubiquitous enzyme in plants and animals and oxidizes phenolic compounds to quinones in the damaged cells. It is named according to the specific substrate specificity such as phenolase, laccase, catecholase, tyrosinase, diphenolase, or monophenolase. Grapes contain tyrosinases and enzymes that oxidize diphenols (catecholase or catechol oxidase, EC 1.10.3.1). Grapes infected with a mold, *Botrytis cinerea*, contain laccase, having a broader substrate spectrum than tyrosinase. The enzyme from Muscat Bailey A (Muscat Hamburg × Bailey) had the specificity not towards monohydroxy groups (cresolase or monophenol monooxygenase, EC 1.14.18.1) but towards dihydroxy groups (catechol oxidase). The mechanisms involved in the oxidation of phenolic compounds such as phenols to quinones has been elucidated. Caftaric acid quinones react with glutathione resulting in 'Grape Reaction Product' (GRP), identified as 2-S-glutathionyl caftaric acid. After completion of glutathionine, the excess caftaric acid quinone can oxidize other must constituents such as GRP and flavonoids, and partially regenerate caftaric acid. PPOs preferably oxidize caftaric and coutaric acids. Flavonoids contribute more to the browning of wine than non-flavonoid phenol compounds. Mechanical harvesting

TABLE 22.1
Enzymes in Wine Production: Source and Applications

Name of the enzyme	Source	Application
• Pectinolytic enzymes	Fungi – *Botrytis cinerea* Yeast – *Saccharomyces* and non-*Saccharomyces* strains Bacteria – lactic acid bacteria	Lowering the must viscosity, improvement of skin maceration and color extraction from grapes
• Proteases	Fungi – *Botrytis cinerea* Yeast – *Saccharomyces* and non-*Saccharomyces* strains Bacteria – lactic acid bacteria	Wine stabilization by prevention of protein haze
• Glycosidases	Fungi – *Botrytis cinerea* Yeast – *Saccharomyces* and non-*Saccharomyces* strains Bacteria – lactic acid bacteria	Improvement of aroma by splitting sugar residues from odorless precursors
• Glucanases	Yeast – *Saccharomyces* and non-*Saccharomyces* strains Bacteria – lactic acid bacteria	Lysis of yeast cell walls, release of mannoproteins Lysis of microbial exopolysaccharide to improve clarification
• Cellulases, hemicellulases and xylanases	Fungi and Lactic acid bacteria	Improvement of skin maceration and color extraction of grapes, quality, stability, filtration of wines
• Phenoloxidases (Laccases & Tyrosinases)	Fungi and Lactic acid bacteria and non-*Saccharomyces* strains	Oxidation of phenolic compounds
• Lipases	Fungi and Lactic acid bacteria	Degrade lipids (*e.g.*, in cell membranes)
• Esterases	Fungi and Lactic acid bacteria	ester formation and degradation
• Tannases	Lactic acid bacteria	Hydrolysis of tannins (polymeric phenolic compounds)
• Lichenases	Lactic acid bacteria	Degradation of polysaccharides
• Ureases	Lactic acid bacteria	Hydrolysis of yeast derived urea, preventing formation of ethyl carbamate

Source: Pogorzelski and Wilkowska (2007); Marchal *et al.* (2010); Comitini *et al.* (2011); du Toit *et al.* (2011); Miklosy and Polo (1995); Sato *et al.* (1997)

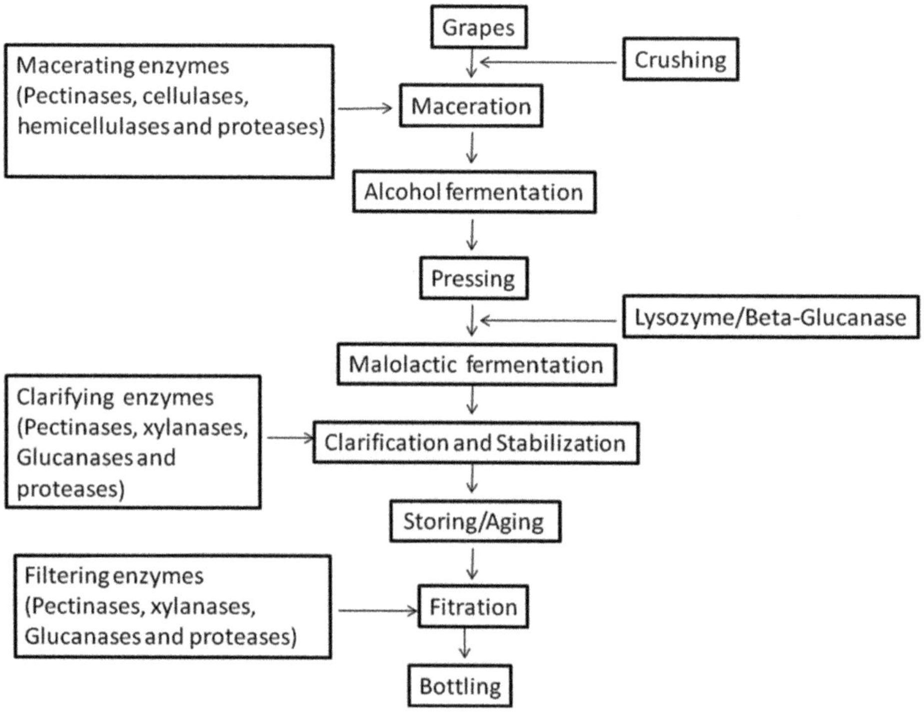

FIGURE 22.1 Use of enzymes in wine production process. *Source:* Romero-Cascales *et al.* (2005); Alexandre *et al.* (2001); Marchal *et al.* (2010); López *et al.* (2015); Hu *et al.* (2016)

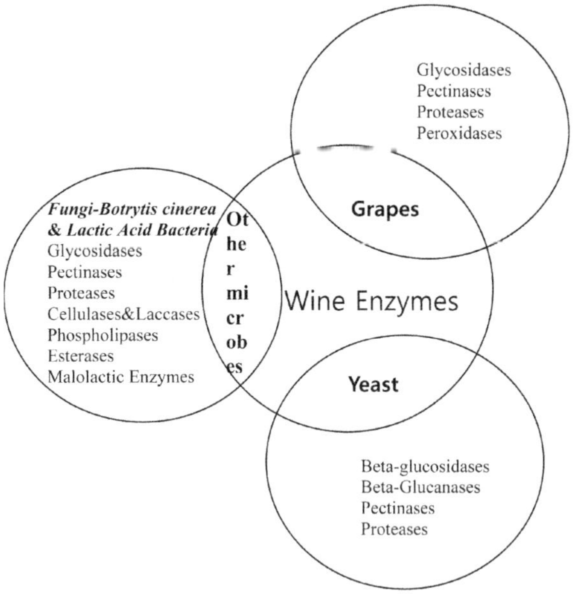

FIGURE 22.2 Major Sources of enzymes that are involved in wine fermentation. *Source:* Gonzales-Pombo *et al.* (2011); Okuda *et al.* (1999); Joshi *et al.* (2011); Guilloux-Benatier *et al.* (2000)

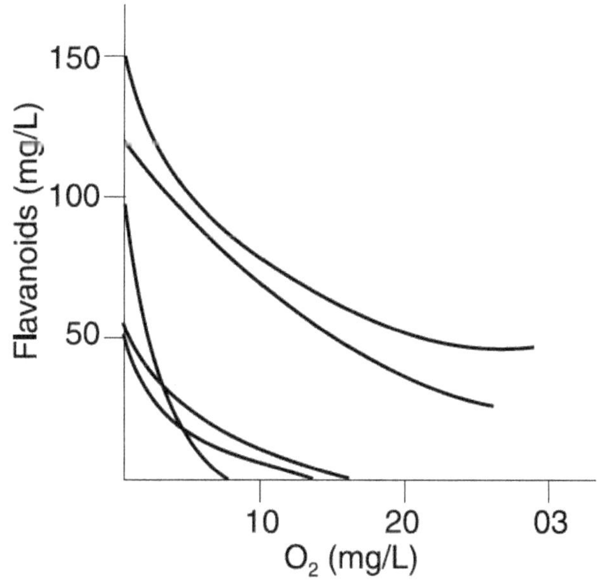

FIGURE 22.3 Flavanoid phenols profile during oxygen consumption of five grape musts

and longer pomace contact time often increase the phenol contents and result in bitter and undesirable wines. PPO is sensitive to SO_2 (tended to be avoided), and also it can be removed with bentonite, but excess fining sometimes results in flat-tasting wines.

For making white wine, hyperoxidation can be done, and for the preparation of red wine, micro-oxygenation can be done. Must hyperoxidation to stabilize white wine against oxidation by the addition of oxygen to the must has also been reported. When oxygen was added to the white grape must (10 mg/L of dissolved oxygen) by pumping oxygen while transferring must from a tank to another tank (*e.g.*,100 g/hr oxygen was added at a flow rate of 10,000 L/hr), flavonoids could be precipitated sufficiently. The relationship of flavonoids and dissolved oxygen levels in five grape musts (Figure 22.3) showed that the 9 mg/L oxygen consumption resulted in flavonoid levels below 100 mg/L. Higher levels of flavonoids by pomace contact require higher levels of oxygen (about 30 mg/L). The phenolic compound levels are different according to the grape varieties and the juice processing procedures. As the excess oxygen addition spoils wine quality, oxygen addition must be carefully practiced. In red winemaking, similar but milder oxygenation than the hyper-oxidation in white wine has been carried out during the last ten years. The method is known as 'micro-oxygenation', and now many wineries in winemaking countries are performing the practice. The oxygen dose ranged from 20 to 80 mL/L/month for four days to one month. As the oxygen-supplying rate is lower than that required, consumption and accumulation of dissolved oxygen does not occur. The wine made with the micro-oxygenation practices had increased color, decreased vegetal aromas, and a change in the type of tannins after 18 months.

22.2.2 Pectinases

Pectinases are induced in grapes along with the ripening of grapes, and grapes to be harvested are softened by the action of pectinases, which are often added to grape juice to clarify and to avoid haze formation in juice and wine. Pectic substances form the middle lamella of plant cell walls and are abundant in the primary cell wall. The major part has highly dehydrated and polymerized galacturonic acid. Galacturonans and rhamnogalacturonans are extracted with hot water containing ammonium oxalate. Other than those acid pectins, the neutral arabinans, galactans, and arabinogalactans are also extracted with the acid substances. The degree of methylesterification and acetylation is varied depending on the kinds of plant. The arabinogalactan-protein (AGP) and pectins have been isolated from grape berry cell walls and pulp tissues by the glycosyl-hydrolase treatment. Homogalacturonan, rhamnogalacturonan I (RG-I), and rhamnogalacturonan II (RG-II) were separated by high performance size exclusion chromatography. Grape berry pulp contains two-fold more AGP and pectins than the cell surface tissues while 75% of the grape cell wall was made from the surface cell tissues that contained three-fold more RG-I and RG-II than the pulp tissues.

Pectin degrading enzymes are contained in grapes and degrade the structures of pectins which hold the rigidity of grape berries and skins. The enzymes are activated along with the grape ripening especially after the *véraison*, the timing of pigmentation of grape berries, and lead to softening of grape berries. The commercial application of pectinases was first reported in 1930 in wine and fruit juice preparation. The effect of pectinases on juice extraction and clarification of plum, peach, pear, and apricot juices has shown the juice recovery of enzymatically-treated pulps increased significantly from 52 to 78 per cent in plum, 38 to 63 per cent in peach, 60 to 70 per cent in pear, and 50 to 80 per cent in apricot. There are various kinds of enzymes that can degrade the rigid structures of pectins by the endo-mode or exo-mode. There are many methods of classification of pectinases, but a recent classification based on the mode of action and substrate used is as shown in Table 22.2. Xyloglucan endo-transglycosylase gene expression has been found closely related to berry softening in Kyoho grape (*Vitis labruscana*) after *véraison*. (Refer to Chapter 9 of this text for more details). The effect of commercial pectolytic enzymes and the time of maceration on wine color and skin cell wall degradation have suggested that commercial pectinase enzyme preparations influence the structure of the skin cell walls of the grapes during winemaking, facilitating the extraction of phenolic compounds and accelerating the skin degradation process that naturally occurs during maceration. The addition of pectinase enzyme has resulted into improvement of quality of wine.

22.2.3 Proteases

Proteases are the enzymes that catalyze the hydrolysis of peptide bonds between the amino acid units of proteins and release peptides and/or amino acids. A broad range of proteolytic enzymes are utilized for the hydrolysis of haze-forming proteins during wine fermentation. The proteolytic activity in fresh wines cannot be detected because the proteolytic activity of grape juice is low and inhibited quickly by low levels of alcohol produced in wine. The proteins as the substrates of proteases are important in winemaking. Grape proteins are important to give a rich taste and body to wine but sometimes they cause haze problems in white wines. The mechanism of protein haze formation has been recently updated. The haze forming proteins in wines are glycoproteins that can be removed by conventional bentonite treatment. The major proteins in grapes are unstable, haze forming proteins and have a molecular weight ranging from 10,000 to 36,000 kDa though a protein with a molecular weight of 28 kDa, and two other proteins having an apparent molecular weight of 35 kDa from a Chardonnay wine have been purified.

Wine from Muscat of Alexandria contained three dominant proteins and separable protein fractions. Surprisingly, the dominant juice proteins of apparent molecular weight ~24 kDa shared extensive homology with thaumatin-like proteins, and the 28 and 32 kDa proteins both had extensive similarity to plant chitinases. Those proteins are classified as the pathogenesis-related (PR) proteins. The protein with the similarity to chitinase had anti-fungal activity against *Botrytis cinerea*, though the thaumatin-like protein did not have a sweet taste like thaumatin. The PR-proteins are induced by stresses such as a fungal attack, physical damages, and physiological conditions except in Shiraz grape under drought stress. The PR-proteins were detected in skins of Pinot noir and Sauvignon blanc grapes. As the PR-proteins are ubiquitous in grapes, the protein profile can be employed as a tool of variety differentiation.

Promising strategies for the degradation of haze-forming proteins with the combination of thermal denaturation of PR proteins with an enzymatic treatment have been proposed. Recently, an acid protease from *Botrytis cinerea* which had proteolytic activity against chitinases at a winemaking temperature of 17 °C and did not need a thermal denaturation step was detected and evaluated.

The proteins from red grapes seemed to get precipitated by making complexes with abundant phenolics and tannins in wine. So, there are few reports on the proteins in wine. But, Muscat Bailey A wines made in 1981 to 1992 have been analyzed, showing the existence of a comparable amount of proteins (33–87 mg/L), which were glycoproteins (molecular weight of 25.5 and 30.0 kDa) in those wines. The data are

TABLE 22.2
Classification of Pectinases

S. No.	Type of pectinase	E C No.	Substrate	Mode of action	Product
1	**Protopectinases**		Insoluble pectin		Polymerised soluble pectin
2	**Esterases**				
	i. PME	3.1.1.11	Pectin	Hydrolysis	Pectic acid + methanol
	ii. PAE	3.1.1.6	Pectin	Hydrolysis	Pectic acid + ethanol
3	**Depolymerases**				
	a. Hydrolases				
	i. Endo PG	3.2.1.15	Pectic acid	Hydrolysis	Oligogalacturonates (reduction in viscosity)
	ii. Exo PG	3.2.1.67	Pectic acid	Hydrolysis	Monogalacturonates
	b. Lyases				
	i. Endo PL	4.2.2.2	Pectic acid	Transelimination	Unsaturated oligogalacturonate
	Endo PNL	4.2.2.10	Pectic acid	Transelimination	Unsaturated methyloligogalacturonates

PME, pectin methylesterase; PAE, pectin acetylesterase; PG, polygalacturonases; PL, pectate lyase; PNL, pectin lyase
Source: Belda *et al.* (2016); Revilla *et al.* (2001); Pérez-Magariño and González-San José (2000); Saranraj and Naidu (2014)

similar to those of white grapes, but the proteins found in red wines were relatively heat stable because of the glycosylation of the proteins. Further juice, fermenting wine, and red wine of Muscat Bailey A grapes contained 17, 18, and 22 protein fractions, respectively. The molecular weight ranged from 12.1 kDa to 30.8 kDa, with pI ranging from 3.7 to 4.9. As the glycoprotein sugars decrease during the winemaking process, the sugar portions of the glycoproteins may be modified through vinification. Proteolytic activity was detected in grape juice, but it gets rapidly inactivated during winemaking. Proteases in grapes are sensitive to alcohol, so the protease activities detected in wine seem to be derived from the yeast used for winemaking rather than from grape berries. Hence, it is very important that the proteases utilized in oenological applications can work under extreme conditions, such as acid pH of 3–4, in the presence of inhibitors (alcohol, sulfur, etc.), and at low temperatures.

22.2.4 Glycosidases

Glycosidases involve glycosyl-bond cleaving enzymes such as α-glucosidases and β-glycosidases. As grapes contain β-glucosyl forms of monoterpenols such as 6-*O*-α-arabinofuranosyl-β-D-glucopyranoside, 6-*O*-α-L-rhamnopyranosyl-β-D-glucopyranoside, 6-*O*-β-apiofuranosyl-β-D-glucopyranoside, and β-D-glucoside as the aroma precursors, corresponding glycosidases (βG, EC 3.2.1.21) are important in winemaking to enhance the aromas by hydrolysis and release of active aromatic compounds. The aroma part of the precursors contains monoterpenols (linalool, nerol, geraniol, citronellol, α-terpineol, etc.), aromatic alcohols, aliphatic residues, and norisoprenoids (Figure 22.4). As the β-glucosylterpenols are the major part of aroma precursors in so-called aromatic grape cultivars such as Muscat, Riesling, Gewürztraminer, etc., β-glucosidase is very important. Grapes contain β-glucosidases and α-glucosidases. The β-glucosidases had optimum activity at pH 5.0 with high sensitivity to glucose but have no specificity to tertiary alcohols such as linalool and α-terpineol which are removed by wine-processing steps (clarification and centrifugation). So, the grape glycosidases practically do not work during juice processing and winemaking practices. Hence, the microbes that secret glycosidases and the commercial enzyme preparations (rich in glycosidase) are suitable for winemaking. Commercial enzyme preparations (glycosidases from *Aspergillus niger*) are suggested to be added at the end of fermentation because of their inhibition by glucose.

22.3 ENZYMES DERIVED FROM MICROORGANISMS

22.3.1 Yeasts and Fungi

Enzymes of *S. cerevisiae* are active during winemaking both at the fermentation period and at the aging time on the lees. Here, the proteases, glycosidases, and various enzymes are not only limited to *Saccharomyces* but also non-*Saccharomyces* yeasts, and even some may have fungal origins. Yeast strains (462) have been isolated from wineries and tested for several enzymatic activities of industrial interest. Considering the seven identified species, only *Aureobasidium pullulans*, *Metschnikowia pulcherrima*, and *Metschnikowia fructicola* showed poly-galacturonase activity. For more information on the production of enzymes using microorganisms, see the literature cited at the end of this chapter.

22.3.1.1 Proteases

The major proteases in the yeast *Saccharomyces cerevisiae* which are involved in the autolysis of *Saccharomyces cerevisiae* are protease A, B, carboxypeptidase Y, S, and aminopeptidase I, II, Co. Only protease A had an optimum acting condition at acidic pH (pH 3.0). Other proteases had an optimum pH at neutral (pH 6 or 7) to alkaline pH (pH 8), so in grape juice and wine (acidic pH values), protease A plays a major role.

22.3.1.1.1 Proteases in 'Sur Lie' and Sparkling Wine

'Sur Lie' and Champagne wine are both stored with the yeast lees for relatively long periods, and high levels of amino acids (nitrogenous compounds) are apparently due to the autolysis of wine yeast. The process results in an increase of amino acid during the aging of Champagne wine on the lees, enhanced its nitrogen compounds during the aging (Figure 22.5 and 22.6). But no increase of nitrogen compounds in wine without lees took place during storage, and the enhancement was very sharp during the first ten-day storage period, and the increase paralleled with the abrupt decrease of protein content in wine (Figure 22.7), amply demonstrating the involvement of yeast proteinase

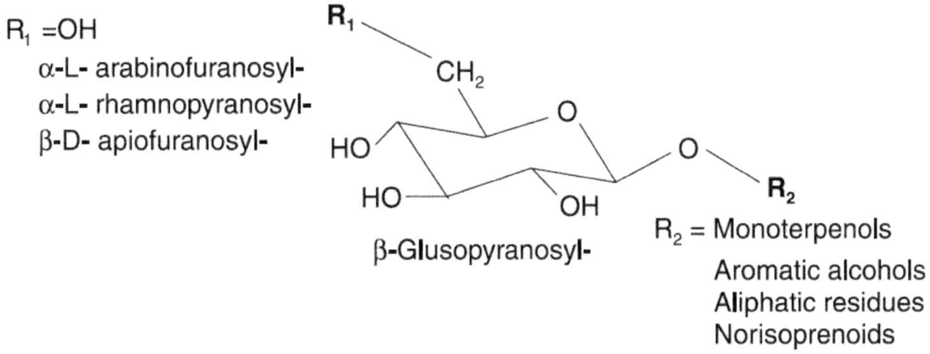

FIGURE 22.4 Structures of glycosidically bound flavor compounds of grapes. *Source:* Hjelmeland and Ebeler (2015)

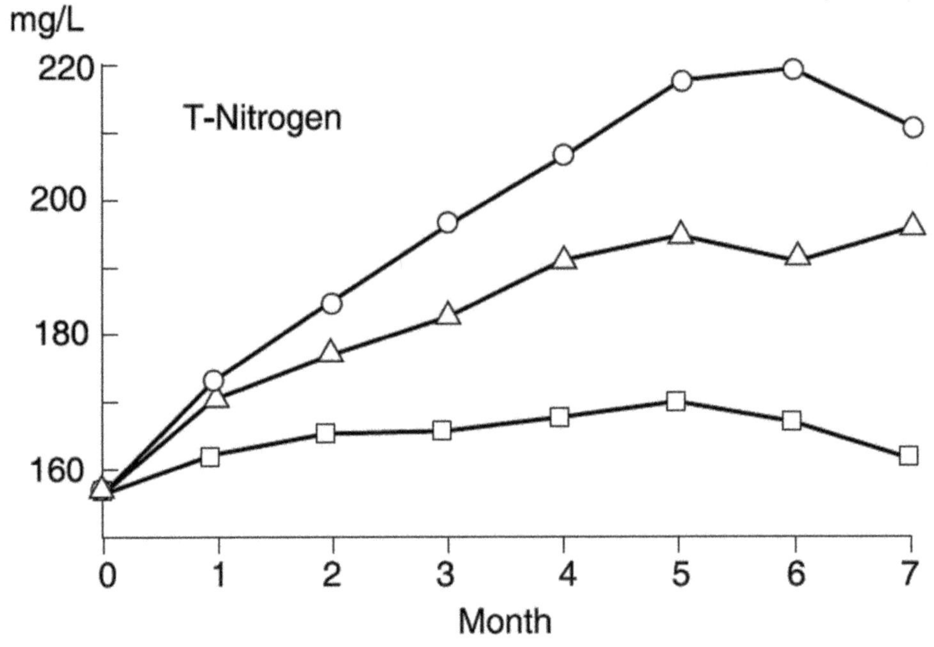

FIGURE 22.5 Changes in total Nitrogen (T-Nitrogen) content in wine without lees (□) with normal volume of less (△) and with double volume of lees (○) during 7-month storage. The yeast stain used was Y-378

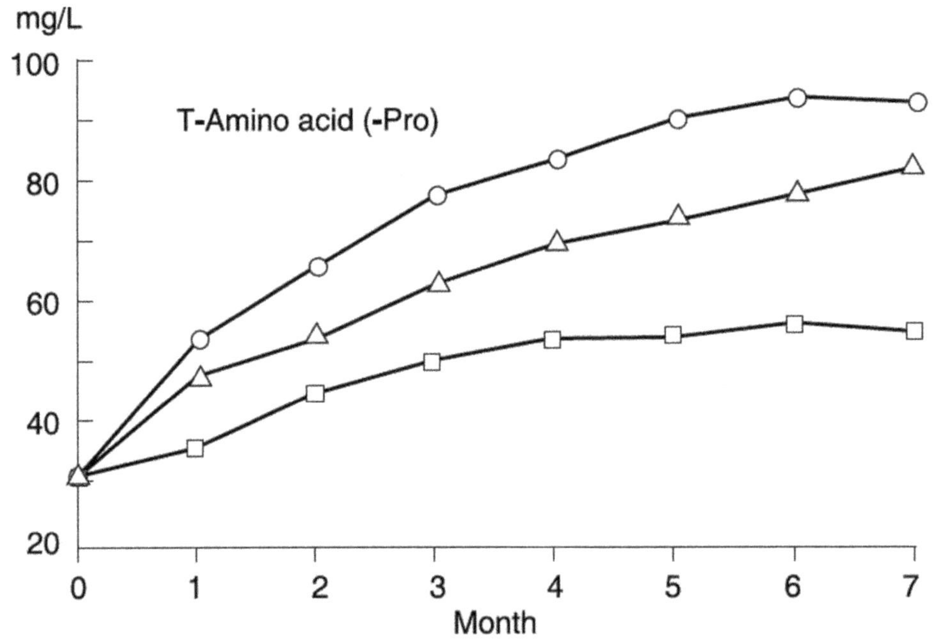

FIGURE 22.6 Changes in total amino acid (T-Amino acid) content in wine without lees (□) with normal volume of less (△) and with double volume of lees (○) during 7-month storage. The yeast stain used was Y-378

during wine aging on the lees. The inhibition of protease activity by the protein A inhibitor Pepstatin showed that the protease A, an acid protease of *S. cerevisiae*, played a major role in the increase of amino acids in wine during yeast cell proteolysis.

Protease A is an endo-type protease, so the increase of amino acids in Champagne or 'Sur Lie' wine could not be explained except when exo-type proteases are involved. Examination of the extra- and intra-cellular protease activities during the storage of wine on the lees revealed that the activity was very low (near the detection limit) in wines during the storage. The intracellular protease activity was detected during the wine storage at 10°C, though it gradually decreased along with the storage (Figure 22.7) and was confirmed in wines with a 100-fold volume of lees at 15°C. The protease activity was clearly detected in wines with and without lees throughout 60-day storage period (Figure 22.8). The extracellular activity increased until the tenth day of storage of wines on the lees, but slightly decreased during the 30- to

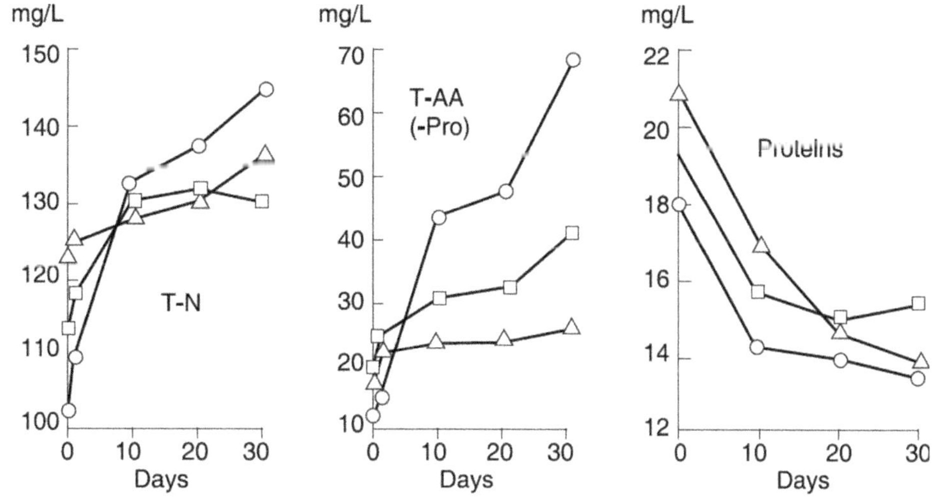

FIGURE 22.7 Changes of total nitrogen (T-N), Total amino acids except for proline (T-AA), and protein content in wines during 30-day storage on the lees. Strain Y-378 (△), *Prise de Mouse* (○) and Lalvin 71-B (□)

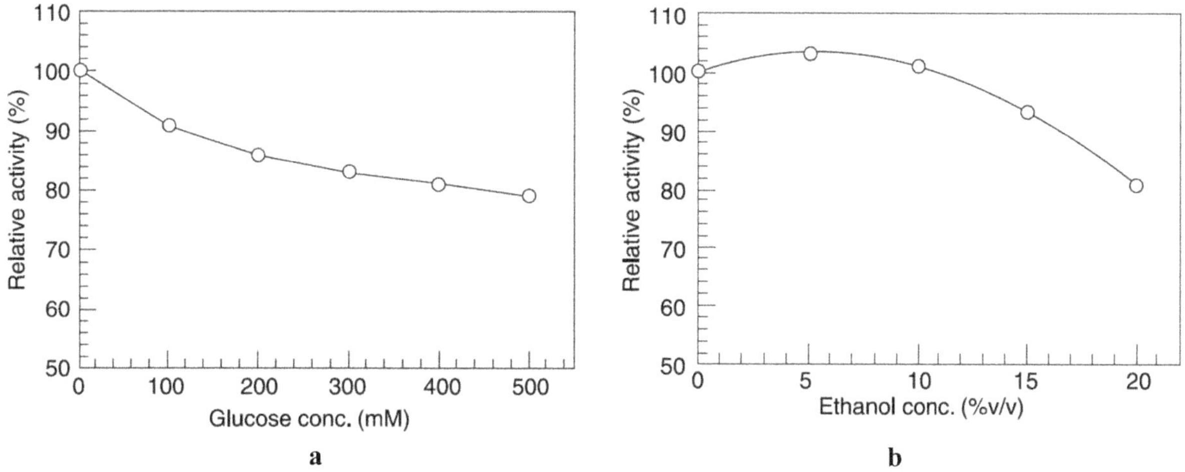

FIGURE 22.8 Effect of (a) glucose concentration (b) on beta-glucosidase activity from *D. hansenii*

60-day storage period. However, carboxypeptidase activity was not detected in wine throughout the storage period of six months at 20°C on the lees, and the intracellular carboxypeptidase activity disappeared at three-month wine storage. The carboxypeptidase was more sensitive than the protease in the wine conditions. The inhibitor experiments showed that the protease contributing to the amino acid increase was protease A. The other proteases mentioned may also be involved in the increase of amino acids in wines on the lees.

22.3.1.2 Glycosidases

Grape glycosidases are sensitive to glucose and have no specificity to tertiary alcohols such as linalool and α-terpineol, and are also easily removed by fining procedures of juice clarification and centrifugation. As we cannot expect aroma liberation from the action of grape glycosidases in wine, yeasts have a major role to produce glycosidases. Wine yeast (*Saccharomyces cerevisiae*) is known to produce glycosidases though these are less sensitive to glucose, and their activities towards glycoside precursors are very low. Non-*Saccharomyces* yeast strains and fungi also produce various glycosidases.

22.3.1.2.1 β-D-Glucosidase from Non-Saccharomyces Yeast

Of glycosidases, β-glucosidase is one of the most important enzymes liberating aromas from grape aroma precursors, because the precursors contain β-glucosyl forms of monoterpenols at the end of their structures. Commercially available β-glucosidase preparations are of fungal origin, mostly from *Aspergillus oryzae*, and the activities are highly inhibited in the presence of glucose. Non-*Saccharomyces* yeasts, which are naturally present in un-inoculated, spontaneous fermentations, can provide a means for increasing aroma and flavor diversity in fermented beverages. Studies reported that glycosidase from non-*Saccharomyces* sp. has remarkable potential to improve aroma complexity and regional characteristics of wine. β-Glucosidases are derived from many non-*Saccharomyces* yeasts (Table 22.3). β-D-glucosidase (Dbg-1) from

TABLE 22.3
List of Non-*Saccharomyes* Yeasts That Produce β-D-Glycosidases

S. No.	Name of the yeast	Property
1	*Candida. molischiana*	Highly tolerant to wine conditions
2	*Candida wickerhamii*	Enhance the wine aroma
3	*Pichia anomala*	Enhance the wine aroma
4	*Hansenula anomala*	Enhance the wine aroma
5	*Debaryomyces castelli and*	Enhance the wine aroma
6	*Debaryomyces hansenii*	High exocellular activity
7	*Debaryomyces polymorphus*	
8	*Kloeckera apiculata*	High aroma producer
9	*W. anomalus*	
10	*Pichia membranifaciens*	
11	*Hanseniaspora uvarum*	High aroma producer
12	*Hanseniaspora vineae*	
13	*Trichosporon asahii*	Works at low temperature
14	*Rhodotorula mucilaginosa*	Works at low temperature
15	*Endomyces fibuliger*	Produce high glucoamylase
16	*Metschnikowia pulcherrima*	
17	*Candida guillermondii*	Highly tolerant to wine conditions
18	*Sporidiobolus pararoseus*	
19	*Issatchenkia terricola*	

Source: Gonzales-Pombo *et al.* (2011); Varela *et al.* (2016); López *et al.* (2015); Baffi *et al.* (2011); Maicas and Mateo (2015); Jolly et al. (2014)

Debaryomyces hansenii, purified 61.1-fold and fully characterized, had the molecular mass of 92 kDa by native- and SDS-PAGE, the pH was at 5.0, had a high tolerance to glucose and alcohol (Figure 22.8a and b), with optimum conditions at pH 7.0 at around 25°C. It was stable at a wide pH (3.0 to 9.0). The aroma precursor glycoside was prepared with a XAD-column, and the aroma liberation was examined using the purified enzyme (Table 22.4). The purified enzyme, when added to the juice of Muscat wine, was made from successfully liberated monoterpenols from the aroma precursor (Table 22.5) and enhanced wine aromas (linalool, geraniol, and nerol), so could be useful in practical winemaking. Cloning of the gene (*dbg-1*) has been successfully made, and hybrid yeast was constructed by cell fusion between *Debaryomyces* and *Saccharomyces* (wine yeast, W-3). Plate 22.1 shows the microphotograph of the parents and the fused cells. As the *Debaryomyces* does not ferment grape juice, a yeast capable of fermenting grape juice with high β-glucosidase activity was constructed. The fused yeast produced about three-fold more terpenols than the parent strain W-3 (Table 22.6), was useful practically, and was not segregated. The progeny was very stable, and it did not change for at least five years. The enzyme from *Debaryomyces hansenii* was purified and also immobilized successfully in one step on hydroxyapatite. As the immobilized β-glucosidase remained active, stable, and reusable, it would be useful practically in winemaking. Interestingly, the strain produced two enzymes (M.W. = 122 kD and 96 kD), and both N-ends of the amino acid sequence were blocked.

TABLE 22.4
Monoterpenols Released by *D. hansenii* β-Glucosidase from the Muscat Glycoside Extract

Monoterpenol (ng)	*D. hansenii* blank	β glucosidase enzyme treatment
Linalool	0	295.7
α-Terpineol	0	78.0
Citronellol	0	17.6
Nerol	0	66.8
Geraniol	0	136.8

Source: Yanai and Sato (1999)

TABLE 22.5
Effect of β-Glucosidase from *D. hansenii* Y-44 on the Concentration of Monoterpenols During the Fermentation of Muscat Wine

Monoterpenol	Untreated[a]	Treated[b]
Linalool	514.2 µg/L	977.2 µg/L
α-Terpineol	46.5	49.6
Citronellol	25.2	38.0
Nerol	70.6	152.6
Geraniol	148.3	189.8

[a] Wine made from must without addition of the enzyme.
[b] Wine made from must initially added with the enzyme.
Source: Yanai and Sato (1999)

22.3.1.2.2 α-L-Arabinofuranosidase

As the aroma precursors have the structure of glycosyl-glucosyl-terpenols, glycosidases such as α-L-arabinofuranosidase (EC 3.2.1.55), α-L-rhamnosidase (EC 3.2.1.40), β-D-apiosidase, and β-D-xylosidase (EC 3.2.1.37) are important along with β-glucosidase. Of aroma precursors, α-L-arabinofuranosyl-β-D-glucopyranosyl-terpenol is one of major precursors, and α-L-arabinofuranosidase will release aromas from grape precursors. Although research on various microorganisms has been carried out, in enological interests the enzyme from *Aspergillus niger* and *Rhodotorula flava* is discussed. As well as this, α-L-arabinofuranosidase was produced from enological yeast stock cultures and isolated the enzyme from *Pichia capsulate*, liberating aromas in winemaking. The α-L-arabinofuranosidase isolated from *P. capsulata* X91 was an intracellular enzyme with a molecular weight of 250 kDa by native-PAGE and 72 kDa by SDS-PAGE; the enzyme seemed to be a homotetramer. The optimal activity of the *Pichia* enzyme was observed at pH 6.0 and 50 °C. In the pH range (3.0–3.8) usually found in winemaking, the enzyme activity

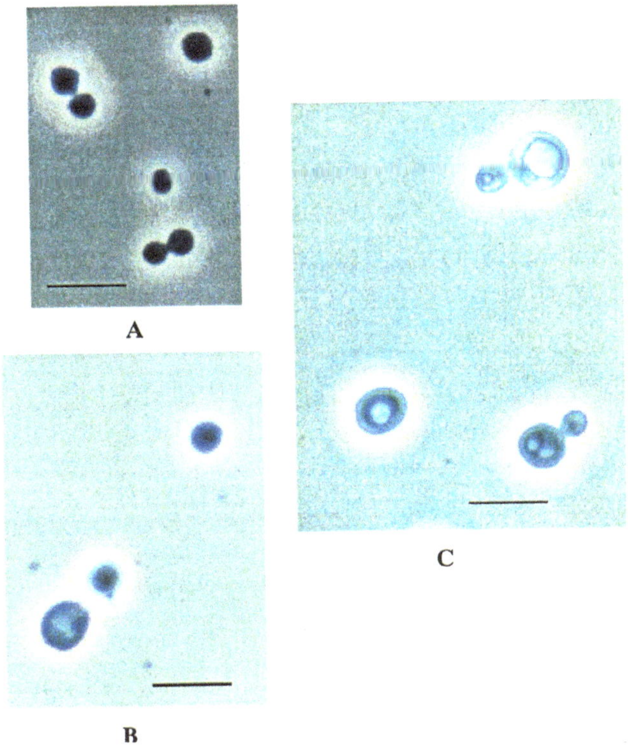

PLATE 22.1 Microphotographs of the parental strains and the fusant F18 Each plate shows: (A) *D. hansenii* CR4; (B) *S. cerevisiae* W-3 (Met$^-$-16); (C) F18, and the bars indicate 5 μm. *Source:* Yanai (2002)

TABLE 22.6
Monoterpenol Contents in Muscat Wine Fermented with *S. cerevisiae* W-3 and the Fusant, F18-1-25

Terpenes analyzed	*S. cerevisiae* W-3	Fusant strain F18-1-25
Linalool	486 μg/L	1916 μg/L
α-Terpineol	162	377
Citronellol	44	106
Nerol	151	405
Geraniol	112	300
Total monoterpenes	965	3104

Source: Yanai (2002)

was 20% of its maximum value while it had 40% of its activity at winemaking temperature conditions (20–30°C). In the presence of alcohol, the α-L-arabinofuranosidase increased activity by 16% with 15% (v/v) ethanol. The enzyme was highly tolerant to glucose, and it maintained more than 60% of its maximal activity in 500 mM glucose, so it will be active in grape juice. The purified enzyme released only arabinose from arabinan and larch-wood arabinogalactan and shortened the arabino-oligosaccharides by one arabinose unit, and is an exo-type. Muscat wine (pH 3.8, 12% ethanol) was treated with the enzyme at 20°C for 24 hours, and the GC-MS analysis of the wine compared to the control wine indicated that the enzyme treatment increased the concentrations of monoterpenols (linalool, citronellol, and geraniol) and the total terpenols by 1.6-fold compared with the control wine. As the aroma precursor is arabinofuranosyl-glucosyl-terpenol, the enzyme must cleave the two glycosyl bonds, but it had no activity on *p*-nitrophenyl β-D-glucopyranoside. It is not clear why the exo-type α-L-arabinofuranosidase can release monoterpenols from the disaccharide aroma precursor.

22.3.1.2.3 *α-L-Rhamnosidase*

Rhamnosides are known to exist as α-L-rhamnopyranosyl residues in grape terpenyl glycosides, the glycosidic precursor of aromatic terpenols, and in citrus bitter flavonoids such as naringin and hesperidin. α-L-Rhamnosidase is an exo-type enzyme and can hydrolyze non-reducing α-L-rhamnosyl groups from rhamnose-containing glycosides and polysaccharides. The enzyme has been industrially used in debittering citrus juice. α-L-Rhamnosidases are produced by a number of mammalian tissues, plants, bacteria, and fungi. Of fungal enzymes, the preparations from *Penicillium decumbens* and *Aspergillus niger* are commercially available. *A. nidulans* produced the enzyme that was tolerant to glucose, SO_2, and ethanol in wine conditions. Low levels of α-L-rhamnosidase activity were detected in *Saccharomyces cerevisiae* Tokaj 7, *Hansenula anomala*, and *Debaryomyces polymorphus*. The enzymes were detected in *Aureobasidium pullulans* and *Candida guillermondii*, but unfortunately, none of them were isolated and fully characterized.

Screening of yeasts producing α-L-rhamnosidase revealed that a strain of *Pichia angusta* produced the highest level of the enzyme that was a monomeric protein, with a molecular mass of 90 kDa, had optimum acting conditions of pH 6.0, temperature of around 40°C, and a Ki for L-rhamnose inhibition of 25 mM. The α-L-rhamnosidase was highly specific for α-L-rhamnopyranoside and liberated rhamnose from naringin, rutin, hesperidin, and 3-quercitrin (Table 22.7). It was also active at the ethanol concentrations of wine, and efficiently released monoterpenols such as linalool and geraniol from the aroma precursor extracted from Muscat grape juice.

22.3.1.2.4 *Other Glycosidases Enhancing Wine Aroma*

As xylose and apiose are also members composing the grape aroma precursors, β-D-xylosidase and β-D-apiofuranosidase could enhance the wine aromas. β-D-Xylosidase (EC 3.2.1.37) has been purified and characterized from *Penicillium wortanni*, *Aspergillus fumigatus*, *A. aculeatus*, *A. niger*, and *Cochliobolus carbonum*, as well as yeasts such as *Cryptococcus albidus*, *Pichia stipites*, *Hanseniaspora*, and *Pichia Candida utilis*. These produced the β-D-xylosidase, that was purified to be a single protein with a molecular mass of 92 kDa (by SDS-PAGE), was most active at pH 6.0, and at around 40°C. A 10% (v/v) concentration of ethanol optimally increased the activity by 57%, but the activity of the enzyme was not inhibited with 300 mM of xylose. The enzyme had a strict specificity towards β-D-xylopyranoside configuration so did not act against α-D-xylopyranoside but was most active against *p*-nitrophenyl β-D-xylopyranoside

TABLE 22.7
Effects of α-L-Rhamnosidase from *P. angusta* on Various Flavonoid Rhamnoglycosides

Substrate	Type of linkage involving rhamnose	Activity (%)[a]
Naringin[b]	α-1,2	100
Rutin[b]	α-1,6	63
Hesperidin[b]	α-1,6	34
3-Quercitrin[b]	α-1,3	31

[a] To measure L-rhamnose-releasing activity towards the rhamnoglycosides, the L-rhamnose released from each substrate was determined by HPLC. Activity using naringin as a substrate was taken as 100%.[b] Naringin, naringenin-7-(α-L-rhamnosyl)-2-β-glucosides; rutin, quercetin-3-(α-L-rhamnosyl)-6-β-lucosides; hesperidin, hesperetin-7-(α-L-rhamnosyl)-6-β-glucoside; 3-quercitrin, quercetin-3-α-L-rhamnoside.

Source: Yanai and Sato (2000)

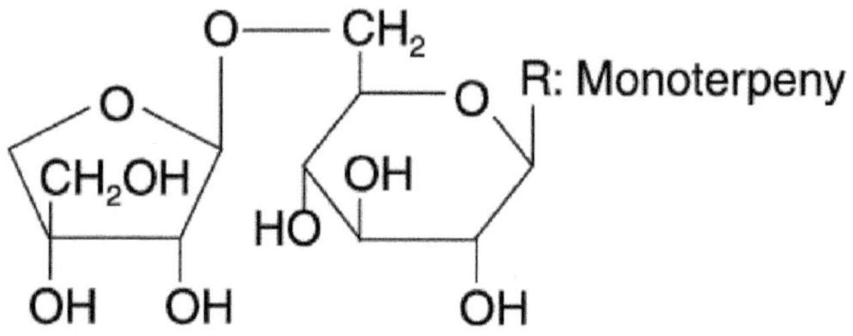

FIGURE 22.9 Structure of beta-apiosyl β-D-glucopyranoside

and showed 23% of activity towards *p*-nitrophenyl β-D-glucopyranoside compared to that against *p*-nitrophenyl β-D-xylopyranoside. It had only weak activity towards *p*-nitrophenyl β-D-cellobioside (6%) and *p*-nitrophenyl α-L-arabinofuranoside (5%). The enzyme was active towards β-1,4-linked xylo-oligosaccharides of the chain length from two to five (Table 22.8). As the enzyme hydrolyzed the oligomers to shorten one xylose unit, it therefore can be called an exo-type enzyme. The enzyme also produced single xylose units and longer xylo-ologosaccharides than the substrate used, so the β-D-xylosidase seemed to have the glycosyltransferase activity too. The performance of aroma liberation was examined using an aroma precursor prepared from Muscat juice. The β-D-xylosidase efficiently liberated monoterpenolse from the aroma precursor (Table 22.6). Muscat wine (pH 3.8, 12% ethanol) was treated with the enzyme. The monoterpenol concentrations were enhanced compared with that of untreated wine. As the β-D-xylosidase from *C. utilis* is tolerant to glucose and shows activity stimulation with ethanol, the enzyme might be useful for technological applications in winemaking. 114 isolates of non-*Saccharomyces* yeasts have been studied and four isolates with both high β-D-xylosidase activities have been identified. From the ribosomal D1/D2 regions sequencing, they were identified as *P. membranifaciens*, *H. vineae*, *H. uvarum*, and *W. anomalus*. An ethanol-tolerant 1,4-β-xylosidase was also purified and characterized from *P. membranifaciens*.

β-D-apiofuranosidase seems to be important to release aromas from the glycoside aroma precursors. It is active towards the precursors of diglycosides, β-D-apiofuranosyl and β-D-glucopyranoside (Figure 22.9). β-apiosidase by *Aspergillus niger* was produced, was partially purified, and had the molecular weight of 38,000 kDa, while the optimum pH and temperature were 5.6 and 50 °C, respectively. The enzyme production was only induced by the addition of apiin and apiosylglucoside in the culture medium and was active towards the grape glycosidic extract, and the sugar linkage (1→6) of apiosylglucosides of nerol, geraniol, and pyran linalool oxide was cleaved by the enzyme. It was inhibited neither by glucose nor by ethanol. The β-D-apiosidase from *A. niger* was purified from an enzyme preparation (Klerzyme 200, Gist-Brocades Co., Delft, The Netherlands), and recently isolation and evaluation of properties of β-apiosidase from the preparation was made. The enzyme preparation contains multi-glycosidases such as β-apiosidase, β-glucosidase, α-rhamnosidase, and α-arabinofuranosidase. The molecular weight of the β-apiosidase was 120 kDa estimated by SDS-PAGE with optimum conditions of pH 5.0 and 40 °C. It cleaved geranyl and linalyl β-D-apiofuranosyl β-D-glucopyranosides and resulted in apiose and the corresponding monoglucosides (geranyl β-D-glucoside and linalyl β-D-glucopyranoside). It was highly resistant to glucose (9%) and was also tolerant to ethanol (84% of the activity remained with 10% ethanol). The pectinase preparation, Klerzyme 200, contains various glycosidases. As the grape aroma precursors have the

TABLE 22.8
Products Accumulated During Reaction of β-D-xylosidase with Xylo-Oligosaccharides

	Products (mol% of total xylo-oligosaccharides)					
Substrate	Xylose	Xylobiose	Xylotriose	Xylotetraose	Xylopentaose	Xylohexaose
Xylobiose	41.5	57.9	0.6			
Xylotriose	4.9	4.3	88.7	2.1		
Xylotetraose	1.9	0.3	6.2	90.7	0.4	0.5
Xylopentaose	2.8	1.1	0.2	10.7	81.9	3.4

Source: Yanai and Sato (2001b)

monoterpenyl-glucosyl-glycoside structures, the enzyme preparation seems to be promising in enhancing wine aromas in practical wine making. (See Chapter 4 of this text.)

22.3.2 Enzymes of *Botrytis cinerea*

Botrytis cinerea is a fungus infecting grapes, which develops into a gray rot as a fungal disease and produces grapes with the noble mold, the material of noble rot wine (See chapter 12 of text). Pectolytic enzymes (polygalacturonase, pectin esterase), cellulases, a laccase, phospholipases, an esterase, N-acetyl glucosaminidase, and proteases had been identified and studied in *B. cinerea* until the year 1990. *Botrytis* laccase is important in the oxidation of grape juice. The laccase activity can be an index of the rotten degree of grapes, and a test-kit is available for this purpose. The activity of β-glucosidase was closely correlated with the pathogenicity of *B. cinerea*, examined near the lesion area of apple fruits. On the contrary, CMCase was negatively associated with the lesion area.

22.3.3 Enzymes of Lactic Acid Bacteria

Various bacteria reside on the surface of grape berries and whole plants, however wine conditions, especially pH, alcohol, temperature, and sulfur dioxide, limit the survival of some kinds of bacteria. Lactic acid bacteria (LAB) such as *Oenococcus*, *Lactobacillus* and *Pediococcus* can survive in the wine conditions. Of LAB, *Oenococcus oeni* (formerly called as *Leuconostoc oenos*) is extremely important in winemaking because of the malolactic fermentation (MLF) that results in deacidification of wine by converting L-malic acid into L-lactic acid and CO_2 (see Chapter 16 of this text for more details).

22.3.3.1 Enzymes of *Oenococcus oeni*

The lactic acid bacterium, *Oenococcus oeni*, plays a major role in conducting malolactic fermentation in wine as stated earlier. The malolactic enzyme had been purified and characterized. The enzyme catalyzes the following reaction:

$$\text{L-Malic acid} \rightarrow \text{L-Lactic acid} + CO_2 \text{ in the presence of } NAD^+ \text{and } Mn^{2+} \quad (21.1)$$

The enzyme was cloned and characterized; it had a molecular weight of 60 kDa and had a 66% similarity with the malolactic enzyme of *Lactobacillus lactis*. The enzyme was expressed in *E. coli* and *Saccharomyces cerevisiae* and revealed that the malolactic enzyme and a malate carrier protein formed a cluster. An exoprotease has been produced by the bacterium in stress conditions (starvation condition), and the enzyme production was enhanced with 60 mg/L of SO_2 and 8% or 12% ethanol. The activity was, however, inhibited with glucose and reverted by cyclic adenosine 3'-5'-phosphate. The purified exo-protease had a molecular weight of 17 kDa by SDS-PAGE and 33.1 kDa by gel filtration. The optimal conditions for activity on grape juice were 25 °C incubation temperature and pH 4.5.

Arginine deaminase (ADI, EC 3.5.3.6), ornithine transcarbamylase (OTC, EC 2.1.3.3), and carbamate kinase (CK, EC 2.7.2.2) of LAB were involved in the arginine metabolism. As the amino acid arginine is one of the major amino acids contained in grape juice, and also could be metabolized to ethyl carbamate precursors, the arginine metabolizing enzymes are important. Arginine is metabolized by the arginine deaminase (ADI) pathway, according to the following scheme:

$$\text{L-arginine} + H_2O \rightarrow \text{L-citrulline} + NH_4^+ (1) \text{ by ADI} \quad (21.2)$$

$$\text{L-citrulline} + \text{Pi} \leftrightarrow \text{L-ornithine} + \text{carbamyl-P}(2) \text{ by OTC} \quad (21.3)$$

$$\text{Carbamyl-P} + \text{ADP} \leftrightarrow \text{ATP} + CO_2NH_4^+ (3) \text{ by CK} \quad (21.4)$$

Examination of various malo-lactic LAB strains for citrulline formation as the ethyl carbamate precursor revealed that all the tested LAB strains excreted citrulline from arginine degradation and degraded citrulline as a sole amino acid. The citrulline excretion rate was linearly correlated to the arginine degradation rate. *Oenococcus oeni* exhibited extracellular β(1→3) glucanase activity, and the culture supernatant effectively lysed viable or dead cells of *Saccharomyces cerevisiae*, though the activity was very low compared with yeast cell wall degrading activities in various other microorganisms, and the optimum temperature (40 °C) of the lytic activity was far from the enological condition. Therefore, the lytic activity of *O. oeni* will not contribute to yeast cell-lysis at the end of alcoholic fermentation.

22.4 COMMERCIAL ENZYME PREPARATIONS

Grapes have enzymes of pectinesterase and polygalacturonase (see Section 22.2), but the activities are insufficient to break down pectic substances. Hence, the addition of commercial enzymes (pectinases, xylanases, glucanases, proteases) is important to improve clarification and filtration. Many of the commercial preparations available are derived from different species of filamentous fungi, especially *Aspergillus* spp., accepted as GRAS (Generally Recognized as Safe) and included in the International Code for Enological Practices of the International Organisation of Vine and Wine (OIV). Pectinase preparations became available in the 1970s, and such preparations have been further developed to date and are used for maceration (mash treatment) to release color and aromas, for clarification to speed up settling, for wine stabilization, and filtration. The preparations usually contain various side enzymatic actions also. The pectinase preparations, especially, have diverse side actions such as hemicellulase, cellulase, glycosidase, or protease activities.

22.4.1 Changes of Wine Composition by Pectic Enzyme Preparations

Pectic enzymes alter the composition and quality of white wine. Treatment of white grapes of Macabeo, Xarel·lo, and Parellada with a pectic enzyme preparation (Citolase 446, Biocatalysis Ltd., UK), increased higher alcohols, free hydroxycinnamic acids, and volatile phenols, whereas most of the esters were reduced. Albillo white grapes (*Vitis vinifera*) were treated with two pectic enzyme preparations, Rapidase CX (Gist-Brocades, France) and Zimopec PX1 (Perdomini SPA, Italy). The enzyme-treated wines had a lower amount of total pectins, higher titratable acidity, and lower acetic acid content than the non-treated wines. The pectinase-treated white wine had higher methanol content than the non-treated wine as was the case with red wine made from grapes of Tinto fino (*V. vinifera*). Besides grapes, efforts have also been made to use pectinase enzyme to extract juice from various stone fruits like plum, peach, and apricot and to improve the physico-chemical and sensory quality of apple wine.

22.4.2 Extraction and Stabilization of Wine Pigments by Pectinase Preparations

Red wines treated with pectinase preparations improved the chromatic characteristics besides increasing the extraction of polyphenols. In the Pinot noir, wine made with pre-fermentation enzyme maceration and extraction of anthocyanins was not increased but enhanced visible color intensity, color density, and polymeric pigment formation. The promotion of the formation of polymeric pigments in enzyme-treated wines continues even after the wines are bottled over 18 months. The wines treated with pectolytic enzymes gave better color characteristics and showed greater stability during two years storage in bottles than those of the control wines. Pectic enzymes help break down the cell walls of the grapes and thus, extract the aromatic precursors. The addition of pectic enzymes during the extraction or fermentation of the must results in an increase in aromatic precursors and are susceptible to beta-glucosidases that the present in the must, from the yeast and bacteria, and enzymes supplemented during fermentation and enhance the aroma of wines.

22.4.3 Glucanase and Lysozyme Preparations in Winemaking

The glucanases (EC 3.2.1) are mainly used in winemaking to avoid the problems caused by glucans (clarification and filtration) of wines produced from *Botrytis*-infected grapes, especially in sweet white wines. The enzyme preparation was made from the culture of *Trichoderma* spp. These enzymatic preparations are a mixture of endo- and exo-β-1,3 glucanases and can also contain hemicellulases and cellulases as secondary activities. It is suggested to add this enzyme to the wine any time between the first racking and filtration but prior to bentonite treatments. *Trichoderma longibrachiatum* endoglucanase also increased aromas in red wine. The other use of β-glucanases is for the maturation of wines.

Lysozyme is a peptidoglycan N-acetylmuramylhydrolase (EC 3.2.1.17), which cleaves the peptidoglycan in the cell wall, leading to fractures in the cell membrane by lysis, after which a fine sediment resembling the texture of fine lees will occur. The main use of lysozyme in winemaking is to prevent or delay malolactic fermentation and to diminish the quantities of SO_2 added to the must or the wine. Lysozyme is also added to increase the foamability of sparkling wines. Pinot noir and Chardonnay wines, originating from lysozyme-treated musts, in addition to bentonite treatment on the wine, showed higher foamability than wines treated only with bentonite. The action of lysozyme may be stopped by adding bentonite or metatartaric acid. If the lysozyme is not eliminated, it can induce a precipitate during aging by enhancing protein instability.

BIBLIOGRAPHY

Alexandre, H., Heintz, D., Chassagne, D., Guilloux-Benatier, M., Charpentier, C. and Feuillat, M. (2001). Protease a activity and nitrogen fractions released during alcoholic fermentation and autolysis in enological conditions. *J. Ind. Microb. Biotech.*, 26(4): 235.

Ariizumi, K., Suzuki, Y., Kato, I., Yagi, Y., Otsuka, K. and Sato, M. (1994). Winemaking from Koshu variety by the *sur lie* method: Changes in the content of nitrogen compounds. *Am. J. Enol. Vitic.*, 45: 312.

Aryan, A.P., Wilson, B., Strauss, C.R. and Williams, P.J. (1987). The properties of glycosidases of *Vitis vinifera* and a comparison of their β-glucosidase activity with that of exogenous enzymes. *Am. J. Enol. Vitic.*, 38: 182.

Baffi, M.A., Bezerra, C.D., Arevalo-Villena, M., Briones-Perez, A.I., Gomes, E. and Da Silva, R. (2011). Isolation and molecular identification of wine yeasts from a Brazilian vineyard. *Ann. Microbiol.*, 61(1): 75–78.

Belda, I., Lorena, B.C., Javier, R., Navascués, E., Marquina, D. and Santos, A. (2016). Selection and use of pectinolytic yeasts for improving clarification and phenolic extraction in winemaking. *Int. J. Food Microbiol.*, 223: 1–8.

Claus, H. and Mojsov, K. (2018).Enzymes for wine fermentation: Current and perspective applications. *Fermentation*, 4(3): 52. doi:10.3390 /fermentation4030052.

Comitini, F., Gobbi, M., Domizio, P., Romani, C., Lencioni, L., Mannazzu, I. and Ciani, M. (2011). Selected non-*Saccharomyces* wine yeasts in controlled multistarter fermentations with *Saccharomyces cerevisiae*. *Food Microbiol.*, 28(5): 873–882.

Dupin, I., Gunata, Z., Sapis, J.C. and, Bayonove, C. (1992). Production of β-apiosidase by *Aspergillus niger*: Partially purification, properties, and effect on terpenyl apiosylglucosides from grape. *J. Agric. Food Chem.*, 40(10): 1886.

Du Toit, M., Engelbrecht, L., Lerm, E. and Krieger-Weber, S. (2011). Lactobacillus: The next generation of malolactic fermentation starter cultures-an overview. *Food Bioprocess Technol.*, 4(6): 876–906.

Feuillat, M. and Charpentier, C. (1982). Autolysis of yeasts in Champagne. *Am. J. Enol. Vitic.*, 33: 6.

Gill, J.V. and Vallés, S. (2001). Effect of macerating enzymes on red wine aroma at laboratory scale: Exogenous addition or expression by transgenic wine yeasts. *J. Agric. Food Chem.*, 49(11): 5515.

Gonzales-Pombo, O., Farina, L., Carreau, F., Batista-Viera, F. and Brena, B.M. (2011). A novel extracellular beta-glucosidase from *Issatschenkia terricola*: Immobilization and application for aroma enhancement of white Muscat wine. *Process Biochem.*, 46: 385–389.

Guilloux-Benatier, M., Pageault, O., Man, A. and Feuillat, M (2000). Lysis of yeast cells by *Oenococcus oeni* enzymes. *J. Ind. Microbiol. Biotechnol.*, 25(4): 193.

Günata, Y.Z., Bayonove, C.L., Cordonnier, R.E., Arnaud, A. and Galzy, P. (1990). Hydrolysis of grape monoterpenyl glycosides by *Candida molischiana* and *Candida wickerhamii* β-glucosidases. *J. Sci. Food Agric.*, 50(4): 499.

Gunata, Z., Brillouet, J.M., Voirin, S., Baumes, R. and Cordonnier, R. (1990). Purification and some properties of an α-L-arabinofuranosidase from *Aspergillus niger.* Action on grape monoterpenyl arabinofuranosylglucosides. *J. Agric. Food Chem.*, 38(3): 772.

Guo, W., Salmon, J.M., Baumes, R., Tapiero, C. and Günata, Z. (1999). Purification and some properties of an *Aspergillus niger* β-apiosidase from an enzyme preparation hydrolyzing aroma precursors. *J. Agric. Food Chem.*, 47(7): 2589.

Hjelmeland, A.K. and Ebeler, S.E. (2015).Glycosidically bound volatile aroma ompounds in grapes and wine: A review. *Am. J. Enol. Vitic.*, 66(1): 1–10.

Hu, K., Zhu, X.L., Mu, H., Ma, Y., Ullah, N. and Tao, Y.S. (2016). A novel extracellular glycosidase activity from *Rhodotorula mucilaginosa*: Its application potential in wine aroma enhancement. *Lett. Appl. Microbiol.*, 62(2): 169–176.

Jolly, N., Augustyn, O. and Pretorius, I.S. (2003). The occurrence of non-*Saccharomyces cerevisiae* yeast species over three vintages in four vineyards and grape musts from four production regions of the Western Cape, South Africa. *South African J. Enol. Viticult.*, 24: 35–42.

Joshi, V.K. and Bhutani, V.P. (1991). The influence of Enzymatic clarification on the fermentation behaviour, composition and sensory qualities of apple wine. *Sci. Aliments*, 11(3): 491–496.

Joshi, V.K., Chauhan, S.K. and Lal, B.B. (1991). Extraction of juices from plum, peach and apricot by the pectolytic enzyme treatment. *J. Food Sci. Technol.*, 28(1): 64–65.

Joshi, V.K., Parmar, Mukesh and Rana, Neerja (2006). Pectin esterase production from apple pomace in solid state and submerged fermentations. *Food Technol. Biotechnol.*, 44(2): 253–256.

Joshi, V.K., Parmar, Mukesh and Rana, Neerja (2011). Purification and characterization of pectinase produced from Apple pomace and evaluation of its efficacy in fruit juice extraction and clarification. *Indian J. Nat. Prod. Resour.*, 2(2): 189–197.

Labarre, C., Guzzo, J., Cavin, J.F. and Diviés, C. (1996). Cloning and characterization of the genes encoding the malolactic enzyme and the malate permease of *Leuconostoc oenos. Appl. Environ. Microbiol.*, 62(4): 1274.

Lao, C., López-Tamames, E., Lamuela-Raventós, R.M., Buxaderas, S. and del Carmen de la Torre-Boronat, M. (1997). Pectic enzyme treatment effects on quality of white grape musts and wines. *J. Food Sci.*, 62(6): 1142.

Le Clinche, F., Piñaga, F., Ramón, D. and Vallés, S. (1997). α-L-Arabinofuranosidases from *Aspergillus terreus* with potential application in enology: Induction, purification, and characterization. *J. Agric. Food Chem.*, 45(7): 2379.

Liu, S.Q. and Pilone, G.J. (1998). A review: Arginine metabolism in wine lactic acid bacteria and its practical significance. *J. Appl. Mirobiol*, 84(3): 315–327.

Lonvaud-Funel, A. (1995). Microbiology of the malolactic fermentation: Molecular aspects. *FEMS Microbiol. Lett.*, 126(3): 209.

López, M.C., Mateo, J.J. and Maicas, S. (2015). Screening of _-glucosidase and _-xylosidase activities in four non-Saccharomyces yeast isolates. *J. Food Sci.*, 80(8): C1696–C1704.

Madhavan, A., Raveendran, Sindhu, Parameswaran, Binod, Sukumaran, Rajeev K. and Pandey, Ashok, (2017) Metagenomic analysis: A powerful tool for enzyme bioprospecting. *Appl. Biochem. Biotechnol.*, 183: 636–651.

Maicas, S. and Mateo, J.J. (2015). Enzyme contribution of non-*saccharomyces* yeasts to wine production. *Univ. J. Microbiol. Res.*, 3(2): 17–25.

Manzanares, P., Ramon, D. and Querol, A. (1999). Screening of non-*Saccharomyces* wine yeasts for the production of β-D-xylosidase activity. *Inter. J. Food Microbiol.*, 46(2): 105.

Marchal, R. and Jeandet, P. (2010). Use of enological additives for colloids and tartrate salt stabilization in white wines and for improvement of sparkling wine foaming properties. In: *Wine Chemistry and Biochemistry*, M.V. Morena-Arribas, Polo, M.C. (Eds.), Springer, New York, NY, pp. 127–158.

Mateo, J.J. and Maicas, S. 2016Application of Non-Saccharomyces yeasts to winemaking process. *Fermentation*, 2(4): 14.

McMahon, H., Zoecklein, B.W., Fugelsang, K. and Jasinski, Y. (1999). Quantification of glycosidase activities in selected yeasts and lactic acid bacteria. *J. Ind. Microbiol. Biotechnol.*, 23(3): 198.

Miklosy, M.V. and Polos, M.C. (1995). *Wine Chemistry and Biochemistry*, M.V. Morena-Arribas, Polo, M.C. (Eds.), Springer, New York, NY: 2010, pp. 127–158.

Morimitsu, K., Koike, S., Yumiko, Y., Sato, M. and Osawa, T. (1997). *Proceedings of Annual Conference of Jpn. Soc. Biosci. Biotech. Biotechnol.*, March 31–April 4, Tokyo, Japan. Patent Kokai H11-60591 (1999), p. 58.

Morosoli, R., Roy, C. and Yamaguchi, M. (1986). Isolation and partial primary sequence of a xylanase from the yeast *Cryptococcus albidus. Biochem. Biophys. Acta*, 870(3): 473.

Nakamura, K., Amano, Y. and Kagami, M. (1983). Purification and some properties of polyphenol oxidase from Koshu grapes. *Am. J. Enol. Vitic.*, 34: 122.

Okuda, T., Pue, A.G., Fujiyama, K. and Yokotsuka, K. (1999). Purification and characterization of polyphenol oxidase from Muscat Bailey A grape juice. *Am. J. Enol. Vitic.*, 50: 137.

Ozcan, S., Kotter, P. and Ciriacy, M. (1991). Xylan-hydrolysing enzymes of the yeast *Pichia stipitis. Appl. Microbiol. Biotechnol.*, 36(2): 190.

Pérez-Magariño, S. and González-San José, M.L. (2000). Effect of pectolytic enzymes on the composition of white grape musts and wine. *Ital. J. Food Sci.*, 12: 153.

Pocock, K.F. and Waters, E.J. (1998). The effect of mechanical harvesting and transport of grapes, and juice oxidation, on the protein stability of wines. *Aust. J. Grape Wine Res.*, 4(3): 136.

Pocock, K.F., Hayasaka, Y., McCarthy, M.G. and Waters, E.J. (2000). Thaumatin-like proteins and chitinases, the haze forming proteins of wine, accumulate during the ripening of grapes (*Vitis vinifera*) berries and drought stress does not affect the final levels per berry at maturity. *J. Agric. Food Chem.*, 48(5): 1637.

Pogorzelski, Eugeniusz and Wilkowska, Agnieszka (2007). Flavour enhancement through the enzymatic hydrolysis of glycosidic aroma precursors in juices and wine beverages: A review. *Flavours Frag.*, 22(4): 251–254.

Riccio, P., Rossano, R., Vinella, M., Domizio, P., Zito, F., Sanserrino, F., D'Elia, A. and Rossi, I. (1999). Extraction and immobilization in one step of two β-glucosidases released from a yeast strain of *Debaryomyces hansenii. Enz. Microb. Technol.*, 24(3–4): 123.

Revilla, I. and González-SanJosé, M. (2001). Evolution during the storage of red wines treated with pectolytic enzymes: New anthocyanin pigment formation. *J. Wie Res.*, 12(3): 183–197.

Río-Segade, S., Pace,C., Torchio,F., Giacosa, S., Gerbi, V. and Rolle, L. (2015). Impact of macerationenzymeson skin softening andrelationship with anthocyanin extraction in wine grapes with different anthocyanin profiles. *Food Res. Int.*, 71: 50–57.

Romero, C., Sanchez, S., Manjon, S. and Iborra, J.L. (1989). Optimization of the pectinesterase/endo-D-polygalacturonase immobilization process. *Enzym. Microb. Technol.*, 11(12): 837–843.

Romero-Cascales, I., Fernández-Fernández, Jose I. , López-Roca, Jose M. and Gómez-Plaz, Encarna. (2005). The maceration process during winemaking extraction of anthocyanins from grape skins into wine. *Eur. Food Res. Technol.*, 221(1): 163–167. doi:10.1007/s00217-005-1144-1.

Rossi, I., Domingo, P., Vinella, M. and Salicone, M. (1995). Hydrolysis of grape glycosides by enological yeast β-glucosidases. In: *Food Flabors: Generation, Analysis and Process Influence*, Charalambous, G. (Ed.), Elsevier Science, Amsterdam, p. 1623.

Saranraj, P. and Naidu, M.A. (2014). Microbial pectinases: A review. *Global J Trad Med Sys*, 3(1): 1.

Sato, M. (2011). Enzymes in wine production. In: *Handbook of Enology: Principles, Practices and Recent Innovations*, V.K. Joshi (Ed.), Asia Tech Publisher, New Delhi, pp. 702–730.

Sato, M., Suzuki, Y., Hanamure, K., Kato, I., Yagi, Y. and Otsuka, K. (1997). Winemaking from Koshu variety by the *sur lie* method: Behavior of free amino acids and proteolytic activities in the wine. *Am. J. Enol. Vitic.*, 48: 1.

Sergi, Maicas and Mateo, José Juan (2016). Microbial glycosidases for wine production. *Beverages*, 2(3): 20; doi:10.3390/beverages2030020.

Schneider, V. (1991). Components des vins obtenus par oxygenation des moûts blancs. *Rev. Fr. Oenol.*, 130: 33.

Schneider, V. (1998). Must hyperoxidation: A review. *Am. J. Enol. Vitic.*, 49: 65.

Sharrel, R., Mohandas, A., Embalil, M.A., Raveendran, Sindhu, Binod, Parameswaran and Ashok, Pandey (2017). Recent advancements in the production and application of microbial pectinases: An overview. *Rev. Environ. Sci. Biotechnol.*, 16(3): 381–394; doi:10.1007/s11157-017-9437-y.

Singleton, V.L. (1987). Oxygen with phenols and related reactions in musts, wines, and model systems: Observations and practical implications. *Am. J. Enol. Vitic.*, 38: 69.

Theron, L.W. and Divol, B. (2014). Microbial aspartic proteases: Current and potential applications in industry. *Appl. Microbiol. Biotechnol.*, 98(21): 8853–8868.

Uesaka, E., Sato, M., Rauju, M. and Kaji, A. (1978). α-L-Arabinofuranosidases from *Rhodotorula flava*. *J. Bacteriol.*, 133(3): 1073.

Varela,C., Sengler,F., Solomon,M. and Curtin, C. (2016).Volatile flavour profile of reduced alcohol wines fermented with the non-conventional yeast species *Metschnikowia pulcherrima* and *Saccharomyces uvarum*. *Food Chem.*, 209: 57–64.

Van Sluyter, S., Durako, M.J. and Halkides,C.J. (2005). Comparison of grape chitinase activities in Chardonnay and Cabernet Sauvignon with *vitis rotundifolia* cv.fry. *Am. J. Enol. Vitic.*, 56(1): 81.

Van Sluyter, S.C., McRae, J.M., Falconer, R.J., Smith, P.A., Bacic, A., Waters, E.J. and Marangon, M. (2015). Wine protein haze: Mechanisms of formation and advances in prevention. *J. Agric. Food Chem.*, 63(16): 4020–4030.

Versari, A., Parpinello, G.P. and Cattaneo, M. (1999). *Leuconostoc oenos* and malolactic fermentation in wine: A review. *J. Ind. Microbiol. Biotechnol.*, 23(6): 447.

Winterhalter, P. and Skouroumounis, G.K. (1997). Glycocojugated aroma compounds: Occurrence, role and biotechnological transformation. In: *Biotechnology of Aroma Compounds, Advances in Biochemical Engineering Biotechnology*, T. Scheper (Ed.), Springer-Verlag, Berlin, Heiderverg, 55: 73.

Yanai, T. (2002). The study of glycosidases produced by yeast and the application in winemaking. Doctoral thesis (in Japanese), The University of Tokyo.

Yanai, T. and Sato, M. (1999). Isolation and properties of b-glucosidase produced by *Debaryomyces hansenii* and its application in winemaking. *Am. J. Enol. Vitic.*, 50: 231.

Yanai, T. and Sato, M. (2000). Purification and characterization of an α-L-rhamnosidase from *Pichia angusta* X349. *Biosci. Biotechnol. Biochem.*, 64(10): 2179.

Yanai, T. and Sato, M. (2000a). Purification and characterization of a novel a-L- arabinofuranosidase from *Pichia capsulata* X91. *Biosci. Biotechnol. Biochem.*, 64(6): 1181.

Yanai, T. and Sato, M. (2000b). Purification and characterization of an a-L-rhamnosidase from *Pichia angusta* X349. *Biosci. Biotechnol. Biochem.*, 64(10): 2179.

Yanai, T. and Sato, M. (2001a). Cloning of β-glucosidase from *Debaryomyces* and the construction of fused yeast cells between *Saccharomyces* and *Debaryomyces*. Jpn. Patent Kokai 2001-10781.

Yokotsuka, K., Nozaki, K. and Takayanagi, T. (1994). Characteristics of soluble glycoproteins in red wine. *Am. J. Enol. Vitic.*, 45: 410.

Young, N.M., Johnston, R.A.Z. and Richards, J.C. (1989). Purification of the α-L-rhamnosidase of *Penicillium decumbens* and characterization of two glycopeptide components. *Carbohydr. Res.*, 191(1): 53.

23 Winemaking
Control and Modeling

M. Remedios Marín-Arroyo, Arturo Esnoz, Antonio López-Gómez and César Elosúa Aguado

CONTENTS

23.1 INTRODUCTION

Winemaking involves a large number of operations but, fermentation is a "key step", especially because of its effect on wine quality and its relevant use of winery equipment including refrigeration systems. There are several reasons why the control of fermentation is very important such as the size and control of the refrigeration systems, the energy consumption management in the winery, to ensure a sustained level in wine quality every campaign and the prevention of sluggish or stuck fermentations. Nowadays, in a great number of wineries, the fermentation is still being controlled traditionally based on daily measurement of density and temperature, or in the best circumstances, it is based on the continuous monitoring and control of just fermentation temperatures. In the future, adequate control of fermentation will be strongly linked with the existence of an accurate fermentation model capable to predict in advance the development of the fermentation process and with the use of suitable sensors permitting to measure the key process variables.

The main objective of the models applicable to enology is the prediction of the kinetic behavior of the fermentation process based on the initial characteristics of the grape juice. Such a method to predict fermentation behavior before the actual fermentation would be a powerful instrument for the enologist, due to technical and economic implications, but its degree of utility would depend upon the capacity for effective prediction. For example, a model that could foresee sluggish fermentation would make it possible to intervene during the initial stages of the process to prevent this from occurring. The possibility of estimating precisely the duration of the fermentation process, at least to the point in which most of the sugar has been depleted, would allow better planning and thus, more efficient use of the resources available in the wine cellar. Similarly, the prediction of the CO_2 production rate would make it possible to control this rate and, consequently, the aroma development in wines as same is the case with respect to energy consumption management or to predict the levels of heat generated during the fermentation: this information would allow the maximum cooling requirements to be calculated for the design of an efficient refrigeration system. Unfortunately, the difficulties in establishing a globally valid model are numerous (large number of parameters involved, wide variability of must composition, poorly understood nature of the process, un-measurability of key process variables, non-linearity of the process), but the importance of the subject has driven many researchers to look for models of practical utility in the field of winemaking.

23.2 DESCRIPTION OF ALCOHOLIC FERMENTATION KINETICS

Alcoholic fermentation usually is carried out under isothermal conditions. The evolution of the main kinetic parameters of a typical isothermal fermentation process under laboratory conditions has been studied, and the fermentation process in these conditions has been well described. The process is characterized by the succession of three principal phases (Figure 23.1).

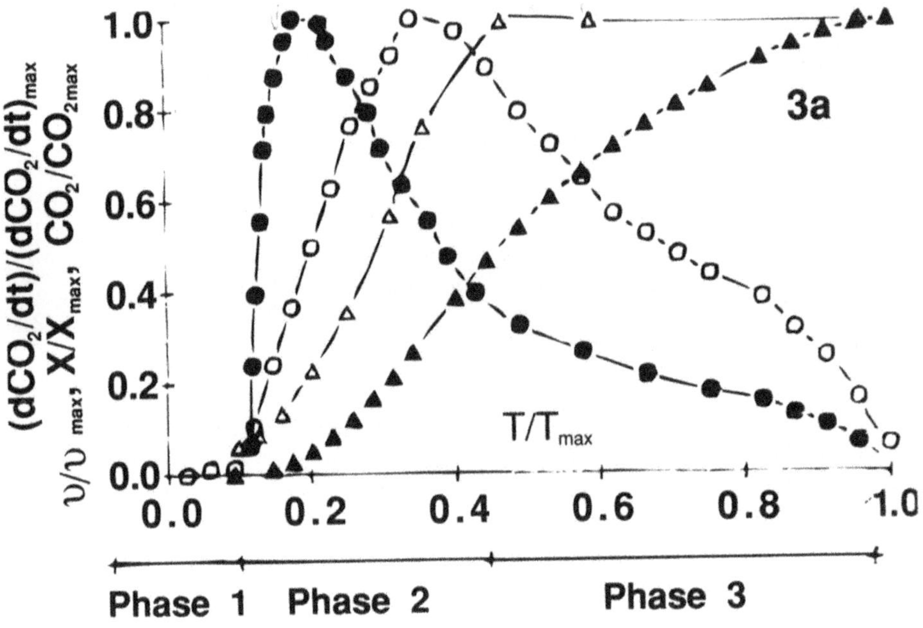

FIGURE 23.1 Description of a fermentation cycle. Evolution of different normalized kinetic parameters during the fermentation: X Δ): cell population. CO_2 (▲): CO_2 production. dCO_2/dt (○): CO_2 production rate. $\upsilon = 1/X \bullet dCO_2/dt$ (●): specific CO_2 production rate. *Source:* Copyright 1990 by the American Society of Enology and Viticulture. Am. J. Enol. Vitic. 41:319-324. Reproduced with permission

Phase 1 (lag phase) corresponds to the progressive saturation of the medium with CO_2. Phase 2 is characterized by a rapid and practically linear increase in the rate of CO_2 production. The second phase continues until the maximum yeast population is obtained. During this phase, the following kinetic parameters occur successively: (a) the maximum CO_2 specific production rate ($\upsilon_{max} = (1/X \bullet dCO_2/dt)_{max}$), and (b) the maximum CO_2 production rate ($(dCO_2/dt)_{max}$). Phase 3: during the third phase, the yeast cells are in a stationary phase and their activity decreases continuously, even when their viability remains at a high level. The shape of the CO_2 production rates curves may be quite variable but seems to be a characteristic of the pair must-yeast strain.

Significant differences exist in the kinetics of enological alcoholic fermentations between isothermal and non-isothermal conditions. But, when kinetics is known for a constant temperature, it is possible to predict the kinetics in non-isothermal conditions with good precision. These observations have practical importance since when a good prediction of the isothermal kinetics is possible, an extension to most enological temperature conditions will be possible, too.

23.3 MATHEMATICAL DESCRIPTION OF THE PRINCIPAL PHYSIOLOGICAL PHENOMENA

The development of alcoholic fermentation is directly related to yeast growth and heat generation during the process, the latter being the most important parameter necessary for an accurate estimation of the required refrigeration power. Heat generation is the direct result of the transformation reaction of the sugar contained in the medium, so the focus of many models has been on the search for mathematical equations, capable of describing this phenomenon based on a series of biologically significant parameters. Several general fermentation kinetic models have been proposed but due to the particular characteristics of alcoholic fermentation, it is necessary to use specific models. The particular characteristics for alcoholic fermentation in the food industry can be summarized in the following points: batch fermentation on natural complex media; anaerobic conditions due to CO_2 generation; inoculation of a selected strain (*Saccharomyces*); low media pH (3.0–3.6); addition of bisulfite ions (50–150 ppm) as an antioxidant and antiseptic; mixed substrate (120–250 g/L) composed of D-glucose and D-fructose; and generation of inhibitory product (80–125 g/L ethanol).

23.3.1 Yeast Growth Kinetics

The kinetics involved in the growth of a microorganism is characterized by its growth rate (μ). The cell growth rate (dX/dt) can be written as follows:

$$\frac{dX}{dt} = \mu X_v \tag{23.1}$$

where X_v represents the viable yeast cell mass and μ represents the specific growth rate of the yeast. The net increase in viable cell concentration can be described by Equation 23.2 which accounts for biomass growth and death:

$$\frac{dX_v}{dt} = \mu X_v - \kappa_d X_v \tag{23.2}$$

The growth rate depends on the concentration of nutrients in the medium. The mechanistic models developed have typically used growth expressions dependent on sugar concentration in a Monod-type relationship. The Monod equation describes the dependence of a microorganism's growth rate on the concentration of a limiting substrate:

$$\mu = \mu_m \frac{S}{K_S + S} \quad (23.3)$$

In this manner, these models have indirectly linked cell growth and ethanol production. However, there are cases where yeast cell growth is clearly complete prior to significant utilization of sugar. So, another juice part must be the growth-limiting nutrient. Even though there is evidence of the influence of the first assimilable nitrogen content in the fermentation medium on the kinetic parameters, this value has only been taken into consideration by the model proposed by Bezenger (1987), but the mechanistic expressions examined were not satisfactory for fitting their kinetic data, and therefore, an empirical model was developed instead. Recently, the nitrogen content has been taken into account, considering that the growth rate follows Monod kinetics with respect to nitrogen as below:

$$\mu = \mu_m \frac{N}{K_N + N} \quad (23.4)$$

The Monod expression is only applicable where the presence of toxic metabolic products plays no role. Alcoholic fermentation can be inhibited by high concentrations of either alcohol or substrate. Many models use the Monod equation in a modified form to take into account the presence of an inhibitor. The analytical form of the function depends on the nature of the inhibitor. In the case of ethanol, it is known that, when the specific growth rate is concerned, it behaves analogously to a non-competitive inhibitor; hence, it affects only the value of μm, but not Ks, and therefore, Equation 23.3 becomes:

$$\mu = \mu_i \frac{S}{K_S + S} \quad (23.5)$$

The principal expressions adopted in order to describe the inhibitory effect of ethanol, substrate and the effect of self-inhibition (Equations 23.6 to 23.13) are given in Table 23.1. While it is considered that inhibition due to ethanol is fully proven, there are contradictory views concerning inhibition due to substrate. In some cases, a significant inhibitory effect of the initial concentration of substrate on the growth rate was observed (the maximum growth rate decreases from 0.4 h^{-1} for $S_0 = 20$ g/L to 0.05 h^{-1} for $S_0 = 250$ g/L). Contrary to this, the data compiled in research involving discontinuous and anaerobic fermentation show that inhibition due to substrate is negligible for sucrose concentrations below 200 g/L.

TABLE 23.1
Equations Proposed to Consider the Different Types of Inhibiting Effect on the Growth Rate

Equation		Influence type	References
		Inhibiting effect of the product	
$\mu_i = (\mu_m - K_p P)$	(Eq. 23.6)	linear	Hinselwood (1946),
$\mu_i = \mu_m \left(1 - \frac{P}{P_m}\right)$	(Eq. 23.7)	exponential	Luong (1985), Holzberg *et al.* (1967),
$\mu_i = \mu_m e^{-K_p P}$	(Eq. 23.8)	hyperbolic	Godia *et al.* (1988), Ghose and Tyagi (1979)
$\mu_i = \mu_m \frac{K_p}{K_p + P}$	(Eq. 23.9)	considers a critical concentration of ethanol (Pc) above which cell growth is impeded; n represents the toxic power of ethanol.	Aiba *et al.* (1968), Boulton (1980).
$\mu_i = \mu_m \left(1 - \frac{P}{P_c}\right)^n$	(Eq. 23.10)		Novak *et al.* (1981), Godia *et al.* (1988), Boulton (1980), Caro *et al.* (1991)
$\mu_i = \mu_m \left(1 - \left(\frac{P}{P_m}\right)^\alpha\right)$	(Eq. 23.11)		Levenspiel (1980), Luong (1985)
		Inhibiting effect of the substrate	
$\mu = \mu_m \left(1 - \frac{P}{P_c}\right) \frac{S}{K_S + S + \frac{S^2}{K_{SW}}}$	(Eq. 23.12)		Ghose and Tyagi (1979)
		Self-inhibition	
$\mu = \mu_m \frac{S}{K_S + S} \frac{K_p}{K_p + P} - K_X X$	(Eq. 23.13)		Godia *et al.* (1984)

Source: Marín, R. 1999. Alcoholic fermentation modelling: Current state and perspectives. *American Journal of Enology and Viticulture*, 50(2): 166

The abilities of yeast strains to inhibit their own development like that of other strains (self-inhibitory and cross-inhibitory effects respectively) have been proven using four yeast strains used by enologists. Both the effects were seen in all cases: the self-inhibitory effect being the weakest of them but more or less pronounced according to the type of yeast strains. Similarly, there is a decrease in the specific growth rate in relation to the increase in the quantity of inoculum.

The possibility of adapting Equations 23.6, 23.8, 23.9, 23.10 and 23.13 (Table 23.1) has been studied in order to describe different fermentation processes and reasonably good adjustments for all of them have been obtained, although with fairly incongruous parameter values in most of the cases. Therefore, apart from Equation 23.13, the same parameter may have very different optimum values when the same equation is adjusted to the data of different experiments. Besides, if one of the parameters is fixed at a reasonable value, in such a way that the rest reach more congruous values, the adjustments are worse. Equation 23.13 is the only one that yields a lower dispersion for the values of the model parameters when it is adjusted to the data from differing experiences, although the values of the constants lack physical sense and μm is increased to a larger than real value. Nevertheless, this equation is more appropriate to describe the results of fermentation at different temperatures and with varying initial sugar concentration levels, then the expressions that only take into account the inhibitory effect of the product.

The effect of pH levels on biomass growth has been considered in very few cases. The following expression takes this effect into account in models that consider growth limitation due to substrate concentration (sucrose in the performed study), and the following has been proposed:

$$\frac{dX}{dt} = \frac{k_1XS}{k_2\left(1 + \frac{k_H}{k_{-H}}\left[H^+\right] + \frac{k_{OH}}{k_{-OH}}\frac{\left[OH^-\right]}{\left[H^+\right]}\right) + S} \quad (23.14)$$

where k_H, k_{-H}, k_{OH} and k_{-OH} are the velocity constants of the reversible reactions:

$$S + \left[H^+\right] \leftrightarrow S - H^+$$

$$S + \left[OH^+\right] \leftrightarrow S - OH^-$$

Although descriptions have been given as to the effect of temperature and must conditions on the fermentation rate and of the inhibitory effect on the metabolism of yeast of other substances produced by the fermentation, such as carbon dioxide, ethyl acetate, isopropanol, propanol, and butanol, octanoic and decanoic acids, this effect has not been considered in fermentation models, since the inhibitory effects of products other than ethanol are normally considered negligible. Alcohol also influences yeast viability. Viability of wine yeast stored at various ethanol concentrations was reduced, and a decrease took place almost linearly with time for a strain of Brewer's yeast. The effect of ethanol on yeast viability has been incorporated into the models in the form of specific expressions for cell death (see Equation 23.15).

$$\kappa_d = \kappa_d' E \quad (23.15)$$

Batch fermentation of ethanol by *Saccharomyces cerevisiae* revealed that lower culture temperatures caused slower growth as well as ethanol production, however, the final cell mass and ethanol concentrations reached levels which were higher than those for higher culture temperatures. To explain these phenomena, a kinetic model based on enzyme deactivation kinetic has been proposed. In the model, growth and ethanol production were expressed as a function of the "integrated ethanol concentration" (Equations 23.12 and 23.13 in Table 23.1). However, the equation that describes the cell mass concentration in the model proposed was similar to Gompertz's curve for population dynamics, in which the final cell mass concentration approaches a finite value, contrary to what happens in equations derived from the Michaelis-Menten enzymatic kinetic models described earlier.

23.3.2 Substrate Consumption Kinetics

The most general equation describing the kinetics of substrate consumption takes into account the consumption of substrate in the formation of biomass, ethanol, and the maintenance of biomass, this last term not being well-defined:

$$\frac{dS}{dt} = -\frac{1}{Y_{X/S}}\frac{dX}{dt} - \frac{1}{Y_{E/S}}\frac{dP}{dt} - mX \quad (23.16)$$

where the coefficient Yx/s represents grams of biomass obtained for every gram of fermented sugar and $Y_{E/S}$ represents grams of ethanol obtained for every gram of fermented sugar. The substrate consumed to maintain the metabolic activity of yeast when cell growth ceases justifies the inclusion of the maintenance term in Equation 23.16. Nonetheless, during the cell growth phase, this term is negligibly facing a specific growth rate.

As far as the global yield of ethanol production based on sugars consumed is concerned, some models make use of the stoichiometrical relationship, which results from the consideration that only a metabolic means is involved in the fermentation ($Y_{E/S}$ = 0.510). Similar mechanistic models have been used to describe sugar use, but modifications in the kinetic equations to consider factors which reduce ethanol yield apart from evaporation have been introduced (*i.e.*, mainly respiration). Accordingly, the percentage of sugar transformed by each process depends on the temperature; under normal temperature and enological conditions; the yield of sugar in ethanol is between 91% and 94% of the theoretical value. However, performance levels of 0.434 ± 0.026, depending on the initial sugar content, have been documented, which means that in the most unfavorable case, performance is 85% of the theoretical value, and the remaining percentage is attributed to the utilization of a part of the sugar content for biomass formation, secondary metabolite production and maintenance, essentially glycerol. Yield levels decrease as the temperature is increased, though they are not significantly affected by the quantity of inoculum. Yield data as a function of the initial sugar concentration for two types of yeast (*Saccharomyces cerevisiae* and *Saccharomyces bayanus*) is offered, wherein values of 0.455 and 0.420 respectively for sugar concentrations of 100 g/L, with a pronounced decrease for concentrations of 220 g/L (0.260 and 0.340 respectively) was recorded. In other cases, the yield varies between 0.463 and 0.485 when the initial sugar concentration is 260 g/L. In the rest of the studies, the yield figures are between 0.350 and 0.490.

23.3.3 Product Formation Kinetics

Alcohol formation is directly related to sugar consumption:

$$\frac{dP}{dt} = a\left(-\frac{dS}{dt}\right) \tag{23.17}$$

and therefore, through Equation 23.16, with the yeast cell population and its growth. The kinetics of product formation have been described, in a general manner, by the Luedeking and Piret equation, originally used to simulate lactic acid production by *Lactobacillus delbrueckii*:

$$\frac{dP}{dt} = \alpha\frac{dX}{dt} + \beta X \tag{23.18}$$

where α and β are constants.

The first term in Equation 23.18 describes the product formation rate when it is coupled with growth, whereas the second term describes the product formation rate when it is totally independent of growth. The application of this equation to explain the kinetics of alcoholic fermentation has been debated at length tough, it is not fully verified for batch alcoholic fermentation. It has only been verified in the case of alcoholic fermentation for sugar content levels of less than 20 g/L, with *Saccharomyces cerevisiae* and is valid only for pure culture fermentation.

In some cases, equations similar to the Luedeking and Piret Equation (eq. 23.18) have been put forward but with α and β as variables: this expression was modified by considering α a variable since the microorganisms dominant in the fermentation medium varies as the fermentation process advances, and alcohol-tolerant microorganisms tend to predominate. The variation of α is related to the concentration of ethanol in the medium as expressed in the following equation:

$$\alpha = \alpha_0 + \alpha_1 P + \alpha_2 P^2 \tag{23.19}$$

Where α_0, α_1, and α_2 are constants whose numerical values depend on the microorganism contributing to the total population. Nevertheless, there are more descriptions of product formation rate by means of equations with only the non-growth-associated term in which the proportionality is variable:

$$\frac{dP}{dt} = \nu X \tag{23.20}$$

TABLE 23.2
Equations Proposed to Describe Specific Growth Rate and Specific Ethanol Productivity in the Different Models

Equations		References
$\mu = \mu_0 e^{-k_1 P}$	(Eq 23.21)	Aiba *et al.* (1968)
$\nu = \nu_0 e^{-k_2 P}$	(Eq 23.22)	
$\mu = \mu_0\left[1-\left(\frac{P}{P_m}\right)^{\alpha}\right]\frac{S}{K_S+S}$	(Eq 23.23)	Luong (1985)
$\nu = \nu_0\left[1-\left(\frac{P}{P_m'}\right)^{\beta}\right]\frac{S}{K_S'+S}$	(Eq 23.24)	
$\mu = \mu_0 e^{-k'\int_0^t P^n dt}$	(Eq 23.25)	Nanba *et al.* (1987)
$\nu = \alpha_0\mu + \beta_0 e^{-k''\int_0^t P^n dt}$	(Eq 23.26)	
$\mu = \mu_0 \frac{S}{(K_S+S)\left(1+\frac{P}{K_{EI}}\right)}$	(Eq 23.27.1)	Boulton (1979) Boulton (1980)
$\mu = \mu_0\left(\frac{G}{K_{SG}+G}+\frac{F}{K_{SF}+F}\right)e^{-0.05P}$	(Eq 23.27.2)	
$\nu = \frac{Y_{E/S}}{Y_m}\mu + mY_{E/S}$	(Eq 23.28)(*)	
$\mu = \mu_g + \mu_d = \mu_0\frac{S}{S+K_S\left(1+\frac{P}{K_1}\right)} - \frac{P}{K_0}$	(Eq 23.29)	Caro *et al.* (1991)
$\nu = f\frac{Y_{E/S}}{Y_g}\mu_g + fmY_{E/S}$	(Eq 23.30)(*)	
$\mu = \mu_m\left(1-\frac{P}{P_m}\right)$	(Eq 23.31)	Ghose and Tyagi (1979)
$\nu = \nu_m\left(1-\frac{P}{P_m'}\right)$	(Eq 23.32)	

(*)Equation that does not appear explicitly in the reference sources.

Source: Marín, R. 1999. Alcoholic fermentation modelling: Current state and perspectives. *American Journal of Enology and Viticulture*, 50(2): 166

where ν represents the specific rate of production.

It has been observed that ethanol production starts just prior to the cessation of exponential cell growth. Table 23.2 (Equations 23.21 to 23.32) is a compilation of the main expressions that describe μ and ν in the different models. Some equations do not appear explicitly as such in the reference sources, although their deduction is directly based on the explicit equations and on the definition of the production rate. Within the exponential growth phase, the rate of the ethanol production stays practically constant and independent of the cellular growth. Since dP/dt is essentially constant, the equations used by Luedeking and co-workers, in which rate of product formation is proportional to both growth rate and cells, do not apply, and therefore, it was proposed to apply the following equation during this exponential growth phase:

$$\frac{dP}{dt} + BP = A \ln \frac{N}{\mu} - C \qquad (23.33)$$

where N is the number of cells per milliliter, A is the proportionality constant between the rate of the ethanol production (modified by BP to take into account the inhibitory effect of the ethanol) and the yeast concentration during the growth phase, B is a proportionality constant and C is a constant which takes into account the delay between the initiation of the cellular growth and the production of ethanol. The following equation has been proposed for the stationary phase, in which the yeast population is constant:

$$\frac{dP}{dt} = B(P_m - P) \qquad (23.34)$$

where P_m is a stoichiometric constant whose value is related to the first substrate concentration. Equation 23.34 shows how the decrease in the production rate is due to the accumulation of ethanol.

23.4 NON-PHYSIOLOGICAL MATHEMATICAL DESCRIPTIONS

Various strategies can be followed to propose models wherein it is unnecessary to give a mathematical description of physiological aspects of the process. The use of a semi-empirical model in which the velocity of sugar consumption is described by a chemical law that depends on substrate and product content is proposed as expressed in the following equation:

$$\frac{dS}{dt} = kS^{\alpha}P^{\beta} \qquad (23.35)$$

where k is a constant to be determined and α and β are pseudo-orders of the reaction. The parameters of the model are adjusted by means of non-linear programming methods, which compare model predictions with experimental data and minimizing errors.

The prediction capacity of the model described by Equation 23.35 has been tested under enological conditions over a period of two vintages and with varying conditions. With a maximum of four measurements after the latency phase, it is capable of predicting the fermented sugar (and thus, thermal planning) within an error of 3.3%. However, this model needs the identification of the parameters (variables depending on the conditions of the medium and variables of the process) throughout the fermentation process. Within the scope of prediction, it is only interesting if it is possible to identify these factors early enough during the fermentation process, which in practice is rarely feasible.

Faced with the difficulty of describing the entire fermentation process by means of only one simple equation, Bezenger has proposed two second degree polynomials which describe the beginning and end of the fermentation process respectively:

$$M = a_0(t - t_0)^2 + b_0(t - t_0) + c_0$$
$$M = a_f(t - t_f)^2 + b_f(t - t_f) + c_f \qquad (23.36)$$

where:

M = loss of mass per liter of medium (as a result of the emission of CO_2 gas)
t_0 = time in which M = 1.5 g/L
tf = time in which M = 50 g/L

The coefficients of the regressive polynomials are a function of the temperature and the initial nitrogen concentration, as well as a linear relationship between a coefficient for the beginning of the fermentation process and a corresponding coefficient for the end of the process, thus fostering the possibility of a prediction model for the end of the reaction as a function of the data compiled during the first few hours of the process. It is also known that there is a linear relationship between the coefficient a_0 and the maximum velocity of mass loss (V_{max}) for the same fermentation, given by the following equation:

$$V_{max} = 6.54a_0 + 0.78 \qquad (r = 0.975) \qquad (23.37)$$

This model makes possible, under certain conditions, the prediction of the maximum fermentation rate based on the initial acceleration figure (eq. 23.37). So, the value of the initial acceleration may yield information about the profile of the end of the reaction and an indication of the accident risk involved in the fermentation. The value of the initial acceleration is related to the maximum reaction velocity, and the maximum rate of fermentation is directly related to the maximum rate of heat loss: it is why the value of the initial acceleration can also indicate a criterion to predict an unexpected increase in temperature within the fermentation recipient. It has to be experimentally confirmed whether the equation parameters are generalized or depend on characteristics of must.

Two modeling trials were performed: the first one consists on the search for polynomial models to predict the time of onset of gas released (T_d), the acceleration of CO_2 production (k_1), the maximal rate of CO_2 production (v_{max}) and the slope of the logarithm of the rate of CO_2 released during the deceleration phase (k_2) as a function of the initial fermentation conditions (fermentation temperature and initial grape juice-sugar concentration). The established models have presented mean relative errors between 20 and 50%. The other modeling strategy consists of polynomial models for the in-line prediction of v_{max} and k_2 as a function of the fermentation temperature and the CO_2 production acceleration, calculated five hours after the onset of gas production (T_d + 5) when the precision of the polynomial models obtained were ± 5.8% for v_{max} and ± 10.4% for k_2.

Based on research involving the estimation of the development of heat production throughout the fermentation process, a mathematical empirical model is proposed, which is successful in describing what occurs under test conditions and at constant temperature:

$$\frac{dQ}{dt} = \frac{K_1}{K_1 - K_2}\left(e^{-K_2 t} - e^{-K_1 t}\right) \quad (23.38)$$

where K_1 and K_2 vary with the temperature according to the following expressions:

$$K_1 = K_{10}\exp\left[-\left(\frac{E_{a1}}{RT}\right) + K_{11}A_t\right] \quad (23.39)$$

$$K_2 = K_{20}\exp\left(-\frac{E_{a2}}{RT}\right) \qquad K_{20} = \frac{k}{S_0} \quad (23.40)$$

23.5 NON-PHYSIOLOGICAL DESCRIPTION BY MEANS OF ARTIFICIAL INTELLIGENCE METHODS

The non-linearities and the changing characteristics of the fermentation process makes it difficult to control the process based on a deterministic model. A different approach is to consider the process to be a "black box" and to use heuristic modeling methods (Group Method of Data Handling – GMDH or Neural Networks) which can be used to obtain unstructured models. The GMDH was proposed in order to establish polynomial behavior prediction models of the basic magnitudes involved in the winemaking process (the volume of CO_2 released during the reaction and the alcoholic fermentation rate). In artificial neural networks, information is stored in the form of weights between artificial neurons. The strategy in using neural networks to solve a given problem, is to choose a network architecture and a learning process whereby representative examples of the knowledge to be acquired are shown to the network which then self-organizes within its structure by adjusting accordingly the synaptic strength of neural connections. Learning usually requires showing a great set of examples to the network. Artificial neural networks systems have the ability to adapt their parameters to match input-output mapping rules of unmodeled non-linear systems. These modeling techniques have recently been applied in the field of winemaking: the neural networks have demonstrated an improved and faster estimation capacity for biological activity than extended Kalman filters as well as to predict sluggish and stuck fermentations. Neural networks have been used for the simulation of alcoholic fermentation, to predict the acidity of Champagne, follow the growth of yeast in a wine-base medium and the yeast fermentation kinetics, as well as chemical and sensory properties of the finished wine, based solely on the properties of the grapes and the intended processing.

The models obtained by means of neural networks seem to be more suited to fermentation modeling, although models based on Group Method of Data Handling seem to be more appropriate and versatile when applied to different types of fermentation. The limited prediction horizon, however, makes both of the methods, in practice, less useful for prediction purposes involving, for example, cooling requirements.

23.6 THE INFLUENCE OF TEMPERATURE

Most of the models have been conceived for isothermal conditions, although some of them have taken into account the effect of temperature variation on the different parameters that determine the development of the fermentation. Since the temperature influences the yeast's metabolic activity, the growth, and maintenance terms must be expressed as functions of the juice temperature. The rise of temperature has a positive effect on the specific growth factor; then, the fermentation rate increases with increasing temperature. However, after a certain value, it induces less favorable yeast growth conditions. At this point, both growth and maintenance cease and "stuck" fermentation results.

The maximum specific growth rate can be expressed as a combination of two terms. The first, an exponential increase in growth with temperature, and the second an exponential decrease in growth (increase in denaturation or death) with temperature:

$$\mu_m = a_1\exp\frac{E_{a1}}{RT}\frac{(T - T_0)}{T_0} - a_2\exp\frac{E_{a2}}{RT}\frac{(T - T_0)}{T_0} \quad (23.41)$$

Besides its influence upon the yeast specific growth factor, the temperature also influences the maintenance coefficient and the saturation constants. The equations proposed by Boulton in order to describe this influence are:

$$m = m_0\exp\frac{-E_{a3}}{RT}\frac{(T - T_0)}{T_0} \quad (23.42)$$

$$\frac{1}{K_S'} = \frac{1}{K_{S0}'}\exp\frac{-E_{a4}}{RT}\frac{(T - T_0)}{T_0} \quad (23.43)$$

The variation of parameter *m* with temperature has an effect opposite to the variation of the specific growth factor with temperature on the rate of sugar consumption. However, for typical values, the effect of the specific growth factor is the most relevant, and the rate of sugar consumption is enhanced with temperature in the range of recommended temperatures for wine fermentation. Equations have been proposed that take into account the variation of the parameters according to the temperature, considering that the substrate's yield coefficients (Ys) vary with temperature in the same way as with the maintenance rate and the saturation constants. The equations are similar to those proposed by Boulton and confirm that the coefficients of Equation 23.42 are largely dependent on the species of micro-organism used.

It is considered that the parameters of their model (deactivation constant of the specific growth rate (k) and the specific growth rate) within the range of physiological temperatures vary as a function of temperature according to the following Arrhenius expressions:

$$\mu_0 = A_1 e^{\frac{-E_1}{RT}} \quad (23.44)$$

$$k = A_2 e^{\frac{-E_2}{RT}} \quad (23.45)$$

23.7 THE CONTROL OF THE FERMENTATION PROCESS IN WINEMAKING

The correct control of the fermentation process is critical in winemaking because it determines the aromatic quality of wine and other sensory attributes. During fermentation, glucose and fructose are metabolized into ethanol, producing CO_2 as a sub-product; therefore, the most relevant parameters that can be used to monitor the process are CO_2, pH, density variations, optical properties of the must (refractive index) and volatile organic compounds, among others; moreover, a key factor is temperature because the quality of the process depends critically on it. As an example, fermentation stops if the temperature is too high, so that it is kept at a value below 20 °C in the case of white wines, whereas for red ones it is set between 30 and 32 °C. In this context, the temperature has to be controlled by refrigeration systems whose efficient use could reduce more than 50% of total energy consumption in a winery. Nevertheless, today the fermentation process control is limited due to the lack of efficient, low-cost sensors and the available measure devices. Furthermore, most of the magnitudes are traditionally measured off-line: a sample from each tank has to be taken to the laboratory and be analyzed there. In the case of wineries with tens or even hundreds of fermentation tanks, this kind of procedure is inefficient, which constitutes the main problem to develop fermentation control systems.

The measuring systems used in bioreactors could be classified by the kind of variable used (physical, chemical or biological) or by the measuring principle. Between all the parameters that could be registered in practice, many fermentation control proposals carried out are based on the measurement of CO_2. A simple linear correlation has been established, which allows, using the measurement of released CO_2 or the must density, the concentration of ethanol and sugars during must fermentation to be determined in real time and with high precision. This gas can be measured directly by specific sensors or indirectly by registering pressure changes inside the tank or even weight changes produced by the mass loss from sugar metabolization: in any case, controlling the CO_2 concentration would allow knowing when the process is over and therefore, estimating if it has been performed during the proper time. Despite having available equipment to measure these variables at an industrial level, the fermentation process is currently carried out with simple control systems fitted, and it is supervised manually (for instance, temperature control).

Off-line methods to measure biomass are not adequate for on-line monitoring. Recent working lines in the sensors field are based on the optical and dielectric properties of the cellular suspensions. The components of the mean (particles in suspension, gas bubble) and of the process (stirring) could interfere with the optical measures. The on-line measures of substrates and fermentation products in a continuous bioreactor lacks adequate sensors too. Its development is limited, in part, by the necessity of robust on-line instrumentation capable to operate in industrial ambient and requiring minimum maintenance. The measures on-line without the necessity of taking samples from the tanks that have to be analyzed in the lab are being studied.

The traditional measuring of biological magnitudes shows the main drawback of offline monitoring, so other physical-chemical properties have been found relevant for fermentation evaluation. Density changes have been measured by aerometers because there is a clear relationship between this magnitude and the evolution of the fermentation process; however, density is typically measured offline in a manner that a sample has to be taken from each tank. There are other possibilities to estimate it based on pressure changes inside the tank that would allow online monitoring. Another approach slightly related to density changes is the measurement of the must's refractive index. It has been reported that its value is modified during the fermentation following a certain trending: it is decreased along the process. This optical property can be evaluated online and in situ by sensors made of optical fibre, which is one of the solutions for fermentation control that shows more potential.

Regarding biological magnitudes, there are three parameters which are essential to follow and to control the fermentation process, although they are not particularly available in real-time: concentration of microorganisms (and their biological activity), substrates and of metabolites. During the last years, a big explosion of devices called biosensor and soft-sensor has occurred. On one hand, it involves the use of biochemical reactions for measurement purposes, and on the other hand, it consists of sensors focused on the on-line software-based state estimators (software sensors). All of them provide batch and an on-line estimate of a large number of variables of wine production. An example of glucose estimation biosensor is illustrated in Figure 23.2.

Most of the biosensors evaluate the ethanol content, which is faster, more selective, and sensitive, than traditional methods;

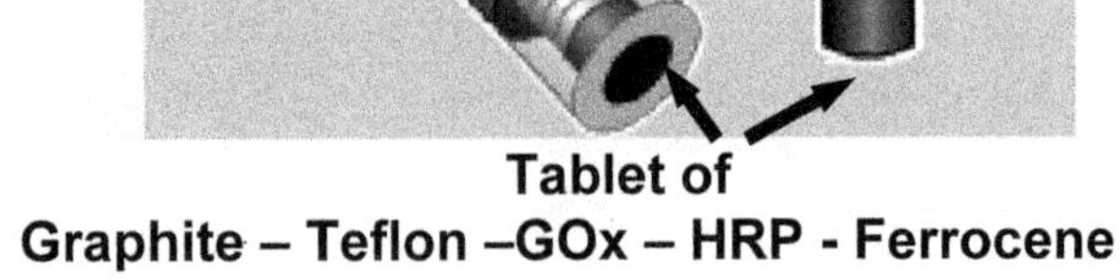

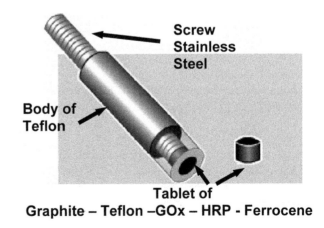

FIGURE 23.2 Electrochemical Glucose Sensor. *Source:* Courtesy of Det. of Analytical Chemistry of *Universidad Complutense de Madrid* (Spain)

biosensors also simultaneously determine other metabolites such as methanol, glucose, fructose, L-malate, lactate, glycerol, sulfite, acetaldehyde, or acetic acid. However, there is a great difference between the know-how of the researchers and the industrial practice. The industrial application of the technical advances obtained in laboratory or pilot plant by the researchers is not often possible because the industrial innovation has to deal with a sector where there is a traditional opposition to innovation. Moreover, the economic impact of the innovation in a short time has a difficult evaluation. There are other issues, such as the use of the same instrumentation employed by researchers in the laboratory or pilot plants (high accuracy and automated performance) in industrial installations, where typically, instrumentation as simple as possible is used to limit the problems of calibration and maintenance.

As the determination of relevant procedure variables have not yet been done at low cost in alcoholic fermentation control, the process' modeling is establishing a link between available data and the variables which are not measured (because it is not possible at the present time or it may be a lot more expensive when compared to the current instrumentation). In this manner, a bio-process dynamic model is used normally to access a number of measured variables through parameters easy to measure. These models could have an interest in the analysis of problems of procedures development or to control the process. Although the work has reviewed fermentation process modeling, the absence of industrial applications is remarkable, due probably to the excessive complexity of the approach used.

23.8 FERMENTATION CONTROL TODAY

The traditional control based on daily measurement of density and temperature is being progressively replaced by continuous monitoring of the process and by automatic control of fermentation temperatures. Actually, the alcoholic fermentation control today in wineries is based on the temperature. The central computers installed in wineries, which could control from 60 to 90 tanks, monitor and control the temperatures in the tanks. The use of the fermentation rate to automatically control the fermentation process still continues without a real industrial application. The brewing industry produces diverse kinds of beers using large outdoor tanks (200 to 1000 m^3). Actually, they are automatically controlled by two main parameters: (1) the fermentation temperature, measured at different levels (two or three) in the tanks, which is controlled by cold water applied in several jackets (from two to four levels); and (2) the pressure in the tanks, which is controlled too, actuating on released CO_2 flow. Low temperatures are used (from 10 to 18 ºC) during fermentation so that the process is controlled by temperature evolution profiles, fitted and related to the kind of beer and the know-how of each brewery. Efforts are being made in improving the process control, following the beer quality evolution during fermentation.

There are several approaches to control temperature, and among them, electronic controllers have been designed through the Programmable Logic Controller (PLC) and microcontroller based on temperature control systems. The PLC and microcontroller offer by far the greatest flexibility in their application. Available features include temperature trending graphs, comprehensive alarming and fault diagnostics, energy-saving systems on cooling, and remote access via LAN. The cost of these systems is also very cheap, which means that systems-based temperature control has been implemented in a large number of wineries.

Approaches based on optical fibre sensors have been also proposed. There is a technology that allows the temperature to be measured in a distributed way: more specifically, the spatial resolution achieved is below one cm, taking advantage of the Brillioun effect. In this manner, optical fibre could be located inside the fermentation tanks attached to the walls, so that the temperature could be monitored at different points in real-time and, what is more, several tanks could be monitored with the same fibre simultaneously. This technology would allow the tank to be refrigerated at different points or areas, which is already performed with electronic sensors although just three of them are placed at different heights inside the tank.

Following on with optoelectronic systems, the refractive index is a parameter to evaluate and control fermentation. As exposed in the previous section, the refractive index is decreased along the fermentation process, and it shows a relationship with other already used parameters, such as the density or sugar concentration. Some prototypes based on refractometry have been developed: they are based on low-cost LEDs and photodetectors; however, some of them do not operate in situ, which is an important drawback. On the contrary, other proposals are based on optical fibre, which allows the refractive index to be measured in real-time: discrete sensors to monitor this parameter could be combined with the distributed temperature systems previously mentioned. Optical fibre refractometers are based on metallic nanocoatings deposited onto the fibre and which are affected by variations in the surrounding refractive index media: the transduction takes place when the light traveling through the fibre is modified by this phenomenon, and the variation is registered by optical detectors.

Some approaches are based on monitoring the fructose/glucose concentration to determine when the alcoholic concentration is over: as the first sugar is an isomer of the second one, it is a real challenge to measure them individually, and even more on real-time. There are electronic sensors that use enzymes which are highly selective to one of the sugars, but the main drawback is that they are not reversible. Following the idea of finding materials with high selectivity to one of the sugars, some molecularly imprinted polymers (MIPs) have been developed. These synthetic molecules are prepared to form a molecular template that holds the target molecule by non-covalent links: therefore, the molecule can be removed just by chemical solvents or even water, allowing the reversible use of the material. The optical or electrical properties of the MIPs depend on the presence of the target molecule so that it can be used to develop electronic or optical fibre sensors. Some prototypes have been already developed and tested at the laboratory.

The detection and control of the volatile organic compounds (VOCs) produced during the process can be also used to determine if fermentation is being performed properly. Initially, the main volatile compound is water (must); as the yeast converts the sugars into ethanol, this compound's concentration is increased. In the case that some VOCs were detected above a certain concentration, the process should be corrected, or, in the worst situation, the must be disposed of. The detection of acetic acid or methanol would be critical in these situations. Although these VOCs are the most relevant ones, there are hundreds of them that determine the final aroma of wine, and it is difficult to find materials with a high selectivity to these volatile molecules. Therefore, it is better to work with several sensors with low selectivity and analyze their responses jointly, which mimics the mammal's olfactory system. As in the case of MIPs, there are materials sensitive to VOCs that suffer reversible changes in their optical and/or electrical properties, so different kind of sensors based on distinct transduction have been developed so far.

Other advanced methods used currently to determine fermentation kinetics involves the on-line measurement of CO_2 production or density changes. Besides this, using fermentation rate evaluation, it is possible to determine in real-time the ethanol and sugar concentrations in must. This represents considerable progress since it has allowed the establishment of the control laws depending on the kind of wine to make. There are several electronic commercially available methods to control CO_2 concentration, although the conditions inside fermentation tanks where hundreds of VOCs are present make necessary the combined use of other sensors to lower cross-correlation problems.

Recent proposals are not focused on a specific parameter but on several of them: in other words, the most promising systems are based on multi-sensors. Not only physical magnitudes (such a temperature) are to be measured simultaneously with chemical or biological ones (pH, VOCs, CO_2, sugars, etc.), but are also employing different technologies (for instance, electronic sensors with optical fibre ones). The big amount of data is to be processed by data mining techniques: among them, the ones that are often used are Principal Component Analysis (PCA) and Artificial Neural Networks (ANNs). PCA is a linear transformation of multidimensional spaces into three or two ones, which allows patterns to be identified. Conversely (as described in Section 23.5), ANN are algorithms based on a nonlinear transformation that learns to identify patterns by a training process. Both approaches can be combined to perform predictive controllers to detect problems during the fermentation process in advance, optimizing the process.

23.9 A NEW CONTROL SYSTEM: THE FUZZY CONTROL IN WINEMAKING

In spite of works carried out during the last 15 years and the advances made in on-line control and fermentation process monitoring, the fermentation process is still controlled (at most of the wineries) in isothermal conditions and using simple control systems fitted and supervised at hand, because fermentation temperature is a parameter easy to control. The main reason is that the offered control systems are too sophisticated, expensive, and few are reliable.

Controlling the temperature and fermentation rate at the same time is generating considerable interest. A procedure to control fermentation rate to avoid fermentation stop at the end of the process has been proposed. As the fermentation rate and its process kinetics determine the cold consumption during fermentation, it is possible to act on the fermentation kinetics, satisfying certain limits of maximum and minimum temperature, to achieve maximum celerity of process and a minimum energy consumption (with rational cold consumption and production in the winery). On this hypothesis, a control system prototype based on fuzzy logic was developed in our laboratory (Public University of Navarra, Spain) to use in alcoholic fermentation process control. This control system is based on a thermal and kinetics fermentation process model. The fermentation temperature and CO_2 production rate, as an indicator of fermentation rate, are the observed parameters. The chosen controller was a double-input single-output fuzzy controller, which uses CO_2 production rate and temperature measurements as inputs and the refrigeration action as an output (Figure 23.3).

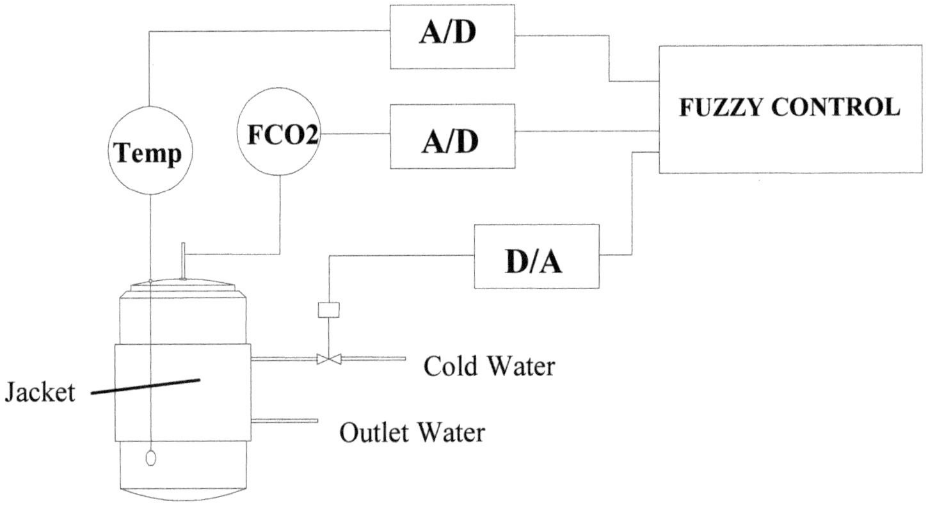

FIGURE 23.3 Fuzzy control system diagram with *Temp* (Temperature Sensor) and *FCO_2* (CO_2 flowmeter). *Source:* Martínez *et al.* (1999)

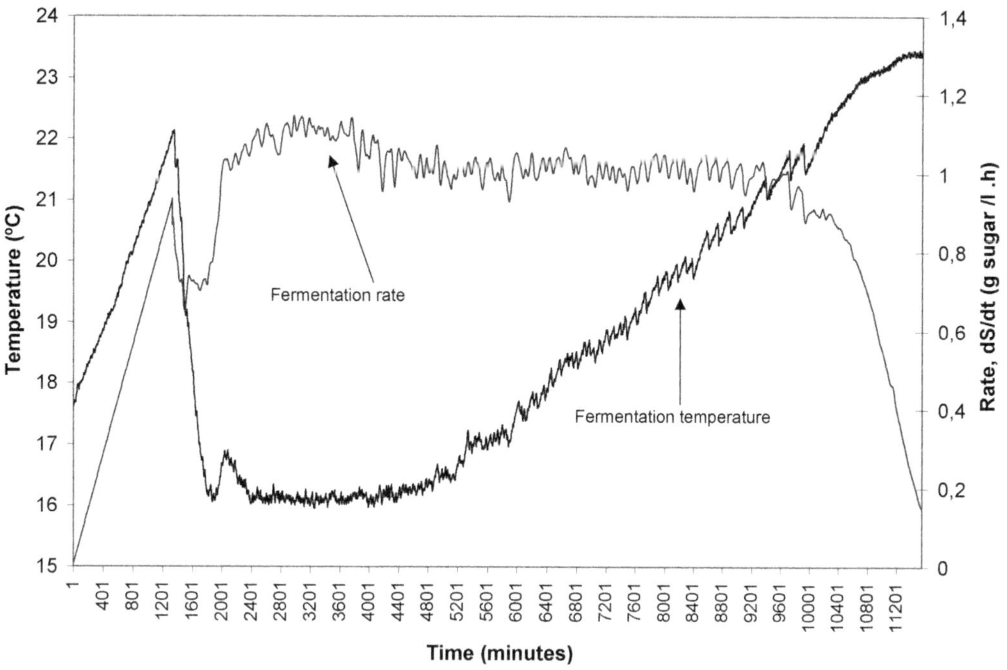

FIGURE 23.4 Fermentation evolution with fuzzy control system

The control objective was to maintain a fermentation rate always lower than a certain level since the higher the fermentation rate is, the higher the flavor loss is (Figure 23.4). First, the control system has been developed using a computer fermentation simulation and control system simulation actuating on a simulated fermentation tank. Thereafter, as a result of validation experiments done at the pilot plant and an industrial level, it was felt there was a need to complete the knowledge about certain parameters of the thermal and kinetics model to fine-tuning of the controller. Nevertheless, fermentation process time and energy saving were observed with the application of the control system developed. Also, the wine quality obtained with this kind of control system was found similar to that of the wine from traditional winemaking control systems (with constant fermentation temperature). The aim was to offer to the wine industry a new control system, advanced and at low cost, allowing efficient use of cold and fermentation tanks, using the temperature and CO_2 flow as real observed variables.

NOMENCLATURE

a_1, a_2: empirical constants in Equation 23.41
f: fermentative conversion coefficient in Caro *et al.* model (dimensionless)
k: decay constant of kinetic parameters in Nanba *et al.* model (h^{-1})
k',k'': decay constants corresponding to μ and β, respectively, in Nanba *et al.* model (L/g/h)
k_1, k_2: empirical exponents in Equations 23.21 and 23.22 (L/g)
m: specific maintenance requirement (h^{-1})
m_0: reference value of m (h^{-1})
t: fermentation time (h)
w: degree of substrate inhibition for growth
A_1, A_2: frequency coefficients in Equations 23.44 to 23.45 (h^{-1} or L/g/h)
A_t: total acidity in initial must (g/L as tartaric acid)
C: specific heat of the fermentation broth (kcal/kg/°C)
C_p: heat capacity (kcal/kg/°C)
E: activation energy in Equations 23.44 to 23.45 (kcal/mol)
E_a: activation energy (kcal/mol)
F: fructose concentration (g/L)
G: glucose concentration (g/L)
N: total nitrogen concentration (g/L)
K_{EI}: ethanol inhibition constant (g/L)
K_P: product inhibition constant (L/g)
K_S: saturation constant for growth (g/L)
K_N: Monod constant for nitrogen (g/L)
K'_S: saturation constant for product formation (g/L)
K'_S: saturation constant for the sugar transfer reaction (g/L)
K'_{S0}: reference value of K'_S (g/L)
K_{SF}: lumped substrate constant for fructose (g/L)
K_{SG}: lumped substrate constant for glucose (g/L)
K_X: self-inhibition constant (h^{-1})
P: product (ethanol) concentration (g/L)
P_m: product (ethanol) concentration above which cells do not grow (g/L)
P'_m: product (ethanol) concentration above which cells do not produce ethanol (g/L)
P^n: ethanol concentration raised to the degree of inhibition
R: universal gas constant (kcal/°C/mol)
S: substrate (sugar) concentration (g/L)
S_0: initial substrate (sugar) concentration (g/L)
T_0: initial temperature (°K)
X: biomass concentration (g/L)

X_v: viable yeast biomass concentration (g/L)
Y_{cal}: metabolic heat generation yield of the microorganism (OD_{650} $kcal^{-1}$)
Y_g: substrate yield for growth in Caro *et al.* model (dimensionless)
Y_m: maximum growth yield
$Y_{x/s}$: Yield of biomass on substrate (dimensionless)
$Y_{E/s}$: Yield of product (ethanol) on substrate (dimensionless)
α: constant defined in Equation 23.11 (dimensionless)
α_0: growth associated constant in Nanba *et al.* model
β: constant defined in Equation 23.25 (dimensionless)
β_0: non-growth associated constant in Nanba *et al.* model
κ: specific death rate (h^{-1})
μ: specific growth rate (h^{-1})
μ_0: initial value of μ (h^{-1})
μ_d: specific death rate (h^{-1}) in Caro *et al.* model
μ_g: specific growth rate (h^{-1}) in Caro *et al.* model
μ_i: specific growth rate considering inhibiting effect (h^{-1})
μ_m: maximum specific growth rate (h^{-1})
ν: specific ethanol production rate (h^{-1})
ν_m: maximum specific ethanol production rate (h^{-1})
ν_0: specific ethanol production rate at zero ethanol concentration (h^{-1})
ρ: density of the fermentation broth (kg/m^3)
dP/dt: rate of product (ethanol) formation (g/L/h)
dQ/dt: rate of heat generation (kcal/L/h)
dS/dt: rate of sugar consumption (g/L/h)
dT/dt: rate of change of medium temperature (deg/h)
dX/dt: cell growth rate (g/L/h)

BIBLIOGRAPHY

Aiba, S., Shoda, M. and Nagatani, M. (1968). Kinetics of product inhibition in alcohol fermentation. *Biotechnology and Bioengineering*, X: 845–864.

Arregui, F.J., Del Villar, I., Zamarreño, C.R., Zubiate, P. and Matías, I.R. (2016). Giant sensitivity of optical fiber sensors by means of lossy mode resonance. *Sensors and Actuators: Part B*, 232: 660–665.

Bely, M., Sablayrolles, J.M. and Barre, P. (1990). Description of alcoholic fermentation kinetics: Its variability and significance. *American Journal of Enology and Viticulture*, 41(4): 319–324.

Bezenger, M.C. (1987). Incidence de l'azote sur le déroulement de la fermentationalcoolique en oenologie. Thèse de Doctorat-Ingénieur, USTL, Montpellier, France.

Boulton, R. (1979). Kinetic model for the control of wine fermentations. *Biotechnology and Bioengineering Symposium*, 9: 167.

Boulton, R. (1980). The prediction of fermentation behavior by a kinetic model. *American Journal of Enology and Viticulture*, 31(1): 40.

Bouyer, F. (1991). *Simulation de cinétiques defermentationalcoolique à l'aide de réseaux connexionnistes.* Rapport de stagepour le Diplôme d'Etudes Approfondies en Génie des Procédés de l'Université de Paris VI. CEMAGREF. Groupement d'Antony. Division ELAN, Laboratoire d'Intelligence Artificielle.

Boveé, J.P., Blouin, J., Maron, J.M. and Strehaiano, P. (1989). Modelisation de la fermentation alcoholique. Actualités oenologiques 89. *Microbiologie et Biologie Moleculaire*, 281– 286.

Boveé, J.P., Strehaiano, P., Goma, G. and Sevely, Y. (1984). Alcoholic fermentation: Modelling based on sole substrate and product measurement. *Biotechnology and Bioengineering*, XXVI(4): 328.

Caro, I., Pérez, L. and Cantero, D. (1991). Development of a kinetic model for the alcoholic fermentation of must. *Biotechnology and Bioengineering*, 38(7): 742–748.

Carrillo, G.E., Roberts, P.D. and Becerra, V.M. (2001). Genetic algorithms for optimal control of beer fermentation. In: *Proceedings of the 2001 IEEE International Symposium on Intelligent Control*, Mexico City.

Cléran, Y., Corrieu, G. and Chéruy, A. (1989). Modèles predictifs de la fermentation alcoolique en oenologie par la méthode MTDG (GMDH). *2ème Congrès de Génie des Procédés. Ed. Lavoisier*, 3(9): 293. Toulouse (France).

Colombié, S., Malherbe, S.. and Sablayrolles, J.M. (2007). Modeling of heat transfer in tanks during wine-making fermentation. *Food Control*, 18(8): 953–960.

Cozzolino, D. (2016). State-of-the-art advantages and drawbacks on the application of vibrational spectroscopy to monitor alcoholic fermentation (beer and wine). *Applied Spectroscopy Reviews*, 51(4): 302–317.

Cramer, A.C., Vlassides, S. and Block, D.E. (2002). Kinetic model for nitrogen-limited wine fermentations. *Biotechnology and Bioengineering*, 77(1): 49–60.

El Haloui, E., Picque, D. and Corrieu, G. (1988). Alcoholic fermentation in winemaking: On line measurement of density and carbon dioxide evolution. *Journal of Food Engineering*, 8(1): 17.

Elosua, C., Bariain, C., Luquin, A., Laguna, M. and Matias, I.R. (2012). Optical fibre sensors array to identify beverages by their odor. *IEEE Sensors Journal*, 12(11): 3156–3162.

Esti, M., Volpe, G., Compagnone, D., Mariotti, G., Moscone, D. and Palleschi, G. (2003). Monitoring alcoholic fermentation of RedWine by Electrochemical Biosensors. *American Journal of Enology and Viticulture*, 54(1): 39–45.

Ferreira, L.S., De Souza, M.B. and Folly, R.O.M. (2001). Development of an alcohol fermentation control system based on biosensor measurement interpreted by neural networks. *Sensors and Actuations*, 75(3): 166–171.

Ghose, T.K. and Tyagi, R.D. (1979). Rapid ethanol fermentation of cellulose hydrolysate. II. Product and Substrate inhibition and optimization of fermentor design. *Biotechnology and Bioengineering*, XXI(8): 1401–1420.

Giovanelli, G., Peri, C. and Parravicini, E. (1996). Kinetics of grape juice under aerobic and anaerobic conditions. *American Journal of Enology and Viticulture*, 47(4): 429–434.

Godia, F., Casas, C. and Solá, C. (1984). Estudio de la cinética de la fermentación alcohólica. Alimentación, equipos y tecnología, Sep–Oct, 259.

Godia, F., Casas, C. and Solá, C. (1988). Batch alcoholic fermentation modelling by simultaneous integration of growth and fermentation equations. *Journal of Chemical Technology and Biotechnology*, 41(2): 155–165.

Goma, G., Strehaiano, P., Uribelarrea, J.L., Mota, M. and Duran, G. (1982). Kinetic considerations on ethanol production. In: *Proceedings of the Second EC Conference on Energy from Biomass*, Berlin, 20–23 Sept.

Harris, M. and Kell, D.B. (1985). The estimation of microbiol biomass. *Biosensors*, 1(1): 17–84.

Henriques, D., Alonso-del-Real, J., Querol, A. and Balsa-Canto, E. (2018). *Saccharomyces cerevisiae* and *S. kudriavzevii* synthetic wine fermentation performance dissected by predictive modeling. *Frontiers in Microbiology*, 2(9): 88. doi: 10.3389/fmicb.2018.00088.

Henriques, D., Minebois, R., Pérez-Torrado, R., Balsa-Canto, E. and Querol, A. (2018). Multi-scale modeling to explain wine fermentation. In: *10th International Conference on Simulation and Modelling in the Food and Bio-Industry, FOODSIM 2018*, Ghent, Belgium, pp. 259–263.

Hinselwood, C.N. (1946). *The Chemical Kinetics of the Bacterial Cell*. Oxford University Press, London.

Holzberg, I., Finn, R.K. and Steinkraus, K.H. (1967). A kinetic study of the alcoholic fermentation of grape juice. *Biotechnology and Bioengineering*, IX: 413–427.

Ivakhnenko, A.G. (1968). The GMDH: a rival method of stochastic approximation. *Automatika*, 3: 129–138.

Jiménez-Márquez, F., Vázquez, J. and Sánchez-Rojas, J.L. (2014). High-resolution low-cost optoelectronic instrument for supervising grape must fermentation. *Microsystem Technologies*, 20(4–5): 769–782.

Jiménez-Márquez, F., Vázquez, F. and Sánchez-Rojas, J.L. (2015). Optoelectronic sensor device for monitoring the maceration of red wine: Design issues and validation. *Measurement*, 63: 128–136.

Jiménez-Márquez, F., Vázquez, J. and, Sánchez-Rojas, J.L. (2016). Optoelectronic sensor for measuring etanol content during grape must fermentation using NIR spectroscopy.*Microsystem Technologies*, 22(7): 1799–1809.

Jiménez, F., Vázquez, J., Sánchez-Rojas, J.L., Barrajón, N. and Úbeda, J. (2011). Multi-purpose optoelectronic instrument for monitoring the alcoholic fermentation of wine. *IEEE Sensors*, 978.

Jiménez-Márquez, F., Vázquez, J., Úbeda, J. and Sánchez-Rojas, J.L. (2016). Temperature dependence of grape must refractive index and its application to winemaking monitoring. *Sensors and Actuators: Part B*, 225: 121–127.

Jones, R.P. (1987). Factors influencing the deactivation of yeast cells exposed to ethanol. *Journal of Applied Bacteriology*, 63(2): 153–164.

Kurtanjek, Z. (1994). Modeling and control by artificial neural networks in biotechnology. *Computers and Chemical Engineering*, 18(Suppl): S627.

Lafon-Lafourcade, S., Geneix, C. and Ribéreau-Gayon, P. (1984). Inhibition of alcoholic fermentation of grape must by fatty acids produced by yeast ghost. *Applied and Environmental Microbiology*, 47(6): 1246–1249.

Lazaro, F., Luque De Castro, M.D. and Valcarcel, M. (1987). Individual and simultaneous determination of ethanol and acetaldehyde in wines by flow injection analysis and immobilized enzymes. *Analytical Chemistry*, 59(14): 1857–1859.

Leão, C. and Van Uden, N. (1985). Effects of ethanol and other alkanols temperature relations glucose transport and fermentation in *Saccharomyces cerevisiae*. *Applied and Environmental Microbiology*, 22: 359.

Levenspiel, O. (1980). The Monod Equation: A revisit and a generalization to product inhibition situations. *Biotechnology and Bioengineering*, XXII(8): 1671.

Locher, G., Sonnleitner, B. and Fiechter, A. (1992). On line measurement in biotechnology: Techniques. *Journal of Biotechnology*, 25(1–2): 23–53.

López, A. and Secanell, P. (1992). A simple mathematical empirical model for estimating the rate of heat generation during fermentation in white-wine making. *International Journal of Refrigeration*, 15(5): 276–280.

Luedeking, R. and Piret, E.L. (1959). A kinetic study of the lactic acid fermentation: Batch process at controlled pH. *Journal of Biochemical and Microbiological Technology and Engineering*, 1(4): 393.

Luong, J.H.T. (1985). Kinetics of ethanol inhibition in alcohol fermentation. *Biotechnology and Bioengineering*, XXVII(3): 280–285.

Luque De Castro, M.D. and Luque García, J.L. (2000). Biosensors in wine production monitoring. *Analytical Letters*, 33(6): 963–969.

Madrid, N., Boulton, R. and Knoesen, A. (2017). Remote monitoring of winery and creamery environments with a wireless sensor system. *Building and Environment*, 119: 128–139.

Marín, R. (1999). Alcoholic fermentation modelling: Current state and perspectives. *American Journal of Enology and Viticulture*, 50(2): 166–168.

Martínez, G., López, A., Esnoz, A., Vírseda, P. and Ibarrola, J. (1999). A new fuzzy control system for white wine fermentation. *Food Control*, 10(3): 175–180.

Matsumoto, K., Matsubara, H., Hamada, M. and Osajima (1991). Determination of glycerol in wine by amperimetric flow injection analysis with an inmobilized glycerol dehydrogenase reactor. *Agriculture and Biological Chemistry*, 55(4): 1055–1059.

Miller, K., Oberholster, A. and Block, D. (2019). Predicting the impact of red winemaking practices using a reactor engineering model. *American Journal of Enology and Viticulture*, 70(2): 162–168.

Miller, K., Noguera, R., Beaver, J., Medina-Plaza, C., Oberholster, A. and Block, D. (2019). A mechanistic model for the extraction of phenolics from grapes during red wine fermentation. *Molecules*, 24(7): 1275–1286.

Mompó, J.J., Urricelqui, J. and Loayssa, A. (2016). Brillouin optical time-domain analysis sensor with pump pulse amplificaction. *Optics Express*, 24(12): 12672–12681.

Moreira, A.P. and Carvalho, J.L. (1997). Intelligent and low cost continuous relative density measurement in batch fermenters. In: *3rd IFAC Symposium in Intelligent Components and Instruments for Control Applications*, Annecy, France.

Mouret, J.R., Farines, V., Sablayrolles, J.M. and Trelea, I.C. (2015). Prediction of the production kinetics of the main fermentative aromas in winemaking fermentations. *Biochemical Engineering Journal*, 103: 211–218.

Nakamura, K., Saegusa, K., Kurosawa, H. and Amano, Y. (1993). Determination of free sulfur dioxide in wine by using a biosensor based on a glass electrode. *Bioscience, Biotechnology, and Biochemistry*, 57(3): 379–382.

Nanba, A., Nishizawa, Y., Tsuchiya, Y. and Nagai, S. (1987). Kinetic analysis for batch ethanol fermentation of *Saccharomyces cerevisiae*. *Journal of Fermentation Technology*, 65(3): 277–281.

Novak, M., Streahiano, P., Moreno, M. and Goma, G. (1981). Alcoholic fermentation: On the inhibitory effect of ethanol. *Biotechnology and Bioengineering*, XXIII(1): 201–211.

Oikonomou, P., Raptis, I. and Sanopoulou, M. (2014). Monitoring and evaluation of alcoholic fermentation processes using a chemocapacitor sensor array. *Sensors*, 14(9): 16258–16273.

Özilgen, M., Çelik, M. and Bozoglu, T.F. (1991). Kinetics of spontaneous wine production. *Enzyme and Microbial Technology*, 13(March): 252–256.

Palleschi, G., Volpe, G., Compagnone, D., Notte, E. and Esti, M. (1994). Bioelectrochemical determination of lactic and malic acids in wine. *Talanta*, 41(6): 917–923.

Sablayrolles, J.M. (1996). Pilotage de la fermentation alcoolique en oenologie. INRA. *Genie oenologique*. Chapitre IV.

Sablayrolles, J.M. (2007). Contrôle de la fermentation alcoolique: Nouveau xoutils - Exemple du simulateur de la fermentation. *Revue Internet de Viticulture et Oenologie*, 26: 1–6.

Sablayrolles, J.M. (2009). Control of alcoholic fermentation in winemaking: Current situation and prospect. *Food Research International*, 42(4): 418–424.

Sablayrolles, J.M. and Barre, P. (1993). Kinetics of alcoholic fermentation under anisothermal enological conditions. I. Influence of temperature evolution on the instantaneous rate of fermentation. *American Journal of Enology and Viticulture*, 44(2): 127–133.

Sablayrolles, J.M. and Barre, P. (1993). Kinetics of alcoholic fermentation under anisothermal conditions. II. Prediction from the kinetics under isothermal conditions. *American Journal of Enology and Viticulture*, 44(2): 134–138.

Shrake, N.L., Amirtharajah, R., Brenneman, C., Boulton, R. and Knoesen, A. (2014). In-Line measurement of color and total phenolics during red wine fermentations using a light-emiting diode sensor. *American Journal of Enology and Viticulture*, 65: 463–470.

Starzak, M., Krzystek, L., Nowicki, L. and Michalski, H. (1994). Macroapproach kinetics of ethanol fermentation by *Saccharomyces cerevisiae*: Experimental studies and mathematical modelling. *The Chemical Engineering Journal and the Biochemical Engineering Journal*, 54(3): 221–240.

Stassi, P., Fehring, J.F., Ball, C.B., Goetzke, G.P. and Ryder, D.S. (1995). Optimization of fermentor operations using a fermentor instrumentation system. *Technical Quarterly, Master Brewers' Association of the Americas*, 32(2): 57–65.

Steinmetz, V., Cauquil, B. and Bocquet, F. (1995). Des neurones dans le champagne. *Ingénierie IEAT*, 3: 23–28.

Strehaiano, P. (1984). *Phénomènes d'inhibition et fermentation alcoolique*. Thèsed'Etat. INP, 80. Toulouse, France.

Strehaiano, P., Moreno, M. and Goma, G. (1979). Fermentation alcoolique: Observations cinetiques. *Connaissance de la Vigne et du Vin*, 13(4): 281–283.

Strehaiano, P., Mota, M. and Goma, G. (1983). Effects of inoculum level on kinetics of alcoholic fermentation. *Biotechnology Letters*, 5(2): 135–140.

Sugden, S. (1993). In line monitoring and automated control of the fermentation process. *Brewers´ Guardian*, 122: 21–26.

Teissier, P., Perret, B., Latrille, E., Barillere, J.M. and Corrieu, G. (1997). A hybrid recurrent neural network model for yeast production monitoring and control in a wine base medium. *Journal of Biotechnology*, 55(3): 157–169.

Turkusic, E., Kalcher, K., Schachl, K., Komersova, A., Bartos, M., Moderegger, H., Svancara, I. and Vytras, K. (2001). Amperometric determination of glucose with an MnO_2 and glucose oxidase bulk modified screen printed carbon ink biosensor. *Analytical Letters*, 34(15): 2633–2647.

Vlassides, S., Ferrier, J.G. and Block, D.E. (2001). Using historical data for bioprocess optimization: Modeling wine characteristics using artificial neural networks and archived process information. *Biotechnology and Bioengineering*, 73(1): 55–68.

Wu, S.S., Ozturk, J.D., Blakie, J.C., Thrift, C., Figueroa and Navesh, D. (1995). Evaluation and application of optical cell density probes ill mammalian cell bioreactors. *Biotechnology and Bioengineering*, 45: 495–502.

Unit 5

Process of Winemaking

INTRODUCTION

Wine has probably inspired more research and publications than any other beverage. In fact, through their passion for wine and wine derived products, a great number of wine scientists have not only contributed to the development of viticultural practices but have also made discoveries in the winemaking process, including in winery equipment, wine fermentations, stabilization and aging processes. Each applied development has led to a better control in winemaking and aging conditions, and consequently to improve the physico-chemical and sensory quality of the different wine categories. Thus, the determined evolution of the wine industry demands persistent advancements in all parts of the production chain.

The goal of this book unit is to summarize in a concise manner the accumulated information about the developments in all phases of winemaking process: the different winery equipment, grape juice extraction and preparation, wine fermentation process, microbiological and biochemical wine stabilization, maturation of wines and brandies, technology of wine packaging and also waste management in wineries and distilleries. All of these topics are written by a group of international researchers, in particular from India, in order to provide up-to-date reviews, overviews and summaries of current research on the different dimensions of wine production.

Therefore, inside this book unit, several chapters provide current research on different topics of recent advances in wine production. In the chapter relating to different aspects of the various operations connected with setting up of wineries, the authors present the main topics associated with machinery and equipment involved in wine production during all phases of the process (winemaking, wine stabilization and aging). The second chapter of this book unit focuses on the principal topics associated with grape juice extraction for wine production, including the main technological operations involved. In addition, general aspects related to the wine yeasts and bacteria activity and also bioreactors in wine fermentation are covered in Chapters 26 and 27. In the chapter regarding the maturation of wines and brandies, the authors discuss several dimensions of the wine and brandy maturation process, such as objectives and theory of maturation, the main factors which affect the maturation of wines and brandies, types of containers used for the maturation process, including the specific characteristics of the different containers which can influence the characteristics of wines and brandies matured in them. Other researchers, in the chapter concerning microbiological and biochemical stabilization of wines, discuss the major causes of the instability of wines and the different tests used to check wine stability. In the last two chapters of this unit, two important topics are covered. One relating to packaging technologies, and the other covers the theme of waste management in wineries and distilleries. In the chapter relating to packaging technologies, several themes are addressed, such as considerations and requirements for packaging, the different materials used for packaging production and also some aspects relating to wine labels. Finally, in the last chapter, the authors approach the thematic of waste management in wineries and distilleries, addressing topics relating to the planning of waste treatment plants, the introduction of clean technologies and in-plant measures and also the most important clean technologies used.

Prof. António Manuel Jordão
Associate Professor of Oenology
Polytechnic Institute of Viseu,
Agrarian Higher School, Viseu. Portugal

24 Winemaking
Fermentation Operations Machinery and Equipment

Wamik Azmi, Kriti Kanwar and Alka Rani

CONTENTS

24.1 INTRODUCTION

A number of unit operations are involved in winemaking. Winemaking was a simple operation during earlier times, but nowadays it is much more controlled, having sophisticated automation using various equipment and machinery. Details of various unit operations (Figure 24.1) connected with wine production in a winery and relevant aspects including machinery and equipment have been discussed in this chapter.

24.2 SETTING UP OF WINERY

The size of a winery is an important consideration in order to select a suitable place for setting up a winery. These can be located near the source of grapes or raw material, so that minimum expenditure is required for the transportation of the raw material to the winery. Sometimes, the crushing and fermenting processes are carried out near the source of the raw material, but the processes like ageing and finishing are carried out in the central part located at some distance from source of the raw material. Other factors such as an availability of labor, insect problems, drainage and waste disposal are also considered for this purpose. The winery should not be located in a city because of the problem of drainage, waste management etc. The winery could be located a little away from the vineyards or orchards in order to avoid insect problems like fruit flies. However, today, the tendency is to locate the winery very close to the vines, and a great number of wineries are located in the middle of the vineyards. Factors like economy, type of fermentation processes, capacity and many functional aspects are also considered in designing a winery, so that the operations are carried out in a beneficial manner.

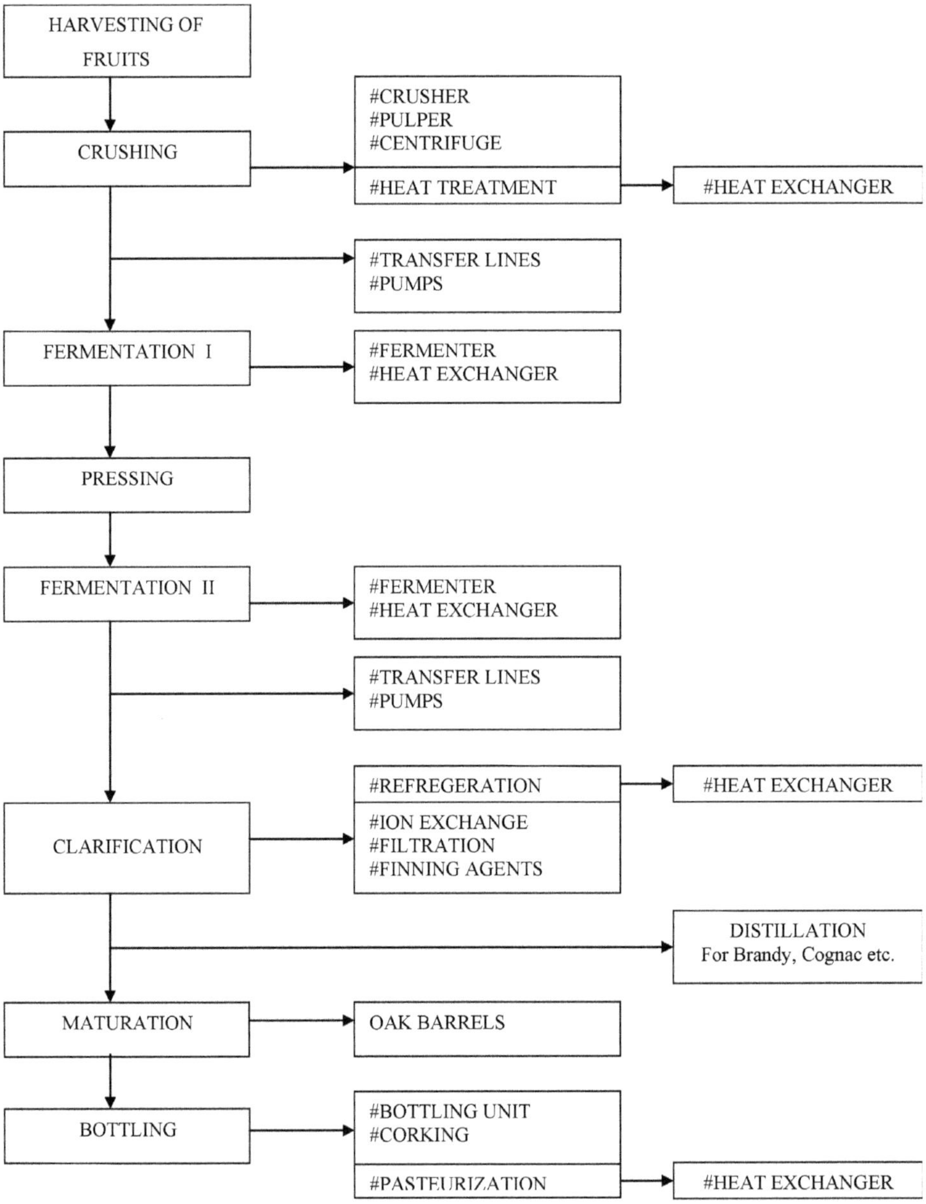

FIGURE 24.1 Various steps involved in red winemaking and brandy making

Generally, four major departments are established in a winery, *viz*., procurement, processing, quality assurance and marketing departments. Due to the growing concern for pollution control measures, the designing of a winery should be such that it provides maximum ease of operation and cleanliness, adequate drainage and a waste disposal system with provision for the effluent treatment of the winery waste including specifying the dumping area. The security and protection system need to be installed at appropriate places. Special protection for the electrical circuits and fixtures is provided with smoke and fire detectors along with suitable alarming systems.

Functionally, the essential components of any winery are the crushing unit, fermentation unit, quality control, sanitation, refrigeration, pumps and transfer lines, along with various utilities such as continuous supply of steam for heating and temperature maintenance, de-mineralized water, chilled water and electricity. Major unit operations of the winery have been depicted in Figure 24.2. The processing of the wine needs various equipment and machinery such as fermenters, crushers, filter presses, heat exchangers, pumps, filters, fining, flow meters etc. in the wineries. Nowadays, most of all the operations are controlled by the automated systems.

24.2.1 Wine Yeast Culture

Saccharomyces cerevisiae var. *ellipsoideus* is the major microorganism involved in winemaking. It becomes important to have wine yeast of good quality, so considerable research has been focused on screening the improved strains and their genetic manipulation. In a winery, the yeast culture in the form of the slants and tablets are obtained from a suitable source like culture collection departments, microbiological laboratories etc. Pure wine yeast in the form of compressed cake, lyophilized yeast or in dry granular form also is widely used. Preparation of yeast starter is initiated at least 2–3 three days prior to the crushing of the grapes using fruit juices or their extracts. The yeast count after 24 hours becomes sufficient so that culture can be used as inoculum within 2–3 days. The traditional method for propagating the pure yeast culture for use in a winery needs no sophisticated instrumentation. It consists of two covered tanks placed one above the other. The upper tank is either jacketed or fitted with stainless steel coil for heating and cooling purposes. The lower tank contains a sparger for aeration, wherein compressed air is passed through an air filter to prevent any contamination. Fresh juice is placed in the upper tank for sterilization and then cooled. This cooled juice is then passed to the pre-sterilized (with steam) lower tank to three-quarters volume. After 3–4 hours, fermenting juice (containing pure yeast) is added to sterile juice in the lower tank and aerated when vigorous fermentation is started, and three-quarters of its contents are used to inoculate the crushed grapes. The remaining fermented juice in the lower tank will act as an inoculum for the next cycle. Modern inoculation propagation scheme uses 1% (v/v) or even lower inoculum, but the most efficient ethanol formation has been obtained with an inoculum size of about 18 g/liter of dry yeast. (For more details see Chapter 26.)

24.3 PRE-FERMENTATION EQUIPMENT

24.3.1 Grape Crushers

The crushing unit is generally located outside the winery. Crushing and de-stemming is done as soon as possible after harvesting the grapes (Plate 24.1). Due to considerations of cleaning and delivery of the raw material, crushing units are generally, but not necessarily, installed outside or at some distance away from the main winery. The crushing unit has conveyers for cleaning the main system as well as a concrete pavement with water that also helps in keeping fruit flies away from the winery. Depending upon the volume of operation, the capacity of the crusher is chosen. Two types of the crushers are generally used, namely Roller Type and Garolla Type. The crushing units should be made up of corrosion-resistant materials. The rollers in the roller crusher are either rubber coated or are of stainless steel. These rollers are adjustable and roll towards each other so that grapes are crushed without breaking the seeds or grinding the stem. The de-stemming

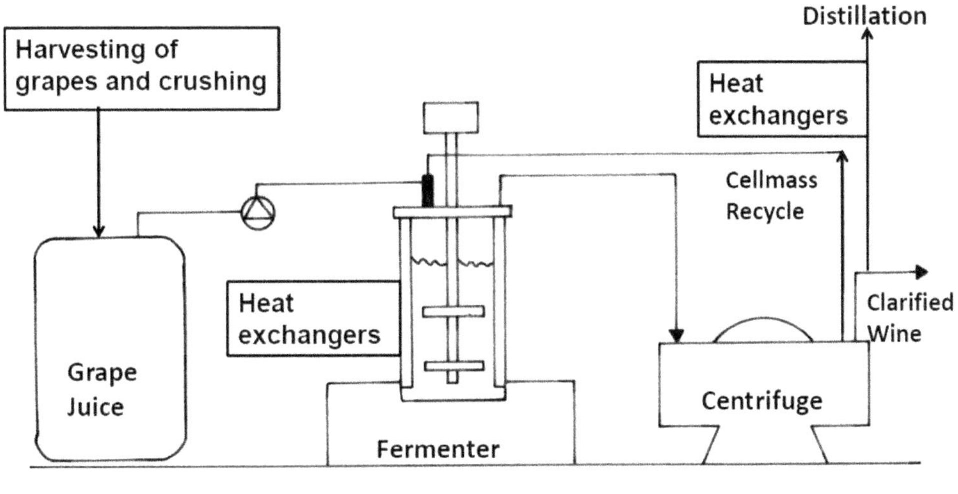

FIGURE 24.2 Major unit operations carried out in a winery to make a wine and brandy

PLATE 24.1 Mechanical crusher used for the release of juice from the grapes. *Courtesy:* Sh. S.G. Chougule, Indage Group of Companies, Mumbai

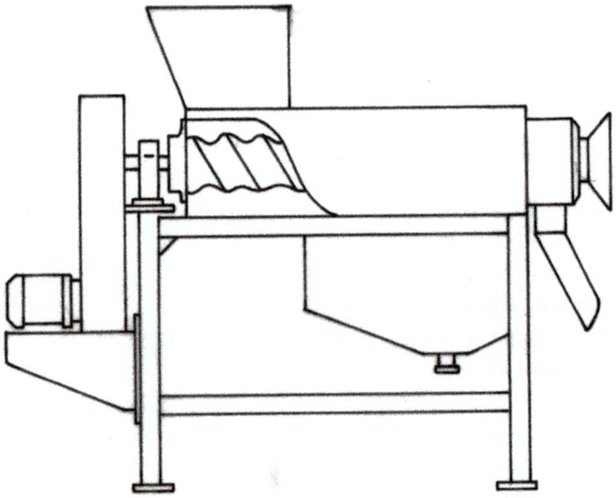

FIGURE 24.3 Schematic diagram of pomace conveyer

is done in conjunction with crushing. The Garolla crusher is made of stainless steel and consists of a larger, horizontal, coarselyperforated cylinder. The rotating blade inside the cylinder crushes the grape bunches with simultaneous removal of the stem.

24.3.2 Conveyers and Pumps

Screw conveyers and continuous chain conveyers are the most common conveyers used in wineries. In screw conveyers, the fruit and pomace move through a trough or tube, a metal helix or a screw (Figure 24.3). The materials are pushed along by the leading face of the helix as the screw turns. A motor and belt is used for the rotation of the screw. The conveyers can be either open or closed, can be operated horizontally, vertically or at an angle. Materials can be loaded and discharged at several points, and heating and cooling of the material is done as they pass through. The average length of screw conveyers varies from 8 to 12 ft, with screw diameter ranging from 3 to 20 inches, while the rotation of the screw varies from 100 to 250 rpm for grapes and pomace. Where it is desired to recover as much wine as possible, the pomace may be discharged through the gate either into a conveyer or a pomace pump. The pomace is usually sliced out with the water into the pomace conveyers where distilling materials are needed. Only limited numbers of pumps have been designed for the transfer of red skin from fermenters to the press. Most of the time the transfer of red skin by pomace pumps results in a significant tearing of the skin and seeds, which increase the amount of tannins during pressing. These pumps operate at a low rotational speed (30–60 rpm) so are not suitable for fluid transfer. A hopper is attached to feed the fluids by conveying screws and has a large outlet tube.

24.3.3 Pumps and Transfer Lines

Pumps are essential components in wineries and used for transfer must, juice and wine. Pumps should be operated at a pressure and speed prescribed by the manufacturer. Transfer by pumps is rapid and convenient for the material, as clean and well-maintained pumps transfer the juice and the wines without leakage or spillage of the material, introduction of air or microbial contamination. Generally, the pumps that are used for the juice transfers can also be used for the wine, while those that are suitable for must have special requirements. The pumps should be suitable with respect to public health and safety. Different types of pumps are piston pump, centrifugal pump, gear rotary and vacuum pump. Vacuum pumps are most frequently used in modern wineries. Such pumps operated by steam are called steam jet pumps, are very efficient and can generate huge amount of pressure. The pumps are mainly classified in two major types on the basis of working: positive displacement pump and centrifugal pump.

The positive displacement pump causes positive displacement of the fluid in the pump cavity, does not require priming and has through-put which is proportional to its speed so it can be used as a metering pump for the on-line addition of material such as SO_2, pectic enzyme solution, bentonite slurries and other oenological additives. The pumps with variable speed drives maintain different flow rates as per requirements. These pumps are of five categories. Reciprocating piston pumps have a reciprocating piston within a cylindrical cavity wherein the fluid is drawn during the reverse stroke and displaced during the forward stroke. These kinds of pumps are generally used for must transfer and have a capacity of 10 to 50 ton/hr. Progressive cavity pumps possess stainless steel helical rotors, which turn within the larger helical cavity within the surrounding casing. The helical cavity is displaced forward as the rotor turns, and a gentle pumping action results. The transfer of must by these pumps generates minimum suspended solid during the transfer. These type of pumps are used where the slurry is shear sensitive with capacity up to 30 to 60 ton/hr. Rotary vane pumps differ from other types of pumps in the type and shape of the vane, which may have two, three or even four segment stars, and their shape varies according to the model. The rotary vanes are geared to rotate at the same speed in opposite directions and are mounted so that as the lobes of one rotor moves into the spaces of the other, during a rotation the fluid is swept and displaced through the outlet. Rotary pumps should not be used when granular material is present in must, juice and pomace. The capacity of this pump varies from 7,200 to 30,000 L/hr. Impeller pumps consist of a rubber star wheel mounted on deformed cylinder housing. The axle of the star is somewhat extended for majority of the rotation that is generally in a counter clockwise direction, has limited outlet pressure conditions, is more suitable for juice or wine transfer over a short distance and has a capacity ranging from 6,000 to 12,000 L/hr. In diaphragm pumps, the two adjacent cylindrical diaphragms are connected to an oscillating shaft; the movement is controlled by the compressed air, with the direction reversing at the end of each stroke. The pumps are air operated and their speed can be controlled (from 3,000 to 60,000 L/hr) by the regulation of air supply with the flow rates.

Other types of widely used pumps in wineries where constant pressure is required are the centrifugal type of pumps. Various modifications of these pumps are used such as self-priming and non-self-priming type. Gear and rotary type of the pumps are the other modifications of the centrifugal type of pumps that provide rotational momentum to the liquid from a rotating disc or an impeller and are in common use in some modern wineries for pumping the must. Such pumps are not usually self-priming, so the pump cavity must be filled with liquid, and a continuous liquid connection exists with the inlet. The liquid intake is at the center of the impeller housing, and the discharge is at the wall in a tangential outlet. The flow can be controlled at the outlet without development of much pressure. The working capacity varies from 12,000 to 60,000 L/hr. All the pumps of this type are fitted with a barometer system or with some digital system for the measurement of the pressure, depending upon the requirements of the winery. Generally, pipes of stainless steel are used to transfer must to the fermenter as well as the transfer of the wine. An aluminum glass pipe may be used, but nowadays wineries prefer to use pipes made of fiberglass, as these are less costly than the glass and stainless steel pipes for the same purpose. However, the transfer line should contain minimum bends to prevent any sort of deposition in the pipeline.

24.4 FERMENTATION EQUIPMENT

24.4.1 Fermenter

The vessels generally used for conducting the alcoholic fermentation in winemaking are called 'fermenters'. A diagrammatic sketch of a fermenter is shown in Figure 24.4. These could be as simple as a wooden vat or sophisticated with automation and controls. These vessels vary considerably in the design and can be very complicated too. The fermenters are designed in such a manner to provide ease in the manipulation of the operation but should be strong enough to withstand the CO_2 pressure generated during the fermentation, and material should be corrosion resistant. Other materials for the construction of the fermentation tanks or inner linings include stainless steel, iron and glass, but stainless steel is preferred. As wine fermentation mostly uses pure culture, the fermenters should have provisions for the control and prevention of any contamination. The recent vessels also have systems to help to increase or regulate the maceration process for the red winemaking process. Suitable provision for the measuring of pH and temperature should be there in the fermenter. It should also be provided with inoculum or seed tanks or smaller sized fermenters, in which inoculum is produced and then aseptically transferred to the main fermenter.

In wine fermentation, the introduction of air to the vessel is not always required except for the initial aeration needed for the yeast cell growth when aerobic conditions are introduced. Generally, impeller or other mixing devices are not required, with certain exceptions like that for red wine production. However, for the production of sparkling wine by 'Charmat Process', CO_2 pressure resistant tanks are needed. (See Chapter 34 for more details.) Depending on the type of fermentation and the volume of the product desired, the size of fermenters is decided upon. In practice, the actual operating volume is always less than the total volume, as some space (generally one-quarter of the total volume) is left, known as the "head space", above the liquid medium for slashing, foaming and aeration, Generally, small laboratory scale fermenters have the volume of one to five liters or higher capacity; pilot

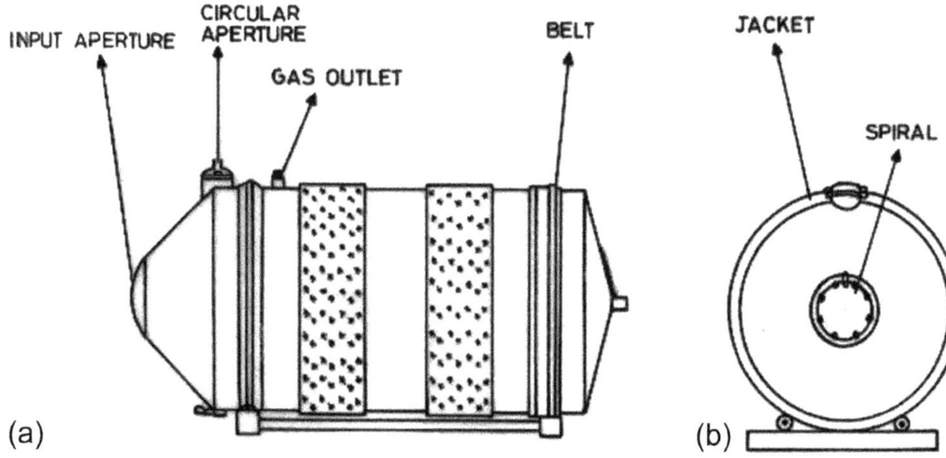

FIGURE 24.4 Horizontal fermentor in a winery (a) and its cross section (b)

plant fermenters have 25 to 100 gallons (up to 2,000 gallons). Fermenters in most commercial wineries are of 20,000 L or even larger in capacity. Both horizontal and vertical types of fermenters are used for winemaking. Horizontal fermenters are cylindrical stainless rotary vats with conical bottoms, having a central input aperture, upper circular aperture and a sifter for straining. Essentially, they contain two valves for leaking out gases. The main vessel is jacketed for cooling or heating, with spirals for mixing the product and unloading the strained substance. The whole operation is automated and controlled by a control panel. Horizontal fermenters are used both for fermentation of red wines and for filtering of white wines. In these type of fermenters, homogenization is excellent because the spiral mixing leads to extremely fast and perfect maceration. Vertical fermenters are cylindrical stainless vessels with conical bottom and are fitted on stainless legs. The main vessel contains a central input aperture for out-flow of CO_2 and has other attachments like a mixer, level meter and test tap. Temperature is continuously monitored and regulated by automatic feeding of a cooling agent in the jackets of the container.

24.4.2 Fermentation

Wine fermentation is carried out at an incubation temperature ranging from 20–25°C. The temperature of fermentation for red wine production is higher than for white wine (see Chapter 26 for more details). During alcoholic fermentation, winemakers control the sugar degradation by an indirect method *i.e.* by the density evaluation of the must. However, a Brix hydrometer or refractometer is also used to monitor sugar content during fermentation. The correction of acidity and pH adjustment of the must prior to fermentation is also done, whenever needed. External factors such as temperature and pH govern the nature of the fermentation process by affecting the growth and the physiology of the cell. In case of red wine, the skin is separated from the must by using a suitable strainer after a few days of fermentation, and its duration depends on the type of red wine to be produced. The liquid after removal of skin is pooled in suitable containers and is filled up to 85–90% of the capacity of the container and is allowed to ferment until all of the sugar is fermented. Yeast and other material starts settling at the bottom of the vessel as sediments after completion of the fermentation. Further, racking is done until the liquid becomes clear.

24.4.2.1 Mode of Fermentation

Two different approaches are possible to culture yeast in liquid medium, namely batch and continuous culture. Most of the fermentation for wine production is of batch type, although continuous fermentation is also practiced in a few wineries.

24.4.2.1.1 Batch fermentation

In general, the batch culture represents a closed system in which cells multiply until some nutrients are consumed completely, or some toxic metabolites accumulate to a certain concentration. The growing culture passes through a number of distinct phases like lag phase, accelerating phase, exponential phase, decelerating phase, stationary phase and decline phase. The exponential phase in the batch culture can be represented by the following equation:

$$dx / dt = \mu x \quad (24.1)$$

Where, x = Concentration of biomass; t = time in hours; μ=specific growth rate in hours^{-1}

The growth-linked product formation can be described with the following equation:

$$dp / dt = q_p x, \quad (24.2)$$

where p = concentration of the product, q_p = specific rate of product formation. Thus, product formation is related to the biomass formation.

In commercial winemaking, the actively growing yeast inoculum is added to avoid any lag phase, as it leads to loss of time. The fermenter is emptied for further processing of the product at the end of fermentation. The fermenter has to be cleaned, refilled, resterilized and then reinoculated. *S. cerevisiae* exhibits double phase growth kinetics (diauxic growth) in batch culture. It is due to the ability of this yeast to metabolize sugar glycolytically during the first phase of the growth, producing ethanol, that means it is subsequently metabolized oxidatively. It is obvious that oxygen must be present for oxidative degradation of sugars. Therefore, limitation of oxygen renders the organism incapable of producing enough energy for growth. Consequently, either the cells cease to grow or they use alternative pathways, most commonly a reductive metabolism of sugar-yielding ethanol to get the required energy. So, the batch growth of *S. cerevisiae* comprises two phases: in the first phase ethanol is formed and is used in the second growth phase as the substrate. The batch type of process produces a better quality wine than the continuous process.

24.4.2.1.2 Continuous Fermentation

A continuous culture represents an open system. Pumps constantly supply grape juice, and an equal fraction of the product mixture including yeast cells is withdrawn. Continuous fermentation contains a perfectly mixed suspension of the yeast cell and juice, in which the grape juice is fed and from which wine is removed at the same time. The rate of microbial growth in this fermentation is not fixed and depends on the availability of the nutrients and other physical factors like aeration, pH, yeast cell density, etc.

When *S. cerevisiae* is grown in a continuous culture, it shows two metabolically different responses. The breakdown of the sugar occurs oxidatively at low dilution rates and products other than biomass and CO_2 are not formed. The rate of CO_2 production equals the rate of oxygen uptake during this period. When the dilution rate is further increased, a particular dilution rate is reached, above which the rate of oxygen uptake no longer increases linearly with increasing dilution rates. Simultaneously, the rate of CO_2 production increases steadily. Cells of *S. cerevisiae* growing at this dilution rates metabolize sugar reductively and produce ethanol accompanied by excessive CO_2 production. The yield of biomass drops owing to the formation of energy-rich ethanol but remains comparable to that found in the first phase of the batch growth. So, the growth of *S. cerevisiae* at high dilution rates resembles that of the first phase in batch culture. But morphology of the cells of *S. cerevisiae* changes with dilution rates when grown in a continuous culture.

In France and Italy, continuous fermentation is used to make wine, is automated and thus can save space and labor, while in countries of the former Soviet Union it has been successful for the production of sparkling wine. Further, yeast continues to grow at 7.6% (v/v) alcohol in continuous fermentation, while in the batch process the yeast cells stop dividing. Although continuous fermentation has many advantages like maintaining the cells at a particular physiological state, it is seldom used for winemaking as the operation demands a continual supply of nutrients, maintenance of sterility of juice and oxygen free conditions for longer duration, which is difficult to achieve and is expensive too.

24.4.3 Heat Exchangers

Heat exchangers are used very frequently in a winery, as alcoholic fermentation is an exothermic process. The amount of heat involved in the process is very small for laboratory fermenters but certainly more significant for large-scale fermenters of a capacity of 10,000 L or more. In a winery, heat exchangers are used in heating/cooling of juice or must wherever required. The process of heat transfer is also used at several points in winemaking such as: in winery setup – air conditioning, cooling of aging cellars; in operation before fermentation – cooling of juice and must, during high temperature short time (HTST) treatment to kill microorganisms and denature the enzymes; during fermentation – temperature control; after fermentation – cooling of wine during storage, cooling of wine to enhance potassium bitartrate, crystallization and precipitation.

The selection of heat exchangers that is best suited to perform a particular heat transfer function is generally dependent on the condition of juice and wine because the efficiency of heating and cooling of any fluid depends largely on the properties, such as viscosity and density, of the fluids. The viscosity of pure ethanol, wine and fruit juice (at 20°C) is 789, 990 and 1090 kg/m^3, respectively, while their densities are 1.2, 1.55 and 2.55 mps, respectively. Different applications of heat exchangers and their types in winemaking include: shell and tube heat exchangers: must cooling, juice heating or cooling, fermentation cooling, wine cooling and heat recovery; spiral heat exchangers: must cooling and juice heating or cooling; plate heat exchangers: high temperature short time treatment, wine cooling and heat recovery; jacketed heat exchangers: fermentation heating or cooling and wine cooling; scraped surface heat exchangers: cold stabilization.

24.4.3.1 Shell and Tube Heat Exchangers

These type of heat exchangers act on the principle of reducing (or increasing) the temperature of a fluid by transfer of sensible heat to a cool (or hot) fluid, the temperature of which is increased (or decreased), thereby heat transfer to and from the flowing fluids takes place with a tubular heater. The function of the heater is to increase the temperature of a fluid by transferring it to the latent heat of the condensation of vapor. The heater consists of a bundle of parallel tubes, and their ends are expanded into tube sheets. These tubes are placed inside a cylindrical shell, which is provided with two channels at each end. Steam or coolant is introduced through a nozzle into the whole shell space, and condensate can be withdrawn through an outlet at the bottom while any non-condensable gas is removed through an opening at the top. The fluid to be heated is pumped into the inlet channel, and then it flows through the tubes into the other channel and then goes out. The two fluids are physically separated but are in thermal contact with the thin metal tube walls supporting them. Sometimes, baffles in the shell are also used to increase the heat transfer efficiency. In the simplest shell and tube heat exchangers, juice enters the tube from one end and comes out at the other end. This single pass heat exchanger is generally referred to as a 1–1 type. In a 1–2 or double pass exchanger, the juice passes through one set of tubes to one end, then passes through a return bend and moves back to the inlet end by way of other sets of tubes. Another type of tube arrangement is exhibited in 2–4 exchangers, in which coolant or steam makes two passes while the juice makes three changes in direction. The reason for such modifications or arrangements is the requirement of a more effective heat transfer. The coolant in the shell can be either a glycol solution or an evaporating refrigerant for cooling. A second type of single heat transfer equipment is the double-pipe heat exchanger or tube-in-tube heat exchanger, wherein a small diameter (25–50 mm) tube is mounted within a large diameter tube. One fluid flows through the inside pipe and the second fluid through the annular space between the outside and the inside pipe in counter direction. This type of arrangement is best suited for the cooling of must. The shell and tube type of heat exchanger is used in wineries for must cooling, heating/cooling of juices and wine and can also be used for cooling or maintaining the temperature during fermentation by the use of an external cooling jacket on the fermenter, internal plates placed within the fermenter or by external shell and tube heat exchangers. However, for cooling of must by this type of exchanger, the tube diameter should be large (>75 mm) to reduce the plugging by skin and fruit fragments. The bends of tube should also have a large radius with outlets to release the blockage and better cleaning.

24.4.3.2 Plate Heat Exchangers

This assembly resembles the plate and frame filter. The plates used in this exchanger are rectangular, with a length of two to three times more than the width. All the plates are fitted with a polymer gasket, which acts as a seal and distributor of the flow. This film of fluid passes down between the two plates alternatively, through a juice inlet. The flow of fluid can be arranged in different ways. In single pass operation, the fluid to be heated passes into every second plate, while the steam passes in the opposite direction in the alternate plates.

Extra plates can be added to increase the effective surface area or the operating capacity of a unit. The juice is distributed through each alternate plate in the plate exchanger while steam flows in the opposite direction through the plates in between; the holding time can be controlled by manipulating the fluid path length in the exchanger. The plate stacks are divided in two sections for HTST treatment of juice. The first section is for the heating of juice by steam, then the cooling of juice is done in the next section of the plate exchanger where steam is replaced by refrigerated glycol. The heat transfer efficiency of the plate exchanger is mainly due to the two sides heating or cooling a thin film of juice. Sometimes, to increase the heat transfer efficiency, a wavy surface on the plates is created, promoting turbulence. The plate exchanger is used for the cooling of wine, with a simultaneous recovery of heat using glycol as a coolant.

24.4.3.3 Spiral Heat Exchangers

Spiral heat exchangers are designed by rolling two adjacent flat jackets into a coil. There are two ports at the top and two other ports at the either side of the center. In spiral heat exchanger, the fluids (juice and coolant) flow in opposite directions. The juice or must for cooling enters at the base and moves through spiral path and goes out through the central port. The coolant enters at the opposite side at the center and passes through second spiral path and is released at the top of unit. The width of fluid path varies from 5–25mm. For turbid juice, spiral heat exchangers are the most suitable. After the clarification of juice by settling or removing the seeds and stem fragments by screens, plate heat exchangers can also be used. A spiral heat exchanger with a wide flow channel is used for the cooling of must and juices.

24.4.3.4 Scraped-Surface Heat Exchangers

The scraped-surface heat exchanger has a short cylinder with two or four scraping blades with axle to prevent the freezing of wine at the exchanger wall. The blades run the entire length of the cylinder with the help of a motor at a speed of 300–500 rpm. The cylinder is chilled from the outside by an annular jacket which contains either expanding refrigerant or glycol solution. For cooling, the flow of wine is maintained so that only 5–10% (v/v) ice is formed and goes in the exit stream. Scraped-surface heat exchangers can be used for the cooling of wine for cold stabilization that is done below freezing point with the formation of ice in wine. Cooling is controlled in such a way that only a thin layer of ice is formed, which during cold stabilization is allowed to melt back to maintain the net ethanol content or the volume of wine. Mixing the cool but ice-free wine generally causes the melting of ice in wine. Subzero cooling of wine enhances the rate of sodium bitartrate crystallization. Care should be taken that the wine does not freeze at the exchanger wall to block the flow.

24.4.3.5 Jacketed Heat Exchangers

Jacketed heat exchangers are conveniently used for the cooling of fermenters and storage tanks. Heat transfer is poor in this type of heat exchanger due to the stationary fluid at the inside surface of the tank. With increases in tank size, heat transfer efficiency is reduced due to a reduction in jacket area per unit volume. If the diameter is increased two fold, the wall area will increase four fold, the volume will increase eight fold, but the area per unit volume will be reduced to half. That is why the jacketed heat exchangers are more suitable for small and medium size fermenters. During the fermentation run, temperature is maintained by removing the heat generated using external cooling jackets. The jacket or internal plate has coolant running through it, but the overall heat transfer is poor. The existence of a temperature gradient in storage tanks leads to the movement of the warmest wine to the top while colder wine descends towards the base of the tank.

24.5 POST-FERMENTATION OPERATIONS AND THEIR EQUIPMENT

24.5.1 Refrigeration

Refrigeration is required for the precipitation of potassium tartrate. Generally, tartrate in wine settles if kept for a long time. Ion exchange resins and refrigeration stabilizes the tartrate. In European countries, winter temperatures chill the wines and cause the excess cream of tartrate to settle down, but in warm places, it is a very slow process. That is why artificial refrigeration up to the freezing point (from –5.5°C to –3.9°C for table wine, from –9.4°C to –7.2°C or slightly more for fortified and dessert wine) of the wines is done. At a low temperature, the wine dissolves more oxygen than at room temperature, resulting in rapid ageing of the wine. It is now an established fact that the solubility of the tartrate not only decreases with the decrease in the temperature but also with increase in the alcohol concentration. The following formula is generally used for the calculation of the temperature required for the solubilization of the tartrate:

$$\text{Temperature }\left({}^{\circ}\text{C}\right)=\left(\%\text{ Alcohol Content}/2\right)-1$$

The response of the wines to the refrigeration temperature varies with the wine. Though the pipes and circulating system employed are relatively inexpensive, the refrigerated room system is the most sophisticated and expensive one.

24.5.2 Ion Exchange

Ion exchangers are used in batch and continuous form for the exchange of hydrogen ions in the removal of potassium and calcium ions and to improve the sensory qualities of wine by lowering its pH. However, not all wine should be subjected to this treatment since the consequences may not be totally positive, due to an excessive decrease in the pH values of the wine. It must be a given percentage of a batch of the wine to be subjected to this treatment. Removal of excess potassium and calcium ions with either sodium or hydrogen ions also takes place. The ion exchanger can be employed. Continuous types of ion exchanger are less expensive than the batch process ones. Fining with bentonite alone or in combination with gelatin is generally done before ion exchange.

24.5.3 Clarification of Wine

Clarification and fining of wine is used to achieve the clarity of the wine and removal of some excessive components present in wine. Sometimes, different enzymes are also added for clarification purposes. Pectic enzymes are used for filtration and clarification of grape must and wine.

24.5.3.1 Natural Clarification of Wine

Natural clarification is the process of the settling of the suspended solids under the action of gravity. The time required for the clarification is the time taken by the smallest particle to fall through the height of the tank and also depends on the interaction between the particle and the presence of the natural convection current or bubble rising from the onset of fermentation within the tank. The average particle size varies from a diameter of 2–4 µm to 16–20 µm. The rate at which the particles denser than the wine settle depends upon the density difference between the particle and the fluid, the particle diameter and the viscosity of the fluid. The particles will rapidly accelerate until they reach a steady falling speed known as terminal velocity. Estimation of the natural settling time for wine is influenced by the size distribution of the particle and the existence of at least three different mechanisms of settling within the tank. These are the laminar settling within the dense layers of particles, the hindered settling within dense layers of the particles and the compaction of heavy solid layers. The solid layers that get collected at the base of the tanks also undergo much slower compaction, which is influenced by the nature of the solids. To improve the settling, settling aids such as bentonite or silica suspension (silica sols) may also be used. (For details see Chapter 29.)

24.5.3.2 Centrifugation of Wine

The most common alternate for wine clarification by settling is the use of centrifugal processes to increase the settling force, one to several thousands times that of gravity. This centrifugal force greatly increases the terminal velocity, decreases the separation time and also provides more consistent clarifying action even in the presence of particle interaction. Sometimes hydro-cyclones are also used to remove the larger suspended particles from juice and wine. Generally, two types of centrifuges are used, the de-sludging type and the decanting type. De-sludging centrifuges consist of a stack of truncated cones that are mounted in the center of a spindle. The spindle is hollow and allows the feed stream to enter from the top and is then distributed at the base of the bowl. The entire bowl, outer wall and disc stack are rotated at higher speeds, producing outward radial force on particle of more than 10,000 times that of gravity. The stream that is distributed at the base is then forced up between the discs, leaving the unit at the top of the conical section. The accumulated solids are removed in an operation known as de-sludging. An optical sensor in the outlet stream detects the overflow of the solid when the bowl is full and usually triggers the de-sludging. The inflow is stopped while the bowl continues to spin, and the base section of the bowl chamber drops down, allowing a series of holes or ports in the circumference to open. The solids are then forced through these ports by the centrifugal action and collected in outer chambers. When the entire contents of the bowl are ejected, that operation is known as total de-sludge. The centrifugation is a very efficient way to increase the clarification of wine and grape musts, but it is only used for low level quality wines because centrifugation induces and increases oxidation and decreases the aromatic characteristics of wines, along with losses of phenolic compounds, etc. In addition, for grape must, a decrease of yeasts is also possible to detect. So, centrifugation could be a potential positive option but only for high wine volumes and with lower quality.

24.5.4 Filtration of Wine

Filtration of wine is employed for the partial removal of the large suspended solids of approximate size ranging from 50 to 200 µm in diameter, grape pulp, bitartrate crystals or some of the yeast. Various grades of the diatomaceous earth or filter sheet are used for the purpose of filtration, and for the complete removal of the microbes, the perpendicular flow polymeric membrane is used. Perpendicular flow filtration is generally done in stages to remove the coarser solid first and then the finer particles by changing the pore size. The membrane-based cross flow filters can be used for the sterile filtration of juices and wines by the removal of solutes. A filtration process is generally defined by its filtration rate, which is the ratio between the driving forces against the cake resistance. as given below:

$$\text{Filtration rate} = \frac{\text{Driving force}}{\text{Cake resistance}}$$

In the filtration process, pressure can be due to two components. One component is needed to pass the constant volume of filtrate through the filter resistance and the second one is the increasing pressure component that is proportional to the resistance from the increasing cake depth. A good filtration system must ensure good or sufficiently fast flow rates of the filtrate and complete retention of the particles. To satisfy these conditions, various types of filters are used in wineries. Filters can be classified according to their porosity, the nature of filter medium and the method of housing the filter medium or the arrangement of fluid flow path. On the basis of filtration media, filters can be divided into three categories: membrane filters, depth filters and adsorptive filters. While adsorptive filtration has some limitations, depth filter can be used very effectively for the removal of suspended solids. Membrane filters are sophisticated and can even separate minute solute from the solution, as they have a mechanism that traps particles larger than the pore size of the filter, while contaminants smaller than the pore size may pass through the membrane or may be removed by some other mechanism. Depth filters are those that entrap the contaminants both within the matrix and on the surface of filter media, contaminants are retained within the tortuous path structure of the filter and characterized by the thickness of the filtration medium which has great particle holding capacity. Adsorptive filters are those that have filter media which exert attractive forces that retain contaminants. However, in modern oenology, filtration has fallen into disuse because it leads to a reduction in the quality of wines, particularly white wines.

24.5.4.1 Filtration Aids

When bacteria and other fine or gelatinous materials are filtered, which prove slow to the filters or partially block a filter, filter aids are used. Diatomaceous earth (Keiselghur) is the most commonly used filtration aid. It provides structural support to the filter cake so that the compressible solid does not form a tight film across the filter surface, thereby reducing the filtration rate to zero (see Chapter 29 of this text for more details). The commercial grades of diatomaceous earth used for wine filters have medium particle size between 14.0 µm in Celite 500 and 36.2 µm in Celite 545.

24.5.4.2 Pressure Leaf Filters

Pressure leaf filters are used for the rough filtration of the tank settlings, as a prefilter for the wine prior to membrane filtration. These filters use diatomaceous earth of fine grade and are employed in large and medium size wineries. These filters are batch filters incorporating several leaves consisting of a metal framework of grooved plates. The leaves are covered with fine wire mesh, or a filter cloth, and pre-coated with a layer of cellulose fibers and thus, steam sterilizable. Leaves can be placed vertically or horizontally.

24.5.4.3 Pad Filters

A simple approach to filtering that does not require the use of pre-coats and body feed is the application of preformed sheets or pads of cellulose and diatomaceous earth. Many wineries use pad filtration in combination with the membrane filtration prior to bottling. Filter pads are designed to collect particles in their interior rather than to develop a cake at the surface. Pads are made of cellulose fibers with the diatomaceous earth trapped in between. Tight pads containing asbestos fibers have been used to remove proteins, other colloids and microorganisms. At present, a range of cellulose acetate and other polymeric filters are used. Cellulose-based depth filtration media are designed to retain contaminants by both mechanical entrapment and electrokinetic adsorption. The depth filters can remove bacteria, particulate matter and haze and cell debris. Generally, the pad filters are classified on the basis of their capacity for water flow.

24.5.4.4 Plate and Frame Filters

These are pressure filters consisting of plates and frames arranged alternatively and held together by means of screws to prevent any leakage between the plate and frame during operation. The plate and frame are generally square in shape and hang from two supporting bars at each side of the filter stack. Usually, pressures gauges are mounted on the front end plate to indicate the pressure of each stream so that pressure drop (which indicates the resistance to flow) across the membrane can be calculated. Plate and frame filters can be used in many ways such as the filtration medium, wherein the filtrate passes through the filter cloth or pad and then get collected at the outlet while the solids are retained within the frame. The degree of clarification is determined by the grade of the diatomaceous earth used in the pad. This filter acts at its best when the solid contents in fermentation broth are low, posing low resistance during filtration. In wineries, it is used to remove the residual yeast cells following initial clarification by centrifugation or by some other means. Seitz and Schenk make the most common types of plate and frame filters used in US wineries.

24.5.4.5 Cartridge and Membrane Filters

Membranes act as a selective barrier and facilitate active separation of components with precision. The whole separation process by membrane depends on the effective pore size (approximate pore size of 0.2–2.0 µm in diameter), which covers a huge range. Pore sizes of 0.65 µm are required for yeast removal, while 0.45 µm is required for the removal of bacteria. Traditional membranes are the perpendicular flow membranes that are used for the filtration of wine just before bottling. The filter cartridges are made of synthetic polymers such as polycarbonate, polysulfonate or polypropylene, with varying pore size ranging from 0.45 to 1.2 µm (Figure 24.5). These are regarded as membrane filters since they collect particles at the surface pores rather than within the torturous path in the body of the pad filters. The membranes are rated on the basis of the size of the largest pore rather than the average or effective pore size that in turn determines the flow condition. Filters of this type have a very limited holding capacity for solids, and the majority of the solids should have been removed by earlier clarification.

24.5.4.6 Cross-Flow Filters

Generally, the wine bottled for commercial purpose passes through a membrane filter, which has virtually replaced heat and chemical treatment because it does not affect sensory

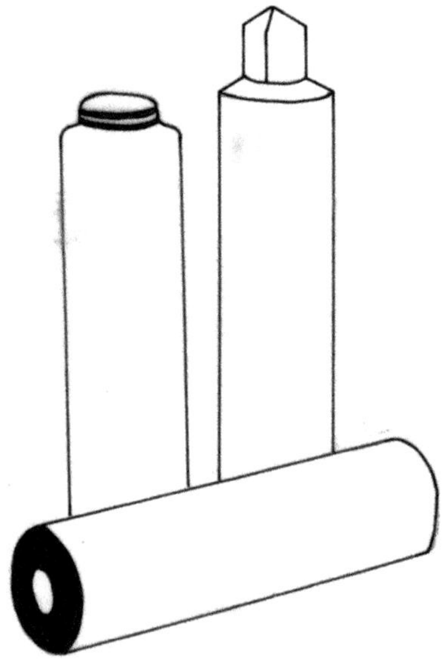

FIGURE 24.5 Cartridge filters used in wine clarification

quality. Preferred membrane pore size is 0.45 μm to prevent the entry of any spoilage microorganisms in passing through. In traditional filters, the flow of fermentation broth was perpendicular to the filtration membrane, termed as axial filters. In this, the entire filtrate passes through the filter medium on the first pass, and suspended solids played a major role in fouling of the filter surface. This blockage of the membrane may lead to lower efficiency of filtration processes. To avoid the plugging of microfiltration membrane, dead end flow filtration devices with built-in pre-filters are generally used. A better option is a cross flow or tangential flow membrane filtration unit. In cross flow filtration, the flow of the medium to be filtered is tangential to the membrane, and no filter cake is built on the membrane. In cross flow filters, most of the fluid passes rapidly across the filter surface, and unfiltered fluid is recycled and passed across the surface many times during the run. Due to this repeated cross flow, no filter cake is built on the membrane and hence, no filter aid is required. In tangential flow filtration, the filtrate is referred to as permeate and the unfiltered fraction is known as retentate. The main drawback of this type of arrangement is the requirement of a large surface area and much larger filtration time. In tangential flow filtration, usually acrylic or stainless steel manifolds are combined with either micro-porous or ultra-filtration membrane. Cross flow filters in winemaking can be employed for microfiltration, ultrafiltration and reverse osmosis. Tangential filtration is a recently developed technology in the oenology sector with good results on wine quality.

24.5.4.6.1 Microfiltration

Microfiltration is a type of filtration device that collects suspended particles in the microbial size range (0.1 to 1.0 μm diameter), and the microporous membrane filters (made of polyvinylidene fluoride-PVDF) function as an absolute screen or sieve, retaining all the particles larger than the pore size. The micro-filtration membrane-based cell separation is efficient, and 99.9% of cells can be retained. Microfilters are widely used in the clarification of juices and wines.

24.5.4.6.2 Ultrafiltration

Ultrafiltration is a process in which solutes of high molecular weight are retained while the solvent and low molecular weight solutes are forced through a membrane of very fine pore size. So, ultrafiltration is the cross-flow separation of large molecular weight solutes rather than the suspended particles. The ultrafiltration membranes are regarded as the filters best suited for retaining a wide range of macromolecules with side-by-side maintenance of high flow rates. The membrane material is generally polysulfone or regenerated cellulose. Factors, other than the molecular weights of the solutes, like concentration polarization or formation of gel layer may affect the passage of molecules through the membrane. Ultrafiltration filters are generally classified by their molecular weight cut-off (MWCO), which is the average size of the range of molecules 1 KD to 100 KD that it will effectively retain. The polysaccharides of 50 KD to 200 KD ranges are found in grapes and produced by yeast. In wineries, 10 KD MWCO membranes are used to separate 20 KD to 40 KD molecular weight proteins to make a more stable wine, but unfortunately they also remove some of the desirable phenolic compounds from the wine. Tannins and red pigment have been removed by using 0.5–2.0 KD MWCO membranes.

24.5.4.6.3 Reverse Osmosis

This is a separation process in which solvent molecules are forced by pressure to flow through a membrane in a direction opposite to that dictated by osmotic forces. Generally, it is used for the concentration of the smaller molecules. The materials of the reverse osmosis membrane are generally polyamide, cellulose or polysulfone, with pore size of varying from 10–100 molecular weight cut-off. The major application of reverse osmosis in wineries is in the production of low ethanol wines using 70–100 molecular weight cut-off membranes and water and a small amount of organics. Most of the ethanol goes out in the filtrate. The rest of the components of wine like amino acids, sugars, organic acids and phenolics are retained. The permeate can be discarded, and water is added to retentate to reproduce wine of low ethanol content. Reverse osmosis can also be used to remove acetic acid in wine, in the production of wine concentrates, to accelerate the crystallization and removal of tartrate from wine.

24.5.5 Distillation of Wine

Wine fermented from fruit juice or pulp is distilled to produce brandy. (See Chapter 38 for more details on brandy.) Distillation is based on the fact that matter can exit in three phases, solid, liquid and gas. As the temperature of pure substances is increased, it passes through these phases, making a transition at a specific temperature from solid to liquid (melting point, t_m) and at a higher temperature from liquid to a gas (boiling point, t_b). It involves evaporating a liquid into a gas phase, then condensing the gas back to liquid and collecting the liquid in a receiver. Substances having a higher boiling point than the desired material will not distill at the working temperature and remain behind in the distillation unit. Applied to the preparation of brandy, alcohol has a lower boiling point than water, and thus, distillation can separate the alcohol from the wine. Since some water also evaporates during this process, liquors are not pure alcohol, and the products are expressed in terms of the alcohol content. In many cases, single distillation is used, but usually the wine is double distilled. Distillation takes place soon after the completion of fermentation. In a two-stage process, the first distillate will contain about 28–30% (v/v) alcohol but after second distillation, alcohol content may rise up to 70% (v/v). Traditional distillation is carried out in a copper still. For marketing purpose, the distillates are generally diluted to an alcohol content of 30–40% (v/v) or blended with fruit juices to yield fruit liqueurs. Batch distillation is generally preferred in order to extract maximum aroma from the wine (see Chapter 35 on brandy for more details). In batch distillation, the still is filled with wine. Flash distillation, usually conducted as a continuous process, consists of vaporizing a definite fraction of liquid, keeping all of the residual liquid and the vapor in intimate contact so that at the end of the operation, the vapors are in equilibrium with the liquid, separating the vapors from liquid and condensing the vapor.

24.5.6 Storage of Wines

The storage of the wine is done in the cool and dry place with proper cleanliness. Sterile conditions are maintained during storage in bottles, glass or plastic containers, so that any contaminant will not spoil the wine. Wood-based containers made of oak as well as containers made of concrete, stainlessness steel and polyester are generally used for this purpose. Containers used for the storage may vary in size: between 225 L (barrel) and 1900 L – oval, puncheons (rounded), pipes (for port), butts (for sherry) – or simply casks appropriate for the specific quality of the wine and the purpose of ageing. Oak cooperage is preferred for red wine and dessert wine and also for white wines. The use of oak wood barrels is a very useful option (during fermentation and/or the ageing process).

24.5.6.1 Filling and Racking of the Wine

The tanks used for table wine are completely filled and sealed, thus air available for the wines is very low or almost negligible. The air in the partially filled containers during filling can cause spoilage by acetification and oxidative degradation of the wine. Containers or bottles with a narrow and small mouth are preferred. For easy and frequent filling, the container size should be small. Racking is a simple but an important process carried out to remove the suspended material and carbon dioxide for clarification. Removing CO_2 raises the redox potential, so this process should not be delayed. One danger of leaving the wine for longer periods without racking is that the sediment of yeast cell autolyses, which lowers the redox potential and forms H_2S. Racking can be done after a longer duration of time if malo-lactic fermentation is desired (to remove total acidity).

24.5.7 Ageing of the Wines

Through the process of ageing, a harsh taste and odor are converted to a pleasant and soothing odor. Wood extracts also change the flavor and give a peculiar aroma. Another important change of wine is the removal of suspended particles, a process termed as 'clarification', wherein the lowering of the temperature, storage in various types of the containers, racking, filtration, fining, centrifugation, pasteurization, refrigeration, passage through ion exchange resins, etc. are involved.

24.5.8 Bottling and Corking

The bottling operation is one of the important operations in winemaking. The bottling operation include cleaning or rinsing units for bottles, filling machines, cork inserting machines, labeling machines and capsulation or foiling machines. The finished wine is generally transported and marketed in glass bottles. The objective behind the bottling of the wines is to prevent its spoilage or deterioration by microorganisms or oxygen. The bottle's shape, size and color varies from brand to brand and from manufacturer to manufacturer (see Chapter 30 for details). The bottling room is a dust-free environment often maintained under a positive pressure of filtered air. The room is generally tiled or lined to allow for ongoing wash down and sanitization on a daily basis. The most critical stage of the bottling operation is the path of the open bottles from the filler to the corker, since these are least protected against airborne contamination at this point. Stainless steel canopies are generally used to cover these distances. The bottling line can involve a manual operation involving several people handling hundreds of bottles per hour to an automated line with several operators handling several hundred bottles per minute. Some lines are in the form of straight lines or U-shaped with materials flowing from one end to the other or returning to a point close to where they were initially dumped from their cartons. The area of the closed bottling hall is such that there is an easy movement of cases of empty and full bottles, and there is a restriction of non-essential foot traffic. The bottle fillers are of various types, such as siphon based, vacuum based, gravity based and pressure based. The gravity fillers range from small hand-operated units (6–10 spouts) to large continuous units (40–120 spouts). The total volume of wine in each bottle must be indicated on the label; overfilling is wasteful and leads to problems with the control of headspace pressures after corking. The customer often dislikes bottles showing different fill heights, so it is important to provide stated volumes, uniform fill levels and normal head space in each bottle. Too much headspace suggests a short fill or leakage since filling, while too little head space leads to wider pressure fluctuations with temperature changes leading to cork movement and leakage. There must be some system to sterilize the fillers. Generally, nitrogen or carbon dioxide is passed into the wine before filling to replace the oxygen present. The fillers on continuous lines take the bottles, which are supplied from the moving trays on one side, positioning them underneath the filling head to initiate the filling procedure. The bottle is usually taken in a complete circle from the point where it was picked up. The base plate on which the bottle stands is raised and filled. Once filled the bottle is lowered, shutting off the valve in the filling spout take place, and bottles are re-positioned on the conveyor and advanced towards the corker. Fillers can usually be modified with interchangeable parts to handle a range of bottle shapes, heights and volumes. The Automatic Bottle Filler allows for quick and easy filling of wine bottles. The filler is easily primed, self-leveling and shuts itself off when the bottle is full. The specially designed tapered tip cascades wine into the bottle. To fill up additional bottles, the automatic filler can be simply removed from the first bottle and placed into the next, while pushing down on the lever makes the wine flow. To achieve sterilization, a filling operation is generally carried out at a temperature of about 60°C.

A cork, used for the sealing the wine bottles, is the outer bark of plant *Quercus suber.* It has good compressibility and is inert in nature to various chemicals. Further, the cork should be smooth and without any defects. Its elasticity ensures continuous pressure on the neck of the bottle for a very long duration of time. It has very little nutrients so is relatively resistant to microbial attack, but it is recommended that cork be sterilized by gamma-radiation to prevent any off-odor formation due to microbial growth. A corking machine does the corking. This machine first compresses cork to a small cylinder and then the compressed cork is driven into the bottleneck by a plunger. Capsules (initially of lead or tin but nowadays of plastic) are used to cover the cork to protect it from cork-boring insects.

24.5.9 Labeling Machines

The traditional labeling operation involves the attachment of a glued front label to the bottle, with options for the attachment of a back label and a neck label, as required. The continuous labeling machine removes a bottle from the line during one cycle of its rotation, picks up a label, applies glue to it, places the label onto the bottle, and then brushes it around the bottle before returning the bottle to the main line. There is now a general trend toward the use of self-adhesive labels, peeled from a backing paper as they are applied, which eliminates the troublesome gluing operation.

24.5.10 Capsulators and Foiling Machines

The final operation is to place a plastic cover or metal foil over the corked bottle and to shrink this tight by heat or to spin-tighten it onto the bottle. The placement is often done by hand even for automated lines, while the spinning is done by machine. The cap or foil is applied for cosmetic reasons and to cover any wine that may have splashed or mold growth that may appear on the outer face of the cork. A number of other operations may be associated with the bottling sequence, but these usually have to do with the logistics of warehousing and shipping rather than bottled wine. These range from stacking cases to coding labels, bottles and cases with bottling dates and timers.

24.6 CLEANING OF THE WINERIES

General cleanliness of the floor, conveyors, floor of tanks, hoses, crushers, etc. is important, otherwise organic matter in these areas attract flies, insects and rodents etc., besides being unhygienic for human health. The dilute solution of hypochlorite is used as a good disinfectant for general cleaning and washing of the floor. Pyrethrin and other oil-based insecticides are also used, but pyrethrin is more common, being more effective for the long term than oil-based insecticides. The dispenser of vapors or DDVP (dimethylo-dichloro-vinylphosphate) in the form of pellets placed in a loading chamber assures effective control in cellars, with one loading, for a full season. Sodium hydroxide is also used. Alkaline phosphates are used as cleaning, dispersing and wetting agents. Usually non-ionic organic compounds like polyethylene glycol at low concentrations are also very effective. The agents, which form complexes with calcium and magnesium ions, are also used as suitable dispersing agents. EDTA is used most commonly as a complexing agent, at a concentration 0.1% removed 260 mg/L (expressed as calcium carbonate) of the hardness of the water. The most common chemicals considered as sterilizing agents are hypochlorite solution, quaternary ammonium salts and a mixture of sodium hydroxide and silicates, trisodium phosphate.

24.7 PROBLEMS ASSOCIATED WITH WINERIES

The growth of various microorganisms (aerobic, facultative anaerobes and anaerobes) that attack wine and deteriorate its quality can be prevented by carrying out fermentation and storage at low temperatures or by using SO_2 (refer to Chapter 41 of this text for microbial spoilage of wine). Vinegar fly, also known as the fruit fly (*Drosophila melanogaster*) is a notorious pest is attracted to the must, particularly the fermenting must. The female population of the fly deposit eggs and larva in the premises of the winery, which carry out another life cycle of the insects. So some method is needed to prevent the development of the eggs, and the larvae feed on the organic matter, pomace fruit etc. if they are not disposed of properly. Many insects are harmful to the winery (red fruit beetle, winged insects, beetles, cockroaches, crickets, houseflies etc.) and breed on human excreta, garbage and other disposed material. Many insects like cockroaches have omnivorous living habits and are dangerous for the winery, while ants and cockroaches can enter any area in the winery and live in cracks, crevices, under moist boards etc, causing nuisance if uncontrolled. Wasps and bees are also very difficult to exclude from areas like wineries. Removal of waste fruit from the winery premises as soon as possible is one of the most effective measures for the control of insects. Alternatively, some powerful insecticide should be used. Seepage and leakage of the tanks, lining, and other apparatus should be prevented. Walls cracks and crevices should be sealed to prevent the movement of cockroaches, ants, beetles and rodents. The use of good insecticides is also a preventative measure. Spraying with 0.1% pyrethrin and 1.0% piperonyl butaoxide repels the insects but does not kill them. The safe removal of wasp colonies can be achieved by professional pest controllers, and the safe removal of bee colonies can be done via bee protection organizations.

BIBLIOGRAPHY

Amerine, M.A., Berg, H.W., Kunkee, R.E., Qugh, C.S., Singleton, V.L. and Webb, A.D. (1980). Winery design, equipment, operation and sanitation. In: *Technology of Wine Making*. 4th ed. AVI, Westport, CT. p. 255.

Bird, D. (2004). *Understanding Wine Technology: The Science of Wine Explained*. DBQA Publishing, UK.

Boulton, R.B., Singleton, V.L., Bisson, L.F. and Kunkee, R.E. (1995). *Principles and Practices of Winemaking*. Chapman and Hall, New York. p. 279.

Butzke, C.E. (2010). Winemaking equipments, maintenance and troubleshooting. In: *Winemaking Problem Solved*. Woodhead Publishing House Ltd., Sawston, Cambridge.

Felder, R.M. and Rousseau, R.W. (2005). *Elementary Principles of Chemical Processes.* John Wiley & Sons, Inc., New Jersey.

Fiechter, A., Kappeli, O. and Meussdoerffer, F. (1987). Batch and continuous culture. In: *Yeast.* Vol. 2. 2nd ed (Eds A.H. Rose and J. S. Harrison), Academic Press Inc., London. p. 99.

Flores, J.H., Heatherbeli, D.A., Hsu, J.C. and Watson, B.T. (1988). Ultrafiltration of white Riesling juice: Effect of oxidation and pre-uf juice treatment on flux, composition and stability. *American Journal of Enology & Viticulture*, 39:180.

Joshi, V.K. (1998). *Fruit Wines.* Directorate of Extension Education, Dr. Y.S. Parmar University of Horticulture and Forestry, Solan, India. p. 255.

Margalit, Y. (2003). *Winery Technology and Operation: A Hand Book for Small Winery.* The Wine Operation Guild, San Francisco, CA.

Marsh, G.L. and Guymon, J.F. (1959). Refrigeration in wine making. *American Society of Refrigerating Engineers*, I. *Data Book.* Vol. I, Chapter, 10.

McCabe, W.L., Smith, J.C. and Harriott, P. (2014). Heat-exchange equipments. In: *Unit Operation of Chemical Engineering.* 7th ed. McGraw-Hill Book Co., New York. p. 440.

Meier, P.M. (1988). Aseptic filling using membrane cartridge filtration. In: *Wine East Buyer's Guide.* p. 15–17.

Moresi, M. (1989). Fermenter design for alcoholic beverage production. In: *Biotechnology Applications in Beverage Production*, C. Canterelli and G. Lanzarini (Eds). Elsiever Applied Science, London. p. 93.

Pirt, S.J. (1975). *Principles of Microbes and Cell Cultivation.* Blackwell Scientific, London.

Rayess, W.E., Albasi, C., Bacchin, P., Thailandier, P., Ranyal, J., Meitton-Peuchot, M.. and Devatine, A. (2011). Cross-flow microfiltration applied to oenology: A review. *Journal of Membrane Science*, 382(1–2):1.

Rupp, W.E. (2007). Diaphram pumps. In: *Pumps Handbook.* 4th ed., I.J. Karassik, W.C. Krutzsch, W.H. Frazer and J.P. Messia (Eds). McGraw Hill, New York.

Schorr, C. (1996). *Filtration Technology for Pharmaceutical and Biological Manufacturing.* Seminar Notes, Cuno Filter System, USA.

Singh, R.S. and Sooch, B.S. (2009). High cell density reactors. in production of fruit wine with special reference to cider-An overview. *Natural Products Radiance*, 84:323–333.

Vine, R.P. (1981). Winery and the laboratory. In: *Commercial Winemaking.* Springer, Dordrecht.

Vine, R.P., Harkeneis, E.M. and Linton, S.J. (2002). Winery design. In: *Winemaking.* Springer, Boston, MA.

Wang, D.I.C., Cooney, C.L., Demain, A.L., Dunnil, P. and Lilly, M.D. (2006). *Fermentation and Enzyme Technology.* Wiley, New York.

25 Preparation of Grape Must for Wine Production

Devina Vaidya, Manoj Vaidya, Swati Sharma and Nilakshi Chauhan

CONTENTS

25.1 INTRODUCTION

Wine is by common usage defined as a product of the normal alcoholic fermentation of the juice of sound ripe grapes. Grape wine is perhaps the most economically important product of alcoholic fermentation of fruit juice because of the commercialization of the product for industry. Accordingly, grape wine has receivedmost of the attention of research and development in oenology. Concerning the preparation of wine, extraction of juice and making must is the most crucial step, therefore in this chapter, we will describe the core aspects relating to juice extraction and the preparation of must inherent to that process.

25.2 FRUIT COMPOSITION, MATURITY AND HARVESTING

Grape clusters are made up of about 3% (2 to 8%) stems, 15% (15 to 20%) skins, 4 per cent (0–6%) seeds, with the flesh/ pulp/ juice accounting for the remaining 78% (74 to 90%). Two parts form a grape bunch: the stem constitutes the body and the berries are composed of skin, pulp and seeds (Figure 25.1). The stem constitutes about three to five per cent of the grape weight. It is rich in water, minerals and contains proanthocyanidins that impart to the wine a certain amount of pungency (stem taste). However, it is important to note that sometimes this is not a good taste for the wines. It is be necessary to have an adequate good stem maturation level, if not, the stem contact is negative. So, usually de-stemming is done. Skin represents tenper cent of the weight of the bunch. A thin white coat called "prune", on which the necessary yeasts deposit to transform grape juice into wine, covers the grape. Red grape berry skin is rich in "phenolic compounds", namely anthocyanins, and this is what colors the wine. Lastly, some aromatic substances, specific to each variety, are present in the skin in the form of aroma precursors.There can be two, three or four seeds per berry, which constitute four to five per cent of the bunch weight. They are rich in tannins and oily substances (lipids). Winemaking techniques do not attempt to extract these substances and contact with the wine is prevented.

The maturity of grapes is an important factor that determines the quality of wines (see Chapter 9 for more details). Fine wine results only from grapes with the proper balance of sugars and acids. Under-ripe grapes may yield thin wines with high acid and little flavor, whereas overripe grapes may result in wines with low acid, high pH and consequently poor quality. The general chemical parameters of grapes during different maturity levels in different cultivars are shown in Table 25.1. At normal fruit maturity, enlargement and physiological accumulation of sugars cease at about 25° Brix in wine grapes. Grape skins are ordinarily the sole source of anthocyanins. Pressure alone at a commercially practical level does not release much red color from fresh skins, but after heating to about 70°C or greater, or with alcohol as in fermentation, the anthocyanins readily leach into the juice.

Harvesting of grapes is done either manually or mechanically. Of the two methods used in harvesting of Cabernet Sauvignon grapes, wines exhibited no significant difference in quality, while mechanically harvested white CV Riesling grapes gave wine a slightly higher quality compared to that obtained from hand harvested grapes. The mechanically harvested lots of both the varieties contained fewer leaves than the hand harvested grapes. The hand harvesting of Chardonnay gave the highest yield and when pressed, yielded the highest volume of juice per ton (Table 25.2), while the

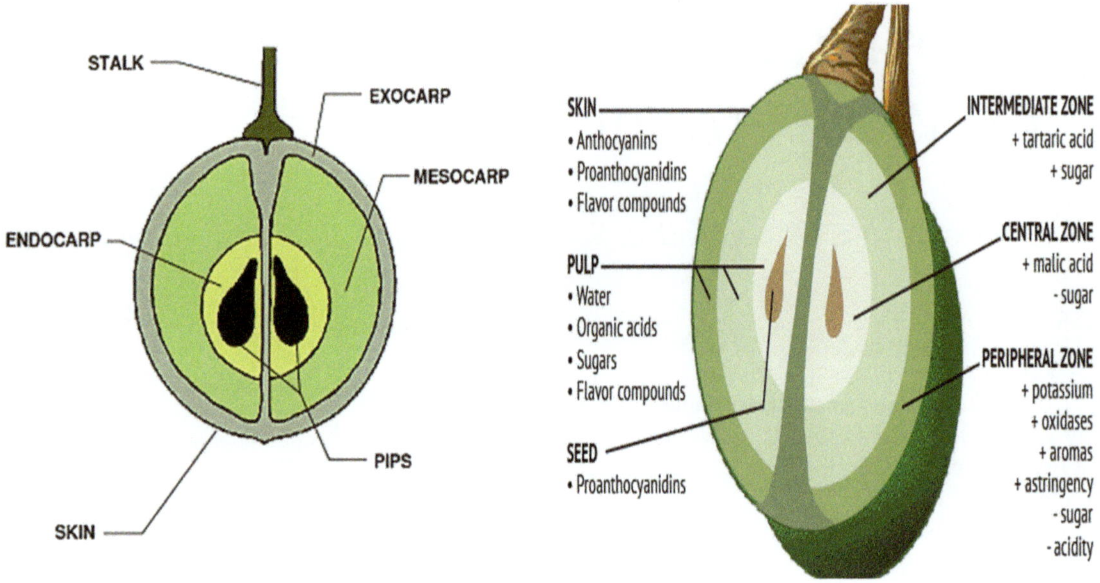

FIGURE 25.1 Grape berry and a look inside of grape

TABLE 25.1
General Chemical Parameters of Grapes During Different Maturity Levels

Variety	Maturity	TSS (⁰Brix)	Total acidity*	pH	Sugar-acid ratio
Arka Kanchan	Early	14.10	072	3.26	19.60
	Mid	14.50	0.54	3.32	28.90
	Late	15.60	0.45	3.46	34.60
Thompson Seedless	Early	17.80	0.84	3.13	21.20
	Mid	19.50	0.68	3.30	28.70
	Late	21.60	0.60	3.48	36.00
Bangalore Blue	Early	16.80	1.28	3.03	13.10
	Mid	20.00	1.13	3.10	17.70
	Late	20.40	0.89	3.25	22.90
Arka Shyam	Early	16.50	0.78	3.18	21.80
	Mid	17.90	0.73	3.30	24.60
	Late	18.80	0.68	3.52	27.70

*g tartaric acid /100ml
Source: Arora and Singh (2008)

TABLE 25.2
Effect of Method of Harvesting on the Grape Must Quality

Treatment	Temp (⁰C)	Yield (L/T)	TA *(g/100ml)	SS (⁰Brix)**
Cane shaker	12.8	782	0.75	22.2
Trunk shaker	16.1	765	0.76	22.2
Hand harvest	20.6	790	0.74	22.6

*g tartaric acid /100ml, ** soluble solids
Source: Jackson and Schuster (2008)

wine made from mechanically harvested grapes exhibited slight oxidation without a significant difference in analytical composition or sensory quality. Mechanical harvesting results in a must containing significantly lower sugar content and higher acid. The reason may be difficulties of mechanical harvesters in distinguishing between ripe, healthy grapes and unripe or rotted bunches which must then be sorted out at the winemaking facility. Another disadvantage is the potential of damaging the grape skins, which can cause maceration and coloring of the juice that is undesirable in the production of white and sparkling wine. The broken skins also bring the risk of oxidation and a loss of some of the aromatic qualities in the wine.

25.3 CRUSHING AND DE-STEMMING

After harvesting, the individual grapes are separated from the stems (which would add unwanted bitterness to the must) by a stemmer (Figure 25.2) and then pumped into one of the four presses to keep up the production according to the quantity

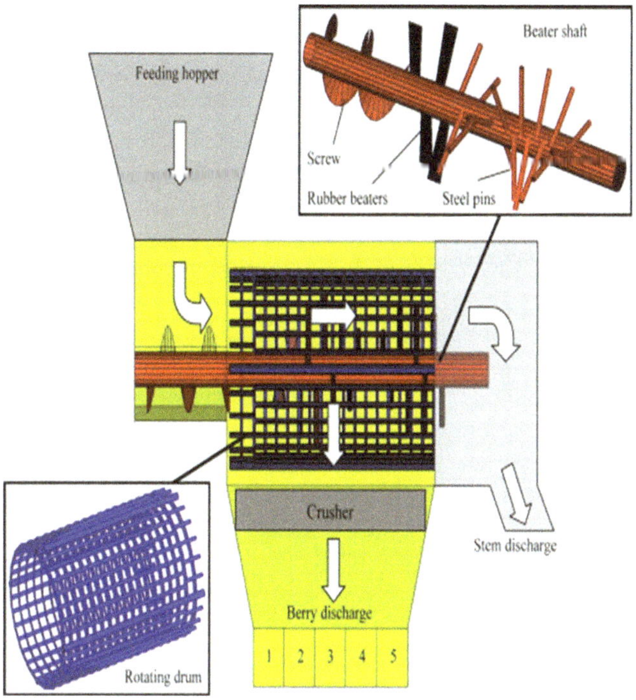

FIGURE 25.2 Grape crusher and de-stemmer

of grapes harvested every day. To press the grapes as soon as possible after being harvested is essential to avoid their oxidation or deterioration in the heat. Crushing is done first to break the grapes skin (normally, almost all the berries are broken) and let the juice escape. Crushing contributes for native yeast release to the grape must and also activates the enzymes liberated from damaged grape cells that begin the autolytic breakdown of cellular constituents and the release of flavor compounds bounded in the grape skins. In addition, cellular enzymes promote the oxidation of lipids and phenolic compounds that favor the early polymerization and loss of easily oxidized phenols. The latter provides protection from the oxidative browning of bottled wine. Oxidation of cellular components also rapidly removes the oxygen normally absorbed during crushing, limiting the growth of aerobic spoilage microbes found in the juice.

In white wine making, before pressing, the crushed grapes are sometimes transferred to a vat to allow the partially freed juice to remain in contact with the skins for a day or two so as to draw out the aromas and flavors components lying just below the skin, a limited maceration in fact for producing white wines with high aromatic compounds, but generally for white wines no skin contact is made. The grapes must be kept very cold to prevent any fermentation starting, and they are usually kept under a blanket of inert gas to prevent oxidation. If contact levels are higher, excessive tannins are also picked up and the wine is likely to be coarse and astringent. However, if the stems, from most varieties, are left in contact with the crushed grapes for an extended time, a "stemmy" off-character or greasy character will be imparted to the wine. Crushed grapes are then generally de-stalked and de-stemmed, thuseliminating the woody part of the bunch. The stemming can be done by several methods. The older type stemmer or crusher has a rapidly spinning blade at the mouth of the crusher, which immediately impacts the cluster and stems it. The stems and berries are thenmoved along an inner perforated drum by a series of blades. The berries are thrown-out through the perforations into a slower-moving outside drum with perforations of smaller diameter. A large sweeping blade pushes the berries out through the perforations, crushing the remaining whole berries. The stems remaining in the inner drum are moved to the end of the drum and discharged. Some crusher stemmer are roller type. In this type of crusher, the grape clusters drop onto the adjustable rollers and are crushed, and then stems and berries are dropped into a perforated drum with several round spinning blades. In this crusher, the juice is moved out through the perforations and stems out the end. The Garolla crusher was introduced from Italy to California and has a large, horizontal, coarsely perforated cylinder. Revolving blades inside the cylinder move the stems toward the exit and hammer the bunches of grapes. The combined effects of the moving parts do a thorough job of crushing and stemming. The major drawback of this type of crusher stemmer is that the excessive quantity of stems are left in the must especially when the paddles are operated at a high speed, but this can be corrected by adjusting the pitch of the blades. The crusher should be made of stainless steel. The differences in the content of juice results from the amount of shearing and tearing force applied to the berry. Berries that are not firm or more shrivelled are more difficult to break open. But application of too much force may cause the skins of the berries to break into many fragments. Strongly buffered cellular juices elevate the pH of the juice as well as of the wine. Since the grape skins are highly macerated, they become a problem in the clarification of the juice. During crushing, the seeds should be kept intact because if the outer shell of the seed is broken, the high level of phenolic material, which the seeds contain, would give a bitter taste to the wine. In addition, prolonged contact between the released juice and stalks encourages the uptake of bitter tannins. Thus, stemming is conducted in conjunction with crushing.

25.4 MUST PREPARATION AND HANDLING

Must preparation is an important operation that constitutes basic differences in the preparation of white or red wine. Inadequate or inappropriate treatment of the original must may cause a number of defects and undesirable features, such as low titratable acidity, high pH, the formation of significant amount of hydrogen sulfide or volatile acidity, the incidence of incomplete fermentation and low intensity of fruit aromas. The grape juice composition is highly influenced by the freshly crushed grapes. A variety of components have different concentrations in red grape juice, pulp and skin, and the most common components are phenolic compounds, minerals and terpene fractions. So in the preparation of red wine, the skin contact is important to promote the extraction of these components, which involves the use of pectic and/or other hydrolytic enzymes and presses which allow the extraction of juice with minimum amount of skin tearing. Coloring matters responsible for the red color of the wines are contained in the red grape skins. However, for tinturier grape varieties, it is also possible to detect high levels of anthocyanins in grape pulp. When pressed, the juice is transparent, so it is necessary to leave the skin in contact with the juice so that the color is extracted in the must. This is called maceration. The standard maceration consists in the simple contact of grape solid parts and must. Maceration period may be extended depending on variety from a few days to months. Limited maceration ($\leq$24 hrs) is favored for the cultivars possessing a distinctive varietal aroma as enhanced extraction of varietal aroma compounds and improved coloration often offset the production of herbaceous odors.

Browning or discoloration of white wines remains a problem in many wines. Types of cultivars, skin content and pressing regime can alter the phenolic composition and thus effect browning. Polyphenol oxidase activity is usually responsible for the browning of non-sulfited juice. The improvement in white wine color seems to be related to the rapid oxidation and polymerization of readily oxidizablephenolics prior to or during fermentation. Hydroxycinnamate derivatives account for most of the non-flavonoid phenol content of grape juices. These compounds serve as the primary substrate for attack by polyphenol oxidase (PPO). The oxidation of juices prior to fermentation was found to reduce cafteric acid content but appeared to increase the wine bitterness. White wines made from oxidized musts were comparable in color with control wines,were not judged as inferior in flavor and were also less susceptible to oxidative browning. To retard the inception of fermentation during maceration, white grape juice is chilled to about 10°C, as the cool temperature favors the production of fruity esters during fermentation and retards the excessive uptake of phenolic compounds. The addition of SO_2 further retards the microbial activity. Nevertheless, the use of SO_2 is declining as it suppresses the action of macerating enzymes and limits the beneficial effects of limited early oxidation. Only if the fruit is partially diseased, the addition of SO_2 is recommended.

Red grapes must is given long maceration at a temperature between 24° to 27°C, which encourages the extraction of pigments and other phenolic compounds essential to the character of the red wines. The duration of maceration can vary from as little as three days to more than three weeks, depending upon the grape cultivars and how quickly the wine is made. Red wines designed to mature quickly tend to have short maceration periods (three to five days) while those intended to age for decades receive a long maceration time (two to three weeks).

25.5 PRESSING

When the grapes are transferred to the press, about a third of the available juice will be free run, and light or gentle pressing will express a further two-thirdsof the juice. This juice is best for white wine making and is often fermented, separately. The Raymond presses are continuous screw presses designed to extract a maximum amount of juice with a minimum amount of pressure, having more clarity without tannins or flavors from the grape skins. Further pressing will yield a juice that is coarser and more astringent, as it extracts more tannins from the skins. After a period of settling that allows the must to clarify, the acidity can be adjusted if necessary, and fermentation can then begin. There is a bewildering array of presses, but all fall into one of the two main categories, batch or continuous. Batch presses require repeat cycles of filling, pressing and dumping. In contrast, continuous presses run without interruption. Juice or fermenting must is added at one end and pressed pomace is received at the other end. Continuous presses are more efficient and rapid than batch presses. Despite this, batch presses are considered more "gentle" with less movement of the grape skins which minimizes the amount of tearing of the skins. The more the grape skins are torn or scoured, the more phenolic compounds and tannins are extracted, which can increase the harshness of the wine. The high pressures usually required in continuous press operation often extract excessive amounts of suspended solids and phenolic compounds that are undesirable for the sensorial characteristics of the wines. Thus, they are currently deactivated for the production of high quality wines. At low levels (about 0.1 to 0.5 %), suspended solids encourage rapid and complete juice fermentation while their higher levels are often associated with unacceptable levels of hydrogen sulfide production and difficult clarification.

To achieve better control over the levels of suspended solids, most premium wine producers use pneumatic batch presses (Figure 25.3) that apply more uniform pressure over a large surface, permitting the use of lower pressures. Pressing does entail, however, several cycles, each separated by a crumbling that breaks up the pressed pomace and permits easier extraction of the entrapped juice/wine. The individual press fractions are often kept separate for independent fermentation. This is particularly common with red wines where the latter pressing contains the highest concentration of pigments and tannins. Thus, it is important to say that pneumatic presses are the most adequate presses for all wines but in particular for white wines. The wine produced from different fractions may be selectively blended to produce the features desired in the finished wine. The juice that is expressed from the skin can be significantly different from that which is not. The extent to which these differences are depends primarily on the type and operation of the press (Table 25.3) and secondarily, on the condition of the grapes.

25.6 CARBONIC MACERATION

The concept of anaerobic fermentation of grape musts, dating back to Pasteure's time, is used in wine production including red wine. Storage of whole grapes before crushing, either for red or white grapes, has been recommended. In red grapes, the skin can be separated and stored under CO_2, and after storage, the color and flavor leaches out with new wine. So the whole red grapes should be held for six to tendays prior to crushing. Fermenting on the skins for one or two days before pressing is indicated where the pressed or drained skins are held until

FIGURE 25.3 Continuous screw presses and pneumatic batch press

TABLE 25.3
White Juice Composition from a Screw Press

Physico-chemical Characteristics	Free run	During filling	Press fraction 1	2	3
Juice density (°Brix)	NA	17.2	17.5	17.5	17.5
PH	3.1	3.1	3.2	3.4	3.5
Titatable acidity (g/L)	8.9	9.9	9.1	8.8	9.1
Tartaric acid (g/L)	4.5	4.1	3.3	3.1	3.1
Total phenol content (mg/L)	306	472	607	1142	1988
Condensed phenols (mg/L)	65	176	211	738	2045
Brown color (absorbance unit at 420 nm)	0.149	0.249	0.412	0.730	0.965
Solids (g/L)	66.2	42	16.8	27.9	23.7

Source: Boulton *et al.* (1999)

the free run juice fermentation is finished. However, for white grapes, there was no gain in quality with the process except possibly with very neutral grapes. Wines resulting from carbonic maceration (CM) (Figure 25.4) have a distinct aroma, in which vague fruity notes predominate as a consequence of the anaerobic metabolism of the grapes. Because these undergo carbonic anaerobiosis, the alcoholic fermentation of these wines is rapid, that is why these are more biologically stable than those wines which undergo traditional vinification processes. Carbonic macerated wines are principally intended for immediate consumption and can be marketed sooner.

The other technique consists of depositing whole grapes, neither crushed nor de-stemmed, in a carbon dioxide enriched atmosphere. An intracellular fermentation then takes place, inside the grape, under the action of enzymes that transform a small quantity of sugar into alcohol (about 2% v/v vol.) (Figure 25.4). There is also production of CO_2, a little glycerol and various secondary products. After this maceration phase, the grapes are then pressed, and the alcoholic fermentation happens normally. When maceration is of short duration, wines obtained are supple and smooth, but a longer maceration (10 to 20 days) can produce wine needing aging.

25.7 GAS BLANKETING

The juice released from the grapes at crushing is saturated with O_2. The taking up of O_2 from the air during the harvesting, transporting, crushing, draining and pressing operations can be minimized by the use of inert gas blanketing. The inert gas case is a desirable storage condition for white grape must but is a cause for poor cell viability in the latter stages of alcoholic fermentation. But when natural micro-flora are used to conduct the fermentation, there is competition for oxygen between the oxidases and the yeast, in which the oxidases would appear to have the advantage. This problem is enhanced by the presence of significant levels of suspended solids, which contain additional oxidase activity in white juices, and incomplete fermentation caused by this continues to be a problem in some countries. This problem can be minimized by inhibiting the oxidation enzymes with the addition of SO_2. Settling of juice also decreases the activity of tyrosinase enzyme, mainly responsible for undesirable effects in wine, because it is largely associated with the solid parts of the grape berry. Bentonite fining has also been found to do this, with 100 g/hL leading to a 30% loss in activity, but it also removes glutathione. Heating of the must to 45 and 65°C will destroy tyrosinase and

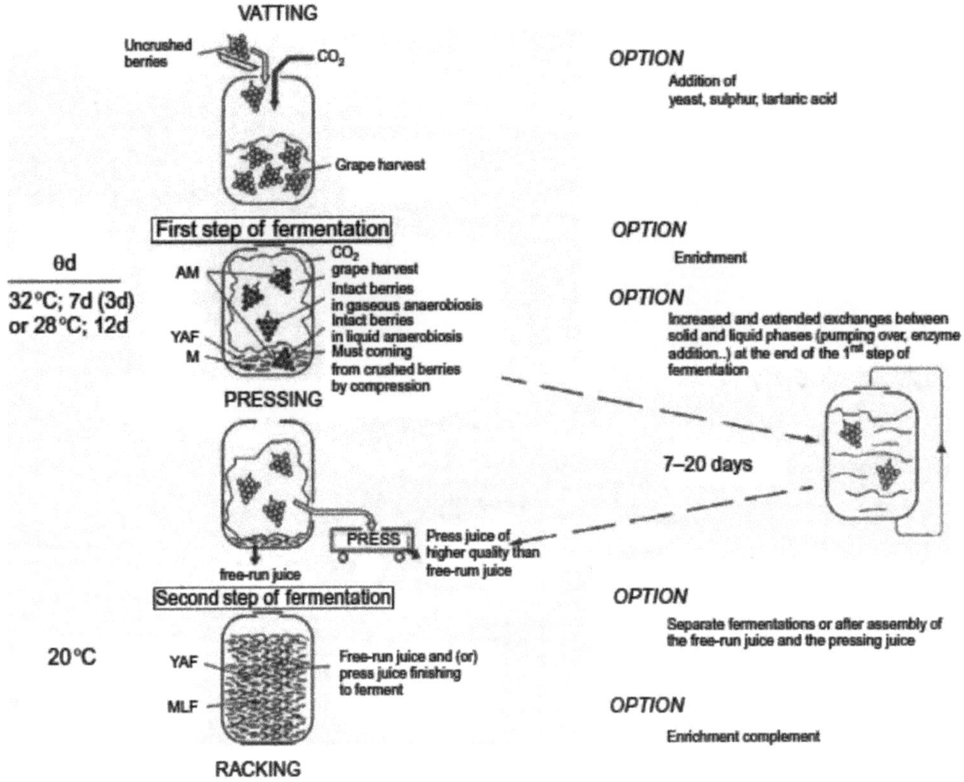

FIGURE 25.4 Scheme of carbonic maceration winemaking. AM, anaerobic metabolism of grape berries; YAF, yeast alcoholic fermentation; MLF, malolactic acid fermentation; M, maceration; temperature (⁰C); action duration (days). *Source:* Tesniereand Flanzy (2011)

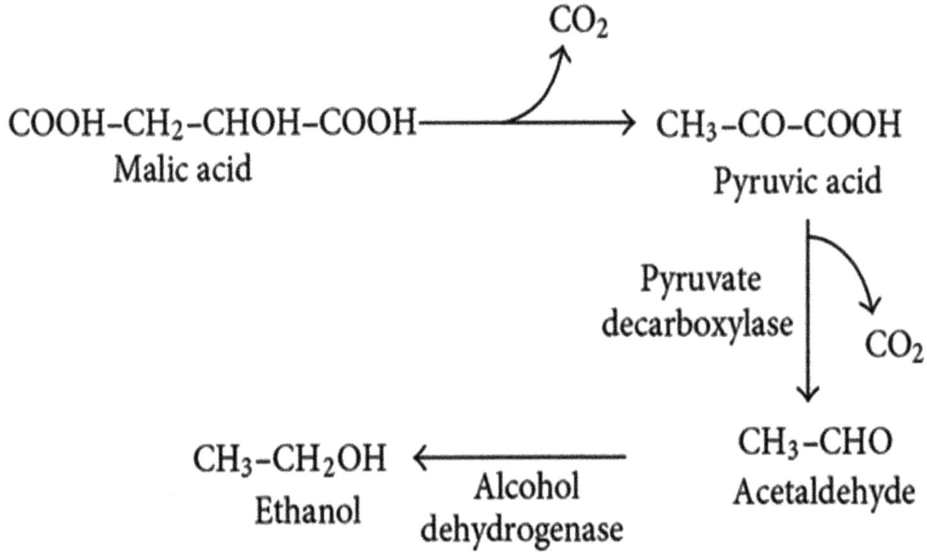

FIGURE 25.5 Enzymatic reactions of carbonic maceration. *Source:* Marquez *et al.* (2014)

laccase, respectively. Another strategy to prevent oxidation is to limit the phenolic substrates available for oxidation, especially the flavanoid content, by soft pressing, no skin contact and removal of stems. A process called hyperoxidation, where large quantities of O_2 are added to the must, can also achieve this. The latter leads to the oxidation of phenolic molecules which settle, and the juice can then be removed from the precipitate by racking, with no SO_2 added to the must at crushing. Juice that did not receive any skin contact can thus be treated with one saturation, but up to three saturations are necessary to remove sufficient flavanoid molecules from juice that did have skin contact. It is imperative that the subsequent clarification is done efficiently before fermentation starts because the precipitate can re-dissolve in alcohol. The reductive conditions during alcoholic fermentation and adsorption to yeast cells reduce the brown color further.

25.8 JUICE AND SKIN SEPARATION FOR WHITE WINES

The juice being prepared for white table wine should be separated from its skins for two main reasons – because the skin represents a major source of natural microflora, coupled with the desire to minimize the extent of phenol extraction from skins. Commonly, the use of gravity settling for separation of skins from juice in vertical tanks is practiced. Drainers are the specially designed equipment, which have a special kind of screen for separation of free run juice from the associated skins. The separation is important for the production of distinctive varietal white wines and may take place immediately after crushing or after a period of juice and skin contact. In some wineries, cylindrical fermentation tanks are used as drainers that allow the skin cap to rise and then drawoff the juice. There are static drainers, in which the skins are not moved across the screen while juice is being removed; others are continuous drainers in which skins move across the screen by sliding over the surface or by an advancing helical screw during juice removal. Large wineries use de-juicers for the process to reduce the time that juice/ must spends in expensive presses. By comparison, de-juicers are inexpensive, simple to operate and have a screw mechanism that lifts the must up a slope, during which juice escapes through a sieve into collecting bays. The de-juiced must is fed into a press where pressure extracts most of the remaining juice. These units do provide considerable juice recovery prior to pressing, but the suspended solids content increases significantly (4% v/v), which requires further clarification commonly by centrifugation or settling prior to fermentation.

25.9 CLARIFICATION OF MUST FOR WHITE WINES

Must is commonly clarified before fermentation, as better wines are produced from clarified juices than from very turbid juices, though the unclarified juices can reportedly make more flavorful wines that those after pectic enzyme treatment. Settlingis by far the slowest method of solid separation, usually improving wine quality. Centrifugation speeds up solid separation and produces better wines than untreated musts do, but it is a not good option for white wine production, in particular for high quality ones. This option introduces an excess of O_2 addition, reduces the yeast population from the grape must and also the nutrient content, etc. So, today it is not a good option for wines of high quality. Pectolytic enzymes help clarify juice but are variable in their sensory effects. The use of fining agents are also an important option for white must clarification (for example by the use of bentonite or other fining agents).

After pressing, the must is filtered using suitable methods. An early clarification of musts followed by a rapid racking causes an almost complete lack of sterols in the clarified fraction as was the case for fatty acids. Fatty acids and sterols in fact are important for yeast growth, fermentation activity and for aromatic volatile compound synthesis.

In producing white wines, juices usually receive some pre-fermentative clarification. Retention of high levels of suspended solids reduces the wine's fruitiness and enhances concentration of higher fusel alcohols. The solids also contain polyphenol oxidases that flavor excessive oxidative browning. Conversely, overzealous removal of suspended solids can cause fermentation to terminate prematurely. This usually is the consequence of the removal of long chain unsaturated fatty acids needed in synthesis of cell membrane sterols. For most white grape juices, simple settling for several hours provides sufficient clarification, but bentonite may be added to promote this process. Filtration and floatation are other options to achieve clarification, whereinfine gas bubbles (commonly N_2) are introduced in slow moving juice, and the suspended pulp becomes attached to the bubbles because of surface tension and starts floating upwards where it can be collected. It can be used as a continuous, relatively fast, low energy process for the clarification of more than 80 per cent of white juice. Wines made from clarified juices are easier to clarify. Musts of moldy grapes should always be clarified prior to fermentation. To obtain a finer wine, fermentation must occur with clear juice. To this end, the must is racked which eliminates vegetal particles (bits of skin, peduncle etc.) and suspended impurities and can be done by gravity, by cooling the wine, by using fining agents and by the use of filtration.

There is a more extensive removal of larger grape pulp particles by centrifugation than by natural settling. Disc centrifuges are used in wine making throughout the world to clarify settled white juices and wines. These centrifuges provide good work capacity, but not good results for wine quality, *i.e.* disc centrifuges yield juices with a lower solid content, while decanting centrifuges are more suitable for handling juices with a higher solid contents. The solids are replaced at one end as a relatively dry paste while the clarified juice leaves at the other end, thus, not providing completely clarified juices. The use of diatomaceous earth filters for the clarification of juice has declined nowadays. Rotary drum filters are still used because of the ease of removal of the filter cake as compared to pressure leaf and plate, and frame filters. But the absorption of amino acids by diatomaceous earth takes place, which is not a major occurrence during normal filtration. The first juice to contact the earth cake may be significantly depleted of these components, but once pre-coat becomes saturated, it cannot adsorb further. Nowadays, cross-flow filtration is commonly used for this purpose because of its ability to perform under a wide range of suspended solid contents. It is best used in combination with other equipment which provides partially clarified juice rather than for use with turbid juices directly.

25.10 JUICE AND MUST TREATMENTS

According to OIV rules, some organic acids and nutrients are permitted in the majority of wine country members but with some limits. The least contentious modification involves correction of total acidity of the must. If the acidity level is undesirably low (>5 g/L), acids such as tartaric or citric acid may be added to develop flavor, color and microbial stability.

More difficult to rectify is the condition of excessively high acidity. Blending with low acid juice can be effective but is often impossible in small wineries with low must volumes. Deacidification may be achieved with the addition of calcium or potassium carbonate salts or using the biological deacidification techniques. The amount of alcohol and sweetness in the finished wine depends on the amount of sugar in the juice. If the grapes are low in sugar, extra sugar may be added before fermentation in the case of sweet wines, but it is not allowed for table wines in many wine-producing countries. The total sugar content, including the natural sugar present in the grapes, should not be more than 20 per cent for good dry table wine. Higher amounts of sugar produces a wine with-harsh flavor because of too much alcohol and could induce problems during alcoholic fermentation, as excess ethanol produced limits the yeast activity. A sugar content of 20 per cent will therefore produce 11–12 per cent alcohol (v/v) in a completely fermented wine. For dessert wines, which are sweeter, the alcohol may be increased to 14–15 per cent (v/v) or more than 15 per cent by using a total sugar content of 24 per cent. Hydrometer or refractometer is used to measure the sugar which should be calibrated in per cent sugar (Balling or Brix scale) and should cover a range from 0 to 35 per cent. A simple calculation for the amount of sugar to add is by Pearson's square method. Newer techniques that can increase sugar concentration without the addition of sugar to do so involve the removal of water. Reverse-osmosis,cryoextraction and entropie concentration are included in such methods. Juice with reduced alcohol-producing potential may be desired for producing low alcohol wines. It is possible without dilution with water, by the addition of glucose oxidase. More details are found in Chapter 36 of this book.

Normally, grapes have enough nutrients to complete fermentation of sugar to alcohol, but sometimes when grapes are moldy or juice has overly been clarified, nutrients, *e.g.* ammonium salts, may be needed to achieve complete fermentation. An amino acid presently permitted in the United States as an additive to juices is glycine. The use of SO_2 to restrict the extent of juice browning and to inhibit or kill most of the natural microflora in the juice has been practiced for many decades. Ascorbic acid can be used as an antioxidant either alone or in conjunction with SO_2. It is uncommon that grape must needs color adjustment before fermentation with moldy grapes, though fungal phenol oxidases may induce browning or the oxidative loss of anthocyanincolor. The addition of SO_2 at normal levels does not inactivate phenol oxidase. The only treatments may be heating the must to over 60°C to inactivate the phenol oxidase and a procedure called thermovinification, which is also used to improve color extraction from weakly-colored grape varieties. This method also favors the rapid fermentation of the treated juice. The must is usually pressed immediately after heating, and only the juice is fermented. A disadvantage is that it may result in production of a cooked flavor and impart a blush tint to the color. A number of enzyme preparations have been proposed for applications in juices and wines. These are important hydrolyzing enzymes such as the pectic enzymes, proteases, glucanases and cellulose glucosidases. The use of pectic enzyme helps to prevent the development of pectin hazes in wines in addition to grape must clarification and also increases the anthocyanin extraction with enhancement in free run yield in draining and pressing operations.

Fruits like apple, peach, plum, pear, strawberry, apricot, mango, litchi etc. are also made into wine. In preparation of wines from these fruits, musts are also made either from their juice or pulp. However, in the must preparation, these are ameliorated, but the addition of additives and their concentrations depends upon the type of fruits used. More details can be seen in the literature cited and in Chapter 37 of this text, which is devoted to these aspects.

BIBLIOGRAPHY

Arribas, M.V. and Polo, M.C. (2005). Wine making biochemistry and microbiology: Current knowledge and future trends. *Critical Reviews in Food Science and Nutrition*, 45(4):265–286.

Blair, R.J., Francis, M.E. and Pretorius, I.S. (2005). *Advances in Wine Science.* Australian Wine Research Institute, Urrbrae, Australia.

Boulton, R.B., Singleton, V.L., Bisson, L.F. and Kunkee, R.E. (1999). *Principles and Practices of Winemaking.* Springer, New York, NY.

Chambers, P.J. and Pretorius, I.S. (2010). Fermenting knowledge: The history of winemaking, science and yeast research. *EMBO Report*, 11(12):914–920.

Clary, C.D., Steinhauer, R.E., Frisinger, J.E. and Peffer, T.E. (1990). Evaluation of machine vs hand harvested chardonnery. *American Journal of Enology and Viticulture*, 41:176–181.

Delfini, C. and Formica, J.V. (2001). *Wine Microbiology: Science and Technology.* Marcel Dekker, New York, NY.

Edwards, C.G. (2006). *Illustrated Guide to Microbes and Sediments in Wine, Beer, and Juice.* WineBugs LLC, Pullman, WA.

Farkaš, J. (1988). *Technology and Biochemistry of Wine*, Vols 1 and 2. Gordon & Breach, New York, NY.

Flanzy, C. (ed.) (1998). *Oenologie: FondementsScientifi ques et Technologiques.* Lavoisier, Paris.

Fleet, G.H. (ed.) (1993). *Wine Microbiology and Biotechnology.* Harwood Academic, New York, NY.

Fugelsang, K.C. and Edwards, C. (2006/7). *Wine Microbiology. Practical Applications and Procedures*, 2nd edn. Springer, New York, NY.

Jackson, D. and Schuster, D. (2008). *The Production of Grapes and Wines in Cool Climates*, 3rd edn. Dunmore Press, Auckland, NewZealand.

Joshi, V.K. (2009). *Production of Wines from Non-Grape Fruit. Natural Product Radiance.* Special Issue July Augest. NISCARE, New Delhi.

Joshi, V.K. (2011). *Handbook of Enology: Principles, Practices and Recent Innovations*, 3 Volumes set, Asia - Tech Publisher and Distributors, New Delhi, p. 1450.

Joshi, V.K. (2017). Pome fruit wines: Production technology 7.1. Specific features of table wine production technology Chapterc. In: *Science and Technology of Fruit Wines*, Eds. Kossovea, Maria, V.K. Joshi and P.S. Panesar, Elsevier, USA, pp. 295–347, 7.

Joshi, V.K., Panesar, P.S., Rana, V.S. and Kaur, S. (2017). Science and technology of fruit wines: An overview chapter 1. In: *Science and Technology of Fruit Wines*, Eds. Kossovea, Maria, V.K. Joshi and P.S. Panesar, Elsevier, USA, pp. 1–72.

Joshi, V.K., Thakur, N.S., Bhat, A. and Chainkya, G. (2011). Wine and brandy: A perspective. In: *Handbook of Enology*, Vol. 1. AsiaTech Publication, New Delhi, p. 1.

Kelebek, H., Selli, S. and Canbas, A. (2013). HPLC determination of organic acids, sugars, phenolic compositions and antioxidant capacity of orange juice and orange wine. *Microchemical Journal*, 9:20–24.

Kocher, G.S., Phutela, R.P. and Gill, M.I.S. (2009). Evaluation of grape varieties for wine production. *Indian Journal of Horticulture*, 66:410–412.

Kosseva, M.R., Joshi, V.K. and Panesar, P.S. Eds., (2017). *Science and Technology of Fruit Wine Production*. Academic Press is an imprint of Elsevier, London, UK, pp. 705.

Marquez, Duenas M., Serratosa, M. and Merida, J. (2014). Identification by HPLC-MS of anthocyanin derivatives in raisins. *Journal of Chemistry*,1: 1-7.

Sachhi, K.L., Bission, L.F. and Adasms, D.O. (2005). A review of the effect of winemaking techniques on phenolic extractionin red wines. *American Journal of Enology and Viticulture*, 56(3):197–206.

Tesniere, P. and Flanzy, C. (2011). Carbonic maceration wines; Characteristics and winemaking process. *Advances in Food and Nutrition Research*, 63:1–15.

Thakur, A., Arora, N.K. and Singh, S.P. (2008). Evaluation of some grape varieties in the arid irrigated region of north-west India. *International Society of Horticultural Science*, 785:79–84.

Vaidya, D., Vaidya, M. and Sharma, S.. and Ghanshyam (2009). Enzymatic treatment of juice extraction and preparation, and preliminary evaluation of Kiwi fruit wine. *Natural Products Radiance*, 8(4):380–385.

26 Culture of Wine Yeast and Bacteria

Tek Chand Bhalla, Navdeep Thakur and Savitri Kapoor

CONTENTS

26.1 INTRODUCTION

The species and strains of *Saccharomyces* specific to wine, apple and pear cider production were first reported by Müller-Thurgauand referred to the addition of yeast cells to beer and bread for enhancing the rate of fermentation. The developments in areas of microbial physiology, biochemistry, genetics, molecular biology, ecology and biochemical engineering have greatly contributed in understanding the extent and behaviour of yeast as a fermentation agent and thus, transforming the art of wine making into a science and technology of wine production. The starter culture or inoculum for conducting wine fermentation may comprise: i) indigenous microfloral yeasts present on the grapes, ii) dry wine yeast, iii) wet yeast (*i.e.* yeast on slants or petri plates), iv) lyophilised yeast, and v) immobilized yeast (entrapped in gels). Besides yeast, lactic acid bacteria, generally *Leuconostoc oenos* is used by wine producers for induction of malolactic fermentation. Here in this chapter, some aspects of the cultivation of wine yeast and lactic acid bacteria have been discussed.

26.2 WINE YEAST: CHARACTERISTICS AND ESTIMATED DEMAND

It often becomes difficult to select a yeast strain from commercial sources for wine fermentation. Wine yeast strains exhibit a lot of variation in their physiology and biochemical characteristics. Some desirable properties of wine yeast important in the selection of a suitable yeast strain for wine production include: (a) fermentation properties: rapid initiation of fermentation; fermentation of variety of carbohydrates; high osmo-tolerance; high ethanol tolerance (up to 15% v/v or more); ability to ferment at low temperature and under pressure; moderate biomass production; (b) flavor characteristics:

production of desirable aldehydes and ketones; reduced sulphide/dimethyl sulphide/thiol formation; low production of higher alcohol and volatile acidity; high glycerol production; potential to produce β-glucosidase to enhance wine flavor; enhanced production of desirable volatile esters; (c) technological properties: high genetic stability; high sulfite tolerance; low foam formation; flocculation properties; compact sediment; resistance to desiccation; resistance to copper; zymocidal (killer) properties; genetic marking; low nitrogen demand; secretion of pectinases, glucanases and proteases; and (d)metabolic properties with health implications; reduced ethyl carbamate (urea production); low biogenic amine formation; low sulfite formation; yeast oxidation of polyphenols.

The desired yeast and bacterial strains for the production of wine are either obtained from commercial sources or propagated by the industry itself for use as inoculum. More than 100 strains of wine yeasts are available in the market from various sources and are usually available in the form of dry powder or tablet, so-called dry wine yeast. Based on the yeast usage of 0.1-0.2 g/L wine, approximately 5,000 tonnes of dry wine yeast is required to produce the total estimated quantity (3.25×10^{10} L) of wine produced in different countries of world. But the actual market for dry wine yeast does not exceed 600 tonnes per annum because many wineries either use the indigenous yeasts or propagate the wine yeast in-house.

26.3 STARTER CULTURES OF YEAST

Wine fermentation is either 'a natural or a pure culture' process. Various forms of starter cultures of yeast for wine production are described here.

26.3.1 Natural Microflora

The yeasts which initiate natural fermentation include species of *Kloeckera*, *Hanseniaspora*, with lesser representation from species of *Candida*, *Metschnikowia*, *Cryptococcus*, *Pichia* and *Kluyveromyces*. *Saccharomyces cerevisiae*, the principal wine yeast, occurs at very low populations (< 50 cfu/g) on healthy grapes.

26.3.2 Pure Culture of Yeast

A pure culture of wine yeast was isolated by Müller-Thurgau from wine ferments in Germany. In 'pure culture' fermentation, selected strain of *S. cerevisiae* is added to the juice at a population of 10^6–10^7 cfu/ml, and this approach is generally favored to get a product of predictable and consistent quality.

26.3.2.1 Wet Yeast

The pure cultures of wine yeast are maintained in the laboratory on agar slants or plates and can also be known as wet yeast.

26.3.2.2 Sparkling Wine Starter

The starter for the production of sparkling wine differs from the starter culture used in still wine production, as the former needs gradual adaptation to increasing ethanol concentration and decreasing temperature incubation.

26.3.2.3 Lyophilized Yeast

Lyophilization, or freeze-drying, involves freezing of a culture followed by its drying under vacuum conditions, which results in sublimation of cell water. The culture is grown to maximum stationary phase, and then cells are resuspended in a protective medium such as milk, serum or sodium glutamate. A few drops of the suspension are then transferred to an ampoule, which is then frozen and is subjected to a high vacuum until sublimation is complete and the culture is dry, after which the ampoule is sealed. The culture may remain viable for up to ten years or more.

26.3.2.4 Immobilized Yeast

For immobilization, yeast cells are grown under aerobic conditions and then entrapped in gels, which permit diffusion of substrates and products. Use of immobilized yeast cells as fermentative agents has an advantage of increasing the overall productivity, due to the high concentration of the cells in the reactor.

26.3.2.5 Dry Wine Yeast

The dehydrated forms of yeast available commercially for use as inoculum to conduct the fermentation of grape juices are known as 'dry wine yeast'. Most of the commercial dry wine yeast preparations have less than 0.8–6.0% residual moisture. The use of dry wine yeast generally ensures quick initiation of fermentation and maintains the quality of wine produced from batch-to-batch.

26.3.2.5.1 First Generation of Dry Wine Yeast

Initially, two active dry wine yeast strains in freeze-dried form of Montrachet and Pasteur strains were commercially produced for a large California winery in 1960. The Montrachet wine yeast strain became a popular reference strain among the yeast scientists involved in the study of the physiological and biochemical characteristics of wine yeasts. Yeast strains like SB1 were developed to restart stuck fermentations although it was sensitive to killer yeasts. All these strains are considered as the first generation of dry wine yeast. However, the first-generation dry wine yeasts could not sustain for long, as their efficiency in wine production in other regions failed due to the differences in grape varieties. Therefore, selection of yeast strains well suited for fermentation of grape varieties in various regions of the world became inevitable.

26.3.2.5.2 Second-Generation Dry Wine Yeast

In 1970, the Lallemand Inc., Canada, in collaboration with other wine research institutions and commercial organizations, developed a second-generation of dry wine yeasts. These dry wine yeasts had technical features which had a predictable influence on the fermentation and quality of wine produced (Table 26.1).

26.3.2.5.3 Third-Generation Dry Wine Yeast

In recent years, the third generation of wine yeasts has emerged, and this includes starter cultures with specific

TABLE 26.1
Some Second-Generation Dry Wine Yeast

Yeast strain	Institution (which developed the strain)	Special fermentation feature
WSK	Wadenswill Institut, Switzerland	Fermentation at low temperature
EC	Institut Oenologique de Champagne Epernay, France	Low foaming and easy settling
K1	Institute Cooperatif du Vin, Montpellier, France	Killer factor
71B	INRA, Narbonne, France	Produce aromatic components

enzyme activities. There are some wine yeast strains like VL1, which have β-glucosidase activity that catalyses the release of bound terpenes in certain grape varieties. The role of yeast in wine production is not limited to fermentation of sugars to ethanol, but it also includes the imparting of desirable characteristics like aroma, either as its own metabolite or helps to release such compounds from grapes. Besides this, wine yeast has to encounter adverse conditions of pH, temperature, pressure and nutrient limitation and has to grow under increasing ethanol concentrations. Thus, during wine fermentation, it should be ensured that there are sufficient nutrients to support the yeast growth (from approximately $2x10^6$ cfu initially to $2x10^8$ cfu during fermentation) and maintain its fermentation activity.

26.4 YEAST PHYSIOLOGY

The wine yeast (*Saccharomyces cerevisiae*) is a unicellular eukaryotic organism and belongs to class ascomycetes. Some morphological characteristics and dimensions of wine yeast cell include shape: round to oval, average cell size: 4x8μm; volume: $10^{-10}cm^3$; and dry weight: $0.3\times10^{-10}g$. The number of viable cells in commercial preparations of dried wine yeast is in the range of $2.0–2.5\times10^{10}$ cell/g. The yeast cells multiply by budding (Figure 26.1), and genetically most of the industrial strains are polyploid, having homologous chromosomes of varying sizes which is the cause of polymorphism among wine yeast strains. The yeast *Saccharomyces cerevisiae* has

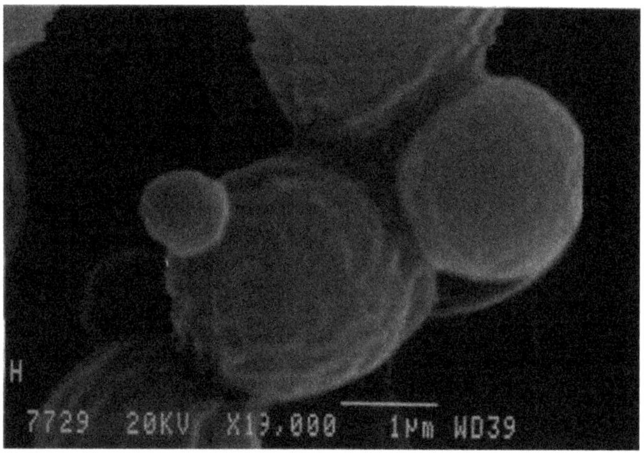

FIGURE 26.1 Scanning electron micrograph of wine yeast *Saccharomyces cerevisiae* showing budding

both the oxidative and fermentative modes of metabolising glucose, *i.e.* it is a facultative anaerobic organism. The following equations summarise these modes of glucose utilization:

i) $C_6H_{12}O_6 + 6O_2 \rightarrow 6CO_2 + 6H_2O\left(\text{Aerobic metabolism}\right)$ Eq. 1

ii) $C_6H_{12}O_6 \rightarrow 2C_2H_5OH + 2CO_2\left(\text{Fermentative metabolism}\right)$ Eq. 2

If all other nutrients are maintained at optimum levels, 100 g of glucose under aerobic condition of yeast growth will yield 54 g of yeast dry mass (YDM), and fermentative metabolism yields about 7.5 g of YDM. The aim of the yeast producer is to achieve higher cell yields per gram of sugar, and thus, the conditions for an aerobic mode of glucose utilization are preferred. The regulation of glucose catabolism in *S.cerevisiae* is quite complicated. At suitable values of glucose and dissolved oxygen concentrations, glucose is utilised for respiration, providing maximum cell yields per unit amount of glucose consumed. If glucose concentration exceeds a certain level (*i.e.* >50mg/ml), the metabolism switches to fermentation even in the presence of oxygen. This condition is termed as aerobic fermentation and is popularly known as Crabtree effect. The catabolite repression and limited respiratory capacity of *S. cerevisiae* are thought to be the causes of this phenomenon. For detailed biochemistry of fermentation, readers may refer to Chapter 17 of this text. In the presence of an excess of glucose, there is a six-fold increase in the enzyme pyruvate decarboxylase (which transforms pyruvate to acetaldehyde and carbon dioxide), and after attaining the optimum capacity to utilise the pyruvate through oxidative metabolism, the surplus pyruvate is diverted to ethanol production. Aerobic fermentation leads to reduced cell yields, and glucose, ethanol and CO_2 are formed as end products. Ethanol inhibits the growth of yeast cells, and therefore, to obtain higher cell yield (as this is the main objective of yeast producers), aerobic fermentation should be avoided; this can be achieved by feeding glucose during cultivation of yeast cells in batch fermentation. It is now well established that glucose transport in *S.cerevisiae* is mediated by facilitated diffusion and a number of glucose carriers (*e.g.* SHF3, HXT1 and HXT2) are involved in this process. HXT2 carrier is of special interest as it is expressed under anaerobic conditions, and wine yeast strains like EC118 which have high fermentation rates are thought to contain multiple copies of the gene coding for this carrier. The type of nitrogen source that is added to the medium also affects the growth of yeast. Ammonium ions are readily assimilated into glutamic acid in *S. cerevisiae*. If urea is the source of nitrogen in yeast cultivation broths, it is first carboxylated by urea carboxylase and then converted to ammonia and carbon dioxide by an enzyme allophanate hydrolase.

26.5 INDUSTRIAL CULTIVATION

Many of the technical features of wine yeasts are affected by conditions of yeast cultivation and drying. That is why yeast producers are very careful in preserving the technical characteristics of wine yeast.

26.5.1 Culture Maintenance

Yeast cultures are commonly maintained on malt agar slants and stored at different locations so that if the desired culture is contaminated or lost in one collection, it can be obtained from the other. The culture slant can be stored for up to six months under sterile mineral oil at 4°C in a refrigerator. From the master/mother culture, a number of slants are prepared, and one of them is used as starter culture for further propagation of wine yeast. The other techniques employed for yeast preservation are cryo-preservation and lyophilization. Cryopreservation at –196°C in liquid nitrogen, freezing at –70°C and lyophilization were compared for brewing inocula over a period of two years. Cryopreservation and freezing were the two maintenance methods that were found to give a lower percentage of variants and respiratory deficients as compared to lyophilization. A high percentage (>95%) of the viable cells was recorded for yeast maintained by cryopreservation and freezing, whereas the viability of the lyophilized yeast cells was <50%. Thus, cryopreservation and freezing (–70 to –80°C) in the presence of 10% glycerol are the methods of choice for yeast preservation.

26.5.2 Fermentation

The initial step in large-scale cultivation of wine yeast is the transfer of culture from the slant to the flask and growing it on a shaker at a suitable temperature under aseptic conditions in the laboratory. The laboratory culture of yeast forms an inoculum for the production plant. Medium components for the propagation of yeast at plant scale includes molasses (as carbon source), ammonium hydroxide (as nitrogen source), phosphoric acid (as phosphorus source), yeast extract (as source of some vitamins) and minerals like magnesium, etc. Use of a mixture of cane and beet molasses as a carbon source is recommended, as they complement each other for vitamins and minerals. The sugar content in molasses is around 50% (w/v), sucrose being the main sugar in the beet molasses while cane molasses contain about 30–50% sucrose. Glucose and fructose comprise the remainder. The molasses used as a carbon source for the cultivation of wine yeast are diluted (1:1 with water) and sent to continuous centrifuges for de-sludging to remove particulate matter. The clear molasses are heat treated with HTST (high temperature short time) or LTLT (low temperature long time). This yields molasses, which are dilute, clear and free from volatile acids, alcohols and microbes.

Various steps are involved in the cultivation of wine yeasts. Initial steps involving transfer of yeasts from slants to flask culture and subsequent growth in small fermenters are performed strictly under aseptic conditions in the laboratory. The yeast produced in the laboratory facility is transferred to first an industrial fermenter (size 20 k litres or 2×10^4 to 10^5 litre) and cultivated as batch fermentation. The contents of the first industrial fermenter are transferred to next larger fermenter (size ~10^5 to 25×10^5 litre), and yeast propagation is done by fed batch mode of fermentation. The feeding rate of molasses to the medium is one of the crucial steps which influences the growth rate of yeast in fed batch fermentation. Aerobic conditions are maintained by supplying one volume of air per volume of culture medium per minute (VVM). The yeast is harvested and washed several times using continuous discharge nozzle centrifuges and stored in the form of cream yeast at 2–4°C. The cream yeast comprises 15–20% solids and is used to seed final fermentation medium from which the cells are harvested for making dry wine yeast. The specific growth rate (μ) for cultivation of yeast under optimum nutrient and culture conditions is in the range of 0.14–0.18h^{-1}. Yield is expressed as Equation 26.3 as given below:

$$\text{Yield}\,(\%) = \frac{\text{Kilogram of yeast biomass formed}\,(27\%\ \text{solids})}{\text{Kilogram of molasses consumed}} \times 100$$

The difference in yields under laboratory and plant conditions are attributed to mixing of media components including oxygen. The requirements for the production of 2600 Kg dry wine yeast (95% solids) have been shown in Table 26.2. The efficiency of oxygen transfer during the propagation of yeast in a large fermenter determines the yield and quality of yeast. Even under an anaerobic mode of fermentation, yeast growth requires oxygen in order to favor the synthesis of sterols and unsaturated fatty acids needed for yeast cell wall synthesis.

TABLE 26.2
Requirements for the Production of Approximately 2600 kg Dry Wine Yeast (95% Solids)

Parameter	Size/volume/ dimensions
I) Fermenter	
• Total volume	10^5 litre
• Working volume	8×10^4 litre
II) Raw materials	
• Molasses (as carbon source)	9.5×10^3 kg
• Aqueous ammonia (27% as nitrogen source)	670 litre
• Phosphoric acid (75%)	67 litre
• Air (at 1VVM)	7.2×10^7 litre
• Yeast as inoculum	435 kg
III) Cultivation time	14 hour

Source: G.H. Fleet (Ed.). *Wine Microbiology and Biotechnology.* Harwood Academic Publ., Chur, Switzerland. p. 421

An aeration system is a very important component of a fermenter and involves energy inputs. To reduce the final cost of the product, fermenters with aeration system that require less energy inputs (kW/kg yeast biomass) are preferred.

During the cultivation of the yeast, biomass production and ethanol formation need to be continuously monitored. If ethanol concentration exceeds certain values, the flow of air is increased and molasses-feeding rate is reduced. Fermentation parameters *e.g.* molasses and ammonia, feed rate, aeration, agitation, oxygen uptake rate, ethanol concentration, respiratory quotient and biomass production rates are monitored by gas chromatography, mass spectrometry, special equipment for measuring respiratory quotient, specific sensors and NADH probe. A number of mathematical models of yeast metabolism are also used to augment yeast biomass yield. Advances in instrumentation, monitoring and control strategies have made it possible to maximize yeast production without affecting its quality. Yeast biomass ranging 3 to 6 kg (27% solids) per m^3 of fermenter per hour for baker's yeast has been achieved. However, wine yeast productivity is much less than the baker's yeast. Even under similar condition of aeration and oxygen transfer, the specific (critical) growth rate for most of the wine yeasts lies in the range of 0.10–0.12 h^{-1} which is less than baker's yeast (0.14–0.18 h^{-1}), and this means the former has a longer doubling time. It seems that wine yeast strains either have an inefficient system for glucose transport or have limited respiratory capacity. The maximum oxygen uptake rate for *S. cerevisiae* is 8 mmol/g/h, while it is 5 mmol/g/h for *S. cerevisiae* strain bayanus. Acclimatization of seed yeast to aerobic conditions has been proposed to increase the respiratory capacity of wine yeast.

The transport of sugar into yeast is a major limiting factor during the cultivation of wine yeast and amongst commercial wine yeasts there are variations for this property. The sugar transport systems in yeast are broadly classified as: i) high affinity system and ii) low affinity system. The strains which possess a high affinity sugar system may exhibit substrate inhibition (*i.e.* rate of sugar transport declines if glucose concentration is more than 10mM) and turn up as poor fermenters, such as the Tokay strain. The strains, which have a low affinity sugar transport system can transport sugar into the cells when sugar concentration in medium is high. The wine yeast strains which ferment the must as efficient fermenter have the latter type of sugar transport system. Nitrogen content is one of the limiting factors of yeast growth and sugar utilization. It has been reported that the specific growth rate would increase if the nitrogen content of the yeast cell is below 32–50% protein. However, the wine yeast cultured under industrial conditions contains 40–45% protein.

26.6 EFFECTS OF INDUSTRIAL CULTIVATION ON PROPERTIES OF WINE YEAST

26.6.1 Microbiological Quality

The main source of contamination in industrial plants is molasses (the carbon source for the industrial cultivation of wine yeast), as the temperature used for its processing does not sterilize it completely. In order to produce a yeast product of high microbiological quality, stringent measures in cleaning, sanitation and sterilization have to be followed at all steps of large scale cultivation, harvesting, drying and packaging of wine yeast, though this may decrease productivity (gram yeast/l/h).

26.6.2 Fermentation Rate

The fermentation rate of a yeast depends upon its protein and phosphate contents. Wine yeast normally contains 40–45% protein (based on N × 6.25) and 2.0–2.45% phosphates (as P_2O_5). The optimum ratio of P_2O_5/N is 1:3, for yeast to be used for carrying out effective fermentation. An increase in this ratio would mean an excess of water linked to yeast protein, which affects the drying, viability and fermentation activity of the yeast.

26.6.3 Production of Hydrogen Sulphide (H_2S)

Under conditions of nitrogen scarcity, H_2S accumulates and also diffuses out of the yeast cells. If the grapes to be used for wine making are deficient in metabolizable nitrogen, the proteolytic activity will be stimulated in the yeast, resulting in H_2S formation as a consequence of breakdown of proteins rich in sulfur-containing amino acids. Yeasts containing proteins of up to 38% are suitable to limit their nitrogen demand in the event of nitrogen deficient grapes/musts. Further, yeast deficient in pantothenate produced more sulfide; that is why molasses used for cultivation of wine yeasts are supplemented with calcium pantothenate.

26.6.4 Ethanol Yield and Tolerance

Sterols and unsaturated fatty acids (C-16: 1 and C-18: 1) play a very crucial role in influencing the fluidity and functions of cell membrane in yeasts. Alcohol tolerance of *S. cerevisiae* has been related with the lipid content and composition. Ethanol tolerance, ethanol yield and cell viability are strongly affected by the fluidity of yeast cell membranes. The yeast cells produced through aerobic and anaerobic modes of yeast cultivation exhibit differences in their chemical composition especially (lipids) of the cell membranes. *S. cerevisiae* does not synthesize sterols under anaerobic conditions, and such yeast cells have defective membranes and are metabolically inefficient. Yeast cells cultivated in the presence of ethanol (14%) have very low levels of free fatty acids. Propagation of wine yeast under aerobic conditions results in having sufficient levels of unsaturated fatty acids and sterols to keep cell membranes integrated during wine fermentations. The practice of introducing a little oxygen into the media during wine production is beneficial for the yeast in maintaining its cell membrane constituents essential for carrying out optimum anaerobic fermentation. Growth factors like biotin, inositol, pantothenic acid and thiamine are essential for yeasts during industrial cultivation and wine fermentation and so are added to the industrial yeast propagation media to meet the variation in their content in molasses. The yeasts cultivated perform satisfactorily under the conditions of wine fermentation, even if grape must has lower level of growth factor than desired. It is known that by altering the nutritional conditions, it is possible to increase the ethanol yield as well as the survival of yeast at high concentrations of ethanol. The dry yeast to be used for initiating fermentation in wineries is produced under conditions which generate yeast cells with sufficient reserves of vitamins and cells membranes with normal structure and function. However, recycling of the residual yeast or spent yeast in the winery or brewery is not advisable; as such yeast does not yield the desired product after fermentation.

26.6.5 Resistance to Sulfur Dioxide

Sulfur dioxide (SO_2) acts as an antioxidant and is used in wineries to curtail microbial contamination. The yeast strains vary in their sensitivity to SO_2, which is related to the level of saturated lipids in the yeast cell membranes. The mechanism of action of SO_2 is considered to be similar to that of propionic acid, as both are more active and less dissociated at low pH. The antimicrobial action of propionic acid is linked with its properties to retard intracellular pH. The yeast strains with higher ATPase activity can tolerate higher levels of SO_2, since this enzyme will facilitate the removal of proton (H^+ ions) from the cells. The selection of yeast strains resistant to SO_2 can be made by growing them in the presence of SO_2 at low pH.

26.6.6 Resistance to Drying and Rehydration

The viability and physiology of microbial cells is affected by drying and rehydration, so these organisms, which remain stable after these operations, are preferred for industrial use. As well as inherent properties of the yeast may be responsible for such stability, trehalose content in the yeast cells has been considered to be one of the very important factors that helps yeast to sustain the effects of drying and rehydration. It also inhibits phase transition of the membranes by binding to polar groups of the phosphate lipids, thus providing protection to the structure of membrane lipids during dehydration or water stress conditions. Trehalose also serves as a source of reserve carbohydrate that prolongs the shelf-life of fresh yeast. Therefore, wine yeast should be cultivated under conditions to yield yeast cells with higher trehalose content, which tolerate the stresses of dehydration and rehydration.

26.7 DRYING OF WINE YEAST

After the propagation of yeast in the large fermenters, it is harvested from broth using continuous discharge nozzle centrifuges. The resulting yeast mass is known as cream yeast (15–20% solids). It is passed through a rotary vacuum filter or filter presses to yield a semi-solid form of yeast mass with 30–35% solids. The semi-solid yeast mass is passed through perforated plates, and noodle-like particles formed are further dehydrated using a dryer. Dry yeast has 95 % solids. The temperature of yeast during drying is not allowed to exceed 35°C and the mode of operation of dryers can be continuous or batch wise. The drying time varies between 15 to 60 minutes, which depends on the type of dryer and method used. The moisture content in the dry yeast is generally brought below 6%. Yeast which was dried using the techniques of freeze-drying and overnight drying exhibited certain biochemical changes that included an increase in activities of dehydrogenases, decline in polyphosphates and pyrophosphatase activities.

26.8 YEAST REHYDRATION

In order to ensure good fermentation, proper rehydration of the yeast is essential. It should not take more than 30 minutes for yeast rehydration and addition of rehydrated cells to must, as yeast will start to utilize reserve material after its activation upon rehydration. The optimum temperature for yeast rehydration is 38–45°C, which yields yeast cells with maximum viable counts and fermentation rates. Although rehydration of dry wine yeast in 1% potassium chloride solution leads to minimum changes in yeast structure and physiology, the most commonly followed practice in wineries is to use water at a temperature between 35–40°C.

26.9 QUALITY CONTROL OF WINE YEAST CULTURE

The parameters used to assess the quality of wine yeast preparations include viability, fermentative activity and contamination with bacteria or other yeasts. Traditional approaches for assessing the quality of yeasts are to evaluate their fermentation/ assimilation profile, specificity of killer character and resistance to mitochondrial inhibitor. New techniques are available for identification of yeasts at a molecular level. Restriction fragment length polymorphism (RFLP) analysis has made it easy to compare and identify the test strain to that of known yeast, and so it is used in authentication, validation and identification of commercial preparation of yeasts for quality control purposes and selection of new strains.

26.10 MALOLACTIC BACTERIA

After the traditional alcoholic fermentation, the high acid wines undergo malolactic fermentation (MLF) conducted by lactic acid bacteria (also called malolactic bacteria), which decreases wine acidity and improves the flavor profile by decarboxylation of L-malic acid to L-lactic acid. It is highly desirable in wines produced from grapes cultivated in cool climates that have a higher content of malic acid and lower pH (3.0–3.5). The wines that undergo MLF exhibit greater microbial stability and are less susceptible to spoilage by other bacteria. However, MLF is not beneficial to wines produced from grapes cultivated in warmer climates, as these are less acidic (pH > 3.5), and further increase in pH may make wine prone to bacterial spoilage. For more details on the effect of MLF on wine quality, see Chapter 18 of this text. The malolactic bacteria which bring about the conversion of malic acid to lactic acid and carbon dioxide are species of *Leuconostoc*, *Lactobacillus*, *Pediococcus* and *Oenococcus oeni*, however wine makers generally prefer this

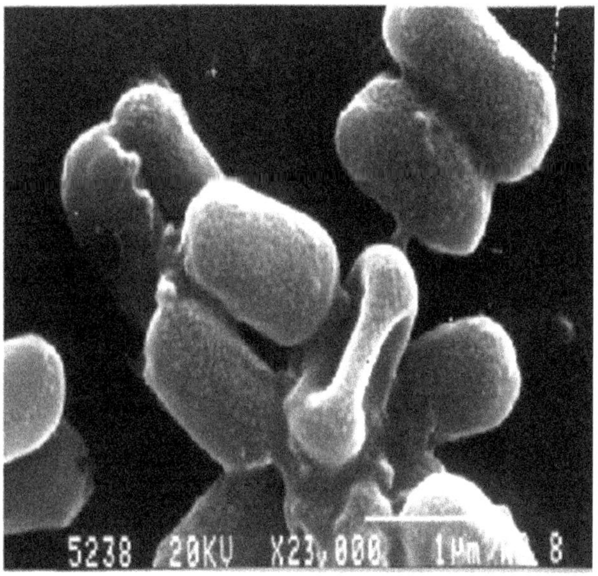

FIGURE 26.2 Scanning electron micrograph of *Leuconostoc* sp. (isolated from traditional fermented beverage of Himachal Pradesh, India)

reaction by inoculation of wine with strains of *Leuconostoc* especially *Leuconostoc oenos*; This has been recognized as the bacterium most tolerant to wine conditions such as low pH, high SO_2 and ethanol content.

26.10.1 Physiology

Leuconostoc oenos (formerly known as beta cocci) is a mesophilic, facultative anaerobic, non-motile, non-spore forming and gram-positive bacterium. The cells are short rods (0.5–0.7 × 0.7–1.2µ in size), generally arranged in pairs or short chains. *Leuconostoc* sp. was isolated from a traditional fermented beverage (*Lugri*) of Himachal Pradesh, India (Figure 26.2). It has an obligate requirement for fermentable carbohydrates and requires nutritionally rich media for its cultivation. The optimum growth temperature is 20–30°C and pH 4.5, and its growth is stimulated in media containing tomato juice. *Leuconostoc oenos* prefers to degrade malic acid instead of metabolizing glucose under non-growing conditions. The transformation of malic acid into lactic acid is enhanced by the growth of this organism at low pH. The malolactic bacteria have the ability to tolerate acidic pH, ethanol, SO_2, low temperature and concentration of nutrients.

26.10.2 Selection and Identification

Leuconostoc strains are generally isolated from wines undergoing active malolactic fermentation. The malolactic acid bacteria are identified by conventional and molecular taxonomic methods and are selected if they successfully bring about the desired changes in wine.

26.10.3 Industrial Production

Large-scale cultivation of *Leuconostoc oenos* is more simple than that of yeast and it can be easily propagated in batch fermenters. *Leuconostoc oenos* is a fastidious bacterium, so it requires supplementation of conventional media with vitamins, etc. The cultures of *Leuconostoc oenos* are normally maintained on agar slants or as frozen or lyophilised forms that are used for inoculation in small flasks, incubated at 20–25°C for three to five days. It in turn is used to inoculate a fermenter of 500–1000 L capacity, keeping the same incubation temperature and period. The number of transfers from one vessel to another during the propagation of this bacterium should be kept at a minimum to reduce the chances of contamination. The last propagation in the larger vessel/fermenter is carried out under the conditions of limited oxygen supply. The cells are harvested from the broth either by centrifugation or micro-filtration, and the concentrated cells (20–100 fold) are subjected to freezing or lyophilization. These are tested for viability and malolactic activity. DNA fingerprinting of *Leuconostoc oenos* is used for its identification. The commercial preparation of malolactic bacteria meant for malolactic fermentation is either pure cultures of *Leuconostoc oenos* or a mixture of strains. Alternatively, the passage of wine is made through a biocatalyst-containing reactor consisting of high concentrations of malolactic bacteria (10^9–10^{10} cfu/ml) immobilized in beads of alginate or carrageenan or entrapped between membranes. Although such systems have proved to be very effective in rapid deacidification of wines, it cannot not become commercially viable due to the long-term instability of malolactic activity of the biocatalyst.

26.10.4 Rehydration

The dry bacterial cell preparation should be properly rehydrated to achieve satisfactory induction of malolactic fermentation. Most of the protocols for the rehydration of freeze-dried *Leuconostoc oenos* preparation involve adaptation of the bacterium to pH, temperature, ethanol and SO_2 concentration at which it has to conduct MLF of wine. A simple and a straightforward procedure for rehydration of *Leuconostoc oenos* for initiation of malolactic fermentation in Champagne has been suggested. This protocol considers SO_2 concentration, pH and temperature as important parameters during rehydration and adaption of the seed culture of *Leuconostoc oenos* for proper induction of MLF in wine. Various steps involved in this procedure were the addition of 2.5 g lyophilised *Leuconostoc oenos* cells to one liter of 50% diluted wine having SO_2<20mg/L, pH 3.2 and a temperature of 20°C for 8–10 days for adaptation. These were then added to 500 L wine for malolactic fermentation that had <1 g sugar/L at 18–20°C for 15–40 days. Any protocol employed for conducting malolactic fermentation of wine should be cost effective.

BIBLIOGRAPHY

Bailey, J.E. and Ollis, D.F. (1986). *Biochemical Engineering Fundamentals*. McGraw Hill Book Co., Singapore, p. 658.

Beker, M.J. and Rapoport, A.I. (1987). Conservation of yeasts by dehydration. *Advances in Biochemistry and Engineering Biotechnology*, 35: 127–171.

Beudeker, R.F., Van Dam, H.W. and Van Der Plaat, J.B. (1990). Developments in baker's yeast production. In: *Yeast Biotechnology and Biocatalysis*, H. Verachtert and R. De Mot (Eds.), Marcel-Decker Inc., New York, p. 103–146.

Davis, C.R., Wibowo, D., Eschenbruch, R., Lee, T.H. and Fleet, G.H. (1985). Practical implications of malolactic fermentation- a review. *American Journal of Enology and Viticulture*, 36: 209–301.

Degre, R. (1993). Selection and commercial cultivation of wine yeast and bacteria. In: *Wine Microbiology and Biotechnology*, G.H. Fllet (Eds.), Harwood Academic Publ., Switzerland, p. 421–448.

Ghareib, M., Youssef, K.A. and Khalil, A.A. (1988). Ethanol tolerance of *Saccharomyces cerevisiae* and its relationship to lipid content and composition. *Foila Microbiol. (Praha)*, 33(6): 447–452.

Joshi, V.K., Sandhu, D.K. and Thakur, N.S. (1999). Fruit based alcoholic beverages. In: *Biotechnology: Food Fermentation Vol. II*, V.K. Joshi and Ashok Pandey (Eds.), Educational Publisher and Distributors, New Delhi. p. 647–744.

Reguant, C., Carrete, R., Constanti, M. and Bordons, A. (2005). Population dynamics of *Oenococcus oeni* strains in a new winery and the effect of SO_2 and yeast strains. *FEMS Microbiology Letters*, 246(1): 111–117.

Rose, A.H. (1977). History and scientific basis of alcoholic beverage production. In: *Economic Microbiology*, A.H. Rose (Ed.), Academic Press, London, p. 1–41.

Rosen, K. (1987). Preparation of yeast for industrial use in production of beverages. In: *Yeast Biotechnology*, D.R. Berry, I. Russel and G.G. Stewart (Eds.), Allen and Unwin Inc., Winchester, MA, p. 471.

Thakur, N., Kumar, D., Kapoor, Savitri and Bhalla, T.C. (2003). Traditional fermented foods and beverages of Himachal Pradesh. *Invention Intelligence*, 38: 173–178.

Van Vuuren, H.I.I. and Dicks, L.M.T. (1993). *Leuconostoc oenos*: A review. *American Journal of Enology and Viticulture*,44: 99–112.

27 Bioreactors in Wine Fermentation

R.S. Singh

CONTENTS

27.1 INTRODUCTION

Winemaking is one of the world's oldest professions. Over the last 20 years, winemaking technology has seen remarkable developments which have improved the quality of wines besides making it possible to prepare wines with a large range of characteristics. The technological innovations are one of the cornerstones with which the successful wine industries of the 21st century can be assured of winning global influence and sustainable profitability. These include biotechnological innovations aimed at improving the efficiency of the production process. This chapter is slanted to highlight the new technologies developed in fermentation vessels, bioreactor designs, continuous fermentation, high cell density reactors and immobilized systems used in winemaking.

27.2 WINE PRODUCTION: CONVENTIONAL BIOREACTOR SYSTEMS

Ancient Greeks started wine fermentation in simple jars. The considerable diversity in bioreactor systems in the wine industry has led to the production of a wide variety of wines with cost-competitiveness, improved quality and low environmental impact.

27.2.1 White and Red Wine Production

White wines are made from free-run juice after crushing the grapes that were de-stemmed and then pressed. The must is clarified before fermentation. The lack of maceration in white wine is not an absolute factor and in some cases, even short maceration of the skin is carried out. Use of SO_2 and heating, chaptalization, amelioration, reverse-osmosis, cryo-extraction, entropy-concentration, double salt precipitation are the steps carried out during white wine production. The fermentation temperature is kept at 15–24°C during white winemaking. For more details, refer to Chapter 29 of this book. Red wine is a macerated wine made from the fermentation of red or black grape juice along with their skins and seeds at a higher temperature than white wine (24–27°C). In red wines, anthocyanin pigments stimulate fermentation. When the active fermentation starts, the solids (skins) rise to the top of the must and form a skin cap (Figure 27.1). The longer skin maceration time produces wines with higher colour intensity,

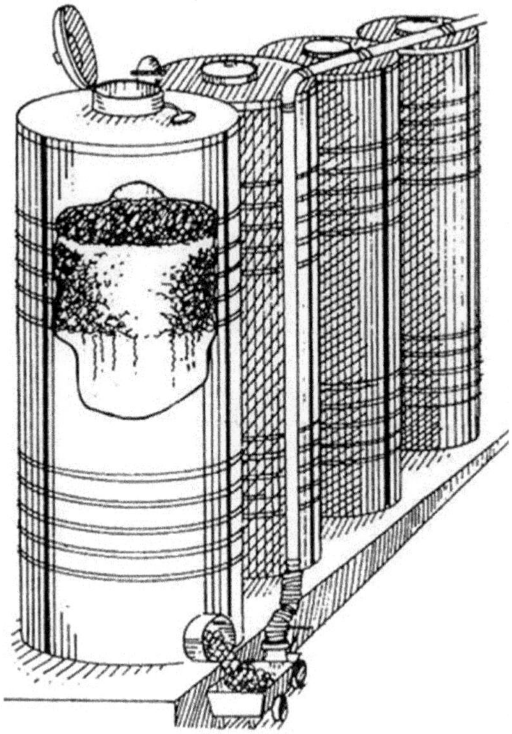

FIGURE 27.1 Jacketed red wine fermenting tanks with floating "cap" of skins. *Courtesy:* Sebastiani Vineyard

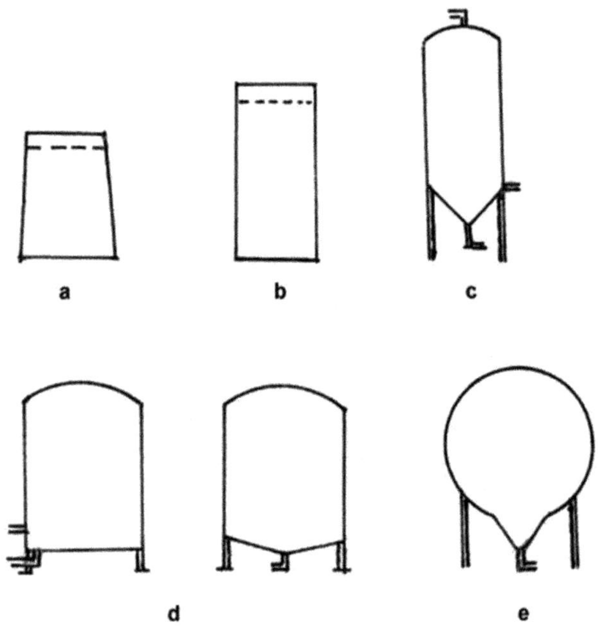

FIGURE 27.2 Shapes of the main fermentation vessels (bioreactors) used in the beverage industry (a) Barrel, (b) Vat, (c) Cylindroconical, (d) Cylindrical, (e) Spheroconical. *Source:* Diviès (1993)

phenolic compounds and higher sensory ratings which preserve wine characteristics during storage. To bring the skins in contact with the colour-extracting liquid, use of submerged caps, pumps, agitators, mechanical removal of the floating head by screw conveyors, rakes or sudden decompression of the compressed CO_2 in the tank can be made. Sometimes carbonic maceration and thermovinification techniques are also used for extraction of colour in red wine making. The addition of montmorillonite clay bentonite to the fermentation medium has been recommended to facilitate the even fermentation of juice low in solids and ensure rapid clarification at the end of fermentation. (For details, Chapter 32 of this text).

27.2.2 Fermentation Vessels or Bioreactors

27.2.2.1 Shape and Size

There is a considerable diversity in size, shape, design and construction materials used in fermentation vessels in winemaking, leading to a wide variety of wines. Almost any non-porous and non-toxic vessel can be used as a fermenter. These can be categorised into two basic categories, open-topped vats and the tanks that have sealed tops. Historically, vats were used in red wine production because direct access to the cap of seeds and skins is required during fermentation. White wines can be vinified in tanks, where oxygen can be excluded from the fermenting juice. Most of the fermenters are of simple design, while complexity in design occurs as its volume is increased, but it also reduces the relative surface area for heat transfer. In red wine production, the complexity of the vessel design depends upon the technique used for submersion of the cap. Conventionally, the batch system of fermentation is used in wineries. Previously, fermentations were carried out in 225–228 L barrels or 6–12 hL vats. However, these have now been replaced with well-designed stainless steel fermenters. Small fermenters are rarely used in red wine fermentation due to the difficulty in achieving adequate cap submersion. Commercial wineries are using fermenters of 200 hL or more capacity, which are cost-effective in terms of capital outlay, mechanization and automation.

Fermenters are of a wide variety of shapes, such as upright-cylindrical, horizontal-cylindrical, V-shaped, square tanks etc. (Figure 27.2), are used in wine production. In most of the fermenters, floor sloping is inclined towards the front. Fermenters with hemispherical or domed bases are also used in winemaking. In spite of the advantages in pomace discharge in red winemaking, the use of conical-based fermenters has not been practised. One of the newest, efficient in the amount of time and energy and one of the most expensive innovations in red wine fermentation, is the rotating stainless steel fermenter (Plate 27.1) that gives optimal skin exposure to the fermenting wine and oxygen exposure to the spoilage organisms. Even then, the use of rolling and cylindrical vessels as an alternative to conventional fermenters has found only limited acceptance in red wines. Various types of '*Defranceschi*' red wine fermenters have the option of automatic discharge, pumping over, plunging system etc. and are used in wineries throughout the world. In South Africa, Grotto defranceschi now produces wine tanks as well as the 'Fermentamatic^{R}' with pump-over and automatic discharge facility (Plate 24.2) and 'CP^{R}' (plunging models). These fermenters have become well known in the wine industry and are used extensively by producers of

PLATE 27.1 Rotating stainless steel fermenter. *Courtesy:* Nigel Dolan, Beringer Blass Wine Estates, Nuriootpa, South Australia

PLATE 27.2 Defranceschi's Fermentamatic[R] wine tank: (a) External view, (b) Internal structure showing pump-over facility. *Courtesy:* Alessandro Furlanetto, Defranceschi S.p.A., Italy

high-quality wines. Both types can be fitted with a central filter column.

27.2.2.2 Construction Material

The fermentation vessels are primarily made-up of stainless steel, fibreglass, cement, plastics and non-aromatic wood. Generally, stainless steel, though expensive, is preferred to other materials because of long-lasting strength, impervious nature, corrosion-proof, rapid heat transfer, aromatic neutrality, formation into any size or shape and because it is easy to clean, a suite of abilities suitable for all forms of temperature control with a capacity to produce the quality wines. The fermenters may be constructed with a jacket in which coolant is circulated for temperature control. Fermenters made of materials other than stainless steel require periodic pumping of the fermenting juice through external cooling coils, or the cooling coils are inserted into the fermenting juice. Fibreglass

PLATE 27.3 A view of Fermentabags™ being used in a wine factory and an individual fermentabag in inset. *Courtesy:* Dieter Sellmeyer, Viticor (Pty) Limited, South Africa

is considerably less expensive, with most of the properties of stainless steel except the heat-conducting properties, and should be properly cured during manufacturing. Although cement is less expensive, it has poor heat conductivity and no mobility. The coating of the inner surface of cement vessels with wax or tiles is essential to prevent calcium uptake during fermentation.

Wooden vessels are expensive, have a comparatively limited life, are not easy to clean, cannot be sterilised, possess poor heat conductivity making temperature control very difficult, as well as imposing restrictions in shaping and design. They are neither impervious nor inert but allow access to the air, which causes faster oxidation and evaporation, which concentrates the flavour in the wine. Wines fermented in barrels has higher quantities of alcohols but lower in fatty acid esters, resulting in a less fruity character in wine besides more occurrence of hydrogen sulphide. Furan aldehydes (with the aroma of grilled almonds) are reduced by yeast during barrel fermentation, resulting in the formation of furfuryl alcohols. Oak imparts wood tannins to low tannin wines, absorbs tannins from very high tannic wine or can simply exchange tannins with other wines besides making wines more complex in sensory properties. Wooden or cement vessels are now replaced by well-designed stainless steel fermenters. Plastic containers are an inexpensive alternative to the traditional red wine fermentation vessels. "A plastic container that collapses as a liquid is dispensed" was used in Europe for at least twenty years before its adoption in Australia. Hickinbotham Jr. has invented a disposable 'Fermentabag™' (Plate 27.3). This flexible fermentation cell is made up of highest food grade specialised plastics (multi-layered polyurethane) and is of 1000 L capacity with a breather pipe in the middle. This material allows for specific amounts of air to be passed through the film to the wine and allows for a 'micro-oxygenation' effect similar to that of oak barrel and is designed for fermentation of red wine only, but can also be used for storage of red wines for 4–6 weeks after fermentation is complete to allow for maturation. The system is also suitable for carbonic maturation has to be filled up to 90–97%. Warm water or brine can be added via the auxiliary ports to speed up or kick start the fermentation in cool climates. Besides this, wine can be continuously circulated through the cap during fermentation instead of the plunge or "punch down" cap. These bags are placed in specified rigid containers so can have mobility, as required. Softer tannined and better-coloured wine can be produced using plastic containers.

27.2.3 Continuous Wine Making

Besides conventional batch fermentation, continuous fermentation processes have been developed for red wines for operations in regions having a large and readily available supply of grapes for continuous maceration to feed into the fermenter. It is not suitable for vintage wine. The technology was firstly installed in Argentina by Cremaschii in 1948, followed by fermenter installations in France in 1955 by Ladousse, with an attempt to produce red wine using fermenters of 5×10^5 L capacity, processing 15×10^4 kg of grapes per day with a corresponding amount of wine generated with an average residence time of the wine of 3 days at 26–28°C in the fermenter. Continuous wine fermenters in Italy and other European countries are used to increase the extraction of pigments from red grape skins. Various advantages of continuous winemaking include: lower operating costs, higher production rates, uniform product quality, more efficient use of raw materials, uninterrupted operation for longer periods, minimum downtime and greater overall productivity, while disadvantages are: more expensive system, complex bioreactor design, lack of food-grade and low-cost support materials, lower viability of the immobilised systems, substandard quality of wine and unsuitability for small scale wineries.

Generally, the maceration and fermentation is carried out in cylindrical or cylindro-conical metallic tower fermenters of 200–4000 hL capacity, where feeding of the grapes is done through the bottom of the fermenter and wine is collected from the top by pre-determined overflow, while the skins and seeds form a cap at the top of the fermenter like in traditional vessels. The cap is then extracted mechanically and sent to a continuous press after draining. The rate of feeding is calculated as a function of maceration time keeping in mind the quality of grapes at harvest. To avoid contamination by lactic acid bacteria, a higher quantity of SO_2 (80–100 mgL^{-1}) is used.

The use of immobilized cells for continuous winemaking is an attractive process including that for ethanol fermentation. Immobilization has clear advantages over the conventional, suspended cell systems. Of all the described immobilized cell reactors, packed bed reactor using gel beads are the most popular, though this has a problem of gas diffusion out of the reactor besides the problems of the accumulation of CO_2 and cells in the void volume of the reactor, the plugging of the bed which increases the backpressure, the channelling of the nutrients and the rupture of the gel beads during the continuous fermentation. Improvement of the productivities and operational stabilities of these reactors are being made mainly in alcohol fermentation with only wine fermentation. Nevertheless, a horizontal multistage bioreactor with replaceable immobilized plates (bio-plates) was constructed and used for the continuous fermentation of Koshu grape must. The bio-plates were made by immobilizing viable cells in the form of membranes onto sintered glass plates, using calcium alginate as the carrier. Five bio-plates were inserted vertically along a rectangular frame constructed from acrylic sheets, thus separating it into six compartments. The fermentation conditions in this bioreactor were similar to those of the traditional method of wine fermentation. There was no significant difference in the concentration of the major components between the wine from the batch and continuous fermentations. Continuous winemaking in a tower bioreactor (1400 mL working volume) by an immobilized alcohol-resistant strain of *S.cerevisiae* AXAZ-1 on delignified cellulosic (DC) material was made continuously for two months. Wine productivity was three- to five-fold higher than that obtained by natural fermentation. The use of β-glucosidase in the wine industry has the potential for upgrading the aroma in wines at the end of the fermentation stage by hydrolysing the monoterpene glycosides which naturally occur in wine. To meet this goal, the decay of the catalytic activity of immobilized β-glucosidase should be limited. Consequently, the preliminary inspection of biocatalyst operational stability is of paramount importance for optimal process design. The chitosan used for immobilization of the enzyme was found suitable as a support and could be easily separated from the finished wines and reused in continuous operations. A freeze-dried immobilized biocatalyst was developed by immobilizing *Saccharomyces cerevisiae* cells. The developed immobilized biocatalyst was used in a multi stage fixed-bed tower (MFBT) bioreactor for the batch and continuous wine making. The MFBT bioreactor showed higher alcohol productivity and good operational stability. The successful implementation of continuous immobilized cell fermentation processes in wine production demands more investigations on the final quality of the wine produced. Continuous systems are not employed at present, as they require a complete reversal of traditional winemaking procedures. Nevertheless in the future, continuous systems probably utilising immobilized yeast, may be introduced.

27.2.4 Microbial Technology

Use of *Saccharomyces cerevisiae* culture to achieve the ethanol fermentation is a common practice now as it helps in maintaining the consistency of fermentation pattern, earlier onset of active fermentation and the minimisation of undesirable by-product formation. The oxygen is required to ensure the growth and survival of yeast during alcoholic fermentation. The optimal period of aeration is at the end of the cell growth and can be made by the use of gas-permeable membranes. The micro bubbling of oxygen into the must is made to reduce CO_2 whose retention is not desired in wines except in sparkling wines so it is allowed to escape freely. The effect of CO_2 has been studied little in the winemaking in contrast to beer fermentation. The yeast growth decreases to zero above seven atm pressure. Temperature is also known to have many subtle effects on yeast metabolism such as when its range is

10–15°C, this results in the production of fruity esters, but around 20°C produces higher alcohols such as isoamyl alcohol and hexanol. Cell sensitivity to the toxic effects of alcohol increases with temperature, maybe due to increased membrane fluidity. The implementation of kinetic models in the wine industry would make on line controlling and command of fermentation tanks more effective. (For more details, see Chapter 23 of this text).

27.3 HIGH CELL DENSITY REACTORS

The productivity in wine fermentation can be increased by using a high density of yeast cells by increasing the effective size or density of the cells by aggregation or immobilizing the cells on some kind of support: such systems are termed as high cell density reaction systems. Various advantages of these systems are: higher fermentation rates due to high cell densities per unit bioreactor volume; repeated use of the same biocatalyst for extended periods of time; high flow rates in continuous processes can be used without the risk of cell washout; high dilution rate in continuous operation decreases the risk of reactor shut-down due to contamination adaptation to continuous processes which can be better optimised and controlled; easy cell/liquid separations, thus minimising down-time and separation costs; a wide variety of microbial strains/genetically modified strains can be used; improved tolerance or protection of cells from inhibitory products; sequential reactions can be carried out by connecting various reactors in series; continuous operation and system control is easy; efficient gas–liquid mass transfer rate; better product uniformity in continuous systems and reduced capital cost for small scale wineries. Innovations in bioreactor technologies during the past few years have been mainly in three areas, *i.e.* bioreactor designs, bioreactor for two phasic reactions and environmental bioreactors.

There are two major systems, *i.e.* homogeneous and heterogeneous, for confining or immobilizing cell biomass. The homogeneous system contains a uniform distribution of microbial biomass in the form of free cells in the medium. The repeated use of biomass can be carried out by centrifugation, flocculation of yeast cells with external or internal decanter or retaining the cells in a membrane reactor. The heterogeneous system has two separate phases, *i.e.* liquid medium, which is to be transformed, and a solid phase, containing the microbial cells and where the biomass is confined using support, auto-flocculation and entrapment in gels.

27.3.1 Homogeneous Reactors

The cell recycle batch fermentations (CRBF), whose principal innovation is the multiple successive uses of the same yeast starter in different batch fermentation, is the more acceptable non-conventional technique in winemaking and does not require radical changes in the winery procedures nor does it imply capital investment in new equipment. As well as yeast cells to be recycled, they can be recovered through natural sedimentation. A single reactor with a centrifugation step for recycling of yeast cells is mostly used. An increase in cell mass as well as in productivity has been achieved by using a partial vacuum system. The centrifugation techniques decrease the viability of the yeast biomass due to the stress to which they are subjected. An alternative method to centrifugation in homogeneous reactors is membrane bioreactors, where the yeast cells are retained in the reactor having a membrane of pore size of less than 0.45 μm, where the material is passed through the membrane, and the transformed product comes downstream from the membrane. The efficiency of the transformation can be increased by recycling the product through the reactor is the main limitation. An off-skin fermentation of clarified *Trebbiano toscano* grape juice was carried out using a non-conventional cell recycle batch fermentation process and revealed that the processes caused a reduction in fermentation time, an improvement in ethanol productivity and yield and thus can be applied to the production of ordinary table wines.

27.3.2 Heterogeneous Reactors

In heterogeneous reactors, microorganisms are confined by immobilization. The basic principle is to increase the density of cells and to maintain their viability for longer periods in repeated use or a continuous system.

27.3.2.1 Immobilization Techniques

Whole-cell immobilization is a process by which "cells are physically confined or localized in a certain defined region of space with retention of their catalytic activities." The success of an immobilization system in large scale industrial applications depends upon the various factors, such as the support material that must be readily available and affordable, the system should be efficient, easy to operate and give good yields, the cells should have prolonged viability in the support, which should not be severely toxic to the cells, the support material should allow for high cell loading (weight of cells/weight of support) and the kinetic behaviour of the loaded support should be understood and not hinder the fermentation. This includes diffusional limitations, local pH and inhibitor accumulation, and any modification of metabolic processes associated with the carrier should be realized and accounted for. Cell mobility can be restricted both by passive and active immobilization techniques (Figure 27.3).

27.3.2.1.1 Self-aggregation

Flocculation is a natural phenomenon resulting in a cell aggregation, but all the cells don't flocculate, and natural cell aggregates are generally unstable and sensitive to shear. Thus, aggregates are promoted by the addition of cross-linking agents. The flocculation of different yeast strains to form a dense concentration of biomass can be made and is a very attractive method of biomass retention involving decantation. As no solid support is required, the higher cell density can be achieved. Flocculated cells are particularly compatible for use in fluidized bed reactors. It is possible to use flocculating yeasts in wine, when fermentation had stopped by using a laboratory reactor equipped with a simple decanter.

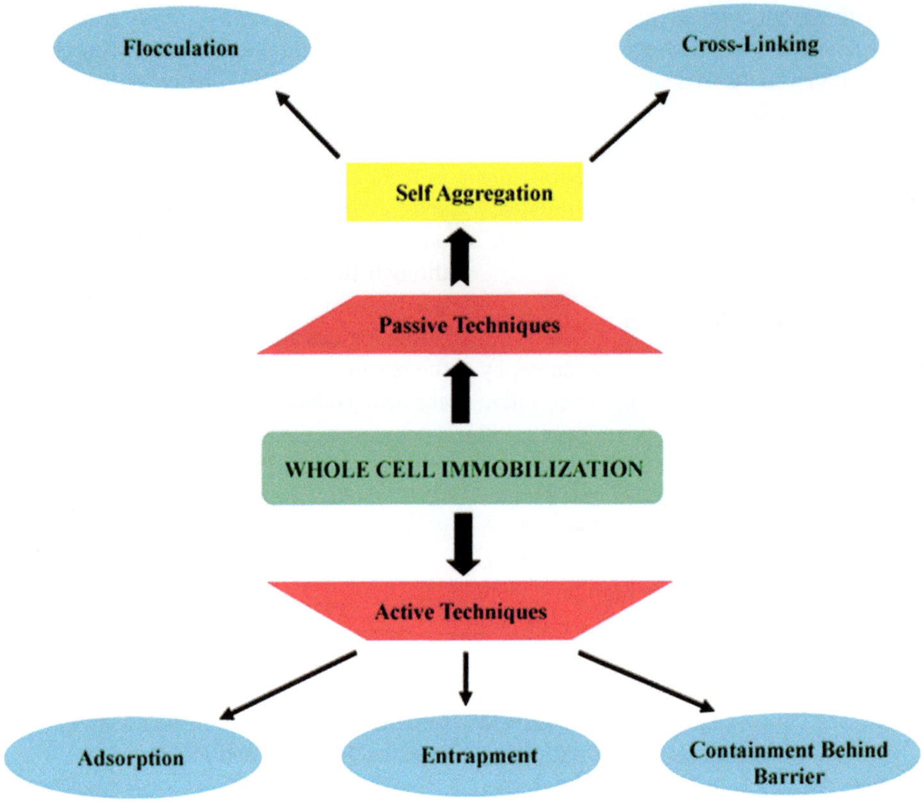

FIGURE 27.3 Techniques of whole cell immobilization

27.3.2.1.2 Adsorption

It is one of the simple and quick immobilization techniques. The affinity of certain microorganisms for growth on solid surfaces is an established phenomenon of different microorganisms on various supports, and the choice of an appropriate resin may be an expensive and time-consuming task. A suitable absorbent should display high affinity towards the biocatalyst with minimal denaturation. The basic reaction of adsorption is an electrostatic interaction between charged support and charged cell and depends on the chemical nature of the surface of the cell wall, where constituents such as peptides, hexosamines and diaminopimelic acid provide the necessary ionic sites for attachment to a charged support. Carriers and biocatalysts are kept together through van der Waals bonding forces, inorganic bonds or H-bridges. Several steps like an adaptation of the polymer (polysaccharide, protein etc.) to serve as the support conditions to promote adsorption of the microbial cells, reversible adsorption and irreversible adsorption are considered for this technique. The use of adsorbed cell reactors has been proposed for deacidifying wines with *Schizosaccharomyces pombe*. However, the desorption of the biofilms occurs due to the gases produced during alcoholic fermentation, an increase in the turbulence of the liquid passing over its surface and autolysis of the underlying cells besides and limitation of the limited access of oxygen to underlying cells and its dependence on pH. But the simplicity of this technique has resulted in a large number of applications in bioprocesses.

27.3.2.1.3 Entrapment

Entrapment is the widest spread cell immobilization technique. It involves the confinement of the living cells in a rigid network which permits the diffusion of the substrates and the products and ensures the growth and maintenance of viability of yeast cells. The typical examples of polymers used for this purpose are polyacrylamide, collagen, cellulose acetate, alginate, *k*-carrageenan, chitosan, gelatine, agar, agarose, epoxy resin and silica. But, alginate is the most common polymer used for entrapment of yeast cells in winemaking. The higher productivity with alginate beads than polyacrylamide cubes is due to the ratio of specific surface areas based on gel particle size. The use of entrapped cells for conducting the alcoholic fermentation of wine and beer being made though no industrial installation is already operational.

27.3.2.1.4 Containment behind a Barrier

Membrane supports can maintain ten times more cell densities than alginate and allows greater volumetric productivity. However, diffusional limitations involving the build-up of inhibitory products are one of the major limitations of these membrane supports. The membrane should be hydrophilic for easy exchange of nutrients and products, and mechanically strong enough to withstand the pressure differentials. The membranes are typically made up of polymers like polyvinyl chloride, polypropylene or polysulfone. Sheets and hollow fibre cartridges are the two basic configurations used for this technique. An interesting variation of standard containment methodology is called encapsulation. Usually, cells are first entrapped in spherical gel, and then the coating of the sphere is carried out with a polymer such as a polyethyleneimine.

27.3.2.2 Immobilized Bioreactor Design

The main function of the reactor is to retain the immobilized complex and to ensure efficient and controlled contact between the catalysts to maximize product formation and give the engineer more processing options since the biocatalyst is in a concentrated and easily handled form. The choice of the reactor depends upon the type of immobilization, metabolism of the cells, the resistance of the matrix to shear stress and the mass transfer requirement. Immobilization system used for the yeast cells would prove a revolutionary perspective for the next future in the production of wines. Typical reactors used for immobilized cells are discussed below.

27.3.2.2.1 Continuous Stirred Tank Reactor (CSTR)

Stirred tank reactor consists of an agitated tank in which fresh substrate is continuously fed and well mixed by the use of impellers, and the corresponding volume of the liquid is removed. The liquid component of the reactor is homogeneous in composition, similar to the concentration of the outflow. With immobilized cells, high fluid velocities are needed to achieve a constant supply of substrate and product removal. CSTRs or back-mix reactors are cheap and versatile, suitable for carrying out liquid phase reactions, give an easy supply of gas and control of pH and temperature, allow easy addition of fresh catalysts to the reactor and have tolerance of substrates containing particulate materials without causing fouling. However, the relatively high power input required to give efficient agitation in CSTR is a disadvantage, resulting in abrasion damage to the immobilized catalyst due to high shear forces at the impeller surface. The substrate concentrations in a CSTR are typically lower than the packed bed and fluidized bed reactor, resulting in lower average reaction rates. A typical example is β-glucosidase immobilized on chitosan pallets. Such reactors promote good mixing within the fermentation, but are however not suitable for heavy immobilized cell preparations unless there is no requirement to prevent settling. Fast stirring rates in CSTR result in high shear stresses, thus increasing cell leakage from alginate, or carrageenan beads, cell detachment from ion exchange resins and floc disruption.

27.3.2.2.2 Packed (fixed) Bed Reactor (PBR)

Packed bed reactor (PBR) is the most frequently used type of immobilized cell bioreactor, where the particles are packed into a column at its maximum density through which the substrate solution passes. If the fluid velocity profile is perfectly flat, the PBR operates as a plug flow reactor (an ideal behaviour). It is largely used for ethanol production. The biocatalyst is packed at its maximum density in the reaction. The degree of conversion of the substrate increases with the length of the column, while the productivity for a given biocatalyst depends on the type of immobilization (entrapment or surface attachment). High cell loadings are often achieved by entrapment, resulting in improved productivity. In principle, it is possible to achieve total conversion into the product so that such reactors are ideal, where total removal of a substrate is essential (detoxification). Continuous fermentation of grape must by psychrophilic yeast immobilized on apple cuts was used in a packed bed reactor (2 L), which produced wines of good quality with a distinctive flavour profile and a pleasant taste. Similarly, the use of immobilized β-glucosidase improved the aromatic quality of Muscat wine in a packed bed reactor. The PBR is simple in operation and of low cost, but ensuring good flow throughout the bed is the main difficulty besides being unstable during long-term operations because of continuous biomass accumulation, mass transfer limitations and CO_2 holdup. This results in channelling and creation of dead spaces, back mixing and matrix disruption. Horizontal packed bed reactors have been used to aid gas removal, thus reducing channelling and gas hold up problems in the continuous system.

27.3.2.2.3 Fluidized Bed Reactor (FBR)

Fluidized bed reactors provide conditions that are intermediate to those of the CSTR and PBR, but mixing is better in the PBR and has lower levels of shear than CSTR. The reactor consists of a column, wherein the biocatalyst particles are maintained in suspension relative to each other by a continuous flow of the substrate or gas at the highest flow rates. The pressure drop of the fluid flow supports the weight of the bed. The reactors offer higher productivity than CSTR because liquid approximates plug-flow is similar to the PBR. However, the FBR is more advantageous for fermentation with substrate inhibition than the PBR and promotes good mass transfer, removes the dead cells from the system, releases large volumes of CO_2 without channelling and minimizes pressure drop. Fluidization avoids such problems as contamination, shear damage and limits to scale up, associated with impeller shaft and blades in stirred tanks. Use of immobilized β-glucosidase in a fluidized bed reactor to improve the aromatic quality of Muscat wine is an example of FBR.

27.3.2.2.4 Rotating Disc Reactor (RDR)

This consists of immobilized cell units such as polyurethane foam sheets or fibre discs attached to a rotating shaft. It is slowly stirred, thus allowing good mixing and removal of dead cells, debris and evolved CO_2. The energy required for RDR is less than that for STR because of its slow mixing speed, thus, it has the potential to accept industrial substrates containing suspended solids to achieve high productivity. No difficulty with high solid media has been encountered in this type of bioreactor.

27.3.2.2.5 Air (gas) Lift Reactor (ALR)

In an air lift reactor, the fluid volume of the vessel is divided into two interconnected zones using a baffle or draft tube. One zone (*riser*) is sparged with air or gas while another zone (*downcomer*) receives no gas. Gas bubbles carry the liquid, which reduces in liquid bulk density, and escapes from the top, while the liquid cascades down in a downcomer. An external loop system may replace the inner draft tube for the recirculation of liquid in some bioreactors. Agitation in the ALR due to gas flow results in low shear with efficient mixing and mass transfer. The size of the draft tube influences the hydro-dynamics of the fermenter, such as the gas hold up. These are highly energy-efficient, relative to stirred fermenters, though productivities are comparable.

27.4 KINETIC CONSTRAINTS AND BEHAVIOUR OF IMMOBILIZED MICROORGANISMS

Alteration of physiological/metabolic properties of immobilized cells or alteration of local micro-environment near immobilized cells may cause changes in the kinetic behaviour of cells during fermentation. Physiological reactions of the cell will vary with the method of immobilization used. Direct cross-linking of the cells chemically modifies the cell membrane affecting cell metabolism and growth, adsorbed cells may have different membrane properties, cell accumulation in a biofilm induces transfer limitations and entrapment techniques change the physical properties of the microenvironment influencing cell metabolism. These include the presence of ionic charges, altered water activity, modified surface tension, cell confinement and the mass transfer limitation (described later), resulting in gradients of oxygen, substrate and product that may enhance or reduce cell metabolism.

The local micro-environment near immobilized cells may be described in terms of solute partitioning effects between the bulk liquid phase and the solid immobilization matrix, external film mass transfer resistance and internal mass transfer resistance. The effect of immobilization on mass transfer and cell metabolism is critical. The external mass transfer involves the transfer of nutrients from the bulk liquid medium to the carrier surface, while internal mass transfer describes the transfer of substrates and products within the carrier, and in and out of the cell. In the immobilized cell systems, due to both internal and external mass transfer limitations, cells are forced to alter their metabolic states and thus, impact the efficiency and quality of a fermentation process. Both the concentration of matrix used for immobilization and the size of the beads influence the productivity in bioprocesses due to this effect. The properties of alginate resulting in weaker gels, probably facilitating a higher degree of substrate and product transport through the gel network, thus enhance fermentation performance. Satisfactory external transfer of reactants has been achieved by the use of fluidized bed reactors. The agitation in continuous stirred tank reactor changes the hydrodynamic conditions of the medium and directly affects the thickness of the diffusional layer.

The physical environment of the cells in the immobilized state, particularly with organisms encapsulated in biocatalyst beads, affects the distribution and morphology within the gel. Various methods have been used to estimate cell density within the gel matrix, including microscopic techniques (also electron microscopy), bead sectioning or dissolution of beads in case of alginate. High cell concentrations have been found near the outer surface of large beads attributable to the poor diffusion of dissolved oxygen (DO) and an inhibitory ethanol concentration, which is transitorily higher at the centre of the beads. The cell density within the matrix can also affect the mass transfer properties of the beads, while the prolonged use of immobilized viable cells often involves potential continuous cell growth along with parallel cell death and lysis. In some cases, the cells appear to propagate in small vacuoles/cavities within the beads, ultimately filling up the available space. Generally, the growth of cells is not uniform throughout the gel particle, but often occurs in the surface region. To obtain microorganisms with good performance, these should be grown directly *in situ* in the gels, instead of preparing the high cell density gels.

The mechanical strength of the gel matrix is important for minimizing gel splitting of yeast cells from the matrix due to the evolution of carbon dioxide by the immobilized yeast during fermentation. Cell-supporting matrices should be permeable to exchange whole cells and/or cell debris with the surrounding medium without losing their mechanical stability. The concentration of sodium alginate, calcium chloride and the curing time of the beads have a pronounced effect on the stability of the gel. Sensitivity of alginate towards chelating compounds such as phosphate, citrate, lactate or anti-gelling cations (Na^+ or Mg^{2+}) is the major limitation, which can be overcome by keeping the gel beads in a medium containing a few millimolar free calcium ions and to keep the Na^+:Ca^+ ratio less than 25:1 for high-G alginates and 3:1 for low-G alginates. Operational stability of immobilized cells in algal polysaccharides can be improved by treatment with hardening agents such as periodate, glutaraldehyde or hexamethylenediamine. The stabilization of calcium alginate gels by adding divalent (Sr^{2+}, Br^{2+}) or trivalent (Ti^{3+}, Al^{3+}, Fe^{3+}) ions can also be made. Abrasion caused by particle-to-particle contact, particularly in a fluidized bed or stirred reactors or particle compression (as in packed bed reactors) may also lead to immobilized cell aggregate breakdown and could be problematic. The rapid growth near the gel surface induces a decrease in the rigidity and mechanical strength of the gel, consequently cavities near the surface rupture, releasing yeast cells into the wine, which causes turbidity. It can be overcome by covering the gel beads with an external gel coating without microorganisms, which has been successfully used in sparkling wine production. Better conversion of glucose to ethanol with better ethanol productivity takes place for immobilized cell system than free cells as the entrapment of yeast cells protects them against ethanol toxicity. The tolerance of immobilized cells to toxic compounds such as alcohols is often enhanced, possibly because of the effect on the cell membrane.

Macromolecular composition of the cells showed higher carbohydrate and DNA content along with faster cell multiplication in immobilized cells than free cells. Scanning microphotometry and microfluorimetry revealed that relative RNA content per cell varied with the radial position of the cells within the beads, which vary almost linearly with the specific growth rate for several species. The specific RNA levels of cells in the rapidly growing colonies of the outer cell layer were two times higher than the values obtained from cells of much smaller colonies found in the bead centre. The consequence to cell captivity was that immobilized recombinant cell populations exhibited better plasmid stability than suspended cells due to attachment of cells to cross-linked gelatine beads decreased for cells near the bead surface. The improvement of plasmid stability is caused by restricted growth in the gel beads that prevent plasmid loss. A limited number of cell divisions occur within each clone of cells before the clone is released from the gel bead, and a low number of generations does not allow the apparition of plasmid-free cells. Thus, the number of cells without plasmids is decreased with immobilization, consequently, immobilization of cells with plasmids significantly enhanced the volumetric productions due to high cell concentrations.

27.5 APPLICATIONS OF IMMOBILIZED BIOREACTORS IN ENOLOGY

27.5.1 Wine Deacidification

27.5.1.1 Malolactic Fermentation

Different strategies to improve the malolactic fermentation (MLF) have also been developed. The use of immobilized cells in fermentation processes over the use of free cells offers several advantages including the recovery of the immobilized cells and their reuse. The entrapment and adsorption/attachment are the two main immobilization techniques used to induce MLF in wine, and entrapment is a popular method of cell immobilization. Improvement of immobilization techniques for wine deacidification has also been achieved using alginates, polyacrylamide, κ-carrageenan, oak chips, cellulose sponge and polyvinyl alcohol hydrogel, while the possibility of controlling MLF by using a reactor with immobilized cells or enzymes has been demonstrated (Table 27.1). It has advantages over the traditional deacidification of wines, as starter cultures can be reused, the decreased formation of secondary products and fermentation can be induced and halted as desired. However, phage infection may decrease the activity rapidly and lead to a slight modification of the sensory qualities of the treated wine. Under natural conditions, the MLF is developed in weeks or months. So it is generally carried out by a small bacterial population. *Oenococcus oeni* must reach a cell concentration of 10^6 cfu mL^{-1} to start MLF. The rapid and complete MLF can be obtained by inoculating the wine with high cell densities of selected strains. The complete breakdown of malic acid is achieved by using an inoculum containing up to 10^7 cells mL^{-1} that has been allowed to develop previously for 24 hours in either a synthetic substrate or in a grape juice based substrate. By inoculating a high density ($<10^6$ cfu mL^{-1}) of *Leuconstoc oenos*, malic acid degradation is uncoupled from the growth of the bacterial cells,

TABLE 27.1
Comparison of Methods Studied to Induce Malolactic Fermentation in Wine

Bacterium or enzyme	Immobilization support	Reactor (Working volume)	Initial L-malic acid (gL^{-1})	Bioconversion rate (%)	Operation period (h)
Lactobacillus sp.	–	CSTR (350 hL)	4.0	62.0–750	24
Lactobacillus sp.	Calcium alginate	FBR (1.4 L)	0.9	45.0	72
Lactobacillus sp.	κ-Carrageenan	CSTR (0.5 L)	9.0	64.0	200
Lb. brevis	κ-Carrageenan	Shake flask (5 mL)	5.00	71.4	1
Lb. casei	Polyacrylamide	Air flow (nr)	48.0	80.0–100.0	360
Oenococcus oeni	–	350 hL	4.0	75.0–100.0	24
Oenococcus oeni	–	Screw tubes (5 mL)	1.0–6.0	94.0–99.4	6
Oenococcus oeni	–	CCR (0.3 L)	0.8–4.6	25.0–100.0	125
Oenococcus oeni	–	Screw tubes (10 mL)	5.0	77.0	5–48
Oenococcus oeni	–	CSTR (0.3 L)	4.2	92.0–95.0	500
Oenococcus oeni	–	Screw Tubes (10 mL)	3.5–7.0	41.0–98.0	24–500
Oenococcus oeni	Calcium alginate	CSTR (2.7 L)	8.0	97.0–100.0	17
Oenococcus oeni	Calcium alginate	FBR (1.4 L)	0.9	82.0	72
Oenococcus oeni	Calcium alginate	CSTR (30 mL)	0.3	100.0	12
Oenococcus oeni	Calcium alginate	CSTR (60 mL)	1.5	60.0	864
Oenococcus oeni	Calcium alginate	CSTR (nr)	4.5	98.7	360
Oenococcus oeni	Cellulose sponge	Shake flask (0.1 L)	3.5	50.0	96
Oenococcus oeni	κ-Carrageenan	CSTR (0.5 mL)	9.0	36.3	48
Oenococcus oeni	κ-Carrageenan	Shake flask (5 mL)	5.0	100.0	1
Oenococcus oeni	Oak chips	CSTR (0.1 L)	8.0	20.0–58.0	264
Oenococcus oeni	Polyacrylamide	Shake flask (5 mL)	2.3–4.5	71.0	1
Malolactic enzyme *(O.oeni)*	–	Membrane reactor (5 mL)	18.6–23.9	62.0–75.0	168

CSTR: Continuous stirred tank reactor; CCR: Cell recycle continuous stirred reactor; FBR: Fluidized bed reactor; nr: no reported dat
Source: Maicas (2001)

and a rapid MLF can be achieved. Thus, by separating the bacterial growth phase from the phase for the conversion of malic acid into lactic acid, better controlled conditions leading to improved results could be obtained easily for MLF. The use of cell-recycle fermentations to reach high cell densities and strategies for controlling continuous fermentation can be made. To induce the MLF, tangential flow or hollow fibre filler is used to separate the cells from the wine in cell recycle bioreactors. In this continuous process, wine after MLF is constantly removed, however cells remain in the vessel and reach high cell densities, which can delay or prevent inhibition of cell growth by lactic acid production and low pH. The shear stress on cells entering the filtration unit and difficulties in scale-up due to the filtration system are the major limitations of recycling bioreactors. An efficient membrane bioreactor has been employed for MLF involving high cell densities, where a complete degradation of malic acid took place during the first day of continuous operation as well as on the second day of operation after storage of the reactor overnight at 4°C. However, the activity declined after 48 h. A stirred tank reactor was successfully used continuously for two to three weeks using free *O. oeni* cells to induce MLF in red wine with 92–95% degradation of the malic acid in wine.

The choice of the immobilization matrix has to be made by the long term conservation of cell viability and for beverage production, its acceptance as GRAS (Generally Recognized As Safe). Alginate is a suitable matrix in both respects. Bioreactor systems consisting of high concentration of malolactic bacteria immobilized in beads of alginate or carrageenan or entrapped between membranes have been developed. κ-carrageenan is another matrix used for immobilization of cells to induce wine deacidification with better perspectives for its industrial applications. Due to enhanced operational stability, immobilization of *Lactobacillus* cells in this matrix was achieved, and the conversion ratio and deacidification acidity were equivalent to 53.9% and 64.0%, respectively. Nevertheless, a continuous flow bioreactor filled with *Lactobacillus brevis* immobilized on 4 types of alginate and κ-carrageenan for the control of MLF in wine has been attempted. Polyacrylamide gel-entrapped *Leuconostoc oenos* cells converted 71% of L-malic acid into L-lactic acid after ten minutes of incubation in the presence of L-malic acid (2.2–4.5 g L^{-1}). Continuous MLF of wine using *O. oeni* cells immobilized by adsorption onto the oak chips gave two to three times higher conversion in a continuous stirred tank reactor than that by batch fermentation. The cells of *Leuconostoc oeni* immobilized by cross-flow filtration through polypropylene capillary membranes carried out MLF in white wine. The use of positively charged cellulose sponges for the immobilization of *O. oeni* cells to carry out MLF in a semi-continuous system has also been made.

27.5.1.2 Deacidification by Yeast

Yeasts like *Schizosaccharomyces pombe* and *Schizosaccharomyces malidevorans* completely metabolize malic acid under anaerobic conditions could also be used instead of malolactic bacteria to bring about wine deacidification. These could be cultured in grape juice either with or before inoculation with *S. cerevisiae*, or juices/wines could be passed through reactors containing immobilized cells of *Schizosaccharomyces* spp. Such use of this yeast immobilized in alginate deacidified red must and partially fermented red wine. High malate fermentation activity was obtained in both continuous and batch systems, and neither the yeast nor the alginate support system exerted any undesirable effect on quality of wine. Chemical and sensory properties of the wine deacidified by *Schizosaccharomyces pombe* biocatalyst have been found to be satisfactory except for the development of off-flavours, which needs to be overcome. Since yeast converts malic acid into ethanol and CO_2, no lactate will be formed in contrast to MLF by lactic acid bacteria. This may be favourable for white wines, but not for red wines because the characteristic taste of lactic acid is desirable for good red wine. *Zygosaccharomyces bailii* also gives good degradation of malic acid, and its use in wine deacidification is worthy of exploration. A multistage bioreactor with replaceable two bio-plates of *Schizosaccharomyces pombe* and five bio-plates of *Saccharomyces cerevisiae* 2HY-1 was used for simultaneous alcoholic and deacidification fermentation in a continuous system. The rate of fermentation and deacidification was recorded highest in the treatment consisting of immobilized *Schizo. pombe* followed by *S. cerevisiae* and combination treatment of both the yeasts in immobilized form. The use of several pilot-scale bioreactors has been made. Successful scaling-up means a short cycle, competitive advantage and cost-effectiveness.

27.5.2 Sparkling Wines

The uses of yeast cells immobilized in natural non-toxic polymers or contained in a micro-porous membrane and of the cell recycle batch fermentation (CRBF) process to conduct the secondary fermentation for the preparation of sparkling wines are exciting. Essentially, the immobilized cells are more readily removed from the bottle, giving significant savings in processing costs, and they conduct satisfactory fermentation and ageing reactions, producing a product with sensory quality comparable with that made by the traditional process.

27.5.2.1 *Prise de Mousse* or Champagne Method

Traditionally, secondary fermentation of wine is carried out in the bottle by inoculation of freely suspended yeast cells. The bottles thus prepared are stacked horizontally in the thermo-regulated premises (12–18°C), where the secondary fermentation is conducted. Once the fermentation has terminated and the wine has been in contact with the yeast for a year, the bottles are turned frequently and inclined to slide the yeast sediment toward the bottle-neck and settles on the stopper. The sediment is then eliminated by "degorgement", and the process is referred to as "remuage" which is both long and arduous, requiring a large amount of labour over a long period (1–4 months) and the occupation of large wine cellars. The use of immobilized yeast cells for this secondary fermentation would greatly simplify "remuage". There are three possibilities of using immobilized yeast for bottle fermentation: mechanical "remuage" using yeast cultures that readily flocculate and agglomerate, entrapped yeasts and "Millipark" membrane cartridges. The limitations of agglomerating cells are that the "remuage" is not avoided, and the choice of the yeast strain is also limited. Entrapped or retained cells have been found to start their activity more rapidly. Yeast cells entrapped in alginate beads of two mm diameter were used in this process, and at the end of fermentation, the kinetics of "prise de mousse" had slowed down for the immobilized cells. The onset of champagne fermentation depends to a great extent on the yeast strains used, size of the beads for the same biomass loaded and high cell concentrations in gel beads. A device has been developed consisting of a small cartridge containing two microporous membranes (one hydrophilic and the other hydrophobic), which are filled with yeast and placed in the neck of the bottle. The two membranes hold the yeast captive, enabling free and total exchange between the yeasts and the wine, and upon opening the cap, the cartridge is automatically ejected. The production of sparkling wine with the immobilized yeasts occurred like those using traditional methods with free cells, except for differences in protein and polysaccharides, and insufficient surface area for reaction exchange, which creates local super-saturation of CO_2 thus limiting further exchange. The experiments conducted in France since 1978 have shown that the industrial reliability of entrapped yeast cells and their use does not cause any significant change in the composition and sensory characteristics of the bottled sparkling wine in comparison with traditionally produced sparkling wines except for differences in certain amino acids, some aroma components and chemical parameters.

The main drawback of the immobilized yeast cells, however, is the leakage of cells from the beads to the wine, thus causing turbidity that should be completely avoided for a safe application of this technology. To avoid cell release, coating alginate/yeast beads with a protective cell-free alginate gel layer for the bottled sparkling wine has been successfully attempted. A minimum critical thickness equal to 0.2 mm in order to avoid cell leakage for the double layer has been suggested. The possibility of supplying wineries with ready to use immobilized cells could help in the implementation of this technique. It is now possible to dry and rehydrate alginate beads coated with a layer of sterile external alginate, and is now possible to freeze dry without changing bead shape using either κ-carrageenan or κ-carrageenan plus calcium alginate, but the problem of cell release still needs to be solved. An industrial machine for the delivery of beads at the rhythm of 20,000 bottles per hour has been developed, and an economic survey for a large champagne plant has proved its competitiveness.

27.5.2.2 Charmat Process or Bulk Method

In the "Charmat Process" for sparkling wine production, sugars, other nutrients and yeasts are added to the base wine, and the secondary fermentation is carried out in large stainless steel sealed tanks, followed by filtration, fining and bottling. But the process requires a considerable number of expensive tanks. Introduction of oxygen during finishing in the tank method increases the aldehyde content and darkens the colour of the wine. In the continuous process, a higher reaction rate can be obtained by using a high cellular concentration for this secondary fermentation. A continuous fermentative system with calcium alginate immobilized yeast bioreactor for the production of sparkling wine in tanks was tested. The experiment set-up used is shown in Figure 27.4. The idea was to run the second fermentation outside the principal stainless steel tank (B) using equipment that could be easily moved from one tank to another. A packed bed (2 L) of beads obtained from 1.5% alginate and 10% yeast (yeast inoculum was 250 g dry weight) or 3 L of 1.5% alginate and 10% yeast (yeast inoculum was 500 g dry weight) was placed in a 10 L bioreactor (A) that could withstand the CO_2 pressure developed there during the secondary fermentation. The flow of nutrient syrup (0.14 L h^{-1}) for yeast was regulated by a pump from a separate vessel (C). The base wine having 9.95% (v/v) ethanol, 5.61 g L^{-1} titratable acidity and pH 3.04 was taken at a rate of 5 L h^{-1} from tank B by a second pump and was mixed with the syrup before entering the bioreactor. The input flow was regulated in such a way that the yeasts were able to use all the sugars, and then the fermented wine was replaced in tank B. By this method, the pressure in the system bioreactor-principal tank was increased constantly up to the set value and the sparkling wine obtained was free of suspending yeasts. The sparkling wine produced by this method had a composition and sensory quality comparable to that of traditional sparkling wines. The continuous process for the production of sparkling wine proposed by Muller-Spath can also be simplified. Moreover, the

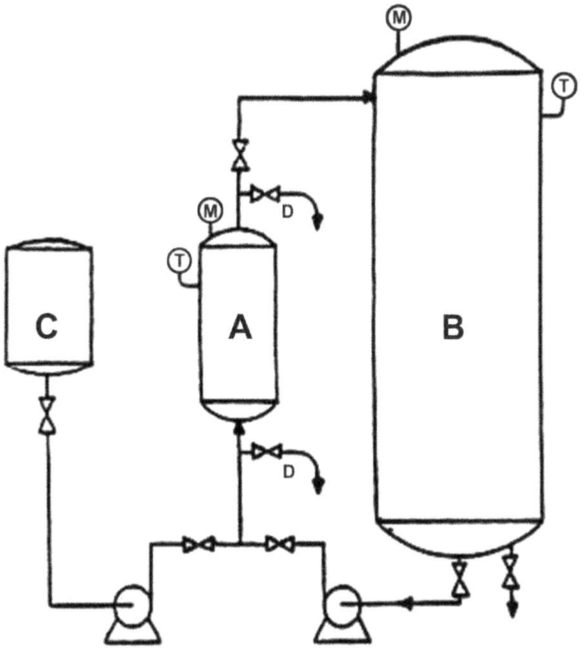

FIGURE 27.4 Scheme of the system used in the production of sparkling wine in tanks with immobilized yeast (A) Bioreactor with immobilized yeast, (B) Sealed tank, (C) Syrup tank. *Source:* Fumi *et al.* (1989)

repetitive continuous fermentation under CO_2 pressure can be carried out easily by cellular entrapment, wherein aromatic characteristics of the products, especially the decrease in higher alcohols and increase in esters, were controlled under these conditions. Fermentation can be carried out at higher pressure, and it is possible to produce sparkling wine or semi-sparkling grape juice with remarkable performance.

27.5.3 Cider Production

Traditional cider (alcoholic beverage from apple) is often uncarbonated and may not even be clarified, but most ciders are clear and sterile filtered or flash pasteurized products. They are artificially carbonated and analogous to bottled and keg beer. In France, carbonation may be achieved via the "Charmat process". Modern commercial vessels are usually of large-sized (2,000–9,000 L) capacity, constructed of lined concrete, lined mild steel or stainless steel, with fermenter depths above 14.5 m, which produce hydrostatic pressure of 1.5 atm, and have been shown to impair cider yeast performance. In recent years, the use of immobilized/co-immobilized cell systems in cider production has successfully been made. Calcium-alginate matrix was used to co-immobilize *Saccharomyces bayanus* and *Leuconostoc oenos* in one integrated biocatalytic system to simultaneously perform alcoholic and malolactic fermentation of apple juice to produce cider in a continuously packed bed bioreactor (2.5 L). A bioreactor loaded with co-immobilized alginate beads (0.985 L volume of liquid and 1.225 L volume of beads) was operated at 30°C, and up-stream feed-flow of apple juice through the bioreactor was adjusted by a peristaltic pump with normal flow rates between 0.05 and 0.55 L h-1 (Figure 27.5). The bioreactor was connected to a stainless steel cylindroconical vessel (150 L), in which the maturation process of the partially fermented juice was accomplished at 10°C. The continuous process permitted much faster fermentation compared with the traditional batch process with better controlled flavour formation. By adjusting the flow rate of feeding substrate through the bioreactor *i.e.* its residence time, it was possible to obtain either "soft" or "dry" cider. However, the profile of volatile compounds in the final product was modified compared to the batch process, especially for higher alcohols, isoamyl acetate and diacetyl linked to different physiological states of yeast in both processes. The feasibility of immobilized/co-immobilized *S. bayanus* and *L. oenos* was studied in different bioreactor configurations for cider making. Four of these reactor configurations were studied concerning the spatial distribution of the two microorganisms. Two bioreactors were very similar for their capacity to use glucose or malic acid. In a reactor with *S. bayanus* in the first stage more ethanol production was shown, while in the second stage low lactic acid was produced by *L. oenos*. In the fluidized bed reactor, the inhibition of glycolytic flow in the yeast has been done. In the mono-reactor configurations, fresh matrix or rehydrated beads of *L. oenos* and *S. bayanus* were used. The volumetric productivities of alcoholic and MLF were twice when the rehydrated matrix was used in reactors. A thermostable tubular APV bioreactor

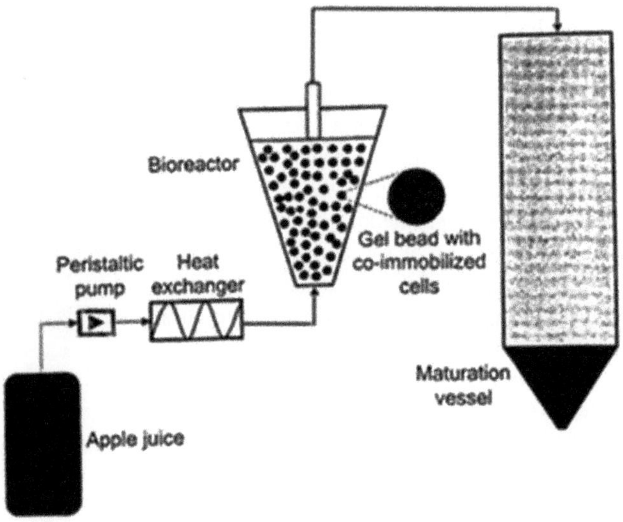

FIGURE 27.5 Schematic presentation of the experimental set-up for continuous cider fermentation. *Source:* Nedovic *et al.* (2000)

(total volume 250 mL) has also been successfully used for cider making using co-immobilized *S. bayanus* and *L. oenos*. A sponge-like material has been used to immobilize both *S. cerevisiae* and *L. plantarum* for carrying out fermentation and partial maturation of cider. Traditional ciders are very frequently subjected to malolactic fermentation. In French cider making, where the primary fermentation is very slow, the malolactic change may occur concurrently with the yeast fermentation, whereas in UK cidermaking, it is most likely to occur once the yeast fermentation has finished and the cider is in bulk store. The MLF can be encouraged by the addition of an appropriate inoculum of *Leuconostoc oenos* into maturing ciders. A controlled MLF in cider using *Oenococcus oeni* free and immobilized cells in alginate beads in flask culture showed similar rates of malic acid consumption. But lower ethanoic acid content and a higher concentration of alcohols were detected with immobilized cells, which have beneficial effects on the sensory quality of cider. A lab-scale continuous process to deacidify apple juices and cider was developed by entrapping *Oenococcus oeni* in a new type of polyvinyl alcohol hydrogel (Lentikats) in a tubular glass reactor (170 mL). The Lentikats bioreactor gave better performance than alginate beads bioreactor.

27.6 ECONOMICS OF IMMOBILIZED BIOREACTORS

Continuous winemaking using immobilized cell systems gives a high production rate with uniform product quality and reduces the down-time, thus lowering operating costs. These systems reduce the production and investment costs by 50% and 75% respectively than batch systems. The use of several pilot-scale bioreactors have also established the feasibility of immobilized bioreactors for malo-lactic fermentation and proved to be cost-effective. In sparkling wine production, immobilized cells make it more cost-effective, as is the case with the use of co-immobilized bioreactors in cider making.

BIBLIOGRAPHY

Amerine, M.A., Berg, H.W., Kunkee, R.E., Ough, C.S., Singleton, V.L. and Webb, A.D. (1980). *The Technology of Wine Making.* AVI Publishing Co., Inc., Westport, CT.

Benda, I. (1982). Wine and brandy. In: *Prescott & Dunn's Industrial Microbiology*, G. Reed (Ed.). AVI Publishing Co., Westport, Inc., USA, p. 293.

Benucci, I., Lombardelli, C., Cacciotti, I., Liburdi, K., Nanni, F. and Esti, M. (2016). Chitosan beads from microbial and animal sources as enzyme supports for wine application. *Food Hydrocoll.*, 61: 191–200.

Benucci, I., Liburdi, K., Cacciotti, I., Lombardelli, C., Zappino, M., Nanni, F. and Esti, M. (2018). Chitosan/Clay nanocomposite films as supports for enzyme immobilization: An innovative green approach for wine making applications. *Food Hydrocoll.*, 74: 124–131.

Berbegal, C., Polo, L., Garcia-Esparza, M.J., Lizama, V., Ferrer, S. and Pardo, I. (2019). Immobilization of yeasts on oak chips or cellulose powder for use in bottle-fermented sparkling wine. *Food Microbiol.*, 78: 25–37.

Boulton, R. (1995). Red wines. In: *Fermented Beverage Production*, A.G.H. Lea and J.R. Piggott (Eds.). Blackie Academic & Professional, Glasgow, UK, p. 121.

Boulton, R.B., Singleton, V.I., Bisson, L.F. and Kunkee, R.E. (1996). *Principles and Practices of Winemaking.* Chapman and Hall, New York, NY.

Brodelius, P. and Vandamme, E.J. (1987). Immobilized cell systems. In: *Biotechnology*, Vol. 7a, H. Rehm and G. Reed (Eds.). VCH, Weinhein, Germany, p. 405.

Cantarelli, C. and Lanzarini, G. (Eds.) (1989). *Biotechnology Application in Beverages Production.* Elsevier Applied Science Publishers, London, UK.

Colagrande, O., Silva, A. and Fumi, M.D. (1994). Recent applications of biotechnology in wine production. *Biotechnol. Prog.*, 10(1): 2.

de Lerma, N.L., Peinado, R.A., Puig-Pujol, A., Mauricio, J.C., Moreno, J. and García-Martínez, T. (2018). Influence of two yeast strains in free, bioimmobilized or immobilized with alginate forms on the aromatic profile of long aged sparkling wines. *Food Chem.*, 250: 22–29.

Deshusses, M.A., Chen, W., Mulchandani, A. and Dunn, I.J. (1997). Innovative bioreactors. *Curr. Opin. Biotechnol.*, 8: 165–168.

Diviès, C. (1993). Bioreactor Technology and wine fermentation. In: *Wine Microbiology and Biotechnology*, G.H. Fleet (Ed.). Harwood Academic Publishers, Chur, Switzerland, p. 449.

Diviès, C., Cachon, R., Cavin, J.-F. and Prévost, H. (1994). Immobilized cell technology in wine fermentation. *Crit. Rev. Biotechnol.*, 14: 135–153.

Djordjević, R., Gibson, B., Sandell, M., de Billerbeck, G.M., Bugarski, B., Leskošek-Čukalović, I., Vunduk, J., Nikićević, N. and Nedović, V. (2015). Raspberry wine fermentation with suspended and immobilized yeast cells of two strains of *Saccharomyces cerevisiae. Yeast*, 32(1): 271–279.

Erickson, L.E. and Fung, D.Y.C. (Eds.) (1988). *Handbook on Anaerobic Fermentations.* Marcel Dekker, New York, NY.

Fleet, G.H. (Ed.) (1993). *Wine Microbiology and Biotechnology.* Harwood Academic Publishers, Chur, Switzerland. p. 165.

Fleet, G.H. (1998). The microbiology of alcoholic beverages. In: *Microbiology of Fermented Foods*, B.J.B. Wood (Ed.). Blackie Academic & Professional, London, UK, p. 217.

Freeman, A. and Lilly, M.D. (1998). Effect of processing parameters on the feasibility and operational stability of immobilized viable microbial cells. *Enz. Microb. Technol.*, 23: 335–345.

Fumi, M.D., Bufo, M., Trioli, G. and Colagrande, O. (1989). Bulk sparkling wine production by external encapsulated yeast bioreactor. *Biotechnol. Letts.*, 11: 821–824.

García-Estévez, I., Escribano-Bailón, M.T., Rivas-Gonzalo, J.C. and Alcalde-Eon, C. (2017). Effect of the type of oak barrels employed during ageing on the ellagitannin profile of wines. *Aust. J. Grape Wine Res.*, 23: 334–341.

García-Martínez, T., Moreno, J., Mauricio, J.C. and Peinado, R. (2015). Natural sweet wine production by repeated use of yeast cells immobilized on *Penicillium chrysogenum. LWT Food Sci. Technol.*, 61: 503–509.

Genisheva, Z., Teixeira, J.A. and Oliveira, J.M. (2014). Immobilized cell systems for batch and continuous winemaking. *Trends Food Sci. Tech.*, 40: 33–47.

Genisheva, Z., Mussatto, S.I., Oliveira, J.M. and Teixeira, J.A. (2013). Malolactic fermentation of wines with immobilised lactic acid bacteria - Influence of concentration, type of support material and storage conditions. *Food Chem.*, 138: 1510–1514.

Gerbsch, N. and Buchhold, R. (1995). New processes and actual trends in biotechnology. *FEMS Microbiol. Rev.*, 16: 259.

Giorno, L. and Drioli, E. (2000). Biocatalytic membrane reactors: Applications and perspectives. *TIBTECH*, 18: 339–349.

Godia, F., Casas, C. and Sola, C. (1991). Application of immobilized yeast cells to sparkling wine fermentation. *Biotechnol. Prog.*, 7: 468–470.

González-Pombo, P., Fariña, L., Carrau, F., Batista-Viera, F. and Brena, B.M. (2014). Aroma enhancement in wines using co-immobilized *Aspergillus niger* glycosidases. *Food Chem.*, 143: 185–191.

Groboillot, A., Boadi, D.K., Poncelet, D. and Neufed, R.J. (1994). Immobilization of cells for application in the food industry. *Crit. Rev. Biotechnol.*, 14: 75–107.

Jackson, R.S. (1994). *Wine Science: Principles and Applications.* Academic Press Inc., Cambridge, USA.

Jagtap, U.B. and Bapat, V.A. (2015). Wines from fruits other than grapes: Current status and future prospectus. *Food Biol. Sci.*, 9: 80–96.

Joshi, V.K. and Pandey, A. (Eds.) (1999). *Biotechnology: Food Fermentation, Microbiology, Biochemistry and Technology*, Vol. I and II. Educational Publishers and Distributors, N, Delhi, India.

Kang, W., Niimi, J., Muhlack, R.A., Smith, P.A.A. and Bastian, S.E.P. (2019). Dynamic characterization of wine astringency profiles using modified progressive profiling. *Food Res. Int.*, 120: 244–254.

Karel, S.F., Libichi, S.B. and Roberston, C.R. (1985). The immobilization of whole cells: Engineering principles. *Chem. Engg. Sci.*, 40: 1321–1354.

Klein, J. and Vorlop, K.-D. (1985). Immobilization techniques - Cells. In: *Comprehensive Biotechnology*, Vol. 2, M. Moo-Young (Ed.). Pergamon Press, New York, p. 203.

Kunkee, R.E. and Goswell, R.W. (1977). Table wines. In: *Alcoholic Beverages - Economic Microbiology*, Vol. I, A.H. Rose (Ed.). Academic Press, London, UK, p. 315.

Lafon-Lafourcade, S. (1983). Wine and brandy. In: *Biotechnology*, Vol. 5, H.J. Rehm and G. Reed (Eds.). Verleg Chemie, Weinheim, Germany, p. 81.

Lea, A.G.H. (1995). Cidermaking. In: *Fermented Beverage Production*, A.G.H. Lee and J.R. Piggott (Eds.). Blackie Academic & Professional, London, UK, p. 66.

Maicas, S. (2001). The use of alternative technologies to develop malolactic fermentation in wine. *Appl. Microbiol. Biotechnol.*, 56: 35–39.

Moreno-García, J., García-Martínez, T., Mauricio, J.C. and Moreno, J. (2018). Yeast immobilization systems for alcoholic wine fermentations: Actual trends and future perspectives. *Front. Microbiol.*, 9: 241.

Nedovic, V.A., Durieux, A., Van Nedervelde, L., Rosseels, P., Vandegans, J., Plaisant, A.-M. and, Simon, J.-P. (2000). Continous cider fermentation with co-immobilized yeast and *Leuconostoc oenos* cells. *Enz. Microb. Technol.*, 26: 834–839.

Nikolaou, A., Tsakiris, A., Kanellaki, M., Bezirtzoglou, E., Akirda-Demertzi, K. and Kourkoutas, Y. (2019). Wine production using free and immobilized kefir culture on natural supports. *Food Chem.*, 272: 39–48.

Ogbonna, J.C., Amano, Y., Nakamura, K., Yokotsuka, K., Shimazu, Y., Watanabe, M. and Hara, S. (1989). A multistage bioreactor with replaceable bioplates for continuous wine fermentation. *Am. J. Enol. Vitic.*, 40: 292–298.

Panesar, P.S., Joshi, V.K., Bali, V. and Panesar, R. (2017). Technology for production of fortified and sparkling fruit wines. In: *Science and Technology of Fruit Wine Production*, M.R. Kosseva, V.K. Joshi and P.S. Panesar (Eds.). Academic Press., Cambridge, USA, p. 487c530.

Pilkington, P.H., Margaritis, A. and Mensour, N.A. (1998). Mass transfer characteristics of immobilized cells used in fermentation processes. *Crit. Rev. Biotechnol.*, 18: 237–255.

Pozo-Bayón, M.A. and Moreno-Arribas, M.V. (2016). Sherry wines: Manufacture, composition and analysis. In: *Encyclopedia of Food and Health*, B. Caballero, P.M. Finglas and F. Toldra (Eds.). Academic Press., Cambridge, USA, p. 779–784.

Pretorius, I.S. (2000). Tailoring wine yeast for the new millennium: Novel approaches to the ancient art of wine making. *Yeast*, 16: 675–729.

Rolle, L., Englezos, V., Torchio, F., Cravero, F., Río Segade, S., Rantsiou, K., Giacosa, S., Gambuti, A., Gerbi, V. and Cocolin, L. (2018). Alcohol reduction in red wines by technological and microbiological approaches: A comparative study. *Aust. J. Grape Wine Res.*, 24: 62–74.

Romo-Sanchez, S., Arevalo-Villena, M., Garcia Romero, E., Ramirez, H.L. and Briones Perez, A. (2014). Immobilization of β-glucosidase and its application for enhancement of aroma precursors in Muscat wine. *Food Bioprocess Technol.*, 7: 1381–1392.

Servetas, I., Berbegal, C., Camacho, N., Bekatorou, A., Ferrer, S., Nigam, P., Drouza, C. and Koutinas, A.A. (2013). *Saccharomyces cerevisiae* and *Oenococcus oeni* immobilized in different layers of a cellulose/starch gel composite for simultaneous alcoholic and malolactic wine fermentations. *Process Biochem.*, 48: 1279–1284.

Sevda, S.B. and Rodrigues, L. (2015). Preparation of guava wine using immobilized yeast. *J. Biochem. Technol.*, 5: 819–822.

Shin, H.M., Lim, J.W., Shin, C.G. and Shin, C.S. (2017). Comparative characteristics of rice wine fermentations using *Monascus koji* and rice nuruk. *Food Sci. Biotechnol.*, 26: 1349–1355.

Sipsas, V., Kolokythas, G., Kourkoutas, Y., Plessas, S., Nedovic, V.A. and Kanellaki, M. (2009). Comparative study of batch and continuous multi-stage fixed-bed tower (MFBT) bioreactor during wine-making using freeze-dried immobilized cells. *J. Food Eng.*, 90: 495–503.

Sirisha, V.L., Jain, A. and Jain, A. (2016). Enzyme immobilization: An overview on methods, support material, and applications of immobilized enzymes. *Adv. Food Nutr. Res.*, 79: 179–211.

Webb, C. (1989). The role of cell immobilization in fermentation technology. *Aust. J. Biotechnol.*, 3: 50–55.

28 Wines and Brandies
Maturation Aspects

Shashi Bhushan, Ajay Rana and Somesh Sharma

CONTENTS

28.1 INTRODUCTION

A sequence of biochemical processes that happen with the passage time, lead to the development of various positive qualities in wine. The wine acquires new characteristics with respect to taste and organoleptic properties; this is called aging or maturation. During the process, wine kept in casks or in bottles gets a more balanced and stable structure. Aging plays significant role in improving the physical, chemical, and sensory quality of wine or brandy. Certainly, matured or aged wines are of superior quality and highly priced. Higher priced wine is likely to have been aged for a longer period to achieve the desired attributes. Hazard analysis and critical control point (HACCP) established for each stage of the process followed by monitoring systems and corrective actions have considerably improved wine quality. Primary quality characters – flavor and aroma of wine or brandy – mainly depends upon the type and quality of their must and on the processing conditions, especially the maturation, aging, or biological aging. It is one of the most important as well as the most complex processes in winemaking. Newly fermented wine is cloudy, harsh in taste, yeasty in odor, and without a pleasing bouquet. Thus, during maturation, it cleared from suspended material, yeasty odors, and condensed tannins by different treatments known as wine fining, which leads to improvement in sensory and organoleptic properties. The wines that benefit from aging become less harsh, less tannic, smoother, and more complex. Once wines complete fermentation, they begin to change, mainly because of air contact and due to this, the natural components of the new wine begin interacting with one another. Wines begin the aging process in tanks or vats where they go through fermentation. After that, most high-quality wines receive some sort of wood aging and then bottle aging. Wood aging is the process of maturing wine in barrels or casks prior to bottling. When the young wines soften and absorb some of the flavors and tannins from the wood, the wine's flavors become concentrated because of slight evaporation. The length of maturation, type of cooperage, treatment of wood, and type of wine all influence the maturation process. The effect of oak wood on the maturation of alcoholic beverages, variability in the anatomy of oak wood, properties of casks and their influence on sensory qualities of the beverages, and the anatomical features of wood and correlation with chemical properties affect the maturation process. Hence, the selection of oak for cask manufacture or in determining its suitability for particular alcoholic beverages has been focused on in this chapter. In recent years, new technologies have been developed for maturing wines, employing wood fragments, gamma irradiation, ultrasonic waves, AC electric field, lees, and micro-oxygenation etc. to accelerate aging and improve quality. Treatments using physical methods (electric field 600 V/cm and duration time three min) shorten the length of the aging process and were found to be equivalent to six months aging in oak barrels.

28.2 MATURATION AND AGING OF WINE AND BRANDY

28.2.1 Maturation and Aging

Maturation or aging denotes the reactions and changes that occur after the first racking that lead to improvements, rather than spoilage, at some stage. Maturation is a separate process from other post-fermentation processes such as clarification, tartrate stabilization, and malolactic fermentation. Aging enriches the wine with aroma active compounds, stabilizes the color, and improves the complexity of mouthfeel (Figure 28.1). Wine aging can be sub-categorized into two major stages, *i.e.* bulk storage and bottle storage or aging. During bulk storage period, the wine may be relatively exposed to oxidation in large containers (cooperage) termed as *maturation*, and in bottle storage or *bottle aging*.

28.2.2 Objective of Maturation

Generally, aging is done for the modification, improvement, and development of attractiveness in wine and to extend its stability and preservation. It affects mature wine by the extraction and interaction of volatile and non-volatile compounds, producing simpler and more stable wine components. The internal environment of the cask provides the conditions for further reactions such as oxidation, hydrolysis, and polymerization, which impart wood related aromas. Besides the significant contribution of wood in the extraction of volatiles and interaction between different phytoconstituents, it also influences the aroma of matured beverages (chemical

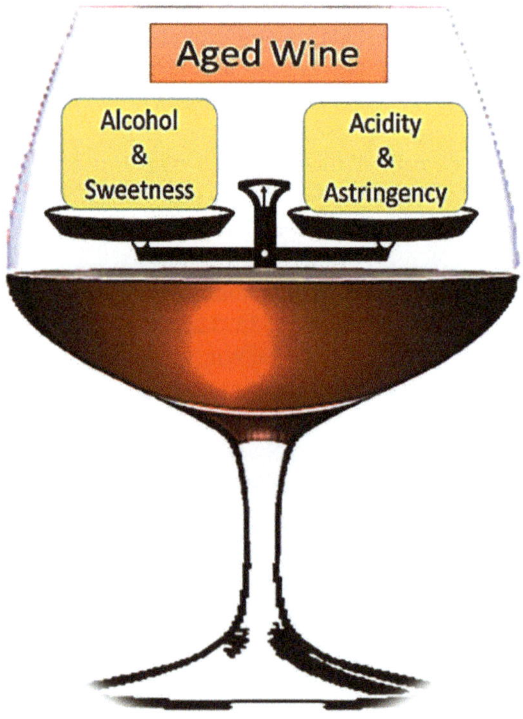

FIGURE 28.1 Attributes of aged wine

transformations of vanillin and furfural to vanillic alcohol and furanic).

These goals are achieved either by subtraction (removal of gassiness or green flavors/harshness), addition (extraction of flavor from oak barrels, development of color and flavor from oxidation and development of bottle bouquets), carry over (retention and carrying forward the varietal aromas and flavors from the wine fermentation during maturation and aging), and multiplication or complexity (small and subtle changes in aroma, for complexity and multiplication). During aging/maturation, changes are caused by three types of reactions occurring simultaneously and continually in the barrel: i) Extraction of complex wood constituents by the liquid, ii) Oxidation of components originally in the liquid and of material extracted from the wood, and iii) Reactions between the various organic substances present in the liquid, leading to the formation of new congeners. These aspects will be developed later in this chapter.

28.3 CONTAINERS/COOPERAGES FOR MATURATION

All winery tanks, vats, and barrels are called *cooperage* whether made of wood or from some other material (Figure 28.2). A wine container ideally should be impervious, inert, durable, strong, easy to clean and maintain, convenient, cheap, and not impart any undesirable flavor. Materials which are used for containers are made from ceramic, concrete, metals, wood, and more recently plastics, preferably HDPE. Different containers/cooperages have set advantages and disadvantages. Concrete cooperage can be used if properly coated with wax or other impervious coating, but it may result in acidic wine due to the presence of calcium and other minerals. At present, the use of concrete cooperage has fallen drastically. Metal cooperage contributes iron to wine, ferrous or ferric ions, which catalyze undesirable reactions in wine. Brass and copper contribute Cu^+ ions, which cause unwanted quality characteristics in wine even at low concentration. Plastic cooperage is made of different type of materials such as fiber glass polystyrene tanks, where light can pass through tanks and can affect the wine detrimentally and may contribute monomers of plastic itself or extractable plasticizers. Some plastics like HDPE transit oxygen rather readily, than others. Thus, this leaves wooden and stainless steel cooperage for modern wine storage and maturation.

Oak barrel is the oldest type of storage container used for the maturation and aging of wine. In France, two species of oak are used for the construction of oak barrels: *Quercus robur* and *Q. petraea*. However, in the United States, *Quercus alba*, *Q. bicolor*, *Q. prinus* etc. are some of the species used for cooperage. Certain Japanese and East Asian oaks such as *Q. deutta*, *Q. mongolica*, and *Q. crispula* may also be suitable for this purpose. The wood structure of casks influences the quality of alcoholic beverages through various mechanisms (see a later part of this chapter). Barrels used for the aging of wines and brandies are constructed from oak heartwood,

FIGURE 28.2 Stainless steel cooperage (a) and oak wood barrel (b) used for aging of wine

which contains the most abundant constituents that are potentially extracted into the wine during aging (Figure 28.3). The distribution of extractives in the tissue and factor(s) that influence the permeability of wood are likely to affect the movement of these extractive liquids and gases through the staves. The type of oak barrel employed also influences polyphenolics (ellagitannins) profile of aged red wines and it is distinguishable whether wine is matured in new barrels or second-fill French oak barrels or American oak barrels.

In order to achieve barrel-aged type wines, the oak wood chips, oak infusion spirals, or oak staves were used in stainless steel tanks (Figure 28.4). These practices are suggestive due to many advantages such as short term storage, greater protection against oxygen exposure, better utilization of cellar space, and smaller evaporative losses. However, wooden barrels are often desired in winemaking because of the release of wood components into the wine and increased oxygen exposure. Adding a toasted section of oak wood into the fermenter provides compounds which aid in the stabilization of tannins

FIGURE 28.3 Varieties of extractive and structural compounds extracted from oak heartwood

FIGURE 28.4 Different alternative products of oak wood used in wine maturation

and colors. A system developed by Stavin and discussed in Stavin's book regarding oak infusion with tanks could help in tannin management, color enhancement and stabilization, and palate development of red wine. The oak chips may be added directly to the stainless steel tank to get the advantages of both wood and stainless steel tanks.

28.4 THEORY AND PROCESS OF MATURATION

28.4.1 Theory of Maturation

During aging, the principal changes in flavor and bouquet of wine are mainly caused by slow oxidation, which could be beneficial or injurious to the quality of the wine. Oxidation results in the rapid accumulation of acetaldehyde and possibly of ethyl acetate and acetic acid and thus, affects the aging process if wine remains exposed to air. As the oak casks are porous, air can enter slowly, and the oxygen is absorbed by the wine and brings about a series of oxidation reactions that may change the sensory qualities as well as storage life of wine. Oxygen also enters during pumping, filtering, and racking, as well as from the head-space in the tank or cask. Oxygen firstly makes the wine flat in flavor, due to the formation of aldehyde. Properly sealed and filled wine will gradually improve in flavor, provided oxidation has not been too severe. The formation of a reddish brown layer of coloring material at the bottom or sides of bottles is the result of the oxidation of tannins, while some of the coloring matter is precipitated later on. Earlier, it was thought that rapid oxidation gives different by products and less desirable aging. Though air contact leads rapidly to the production of a strong oxidant, the effects of both systems should be the same, provided wine is kept well mixed and same amount of oxidation occurs. Aldehydes and acetals are said to be responsible for the development of a new taste of immature table wines. Some of the higher alcohols of the wine may be converted to acids which in turn may form esters with ethyl and other alcohols to contribute to the flavor and bouquet of aged wine.

28.4.2 Maturation Process

Maturation and aging are characterized by changes in the overall bouquet of wine or brandy besides changes in the color, flavor, and taste of the matured product. Oxygen is absorbed during different operations of winemaking. A lesser extent of alcohol is lost by evaporation through the pores of the wood, and results in headspace beneath the bung. Air diffuses in to fill these spaces, and wine then absorbs the oxygen from this air. The cellar worker replaces this space with wine but unavoidably aerates the added wine, thus introducing more oxygen. So, the prevalent practice of topping barrels in order to prevent or reduce oxidation may in fact do more harm than good. The saturation level of oxygen in wine is considered to be 6–7 mL/liter. To control this, a nitrogen-stripping column, which on a single pass reduced oxygen to 2 mL, has been used. However, CO_2 was found to be less effective for this purpose. The presence of high contents of sulfur dioxide (SO_2) in wines retards aging, since much of the O_2 is consumed in oxidizing the sulfurous acid, so controlled oxidation and reduction of various wine components provide opportunities for improvement. That said, SO_2 plays a significant role in wine maturation and stabilization. Even though wine yeast produces an adequate content of SO_2 during fermentation, the exogenous addition of SO_2 has become a routine practice in order to control the pH of wine.

A number of alcohols and acids in the wine may unite to form esters during aging. There are a number of fixed acids, namely tartaric acid, malic acid (present in fruit juice), succinic acid, and lactic acid (present and formed during fermentation). Other acids which are formed during fermentation are acetic acid, propionic acid, formic acid, along with traces of volatile acids. Acetic acid is in abundance, so important volatile esters, largely those of acetic acids, can affect spoilage but in small amounts contribute to the fragrance of the wine. Aldehydes formed by oxidation from various alcohols in the wine affect the flavor and bouquet and sometime also form insoluble compounds with the tannins and coloring matter or with other aldehydes. It is observed that different wines require different amounts of oxygen. The contribution of non-volatile esters to the wine odor has not been well established, however, these compounds provide uniqueness to the final product. The effect of varietal characteristic aromas and flavors (see Chapters 4 and 10 of this book) has also been observed, which are carried over to a large degree in the wine and form an important aroma component of aged wine.

28.5 FACTORS EFFECTING MATURATION OF WINE

Maturation of wine and brandy is affected by a number of factors such as time and temperature, oxygen, dissolved oxygen, substrates etc.

28.5.1 Effect of Time and Temperature

The maturation period has a great influence on the quality of wine, though the type of wine and conditions also effect the bouquet production in the final product. Temperature fluctuations are not good for flavor and the chemistry of wine/brandy maturation. A low temperature for a long time slows down the efficiency of the aging process, while an unnecessarily high temperature cooks the wine. A light white wine, such as a Riesling, will age more rapidly than a heavy red wine and would become ready for bottling in one or two years only. Table wine has been found at its optimum quality in the tank, and properly aged wines placed in bottles continue to improve for several years. Fortified wines also undergo a rapid initial aging and within a few months after the vintage, are generally sufficiently aged for consumption. Ageing of strawberry wine for nine months improved the palatability and quality characteristics. Port wines require several years of wood aging, while others improve with bottle aging only. Sherry is cooked for several months at a relatively high temperature in order to hasten aging and develop the desired flavor and color. In addition to adverse changes, an increase in the intensity of bottle bouquet is associated with aging.

The temperature coefficient (Q_{10}) is doubled with every 10°C rise in temperature for all the chemical reactions. The optimum temperature is fairly low because of the destruction of enzymes highly ordered but weakly bonded. Maturation and aging involve quite a few major reactions and a decreasing flux of some components including desirable ones, while others increase or are formed. Shipping and storage of wines at elevated temperatures can cause a rapid deterioration in wine quality due to negative effects on color and flavor, but these changes vary with the type of wine, the presence or absence of O_2, and time and temperature during storage. Slight to severe browning in heated white wines is accompanied by the development of sherry-like oxidized characters; the increased intensity of bottle bouquet is associated with aging. For varietal white wines, maturation temperature ranged between 7–21°C, with optimum at 13°C. Heating table and dessert white wines at 53°C for up to 20 days under anaerobic conditions produced wines with enhanced bottle bouquet and improved quality. The concentration of several compounds associated with aging under normal conditions, notably vitispirane, 1,1,6-trimethyl-1,2-Dihydronaphthalene (TDN), and diamascenone, increased in Riesling juice heated in the absence of air at 50°C for 28 days. These volatiles may arise from hydrolysis of glycoside precursors which are accelerated at the elevated temperatures.

Hydrolysis of terpene glycosides initially produces an increase in varietal aroma in aromatic wine, such as Muscat. Storage of Chardonnay and Semillon wines under anaerobic conditions for three weeks at 45°C increased the intensity of oak, honey, and smoky attributes, which were speculated to be the result of glycoside hydrolysis. Recently, an increased intensity of lime, honey, oak, toasty, and nutty aromas in Semillon wines were found correlated with glycosyl glucose levels. The most significant effect of short term storage at elevated temperatures on wine sensory properties was the aroma. A significant difference in Chardonnay wine aroma was observed when stored at 40°C for five to nine days. The older oaked wines are more resistant to heat-induced changes than younger ones, which would be expected to have higher levels of yeast-synthesized esters and unhydrolyzed glycosides.

For varietal red table wines, cellaring temperature averaged at 15°C with a range of 7°C–23°C. Elevated temperatures and inhibitory ethanol concentrations have a lethal effect on yeast cells, and these factors target the yeast mitochondria. Ethanol is a powerful mutagenic agent, and the effect is enhanced by incubation at high temperatures. The ability of yeasts to resist the high ethanol concentration decreases with the increased temperature, explaining the significant deterioration of the yeast film as the temperature of the wine rises in sherry maturation.

Bottle aging is one of the most important steps in Champagne making, and during this period wine remains in contact with the yeast (lees) for one year for the complete maturation process. However, a contact period of at least four months of bottle fermented wines with lees had a significant effect on the final quality and style of wine. During this period, yeast autolysis took place, which is known to play an important and subtle role in producing aroma and flavor. In autolysis, the cell constituents are broken down by their own enzymes, releasing proteins, peptides and amino acids, lipids, and carbohydrate complexes by the action of proteases, and is considered to be the most important aspect of yeast autolysis. Temperature during autolysis plays an important role, as a 10°C increase in temperature corresponds to a six to seven per cent increase in the rate of autolysis, and after 45°C and above, it denatured the protease enzyme, preventing further autolysis. However, temperatures not exceeding 10°C are preferred for the aging of wine on yeast maturation. Ethyl acetate and acetaldehyde increased, while the concentration of other volatile compounds remained unaffected during maturation on the lees. The temperature for bottle aging is generally recommended at about 13°C but not less than 10°C. A constant temperature is desirable for bottle aging to avoid pressure change and cork movement but also in bulk to control headspace variation.

28.5.2 Effect of Oxygen

Oxidation and associated changes, including browning, are desirable in a few wine types such as Madeiras, Malagas, certain Sherries, and Tawny Ports, but undesirable in many white and most blush and rosé table wines. There is an optima for such wines beyond which oxidation adversely affects the quality. Controlled amount of oxygen is added in the wines to enhance the aroma and mouthfeel of wines especially in the case of red wines. Wine is inevitably exposed to oxygen until it has been bottled. However, it has been found that synthetic plastic corks still allow entry of oxygen when compared to natural corks.

28.5.2.1 Oxygen Dissolved in Wine

Solubility of O_2 from air into wine, saturated at room temperature and atmospheric pressure is roughly 6 mL/L to 8 mL/L. The solubility increases by about ten per cent at a temperature of 5°C or lower. If the O_2 is replenished in the wine by diffusion from headspace or renewed contact with air, the consumption reaction will continue to repeat. The total amount of O_2 which a wine can take up is large. Ultimately as a result of oxidation, the color is shifted toward amber, oxidized flavors are produced, and precipitations are probably caused, especially in red wines. Oxidation, beyond the unavoidable minimum, rarely improves white table wines, but generally improves rich red table wines initially. The transition between an over-oxidized white table wine and a minimally oxidized sherry-type wine appears to take place near ten saturations or 60 mL O_2/L of wine. Red wines generally seem to improve up to about 10 saturations, and after about 30 saturations (180 mL/O_2/L) begin to show definite impairment in some characteristics.

28.5.2.2 Major Substrates in Wine for Oxidation

A number of constituents (ferrous ions, sulfite, ascorbic acid, phenols, and ethanol) are present in wines and can be oxidized more readily than others by serving as the substrates for oxidation and affecting the quality of the final product. While individual phenols can react very differently, the total content of phenols in a wine is a rough measure of its capacity to take up oxygen, capability to withstand oxidation, or to change when exposed to oxygen. It is evidenced in many ways by comparing white, pink, or red wines; by observing increased pomace contact or pressing force within or between the classes of wines; by studying the effect of total or selective removal of phenols by adsorbents such as active carbon; or by comparing a model solution with similar wines. The older methods of determining plant phenols mostly depend on their oxidation. Oxidation diminishes phenol content of wines. Considering the amount of O_2, a wine can take up 60–600 mL/L from light white wine to heavy red wine. There are no other autoxidizable substances present in sufficient amounts in wine to react with that much oxygen except ascorbic acid, unless added in large amounts. SO_2 in white wines unlikely in red, is a noteworthy scavenger O_2 but it does not react unless in free form and already present in wine. Conversion of tartaric acid to dihydroxymaleic acid and its further oxidation has been suggested as a major factor but without evidence.

28.5.2.3 Reactions Coupled to The Oxidation of Phenols

The oxidation of phenols in wine has a significant impact on the sensory and chemical profile of wines. Generally, two types of phenolic oxidation reactions are reported in wines – enzymatic and non-enzymatic. Enzymatic oxidation occurred mostly in the presence of polyphenol oxidase enzymes that are present in the majority of fruits especially grapes. These reactions mainly rely on the content of flavan-3-ols and hydroxycinnamates, such as caffeoyl tartaric acid and para-coumaroyl tartaric acid. Non-enzymatic oxidation reactions, which are also known as chemical oxidation reactions, occurred by oxidation of the catechol or galloyl phenolic groups (Figure 28.5). Reactions of vicinal dihydroxy phenol derivatives (cafferic acid, catechins) and gallic acid present in wines generate quinone and hydrogen peroxide after oxidation. In the process, hydrogen peroxide converts ethanol to acetaldehyde a major product of oxidation. It justifies the source

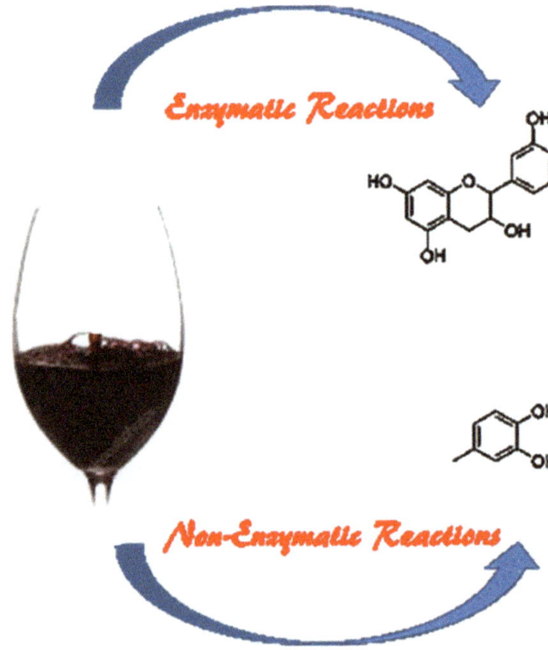

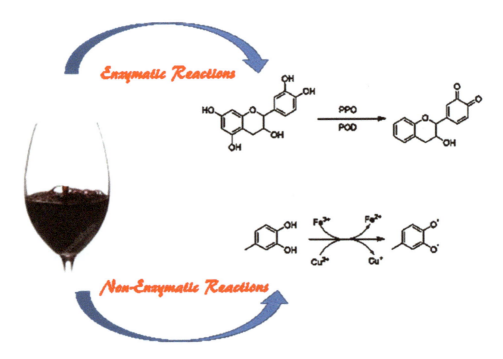

FIGURE 28.5 Oxidation of wine phenolics

of acetaldehyde production during non-microbial oxidation, which plays an important role in red wine pigment development during aging.

28.5.2.4 Phenols Other Than Vicinal-diphenol in Wine

Decline of total phenols upon oxidation more than accounted for is by a separate determination of the vicinal diphenols. Phenols such as phloroglucinol increase the O_2 consumption and browning when added to oxidizable phenols such as caffeic acid. In white wine, browning is related more to flavonoid content than to flavonoid phenols. Oxidation is auto-catalytic, particularly as evidenced from browning. The golden color of dry wines appears to be almost exclusively of phenolic origin. Dry wines thoroughly protected from oxidation ordinarily do not brown. A similar but much more autocatalytic-pronounced effect is observed with forced oxidative browning. In fact, with browning there is a lag before it begins, and sometimes even a temporary drop in color at the early stages of oxidation takes place.

28.5.2.5 Effect of pH on Wine Oxidation

The pH plays significant role in oxidation, as high pH makes wine more prone to oxidative deterioration. Phenolics are weak acids, are at least a million times more likely to lose protons than alcohols, and are about ten thousand times more acidic than water. The optimum pKa of phenols is at pH 9–10, above which 50 per cent or more of phenol exists as phenolate ions. Auto-oxidation is rapid, reaching completion in 30 minutes at room temperature. Apart from prevention of oxidation, acidic pH provides biological stability to wine by inhibiting growth of spoilage microorganisms and stabilizes wine color. As discussed earlier, SO_2 is effectively used in wine to control pH.

28.5.2.6 Increased O_2 Consumption

Wine phenols take up a large amount of O_2, and fast oxidation decreased O_2 consumption, but highly oxidized dry Sherries and Madeiras can rapidly take up additional O_2 if made alkaline. The amount of O_2 consumed per unit phenol, lost under slow acidic conditions, is high and is estimated to be 1.4–1.8 times as much as consumed under alkaline conditions. Slow oxidation does not exhaust the original oxidizable substrates of wine as rapidly as does fast oxidation and suggests an additional slower reaction taking place which augments the pool of oxidizable substrates. However, it was found that an oxygen molecule in its original triplet state does not directly react with the wine phenolics due to unpaired outer electrons. Oxygen accepts electrons from free radicals and reduced metals.

28.5.2.7 Regenerative Polymerization

Phenols are well known to polymerize as a result of oxidation, and the reactions can be complex and can take several forms, but one to illustrate here would be the interaction of two semi-quinone free radicals. The unpaired electrons readily combine to generate a new shared pair covalent bond.

28.5.2.8 Oxygen as the Oxidant in Wine

Oxygen has both positive and negative contribution to overall wine quality. The major activation route in wine appears to be substrate activation via production of free radicals in the form of semi quinones from phenols. Oxygen brings about oxidation by itself being reduced, and the full reduction of O_2 to $2H_2O$ requires the addition of four electrons. Autoxidation of wine catechol to ortho-quinones is the common example.

28.5.2.9 Alkaline Oxidation of Gallic Acid

It was found that autoxidation of pure gallic acid in water at pH 14 produced an uptake of nearly five atoms of O_2 per mole of gallic acid. Regenerative dimerization of gallic acid is about 98 per cent and remaining 2 per cent is completely oxidized to monomer. The hexahydroxy diphenic acid dimer is then re-oxidized to an identical product as obtained by similar treatment of ellagic acid.

28.5.3 Effect of Cooperage

For the past several years, the percentage of wines stored in stainless steel tanks has increased due to its many advantages over oak barrels for wine storage such as: lower long term costs, greater protection against O_2 exposure, and better utilization of cellar space and smaller evaporative losses. However, oak barrels are preferred by wine makers in aging certain wines for extraction of wood components and increased O_2 exposure, as desired in dry table wines. In an effort to achieve barrel-aged style wines using stainless steel tanks, the use of oak chips has also been suggested. During barrel aging, the phenolic composition of the wine changes due to an increased extraction from the wood, including non-flavonoids such as lignins, hydrolyzable tannins, gallic acid, ellagic acid, aromatic acids, and aldehydes, depending on the aging time, the oak type, size, toasting level and previous use of barrel. Thus, the main effects of maturation in wooden cooperage which are not found with stainless steel tanks are that extractives are furnished from the wood to the wine, and because evaporation produces ullage, air contact and oxidation may be encouraged – evaporation occurs through the wood also.

28.5.4 Effect of Oak Wood Chips

Traditionally, grape brandy is matured in oak wood cooperage, and its flavor is based on compounds which increase during aging as a result of the oxidation process and extraction from wooden barrels and wood chips. Slightly roasted wood chips of different trees viz. *Quercus*, *Albizia*, and *Bombax* were added to peach brandy. The inclusion of these wood chips in the distillate significantly affected the titratable acidity, aldehyde, fusel oil, furfural, and ester contents. Aging of peach wine with wood chips also significantly changed the biochemical characteristics of wines (Table 28.1). The maturation of wines with wood chips gave higher phenols due to the extraction of phenolic compounds from wood during maturation. As extraction is a surface-related phenomenon, the shape and size of the wooden container have significant roles to play.

TABLE 28.1
Effect of Addition of Different Wood Chips on Total Phenols, Esters, Higher Alcohols of Peach Wine of Different Cultivars

	Total phenols (mg/L)				Total esters (mg/L)				Higher alcohols (mg/L)			
Cultivar	W_0	W_1	W_2	W_3	W_0	W_1	W_2	W_3	W_0	W_1	W_2	W_3
Richahaven	253	264	256	251	115	122	121	120	125	123	129	127
Redhaven	315	320	315	313	115	122	191	119	144	139	150	164
Sunhaven	290	307	295	291	120	127	124	125	138	130	142	156
July alberta	214	230	221	218	108	113	111	111	215	250	245	256

W_0 = Control (No chips) W_1 = *Quercus* chips
W_2 = *Albizia* chips W_3 = *Bombax* chips
Source: Joshi and Shah (1998)

Extractives are the compounds found in oak wood that are soluble in either water or organic solvents.

28.5.5 Changes in Tannins and Their Derivatives

The most important group of phenolic compounds are the tannins, being a loosely defined group of water-soluble polyphenols responsible for their biological activity and astringency with the ability to bind proteins. The principal group in oak is the hydrolyzable tannins (gallotannins and ellagitannins), and the second category is the non-hydrolyzable tannins (condensed tannins or proanthocyanidins). These are oligomeric, linked by C–C bonds, which are not susceptible to hydrolysis. In oak heartwood, hydrolyzable tannin concentration is higher. A variety of other phenolic compounds are also found in oak wood. The fluorescent cumarin, compound scopoletin, is used as an indicator of alcohol maturity in casks. A wide range of volatile phenolics, particularly aromatic aldehydes and acids derived from lignin, contribute to wine maturation.

28.5.6 Changes in Other Extractives from Oak

Many compounds found in matured wine are derived from oak lignin. The possible mechanism of their formation include: degradation of lignin to aromatic compounds due to toasting or charring of casks; extraction of monomeric compounds and of lignins from the wood; and formation of aromatics by ethanolysis of lignin and further conversion of compounds into alcohol. Two γ-lactones isomers (*cis* and *trans*), as being major components of the volatile fraction of oak wood extractive, might be formed from the oxidation of lipids. It was found that both the volatile and fixed acids may increase during maturation. Acetic acid and dicarboxylic acids are identified as aroma-generating compounds and catalyzing reactions, forming lactones, esters, and other compounds. Both polyphenol oxidase and peroxidase activity have been found in heartwood of oak. Pyrazines and pyridine are also found in low concentrations. But these might not contribute to any type of flavor. The information available on this group of volatile compound is very scarce, despite being important flavor and coloring compounds.

28.5.7 Changes in Polysaccharides

Polysaccharides are one of the main macromolecules of wine that have an important role in filterability, the interaction of fermentation flora with aromatic compounds, a protective effect on protein haze formation, or on crystallization of tartrate salts. In wines, the polysaccharides primarily come from grape berries and wine yeast during fermentation. Generally grape berries release rhamnogalacturonan- and arabinan-type polysaccharides, whereas wine yeast adds mannans and glucans. Both types of polysaccharides act different while altering the quality of the organoleptic properties of wines. Aging of Carignan Noir red wines is highly influenced by polysaccharide content (Arabinogalactans, rhamnogalacturonans II, and galacturonans). Mannans and mannoproteins were apparently stable with values in the range of 120 to 150 mg/L, although a slow evolution was towards low molecular weight fraction. However, type II Arabinogalactans and rhamnogalacturonans II, the two main pectic polysaccharides present in wines, were stable for approximately ten years but decreased in the older wines. The oligomers of homo and rhamnogalacturonans II decreased dramatically during aging from 230 mg/L to 50 mg/L in wine. These slow hydrolytic phenomena were accompanied by a decrease in total polysaccharides during the aging.

28.6 WOOD SELECTION, TOASTING, AND CHANGES IN OAK STRUCTURE

28.6.1 Criteria for Wood Selection

The wood is selected based on the cost of wood, ease of construction of the barrel, and the impact on wine or brandy flavor. The wood supply must be adequate to meet demand with affordable economic price. Economic constraints demand the re-use of casks within the industry, despite casks decreasing in viability with every use. Although their previous use is often claimed to contribute to the taste of mature distillate, the purchase of used casks is primarily due to their low cost. The wood must allow the construction of a tight cask, allowing minimum leakage and being of suitable strength. Low wood permeability is necessary in order to ensure a light cask with little leakage. Oak wood has low permeability together with wide availability and thus is the main source for cooperage making. The wood must have the necessary properties to produce mature wine or brandy with the desired flavor. Comparison of Seyval blanc wine aged in barrels and stainless steel tanks with oak chips revealed that greater O_2 exposure during aging in new barrels made significant differences in the sensory score. The main differences in chemical composition among the wood treated wines were the increase in volatile acidity and decrease in pH for new and used barrels, when compared to wines matured with oak chips. The changes are caused by direct extraction of wood compounds, decomposition of wood macromolecules and their extraction into the beverage, the reaction between wood compounds and the raw product, and reactions involving only wood extractives or only the beverage component.

When Seyval blanc wine was aged in different oak barrels (American and French oak), 33 different compounds from the ethanol extract of wine were separated by HPLC. In addition, six volatile compounds were identified and quantified as gallic, protocatechenic, vanillic, caffeic, syringic, and P-coumaric acids. Aging increased gallic acid for all oak-aged wines, 7 per cent of the total increase in non-flavonoid phenols. Aging in different types of oak effected primarily quantitative values rather than qualitative, as shown by the chromatographic profile of Seyval blanc wine in different types of oak and without barrel aging. Variability of oak wood, characteristics (ring width, early wood width, texture, large vessel tyloses, multiseriate rays, total porosity etc.) of oak wood resulted in strong variability, with the abundance of tyloses in the large ring wood vessels influencing the risk of cask liquid leakage.

28.6.2 Method of Wood Toasting and Its Effect

28.6.2.1 Method of Toasting

Wood is acknowledged to be an essential element in the aging of wines, brandies, or spirits. Since ancient times, different woods have been employed for making wine barrels, which include red oak, chestnut, red or sweet gum, sugar maple, beech, black cherry, acacia, mulberry, and the wood used almost exclusively today, white oak. The different factors which influence the qualities of the oak used for cooperage include: origin, species, method used to obtain staves, the stave-drying technique (natural or artificial), and the treatments used during the cask making process. The toasting phase is particularly crucial in influencing the compounds in the wood. In the case of barrel production, the oak used for barrels pass through several processing stages important to wine flavor, principally seasoning and toasting. In the commercial production of wine barrels, toasting is normally carried out over oak fires, but the time of toasting varies according to the toast required. The designations of light, medium, and heavy toast are the traditional terms based on visual appearance of the inside face of the staves. Light toasting implies visual darkening of inside face of staves whereas medium means toasted at an intermediate level, like toasted bread. Heavy toasted wood is very dark and has a chocolate-like appearance. According to the level of toasting, the concentration of different compounds varied (Table 28.2).

TABLE 28.2
Level of Selected Compounds in Light, Medium, and Heavy Toast in the Top 4 mm of Surface Wood (in mg/kg)

	Surface toast		
Compounds	**Light**	**Medium**	**Heavy**
Ellagitannins	6462	5035	4436
Ellagic acid	848	1900	1487
5-HMF	92	208	201
Furfural	97	57	142
Vanillin	25	55	50
Syringaldehyde	94	290	345

Source: Hale *et al.* (1999)

28.6.2.2 Effect of Toasting

The volatile compounds and odors formed by toasting the oak wood can greatly influence the composition and quality of wines and brandies matured in casks. During the toasting process, the chemical changes are much more extensive. Basically, the toasting of wood causes a thermal degradation of wood components, resulting in many aromatic compounds by pyrolysis and hydrothermolysis. In addition, the major changes take place in the structure as well as in the amount of oak tannins. Chemical bonds between the three polymers (cellulose, hemicellulose, and liginin) are disrupted, while the hemicellulose and lignin in particular are degraded. The structure and amount of oak tannins also change. During toasting of American oak, the extent of thermal breakdown of the major constituents became greater as seasoning time was increased, even though the temperature of the wood was lower during toasting. The greatest difference was in polyose caramelization products, such as 5-hydroxymethyl furfural, furfural, and 5-methyl furfural, which were found highest on the surface and decreased rapidly into the depth of the stave. However, tannins were lowest at the toasted surface, and ellagic acid was formed at the expense of ellagitannins. Oak lactones occurred in their highest concentration halfway into the depth of the stave. So the wine that was matured in contact with only the surface layer of toasted wood had a dominant toasted character, higher total intensity but low complexity. Furthermore, the toasting level corresponding to the maximum chemical concentration level depended on the chemical nature of the compounds. Phenyl ketones, resulting from the thermal breakdown of lignin, were the only compounds that seemed to increase steadily with the toasting level. However, these compounds only appear in significant quantities at temperatures above 200°C, and to measure accurately the toasting temperature of less than 200°C is impossible. So each cooperage has its own toasting methods.

28.6.2.3 Changes in Oak Structure Due to Toasting

The main microscopic feature of heavy toasting was the production of transverse cracks in the wood surfaces which were not observed in medium and light toasted wood. These cracks were particularly straight, even when they crossed cell walls. At the toasted surface also, these cells showed thinning and possible volatilization of material from the ray cell lumina. The effect of heat on the physical, structural, and chemical properties of wood, along with other factors such as time of treatment, atmosphere pressure, water content, and changes in wood constituents has been investigated. Heat has been shown to cause degradation of wood polymers such as hemicellulose, cellulose, lignin, and extractive tannins. In addition to the formation of new compounds, a disorganization of the wood structure, the accessibility of solvents, and the rate and amount of extraction also increased.

28.7 AGING REGIME TYPICAL FOR WINES

The aging program selected for a specific wine depends upon the type of wine, the style within type, the price category, the marketing approach, and above all, the effects of the maturation and aging on that particular wine. The variables listed are applicable in every type and most styles of wines. *Style* refers to, for example, dinner wines versus more simple fresh 'picnic' wines. Wines made for instant marketing and consumption, are not aged for longer. Maturation process is different for each type of wine. For table wines, being a more common and important class, maturation and aging is highly emphasized. Difference in maturation levels are well noted in white versus red wine. The dessert wine requires a longer maturation and aging period in contrast to dry wines with a shorter maturation process (Table 28.3). Whereas, the production of sparkling wines usually prepared from grapes that are less ripe with higher acidity that results in wine having lower alcohol content. Generally, keeping the wine for some time with lees is the speciality of sparkling wine maturation.

In sherries or similar wine products, oxidation in one form or another is much more necessary and extensive. Oxidation may be microbial as in flor-type sherries, chemical reaction with air as in Oloroso-type sherries, and a combination of the two as in Amontillado types. Accelerated air oxidation by heating is done in Madeira-types, including California baked sherries. Bottle aging is particularly done for the Fino-flor types. The total time suggested for the Fino types is 4 years and for Olorso and Amontillado 12 years and 16 years (Solera system), respectively.

TABLE 28.3
Recommended Aging Regimes for Different Types of Wines

Wine type	Barrel	Bottle
	White table	
Dry	0–6 months	4 years
Sweet	6 months	4–10 years
	Red table	
Dry	0–6 months	0–12 months
Sweet	> 1 year	> 5 years
	Port	
Rubby	0–2 years	0–2 years
Tawny	> 4 years	0–2 years
Vintage	6 months–2 years	> 10 years
Moscatel	1–10 years	0+ years
	Sherry	
Flor	> 3 years	0+ years
Amontillado	> 3 years	0+ years
Oloroso	1–3 years	0+ years
Madeira	0–3 years	0+ years
	Sparkling wine	
Cuvee	0–12 months	–
Bulk	0–12 months	0
Bottle fermented	–	> 1–3 years

Source: Boultan *et al.* (1995)

28.7.1 Maturation and Aging of Table Wines

As the fermentation takes place, wines are kept in barrels/cooperages for maturation, and then mature ripe wine is allowed to age in bottles to develop the characteristic flavors and bouquet for the type of wine. These specific characteristics particular to a wine are affected by various factors such as varietal characters of fruits, type of fruit wines, constituents of fruits, style of wine, aging system, and other additional characteristics involved as the wine passes through the various production stages. Different aspects with respect to quality and safety during maturation and aging are shown in Figure 28.6. During the first year of aging, clarification treatment of the new red wine is customary with gelatin to remove excess tannins or bentonite after the first racking. It is a desirable procedure and hastens the development of the wine. During aging, the wine should be racked at least twice a year to get rid of sediment and to aerate it, thereby promoting normal aging. Aging of these wines is hastened by aerating and pasteurizing in smaller containers because the aging process proceeds slowly when in large containers. These processes are not considered suitable for the preparation of fine quality red wines, as the development of a fine bouquet and flavor is so far attained only by a slow aging process. For a fine Cabernet Sauvignon, one should age for about two to three years in oak barrels, and after bottling, such wines should be held at least for a year. A red wine of light body and tannin content requires a shorter aging period than a Cabernet Sauvignon with heavy body and high tannin content, and it might be best after one year in barrels and one year in bottles.

AGING		**Critical factors**	**Effect on quality**
	Quality	• Control of SO2 amount (used as preservative) • Control of air absence/ temperature • Control of foul smell, spoilage microorganisms and barrel cleaning	• Changes in sensorial characteristics • Barrel flavour and wine oxidation • Growth of *Candida*, *Pichia*, *Brettanomyces*, *Dekkera* and *Acetobactor*
		Critical factor	**Safety reasons**
	Safety	• Ethyl carbamate concentration (new barrel effect) • Control of barrel cleaning / winery/ barrel suitability	• Residues • Ethyl carbamate • Growth of undesirable microorganisms/ contaminations from dirty winery

FIGURE 28.6 Aging with respect to quality and safety aspects

The off-flavor, described as the 'phenolic character', appears in all red wines at various stages during fermentation and aging. Vinyl and ethyl phenols result from the microbiological transformation of *trans*-ferulic and *trans-p*-coumaric acids, the non-volatile odorless precursors, present in all wines, believed to be produced by lactic acid bacteria in wines. Certain lactic acid bacteria possess the enzymatic capacity needed for the decarboxylation of cinnamic acids, and in some cases, for the reduction of the vinyl phenols to the corresponding ethyl phenols present in wine. Yeasts of the genus *Brettanomyces* sp. were responsible for the phenolic character of red wines.

Changes in volatiles present before maturation on the development of the mature (20 weeks) character of red wine showed that the increased contact time of wine with the wood decreased ethyl esters of hexionic octanoic and decanoic acids. This is due to hydrolysis favored by the more aqueous environment in wine and the low proportion of isoamyl acetate. However, the increase of isoamyl and hexyl acetate after 20 weeks in wood was related to changes in equilibrium for higher levels of acetic acid, isoamyl alcohol, and hexanol. Smaller wooden casks produced oxidation conditions which were too severe, possibly due to the higher surface contact of wine and the presence of porous staves in dried wood which increased the evaporation losses similar to oxygen penetration; this was however not a major problem. Differences in phenolic compounds found during aging may contribute to variations in the evolution and volatility of some compounds during wine aging. The practice of topping up to reduce evaporation losses might not be recommended specifically in small oak containers because exposing even to such low levels (1%) of oxygen in headspace might cause fast oxidation and growth of acetic acid bacteria.

With evolution of new technologies, the solicitation of physical processes using ultrasonic waves (20 KHz) to accelerate the aging process and speedup the extraction of aroma compounds in must and wine. It subsequently, hastens the cell growth and esters formation. Earlier ultrasonic waves, having power between 20–100 KHz, were considered sufficient for efficient chemical reactivity. The waves produce their effect via cavitation bubbles, which are generated during the rarefaction cycle of the wave, when micro-bubbles are formed which collapse in the compression cycle. The ultrasonic equipment, working in the range of 20–100 KHz, was found relatively inexpensive and readily available. The treatment enabled the aging of wine with a similar quality of taste as the conventional method of aging rice wine. It is a faster process that reduces aging time to one week compared to a few months but it needs further validation in wines wines made from other sources apart from rice. It is necessary to filter the wine at least once during aging and give it a finishing filtration before bottling. For more information on the stabilization of wine, see Chapter 29 of this text.

28.7.2 Sherries

The aging system used in Sherry-making is not static and takes on dynamic qualities. Products at different stages of aging are mixed as per a set sequence to reach the final product, and aging may be biological (Jerez fino), physical-chemical (Jerez olorosa), or may involve both types of aging successively (as in Amontillado). The production of 'Fino-Sherry' wine involves film-forming yeasts (flor yeasts), whose contribution to flavor development is essential. The yeast population evolved during a semi-continuous culture system and Solera aging system makes it unique, while the film formation and oxidative metabolism might reflect an adaptive mechanism that allows the cells to survive under these conditions. The process has two phases; first there is anadae which is *static* since the wine stays in a butt as a vintage for a variable number of years. The second phase, called the solera system, is *dynamic* and consists of several scales with a different aging system. The '*velum*' formed on the surface of 'Fino-Sherry' wines in different systems and different steps of biological aging is also characterized. About 95 per cent of the whole population belongs to *S. cerevisiae* [*beticus* (75%), *S. mountuliensis* (15%); *S. cheresiensis* (5%) and *rouxii* (1%)] in both *static* and *dynamic* aging systems. The solera system was used as an aging process for four different wines, divided into five stages of aging in ascending order of age, *i.e.* Sobretablas, 3rd criadera, 2nd criadera, 1st criadera, and solera, respectively. The age that results for the Oloroso is 12 years and for Amontillado 16 years, which is the longest one.

28.7.2.1 Solera System

Each container in every stage is never completely emptied, and the maximum removal at any one time is about 25 per cent or less. When wine is withdrawn from the final (solera) stage, it is replenished by an equal amount of wine from the next younger stage in that particular solera system, and so on, throughout the stages. When a wine is drawn from each container of the next younger stage, some of that wine is placed in proportion in all the containers of the older stage. Variation of the dry extract of sherry during its aging process depends on the type of aging process. During biological aging, a series of enzymatic reactions take place due to the metabolism of the yeast acting in the medium causing reductions in volatile acidity, alcohol, and concentration of esters. In the biological aging of fino sherry, the formation of aroma compounds takes place as a result of the oxidative metabolism of the *Saccharomyces cerevisiae* 'flor' yeast, consuming some compounds such as ethanol, glycerol, organic acids, and amino acids including proline, and at the same time producing other compounds namely higher alcohols.

28.7.3 Sparkling Wine/Champagne

From a maturation and aging point of view, the flavor of aged-on-yeast sparkling wine increased when kept at least five years on the yeast. But, mostly, maturation is done for six months to a year between starting the re-fermentation and disgorging the wine. Champagnes are aged for one year minimum and *vintage* are for three years, whereas Spanish Cavas requires nine months aging on lees. During tank fermentation, wines are not usually aged for so long on the yeast but stirring or other treatments may be used to hasten yeast autolysis. A change in phenolic compounds during second fermentation, *i.e.* aging in contact with lees, of the Cava takes place. Due to the formation of pigments during aging (oxidative browning), its color tends to go from pale yellow to brown, which further increases after 15 months. But the hydroxycinnamate group decreases during aging up to 12 months, which inversely correlates with an increase in oxidative browning, caused by quinones and polymers. Hydroxycinnamates are most susceptible to oxidation, especially *cis* and *trans* caftaric and 2-5-glutathionyl caftaric acids. The release of phenolic compounds during the autolysis period (9–18 months) into the medium, along with different kinds of cytoplasmic enzymes from yeasts, changes the *trans* coutaric acid to *cis* coutaric acid.

28.8 MATURATION OF BRANDY

Historically, the aging of brandy is done in oak barrels for a number of years to develop strong sensorial characteristics and formation of hundreds of volatile compounds derived from different stages of the production process. The chemical modification occurs either from compounds already present in grapes and wines or from the distillation process. The majority are oak derived. The maturation process is affected by many factors including the type of oak species used and toasting of wood. Substantial information has been obtained and focused on the physico-chemical characteristics of brandy, oak aging, and other processes. The effect of maturation is quite distinct in brandy. Satisfactory maturation times may vary from one to more than fourteen years, there is normally a significant flavor distinction, volume and strength are lost due to evaporation of water and alcohol through the porous wood of the casks, so the quality varies according to the type of brandy, the size of cask, wood type, and environment of maturation. The presence of oxygen and the oxidative reactions related to the extraction of some substances from the oak wood exert a significant impact on the oxidation of polyphenolic fraction of lignin and tannin substances, but the main products are aromatic aldehydes, sugars, and other compounds. Lignin was found to be one of the most valuable compounds in oak wood maturation of brandy, which undergoes ethanolysis during aging and where O_2 acts as an oxidizing agent. The results so obtained led to the development of methods for thermal treatment, oxidation, and hydrolysis of oak wood in the production of brandy with 60 per cent ethanol.

28.8.1 Brandy Maturation and Type of Compounds

The sensorial attributes modified during the aging of brandies in oak wood are the most pronounced and significant, ultimately for acceptance by consumers. Aging is usually performed in oak barrels for many years, developing hundreds of volatile compounds which contribute to the aroma of brandies. During brandy aging in oak casks, low molecular weight phenolics such as cinnamic aldehydes, benzoic acid, and gallic acid were accumulated primarily as a result of the breakdown of large molecular weight polyphenols. The formation of numerous volatile and non-volatile compounds would depend on how long physical and chemical interactions take place inside the barrel. These phytochemicals are mainly derived from grapes (from the berries themselves), wines developed by distillation process, or induced by oak contact etc. These compounds can be classified as having their origin from one of three general sources: initially present in the distillate, extracted from the oak wood, or as a product of chemical reactions taking place during aging.

28.8.2 Effect on Flavor

Factually, the flavor of brandy is based on number of compounds such as carbonyls, higher alcohols, esters, acetals, and phenolic compounds. The higher the interaction and chemical conversion, the more the role of these compounds in the development of brandy flavor is strongly expressed. The aroma profile by sensorial descriptive analysis of brandies aged in different wood barrels along with the studies on modification of the aroma profile due to toasting levels was compared using seven different woods: one Portuguese chestnut wood, three Portuguese oak woods from three different locations, two French oak woods of different species, and one American oak wood. The barrels were subjected to three levels of toasting (light, medium, and strong) by heat treatment. Aging the brandies for four years showed a significant effect of heat treatment on aroma properties. Strong toasting resulted in a complex aroma profile with highest significant intensities of vanilla, woody, spicy etc. and lowest of fruity, green, taints, and glue. Similar was the aroma description profile for brandies aged in chestnut and barrels to those of oak wood aged brandies. Flavor congeners affect the type of brandy also. Naturally, changes in taste or aroma will be due to changes in flavor congeners (esters, carbonyls, sulfur compounds, lactones, phenols, and nitrogen base). A review of various compounds extracted from oak wood and present in distillates including brandy revealed that about 81 components were identified in a sample of genuine Cognac, and evidence for 16 additional compounds was found. Analysis of concentrated methylene chloride extracts of various experimental brandies beverages, unaged and aged in French and Americans oak barrels for higher boiling volatile compounds, showed that diethyl succinate, 5-methyl furfural, and β-methyl-γ-octalactone are the compounds derived from oak during aging, and the higher amount was in American oak barrels However, the ethyl esters of fatty acids present in unaged brandy distillates diminished during aging, particularly ethyl laurate and ethyl caprate.

28.8.3 Influence of Oxidation Processes on Maturation of Brandy

The mechanism and nature of non-enzymatic oxidation occurring during aging showed that the oxidation process of lignin and the formation of aromatic aldehydes has a predominant role to play in the flavor development of wine distillates. The content of aromatic aldehydes in Preslav brandy increased with an increase in time of aging, which suggests a corresponding reduction in free oxygen content of the wine distillate (Table 28.4). The mechanism of oxidation of wine distillates has been summarized as follows: the alcohol is first dehydrated to acetaldehyde by Cu dissolved in the distillate, where Me is the ion of the microelement in balance state and Ln is an organic ligand (tannin, lignin, and hydrocarbon etc.) followed by hydration and dehydration reactions. H_2O_2 reacts with copper oxide to form copper hydroxide, which possesses an even higher catalytic activity. Copper hydroxide further reacts to form Copper tannate or Copper phenolate or CuL_2-Copper carboxylate etc. It has been established that the peroxide reaction of copper hydroxide is detectable at a concentration as low as 10^8 M. The non-enzymatic oxidation of wine distillates during maturation in oak barrels also takes place. The oxidation reactions during aging of wine distillates in oak barrels proceed with the participation of tannin matter, lignin, metal catalysts, and peroxy compounds. During oxidative degradation, lignin is converted into coniferyl alcohol, guayacyl glycerol, 3-methoxy,-4-hydroxy phenyl pyruvic acid, and other products, which through the oxidation of the hydroxyl group, form coniferyl aldehyde. Coniferyl aldehyde is oxidized at the double bond and turns into vanillin. It is one of the paths of formation of flavoring substances in brandy. Coniferyl, sinap, and *p*-hydroxy cinnnamic alcohols are also oxidized to their respective aromatic aldehydes residues.

TABLE 28.4
Variation in the Content of Aromatic Aldehydes in Preslav Brandy During Maturation

	Maturation years			
Aldehydes (mg/L)	**3**	**5**	**9**	**14**
Syringaldehyde	2.7	3.3	4.7	5.5
Vanilline	1.2	2.6	4.8	6.3
Conileraldehyde	3.6	5.1	7.8	9.5
p-hydroxybenzaldehyde	1.4	2.8	3.3	3.6
Total aromatic	9.4	15.8	23.1	28.6

Source: Litchev (1989)

BIBLIOGRAPHY

Amerine, M.A., Kunkee, R.E., Ough, C.S., Singleton, V.L. and Webb, A.D. (1980). *Technology of Wine Making*. AVI Publishing Company, the University of Michigan.

Boultan, R.B., Singleton, V.L., Bisson, L.F. and Kunkee, R.E. (1995). The maturation and aging of wines. In: *Principles and Practices of Winemaking*, Springer, Boston, MA.

Caldeira, I., Belchior, A.P. and Canas, S. (2013). Effect of alternative ageing systems on the wine brandy sensory profile. *Cienc. Tec. Vitiv*, 28(1): 9–18.

Cano-Lopez, M., Bautista-Ortin, A.B., Pardo-Mingiez, F., Lopez-Rocha, J.M. and Gomez-Plaza, E. (2008). Sensory descriptive analysis of a red wine aged with oak chips in stainless steel tanks or used barrels: Effect of the contact time and size of the oak chips. *J. Food Qual.*, 31(5): 645–660

Chang, Y.L. (1992). Enzymatic oxidation of phenolic compounds in fruits. In: *ACS Symposium Series, Vol. 506, Phenolic Compounds in Food and Their Effects on Health I, Chapter 24*, pp. 305–317. ACS Symposium Series, Vol. 506.

Chatoneet, P. (1998). Volatile and odoriferous compounds in barrel aged wines: Impact of co-operage techniques and aging conditions. In: *Chemistry of Wine Flavour.* Chapter 14, pp 180–207. ACS Symposium Series, Vol. 714.

Feuillat, F. and Keller, R. (1997). Variability of oak wood (*Quercus robur* L., *Quercus petraea* Liebl.) Anatomy relating to cask properties. *Am. J. Enol. Vitic.*, 48(4): 502.

García-Estévez, I., Escribano-Bailón,M.T., Rivas-Gonzalo, J.C. and Alcalde-Eon, C. (2017). Effect of the type of oak barrels employed during ageing on the ellagitannin profile of wines. *Aust. J. Grape Wine Res.*, 23(3): 334–341.

Guadalupe, Z. and Ayestaran, B. (2007). Polysaccharide profile and content during the vinification and aging of Tempranillo red wines. *J. Agric. Food Chem.*, 55(26): 10720–10728.

Hale, M.D., McCafferty, K., Larmie, E.D., Newton, J. and Swan, J.S. (1999). The influence of oak seasoning and toasting parameters on the composition and quality of wine. *Inter. Symp. Oak Wine Mak./Am. J. Enol. Vitic.*, 50(4): 495.

Hankerson, F.P. (1947). *The Cooperage: Hand Book.* Chemical Publishing, New York.

Harbertson, J.F. (2008). A guide to the fining of wine. Washington State University Extension Manual EM016. http://winegrapes.wsu.edu/Newsletters/vol18-1-2008.pdf.

Jaarsveld, F.P. and Hattingh, S. (2012). Rapid induction of ageing character in brandy products. Ageing and general overview. *S. Afr. J. Enol. Vitic.*, 33(2): 225–252.

Joshi, V.K. and Shah, P.K. (1998). Effect of wood treatment on chemical and sensory quality of peach wine during aging. *Acta Alimentaria*, 27(4): 307.

Kilmartin, P.A. (2009). The oxidation of red and white wines and its impact on wine aroma. *Chem. N. Z.*, 73: 18–22.

Litchev, V. (1989). Influence of oxidation processess on development of the taste and flavour of wine distillates. *Am. J. Enol. Vitic.*, 40: 31.

Marko,S.D., Dornedy, E.S., Fugelsang, K.C., Dormed, D.F., Gump, B. and Wample, R.L. (2005). Analysis of oak volatiles by gas chromatography –mass spectrometry after ozone sanitization. *Am. J. Enol. Vitic.*, 56(1): 46.

Martinez-Lapuente, M., Guadalupe, Z., Ayestaran, B., Ortega-Heras, M. and Perez-Magarinno, S. (2013). Changes in polysaccharide composition during sparkling wine making and aging. *J. Agric. Food Chem.*, 61(50): 12362–12373.

Mosedale, J.R. (1995). Effect of oak wood on the maturation of alcoholic beverages with particular reference to whisky. *Forestry*, 68(3): 203.

Mosedale, J.R. and Puech, J.L. (1998). Wood maturation of distilled beverages. *Trends Food. Sci. Technol.*, 9(3): 95.

Nishimura, K. and Ohnishi, M. (1983). Reactions of wood components during maturation. In: *Flavor of Distilled Beverages: Origin and Development*, J.R. Piggot (Ed.), Ellis Horwood Series, London, p. 241.

Oliveira, C.M., Ferreira, A.C.S., De Freitas, V. and Silva, A.M.S. (2011). Oxidation mechanisms occurring in wines. *Food Res. Int.*, 44(5): 1115–1126.

Owades, J.L. (1992). Distilled beverages and spirits. In: *Ensyclopedia of Food Science and Technology*, Y.H. Hui (Ed.), A Wiley Interscience Publ, New York, p. 601.

Quinn, M.K. and Singleton, V.L. (1985). Isolation and identification of ellagitannins from white oak wood and an estimation of their roles in wine. *Am. J. Enol. Vitic.*, 36: 148.

Ribereaus Gayon, P. (1978). *Flavour of Food and Beverages*, G. Charalambous and G.E. Ingle (Eds.). Academic Press, New York.

Shah, P.K. and Joshi, V.K. (1999). Effect of different sugar sources and wood chips on the quality of peach brandy. *J. Sci. Indus. Res.*, 58: 9952.

Sharma, Somesh and Joshi, V.K. (2003). Effect of maturation on the physiochemical and sensory quality of strawberry wine. *J. Sci. Ind. Res.*, 62: 601.

Shubert, W. (1965). *Lignin Biochemistry*. Academic Press, New York.

Singleton, V.L. (1974). Some aspects of the wooden container as a factor in wine maturation. *Adv. Chem.*, 137: 254.

Singleton, V.L. (1995). Maturation of wines and spirits: Comparisons, facts and hypotheses. *Am. J. Enol. Vitic.*, 46: 98.

Spillman, P.J., Pollnitz, A.P., Liacopoulos, D., Pardon, K.H. and Sefton, M.A. (1998). Formation and degradation of furfuryl alcohol, 5-Methylfurfuryl alcohol, vanillyl alcohol, and their ethyl ethers in barrel-aged wines. *J. Agric. Food Chem.*, 46(2): 657–663.

Tao, Y., García, J.F. and Da-Wen, Sun (2013). Advances in wine ageing technologies for enhancing wine quality and accelerating wine ageing process. *Crit. Rev. Food Sci. Nutr.*

Waterhouse, A.L. and Laurie, V.F. (2006). Oxidation of wine phenolics: A critical evaluation and hypotheses. *Am. J. Enol. Vitic.*, 57: 306–313.

Yıldırım, H.K. and Dündar, E. (2017). New techniques for wine aging. *BIO Web Conf.*, 9: 02012. doi:10.1051/bioconf/20170902012.

29 Chemical and Microbiological Stabilization of Wines

Nivedita Sharma, Bhanu Neopaney and Poonam Sharma

CONTENTS

29.1 INTRODUCTION

One fine day you buy a beautiful clear bottle of wine to celebrate your birthday falling in a few months. On the day of celebration, you suddenly go panicky on finding the clear bottle liquid turning cloudy when the bottle is opened. This is a typical example of what can happen when a wine is unstable. Instability of wine refers to the still changes that may occur in wine when conditions are favorable. These changes which take place during the production of wine can be organic/chemical or desirable/undesirable. Instability could be the consequence of more than one cause, and its determination is of paramount significance in influencing the wine's acceptability and finally sale. Why there is instability in wine and how it is rectified forms the subject matter of this chapter.

29.2 TESTING OF WINE FOR STABILITY

It is very simple to determine if the wine is clear or cloudy. In clear wine, the light passes through the wine bottle as a bright light, while the light is dim in cloudy wine. There are simple tests that can be performed on the wine sample to find out if it is unstable and has the potential for precipitation. These are of two types viz., cold stable test and heat stable test (Table 29.1). A little bit of precipitation in either case is normal, as wine is exposed to extreme temperature conditions that are unusual for their storage. Some metals ions like copper and iron are present in wines naturally. But the presence of copper at more than 0.2 mg/L and iron value higher than 5 mg/L can cause wine to be unstable.

29.3 INSTABILITY OF WINE – MAJOR CAUSES

Both chemical and microbiological causes are responsible for the instability of wine. The different factors causing instability in wine and their remedial steps are discussed here.

29.3.1 Wine Oxidation

The eminent French Scientist, Pasteur observed that "oxygen is the ardent enemy of all wines". Air is omnipresent and oxygen present in it is highly reactive with unprotected grape must. A variety of materials is present in juice or wine and oxidation affect many of these adversely, resulting in bitter unpleasant off-odor and off-tastes. The activity of several enzymes like polyphenol oxidase is responsible to bring out oxidative changes in grapes. Laccase is another important oxidation-causing enzyme, found only in grapes infested with molds. Brown color is produced in the juice when oxygen reacts with certain phenolic compounds in the presence of these enzymes. During alcoholic fermentation, these enzymes, because of their inability to tolerate the alcohol produced, do not generate brown compounds unlike juice. The oxygen is exhausted in freshly crushed grapes by polyphenol oxidase. But to have the activity of such enzymes there is a need to have the precursor for oxidation *i.e.* enzyme, substrate, and oxygen. The latter is the single most important factor influencing the enzymatic activity. To measure the quantity of dissolved oxygen in grape juice, the following standard equation is used:

$$\text{Dissolved oxygen}(\text{DO}),\text{In}(\text{mg/L } O_2) = 2.63 - 0.0179^{\circ}\text{S} - 0.0190^{\circ}\text{T} \quad 29.1$$

TABLE 29.1
General Wine Stability Tests

Component	Stability test
Tartrates • Potassium Hydrogen Tartrate • Calcium Tartrate	• Cold stability test: Store at 3°C for 14 days – precipitation in crystal from crystalline ppts (shows instability)
Protein	• Heat stability test: Heat at 60°C for 48 hrs – haziness shows wine is unstable.
Copper	• Expose the bottle to indirect sunlight for 7 days – cloudiness is an indication of instability.
Iron	• When stored at 0°C for a week, aerate vigorously – precipitates formation show instability.
Brix	• Brix (°Bx) is one measure of the soluble solids in the grape juice and represents many soluble substances such as salts, acids and tannins, sometimes called total dissolved solids (TDS).
Residual sugars	• Using Hydrometer: A brix hydrometer having a scale of +5.0 to –5.0 to estimate the residual sugar content and evaluate the completion of fermentation. A wine with residual sugar over 0.2 to 0.3 percent, a high pH (>3.5), and low $S0_2$ condition, is also susceptible to spoilage by lactic acid bacteria.
Total sulfur	• Riper titration method: this method is based on the redox reaction where sulfur dioxide is in the form of the bisulphite ion. Unreacted iodine forms a blue complex with starch indicator to signify the endpoint. The addition of sodium bicarbonate prior to commencing the titration creates an inert blanket of carbon dioxide gas to prevent interference caused by oxygen in air. Red wines may require decolorizing with activated carbon prior to performing the titration in order for the endpoint color change to be observed. Color pigments in red wines may also interfere with the result, so this method is best suited to pale colored or white wines only.
Titratable acidity	• Titratable acids represent the sum of all acids in wine, except carbonic acids (H_2CO_3 or H2O + CO2). Titratable acidity (acidity) is determined by wine titration (after removal CO2) until the end point of titration by a strong base and is expressed by proton number received as equivalent concentration of selected acid.
Per cent alcohol	• Dichromate oxidation test: Potassium dichromate oxidizes primary alcohols to the corresponding carboxylic acid. The intermediate product is the aldehyde. The reaction is critically dependent upon hydrogen ion concentration for the complete oxidation to occur, rather than to a mixture of aldehyde and acid. The reaction conditions most favorable to reaction completion are 60–65 minimum of 30 minutes. Reduction of chromium from the VI oxidation state to the III oxidation state as a result of the oxidation reaction can be observed as the color change from orange to green.

Source: Rajkovic *et al.* (2007); Zoeckein *et al.* (1990)

where °S = degree Brix (°B), T = degree centigrade (°C), In = natural logarithm. This equation is applicable to other sugar solutions besides grape must.

The oxidation of white wine is never desirable, while in some cases, red wine is oxidized deliberately to remove traces of hydrogen sulfide or hasten maturity. "Pinking of wine" is the term used for the color that may appear in white wines (salmon-red blush color) produced exclusively from white grape varieties due to the presence of anthocyanins, which are located both in the pulp and in the skin of the grape. The minimum amount of anthocyanins required for the pink color visualization in wine is 0.3 mg/L. The pigmented material can be removed during early filtration and will not exist after bottling. Addition of ascorbic acid (25 mg/L) along with sufficient SO_2 is a recommended remedy to prevent this reaction. "Browning" is another problem in wine caused by oxidative changes brought in primarily by the presence of molecular oxygen. The reaction of oxygen with wine is slow, resulting in the formation of hydrogen peroxide, which being highly reactive may cause wine to oxidize rapidly. It leads to browning besides changes in flavor. Addition of SO_2 proves as an efficient antioxidant and removes hydrogen peroxide as:

$$H_2O_2 + SO_2 \rightarrow H_2O + SO_3 \qquad 29.2$$

When SO_2 is added to wine, approximately half of it rapidly combines with the components of wine, while the other half is left in the wine as a free form which is effective against oxidation. Enough SO_2 is added when the grapes are crushed, to achieve 30–50 mg of SO_2/L. Vigorous oxidation occurs when wine is bottled, and oxidation at this stage is highly detrimental leading to shortened life of bottled wine unless adequate SO_2 is present. Thus, winemakers raise the free SO_2 content of their wines to about 30 mg/L just before bottling. Addition of polyvinyl polypyrrolidene (PVPP) is also advised in white wine especially if the phenolic content is very high. The former absorbs the brown color and phenols. Binding action of PVPP preferably is as follows:

> Leucoanthocyanine> catechins >flavonols> phenolic acids

This binding is through a hydrogen bond to the oxygen of the ketoamide group of the five membered ring. Normally, PVPP is added at the rate of 6 lb/1000 gal (700 mg/L) to the wine. Casein also combines with browning, causing phenols and browned complexes and results in fresher tasting wines upon fining. The recommended dose of casein varies from 50–1000 mg/L. The exclusion of oxygen air from wine during racking is another important step to avoid oxidation.

29.3.2 Protein Precipitation

Haziness of wine is a result of protein instability. The major source of protein in wine is grape, but there are half a dozen or so proteinaceous compounds that are produced in the wine. Variety, vintage, maturity, condition of fruit, pH, and processing methodology all affect the must composition and subsequent wine protein content. The two major groups of proteins responsible for protein instability are chitinases and thaumatin-like proteins (TLPs). Beta-glucanases also cause instability in wine, but due to their less quantities, they do not make a significant contribution.

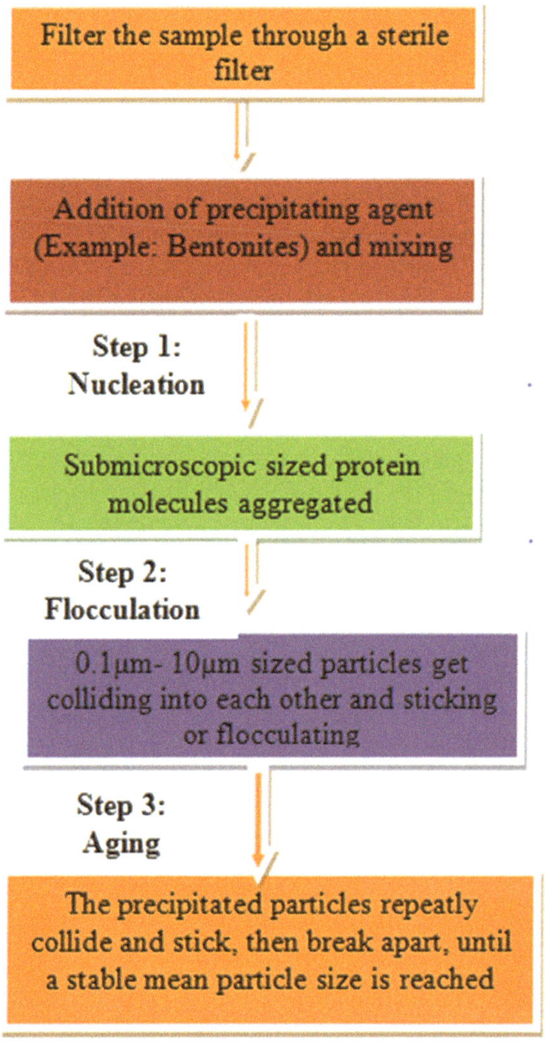

FIGURE 29.1 Schematic representation of protein precipitation

Protein precipitate formation occurs in a stepwise process as shown in Figure 29.1.

The range of protein nitrogen content of the wine is between 10–275 mg/L and can be characterized based upon size and electric charge and has 11,000 to 28,000 Dalton molecular weight. However, the actual protein levels at which wine will remain stable is not yet known despite of several studies conducted in this field. Wine proteins bind to a minor quantity of grape phenolics (phenolic acids, stilbenoids, flavanols, and flavonoids), and these compounds are thought to result in the formation of protein haze. However, protein (short peptide) derived from yeast during fermentation and lees contact are not involved in protein instability. The protein is least soluble at its isoelectric point (at a certain pH, the +ve and -ve charges of each protein fraction are equal). The isoelectric properties of protein affect precipitation and its affinity to get removed with various fining agents. Different protein stability tests have been devised to accurately predict how long the wine can stay stable after bottling.

29.3.2.1 Heat Tests

In this easy to perform heat test, wine is exposed to elevated temperatures for varying time periods (Figure 29.1). Both the temperature of heating and duration of heating influence the precipitation of protein. Heating a wine sample at 40°C for 24 hrs precipitates about 40% of protein, whereas at 60°C for the same time precipitates about 95–100% of protein, and each wine exhibited its own characteristic pattern of turbidity formation when exposed to increasing temperatures (30–80°C). Different protein stability tests were compared, such as i) 5 min heating at 80°C in double boiler, ii) 10 days at 35°C in oven, iii) Addition of 5 g/L tannin, iv) Addition of phosphomolybdic acid (Bento test), v) Addition of trichloroacetic acid (TCA), and vi) Chromatographic separation. It was found that heating for 30 min at 80°C was more efficient, but the cloudiness determined after 10 days at 35°C was more important. Trichloroacetic acid and Bento test acted on all protein, not just the heat-unstable ones. Adding bentonite to fermenting grape juice removed protein effectively, and it was more efficient than ultrafilteration or heating the wines.

29.3.2.2 Cold Stability Test

Grape juice generally composed of potassium and tartaric acid which join together to form the salt potassium hydrogen tartrate (KHT). KHT is highly soluble in grape juice when compared to ethanol. After fermentation when wine becomes saturated with KHT, it precipitates out of the solution. At low temperatures, KHT solubility decreases, and thus if an unstable wine is bottled and chilled, crystals form in the bottle. Therefore the cold stability test gives an indication that precipitation of KHT will occur after bottling. The most commonly used techniques and methods to test cold stability in wines are given below:

- Refrigeration test
- Freeze test
- Conductivity test
- Concentration of product
- Saturation point (Tsat)
- Cold stabilization in the cellar

29.3.2.3 Precipitation Test

Precipitation of wine protein is done by a number of chemical methods such as precipitation of protein using ethanol, phosphoglutaric acid, tannic acid, ammonium sulphate, trichloroacetic acid, and phosphomolybdic acid. Mostly the precipitation test is stronger than the heat test, which causes denaturation and precipitation of all protein fractions. Ethanol precipitation tests are based on the fact that increasing amounts of ethanol will decrease protein solubility, causing the proteins to precipitate out of the solution as a haze. The least soluble protein fractions are expected to precipitate first. Tannin precipitation tests are based on the assumption that proteins precipitate in wine by linking to high molecular weight compounds, like tannins. Oversimplifying a bit, the more prone in, the more precipitation, if there is not a shortage of tannin.

29.3.2.4 Bento Test

In this, the wine proteins are denatured and precipitated by adding a solution of phosphomolybdic acid in hydrochloric acid. Precipitation is directly proportional to the quality of wine protein. Bento test reagents are available commercially under different trade names. The selectivity of bentonite for removal of proteins showed that those having higher molecular weight were found most difficult to remove, and ultrafiltering the proteins using 10,000–30,000 cut-off filter held back 3–20 mg/L of protein. The 10,000 cut-off filter retained 99% of the total protein under a molecular weight of 30,000 daltons (Table 29.2). Bentonite was found more sensitive than a heat test technique using 70°C and 15 min exposure.

29.3.2.5 TCA Acid Test

After adding 1 mL of reagent to 10 mL of wine and heat in boiling water for 2 min. and then cooling to room temperature, the presence of haze indicates the occurrence of proteins. Haziness of protein can be measured by a nephlometer in terms of nephelos units. This test is stronger than a heat test and is economical and quick. The principle of nephelometric measurement is a passage of a light ray through a turbid medium, resulting in scattering which can be measured at any angle (*i.e.* 90°) relative to the plane of incident light.

29.3.2.6 Flash Pasteurization

In this case, wine is flash pasteurized at 160°F (71°C) then chilled and filtered. It is occasionally used for dessert wines,

TABLE 29.2
Protein Removal by Different Methods

Treatment	Molecular weight	Result
Bentonite	12,000 and 20,000–30,000 Dalton	Removed 1st fraction
	60,000–65,000 Dalton	Difficult to remove
Ultrafiltration	10,000–30,000 Dalton, cut-off filter	Held back 3–20 mg/L of protein
	10,000 Dalton cut-off filter	Retained 99% of total protein under 30,000 Dalton

Source: Lucchetta *et al.* (2013)

especially muscat types which require excess bentonite treatment.

29.3.2.7 Silicon Dioxide (Kieselsol)

Kieselsol is a heavy, liquid silica colloid. This material reacts with protein in the wine and precipitates out quickly. Kieselsol is sometimes used to remove excess protein material from white and blush wines. Kieselsol is often used in combination with gelatins to clarify white and blush wine, and gelatin-kieselsol fining often produces excellent clarification. Gelatin should be added to the wine first. Then, the kieselsol should be added a day or two later. Kieselsol is found in Super Kleer KC Finings. Steps involved during kieselsol test:

- De-gas wine vigorously for five minutes by stirring with the handle of a spoon or with a drill mounted stirring device.
- Add kieselsol to wine and stir for one minute. Then, add chitosan and stir well.
- Top up to within two inches of the airlock. Attach bung and airlock.
- Let wine stand until day 42–45 in an elevated cool area (15–19°C of temperature).

29.3.3 Color Stabilization

Color stabilization in white and red wine is altogether a different phenomenon. In white wines, the color stability is directly related to oxidation problems such as pinking. Young red wine pigments are essentially those of red grapes. The pH of wine and the addition of SO_2 cause instability/change in the color of red wine. The red pigments are anthocyanin and related polymers. The pH of wine changes the percentage of ionized anthocyanin compounds and consequently the color of the wine. Sulfur dioxide partly causes the discoloration of anthocyanins, thus affecting the overall intensity of the color. There are also many colorless phenolics present in red wine such as phenolic acids, tannins, and various classes of flavonoids. During the aging of wine, these compounds can interact to produce changes in color and taste characteristics.

The color of red wine also changes upon maturation, and no red wine will remain color stable for a long time. Anthocyanin molecules polymerize among themselves and other phenolics compounds, causing precipitation. During cold stabilization, all red wines have some of their color precipitated. The role of some of the key compounds responsible for wine color during aging using model solutions showed that new highly colored dimeric compounds catechin-anthocyanin are formed during the initial stages of the aging of red wine when anthocyanins, flavan 3-ols, and acetaldehyde are present. As a result, anthocyanins and flavan 3-ols get linked by an acetaldehyde bridge. The reaction is favored and is fast at low pH. These reactions also proceed in the presence of SO_2, but the rate of the reaction is slow here, so an appropriate amount of SO_2 helps to stabilize the color of red wine. Cold precipitation near the freezing point of wine *i.e.* –2 to –3°C followed by filtration may remove the polymerized color pigments, thus reducing the "mask" which otherwise is formed on the bottle of any red wine within a span of five to ten years.

29.3.4 Hydrogen Sulfide

The main source of hydrogen sulfide in grape juice is the improper use of dusting sulfur on the grapes to prevent the occurrence of mildew. Hydrogen sulfide is a gas formed by the reduction of sulfur. Volatile sulfur compounds of wines are classified in two categories according to their boiling points, *i.e.* light sulfur compounds (bp. <90°C) and heavy sulfur compounds (bp. >90°C). The sulfur-containing compounds like sulfur dioxide, hydrogen sulfide, mercaptans, and dimethyl sulphide present during fermentation play a positive role in overall aroma profile. When these compounds are present in a concentration five-fold higher than their odor threshold, they degrade wine aroma, but when present in a concentration three-fold lower, can cause beneficial effects. However, the majority of sulfur compounds like sulfur dioxide, hydrogen sulfide, mercaptans, and dimethyl sulfide also have detrimental effects on wine quality, and terms such as cauliflower, cabbage, onion, raw potato, garlic, rubber, burnt and rotten meat have been used to describe their aroma. The mechanisms of synthesis of these compounds in wine are not fully defined, though amino acids in grape juice especially methionine act as their precursor (For details on biochemical basis of the formation of aroma compounds and off-flavour, see Chapter 17 of this text.)

Low temperature of fermentation (18°C) also substantially limits the formation of volatile sulfur compounds like *S*-methyl thioacetate, 2-mercaptoethanol, acetic acid-3-(methylthio) propyl ester, 3-mercapto-1-propanol, 4-(methylthio)-1-butanol, 3-(ethylthio)-1-propanol, 3-methylthiopropionic acid, and *N*-3-(methylthiopropyl)acetamide in wines, besides the use of a suitable yeast strain for each fermentation. The technique of prolonged contact of new wines with their lees for aroma enrichment may lead to an over-production of hydrogen sulfide. Addition of 4g copper sulphate in 1000 liters of wine can remove the H_2S. The addition of 25 mg/L of SO_2 letting the wine stand for 7–10 days in a closed container will then, filter the wine. This will usually remove the H_2S but not mercaptans or disulfides as per the reactions shown below:

$$2S^{=} + SO_2 + 4H^{+} \rightarrow 3S^{o} + 2H_2O \qquad 29.3$$

$$S^{=} + ½O_2 + 2H^{+} \rightarrow S^{o} + H_2O \qquad 29.4$$

The elemental sulfur settles and precipitates and can be removed by tight filtration. Still if it is not removed, the wine becomes anaerobic and the $S^{=}$ will be found again. Yeast under anaerobic conditions in wine reduces SO_2 to $S^{=}$. If the H_2S is not removed from the wine, it may be converted to ethyl mercaptan which is extremely difficult to remove, and its removal generally decreases the wine quality to a greater extent. Ethyl mercaptan has a cabbage like, skunk-like, or treated natural gas smell which is very disagreeable. The threshold level of various volatile sulfur compounds has also been determined.

29.3.5 Tartrate Stabilization/ Cold Stabilization

Cold stabilization is often utilized to avoid the precipitation of tartrate crystals, which are common in unstable wines at cooler temperatures. Prior to putting a wine through cold stabilization, it is worth the time and effort to analyze the wine for cold stability. Not all wines end up having cold stabilization problems. For those wines that do not, going through the cold stabilization process can actually minimize wine quality by stripping out delicate aromas and flavors, or altering taste or mouthfeel attributes of the wine. This doesn't touch upon the amount of wasted time and effort to cold stabilize wines that are otherwise cold stable. In wine, tartrate deposits are generally crystalline and can be recognized easily. Though the quality of wine is not affected by its presence, in commercial wines it is a prerequisite to remove the excess tartrates. Potassium bitartrate (KHT) is present in abundance in grape juice. It is insoluble in water as well as water alcohol solutions. Potassium bitartrate is a function of temperature, the number of points of crystallization, and the amount of bitartrate present. The age-old practice to eliminate tartrate was to cool the wine near its freezing point for a few days which causes the potassium hydrogen tartrate (KHT) to settle out. Here, one precaution is the careful exclusion of air from the cold wine as the solubility of oxygen increases with decreasing temperature, and when it returns to cellar temperature, oxidation can become a problem. The filtration/racking should be done with cold wine itself as these precipitates may re-dissolve with a rise in temperature. Certain cold insoluble proteins, colored pigments; and calcium tartrate also get precipitated along with KHT. As crystallization of tartrate salt is a major source of instability in wines, it has led to several studies for better clarification and in turn imparting better stability in wine. Both tartrates, *i.e.* potassium hydrogen tartrate (KHT) and calcium tartrate (CaT), may crystallize, KHT being the most common. Tartrate salt crystallization occurs spontaneously during both alcoholic fermentation and wine storage. The crystallization of potassium hydrogen tartrate, however, is dependent on wine composition *i.e.* polysaccharides and polyphenols. Seeding cold wine with finely ground KHT crystals leads to a very rapid clarification of wines. Constant agitation of crystals and continuous filtration of tartrate is required in this method. Modern machinery is installed in many wineries based on this method for continuous removal of tartrates.

When chilled wine is passed through a bed of finely ground KHT, a rapid tartrate stabilization of wine can be achieved. The removal of tartrate can be hastened by first ultrafiltering the wine, thus leading to complete stabilization. Some colloids also act as effective stabilizing agents and can stabilize tartrates for a long time. The CMC (carboxymethyl cellulose) can be used as a stabilizing agent. It can react as other colloids do to block the nuclei and prevent the crystal growth. Yeast cells also act as a source of KHT crystal formation. Colloids from wines interacted with KHT crystal faces and affected their growth, where the polyphenols strongly inhibited crystallization and resulted in small crystals with a one-dimensional growth (Table 29.3). In contrast with polyphenols, cubic crystals were obtained when the wine polysaccharides were associated with yeast cells.

Electrodialysis is the latest concept used for the tartaric stabilization (potassium bitartrate) of wine (Figure 29.2). Formation of tartrate crystals is prevented in the bottles by running the wine through an electrodialysis unit run in a batch system. In this technique, the proportion of ions, *i.e.* potassium and tartrate, causing undesirable instability are first determined. However, the wine is coarsely filtered before electrodialyzing it. An electric field is applied to the ions, which puts them in motion, and they pass through the electrodialysis membranes selectively, up to the set point determined earlier. Selective tartrate removal systems in electrodialysis have allowed the commercialization of this process. Electrodialysis has been recognized by the International Organisation of Vine and Wine (OIV) and has been approved for commercial use by the European Union for all types of wines, as well as in the United States and other countries.

Calcium tartrate is usually not a problem in table wines unless calcium has been added to the wine either by adding calcium carbonate to lower the excessive acidity or by not washing the filter pads. Calcium present in the range of 50–100 mg/L is a safe stability level. Complete and rapid stabilization of calcium can be achieved by the addition of salt D, L potassium tartrate. The practical advantages of this system are quality preservation, checking oxidation hazards,

TABLE 29.3
Parameter Associated With Precipitation Kinetics and KHT Crystals

	Solution	Induction time (d)	Rate of conductivity decrease (% d^{-1})	Maximal conductivity decrease (%)	Crystal size after 24 h (μm)	Crystal size after cold stabilization (μm)	Crystal shape
[1]	KHT	2	1.4	4.8	300	700–1000	Perismatic
[2]	KHT+Yeasts	<1	1.2	6.8	100–200	150–350	Perismatic
[3a]	KHT+Yeasts+ Polyphenols	>10	0	0	–	–	–
[3b]	KHT+Yeasts+ Polyphenols	5	2.7	2.7	50	50–50	Thin plates
[4]	KHT+yeasts+ polysaccharides	<1	3	12	300–400	400–600	Cubic

In 3b, wine the concentration of tartaric acid and potassium was increased to twofold to obtain crystal
Source: Petermann *et al.* (1999)

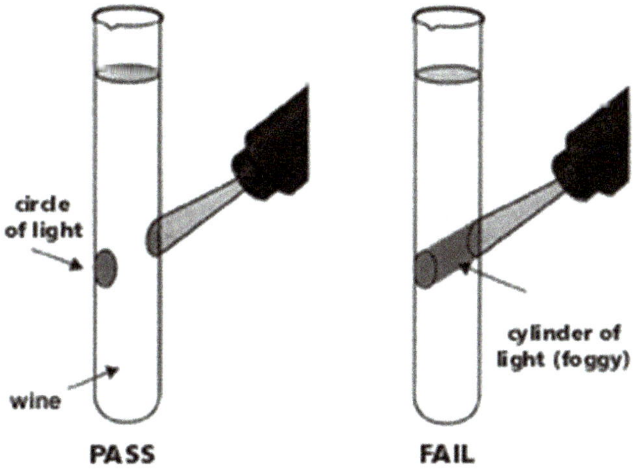

FIGURE 29.2 Heat stability test of wine (www.awri.com)

more efficiency, more reliability (than the cold treatment), and being more economical (Figure 29.3).

29.3.6 Stabilization by Ion Exchange

The wine can be stabilized without significantly affecting its quality. In this process, wine is passed through a column containing resin in cationic or anionic form. In cation form, the resin may be charged with sodium (Na^+) or hydrogen (H^+), or a mixture of Na^+ and H^+. When the wine is treated with cationic resin in sodium form, the Na^+ of the resin is exchanged with K^+ (and other cations such as Ca^{++} and Mg^{++}) from the wine. This results in the formation of sodium bitartrate which is more soluble. There is a slight reduction in acidity. The increase in sodium content of the wine could be undesirable. In such a situation, a mixed resin in Na^+ and H^+ form could be used. This would limit the amount of Na^+ in the wine. However, the acidity of the wine would increase due to the exchange between the H^+ ion (from resin) and the K^+ ion (from wine). This may be suitable for treating low acid wine, which would benefit from increased acidity while being stabilized.

When a wine is treated with anionic resin in hydroxyl (OH^-) form, the OH^- ion is exchanged for the tartrate anions (and other anions). This lowers the tartrate content of the wine. By passing the wine through both cation (H^+ form) and anion (OH^- form) exchange resin, one exchanges H^+ and OH^- ions for potassium and tartrate ions. Thus the net result is the exchange of bitartrate for water.

29.3.7 Heavy Metal Stabilization

The metal ions (molecules) most frequently found in the wine are lead, vanadium, zinc, chromium, nickel, and manganese. The three most common metals which can cause problem with wines are copper, iron, and aluminum. Tartaric acid present in grape must and wine dissolves most of metals. In a process known as blue fining, potassium ferrocyanide is used to remove any copper and iron particles that have entered the wine from bentonite, metal winery and vineyard equipment, or vineyard sprays such as Bordea.

29.3.7.1 Aluminum

This causes the reduction of SO_2 in the wine to H_2S as well as dissolved it in the wine. After short exposure of wine to aluminum, it develops H_2S, and the wine stored in this manner is usually beyond recovery.

29.3.7.2 Iron

This reacts differently with red and white wines. In red wines, ferric tannate gives red wine a blue color, a phenomenon known as "blue casse", while in the case of white wine it is called "white casse" and which takes place due to the ferric phosphate. The latter is formed within a pH range of 2.9–3.6. The safe and preferable amount of iron in wine is less than 5 mg/L. An average of between 2.5–5.0 mg/L of iron in wine is safe, and no further attention is given to its removal. Some of the tests used to determine the iron stability are: hydrochloric test (HCl), hydrogen peroxide (H_2O_2) test where a few drops are added, and citric acid test. Casein-fining is only

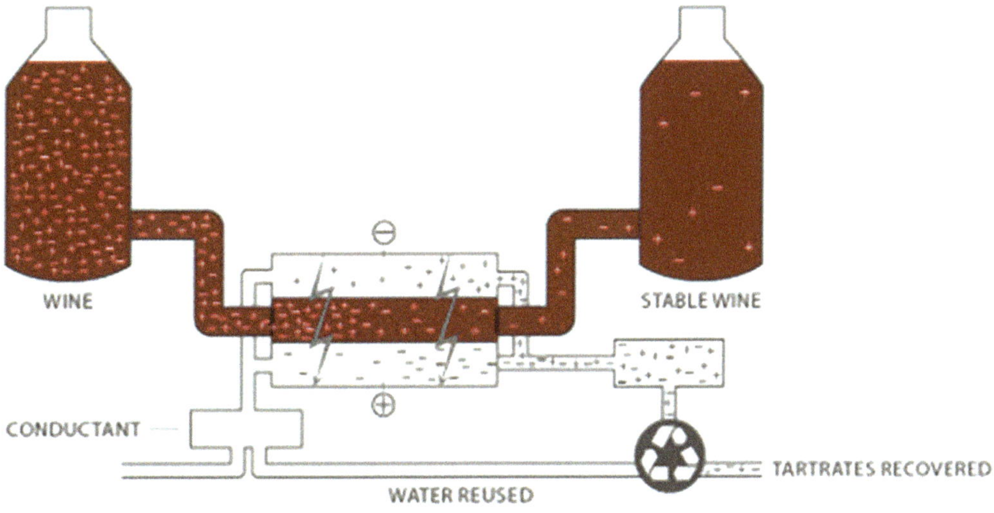

FIGURE 29.3 Tartrate removal in wine by electrodialysis process (www.rathfinnyestate.com)

used for white wine and is the easiest treatment. (For more details, see the section on fining in this chapter.)

29.3.7.3 Copper

Reaction of copper with proteins, tannins, or both causes very disagreeable hazes in wine. It mainly comes from grape sprays, addition of $CuSO_4$ for H_2S removal, etc. The presence of copper in wines can be checked by adding a few drops of sodium sulfide solution to 20 mL of wine and comparing the content of the tube with one without the addition of sodium sulfide. If more than 0.5 mg/L of copper is present, a turbidity is formed in the sodium sulfide-treated tube. The problem of copper in wine can be overcome by using yeast cells which adsorb copper, chelating agents like 8-hydroxyquinoline, as well as blue fining agents.

29.3.8 Polysaccharides and Polyphenols

29.3.8.1 Polysaccharides

Polysaccharides are the main macromolecules of colloidal nature in wines and play a very important role to maintain the basic properties of wines. Their roles depend upon their concentrations/quantity and chemical composition. Polysaccharides and polyphenols can modify the efficiency of fining treatments, participate in protein haze formation, or hinder the precipitation of potassium tartrate crystals. Microporous membranes used for dead ends or crossflow microfiltration may also get plugged by polysaccharides and polyphenols during wine filtration, thus affecting the overall filterability. These are also called "protective colloids", and instability of wine is enhanced upon their removal from wine. Moreover, these are found essential for the sensory properties of wine. Knowledge of the structure, composition, and physico-chemical properties of these polysaccharides and polyphenols can help to understand these mechanisms better. Wine polysaccharides come from both the cell walls of the grape itself and yeasts and other microorganisms that act during wine making. Wine polysaccharides are divided into three main categories:

i) Arabinose and galactose rich polysaccharides (derived from the grape cell wall)
ii) Rhamnogalacturonan-rich polysaccharides (also released from the grape cell wall)
iii) Mannoproteins released by yeasts

Polysaccharides are extracted during the mechanical operations applied to the grapes (crushing, pressing, and pumping) during the winemaking process. During the filtration stage of winemaking, the content of polysaccharides can be decreased.

29.3.8.2 Polyphenols

The main phenolics compounds of red wines are anthocyanins and proanthocyanidins (condensed tannins) that are found in grape skin (procyanidins and prodelphinidins), stems and seeds (procyanidins only) but are extracted during wine making. Anthocyanins are responsible for the red, purple, and blue pigmentation of grape berries and, consequently, red wine. During wine aging, physical, biochemical, and chemical reactions lead to the modification of the polyphenolic content of wine, color, astringency, and stability. The colloidal behavior of tannins is of paramount importance as it enlightens the properties of tannins during wine clarification and stabilization. Tannins can form hazes and precipitates, interact with proteins, as well as impact on membrane fouling.

Quercetin, a phenolic compound present in grape leaves, causes yellow haze and yellow precipitates in white wine. This is a compound incorporated in wine due to poor mechanical harvesting and gets hydrolyzed during fermentation. In cold stabilization of white wines, tartrate esters of phenolic acids and polysaccharides accounted for 0.2–0.8% of KHT crystals on a dry weight basis. 2-3-glutathionyl caftaric phenolic acids have shown specific affinity toward KHT crystals and are attributed to its glutathionyl moiety. Rhamnogalacturonan I (RG-I) and rhamnogalacturonan II (RG-II) had a peculiar behavior as the RG-I is adsorbed on crystal surfaces, whereas RG-II was not detected. Arabinogalactans and mannoprotein were also associated with tartrate crystals and may thus hamper crystal growth. In red wines, due to the presence of anthocyanin and tannins, organic compounds were associated with a higher proporation of KHT crystals than white wine.

Tannins are the major wine components associated with KHT crystals. In white wines, it is in lower proportion, *i.e.* about 2% as compared to 20% in red wines. The physico-chemical properties of these polysaccharides and polyphenols viz., their net charge, nature of their hydrophilic or hydrophobic character greatly contribute to the stability of wine. The net charge densities of many main polysaccharides has also been determined. Noticing the importance of net charges present on polysaccharides and polyphenols on the stability of wine, a method developed by Miitek (Germany) to assess the net charges on them was developed.

29.3.9 Microbiological Stability

Yeast and fermentation conditions are claimed to be one of the most important factors influencing the stability of wine. Wine itself acts as a good medium for the growth of many types of yeasts, bacteria, or molds, if not handled hygienically or protected properly, thus leading to instability/spoilage of wine. The winemaking process includes multiple stages at which microbial activity can occur, altering the quality of wine including sensory properties and rendering it unacceptable. The microscopic examination of wine is thus a pre-requisite for having a glimpse of these tiny creatures so as to determine their significance. The main spoilage organisms include yeast and bacteria, which play a vital role in the microbiological stabilization of wine. The values of pH found in wine are more vulnerable to wine yeast contamination while bacteria multiply rapidly in high pH wines. The different microorganisms seriously affecting the biological stability of wine include species and strains of yeast genera *Brettanomyces*, *Candida*, *Hanseniaspora*, *Pichia*, *Zygosaccharomyces*, etc., the lactic acid bacterial genera are *Lactobacillus*,

TABLE 29.4
Microorganisms Affecting the Stability of Wine

Genus	Changes caused in wine
Brettanomyces sp.	Off-taste with small yeasts in wine
Candida sp.	Surface film
Torulospora sp.	Crinky film
Dekkera sp.	Mousy off-flavor
Acetobacter sp.	Surface film
Bacillus sp.	Cloudiness/loss of color
Leuconostoc sp.	CO_2 evolution
Lactobacillus sp.	Turbidity/Ropiness
Chrysoniliasitophila	Cork taint

Source: Fugelsang and Edwards (2010)

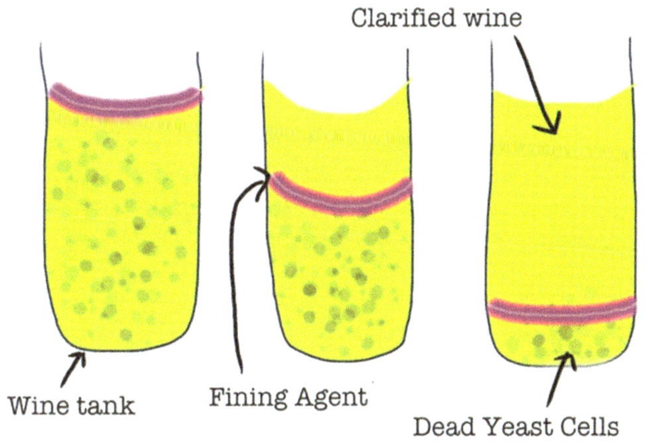

FIGURE 29.4 Fining of wine (www.enofylzwineblog.com)

Leuconostoc, and *Pediococcus*, while acetic acid bacterial genera includes *Acetobactor* and *Gluconobacter.* Two species of *Dekkera*, *i.e. D. bruxellensis* and *D. anomala*, are involved in wine spoilage. Besides these, *Torulospora*, *Debaromyces*, *Saccharomycodes*, *Kluveromyces*, *Kloeckera*, *Schizosaccharomyces*, and *Rhodotorula* are also found to cause undesirable changes in wine. Cork taint in wines is mostly due to the cork stoppers being contaminated by molds like *Chrysoniliasitophila*, *Penicillium* spp., *Cladosporium* spp. etc. Some of the important microorganisms involved in the biological instability of wines and the changes caused by them are listed in Table 29.4. These microorganisms when growing in wine generally result in bitterness and off-flavors – mousiness, ester taint, phenolic, vinegary, or buttery taste and cosmetic problems such as turbidity, viscosity, and sediment and film formation. The wholesomeness of wine is also affected by these microorganisms producing biogenic amines and precursors of ethyl carbamate. The judicious use of chemical preservatives like SO_2 is recommended to keep wine contamination-free, adequate pH values and adequate hygienic conditions will also help.

Stuck fermentation is a common problem in winemaking in which yeast prematurely shuts down instead of converting all grape sugars into alcohol and carbon dioxide. The remaining sugars are used by bacteria to cause the spoilage of wine.

29.4 FINING AGENTS

Fining is the deliberate addition of an adsorptive component that is followed by the resettling or precipitation of partially soluble components from the wine. Fining is an important operation (Fig 29.4) for wine clarification to get rid of undesirable substances from the wine which lead either to wine instability, softening and reducing the astringency of the wine, and other sensorial defects. The soluble components are removed by adding fining agents which render them insoluble (Table 29.5).

The type of fining treatment given to the wine depends upon the wine sample to be clarified. The fining agents can be classified into various groups like electrostatic, adsorbent, ionic, and enzymatic. These agents include activated carbon, fining yeast, PVPP, copper sulfate, pectinase, and pectolase. The use of proteins as fining agents is to reduce the astringency of wine or reduce its color by the adsorption and precipitation of polymeric phenols and tannins. These fining agents remove polyphenols from wine, and differently hydrolyzed gelatins are used to treat red wine.

TABLE 29.5
The Various Fining Agents and Their Concentration Generally Used for Fining White and Red Wine

Fining agents	Concentration used in wine (mg/L)	
	Red wine	White wine
Casein	60–240	60–120
Albumen	30–240	–
Gelatin	30–240	10–120
Bentonite	–	120–720
Silica sol.	40–200	40–200
PVPP	120–480	120–240
Activated carbon	120–600	120–600
Agar	120–480	120–480
Acacia	120–480	120–480
*Ferrocyanide salt	*0.02	*0.02
**Copper sulfate	**0.2–0.5	**0.2–0.5

* ,** The range shows the residual as the salts are sparingly soluble
Source: Fugelsang and Edwards (2010)

Bentonite is widely used for the adsorption of proteinaceous materials from wines. Bentonite is a gray, clay granule that is used in wines as a clarifier and also as a potential stabilization agent to remove, for example, proteins. It possesses a negative charge and thus causies suspended particles in wines to cling to it as these settle to the bottom of the container. It is very effective in dragging out tannins, yeast, and other unwanted proteins that may linger on after fermentation. By using bentonite, the treated wine has a crystal clear appearance and a radiant color. Besides this, it also helps to reduce the oxidation capacity of wine as well as certain off-flavors from it. The synthetic polymers like polyglycine, polyamide, and polyvinyl-polypyrrolidone (PVPP) are able to absorb monomeric phenols such as catechin and other flavonoids at high levels, thus preventing the browning of wine. The polysaccharides, agar, and gum arabic are protective colloids which can partially neutralize surface charges on other naturally dispersed colloids, thereby allowing them to coagulate, giving a haziness to the wine.

White wines are mostly stabilized by removing the heat unstable proteins through adsorption by bentonite, but bentonite fining is not an efficient step and can also remove other wine components. Therefore to avoid this, activated charcoal is used because it acts by non-selective adsorption and is useful for making a simple "base" wine. It removes wine faults and cleans up bad fruit before fermentation. These days, zirconium dioxide (zirconia) has been recognized as a potential candidate for fining. In this process, zirconia pellets are enclosed into a metallic cage submerged in the wine. By using these pellets, the wine can be treated for the time required for the stabilization of proteins and finally the pellets can be removed without further manipulation.

29.5 FILTRATION

Filtration as applied to wine is a very common operation which removes the suspended solids and retains the microorganisms completely. Physical stabilization removes the formation of hazes and deposits after packaging, while microbiological stabilization removes yeasts and bacteria. Filtration of contaminants occurs in three steps: primary, intermediate, and final (terminal) filtration steps. While fining clarifies wine by binding to suspended particles and precipitating out as larger particles, filtration works by passing the wine through a filter medium that captures particles larger than the medium's holes. Complete filtration may require a series of filtering through progressively finer filters. Many white wines require the removal of all potentially active yeast and/or lactic acid bacteria if they are to remain reliably stable in bottle, and this is usually now achieved by fine filtration. Most filtration in a winery can be classified as either coarser depth filtration or finer surface filtration.

29.5.1 Depth Filtration

In depth filtration, often done after fermentation, the wine is pushed through a thick layer of pads made from cellulose fibers, diatomaceous earth, or perlite.

29.5.2 Surface Filtration

In surface filtration, the wine passes through a thin membrane. Running the wine parallel to the filter surface, known as cross-flow filtration, will minimize filter clogging. The finest surface filtration, microfiltration, can sterilize the wine by trapping all yeast and, optionally, bacteria, and so is often done immediately prior to bottling. An absolute rated filter of 0.45 μm is generally considered to result in a microbially stable wine and is accomplished by the use of membrane cartridges, most commonly polyvinylidene fluoride (PVDF). Certain red wines may be filtered to 0.65 μm, to remove yeast, or to 1.0 μm to remove viable *Brettanomyces* only.

29.5.3 Earth Filtration

In this type of filtration, a layer of coarse grade earth is deposited on a supporting screen made of nylon or stainless steel as the first step. In the second step, earth is mixed into the wine, and the mixture is passed through the screen. This leads to a continuously replenished filtration surface.

29.5.4 Pad Filtration

In this process, wine is passed across a filter pad made of cellulose, cotton, or synthetic fibers such as polyethylene.

29.5.5 Membrane Filtration

This form of filtration uses a cartridge made of cellulose esters, polysulfonate, nylon, polypropylene, or glass fibers (Figure 29.5).

29.5.6 Cross-Flow Filtration

The wine is passed along the surface of a porous membrane with sufficient pressure to force liquid through the pores.

29.5.7 Ultrafiltration

This is a process where a semi-permeable membrane separates the components of liquid/solute mixture according to their molecular size. In ordinary filtration, the process liquid flows perpendicular to the filter, in ultrafiltration (UF) the process liquid flows tangential to the membrane. When the solution containing two solute flows tangentially through a semi-permeable membrane, the solute having smaller molecular size permeates easily while the solute having larger molecular size is retained. When hydrostatic pressure is applied to the membrane, the smaller molecules permeate through the membrane and larger molecules do not and so would be retained by the membrane. UF is generally classified according to molecular weight cut-off (MWCO), *i.e.* average size of the range of molecule that they will retain, and 10,000–40,000 daltons MWCO is most commonly used for the removal of proteins from milk and apple juice; 50,000–200,000 daltons are used to remove the polysaccharides that have been produced by the grape and yeast; and 10,000–50,000 daltons are mostly used for the removal of protein fractions in white wines. UF has a unique feature among filtration processes such that retentate material always gets concentrated at the membrane-solution interface and swept by fluid dynamic forces. UF can be operated in the absence of external means.

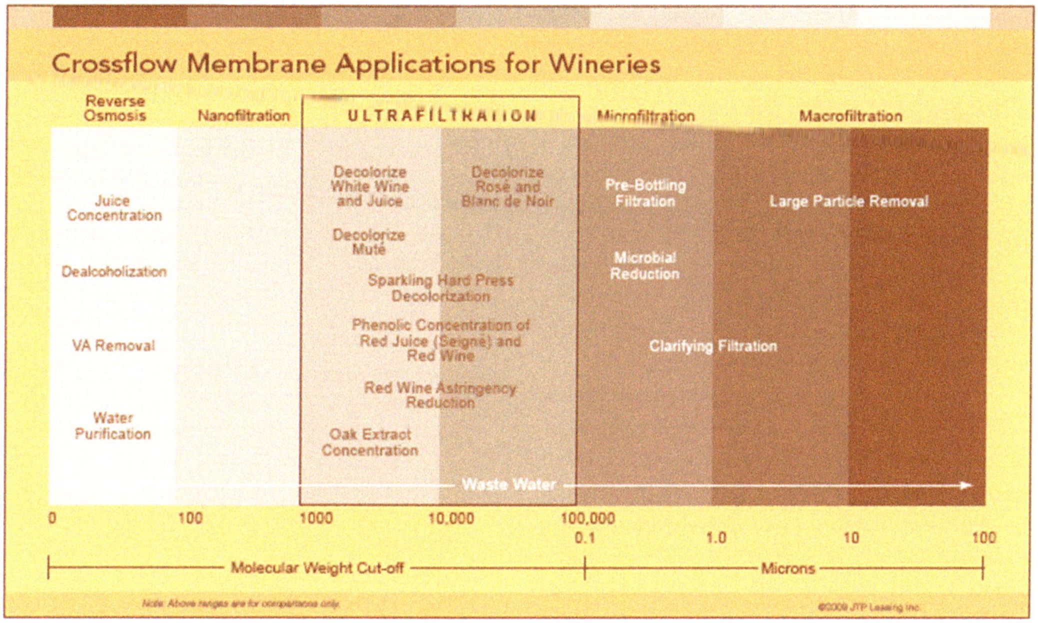

FIGURE 29.5 Different types of membrane systems for wine filtration (www.armourtech.co.nz)

29.5.7.1 Removal of Harsh and Astringent Compounds

Prior to UF, it is most important to study the molecular weight composition in comparison to percentage composition of the pressed wine. The pressed juice wine has a higher concentration of harsh and astringent phenolic compounds emanated from the skin and seed, while in the free juice wine these components are present in lower concentrations due to minimum contact with seed and skin. These have their molecular weights in the range of 600–2,000 and are separated from other phenolics ranging in molecular size between 200–500 and from sugars, acids, and the alcohols of aroma compounds which are in the vicinity of 200 molecular weight. It was postulated that UF with a molecular membrane cut-off of about 1000 could selectively remove these harsh and astringent phenolic compounds without affecting other constituents of the wine. To achieve the main objective, *i.e.* removal of harsh and astringent phenolic compounds from the pressed wine, the concept of UF was developed in the wine industry. After verification of the concept, it is now more important to select the proper UF system to ensure that: the membrane permits maximum retention of fruitiness in wine and rejects the undesirable harsh and astringent components; the system produces steadily high fluxes without fouling of the membrane; and the cleaning efficiency of the membrane in repeated use is maintained.

29.5.7.2 Protein Stabilization

To provide heat or protein stability in wine, a number of 10,000 MWCO membranes have been tested to remove the protein. This application serves as an alternative to the removal of bentonites and hence, has received considerable research attention, but it is not widely used commercially as it removes the phenolic components which are retained by 20,000 MWCO membrane for reasons other than haze. It is difficult to remove the haze precursors as they (peptides MW <10,000) form complexes with phenolics or alone are often involved in these stabilities. Wines should undergo protein (heat) stability after they are cold stabilized due to the fact that cold stabilization will affect the acidity (pH and TA) of the wine and therefore alter the protein stability properties of the wine. Again, winemakers are encouraged to check the wine for protein stability prior to treating a wine with bentonite. Bentonite is a fining agent used to bind any proteins in a wine that would otherwise be considered unstable. However, if the addition of bentonite is unnecessary (*i.e.*, the wine is protein stable and does not provide a component for bentonite to bind to), bentonite can bind to other components in the wine, most specifically aroma and flavor active compounds. While this has been shown in the research literature, it is unclear how detrimental the loss of aromatic compounds is to the wine. Additionally, bentonite additions have been noted to strip color out of rosé and red wines.

29.5.7.3 Tannins and Color Removal

Various experiments have been conducted using 500–2,000 MWCO membrane to remove red pigment to astringent tannins from press fractions. BATF considers this to be a significant alteration in wine composition and hence, it is not approved in the US winemaking industry. Fining agents and ultrafilteration help overall in retaining wine's color and clarity and thus, maintaining the stability of wine.

29.5.7.4 Filtration for Microbiological Stabilization

In wine production it is important that the wine is microbiologically stabilized for prolonged shelf-life. Filtration of organisms is performed to stabilize wine by removing or controlling the proliferation of foreign and cultured organisms. A wine that has not been sterilized might still contain live yeast cells and bacteria. If both alcoholic and malolactic fermentation have run to completion and neither excessive oxygen nor *Brettanomyces* yeast are present, this will not cause any problem. Modern hygiene has largely eliminated spoilage by bacteria such as acetobacter, which turns wine into vinegar. If there is residual sugar then it may undergo secondary fermentation and create dissolved carbon dioxide as a by-product. A wine that has not been put through complete malolactic fermentation may undergo it in the bottle, reducing its acidity, generating carbon dioxide, and adding a diacetyl butterscotch aroma. *Brettanomyces* yeasts, which add 4-ethylphenol, 4-ethylguaiacol, and isovaleric acid give a horse-sweat aroma. These phenomena may be prevented by sterile filtration by the addition of relatively large quantities of sulfur dioxide and sometimes sorbic acid mixed in an alcoholic spirit to give a fortified wine of sufficient strength to kill all yeast and bacteria. The phenomena may also be prevented by pasteurization.

Pasteurization: In this process, the wine is heated to 185°F (85°C) for one minute, then cooled to 122°F (50°C) at which it remains for up to three days, killing all yeast and bacteria. It may then be allowed to cool or be bottled "hot" and cooled by water sprays. A gentler procedure known as flash pasteurization involves heating to 205°F (95°C) for a few seconds, followed by rapid cooling.

Filters used for the Removal of Microorganisms: The SupaPore VPW filter, which contains an advanced single layer PES membrane, offers the highest flow rates whilst effectively removing all spoilage microorganisms. Therefore, different filters are used for the reliable removal of larger organisms, while the VTEC and ZTEC-B filters have been used for the removal of various bacteria and yeasts.

BIBLIOGRAPHY

Benjamin, W.R., Jablonski, J.E., Halverson, C., Jaganathan, J., Md. Abdul, Mabud and Jackson, L.S. (2019). Factors affecting transfer of the heavy metals arsenic, lead, and cadmium from diatomaceous-earth filter aids to alcoholic beverages during Laboratory-Scale Filtration. *Journal of Agricultural and Food Chemistry*, 67(9): 2670–2678.

Butzke, C. (2010). What Should I use: Sodium or calcium bentonite? In: *Winemaking Problems Solved*, Christian E. Butzke, Ed. Woodhead Publishing Limited and CRC Press, Boca Raton, FL.

Cabello-Pasini, A., Victoria–Cota, N. and Macias-Carrnza, V., Harnande-Garibay, E. and Muriz-Salazar, R. (2005). Clarification of wines using polysaccharides extracted from sea weeds. *American Journal of Enology and Viticulture*, 56(1): 52–59.

Castillo-Sánchez, J.X., García-Falcón, M.S., Garrido, J., MartínezCarballo, E., Martins-Dias, L.R. and Mejuto, X.C. (2008). Phenolic compounds and colour stability of Vinhão wines: Influence of wine-making protocol and fining agents. *Food Chemistry*, 106(1): 18–26.

Fugelsang, K. and Edwards, C. (2010). *Wine Microbiology*. 2nd Edition, Springer Science and Business Media, New York, pp. 29–44, 88–91, 130–135, 168–179.

http://www.gravertech.com/PDF/Product_Sheets/LPF/Wine app_806.pdf; https://phys.org/news/2014-08-prions-trigger-stuck-wine-fermentations.html, and https://www.awri.com.au/industry_support/winemaking_resources/laboratory_methods/chemical/cold_stab/.

https://phys.org/news/2014-08-prions-trigger-stuck-wine-fermentations.html.

https://www.awri.com.au/industry_support/winemaking_resources/laboratory_methods/chemical/cold_stab/ (2015).

Iland, P., Bruer, N., Ewart, A., Markids, A. and Sitters, J. (2012). *Monitoring the Winemaking Process from Grapes to Wine: Techniques and Concepts*, 2nd Edition, Patrick Iland Wine Promotions Pty. Ltd., Adelaide, Australia.

Jenny, A.S., Fernanda, C., Luis, F.R., Ana, S.P.M.,Aureliana, C.M., Manuel, A.C., Domingues, M.R.M. and Fernanda, M.N. (2014). Origin of pinking phenomenon of white wines. *Journal of Agricultural and Food Chemistry*, 62(24): 5651–5859.

Lucchetta, M., Pocock, K.F., Waters, E.J. and Marangon, M. (2013). Use of zirconium dioxide during fermentation as an alternative to protein fining with bentonite for White Wines. *American Journal of Enology and Viticulture*, 64(2): 404–409.

Marchal, R. and Waters, E.J. (2010). New directions in stabilization, clarification and fining of white wines. In: *Managing Wine Quality*, Volume 2, Andrew G. Reynolds, Ed. Woodhead Publishing Limited, Great Abington, UK.

OIV. (2012a). *Compendium of International Methods of Analysis of Wines and Musts*, Volume 2. Organisation Internationale de la Vigne et du Vin, Paris.

OIV. (2012b). *International Code of Oenological Practices*. OrganisationInternationale de la Vigne et du Vin, Paris.

Petermann, L.B., Vernhet, A., Dupre, K. and Moutounet, M. (1999). KHT stabilization: A scanning electron microscopy study of the formation of surface deposits on stainless steel in model wines. *Vitis*, 38(1): 43.

Pocock, K.F., Salazar, F.N. and Waters, E.J. (2011). The effect of bentonite fining at different stages of white winemaking on protein stability. *Australian Journal of Grape and Wine Research*, 17(2): 280–284.

Rajkovic, D.N., Ivana, M.B. and Petrovic, A. (2007). Dtermination of titeratable acidity in white wine. *Journal of Agricultural Sciences*, 52(2): 169–184.

Riou, V., Vernhet, A., Doco, T. and Moutounet, M. (2002). Aggregation of grape seed tannins in model wine effect of wine polysaccharides. *Food Hydrocolloids*, 16(1): 17.

Suhong, L., Yingfeng, A., Weina, F., Xioas, S., Wenjie, L. and Tuoping, L. (2017). Changes in anthocyanins and volatile components of purple sweet potato fermented alcoholic beverage during aging. *Food Research International*, 100(2): 235–240.

Tschiersch, C., Nikfardjam, M.P., Schmidt, O. and Schwack, W. (2010). Degree of hydrolysis of some vegetable proteins used as fining agents and its influence on polyphenol removal from red wine. *European Food Research and Technology*, 231(1): 65–74.

Vernhet, A. (2019). Red wine clarification and stabilization. *RED Wine Technology*. https://doi.org/10.1016/B978-0-12-814399-5.00016-5

Vernhet, A. and Moutounet, M. (2002). Fouling of organic microfiltration membrane by wine constituents: Importance, relative impact of wine polysaccharides and polyphenols, and incidence of membrane properties. *Journal of Membrane Science*, 201(1–2): 103–122.

Zoeckein, B., Fugelsang, K.C., Gump, B.H. and Nury, S. (1990). *Production Wine Analysis*. Van Nostrand Reinhold. Springer US.

30 Packaging Technology of Wines

S.S. Marwaha, S.K. Soni and U. Marwaha

CONTENTS

30.1 INTRODUCTION

Wine production and consumption have diversified during the last several decades with innovations in packaging from barrels to glass bottles with different types of closures. As the bottling process has an influence on the appearance at the point of sale and on the final consumption, it is necessary to consider effects on the wine for quality, product safety, and respecting economic limits. Modern filling technologies aim at minimizing the irregular effects that can be caused by oxygen or microorganisms. The choice of the filling method has a great impact on the level of oxygen uptake before bottling and the migration of oxygen passing through the closure after bottling. Shelf life can be extended without needing to undergo a ripening process in the bottle. Product stability, behavior, and development can be planned and finally meet the consumer's expectations.

With the developments in packaging technology and advances in material sciences, there appears to be a distinct possibility of innovations in wine packaging in the near future. Modern packaging provides all the information about the contents of a product and also educates the consumer about use and shelf-life besides protecting and preserving the product. The need for packaging of wines, packaging materials including glass and plastic containers, the requirements for packaging, declarations about the product and Government health warnings with respect to wines have been discussed hereafter.

The main purpose of packaging for different products is to i) contain, preserve and to protect the food product, ii) dispense the food product, iii) communicate details of the product to the consumer, iv) provide aesthetic appeal and iv) withstand storage and transport hazards. At the same time, packaging is a marketing tool and is definitely an element of cost, but the same is usually discarded after the product reaches the consumer. Traditional packaging does not appear to be sufficient in view of increasing customer expectations, increasing product complexity and national and international initiatives towards the economic system aimed at eliminating waste and the continual use of resources and minimising the carbon footprint of manufactured products. Innovative packaging with enhanced functionality is thus required to accommodate a variety of additional consumer needs. These include offering foods processed with fewer preservatives, products that meet increased regulatory requirements, and packaging that allows for cradle-to-grave tracking and thus may serve as a protection against lawsuits. Over the past two decades, terms such as active packaging, intelligent packaging and smart packaging have emerged in literature and are often used interchangeably. These all refer to packaging systems used for foods, drinks, pharmaceuticals, cosmetics and many other perishable goods. Strictly speaking, one should actually distinguish between intelligent, smart and active packaging. Active packaging is generally defined as "incorporation of certain additives into packaging systems with the aim of maintaining or extending product quality and shelf-life". Intelligent packaging is known as the packaging system that is capable of carrying out intelligent functions (such as sensing, detecting, tracing, recording and communicating) to facilitate decision-making to extend shelf life, improve quality, enhance safety, provide information and warn about potential problems. Smart packaging is the packaging that possesses the capabilities of both intelligent and active packaging. Smart packaging provides a total packaging solution that on the one hand monitors changes in the product or the environment (intelligent) and on the other hand acts upon these changes (active). Smart packaging serves as a means of expanding markets in the context of globalization, helps to accommodate stricter national and international food safety regulations and even serves as a protection against potential threats of food bioterrorism. The digital transformation has also been captured in packaging. However, the use of smart packaging should not be primarily oriented towards trends and image cultivation, but should focus on the benefits for the company itself and the customer. The potential and added value of innovative packaging should be profitable for consumers, manufacturers and retailers alike.

With the advent of new technologies used in manufacturing and converting a variety of basic materials into various packaging forms, the packaging designer has a number of options available such as i) cellulosic materials like paper and cellulosic films, ii) glass, iii) metals like aluminum, steel, tin plate and iv) plastics with a host of polymers. Most packaging today is derived from one or more of the basic materials listed earlier, catering to specific requirements of the products and markets. Some of the prominent packaging forms are paper board cartons and boxes, paper wrappers and sacks, glass used as ampoules, vials, bottles and carboys, metals used as cans, drums, etc. Aluminum metal is also used in the form of foil either by itself or in combination with other materials like paper and plastics. Figure 30.1 illustrates the common packaging materials used for packaging various food products. Appropriate and adequate packaging is necessary for wine and brandy so as to reach the relevant consumers.

The choice of packaging material depends upon a number of factors including i) product-package compatibility, ii) shelf-life desired, iii) packaging process adopted, iv) volume of business, v) unit quantity packed, vi) type of market and vii) cost of packaging. Protection from contamination is the most important aspect of food packaging that is ensured by hermetically sealing the package. Secondly, the packaging material used should have adequate barrier properties to protect the contents from moisture and ingress of gases like oxygen. In different forms, a variety of packaging materials are used for the packaging of fermented products, such as glass bottles with appropriate closures or plastic containers used for packaging alcoholic beverages.

FIGURE 30.1 Common materials used for packaging of various foods products

30.2 GLASS CONTAINERS FOR PACKAGING

Traditionally, glass containers of different shapes and sizes have been used for packaging wine and brandy. Owing to its inertness, glass seems to be one of the most favored packaging media, particularly for alcoholic beverages. Glass containers are used for packaging in the form of ampoules, vials, besides bottles of different shapes and sizes. But the effectiveness of glass containers as a package is achieved only with an equally effective closure system. There are three types of glass in terms of quality in relation to surface alkalinity. The glass of type I, commonly known as neutral glass or borosilicate glass, is used for packaging of very sensitive products. Type II has the inner surface alkalinity neutralized by a special treatment to make the glass inert to products which are sensitive to changes in pH value. Glass of type III with limited alkalinity is used for packaging of products which are not very sensitive to changes in pH value.

Glass is transparent, therefore its contents can be clearly seen for any impurities or suspended particles, but this property sometimes becomes detrimental to its content as some of the products tend to deteriorate on exposure to sunlight. In such cases, amber colored glass or green color glass is used. These colors filter the detrimental wavelength of sunlight. Though glass containers are considered to be very fragile, they are one of the toughest containers used for packaging. Nevertheless, efforts are made to reduce the fragility of glass containers by coating the outer surfaces. As the glass containers are at almost room temperature, these coatings are also known as cold end coatings. Another method of preventing breakage of the glass is to pack them with techniques such as shrink packaging to ensure minimum movement and rattling of glass containers during their journey. One of the most important characteristics of glass containers is their top to bottom compression resistance. Some of the parameters that are often evaluated for deciding suitability of glass containers for a particular application are discussed here.

Dimensions refer to height, body diameter, neck diameter, etc. and are critical, as they affect productivity on the high-speed machines. Volume is an important parameter, as an oversize bottle would give an impression of less volume to the consumer. On the other hand, an undersize bottle would overflow with the correct fill-volume. Many machines are designed based on the level filling principle, where the filling machine automatically stops as soon as filling touches the liquid in the bottle. Neck dimension and finish become critical, particularly on high-speed machines for application of closures, and therefore their tolerances and finish become very important. Many times, the formulation of ingredients, if not uniform, produces glass containers without proper homogeneity. For testing the homogeneity of the glass, section analysis is done wherein a small ring is cut out from a glass body and seen against polarized light in polariscope.

Glass containers during their use often receive thermal shock due to sudden exposure to very high or low temperatures. To evaluate whether the glass container has adequate strength to withstand such thermal shock, a thermal shock test is conducted. In the test, glass containers are filled with boiling water and then, these containers are lowered in a water bath at room temperature up to the neck. As a result of this, while the inner surface of the container is at a high temperature, the outer surface is exposed to room temperature and thus, the container receives thermal shock. Glass containers of type I should be able to withstand thermal shock of 45°C. The shape of the container also influences its ability to withstand shocks. Generally, cylindrical bottles and containers are common as they have better strength than other, non-cylindrical, shapes. It is because when the containers are under stress during storage, the load of the upper layers of bottle is distributed evenly in a cylindrical shaped container, whereas in glass containers of rectangular, square or other angular shapes, stresses tend to get concentrated in the corners, at the shoulders and at other such points so that with little impact, the glass breaks. In the emerging scenario, however, glass appears to be getting replaced with alternate packaging such as plastic containers, composite containers, carton system, flexible pouches, etc. However, glass containers are the most eco-friendly packaging, as the container itself can be reused a number of times. Even after breakage of the container, the broken glass can be recycled to produce a fresh new glass container and this offers an advantage for the environment. Thus, glass containers will continue to dominate the packaging scene even in the 21st century.

30.2.1 Bottles

The history of the wine bottle is fascinating and dates back to 6,000 BC with invention of "kvevri", large earthenware vessels used for the fermentation, storage, ageing and transportation of traditional Georgian wine (Figure 30.2). Next came the "amphora", invented by the Egyptians and reaching peak usage during the time of the Greek and Roman empires. These ceramic containers came in a cornucopia of shapes and sizes but always had two handles for ease of transport and a long slim neck that reduced the amount of wine being exposed to oxygen (Figure 30.2). Originally sealed with clay stoppers, the Greeks and Romans settled on cork as the best way to seal their amphorae in order to prevent wine spoilage. The ancient nation of Gaul which encompassed most of modern-day France and Switzerland, in addition to parts of Belgium, Luxembourg, Germany and Northern Italy, is responsible for the technological innovation of storing wine in wooden barrels. Glass bottles began to be used in the 17th century, although they were different shapes to the wine bottles of today. They were squat, with large bases and short necks. By the 1820s, wine bottles resembled the traditional ones we use today. These bottles needed stoppers, and cork was found to be ideal for the purpose, although it does allow a small amount of oxygen into the bottle. Plastic wine bottles made their debut in the 21st century, and clearly, they have advantages over glass bottles. These are lighter to carry than glass ones, and they can be recycled.

The Romans were the first to use glass for wine storage. They developed the art of glass blowing and found that glass bottles were great for the storage of wines. They could seal the wines completely and ensure that the storage utensil didn't affect the flavor of the wine. It was also easier to check just how much wine was left in the bottle without opening it.

FIGURE 30.2 Kvevri, the large earthenware (upper) and Amphora, the ceramic containers (lower) traditioally used for storage of wine

Naturally, this would have been considered a great boon to the ancient wine trade.

Glass bottles were hand-made and the sizes varied from bottle-to-bottle. Because of this, no one could be absolutely certain of how much wine was stored in the bottles. Because the customers didn't know just how much wine they were getting, the trade wasn't fair and therefore Romans made selling wines in glass bottles illegal.

The glass bottle with which we are familiar now was introduced in the 17th century. While the shapes varied from manufacturer to manufacturer, the general idea was the same. These wine bottles were made from dark, resilient glass, had a long, slim neck, and a cork sealing the mouth. No matter what the shape of the wine bottle was, these characteristics remained the same.

The shape of the wine bottle evolved over time. The very first bottles introduced in the mid 17th and used during the 18th century were broad-bottomed and had short necks. They were still experimenting with the size and shape during those early years. There were all kinds of designs, some elegant and artistic while others were downright ridiculous.

There were several shapes that become quite popular. The Belgian shape or the Onion Shape was common during those years. Eventually, the bottle shape started to take on the more familiar design. The modern bottle shapes started to emerge in the early 19th century. The earliest bottles of modern design started appearing in the markets in the 1820s. There are five distinct bottle styles that are common in the modern wine industry. They are the Bordeaux Shape, Burgundy Shape, Rhine Shape, Chianti Shape and Champagne Shape (Figure 30.3). In addition, there are some more shapes used for the packaging of wines. Historical credit for invention of glass blowing goes to the Romans, although glass was essentially a luxury item for centuries. In 1821, an English company patented a machine to mould bottles that were uniform in size and shape. Selling wine already bottled, however, was illegal in England until 1860, due to both the political influence of pub owners and the lack of labeling standards and means of authenticating the fill volume. For non-traditional bottle shapes, there is not much consumer acceptance. There is plenty of evidence to show that packaging Cabernet Sauvignon in a Burgundy bottle, or Pinot Noir in a Bordeaux bottle just does not work. One pathway to attention-getting bottles is *via* attractive colors.

30.2.1.1 Wine Bottle Shapes

Shapes for wine bottles evolve primarily from area tradition. There are 'classic' shapes in general use by the majority of producers from any given area and 'modern' shapes that are essentially more variations of the classics. The Italians seem to have the most variations, such as the tall bottles of fanciful shapes that sometimes hold Chianti. There are also fairly wide variations in glass colors, from crystal clear through various shades of green and brown to nearly opaque, occasionally blue. Light, whether natural or artificial accelerates the wine spoilage. Darker colored bottles and certain shades protect wine from light, but producers generally select glass color based upon packaging appeal, rather than solar security. Every bottle has a bottom, which may be flat or 'punted'. The punt evolved as a pushed-up section of varying depth in the center of the bottom. This indentation was formed as a "handle" for glass blowers to turn their creation and to enable the bottles to stand upright and consequently, was much easier to form an even plane by pushing up on the center of the bottom, rather than turning one that was perfectly flat. The punt forms a handle for Champagne bottles for racking, also known as riddling, a process of moving the sediment remaining in the bottle from the second fermentation to rest in the neck of the bottle for easy removal and strengthens and spreads the pressure over more surface area to prevent sparkling wine bottles from bursting. On the classic shapes, necks vary slightly in length, which enables wine bottle handlers to observe level of fill or 'ullage'. A high fill is desirable, because this means there is less oxygen trapped in the bottle to hasten spoilage. A fill which is too high can be too sensitive to small changes in temperature and be prone to leakage.

30.2.1.1.1 Bordeaux

This is probably the most used ever, both for red and white wine. Originally from Bordeaux, the port city in southwestern France, the bottle has a very pronounced shoulder to hinder the escape of any deposits while pouring the wine. These are the straight sided bottles with steep, tall shoulders, are typically made with dark green glass for red wines, light green glass for white wines and clear for dessert wines. This bottle

FIGURE 30.3 Common shapes of wine bottles

houses the two most popular red wines in the world, Cabernet Sauvignon and Merlot.

30.2.1.1.2 Burgundy

This is sturdier, heavier and wider than the other types, with gently sloping shoulders and was invented in the 19th century. Within a few decades, the bottle became ubiquitous as the bottle used to house good Pinot and Chardonnay wines, the two most representative wines of Burgundy. Nowadays, most red wines with a flavor profile similar to Pinot Noir – light, bright and complex – such as Nebbiolo, Gamay and Etna Rosso are also be found in this style bottle. It is made of good thickness and is transparent in color for white wines and green or brown for red wines.

30.2.1.1.3 Rhine

Originally from the Rhine area, this is often green in color for wines from the Mosel, or brown, and it is used for white wines from Germany and many other areas of the world. Its main feature is the lack of the shoulder. In Alsace this bottle is called the Alsatian and differs slightly in shape.

30.2.1.1.4 Chianti

Though this has fallen out of general use, a traditional Chianti bottle is round, and is distinguished by the raffia basket woven around the base. This is used to store a Chianti wine, produced in the Chianti region of central Tuscany.

30.2.1.1.5 Champagne

This is a bottle design that was born out of necessity. Very thick glass with gentle sloping shoulders and a long neck with a rather large punt. Champagne bottles need to withstand the high pressures (exceeding 90 psi) exerted by the carbonation development after bottling. The punt is needed to help reduce the pressure felt along the bottom of the bottle to prevent blowing out and frequent breakage.

30.2.1.1.6 Alsace and Mosel

A narrow, thin tall bottle with a very gentle sloping shoulder is typically light green, but may occasionally be brown glass, typically used for wines, such as Riesling, Gewurztraminer and Pinot Gris.

30.2.1.1.7 Fortified Wines

Sturdy bottles are typically used for fortified wines including Port, Madeira and Sherry. These have a larger bulge in the neck to help capture the sediment when it is decanted. Usually, they have a cork stopper rather than the larger corks typically used for other wines. Unopened bottles of fortified wine can be stored in a cool, dark location. Once opened, it is best to drink fortified wines as soon as possible. However, Vermouth can retain its flavor for up to three months. All open bottles of fortified wine should be stored upright in the refrigerator.

30.2.1.2 Evolution of Wine Bottle Shapes

The shapes of different glass bottles evolved with different glass houses over a period of time. Free blown English wine shapes are normally pre-1820 and range in colour from dark olive to dark amber. Time spans for evolution of various bottle shapes are represented diagrammatically in Figure 30.4.

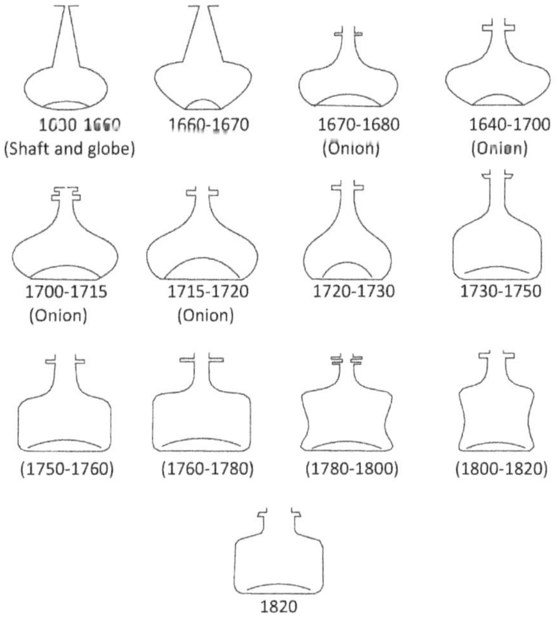

FIGURE 30.4 Time spans for the origin of wine bottle shapes

30.2.1.3 Wine Bottle Sizes

Wine bottle dimensions and wine bottle shapes vary. Wine bottles typically measure 3 to 3.2 inches in diameter and are about 12 inches tall. Champagne comes in slightly larger containers that measure up to 3.5 inches in diameter and closer to 12.5 inches tall in 750-milliliter amounts. Various commonly used wine bottle sizes and designations are shown in Table 30.1 and Figure 30.5.

30.3 PLASTICS FOR WINE PACKAGING

In recent years, moves have been made to replace glass, which while being excellent at protecting wine from the ingress of oxygen, is heavy and has a tendency to break. The latest development is the appearance of plastic bottles made of polyethylene terephthalate (PET) for several wines, which started with a New Zealand Sauvignon Blanc and an Australian Shiraz Rosé.

The plastics, being less expensive, are finding more and more use in food packaging, due to their light weight and convenience of use, and rise of flexible packaging for food is contributing to its popularity. Only food grade plastics are used for this purpose. PET belongs to the polyester family, and has been developed for the preservation of foods and beverages (sodas, juices, water, and now wine). Chemically, it is a combination of ethylene glycol and terephthalic acid, which form a polymer chain. This chain is then broken down into small pellets, then heat applied to create a liquid which may then be formed into any desired shape. Advantages of this type of material are cited as it being transparent, low cost and strong. As a result of this technology, PET has been considered

TABLE 30.1
Wine Bottle Sizes and Designations

Measure	Size equivalence	Servings	Popular name
187 milliliters	quarter bottle	1	Split
375 milliliters	half bottle	2	Tenth (often wrongly referred to as a "Split")
500 milliliters	two-thirds bottle	3 -	Half Liter
750 milliliters	standard bottle	4 +	Fifth
1.5 liter	two bottles	8 +	Magnum
3 liter	four bottles	17 +	Double Magnum, Jeroboam (sparkling wine only)
4.5 liter / 5 liter	six / six + bottles	25 / 28	Jeroboam (claret shape) Jeroboam (burgundy shape)
6 liter	eight bottles	34	Imperial Magnum (claret shape) Methuselah (sparkling or burgundy shape)
9 liter	twelve bottles	50 +	Salmanazar
12 liter	sixteen bottles	67 +	Balthazar
15 liter	twenty bottles	112 +	Nebuchadnezzar (sparkling wines)
± 12 liter to 16 liter	± sixteen to twenty bottles	90 to 112 +	Nebuchadnezzar (table wines)

Source: http://www.westcoastwine.net/index.html

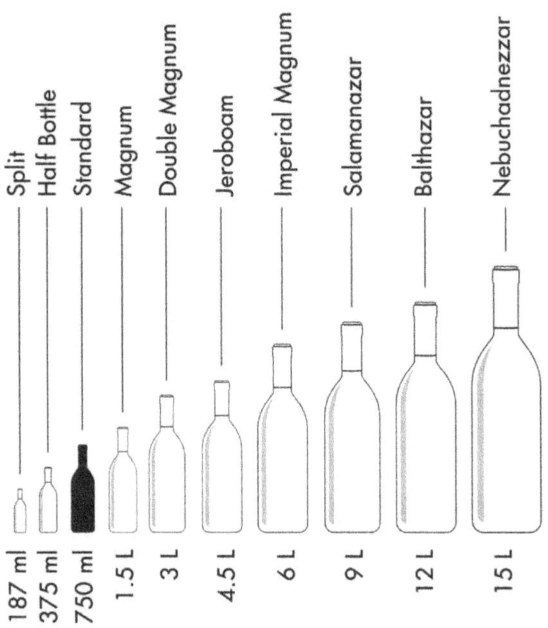

FIGURE 30.5 Various wine bottle sizes and their names. *Source:* https://www.bmj.com/content/359/bmj.j5623

a potential acceptable replacement for glass, available in both single-layer and multi-layer forms. PET has excellent gloss and clarity with high mechanical properties, and good resistance to gases, water vapor and chemicals besides dimensional stability over a wide range of temperature.

30.4 BAG-IN-BOX SYSTEM

Efforts have been made throughout the history of wine cask packaging to upgrade the packaging and presentation of the product at the point of sale, besides controlling the costs. These include bag and box manufacture, filling systems, component materials, improvement of oxygen barrier properties, understanding and avoiding product taints, improving the physical properties of the cask for handling and distribution process to which it is subjected, developing more reliable quality control and assurance system, and catering to the needs of the different market segments. The bags (capacity 500 mL to 20 liters or more) are fabricated using laminates (polyester, nylons, HDPE, LDPE, aluminum foil and specialty coatings like EVOH, ionomer) which are tailored for compatibility with the product and have a specially designed valve for filling and dispensing the wine. The bags are generally pre-sterilized, and sterilized products are filled under sterile conditions, which helps in giving a longer shelf-life. Two plies of polymeric film are used to make bags for wine casks and have a dispensing fitment attached to one face of the bag. The inner ply is generally a low-density polyethylene, ethyl vinyl acetate of linear low density polyethylene, used to avoid the occurrence of flex cracking during shipment. The outer ply (or gas barrier) gives strength to the bag while providing protection for the contents by reducing the ingress of oxygen and the egress of sulfur dioxide and volatiles which are associated with product quality. When designing boxes, there are several important factors to consider, such as customer related factors, design of graphics, convenience for carrying and storing and dimensions to allow easy handling. The cask has to be free of bulge and warp, yet resist bulge on the filling line, fold crisply along score lines, fit within the outer box on high speed palletizing lines, and the performance around the spiggot hole must open cleanly but not burst open when traveling along the filling line (Figure 30.6).

30.5 WINE BOTTLE CLOSURES

Keeping wine in the bottle until consumption necessitates the use of closure with dual purposes of wine containment and its preservation. Wine is sensitive to oxygen and will spoil before

FIGURE 30.6 Typical bag-in-box systems for wine packaging

it has time to evaporate, so the latter purpose is the more critical. In ancient times, when little was known of wine chemistry (or general hygiene for that matter), devices such as tightly bundled straw or oil-soaked twisted rags may have been stuffed into wine bottles to prevent spilling. Although glass bottles appear smooth-surfaced, they are actually imperfect especially inside the neck, so these closures were only marginally effective for containment and not at all for preservation. Stoppers made from bark of the cork oak, *Quercus suber*, were discovered to be excellent closures, because of their elasticity and their apparent impenetrability to both moisture and oxygen.

30.5.1 Corks

Wine consumers enjoy the ritual of carefully removing the cork from a bottle. The industry has long promoted the idea that superior wine comes only in cork-finished bottles. Natural cork remains the closure of choice throughout most of the wine world, but the advancing technology of synthetic corks made of various ethylene-vinyl acetate compounds is encouraging many vintners to take a closer look. In 1994, the sale of more than several million synthetic corks was reported as hardly significant, but that level was surpassed in just the first month of 1995. For the venturesome vintner, synthetic corks may be worth a try, but still the natural corks are predominant.

30.5.2 Capsules

Traditionally, capsules have been a very important part of packaging, primarily for consumer appeal. Manufacturers have made much about the ability of capsules to "seal" bottles, though they can in fact be removed and replaced rather easily. In the case of most hard poly types, it is simply a matter of pulling them off and putting them right back on again. Until the late 1980s, lead and lead-alloy capsules were the epitome of fine wine capsules – durable, efficient, attractive in appearance but expensive. When several controversial studies indicated the possibility of trace amounts of lead migrating from the capsule into poured wine, lead was banned by the Bureau of Alcohol, Tobacco, Firearms, and Explosives, (ATF, USA). Great advances have been made in the appearance of heat-shrink PVC film capsules, and they offer great resistance to tampering. Even more improvements have been achieved in replacing lead with tin-aluminum and aluminum polylaminate roll-on capsules. As with corks, capsule diameter must be precisely fitted to the bottle neck in order to avoid folding, wrinkling or tearing when applied. The revolutionary B-Cap® closure system features a custom-designed, functional, economical, attractive, user friendly, drip-resistant, flange-lip bottle mouth, in which the corks are driven as usual, but instead of a capsule, a small decorative paper circle is applied to cover the top of the cork.

30.6 WINE LABELS

The wine may, indeed, be very good; the question is whether or not it is saleable. Is the label something which can exude pride in the consumer's cellar? Is the packaging something serious wine imbibers want to present at their dinner table with guests? If the answer to any of these questions is in doubt, then results are likely to be a bottle or two bottles of "courtesy" sales rather than full-case sales and repeat orders. The label rationale includes tags, stickers and other supplemental attention-getters.

There are many shapes of labels in use, which radically depart from the traditional square or rectangle. Design and printing of odd shapes generally requires custom dies for cutting. Non-traditional shapes cause greater paper waste. Labeling machines usually require expensive custom parts fabricated in order to handle custom shapes.

There can be little question that color and its integration in wine packaging should be placed in the hands of professionals, though expensive but absolutely essential to properly address each wine to its target market. Color psychology seems to differ with each psychologist who feels qualified to write about it. Several references have been distilled down to this general set of consumer truisms. Examples can be cited of

White: Clean-winter-delicate; Black: Formal-death-night; Brown: Heavy-religion-obscure; Red: Hot-danger-stop; Blue: Cool-thin-distant; Green: Life-summer-proceed; Yellow: Floral-spring-fresh; Orange: Citrus-autumn-ripe; Violet: Royal-soft-expensive. Obviously, wine labels should be totally representative of the product and the producer. There are some wine labels which are funny and others set out to generate label smile appeal. Consumers may not take any wines in the line seriously, and a solid reputation may erode to novelty. Local laws dictate label information for the point of sale where the wine is marketed, rather than where the wine is made.

30.6.1 Vintage Dating

A vintage year may be used on labels of wine produced from grapes harvested and fermented within that calendar year and which are labeled with an appellation more specific than a country name.

30.6.2 Declaration of Sulfites

Bottled wines must have a label affixed that is a declaration of sulfites. The label may be front, back, strip or neck, but it must be on every bottle.

30.6.3 Government Health Warning

The wines bottled for sale or distribution must have a health warning statement on the label indicating: "GOVERNMENT WARNING". All the wine labels can be classified into four basic categories: 1) APPELLATION: named for the place the grapes are grown, 2) VARIETAL: named for the predominate type of grape used, 3) GENERIC: named for a commonly recognized style of wine and 4) PROPRIETARY: name created and owned by the brand.

BIBLIOGRAPHY

Aidan, C. (2018). Wine packaging: Is it time to can the classic 750mL bottle?: Alternative wine packaging could be on the rise. *Australian and New Zealand Grape Grower and Winemaker.* 90: 92.

Boulton, R.B., Singleton, V.L., Bisson, L.F. and Kunkee, R.E. (1997). *Principles and Practice of Winemaking.* Chapman and Hall, New York, 427.

Bureau, G. and Multon, J.L. (1996). *Food Packaging Technology,* Vol. 1 & 2. VCH Publishers Inc., New York.

Cooper, J. (2019). Briefing: Developing a more circular economy model for wine packaging and delivery. *Proceedings of the Institution of Civil Engineers - Waste and Resource Management.* 172(2): 40–41.

Eldred, N.R. (1992). *Package Printing.* Publishing Plainview Co., Inc, Jelmar.

Joshi, A.A. and Mokashi, N.G. (1999). Packaging of fermented food products. In: *Biotechnology: Food Fermentation: Microbiology, Biochemistry and Technology,* Vol. 1, V.K. Joshi and A. Pandey, eds. Educational Publishers & Distributors, New Delhi, 477–500.

Lai, M.B. (2019). Consumer behavior toward wine products. *Case Studies in the Wine Industry.* Woodhead Publishing Series in *Food Science, Technology and Nutrition,* 33–46.

Lee, T.H. and Simpson, E.F. (1993). Microbiology and chemistry of cork taints in wine. In: *Wine Microbiology and Biotechnology,* G.H. Fleet, ed. Harwood Academic Publishers, Chur, Switzerland, 77–164.

Margot, S. (2019). Labelling and packaging: The first taste is almost always with the eye. *Wine and Viticulture Journal.* 34: 59–60.

Osborn, K.D. and Jenkins, W.A. (1992). *Plastic Films: Packaging Applications.* Technomic Publishing, Lancaster.

Paine, F.A. and Paine, H.Y. (1992). *Handbook of Food Packaging.* 2nd edn. Blackie Academic & Professional, London.

Rapp, B. (2005). Packaging of wine. *Materials Today.* 8(3): 6. doi:10.1016/S1369-7021(05)00720-0.

Schaefer, D. and Cheung, W.M. (2018). Smart packaging: Opportunities and challenges. *Procedia CIRP.* 72: 1022–1027.

Strobl, M. (2019). Red wine bottling and packaging. In: *Red Wine Technology,* Morata, A., ed. Academic Press, London, 323–339.

31 Technology of Waste Management in Wineries and Distilleries

Chetan Joshi

CONTENTS

31.1 INTRODUCTION

Increase in population, rapid industrialization especially in the food processing and brewing industries and depletion of natural resources, are some of the major issues confronting most of the countries in the world. The food processing industries are major generators of food waste during postharvest production and during processing of products. All over the world, large quantities of waste are generated from the food industries. India is one of the largest producer of food and food products in the world and is also the leading generator of food waste due to several factors, including the postharvest losses of various crops. The solid waste generated from food production and fruit processing is about 4.5 million tons per annum. These activities are continuously destroying the purity and natural beauty of air, water, soil, plants and animals, and other constituents of our environment, which are essential for our living on the earth.

From time immemorial, the biosphere has acted as a sink for all the waste products and recycles them to make good the loss, so that every generation finds the biosphere to be the same as the one preceding it. But, of late, this self-cleaning and equilibrated maintenance of the biosphere is disastrously disturbed as the waste produced from man's activities is continuously exceeding the Earth's capacity to purify herself; consequently, our biosphere is becoming more and more poisonous and days are not far away when this planet will become uninhabitable.

Feeling alarm over the growing pollution of the natural resources of Earth, the United Nations focused on environmental problems in a conference on Human Environment in 1972 in Stockholm and emphasized the need for greater efforts for environmental protection world-wide. In the United States, the Environmental Protection Agency (EPA) was created in 1970 to administer the Environment Protection Program. In India, after the enactment of the Water (Prevention & Control) Act, 1974, came another comprehensive act, Environment Protection Act, 1986. At present, all the developed and developing countries have strict environmental legislation to control environmental pollution.

The wineries (for making wine) and distilleries (to distill wine to make brandy) are no exception. Nevertheless, the waste is a precious source of several useful materials, as such or after fermentation. The pollution load of biodegradable waste has increased many-fold, including that from the wineries and distilleries. In fact, turning winery waste into valuable products should be an essential component of good winemaking practices. Recycling of waste is related to microbiological degradation of the waste. What are the sources of waste, what is the nature, composition and quantity of waste generated? Is this waste useful or just suitable for treatment to achieve reduction of pollutants? Can and how can these be utilized profitably? All of these aspects, including the utilization of waste, are described in this chapter.

31.2 POLLUTION LOAD AND TYPE OF WASTE GENERATED

Preparation of wine and brandy basically involves fermentation and involve a series of steps, as shown in Figure 31.1. Typical wine industry effluent waste includes unconsumed inorganic and organic acids, microbial cells, filter aids, waste wash water from cleaning of equipment containing traces of solvents, alkali etc. having high BOD (Bio-chemical Oxygen Demand) and large quantities of solid waste. The wine distilleries too produce large quantities of waste called `vinasse' that is acidic and has organic content. Effluents from a wine distillery primarily consist of organic acids with a high level of soluble biodegradable fractions, besides various phenolic compounds. Table 31.1 presents pollution load of effluents generated from wineries and distilleries. The main characteristics of wastewater from wineries are: temperature – 20–29°C; color – brownish yellow; pH – 4.0; BOD – 1,500–2,000 mg/L; dissolved solids – 6,800–9,400 mg/L; suspended solids – 15–17 mg/L. The chemical oxygen demand (COD) of a distillery in Wellington, South Africa was found to range from 20,000 to 30,000 mg/L, with a low pH (3–4), though COD values of grape-based distillery effluents have been documented to be between 22,000 and 48,000 mg/L.

31.3 IMPACT OF POLLUTION AND THE NEED FOR CONTROL

Table 31.1 describes the general composition of effluents produced from wineries and a distillery. The potential environmental impact of winery and/or distillery wastes is summarized in Table 31.2.

Upon discharge into natural stream or clean water sources, the organic waste normally undergoes aerobic decomposition, mainly accomplished by microorganisms. Consequently, the oxygen content of natural streams gets depleted and sometimes even completely exhausted to the extent that anaerobic

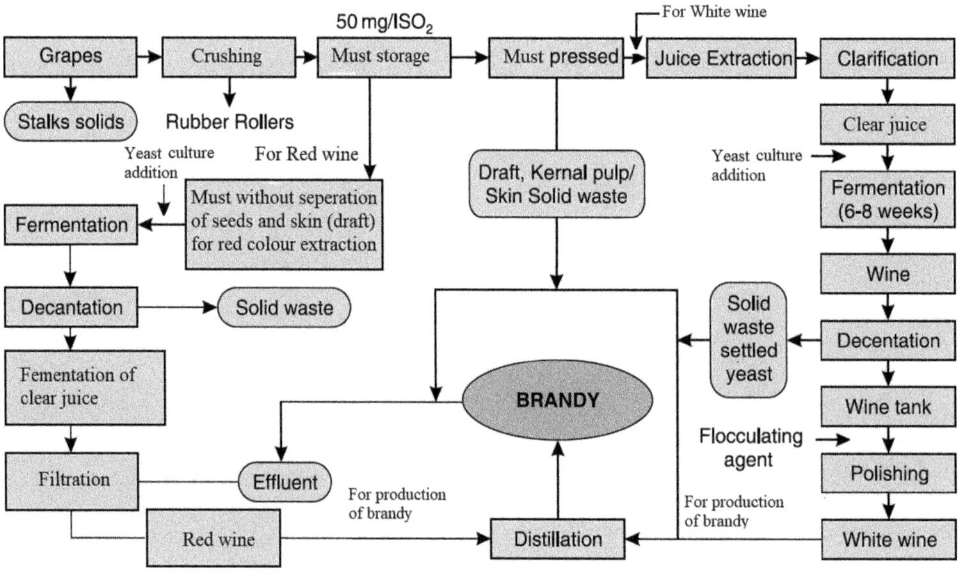

FIGURE 31.1 General outline of the process to produce wine and brandy

TABLE 31.1
General Characteristics of Untreated Winery and Distillery Effluent

Analysis	Distillery	Winery Vintage	Winery Non-vintage
Suspended solids (mg/L)	5,000–30,000	100–1,300	100–1,000
pH	3–5	4–8	6–10
Total dissolved solids (mg/L)	1,100–4,500	<550–2,200	<550–850
Biochemical oxygen demand (mg/L)	13,000–35,000	1,000–8,000	<1,000–3,000
Total organic carbon (mg/L)	1,000–15,000	1,000–5,000	<1,000
Total Kjeldahl nitrogen (mg/L)	500–1,700	5–70	1–25
Sodium (mg/L)	260–540	110–310	250–460
Total phosphorus (mg/L)	100–400	1–20	1–10
Carbon: nitrogen: phosphorus	10–50:4:1	30–100:4:1	15–30:5:1
Calcium (mg/L)	90–140	13–40	20–45
Magnesium (mg/L)	70–100	6–50	10–20
Sodium absorption ratio (SAR)	8	4–8	7–9
Potassium (mg/L)	1,300–2,100	80–180	40–340

Source: NWQMS (1998). Agriculture and Resource Management Council of Australia & New Zealand, and the Australian & New Zealand Environment and Conservation Council (1998) National Water Quality Management Strategy. Effluent management guidelines for Australian wineries and distilleries. Commonwealth of Australia. www.environment.gov.au/water/policy-programs/nwqms/

TABLE 31.2
Potential Environmental Impact of Winery and/or Distillery Wastes

Constituent	Indicators	Effects
Organic matter	Biochemical oxygen demand, total organic carbon and chemical oxygen demand.	Depletes oxygen when discharged into water leading to the death of fish and other aquatic organisms. Odors are generated by anaerobic decomposition if waste is stored in open lagoons or applied to land. Phenolic compounds may reduce light transmission in water.
Acidity	pH.	Kills aquatic organisms at extreme values. Affects crop growth, microbial activity in biological wastewater treatment processes and heavy metal solubility.
Nutrients	Nitrogen, phosphorus and potassium.	Can lead to eutrophication or algal blooms when discharged into water or stored in lagoons; algal blooms can cause undesirable odors. Toxic to crops in large doses. Nitrogen as nitrate and nitrite in drinking water can be toxic to infants.
Salinity	Electrical conductivity and total dissolved salts.	Imparts undesirable taste to water, is toxic to aquatic organisms and affects water uptake by crops.
Sodicity	Sodium adsorption ratio and exchangeable sodium %.	Affects soil structure (hard and dense subsoil) causing surface crusting, low infiltration and hydraulic conductivity.
Heavy metals	Cadmium, chromium, cobalt, copper, nickel, lead, zinc, mercury.	Toxic to plants and animals.
Solids	Total suspended solids.	Reduces soil porosity and leads to reduced oxygen uptake. Reduces light transmission in water. Anaerobic decomposition generates odors.

Source: Adapted from EPA Guidelines for Wineries and Distilleries. 2004. Environment Protection Authority. Government of South Australia. http://www.environment.sa.gov.au/epa/pdfs/guide wineries.pdf

conditions start emerging in the natural water streams. Such conditions are undesirable in a natural stream as these make the streambed black, with off-odors, rendering the water unfit for any human consumption as well as for the living system. The process of depletion of the oxygen content of a natural water stream due to discharge of highly polluting organic matter is called de-oxygenation of the stream. However, in course of time and flow of the river, due to utilization of organic matter by microbial population, food supply gets exhausted, significantly reducing the microbial activities and biological demand in the stream. Thus, the process of re-aeration of a natural stream gets under way, with replenishment of oxygen in the natural stream occurring due to absorption of oxygen from the atmosphere and release of oxygen by green plants during photosynthesis, a process known as re-oxygenation of the stream.

The inter-play between the de-oxygenation and re-aeration produces a well-defined profile of dissolved oxygen, known as the oxygen sag curve. The sag curve possesses two characteristic points, *i.e.*, the critical point where there is maximum dissolved oxygen (DO) and dissolved oxygen deficit (DC), where the rate of recovery of DO is maximal. Both the characteristics are critical to avoid anaerobic decomposition of organic waste, to support aquatic life in the stream and also help in determining the maximum allowable organic loading of the stream. Alcohol (one form of organic carbon) is also toxic to many freshwater organisms. Some of the aquatic species will die off in specific stretches of stream due to depleted DO levels as oxygen is essential for survival of most of the aquatic life. So, it becomes essential for wine industry to treat the wastewater generated to the extent where impact on natural streams/water bodies is minimum.

Dissolved oxygen (DO) acts as an indicator of the health and purity of a natural stream. Ideally, the oxygen concentration should be at least 90% of the saturating concentration at the ambient temperature and salinity of the water. As soon as any industry discharges an organic nature effluent, DO starts depleting and instant demand for oxygen, is called Biochemical Oxygen Demand (BOD) and its measurement, becomes one of most important parameters in water pollution control and wastewater treatment plant design. The organic content of wastewater can be measured directly, as total organic carbon (TOC) but this does not indicate whether the organic matter is bio-degradable or not. BOD is the measured amount of oxygen required by acclimated microorganisms to degrade the organic matter in the wastewater to CO_2 and H_2O in a closed system in five days at a temperature of 20°C. BOD is proportional to the organic matter degraded.

The oxidation process of organic matter proceeds in two stages. In the 1st stage, carbonaceous matter is oxidized within 7–10 days, while, in the 2nd stage, oxidation of nitrogenous matter takes place. Accordingly, the expression for BOD is derived as:

$$\mathrm{BOD}_5 = \mathrm{Y}_5 = \mathrm{L}\left(1 - \mathrm{e}^{-5\mathrm{k}}\right) \qquad 31.1$$

where Y is BOD at the end of 5 days (BOD_5), and L= ultimate BOD

The expression is called the first-order BOD equation. Reaction rate constant K in the above equation varies significantly with the type of waste and temperature. For typical organically polluted water, the value of K (base 20) is 0.23 per day. For other temperature ranges, K can be determined using the van't Hoff Arrhenius relationship:

$$\mathrm{K}_{\mathrm{t}} = \mathrm{K}_{20.\mathrm{Q}}{}^{(\mathrm{T}-20)} \qquad 31.2$$

where K_t = 1.047, a dimensionless temperature coefficient; T = temperature (°C)

Increases in pollution load, together with public awareness about the damage caused to natural resources, have made regulatory authorities active all over the world. So, legislation and policies have been framed and water quality standards and limits have been defined. Streams in excess of limits provided by the regulatory agencies have been laid out (Table 31.3) which restrict the polluters from discharging the wastewater into the streams in excess of limits provided by the regulatory agencies. It is permissible to discharge wastewater only after treatment to the desired standards, making this requirement to be legally binding on the polluter.

TABLE 31.3
Stream Water Quality Standards

Parameter	Wastewater Effluent	Stream Quality
1. Canadian objective		
BOD_5	15 mg/L max.	4 mg/L max.
SS	15 mg/L max.	–
DO	2 mg/L min.	4 mg/L min.
Total Coliforms	–	5,000/100 ml max. (water supply) 1,000/100 ml max. (swimming area)
Fecal Coliforms	200/100 ml max.	500/100 ml max. (water supply)
2. United States standards (typical)		
BOD_5	30 mg/L max.	4 mg/L max.
SS	30 mg/L max.	–
DO	–	4 mg/L mix.
Total Coliforms	–	5,000/100 ml max. (water supply) 1,000/100 ml max. (water supply)
Faecal Coliforms	200/100 ml max.	500/100 ml max. (water supply)
3. Japanese standards		
BOD_5	20 mg/L max.	2mg/L max.
SS	70 mg/L max.	25 mg/L max.
DO	–	7.5 mg/L min.
Total Coliforms	–	5,000/ 100 ml max. (water supply) 1,000/100 ml max. (swimming area)
Faecal Coliforms	30/100 ml max.	–

Source: Henry and Heinke (1989)

31.4 PLANNING A WASTEWATER TREATMENT PLANT

Selection of the least-polluting process from manufacturing out of the various process options should be explored. The least polluting process, even if available at a high cost, should be given preference as it pays in the long run. A Process Flow Sheet should be drawn up and the source of pollution indicated on it (Figure 31.1). The wastewater from a winery is generated intermittently in nature and daily/ annual variation can fluctuate greatly. The determination of flow characteristics may include average daily flow, maximum daily flow, peak hourly flow, minimum daily flow, minimum hourly flow and sustained flow, and their relative importance can be seen from Table 31.4. Wastewater flow can be measured with the help of flow-measuring devices like rectangular or V-notch weirs. The cross-section of the wastewater channel is designed for self- cleaning velocity. The wastewater collection system should be reliable, non-clogging, accurate and accessible for maintenance and cleaning. The wastewater is subjected to detailed analysis for polluting parameters like pH,

TABLE 31.4
Flow Characteristics – Importance in Wastewater Design

Flow Characteristics	Design Importance
Average daily flow	Treatment plant capacity, pumping and chemical sludge, solid and organic loading rates.
Maximum daily flow	Design facilities involving retention in flow equalization tank.
Peak hourly flow	Design of wastewater collection system, pumping station design, wastewater flow meters, grit chambers, sedimentation tank or channels design.
Minimum daily flow	Sizing of pipes/conduit where solid deposition or clogging might occur.
Minimum hourly flow	Sizing of flow meters that pace chemical feed system, recirculation of wastewater in trickling filter design etc.
Sustained flow	Sizing/equalization basin and other plant hydraulic components.

Source: Moosbrugger *et al.* (1993)

TDS (Total Dissolved Solids), TSS (Total Suspended Solids), BOD (Biochemical Oxygen Demand), COD (Chemical Oxygen Demand), etc. in a wastewater testing laboratory. The common measurements of organic carbon are defined here as:

- COD – Chemical Oxygen Demand. The amount of oxygen consumed during the chemical breakdown of organic materials and the oxidation of inorganic chemicals in water.
- BOD_5 – Biochemical Oxygen Demand. The amount of oxygen consumed over five days by microbes as they break down organic materials in water.
- TOC – Total Organic Carbon. The amount of carbon present in organic compounds in a water sample, regardless of their biodegradability. The TOC and TIC (Total Inorganic Carbon) values present (*e.g.* carbonates and dissolved carbon dioxide) make up Total Carbon (TC).
- DOC – Dissolved Organic Carbon. The amount of organic carbon remaining in a sample after fine filtration.
- POC – Particulate (or Purgeable) Organic Carbon. The amount of organic carbon able to be filtered from a water sample.

The accuracy of test results is very important in designing a wastewater treatment plant. Finally, selection of a specific wastewater treatment process depends upon the nature of the product manufactured, the capacity of the manufacturing plant, the quantity of the wastewater generated, the availability and cost of land for the wastewater treatment plant, climatic factors like temperature, etc., at the site of construction, regulatory requirements of pollution control authorities and the existing status of the water quality at the final disposal point of the natural stream or river. So, wastewater stabilization ponds, oxidation ditches and aerated lagoons are suitable where land is cheap and the climate is moderately hot, whereas a fluidized bed anaerobic system is more suitable where land is scarce and costly.

31.5 ENVIRONMENTAL MANAGEMENT OF WINE WASTE

31.5.1 Solid Waste – Process

Over the decades, the utilization of grape wine waste for alternative uses has been environmentally inefficient, with a large proportion of the waste discarded in landfills not suitably planned and executed, resulting in serious damage to water sources and pollution of land from the leaching of toxic compounds like heavy metals (heavy metals equal to more than 7 mg/L are considered toxic). Public health is also affected due to foul smells and the risk of spreading diseases by flies and pests. Many factory managements throw the solid waste from a winery into effluent-carrying drains, thus increasing the pollution load of the waste water manifold. Hence, removal and recovery of the solid waste at the source is very important.

31.5.2 By-Product Recovery From Winery and Distillery Waste

The grape wine waste is called grape pomace and is the major solid waste, consisting of skin, seeds and stems. The waste from a winery is composed of mainly the parts of berries, lees (lees is comprised of sediments which mainly consists of sedimented yeast, and parts of fruit skin), seeds, tannins, tartrate, etc. In Figure 31.2, the individual parts of a grape berry and their composition is shown. The composition of some of the grape pomace and the detailed composition of grape pomace shows it to be a rich source of several useful products (Table 31.5a and b). Recent findings have shown that grape pomace is a significant source of dietary fiber, phenolic compounds and soluble sugars. Red grape pomace contains a higher quantity of phenolics, whereas white grapes have higher concentrations of sugars. Several products can be recovered (Table 31.6). The grape pomace mass generated has been estimated to be 25–30% of that of the grapes processed. Thus, dumping of such a waste into natural resources is like throwing gold coins into the sea. The by-product recovery potential of winery waste includes grape seed oil, tannins, tartrates, stock feed and fertilizer from solid waste besides preparation of vinegar, concentrates etc., as described here.

31.5.2.1 Recovery of Tartrates

Tartrates, present as potassium bitartrate, are recovered as calcium tartrate from winery solid waste (pomace still slops from brandy distillation, from lees that settle in wine tank and in the argoles that separate in the wine storage tanks). On an industrial scale, tartrate is produced from winery by- products. For recovery of tartrates from pomace, all stems are removed, and the pomace is compacted.

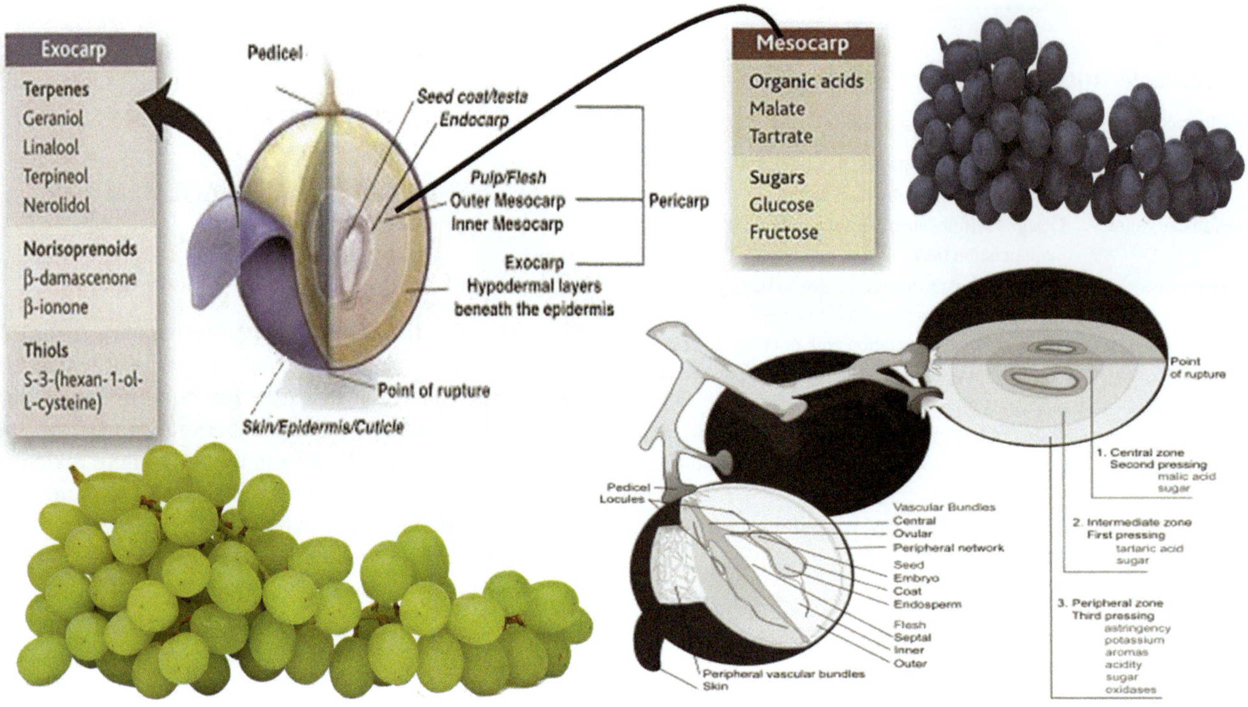

FIGURE 31.2 Parts of grape berries and their composition. *Source:* Adapted from Lund and Bohlmann (2006)

TABLE 31.5(a)
Basic Chemical Composition of *Vitis vinifera* Fresh Weight Grape Pomace From Five Varieties

Grape Variety	Skin (%)	Seed (%)	Stem (%)	TSSC* (%)
Morio Muscat	85.99	12.77	1.25	83.9
Muller Thurgau	90.67	7.84	1.49	72.4
Cab. Sauvignon	77.41	20.91	1.68	28.2
Pinot Noir	73.35	12.34	0.54	27.7
Merlot	83.18	14.98	1.84	22.2

* Total soluble solid content
Source: Jiang *et al.* (2010)

The three extraction steps include (i) production of calcium tartrate from potassium bitartrate by the use of calcium hydroxide/calcium carbonate; (ii) the production of tartaric acid by reaction of calcium tartrate with sulfuric acid; and finally (iii) the de-colorization of tartaric acid with activated charcoal and its further crystallization under reduced pressure. Recently, the wine lees and pomace have been used to extract tartaric acid in Portugal. The amount of calcium tartrate recovered ranged from 120–145 kg/ton and 70–85 kg/ton from wine lees and grape pomace, respectively.

31.5.2.2 Recovery of Grapeseed Oil

The recovery of oil from grape seeds involves drying of pomace in rotary dryers. Seeds are separated by thrashing and sieving. The oil is extracted by pressing the seeds or by grinding the seeds with the help of solvents. A comparison of the seed oil contents of red and white grape varieties has been made in Table 31.7a, and the composition of the seed oil is shown in Table 31.7b.

31.5.2.3 Recovery of Tannins

Wine pomace phenolics have antioxidant activity, are linked to the prevention of heart-related diseases and have anti-cancer properties. Tannins extracted from grape seeds are reused in wine making in France and Italy. The quantities of total phenolics, total flavonoids and proanthocyanidins in grape extract and grape seeds are shown in Table 31.8.

31.5.2.4 Bioactive Compounds and Their Applications

The increased awareness of the compounds contained within grape pomace led to associated research and development in many industries. The grape pomace contains polyphenols, dietary fibers and oil with bioactive characteristics as described in Section 31.5.2 for alternative applications as bioactive compounds. The bioactive compounds found in grape pomace flour are shown in Table 31.9.

An organic grape pomace product is 'Bioflavia', is commonly sold in 300 g bottles. The product is made from red grape wine pomace. Bioflavia powder can be used in smoothies or alternative beverages. A recent study has found grape pomace to be a possible input for biosurfactant production.

Biosurfactants are important for food processing applications, due to their emulsifying abilities. There is also an increased demand for natural biosurfactants compared with the synthetic agents, due to their lower toxicity, ease of biodegradability and lower cost.

TABLE 31.5(b)
Physicochemical and Microbiological Characteristics of Grape Pomace (*Vitis vinifera* L.) Flour

Parameters (% Dry Basis)	Results (Mean ±SD)	
Moisture (g/100g)	3.33 ± 0.04	
Ash (g/100g)	4.65 ± 0.05	
Total Lipids (g/100g)	8.16 ±0.01	
Protein (g/100g)	8.49 ± 0.02	
Carbohydrate (g/100g)	29.2	
Pectin (g/100g)	3.92 ± 0.02	
Fructose (g/100g)	8.91 ±0.08	
Glucose (g/100g)	7.95 ± 0.07	
Total dietary fiber (g/100g)	46.17 ±0.80	
Total Calories (Kcal/100g)	224	
Minerals	**Results (Mean ±SD)**	
Calcium (mg/100g)	0.44 ±0.715	
Magnesium (mg/100g)	0.13 ±0.255	
Sodium (mg/100g)	0.044 ± 0.056	
Potassium (mg/100g)	1.40 ±0.313	
Iron (mg/100g)	18.08 ±0.03	
Manganese (mg/100g)	0.817 ±0.550	
Phosphorus (mg/100g)	0.183 ±0.255	
Sulfur (mg/100g)	0.089 ±0.336	
Zinc (mg/100g)	0.98 ± 0.702	
Microorganism	**Result**	**Tolerance***
Salmonella (in 25g)	Absent	Absent
Fecal Coliforms (MPN/g)**	<3	10^3
Bacillus cereus (CFU/g)***	<100	3×10^3

According to Resolution No. 12 of January 2, 2001, the National Health Surveillance Agency (2001), Food Group 10, item a. **MPN/g = Most Probable Number per gram. ***CFU/g = Colony Forming Unit per gram. SD = standard deviation

Source: Sousa *et al.* (2014)

31.5.2.5 Cosmetology and Health-Related Product Recovery

Grape seeds are the waste of wineries and, nowadays, many cosmetics use grape seed oil, grape seed extracts and grape juice as a main ingredient for the production of creams, shampoos, body lotions and hair treatment products. The grape seed oil is used for the treatment of skin problems like acne, skin lightening and healing, reduction of dark circles around eyes, hydration, protection of skin from aging and UV radiation and for hair treatment. Grape seed oil can accelerate the healing process of wounds on human skin and is able to increase the amount of antioxidant known to give resistance to free radicals, which damage the tissue.

31.5.2.6 Pollulan

Apart from grapes, wine and brandy are prepared from other fruits such as apples. In this production process apple pomace is generated. Pollulan, is a product of apple pomace, with applications in the food and cosmetics industries to improve texture of the end products, in healthcare products like lotions, shampoos, denture adhesives, and capsules for supplements. A Pollulan-based antimicrobial active packaging system is also being developed.

31.5.2.7 Recovery of Pigments

Red grape skin extract is a purplish-red liquid prepared by aqueous extraction of fresh de-seeded marc after grapes have been pressed to produce grape juice or wine. The extract is concentrated and was found to be suitable in color additive mixtures for coloring food. Anthocyanins are other useful products which can be recovered from grape pomace. The anthocyanin composition of grape pomace is shown in Table 31.10.

Anthocyanin pigments have been successfully extracted from both wine lees and grape pomace, with quantities ranging from 3–5 kg/t and 2–4 kg/t, respectively.

31.5.2.8 Recovery of Other Useful Products

Other recoverable products from the waste from wineries and distilleries include yeast as a single-cell protein, baker's yeast and sludge from the distillery for use in the pharmaceutical industry and for yeast hydrolysate preparations for the food industry. The yeast, after drying into flakes or powder, has been used to make yeast tablets, after debittering, yeast extracts and peptone preparations. Being rich in protein, fats, minerals, fibers and B vitamins, the dried yeast is excellent for use in poultry, pigs and cattle feed. Carbon dioxide is also recovered from the wine industry for various applications, including carbonation of soft drinks or wine. The waste from apple juice extraction is pomace (Plate 31.1) that has been used to extract pectin (used in jams and jelly preparations) and as an ingredient of medium to produce biocolors, using *Rhodotorula, Sarcina, Achromobacter, Micrococcus*, etc.

31.5.2.9 Use as a Fertilizer – Composting of Wine Pomace

Grape pomace composting is beneficial as demands for organic fertilizers are increasing rapidly. Currently, most of grape pomace produced by the wine industry is composted.

Composting is based on aerobic decomposition, whereby organic compounds are broken down into inorganic forms of elements like carbon and nitrogen. A comparison of the composition of grape pomace before and after composting is shown in Table 31.11a and b.

For optimum composting, the material being composted must have high moisture content and sufficient carbon-to-nitrogen ratio to provide nutrients for the microbes to survive and continue biodegradation. The optimum C:N ratio varies from 25:1 to 35:1. Typical grape pomace has a carbon content of 46.6 (g/100g), nitrogen 173 (g/100g) and a C:N ratio of 26: 94, which qualifies as an optimal substrate for composting. The compost generated from grape waste from a winery still needs more nitrogen to meet the optimum requirements of 22.267 kg N/acre and, thus, addition of fertilizer, *e.g.*, in the

TABLE 31.6
Treatment of Grape Wastes, Physicochemical Properties and Their Use

Final Products	Treatment	Physicochemical Characteristics	Uses	References
• Grape waste	Composting of grape waste and hen droppings	Organic matter content	Fertilizer for corn seed	Ferrer *et al.* (2001)
• Grape seed and skin extracts	Fractionation of grape seed and skin extracts from grape waste	Phenol content	Dietary supplements for disease prevention	Shrikhande (2000)
• Pressed grape skin	Composting of solid waste and wastewater	Organic matter content	Fertilizer	Manios (2004)
• Wine pomace and grape seeds	Lyophilization and extraction of flavanols	Flavanol content	Dietary supplements, production of phytochemicals	Gorualez- Paramas *et al.* (2004)
• Grape marcs, stalks and dregs	Lyophilization and extraction of polyphenols	Polyphenol content	Dietary supplements	Alonso *et al.* (2002)
• Grape skins, seeds and stems	Acidolysis of a polymeric proanthocyanidin fraction of grape pomace in the presence of cysteamine	Flavanol content	Source of flavonols	Torres & Bobet (2001)
• Grape seed extract (GSE)	Pre- and post-mortem use of grape seed in feeding experiment	Phenol content	Feedstuff for dark poultry meat	Lau & King (2003)
• Grape skin pulp	Fermentation by *Aureobasidium pullulan*	Ethanol precipitate	Pullulan production	Israilides *et al.* (1998)
• Grape seeds	Solid-state cultivation by *Trametes hirsuta*	Lignocellulosic content	Laccase production	Moldes *et al.* (2003)
• Grape pomace	Solid-state cultivation by *Pleurotus* sp.	Pruning content, high phenolic components and total sugars	Feedstuff for animals	Sanchez *et al.* (2002)
• Wastewater	Electrodialysis	Tartaric acid content	Additive in medicines and cosmetics, acidulant compound in soft drinks	Andres *et al.* (1997)
• Wastewater	Electrodialysis at 60 °C	Tartaric acid and malic acid content	Food and pharmaceutical industries	Smagge *et al.* (1992)

Source: Ioannis *et al.* (2006)

TABLE 31.7(a)
Seed Oil Contents

Grape Seed Cultivars	Oil Content (% v/w)*	Degree of Unsaturation (%)	Tocopherol (mg/kg)
Red	13.1–19.6	86.6–89.3	357–578
White	14.7–17.8	86.5--88.1	364–486

*Avg. (red & white) = 16.3%.
Source: Adapted from Baydar and Akkurt (2001)

TABLE 31.7(b)
Average Composition of Grape Seed Oil Fatty Acids

Common Name	Acid Name	Average Percentage Range
Omega-6	Linoleic acid	69–78
Omega-9	Oleic acid	15–20
Palmitic acid	Hexadecanoic acid	5–11
Stearic acid	Octadecanoic acid	3–6
Omega-3	Alpha-linolenic acid	0.3–1
Palmitoleic acid	9-hexadecenoic acid	0.5–0.7

Source: Ioannis *et al.* (2006); http://en.wikipedia.org/wiki/grape.seed.oil

form of ammonium nitrate is required to be made before the same is applied to the vineyards. Some other nutrients, *e.g.*, phosphorus, potassium and magnesium are needed to make the compost suitable for plant growth.

As there is growing demand for organic fertilizers, small wineries are using 100% of their grape pomace to produce compost on-site, for reuse in the vineyard, to enhance the economy of wine production.

The large wineries may own treatment facilities for on-site recovery of useful products, including the chemicals and energy production, using the bio-refining concept or, if available, to supply to other industries for treatment and by-product recovery or energy production facilities, as composting may need a large area for disposal of the composted waste.

TABLE 31.8
Quantity of Total Phenolic Substances, Total Flavonoids and Proanthocyanidins

Total Phenols (GAE)	
Grape Extract (g/L)	2.86 ± 0.01 g/L^{-1}
Grape Seeds (g/100g)	8.58 ± 0.03 g/100 g d.m.
Total Flavonoids (CE)	
Grape Extract (g/L)	2.79 ± 0.01 g/L^{-1}
Grape Seeds (g/100g)	8.36 ± 0.04 g/100 g d.m.
Proanthocyanidins (CyE)	
Grape Extract (g/L)	1.38 ± 0.06 g/L^{-1}
Grape Seeds (g/100g)	5.95 ± 0.17 g/100 g d.m.

GAE, gallic acid equivalent; CE, (+)-catechin equivalent; ME, cyanidin equivalent; d.m., drymatter
Source: Negro *et al.* (2003)

TABLE 31.9
Bioactive Compounds in Grape (*Vitis vinifera* L.) Flour

Bioactive Compounds	Results (Mean ±SD)
Vitamin C (mg ascorbic acid/100g)	26.25 ± 0.01
Total anthocyanins (mg/100g)	131 ± 0.4
Soluble dietary fiber (g/100g)	9.76 ± 0.03
Insoluble dietary fiber (g/100g)	36.40 ± 0.84

SD, standard deviation
Source: Sousa *et al.* (2014)

31.5.2.10 Biogas/Energy By-Products

Biogas production as a by-product from grape pomace is another attractive use of grape pomace. Biomethane is produced during anaerobic digestion of grape pomace in anaerobic bioreactors by oxygen-free fermentation of biomass, which leads to the production of biogas, containing up to 70% methane. From dried milled red grape pomace, biogas with a methane content of 62.5% can be obtained. Bioenergy can be a highly attractive by-product, with the opportunity for income from waste compared to the disposal cost.

TABLE 31.10
Anthocyanin Content (mg/kg) d.m.) of Grape Skin

Compound	Value (mg/Kg^{-1} d.m.)
Delphinidin 3-*O*-glucoside	68–5,552
Cyanidin 3-*O*-glucoside	37–1,903
Petunidin 3-*O*-glucoside	65–6,680
Peonidin 3-*O*-glucoside	515–12,450
Malvidin 3-*O*-glucoside	1117–50,981
Delphinidin 3-*O*-acetylglucoside	392–956
Petunidin 3-*O*-acetylglucoside	545–1,375
Peonidin 3-*O*-acetylglucoside	1,371–1,484
Peonidin 3-*O*-acetylglucoside	45–8,688
Cyanidin 3-*O*-coumaroylglucoside	374–1,071
Petunidin 3-*O*-coumaroylglucoside	974–2,458
Peonidin 3-*O*-coumaroylglucoside	68–6,828

Source: Adapted from Kammerer *et al.* (2004)

31.5.2.11 Biorefining Concept of Fruit Winery Waste

The concept of a biorefinery was recently introduced along the lines of a petroleum refinery, due to the depletion of natural resources of fossil fuels, leading to the severe energy crisis as well as pollution of environment by the combustion of fossil fuels, resulting in global warming and climate change. As an alternative, the biorefinery concept uses biomass waste, waste from the food industry, including wineries, to recover useful products, such as fuel, heat and energy as by-products, employing processes similar to those operating in the petroleum industry.

All over the world, including India, governments are encouraging the use of waste biomass from agricultural sector

PLATE 31.1 A mechanical sludge drier for apple pomace for utilization of solid waste for cattle feed. *Courtesy:* HPMC, Parwanoo, Himachal Pradesh, India

TABLE 31.11(a)
Chemical Composition of Compost Derived from Winery Waste

Element	Values
N (%)	2.14–3.74
P (%)	0.18–0.52
Ca (%)	3.17–14.3
Mg (%)	0.3–0.61
Fe (%)	0.5
Zn (mg/Kg^{-1})	77–109
Cu (mg/Kg^{-1})	30–46
Ni (mg/Kg^{-1})	9.1–17.6
Cr (mg/Kg^{-1})	23.4–147
Pb (mg/Kg^{-1})	8–19
Cd (mg/Kg^{-1})	0.2–0.4

Source: Ioannis *et al.* (2006)

TABLE 31.11(b)
Physicochemical and Chemical Characteristics of Compost Derived from Winery Wastes

Parameter	Values
pH	6.5–8.5
EC (MS cm^{-1})	1.57–4.1
Volatile solids (%)	46.8–67.5
C/N ratio	11.9–19.5
Moisture (%)	47--66
CEC (Cmol kg^{-1})	108.65
OM (%)	84.15–89.1
C (%)	40.5–51.5

OM = Organic matter
CEC = Cation exchange capacity
C = Carbon
C/N = Carbon/nitrogen
EC = Electrical conductivity
Source: Ioannis *et al.* (2006)

to produce energy with minimal impact on the environment, and to avoid huge wastage of food/fruits and agricultural produce and the negative consequences of its disposal. Accordingly, the Government of India has developed a policy to promote Mega Food Parks or Zones to facilitate the establishment of food processing units with all the associated infrastructure. These Mega Parks can operate waste-management systems based on the biorefinery concept for better recovery of the useful chemicals and energy for economic benefits. In terms of pollution control, the biorefinery concept is a central common recovery and waste-treatment system, and is a better pollution control system than conventional technologies. Both solid and liquid waste can be treated in the biorefinery concept. The wineries and other food industries located in such Parks can benefit from the use of common facilities. Presently, the cost of the biorefinery concept is very high but it is the future technology for waste management.

Outsourcing the waste to common by-product recovery and generation facilities, designed according to the biorefinery plan, may be more beneficial to the wine industry instead of having of its own recovery and generation infrastructure on the site.

31.5.3 Winery Solid Waste

31.5.3.1 Use as a Fertilizer

Some large wineries dehydrate the winery solid waste (pomace) to a low moisture content and grind it for use in feeding livestock, particularly dairy cows. The main hurdle in the use of grape pomace as stock feed, however, is its high fiber content, largely the indigestible hulls of seeds. So, at best, it can be used as a feed supplement. In addition, grape pomace was found to be better as stock feed than good-quality wheat straw. Grape pomace can be mixed with bran and molasses for use as a supplementary feed for livestock. It can also be improved as a cattle feed if lime is added to raise the pH. Apple pomace has successfully been made into animal feed after drying, or after fermentation, following removal of the ethanol and drying of the leftover material.

31.5.3.2 Use as Stock Feed

Winery pomace has almost the same fertilizing value as farmyard manure, with nitrogen content of 1.5–2.5% and 0.5%–2.5% potassium on a dry weight basis. But application to crop land of grape pomace created temporary toxic conditions; if applied through a fertigation system with trickle irrigation and liming, to recover tartrate before sewage treatment, it could be useful and safe for irrigation.

31.6 PROCESS MODIFICATION/CLEAN TECHNOLOGY IN PLANT-MEASURES

To reduce the pollution load in the streams from the food industries, including wine and distillery waste, several methods are employed, such as the use of process modification, clean production methods and better in-house keeping. These also result in savings in treatment costs for waste disposal and improvement in the quality of the final product.

31.6.1 Process Modification

Water consumption for various processes in wine production, include the use of rinsing and washing water for bottle washing, cleaning water from machines, pipes and tanks, etc. The increase in container/barrel size resulted in a drastic reduction in the volume of wastewater used during the rinsing process, which also reduces the pollution load significantly, and consequently the cost of the treatment of waste water is decreased.

31.6.2 In-Plant Measures

In-plant measures can be taken within the factory that can significantly reduce the pollution load in a fermentation industry. Segregation of polluting and non-polluting streams in a winery gives an overview of the situation and decides the extent of the treatment needed for each polluting stream. It is a cost-effective method, as addition of non-polluting streams only increases the hydraulic load, requiring larger treatment tanks. In a winery, cooling water should be recycled, whereas minor polluting streams could be discharged following pre-treatment, instead of joining the polluting streams.

31.6.3 In-House Keeping and Clean Production Methods

In the wine industry, some clean-production measures and in-house keeping can be adopted to reduce the hydraulic load as well as the pollution load drastically, thus saving the cost of the treatment of wastewater. In wine production, after extraction of the juice, the mash is pressed and draff (consisting of pulp and skin) is separated and retained. After first racking of white wine, 2–3.5% by volume of settled yeast and 1% of separator yeast is left. Removal of this yeast squash helps in reduction of the pollution load. During the water-polishing process, substances settle out, equivalent to 2.5 % by volume of the wine produced and having a high BOD (100,000 mg/L) and hence, their removal reduces the pollution load to a great extent. Some in-plant measures, like the use of small-diameter hose pipes, cleaning of containers once a day, avoidance of product loss/spillage, use of water-saving valves and use of pulsating jets in bottle washing section can reduce the generation of wastewater significantly.

31.7 WASTEWATER TREATMENT

Process modification, clean technology, in-plant measures, and the reuse and recovery of material of immense economic importance from winery waste, aimed at reduction of hydraulic and pollution load in the winery are cost effective and help in reduction of the pollution load at the source itself, resulting in huge cost savings. But still these are not enough to meet the stringent effluent discharge standards set by various regulatory agencies/authorities for prevention and control of pollution. Wastewater still needs further treatment, normally called end-of-pipe treatment, using an appropriate technology, as discussed in the subsequent sections.

31.7.1 Treatment Process – Wastewater Treatment

In wastewater engineering, treatment processes include primary treatment, secondary treatment and tertiary treatment.

31.7.1.1 Primary Treatment

The primary treatment can include the following:

(i) **Removal of Solids**
The main objective of solids removal is to prevent floating and suspended matter present in wastewater from entering secondary treatment and causing problems in pumping the wastewater. For removal of coarse solid particles, 20- to 40-mesh screens are commonly used. Filtration, centrifugation and ultrafiltration (UF) are used for removal of solids from winery wastewater.

(ii) **Neutralization**
The low pH of wastewater can upset the working of secondary biological treatment, resulting in poor settling of primary sludge. So, the wastewater is neutralized in the pH range of 6–9 with lime and caustic soda in a mixing tank with wastewater retention capacity of about 3 minutes with a high-speed mechanical flash mixer.

(iii) **Equalization Tanks**
The objective of a flow equalization tank is to provide constant organic loading to the secondary biological treatment and to prevent shock loads which results from frequent flow variations. It also helps to maintain the biological oxidation activity in the shut-down periods of the plant. Such tanks are provided prior to biological oxidation, with a retention period of 4–5 hours at an average flow rate to be determined by the specific plant.

(iv) **Primary Clarifier or Primary Settling Tank**
The purpose of the primary clarifier is to achieve solid–liquid separation. The solid particles fall to settle on the bottom of tank under gravity. The depth of the clarifier may be designed to permit accumulation of solids. The primary clarifier normally has a peripheral drive mechanism to remove sludge. A clarifier with chemical flocculation is used for settling and removal of solids in the winery industry. Circular clarifiers are more effective in solid–liquid separations. The primary treatment of wastewater is quite effective in neutralization, 50–60% TSS (Total Soluble Solids) reduction and removal of almost all the settleable solids, with 25–30% reduction in BOD.

31.7.1.2 Secondary Treatment

The primary treated wastewater still has about 70–80% BOD left in it and secondary treatments (biological treatments or bio-oxidation) are basically designed to remove BOD present in the winery wastewater.

(i) **Secondary Treatment Method**
Wastewater from a winery could be treated in secondary biological treatments, which basically consist of aerobic biological treatment, anaerobic biological treatment and combined aerobic and anaerobic biological treatment.

(ii) **Aerobic Biological Treatment**
In aerobic biological treatment, microorganisms utilize the dissolved starches, sugars and other carbohydrates as food and, in the process, the bacterial population is increased many-fold, with conversion of waste into bacterial solids. To complete this process, the microorganisms need oxygen that must

be supplied through a mechanical aerating system in an aeration tank. Proper conditions need to be maintained for bacterial growth by supply of oxygen and nutrients in the form primarily of nitrogen and phosphorus. The waste load is expressed as organic matter (BOD) available as food in wastewater and divided by the number of microorganisms available in the aeration tank, referred to as the food-to-microorganism (F/M) ratio or mixed liquor-suspended solids (MLSS). BOD load and MLSS form the basis for the design of an aerobic biological treatment system.

(iii) **Activated Sludge Process**
Due to the flexibility of this system (Plate 31.2), it can be used for high-strength intermittently produced wastewater, such as those generated by a winery. A pure oxygen-activated sludge system with a BOD of 2,269 mg/L has successfully been used. A satisfactory treatment was achieved with mechanically aerated biological plant, when winery effluent in combination with municipal sewage. An activated sludge treatment plant in West Germany produced effluent with BOD 3–15 mg/l when winery and sewage waste was treated at sludge loading (SL) of 0.2 kg BOD/ m^3/day.

(iv) **Trickling Filters**
Trickling filters have been used to provide biological wastewater treatment for organic waste for nearly 100 years. These are circular beds of coarse aggregates, about 1.8 m deep with surface rotating arms for a wastewater spray system called liquid distributors. A microbial film is formed on the surface of the aggregates, which serves to oxidize the organic matter present in the wastewater. These filters are rugged and have the capacity for shock loads. But concentrated wastewater application to a trickling filter causes excess growth of biofilm that can be prevented by recirculation of finally-treated effluent. A typical trickling filter can be designed for organic or sludge loading of 0.1 kg BOD/m^3/day. A BOD_5 reduction of 70–85 % can be obtained with a properly designed trickling filter. A sludge loading of 0.3–0.4 kg BOD/m^3 for treatment of winery waste has been recommended.

(v) **Aerated Lagoons**
These lagoons are simple holding basins with a depth of 2–4 m. Biodegradation of organic wastewater is carried out by aerobic microorganisms, which need a continuous supply of oxygen. The oxygen is supplied by means of floating surface mechanical aerators. The typical sludge retention period is four days. A typical layout plan is shown in Figure 31.3. Aerated lagoon is designed as a complete-mix biological reactor without sludge recycling. For the treatment of brewery waste effluent, two aerated ponds/lagoons are used in series and the BOD is reduced to less than 20 mg/L at the final outlet. A BOD_5 reduction of 80–95 % can be achieved in a properly designed aerated lagoon.

(vi) **Waste Stabilization Ponds**
These are called the natural processes of wastewater treatment for organic waste and can be classified as aerobic, anaerobic or facultative ponds, in which wastewater is treated and stabilized by a combination of aerobic, anaerobic and facultative bacteria. Organic loading of 100 kg BOD/acre/day and a depth of 1.2–2.5 m can be assumed. A properly designed waste stabilization pond (where land cost is cheap, and the temperature is moderate) can remove BOD by 80–95 %. Table 31.12 briefly describes the design and BOD-removal performance of various types of waste stabilization ponds.

(vii) **Oxidation Ditch**
It is an oval-shaped closed channel. Cage-rotors are used for intensive aeration of the wastewater. The oxidation ditch can be designed with channel depth of 1–1.5 m, BOD loading of 0.2 kg BOD/kg MLSS when

PLATE 31.2 An activated sludge process with mechanical surface aerators primary clarifier is also visible. *Courtesy:* HPMC, Parwanoo, Himachal Pradesh, India

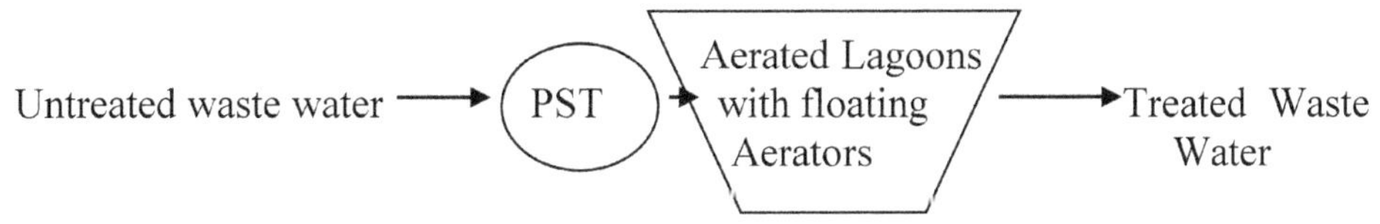

FIGURE 31.3 Aerated lagoons process. *Source:* Adapted from Joshi (2000)

TABLE 31.12
Design and Performance Characteristics of Waste Stabilization Ponds

Design/Performance Parameter	Types of Ponds			
	Aerobic	Anaerobic	Facultative	Aerated Lagoon
Pond size (acres)	Less than 10	0.5–2.0	2–10	2–10
Wastewater residence time (days)	10–40	20–50	5–30	3–10
Depth of pond (ft)	3–4	8–16	4–8	6–20
Temperature (range °C)	0–30	6–50	0–50	0–30
BOD_5 removal (%)	80–95	50–85	90–95	80–95

Source: Moosbrugger *et al.* (1993)

a variable oxygen requirement of 1.5–2 kg/kg BOD is applied. Reduction of 90–95 % BOD can be achieved in a properly designed oxidation ditch system.

(viii) **Rotating Biological Contactors (RBC)**
The Rotating Biological Contactors (RBC) system consists of closely-spaced polymeric-material discs mounted on a shaft which is rotated at a very low speed (2–5 rpm). The discs are submerged in a tank, through which wastewater passes continuously. A biofilm develops on the discs' surface with alternative exposure to wastewater and atmospheric oxygen. For brewery wastewater, 85% COD removal is achieved when treated in an RBC system.

(ix) **Combined Aerobic Process**
By combining aerobic treatment processes, best performance and economics can be achieved in wastewater treatment. The use of the combined aerobic process has increased recently due to significantly higher organic loading rate and greater reduction in BOD than a conventional aerobic system, besides reducing the considerable cost of construction and treatment.

31.7.1.3 Anaerobic Biological Treatment

Anaerobic digestion of organic matter can also be carried out. Anaerobic biological treatment is also employed for treatment of effluent from a distillery. Since distillery waste contains, in addition to other components, phenolic compounds (inhibitory to microorganisms) that make the anaerobic digestion process difficult, so partial removal of such compounds can be achieved by pre-treatment with the fungus *Geotrichum candidum*. The pretreated effluent was anaerobically degraded more rapidly than the untreated original wine distillery wastewater for the same COD loading level. The anaerobic bioreactor had a suspension of micronized clay (saponite) to which the microorganisms responsible for the process get adhered. The treatment of organic waste proceeds in two steps: i) the degradation of organic matter or waste into volatile acids and ii) the conversion of these volatile acids into methane gas under anaerobic conditions (absence of oxygen) by anaerobic microorganism. Besides methane (CH_4), CO_2 is also generated in the process. But the process is quite slow, and a conventional anaerobic biological system may take 30–60 days to complete the treatment process. Process modifications have been made in the design of anaerobic biological reactor with two- stage anaerobic reactors for enhanced energy production. The sludge recycling was introduced to reduce hydraulic retention time to 1–4 days, which resulted in smaller reactor constructions, besides improving the performance of the system (Figure 31.4). These are called stirred-tank reactors or anaerobic contact process (ACP).

31.7.1.4 Immobilized Anaerobic Biological Reactors

(a) **Up-Flow Anaerobic Sludge-Blanket Process (USABP)**

In USAB, the organic waste is introduced at the bottom of the reactor and the wastewater flows upward through a blanket composed of biologically formed granules. Treatment occurs as the wastewater comes in contact with the immobilized granules. Methane is produced under anaerobic conditions from the biodegradation of the organic waste, which can be used as an additional source of energy. The hydraulic retention time (HRT) is reduced to 4–12 days due to the large surface area available for biological oxidation as granules are immobilized with anaerobic bacterial culture. The sludge produced is negligible due to direct conversion of organic load into biogas and CO_2. A properly designed USAB reactor (Plates 31.3 and 31.4) can reduce organic loading (in terms of COD) to about 75–85%. The performance of USAB reactor with brewery waste, winery waste and distillery water was

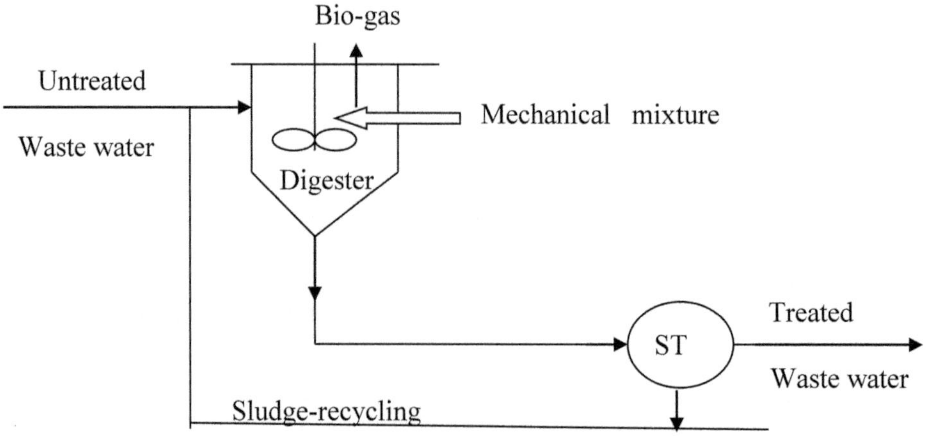

FIGURE 31.4 Anaerobic contact process (ACP). *Source:* Adapted from Joshi (2000)

PLATE 31.3 Up-flow sludge blanket bioreactor (USAB) with activated sludge process – a combination of anerobic and aerobic system for high strength waste from food processing industries. *Courtesy:* Mohan Meakins, Solan, India

found to be satisfactory. Recently, a USAB effluent plant on distillery wastewater was evaluated and gave a performance of 90% removal of COD, indicating its suitability to pre-treat the high-strength distillery effluents. The system was used to treat grape wine distillery waste, which was converted into a palletized sludge bed with COD removal of greater than 94 per cent at a maximum loading rate of 15 kg/m^3 sludge bed.

(b) **Fluidized Bed Anaerobic Filter Process**

Fluidized bed anaerobic filters are used for treatment of fermentation industry waste and consists of vertical columns filled up with various types of media, like PVC granules and/or charcoal, kept in the fluidized state with an upward current of wastewater with suitably designed wastewater distributors (Plate 31.4). The anaerobic microorganisms are allowed to grow on filter media, and wash-off from the filter is prevented. Due to the large surface area available for biodegradation, a hydraulic retention time as low as three hours could be achieved, compared to 20–30 days in a conventional anaerobic bioreactor. These bioreactors can also be operated at temperatures between 5 and 55^0 C, which is a limitation in the USAB reactor system. Properly designed systems can reduce COD by about 85–90% of waste from the wine industry, with the additional benefit of harnessing bioenergy production. These filters treated the wastewater from breweries and wineries effectively (Figure 31.5).

(c) **Combination of Aerobic and Anaerobic Biological Process**

Although anaerobic biological treatment processes are ideally suited to treating high- strength organic waste, the biodegradation

PLATE 31.4 Fluidized bed biological reactor (FBBR) with activated sludge process – a combination of anerobic and aerobic system. *Courtesy:* Mohan Meakins, Solan, India

is not complete, so the wastewater needs further processing. As a result, combinations of anaerobic–aerobic biological processes are designed to bring polluted wastewater pollution to the acceptable standards (Figure 31.6). The aerobic biological oxidation is carried out in aerobic treatment systems like the activated sludge process, oxidation ponds and biofilters.

31.8 WASTEWATER TREATMENT TECHNOLOGIES

The reader may realize that, after adopting technologies, like process modifications, clean production methods in the plants, reuse and recovery of useful products from the winery waste, aimed at reduction of hydraulic and pollution waste in wine industry are cost effective and help in the reduction of pollution load at the source itself, resulting in huge cost savings in further treatment in the conventional wastewater treatment.

31.8.1 Conventional Wastewater Treatment

The wastewater still needs conventional wastewater treatment as wastewater is not good enough to be discharged into natural water bodies in compliance with the discharge standards or permit conditions notified by the Environment Protection Agencies/Authorities of the concerned nation. The Environment Protection Acts provide for stringent penal actions including closing down the industry

31.8.2 Wastewater Treatment Process

Please refer to Section 31.7 for full details.

31.9 TREATMENT OF SOLID WASTE AND EFFLUENT DISPOSAL

Secondary sludge may be dried on conventional sludge-drying beds and used for composting or landfills; the waste water is disposed of in the environment as per the regulatory frame work of pollution control acts of the specific country. The discharge is permitted on land or into natural water bodies like river, lake or sea, where a minimum of ten times dilution is available with respect to the treated effluent. In general, civic bodies like the municipal authority accept only the treated wastewater (free of any toxic matter) of any industry if it is within the specified standard for disposal in a public sewer. The treated effluent is used for irrigation through flood irrigation, channeling or sprinkler irrigation with nozzles, on pervious soil, where filtering, biological decomposition and adsorption is achieved by the soil to remove the contaminants.

31.10 ADVANCED WASTEWATER TREATMENT

Advanced wastewater treatment is the additional treatment needed to remove the substances remaining in the wastewater after secondary biological treatment. The secondary treated wastewater may still contain considerable concentrations of pollutants like suspended solids, biodegradable organics, volatile organic compounds, nutrients, in the form of ammonium, nitrate, phosphorus, calcium and magnesium, chloride, sulphate, surfactants, toxic and carcinogenic compounds or elements. The need for advanced treatment is being felt due to increasing pollution of our water resources, resulting in serious environmental and health concerns. The environmental regulatory authorities are imposing stringent limiting concentrations for pollutants mentioned earlier, where wastewater is discharged into surface water bodies like rivers and lakes, to safeguard water supply schemes, tourism or fisheries. To meet the stringent standards, secondarily treated wastewater needs to be retrofitted with additional advanced water treatment facilities.

Depending upon the regulatory requirements, advanced wastewater treatment technologies can be selected. Filtration and micro-strainers are used for removal of fine suspended solids, biological nitrification for denitrification and air stripping of nitrogen removal, chemical precipitation of metal salts for phosphorus removal; carbon adsorption for toxic and refractory organics removal and volatile organic compounds with volatilization and gas stripping.

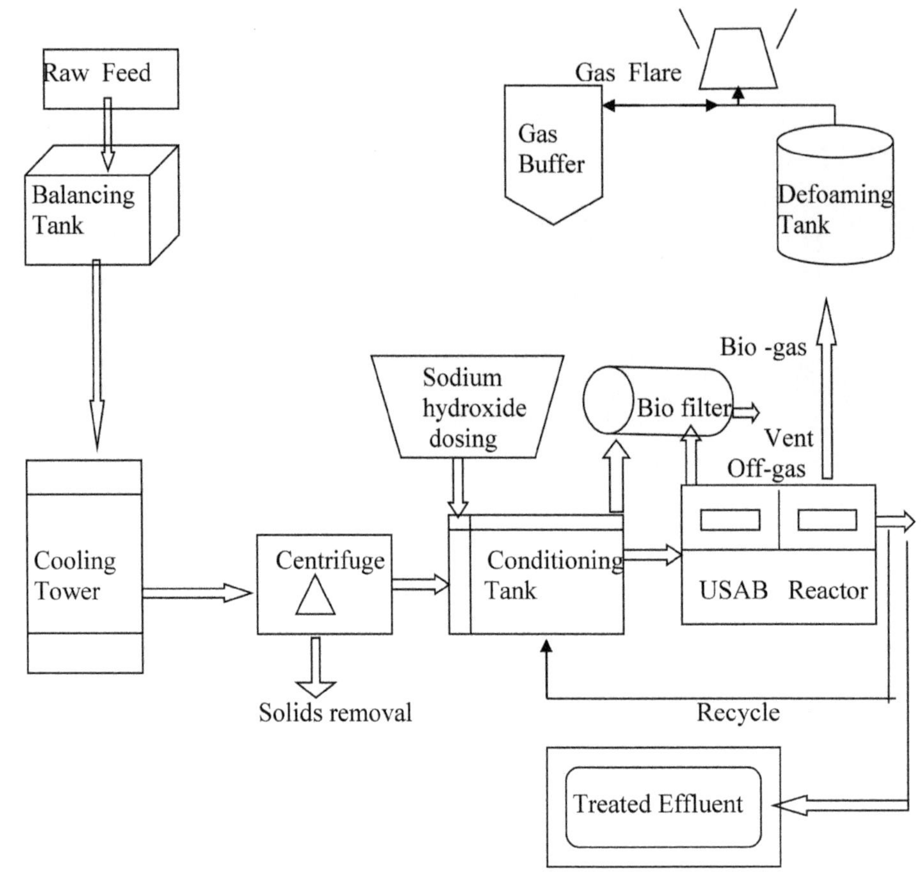

FIGURE 31.5 Process flow diagram of a UASB effluent treatment plant in a distillery. *Source:* Adapted from Wolmer and Villiers (2002)

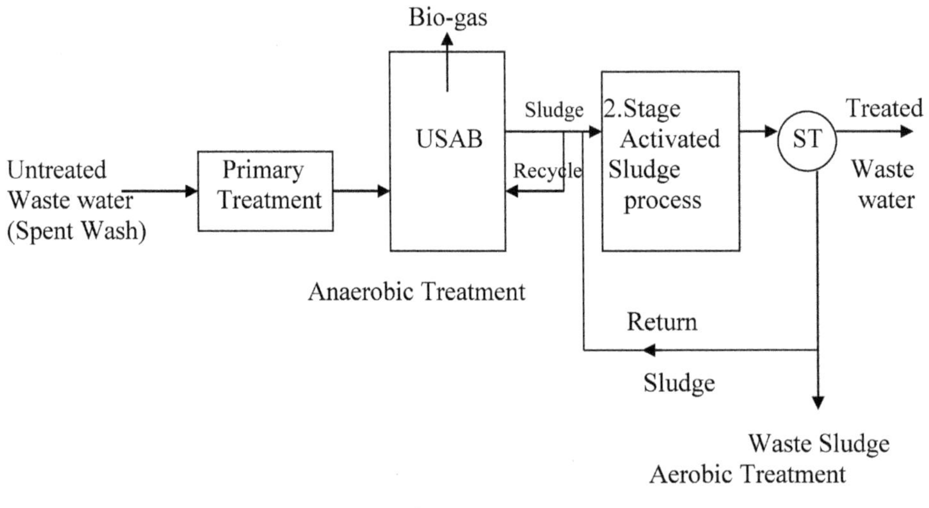

FIGURE 31.6 Combined anaerobic and aerobic treatment for distillery waste. *Source:* Adapted from Joshi (2000)

With a combination of various advanced wastewater treatments following secondary treated wastewater, treatment levels achievable have been reported for various parameters: fine suspended solids: 3–10 mg/L; BOD 1.0–20 mg/L; COD 5–0 mg/L; total nitrogen 30–35 mg/L; NH_3-N: 1.0–25 mg/L; PO_4 as P 1.0–10 mg/L; turbidity 0.3–15 mg/L. Advanced treatment of secondary treated wastewater is useful to protect our lakes and reservoirs from eutrophication. Nitrate removal becomes quite significant in terms of health concerns, when wastewater disposal is made into natural streams being used as drinking water supplies.

31.11 WASTEWATER REUSE APPLICATION

Due to increasing pollution of natural water resources, the regulatory agencies are not only imposing stringent limits for specific pollutants but also limiting the quantity of wastewater permitted to be disposed of, even for the highly treated wastewater with advance treatment. The permitted conditions of regulatory agencies, may direct the industry to work out a plan for reuse or recycling of highly treated wastewater, to, say, the extent of 20–30% of total wastewater generated, depending upon the site conditions and the volume of the receiving water body.

Constructed wetlands with floating aquatic plants to a depth of 0.5–1.8 m, with a residence period sufficient to remove the residual contaminants like suspended solids, organic matter, nitrogen, ammoniacal nitrogen, nitrate nitrogen, biological denitrification, phosphorus and heavy metals. The treated wastewater is not recommended to be used in crop production. The treated wastewater from a constructed wetland can be used for landscaping and green belt development. The treated wastewater can also be used as raw water for toilet flushing, lawn sprinkling, cooling towers, boiler feed, road washing after suitable treatment, if necessary. In 2000, the US Water Resources Council reported 865 Bgal/day treated wastewater was recycled and reused in the US from the industrial sector alone.

The environmental management practices are still limited to the conventional treatment, which are not cost effective due to poor by-product recovery and the high cost of energy utilization in operating the systems. Overall, poor environmental practices result in pollution of our natural water resources and land, incurring additional costs with respect to health- related problems. However, the conventional technology for environment management has not been found to be effective in protection of our natural resources, streams and rivers, in generating energy from waste, including biomass-based thermal plants with central steam-generating system for large-scale food processing need to be given priority over the energy-intensive conventional treatment system. This integrated approach may maximize the benefits and reduce waste disposal to a minimum.

BIBLIOGRAPHY

Alonso, A., Guillean, D., Barroso, C., Puertas, B. and Garcia, A. (2002). Determination of antioxidant activity of wine byproducts and its correlation with polyphenolic content. *Journal of Agricultural Food and Chemistry*, 50, 5832–5836.

Amerine, M.A., Kunkee, R.E., Ough, C.S. and Singletion, V.L. (1980). Winery by-products. In: *Technology on Wine Making.* 4th edn. AVI. Publishing Co., Inc, Westport, CT.

Andres, L., Riera, F. and Alvarez, R. (1997). Recovery and concentration by electrodialysis of tartaric acid from fruit juice industries waste waters. *Journal of Chemical Technology and Biotechnology*, 70, 247–252.

Barnes, D., Forester, C.F. and Hrudly, S.E. (1984). *Survey in Industrial Waste Treatment*, Vol 1. Food and Allied Industries. Pitman Pub, Co., London.

Bartrain, J. and Balance, R. (1996). *Water Quality Monitoring.* E and FN SPON, London.

Baydar, N.G. and Akkurt, M. (2001). Oil content and quality parameters of some grape seeds. *Journal of Agriculture and Forestry*, 25(3): 163, 168.

Borja, R., Martin, A., Martín, A., Maestro, R., Luque, M. and Durán, M.M. (1993). Enhancement of Anaerobic digestion of wine–Distillery waste water by removal of phenolic inhibitor. *Bioresource Technology*, 45(2): 99–104.

Braga, F.G., Lencarte, Silva, F.A. and Alves, A. (2002). Recovery of winery by-products in the Douro Demarketed Regions: Production of calcium tartarate and grape pigments. *American Journal of Enology and Viticulture*, 53(1): 41–45.

Burgess, S. and Morres, G.G. (1984). Two phase anaerobic digestion of distillery effluent. *Process Biochemistry (Supplement)*, 19(5): IV–V.

Cheng, S.S., Lay Wet, Y.T., Vin, M.H., Roam, G.D. and Chang, T.C. (1990). A modified USAB process treating winery waste. *Water Science and Technology*, 22(9): 167–174.

Dwyer, K., Hosseinian, F. and Rod, M. (2014). The market potential of grape waste alternatives. *Journal of Food Research*, 3(4), 91–106.

EPA Guidelines for Wineries and Distilleries. (2004). Environment protection authority. Government of South Australia. http://www. environment.sa.gov.au/epa/pdfs/guide wineries.pdf.

Ferrer, J., Paez, G., Marmol, Z. *et al.* (2001). Agronomic use of biotechnologically processed grape wastes. *Bioresource Technology*, 76, 39–44.

García-Lomillo, J. and González-SanJosé, M.L. (2017). Applications of wine pomace in the food industry: Approaches and functions. *Comprehensive Reviews in Food Science and Food Safety*, 16(1): 3–22.

Gonzalez-Paramas, A., Esteban-Ruano, S., Santos-Buelga, C., Pascual-Teresa, S. and Rivas-Gonzalo, J. (2004). Flavanol content and antioxidant activity in winery byproducts. *Journal of Agricultural Food and Chemistry*, 52, 234–238.

Grayson, M. and Eckbroth, D. (1995). *Kirkothmer Concise Encyclopedia of Chemical Technology.* Wiley and Sons, New York, 163.

Green, J.H. and Kramer, A. (1984). *Food Processing Waste Management.* AVI Pub. Co Inc, Westport, CT.

Gupta, K. and Joshi, V.K. (2000). Fermentative utilization of food processing waste. In: *Postharvest Technology of Fruits and Vegetables*, L.R. Verma and Joshi V.K., eds. The Indus Publ, New Delhi, 1171–1193.

Henry, J.G. and Heinke, G.W. (1989). *Environmental Science and Engineering.* Prentice Hall, Englewood Cliffs, New York.

Ioannis S.A., Ladas, D. and Mavromatis, A. (2006). Potential uses and applications of treated wine waste: A review. *International Journal of Food Science and Technology*, 41(5): 475–487.

Israilides, C., Smith, A., Harthill, J., Barnett, C., Bambalov, G. and Scanlon, B. (1998). Pullulan content of the ethanol precipitate from fermented agro-industrial wastes. *Applied Microbiology and Biotechnology*, 49, 613–617.

Jiang, Y., Simonsen, J. and Zhao, Y. (2010). Compression-molded biocomposite boards from red and white wine grape pomaces. *Journal of Applied Polymer Science*, 119(5): 2834–2846.

Joshi, C. (2000). Food processing waste treatment technology. In: *Postharvest Technology of Fruits and Vegetables*, Vol. 1, L.R. Verma and Joshi V.K., eds. Indus Publishing Co., New Delhi, 440.

Joshi, C. and Joshi, V.K. (1990). Food processing Waste Management Technology- need for an integrated approach. *Indian Food Packer*, 44(5): 56–61.

Joshi, V.K. (2002). Food processing waste management - Oppurtunities and challenges. In: *Biotechnology in Agriculture and Enviornment*, J.K. Arora, Marwaha S.S. and Grover R., eds. Asia Tech Publishers and Distributors, New Delhi, 129–148.

Joshi, V.K. and Sharma, S.K. (2011). *Food Processing Wastes Management*. New India Publishing Agency, Pitam Pura, New Delhi, 472.

Joshi, V.K. and Pandey, A. (Eds) (1999). *Biotechnology: Food Fermentation*, Vol.1 & 2. Educational Publishers and Distributors, New Delhi.

Joshi, V.K., Devender, Attri and Rana, Neerja S. (2013). Solid state fermentation with reference to apple pomace utilization. In: *Food Processing and Preservation*, A.K. Bakshi, Joshi V.K., Vaidya D. and Sharma S., eds. Jagmander Book Agency, New Delhi, 647–657.

Joshi, V.K., Pandey, A. and Sandhu, D.K. (1999). Food factory waste management technology. In: *Biotechnology: Food Fermentation*, Vol. II, V.K. Joshi and Pandey Ashok, eds. Educational Publishers and Distributors, New Delhi, 1291–1348.

Joshi, V.K., Raj, Dev and Joshi, C. (2011). Utilization of wastes from food fermentation industry. In: *Food Processing Wastes Management*, New India Publishing Agency, Pitam Pura, New Delhi, 295–356.

Joshi, V.K., Devi, M.P., Attri, D. and Sharma, R. (2012). Biocolour: Chemistry, production, safety and market potential. In: *Food Biotechnology: Principles and Practices*, V.K. Joshi and Singh R.S., eds. Jagmander Book Agency, New Delhi, 641–689.

Kammerer, D., Claus, A., Carle, R. and Schieber, A. (2004). Polyphenol screening of pomace from red and white grape varieties (*Vitis vinifera* L.) by HPLC-DAD-MS/MS. *Journal of Agricultural and Food and Chemistry*, 52(14): 4360–4367.

Lau, D. and King, A. (2003). Pre- and post-mortem use of grape seed extract in dark poultry meat to inhibit development of thiobarbituric acid reactive substances. *Journal of Agricultural Food and Chemistry*, 51, 1602–1607.

Librán, C.M., Mayor, L., García-Castellό, E.M. and Vidal Brotons, D.J. (2013). Polyphenol extraction from grape wastes: Solvent and pH effect. *Agricultural Sciences*, 4(9B), 56–62. doi:10.4236/as.2013.49B010. Copyright © 2013 SciRes. OPEN ACCESS.

Lund, S.T. and Bohlmann, J. (2006). The molecular basis for wine grape- quality—A volatile subject. *Science*, 311(5762): 804–805. www.sciencemag.org.

Manios, T. (2004). The composting potential of different organic solid wastes: experience from the island of Crete. *Environment International*, 29, 1079–1089.

Metcalf and Eddy. (1991). *Waste Water Engineering Treatment Disposal-Reuse* (3rd edition). McGrawHill, Inc., New York, USA.

Moldes, D., Gallego, P., Rodrıguez-Couto, S. and Sanroman, A. (2003). Grape seeds: the best lignocellulosic waste to produce laccase by solid state cultures of *Trametes hirsuta*. *Biotechnology Letters*, 25, 491–495.

Moosbrugger, R.E., Wentzel, M.C., Ekama, G.A. and Marais, G.V.R. (1993). Treatment of wine distillery waste in USAB systems. Feasibility, alkalinity, requirements and pH control. *Water Science and Technology*, 28(2): 45.

Negro, C., Tommasi, L. and Miceli, A. (2003). Phenolic compounds and antioxidant activity from red grape marc extracts. *Bioresource Technology*, 87(1): 41–44.

NWQMS. (1998). Agriculture and Resource Management Council of Australia &New Zealand, and the Australian & New Zealand Environment and Conservation Council (1998) National water QualityManagement Strategy. Effluent management guidelines for Australian wineries and distilleries. Commonwealth of Australia. www.environment.gov.au/water/policy-programs/nwqms/.

Panday, G.N. and Carney, G.C. (1989). *Environmental Engineering*. Tata McGraw, Hill Publishing Co., New Delhi.

Rao, M.N. and Datta, A.K. (1987). *Wastewater Treatment*. Indian edn. Oxford & IBH Publishing Co., New Delhi.

Rebecchi, S., Bertin, L., Vallini, V., Bucchi, G., Bartocci, F., Fava, F., Totaro, G. and Gavrilescu, M. (2013). Biomethane production from grape pomaces: A technical feasibility study. *Environmental Engineering and Management Journal*, 582-9596(12): S11, 105–108.

Rice, A.C. (1976). Solid waste generation and by product recovery potential from winery residue. *American Journal of Enology and Viticulture*, 2: 21–26.

Sanchez, A., Ysunza, F., Beltran-Garcia, M. and Esqueda, M. (2002). Biodegradation of viticulture wastes by Pleurotus: a source of microbial and human food and its potential use in animal feeding. *Journal of Agricultural Food and Chemistry*, 50, 2537–2542.

Seluy, L.G., Comelli, R.N., Benzzo, María T. and Isla, M.A. (2018). Feasibility of bioethanol production from cider waste. *Journal of Microbiology and Biotechnology*, 28(9): 1493–1501. doi:10.4014/jmb.1801.01044 Research.

Shrikhande, A. (2000). Wine byproducts with health benefits. *Food Research International*, 33, 469–474.

Silvero, C.M., Anglo, P.G., Montero, G.V., Pacheco, M.V., Alamis, M.L. and Luis, V.S. Jr. (1986). Anaerobic Treatment of distillery slops using as up-flow anaerobic filter reactor. *Process Biochemistry*, 21(6): 192.

Smagge, F., Mourgues, J., Escudier, J., Conte, T., Molinier, J. and Malmary, C. (1992). Recovery of calcium tartrate and calcium malate in effluents from grape sugar production by electrodialysis. *Bioresource Technology*, 39, 85–189.

Sousa, E.C., Uchôa-Thomaz, A.M.A., Carioca, J.O.B., Morais, S.M.D., Lima, A.D., Martins, C.G., Alexandrino, C.D., Ferreira, P.A.T., Rodrigues, A.L.M., Rodrigues, S.P., Silva, J.D.N. and Rodrigues, L.L. (2014). Chemical composition and bioactive compounds of grape pomace (*Vitis vinifera* L.), Benitaka variety, grown in the semiarid region of Northeast Brazil. *Food Science and Technology*, 34(1): 135–142.

Thakur, N.S., Joshi, V.K. and Slalthia, S. (2011). Waste Utilization – Fruit and vegetable processing industry. In: *Food Processing Wastes Management*. New India Publishing Agency, Pitam Pura, New Delhi, 73–96.

Torres, J. and Bobet, R. (2001). New flavanol derivatives from grape (*Vitis vinifera*) byproducts. Antioxidant aminoethylthio-flavan-3-ol conjugates from a polymeric waste fraction used as a source of flavanols. *Journal of Agricultural of Food and Chemistry*, 49, 4627–4634.

Wolmar, B. and Villiers, H. de (2002). Start up of a USAB effluent Treatment plant on distillery wastewater. *Water S.A.*, 28(1): 63.

Unit 6

Technology for Production of Wine and Brandy

INTRODUCTION

Unit 6 deals primarily with the production specifics of the various, major types of wine and their by-products. As used here, wines include not only those derived from grapes, but also other fruits.

The section begins with a general discussion of table wine production. Before delving into the production specifics of the various types of table wine, they need to be classified. This is typically based on those features that give clues as to their sensory characteristics – notably their color, aging potential, flavor intensity and varietal origin. Were wines living objects, as could be implied from the frequent use of anthropomorphic terms (*e.g.*, "maturing," "aging," "breathing," "nose," "legs"), one would attempt to classify wines by evolutionary origin. However, many technologically and sensorially similar wines have distinctly different origins, both geographically and temporally. What subsequently follows is an outline of the unique production demands for producing these major, still, table wine styles: white, rosé and red wines. These often come in a variety of dry through sweet variants and may involve distinctive production procedures, unique to each style.

The second in this series of chapters discusses the production of the various major types of fortified wines. Many of these had distinctive regional origins, as well subtypes with recognizable sensory difference, based on decisions made during production. Some are dry, functioning primarily as aperitifs, whereas others are sweet, functioning as dessert substitutes. Most are oxidized, appropriate for wines typically consumed in small amounts, permitting their characteristics to remain stable for weeks after opening. All fortified wines are comparatively 'young' in terms of origin, as related to when wines were first made (some 7000 years ago). This is due to the slow development of adequately effective distillation equipment (late Middle Ages), and the realization that fortification could microbially stabilize wine. This was particularly valuable in hot, Mediterranean climes, where the wines were more susceptible to early spoilage. The fortification also provided consumers with additional 'warmth', definitely appreciated during cold, damp, winter weather in northern Europe before central heating.

The third in the series deals with another late entry into the vinous pantheon – sparkling wines. Their development is similarly predicated on a series of, often serendipitous or independent, discoveries and inventions. The chapter dealt with the first, and often considered, most iconic type of sparkling wine, champagne. The first sections deal with the modern production procedures and aspects of its microbiology. The latter portion concentrates on the production, stability and techniques for studying an effervescence – the essential feature that most distinguishes sparkling wine. The latter demonstrates the value of disparate disciplines (in this case, physics) applied to the wine investigation.

The next chapter shifts away from grape-based wines, to investigate alcoholic beverages derived from two major, temperate-zone, orchard crops – cider (from apples) and perry (from pears). When such beverages were first produced appears unknown, but most likely postdates developments in agriculture and a settled lifestyle (only rational if one plants and establishes orchards and provides an incentive to select varieties appropriate for cider and perry production). Although many production procedures are similar to those involved in grape-wine generation, the structural characteristics of apples and pears, and their much lower fermentable sugar content, imposes unique demands on cider and perry production. Current production procedures are highlighted for the production of hard (alcoholic) versions of both ciders and perries.

Chapter 36 is a unique contribution as it deals with a topic seldom discussed in wine references – the production of reduced-alcohol wines. This is not surprising, as most consumers appreciate the wine's alcoholic content as much as the product's flavor characteristics (those that make wine unique). Nonetheless, there is a niche market for wines that are both dry and of low alcohol content. This may be easy with wines made from other fruits (which are inherently low in fermentable sugar content), but difficult with ripe wine grapes, which are remarkable for their high, fermentable sugar content. Thus, grape-wines fermented dry are almost inherently alcoholic (greater than 9% and can reach 15% without fortification). The difficulty of producing flavorful, dry, low-alcohol wines relates to problems associated with limiting or reducing their alcohol content. The chapter discusses the modern procedures by which low alcohol content wines can be produced, with a minimum of flavor modification or loss. For low-alcohol-content wines to achieve more than restricted market appeal, they must have flavor contents approaching those of the source wine.

Chapter 37 returns to non-grape wines, but this time the wide range of alcoholic beverages that can be produced from fermenting the juice of other, temperate and tropical zone fruits. Although wine production is a minor component of their agricultural use (most being consumed as fresh fruit), it permits the conversion of a perishable fruit crop into a semi-stable, value-added, agricultural product. In tropical countries, ill-adapted to the cultivation of grapes, apples, and pears, it permits the generation of local employment, and can function to reduce the financial drain induced by the importation of alcoholic beverages from abroad (especially for those customers not already accustomed to, and preferring, wines produced elsewhere).

With the next chapter, we come to a product derived from distilled wine – brandy. Although the most famous brandies have *appellation contrôlé* designations (*e.g.*, cognac and armagnac), there are multiple other grape-based brandies. Brandies may also be derived from the pomace (after fermentation), or other fruit-based wines. The characteristics of a brandy depend primarily on the type of still, the decisions made during distillation, and the type and duration of maturation. Base wines are usually chosen to be of neutral character, so that the flavor of the beverage is derived principally from the distillation process and the distillate's maturation. Unrectified wine distillates are usually termed 'wine spirits' and may be preferred to brandy in fortifying wine (notably ports).

Recent technologies have advanced winemaking prowess and quality to a point that could have only been dreamt of in the past. Thus, fine wines are now commonplace, and the likelihood of purchasing a microbially spoilt wine now thankfully rare.

Dr Ron Jackson
Fellow of the Cool Climate Oenology and Viticulture Institute,
Brock University, Canada

32 Production of Table Wines

Ronald S. Jackson

CONTENTS

32.1 INTRODUCTION

Red and white wines constitute the most important class of wines. Most wines follow similar production procedures (Figure 32.1). The principal variant, however, relates to anthocyanin and tannin extraction for the production of red wines. Further differentiation depends on sugar retention or on the type of processing before or after alcoholic fermentation. In this chapter, the basic production features of most wines are discussed initially, followed by those aspects that differentiate red and white wines.

32.2 WINE PRODUCTION BASICS

32.2.1 Sugar and Potential Alcohol Estimates

An assessment of juice soluble solids (a fairly accurate measure of sugar content) is typically one of the first tests

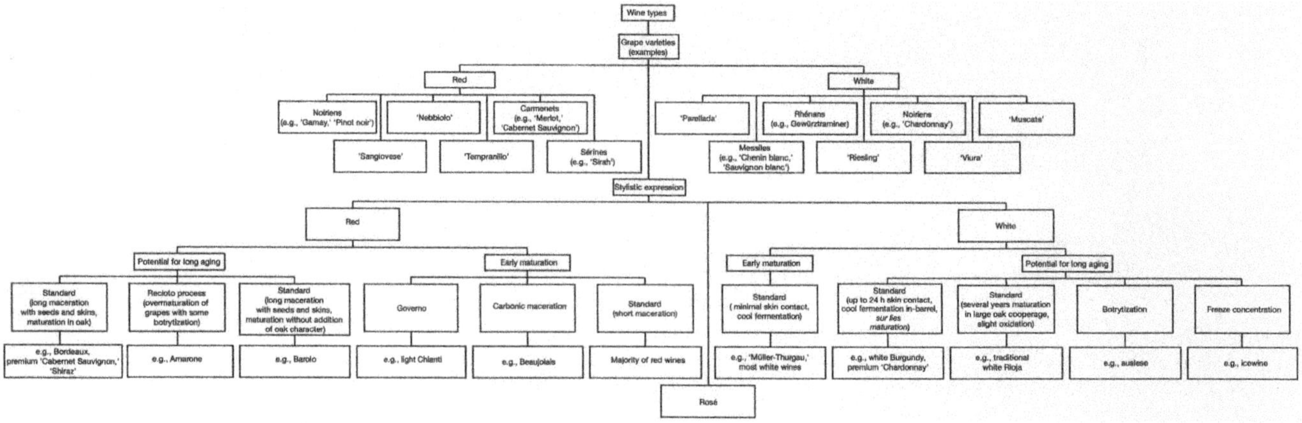

FIGURE 32.1 Categorization of table wines based on their general sensory attributes, with some common examples (Reproduced by permission from Jackson, R.S., 2016, Wines: types of table wines. pp. 556–561. In: *Encyclopedia of Food and Health*. 3rd ed. (B. Caballero, P. Finglas and F. Toldra, eds.), Academic Press, Oxford, UK)

conducted on grapes reaching the winery. This is measured either in°Brix, Baume or Oeschle. Brix estimates the sugar content, Baume approximates to the juice's alcohol-producing potential, whereas Oeschle gauges specific gravity. Because these estimates are good indicators of grape maturity and flavor, they are often used in assessing the crop's financial worth. Nonetheless, as soluble sugar content can increase during a long maceration (as in red wine production), initial assessments may underestimate alcohol potential. Additional sources of conversion error can involve the fermentation temperature, juice amino acid content, and whether fermentation occurs spontaneously or follows inoculation. Because spontaneous fermentation involves multiple cycles of yeast growth and division (increased biomass), less sugar is left for conversion to alcohol. Measuring the sugar content is particularly important with immature grapes, as sugar may need to be added (chaptalization) to reach a desired or legislated minimum alcohol content. Specific gravity measurements may also be used during fermentation as an indicator of its progress. This was essential in large fermenters before refrigeration became available to avoid overheating. The latter could induce the premature termination of fermentation, thus increasing the likelihood of microbial spoilage. Assessing sugar content is also essential if retention of a particular residual sugar content is desired.

32.2.2 Additions and Adjustments

32.2.2.1 Sulfur Dioxide

Until comparatively recently, standard winemaking practice included the automatic addition of sulfur dioxide (SO_2) to the grapes before or immediately after crushing. The amount (30–100 mg/L) varied, depending on the maceration temperature and the health of the grapes. This was predicated on the view that it controlled both contaminant yeasts and bacteria and limited potential oxidation of the bottled wine (especially with white wines). In situations where the grapes are healthy and maceration occurs at cool temperatures, there is little need for the antimicrobial action of SO_2. In addition, inoculation with high doses of *Saccharomyces cerevisiae* obviates the need for SO_2 addition to suppress the activity of indigenous yeasts. Even in spontaneous fermentations, *S. cerevisiae* usually comes to rapidly dominate fermentation. Whether indigenous yeasts contribute favorably to the wine's sensory appeal is a contentious issue – there being proponents on both sides of the issue (See Chapter 15 of this volume for details on the role of non-*Saccharomyces* yeasts). It is only with moldy grapes that there is general consensus that SO_2 addition is beneficial. While effective in inhibiting the action of contaminant microbes, the usually applied doses are ineffective against fungal polyphenol oxidases, laccase (see Chapter 22 of this volume for details).

Current views are that adding sulfur dioxide to the juice may be detrimental. It can delay the oxidation of readily oxidizable phenolics, their subsequent precipitation during fermentation, reducing the likelihood of the early browning of white wines. In addition, sulfur dioxide can suppress the early onset of malolactic fermentation. Nonetheless, its addition could be useful if there was reason to believe that the juice was contaminated with bacteriophages (pathogens of *Oenococcus oeni*).

After the completion of alcoholic and malolactic fermentation (if the latter is desired), the addition of sulfur dioxide, especially to white wines, is still generally recommended for its joint antimicrobial and antioxidant effects. Similar beneficial actions promote sulfur dioxide's addition just before bottling. The antimicrobial effect of SO_2 during maturation is clearly beneficial as is its antioxidant activity functions in limiting in-bottle oxidation.

32.2.2.2 Acidity and pH Adjustment

Adjustment of juice low in acidity typically occurs before vinification (limiting undesirable microbial activity and enhancing color stability), whereas juice high in acidity is generally de-acidified after fermentation. Regrettably, adjustments before fermentation are only approximate. Limited de-acidification often occurs during fermentation due to yeast or bacterial metabolism. Although uncommon, acidity can rise during fermentation, due to the synthesis of malic acid by some strains of *S. cerevisiae*. Typically, the winemaker aims for juice with a total acidity in the range 7–9 g/L (tartaric acid equivalents) and a pH of 3.1–3.4 for white wines.

Typically, red wines have (and are preferred at) a higher pH than white wines. This partially results from red grapes maturing more effectively in warmer regions, conditions that favor the selective metabolism of malic acid. In addition, fermentation in contact with the skins promotes greater potassium extraction, favoring lower acidity and a higher pH (by favoring the conversion of tartaric acid to insoluble salts). Although stylistic and regional preferences exclude precise recommendations, red wines are often preferred within a range of 5.5 and 8.5 g/L total acidity, and a pH range in the range 3.1–3.5.

Wine acidity is of primary importance through its effect on pH. A relatively low pH gives wine a fresh taste, enhances microbial stability, reduces browning, augments SO_2 activity, increases the production and stability of fruity esters, and favors vibrant and stable coloration in red wines. Nonetheless, there is no simple correlation between total acidity and pH, depending as it does on both the amounts of the various acids and the buffering capacity of the juice and/or wine.

If the pH needs to be lowered, experimentation with small lots of must or wine is usually required to determine the amount of acid required. Because tartaric acid is more ionized than malic acid, within the normal range of wine pH values, its adjustment has a greater effect on the pH than an equivalent change in malic acid content. In addition, tartaric acid is relatively resistant to microbial metabolism.

However, where de-acidification – pH adjustment is desired, it can be achieved by either physico-chemical or biological means. Physico-chemical de-acidification may involve either acid precipitation or ion-exchange, while biological de-acidification occurs *via* malolactic fermentation. The addition of inorganic salts (usually calcium carbonate) favors ion exchange, between hydrogen ions from tartaric acid and salt cations. This reduces tartrate solubility, initiating crystal formation and precipitation. To speed the reaction and avoid subsequent formation of calcium malate crystals in bottled wine, the double-salt procedure is preferred. In this process, calcium carbonate is added to a small proportion of the wine (approximately 10%), raising its pH to above 5.1. This assures rapid formation and precipitation of tartrate salts. After about 60 minutes, the remainder of the wine is slowly added to bring the pH back to an acceptable value. Subsequent filtration, centrifugation, or settling removes the salts, making the reaction irreversible.

The double-salt procedure typically works well with wines of medium to high total acidity (6–9 g/mL) and medium to low pH (< 3.5), but can induce an excessive pH rise in wines showing both high acidity and high pH. Ion exchange can be used in this situation, or tartaric acid added to the wine before incorporation of calcium carbonate. Subsequent precipitation of potassium tartrate removes the excess potassium, while the added tartaric acid lowers the pH. Ion exchange involves passing the wine through a resinous column saturated with a particular anion or cation. During passage, ions from the column exchange with those in the wine. For de-acidification, an anion-exchange resin is used. The principal limiting factor in ion-exchange use, other than legal restrictions and cost, is its tendency to reduce wine quality by removing flavorants and pigments.

In most situations, though, malolactic fermentation is the preferred method of de-acidification. It has the potential added advantage of desirable flavor enhancement (see also Chapter 18 in this volume).

32.2.2.3 Flavor Enhancement

Because many varietal aromatics are bound in nonvolatile glycosidic complexes, there has been considerable interest in releasing this unexpressed flavor potential. Liberation can involve either acidic or enzymatic hydrolysis. Because acidic hydrolysis occurs rapidly only at high temperatures (that induces cooked flavors), early liberation of bound aromatics is only enologically acceptable *via* enzymatic hydrolysis. For this, β-glycosidases are preferred, due to their ability to break bonds with a range of sugars, not only glucose. Although generally enhancing a wine's fragrance, enzyme preparation should be experimented with prior to use. It is also a moot point whether such enzymes should be used with premium wines designed for long aging – by favoring early liberation, they may dissipate or degrade before the wine is opened. In addition, some commercial preparations possess additional enzymatic activities, such as cinnamoyl esterases that can release hydroxycinnamic acids. If these acids are subsequently metabolized to vinyl phenols, they may generate spicy, smoky, clove, phenolic, or stable-like off-odors. Enzymes are typically added at the end of fermentation to avoid inhibition of their catalytic action by sugars in the juice. Typically, enzymatic additions are used only with white grapes. Enzymatic addition to red wines has been less common due to the former view that most red wine flavorants came from non-glycosidically bound constituents, notably phenolics. In addition, the enzymes break the glycosidic bonds associated with anthocyanins, initially leaving them more susceptible to oxidation and early color loss (see also Chapter 22).

32.2.2.4 Volatile Loss

During fermentation, large amounts of carbon dioxide (CO_2) are released (approximately 2.3 L/g sugar). This could create toxic working conditions if the winery were not adequately ventilated. The gas also acts as a sparging agent, removing volatile compounds from the fermenting juice. From a sensory aspect, the loss of water and alcohol is insignificant. Of greater impact could be the dissipation of aromatics, such as terpenes and esters. Up to 25% of all the fruit esters may be lost during fermentation. Where considered of sufficient significance, they can be trapped as they escape, and added back to the wine.

32.2.3 Fermentation

32.2.3.1 Fermenters

Demands for the efficient production of clear stable wines have produced a new generation of fermenters. The unique requirements of different wines have also contributed to this diversity. Some features of fermenter design apply to both red and white wines, such as the avoidance or limited access to oxygen, prevention of exposure to spoilage microorganisms, adequate temperature control, efficient means of gas escape, effective procedures of pumping over or submerging the pomace cap in red wines, and facilitation of emptying, cleaning and maintenance. It is also desirable if the fermenter can also act as a storage cooperage after fermentation (see also Chapters 24 and 27 in this volume).

32.2.3.2 Yeast Inoculation

With the development of reliable and inexpensive means of preserving yeasts, it became possible to supply winemakers with a wide selection of yeast strains. Thus began a transformation in winemaking, where inoculation has become standard practice (see also Chapters 14, 15 and 27 in this volume). With inoculation, it is typical to add sufficient yeast to bring the population up to about 10^6 cells/mL juice. Because this usually involved active dry yeasts, which do not have fully functional membranes, it is essential to reactivate the cells before inoculation. This typically occurs in water or dilute grape juice at about 40°C.

Nonetheless, some winemakers are reverting to spontaneous fermentation. It is viewed as a means of retaining regional 'typicity,' or as a means of donating distinctiveness to a producer's wines. That epiphytic grape yeasts can influence the aromatic attributes of a wine is without question. More to the question is whether these influences are desirable, but then this is for the winemaker and consumer to decide. Alternatively, joint inoculation with a series of yeast strains and/or species (commercial and/or locally isolated) may be used.

32.2.3.3 Inoculation with Lactic Acid Bacteria

Malolactic fermentation (MLF) is normally, and preferentially, induced by *Oenococcus oeni*. It is not only the principal means by which the acidity (sourness) of a wine may be mellowed, but also is a means of donating desired flavorants. In addition, once completed, MLF limits the growth of spoilage-inducing lactic acid bacteria. Because various strains affect the flavor differentially, inoculation with a desirable strain is now generally preferred *versus* spontaneous MLF. With most inoculum preparations (lyophilized), the bacteria need to be reactivated and the population increased before addition to the juice or wine. Inoculation aims to achieve a viable population of about 10^6 to 10^7 viable cells/mL. Just as there are varying opinions on the desirability of MLF, there are differing views about when it is best to inoculate. Typically, joint alcoholic and malolactic fermentation may be preferred, so that the wine can be racked early and stored safely for maturation. Nonetheless, MLF is usually encouraged to occur after the completion of alcoholic (yeast-induced) fermentation. Where the process is not desired, notably with wines low in acidity (often from grapes grown in hot climates), various methods are available to limit or prevent its spontaneous occurrence (see also Chapter 18 in this volume).

32.2.4 Maturation

32.2.4.1 Conditions

Except for wines produced by carbonic maceration, wines are matured for several months to years before bottling. During this period, particulate material tends to settle, the supersaturated state of CO_2 equilibrates to ambient, malolactic fermentation occurs (if desired at the end of fermentation), the color of red wines begin to stabilize, and liberation of varietal aromatics bound in nonvolatile complexes begins (see also Chapter 28 in this volume). Typically, maturation occurs at relatively stable, cool temperatures. This not only aids the precipitation of particulate matter but retards the growth of potential spoilage organisms.

For economy, large storage vessels are preferred. It facilitates the mechanization and automation of cellar activities. However, the economies of scale may be sacrificed to maintain the individuality of specific lots of wine, maturation typically occurring in small oak cooperage (barrels, casks, pipes). The specific attributes supplied by maturation in oak depend on the species, its growing conditions, the type of stave seasoning, the degree of toasting (during barrel construction), repeat usage, and the duration of maturation. In situations where an oak character might clash or overpower a wine's varietal or fruity attributes, stainless steel is the standard cooperage construction material (see also Chapter 24 in this volume).

At the end of maturation, the wine is typically fined and/or filtered, and a dose of sulfur dioxide added (notably to white wines) before bottling.

32.2.4.2 Color Adjustment

Occasionally white wines may develop a pinkish coloration. Typically, this is removed with PVPP (polyvinylpolypyrrolidone). Ultrafiltration can also be used but has the disadvantage that desirable flavorants may simultaneously be removed.

32.2.4.3 Blending

To some degree, all wines are blended, even if it only involves the combination of free- and press-run fractions, or the same wine matured in different cooperages. Nonetheless, blending may involve the combination of wines from different varieties, regions, countries, and/or vintages. The type and extent of blending permitted are often dictated by the region's *appellation contrôlé* laws. Blending is a critical component is generating and maintaining the brand consistency in most sparkling and fortified wines. One of the principal advantages of blending is flavor improvement. In an experiment where similarly ranked wines were blended, the combination was usually ranked higher (but never lower) than that of the component wines.

The timing of blending depends primarily on the type of wine produced. For example, the increased use of press wine is often of benefit in poorer vintages, or the incorporation of a similar or older wine. For sherries, fractional blending occurs periodically throughout processing, whereas, for ports, with grapes at pressing, and/or the wines after maturation are blended, whereas, with sparkling wines, the *assemblage* is formulated just before the second, in-bottle fermentation.

32.3 TABLE WINE CLASSIFICATION

Table wines are primarily classified using features such as color, provenance, relative sweetness and varietal origin. Another attribute by which wines are occasionally categorized is style, notably the use of carbonic maceration or partial dehydration. Features such as vineyard origin, producer, vintage or viticultural procedure (*e.g.*, organic or biodynamic) may be used, but often are so sensorially variable as to be essentially of little value (at least consistently). An example of a table wine classification, illustrating some of these factors, is presented in Figure 32.1. Much of what follows relates to those factors influencing wine style.

32.4 WHITE WINES

White wines are usually grouped relative to residual sugar (dry versus sweet) or carbon dioxide (still *versus* sparkling) content. Many features of their production are similar, but several are unique (Figure 32.2). Those aspects that apply to all categories will be discussed first, with those applying specifically to sweet and sparkling wines noted subsequently.

32.4.1 Oxygen Exposure

White wines are generally viewed as being more susceptible to oxidation than are red wines, presumably due to their reduced ability to inactivate (consume) oxygen. However, this may be partially an illusion, due to their pale color showing the effects of oxidative browning more readily than red wines. Either way, white wines generally are treated with more sulfur dioxide before and after fermentation than are red wines. For a period, grapes were blanketed with carbon dioxide or nitrogen, during or immediately after crushing, on the assumption that it would limit subsequent in-bottle browning. However, with some cultivars, the practice exacerbated the problem, as well as enhancing the tendency of the wine to 'stick' and produce excess amounts of hydrogen sulfide (H_2S). The reasons for these counterintuitive results became clear when it was realized that yeasts (especially during spontaneous fermentation) benefitted from the uptake of oxygen during crushing. Oxygen favored the production of unsaturated fatty acids and sterols, essential for the production of new membranes during cell division. Oxygen uptake also promoted the early oxidation and precipitation of readily oxidized phenolics (limiting post-bottling browning). This has now encouraged some producers to bubble air through the juice of some cultivars immediately after crushing (hyperoxidation). The enhanced straw-to-golden colors associated with hyperoxidation may or may not be considered of significance.

32.4.2 Maceration (Skin Contact)

Maceration refers to the contact time between the seeds and skins (pomace) and the juice after crushing. It facilitates the release of flavorants and nutrients from the skin as well as the activation of several cellular enzymes. However, it also favors the release of bitter/astringent, oxidizable phenolics. Correspondingly, maceration is often limited, a tendency enhanced by the widespread adoption of mechanical harvesting (notably with cultivars whose grapes rupture easily during harvesting or have high levels of soluble phenolics). Unfortunately, short maceration periods can also limit the uptake of varietal flavorants from the skins. Flavorant *versus* phenolic uptake can be also influenced by the use of gentler

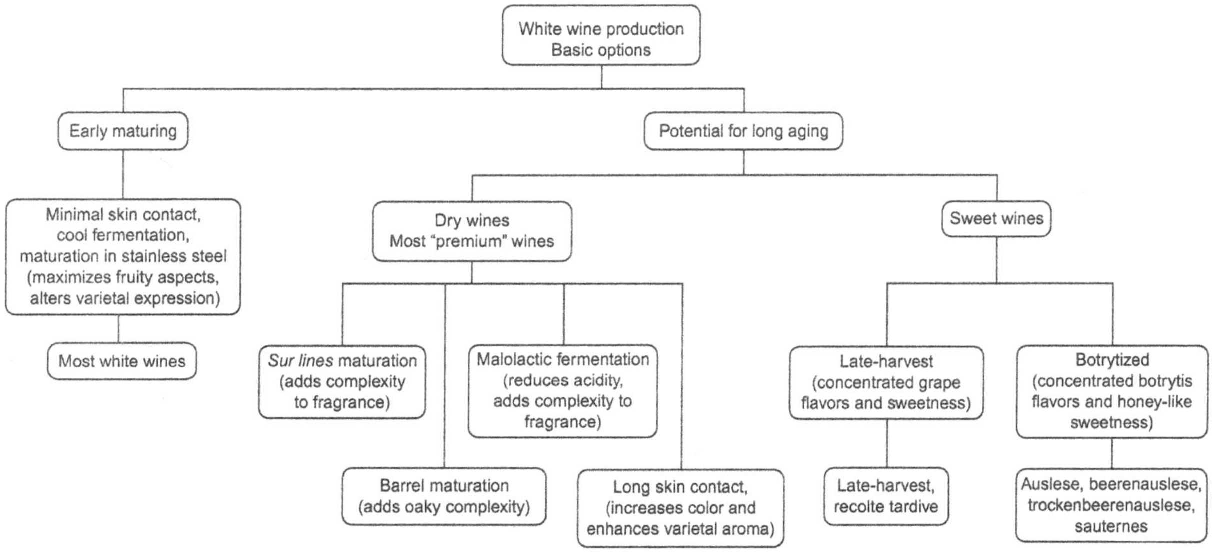

FIGURE 32.2 Classification of still white table wines based on basic production options (Reproduced by permission from Jackson, R. S. (2017). Wine Tasting: A Professional Handbook. 3rd ed. Academic Press, London, UK)

pressing (*e.g.*, pneumatic presses), and adjustment in the amount of press-run wine used in the final blend. Maceration also influences the uptake of essential nutrients (*e.g.*, amino acids and fatty acids) required for yeast and bacterial growth and metabolism. For example, extended skin contact enhances yeast synthesis of extracellular mannoproteins, reduces the production of toxic C_{10} and C_{12} fatty acids, and favors malolactic fermentation.

32.4.3 Suspended Solids

Another feature that tends to distinguish white from red wine production is the clarification that occurs after pressing and before fermentation. This may entail spontaneous settling (12–24 hours), settling combined with a fining agent, centrifugation or filtration. With the use of pneumatic presses, settling may be adequate to reduce the suspended solids in the juice to within an acceptable range (0.1–0.5% for most cultivars). Alternatively, suspended solids can be removed by flotation (micro-bubbles of air, nitrogen or oxygen passed through the juice) (see also Chapter 21 in this volume). Within the acceptable range, suspended solids favor quick and complete fermentation, the production of fruit esters, while avoiding the generation of undesirable concentrations of fusel alcohols and H_2S.

32.4.4 Fermentation Temperature

In addition to deciding how fermentation should commence (spontaneous *versus* induced, and the choice of species and strain), regulating the fermentation temperature provides the winemaker with one of their major tools in directing the wine's sensory character. Spontaneous fermentations may be started at about 20°C (to promote early initiation of yeast growth), followed by cooling to favor the synthesis and retention of fruit esters. If a particularly cool fermentation is desired (down to 10°C), inoculation with *Saccharomyces uvarum* is often preferred. In addition, by slowing ethanol production (and its associated suppression of most of the grape flora), cool temperatures can prolong the activity of indigenous species such as *Kloeckera apiculata*.

32.4.5 Malolactic Fermentation

As noted, malolactic fermentation (MLF) is being increasingly used to add flavor complexity and to enhance the perception of body. The de-acidification and microbial stability benefits of MLF are often now regarded more as incidental benefits, rather than the primary reason for encouraging its occurrence. Even in warm climatic regions, where de-acidification is unnecessary or detrimental, the juice may be acidified to permit the use of malolactic fermentation.

Although generally viewed as beneficial, the frequent production of diacetyl during MLF may reduce or mask the subtle character of some fruity wines. This may, however, be avoided by the appropriate choice of bacterial strain (those that produce little diacetyl), extended wine/bacterial contact (favoring the metabolism of diacetyl to acetoin), or by the addition of a fresh yeast culture (yeasts metabolize diacetyl). Although some white wines are viewed as benefitting from MLF (*e.g.*, 'Chardonnay'), others often are not (*e.g.*, 'Riesling'). In the latter case, MLF is avoided by a combination of cooling after fermentation, early and frequent racking, the addition of sulfur dioxide after the completion of fermentation, or the addition of lysozyme or bacteriocins, such as nisin and pediocin.

32.4.6 In-Barrel Fermentation

Beginning in Roman times, wine shifted from fermentation in clay pithoi (large amphoras) to barrels. However, in-barrel fermentation is now generally restricted to small lots of special juice, or to give the wine a distinctive character not generated otherwise. Because temperature control is difficult, fermentation temperatures may rise significantly above ambient, reducing the production fruit esters. For cultivars such as Chardonnay, this is thought to produce a purer expression of the wine's varietal aroma. In addition, the uptake of oak flavorants is favored, while limiting the incorporation of oak ellagitannins (they precipitate with yeast cells at the end of fermentation). In addition, wines given in-barrel fermentation are usually racked late, favoring the uptake of flavorants from the lees (see below). Delayed racking also favors spontaneous MLF. Although the increased labor involved in using small fermenters is costly, the relative expense of barrel purchase and upkeep is reduced if they are equally used for maturation.

32.4.7 *Sur Lies* Maturation

Sur lies maturation, formerly used locally in a few French regions to enhance wine flavor, has now spread overseas. At the end of fermentation, the wine is left in (or transferred to) barrels containing lees for a period generally lasting 3–6 months. The large surface area/volume ratio favors the uptake of flavorants from the dead and dying yeast cells. Although constituents such as ethyl octanoate, ethyl decanoate, various amino acids, and mannoproteins are known to diffuse into the wine, their relative sensory impacts are uncertain.

During *sur lies* maturation, it is important to avoid the development of a low redox potential in the lees (and the production of reduced-sulfur off-odors). For this purpose, the wine is periodically stirred (*battonage*). Although protecting against thiol off-odor production, it runs the risk of promoting the activity of acetic acid bacteria. To limit this occurrence, oxygen uptake must be limited, and sulfur dioxide is added as an antimicrobial.

32.4.8 Sweet Wines

Sweet wines have been produced for millennia, their preparation being almost as varied as the peoples who have produced them. Currently, most sweet wines are produced by either juice concentration prior to fermentation (*e.g.*, partial dehydration), or the addition of unfermented grape juice just prior to bottling (Figure 32.3). Adding unfermented grape juice has

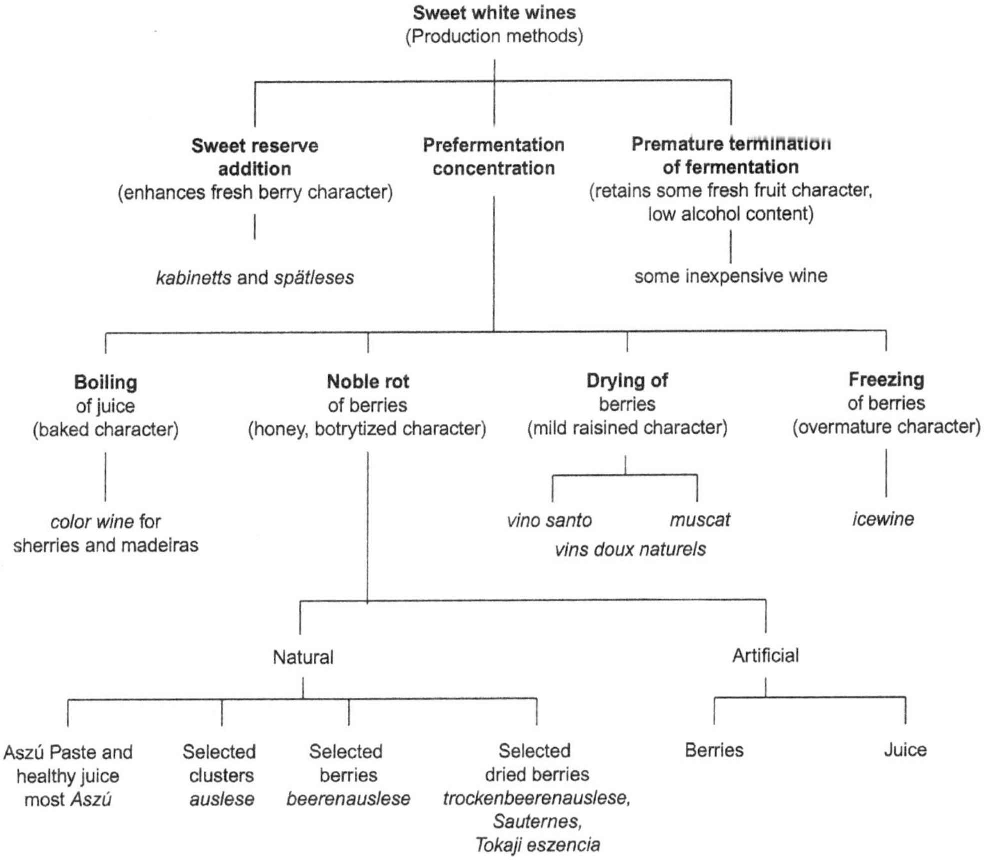

FIGURE 32.3 Classification of sweet white wines based on production method (Reproduced by permission from Jackson, R. S. (2017). Wine Tasting: A Professional Handbook. 3rd ed. Academic Press, London, UK)

the advantage of generating no extraneous flavors. Before use, the grape juice is usually stored at slightly below 0°C (after clarification), with sufficient SO_2, or is sterile filtered.

Because sweet wines contain readily metabolized sugars, and are therefore inherently unstable, stringent measures must be taken to prevent microbial spoilage. This is typically achieved by sterile filtration (or pasteurization), followed by bottling under aseptic conditions and the addition of sulfur dioxide.

An entirely different method for generating sweet wines involves grapes that have undergone what is called noble rot. This involves the growth of *Botrytis cinerea* on grapes late in the season, under marked, cyclical fluctuations in humidity (foggy nights followed by dry, sunny days). Under these conditions, the infected grapes become partially dehydrated (concentrated). In addition, varietal aromatics tend to be degraded while *Botrytis*-generated flavorants accumulate. Harvesting is done manually to avoid rupturing the infected fruit and the grapes are pressed whole. This limits the uptake of β-glucans produced by *B. cinerea*, that can disrupt fermentation as well as seriously complicating filtration (see also Chapter 12 in this volume).

An alternative method of producing sweet wines involves freezing. In the production of ice wines, the grapes are left on the vine until vineyard temperatures fall to –7 to –8°C. Crystallization of much of the grape's water content concentrates the sugar content. The concentrated juice is extracted by pressing the grapes while they are still frozen. Subsequent slow, incomplete fermentation produces a wine high in residual sugar content. Although overmaturation of the grapes during the protracted stay in the vineyard (up to 3–4 months) may result in a loss in varietal aroma, concentration appears to partially compensate for this loss. Freezing also induces sufficient tartrate crystallization to avoid excessive wine acidity.

32.4.9 Base Wine Production

Most white grapes, and a few red grapes, are used in the production of dry and sweet white table wines. Nevertheless, some are used as base wines for the production of other wine styles, notably sparkling wines and sherries, as well as for distillation to produce brandy (see also Chapters 33, 34 and 38).

32.5 RED WINES

Unlike white wines, most reds fall within a single category – dry and still – with ports being the major exception. Red wines are also generally more aromatically distinctive than their white counterparts. This presumably relates to their longer maceration (and corresponding greater uptake of flavorants from the skins), and possibly the larger number of

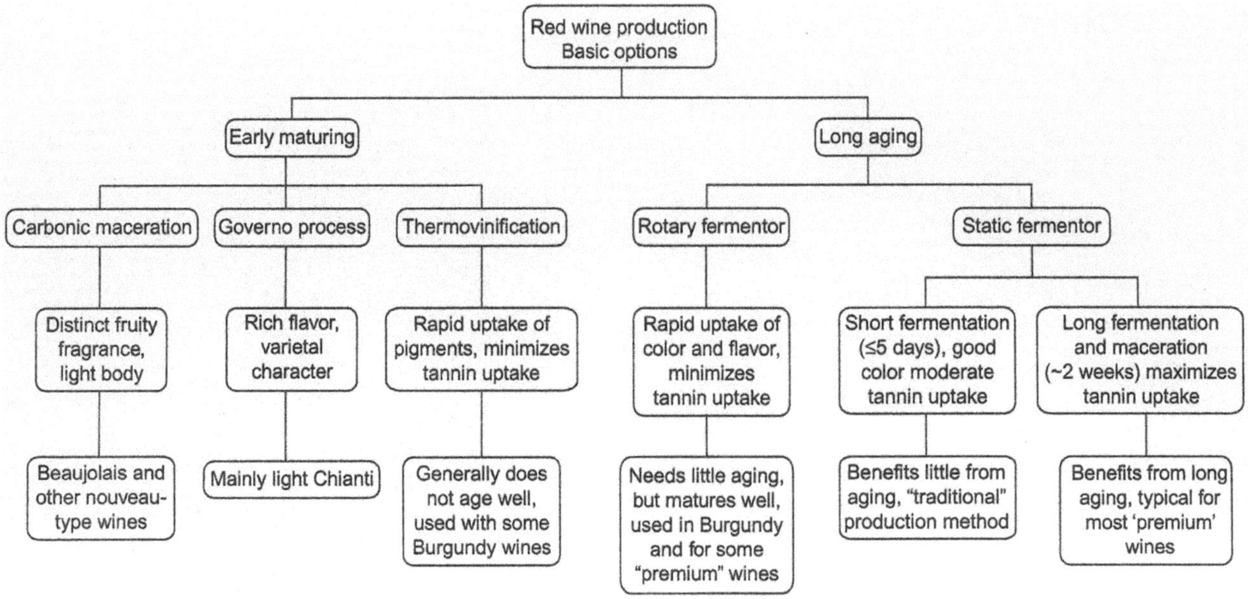

FIGURE 32.4 Classification of red table wines based on basic production options (Reproduced by permission from Jackson, R. S. (2017). Wine Tasting: A Professional Handbook. 3rd ed. Academic Press, London, UK)

aromatically distinct grape varieties. Several unique wine production processes also enhance the sensorial character of red wines (Figure 32.4). As with white wines, stylistic differences can be arranged along a spectrum relative to extent to which they express their varietal or stylistic origins. The remainder of Section 32.5 will discuss how pre-fermentative, fermentative and subsequent processing affect the wine's sensory attributes.

32.5.1 Maceration

Fermentation in the presence of the pomace is a universal feature of red wine production. Without it, the wine would have neither its color nor its flavor characteristics. Despite maceration, only about 20–30% of the grape's anthocyanins and tannin constituents tend to be extracted. Many are complexed with grape cell remains, or tightly bound within the seed coats. Correspondingly, they remain with the pomace, once the wine is pressed.

For cultivars from which color extraction is comparatively poor (*e.g.*, 'Pinot noir'), the crushed grapes may also be given a cold, pre-fermentative soak (maceration) in the juice. This may last from two to six days. The procedure improves color depth, with shifts into the red-purple range. A fresh berry fragrance may also result, possibly due to improved solubilization of some aromatics in the absence of ethanol, the enzymatic release of bound aromatics and/or the action of indigenous yeasts. Cold maceration is also correlated with an increase in the population of indigenous yeasts, as well as a selection for strains of *S. uvarum* and *Saccharomyces bayanus*. There is little data on the optimal temperature conditions for cold maceration, but recommendations can range from 4 to 15°C. Color intensification may be further enhanced by maturation of the wine in oak cooperage.

Opinions relative to SO_2 addition vary widely across the industry. Besides the antioxidant and antimicrobial effects of sulfur dioxide, its addition can enhance pigment and tannin extraction. Regrettably, the binding of sulfur dioxide to anthocyanins delays their polymerization with proanthocyanins, and, correspondingly, may result in reduced long-term color stability.

Maceration provides winemakers with one of the principal means by which they can influence the wine's sensory attributes and aging potential. For example, short macerations (< 24 hours) are used to produce rosé wines, with softer reds usually being pressed within 3–5 days, whereas many premium wines are given extended macerations. Peak anthocyanin extraction usually occurs within the first 2–5 days (depending on the variety and fermentation temperature). In contrast, most skin proanthocyanidins (tannins) take 5–10 days to be solubilized. Another extraction peak usually begins after 15 days, principally involving seed tannins. Thus, a relatively short fermentation supplies adequate coloration, without the incorporation of marked bitter/astringent tannins that take prolonged aging to mellow. Extended maceration (as long as 3 weeks) has usually been considered essential for long aging potential, but without experimental verification.

The extraction of anthocyanins and other flavonoid phenolics during maceration is regulated primarily by its duration, temperature and the frequency and vigor of pumping over. Pumping over is usually required to submerge the pomace within the fermenting juice. Carbon dioxide, released during fermentation, trapped within skin fragments, causes them to float to the surface. This isolates most of the fermenting juice from the principal source of anthocyanins and most flavorants. Submersion of the pomace may involve manually punching the cap down into the juice, as was once done in open vats. However, mixing the fermenting juice with the pomace is now almost exclusively automated, permitting fermentation to occur in tanks. Before tank refrigeration became common, pumping over was incorporated with systems to regulate the temperature fermentation (see also Chapters 24 and 25 in this volume)

With the widespread adoption of efficient stemmer-crushers, it is no longer necessary for grape stems to be included in the ferment (to provide escape channels during pressing). However, if deemed necessary, mature (brown) stems from grape clusters may be included in the ferment, to augment the phenolic content of the wine. Stems may also act as a source of oleanolic acid to facilitate the proper function of yeast cell membranes.

32.5.2 Fermentation Temperature

Fermentation temperatures for red wines are typically higher than those preferred for white wines. Temperatures above 20°C favor the extraction of anthocyanins and other phenolics. Higher temperatures also affect yeast metabolism, resulting in an increased production of acetic acid, acetaldehyde, glycerol and acetoin, as well as reduced production of esters. The higher glycerol contents may reduce the perceived bitterness and astringency of tannins, providing a smoother mouthfeel. Higher rates of metabolism also enhance the sparging effect of CO_2 production. For example, esters are lost at a more rapid rate than ethanol. Nonetheless, sparging is not known to affect significant varietal aroma expression.

With automatic pumping over, the formation of a temperature differential of up to 10°C between the cap and the juice is avoided. Although periodic pumping over temporarily disrupts this differential, it can rapidly redevelop due to the insulating properties of pomace. Thus, depending on the frequency of pumping over, red wines may experience fermentation at two distinct temperatures, the lower major phase in the juice and a periodic, hotter phase in the semi-solid cap.

32.5.3 Malolactic Fermentation

Malolactic fermentation can occur spontaneously, days to months after the completion of alcoholic fermentation. In wines with excess acidity, de-acidification improves the wine's mouthfeel (by softening the perceptions of bitterness and astringency). It was also frequently encouraged in wines of marginal to low acidity, in the belief that it would prevent the subsequent growth of spoilage lactic acid bacteria. Current thought, though, is that the associated improved microbial stability comes primarily from the application of antimicrobial actions after the completion of MLF, such as cooling, the addition of SO_2 and early racking and fining.

To favor the onset of MLF, wines were traditionally transferred to small cooperage, loosely bunged, placed in warmer parts of the cellar and racking and the addition of SO_2 delayed. Because of the unreliability of its start, MLF is now typically induced by inoculation with a desirable strain of *Oenococcus oeni*. Occasionally, joint MLF and alcoholic fermentation is induced to permit the removal of the wine from the lees upon the completion of alcoholic fermentation. This avoids the expense of frequently checking individual barrels for the completion of MLF (and the associated topping up of loosely bunged barrels).

32.5.4 Maturation

Upon the completion of alcoholic fermentation, the wines may be matured for two or more years before fining and bottling. During this period, clarification occurs, and the anthocyanin content progressively becomes stabilized. Stabilization primarily occurs between anthocyanins and other flavonoids (tannins). This involves both direct and indirect polymerization, involving other constituents (notably acetaldehyde). Within about 8–12 months of completing fermentation, most free anthocyanins have bound to catechin (tannin)-based polymers. Subsequent polymerization continues until essentially all free anthocyanins have combined with tannins. This is an important process because, unlike free anthocyanins, the polymers are largely unaffected by SO_2 bleaching or oxidative browning. Because the polymers are often less intensely colored than free anthocyanins, and have more brickish tints, the intensity and color of the wine changes. Polymerization can occur anaerobically but is facilitated by the slight uptake of oxygen (formerly a consequence associated with racking). Hydrogen peroxide, a by-product of phenolic oxidation, primarily oxidizes small amounts of ethanol to acetaldehyde. It does not accumulate to levels with sensory significance as it binds quickly with other constituents, notably forming a 'bridge' between terminal tannin subunits and anthocyanins.

Traditionally, most red wines were matured in small oak cooperage (barrels). However, for wines designed for early consumption, the flavor complexity donated by oak exposure is considered unnecessary or undesirable and is consequently omitted. Alternatively, sources of oak flavor, less expensive than maturation in barrels, may be used, notably oak chips or slats to which the wine is exposed. Nonetheless, the maturation of premium wine is still principally conducted in oak cooperage, typically 250-L barrels. Their large surface area-to-volume ratio promotes the effective extraction of oak flavorants. Any concurrent oxidation that occurs comes primarily from loosely bunged barrels, during racking, or through barrel ends (junction of the head and side staves) (Moutounet *et al.* 1998). Historically, most maturation took place in larger barrels or small tanks (1000–10,000 L). The duration of aging was also longer (approximately 5 y), as compared to the standard 2 years in 250-L barrels.

The flavor imparted by oak depends largely on the degree of toasting given the cooperage during raising (construction) and the frequency of barrel reuse. More subtle differences arise from the species and regional origin of the oak (American or European), the conditions under which the trees grew and how the wood was seasoned (see also Chapter 28 in this volume).

32.5.5 Fining

When to press the wine off the pomace is usually based on past experience and the style of wine desired. Such a pragmatic approach is required as there is no known direct correlation between a wine's phenolic content (consisting potentially of more than 100 compounds) with the wine's perceived bitterness and astringency, flavor or aging potential. Insufficiency with respect to tannin content may be adjusted by maturation in oak (hydrolyzable *versus* condensed tannins), or by the addition of commercial tannin sources (usually hydrolyzable). Excessive astringency may be reduced by the judicious addition of proteinaceous fining agents (they primarily remove polymeric tannins versus their monomers) early during maturation. This is designed to avoid removal of polymeric pigments that begin to form progressively during maturation. However, if limiting phenolic-based bitterness is desired, fining is better delayed, and products such as polyvinylpolypyrrolidone (PVPP) should be used. It tends to preferentially remove mono- and dimeric phenolics. By delaying PVPP fining, adequate time is provided for the formation of more color-stable anthocyanin-tannin polymers.

32.5.6 *Recioto* Process

The *recioto* process is an ancient procedure, used in Veneto and Lombardy. It involves selecting fully mature clusters of grapes and placing them in shallow, slatted trays in well-ventilated storage areas. During the several months of cool storage, the grapes partially dehydrate, a process termed *appassimento.* During this period, the grapes lose about 25 to 40% of their moisture content. The cool temperatures and low humidity levels (less than 90%) limit microbial activity, although a portion of the grapes undergo a slow, limited infection by the noble rot fungus, *Botrytis cinerea.* The chemical effects produced are similar to those found during the noble rotting of white grapes (described in Section 32.4.8), *e.g.*, enhanced glycerol and gluconic acid contents, reduced tartaric acid values, and an increased fructose/glucose ratio. Although the infected grapes turn a pale purple brown (healthy grapes turn bluish), the wine produced shows a brickish red color. After the *appassimento* period, the grapes are stemmed, crushed, and the must raised to about 20°C, for the commencement of fermentation, if the dry *amarone* style is desired. Fermentation is traditionally conducted by indigenous yeasts, the dominant strains by the end of fermentation usually being those of *S. bayanus.* Fermentation is slow, often lasting about 40 days. The wine is usually matured for several years in used oak cooperage (to minimize the uptake of a marked oaky character). The wine is characterized by a velvety, smooth taste, atypical for a young wine possessing such flavor and aging potential. The wine also possesses a unique, sharp, oxidized, phenolic fragrance, resembling the pungency of some tulip and daffodil flowers. Besides the common, dry, *amarone* style, the wine may also be produced in sweet (*amabile*) and sparkling (*spumante*) versions.

32.5.7 *Governo* Process

Another traditional Italian (in this case Tuscan) winemaking procedure is called the *governo* process. It also involves partially dried grapes, but produces a light, early-drinking wine – what used to characterize Chianti wines. In the process, about 3–10% of the harvested grapes are separated from the main crop and placed under cool, dry conditions. During the ensuing partial drying, the dominance of *Kloeckera apiculata* in the indigenous grape flora declines, while the population of *S. cerevisiae* increases. Following partial dehydration, the grapes are crushed, and fermentation commences. The fermenting must is then added to the wine which had previously been produced from the major portion of the crop, using traditional methods. Usually, the cellar is heated to promote the slow fermentation of the wine/fermenting must mixture.

32.5.8 Carbonic Maceration

Carbonic maceration is a modern version of an old technique, receiving renewed attention and investigation. Its advantage is the production of a distinctly fruity wine that can be consumed shortly after fermentation. Versions made primarily from the free-run press fraction tend to be light in style, but do not age well (about one year). Such wines are typically designated *nouveau* (*novello*) wines. However, wines made from the more flavorful, more intensely colored, press-run fractions may age well for several years.

Along with the *recioto* and *governo* processed wines, those using carbonic maceration derive most of their distinctive attributes from the grape processing they undergo before alcoholic fermentation.

For carbonic maceration, the grapes are harvested manually (to minimize grape breakage). Harvesting is preferred in the heat of the day, so that the grapes reach the winery distinctly warm. In Beaujolais, the clusters are dumped into shallow, broad vats, and the tops covered. Carbon dioxide may be flushed through the piled grape clusters to eliminate air, promoting an earlier initiation of grape-cell fermentation. In other regions, crates of grapes may be placed in special structures that are flushed with CO_2. Berry auto-fermentation is usually encouraged to occur at above 30°C. This favors the development of flavorants that characterize carbonic maceration.

During grape-cell fermentation, a small quantity of alcohol is produced (about 2%). Of greater significance, however, is the modification of the shikimic acid pathway. The result is the production of several volatile phenolics, such as ethyl cinnamate, benzaldehyde and vinylbenzene. These may partially generate the unique carbonic maceration aroma, along with other volatile phenolics that accumulate at higher concentrations than during traditional fermentation. Nonetheless, the precise chemical nature of the typical carbonic maceration fragrance (strawberry–raspberry, cherry–kirsch) remains to be established.

During carbonic maceration, the grape structure weakens and progressive berry rupture releases juice. By the end of carbonic maceration, up to 35–55 % of the juice may have been liberated, becoming colonized by indigenous yeasts and lactic acid bacteria.

Once grape-cell fermentation has ceased (about 6–8 days), the grapes are crushed and pressed. After pressing, the juice is usually cooled to between 18–20°C for completion of alcoholic fermentation. Fermentation is typically conducted by the same yeasts that were already active in the juice released before pressing. However, to minimize the production of ethyl acetate, inoculation with a specific strain of *S. cerevisiae* may be employed.

The free- and press-run fractions may be fermented separately (for possible selective combination later) or may be combined before fermentation. If malolactic fermentation has not already taken place by the end of fermentation, the wine is usually inoculated with *Oenococcus oeni* to induce its early completion.

Although treating the whole crop to carbonic maceration is standard practice, the grapes may be divided into lots – one treated to carbonic maceration and the other to standard practices. The two are subsequently blended to combine the features of both. Although much less common, rosé and white versions of carbonic maceration wines are produced.

32.5.9 Thermovinification

Thermovinification is a modern treatment designed to enhance the extraction of pigments, primarily from diseased or poorly colored grapes. Whole or crushed grapes are exposed to short periods of high temperatures to inactivate laccase enzymes (in diseased grapes) and/or to enhance anthocyanin extraction. In most versions, intact or crushed grapes are heated to between 50 and 80°C. Occasionally, whole grapes may be exposed to steam or boiling water (flash heating). Such treatments are typically short (approximately 1 minute), heating only the outer layers of the skin to approximately 80°C. Killing the skin cells facilitates the rapid release of anthocyanins during subsequent maceration (approximately 45°C for 6–10 hours). Heating may be supplied with or without continuous stirring of the must.

Maceration time varies inversely with the temperature – the higher the temperature, the shorter the duration (Wiederkeher, 1997). Temperatures as low as 32°C may be used for varieties such as 'Pinot noir' (Cuénat *et al.* 1991), from which the varietal aroma would be destroyed at higher temperatures. Only mold-free grapes can safely be treated at temperatures below 60°C (the fungal polyphenol oxidase, laccase, is denatured only above this temperature). Subsequent vinification may be conducted in the presence or absence of the seeds and skins. Thermovinification dramatically increases anthocyanin extraction, generating wines with a rich red color. The bluish colors and cooked flavors occasionally noticed can usually be avoided by strict exclusion of oxygen during thermovinification and limiting the duration. Careful use of the technique can also generate wines low in astringency and with reduced vegetal odors and reduce the varietal aroma characteristic of certain *Vitis labrusca* cultivars and hybrids.

BIBLIOGRAPHY

Boulton, R. (1979) The heat transfer characteristics of wine fermentors. *Am. J. Enol. Vitic.* 30, 152–156.

Boulton, R.B., Singleton, V.L., Bisson, L.F. and Kunkee, R.E. (1996) *Principles and Practices of Winemaking.* Chapman & Hall, New York, NY.

Bucelli, P. (1991) Il governo del vino all'uso del Chianti, note storiche e aspetti tecnici. *Vini d'Italia* 33(4), 63–70.

Cheynier, V., Remy, S. and Fulcrand, H. (2000) Mechanisms of anthocyanin and tannin changes during winemaking and aging. In: Rantz, J.M. (Ed.), Proceeding of the ASEV 50th Anniversary Annual Meeting, *Seattle, WA*, pp. 337–344. American Society of Enology and Viticulture, Davis, CA.

de Revel, G., Martin, N., Pripis-Nicolau, L., Lonvaud-Funel, A. and Bertrand, A. (1999) Contribution to the knowledge of malolactic fermentation influence on wine aroma. *J. Agric. Food Chem.* 47(10), 4003.

du Toit, W.J., Marais, J., Pretorius, I.S. and du Toit, M. (2006) Oxygen in must and wine: A review. *S. Afr. J. Enol. Vitic.* 27(1), 76–94.

Flanzy, C., Flanzy, M. and Bernard, P. (1987) *La vinification par macération carbonique.* Institute National de la Recherche Agronomique, Paris, France.

Fugelsang, K.C. and Edwards, C. (2007) *Wine Microbiology. Practical Applications and Procedures*, 2nd ed. Springer, New York, NY.

Guymon, J.F. and Crowell, E.A. (1977) The nature and cause of cap-liquid temperature differences during wine fermentation. *Am. J. Enol. Vitic.* 28, 74–78.

Jackson, R.S. (2014) *Wine science. Principles and practice*, 4nd ed. Academic Press, San Diego, CA.

Kennedy, J.A. (2010) Wine colour. In: Reynolds, A.G. (Ed.), *Managing Wine Quality. Vol. I: Viticulture and Wine Quality*, pp. 73–104. Woodhead Publishing Ltd., Cambridge, UK.

Konig, H., Unden, G. and Frohlich, J. (Eds.) (2009) *Biology of Microorganisms on Grapes, in Must and in Wine.* Springer, New York, NY.

Lerm, E., Englebrecht, L. and du Toit, M. (2010) Malolactic fermentation: The ABCs of MLF. *S. Afr. J. Enol. Vitic.* 31(2), 186–212.

Magyar, I. (2011) Botrytised wines. *Adv. Food Nutr. Res.* 63, 147–206.

Mencarelli, F. and Tonutti, P. (Eds.) (2013) Sweet, reinforced and fortified wine. *Grape Biochemistry, Technology and Vinification.* Wiley-Blackwell, Chichester, UK.

Moreno-Arribas, M.V. and Polo, C. (Eds.) (2010) .*Wine Chemistry and Biochemistry.* Springer Science + Business Media, New York, NY.

Moutounet, M., Mazauric, J.P., Saint-Pierre, B., and Hanocq, J.F. (1998) Gaseous exchange in wines stored in barrels. *J. Sci. Tech. Tonnellerie* 4, 131–145.

Paronetto, L. and Dellaglio, F. (2011) Amarone: A modern wine coming from an ancient production technology. *Adv. Food Nutr. Res.* 63, 285–306.

Razungles, A. (2010) Extraction technologies and wine quality. In: Reynolds, A.G. (Ed.), *Managing Wine Quality, vol. II.Viticulture and Wine Quality*, pp. 589–630. Woodhead Publ. Ltd., Oxford, UK.

Reynolds, A.G. (Ed.) (2010) *Managing Wine Quality. Vol. I and II. Viticulture and Wine Quality.* Woodhead Publishing Ltd, Oxford, UK.

Ribereau-Gayon, P., Doneche, B., Dubourdieu, D. and Lonvaud, A. (2006) *Handbook of Enology Vol. I: (The Microbiology of Wine and Vinifications)*, 2nd ed. John Wiley and Sons, Chichester, UK.

Sacchi, K.L., Bisson, L.F. and Adams, D.O. (2005) A review of the effect of winemaking techniques on phenolic extraction in red wines. *Am. J. Enol. Vitic.* 56, 197–206.

Schäfer, T. and Crespo, J.G. (2007) Study and optimization of the hydrodynamic upstream conditions during recovery of a complex aroma profile by pervaporation. *J. Membr. Sci.* 301(1–2), 46–56.

Schneider, V. (1998) Must hyperoxidation: A review. *Am. J. Enol. Vitic.* 49, 65–73.

Singleton, V.L. and Ough, C.S. (1962) Complexity of flavor and blending of wines. *J. Food Sci.* 12(2), 189–196.

Tesniere, C. and Flanzy, C. (2011) Carbonic maceration wines: Characteristics and winemaking process. *Adv. Food Nutr. Res.* 63, 1–15.

Usseglio-Tomasset, L., Bosia, P.D., Delfini, C. and Ciolfi, G. (1980) I vini Recioto e amarone della valpolicella. Vini Ital. 22, 85–97.

Waterhouse, A.L. and Caputi, A. Jr. (Co-chairs). (1999) Oak in winemaking. International symposium proceedings. *Am. J. Enol. Vitic.*, 50, 469.

33 Fortified Wines
Production Technology

Vandana Bali, Parmjit S. Panesar, V.K. Joshi and Somesh Sharma

CONTENTS

33.1 INTRODUCTION

Fortified wines originated in the 17th century, when wines fortified with brandy were used by the sailors on long voyages. To protect the wine from being spoiled, addition of high-alcohol brandy acted as a preservative. The word 'fortified' is derived from the Latin word '*fortis*' meaning 'strong', so that addition of spirit to the wine to strengthen it not only boosted its taste but also provided a high alcohol content which protected it from spoilage when exposed to the air. The fortified wine could be dry or nearly dry (as with most sherry styles) or elegantly sweet. An added lift to the aroma and taste to fortified wines is provided by the added spirit (port styles). The major fortified wines are sherry, port, vermouth, aperitif and Madeira.

European Union regulations define fortified wines generally as those having an acquired alcohol content by volume of between 15% and 22% (v/v), and a total alcohol content of at least 17.5% (v/v); within these rules, allowance is made for *vino generoso*, wines with 15.0% (v/v) alcohol and less than 5g/l sugar, produced in demarcated areas (Council Regulation (EC) No. 822/87, 1987). Fortified wines have been classified into two main categories: (i) those having their own characteristic flavor because of whole or partially fermented grapes, or distilled spirits, including sherry, port and Madeira; and (ii) those in which, in addition to the above products, flavor from non-grape products is added, which are normally known as flavored fortified wines, such as vermouth and aperitifs. The different processes involved in the production of sherry, port and vermouth are discussed in this chapter.

33.2 SHERRY

33.2.1 Definition and Nomenclature

Sherry is a fractionally blended wine made from white grapes, which is subjected to prolonged oxidative maturation and has considerable finesse. Sherry is one of the oldest wines in the world but not as it exists today. The name "sherry" is given to a number of related dessert wines types with 15–20% alcohol, originally developed in the South of Spain in the area around *Jerez de la Frontera*, in the province of Cadiz. The oxidative reactions occurring during their aging, which may be biological, as in *Fino* sherries and *Manzanilla* sherries, physico-chemical as in *Oloroso* sherries, or first biological and then, physico-chemical, as in *Amontillado* sherries. *Finos* are dry and consumed primarily as aperitifs before meals, whereas *Olorosos* are often sweetened and consumed after or with dessert. *Amontillado* sherries begin their development as a *Fino*, but end as an *Oloroso,* and are, typically, treated as a type of the *Fino* sherry wine.

33.2.1.1 Types of Sherry

- ***Fino de Jerez***: This type of sherry is the driest and youngest of all sherries, approximately 3–5 years old. After fortification of newly pressed wine (up to 15% volume), a layer of protective yeast is introduced to the three-quarter-filled barrel. This layer of yeast protects the wine from becoming oxidized and also provides it with the bone dryness by consuming all the sugars present in the wine.
- ***Manzanilla***: This type of sherry constitutes young and dry forms of the *Fino* type.
- ***Amontillado*:** This type of sherry is produced by the fortification (up to 18%) of *Fino* or *Manzanilla* sherry, followed by further oxidative aging which results in imparting the brownish color to the sherry.
- ***Oloroso***: This type of sherry is produced by fortification (up to 18%) after a second pressing of grapes and then, transferring the wine into barrels where it remains exposed to the air at up to 17 or 18°C for approximately 40 years.
- ***Palo Cortado***: This type of sherry is the result of oxidation and further fortification of a *Fino* or *Manzanilla* sherry. It constitutes the structure and body of an *Oloroso* and the aromatic refinement of an *Amontillado* sherry.
- ***Pedro Ximenez* (PX):** This type of sherry is the sweetest of all the sherries. The grapes are either picked very ripe and/or are sun dried before pressing.
- **Cream Sherry:** It is a blend of *Oloroso* and *Pedro Ximenez* sherry.

33.2.2 Areas of Production, Climate and Soil

Spanish sherry is produced in a restricted area, in southern Spain, close to the town of *Jerez de la Frontera*, known as the sherry triangle. The triangle consists of cities of *Jerez de la Frontera*, *Sanlúcar de Barrameda* and *El Puerto de Santa María*. A committee, known as the *Consejo Regulador de la Denominacion de Origen 'Jerez-Xerex-Sherry'*, appointed under Government authority, strictly controls the materials and techniques used in the manufacture of sherry, the vineyards, and the places where it may be fermented and matured. Sherry wine is also produced in Australia, California, Canada, France, the Soviet Union and South Africa. Almost all wine-producing areas have made legal regulations for protection of their own wines. The factors affecting the quality of sherry include the climate of the overall area and the microclimate of the vineyards, such as slope aspect and drainage. The best vineyards are located over hills composed of a porous and chalky soil (alkaline pH with a high active lime content). The type of soil which occurs in a broad area immediately to the west of *Jerez* is locally known as *albariza*. The albariza soil retains water from autumn and winter rains and provides it during the arid growing season. The porosity of the soil allows vine roots to penetrate deep into the soil and collect the trapped water below the impermeable surface during the arid season. The climatic conditions of the sherry-producing area are markedly constant from year-to-year, with temperature ranging from 10.5°C in winter to 25°C in summer, to a maximum temperature in July and August between 35°C and 38°C. The rainfall remains moderate but quite variable within the year, being heavy and prolonged from the end of November to February, with very little from May to August.

33.2.3 Grape Varieties

In Jerez, a number of varieties of *Vitis vinifera* are used, but, at present, Palomino type grapes are mostly (90%) used for dry wine production. The two types of Palomino type grape variety are *Palomino Fino* and *Palomino de Jerez*. The sweet wines are made from the *Pedro Ximenez* (PX) grape, which are not planted in the Jerez area, as it grows better in the Montilla district. Furthermore, for all practical purposes, the high-yielding *Palomino Fino* is preferred to *Palomino de Jerez* for plantations of sherry vineyards. *Moscatel* grapes are less commonly used for producing sherry as compared with the other two varieties.

33.2.4 Technology of Production

Different steps involved in the production of sherry wines of different styles are depicted in Figure 33.1.

33.2.4.1 Crushing and Pressing

Depending upon the types of equipment available in different areas, different processes are used for crushing and pressing of grapes. Traditionally, in Jerez, crushing of grapes was done by treading in a low wooden trough known as a *lagar*. The crushing of grapes was done earlier by treading with nailed boots, with the grapes then being pressed by a screw covered with bands made from esparto grass. This process is now performed only as a demonstration on a small scale. Nowadays, the grapes are crushed using crushers. After the first pressing, the free-run juice obtained accounts for approximately 85% of the total juice. Another 5% of the juice is extracted by a second pressing, whereas further pressings result in the production of material appropriate for vinegar production or for distillation purposes. For increasing the total acidity of the must and supporting the mechanical pressing, calcium sulfate is added in the form of gypsum. In California, during sherry making, Garolla-type crushers and stemmers are used for crushing and stemming the grapes, and at the winery in the same manner as previously mentioned as for other wines. Since the quality of the finished wine is influenced by the mechanical treatment of the grapes, alternative continuous systems are being investigated. Traditionally, prior to pressing, *yeso* (plaster) was used to aid clarification and the production of tartaric acid (when combined with tartar cream). Nowadays, tartaric acid is directly added (*desfangado*) to clear the must before fermentation.

33.2.4.2 Fermentation

The fermentation is carried out in casks or oak butts for the production of characteristically flavored blended wines. However, currently most of the fermentation process of wine takes place in tanks on the same site as the vintage equipment. The factors affecting flavor and aroma of wine are primarily the type, quality of the must, and the processing conditions. Most sherry fermentations are carried out by inoculating with strains of *Saccharomyces cerevisiae*. The temperature during fermentation should be 25–30°C. The first phase of sherry fermentation, *i.e.*, a tumultuous fermentation constituting a vigorous initial phase lasting up to a week, is dominated by *Candida pulcherrima*, *Kloeckera apiculata*, *Saccharomyces cerevisiae* var. *ellipsoideus*, *Saccharomyces rosei*, *Saccharomyces Italicus* and *Hanseniaspora guilliermondii* present in significant proportions. As the alcohol concentration increases, the population of *Candida* spp. declines and that of *Saccharomyces chevaliers* increases. During the second phase of fermentation, *i.e.*, the lenta, or slow fermentation, *S. cerevisiae*, *S. chevalier* and *S. italicus* predominate. *Hansenula, Hanseniaspora, Kloeckera, Pichia, Candida, Debaryomyces, S. rosei, Saccharomyces uvrum, Kluyveromyces* and many other molds similar to *Mucor* spp. constitute the microflora of the unfermented must.

33.2.4.3 Settling and Racking

After fermentation, it is advantageous to allow the low-acid, delicate base wine (11 to 12.5% alcohol) to settle for a few days. The settling of the base wine before fortification helps to get rid of the yeast lees which further prevent the loss of fortifying spirit in the lees. However, in many cases, wine from the fermenter is directly pumped into the fortifying room, as settling time for the wine is not always practicable to allow.

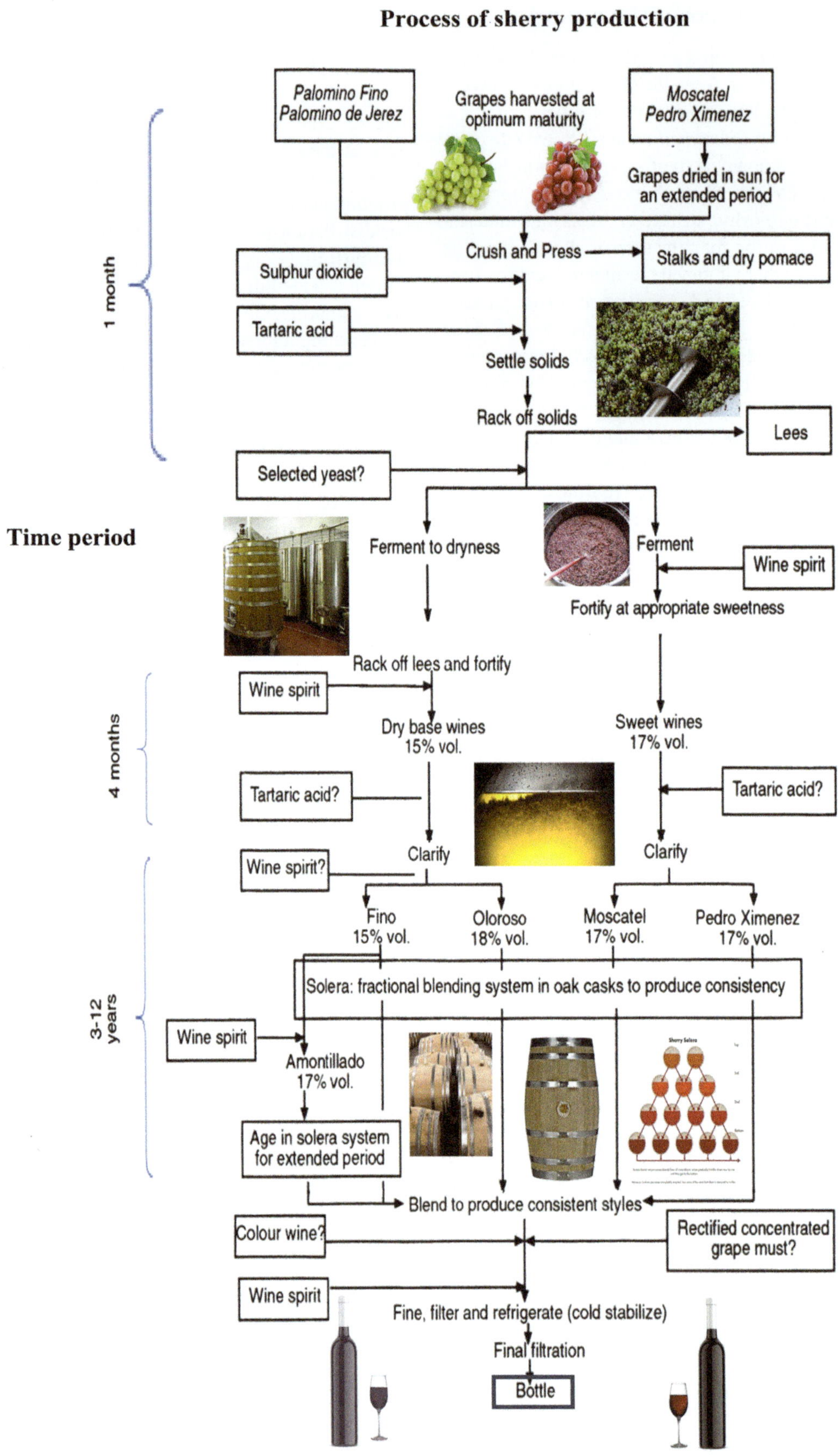

FIGURE 33.1 Flow sheet showing the main processes involved in the preparation of sherry wine. *Source:* Adapted from Goswell and Kunkee (1977)

33.2.4.4 Fortification

The wine is pumped into a special tank for fortification purposes from the fermenter or settling tank, where its alcohol content is determined. The Federal Gauger supervises the actual addition of high-proof grape spirits to the wine, and the alcohol content is increased to between 14.5 and 15.5% (v/v). To obviate the heavy clouding which takes place when high-strength alcohol is mixed with new wine, the alcohol is pre-mixed with an equal quantity of a suitable wine of the same type, and the cloudiness is allowed to settle for about three days. To secure the uniform mixing of high-proof spirits and wine, the wine is pumped over strongly during and after addition of the spirits until it is found by analysis that the fortified wine is of uniform alcohol content throughout the fortifying tank. In most plants, the mixture is agitated vigorously using compressed air but, in others, mixing is done by a propeller, or by a combination of both compressed air and a propeller.

After fermentation, the wine may be classified as *palo* (marked with a vertical slash), or as *gordura* (marked with a circle). The *palo* wines are fortified to 15–15.5%, and result in the production of delicate *Fino* or *Manzanilla* styles, whereas *gordura* wines are fortified to comparatively high alcohol levels (17–18%), thereby limiting the growth of the microflora, and resulting in *Oloroso* sherries.

33.2.4.5 Aging

Once the wine has been fortified (to 15–15.5%), a biofilm forms on the surface of the wine by 'flor yeast' (or *flor del vino*: the "flower" or flor velum) during biological aging. The yeast involved in flor velum formation mainly consists of different subspecies of *Saccharomyces* (mainly *Saccharomyces cerevisiae beticus, Saccharomyces cerevisiae cheresiensis, Saccharomyces cerevisiae montuliensis* and *Saccharomyces cerevisiae rouxii*), according to their abilities to ferment different sugars. Biological aging is one of the most important phases in the production of sherry. During biological aging, over a long time, the wine is subjected to intensive action of microorganisms. At different wine maturation stages and using different aging systems such as static (*anadas*) and dynamic (*solera*) systems, there are significant differences in the flor yeast population observed, integrating the velum in sherry. There is a predominance of *S. cerevisiae beticus*, followed by *cheresiensis, montuliensis* and *rouxii* subspecies. The sensory properties of sherry wines are dependent on the subspecies succession during biological aging. However, the progressive decrease in the population of *S. cerevisiae beticus* takes place in the older stages of maturation, being substituted by S. *cerevisiae montuliensis*. Consequently, after fortification and short-term maturation, the base wines (known as *sobretabla*) are further classified. The biological aging of base wine with a good flor film will result in a *finos*, while oxidative aging will produce *Oloroso. Oloroso* are rather dark wines of a full-bodied type. The *finos*, which gradually shift towards oxidative maturation (due to lack of micronutrients resulting in the depleted flor film) produces the intermediate types of sherry classified as *Palo Cortado* and *Amontillado.*

The young *finos*, after the first year of maturation, are allowed to mature as unblended wines, usually for about a year. The collection of butts of such wine is called an *anada*. The casks are filled to approximately 80% of their capacity for the free development of the *flor*, and the concentration of alcohol is kept between 14.5 and 15.5% by volume. *Fino* sherries remain in an *anada* until they are required for a highly structured fractional blending system known as the 'Solera System'.

33.2.4.5.1 The Solera System

The solera system is essential to the production of sherries as it allows a unique and complex system of fractional blending during aging and the consistent wine mixture from different *soleras* give a uniform end product. The frequent transfer of wine *via* many stages of fractional blending is required for the production of *fino* sherry. An array of barrels that are arranged in hierarchical way, as tiers or scales, is called a solera system. The last tier in a collection of butts contains the wine which is mature and ready to use, also known as the *solera,* whereas the rest of the tiers are called *criadera*. Approximately, one-quarter of the wine in the last (oldest) *criadera* is removed (a process called *saca*) up to three times per year in this system. Furthermore, the wine which is removed is then fined and bottled. The quantity of the wine removed from the *solera* is replaced with the same quantity from the next oldest *criadera*, which again is filled by an equivalent volume from the next *criadera* and so on. This process of replacing the old *criadera* with the next-oldest one continues until the youngest *criadera* is reached. The volume in the youngest *criadera* is stocked up by the addition of new wine named *sobretabla*. This process of taking away a fraction of the wine and replacing it with wine from the younger scale is called a *rociar* or 'to wash down'. The age of the wine at a particular stage cannot be determined accurately due to the complexity and nature of the solera system, although the average age can be estimated mathematically.

Aging (maturation) is always achieved in two stages. The first *criadera* in the system is the *anadas*, which is static (matured for one year without blending), and the second phase is said to be dynamic, consisting of several scales with aging stages, the oldest blend being called *solera,* followed by first, second and third *criadera,* in the case of a four-scale system. In both the static and dynamic phases, wine is kept isolated from the air by the surface film of yeasts, which develops an oxidative metabolism, resulting in most of the sensory properties of sherries. The yeast film formed plays a significant role in the biological aging of sherry wine in the *solera* system. The blending with young wine provides necessary micronutrients to support the yeast. The absence of these micronutrients will result in depletion of the flor, and, hence, oxidative maturation of the wine. Also, the flor yeast metabolism contributes greatly to the organoleptic characteristics of the sherry. Low oxygen conditions produced by the yeast, due to the biofilm, together with inhibitory concentrations of ethanol in the sherry wines prevent the development of contaminating microorganisms. *Saccharomyces* yeasts, being most ethanol tolerant, can proliferate during sherry maturation. Temperature and ethanol strongly influence the state of the flor yeast on the surface of *fino* sherry wine. Higher temperature and ethanol content deteriorate the *flor* film and maturation process.

33.2.4.5.2 Physico-Chemical Characteristics of Sherry Wines During Aging

The oxidative (respiratory) metabolism of velum yeasts, which use and metabolize several substances, such as ethanol or glycerin (glycerol), change both the sensory characteristics and the general composition of the wine. The development of acetaldehyde during maturation of sherry wine is generally an indicator of overall flavor development. Once the flor formation takes place on the wine surface, the film is allowed to grow relatively undisturbed in all stages of the solera. The butts are seldom emptied, hence the sediment, composed of the film that has settled to the bottom of the *butts*, is allowed to accumulate. The sherry clarification process is carried out using bentonite, followed by filtration.

The decolorization of wine may be achieved using activated charcoal. The consecutive maturation takes place in large wooden cooperage that ensures that the wine takes on slight oaky characteristics. The fine quality wine is aged for twenty or more years. The esters, ethyl acid succinate and a few alcohols noticeably increase during aging. Thus, aldehydes, esters, acetals and higher alcohols are all involved in the *flor* sherry bouquet. The partial oxidation of ethanol to acetaldehyde takes place using alcohol dehydrogenases, while partial degradation of the glycerol also occurs (from 7–8 g/L to 2 g/L in 8–10 years). During aging of the wine, the amount of different amino acids, mainly proline, leucine, valine and phenylalanine, decreases, whereas the amount of acetoin, 2,3-butanediol increases. The diallyl acetate formation, along with the acetaldehyde content, contributes to the focal aroma of the sherry.

33.2.4.6 Sweetening and Blending

The fermentation and maturation of each of the sherries, whether it is *Fino, Amontillado* or *Oloroso*, takes place in the *anada* and *solera*. *Fino* sherries are consumed dry or very slightly sweetened, but, outside the Jerez area, *Amontillados* and *Oloroso*, particularly blends of more than one basic type, are usually sweetened. The wine is blended by a process known as *cabeceo*, in which the final blend is assembled firstly on a small scale and later applied proportionally to the wine on a large scale. Traditionally, the sweetened wines are also darkened in color due to the use of coloring matter. All but the very cheapest sherries are sweetened by the use of special sweetening wines, *PX* and/or *Mistela*. The *PX* wine has an alcohol concentration of about 9% (v/v) and a Baume value of 22°, with a full color and the strong characteristic flavor of variety of the grape from which it is made, whereas *Mistela* contains about 13.5% (v/v) alcohol and has a Baume value of about 8°, and is lighter with a less intense varietal flavor. Coloring is adjusted with a very intensely darkened caramelized grape must.

33.2.4.7 Clarification and Stabilization

The unstable coloring matter and proteins are removed and a high degree of stabilization is achieved using individual dry-sherry components and the *mistelas* after passing through *criaderas* and *soleras*. However, this equilibrium is again disturbed by wine blending, resulting in the wine becoming cloudy. Therefore, the clarification and filtration of blended wine is usually achieved generally using clarifying agents, such as gelatin, bentonite, Spanish clay, isinglass or egg whites. Furthermore, the use of charcoal can also make the color lighter. Wines are refrigerated for stabilization to avoid tartrate deposition and the clouding of bottled wine. The precipitated matter can also be removed by using diatomaceous earth filtration.

33.2.4.8 Addition of Sulfur Dioxide

The SO_2 content of the final wine blend should be around 100 ppm to avoid spoilage by *Lactobacillus trichodes*.

33.2.4.9 Finishing

Before bulk bottling or shipment of the wines, sherry undergoes a polishing filtration. The wine is filtered through filter pads which do not add calcium or asbestos to it.

33.2.4.10 Bottling

Greenish- or brown-colored bottles are utilized to pack sherry and are generally closed with screw caps. Mainly, the corks which can be removed without using a cork puller and are basically used for re-closure, are used for premium sherries. The *fino* sherries require special handling care to prevent problems related to its browning, therefore bottling and corking is done under inert gases.

33.2.5 Quality of Sherries

Analysis of Spanish sherries and California sherries for quality characteristics is carried out extensively and a comparison of different sherries is shown in Table 33.1. The Spanish and the California sherries have similar alcohol contents. The ash content of the Spanish sherries is higher than that of the Californian sherries, due to the use of gypsum (plaster) in their production. The Spanish sherries have higher aldehyde content than the American sherries, owing to the content of *flor* yeast used in their production.

33.3 PORT

33.3.1 Definition and Characteristics

Port is a sweet fortified wine which is prepared from grapes, red or white, growing in northern Portugal in the deep valley walls slopping down to the River Douro. Port wines were originally shipped out of the Portuguese town of Oporto and the true ports that are produced in this region are labelled 'Porto' rather than 'Port'. The production and export of port is mainly regulated by the *Insitituto do Vinho do Porto*. The area in which grapes for port are produced and vinified

TABLE 33.1
Composition of Various Sherries

Type and Source	Alcohol (%)	Ash (g/100 mL)	Total Acid as Tartaric Acid (g/100 mL)	Volatile Acid (g/100 mL)	Total Solids (g/100 mL)
California dry	19–20.64	0.228–0.400	0.332–0.497	0.042–0.079	1.58–4.83
Spanish, fino or amontillado	17.74–20.72	0.37–0.49	0.36–0.53	0.48–0.124	1.45–4.92
Spanish, oloroso	18–20	0.37–0.52	0.42–0.57	0.073–0.120	5.3–12.7
Eastern USA, sherry	15–20	0.17–0.30	0.31–0.480	0.046-0.110	2–7.5

Source: Amerine *et al.* (1980)

constitutes the world's oldest demarcated wine area and the demarcated Region of Douro, the Alto Douro, stretches some 100 km upstream from Oporto along the River Douro's valley from *Barqueiros*, to *Barca de Alva* and along the valleys of its tributaries, the Corgo, Jorto, Pinhão, Tua, etc. The wines must be aged, either within the territory or in Vila Nova de Gaira, the city opposite Oporto on the mouth of the Douro, in order to qualify for a certificate of origin. Various types of ports, such as white port (sweet or dry), ruby and tawny ports, are available on the market. A great dessert (aperitif) wine, the 'tawny', is the most versatile of ports. All of these type of wines do not show any date on their label. The port wines which have an indication of age are usually tawnies of very good quality. After getting special approval from the Port Wine Institute, port wine can state one of the indications of age (*i.e.* 10, 20, 30 and 40 years of average age) on their label, besides the date of bottling.

33.3.2 Viticulture

Port is made in an approved area, and its viticulture is controlled by an organization known as *Casa do Douro*, a guild of all the farmers with a government-appointed directorate. The vineyards extend between two and five kilometers from the riverbank to the North and South and also into the lower valleys of the main tributaries of the Douro. Every year, in accordance with the location, the nature of the soil and slope, viticultural practices, varieties etc., the Casa do Douro assign licenses to all the registered farmers for producing a fixed quantity of fortified wine, as per their classification (from A to F, *i.e.*, from the best to the worst), called the *benefício* system. The A and B graded vineyards are authorized to convert a comparatively larger proportion of their fruit into port than the E and F grades. The ranking also affects the set price. Out of more than 30 varieties growing in Douro region, only few varieties, such as '*Tinta cao*', '*Touriga nacional*', '*Alvarelhao*', '*Souzao*', '*Bastardo*', '*Tinta barroca*', '*Tinta roriz*', etc., are utilized commonly for port production.

33.3.2.1 Soil and Climate

In the Douro Valley, the best quality must is produced from the grapes on the slopes of the hills above the river. The valley floor and lower slopes are often composed of alluvial sand, grapes from which give a rather low-grade must while those from the hill-top give a 'green' (*i.e.*, an acid) must with a lower alcohol content. The preferred aspect is facing South-West, where the grapes are exposed to sun until late in the evening. In the Douro, the favored soil is based on schistous rock, which absorbs and holds moisture well, and is found on the lower river valley slopes.

The unique nature of the climate in the Douro is due to its location where the mountains protect the region from the humid westerly winds that blow in from the Atlantic. It is characterized by very cold winters and hot dry summers, with variable rainfall throughout the region but in regular amounts all the year. Rainfall is fairly heavy from November to the end of February, with frequent showers from February to April and occasional showers from May to September. Frost is common and frequently severe during January and February. The mean maximum temperatures vary between 33°C at Tua and about 30°C at Regua. A drying wind, known as 'Suao', sometimes blows, which is harmful to small and green grapes.

33.3.2.2 Vineyards and Grape Varieties

Mostly, vine planting is carried out in either horizontal rows or on vertically aligned rows. The horizontal-row plantations are carried out where the slope permits, for instance, on hand-built stone terraces, bulldozed earth terraces, or in unterraced plots, whereas the unterraced slopes are used for vertically aligned row plantations. Modern vineyards are often on wider terraces with about nine rows of vines. The type and blend of grape varieties play an essential role in the quality of the wine. Generally, most vines consist of an approved red port variety of *Vitis vinifera*: '*Bastardo*', '*Mourisco tinto*', '*Tinta amarela*', '*Tinta barroca*', '*Tinta francisca*', '*Tinta roriz*', '*Tinta cao*', '*Touriga francesa*' and '*Touriga nacional*', whereas white grape varieties include '*Esgana cao*', '*Folgosao*', '*Gouveio*' (Verdelho), '*Malvasia fina*', '*Rabigato*' and 'Viosinhoo'.

33.3.3 Technology of Port Production

The production of port involves several operations, as depicted in Figure 33.2 and described here.

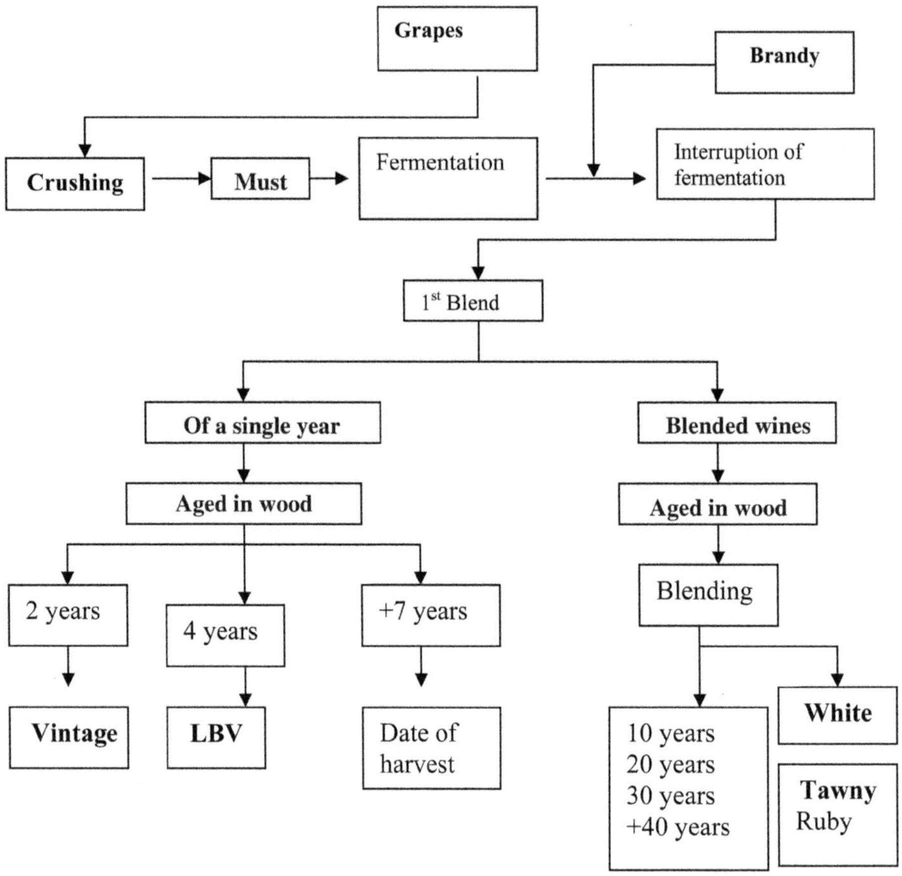

FIGURE 33.2 General outline of preparation and aging of port wine

33.3.3.1 Grape Crushing and Must Preparation

The conventional methods used for preparing certain types of port wine comprise the crushing of grapes in *lagares* after their destemming (separation from the stalks). Both the color and the flavor are completely and rapidly extracted from the berry skins for the production of good-quality port. Traditionally, the treading operation was carried out by men and women, whereas, nowadays, it is carried out using mechanical equipment that simulate the action of the feet. Basically, the centrifugal type and the roller crushers are two main mechanical crusher types used for crushing purposes. Both types of crushers are electrically driven or powered by an internal combustion engine when electricity is not available. But, small installations, manually operated roller crushers are still employing for this purpose. During the transfer of crushed grapes into the fermentation tanks, sulfur dioxide is introduced as potassium metabisulphite (KMS) or as a solution of sulfur dioxide (SO_2) gas (50–150 mg/kg grapes). Before pressing and inoculation, heating of must to 70–75°C for 15 min is done, *i.e.*, thermovinification.

33.3.3.2 Fermentation

The selected wine yeast strain may be inoculated into the fermentation tanks, but most producers still rely on the selection of desirable strains from the grape's natural microflora and on equipment used in the winery by additions of SO_2. The favorable temperature for fermentation is maintained between 26 and 28°C. Initially, *lagares*, *i.e.* shallow, granite troughs were used for carrying out fermentation, and barelegged workers were employed for regular treading sessions for maceration. Some wineries still practice these traditional methods. But nowadays, various forms of cement or steel fermentation tanks are used for wine preparation. Also, the use of autovinifiers of an Algerian design is widespread. Autovinifiers with certain modifications for allowing the operation of a mechanical helical device, named a *remontadore*, have also been employed in some installations. Greater extraction of anthocyanins can be achieved by fermenting musts at temperatures above 28°C for darker and highly regarded young wines. Some wineries continue to advocate *lagares* and treading to vinify some of their best fruit. The fermentation is generally checked by estimation of the specific gravity of the juice. The fermenting liquid is separated from the solids when about 4–5% (v/v) alcohol is produced, and immediately mixed with spirit in a wooden, steel or cement vessel. For pressing the grape solids wet mass, either a continuous screw press, a pneumatic press, a horizontal piston or the widely employed continuous press is used.

33.3.3.3 Fortification

The fortification is carried out using unique brandy (about 75% v/v alcohol), prepared by the *Casa do Douro* or the *Junta*

Nacional dos Vinhos to make the port wine using either casks or vats. In either case, proper mixing is done. Addition of brandy to grape juice, *i.e.*, must, also stops the fermentation process. The brandy, being non neutral, provides a different character to the finished wine as compared to the one produced by using neutral alcohol. It imparts specific sensory characteristics, improved chemical stability and helps in controlling the appropriate degree of sweetness in the wine.

33.3.3.4 Blending

The shipper gradually blends these wines together to produce fewer types of wine, unless the wine is to be marketed as a vintage wine. Blending is fundamental to the quality and style of port. During blending, a fraction of the prior lot is incorporated for consistency. The sweetness of wine is adjusted, using *geropigas* and dry wines. The reserves of previous lots, consisting of blends of different previous years, comprise the backbone of the future lot. However, the reserves can be noticeably different from the final blend in terms of style, and this type of system is referred to as a partial blending system.

33.3.3.5 Clarification

After fortification, the port undergoes the natural process of clarification for three to four weeks, or longer, in most of the wineries. Although port usually clears rather rapidly, as compared to other fortified wines, during the extraction of color using heat, adding pectic enzymes is crucial. Fining using proteinaceous agents, like gelatin, casein, egg white, or the clay bentonite, are used for clarification purposes. After fining, the wine usually settles down clear or brilliantly clear, although further filtration may be needed to remove small flocks of bentonite or pulp etc. Blending follows the filtration process so as to obtain a standard port with desired sugar content. The final blend is fined after 4 weeks to 6 months before bottling and shipment. Traditionally, egg white has been employed to fine, although isinglass is also popular as a fining agent.

33.3.3.6 Stabilization

The cold stabilization of port like ruby wines, is generally done for the removal of unstable tartrates, colloidal material and coloring matter. The two common systems used for this purpose are channeling the wine *via* a heat exchanger and an ultracooler to achieve a temperature reduction of up to 8°C, followed by holding the wine for about 1 week or so inside an insulated or cooled tank. The second system, *i.e.*, the continuous system, comprises chilling and passing of the wine continuously *via* a crystallizing tank. Furthermore, diatomaceous earth, followed by sheets and cartridge membrane, are used for the depth filtration of wine.

Excessive refrigeration may cause loss of color and body, so careful control is necessary at all stages. The use of gum arabic or gum acacia in the wine to act as a colloidal protector is common. But the use of SO_2 (50 or100 mg/L) has a marked stabilizing effect if added to a finishing blend just before bottling. The flavor of the wine may be impaired so care must be taken during stabilization. Wine storage at a temperature of −9°C for 6 to 7 days, along with filtration and flash pasteurization, solves the major hurdle of prevention of deposits in the port wine bottle during its stabilization process.

33.3.3.7 Aging and Maturation

The duration of the aging process varies from a few weeks to several years, depending on the type of wine to be prepared. The aging of the wine is characterized by earlier operations of fining, filtration and refrigerating, along with pasteurization. For rapid aging of the wine, numerous repeated cycles may be performed, particularly when a wine is in high demand. Aeration generally speeds up aging. Of the processes involved in aging, only oxidative changes can be intensified by quick aging. Thus, quickly aged wines are usually slightly oxidized but generally weaker in color.

The deep purple-red-colored young red port is usually astringent and harsh. The water- soluble pigments obtained from the grapes, the anthocyanins, principally impart color to the port, just like with other red table wines. Afterwards, leaving young wines undisturbed for two to three months usually leads to an increase in their visible color. This phenomenon is known as 'closing up'. The formation of aldehyde-bridged polymers between anthocyanins and other phenolics causes this effect. The maximum levels of acetaldehyde can be gained by stopping the process of fermentation at its most active phase, which otherwise would later be reduced to ethanol. The wine assumes a characteristic tawny color after several years of its wood storage. These color alterations are accompanied by a softening of astringent and fiery characters and the development of enhanced complex nutty and related flavors associated with wood extraction. The increase in acetate and other esters with time takes place by esterification and formation of ethyl lactate, diethyl malate and diethyl succinate that can be used as indicators of wine age. The amount of 5-(hydroxymethyl)furfural produced by fructose degradation under acidic conditions is an indicator of wine age, though it makes only a slight contribution to flavor. There are two types of port which depend upon their method of aging, *i.e.*, bottle aged (Vintage and Single Quinta Ports) and barrel aged (all other ports).

33.3.3.8 Bottling

The membrane filtration of the wine is performed prior to its bottling or bulk shipment. Apart from vintage ports, stopper corks are generally used for all styles; for crusted and some late-bottled vintage ports, driven corks are employed. The process of aging is carried on inside the bottle. Filtration and cold stabilization are not employed for the bottle-matured styles, but the formation of heavy deposits mean that decanting of these ports is considered essential to final product quality.

33.3.4 Types of Port

33.3.4.1 Vintage Port

Vintage port is produced from a single harvest of excellent quality grapes with exceptional sensory characteristics. The wine is bottled between the 2nd and 3rd year of maturation and is matured in the bottle for many years.

33.3.4.2 Late Bottled Vintage (LBV) Port

LBV port is also produced from a single harvest of good quality grapes with good sensory characteristics. The wine is matured firstly in wooden barrels and then, later in bottles for four to six years.

33.3.4.3 Tawny Port

Tawny port is produced from a blend of good-quality red or white grapes. These undergo small barrel aging in a controlled oxidative process.

33.3.4.4 White Port

This type of port is prepared from good-quality white grapes. They vary from dry to very sweet.

33.3.4.5 Ruby Port

Ruby port is produced from red grapes. It possesses a deep ruby color and is bottled and consumed young.

33.4 VERMOUTH

33.4.1 Definition and Origin

Aperitif wines, known as 'vermouths', are produced using grape wine by addition of a blend of herbs and spices or their extract, and has been quite popular in USA and European countries, including Russia and Poland. The term 'vermouth' is derived from the German word 'Wermut' (English 'worm wood' which is the English name for the plant *Artemisia absinthium*, a herb used to produce absinthe). Vermouth is a fortified wine (alcohol 15 to 21 %, (v/v)) flavored with a characteristic mixture of herbs and spices, some of which provides an aromatic flavor whereas others impart a bitter flavor. It can be the sweet Italian type or the dry French type. The Italian type of vermouth has an alcohol content between 15 and 17% (v/v) with 12 to 15% (w/v) sugar whereas the French vermouth has an alcohol content of 18% (v/v), with 4% (w/v) reducing sugar. The dry type vermouth has a smaller amount of herbs and spices. Since early Roman and probably early Greek times, wormwood has been added to wine.

33.4.2 Herbs and Spices

A number of herbs and spices (Table 33.2) and their different parts, like the seeds, wood, leaves, bark, or roots in dry form, are utilized in vermouth manufacture. The categories classifying the important herbs and spices utilized for the production of vermouth are bitter, aromatic or bitter/aromatic.

33.4.3 Technology of Preparation

The main steps for vermouth production are depicted in Figure 33.3, and are discussed here.

33.4.3.1 Preparation of the Base Wine

The grape juice or its concentrate, according to the routine method (Chapters 32 and 37 of this volume), are used for the production of base wine. The base wine must possess characteristics of being sound, neutral and cheap for its conversion into vermouth. In France, fortified grape musts, referred to as *mistelas*, are favored for the preparation of vermouth but refined beet sugar is preferred by many Italian producers. Caramel is an essential ingredient in base wine, therefore must be prepared carefully. Wine with naturally higher acidity is utilized for the production of American vermouth. Citric acid is used as a total acidity regulator. For balancing the dilution of base wine with botanical extracts, the alcoholic content is kept higher. The fortification is done using neutral high-proof brandy.

33.4.3.2 Brandy Distillation

After fermentation, the wine is separated into different fractions. The base wine goes through the distillation process and the different fractions, referred to as 'head', 'heart' and 'tail', are separated. The 'heart' fraction of the distillate is retained and undergoes double distillation which increases the ethanol content in the brandy. The 'head' and 'tail' fractions of the distillate are discarded.

33.4.3.3 Methods of Flavoring the Base Wine

Extracts of different plant parts (Table 33.2), with medicinal, antimicrobial or antioxidant properties, are used for flavoring the base wine, which can provide health benefits to the consumer. The quality of these botanicals is affected by the climatic conditions under which the plants are grown, and the harvesting conditions. Separate extraction methods are required and employed for various herbs and spices. The extract can be made using a direct extraction method, preparation of flavoring concentrates, extraction and distillation. A type of fractional blending system is used by some companies to maintain the consistency in composition of the herb and spices extract. Brandy or alcohol extracts of spices and herbs are also available for flavoring the vermouth. Direct extraction is the simplest procedure for flavoring the base wine, wherein defined quantities of botanicals are added to the wine and left undisturbed till absorption of the desired flavors and aromas is achieved. The botanicals may be finely ground (sometimes giving undesirable flavoring) to accelerate the extraction. Although higher temperatures make the extraction process more rapid, the wine is generally stirred at regular intervals and may be heated or kept at room temperature. The extraction tank is usually covered to reduce excessive loss of volatile flavors and aromas. The partial extraction process is favored over complete extraction, as the latter may result in the extraction of undesirable flavors. Pressing of the spent material is also not preferred for similar reasons.

The botanicals are placed in a special vessel, separate from the extraction tank, for the preparation of the concentrated extract and, then the wine is circulated from the tank into the extraction vessel containing the herbs, until most of the desired flavors and aromas are extracted. It is then, used to

TABLE 33.2
List of Herbs and Their Plant Parts Used in the Production of Vermouth and Related Wines

Common Name	Scientific Name	Portion of Plant Commonly Used
Allspice	*Pimenta dioica or Pimenta officinalis*	Berry
Aloe (socotrine)	*Aloe perryi*	Plant
Angelica	*Angelica archangelica*	Root (occasionally seed)
Angostura	*Cuspar febrifuga or Cuspar galipea*	Bark
Anise	*Pimpinella anisum*	Seed
Benzoin, gum benzoin tree	*Styrax benzoin*	Gum
Bitter almond	*Prunus amygdalus*	Seed
Bitter orange	*Citrus aurantium* var.*amara*	Peel of fruit
Blessed thistle	*Cnicus benedictus*	Aerial portion + seeds
Calamus, sweet flag	*Acorus calamus*	Root
Calumba	*Jateorhiza columbo*	Root
Cascarilla	*Croton eleuteria*	Bark
Cinchona	*Cinchona calisaya*	Bark
Cinnamon	*Cinnamomum zeylanicum*	Bark
Clammy sage, common clary	*Salvia selarea*	Flowers and leaves
Clove	*Syzygium aromaticum*	Flower
Coca	*Erythroxylon coca*	Leaves
Common horehound	*Marrubium vulgare*	Aerial portion
Common hyssop	*Hyssopus officinalis*	Flowering plant
Coriander	*Coriandrum sativum*	Seed
Dittany of Crete	*Amaracus dictamnus*	Aerial portion+ flowers
Elder	*Sambucus nigra*	Flower (also leaves)
Elecampane, Common inula	*Inula helenium*	Root
European centaury	*Erythraea centaurium*	Plant
European meadowsweet	*Filipendula ulmaria*	Root
Fennel	*Foeniculum vulgare*	Seed
Fenugreek	*Trigonella foenum-graecum*	Seed
Fraxinella, gasplant	*Dictamnus albus*	Root
Galangal, galingale	*Alpinia officinarum*	Root
Gentian	*Gentiana lutea*	Root
Germander	*Teucrium chamaedrys*	Plant
Ginger	*Zingiber officinale*	Root
Hart's tongue	*Phyllitis scolopendrium*	Plant
Hop	*Humulus lupulus*	Aerial Portion+ flower
Lemon balm, common balm	*Melissa officinalis*	Flowering plant
Lesser cardamon	*Elettaria cordamomum*	Dried fruit
Lung wort, sage of Bethlehem	*Pulmonaria officinalis or Pulmonaria saccharata*	Aerial Portion+ flower
Lungwort lichen, lung moss	*Styeta polmonacea*	Plant (a lichen)
Marjoram	*Origanum vulgare*	Aerial Portion+ flower
Masterwort, hog's fennel	*Peucedanum ostruthium*	Root
Nutmeg and mace	*Myristica fragrans*	Seed
Orris, Florentine iris	*Iris germanica* var. *florentina*	Root
Pomegrante	*Punica granatum*	Bark of root
Quassia	*Quassia amata*	Wood
Quinine fungus	*Fomes officinalis*	Plant
Rhubarb	*Rheum rhapanticum*	Root
Roman chamomile	*Anthemis nobilis*	Flowers
Roman wormwood	*Artemisia pontica*	Plant
Rosemary, old man	*Rosmarinus officinalis*	Flowering plant
Saffron, crocus	*Crocus sativus*	Portion of flower
Sage	*Salvia officinalis*	Aerial portion+ flowers
Savory (summer)	*Satureja hortensis*	Aerial portion of plant

(*Continued*)

TABLE 33.2 (CONTINUED)
List of Herbs and Their Plant Parts Used in the Production of Vermouth and Related Wines

Common Name	Scientific Name	Portion of Plant Commonly Used
Speedwell	*Veronica officinalis*	Plant
Star anise	*Illicium verum*	Seed
Sweet marjoram	*Marjorana hortensis*	Aerial portion + flower
Thyme, garden thyme	*Thymus vulgaris*	Leaf
Valerian	*Valeriana officinalis*	Root
Vanilla	*Vanilla fragrans*	Bean
Wormwood	*Artemesia absinthium*	Plant
Yarrow	*Achillea millefolium*	Plant
Zedoary, setwell, curcum	*Curcuma zedoaria*	Root

Source: Joslyn and Amerine (1964)

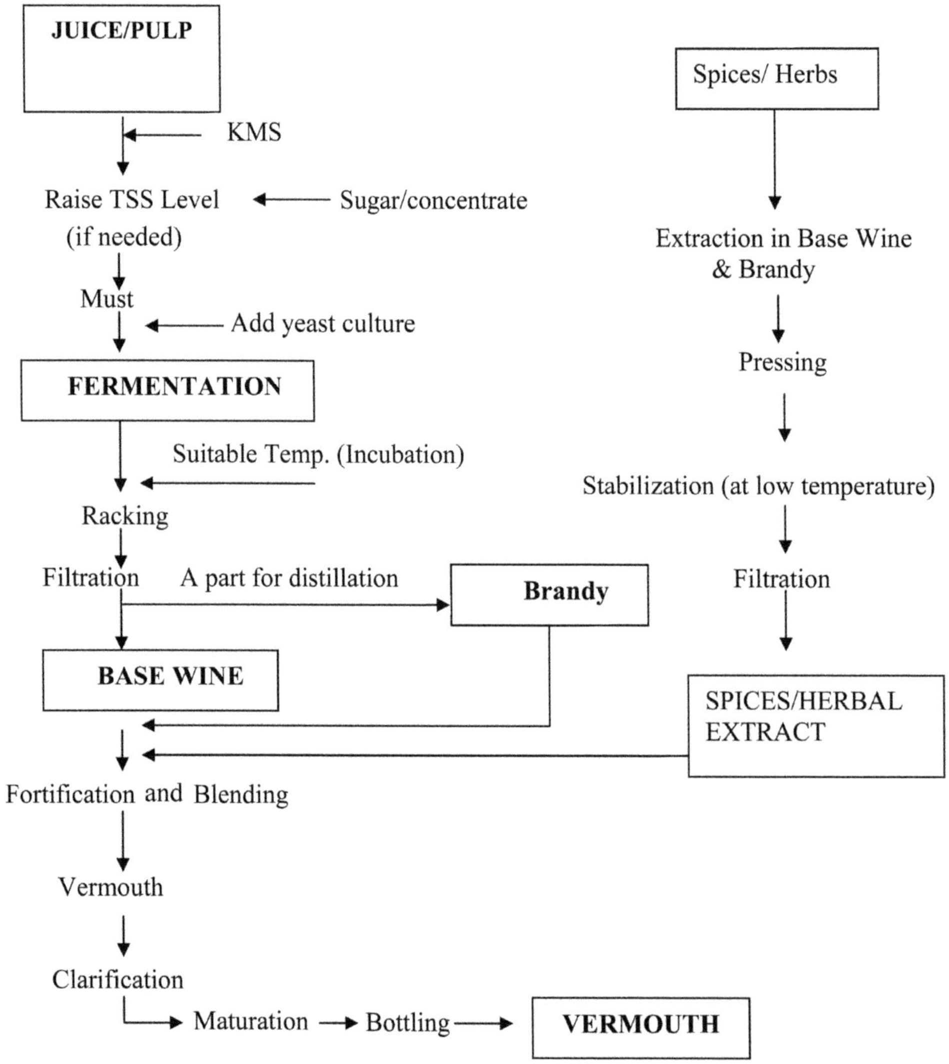

FIGURE 33.3 Flow diagram for the preparation of vermouth. *Source:* Joshi *et al.* (1999)

flavor a comparatively large volume of base wine. An amount of 0.5 to 1.0 oz of mixed dry flavoring materials per gallon of base wine is enough for flavoring the sweet vermouth (Italian style), while 0.5 oz of the botanical mixture is needed for the dry (French style) vermouth. The extract concentrate used in flavoring of the base wine can also be prepared using hot water. Nevertheless, when the initial softening of botanicals has been carried out using hot water, then the subsequent extraction by wine or brandy would be easy. The commercially available brandy or alcohol extracts can be supplemented to flavor a commercial lot of base wine for vermouth. For balancing the flavors in the base wine, which is flavored earlier using the direct extraction method or by adding wine extract concentrate, small volumes of either of these extracts are used.

33.4.3.4 Sweet Vermouth

Sweet vermouth contains high sugar levels (around 150 g/L; 12–15% reducing sugar), with alcohol content varying from 15 to 17% (v/v), total acidity about 0.45%, and tannic acid about 0.04%. Typical Italian vermouth is sweet, dark, amber-colored, having a light Muscat and a sweet nutty flavor, with a pleasing aroma and a generous and warming flavor, along with a slightly bitter but agreeable aftertaste. These wines should contain around 15.5% (v/v) of alcohol and 13% (w/v) or greater reducing sugar level. American vermouths usually have greater alcohol level and comparatively less sugar, compared with the Italian vermouth. Earlier, a fortified wine with a muscat flavor, such as that of the Muscat Blanc produced from grapes grown in Piedmont in northern Italy, was the preferred base wine but for other white wines are also used nowadays. An alcohol extract of different botanicals is generally used for flavoring of the base wine in Turin. In Italy, the liquid invert sugar or sucrose is added to the sweet vermouths for sweetening purposes, while caramel is used to provide the dark color. In France, direct mixing of the botanicals in wine for 7 to 14 days, with intermittent stirring and tasting of the wine during extraction, is normally done to flavor the base wine. The wine is drawn-off and filtered, if there is excessive bitterness or "herbaceous" character.

In California, sweet vermouth is produced using base wine, consisting of a light-colored fortified sweet wine, such as new angelica or white port. Grape concentrate or sucrose is used to maintain the appropriate sugar level of the wine, whereas acidity is regulated by citric acid. Sweet vermouth has a final alcohol content of 17% (v/v), total soluble solids (TSS) of about 13 to 14%, total acidity of about 0.45% and tannins about 0.04%. Caramel concentrate, darkened by heating, can be used for coloring purpose. Techniques such as pasteurization, refrigeration and filtration are generally used for vermouth stabilization, but cloudiness may develop on refrigeration, if licorice or catechu is used. Prolonged aging of vermouth is also undesirable.

33.4.3.5 Dry Vermouth

French-type (dry) vermouths are generally more bitter, with more alcohol and less sugar but are lighter colored than sweet vermouths. Comparatively lower levels of botanicals are required in dry vermouth than in sweet vermouth, although larger amounts of wormwood, bitter orange peel and *Aloe* (a bitter herb) are required as supplementary ingredients. In typical French dry vermouth, the alcohol content is 18% by volume, reducing sugar 4%, total acidity (as tartaric acid) 0.65% and volatile acidity (as acetic acid) 0.053%. In European methods, the recipe for dry vermouth contains more wormwood and bitter orange peel than does sweet vermouth. The bitter botanical, *Aloe*, is used as an extra ingredient. A white wine of Herault region is favored, which can be used after blending with white wine prepared using the Grenache variety of grapes. The wine must possess the characteristic of being sound, light- colored, with moderate acidity. The fortification of wine with superior quality high-proof brandy to approximately 18% alcohol content is preferred.

The French methods employ smaller number of individual botanicals than are used in the Italian vermouth (to a total of about 0.5 oz. per gallon). In one of the French methods, the fortified wine, of about 18% alcohol content, is left covered, with herbs and spices being placed in the extraction tank. The wine is then drawn-off after 30 to 40 days. This process of extraction is repeated a number of times, using fresh wine. Then, base wine (in an appropriate volume) is combined with the extracts to attain the preferred flavor level. After blending, refrigeration and filtration of vermouth is done. Then, grape juice, preserved with alcohol, also referred to as *mistelle,* or grape juice preserved with SO_2, also referred to as *mute*, is supplemented to achieve the preferred reducing sugar level. In California methods for dry vermouth, a neutral sauterne-type wine is usually favored as the base wine. It is mostly prepared by using a fortified sauterne lot having lower SO_2 content and 24% (v/v) alcohol content, with mixing of fortified wine with the 12–14% (v/v) alcohol content sauterne to give a blend containing 18–18.5% (v/v) alcohol. Use of carbon as a decolorizing agent in vermouth is not favored as the carbon absorbs flavoring and aromatic compounds.

33.4.3.6 Fortification and Blending

According to a proprietary formula, the base wine, brandy (to raise the alcohol content), extract of botanicals (flavoring) and sugar as caramel (coloring) are blended for every vermouth type. Both the alcoholic content (ethanol) and the concentration of dry extract affect the wine's viscosity.

33.4.3.7 Aging and Finishing

The sweet vermouths are aged for approximately 4.5 years or more after flavoring in Italy and France, but the time between the mixing of herbs and spices, and the bottling of vermouth is usually three to five years. Cold stabilization by refrigeration of the new vermouth is carried out, which is followed by filtration and then, aging. A longer aging period contributes to lowering the vermouth's quality. For maturation of the wine, generally oak barrels are used but other wooden barrels can also be utilized. Wooden barrels provide vanillin and other organic molecules, while absorbing some phenolics and tannins during aging. For prevention of spoilage of the wine, the pH should be maintained to a low level with an appropriately high SO_2 content.

33.4.3.8 Bottling

Before bottling, the sweet vermouth is filtered again. Spoilage of sweet vermouths can be restricted by increasing the sulfur dioxide content to more than 75 ppm. The dry vermouths are finished and bottled young, and a long aging period is not preferred.

33.4.4 Vermouth from Other Fruits

Fruits other than grapes, such as mango, plum, apple, pear, etc., have been used effectively for production of vermouth.

33.4.4.1 Mango Vermouth

Aromatized wines from mango, referred to as mango vermouth, have been produced. The base wine was prepared using 'Banganpalli' mangoes by increasing the total soluble solids (TSS) content to 22°Brix, adding 100 ppm SO_2, and 0.5% pectinol enzyme, and fermenting it at 22±1°C, using Montrachet strain 522 of *S. cerevisiae*, blending the wine with a herbs and spices mixture.

33.4.4.2 Apple Vermouth

A method for the production of apple vermouth has been developed. Almost all the preferred characteristics required to produce vermouth are present in apple base wine. Apple vermouth with various ethanol concentrations (12, 15 or 18%, v/v), sugar content (4 or 8%, w/v) and botanical extract (2.5 or 5.0%) has been produced and assessed.

33.4.4.3 Plum Vermouth

Plum vermouth with commercial acceptability has also been prepared, using the botanicals. The addition of a herbs/spices extract raised the total phenols, aldehyde and ester content of the plum vermouth. The sensory assessment described sweet plum vermouth with 15% alcohol content as the best product.

33.4.4.4 Sand Pear Vermouth

The sand pear vermouth was prepared by extracting the fruit juice and converting it into vermouth. Both types of vermouth, dry and sweet, along with different alcohol levels, can be produced using sand pear base wine. The addition of botanical extracts during conversion of base wine into vermouth raised the levels of TSS, acidity, aldehydes, phenols and esters. The greatest acceptability of sweet vermouth was with 15% (v/v) alcohol content.

33.4.4.5 Tamarind Vermouth

Tamarind (*Tamarindus indica*) fruits can be made into vermouth of satisfactory quality. To produce vermouth, a base wine is prepared using tamarind fruit (50 g/L), with 0.9% acidity achieved by increasing the TSS to 23°Brix, SO_2 to 150 ppm and fermenting it using *S. cerevisiae* var *ellipsoideus* at a temperature of 27±1°C. Both dry and sweet tamarind vermouths with 17% alcohol content are acceptable.

33.4.4.6 Pomegranate Vermouth

Various spices (clove, cardamom and ginger) are used for the preparation of sweet vermouth using pomegranate ('Ganesha') wine base. Out of these, the pomegranate vermouth using cardamom had the greatest acceptability in terms of total acidity, sweetness, body and general quality, followed by ginger and then, clove vermouth. A three-month maturation period is required.

33.4.4.7 Apricot Vermouth

Both apricot base wine and vermouth using cultivated and non-domesticated apricot varieties have been produced with varying levels of sugar (8–12°Brix), alcohol (15–19%) and spices (2.5–5%). The apricot distillate aroma has a high alcohol concentration with specific terpenes such as linalool, ocimenol, α-terpineol, nerol and geraniol.

BIBLIOGRAPHY

Alexandre, H. (2013). Flor yeasts of *Saccharomyces cerevisiae*-their ecology, genetics and metabolism. *Int. J. Food Microbiol.*, 167(2): 269–275.

Amerine, M.A., Kunkee, K.E., Ough, C.S., Singleton, V.L. and Webb, A.D. (1980). *The Technology of Wine Making*. 4th edn. AVI Publishing Co. Inc, Westport, CT.

Bakker, J. (1992). The sensory and chemical characteristics of the colour of port. *Food Technol. Int. Eur.* 121–125.

Caregnato, M. (2006). Technology and tradition in port wine making and marketing terminology. In (thesis): *Il Vino di Porto fra Tradizione e Modernità – Un Contributo Terminologico in Portoghese e Italiano, SSLMIT – University of Trieste*, pp. 123–143.

Ferreira, A.C.S. and de Pinho, P.G. (2004). Nor-isoprenoids profile during port wine ageing-influence of some technological parameters. *Anal. Chim. Acta*, 513(1): 169–176.

Fonseca, A.M. da, Galhano, A., Pigmental, E.S. and Rosas, J.R.P. (1987). *Port Wine-Notes on its History*. Production and Technology. *Instituto do Vinho do Porto*, Porto.

Gonzalez Gorden, M.M. (1990). *Sherry: The Noble Wine*. 2nd edn. Quiller Press Ltd., London.

Goswell, R.W. and Kunkee, R.E. (1977). Fortified wines. In: *Alcoholic Beverages*, A.H. Rose (Ed.), Academic Press, London, pp. 477–534.

Genovese, A., Ugliano, M., Pessina, R., Gambuti, A., Piombino, P. and Moio, L. (2004). Comparison of the aroma compounds in apricot (*Prunus armeniaca* L. cv Pellecchiella) and apple (*Malus pumila* L. cv Annurca) raw distillates. *Ital. J. Food Sci.*, 16: 185–196.

Ibeas, I., Lozano, I., Perdigones, F. and Jimenez, J. (1997). Dynamics of flor yeasts populations during the biological ageing of sherry wines. *Am. J. Enol. Vitic.*, 48: 75–79.

Jackson, R.S. (1999). Grape based fermentation products. In: *Biotechnology: Food Fermentation (Microbiology, Biochemistry and Technology)*, Vol. II, V.K. Joshi and A. Pandey (Eds.), Educational Publishers and Distributors, New Delhi, pp. 583–647.

Jackson, R.S. (2000). *Wine Science – Principles, Practices, Perception*, 2nd edn. Academic Press, San Diego, CA.

Jackson, R.S. (2014). Specific and distinctive wine styles. In: *Wine Science: Principles and Applications*. R.S. Jackson (Ed.), 4th edn. Academic Press, Cambridge, Massachusetts, Chapter 9, pp. 628–677.

Jackson, R.S. (2017). Styles and types of wine. In: *Wine Tasting: A Professional Handbook*. R.S. Jackson (Ed.), 3rd edn. Cambridge, Massachusetts, Chapter 7, pp. 293–335.

Joshi, V.K., Sandhu, D.K. and Thakur, N.S. (1999). Fruit based alcoholic beverages. In: *Biotechnology: Food Fermentation*, Vol II. V.K. Joshi and A. Pandey (Eds.), pp. 647–744. Educational Publishers and Distributors, New Delhi.

Joshi, V.K., Chauhan, S.K. and Shashi, B. (2000). Technology of fruit based Alcoholic Beverages. In: *Post Harvest Technology of Fruits & Vegetables*, L.R. Verma and V.K.Joshi (Eds.), Indus Publishing Co., New Delhi, pp. 1019–1101.

Joslyn, M.A. and Amerine, M.A. (1964). *Dessert, Appetizer and Related Flavored Wines*. University of California. Division of Agricultural Sciences, Berkeley, CA.

Martinez, P., Perez, L. and Benitez, T. (1997). Evolution of flor yeast population during the biological ageing of sherry wines. *Am. J. Enol. Vitic.*, 48: 160–168.

Martinez, P., Valcarcel, M.J., Perez, L. and Banitez, T. (1998). Metabolism of *Saccharomyces cerevisiae* flor yeasts during fermentation and biological ageing of fine sherry: By products and aroma compounds. *Am. J. Enol. Vitic.*, 49(3): 240–250.

Mesa, J.J., INFante, J.J., Rebordinos, L., Sanchez, J.A. and Cantoral, J.N. (2000). Influence of the yeast genotypes on enological characteristics of sherry wines. *Am. J. Enol., Vitic.*, 51: 15–21.

Moreno, J.A., Zea, L., Moyano, L. and Medina, M. (2005). Aroma compounds as markers of the changes in sherry wines subjected to biological ageing. *Food Control*, 16(4): 333–338.

Muñoz, D., Peinado, R.A., Medina, M. and Moreno, J. (2007). Biological aging of sherry wines under periodic and controlled microaerations with *Saccharomyces cerevisiae* var. *capensis*: Effect on odorant series. *Food Chem.*, 100(3): 1188–1195.

Panesar, P.S., Joshi, V.K., Panesar, R. and Abrol, G.S. (2011). Vermouth: Technology of production and quality characteristics. In: *Advances in Food and Nutritional Research*, vol. 63,. Elsevier, Inc., London, UK, pp. 253–271.

Panesar, P.S., Joshi, V.K., Bali, V. and Panesar, R. (2017). Technology of production of fortified and sparkling fruit wine. In: *Science & Technology of Fruit Wine Production*, M.R. Kosseva, V.K. Joshi and P.S. Panesar, (Eds.), Academic Press, Cambridge, Massachusetts, pp. 487–530.

Peinado, R.A., Mauricio, J.C. and Moreno, J. (2006). Aromatic series in sherry wines with gluconic acid subjected to different biological aging conditions by *Saccharomyces cerevisiae* var. *capensis*. *Food Chem.*, 94(2): 232–239.

Pozo-Bayón, M.Á. and Moreno-Arribas, M.V. (2016). Sherry wines: Manufacture, composition and analysis. *Encyclopedia of Food and Health*, pp. 779–784.

Reader, H.P. and Dominguez, M. (1995). Fortified wines: Sherry, port and Madeira. In: *Fermented Beverage Production*, A.G.H. Lea and J.R. Piggott (Eds.), Blackie Academic and Professional, London, UK, Chapter 8, pp. 157–194.

Real, A.C., Borges, J., Cabral, J.S. and Jones, G.V. (2017). A climatology of vintage port quality. *Int. J. Climatol.*, 37(10): 3798–3809.

Siebert, T. (2017). Agro-processing of fermented beverages: From the vine to farmlink. *FarmacoBiz*, 3(5): 40–42.

Simal-Gándara, J. (2015). Sweet, reinforced and fortified wines: Grape biochemistry, technology and vinification. *J. Wine Res.*, 26(1): 64–66.

34 Sparkling Wine Production

Philippe Jeandet, Yann Vasserot and Gérard Liger-Belair

CONTENTS

34.1 INTRODUCTION

A wine is called sparkling if it is surcharged with carbon dioxide (not less than 5 g/L at 20°C). There are four types of sparkling wines, of which type I includes wines with excess carbon dioxide (CO_2) produced by fermentation of residual sugar from primary fermentation. Type II wines include those with excess CO_2 produced from malolactic fermentation (MLF), whereas type III wines are characterized by excess CO_2 produced from fermentation of sugar added after the primary fermentation, with most of the sparkling wines of the world being produced by this method. In the type IV wines, excess CO_2 is added and these wines include the carbonated and crackling wines. Amongst the sparkling wines, champagne is the most famous. The focus in this chapter will specifically be on various aspects of champagne production, such as the difficulty in initiating MLF in Champagne wines, the magical presence of bubbles, effervescence (related to gas discharge from a liquid by bubbling), quality of the product (related to the size of the bubbles formed in the flute) from its appearance when and after pouring in the glass, and the formation of the foam ring (the collar) on the liquid surface as a result of bubbles formed in the glass.

34.2 PRODUCTION METHODS OF SPARKLING WINES

Several methods are available to transform the base wine into a sparkling wine, such as the *méthode champenoise*, the *Crémants* from France and Luxembourg, the *méthode traditionnelle* (formerly the *méthode champenoise*), *e.g.*, used for Cavas, the transfer method, the *méthode ancestrale* (Limoux, Gaillac), also including the *Dioise* method and the bulk method (*Cuve Close*).

34.2.1 Champenoise Method

Production of sparkling wines from France and Luxembourg uses the *champenoise* method for both the base wine and the *prise de mousse*. Differences from Champagne winemaking occur, especially when considering the separation of juices after pressing the whole grapes. The second phase or second fermentation and aging in the bottle is very similar to that of Champagne wines, except for the fact that *Crémants* have only twelve months aging in contact with lees. Theoretically, Cava production uses the *méthode traditionnelle* (formerly the *méthode champenoise*), but the base wine is rarely obtained from pressing whole grapes, which is different from the Champagne method. The most commonly used technique for making the base wine is similar to that of a traditional white wine. The second phase is exactly the same as for Champagne or *Crémants*. Aging on lees usually takes nine months for Cava wines.

34.2.2 The Transfer Method

In the transfer method, the base wine (obtained by a traditional white winemaking method) is fermented and aged on lees in the bottle (as for the *méthode traditionnelle*) but there are no constraints on riddling (a crucial, expensive and lengthy operation) and disgorging. After bottle fermentation and proper aging, the bottles are automatically emptied into a steel tank without degassing, since the wine is maintained at an isobarometric pressure. At this stage, the dosage (addition of sugar) can be directly added in the tank (scheme 1) but winemakers generally prefer adding the dosage in another steel tank after having filtered the wine (scheme 2). After standing for several days, the wine is filtered and bottled (scheme 1) or just bottled (scheme 2). All operations are carried out under a carbon dioxide atmosphere using isobarometric bottling. Advantages of the transfer method are the suppression of the lengthy operation of *remuage* and the fact that the dosage applied is more uniform. However, this method is expensive and energy-consuming, and a real risk of oxidation of the wine exists.

34.2.3 *Méthode Ancestrale*

The *méthode ancestrale*, which is a production method but very difficult to control, has been developed in the vineyards of Limoux and Gaillac. The base wine is made with whole grapes (mainly of the Mauzac variety) or through a traditional white winemaking step, using a semi-fermented wine. In fact, sugars are used for both the primary alcoholic fermentation and the second fermentation. At various steps of the process, it is essential to stop fermentation each time it starts to accelerate. In this way, refrigeration (down to 0°C), sulfiting, depletion of yeast nutrients (using settling, fining, filtration or centrifugation) are repeated as many times as necessary to regulate or stop the activity of the yeast. The wine is then filtered and kept at 0°C until springtime. At this step, the second fermentation takes place (2–3 months) in the bottle (at rigorously controlled temperatures) with yeast and the remaining sugars of the semi-fermented wine. Riddling and disgorging (without dosage) then take place, but the wine can also be sold with a slight yeast deposit at the bottom of the bottle. These wines receive the *Blanquette Méthode Ancestrale* Appellation of Controlled Origin.

For the *Dioise* method, the same principles of winemaking as for the *ancestrale* method are used, wherein a semi-fermented wine from the '*Muscat à petits grains*' variety is filtered and refrigerated at 0°C, and the second fermentation in the bottle (using sugars remaining after the first fermentation) is carried out. To increase the extraction of aroma compounds, pectinolytic enzymes are added to 'Muscat' grape berries in the crusher which requires fining treatments since the must obtained exhibits high turbidities ranging from 1000 to 1500 NTU. The flotation technique is used to clarify the must. After the second fermentation is stopped by refrigerating the cellar, bottles are emptied into a steel tank maintained at an isobarometric pressure under a CO_2 atmosphere to avoid degassing (as for the transfer method). After filtration, the wine is bottled using isobarometric bottling. The final alcoholic content of the *Clairette de Die* is of approximately 7.5° with 40 to 50 g/L residual sugars. The production of *Asti spumante* follows the same process, with this wine being made from the same grape variety.

34.2.4 Bulk Method

The bulk method (*cuve close*), otherwise known as the tank method or charmat process, is a simpler and more cost-effective technique that has been developed to obtain ordinary low-cost wines. The second fermentation does not take place in the bottle, but the base wine is sent to a reinforced steel fermentation tank able to contain several hundreds of hectoliters of wine. Yeast and sugars are added, and the wine is maintained at a temperature of 20 to 25°C. The *prise de mousse* duration does not exceed ten days. The second fermentation is stopped by a light sulfiting and by refrigerating the wine at –2°C. After having been cold-stabilized at –5°C for several days, the wine is filtered at a low temperature and then bottled using isobarometric bottling. One disadvantage of this method is that there is no aging of the wine, as the contact of the wine with the lees is insufficient.

34.3 CHAMPAGNE: PREAMBLE AND PRODUCTION TECHNOLOGY

34.3.1 Champagne

Champagne is undoubtedly the most prestigious effervescent wine in the world and up to 300 million bottles of this wine type are produced each year. Tradition has made this sparkling wine unique, authentic and a symbol of lifestyle and any form of celebration. Champagne would not be Champagne without the magical presence of the bubbles linked to the CO_2 formed during the secondary fermentation, which is a characteristic of the *méthode champenoise*. It is a bottle-fermented sparkling wine, where its secondary fermentation takes place in the bottle. Champagne winemaking is a long and complex process. Quality regulations have built up this very famous wine throughout three centuries. The various steps involved in champagne production include base wine production, sugaring, yeasting, bottling, blending stabilization, remuage and disgorging. A proper blending of base wines is necessary before secondary fermentation is carried out.

34.3.2 The Grape Varieties for Champagne

Champagne is made from three main varieties: 'Pinot noir', a black grape variety that gives Champagne wines their aromas of red fruits, as well as their strength and body; 'Pinot meunier', another black cultivar characterized by its suppleness and spiciness, giving Champagne wines their roundness and fragrance; and 'Chardonnay', a white grape variety, providing Champagne wines with their finesse, floral and, in some instances, mineral notes very typical of the *Blancs de Blancs* Champagnes (Champagnes made with only Chardonnay grapes). The Champagne vineyard belongs to the so-called *septentrional* (northern) vineyards and this particular location leads to cold growing conditions, making the maturation process (ripening of the grape berries) difficult. The stage of maturity of the grape berries is a fundamental parameter which determines the quality of the wines. Thus, a detailed understanding of the physiological and biochemical events of the maturation process in grapes is of great interest, so that determining the changes in sugar and organic acid contents of grape berries during ripening in the Champagne vineyards would be valuable for estimating the harvest date (See Chapter 9 of this volume for more details).

34.3.3 Base Wine Preparation

Champagne wines are produced in two steps: the base wine is made first, after which a second fermentation is initiated in the bottle (*prise de mousse*). Champagne is typically a white wine whose grapes are hand-harvested and pressed very lightly so as to limit the extraction of phenolic compounds from the skins and the stems, *i.e.*, to avoid the diffusion of color (anthocyanins) into the must. Some three-quarters of this wine are obtained from two black varieties, 'Pinot noir' and 'Pinot meunier', the other quarter being obtained from the white variety, 'Chardonnay'. In order to obtain high-quality musts, proper picking and pressing conditions are required. For example, sorting to remove rot-attacked grape clusters is needed during picking, since it has been shown that gray mold, caused by the phytopathogenic fungus, *Botrytis cinerea* Pers., can have deleterious effects on the foaming properties of Champagne wines (See Chapter 12 of this volume for details). Grapes must be transported as intact as possible to avoid skin maceration in the juice and the oxidation phenomenon. The entire grape clusters are pressed lightly (with pressures ranging from 1.5 to 2 bars) without prior crushing. Strong regulations enacted in 1993 have specified the extraction yield of juice: 4000 kg of grape bunches produce 25.50 hL of must, the first 20.50 hL constituting the *cuvée* and the last 5 hL constituting the *taille*. Only the *cuvée* will be used for producing quality Champagnes; the *taille* is destined for distillation or is vinted apart from the *cuvée* to be possibly used for blending. The *taille* is characterized, in comparison with the *cuvée*, by a decrease in the total acidity (tartaric and malic acids), an increase in the mineral and phenolic concentrations, together with an increase in the pH. Finally, the *taille* is a heavier and less fine product than the *cuvée*, explaining why it is rarely used for winemaking. As soon as the juice is extracted, SO_2 is added at doses varying from 3 to 8 g/hL, depending upon the pressing fraction (*cuvée* or *taille*) and the level of gray mold in the vineyard. Musts with turbidities of 200 to 400 Nephelometric Turbidity Units (NTU) are clarified by centrifugation or after static settling, *débourbage* (18 to 24 hours at temperatures ranging from 6 to 15°C), using tannins, bentonite or casein (+ bentonites) as fining agents. During static settling, some must proteins and pectins, together with mineral cations and phenolics are removed. The mechanism by which proteins are removed from musts or the wine by the use of bentonites, *i.e.*, interactions between bentonites and proteins, are beginning to be precisely understood.

34.3.4 Primary Fermentation

Once musts are clarified, the alcoholic fermentation occurs after inoculation (10–15 g/hL) with selected strains of *Saccharomyces cerevisiae bayanus* (mainly strains DV 10, IOC and Levuline CHP). Champagne wines are generally, though not always, chaptalized (*i.e.*, sugar is added). Alcoholic fermentation usually takes place in stainless steel tanks at temperatures ranging from 15 to 18°C (sometimes fermentation is conducted at lower temperatures, *e.g.*, 13°C). Since musts or wines may be stained by anthocyanins in the case of 'botrytized' or overmature grapes, charcoal is commonly used as an agent for reducing color in colored musts and wines. Since charcoal has detrimental effects on the organoleptic and foaming properties of Champagne wines, it could advantageously be replaced by yeast lees recovered after alcoholic fermentation of 'Chardonnay' musts. Yeast lees are also able to adsorb some of the volatile sulfur compounds found in wines, as well (Refer to Chapter 32 of this volume, for more details on red and white wine production).

34.3.5 Malolactic Fermentation

Malolactic fermentation (MLF) of base wines is now widely used in Champagne to avoid the occurrence of this fermentation during bottle fermentation (*prise de mousse*) or during the aging of wines on lees, which may lead to an increase in the volatile acidity and the appearance of lactic notes in wine tasting. It is achieved by inoculating lactic acid bacteria (*Oenococcus oeni*), using a fermenting starter containing 10^7 cells/mL. In some Champagne houses, MLF is not practiced, in order to retain the fresh and fruity characteristics in the wine, and could be avoided by using high SO_2 concentrations (8 to 10 g/hL). MLF-inhibiting capabilities of lysozyme (added at the must or the wine level) and its possible effects on Champagne wine foamability have also been determined (See Chapter 18 of this volume for details). Because of the difficulty to readily initiate MLF in highly acidic wines like Champagne, the nutritional requirements and the metabolism of lactic acid bacteria under wine conditions will be discussed in Section 34.4.

34.3.6 Clarification

After completion of MLF, wines undergo clarification treatments (static settling or centrifugation + fining, including the use of bentonite, gelatin + [tannins or silica gel], casein + bentonite, charcoal + bentonite, fish gelatins or wheat gluten). Because of the concern associated with bovine spongiform encephalopathy, leading the public and winemakers to lose their confidence in the use of animal proteins in enology (gelatins, egg proteins, casein, etc.), efforts have been made to study the possibility to replace animal proteins with those originating from plants in fining treatments (see Chapter 29 of this volume). Wines at this stage are called vins clairs.

34.3.7 Blending

The practice of blending wines from different grape varieties, different origins (*crus*) and different years (reserve wines) is essential to maintain the quality of Champagne and the House style unique to every Champagne producer. Reserve wines are kept for two or three years (sometimes on lees) in tanks at 12–13°C and protected from oxygen; reserve wines may also be kept for a longer time (*e.g.*, ten years). The Brut Non-Vintage and the Demi-Sec Champagnes are usually a blend of wines from several years, from different grape varieties and a number of *crus*. Vintage Champagnes are produced exclusively from the wines of a single harvest.

34.3.8 Stabilization

Stabilization of wine with respect to potassium hydrogen tartrate (KHT) is a critical point in winemaking in Champagne. Before being stabilized, wines may be filtered on a simple, continuous earth filter. The stability usually required in Champagne wines corresponds to the temperature of –4°C. Briefly, KHT stabilization is obtained by treating the wine with artificial cold, using different technologies, such as slow cold stabilization without KHT crystal seedling, rapid cold stabilization including KHT crystal seedling by the static contact process or by the dynamic continuous process. The very expensive treatment of wines with artificial cold could be advantageously replaced by the addition of inhibitors of the crystallization process of KHT, such as metatartaric acid or carboxymethylcellulose (CMC). Such inhibitors increased the width of the supersaturation field of KHT in the wine, thus delaying tartrate salt precipitation in the bottle. Once the wine is stabilized, with respect to both KHT and colloids, it is filtered (on earth filters or cellulose cartridges). Wines obtained at this stage are called *vins de base champenois*.

34.3.9 Secondary Fermentation and Bottle Aging

The secondary fermentation in the bottle (the so-called *prise de mousse*) may occur by adding the *liqueur de tirage*, that is, saccharose (18–24 g/L, according to the CO_2 concentration desired), immobilized or freely suspended yeast (inoculation with 10^6 cells/mL), diammonium phosphates and riddling adjuvants (bentonite, alginate). The yeast used is acclimatized for lower temperature fermentation and with alcohol tolerance in the base wine. After capping with a crown stopper, the bottles are placed horizontally on laths (*sur lattes*) or directly in crates. This secondary fermentation is a slow (six to eight weeks) and steady low-temperature fermentation (11–12°C) characteristic of the so-called *prise de mousse*. The time interval between addition of the *liqueur de tirage* and expedition (completion of fermentation) must be, for Champagne, of at least 15 months. After the secondary fermentation is completed, Champagne wines undergo a long maturation (1.5 to 8 years, or even more!) on yeast lees. There is a wine enrichment in terms of compounds synthesized and released by the yeast. Autolysis, an enzymatic self-degradation process of yeast cell constituents, that takes place immediately after cell death at the end of fermentation, also occurs during aging. It consists of the release of some yeast constituents, such as amino acids, peptides, nucleotides, mannoproteins and aroma compounds that would influence the wine's sensory properties. Autolysis has a unique occurrence and significance during production of Champagne wines, that must have at least 12 months in contact with the yeast in the bottle. Gas exchanges may also occur during aging; specifically, there is a loss of CO_2 while O_2 is able to penetrate the bottle through the crown stopper, leading to a slight oxidation. But during the process, some sensory defects could occur especially when wines are submitted to light exposure (even in the bottle). Light may be responsible for the vitamin B_2-related photodegradation of methionine, leading to the formation of volatile sulfur compound,s such as methanethiol and dimethyldisulfide (DMDS), which give the wine the cooked cauliflower or wet wool smells (the so-called *goûts de lumière*).

34.3.10 Remuage

Subsequent sedimentation and removal of yeast requires the time-consuming and expensive procedure of riddling (*remuage*), that is, frequent but controlled turning of the slanted, inverted bottles to bring the yeast sediment together with adjuvants (bentonites or alginates) down to the cork. Nowadays, economic pressures have shortened times of *remuage* down to three to four days with freely suspended yeast and even two days with agglomerating cells. Otherwise, the use of immobilized yeast cells instead of inoculation with freely suspended yeast for this secondary fermentation would considerably simplify the riddling process. The disadvantage of shortening times of *remuage* is that the wines obtained are sometimes not crystal clear, containing particles in suspension termed as *voltigeurs*, which have been shown to be of mineral origin (*i.e.*, bentonites added in the *liqueur de tirage*).

34.3.11 Disgorging

Once the deposit (yeast + adjuvant) is concentrated against the crown cap, disgorging can occur. This operation, together with dosage, corking and wire-capping, is done automatically only for 750-mL bottles or *demie*-bottles. Flasks of higher capacity (Magnum, Jeroboam, Mathusalem) are still disgorged manually. During the process, the inverted bottle is partially immersed in a low-temperature brine, freezing the top of the bottle and entrapping the yeast + adjuvant deposit, which is removed, embedded in ice. The *liqueur d'expédition* or dosage consisting of sugar (for a final concentration in the bottle of 6–50 g/L saccharose) and antioxidants (SO_2, citric acid or ascorbic acid) is then added. During this operation, there is a loss in the CO_2 pressure ranging from 0.5 to 0.8 bar. Finally, the bottle is corked. As with non-effervescent wines, Champagnes are affected by corkiness, estimated to be at least 0.5 to 2% of bottled Champagnes, a percentage which is of real concern to the wine industry. The protective effect of a composite cork stopper on Champagne wine pollution with 2,4,6-trichloroanisole, the main component responsible for corkiness, is now being studied.

34.4 MALOLACTIC FERMENTATION IN SPARKLING WINE PRODUCTION

Malolactic fermentation (MLF) is nowadays a very important stage in sparkling wine production. The conversion of malate into lactate by lactic acid bacteria decreases the acidity of the wine, increases its pH and makes these wines microbiologically more stable, besides resulting in favorable changes in the flavor of the wine (See Chapter 18 of this volume for details). But it should occur before bottling in order to prevent subsequent bacterial growth which would form undesirable deposits. For these reasons and because of improved sanitary conditions in winemaking, wherein MLF may not occur spontaneously, many winemakers tried to induce MLF with starter cultures developed from freeze-dried or frozen *Oenococcus œni* biomass. Nevertheless, it often remains very difficult to induce MLF, especially in Champagne wines, and difficulties are usually attributed to the cumulative inhibitory effects of the low pH, high alcohol and SO_2 contents of wines. At the same time, these difficulties could also have, as their origin, the deficiency or the imbalance of Champagne wines with respect to some nutrients such as the free amino acids. In fact, amino acids are, along with short peptides, the most important source of nitrogen for the growth of wine lactic acid bacteria and most of them are either required or are stimulatory to the growth of *Oenococcus œni*.

34.4.1 Effects of Amino Acids

34.4.1.1 Effect on Bacterial Growth

Amino acids have been classified into three groups, based on the degree of bacterial growth in each amino acid-deficient medium. Where 0% to 10% of normal growth is observed in the amino acid-deficient medium, the amino acid is considered to be 'essential'; from 10% to 50%, 'favorable'; and over 50%, 'indifferent'. The number of essential amino acids for *Oenococcus œni* varies from five (Strain MC1) to 11 (Strain NCBF 1707), as shown in Table 34.1. The indigenous (Moët et Chandon; MC) *Oenococcus* strains have lower requirements than the reference strains, indicating that the indigenous strains are better able to grow under stringent conditions, as in wine. Whenever *Oe. œni* strains are used, L-arginine, L-isoleucine, L-glutamic acid and L-tryptophan are essential for the growth of *Oe. œni*. *Oe. œni* also did not have an absolute requirement for tyrosine. Although L-valine is not an essential amino acid, its absence markedly limited bacterial growth.

34.4.1.2 Effect on D-Glucose Fermentation

Between 80 and 100% of the D-glucose was consumed in the complete medium (*i.e.*, containing all amino acids) by every strain. Absence of an essential amino acid, such as L-arginine, L-glutamic acid, L-isoleucine, L-tryptophan or L-methionine, inhibits D-glucose utilization. When the amino acid deficiency limited *Oenococcus* growth, as was the case for L-histidine, L-lysine and L-valine, D-glucose was only partially metabolized. In most of the media used, consumption of D-glucose was correlated with the amount of growth, confirming D-glucose as the main carbon source for growth of *Oe. œni*. As was established earlier, *Oe. œni* is hetero-fermentative and, from 1 g of D-glucose, produced 0.5 g of D-lactic acid, 0.33 g of ethanol and acetate, 0.09g of carbonic gas and 0.08 g of biomass. D-glucose fermentation and end-product formation showed that, in the complete medium, D-glucose was entirely fermented, achieving almost the theoretical yield of D-lactic acid production, so that the fermentation appears to be mainly realized by the hetero-fermentative pathway. Following amino acid deficiency, D-glucose metabolism is modified, and the modifications obtained can be classified into four types (Figure 34.1) as given below:

Type I: medium without L-threonine or L-tyrosine: D-lactic acid production was nearly the same as in the complete medium.

TABLE 34.1
Growth of Six *Oenococcus œni* Strains in Different Synthetic Media after Incubation for 9 Days. Percentages of the Ratios of Absorbance at 650nm Observed for the Deficient Medium and the Complete One

Amino acid omitted in the culture medium	Strains					
	MC1	MC2	MC4\	NCFB 1707	NCFB 1823	NCFB 1674
None	100	100	100	100	100	100
L-alanine	101	92	91	90	94	109
L-arginine	3	3	3	0	2	0
L-aspartic acid	61	71	64	52	77	74
L-cysteine	7	7	15	97	2	88
L-glutamic acid	1	8	2	0	1	0
Glycine	102	104	108	103	115	102
L-histidine	93	77	84	1	92	91
L-isoleucine	4	5	5	0	0	0
L-leucine	71	69	65	0	2	0
L-lysine	57	98	89	2	59	43
L-methionine	16	8	9	2	6	4
L-phenylalanine	100	91	99	1	109	91
L-proline	113	108	101	92	116	106
L-serine	95	111	111	85	102	89
L-threonine	87	111	77	95	94	93
L-tryptophan	7	6	2	0	5	0
L-tyrosine	92	86	89	0	96	86
L-valine	30	35	32	1	20	15

Strains MC1, MC2 and MC4 were isolated from champagne wines by Möet & Chandon laboratory. *Source:* Fourcassie *et al.* (1992)

Type II: medium without glycine, L-alanine or L-phenylalanine: D-lactic acid production was less extended and an increase in the synthesis of other products was noted.

Type III: medium without L-proline or L-lysine: D-lactic acid production was markedly reduced and the amount of other products was higher than in the complete medium.

Type IV: medium without L-aspartic acid, L-leucine or L-serine: an overproduction of D-lactic acid was observed. It was accompanied by an underproduction of biomass in the case of L-aspartic acid or L-leucine deficiencies.

34.4.1.3 Effect on L-Malic Acid Consumption

In the complete medium, the consumption of L-malic acid varied with the strain used, but such differences between *Oe. œni* strains had already been extensively described. In most cases of amino acid deficiency, the amount of L-malic acid consumed was, however, reduced. When no growth occurred (in a medium without L-arginine, L-glutamic acid, L-isoleucine or L-tryptophan), only a weak degradation activity was noted, which may be due to the residual malolactic. When the growth was largely reduced, as was the case when L-histidine, L-methionine, L-valine or L-cysteine were absent, only a small amount of L-malic acid was degraded. Thus, L-malic acid could be entirely consumed, though bacterial growth was reduced following amino acid deficiency. More surprising is the fact that L-proline, glycine, L-phenylalanine or L-tyrosine deficiencies reduced L-malic acid consumption although growth occurred, as in the complete medium and in resting cells; some amino acids were able to stimulate L- malic acid consumption specifically *in Oenococcus œni* strains.

34.4.2 Effect of an Excessive Concentration of One Amino Acid

34.4.2.1 Effect on Bacterial Growth

When cultures were performed with L-malic acid, most of the amino acids tested had no effect on bacterial growth. Nevertheless, L-aspartic acid and L-isoleucine showed an inhibitory effect on all the bacterial strains used. It is interesting to note that, despite their inhibitory effect at high concentrations, L-isoleucine and L-tryptophan are essential amino acids for *Oenococcus œni* NCFB 1707. However, some amino acids, like L-arginine and L-threonine, are able to slightly stimulate the growth of the two indigenous *Oe. œni* strains. When L-malic acid is omitted from the culture medium, the inhibitory character of L-aspartic acid and L-isoleucine is intensified and all the other amino acids, with the exception of L-glutamic acid, showed a stimulatory effect on bacterial growth.

34.4.2.2 Bacterial Growth Stimulation by Amino Acids

Bacterial growth stimulation by high concentrations of L-arginine can result from the susceptibility of this amino acid to ATP synthesis through its deamination *via* the arginine deaminase pathway. Nevertheless, it is clear that high concentrations of L-threonine modify D-glucose metabolism with an underproduction of D-lactic acid, with more D-glucose probably being used for biomass production.

34.4.2.3 Bacterial Growth Inhibition by Amino Acids

The D-form of aspartic acid, formed from L-aspartic acid by an aspartate racemase, is an important component in the peptidoglycan layers of bacterial cell walls. Then, L-aspartic acid is an important amino acid for lactic acid bacteria and is, with L-glutamic acid and L-arginine, one of the most consumed amino acids. Nevertheless, as illustrated in Figure 34.2, for the reference strain *Oe. œni* NCFB 1707, high concentrations of L-aspartic acid inhibit bacterial growth, resulting in a decrease in both the maximum biomass production and the maximum growth rate. Since fumaric acid is known for its bactericidal activity against lactic acid bacteria (LAB), the inhibitory effect of high concentrations

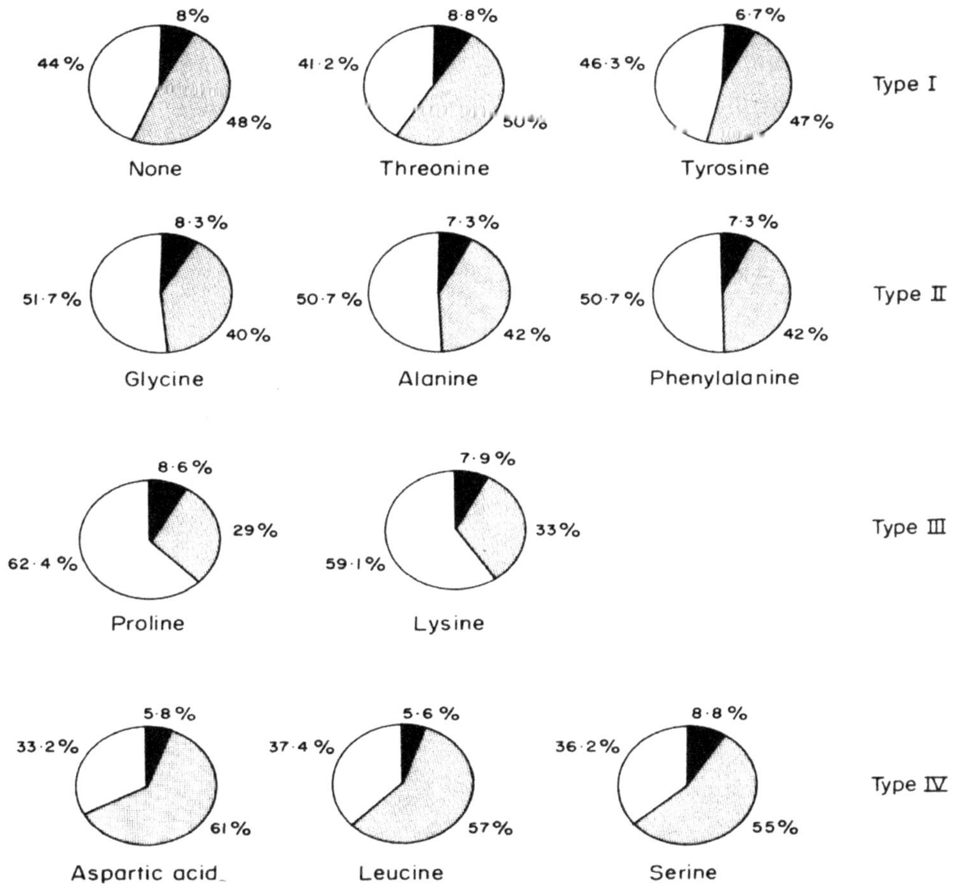

FIGURE 34.1 End-product distribution of heterolactic metabolism of D-glucose for *Oenococcus oeni* strain MC2. The amino acid deficiency is indicated at the bottom of the disk: (■), percentage of biomass production; (□), percentage of D-lactic acid production; (⊟), percentage of other products

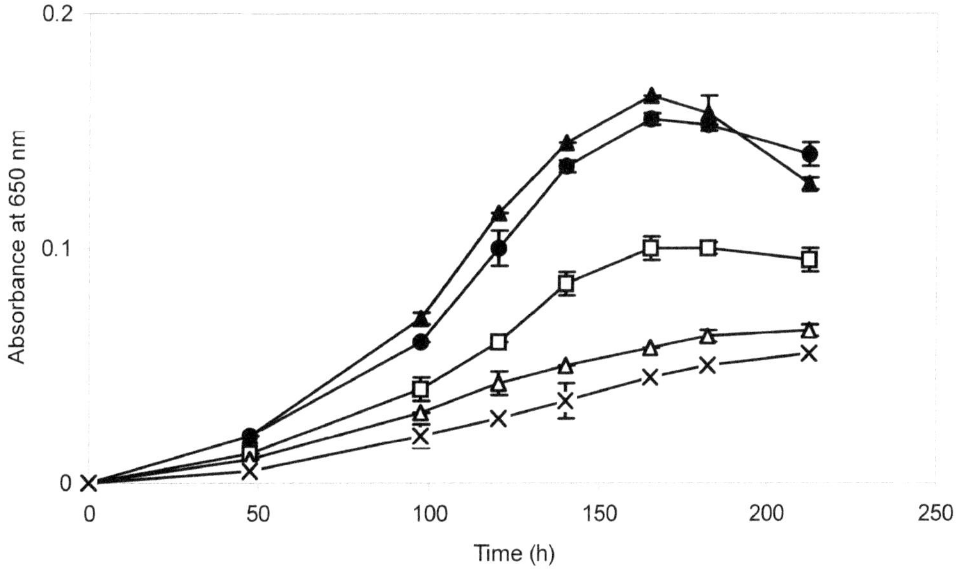

FIGURE 34.2 Effects of varying L-aspartic acid concentration on the growth of *Oenococcus œni* NCFB 1707. Basal medium with L-aspartic acid mmol/L: (□), 0; (•), 0.15; (▲), 0.3;), 6; (x), 10. *Source:* Vasserot *et al.* (2001)

of L-aspartic acid could be due to the deamination of this amino acid, through an aspartase activity, into fumaric acid and ammonia. Occurrence of an aspartase activity has been reported in *Escherichia coli*, although, presently, no evidence of the presence of such an activity in LAB has been presented. The inhibitory effects of L-aspartic acid could also be due to its involvement in antagonistic interactions with the consumption of essential amino acids, as reported for *Leuconostoc mesenteroides*. Further inhibitory effect on bacterial growth was all-the-more reduced since the global amino acid concentration in the culture medium is increased. The same phenomenon has also been observed for the indigenous strains *Oe. œni* 8403 and *Oe. œni* 8406. It can be concluded that the inhibitory effect of this amino acid is due not to its high concentration but to it being in excess with respect to other amino acids, supporting the antagonistic interactions hypothesis.

Only L-glutamic acid is able to significantly reduce the inhibitory effect of L-aspartic acid on bacterial growth with almost the same effectiveness as that obtained in a culture medium in which all the amino acids are used at 1.2 mmol/L. Since L-glutamic acid is an essential amino acid for *Oe. œni* NCFB 1707 and since the transport of this amino acid was found to be inhibited competitively by L-aspartic acid in other strains of LAB, it was thus hypothesized that an excess of L-aspartic acid possibly inhibits *Oe. œni* growth by preventing the bacterial use of L-glutamic acid, which has been confirmed by transport assays with labelled (^{14}C)-L-glutamic acid. So, when the L-aspartic acid concentration increased from 0 to 0.033 mmol/L, the initial rate of L-glutamic acid transport was reduced from 0.47 pmol/min/mg dry mass to 0.06 pmol/min/mg dry mass, as was observed with the indigenous strain *Oe. œni* 8403. Thus, L-aspartic acid and L-glutamic acid share the same transport system.

34.4.3 Effect on L-Malic Acid and D-Glucose Consumption

L-malic acid consumption is hardly affected by an excessive amount of one amino acid. In fact, whichever amino acid is considered, L-malic acid remains entirely metabolized at the end of the bacterial growth even if its consumption rate decreases, as can be obtained with high concentrations of L-aspartic acid and L-isoleucine. The effect of an excessive amount of one amino acid on D-glucose fermentation is more evident and varies with the presence or absence of L-malic acid. When cultures are performed with L-malic acid, D-glucose metabolism is not modified by an excessive amount of one amino acid and this is still in good agreement with Cavin's empirical equation. Nevertheless, in the case of an excessive amount of L-aspartic acid, L-isoleucine or, to a lesser extent, L-glutamic acid (amino acids which are all potent inhibitors of bacterial growth), D-glucose fermentation is not complete. When L-malic acid is omitted from the culture medium, an excessive amount of L-arginine and L-threonine modifies D-glucose metabolism, with an underproduction of D-lactic acid.

34.5 Bubble Dynamics in Champagne Wines

Upon pouring Champagne into a glass and looking to as what is happening in the confined space, it would be interesting to see the bubbles being born on several spots of the glass wall, detaching and then rising in-line toward the free surface in the form of elegant bubble trains, as in so many tiny hot-air balloons. While collapsing at the free surface, bubbles would also emit a typical crackling sound and produce a cloud of tiny droplets that pleasantly tickle the taster's nostrils. Without bubbles, Champagne and sparkling wines are unrecognizable. Small bubbles rising through the liquid, as well as a bubble ring (the so-called *collerette*) at the periphery of a flute poured with champagne, are the hallmark of this traditionally festive wine, and, even if there is no scientific evidence yet to correlate the quality of a Champagne with the fineness of its bubbles, people nevertheless often make a connection between them. It has now become an important stake in the Champagne research area to achieve the perfect petite bubble and to examine more closely the behavior of bubbles in carbonated beverages, illustrate and finally understand better the role played by each of the numerous parameters involved in the effervescence. For example, it is important to distinguish between ascending Champagne bubbles from their beer counterparts. The simple but close observation of a glass poured with carbonated beverages also recently revealed an unexplored and visually appealing phenomenon. Three main steps of a bubble's life include the bubble birth, the bubble ascent and the bursting of a bubble at the free surface of the liquid.

34.5.1 The Bubble Genesis

In the case of Champagne and other sparkling wines, the main gas responsible for bubble production is CO_2, which is produced by yeast during the fermentation in the closed bottle. Yeast convert sugars to alcohol and CO_2 molecules (among other compounds). According to Henry's law, an equilibrium progressively establishes between CO_2 molecules dissolved in the liquid and CO_2 molecules into the vapor phase in the headspace under the cork. When the bottle of Champagne is opened, the CO_2 pressure in the vapor phase suddenly falls. The thermodynamic equilibrium of the closed container is broken, and the liquid becomes supersaturated with CO_2 molecules. To recover a new stable thermodynamic state corresponding to the atmospheric pressure, CO_2 molecules must escape from the supersaturated liquid. When poured into a glass, two mechanisms enable dissolved CO_2 molecules to escape from the supersaturated liquid medium: diffusion through the free surface of the liquid, and bubble formation. However, to cluster into the form of bubble embryos, dissolved CO_2 molecules need to diffuse their way through the liquid molecules strongly linked by the so-called van der Waals attractive forces. Henceforth, bubble formation is characterized by an energy barrier to overcome. It requires very high supersaturating ratios, which are totally unrealistic in the case of carbonated beverages. In weakly supersaturated liquids, such as Champagne, sparkling wines and carbonated beverages in general, bubbles need pre-existing gas cavities with radii of curvature greater than a critical radius in order to overcome the nucleation energy barrier and grow freely. In 'bubbly', the critical radius below which bubble production becomes impossible is sub-micrometric, around 0.2 μm. As a result, bubble formation spots on the wall of a glass poured with carbonated beverage necessarily reveals tiny pre-existing gas cavities greater than this critical radius.

To have access to bubble production sites, a high-speed video camera fitted with a microscope objective was pointed at the base of each investigated bubble train, observing carefully hundreds of different "bubble nurseries". Contrary to general assumption, nucleation sites are not located on irregularities of the glass itself. The length-scale of glass irregularities is far below the critical radius of curvature required for bubble nucleation. Nucleation sites are located on impurities stuck on the glass wall and most of the sites were hollow and roughly cylindrical exogenous cellulose fibers coming from the surrounding air or remaining from the wiping process. Because of geometrical properties, such particles cannot be completely wetted by the liquid and are able to entrap gas pockets during the filling of a glass. Four typical nucleation sites found in a glass poured with champagne are displayed in Figure 34.3. Gas pockets trapped inside the particles are clearly apparent. Dissolved CO_2 molecules migrate into the gas pocket. A bubble appears and grows rooted to its nucleation site because of capillary forces. Finally, the increasing buoyancy induces its detachment, thus providing an opportunity for a new bubble to nucleate, grow rooted to its nucleation site and detach at the same size, and so on, until bubble production stops through lack of dissolved gas. Pre-existing gas cavities trapped inside particles stuck on the glass wall provide bubble production in carbonated beverages. This cycle of bubble production at a given nucleation site is characterized by its "bubbling" frequency, that is, the number of bubbles produced per second. This clockwork and repetitive bubble production from nucleation sites can be easily underscored by the naked eye by lighting bubble trains with a stroboscope. By equaling the flash frequency of strobe lighting with the frequency of the cycle of bubble production, the corresponding bubble train appears 'frozen'. The time needed to reach the moment of bubble detachment depends on the geometrical properties of the given nucleation site. Because a collection of particle shapes and sizes exists on the glass wall, the bubbling frequency may also vary from one site to another. Since the kinetics of the bubble growth also depends on the dissolved CO_2 molecule content, differences in the bubble formation frequencies may be observed from one beverage to another. For example, in Champagne wines, where the dissolved gas content is approximately three times higher than in beer, the most active nucleation sites emit up to about thirty bubbles per second, whereas in beer, nucleation sites emit up to only about ten bubbles per second.

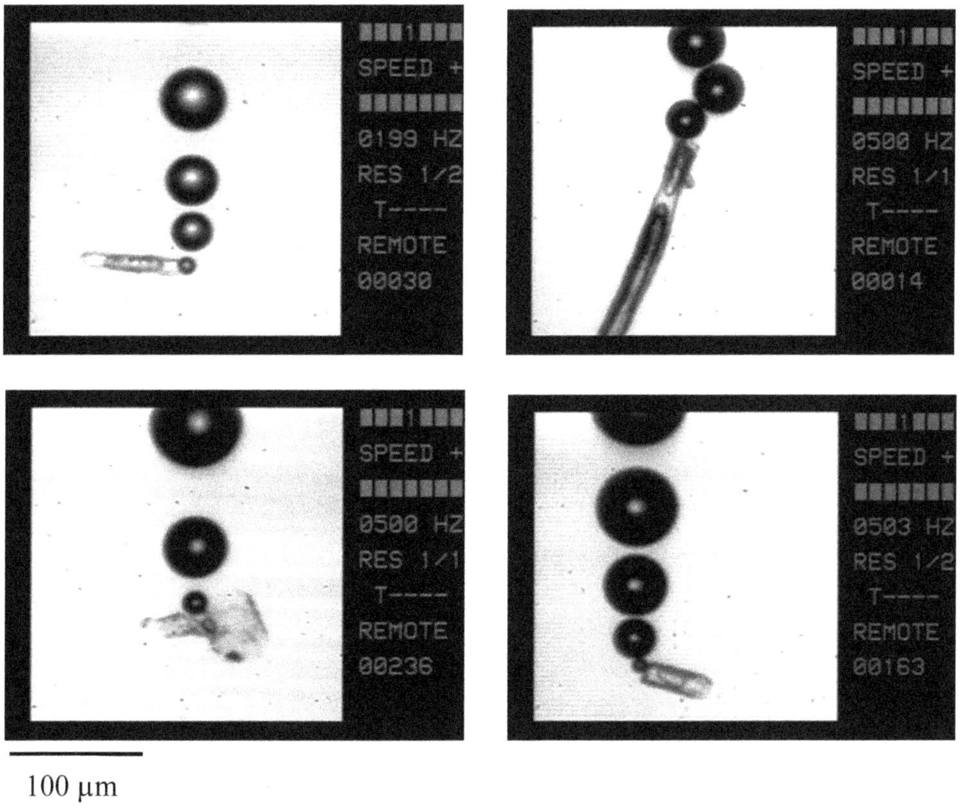

FIGURE 34.3 Close-ups of four particles acting as nucleation sites on the wall of a glass poured with champagne. Gas pockets trapped inside the particles appear. Copyright (2002), reproduced by permission of the American Chemical Society

34.5.2 The Bubble Rise

A short examination of the typical regular bubble train presented in Figure 34.4 clearly shows that bubbles grow during ascent. This bubble growth during ascent is caused by a continuous diffusion of dissolved CO_2 molecules through the bubble interface. While expanding, bubbles increase their buoyancy, causing them to continuously accelerate and separate from one another on their way up. Beers and sparkling wines are obviously not pure liquids. They contain, in addition to alcohol and dissolved CO_2 molecules, many other organic compounds which may show surface activity (mostly composed of proteins and glycoproteins in Champagne wines and beers). Like soap molecules, such molecules have a water-soluble and a water-insoluble part, called 'surfactants'. Surfactants prefer to gather around the surface of a bubble, the water-insoluble part sticking out of the liquid into the gas-filled bubble, rather than staying in the liquid bulk. The role of this surfactant coating around gas bubbles becomes crucial when buoyancy induces their detachment from the wall impurities and forces them to plough their way through the liquid molecules. Adsorbed surface-active molecules stiffen a bubble by forming a sort of shield on its surface. Actually, surfactants encountered along the bubble path progressively adsorb at the surface of a rising bubble, thus increasing the immobile area of the bubble surface. According to fluid dynamics, a bubble rigidified by surfactants and rising through a liquid runs into more resistance than a bubble presenting a more flexible skin free from surface-active materials. Therefore, the drag coefficient experienced by a bubble of fixed radius rising in a surfactant solution progressively increases, thus decreasing its velocity of rise to a minimal value when the bubble interface gets completely contaminated. In ultra-pure water free from surfactants, a millimetric bubble rises at a velocity close to 30 cm/s but with small concentrations of proteins (only of order of several mg/L), the bubble velocity is progressively decreasing as soon as the bubble interface traps proteins, to ultimately reach a final velocity of the order of 15 cm/s (about half that in pure water) when the bubble interface is completely rigidified by a protein coating.

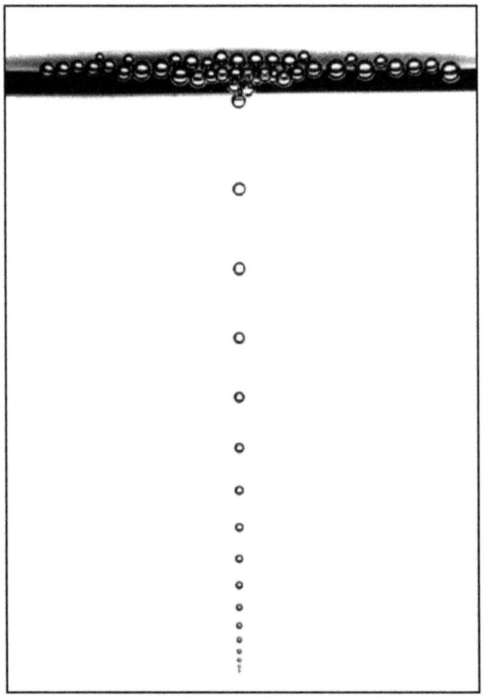

FIGURE 34.4 Typical photograph of a regular bubble train. The dark line running horizontally through the picture is due to the liquid meniscus between the free surface and the glass wall (photograph by Gérard Liger-Belair)

The case of rising and expanding bubbles is a little bit more subtle than that of bubbles of fixed radius, since the former type of bubble expands during their rise through the supersaturated liquid, the bubble interface continuously increases, and so continuously offers newly created surface to the adsorbed surface-active materials. Expanding bubbles, therefore, experience two opposing effects. If the rate of dilation of the growing bubble overcomes the rate at which surface-active molecules stiffen the bubble surface, a bubble progressively cleans its interface. So, the ratio of the bubble surface covered by surfactants to the bubble surface free from surfactants decreases. But if this ratio increases, the bubble surface inexorably gets contaminated by a surfactant monolayer and progressively becomes more rigid.

By measuring the drag coefficients experienced by expanding Champagne and beer bubbles during their way up and by comparing them with data found in the huge scientific literature dealing with bubbles of various sizes ascending through various liquids, evidence has been provided that beer bubbles showed a behavior very close to that of rigid spheres. On the contrary, bubbles of Champagne, sparkling wines and soda were found to present a more flexible interface during ascent, which is not a surprising result, since beer contains much higher amounts of surface-active macromolecules (of the order of several hundred mg/L), more likely to be adsorbed at a bubble interface than Champagne (only several mg/L). Because of the lower gas content in beer, the growth rates of beer bubbles are also lower than those of Champagne. As a result, the "cleaning" effect due to the bubble's expansion may be too weak to avoid the rigidification of the beer bubble interface. In Champagne, sparkling wines and sodas, there is insufficient surface-active material and bubbles grow too quickly to succumb to the same tragic fate. In fact, because the drag experienced by a rising bubble is all the higher because the coating of surfactants is well developed, bubbles of the same size are rising slower in beer than in the other carbonated beverages.

34.5.3 The Bubble Collapse At the Free Surface

A few seconds after being born on a glass wall impurity, a few centimeters below the liquid surface, bubbles reach the free surface with a nearly millimetric diameter. Similar to an iceberg, a gas bubble only emerges slightly above the free surface of the wine. Most of the bubble volume lies below the liquid surface. The emerged part of the bubble, the bubble-cap, is a spherical portion of liquid film which progressively gets thinner due to drainage. Actually, a bubble-cap which has reached a critical thickness becomes sensitive to vibrations and thermal gradients, and finally ruptures. The disintegration process of thin liquid sheets was examined, and it was found that a hole appears in the bubble-cap, which very quickly propagates, propelled by surface tension forces. After the disintegration of the bubble-cap, which occurs on a time scale of 10 to 100 μs for millimetric bubbles, a complex hydrodynamic process ensues which induces the collapse of the submerged part of the bubble. This process, for bubbles collapsing at the surface of a glass poured with Champagne, was studied by using high-speed macrophotography. A reconstructed time sequence illustrating six stages of the collapse of a Champagne bubble is presented in Figure 34.5 and described here. Between frames 1 and 2, the thin liquid film, which constitutes the emerged part of the bubble, has just ruptured. During this extremely brief initial phase, the bulk shape of the bubble has been frozen by the flashlight and a nearly millimetric open cavity remains in the liquid surface. As it collapses, the bubble cavity gives rise to a high-speed liquid jet above the free surface (frames 3 and 4). Near the base of the liquid jet in frame 4, one can distinguish an extremely small bubble (around 100 μm), probably entrapped during the collapsing process. Due to its own velocity, this upward liquid jet becomes unstable. A capillary wave develops along the jet (the well-known Plateau-Rayleigh instability which appears clearly in frame 5). In frame 6, the liquid jet finally breaks up into droplets called 'jet drops'. The combined effects of inertia and surface tension give detaching jet drops various and often amazing shapes. Finally, in frame 7, droplets ejected by the parent bubble recover a quasi-spherical shape. Due to surface excitations following bubble collapse, capillary wave trains centered on the bursting bubble are propagating at the free surface. On the right side of the central bubble in frame 7, the tiny bubble entrapped during collapse can be observed. Since hundreds of bubbles are bursting every second during the first minute that follows the pouring of some carbonated beverage, one can conclude that the liquid surface is literally spiked with such cone-shaped structures, unfortunately too short lived to be observed by the naked eye. At a millimetric scale, such a violent hydrodynamic phenomenon, which leads to the projection of an high-speed liquid jet, is driven by the capillary pressure gradients arising around the open cavity frozen at the free surface after the disintegration of the bubble-cap. Immediately after the rupture of the bubble-cap, sides of the cavity become a region of positive curvature. A ring of high pressure ensues on the sides of the open cavity. At the same time, due to a negative curvature, a low-pressure zone exists around the underside of the cavity. As a result, fluid is rapidly drawn from the sides to the axis of symmetry. The underside of the cavity becomes a region of high pressure which pushes fluid upward to produce the liquid jet.

The liquid jet that follows a bubble collapse strikingly resembles, in miniature, what one can observe as a drop impacts the surface of a pool of liquid. Harold Edgerton (1903–1990), the twentieth century master of stop-action photography, invented and developed the electronic flash and popularized "high-speed events" by photographing everything, from flying bullets and instants of sport actions, to drops of liquid impacting surfaces. Despite noticeable differences in time and length scales, hydrodynamic structures arising after a drop impact are clearly very close to those that follow a bubble collapse. In addition to esthetic considerations, bubbles bursting at the free surface impart feel to Champagne wines, beers and many other beverages. Jet drops are ejected up to several centimeters above the free surface with a velocity of several meters per second. Nociceptors (pain receptors) of the nose are thus stimulated during tasting, as are receptors in the mouth when bubbles burst over the tongue. Furthermore, in addition to these mechanical stimulations, bubbles bursting at the free surface are also expected to play a major role in flavor release. Due to their molecular structure, many aromatic compounds of carbonated beverages show surface activity. Bubbles rising and expanding in the liquid bulk act as aromatic molecules, trapping and dragging such surface-active molecules along their way up. Such molecules progressively concentrate at the free surface at much higher concentrations than those found in the liquid bulk. Bubbles collapsing at the free surface are thus expected to spray in the air a cloud of tiny droplets, over-concentrated with respect to potentially aromatic molecules, thus highlighting the flavors of Champagne, sparkling wines etc.

Moreover, despite the large body of research concerned with collapsing bubble dynamics, the close-up observation of bubbles collapsing at the free surface of Champagne recently revealed an unexplored and visually appealing phenomenon. Most of the previous studies were conducted with single bubbles collapsing at free surfaces. Now, for a few seconds after pouring, the free surface is completely covered with a monolayer composed of quite monodisperse millimetric bubbles collapsing close to each other. Effervescence in carbonated beverages thus ideally lend themselves to a preliminary work with bubbles collapsing close to each other. Since the collapsing process of a millimetric cavity at the free surface of a liquid air liquid interface is very short (2–3 ms), very few photographs freeze snapshots of the collapsing process in a bubble monolayer. Photographs displayed in Figure 34.6, for example, were taken immediately after the rupture of a bubble-cap in the bubble monolayer. Clusters of bubbles can be observed which are strongly deformed towards a bubble-free central area, leading to unexpected and visually appealing flower-shaped structures. A few time sequences of the whole process have also been captured with the high-speed video camera. One is presented in Figure 34.7. Between frame 1 and frame 2, the bubble-cap of a bubble ruptured and left an open cavity at the free surface (the submerged part of the bubble). Paradoxically, adjacent bubble-caps are sucked but

FIGURE 34.5 Reconstructed time sequence illustrating six stages of the collapse of a single bubble at the free surface of a glass poured with Champagne. The time interval between each frame is around 1 ms (Photograph by Gérard Liger-Belair). Copyright (2001), reproduced by permission of the American Society for Enology and Viticulture

not blown-up by bursting bubbles, contrary to what would have been thought at first glance. Capillary pressure gradients around the open cavity left by the ruptured bubble-cap are, once more, the driving forces of this violent sucking process. But, contrary to the single bubble collapse case, liquid flows appear in the thin film of adjacent bubble-caps.

During this sudden stretching process, adjacent bubble-cap areas significantly increase. A systematic image analysis of numerous time sequences conducted with a high-speed video camera demonstrated an average increase of approximately 15% of bubble-cap areas adjacent to a central collapsing bubble. Stresses in the bubble-caps of adjoining bubbles were

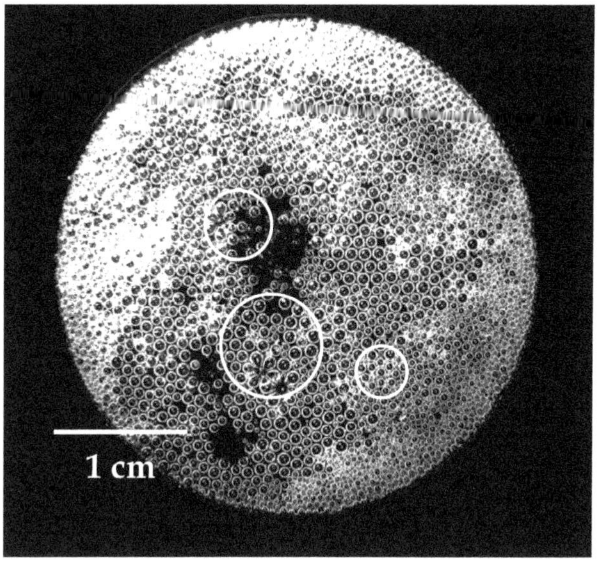

FIGURE 34.6A Top view of the bubble monolayer at the free surface of a glass of Champagne a few seconds after pouring

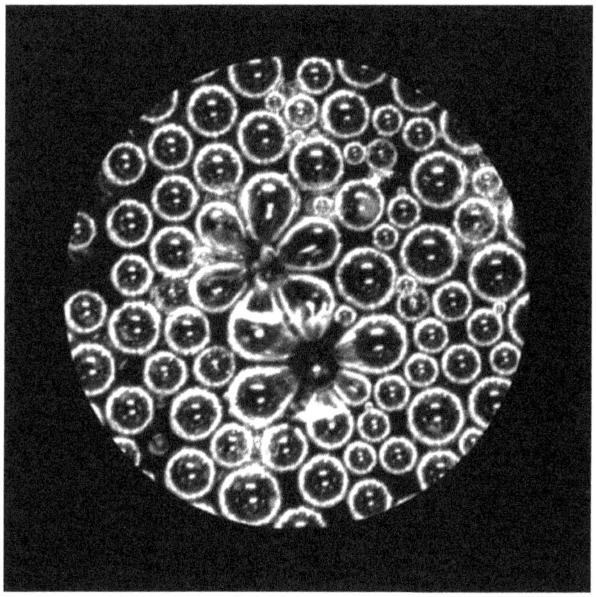

FIGURE 34.6B Detail of the double flower-shaped structure circled in white. In the center of the picture, one can observe two bubbles entrapped at the same time into the wake of the two collapsing cavities. (Photograph by Gérard Liger-Belair.) Copyright (2003), reproduced by permission of the American Chemical Society

evaluated during the sucking process and found to be of the order of 10^5 to 10^6 dyn/cm. By comparison, in previous studies, numerical models conducted to stresses of the order of (only) 10^4 dyn/cm^2 in the boundary layer around single millimetric collapsing bubbles. Therefore, stresses in the bubble-caps of bubbles adjacent to millimetric collapsing cavities should be, at least, one order of magnitude higher than those observed around single collapsing cavities. While absorbing the energy released during collapse, as tiny "air-bags" would do, adjoining bubble-caps store this energy into the thin liquid film of emerging bubble-caps, leading finally to stresses higher than those observed in the boundary layer around single millimetric collapsing bubbles.

34.6 TRENDS IN CHAMPAGNE RESEARCH

34.6.1 Application of Metabolomics to the Characterization of the Chemical Fingerprint of Champagne Bubbles

Once Champagne is poured into a glass, bubbles, originating from the glass cavities or particles left on the glass surface (see above), drag tension-active compounds along their ascension through the liquid surface where they burst, leading to a myriad of tiny droplets in the form of aerosols above the free surface. The presence of these bursting-aerosol-forming bubbles was revealed using laser tomography techniques. Bubbles bursting at the free liquid surface, which are enriched in amphiphilic molecules able to adsorb at the CO_2-liquid interface, lead to hundreds of tiny liquid jets as revealed by high-speed photography and laser tomography (Figure 34.8). Starting from the assumption that a parallel can be made between the chemical composition of the bubble interface and the droplets which originate from the bubble burst, a metabolomics study was carried out to determine which tension-active compounds can adsorb and concentrate at the bubble–liquid interface and what role can be attributed to this phenomenon in tasting Champagne.

The application of ultra-high-resolution mass spectrometry (Fourier Transform Ion Cyclotron Resonance Mass Spectrometry) revealed that some compounds can concentrate in Champagne aerosols by concentration factors reaching up to 30-fold *versus* the Champagne bulk. These compounds, the concentration of which increased in the droplets, were found to belong to a nearly complete series of saturated fatty acid structures (from C10 to C24) as well as unsaturated fatty acids (from C14 to C18). These observations are of particular relevance as fatty acids and their ethanolic esters are the main aroma-giving compounds formed upon fermentation of the white wines. In addition to fatty acids, norisoprenoid compounds, which are also important components of the wine aroma, were found to be overconcentrated in the aerosols. Thus, the bubbles, upon bursting, consistently drive aroma molecules at the free surface of a glass poured with Champagne, thus playing an important role in the tasting of this wine.

34.6.2 Application of Metabolomics To the Chemical Composition And the Age of Champagne Bottles

In July 2010, divers discovered 168 Champagne bottles in a shipwreck off the Finnish Aland archipelago in the Baltic Sea. Some of them were identified thanks to branded engravings on the surface of the cork as Champagnes from the Veuve Clicquot–Ponsardin house. According to historical records found in the archives of the Veuve Clicquot--Ponsardin

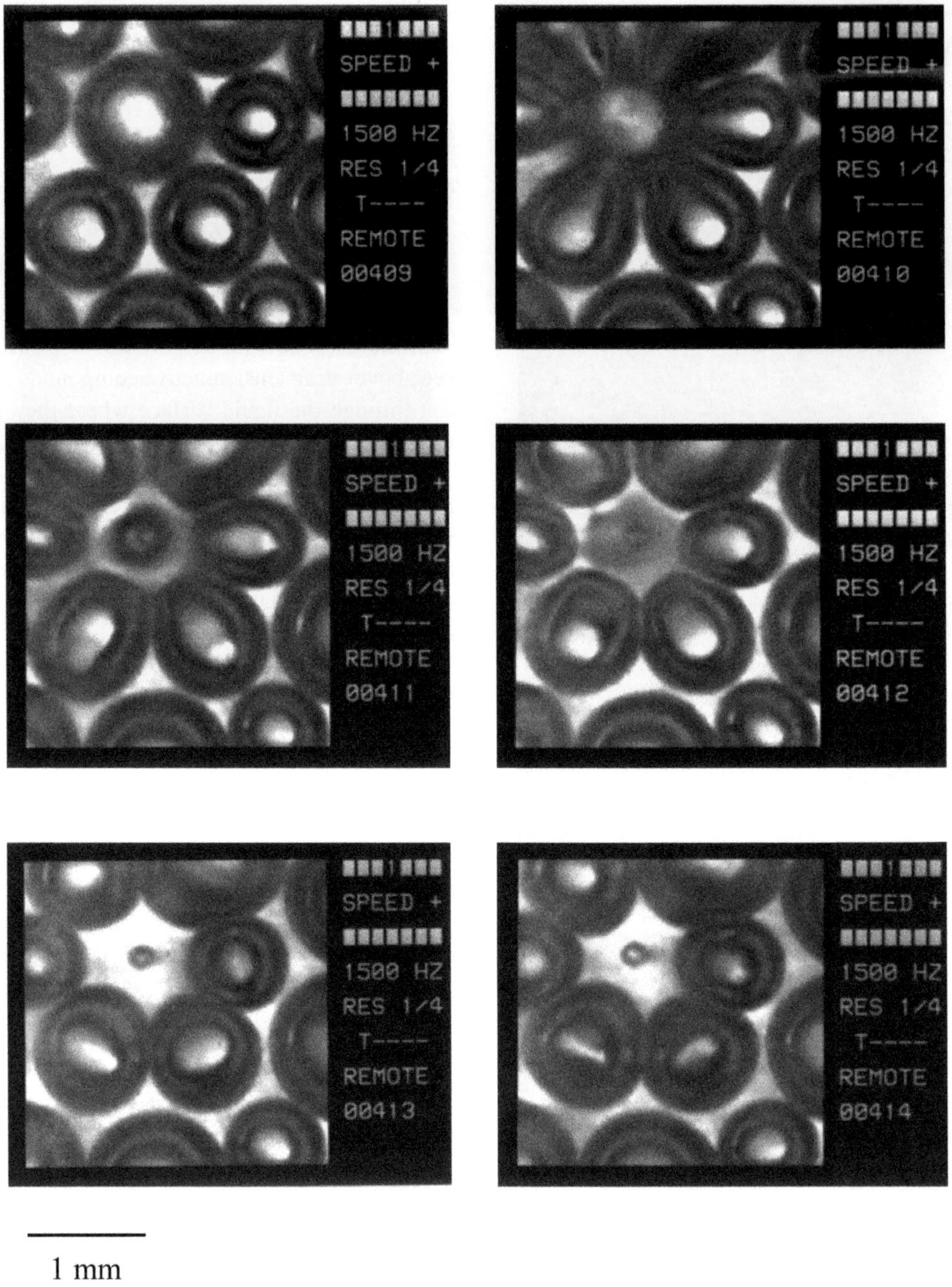

FIGURE 34.7 Time sequence of the collapsing process taken with a high-speed video camera filming at 1500 frames per second. Paradoxically, adjoining bubble-caps are sucked and not blown-up by the central bursting bubble. Copyright (2003), reproduced by permission of the American Chemical Society

house, especially letters from Madame Veuve Clicquot, as well as our own analyses, estimate the age of these enigmatic Champagne bottles to be around 170 years, that is, one of the oldest Champagnes to have made it to our time. A combined organic spectroscopy-based metabolomics and metallomics approach led us to reveal the detailed composition of these 170-year-old Champagnes from the Baltic Sea. Applying Inductively coupled plasma–atomic emission spectroscopy and sector field mass spectrometry identified high concentrations of metals in those Champagnes, such as iron and copper, probably as a result of the winemaking practices in use at the time (use of iron-containing vessels for the winemaking process and copper to protect the grapevine against fungal pathogens). Unexpectedly high sodium chloride amounts were also found in the Champagne samples from the Baltic Sea, probably as a consequence of the fining process, the so-called *collage*, used at this time. Other observations confirmed that sea water contamination of the bottles cannot account for this high sodium chloride content. High-resolution mass spectrometry also revealed the presence of wood markers such as 5-carboxyvanillic acid and castalin, as well, attesting that the wines had been vinified in barrels.

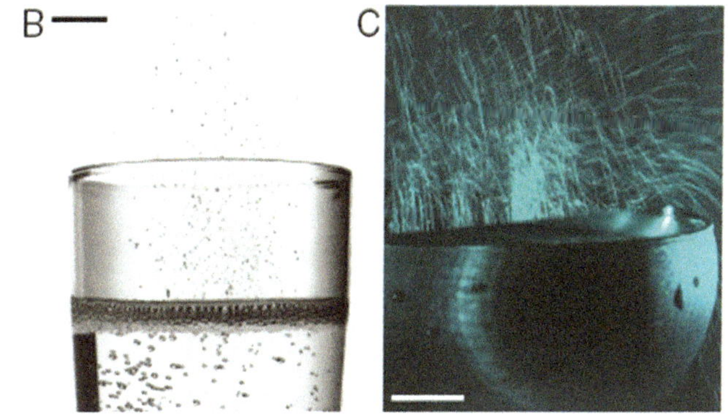

(A) (B)

FIGURE 34.8 Observation of the aerosols formed at the liquid surface of a glass poured with Champagne using high-speed photography (Figure 34.8A) or laser tomography (Figure 34.8B). *Source:* Liger-Belair, G., Polidori, G. and Jeandet, P. *Chemical Society Reviews* (2008)

Possibly the most striking feature of these Baltic Champagne bottles was their extraordinarily high sugar content, measured as over 140 g/L when, today, Brut Champagnes only contain a few grams of sugars. These impressive sugar amounts were added in the *liqueur d'expédition* in the form of a grape syrup with over 700 g/L of sugar! This was a common feature of this time as Russian consumers were reputed to drink Champagnes (or any other wines) with sugar. The agent of Madame Veuve Clicquot in Saint Petersburg indeed reported the following fact: 'Here, they (Russians) always have some sugar on any table close to their wine glass, for they add sugar not only to red wine but also to Champagne'. For these consumers, Madame Veuve Clicquot created a Champagne known as *Champagne à la Russe* whose sugar content approached 300 g/L!

Finally, the aroma composition of the Champagnes from the Baltic Sea was first revealed during tasting sessions and confirmed by a set of aroma analysis techniques. The first nose was rather negative, as the wines were described with cheesy and animal notes, due to the presence of butyric acid and volatile phenols, respectively. Upon swirling the wine in the glass, more pleasant aroma took place with fruity and floral notes (presence of ethyl esters of fatty acids, isoamyl acetate, phenyl-2-ethanol, etc.). The acetic acid content of these wines was very low and comparable to that of modern Champagnes, showing that the microbiological process was perfectly controlled at the time. Use of modern metabolomic approaches has revealed the ultimate composition of Champagne wines with potent applications to many other old wines and alcoholic beverages.

BIBLIOGRAPHY

Beelman, R.P. and Gallander, J.F. (1979). Wine deacidification. *Adv. Food Res.*, 25:1.

Brissonnet, F. and Maujean, A. (1991). Identification of some foam-active compounds in champagne base wines. *Am. J. Enol. Vitic.*, 42:97.

Brissonnet, F. and Maujean, A. (1993). Characterization of foaming proteins in a champagne base wine. *Am. J. Enol. Vitic.*, 44:297.

Cilindre, C., Castro, A.J., Clément, C., Jeandet, P. and Marchal, R. (2007). Influence of *Botrytis cinerea* infection on Champagne wine proteins (characterized by two-dimensional electrophoresis/immunodetection) and wine foaming properties. *Food Chem.*, 103(1):139.

Cilindre, C., Jegou, S., Hovasse, A., Castro, A.J., Schaeffer, C., Clément, C., Van Dorsselaer, A., Jeandet, P. and Marchal, R. (2008). Proteomic approach to identify Champagne wine proteins as modified by *Botrytis cinerea* infection. *J. Proteome Res.*, 7(3):1199.

Cox, D.J. and Henick-Kling, T. (1990). A comparison of lactic acid bacteria for energy yielding (ATP) malolactic enzyme systems. *Am. J. Enol. Vitic.*, 41:215.

Dambrouck, T., Marchal, R., Cilindre, C., Parmentier, M. and Jeandet, P. (2005). Determination of the grape invertase content (using PTA-ELISA) following various fining treatments *vs* changes in the total protein content of wine. Relationships with wine foamability. *J. Agric. Food Chem.*, 53(22):8782.

Dambrouck, T., Marchal, R., Marchal-Delahaut, L., Parmentier, M., Maujean, A. and Jeandet, P. (2003). Immunodetection of proteins from grape and yeast in a white wine. *J. Agric. Food Chem.*, 51(9):2727.

Fleet, G.H. (Ed.) (1993). *Wine Microbiology and Biotechnology*, Harwood Academic Publishers, Chur, Switezerland.

Fourcassié, P., Makaga-Kabinda-Massard, E., Belarbi, A. and Maujean, A. Growth. (1992). D-glucose utilization and malolactic fermentation by *Leuconostoc oenos* strains in 18 media deficient in one amino-acid. *J. Appl. Bacteriol.*, 73(6):489.

Gerbaux, V., Villa, A., Monamy, C. and Bertrand, A. (1997). Use of lysozyme to inhibit malolactic fermentation and to stabilize wine after malolactic fermentation. *Am. J. Enol. Vitic.*, 48:49.

Gougeon, R.D., Reinholdt, M., Delmotte, L., Miéhé-Brendle, J., Chézeau, J.M., Le Dred, R., Marchal, R. and Jeandet, P. (2002). Direct observation of polylysine side-chain interaction with smectites interlayer surfaces through ^{1}H-^{27}Al heteronuclear correlation NMR spectroscopy. *Langmuir*, 18(8):3396.

Gougeon, R.D., Reinholdt, M., Delmotte, L., Miehe-Brendle, M. and Jeandet, P. (2006). Solid state nuclear magnetic resonance investigations on the interactions between a synthetic montmorillonite and two homopolypeptides. *Solid State Nucl. Magn. Reson.*, 29:322.

Gougeon, R.D., Soulard, M., Reinholdt, M., Miéhé-Brendle, J., Chézeau, J.M., Le Dred, R., Marchal, R. and Jeandet, P. (2003). Polypeptide adsorption onto a synthetic montmorillonite: A combined solid-state NMR, X-ray diffraction, thermal analysis and N_2 adsorption study. *Eur.J. Inorg. Chem.*, 7:1366 –1372.

Jeandet, P. and Charters, S. (2011). Producing Champagne. In: *The Business of Champagne, a Delicate Balance*, S. Charters (ed.), Routledge Studies of Gastronomy, Food and Drink, Great Britain and USA, pp. 15–24.

Jeandet, P., Lenfant, J.C., Caillet, M., Liger-Belair, G., Vasserot, Y. and Marchal, R. (2000). Contribution to the study of the *remuage* procedure in champagne wine production: Identification of *voltigeurs* (particles in suspension in the wine). *Am. J. Enol. Vitic.*, 51:418.

Jeandet, P., Clement, C. and Conreux, A. (2007). *Macromolecules and Secondary Metabolites of Grapevine and Wine*, Intercept / Lavoisier, Paris, London, New York.

Jeandet, P., Heinzmann, S., Roullier Gall, C., Cilindre, C., Aron, A., Deville, M.A., Moritz, F., Karbowiak, T., Demarville, D., Brun, C., Moreau, F., Michalke, B., Liger-Belair, G., Witting,

M., Lucio, M., Steyer, D., Gougeon, R.D. and Schmitt-Kopplin, P. (2015). Chemical messages in 170-year-old champagne bottles from the Baltic Sea: Revealing tastes from the past. *Proc. Natl Acad. Sci. U. S. A.*, 112(19):5893–5898.

Jegou, S., Conreux, A., Villaume, S., Hovasse, A., Schaeffer, C., Cilindre, C., Van Dorsselaer, A. and Jeandet, P. (2009). One step purification of the grape vaculoar invertase. *Anal. Chim. Acta*, 368:75–78.

Liger-Belair, G. (2005). The physics and chemistry behind the bubbling properties of Champagne and sparkling wines: A state-of-the-art review. *J. Agric. Food Chem.*, 53(8):2788.

Liger-Belair, G., Cilindre, C., Gougeon, R., Lucio, M., Gebefugi, I., Jeandet, P. and Schmitt-Kopplin, P. (2009). Unraveling different chemical fingerprints between a champagne wine and its aerosols. *Proc. Natl Acad. Sci. U. S. A.*, 106(39):16545–16549.

Liger-Belair, G. and Jeandet, P. (2003). Capillary-driven flower-shaped structures around bubbles collapsing in a bubble raft at the surface of a liquid of low viscosity. *Langmuir*, 19(14):5771.

Liger-Belair, G. and Jeandet, P. (2003). More on the surface state of expanding Champagne bubbles rising at intermediate Reynolds and high Peclet numbers. *Langmuir*, 19(3):801.

Liger-Belair, G., Lemaresquier, H., Duteurtre, B., Robillard, B. and Jeandet, P. (2001). The secret of fizz in Champagne wines: A phenomenological study. *Am. J. Enol. Vitic.*, 52:88.

Liger-Belair, G., Marchal, R., Robillard, B., Dambrouck, T., Maujean, A., Vignes-Adler, M. and Jeandet, P. (2000). On the velocity of expanding spherical gas bubbles rising in line in supersaturated hydroalcoholic solutions: Application to bubble trains in carbonated beverages. *Langmuir*, 16(4):1889.

Liger-Belair, G., Parmentier, M. and Jeandet, P. (2006). Modeling the kinetics of bubble nucleation in champagne and carbonated beverages. *J. Phys. Chem. B*, 110(42):21145.

Liger-Belair, G., Polidori, G. and Jeandet, P. (2008). Recent advances in the science of Champagne bubbles. *Chem. Soc. Rev.*, 37(11):2490–2511.

Liger-Belair, G., Vignes-Adler, M., Voisin, C., Robillard, B. and Jeandet, P. (2002). Kinetics of gas discharging in a glass of champagne: The role of nucleation sites. *Langmuir*, 18(4):1294.

Liger-Belair, G., Voisin, C. and Jeandet, P. (2005). Modeling non – classical heterogeneous bubble nucleation from cellulose fibers: Application to bubbling in carbonated beverages. *J. Phys. Chem. B*, 109(30):14573.

Liger-Belair, G., Voisin, C., Topegaard, D. and Jeandet, P. (2004). Is the wall of a cellulose fiber saturated with liquid whether or not permeable with CO_2 dissolved molecules ? *Langmuir*, 20:4132.

Liger-Belair, G., Tufaile, A., Robillard, B., Jeandet, P. and Sartorelli, J.C. (2005). Period-adding route in sparkling bubbles. *Phys. Rev. E*, 72(3 Pt 2):037204.

Machet, F., Robillard, B. and Duteurtre, B. (1993). Application of image analysis to foam stability of sparkling wines. *Sci. Aliments*, 13:73.

Maujean, A. (2000). Organic acids in wine. In: *Handbook of Enology Vol. 2. The Chemistry of Wine, Stabilization and Treatments*, Pascal Ribéreau-Gayon, Yves Glories, Alain Maujean, Denis Dubourdieu (Eds.), John Wiley & Sons, LTD, Chichester, p. 3.

Okada, H., Yohda, M., Giga-Hama, Y., Ueno, Y., Ohdo, S. and Kumagai, H. (1991). Distribution and purification of aspartate racemase in lactic acid bacteria. *Biochim. Biophys. Acta*, 1078(3):377.

Robillard, B., Delpuech, E., Viaux, L., Malvy, J., Vignes-Adler, M. and Duteurtre, B. (1993). Improvements of methods for sparkling base wine foam measurements and effect of wine filtration on foam behavior. *Am. J. Enol. Vitic.*, 44:387.

Rudin, A.D. (1957). Measurement of the foam stability of beers. *J. Inst. Brew.*, 63(6):506.

Tracey, R.P. and Britz, T.J. (1989). The effect of amino acids on malolactic fermentation by *Leuconostoc œnos*. *J. Appl. Bacteriol.*, 67:589.

Valade, M., Moulin, J.P. and Laurent, M. (1984). Réactivation d'une biomasse de bactéries lyophilisées: Utilisation pour le déclenchement de la fermentation malolactique. *Le Vigneron Champenois*, 4:183.

Vasserot, Y., Dion, C., Bonnet, E., Maujean, A.. and Jeandet, P. (2001). A study into the role of L-aspartic acid on the metabolism of L-malic acid and D-glucose by *Oenococcus oeni*. *J. Appl. Microbiol.*, 90(3):380.

Vasserot, Y., Dion, C., Bonnet, E., Tabary, I., Maujean, A. and Jeandet, P. (2003). Transport of L-glutamate in *Oenococcus oeni* 8403. *Int. J. Food Microbiol.*, 85(3):307.

Vasserot, Y., Caillet, S. and Maujean, A. (1997). Study of anthocyanin adsorption by yeast lees. Effect of some physicochemical parameters. *Am. J. Enol. Vitic.*, 48:433.

Vasserot, Y., Pitois, C. and Jeandet, P. (2001). Protective effect of a composite cork stopper on Champagne wine pollution with 2,4,6-trichloroanisole. *Am. J. Enol. Vitic.*, 52:280.

Vasserot, Y., Steinmetz, V. and Jeandet, P. (2003). Study of thiol consumption by yeast lees. *Ant. Leeuwenhoek Int. J. Gen. Mol. Microbiol.*, 83(3):201.

Viaux, L., Morard, C., Robillard, B. and Duteurtre, B. (1994). The impact of filtration on Champagne foam behavior. *Am. J. Enol. Vitic.*, 45:407.Jeandet, P.

35 Production of Cider and Perry

V.K. Joshi, Laura Fariña, Somesh Sharma and Naveen Kumar

CONTENTS

35.1 INTRODUCTION

Cider is a low-alcohol drink produced from apples (*Malus domestica* Borkh.). It is a popular beverage, especially in those countries where grapevine cultivation is not practiced due to unsuitable agroclimatic conditions. It is obtained by the complete or partial fermentation of fresh apple juice or from a mixture of the juice from fresh apples and fresh pears, with or without the addition of potable water. 'Cider' is an alcoholic beverage, whereas 'cyder' (used in Australia) is usually a non-alcoholic beverage (apple juice).The fermented juice called 'cider' in England is known as 'hard cider' in the USA because, in that country, 'cider' refers to cloudy unpasteurized apple juice. In Europe, fermented apple juice is known as 'cider' (France), 'Sidre' (Italy), 'Sidra' (Spain), and 'Appelwein' (Germany and Switzerland), where the name for the corresponding unfermented product is clearly distinguished as apple juice. Depending upon its alcohol content, cider can be a soft cider (1–5% alcohol) or hard cider (5–8%), and it may be made from fresh juice or from the juice of a single cultivar, which is then classified as vintage cider. White ciders are made from decolourised apple juice or pale-coloured juice. According to the sugar level, cider can be divided into four categories: dry (without sugar), with sugar (invert sugar) content up to 15 g/dm^3, without sugar, semi-dry (15–30 g/dm^3), or semi-sweet (30–60 g/dm^3). The cider produced by the "*Methode Champenoise*" is called champagne cider. Sparkling sweet cider is made by fermenting apple juice, contains not more than 1% alcohol (v/v), and retains the natural CO_2 formed during fermentation. Sparkling cider has lower sugar content and a higher alcohol content of 3.5% but with only partial retention of CO_2 formed during fermentation. Carbonated cider is charged with commercial CO_2 to produce effervescence.

The beverage obtained by the complete or partial fermentation of the juice of fresh pears or a mixture of the juice of fresh pears and apples, with or without the addition of drinking water, is called 'perry'. The word 'perry' is derived from the word 'Pirrium', for a pear, which, in old Latin, was Pera, and, in old French, it was '*pere*' or 'perey'. Both cider and perry are believed to have been produced for over 2000 years. The first record of cider was documented in 1205, but perry has a less ancient history. In recent years, cider has become an increasingly important commercial product. Cider and perry involve many similar steps in their preparation, which is why these two drinks have been grouped together in this chapter.

35.2 CIDER PRODUCTION TECHNOLOGY

35.2.1 Methods of Making Cider

The traditional procedure for making cider is a complex process that combines two successive biological fermentations: the first one is the classical alcoholic fermentation of sugar into alcohol conducted mainly by strains of the yeast

Saccharomyces cerevisiae, whereas the second one is malolactic fermentation by lactic acid bacteria, that occurs during the maturation process. The latter is an important manufacturing step, to reduce the acidity of the cider and stabilise it with respect to microbial spoilage through the bacteriostatic effect of the lactic acid produced.

Flavour is one of the important sensory attributes that determines the overall quality of cider. The various factors that influence the flavour of cider are the raw material of the apple fruit used for extracting the apple juice, the yeast used in fermentation, fermenter design and operation, secondary fermentation, maturation, etc. There are two methods of cider production, namely the traditional method and the modified method. In the traditional method, unwashed apples of different varieties are used. In the modified method, solely acidic varieties of apples are used, with temperature control during fermentation. The different methods used to produce cider have been reviewed in a systematic way in Table 35.1 while the flow diagram of the process of cider making is depicted in Figure 35.1.

35.2.2 Raw Materials

The quality of cider depends upon various factors, such as the cultivars, the condition of the fruit, and the method of preparation. Theoretically, it can be made from any apple, but the choice of the right cultivar is one of the important factors influencing the quality of cider. Apples for cider making can be separated into two main classes, depending on their use: dessert and cider apples. The latter are generally bitter, astringent, and more rustic. Many cider varieties are more resistant to the major pathogens of apples. Phenolic compounds are

TABLE 35.1
Different Methods Used in Cider Preparation

Type of Method	Fruit	Juice	Parameters: Additive	Fermentation	Maturation	Others
			European			
Method 1	(a) Some stored for 3–4 days and others macerated	Extracted as usual, cold stabilized at 0–7.8°C	SO_2 50–100 mg/L	Temperature 4.4–10°C, pure yeast added in some, mixed in others	Secondary fermentation in casks for several months	Malolactic fermentation, produces CO_2 in bottles
	(b) –	–	Pectic enzymes added for clarification			
Method 2	Lower sugar, higher acidity	Juice extracted, no maceration, juice centrifuged for bacteria and yeast removal	Lactic acid added to increase the acidity, if needed	Pure yeast, such as Steinberg, added	–	–
Method 3	Sound fruits separated by flotation	Juice extracted in a hydraulic press	–	Natural fermentation has specific gravity ranging from 1.008 to 1.005	Storage in concrete tanks lined with a coating	Before delivery, cider is sweetened with syrup
Method 4	–	–	–	Fermentation allowed up to specific gravity of 1.025–1.030 (5–7.5°B), filtered or centrifuged	Stored in wooden casks	Carbonated and bottled
			American			
Method 5	(a) Sound apples are used for cider making	Juice is extracted in usual press after crushing in a mill.	Sulphur dioxide 100–125 ppm added, glucose added to give 13% alcohol Sweetened	Spontaneous fermentation may begin during settling.	–	Clarified by bentonite treatment, mixed with apple juice, blended to give 10 °B, filtered
	(b) Instead of juice extraction, apple juice concentrate is used			Champagne yeast, 24.4°C temperature was the best	–	–

Summarised from: Amerine *et al.* (1980); Joshi *et al.* (1999b); Joshi *et al.* (2011)

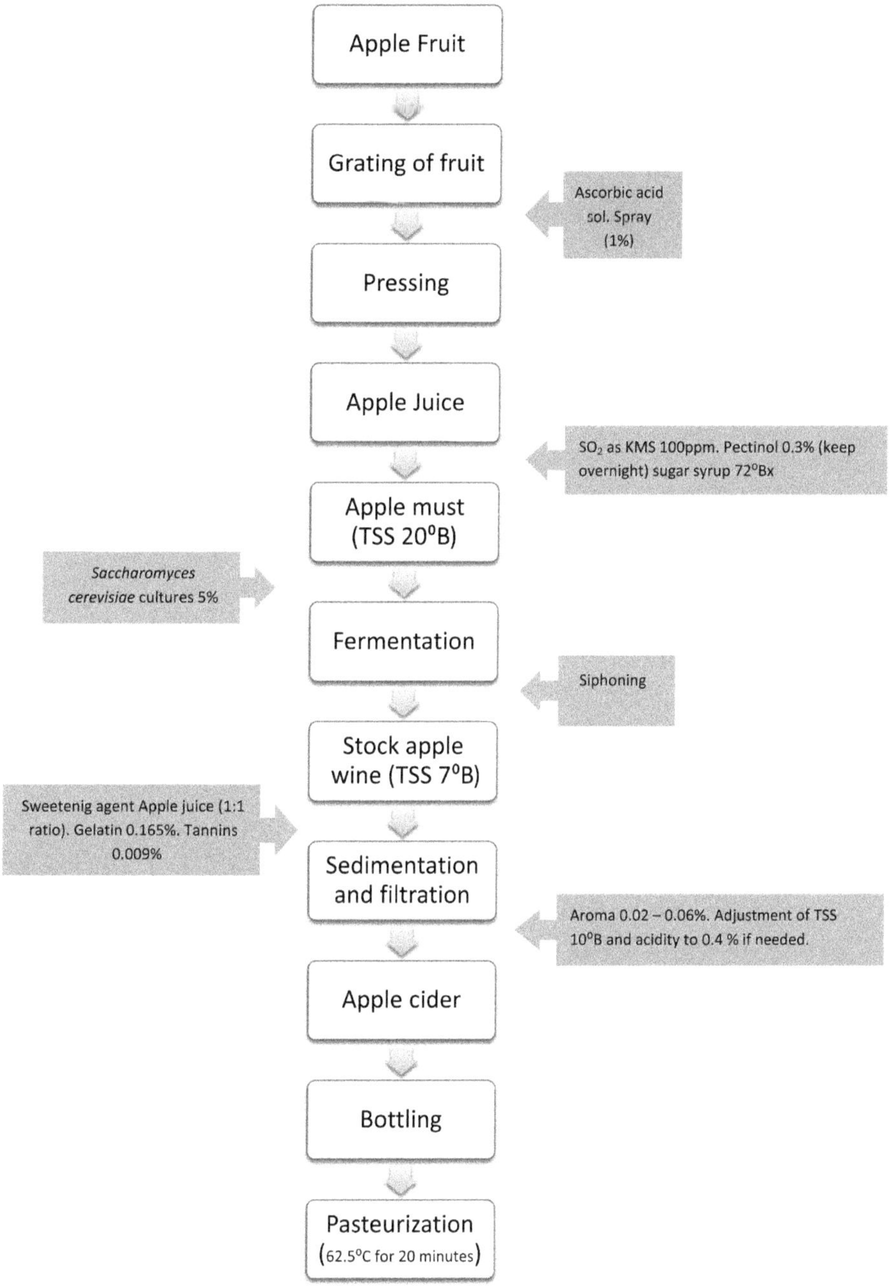

FIGURE 35.1 Flow sheet of preparation of sweet cider. Abbreviations: KMS, potassium metabisulfite; solution; TSS, Total Soluble Solids. *Source:* Based on Amerine *et al.* (1980); Rana *et al.* (1986); Joshi (1997)

responsible for the bitterness, astringency, and colour, and they may also partly contribute to the aroma of cider, as well as to disease resistance. Different varieties of apples suitable for cider making have been recommended, especially those belonging to four different groups: bittersweet (low in acidity but high in tannins), bitter sharp (high acidity and high tannins), sharp varieties (high acidity but low tannins), and sweet varieties (low acidity and tannins). Apples from the different groups are mixed in order to obtain the most appropriate combination of acidity, sweetness, and tannin. The best fruits for cider production are the apples at the early stages of ripeness due to their excellent aroma, whereas overripe fruit are less suitable. The sweet, low-acid cultivars, such as 'Delicious', 'Cortland', 'Ben Davis', and 'Rome Beauty' are recommended for basic apple juice. The juice of 'Jonathan', 'Stayman Winesap', 'Northern Spy', 'Rhode Island Greening', 'Wayne', and 'Newtown' possesses higher acid levels and adds tartness to the cider. 'MacIntosh', 'Gravenstein', 'Ribston Pippin', 'Golden Russet', and 'Delicious' are aromatic and add flavour and bouquet to the cider. The body and flavour can be improved by using astringent apples such as 'Red Astrakhan', 'Lindel', or crab apples (see Chapter 7 of this volume for more details). The use of apple juice concentrate to produce this beverage has increased considerably due to a number of advantages offered by it, such as price stabilisation, quality maintenance, and prolonged storage of concentrate without spoilage. However, this leads to the loss of the development of specific cultivars for cider making. Since the fermentation conditions affect the quality of cider more than do the cultivars, the varietal effect is not given as much importance these days.

In India, cider production is in its infancy and the following varieties used in India have been found to be suitable for cider production: 'Ambri-Kashmiri', 'Red Delicious', 'Golden Pippin', 'Maharaji','Golden Delicious', and 'Rus Pippin', and crab apples. A comparative study of scabbed fruits *vs* non-diseased fruits showed that the juice from fruits with fewer than 15 spots did not affect the fermentation behaviour, nor the physico-chemical or sensory qualities of the cider produced. In traditional apple orchards, fruits are generally allowed to fall naturally from the trees or are shaken from the trees using long poles (lugs). Then, they are picked up either by hand or by machine, but, in intensive bush orchards, mechanically shaking of the tree causes the fruits to fall. Fresh apple fruits from which the juice is to be extracted should be fully ripe, and they are generally stored for a few weeks after harvest so that all of the starch can be converted into sugar. Dessert and culinary apples lose more body and flavour during fermentation than do the cider apples. Blending has always been an important step in controlling the uniformity of the finished product. Juice from apples in the sweet group is considered good for blending with strong-flavoured juices, while those in the bittersweet group give the cider a tangy sensation. Juice for making sparkling sweet cider should not be too sweet or too heavy in body. More acid apples, often with unrepresentative looks, are good for the production of apple cider. The characteristic acrid-bitter taste of good ciders is obtained from the high content of tannin. However, to obtain a flavourful apple cider, apples are required that contain 200–400 mg, or even more, the tannin in 100 g of apple juice. Fruits of varieties used for production of apple cider are totally inedible, being sour, and sometimes even bitter.

Juice used for fully fermented and sparkling ciders are high in sugar, of moderate acidity, and fairly astringent. Certainly, the apple composition is an important factor influencing the quality of cider. Carbohydrates are the principal food constituent in apples, but apples are a poor source of protein. Among the minerals in apple fruits, K, P, and Ca are present in significant amounts. The amino acid composition of cider apples includes asparagine, aspartic acid, glutamic acid, serine, and alanine, while others are present in only traces. Phenolic compounds constitute a significant component of apples, where they also act as antioxidants.

35.2.3 Milling and Pressing

Apples selected for juice processing are washed and inspected for the presence of any foreign materials and decay, because these will have adverse effects on the microbiological status and on the ultimate cider quality (Table 35.1). An early practice is to empty bulk truckloads or bins of apples onto a deleafing screen and into a tank of water. A circulating pump is used to direct the apples to an elevating conveyor which discharges the fruit to an inspection belt. At this point, any damaged or decayed fruit and extraneous material are removed. Rinsing with clean water is accomplished at the scrubber or after inspection. The routine replacement of the holding water is necessary. Fruits are also transferred into the mill using a water flume that provides the additional advantage of washing the fruit. In preparation for pressing, the apples are ground to a mash using either hammer or grating mills, and slicers are required for difficult extractions. Several types of equipment have been developed for pressing and extracting apples juices, such as hydraulic presses, screw presses, basket presses, and pneumatic presses. The choice of equipment depends on the production capacity, product yield, ease of cleaning and sterilisation, and the duration of the production season. For many years, rack and frame presses were used in the apple juice industry. These presses have a frame containing a slatted board covered by a cloth into which a measured amount of mash is transferred and the corners of the cloth are folded over to form an envelope. The frame is removed, and the next slatted board is added together with the frame and another cloth, and this procedure is repeated 10–20 times. Then, it is pressed hydraulically to remove the juice (achieving 80% recovery). In modern plants, the apples are crushed in a grater-type mill made of stainless steel. Next, the pulp is crushed to extract the juice using a cider press. Furthermore, mechanical, hydraulically operated plate presses are also used.

35.2.4 Controlling Microorganisms before Fermentation

For proper fermentation, the microflora of the juice must be controlled before inoculation with yeast, to avoid off-flavours or similar defects in the cider. There are several methods to accomplish this. In Northern France, centrifuging or fining of the juice with gelatin and tannins, followed by filtration to reduce the rate of fermentation, is practiced. Another approach is to treat the juice with pectin-hydrolysing enzymes and then filter before adding the yeast. The danger of bacterial spoilage is still present with these methods, and sulphur dioxide (SO_2) is used extensively to prevent spoilage. The natural fermentation of apple juice depends upon the ability of naturally occurring yeasts in the juice to convert the fruit sugars to ethanol. These yeasts are native to the fruit or they are natural contaminants on the pressing equipment. The typical microflora of freshly pressed apple juice include the following kinds of yeasts: *S. cerevisiae, Saccharomyces uvarum, Saccharomycodes ludwigii, Kloeckera apiculata, Candida pulcheriima, Pichia* spp., *Torulopsis famata, Aureobasidium pullulans,* and *Rhodotorula* spp.; as well as bacteria, such as *Acetobacter xylinum, Pseudomonas* spp., *Escherichia coli, Salmonella* spp., *Micrococcus* spp., *Staphylococcus* spp., *Bacillus* spp., and *Clostridium* spp. Since the effectiveness of SO_2 is dependent on the undissociated form (so-called molecular SO_2), the apple juices should always be brought below a pH of 3.8 by the addition of malic acid before SO_2 addition. Accordingly, the amount of SO_2 added at the pH range 3.0–3.3 should be 75 ppm, 100 ppm at pH 3.3–3.5, and 150 ppm at pH 3.5–3.8. However, it has been noted that juices with a pH greater than 3.8 could not be satisfactorily treated within the legal limit of 200 ppm SO_2. After sulphiting, the juice should be allowed to equilibrate for a minimum of 6 hours before free SO_2 is determined. The addition of SO_2 in cider making has some advantages. It inhibits the development of the native bacteria of the must, decreases the risk of losing a batch, and allows for a better fermentation. SO_2 is also used as an antioxidant and an inhibitor of oxidizing enzymes (to prevent darkening of the must).

35.2.5 Amelioration

Correction of the raw material to make a product of consistent quality is referred to as amelioration (*i.e.*, adjustment of the sugar and/or acid content of the juice as regulated by the respective standards). Sugar and acid levels are very important in cider. During the fermentation, the sugar is converted into alcohol. The amount of sugar in the fresh juice reflects the alcoholic strength of the final product while the acid level should decrease during fermentation. If the cider is contaminated by *Acetobacter*, fermentation is the net result and the final product is not cider but cider vinegar. The initial sugar concentration (ISC) influences the quality of the cider and a TSS value of 20°B (degrees Brix) was found to be optimal (Figure 35.2). The nitrogen-containing compounds of must are important for the growth of yeast and, hence, for the fermentation rate and the production of aroma compounds. Supplementation of nitrogen (as a nutrient for yeast) is in the form of diammonium hydrogen phosphate (DAHP) and a concentration of 0.1% has been used. The addition of DAHP is essential for rapid fermentation.

35.2.6 Inoculation

The traditional method of cider making does not employ any external source of yeast. The indigenous microflora of apples (of the order of 5×10^4 CFU/g of stored fruits) carries out the spontaneous fermentation. After sulphiting, the juice is inoculated with the desired yeast culture in cases where inoculated fermentation is employed. The growth of yeasts, acetic acid bacteria (AAB), and lactic acid bacteria (LAB) can be reduced by washing and sorting of apples before milling and pressing. High counts of bacteria, including LAB, were observed during alcoholic fermentation and storage of cider. The levels of LAB found in musts fermented in small vessels using acid-washed apples were lower. However, the must fermented by using unwashed apples, blended from different varieties, had only a limited number of microorganisms. The growth of microorganisms can be limited by fermentation and storage at a temperature of 10^0C,in combination with low pH. The desirable characteristics of yeasts used in cider making include their ability to produce polygalacturonase to breakdown soluble pectins and to produce rapid onset of fermentation. They should be relatively resistant to SO_2, low pH values, and high ethanol concentration, and have low requirements for vitamins, fatty acids, and oxygen. Fermenting to 'dryness' should not produce excessive foam. They should efficiently use the sugars with minimal production of SO_2. They should not produce hydrogen sulphide (H_2S) or acetic acid, but should produce the required aroma components, organic acids, various esters, and glycerol. In a traditional cider fermentation, where no yeast is added and no sulphite is used, the first few days are dominated by the non-*Saccharomyces* yeast species (*Candida pulcherrima*, *Kloeckera apiculata*, and species of *Pichia, Torulopsis*, and *Hansenula*), which multiply quickly to produce a rapid evolution of gas and alcohol and generate a distinctive range of flavours characterised by ethyl acetate, butyrate, and related esters. As the alcohol level rises (to 2–4 percent), these initial fermenters begin to die out and the microbial succession is taken over by *S. uvarum*, which completes the fermentation. The yeast cells become sub-lethally damaged by the

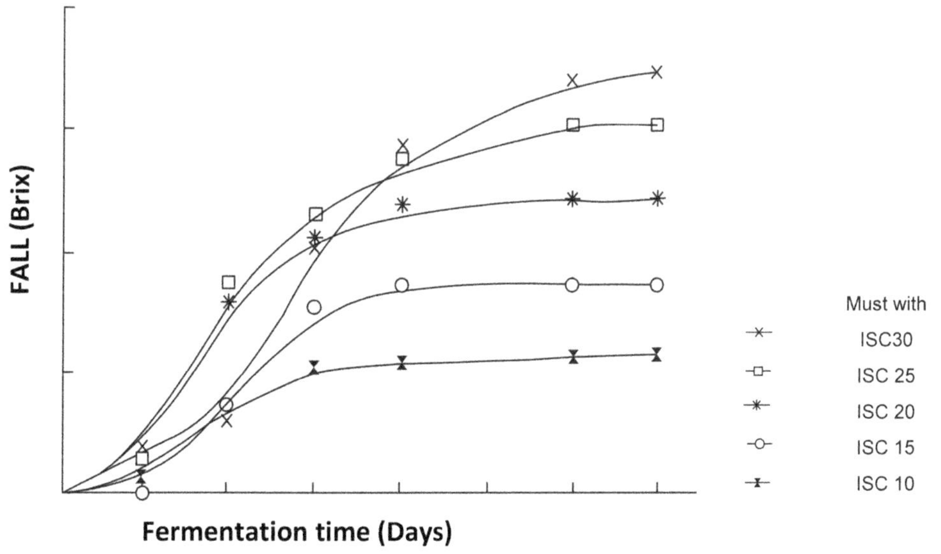

FIGURE 35.2 Effect of initial sugar concentration on the rate of fermentation in apple must made from apple juice concentrate. *ISC*: initial sugar concentration. *Source:* Joshi and Sandhu (1997)

increasing concentration of alcohol, but death does not occur until the concentration exceeds 9% alcohol (v/v). The number and type of yeasts in the juice is also influenced by the method of juice extraction; consequently, the quality of cider is also affected.

A suitable yeast strain is selected for the preparation of cider. It could be an in-house culture propagated and maintained in flasks or larger containers, or it could just be a commercially produced dried yeast, which is a simpler method. The use of a mixed inoculum of *S. uvarum* and *Saccharomyces bayanus* is a widespread practice on the grounds that the first species provides a speedy start, and the second copes better with fermentation to dryness to produce a high alcohol content. These dried yeasts require no pre-propagation and are simply hydrated in warm water before pitching them directly into the juice. A small quantity of heat-sterilised juice is inoculated with a dry culture or a liquid-nitrogen-frozen culture and, after fermentation, the inoculum is added to a larger volume of sulphited juice. The procedure is continued until a final inoculum of 1% or greater by volume is obtained.

Various yeast species have been isolated during cider making at different periods, *i.e.*, at the beginning of fermentation(A),during active fermentation (B), and in the third period (C), when the fermentation is slow. The categories of yeast present during fermentation are the yeast species with strong fermentative metabolism, apiculate yeast, and the species with an oxidative metabolism. During initial fermentation, *K. apiculata* appeared to be the dominant species. Later, during active fermentation, the numbers of *K. apiculata* decreased, and the rapid growth of *S. cerevisiae* took place. Interestingly, alcoholic fermentation was carried out by *K. apiculata* and *S. cerevisiae*, and their distribution was found to be similar in both the fermentation methods. In the traditional method, the malolactic fermentation proceeded at the same time as alcoholic fermentation, but, in the modified method, no malolactic fermentation occurred, but this produced cider with a lower content of volatile acids. Controlled malolactic fermentation in cider, using the LAB, *Oenococcus oeni*, immobilized in alginate beads as a starter culture, results in malic acid degradation similar to that achieved with the free cells of *Oenococcus*. Another interesting approach for continuous production of cider was attempted with the addition of Ca-alginate material to co-immobilized *Saccharomyces bayanus* and *O. oeni* in one integrated biocatalyst system. This permitted much faster fermentation than could be achieved from traditional cider making, with better flavour from the fermentation.

As in other fermented beverages, the tendency in recent years has been the utilisation of non-conventional, non-*Saccharomyces* yeasts in the production of cider. The screening of 163 non-*Saccharomyces* yeast species revealed that *Candida zemplinina, Hanseniaspora vineae* and *Torulaspora delbrueckii* were found to produce relatively more volatiles than the other species that were evaluated. The use of these non-conventional yeasts produces ciders with notes like tropical fruits, kernel fruits, and stone fruits, with floral aromas.

35.2.7 Fermentation

Traditionally, barrels of oak would have been used for the fermentation of cider. Wooden barrels or vats of mild steel with a ceramic or resin lining, bitumen-lined concrete vats, or, more recently, stainless steel and even lined fibreglass-resin vats or tanks have been commonly employed for cider fermentation. Predominantly, stainless steel tanks (vertical, conico-cylindrical) are now used for the fermentation of cider. These tanks may be equipped with temperature control systems, level indicators, and carbon dioxide venting and blanketing systems. Correct sulphiting of the juice and proper cleaning of all the equipment will help ensure a good start to fermentation. The best procedure for assuring a good fermentation is to employ a large inoculum, as all yeast strains perform in the same manner at the same concentrations. If exposed to air during fermentation, either juice or cider will usually develop a surface film of acetic acid bacteria or yeast. This aerobic spoilage can be prevented by excluding air from the vats by properly sulphiting the juice. Heated juice ferments faster than unheated juice but sulphited juices ferment more slowly than those without sulphite. The availability of soluble nitrogen in the juice can affect the rate of cider fermentation.

Fermentation with *Schizosaccharomyces pombe* reduced malic acid concentration in several fruits, including apple, though with low rates of alcohol production. *Leuconostoc* spp. have also been employed to reduce the acidity of the fermented product. Simultaneous inoculation of apple juice with *S. cerevisiae* and *Schizo. pombe* produced cider with acceptable levels of alcohol and acidity. Ion-exchange sponges, with a tailored surface charge for immobilisation of *S. cerevisiae*, encouraged yeast growth and reduced the fermentation rate. Temperature affects both the rate of fermentation and the nature of the metabolites formed. It takes 3–4 weeks to attenuate cider fermentation at temperatures within the range of 20–25°C although a temperature of 15–18°C is preferred in Germany and France for flavour development. Higher temperatures increase the rate of fermentation but enhance the risks of contamination with undesirable thermophilic microorganisms. The best-flavoured cider is generally produced from a slow fermentation process by maintaining the temperature at around 16°C and reducing the yeast population by racking or by the addition of sulphur dioxide. However, if the cider fermentation is too slow, it may be susceptible to cider sickness, which imparts a milky white appearance and a sweet pungent odour to the cider. Occasionally, a fermentation may be slowed down or even stalled. Aeration has been found to be useful for restoring yeast activity in such cases. Stuck fermentations could be restarted if the temperature was 12–13°C, some fermentable sugar remained and at least 10,000 yeast cells per mL were present in the fermenting must (for details on this aspect, see Chapter 18 of this volume).

35.2.8 Clarification

Once the fermentation is finished, the cider is stored for a few days to allow for the liberation of amino acids and enzymes produced by the autolysis of yeast.

The objective of the clarification process is to make the cider clear. This can be done by using a closed filter system, which avoids exposing the cider to air (with the concomitant risk of acetic acid bacteria contamination), or by mixing gelatin, bentonite, and pectic enzymes into the cider. After this clarification process, the cider is transferred into storage vats.

35.2.9 Ageing/Maturation and Secondary Fermentation

After clarification, the cider may be stored in recipient vessels for large volumes, or bottled. In all the cases, extreme care in the sanitation of the storage vessels is needed in order to prevent contamination with undesirable microorganisms. One fundamental factor in storage is the temperature, which can be as low as 4°C but not higher than 10°C. Sparkling or charged ciders have to be stored in pressure tanks to avoid loss of CO_2. Uncharged cider should be kept in a full, closed tank with an air trap or under a blanket of carbon dioxide, nitrogen, or a mixture of the two. If air is not excluded from the tanks, the risk of spoilage occurring from acetic acid bacteria or film yeast is high.

Maturation is an important step in cider making. Like other alcoholic beverages, ageing in oak barrels is a very old tradition, and it is used to improve the quality and the sensory characteristics of wine, such as important improvements in aroma, colour and taste. However, the oak composition and the conditions under which wine/cider maturation take place are of primary importance for the impact of oak on wine cider quality. Several aroma compounds which have been identified in apple wine include *cis*- and *trans*-oak lactones (*cis*- and *trans*-methyl-octalactone), vanillin (4-hydroxy-3 methoxybenzaldehyde), eugenol (4-allyl-2-methoxyphenol), and guaiacyl derivatives (Table 35.2). These are derived principally from the oak barrels during the process of oak barrel maturation of wine and cider. Enumeration, isolation, and identification of LAB during processing and storage of Australian cider revealed *O. oeni* as the predominant bacterium. During maturation, the growth of LAB can occur extensively, especially if wooden vats are used. The ensuing malolactic fermentation (MLF) would convert malic acid into lactic acid, reduce acidity and impart subtle flavours, which generally improve the flavour of the product, though sometimes overproduction of LAB metabolites damages the cider flavour(*e.g.*, excessive production of diacetyl, the butterscotch-like taste). Production of aldehydes, among the flavour compounds, takes place, the most relevant one (acetaldehyde) being produced by the oxidation of ethanol (in a direct chemical reaction with air). Alcohols in wine react with organic acids (tartaric, malic, succinic, and lactic acids) to form esters which increase with the ageing of wine, as happens with the increasing total concentration of volatile compounds. Higher alcohol formation was closely related to the aroma and taste of wine. During maturation, there is a decrease in the tannins, due to their complexing with proteins and polymerization, and subsequently, precipitation takes place.

35.2.10 Final Treatment and Packaging

Prior to packaging different batches of cider, which are generally made from mixtures of different juices, are blended to achieve a specific flavour. Cider can be sold as a still or a sparkling beverage, with varying degrees of sweetness and clarity. The amount of carbonation ranges from saturation for ciders in jars to 2–2.5 volumes of CO_2 in most bottled ciders and up to 5 volumes in champagne cider. Carbonation pressure ranges from 2.5 to 3.5 bar, with higher pressure being used in the case of polyethylene terephthalate (PET) plastic bottles. The sweetening agent may be from unfermented juice, sugars, or concentrate, depending upon the appropriate regulations. The clarity of ciders can range from turbid farm cider to brilliantly clear commercial ciders. The majority of commercial ciders are decanted into kegs, bottles, or cans. Keg cider is carbonated and pasteurised and filled in the plant into stainless steel kegs, which are rinsed, washed and sterilised prior to filling. It can also be filled into glass bottles that may be carbonated and pasteurised after filling. Common container closures are crown caps and roll-on or plastic stoppers that have replaced corks. The cider is then pasteurised at 60°C for 20–30 min or preserved with SO_2 as the best practical solution. Vinyl spotted caps are satisfactory for bottling ciders. Inert gases like CO_2, N_2, or mixtures thereof, can also be used for the storage of cider. Locally sold still cider may be sold in plastic containers after being dispensed from a refrigerated bulk container. Carbonated cider is either sterile-filtered or flash-pasteurised before packaging.

35.3 QUALITY OF CIDER

35.3.1 Chemical Composition of Cider

The apple variety used is the primary factor influencing the composition of the volatiles of apple cider, while the effect of yeast strains and the maturity of apples is highly variety-specific. The most important compounds formed during fermentation, which are considered as key products affecting the quality of cider, are higher alcohols, esters, organic acids, and carbonyl compounds, jointly with sugars and tannins (provided by the apples) (Table 35.3). Different types of ciders are classified according to their ethanol content, which varies from 0.05 to 13.6%. Methanol is also produced in small quantities (10–100 ppm) as a result of demethylation of pectin in the juice. Like apple must, cider contains a variety of organic acids, the concentrations of which depend on the maturity and fermentation conditions. Dry ciders, made by the traditional method of fermentation in which the apples are not washed, have a volatile acidity higher than the ciders made after washing and blending the apples. In storage, the acetic acid content increases. During the traditional method of fermentation, malic acid in the must is found in lower amounts (3.0–3.8 g/L) than that made by modern fermentation (4.8 g/L).

The formation of higher alcohols is an important criterion to determine the quality of the alcoholic beverages, depending upon yeast strain, cultivars used, and the fermentation conditions employed. The biosynthesis of higher alcohols is

TABLE 35.2
Concentration of Alcohols, Acids, and Esters in Cider Aged With Oak Chips (µg/L)

			Ciders Aged With		Ciders Aged With French Oak			Ciders Aged With American Oak		
Volatile compounds	**Control**	**Chinese**	**American**	**French**	**Light**	**Medium**	**Heavy**	**2 g/L**	**4 g/L**	**8 g/L**
Ethyl acetate	23.4	23.8	27.8	25.1	23.73	25.1	24.1	22	27.8	25.3
Ethyl butanoate	0.13	0.18	0.214	0.192	0.172	0.192	0.268	0.173	0.214	0.138
3-Methylbutyl acetate	0.917	1.36	1.65	1.28	1.75	1.28	2.25	4.76	1.65	4.26
Butyl acetate	0.209	0.057	0.018	0.717	0.301	0.717	0.29	ND	0.018	0.008
Ethyl hexanoate	0.436	0.43	0.458	0.501	0.6	0.501	0.906	0.29	0.458	0.323
Hexyl acetate	0.009	0.012	0.011	0.032	0.009	0.032	0.011	0.021	0.011	0.012
Ethyl heptanoate	0.003	0.006	0.006	0.041	0.004	0.041	0.002	0.004	0.006	0.029
Ethyl octanoate	0.508	0.302	0.378	0.299	0.547	0.299	0.897	0.657	0.378	0.713
Ethyl dodecanoate	0.1	0.068	0.063	0.063	0.111	0.063	0.136	0.071	0.063	0.117
2-Phenylethyl acetate	0.234	0.567	0.361	0.233	0.241	0.233	0.291	0.184	0.361	0.161
Total Esters	**25.946**	**26.783**	**30.958**	**28.458**	**27.465**	**28.458**	**29.151**	**28.159**	**30.958**	**31.061**
1-Propanol	94.5	82.1	90.4	82.8	67.4	82.8	128	73.8	90.4	71.5
2-Methylpropanol	60.3	50.8	49.7	47	59.7	47	42.4	61.6	49.7	68.4
1-Butanol	0.339	ND	ND	0.28	0.228	0.28	0.528	ND	ND	0.26
3-Methylbutanol	117	106	113	104	109	104	92	94.8	113	114.2
1-Hexanol	0.055	0.128	0.116	0.12	0.076	0.12	0.061	0.038	0.116	0.027
2-Phenylethanol	6.9	7.97	7.12	12.5	6.3	12.5	4.89	20.9	7.12	16
Total alcohols	**279.094**	**246.998**	**260.336**	**246.7**	**242.704**	**246.7**	**267.879**	**251.138**	**260.336**	**270.387**
Hexanoic acid	0.987	1.76	1.675	1.86	1.94	1.86	1.762	2.21	1.675	1.08
Octanoic acid	2.37	2.61	2.375	4.23	3.05	4.23	1.39	2.31	2.375	2.871
Decanoic acid	0.198	0.274	0.268	0.382	0.206	0.382	0.43	0.089	0.268	0.2
Total acids	**3.555**	**4.644**	**4.318**	**6.472**	**5.196**	**6.472**	**3.582**	**4.609**	**4.318**	**4.15**

Source: Chatonnet *et al.* (1999); Jarauta *et al.* (2005)

TABLE 35.3
Mean Concentration (mg kg–L of Fresh Weight) of Phenolic Compounds Present in the Whole Fruit (2008 and 2009 Harvest Years)

Parameter	Total Catechins	Total PCA	Total Flavanols	DPn	Total HA	Total DHC	Total Flavonols	Ideain	Total Polyphenols
Average	253	1508	1761	30	715	131	94	13	2707
Median	237	1394	1644	29	654	127	88	11	2488
Minimum	33	376	592	21	86	35	20	3	1058
Maximum	656	4112	4769	56	2000	332	274	58	6418

Source: Verdu *et al.* (2014)

generally linked to amino acid metabolism. It is also known that higher fusel alcohols (mixtures of several alcohols (chiefly amyl alcohol) produced as a by-product of alcoholic fermentation) are generated from cloudy rather than from clear juice fermentation.

The phenolic concentration and profile have important effects on the sensory properties of apple ciders, such as colour, bitterness, and astringency. The presence of procyanidins (high molecular-weight phenolics) is known to contribute to astringency, whereas the lower molecular-weight compounds contribute to the bitter taste. Simultaneously, they influence the sweetness and sourness of cider, thus further highlighting their importance in overall flavour development. Volatile phenolics, mainly formed by enzymatic decarboxylation during fermentation, contribute to aroma.

The cultivar and ripening stage of apples (the latter to a lesser degree) have an important role to play to determin the phenolic composition of cider. Unripe apples contain higher contents of the phenolic compounds than ripe apples, although this is dependent on the cultivar.

As an example, Figure 35.3 shows the concentrations (mg/100 mL) of the different types of polyphenol found in bittersweet English cider: chlorogenic acid (98), epicatechin (38), dimeric procyanidin (79), trimeric procyanidin (26), and tetrameric procyanidin (21).

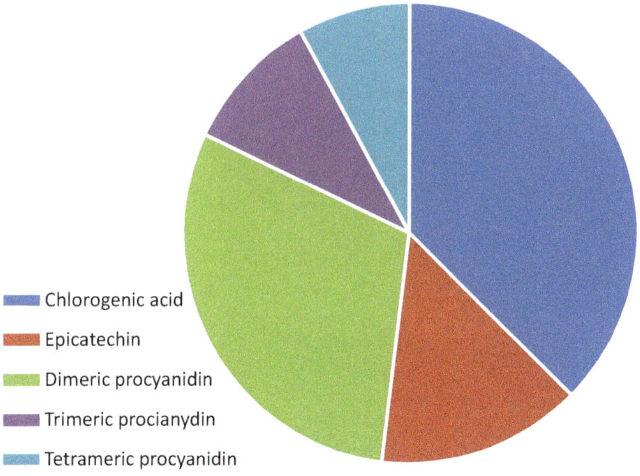

FIGURE 35.3 Polyphenols in bittersweet English cider. The concentrations of polyphenols (mg/100 mL) in cider are shown. *Source:* Based on Lea *et al.* (1984)

No significant change in phenolic content takes place during fermentation, although the chlorogenic, caffeic, and *p*-coumaryl acids may be reduced to dihydroshikimic acid and ethyl catechol. The chlorogenic and caffeic acids contents in apple juice and cider apple cultivars correlated very closely with total phenols.

Chlorogenic acid constitutes 6.2–10.7% of total phenols, and this acid is responsible for non-enzymatic, auto-oxidative, browning reactions. The most important carbonyl compounds formed during cider fermentation are acetaldehydes, diacetyl, and 2,3-pentanedione. Aldehydes, having very low flavour thresholds, tend to be considered as off-flavours (green leaf-like flavours). As intermediates in the formation of ethanol and higher alcohols from amino acids and sugars, the formation of small quantities of aldehydes takes place, which are excreted and then, reduced to ethanol during the later stages of fermentation. Diacetyl makes an important contribution to the flavour of cider and its presence is considered essential for the correct "cider flavour". This compound is mainly formed during malolactic fermentation. Generally, esters are considered desirable, and they are present in smaller concentration than alcohols, with the notable exception of ethyl acetate and 2-and 3-methyl butyl acetates which, in 'Yarlington Mill' juice, increases 200-fold during fermentation. Esters constitute a major group of desirable flavour compounds. Among the esters that can be formed, the most significant in fermented beverages are ethyl acetate (fruity flavours), isoamyl acetate (pear drops), isobutyl acetate (banana-like), ethyl hexanoate (apple-like) and 2-phenyl acetate (honey, fruity, flowery flavours). They are formed by yeasts during fermentation in a reaction between alcohols, fatty acids, co-enzyme A, and an ester-synthesising enzyme.

35.3.2 Sensory Qualities

Appearance, colour, aroma, taste, and other factors constitute the sensory quality of cider. The aroma and taste are determined by a number of factors, such as the apple varieties, vinification practices, fermentation, and maturation. The taste of cider depends more on the composition of the apples, whereas cider odour is governed more by technological factors and the yeast employed more than the apple varieties used to make the cider. Cider with higher juice content is preferred in terms of sensory quality characteristics. The influence of cidermaking technology on low-boiling-point compounds can also be clearly seen in the preparation of semi-sweet cider. The cider flavour is assessed using both subjective and objective approaches. In the subjective approach, panels of trained or untrained judges can be employed to recognise specific flavours.

In human beings, flavour sensation by taste is limited to sweetness, sourness, bitterness, and astringency, together with such tastes as metallic and pungent. Quantitative descriptive analysis has also been applied to cider in order to profile their flavours analytically. Out of 86 descriptors used, 33 descriptors made the greatest contribution towards characterising the cider aroma and perry essence. A cider flavour wheel, like that for the beer flavour profile, is employed in the cider industry (Figure 35.4). At a simple level, a number of general descriptors can be used (fruity, acidity, sweetness, astringency, alcohol, body, bitterness, and sulphury). But, at an analytical level, the number of descriptors is kept large to differentiate the ciders of different types. Another approach which has been applied to flavour profiling is sniff analysis, where the fractionated effluent from a gas chromatograph (GC) is assessed by specially trained judges. Capillary gas chromatography head-space samples of cider has been performed to characterise the aroma compounds. As many as 200 compounds contribute to the flavour of ciders, but the key compounds are alcohols, acids, aldehydes, esters, and sulphur compounds. The spicy, aromatic, and apple notes differentiate the cider from other fermented beverages. Changes in the volatile profile over time have been described, as was reported in Spanish ciders (originating from Asturian and Basque regions). The content of 4-ethylcatechol was found to be closely associated with older samples, whereas 4-vinylguaiacol was linked with young ciders. The colour of cider is determined by the extent of juice oxidation or degradation, and it is possible to make water-white high-tannin ciders if oxidation is completely inhibited. During fermentation, however, the initial colour diminishes by around 50%. Traditional English and French ciders, made from bittersweet fruit, have been distinguished by relatively high levels of bitterness and astringency caused by the procyanidins (tannins).

35.3.3 Spoilage of Cider

Some sweet ciders, with a residual sugar content, with a pH above 3.8 and stored at ambient temperature, develop a defect called ropiness or oiliness. This is caused by certain strains of lactic acid bacteria (*Lactobacillus* and *Leuconostoc* spp.)

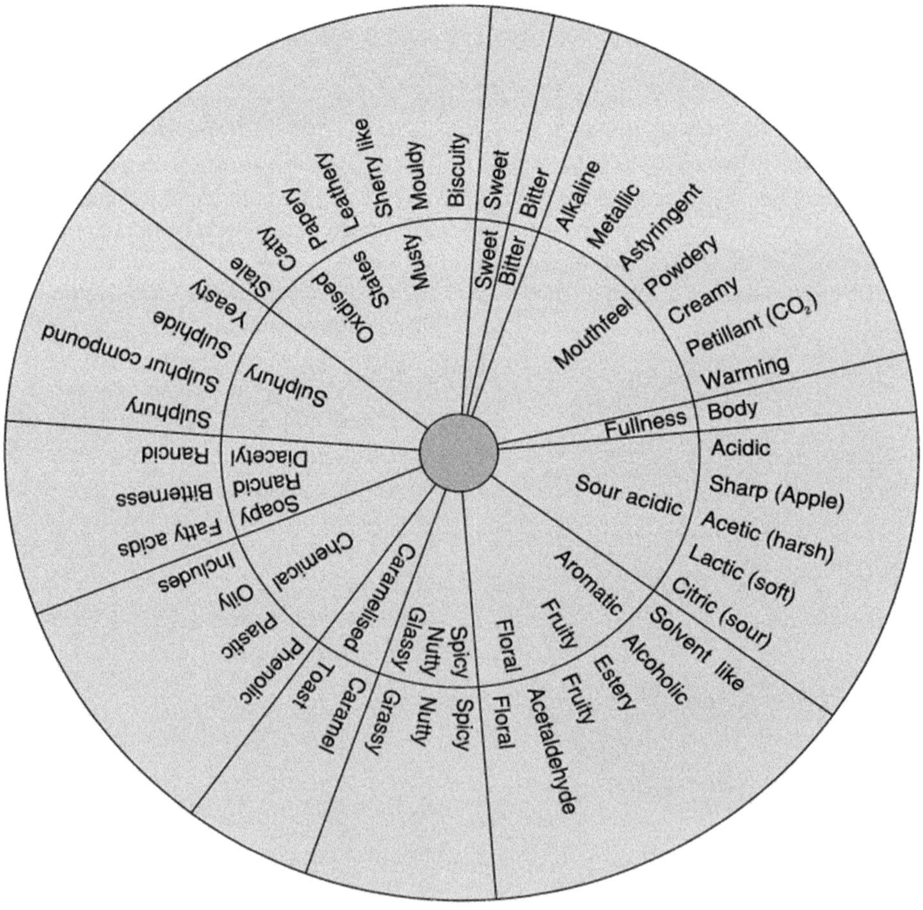

FIGURE 35.4 Cider flavour wheel. *Source:* Anon. (1994); Joshi *et al.* (2011c)

that produce a polymeric glucan. It thickens the cider's consistency, and, when poured, it appears oily in texture with a detectable sheen. Consequently, it moves like a slimy 'rope' when poured from a bottle. Properly sulphited juice with a low pH can correct the defect. However, the treated cider would need to be blended before use. A much simpler approach would be to pasteurise the affected juice. Another defect is referred to as mousiness which occurs in un-sulphited cider with a high pH that has necessarily been exposed to air during fermentation. The growth of film-forming yeasts, such as *Brettanomyces* spp., *Pichia membranefaciens*, and *Candida mycoderma* also produce a 'mousy' flavour (based on 1,4,5,6-tetrahydro-2-acetopyridine). *Saccharomycodes ludwigii* is often resistant to SO_2 levels (1000–1500 ppm) and can grow slowly during fermentation and maturation, resulting in the production of a butyric acid flavour (vomit, rancid cheese) and the formation of flaky particles which spoil the appearance of the cider. Contamination of the final product with *S. cerevisiae*, *Saccharomyces bailli or S. uvarum* increases the concentration of CO_2.

Ciders low in acidity, tannins, and nitrogen but high in mineral matter occasionally develop an olive-green colour, the fermentation ceases, and starch is deposited. The combination of iron with tannins in ciders produces a black or greenish-black colour. Bottled cider stored at high temperature produces a sediment called casse. This is caused by the action of peroxidase on tannins. It can be prevented by the addition of SO_2 after completion of fermentation. The classical microbiological disorder of stored bulk ciders is called as cider sickness or 'Framboise' in French which is caused by the bacterium *Zymomonas anaerobia*, fermenting sugar in bulk sweet ciders stored at pH values greater than 3.7. The features include a renewed and 'almost explosive fermentation', accompanied by a raspberry or banana-peel aroma and a dense white turbidity in the beverage, which is due to the production of acetaldehyde at high levels by *Zymomonas*. The acetaldehyde reacts with the tannins to produce an insoluble aldehyde–phenol complex and, consequently, turbidity. Flavour taints in ciders may arise from the presence of naphthalene and related hydrocarbons, where tarred rope might have been stored adjacent to a cider keg. A new taint in ciders is caused by indole, and this is derived from tryptophan breakdown at concentrations in excess of 200 ppb, where its odour becomes increasingly faecal and unpleasant.

35.4 PERRY

Pears (*Pyrus communis* L.) are grown in the temperate areas of the world and are consumed fresh, but some of them are used in processing perry (pear cider). The pear cultivars fall into two groups: European pears (*i.e.*, pears having soft-fleshed fruit with inconspicuous grit cells) and oriental (Asian) pears (*Pyrus pyrifolia*). 'Bartlett' is the most commonly grown European pear cultivar. A few other cultivars of this type include 'Giffard', 'Precoce', 'Marettini', 'Sekel', 'Anjou', 'Bose', 'Conference', 'Easter Winter', 'Nelis', 'Forelle', 'Kieffer', 'Flemish Beauty', etc. Asian or oriental pears cultivars include 'A-Ri-Rang', 'Chojuro', 'Ichiban Singo', 'Seuri', 'Yokumo', and 'Kosue'. Nutritionally, pears are considered to be a fairly good source of fibre and malic acid. The alcoholic beverage made from pear is called perry. The National Association of Cider Makers has drafted standards for British Cider and Perry. It reads that perry is: 'the beverage obtained by the complete or partial fermentation of juice of the pear or a mixture of the juice of pears and apples, with or without the addition of water, sugars, or concentrated pear or apple juice, provided that not more than 2.5% of the juice shall be apple juice'. The production of perry can be a promising alternative for the utilization of sand pear fruit, which has a very limited outlet for its direct consumption. Production of perry from sand pear has been successful.

35.4.1 Composition of Pears

Asian pears have a sweet to sweet-tart taste and a fragrant aroma. They contain up to 15% natural sugar, although 9–12% is more typical. Pears contain three major sugars, namely sucrose, glucose, and fructose. Fructose occurs in the greatest concentration, at about 7.0% (w/v), whereas the glucose content is usually low (2–2.5 w/v), with sucrose at about 1% (w/v). Sorbitol (a natural sugar alcohol, which acts as a sweetener) is also found in perry juice (1–5% (w/v)), but it is not fermented by yeasts so it remains after fermentation and increases the specific gravity of dry perry. The organic acids present in perry pear include predominantly malic acid (0.45–0.88% (w//v)) and quinic acid (0.05–0.13% (w/v)), while citric, citramalic, lactic, and succinic acids are found in traces. Pears produce juice with pH values within the range of 3.6–3.8. The nitrogen content in perry pears does not exceed 10mg/100mL, and the amino acid that occurs in the greatest quantity in pears is proline. The other amino acids include aspartic acid, asparagine, and glutamic acid. Only one group of tannins is capable of combining with protein and should, more precisely, be called procyanidins, containing within their molecules a phenol structure which is associated with bitterness and astringency.

35.4.2 Process of Making Perry

Perry of good quality can be made from the pears with high tannin content, such as the 'Bartlett' pear. The fruit used to make perry should be astringent to taste and usually full of stone cells or "grit", but ripe dessert pears are not successful in perry making, unless the natural esters of the fruits are removed. Many of the steps to made perry are similar to those for cider. For the preparation of perry, the fruits are grated and pressed in a rack and cloth press. The juice with an original total soluble solids (TSS) of 14°B and 0.25% acidity is ameliorated to 21°B and 0.5% citric acid along with 100 ppm SO_2. For rapid fermentation of sand pear juice, an exogenous addition of a nitrogen source is made, and the TSS is raised to 20°B with the addition of 0.1% DAHP as a nitrogen source. Due to inversion, sucrose is unlikely to be present in any significant quantity at the beginning of fermentation, and the glucose and fructose are fermented by the yeasts. The sugar and sugar-like compounds that the yeast does not metabolise are left in dry perry, so that xylose can be present in quantities of about 0.05% (w/v). In dry perries, the sorbitol content can be in excess of 4% (w/v). Probably, the most important change that takes place during fermentation or storage is the breakdown of organic acids. One change in the organic acids of perries is the metabolism of citric acid. To make perry, the alcoholic fermentation is carried out at 22±1°C by addition of a yeast culture at a rate of 5%. The effect of the addition of different sugar sources (sucrose, fructose, glucose, honey, molasses, or jaggery) on the fermentation behaviour and the sensory quality of perry showed that the highest rate of fermentation was recorded in must made with jaggery (cane sugar), followed by honey, glucose, fructose and molasses, with the lowest rate being in the case of sucrose. Temperature affected the fermentation of sugar to ethanol by the yeast. The maximum ethanol yield, with a corresponding yeast fermentation efficiency of 87.79%, was 8.85% (v/v) in juice fermented at 30°C.

Perry with 5% alcohol, 10°B TSS, and 0.5% acid content was found to be acceptable in terms of sensory quality. The sensory evaluation of perry from different treatments, after six months of maturation, showed that the product made with jaggery had a more attractive colour than the others, whereas all the products were comparable in taste, except for the perry made from molasses, which was unacceptable. Except for a little after-taste, the product made from must with jaggery had the best overall acceptability.

BIBLIOGRAPHY

Amerine, M.A., Kunkee, R.E., Ough, C.S., Singleton, V.L. and Webb, A.D. (1980). *Technology of Wine Making*. AVI Publ. Comp. Inc, Westport, CT.

Anon. (1994). National association of cider maker, 1994.

Beech, F.W. (1993). Yeasts in cider making. In: *The Yeasts*, 2nd edition, Vol. 5, *Yeast Technology*, A.H. Rose and J.S.Harrison (eds.). Academic Press, London, p. 169.

Beech, F.W. and Carr, J.G. (1977). Cider and Perry. In: *Economic Microbiology*, A.H.Rose (ed.). *Vol.VI, Alcoholic Beverages*. Academic Press, London, p. 139.

Carr, J.G. (1987). Microbiology of wines and ciders. In: *Essays in Agricultural and Food Microbiology*, J.R. Norris and G.L. Pettipher (eds.). John Wiley, London, p. 291.

Chatonnet, P., Cutzach, I., Pons, M. and Dubourdieu, D. (1999). Monitoring toasting intensity of barrels by chromatographic analysis of volatile compounds from toasted oak wood. *Journal of Agriculture and Food Chemistry*, 47(10): 4310–4318.

Downing, D.L. (1989). *Processed Apple Products*. AVI Publishing Company, New York.

Frącek, G. (2012). *Dajmyszansęcydrom w Polsce*. Przemysł FermentacyjnyiOwocowo-Warzywny, 2012, no. 11–12.

Herrero, M., Laca, A., Garcia, L.A. and Diaz, M. (2001). Controlled Malolactic fermentation in cider using *Oenococcus oeni* immobilized in alginate beads and comparison with free cell fermentation. *Enzyme and Microbial Technology*, 28(1): 35.

Jarvis, B. (1993). Chemistry and microbiology of cider making. In: *Encyclopedia Food Science and Nutrition*, 1st edition. Caballero, B. (ed.). Academic Press, US, Massachusetts.

Jarvis, B. (1993). Cider: Cyder; hard cider. In: *Encyclopedia Food and Nutrition*, 1st edition. Caballero, B. (ed.). Academic Press, US, Massachusetts.

Jarvis, B., Foster, M.J. and Kinsella, W.P. (1995). Factors affecting the development of cider flavour. *Journal of Applied Bacteriology Symposium Supplement*, 79: 55.

Jarauta, I., Cacho, J. and Ferreira, V. (2005). Concurrent phenomena contributing to the formation of the aroma of wine during aging in oak wood: An analytical study. *Journal of Agriculture and Food Chemistry*, 53(10): 4166–4177.

Joshi, V.K. (1997). *Fruit Wines, Directorate of Extension Education*, 2nd edition. Dr. YS Parmar University of Horticulture and Forestry, Nauni, Solan (HP), p. 255.

Joshi, V.K. and Attri, B.L. (2017). Pome fruit wines: Production technology 7.1. Specific features of table wine production technology Chapter 7. In: *Science and Technology of Fruit Wines*, Maria Kossovea, V.K. Joshi and P.S. Panesar (eds.). Elsevier, UK, London, pp. 295–347.

Joshi, V.K. and Sandhu, D.K. (1997). Effect of different concentrations of initial soluble solids on physico-chemical and sensory qualities of apple wine. *Indian Journal of Horticulture*, 54(2): 116–123.

Joshi, V.K., Sandhu, D.K., Attri, B.L. and Walia, R.K. (1991). Cider preparation from apple juice concentrate and its consumer acceptability. *Indian Journal of Horticulture*, 48(4): 321–327.

Joshi, V.K., Sandhu, D.K. and Thakur, N.S. (1999). Fruit based alcoholic beverages. In: *Biotechnology: Food Fermentation*, Vol. II, V.K. Joshi and Ashok Pandey (eds.). Educational Publishers and Distributors, New Delhi, p. 647.

Joshi, V.K., Sharma, Somesh and Parmar, Mukesh (2011). Cider and Perry. In: *Handbook of Enology*, Vol 1II, V.K. Joshi (eds.). Asia Tech Publishers, INC, New Delhi, pp. 1116–1151.

Kosseva, M.R., Joshi, V.K. and Panesar, P.S. (eds.) (2017). *Science and Technology of Fruit Wine Production*, Academic Press is an imprint of Elsevier, London, UK, p. 705.

Labelle, R.L. (1980). *Apple Cultivars Tested as Naturally Fermented Cider at Geneva*. Memo. N.Y. State Agric. Exp. Stn, Geneva, New York.

Laksonene, O., Kuldjärv, R., Paalme, T., Virkki, M. and Yang, B. (2017). Impact of apple cultivar, ripening stage, fermentation type and yeast strain on phenolic composition of apple ciders. *Food Chemistry*, 233: 29.

Lea, A.G.H. (1984). Colour and tannins in English cider apples. *Flussiges Obstetrics*, 51: 356.

Lea, A.G.H. (1995). Cider making. In: *Fermented Beverage Production*, A.G.H. Lee and J.R. Piggott (eds.). Blackie Academic and Professional, London, UK, p. 66.

Lea, A.G.H. and Arnold, G.M. (1978). The phenolics of ciders: Bitterness and astringency. *Journal of the Science of Food and Agriculture*, 29(5): 478–483.

Lea, A.G.H. and Drileau, J.F. (2003). Cider making. In: *Fermented Beverage Production*, A.G.H. Lea and J.R. Pigott (eds.). Kluwer Academic/Plenum, Publisheres, New York, pp. 59–87.

Nedovic, V.A., Durieux, A., Van Nedervelde, L., Rossels, P., Vandegans, J., Plainsant, A.M. and Simon, J.F. (2000). Continuous cider fermentation by co-immobilized yeast and *Leuconostocoenos* cells. *Enzyme and Microbial Technology*, 26: 834–839.

Picinelli, A. Antón, Mangas, M.J., J. and Suárez, B. (2016). Characterization of Spanish ciders by means of chemical and olfactometric profiles and chemometrics open overlay panel. *Food Chemistry*, 213: 505–513.

Plainsant, A.M. and Simon, J.F. (2000). Continuous cider fermentation by co-immobilized yeast and *Leuconostocoenos* cells.*Enzyme and Microbial Technology*, 26: 834–839.

Proulx, A. and Nichols, L. (2003). *Cider. Making, Using & Enjoying Sweet & Hard Cider*, 3th edition. Storey Publishing, North Adams, pp. 3–109.

O'Reilly, A. and Scott, J.A. (1993). Use of an ion-exchange sponge to immobilise yeast in high gravity apple based Cider alcoholic fermentation. *Biotechnology Letters*, 15(10): 1061–1066.

Rana, R.S., Vyas, K.K. and Joshi, V.K. (1986). Studies on production and acceptability of cider from Himachal Pradesh apples. *Indian Food Packer*, 40: 56–66.

Rosend, J., Kuldjärv, R., Rosenvald, S. and Paalme, T. (2019). The effects of apple variety, ripening stage, and yeast strain on the volatile composition of apple cider. *Heliyon*, 5(6): e 01953.

Salih, A.G., Le Quere,J.M., Drilleau, J.F. and Fernandez,J.M. (1990). Lactic acid bacteria and malolactic fermentation in manufacture of Spanish cider making. *Journal of the Institute of Brewing*, 96: 369–372.

Ścibisz, I., Bonin, S., Mitek, M. and Ziarno, M. (2015). *Porównanieskładucukrów w cydrachpolskichiimportowanych.* PrzemysłFermentacyjnyiOwocowoWarzywny, no. 5.

Sharma, R.C. and Joshi, V.K. (2005). Apple processing technology. In: *The Apple*, K.L.Chadha and R.P.Awasthi (eds.). Malhotra Publish. Co., New Delhi, pp. 455–498.

Verdu, C.F., Childebrand, N., Marnet, N., Lebail, G., Dupuis, F., Laurens, F., Guilet, D. and Guyot, S. (2014). Polyphenol variability in the fruits and juices of a cider apple progeny. *Journal of the Science of Food and Agriculture*, 94(7): 1305–1314.

Wei, J., Wang, S., Zhang, Y.,Yuan, Y. and Yue, T. (2019). Characterization and screening of non-Saccharomyces yeasts used to produce fragrant cider. *LWT*, 107: 191–198.

Wright, J.R., Sumner, S.S., Hackney, C.R., Pierson, M.D. and Zoecklein, B.W. (2000). A survey of virginia apple Cider producers practices. *Dairy, Food and Environmental Sanitation*, 20(3): 190.

Yulianti, F., Retimeire, A.C., Glatz,A.B. and Boylston,T.D. (2005). Sensory,flavor and microbial analyses of raw,pasteurized and irradiated apple cider. *Journal of Food Science*, 70(2): S153.

36 Technology of Reduced-Alcohol Wine Production

Leigh M. Schmidtke and Rocco Longo

CONTENTS

36.1 INTRODUCTION

During the past three decades, a fundamental shift in the attitude of consumers towards alcohol (ethanol) consumption has occurred. Changes in legal blood alcohol limits for driving and increased awareness amongst consumers of the adverse health effects of excessive alcohol intake have resulted in decreasing wine consumption in traditional market segments. Beverages with reduced ethanol concentrations are also more favourably excised in most countries, thereby providing some competitive advantages for wine sales, compared with full alcoholic strength wines. Higher ethanol levels can also affect the sensory profile of wine, in part due to altered solubility and volatility of aroma compounds and their interaction with protein complexes. Consequently, demand for reduced alcoholic strength beverages has encouraged several manufacturers to produce wine and wine products with lower-than-normal alcohol content, though both the manufacture and sale of reduced alcoholic strength beverages is fraught with legal complexities.

Definitions of wine and wine products specify minimum ethanol concentrations of varying quantity, applicable to different countries, such as in Australia, where products labelled as wine must contain greater than 4.5% (v/v) ethanol (alcohol). Fermented grape products with an alcohol concentration of less than this amount may not fit this definition. Thus, the use of the term 'Reduced-Alcohol Wine' to describe a product destined for the Australian market may not be permitted, although the product could be branded as a 'Reduced-Alcohol Wine Product'. The term 'Low alcohol' may be applied to beverages derived from fermented grape juice that contains less than 1.15% (v/v) ethanol and 'Non-intoxicating' implies that the beverage contains less than 0.5% (v/v) ethanol. Further confusion arises from wine production techniques using early-harvested grapes, with naturally occurring low levels of fermentable sugars, and, consequently, naturally low alcohol concentrations. These products may not have undergone any post-fermentation process to modify alcohol levels but may still contain less than the minimum specified ethanol concentration required to be considered a wine. Some countries have legislative complexity arising from different government authorities regulating wine and reduced-alcohol wine. In the United States of America, wine, as defined within the Federal Alcohol Administration Act, must contain between 7 and 24% (v/v) (alcohol), while the Bureau of Alcohol, Tobacco and Firearms regulates labelling requirements. As de-alcoholised wine products usually contain less than 7% (v/v) ethanol, labelling provisions are covered by a separate federal act administered by the Food and Drug Administration. The use of the term 'Partially Fermented Wine Product' may be suitable for some stockkeeping units; however, the use of thermal distillation or membrane processes for reduction of alcohol from wines made from grape juice that has been fermented to dryness (<2.0 g/L sugar) could render this term unsuitable. Further confusion may also arise based on the technique for reducing the alcohol concentration. The use of glucose oxidase enzymes for the reduction of fermentable sugars may not be permitted as processing aids for wines in some countries or regions, but their use in the manufacture of wine products or food products may be permissible. Such definitions vary according to the specific labelling laws for individual countries or economic trading zones, and exceptions based upon historical production of specific wine styles are notable. The terms 'alcohol-reduced', 'de-alcoholised' and 'low-alcohol' are used interchangeably throughout this chapter in the context of wine production, although it is acknowledged that specific legal definitions of these terms may exist.

TABLE 36.1
Technologies for Reducing Ethanol Concentration in Wine and Fermented Beverages

Stage of Wine Production	Principle	Technology
Pre-fermentation	Reduced fermentable sugars	Early fruit harvest
		Juice dilution
		Glucose oxidase enzyme
Concurrent with fermentation	Reduced alcohol production	Modified yeast strains
		Arrested fermentation
Post-fermentation	Alcohol removal	Spinning cone column
		Wine blending
		Evaporative perstraction
		Reverse osmosis
		Solvent extraction
		Ion exchange

Source: Duerr and Cuenat (1988); Pickering (2000); Smith (2002); Schmidtke *et al.* (2012); Longo *et al.* (2017)

36.2 PRODUCTION TECHNIQUES: LOW- OR REDUCED-ALCOHOL WINE

Production techniques for manufacturing low-alcohol or reduced alcoholic strength beverages have been developed over the past 15 to 20 years in order to satisfy a consumer demand for healthier alcoholic products. Producers may also wish to marginally reduce alcohol concentrations in wines to correct balance and maintain consistency of style between vintages or blends. Several engineering solutions and wine production strategies that focus upon pre- or post-fermentation technologies have been described and patented for the production of wines with ethanol concentrations lower than would naturally arise through normal fermentation and wine production techniques. However, consumer perception and acceptance of the sensory quality of wines manufactured by techniques that utilise thermal distillation for removal of alcohol is generally unfavourable. Attention has focused on non-thermal production processes and the development or selection of specific yeast strains with down-regulated or modified gene expression to achieve alcohol production. The production methods that are of most relevance to the wine industry, though not exhaustive, are summarised in Table 36.1, and generally form the basis of the discussion in this chapter.

36.3 PRE-FERMENTATION TECHNOLOGIES FOR LIMITING ALCOHOL PRODUCTION

Limiting alcohol production during fermentation by reducing fermentable sugars in the juice through early grape harvest, juice dilution or arresting fermentation while significant levels of unfermented sugars remain in the wine are some of the options that enable wines with reduced alcohol levels to be produced. Early grape harvest may result in wines that are sensorially undeveloped due to reduced flavour precursor development in the grapes prior to harvest, high acidity levels and a relative lack of yeast-contributed flavour compounds. Blending of a wine produced from early-harvest grapes with that made from more mature fruit may be a solution to minimise the perception of acidic and herbaceous sensory attributes, while reducing alcohol content. Comparatively, arrested fermentation, leaving high residual sugar levels in wines, may dictate that the finished product will require Pasteurisation for microbial stabilisation, thereby leading to potential loss or alteration of volatile flavour and aroma compounds. Juice dilution can only be performed using a low Brix grape adulterate, as addition of water is not a permitted process for wine production. Low Brix grape juice, a by-product of grape juice concentrate, is more commonly used to maintain wine concentration during reverse osmosis or thermal distillation techniques for alcohol removal. Recent efforts that target pre-fermentation production strategies for reducing alcohol content in wines have, therefore, focused upon technologies that minimise loss or alteration of desirable sensory qualities and minimise off-flavour development. The use of enzymes for reduction of fermentable sugars in grape juice prior to fermentation is one such method and will form the basis of discussion for pre-fermentation production options.

36.3.1 Glucose Oxidase: Biochemical Principle

Glucose oxidase (EC 1.1.3.4) (GOX) is a glycoprotein with dehydrogenase activity that catalyses the oxidation of ß-D-glucose to D-glucono-1,5-lactone (D-gluconic acid δ-lactone). This reaction requires the presence of molecular oxygen, and a flavin adenine dinucleotide (FAD) cofactor to participate in electron donation to form hydrogen peroxide. A second enzyme, catalase, is frequently present in commercial GOX preparations to degrade the unwanted hydrogen peroxide that is formed as a by-product during the oxidation of the glucose substrate. D-gluconic acid δ-lactone spontaneously hydrates to form gluconic acid. The biochemical basis for these reactions is illustrated in Figure 36.1.

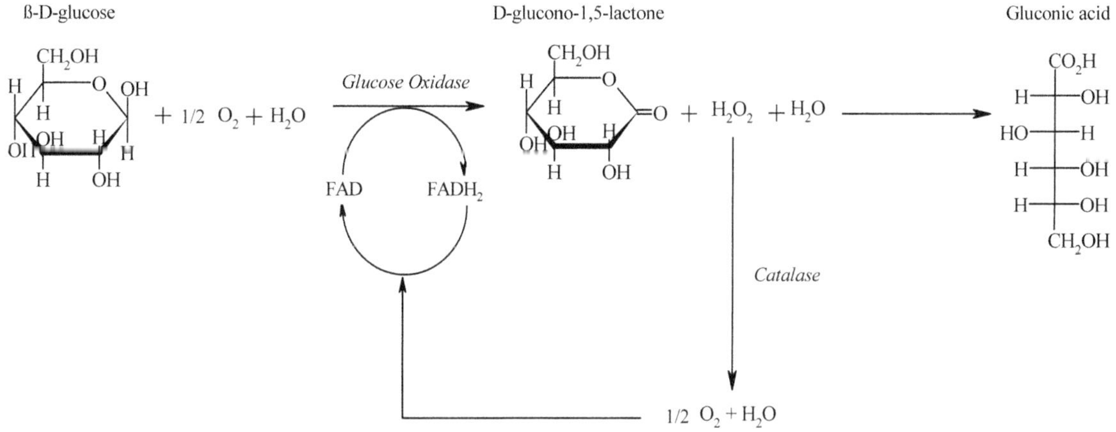

FIGURE 36.1 Biochemical oxidation of ß-D-glucose to D-glucono-1,5-lactone (gluconic acid δ-lactone) by glucose oxidase and subsequent hydration to gluconic acid. Catalase enzyme degrades the hydrogen peroxide formed during glucose oxidation. *Source:* Pickering *et al.* (1998)

36.3.1.1 Treatment of Grape Juice with GOX

As gluconic acid is not fermented by yeasts, a decrease in alcohol production can be achieved in wines prepared from GOX-treated juice. Glucose reduction in grape juice using GOX is presently limited to white grape varieties, as a period of clarification followed by enzyme reaction must first occur prior to yeast inoculation for fermentation to commence. As the glucose fraction represents approximately 50% of the total fermentable sugars, the theoretical maximum reduction in alcohol production is 50%, compared to wines made from untreated juice. In practice, some inefficiency in glucose oxidation arises and reported alcohol reductions in wines produced using GOX-treated juice range from less than 4% to 40%. The efficiency of glucose oxidation is dependent upon enzyme concentration, juice pH, dissolved oxygen concentration, processing time and temperature. The rate of formation of gluconic acid and its contribution to titratable acidity can be used for process monitoring, along with measurements of glucose concentration.

The most efficient conversion of glucose to gluconic acid in grape juice is reported to occur in a pH range of 5.5 to 6.0, corresponding to the reported pH range for optimum GOX activity. At normal grape juice pH, the rate of gluconic acid production is reduced by up to 75%, due to acid inhibition of enzyme activity. De-acidification of the juice with calcium carbonate may, therefore, be required to ensure a reduction in glucose concentration sufficient to achieve an adequate reduction of alcohol in the finished wine. Catalase, for efficient removal of H_2O_2, is present in most GOX preparations, and is not affected by the high acidity of grape juice.

Molecular oxygen is an essential requirement for GOX activity and must be sparged into the juice during enzyme treatment. Agitation will assist dispersion of oxygen bubbles, and enhance GOX activity, probably by limiting bubble size and maximising bubble surface area-to-volume ratio, thereby increasing dissolved oxygen (DO) concentrations. Different rates of air sparging, bubble size, sparger design and mixing rates are important considerations that influence the rate of gluconic acid production. Problems with excessive foaming and evaporation have been reported with high aeration levels. Sparging of juice and peroxide formation from glucose oxidation will also result in oxidation of polyphenolic constituents of the grape juice and development of a brown colour. Oxidised phenolics precipitate during fermentation and, consequently, the resulting wine is reported to have a more developed golden yellow colour than wines produced with reductively handled juice. GOX wines do have less susceptibility to browning and 'pinking' colour reactions during short- to medium- term storage than do control wines. Loss of grape volatile precursors and components may also potentially arise from air or oxygen sparging at excessive rates for prolonged periods.

Reports on optimal temperature for GOX activity are ambiguous. More rapid glucose oxidation takes place at a temperature of 20°C than at 30°C, which contrasts with the optimum temperature of between 30°C and 35°C for GOX activity. Other reports, however, have not demonstrated any significant difference in gluconic acid production rates in GOX-treated juice at temperatures of 20°C and 30°C. Several advantages are apparent for lower processing temperatures, such as the achievement of high DO levels in the juice but high oxygen concentration is an important rate-limiting reactant for GOX activity. Undesirable microbial growth may also be decreased at 20°C, although many wine spoilage organisms are quite capable of growth at this temperature.

36.3.1.2 GOX-Produced Wines and Their Composition

A summary of glucose reduction, gluconic acid concentrations, wine composition and resulting alcohol reduction in juice and wines arising from GOX treatment is shown in Table 36.2 and 36.3. High levels of gluconic acid, produced from the conversion of the glucose fraction of the total fermentable sugars, are a significant problem for finished wine quality. The contribution of gluconic acid to total acidity in wines made by GOX-treated juice may render the wines out of balance. To moderate this problem, and to optimise GOX activity, de-acidification of the grape juice by addition of calcium carbonate prior to GOX treatment may be necessary.

TABLE 36.2
Glucose Oxidation Rates in Grape Juice and Alcohol Reduction Percentages in Corresponding Wine Made From Glucose Oxidase-treated Grape Juice

Grape Variety	Glucose Oxidase/Catalase Concentration (mg/L)	Glucose Reduction (%)	Gluconic Acid (g/L)	Alcohol Reduction (%)
Chasselas	500	NS	NS	23
Müller Thurgau	2000	87	73	40
Muscat Gordo Blanco	50	13	11	3.6
Muscat Gordo Blanco	200	31	25	13
Muscat Gordo Blanco	1000	56	53	33
Riesling	500	36	10-52	23
Riesling	2000	NS	NS	36

NS: not stated
Source: Heresztyn (1987); Pickering *et al.* (1999a); Villettaz (1987)

TABLE 36.3
Composition of Müller-Thurgau Wine Prepared From GOX-treated De-acidified Juice

Component		De-acidified Juice	GOX Treated Juice	Control Juice	Decreased Alcohol Wine§	Control Wine§
Ethanol	% (v/v)				6.2	10.5
Glucose	g/L	84.7	10.7	84.7	<1.0	<1.0
Fructose	g/L	89.8	87.2	89.8	<1.0	<1.0
Gluconic Acid	g/L	<0.3	72.7	<0.3	66.7	<0.3
Tartaric Acid	g/L	1.9	1.7	4.3	1.8	2.9
Titratable Acidity†	g/L	3.2	26.7	7.1	27.8	8.1
pH		4.89	2.93	3.25	3.05	3.13

§Analysis at time of bottling; †As tartaric acid. *Source:* Pickering *et al.* (1999b)

With the production of wines from GOX-treated grape juice, the formation of substantial quantities of carbonyl compounds occurs, resulting in significantly higher sulphur dioxide (SO_2) binding than in control wines; consequently, the total concentration of SO_2 necessary to achieve microbial stabilisation in such wines may approach or even exceed legal limits. A further outcome of GOX treatment is an increased susceptibility of these wines to undergo premature yellow colour development consistent with increased flavonoid production.

36.3.2 Fermentation Technologies For Limiting Alcohol Production: Use of Novel Yeast Strains

Selection of specific yeast strains and their use as starter cultures for the consistent manufacture of a particular wine style is a common winemaking practice, that also ensures greater fermentation reliability and predictability than natural fermentations. The requirement of some producers to harvest grapes at high levels of fermentable sugar, in order to achieve typical varietal characters, produces wines with excessive alcohol concentrations and undesirable palate hotness. Selection and use of yeast strains with lowered ethanol production during fermentation could be one strategy. The focus on this field is the development of *Saccharomyces cerevisiae* yeast strains capable of lower yields of ethanol through the application of metabolic engineering or the combination of non-*Saccharomyces* yeasts, able to divert the carbon flux towards multiple metabolites rather than to ethanol production, with the high fermentative ability of *S. cerevisiae* strains. A problem associated with fermentation using novel or wild yeast species is potential off-flavour development and undesirable organoleptic characters, hence the development of genetically modified *Saccharomyces* strains for wine production.

36.3.3 Genetically Manipulated Saccharomyces cerevisiae

The potential to genetically engineer specific strains of *S. cerevisiae* that divert grape carbon compounds from ethanol to production of other metabolites, or which exhibit increased biomass, has been identified as a strategy to decrease the final ethanol concentration of wines. Such endeavours have chiefly focused upon alterations in glycerol production at the expense of ethanol. Some 4–10% of grape juice carbon is normally directed to glycerol production during fermentation by *S. cerevisiae*, with the majority being produced during the initial stages of biomass formation. The final concentration of glycerol in finished dry wines is normally in the range 4–9 g/L and is dependent upon the yeast strain and a range of environmental signals, including temperature, pH, sugar concentration, nitrogen source and SO_2 levels. The major benefits of glycerol formation for yeast cells during fermentation are two-fold. Glycerol is normally produced by *S. cerevisiae* during biomass formation at the beginning of fermentation, to protect cells from the high osmolar concentrations of sugars, thereby preventing cellular dehydration. Also, as the reactions for glycerol formation involve oxidation of nicotinamide adenine dinucleotide hydride (NADH), the redox imbalance, that arises from anaerobic glycolysis and glucose repression of respiration, is corrected. It is the correction of the redox balance that is considered to be the most important biological function of glycerol formation. Glycerol formation arises from reduction of the glycolytic intermediate dihydroxyacetone-phosphate to glycerol-3-phosphate and subsequent dephosphorylation. These reactions are catalysed by an NADH-dependent glycerol-3-phosphate dehydrogenase (GPDH) and a specific glycerol-3-phosphatase (GPP). Two isoforms of GPDH have been described and designated GPDH-1 and GPDH-2, with expression of the genes encoding these isoenzymes being regulated by the yeast requirement for osmoprotection and redox balance, respectively.

As glycerol and ethanol production by yeasts during fermentation are important regulators of cellular redox balance through the regeneration of NAD^+, any influence upon the flux of these compounds will alter the concentration of a range of other metabolites, that are also involved with redox balance. Acetaldehyde, acetate, succinate, acetoin and 2,3-butanediol appear to be the most important of these metabolites. Genetic manipulation of yeast strains to overexpress either gene *gpdh-1* or *gpdh-2* has resulted in 19–22% decreases in ethanol production in model solutions but significantly less in grape juice fermentations. Concomitant with increased glycerol formation and decreased ethanol production were increases in acetate, succinate, acetoin, acetaldehyde and 2,3-butanediol. The changes in concentration of these and other metabolites that arise in wines fermented with yeast strains with overexpressed GPDH genes may render the product unacceptable, particularly if the level of acetate is excessively high. Acetate formation by yeasts during fermentation may occur either by hydrolysis of acetyl-CoA or *via* the pyruvate dehydrogenase (PDH) by-pass pathway, in which pyruvate is decarboxylated to acetaldehyde followed by oxidation to acetate. The enzymes involved in the PDH by-pass are pyruvate decarboxylase and an acetaldehyde dehydrogenase that belongs to the aldehyde dehydrogenase (ALD) group of enzymes. The ALD group of enzymes in *S. cerevisiae* has been extensively characterised, with five isoforms (ALD2–6) being designated, with the most important of these during fermentation being ALD-5 (mitochondrial) and ALD-6 (cytosolic), as both are constitutive enzymes. The expression of isoforms ALD2–4 is glucose repressed and therefore, these enzymes do not play any role in cellular redox balance during grape juice fermentation.

A novel method to overcome the deleterious effects of high acetate levels in fermentations conducted by *gpdh-2*-overexpressed yeasts has been developed by deletion of the acetaldehyde dehydrogenase-6 (ALD-6) gene. A substantially lower ethanol and an acceptable acetate concentration was then achieved. A summary of metabolite concentrations involved in cellular redox balance, in finished fermentations, using genetically modified yeast strains targeting expression of the GPDH and ALD enzymes can be seen in Table 36.4. For limitations of genetic technology for yeast strain manipulation, the reader is directed to Chapter 19 of this volume.

36.4 POST-FERMENTATION TECHNOLOGIES FOR REMOVING ALCOHOL

36.4.1 Application of Reverse Osmosis in Low-Alcohol or De-Alcoholised Wines

36.4.1.1 Theoretical and Historical Background

Membrane filtration was first recognised in 1748, when milk was separated by passing through a biological membrane (goat skin bag). However, it was neither understood nor explained for another 200 years. In 1950, Westinghouse Corporation developed the first successful membrane, known as a "symmetrical membrane", which was used for desalination of sea water (still a major application of reverse osmosis). Unfortunately, this symmetrical membrane was not commercially successful as it fouled rapidly and, thus, exhibited very low liquid permeation rates. This problem was solved by the development, in 1960, of an asymmetrical membrane that led to the successful commercialisation of the technology shortly afterwards. Compared to the symmetrical membrane, asymmetrical membrane material is supported by one or more layers of micro-porous polymeric substance to create a composite membrane configuration. It may have (1) a thin skin of porous membrane, (2) a polymeric micro-porous support and (3) a polyester support.

Membrane filtration was introduced commercially into the food and beverage industry in the early 1970's, being quickly embraced by the dairy industry. While ultrafiltration is more common in the dairy industry for whey processing and the preparation of pre-cheese, it is also used in the beverage industry for filtration, cold stabilisation and solids concentration. Reverse osmosis (RO), on the other hand, is usually applied for the production of de-alcoholised or reduced- alcohol wine and beer. In fact, the first patent for the application of RO in alcoholic beverages was obtained by the West German brewing company Lowenbrau in 1975 for the de-alcoholisation of beer. However, there are other applications of reverse osmosis in wine production, including the removal of colours and flavours, for must concentration (as an alternative to chaptalisation), for the development of new products such as aperitifs, for wine stabilisation against tartrate precipitation and for de-acidification (removal of volatile acids) of grape juices.

TABLE 36.4
Changes in Metabolite Concentrations of Yeast Cultures With Modified Gene Expression for Glycerol Production

				Concentration of Metabolites						
Yeast Strain Designation	**Modified Gene Product Expression**	**Growth Medium**	**Ethanol reduction (%)**	**Ethanol (g/L)**	**Glycerol (g/L)**	**Acetate (g/L)**	**Succinate (g/L)**	**Acetoin (g/L)**	**2,3-butanediol (g/L)**	**Acetaldehyde (mg/L)**
Michnick *et al.* (1997)		glucose 100 g/L, pH 3.3								
V5/pVTU	Control strain			46.8	4.3	NS	NS	NS	NS	NS
GPD1 V5/GPD1	GPDH-1 overexpressed		22	36.6	14.0	NS	NS	NS	NS	NS
V5/pVTU	Control strain	glucose 200 g/L, pH 3.3		89.2	7.1	0.52	0.25	<0.1	0.90	<100
GPD1 V5/GPD1	GPDH-1 overexpressed		19	72.5	28.6	1.60	0.54	6.10	1.30	220
Remize *et al.* (1999)		glucose 200 g/L								
pvt100-U-ZEO R	Control strain			88.4	7.4	0.42	0.40	0.00	0.24	0.01
pvt100-U-ZEO-GPD1 R	GPDH-1 overexpressed		3	85.7	16.5	1.18	1.11	0.06	1.92	0.04
de Barros Lopes *et al.* (2000)	Chardonnay juice 21.8°Brix, pH 3.16									
AWRI 838	Control strain			129.8	7.9	0.58	0.39	NS	NS	NS
AWRI 838 GPD2-OP	GPDH-2 overexpressed		4	124.0	16.5	1.02	0.65	NS	NS	NS
Eglinton *et al.* (2002)		glucose 80g/L								
GPD2 ALD6	Control strain			34.1	5.1	0.66	0.59	NS	NS	0.64
GPD2-OP ALD6	GPDH-2 overexpressed, ALD6 normal expression		24	26.0	13.4	1.42	0.83	NS	NS	8.36
GPD2 ald6Δ	GPDH-2 normal expression, ALD6 deletion		9	31.1	6.0	0.20	0.59	NS	NS	0.79
GPD2-OP ald6Δ	GPDH-2 overexpressed, ALD6 deletion		20	27.3	16.3	0.36	0.88	NS	NS	8.79

GPDH-1: glycerol-3-phosphate dehydrogenase-1; GPDH-2: glycerol-3-phosphate dehydrogenase-2;
ALD6: aldehyde dehydrogenase-6, NS: not stated, *Source:* de Barros Lopes *et al.* (2000); Eglinton *et al.* (2002); Michnick *et al.* (1997); Remize *et al.* (1999)

As a membrane technology, RO requires low energy input, operates at ambient temperatures, allows reproducible control over separations, requires no disposable filtration media or other additions and is easily automated for continuous operation. Specifically, in comparison with other methods for producing low-alcohol wines, such as distillation, spinning cone technology or arrested fermentation, the reduced alcohol wines produced by RO usually have flavour and aroma profiles comparable to those of the regular wines from which they were obtained.

36.4.1.2 Membrane Types and Configurations

Reverse osmosis membranes can be made up of several different materials, including cellulose acetate, regenerated cellulose, synthetic polymers and ceramics. The cellulosic materials are not as durable as and give lower flux rates than the synthetic polymers, which are also more selective. Ceramics, while very strong and durable, are also expensive, being designed originally for separation of uranium isotopes. The most successful membranes are the asymmetric (heterogenous) types, which are thin skins of membrane material bonded to one or more layers of polymeric support material. In RO, thin-film composite (TFC) membranes are very commonly used where a polymer with high strength and porous structure is chemically bonded to a very thin film of polymer (membrane material) with the required permeation selectivity. Such membranes give very good flux characteristics, are very durable, especially under the high-pressure application of RO, are cleanable and allow back flushing to restore initial flux rates by destabilising any build-up of materials on the membrane surface.

Reverse osmosis, like other membrane techniques, operates under the principle of cross or tangential flow, whereby the liquid flows in parallel with or tangential to the membrane surface at high velocity under pressure. Some liquid passes through the membrane but the solids or materials with molecular weight higher than the nominal molecular weight cut-off (NMWCO) of the membrane will be swept along in the stream of feed across the membrane. Recycling will ensure that more permeate will pass through the membrane during each cycle until the desired concentration of the feed is achieved. In order to achieve this effectively, several module configurations have been developed. These include the flat sheet (also known as plate and frame), tubular, hollow-fibre and spiral-wound configurations. The spiral-wound configuration makes the most economic use of space for a given membrane area, being in the form of flat membranes rolled up together like a cigar. The original space between the membranes serves as permeate collection channels and the new space generated from winding the membranes becomes the feed channel. This configuration is very common for RO used in wines.

36.4.1.3 Applications and Limitations

In a reverse osmosis process for alcohol reduction in wines, the feed is the regular wine with normal alcohol content. This wine is pumped at pressures of up to 4 MPa (40 atm) through a membrane module. Water and ethanol, being small molecules (molecular weight <200 Da), pass through the membrane into the permeate stream. The retentate is redirected to the feed tank and the wine is continuously de-alcoholised and concentrated. Wine may be restored to the original water content by either the addition of low-Brix juice, a by-product of grape juice concentrate, or more commonly, the permeate is distilled to separate the ethanol from the water component which is then added back to the wine. Addition of low-Brix juice to the feed (wine) may be continuous to keep the volume constant during the process or may be begun well before the wine's concentration (and osmotic pressure) reaches a high-enough level to stop the permeation. Basically, the more low-Brix juice which is added, the lower the alcohol content in the feed tank, which also increase permeation, since the osmotic pressure is also reduced. The simultaneous production of low-Brix juice or alcohol-enriched wine, with de-alcoholised wine, is possible by using two reverse osmosis units in parallel, one with an ethanol-impermeable membrane, redirecting the filtrate from this unit to the feed supplying the ethanol-permeable unit. The whole process is usually carried out in a closed loop system, under an inert gas blanket to discourage the incidence of oxidation or other undesirable chemical reactions. Reverse osmosis can be used to reduce alcohol content in wine from about 12–15% (v/v) to less than 0.5 % (v/v), producing a wide range of low-alcohol wines, thereby allowing production flexibility.

36.4.2 Evaporative Perstraction

Evaporative perstraction, also called osmotic distillation and isothermal membrane distillation, is a common technology for the ethanol reduction of wine and seems to have minimal effects on product sensory profile when up to 2% (v/v) ethanol is removed. Evaporative perstraction is a process in which two aqueous solutions – feed (wine) and stripping (water) solutions – flow past either side of a porous hydrophobic membrane. The driving force of the mass transfer process is the partial pressure or vapour pressure difference of the volatile solute in the feed and stripping solutions. Indeed, aroma volatile compounds are more soluble in ethanol/water solution than in water solution and so their vapour pressure is low. Aroma compounds are retained in the final wine, while ethanol, which is the most volatile component in wine, rapidly moves through the membrane from the feed solution (wine) to the stripping solution. Changing some working conditions, such as the feed and stripping flowrates, the temperature and the pH, it is possible to vary the rate of ethanol removal. However, an increase in working temperature and time causes the losses of some aroma volatile compounds and may affect the wine sensory properties. Therefore, it is advisable to maintain a low temperature and to change other operating conditions to increase the rate of ethanol removal, using a higher membrane area.

36.4.3 Application of the Spinning Cone Column for Alcohol Removal

36.4.3.1 Spinning Cone Column

The spinning cone column (SCC) is a device used to extract volatile flavour components from a liquid or slurry. The column consists of a vertical shaft rotating at approximately

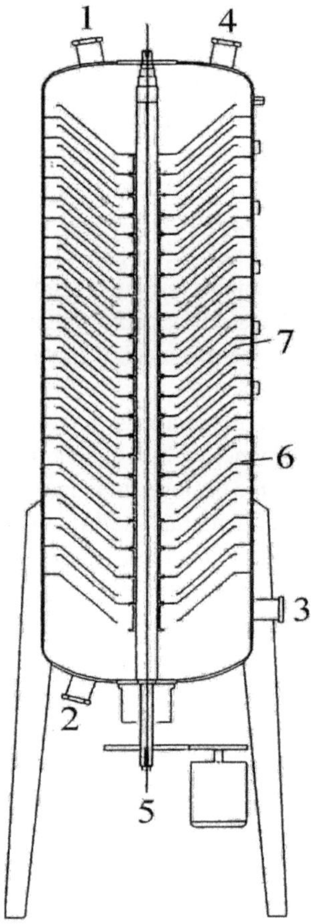

FIGURE 36.2 Mechanical layout of the spinning cone column (SCC). 1: Product in; 2: Product out; 3: Gas in; 4: Gas out; 5: Rotating shaft; 6: Stationary cones; 7: Rotating cones. *Source:* Courtesy of Flavourtech, Lenehan Rd, Griffith, Australia

350 rpm, supporting up to 22 inverted (pointing downwards) cones. Between each pair of cones, there is a fixed inverted cone, attached to the casing of the column (Figure 36.2). The liquid feed is fed to the top of the column, into the first spinning cone. A film of liquid is flung outwards by centrifugal action onto the inside surface of the casing. The liquid will then drop onto a fixed cone and migrates as a thin film downwards and towards the centre of the cone under the influence of gravity. From here, the liquid will pass onto the second spinning cone, and the movement is repeated several times until the liquid finally reaches the bottom of the column. As the liquid film is quite thin, the liquid hold-up volume is low and the residence time is typically around 20 seconds. The SCC can handle a range of different materials as feed, from low-viscosity products (*e.g.*, wine) to more viscous materials (*e.g.*, coffee extract).

A stripping gas, such as nitrogen, is admitted to the base of the column and passes through the voids between the rotating and fixed cones. An alternative stripping vapour can be generated by re-directing a portion of the product discharge through a heater, prior to re-injection. The gas, along with volatile components it has picked up, is collected at the top of the column. Along its tortuous path upward, the stripping gas is exposed to considerable turbulence, caused by fins attached to the underside of the spinning cones (Figure 36.3). It is this turbulence and the fact that the liquid is spread out as a thin film on the upper surfaces of the rotating and fixed cones that enhances a mass transfer of volatile product into the stripping gas. The considerable number of cones also ensures an adequate path length for both the liquid and the stripping gas within the column. The column operates under a negative pressure, so that volatile components will be evaporated-off at a reduced temperature. Typical feed and column temperatures are approximately 30°C. Reasonable clearances between rotating and fixed cones ensure that pressure drops are minimised, which, in turn, enables the mass transfer process to occur at almost constant pressure (and hence constant temperature) within the column. Other ancillary items are required for the process to function efficiently (Figure 36.4). After leaving the feed tank (1; Figure 36.4), the product is warmed in a regenerative heat exchanger (3) and fed into the spinning cone column (5). Stripping gas is obtained from treatment of a portion of the product discharge through a re-injection heater (7), with the remainder of the product discharge being recovered, once passed back through the heat exchanger (3).

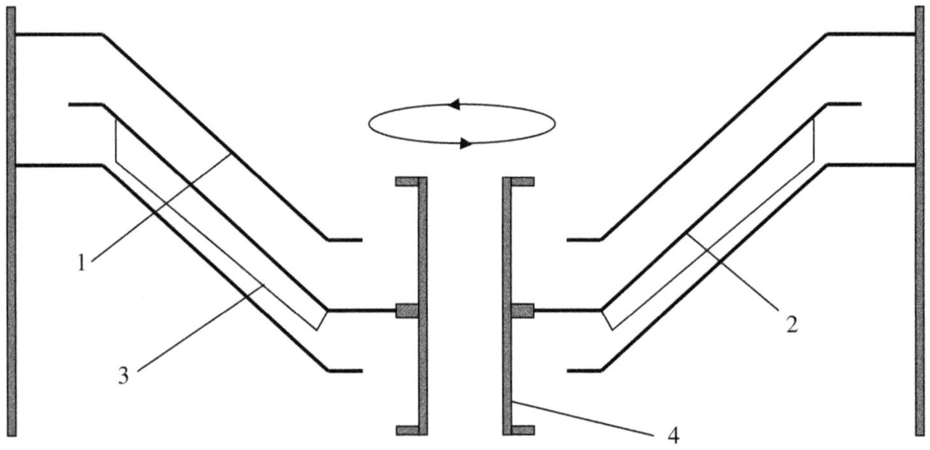

FIGURE 36.3 Cross section of cone showing fins to create turbulence. 1: Stationary cone; 2: Rotating cone; 3: Fin; 4: Rotating shaft. *Source:* Courtesy of Flavourtech, Lenehan Rd, Griffith, Australia

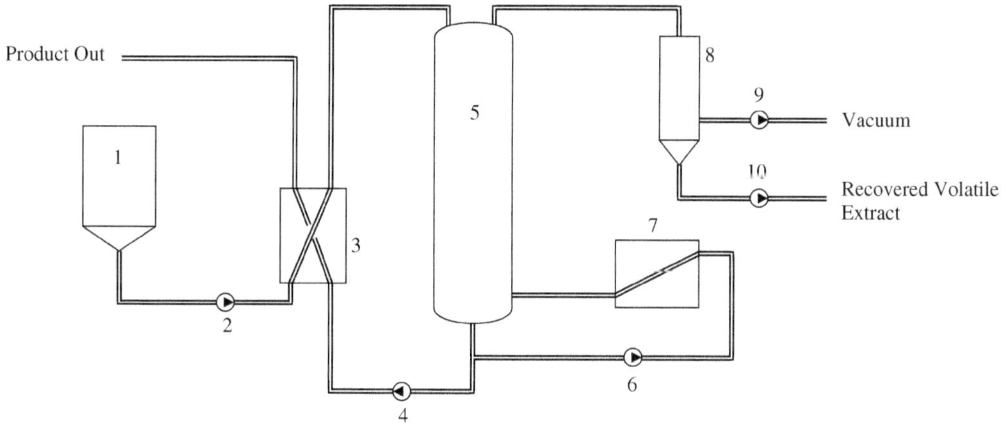

FIGURE 36.4 Typical layout of a spinning cone column and ancillary items. 1: Product feed tank; 2: Product feed pump; 3: Product heat exchanger; 4: Product discharge pump; 5: Spinning cone column; 6: Product re-injection pump; 7: Product re-injection heater; 8: Condensate cyclone; 9: Vacuum pump; 10: Recovered volatile extract pump. *Source:* Courtesy of Flavourtech, Lenehan Rd, Griffith, Australia

On leaving the column, the stripping gas/vapours are fed to a condensate cyclone (8) and volatile components are recovered separately from the treated product. The SCC can be sealed to operate under aseptic conditions and can be a component of a pasteurising or sterilising process.

The SCC offers an alternative to the more traditional method of using evaporators in the removal of alcohol from wine. The general process for adjusting alcohol concentration in finished wine, using SCC technology, consists of a two-stage process. The first pass of wine through the SCC occurs at low temperature (~28°C) under vacuum to recover volatile wine aromas in approximately 1% of the total product volume. The second pass of the product occurs with the de-aromatised wine at slightly higher temperature (~38°C) and under vacuum conditions to remove the alcohol. The final de-alcoholised wine is constructed by re-blending the recovered aroma with the de-alcoholised and de-aromatised base. Blending with full-strength wine, juice or juice concentrates to final product specifications, prior to filtration and packaging, enables a range of product styles to be developed.

36.4.4 Supercritical Solvent Extraction

Compression of a gas at temperatures above its critical point will result in the formation of a supercritical fluid with increased solvent properties, that can be exploited for separation or liquid extraction. The use of CO_2 for supercritical extraction in the food industry is gaining popularity and offers several advantages as the critical temperature for this gas is relatively low, at 31°C, no toxic substances are required for use, and it is relatively inexpensive and easily handled. Furthermore, the use of carbon dioxide in wine production does not pose any legal difficulties and is ideally suited for extraction of alcohol from either wine or beer. One process for the removal of alcohol from wine or beer, using supercritical CO_2 extraction, and the production of low-alcohol beverages subjects the high alcoholic strength beverage to low temperature/high vacuum distillation. The captured volatile fraction, containing alcohol and aroma compounds, is then, subjected to supercritical extraction at 80 to 100 bars pressure. Partial expansion of the supercritical fluid by pressure drop to 18 to 25 bars extracts the aroma portion of the distillate, which, after CO_2 scrubbing, is returned by sparging into the de-aromatised wine base remaining after distillation. A flow diagram, illustrating the production of low-alcohol wine, is shown in Figure 36.5. The critical extraction process is conducted in a counter-current column in which the solvent (carbon dioxide) is pumped as a liquid into the bottom of the column and the volatile alcohol mixture is fed into the top. The extracted volatile aromas and CO_2 gas are recovered from the column head and ethanol-water component is drained away as a liquid from the bottom. Whilst technically feasible, supercritical extraction using CO_2 for the production of low-alcohol beverages is not commonly employed within the beverage industry. High capital costs, a requirement for high-vacuum distillation and inflexibility of the plant remain significant barriers for the uptake of this technology within the wine industry.

36.5 SENSORY QUALITY OF LOW-ALCOHOL WINES

Although no difference between reduced-ethanol and untreated wines could be detected up to certain levels of ethanol reduction (–2% v/v), a general consumer perception is that reduced alcoholic wines lack body and flavour. The removal of ethanol from wines will have an obvious effect upon the sweetness and palate weight of a wine, due to the sensory characteristics of alcohol, whilst decreasing the perception of acidity and increasing the perception of bitterness and astringency. The volatility of aroma compounds is reduced in the presence of ethanol due to their non-polar nature and these compounds have greater solubility in full alcoholic strength wines. Thus, volatile components may be more easily lost from de-alcoholised wines than from full-strength wines during processing. Loss of aroma compounds is likely to be greater by thermal distillation techniques than as a result of low-temperature processes. The removal of ethanol from the wine may also increase the binding of aroma compounds

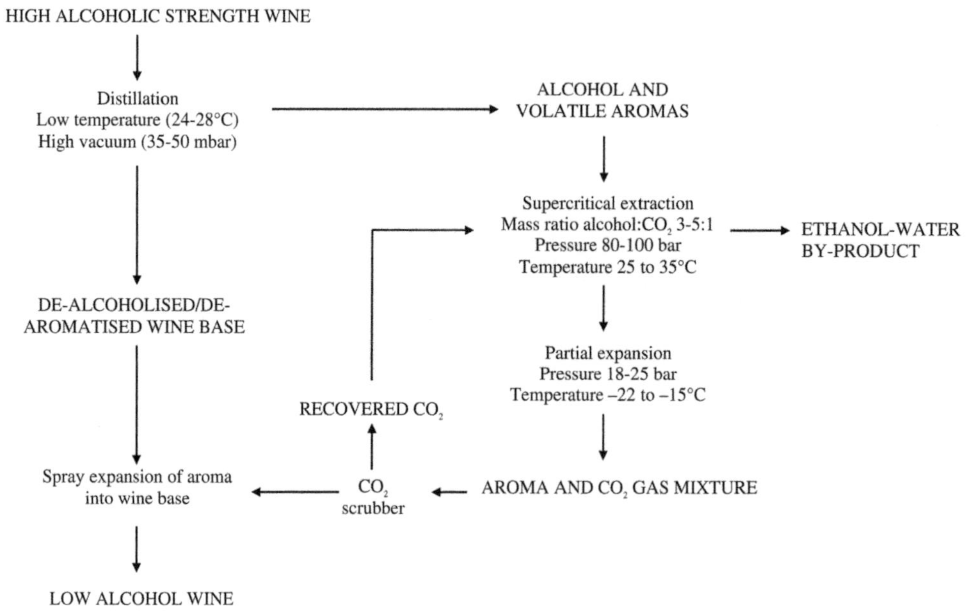

FIGURE 36.5 Processing scheme for production of reduced-alcohol wines, using supercritical carbon dioxide extraction. *Source:* Seidlitz *et al.* (1991)

to proteinaceous materials in wines, a process that leads to diminished volatility and, thus, reduced sensory perception of these compounds. The changes in flavour profile due to ethanol removal are, therefore, a complex interaction of altered volatility of and concentration of aroma compounds, loss of alcohol-related sweetness and changes in the perception of mouth-feel characteristics. The magnitude of sensory changes associated with de-alcoholised wine are dependent upon the quantity of ethanol remaining in the product.

BIBLIOGRAHY

Baldwin, G. (1998). Reverse osmosis and its future relevance to the wine industry. *Aust. Grapegrower Winemaker* 414a: 129.

de Barros Lopes, M., Rehman, A., Gockowiak, H., Heinrich, A. J., Langride, P. and Henschke, P. A. (2000). Fermentation properties of a wine yeast over-expressing the *Saccharomyces cerevisiae* glycerol 3-phosphate dehydrogenase gene (GPD2). *Aust. J. Grape Wine Res.* 6(3): 208.

Belisario-Sánchez, Y. Y., Taboada-Rodríguez, A., Marín-Iniesta, F., Iguaz-Gainza, A. and López-Gómez, A. (2012). Aroma recovery in wine dealcoholization by SCC distillation. *Food. Bioprocess. Technol.* 5(6): 2529–2539.

Duerr, P. and Cuenat, P. (1988). Production of dealcoholised wine. In: *Proceeding of the Second International Symposium for Cool Climate Viticulture and Oenology*, Auckland, 11–15 January 1988, R. E. Smart, R. J. Thornton, S. B. Rodriguez and J. E. Young (Eds.). New Zealand Society for Viticulture and Oenology, Auckland, p. 363.

Eglinton, J. M., Heinrich, A. J., Pollnitz, A. P., Langridge, P., Henschke, P. A. and de Barros Lopes, M. (2002). Decreasing acetic acid accumulation by a glycerol overproducing strain of *Saccharomyces cerevisiae* by deleting the *ALD6* aldehyde dehydrogenase gene. *Yeast* 19(4): 295.

Englezos, V., Rantsiou, K., Torchio, F., Rolle, L., Gerbi, V. and Cocolin, L. (2015). Exploitation of the non-*Saccharomyces* yeast *Starmerella bacillaris* (synonym *Candida zemplinina*) in wine fermentation: Physiological and molecular characterizations. *Int. J. Food Microbiol.* 199: 33–40.

Erten, H. and Campbell, I. (2001). The production of low-alcohol wines by aerobic yeasts. *J. Inst. Brew.* 107(4): 207.

Harders, T. and Sykes, S. (1999). Comparison of a spinning cone column and other distillation columns. *Food Aust.* 51(10): 469.

King, E. S., Dunn, R. L. and Heymann, H. (2013). The influence of alcohol on the sensory perception of red wines. *Food Qual. Pref.* 28(1): 235–243.

Kontoudakis, N., Esteruelas, M., Fort, F., Canals, J. M. and Zamora, F. (2011). Use of unripe grapes harvested during cluster thinning as a method for reducing alcohol content and pH of wine. *Aust. J. Grape Wine Res.* 17(2): 230–238.

Liguori, L., Russo, P., Albanese, D. and DiMatteo, M. (2013). Evolution of quality parameters during red wine dealcoholization by osmotic distillation. *Food Chem.* 140(1–2): 68–75.

Lisanti, M. T., Gambuti, A., Genovese, A., Piombino, P. and Moio, L. (2013). Partial dealcoholization of red wines by membrane contactor technique: Effect on sensory characteristics and volatile composition. *Food Bioprocess Tech.* 6(9): 2289–2305.

Longo, R., Blackman, J. W., Torley, P. J., Rogiers, S. Y. and Schmidtke, L. M. (2017). Changes in volatile composition and sensory attributes of wines during alcohol content reduction. *J. Sci. Food Agric.* 97(1): 8–16.

Longo, R., Blackman, J. W., Antalick, G., Torley, P. J., Rogiers, S. Y. and Schmidtke, L. M. (in press). Harvesting and blending options for lower alcohol wines: A sensory and chemical investigation. *J. Sci. Food Agric.* doi:10.1002/jsfa.8434.

Meillon, S., Urbano, C. and Schlich, P. (2009). Contribution of the Temporal Dominance of Sensations (TDS) method to the sensory description of subtle differences in partially dealcoholized red wines. *Food Qual. Pref.* 20(7): 490–499.

Michnick, S., Roustan, J.-L., Remize, F., Barre, P. and Dequin, S. (1997). Modulation of glycerol and ethanol yields during alcoholic fermentation in *Saccharomyces cerevisiae* strains overexpressed or disrupted for GPD1 encoding glycerol 3-phosphate dehydrogenase. *Yeast* 13(9): 783.

Navarro-Aviño, J. P., Prasad, R., Miralles, V. J., Benito, R. M. and Serrano, R. (1999). A proposal for nomenclature of aldehyde dehydrogenase in *Saccharomyces cerevisiae* and characterization of the stress-inducible *ALD2* and *ALD3* genes. *Yeast* 15(10A): 829.

Pickering, G. J. (2000). Low- and reduced-alcohol wine: A review. *J. Wine Res.* 11(2): 129–44.

Pickering, G. J., Heatherbell, D. A. and Barnes, M. F. (1998). Optimising glucose conversion in the production of reduced alcohol wine using glucose oxidase. *Food Res. Int.* 31(10): 685.

Pickering, G. J., Heatherbell, D. A. and Barnes, M. F. (1999a). The production of reduced-alcohol wine using glucose oxidase-treated juice. Part I. Composition. *Am. J. Enol. Vitic.* 50(3): 291.

Pickering, G. J., Heatherbell, D. A. and Barnes, M. F. (1999b). The production of reduced-alcohol wine using glucose oxidase-treated juice. Part II. Stability and SO_2 binding. *Am. J. Enol. Vitic.* 50(3): 299.

Pickering, G. J., Heatherbell, D. A. and Barnes, M. F. (1999c). The production of reduced-alcohol wine using glucose oxidase-treated juice. Part III. Sensory. *Am. J. Enol. Vitic.* 50(3): 307.

Remize, F., Andrieu, E. and Dequin, S. (2000). Engineering of the pyruvate dehydrogenase bypass in *Saccharomyces cerevisiae*: Role of the cytosolic Mg^{2+} and mitochondrial K^+ acetaldehyde dehydrogenases Ald6p and Ald4p in acetate formation during alcoholic fermentation. *Appl. Environ. Microbiol.* 66(8): 3151.

Remize, F., Roustan, J. L., Sablayrolles, J. M., Barre, P. and Dequin, S. (1999). Glycerol overproduction by engineered *Saccharomyces cerevisiae* wine yeast strains leads to substantial changes in by-product formation and to a stimulation of fermentation rate in stationary phase. *Appl. Environ. Microbiol.* 65(1): 143.

Rizi, S. S. H., Yu, Z. R., Bhaskar, A. R. and Chidambarra Raj, C. B. (1994). Fundamentals of processing with supercritical fluids. In: *Supercritical Fluid Processing of Food and Biomaterials*, S. S. H. Rizvi (Ed.). Chapman & Hall, Glasgow, p. 1.

Röcker, Schmitt, M., Pasch, L., Ebert, K. and Grossmann, M. (2016). The use of glucose oxidase and catalase for the enzymatic reduction of the potential ethanol content in wine. *Food Chem.* 210: 660–670.

Rossouw, D., Heyns, E., Setati, M., Bosch, S. and Bauer, F. Adjustment of trehalose metabolism in wine *Saccharomyces cerevisiae* strains to modify ethanol yields. *Appl. Environ. Microbiol.* 79(17): 5197–5207.

Scanes, K. T., Hohmann, S. and Prior, B. A. (1998). Glycerol production by the yeast *Saccharomyces cerevisiae* and its relevance to wine: A review. *S. Afr. J. Enol. Vitic.* 19(1): 17.

Schmidtke, L. M., Blackman, J. W. and Agboola, S. O. (2012). Production technologies for reduced alcoholic wines. *J. Food Sci.* 71(1): R25–R41.

Scott, J. A. and Huxtable, S. M. (1995). Removal of alcohol from beverages. *J. App. Bacteriol. Symp. Suppl.* 79: 19S.

Seidlitz, H., Lack, E. and Lackner, H. (1991). Process for the reduction of the alcoholic content of alcoholic beverages. *United States Patent Application US 5 034 238.*

Smith, F. (2002). Engineering wine-techniques used to overcome problems in winemaking. *Aust. N.Z. Grapegrower Winemaker* 465: 71.

Tilloy, V., Cadière, A., Ehsani, M. and Dequin, S. (2015). Reducing alcohol levels in wines through rational and evolutionary engineering of *Saccharomyces cerevisiae. Int. J. Food Microbiol.* 213: 49–58.

Van der Horst, H. C. (2001). Membrane processing. In: *Mech. Autom. Dairy Process,* Tamineand, A. Y. and Law, B. A. (Eds.). Sheffield Academic Press, Sheffield, p. 296.

Varela, C., Dry, P. R., Kutyna, D. R., Francis, I. L., Henschke, P. A., Curtin, C. D. and Chambers, P. J. (2015). Strategies for reducing alcohol concentration in wine. *Aust. J. Grape. Wine. Res.* 21: 670–679.

Wright, A. J. and Pyle, D. L. (1996). An investigation into the use of the spinning cone column for insitu ethanol removal from a yeast broth. *Process Biochem.* 31(7): 651.

37 Production Technology of Fruit Wines

V.K. Joshi, Laura Fariña and Ghanshyam Abrol

CONTENTS

37.1 INTRODUCTION

Wine is the drink that results from the complete or partial alcoholic fermentation of grapes, by the natural microflora of grapes or by added yeast culture. The word "wine" signifies the fermented product from grapes; for fruit wines other than the grapes, the name of the fruit is prefixed, *e.g.*, apricot wine. The non-grape fruits are rich in nutrients, especially plant pigments, vitamins, polyphenolics and flavonoids. Fruits like apple, plum, peach, pear, apricot, currants, etc. are used to create diverse sorts of fruit wines, and their distillate is called brandy. These products are specific to the particular fruits. The fruit wines also have nutritive values as well as therapeutic values. Contrasted with the volume of wine produced from grapes and consumed on the planet, the volume of wine created from non-grape natural products is not significant, aside from cider (from apples; Chapter 35) and perry (pears; Chapter 35), whereas plum wines are very well known in Germany and the Pacific Coast States of the US. Here, in this chapter, the characteristics of different fruit wines are described. Since it will not be possible to discuss each and every fruit wine, efforts have been made to discuss a few fruit wines and their production technology in this chapter.

37.2 FRUIT WINE PRODUCTION: BASICS AND PRE-REQUISITES

The processes of fruit wine production are similar to those of wine from grapes, but the production of wine from other fruits has to be modified due to certain physico-chemical characteristics of these fruits. These include the preparation of the must and fermentation. Nevertheless, the alcoholic fermentation basically remains the same, as performed by yeast for any other fruit-based alcoholic beverages, including those from grapes. Some of the characteristics of non-grape fruits have been summarized as below:

- The juices extracted from most of the stone fruits are lacking in the sugar contents required for fermentation and thus, have low fermentability without additions.
- The greater acidity in some of the fruits makes it all the more difficult to prepare a palatable wine.
- Production of quality fruit wine is affected by several factors, namely fruit variety, stage of harvest and maturity of the fruit, total sugar content, acid, total phenolics, pigments, nitrogenous compounds, etc., in the fruit or the must.
- Additives, like nitrogenous compounds, pectin esterase, sulfur dioxide or other preservatives, fruit pulp or juice, yeast strains and other vinification practices, post-fermentation operations, especially the length of maturation and the method of preservation employed, also vary, though there are some similarities.
- Addition of a sweetening agent, like fruit concentrate, sugar, acid, nitrogen source, clarifying enzyme, filtration aid, etc., are almost indispensable for making a fruit wine.
- Spices and herbs or their extract are indispensable raw materials for the preparation of fortified wines, like vermouth.
- The pectin esterase enzyme has also been used to increase extraction of juices from stone fruits, like plum, apricot and peach, which might later be used for the production of wines.
- Several nitrogen sources are used but di-ammonium hydrogen phosphate (DAHP) has many advantages (less shifting in pH of the must and low cost), so it is the most commonly used nitrogen source.

37.3 GENERAL METHOD OF PREPARATION OF FRUIT WINE

In preparation of the table, fortified and sparkling fruit wines, many steps are common, and the following is the general outline (Figure 37.1) for the preparation of a fruit wine.

37.3.1 Preparation of Yeast Culture

The use of *Saccharomyces cerevisiae* strains allows a rapid and reliable fermentation, reducing the risk of sluggish or stuck fermentation, and the risk of microbial contamination.

The ability to achieve a rapid and complete fermentation of juice sugars to ethanol is an essential requirement of a selected yeast strain but there are other properties to evaluate. It is important that the yeast chosen should be sulfur-tolerant, causes minimal foaming, settles out quickly at the end of fermentation and produces a suitable profile of volatile compounds (by the yeast) to complement the typical aromas of the fruit being fermented. A good culture of an appropriate yeast *S. cerevisiae* strain needs to be procured. Yeast in the form of a slope, slant or tablet should be used. For more details, see Chapter 26 of this volume.

37.3.2 Preparation of Must

The fruit from which wine is to be prepared is converted into juice or pulp. For fruits like citrus, pineapple a juice extractor could be used whereas, for apple and pear, the fruit is first grated and then pressed to extract the juice. In stone fruits, like plum and apricot, the fruit is made into pulp by boiling the fruits with 10% (w/v) water and passing it through a pulper, or the required amount of hot water is added to the mashed material followed by addition of the pectinesterase enzyme preparation when the temperature has cooled to about 40°C. Usually, the fruits are good sources of the nutrients required by yeast for satisfactory fermentation into wine. Commonly, DAHP (Diammonium hydrogen phosphate, 0.1%) is added to hasten the fermentation. Fruits, like plum or apricot, are highly acidic and need either to be diluted or have the acidity neutralized by addition of a calculated amount of calcium carbonate. In the preparation of must, sulfur dioxide (SO_2) is added at the rate of 100–150 ppm; (to incorporate 100 ppm of SO_2, approximately 200 mg/L of potassium metabisulfite (KMS) is added). In plum fermentation, addition of sodium benzoate has been found to give a better-quality wine than KMS though there is a need for more elaboration. The material is then kept overnight in a narrow-mouthed container and closed with a lid. Next day, the seeds and skin are removed by filtration.

The sugar contents, as shown by refractometry in terms of % Total Soluble Solids (%TSS) or °Brix (°Bx) have to be increased, if the original sugar content of the juice is not high enough. For most of the musts of the stone fruits, the sugar contents are raised by addition of syrup, with the required sugar content varying depending on the final product to be made. For example, it is 20°Bx for cider and 24°Bx for wild apricot wine (Chuli) and plum wine. Specific TSS values for the wine to be prepared are indicated in the individual Technology of Production of Various Fruit Wines, Section 37.4.

37.3.3 Fermentation

A 24-hour-old active inoculum of yeast, prepared as above (Section 37.3.1) is used for fermentation. The juice for fermentation filled up to three-quarters of the space in the flask/polyethylene barrel or glass carboy with a narrow mouth, fitted with air lock. The active yeast culture is then added, and a cottonwool plug is put back into the mouth of the container. Fermentation will start and can be monitored by the extent of bubbling reduction in degree Brix (°B). In the initial stages, the fermentation is fast but towards the end, it slows down. The fermentation should be monitored for a decrease in °B after a suitable interval of time. When the fermentation is near completion or is completed, there is a risk of acetification of wine by oxidation of ethanol by acetic acid bacteria. To prevent acetification, *i.e.*, conversion into vinegar, air locks or balloons (for small containers) are put into the mouth of the fermentation vessel. With an air lock, excess carbon dioxide passes out without allowing the air to come in contact with the wine.

37.3.4 Finishing of Fermentation

Fermentation is considered complete when no more bubbles come out and it can also be identified by measuring the TSS, using refractometry. If the initial TSS was 24°B, after fermentation, TSS should have fallen to 7 or 8°B for apple, plum or apricot fruit wine at the end of fermentation.

37.3.5 Siphoning and Racking

After the completion of fermentation, the yeast and other material settle to the bottom of the container, with clear liquid separating out above it. The clear liquid is siphoned or racked in the case of pulpy material, is filtered through a cheesecloth or muslin cloth, then siphoned. This is followed by cold storage stabilization for a week, followed by another racking. The fermented liquid is again siphoned, leaving the residue to one side. Two or three rackings are usually carried out during the first 15–20 days after completion of fermentation, followed by subsequent racking once a month. During the inter-racking period, no head space should be kept in the bottle or container, and it should be closed tightly to prevent acetic acid bacteria converting the wine into vinegar.

37.3.6 Maturation

The newly made wine is harsh in taste and has a yeasty flavour. In addition to the clarification, the process of maturation makes the wine tasty and fruity in flavor. The maturation period may extend to a year or may be only 2–3 months, as in the case of cider. During maturation, either oak wood chips should be added to the wine, or it should be kept in oak wood barrels with a suitable quantity of preservative without contact with the air, to prevent spoilage.

37.3.7 Clarification

After racking, if the wine is not clear it should be clarified using a filter aid such as bentonite, celite or using the tannin/gelatin treatment. These treatments usually make the wine crystal clear. In cases where no clarification is achieved, it is clearly due to the use of an unsatisfactory yeast source, to bacterial contamination, or to yeast autolysis, when the only solution is making vinegar from the wine.

37.3.8 Blending

The wines from plum or apricot may need some amount of sweetening with cane sugar. For apricot and plum wine, a total soluble solids (TSS) of 12°Bx (raising the sugar content by 4%) is considered to be optimal. This should again be followed by filtration.

37.3.9 Pasteurization

Wines with low alcohol content, especially cider (soft and hard), are pasteurized at 62°C for 15–20 minutes, after keeping some head space in the bottle, followed by crown corking. Alternatively, the table wines can be preserved by the addition of preservatives like sulfur dioxide, sodium benzoate, sorbic acid etc. The fortified wines need no preservation as alcohol itself act as a preservative, as happens in the case of wines with carbon dioxide as long as the container is not opened. When crown corks are used, the bottle should be stored upright, whereas those with bark corks may be stored in a cool place on their side, preferably after sealing with wax.

37.3.10 Storage and Labeling

After blending and pasteurization, the wines are usually stored until sent to the market for sale. Usually, blending and pasteurization is carried out only for a short time before sale. Any sediment in the bottle after such storage should be removed by fine filtration (Figure 37.1).

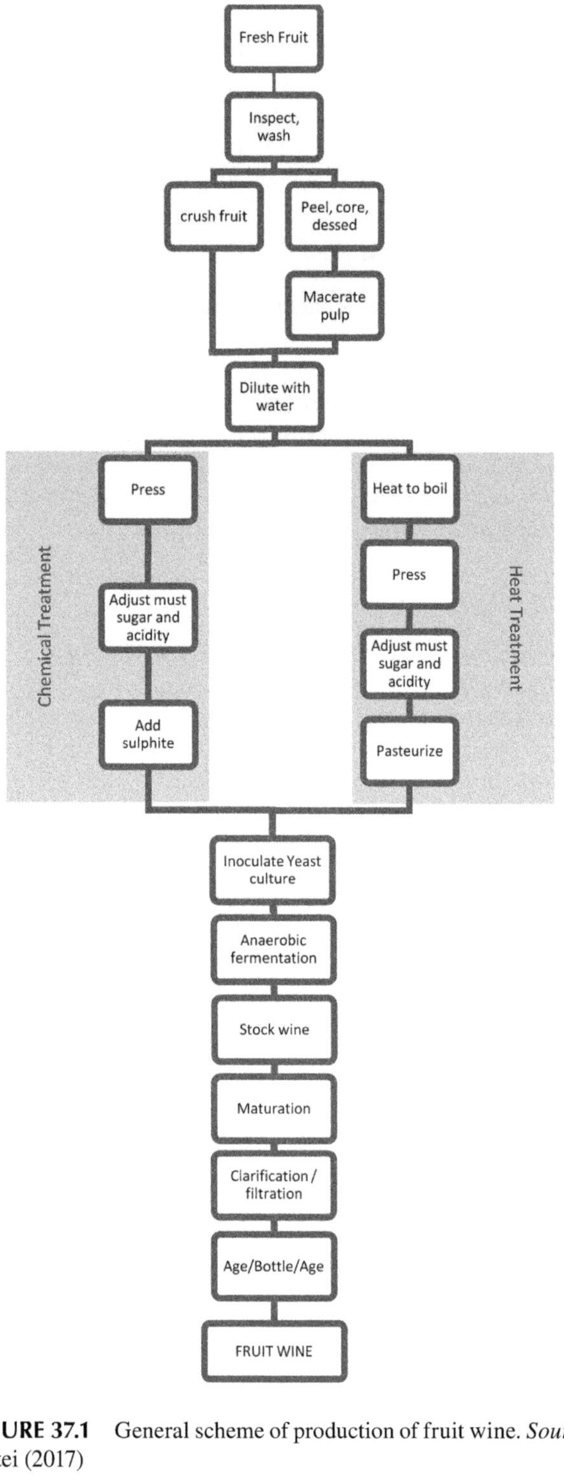

FIGURE 37.1 General scheme of production of fruit wine. *Source:* Matei (2017)

37.4 TECHNOLOGY FOR PRODUCTION OF VARIOUS FRUIT WINES

37.4.1 Apple Wine

Apple fruits are utilized to prepare a table wine. In its preparation, apple juice or concentrate is the raw material but amelioration with sugar or juice concentrate is essential. Natural fermentation is also practiced for making apple wine but pure cultures of the yeast *S. cerevisiae* are preferred as the final quality of wine is predictable and is likewise attractive from the enological and toxicological points-of-view. Addition of ammonium salts to the must reduces the higher alcohol production in wine due to non-degradation of amino acids the must. Bitterness could be the result of fermentation conditions, duration of fermentation or the effect of the variety used in apple wine preparation, particularly in terms of yeast nutrients. Washing and crushing the fruits, adding 50 ppm of SO_2 and 10% water in making apple wine is recommended. Utilization of a nitrogen source, for example, DAHP, and inoculation with a yeast culture is required to deliver a wine with 5.6–7.3% alcohol. For more information, see Chapter 35 in this volume and Figure 37.2.

37.4.2 Pear Wine

Perry or pear wine is produced by using pears, for example 'Bartlett' pear, with a high tannin content by grating and squeezing the organic product in a rack and material press, enhancing the Total Soluble Solids (TSS) of the juice to 21°Bx from 14°B and increasing the acidity to 0.5% from 0.25% by adding citric acid alongside 100 ppm SO_2. The fermentation is carried out with a pure wine yeast culture. Different methodologies followed are as described for apple wine preparation. It can be sweetened, braced or mixed with other fruit wines. For more information, see Chapter 35 in this volume. Printer Degree Brix is written as °B.

37.4.3 Custard Apple Wine

Annona squamosa is known as custard apple, sweetsop or sugar apple. Although cultivated in different tropical areas around the world, it is native to tropical America. This species has been utilized for the preparation of wine. The juice is made by peeling the fruit and expelling the seeds and adding pectinase enzyme to the mash. Both concentration and incubation period influenced the juice recovery. The juice was enhanced to a TSS of 23°B and an acidity of 0.7%, with 0.05% DAHP and 125 ppm SO_2 added. The must was inoculated with wine yeast, *S. cerevisiae*, fermented at 30°C, and finally clarified with bentonite. This wine has low levels of phenols, when compared with red wine, although custard apple wine showed high radical scavenging (or antioxidant) capacity (DPPH), *in vitro* comparable to wines prepared from fruits like pineapple, lime and tamarind. This wine may serve as a good source of antioxidants, providing significant health benefits.

37.4.4 Mango Wine

Mango (*Mangifera indica*) is a tropical fruit distributed worldwide. The shelf-life of mango is 2–4 weeks (when stored at 10–15°C), which limits their availability as fresh fruit. Large-scale production of mango wines could be an alternative to prolonging the shelf-life of mango fruit and reducing post-harvest losses.

To make mango wine, the natural products are made into must and its TSS raised to 20°B by adding unadulterated sweetener (cane sugar), 100 ppm SO_2 and pectinase enzyme

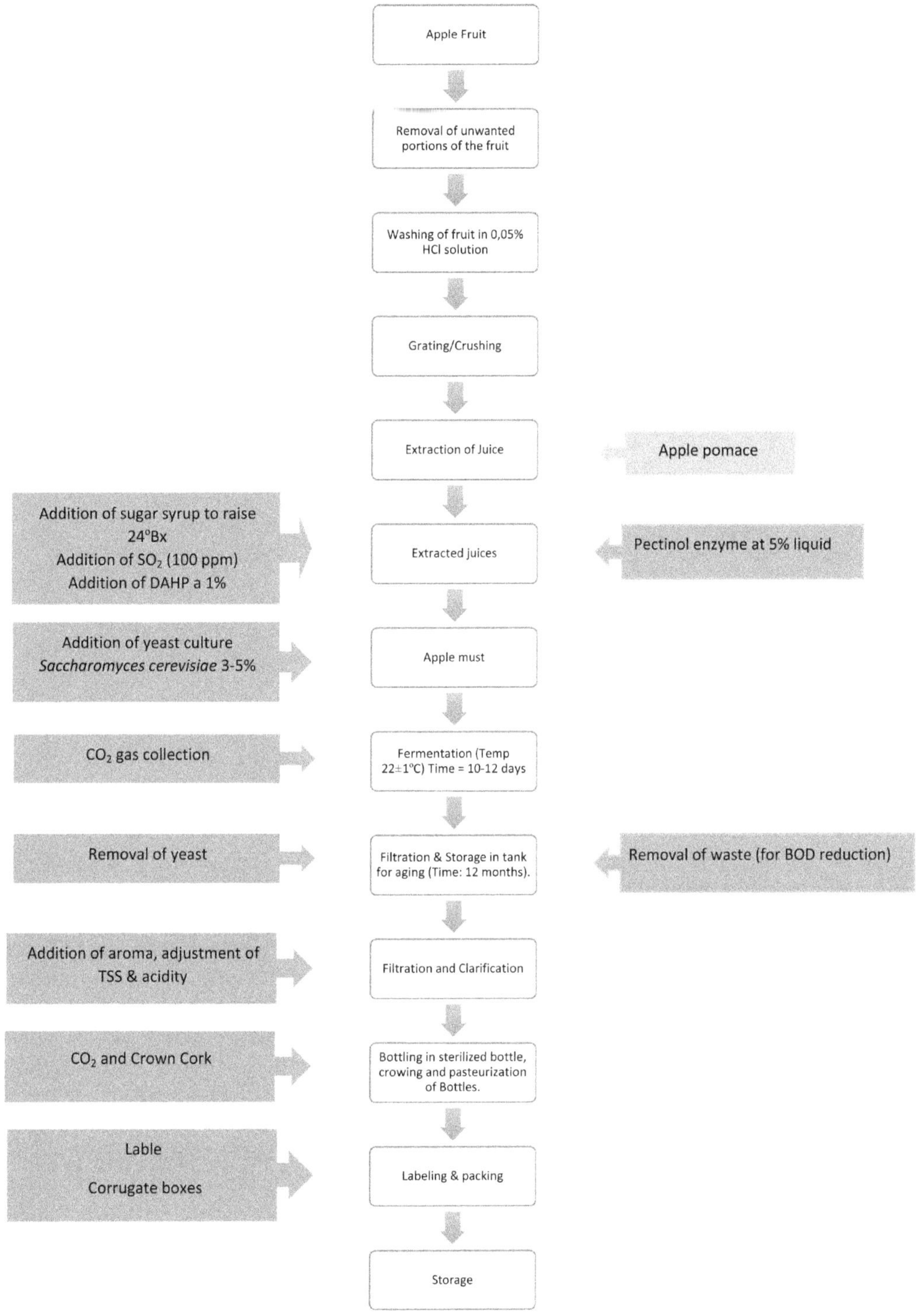

FIGURE 37.2 Flow diagram for apple wine production

at 0.5% to the mash to make the must. *S. cerevisiae* var. *ellipsoideus* (No.522) is added at a rate of 10% and fermentation of the must is carried out for 7–10 days at 22°C. After racking and filtration, the wine is treated with bentonite for clarification and packaged with 100 ppm SO_2 as potassium metabisulfite (KMS). Sweet and dry wine was produced using the fruit of the 'Dashehari' cultivar, where the fermentation was halted by adding 10% (v/v) mango brandy following aging. To make the sweet wine, cane sugar is added at the rate of 5 g/L. The alcohol content of mango wines ranges from 5 to 13%, and they are low in tannins.

37.4.5 Jambal Wine

Dry wine of acceptable quality can be produced using jambal fruit (*Syzygium cumini*). The technique to make the best wine from fruits of cultivar Jamun included dilution of the crushed fruits with water in the proportion of 1:1, improvement of the

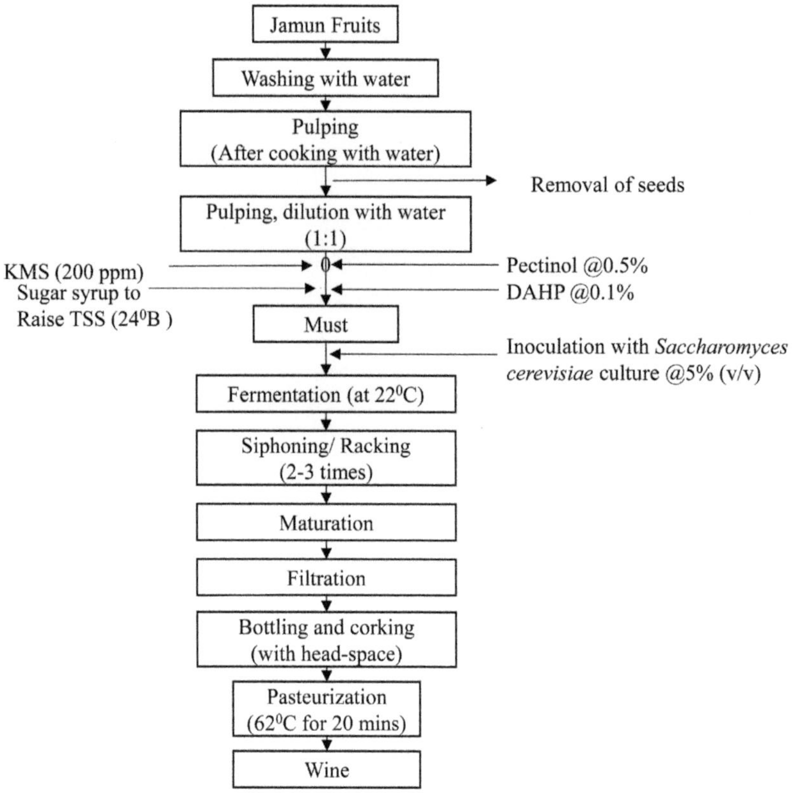

FIGURE 37.3 Flow sheet for the preparation of Jamun wine. *Source:* Joshi *et al.* (2012)

must TSS to 24°B with cane sugar, addition of DAHP at the rate of 0.2%, and addition of 150 ppm SO_2 and 0.25% of the pectic esterase enzyme, with the fermentation completed by the addition of 2% *S. cerevisiae* culture, followed by racking, filtration and packaging. The process of preparation of Jamun wine is shown in Figure 37.3.

37.4.6 Muskmelon Wine

Musk melon (*Cucumis melo*) juice recovered from fruits unfit for table purpose was fermented for 96 hours at 30 ±5°C. The wine showed a decent sensory quality when adjusted to a TSS of 10 to 12°B, with an acid ratio of 34.5 and 41.4, respectively. The dilution of muskmelon pulp (1:1), with the addition of DAHP, pectinesterase enzyme and citric acid can be successfully used for the preparation of good quality muskmelon wine.

37.4.7 Coconut Toddy

Toddy (an alcoholic beverage) is obtained by the natural fermentation of coconut palm inflorescence sap collected in clay pots, sterilized by inverting over a flame for 5 min and being allowed to ferment in open pots for up to two days. During the period of sap collection, microorganisms from the atmosphere enter the clay pots and multiply in the palm sap. The sap contains 15–18% sucrose, which, following fermentation, gives 7% (v/v) ethanol when it is known as *toddy*. Freshly prepared *toddy* has an average alcohol content of 7.9% (v/v), a titratable acidity of 14.36 mM HCl per 100 mL, and a pH value of 3.7, and it is consumed as a soft drink. It is generally stored at ambient temperature and is sold in shops. From the changes observed in tuba (a popular fermented coconut sap in the Phillippines) during one-week storage at various temperatures, it is apparent that toddy should not be sold more than two days after collection to maintain its high alcohol content and its low titratable acidity. The alcohol content in *toddy* can be increased (up to 9% from 6–7 %) if the growth of non-fermenting organisms is inhibited during fermentation. Addition of 0.08% of NH_4^+ ions in the form of ammonium sulfate in fermenting toddy improved ethanol production. The growth of Non-ethanol-producing microorganisms is inhibited by mixing sodium metabisulfite into sterilized coconut sap before the onset of fermentation by a pure culture of *S. cerevisiae.*

37.4.8 Palm Sap Wine

In numerous territories of Africa and Asia, the sap obtained from different types of palm (*Acrocomia mexicana*) is matured to deliver the wine called palm wine, coyol wine or *Vino de Coyol* or toddy. It is basically an overwhelmingly smooth white, opalescent suspension of live yeast and bacteria with a sweet taste and a vigorous fizz. It uses sap from different types of palm, including coconum palm (*Cocos nucifera*), the oil palm (*Elaeis guineensis*), palmyrah palm (*Borassus flabellifer*), the nipa palm (*Nipa fructicans*) and the wild date palm (*Phoenix sylvestris*). There is impressive art to binding the flower spathes, pounding them to cause the sap (600 to

TABLE 37.1
Analysis of a Typical Coyol Wine (From the Sap of Coyol Palms)

Characteristics	Mean value
pH	4.0
Alcohol (%)	12.86
Brix (°B)	16.00
Protein (%)	0.61
Phosphorus (ppm)	38
Sodium (ppm)	28
Calcium (ppm)	142
Magnesium (ppm)	57
Iron (ppm)	2.5
Manganese (ppm)	0.5
Copper (ppm)	0.9
Zinc (ppm)	0.2
Potassium (ppm)	2.540

Source: Balick (1990)

1000 mL per spathe) to flow properly by cutting the spathe tip and collecting the sap.

The palm sap is a colorless liquid containing 10–12% sugar, is neutral in reaction, and contains 4.20±1.4% sucrose, 3.31±0.95% glucose and 0.38±0.15% NH_3. Bacteria related to *Zymomonas mobilis*, yeasts like *S. cerevisiae* and *Schizosaccharomyces pombe* are invariably present but the numbers of lactic acid bacteria (LAB) and other species of bacteria in the palm juice vary. The sap is allowed to ferment naturally for 24 hours before the sale. The bottles have constant foaming/bubbling, due to the fermentation process known as *herviendo* or 'boiling wine'. To continue the fermentation, sugar (4.5 kg of sugar is added to 30 kg of sap) has to be added to the palm juice. Within a few days, coyol wine must be sold, otherwise, it may be converted into vinegar. During the fermentation process, LAB lowers the initial pH of the juice from 7.4 to 6.8, and the wine is consumed after 12 hours of fermentation, by which time the pH is normally between 5.5 to 6.5. The ethanol content in the beverage does not usually rise above 7.0%, though a concentration of 12.86% can be achieved. Coyol or palm sap wine is a source of minerals, especially potassium (Table 37.1).

37.4.9 Pomegranate Wine

To prepare a wine from pomegranate fruit (*Punica granatum*), the entire organic products are squeezed from the fruit without mashing, to avoid unnecessary astringency in the wine. This fruit is useful for manufacturing wines rich in bioactive compounds. Sugar is added to the juice to bring its TSS to 22–23°B, potassium metabisulfite (2 lbs per 1000 gallons) is added to the juice to prevent the growth of undesirable microorganisms and the juice is fermented with a wine yeast. It is aged and finished in the same manner as red grape wine. To make sweet table wine, sugar is added to bring its TSS to 8–10°B after aging and the wine is flash-pasteurized at 60°C and sealed, followed by cooling by spraying with cold water. A port-like wine can be made by fortifying the sweetened wine to about 20% alcohol. A good-quality wine has been prepared from the 'Ganesh' variety of pomegranate.

37.4.10 Banana Wine

To prepare wine from banana (*Musa* × *paradisiaca*), the peel is removed physically and the peeled fruits are passed through a pulper with a 60-mesh screen, and water is added, finally, the juice is centrifugated. The yield is 82 L/100 kg of the natural product. Potassium metabisulfite (100 ppm) is added to avoid the development of undesirable microorganisms. Fermentation is carried out at 18±1°C with 2% yeast added after maturation is carried out at low-temperatures of 5–7°C for ten days. For the greatest recovery of juice, specific concentrations of the pectinase enzyme and various incubation periods were tested at 28±1°C, with incubation with 0.2% pectinase for four hours being the best combination for extraction of juice to generate wine. Addition of sugars and pectinase (0.15%), containing pectolytic enzyme or pectin methyl esterase and polygalacturonase, to banana puree has been used to make a wine.

37.4.11 Plantain Wine

Plantain fruits, like banana, are perishable. The fruits contain a considerable amount of sugar, and they can be utilized to make a wine. Figure 37.4 shows various steps involved in the production of plantain wine.

Mature firm and yellow plantain fruits were stored for five days at 28°C until they were overripe and were used for preparation of plantain wine. Passion fruit (*P. edulis var flavicarpa*) concentrate of 36°B was blended with plantain puree in some wine treatments to improve its quality attributes. Fruits were washed with chlorinated water (100 ppm) peeled, sliced and blended for 1 minute. The recipe for plantain wine consists of plantain pulp with or without peel (24%), water (36%), granulated sucrose (37.5 ± 1.0%), pectolase (0.15%), amylase (0.03%), yeast (0.23%), diammonium phosphate with ammonium phosphate (2.2%) and KMS (0.15%). The pH was adjusted to 3.1–3.5 by the addition of citric acid. Specific gravity of plantain wine was found to be less than 1.0, varying between 0.992 and 0.994 and had < 9°B which was considered to be dry. The tannin content of the wine (mg/100 mL) increased with the addition of peel and was further enhanced by the addition of passion fruit, which were astringent in taste due to the interaction between tannins the salivary proteins and glycoproteins.

37.4.12 Guava Wine

Guava fruits are available in abundance at a low price and can be utilized for production of wine of highly acceptable quality, though it has low sugar content, yet has a characteristic flavor and a golden-yellow colour. Two types of wine have

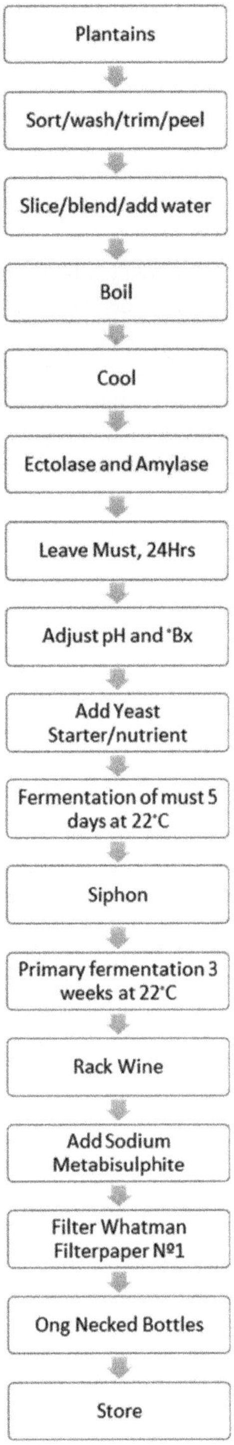

FIGURE 37.4 Flow diagram of unit operation of plantain wine production. *Source:* Alisimon and Badrie (2002)

been prepared from guava fruit: guava juice wine (GJW) and guava pulp wine (GPW). For making wine from pulp, dilution with water is essential and a dilution level of 1:2 was found to be better than 1:3. Sugar (24°B), KMS (to give 125 ppm SO_2), pectinase (0.5%) and yeast (2%) are added to initiate the fermentation, which is carried out at 20±2°C. The treatment of pulp with pectinase increased the final yield of wine to the tune of 18%. Wine prepared from guava juice obtained by treatment with pectinase for juice extraction gave wine with low tannin content, optimum color, flavor and acceptable sensory qualities. However, fermentation of guava pulp in the presence of pectinase reportedly yielded a wine with high tannin, dark colour and astringent taste. The best wine was obtained by fermentation of juice. Guava pulp wine is prepared in the same way as guava juice wine. When the Brix reading reached 10°Bx, the pomace is removed and more sugar (10%) is added to the fermenting materials and the mixture is allowed to ferment further. Rest of the procedure is the same as for any wine preparation. Comparison of chemical characteristics of wine fermented from guava juice or pulp is presented in Table 37.2.

37.4.13 Ber Wine

Ber juice prepared from ber pulp (*Ziziphus mauritiana*) is used to prepare wine. The pulp is diluted with water in a 1:1 ratio and pectinase enzyme is added at the rate of 5 mL/L. After incubation, the juice is filtered through muslin cloth, and sugar is added to the juice to raise its TSS to 23°B. Inoculation of the juice with yeast (*S. cerevisiae*) initiated the fermentation. After eight days, the wine was racked and kept for cold stabilization at 5°C for 10 days. It is then here racked again and clarified, using agar bentonite mixture (2:1) at the rate of 0.1 g/100 mL. Optimized fermentation conditions for preparation of good-quality bar wine includes the use of 50 to 100 ppm SO_2, 2 to 5% yeast inoculum, and 3.5 to 4.0 pH.

37.4.14 Plum Wine

To prepare plum wine, for every pound of plums, 1 L of water is added, followed by addition of the starter culture. The mixture is allowed to ferment for 8–10 days before pressing. Addition of pectolytic enzymes before fermentation facilitates the processing by increasing the yield of juice and accelerating wine clarification. Additional sugar may be added to the partially fermented juice, depending on the type of wine required (table or dessert). Alternatively, fully ripe plum fruits are diluted with water in a 1:1 ratio (on a whole-fruit basis), and 0.3% pectinol, 150 ppm SO_2 and sufficient sugar to raise the TSS of the must to 24°Bx are added and fermented to produce the wine. Aging, filtration, bottling, and processing are similar to that for any fruit wine. A considerable improvement in the quality of the plum wine takes place when the sugar content of the wine is raised to 12°B. The best wine was produced by *S. cerevisiae* strain UCD 595 in must of Santa Rosa plum. In plum fruits of cv. Black Beauty, to make the best sweet table wine, an Initial Sugar Concentration (ISC) of ISC-30 was necessary. The use of sodium benzoate instead of KMS produces wine with better colour and sensory qualities without affecting the physico-chemical characteristics. Some of the physico-chemical characteristics of a plum wine are given in Table 37.3.

Of the cultivars evaluated, Santa Rosa plum made the best-quality wine. The mineral contents of wines from varieties 'Santa

TABLE 37.2
Comparison of the Chemical Characteristics of Guava Wine Prepared From Juice, Pulp and Diluted Pulp

Treatment	Guava juice wine	Guava pulp wine	Guava pulp (1:2 dil.) wine	Guava pulp (1:3 dil.) wine
Total soluble solids °B (refractometer reading)	17.0	17.5	15.0	17.0
pH	3.40	3.50	3.20	3.25
Volatile acidity (Acetic acid/100 mL)	0.033	0.046	0.052	0.063
Total Acidity (g/100 mL TA)	0.901	0.796	0.750	0.627
Brix reading (Hydrometer reading)	9.0	9.0	6.5	10.5
Total SO_2 (ppm)	153.6	115.0	156.8	127.2
Reducing sugar (%)	7.8	10.0	8.3	9.2
Total Tannins (g/100 mL)	0.020	0.130	0.110	0.055
Alcohol (%)	10.0	10.7	11.5	9.8
Total aldehyde (ppm)	58.1	37.4	59.8	50.6
Yield of wine (%)	60.0	60.0	-	-

TA = Tartaric acid.
Dil.= Dilution.
Source: Bardiya *et al.* (1974)

TABLE 37.3
Physico-Chemical Characteristics of Plum Wine

Characteristics	Range
Ethanol (% v/v)	8.5–11.0
Total soluble solids (°B)	8.0–12.0
Titratable acidity (% malic acid)	0.62–0.68
Volatile acidity (% acetic acid)	0.028–0.040
Total esters (mg/L)	104.0–109.0
Colour (tintometer color units):	
Red	6–10
Yellow	10
Total coloring matter and tannins (mg/100 mL)	119

Source: Joshi and Sharma (1995)

TABLE 37.4
Comparison of Some Characteristics of Plum Wine Fermented With and Without Skin

Characteristics	Fermentation with skin	Fermentation without skin
Acidity (% MA)	1.43	1.12
pH	3.68	3.75
Alcohol (% w/v)	8.71	8.65
Sugar utilization	16.2	16.0
Color value (OD 540 nm)	0.80	0.21
Total coloring matter (mg/100 mL)	208	116

OD = Optical Density
Source: Vyas and Joshi (1982)

Rosa', 'Methley' and 'Green Gage' were desirable from enological and nutritional quality points of view. Plum fruit, being highly acidic, is less suitable for the production of a palatable wine. The de-acidification of the plum must by *Schizosaccharomyces pombe* was independent of the sugar content of the must but was adversely affected by a pH of 2.5, higher levels of ethanol and a SO_2 content of more than 150 ppm, whereas addition of ammonium sulphate enhanced the acid degradation by this yeast. The de-acidification of plum must, followed by adjustment of acidity to a desirable level, also improved the quality of plum wine; for details, see the literature cited.

Osmotic techniques were used to make the wine. The sensory quality of wine prepared from water-blanched and osmotically treated fruits was the best. Plum wine can be prepared from plum fruits either with or without the skin of the fruit, but the wines were found to have considerable differences in physico-chemical characteristics. A comparison of such characteristics of plum wine is made in Table 37.4.

Like other fruits, plum has biologically active compounds (like phenolic compounds) and the operations applied in plum processing and in the production of plum wine can significantly affect the content of these compounds. Likewise, the content of total phenols, flavan-3-ols and anthocyanins in wines differs significantly, depending on the plum variety. Among the varieties studied in Serbia, the wine produced with variety 'Čačanskalepotica' was characterized by the highest content of total phenols, total anthocyanins and flavan-3-ols, the greatest color intensity and the strongest antioxidant activity against DPPH free radicals.

For maturation, it is an age-old practice to use wooden casks for storage of alcoholic beverages produced from the grapes. Addition of wood chips from *Albizia* and *Quercus*

TABLE 37.5
Effect of Wood Chips on Physico-Chemical Characteristics of Plum Wine

Treatment	Tannins (mg/100 g)	Methanol (μL/L)
Control	139.00	240.00
Alnus nitida	96.00	236.00
Populus ciliata	107.00	338.00
Celtis australis	122.00	327.00
Toona ciliata	148.00	181.00
Salix tetrasperma	100.00	229.00
Bombax cieba	122.00	180.00
Albizia chinensis	111.00	249.00
Quercus leucotrichohora	139.00	198.00
CD (P=0.05)	20.15	11.29

Source: Joshi *et al.* (1999)

produced wines of the highest sensory qualities, including plum wine. In general, treatment of wines with wood chips increased the methanol content, with wood chips of *Populus, Celtis* and *Salix* increasing the methanol level more than did *Albizia* and *Quercus* (Table 37.5). Wood chips of *Quercus* and *Albizia* were found to be particularly suitable for maturation of plum wine.

37.4.15 Citrus Wines

All citrus wines, except for lemon, are acceptable but have to be sweetened (by addition of 2–3% sugar), followed by pasteurization and maturation, though bitterness remained associated with these wines. The method of orange wine preparation includes sweetening the respective juice, addition of KMS at the rate of 200 ppm, pectinol enzyme at 0.5% and DAHP at 0.1%. A sweet dessert wine has been made out of oranges, and orange varieties grown in Adana, southern Turkey, have been evaluated for wine making. Juice from ripe but not rotten fruit was extracted by an FMC juice extractor after sorting and washing. Crushing of oranges has not been found to be satisfactory for wine making, since oil from the skin also extracts into the juice thus, inhibiting the fermentation process. Addition of KMS to give 150 ppm of SO_2, amelioration of the TSS with sugar to achieve 22–23°B, addition of 0.1% pectic enzyme and fermentation to dryness are carried out. Wines are matured for a period of at least six months. If necessary, the wines could again be clarified before consumption.

The wine is sweetened to 10°B, adjusted to a total SO_2 of 200 mg/L, followed by filtration and pasteurization. However, if it is fortified to 20% alcohol and aged, it can be bottled without pasteurization.

An effective method of making wine without bitterness from 'Kinnow' (a hybrid mandarin grown widely in northern India) involved treatment of the juice with Amberlite XAD-16 ion-exchange resin (to reduce bitterness in wine by removing limonin and naringin), followed by fermentation, maturation etc. Debittering the juice either prior to or during fermentation improved the sensory quality of 'Kinnow' juice.

37.4.16 Peach Wine

Peach fruit has lesser acid than plum or apricot but, being pulpy, has to be diluted with water to prepare the must for fermentation. The method used for making peach wine consists of dilution of the pulp in the ratio of 1:1, raising the initial TSS to 24°B and adding pectinol and DAHP at rates of 0.5 and 0.1%, respectively. To the must, 100 ppm of SO_2 is added to control the activity of undesirable microflora. Differences in fermentation behavior of different peach cultivars were also observed. Amongst the cultivars evaluated, 'Redhaven', 'Sunhaven', 'Flavourcrest', 'Rich-a-haven', 'J.H. Hale' and 'July Elberta' were rated to be suitable for conversion into wine of acceptably sensory quality. Some of the physico-chemical characteristics of wines produced from different cultivars of peach are given in Table 37.6.

Recent experience found that peach wine (extracted from fruits with skin intact) contains significantly higher amounts of total phenolics and flavonoids than white wine, with a similar result being obtained with respect to the antioxidant capacity. Sensory analysis indicated that the peach wine was very well accepted by the regular consumers of wine and can be a very interesting product on the market.

37.4.17 Sea Buckthorn Wine

Sea buckthorn (*Hippophae rhamnoides*), belonging to the Elaeagnaceae family, is a thorny, dioecious bush growing wild in the cold and dry regions of the Indian Himalayas. The berries contain large amounts of vitamins C, essential oils, proteins and bioactive substances. Among the different products that can be prepared from it, is wine. The acid content of sea buckthorn juice is a major factor that affects its fermentation and, hence, the quality of the final product. So, de-acidification of pulp was attempted to reduce the acidity by diluting the pulp with water in the ratios of 1:5, 1:6, 1:7 and 1:8 in one set of trials and with sodium bicarbonate at different concentrations (0.6, 0.8, 1.0 and 1.2%) in the second. The pulp was ameliorated with sugar (24°Bx), 100 ppm SO_2 and 0.5% pectinase enzyme, with or without DAHP (0.1%), and fermented with pure wine yeast culture *S. cerevisiae* var. *ellipsoideus* (5% v/v) at 22±1°C. Sea buckthorn must, prepared by dilution with water, had better fermentation behavior than that prepared by $NaHCO_3$. Addition of DAHP, in general, enhanced both the rate of fermentation as well as the ethanol content. The highest rate of fermentation (rate of fermentation, RF=0.80) was recorded at a 1:5 dilution with 0.1% DAHP. After fermentation, wines prepared by diluting the pulp had ethanol contents of 9.3 to 13.18% (v/v) while wines in which the pulp was treated with $NaHCO_3$ ranged between 8.06 to 10.2% (v/v) in ethanol content. The highest alcohol contents were recorded in must made with 1:6 dilution with 0.1% DAHP followed by 1:5 dilution with 0.1% DAHP. The wines prepared with

TABLE 37.6
Physico-Chemical Characteristics of Wines from Different Peach Cultivars

Cultivars	Ethanol (%v/v)	Higher alcohols (mg/L)	Volatile acidity (% AA)	Total esters (mg/L)	Total phenols (mg/L)	pH	Total sugars (mg/L)
Richahaven	10.73 (3.27)	113.6	0.023 (0.723)	98.23	241.8	3.60	1.33 (1.15)
J.H. Hale	10.67 (3.26)	126.9	0.029 (0.727)	90.92	259.0	3.80	1.24 (1.12)
Redhaven	11.10 (3.33)	121.4	0.023 (0.723)	98.40	301.7	3.71	1.26 (1.12)
Flavorcrest	11.33 (3.36)	134.4	0.029 (0.722)	94.33	306.1	3.61	1.19 (1.89)
Sunhaven	11.67 (3.41)	131.6	0.022 (0.725)	101.50	278.6	3.71	1.22 (1.05)
Starkearly Giant	10.77 (3.28)	144.2	0.026 (0.725)	97.80	206.0	3.81	1.32 (1.15)
Kateroo (wild peach)	11.83 (3.24)	154.3	0.029 (0.727)	92.37	230.4	3.84	1.22 (1.10)
July Elberta	10.87 (3.14)	151.5	0.020 (0.721)	95.50	203.3	0.74	1.36 (1.17)

AA = Acetic Acid; Values within the parenthesis are the transformed values
Source: Joshi *et al.* (2017)

TABLE 37.7
Effect of De-Acidification on TSS, Titratable Acidity, Ascorbic Acid and Total Sugars of Sea Buckthorn Wine

Treatments	TSS (°Bx)			Ethanol Content (% v/v)			Ascorbic Acid (mg/100 g)		
	Without DAHP	With DAHP	Mean	Without DAHP	With DAHP	Mean	Without DAHP	With DAHP	Mean
				De-acidification By Dilution					
T_1 (1:5)	7.2	7.8	**7.5**	9.30	11.62	**10.46**	750	800	**775**
T_2 (1:6)	7.2	7.2	**7.2**	9.61	13.18	**11.39**	700	700	**700**
T_3 (1:7)	7.0	7.0	**7.0**	11.05	12.40	**11.72**	500	530	**515**
T_4 (1:8)	7.0	7.0	**7.0**	9.92	11.60	**10.76**	400	400	**400**
				De-acidification By $NaHCO_3$					
T_5 (0.6%)	9.8	9.8	**9.8**	9.30	9.30	**9.30**	500	500	**500**
T_6 (0.8%)	10.0	10.0	**10.0**	8.06	9.30	**8.68**	400	500	**450**
T_7 (1.0%)	9.0	9.2	**9.1**	8.30	9.60	**8.95**	400	400	**400**
T_8 (1.2%)	9.4	10.2	**9.8**	9.80	10.20	**10.00**	400	400	**400**
Mean	**8.3**	**8.5**		**9.42**	**10.90**		**506.3**	**528.75**	
$CD_{0.05}$									
Treatment(T)	NS		1.53		113.8				
Sub-treatment (S)	NS		0.76		NS				
S×T	NS		2.16		161.0				

DAHP = Di-Ammonium Hydrogen Phosphate
TSS = Total Soluble Solids
$NaHCO_3$ = Sodium Bicarbonate
Source: Joshi *et al.* (2011)

DAHP at 0.1% had higher ethanol contents than those without DAHP. Variations in physico-chemical characteristics and sensory qualities of the wines were also observed (Table 37.7). Among the physico-chemical characteristics, TSS ranged between 6.8 and 10.2°B. The titratable acidity ranged between 0.96 to 2.48 as % citric acid-equivalents, depending on the level of dilution employed or the de-acidification carried out by the use of $NaHCO_3$. The total sugars in different wines ranged from 1.5 to 3.35%, whereas ascorbic acid content was recorded between 400 and 800 mg/100 mL.

37.4.18 Kiwifruit Wine

A wine of good quality can be prepared from varieties of kiwifruit 'Bruno' and 'Hayward', after amelioration. Earlier attempts to make wine using conventional practices did not succeed, as the product was described as grassy, green and stalky in aroma and taste, with unacceptable levels of bitterness and astringency. A process (Figure 37.5) was subsequently developed to prepare a white wine of outstanding character. The wines made from clarified kiwifruit juice, extracted with the use of pectolytic enzymes has been described as possessing an intense, fruity, 'Riesling Sylvaner'-type aroma. The aroma is developed during fermentation for about four days at 15°C. The removal of astringency and bitterness correlated with a reduction in total phenols and the removal of flavonoids. Based on the retention of SO_2 in the free and bound form, a practice of adjusting free SO_2 content gradually over a 6- to 8-week period, which allowed adequate time for settling and clarification, has been made. After 20 weeks of maturation, the wines with SO_2 had only the typical 'Riesling Sylvanar-type' of aroma, while those without SO_2 developed an oxidized aroma and excessive brown color.

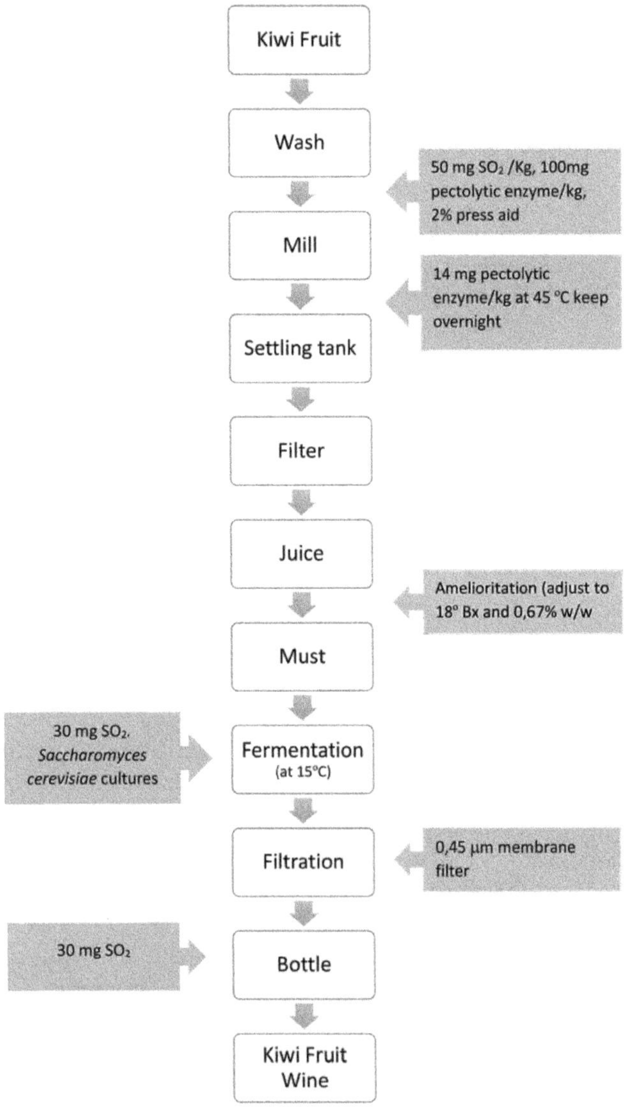

FIGURE 37.5 Flow diagram of kiwifruit wine production. *Source:* Adapted from Lodge (1981)

TABLE 37.8
A Comparison Between Unclarified and Clarified Kiwifruit Juice

Parameters	Unclarified kiwifruit juice	Clarified kiwifruit juice
TSS (°B)	14.40±0.094	14.26±0.004
pH	3.43±0.002	3.50±0.001
Reducing sugar (%)	8.34±0.014	8.17±0.009
Total sugar (%)	10.59±0.005	10.14±0.008
Ascorbic acid (mg/ 100mL)	196.55±0.523	154.59±0.014
Pectin (as % calcium pectate)	0.92±0.007	0.12±0.002
Relative viscosity (cps)	5.43±0.113	1.34±0.001
Total phenols (mg gallic acid/L)	389±0.012	240±0.01

Source: Vaidya *et al.* (2009)

Kiwifruit is a nutritionally rich fruit with high ascorbic acid content (193 mg/100 g), but the extraction of its juice is difficult due to the slimy pulp. To overcome this problem, a combination of enzymes (pectinase 0.025 g/kg + amylase 0.025 g/kg + mash enzyme 0.05 g/kg) were used to macerate pulp (two hours at 50°C), facilitating the extraction of juice. The treatment enhanced the juice recovery (78.46%) compared with the control (58.44%), and the treatment did not affect the TSS, titratable acidity, pH, reducing or total sugars of the clarified juice. A comparison of physico-chemical characteristics of unclarified juice with that of enzymatically clarified juice showed a marked decrease in pectin content and consequently, a decrease in the viscosity of the juice (Table 37.8). The outstanding feature of the juice was its high acidity and high concentration of ascorbic acid, which, however, decreased by 21% after clarification. The recovered juice was ameliorated with sugar (to 22±1°Bx), 100 ppm SO_2 and 0.1% DAHP, and was fermented by a pure culture of *S. cerevisiae* at 22±1°C. After fermentation, a wine of 9.7% alcohol and 7–8°B residual TSS was obtained. Blending with sucrose syrup made the wine palatable. Since, in the present study, the enzyme combinations were used for a period of two hours, high yield and clarity of juice were recorded.

37.4.19 Strawberry Wine

Strawberry wine of good quality has the appealing colour of premium rosé wine but it is often short lived. When frozen berries are employed to make wine, these are first thawed and then, the juice is extracted. The juice is ameliorated to 22°B by the addition of cane sugar added in stages, whereas addition of sugar after fermentation dilutes the alcohol level. The must is mixed with 0.1% DAHP and the fermentation is carried out by addition of 1% of yeast culture at a temperature of 27°C. After the completion of fermentation, the wine is racked, bottled and stored in the dark. The composition and maturity of the fruits and mold contaminants affect the

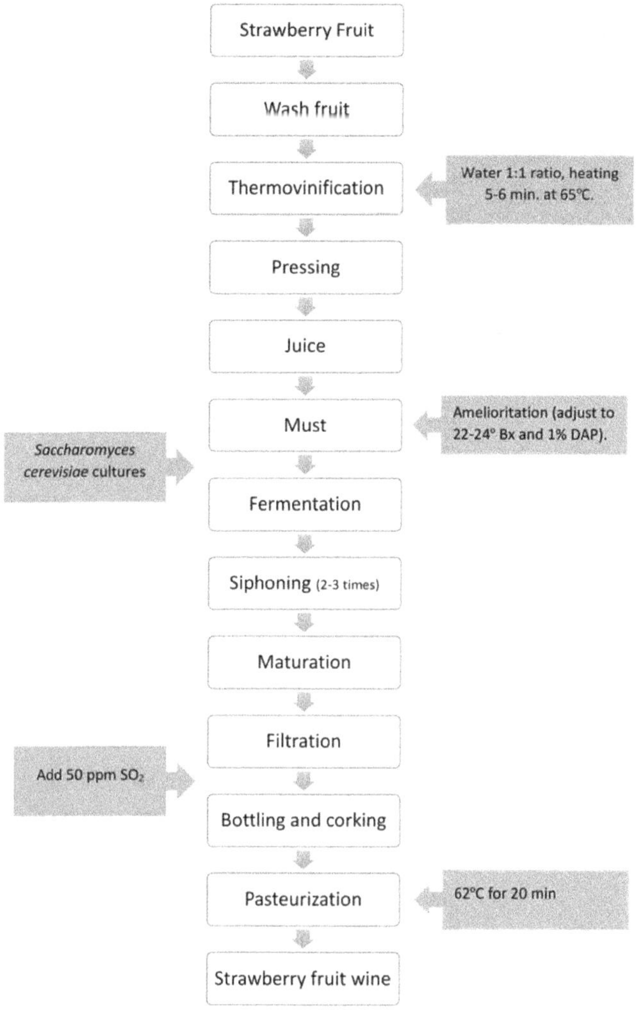

FIGURE 37.6 Unit operation for the production of strawberry wine with the thermovinification method. *Source:* Joshi *et al.* (2005)

quality of the wine. But over-ripe fruits with higher anthocyanin and total phenolics gave wines with better color than did fully ripe fruits. Treatment of juice with enzymes (mainly pectinases) increased polymerization, increased color extraction and color intensity in strawberry wine. Wine from strawberry fruits of three cultivars, namely 'Camarosa', 'Chandler' and 'Douglas' made by three different methods were compared (thermovinification, fermented on the skin and carbonic maceration). Thermovinification imparted many desirable quality characteristics to the wine relative to the control (Figure 37.6). Maturation of wine for nine months made many of the physico-chemical characteristics more desirable and improved the sensory quality of the wine. The flavor profiling of wines from the three cultivars characterized the wine made by different methods by descriptive analysis. Table 37.9 presents the physico-chemical characteristics of strawberry wine.

37.4.20 Red Raspberry Wine

A significant volume of juice of raspberry (*Rubus idaeus*) is used for the preparation of wine. Commercially, block-frozen

TABLE 37.9
Effect of Cultivars on Various Physico-Chemical Characteristics of Strawberry Wine

	Cultivar of strawberry		
Parameters	**Camarosa**	**Chandler**	**Douglas**
TSS (°Bx)	9.7	8.1	8.8
Total sugars (%)	1.7	0.6	1.0
Reducing Sugars (%)	0.135	0.124	0.128
pH	3.18	3.21	3.26
Alcohol (%v/v)	11.2	11.5	9.2
Higher alcohol (mg/L)	155	169	151
Total Volatile acidity (% acetic acid)	0.026	0.032	0.025
Esters (mg/L)	90.9	78.3	102.4
Anthocyanins (OD/mL wine)	0.15	0.145	0.104

TSS = Total Soluble Solids
OD = Optical Density
Source: Sharma *et al.* (2009)

red raspberries ('Meeker') are used for the preparation of wine. The berries are partially thawed at 27°C and are crushed through a hammer mill. Addition of 25 ppm of SO_2 is made. Wines were prepared from must with or without pectic enzymes, with pasteurization or preservation with SO_2 or potassium sorbate. The bottled wines were stored in the dark for six months. Fermentation of pulp, depectinized the juice and pasteurized juice affected the composition and other physico-characteristics of raspberry wine. During fermentation, anthocyanin pigments were degraded to a greater extent (about 50%) after storage. Cyanidin 3-glucoside was the most unstable anthocyanin, disappearing completely during fermentation while cyanidin-3-sophoroside (the major anthocyanin) was the most stable pigment. The pasteurized, depectinized and finned wine had the most stable colour and the best appearance after storage.

37.4.21 Cherry Wines

Like other stone fruits, cherry fruits can also be utilized for preparations of wine, with sour cherries being preferred. But the cherry fruit has been found to be more suitable for preparation into a dessert wine than a table wine. The alcohol content of such wines may range from 12 to 17%. A blend of dessert and table varieties of cherries can also be used for wine making. To prepare a dessert wine of 16% alcohol, each liter of juice is ameliorated with 430 g of sugar. The addition of KMS before fermentation is advisable. The clarity of the wine is improved considerably if pectolytic enzymes are used, but the addition of urea to the cherry must did not improve the fermentability. To enhance the flavor, about 10% of the pits may be broken down while crushing the cherries; production of hydrogen cyanide from hydrolysis of the amygdalin present in the pits of cherries has been detected. The cherry wine does not require a long aging period. To make the wine sweet,

TABLE 37.10
Contents of Total Phenolics and Total Anthocyanins, Total Antioxidant Capacity, and DPPH Radical-Scavenging Activity of Cherry Fruits and Their Products

Cultivar		Total phenolics^{a}	Total anthocyanins^{b}	Total antioxidants^{c}	DPPH-radical-scavenging activity (%)^{d}
Danube	Fruit	188 ± 8.1^{b}	70.1 ± 5.4^{b}	338 ± 14.0^{bc}	60.711.0^{c}
	Wine	79.4 ± 4.4^{d}	29.6 ± 5.9^{d}	289 ± 2.9^{d}	53.0±2.9^{d}
Balaton	Fruit	221 ±2.0^{a}	93.5 ±2.2^{a}	404±8.1^{a}	75.2±4.1^{a}
	Wine	149±1.5^{c}	63.4 ± 4.3^{c}	387 ±4.7^{b}	69.0 ± 0.3^{b}

All mean values are from triplicate determinations.^{a,b,c,d} Values in the same column with different superscript letters differ significantly ($p < 0.01$) by Duncan's multiple-range test.

[a] Total phenolics are expressed in mg of gallic acid equivalents (GAE) per 100 g of fresh cherries or mg of GAE per L of liquor. [b] Total anthocyanins arc expressed in mg of cyanidin 3-glucoside equivalents (CGE) per 100 g of fresh cherries or mg of CGE per L of liquor. [c] Total antioxidant activity, is expressed as mg of vitamin C equivalent antioxidant capacity (VCEAC) per 100 g of fresh cherries or mg of VCEAC per L of liquor. [d] DPPH-radical-scavenging activities of each extract arc 100 μg/mL or 100 μL of liquor.

sugar can be added prior to bottling. The bottled wine may be preserved either by germ-proof filtration or pasteurization. It was found that both the juice- and winemaking processes reduced the total phenolics to about 30–40% or 42–60%, respectively, and total anthocyanin capacity to about 60–77% in juice and 85–95% in wine. However, the radical-scavenging (antioxidant) activity, superoxide dismutase (SOD) and catalase (CAT) enzyme levels were higher in cherry wine than in cherry juice (Table 37.10).

37.4.22 Pineapple Wine

To produce a wine having 12–13% alcohol, the TSS of pineapple juice is raised by the addition of sugar (preferably, on the third day of fermentation), yeast-assimilable nitrogen is adjusted to 150 mg N/L, using DAHP, and the final pH of the juice is corrected to 3.5 with tartaric acid. The wine is preserved by pasteurization but can also be fortified and sweetened. The pineapple wine produced without dilution of juice is characterized by the presence of more quantities of higher alcohols, ethyl and acetate esters, compared to pineapple wines using diluted juice. This could be due to the higher concentration of nutrients of undiluted pineapple juices. The ethyl esters and fatty acids formed enzymatically during the fermentation process constitute an important group of aroma compounds that may contribute to the 'fruity' and 'cream/fatty' notes to pineapple wine's sensory properties.

37.4.23 Date Wine

Wine like Sherbote and Nabit can be prepared from dates by making date syrup followed by quick fermentation, taking 36–48 hours. Dakkai is made from whole dates, with a fermentation period of 96 hours. The method of Dakkai (with a liquid of pale yellow appearance) preparation includes mixing whole date fruits with lukewarm water in an earthenware container, which is allowed to undergo fermentation by natural microflora followed by straining. However, to prepare Nabbit and Sherbote, one part of dates is boiled with three parts of water, followed by straining, giving a sweet brown-colored syrup. In addition, in the production of Sherbote, ground cinnamon and ginger (50 g/L) are tied in a piece of cloth and dropped in the liquor to be fermented for one to three days.

37.4.24 Apricot Wine

Apricot is a delicious fruit grown in many parts of temperate hilly regions of tropical countries, including India. Due to its high flavor, the fruit holds promise for conversion into wine. In India, wild apricot fruits, grown naturally at higher altitude, are used locally to make liquor, which completely lacks nutrients. Further, preparations of apricot wine from the 'Newcastle' variety and wild apricot (*chulli*) have been reported. The physico-chemical characteristics of wine produced from apricot *cv* Newcastle are given in Table 37.11.

A method for the preparation of wine from wild apricots has been developed which consists of diluting the pulp in the ratio of 1:2, addition of DAHP at a rate of 0.1% and 0.5% pectinol, TSS of 24°B and fermentation with *S. cerevisiae*.

TABLE 37.11
Physico-Chemical Composition of Apricot Wine

	Apricot (Newcastle)
Characteristics	Mean ± SD*
TSS (°Bx)	8.20 ± 0.07
Titratable acidity (% MA)	0.76 ± 0.02
pH	3.15 ± 0.02
Ethanol (% v/v)	10.64 ± 0.09
Reducing sugars (%)	0.34 ± 0.01
Total sugars (%)	1.11 ± 0.02
Volatile acidity (% AA)	0.025 ± 0.002
Total phenols (mg/L)	253.60 ± 0.8
Total esters	120.6± 0.6

Source: Joshi *et al.* (1990)

TABLE 37.12
Effect of Different Initial TSS Levels Using Sugar and Honey on Chemical Characteristics of Wild Apricot Wine

	TSS (°Bx)		Ethanol (%v/v)		Higher alcohols (mg/L)		Total phenols (mg/L)	
Treatment	Sugar	Honey	Sugar	Honey	Sugar	Honey	Sugar	Honey
Initial TSS 22°Bx	8.33	8.27	10.23	9.18	113.0	121.0	245.0	238.0
Initial TSS 24°Bx	8.57	8.37	11.70	9.86	131.5	136.5	264.0	255.0
Initial TSS 26°Bx	8.77	8.60	12.17	10.38	153.0	158.0	279.0	270.0

Furthermore, with the increase in the dilution level, the rate of fermentation, alcohol content and pH of the wines increased, whereas a decrease in titratable acidity and volatile acidity, phenols, TSS, color values and mineral contents took place. Addition of DAHP at the rate of 0.1% enhanced the rate of fermentation. The wine from 1:2 diluted pulps was rated to be the best.

In another method, honey was used to ameliorate the wild apricot must, instead of sugar. Comparison of physico-chemical characteristics of various wines produced are made in Table 37.12. Higher alcohols and total phenols in honey wine were found at higher concentrations than those of sugar wine. Depending upon the initial sugar contents, ethanol was produced. Out of three different initial sugar concentrations (ISC), namely 22, 24 and 26°B wood chips, wild apricot wine of 22°B and treated with *Quercus* wood chips was rated the best.

In the maturation of apricot wine cv. Newcastle, wood chips of *Quercus, Albizia, Bombax* and *Toona* were found to increase the tannin contents appreciably, whereas the *Populus, Celtis and Salix* wood chips increased the methanol levels in the treated wine. There were significant improvements in the sensory qualities of all the wines, except those treated with woods of *Toona, Populus* or *Alnus*. Out of the different wood-chips tried, *Albizia* and *Quercus* produced wine of the highest sensory qualities. The effect of the addition of different wood chips viz., *Quercus, Bombax* or *Acacia*, on ethanol, phenols and sensory quality characteristics of wild apricot wine has been documented.

37.4.25 Aloe Wine

The juice of *Aloe barbadensis* (*Ghikumar*), a succulent plant which has been used in the treatment of diabetes and cancer, can be fermented into a wine. The mixed juice of aloe and grapes ('Perlette'), having pH of 4.5, TSS of 100°Bx, 3.5% pomace, KMS 100 ppm, was fermented with 10% *S. cerevisiae* var *ellipsoedeus* at 28±2°C for 24 hours to make a wine with increased amounts of total phenols and catechin, known for their antioxidant role.

37.4.26 Low-Alcohol Bitter Gourd drink

A very low-alcohol (2%) wine named *Karela Somras* has been developed for use by diabetics by fermentation of a mixture of bitter gourd and grapes. The fermentation is conducted at 25–28°C for 6–7 days, followed by separation from the sediment by siphoning and allowing to stand for 3–4 weeks or until clear. Completely filled bottles are corked and stored. The drink reduces the sugar level in diabetics due to the presence of certain components in it which have an action similar to that of insulin or which probably induce the production of insulin.

37.4.27 Mixed-Fruit Wines

Different combinations of fruits can be made to produce wines. Combinations of grape with other fruit is made so as to have the vinosity of grape and the flavor of the specific fruit used. To prepare the must, the juices are mixed, the nutrient salts, sugar and pectolytic enzymes are dissolved in water and then, mixed thoroughly. Details are described in the literature cited in this volume.

37.4.28 Litchi

Litchi has plenty of flavor, minerals and vitamins and is used to prepare alcoholic beverages in China. Low-alcohol high-flavored beverage, using the techniques of partial osmotic dehydration, have been made, containing 5–6% alcohol, 3–4% sugar and 0.35% acid. In this method, the litchi fruits at optimum maturity are washed and peeled, followed by dipping in a sugar solution of 70°B for four hours at 50°C, the treated fruits then, being taken out and drained, followed by pulping in a pulper. The TSS is adjusted to 22°B by dilution with water and inoculated with an active culture of wine yeast to carry out the alcoholic fermentation that is allowed to continue until the TSS reaches 7°B. The wine is matured followed by blending with equal quantities of fresh litchi juice, and filtered and used to fill glass bottles. The bottles are closed with crown corks and pasteurized in water at 62.5°C for 20 minutes.

37.4.29 Mulberry Wine

The mulberry (*Morus indica* berries) were converted into pulp by the hot pulping method and used for the preparation of mulberry wine. The TSS of the must was raised with sugar to 22°B, 24°B or 26°B; to prepare the must, 0.1% DAHP as nitrogen source and 0.5% pectinase enzyme for clarification were added. The respective musts were inoculated with 5% of activated culture of *S. cerevisiae* var. *ellipsoideus* (UCD

595). Fermentation was carried over at room temperature (22–25°C) until a stable TSS was reached. The mulberry wine produced from must having 26°B had higher TSS, total sugar content, pH, volatile acidity, ethanol, total phenols and total anthocyanin.

37.4.30 Mead

Wine called 'mead' is made from honey by alcoholic fermentation in a manner similar to that of any wine and is known for its excellent effect on digestion and metabolism. For fermentation, it is necessary to add enough acid, tannins, nitrogenous and phosphate sources to the diluted honey. To make a dry table wine, honey is diluted with water to reach 22°B, and 5 g citric acid, 1.5 g DAHP, 1 g of potassium bitartrate and 0.25 g each of magnesium chloride and calcium chloride are added, along with 100 ppm of SO_2. Active yeast culture (3–5%) is added to carry out the fermentation, which is continued until TSS is stabilized. The wine is siphoned, racked and matured, and sweetened to 5 to 10°B by addition of honey or other sweetening agents, and pasteurized or sterile bottled. The sweetened wine can also be fortified with high-proof brandy.

Honey has also been used as a source of sugar to make wine from apple, plum and pear. A blending ratio of 8: 5: 3 for pulp/juice, water and honey is maintained, and 0.1% DAHP, 0.3% citric acid, `Pectinol' 0.5% and 100 ppm SO_2 as KMS are added. Active yeast culture of *S. cerevisiae* (5%) initiated the fermentation, wines were siphoned, racked and matured for a year, bottled with a 2.5-cm headspace and pasteurized at 62.5°C for 15 minutes. Out of the combinations tried, apple honey wine was the most acceptable wine.

37.5 SPARKLING WINE

- A wine is called sparkling if it contains significant amounts of carbon dioxide (not less than 5g/L at 20°C).
- Sparkling wine, such as champagne from grape, is produced by secondary fermentation in closed containers such as bottle or tanks to retain the carbon dioxide produced.
- There are four types of sparkling wines that could be made as described below:
  1. Includes those with an excess carbon produced by fermentation of residual sugar from the primary fermentation (Australian, German, Loire and Italian, and the Muscato and Ambila of California);
  2. Includes those with an excess CO_2 from a malolactic fermentation (vinho verela wines of northern Portugal),
  3. Includes those wines with excess of CO_2 from fermentation of sugar added after the process of fermentation. Most of the sparkling wines of the world are of this type,
  4. Includes those wines where excess of CO_2 is added and includes the carbonated wines.
- Different steps involved in the production of sparkling wine are: preparation of a base wine, sugaring, yeasting, bottling, proper blending, secondary fermentation in bottles or tank maturation, finishing and disgorging.
- The maturation is a very important step in the production of sparkling wine and may extend for up to 3 years.
- Bottle aging results in the wine acquiring the physico-chemical and sensory characteristics usually associated with sparkling wines. In the production of sparkling wine, preparation of base wine is the first and foremost step.
- The quality of a sparkling wine depends largely on the characteristics of the base wine, which, in turn, are dictated by a number of factors, such as fruit cultivar, yeast culture, preservatives, temperatures, nitrogen source, etc.

For more details on the production of sparkling wine production, see Chapter 34 of this volume.

37.5.1 Sparkling Plum Wine

Plum fruits have been evaluated for the preparation of sparkling wine. Different preservatives had been used in the elaboration of plum wine *e.g.* potassium metabisulfite or sodium benzoate. It was concluded that either preservative were suitable for conservation of sparkling wine, though sodium benzoate-treated wine was considered better from sensory evaluations.

BIBLIOGRAPHY

Abrol, G.S. and Joshi, V.K. (2011). Effect of Different Initial TSS level on Physico-chemical and sensory quality of wild apricot mead. *Int. J. Food Ferm Tech.*, 1(2): 221–229.

Abrol, G.S., Sharma, K.D. and Kumar, S. (2015). Effect of initial total soluble solids on physico-chemical, antioxidant and sensory properties of mulberry (*Morusindica* L.) wine. *J. Proc. Energy Agricola* 19(5): 228–232.

Agu, R.C., Okenchi, M.U., Ude, C.M.,Onyia, A.I., Onwumelu, A.H. and Ajiwe, V.I.E. (1999). Fermentation and kinetic studies on Nigerian Palm wines-*Elaesguineensis* and *Raphia hookri* for preservation by bottling. *J. Food Sci.Technol.*, 36(3): 205.

Alisimon, R. and Badrie, N. (2002). Utilization of peel in the Plantain wine production. *J. Food Sci.Technol.*, 15(1): 32–34.

Amerine, M.A., Berg, H.W., Kunkee, R.E., Qugh, C.S., Singleton, V.L. and Webb, A.D. (1980). *The Technology of Wine Making.* 4th edn. AVI, Westport, CT.

Atputharajah, J.D., Samarajeewa, U. and Vidanpathirana, S. (1986). Efficacy of ethanol production by coconut toddy yeasts. *J. Food Sci. Technol.*, 23: 5.

Attri, B.L. and Singh, D.V. (2002). Pineapple Wine: An alternate to use culled fruits. *Indian Food Packer*, 56: 79.

Balick, M.L. (1990). Production of coyol wine *Acrocomiamaxicana* (Arecaceae) in Honduras. *Econ. Bot.*, 44(1): 84.

Bardiya, M.C., Kundu, B.S. and Tauro, P. (1974). Studies on fruit Wines- Guava Wine. Haryana. *J. Hort. Sci.*, 3: 140.

Beever, D.J. (1993). Kiwi fruit. In: *Encyclopedia of Food Science. Food Technology and Food Nutrition*, R.Macrae, R.K. Robinson and S.J. Sadler (Eds.), Academic Press, New York, p. 2626.

Brathwaite, R.E. and Badrie, N. (2001). Quality changes in Banana (*Musa acuminata*) wines on adding pectolase and passion fruit. *J. Food Sci. Technol.*, 38(4): 381–384.

Canas, A. and Unal, U. (1994). A study on the evaluation of orange varieties grown in Adana for Wine production. *Turk. J. Agric. For.*, 18(1): 1–7.

Chakraborty, K., Saha, J., Raychaudhuri, U. and Chakraborty, R. (2014). Tropical fruit wines: A mini review. *NPAIJ*, 10(7): 219.

Davidović, S., Veljović, M., Pantelić, M., Baošić, R., Natić, M., Dabić,D., Pecić, S. and Vukosavljević, S. (2013). Physicochemical, antioxidant and sensory properties of peach wine made from Redhaven cultivar. *J. Agric. Food Chem.*, 61(6): 1357–1363.

Dellacassa, E., Trenchs, O., Fariña, L., Debernardis, F., Perez, G., Boido, E. and Carrau, F. (2017). Pineapple (*Ananas comosus* L. Merr.) wine production in Angola:ro Characterisation of volatile aroma compounds and yeast native flora. *Int. J. Food Microbiol.*, 241: 161.

Faparusi, S.I. (1973). Origin of initial microflora of palm wine from oil palm trees (*Elaeis guineensis*). *J. Appl. Bacteriol.*, 36(4): 559–565.

Fowles, G. (1989). The complete home wine maker. *New* Scientist, September, 38.

Heatherbell, D.A., Struebi, P., Eschenbruch, R. and Withy, L.M. (1980). A new fruit wine from Kiwi fruits: A Wine of unusual composition and Riesling Sylvaner character. *Am. J. Enol. Vitic.*, 31: 114–120.

Jagtap, U. and Bapat, V. (2015). Phenolic composition and antioxidant capacity of wine prepared from custard apple (*Annona squamosa* L.) fruits. *J. Food Process. Preserv.*, 39(2): 175–182.

Joshi, V.K. and Bhutani, V.P. (1991). The influence of Enzymatic clarification on the fermentation behaviour, composition and sensory qualities of apple wine. *Sci. Aliments*, 11(3): 491–496.

Joshi, V.K. and Sandhu, D.K. (1994). Influence of juice contents on quality of apple wine prepared from apple juice concentrate. *Res. Ind.*, 39(4): 250–252.

Joshi, V.K. and Sandhu, D.K. (2000). Quality evaluation of naturally fermented alcoholic beverages, microbiological examination of source of fermentation and ethanolic productivity of the Isolates. *Acta Aliment.*, 29(4): 323–334.

Joshi, V.K. and Sharma, S.K. (1995). Comparative fermentation behaviour, physico-chemical and sensory characteristics of plum wine as effected by the type of preservatives. *Cheme. Mikrobiol. Technol. Lebensm.*, 17(3/4): 45–53.

Joshi, V.K., Abrol, G.S. and Thakur, N.S. (2012). Wild apricot vermouth: Effect of sugar, alcohol content and spices level on the physico-chemical and sensory quality. *Indian Food Packer*, 66(2): 53–62.

Joshi, V.K., Bhutani, V.P. and Sharma, R.C. (1990). Effect of dilution and addition of nitrogen source on chemical, mineral and sensory qualities of wild apricot wine. *Am. J. Enol. Vitic.*, 41(3): 229–231.

Joshi, V.K., John, S. and Abrol, G.S. (2013). Effect of addition of herbal extract and maturation on apple wine. *Int. J. Food Ferment. Technol.*, 3(2): 107–118.

Joshi, V.K., John, S. and Abrol, G.S. (2014). Effect of addition of extracts of different herbs and spices on fermentation behaviour of apple must to prepare wine with medicinal value. *Natl. Acad. Sci. Lett.*, 37(6): 541–546.

Joshi, V.K., Panesar, P.S. and Abrol, G.S. (2017). Stone fruit wines. In: *Science and Technology of Fruit Wine Production*, M. Kossovea, V.K. Joshi and P.S. Panesar (Eds.), Academic Press, London, pp. 348–381.

Joshi, V.K., Sandhu, N. and Abrol, G.S. (2014). Effect of initial sugar concentration and SO_2 content on the physico-chemical characteristics and sensory qualities of mandarin orange wine. *Int. J. Food Ferment. Technol.*, 4(1): 37–46.

Joshi, V.K., Sandhu, D.K. and Thakur, N.S. (1999). Fruit based alcoholic beverages. In: *Biotechnology: Food Fermentation* (Microbiology, Biochemistry and Technology), V.K. Joshi and A. Pandey (Eds.), Vol. II. Educational Publisher & Distributors, New Delhi, pp. 647–744.

Joshi, V.K., Sharma, R. and Abrol, G.S. (2012). Stone fruit wine and brandy. In: *Handbook of Plant Based Fermented Food and Beverage Technology*, Y.H. Hui and E.O. Evranuz (Eds.), CRC Press, Boca Raton, FL, pp. 273–306.

Joshi, V.K., Sharma, P.C. and Attri, B.L. (1991). A note on deacidification activity of *Schozosaccharomyces pombe. J. Appl. Bacteriol.*, 70(5): 385.

Joshi, V.K., Sharma, Somesh and Bhushan, Shashi (2005). Effect of Method of Preparation and cultivar on the quality of strawberry wine. *Acta Aliment.*, 34(4): 339–353.

Joshi, V.K., Thakur, N.K. and Kaushal, B.B.L. (1997). Effect of debittering of Kinnow juice on physico-chemical and sensory quality of kinnow wine. *Indian Food Packer*, 5(4): 5–9.

Joshi, V.K., Sharma, Somesh, Bhushan, Shashi and Devender, Attri (2004). Fruit based alcoholic Beveregaes. In: *Concise Encyclopedia of Bioresource Technology*, Ashok Pandey (Ed.), Haworth Inc., New York, pp. 335–345.

Joshi, V.K., Sharma, R., Girdher, A. and Abrol, G.S. (2012). Effect of dilution and maturation on physico-chemical and sensory quality of jamun (black plum) wine. *Indian J. Nat. Prod. Resour.*, 3(2): 222–227.

Joshi, V.K., Sharma, S.K., Goyal, R.K. and Thakur, N.S. (1999). Sparkling plum wine: Effect of method of carbonation and the type of base wine on physico-chemical and sensory qualities. *Braz. Arch. Biol. Technol.*, 42(3): 315–321.

Joshi, V.K., Sharma, S., Sharma, R. and Abrol, G.S. (2011). Effect of dilution and de-acidification on physico-chemical and sensory quality of seabuckthorn wine. *J. Hill Agric.*, 2(1): 47–53.

Kime, R.W. and Lee, C.Y. (1987). The use of honey in apple wine making. *Am. Bee J.*, 270–271.

Kulkarni, J.H., Singh, Harmail and Chadha, K.L. (1980). Preliminary screening of Mango varieties for wine making. *J. Food Sci. Technol.*, 17(4): 218.

Kundu, B.S., Bardiya, M.C. and Tauro, P. (1976). Studies on fruit Wines. Banana Wine. *Haryana J. Hort. Sci.*, 5(34): 160.

Matei, F. (2017). Technical guide for fruit wine production. In: *The Production of Fruit Wines – A Review*, M. Kosseva, V.K. Joshi and P.S. Panesar (Eds.), Academic Press, London, pp. 663–703.

Miljić, U., Puškaš, V., CvejićHogervorst, J. and Torović, L. (2017). Phenolic compounds, chromatic characteristics and antiradical activity of plum wines. *Int. J. Food Prop.*, 20(2): 2022.

Morse, R.A., Stainkraus,K.H. and Patterson, P.D. (1975). *Wines from the Fermentation of Honey. Honey: A comprehensive Survey.* Eva. Crane (ed) Heinmann, London.

Ogodo, A., Ugbogu, O., Agwaranze, D. and Ezeonu, N. (2018). Production and evaluation of fruit wine from *Mangifera indica* (cv. Peter). *Appli. Microbiol. Open Access*, 4: 144.

Okafor, N. (1975). Microbiology of Negerian palm wine with particular reference to bacteria. *J. Appl. Bacteriol.*, 38(2): 81.

Panesar, P.S., Joshi, V.K., Bali, V. and Panesar, R. (2017). Technology for production of fortified and sparkling fruit wines. In: *Science and Technology of Fruit Wine Production*, M.R. Kosseva, V.K. Joshi and P.S. Panesar (Eds.), Academic Press, London, pp. 487–530.

Panesar, P.S., Joshi, V.K., Panesar, R. and Abrol, G.S. (2011). Vermouth: Technology of production and quality characteristics. *Adv. Food Nutr. Res.*, 63: 251–283

Reddy, L.V.A. and Reddy, O.V.S. (2005). Production and characterization of wine from mango fruit (*Mangifera indica* L). *World J. Microbiol. Biotechnol.*, 21(8–9): 1345.

Rommel, A., Heatherbell, D.A. and Wroslad, R.E. (1990). Red Respberry Juice and Wine: Effect of processing and storage on anthocyanin pigment composition, colour and appearance. *J. Food Sci.*, 55(4): 1011.

Sharma, S.K. and Joshi, V.K. (1996). Optimization of some parameters of secondary fermentation for production of sparkling wine. *Indian J. Exp. Biol.*, 34(3): 235–.

Sharma, Somesh and Joshi, V.K. (2004). Flavour profiling of strawberry wine by qualitative descriptive analysis technique. *J.Food Sci Technol*, 41(1): 22–46.

Sharma, S., Joshi, V.K. and Abrol, G.S. (2009). An overview on strawberry [*Fragaria* × *ananassa* (Weston) Duchesne ex Rozier] wine production technology, composition, maturation and quality evaluation. *Nat. Prod. Radiance*, 8(4): 356–365.

Sharma, R., Joshi, V.K. and Abrol, G.S. (2011). Effect of blending and dilution on physico-chemical and sensory quality of seabuckthorn wine. Proceedings of National Conference on Seabuckthorn (Hippophae L.): Emerging Trends in R&D on Health Protection & Environmental Conservation, Palampur, India, December 1–3, 2011, pp. 122–127.

Shukla, K.G., Joshi, M.C., Yadav, S. and Bisht,N.S. (1991). Jambal Wine making: Standardization of methodology and screening of cultivars. *J. Food Sci. Technol.*, 28(3): 142.

Teotia, M.S., Manan, J.K., Berry, S.K. and Sehgal, R.C. (1991). Beverage development from fermented (*S. cerevisiae*) muskmelon (*C. melon*) juice. *Indian Food Packer*, 4: 49.

Tewari, H.K., Sahota, H.K., Chawla, A., Kathuria, S. and Joolka, T.S. (2001). Value added Fermented beverage from *Aloe barbadeusis* and *Vitis vinifera*. In: *Biochemistry- Environment and Agriculture*, A.P.S. Mann, S.K. Munshi and A.K. Gupta (Eds.), Kalyani Publishers, New Delhi, p. 276.

Vaidya, D., Vaidya, M., Sharma, S. and Abrol, G.S. (2009). Enzymatic treatment for juice extraction and preparation and preliminary evaluation of Kiwifruit wine. *Nat. Prod. Radiance*, 8(4): 380–385.

Velić, D., Amidžić, D., Velić, N., Klarić, I., Petravić, V. and Mornar, A. (2018). Chemical constituents of fruit wines as descriptors of their nutritional, sensorial and health-related properties. *Descript. Food Sci.*. doi:10.5772/intechopen.78796.

Velić, D., Velić, N., Klarić, D.A., Klarić, I., Tomina, V.P., Košmerl, T. and Vidrih, R. (2018). The production of fruit wines – A review. *Croat. J. Food Sci. Technol.*, 10(2): 279–290.

Vyas, K.K. and Joshi, V.K. (1982). Plum wine making standardization of a methodology. *Indian Food Packer*, 36(6): 80–85.

38 Technology of Brandy Production

Anju K. Dhiman, Surekha Attri and Preethi Ramachandran

CONTENTS

38.1 INTRODUCTION

When the word brandy is mentioned, most cognoscenti think of 'Cognac'. The name brandy comes from the Dutch word *Brandwiji*, meaning 'burnt wine'. The name is apt as most brandies are made by applying heat, originally from an open flame. The heat drives out and concentrates the alcohol naturally present in wine. Brandy has been called *aqua vini*, *eaux-de-vie*, *weinbrand*, *Cognac*, *Branntwein*, *aquardiente* and *aquavit*. However, the more specific names, Cognac and Armagnac, are often applied to certain brandies from different regions of France. Grapes are most frequently employed to make wine that is distilled to concentrate the alcohol to a higher percentage than in the base wine. Some distillates are aged in oak barrels and some are coloured with caramel to imitate the effect of ageing. Some brandies are produced using a combination of both ageing and colouring. The process of brandy production is described below.

38.2 DEFINITION AND CHARACTERISTICS OF BRANDY

Brandy is a distillate or a mixture of distillates obtained solely from the fermented juice or mash of grapes. It is produced at less than 190° proof in such a manner that the distillate possesses the taste, aroma and characteristics generally attributed to the product. It is bottled at not less than 80° proof, and generally contains 35–60% alcohol by volume (70–120 US proof). In contrast, 'fruit brandies' are distilled from the fermented juice or mash of whole, sound ripe fruit, from grape, citrus or other fruit wine (with or without the addition of no more than 20% by weight of the pomace of such juice or wine) or 30% by volume of the lees of such wine, or both, calculated prior to the addition of water to facilitate fermentation or distillation. It may include mixtures of such brandies with not more than 30% (calculated on a proof gallon basis) of lees brandy. 'Fruit brandies' (other than grape brandy) derived from one type of fruit are designated by the word 'brandy' qualified by the name of such fruit, *e.g.* peach brandy or apple brandy. Cognac is the grape brandy distilled in the Cognac region of France and is entitled to be so designated by the laws and regulations of the French government. 'Geographical designation' of distinctive types of distilled spirits with geographical names that have not become generic are *Eau de Vie Dantzig* (Dantziger Foldwasser), *Ojen* and *Swedish Punch.* Such geographical names are only used to designate distilled spirits conforming to the standards of the region. Geographical names that have not become generic should not be applied to distilled spirits produced anywhere else than the designated region, *e.g.* Cognac, Armagnac, Greek brandy and Pisco brandy.

Readers can refer to Chapters 1 and 2 for the classification and characteristics of brandies and wine regions, respectively.

38.3 HISTORICAL BACKGROUND

The tales of how brandy was first invented are many, but the true origin of brandy may never be known. The preparation of brandy is an ancient process that has lost its origin to antiquity. Additionally, whether it was first invented by the Greeks, the Arabs or the Chinese is also debatable. The derivations of the words 'alcohol' and alambic, from the Arabic word *al-koh'l* and *al-anbiq*, indicate that it was from the Islamic world that the practice of distillation first entered Europe, and Arabic records indicate their use of stills in the 8th century. Until the end of the 15th century, distilled wine appears to have been largely used as a medicine and is now known as *aqua vitae.* It was considered to possess such strengthening and sanitary power that a physician named it 'the water of life' (*L'ean de vie*), a name it retains to this day. However, now it is rendered as one of life's most powerful and prevalent destroyers due to its excessive potations. In 1493, a Nuremburg doctor noted that everyone in the city had formed the habit of drinking *aqua vitae.* Consequently, 3 years later, the city authorities forbade the sale of alcohol on feast days. Due to the high cost of stills and that the process of distillation was a very secret and restricted process, the production of distillation was limited. Thus, within France, the production of brandy only emerged from the control of doctors and apothecaries when Louis XII granted the privilege of distilling to the guild of vinegar makers in 1514 with Francis I granting the same privilege to viticulturists in 1537.

The commercial distillation of brandy from wine originated in the 16th century when Dutch settlers came to the French region of Cognac to buy salt, wood, wine and other goods. The story regarding the development of brandy involves a Dutch shipmaster, who was in the business of transporting wine from France to Holland. Trying to make as much money as possible, he realised that by removing most of the water by distillation, he could concentrate the wine and thus could ship more back to Holland, adding the water once it arrived. Unfortunately, he did not think his friends and crew would start to drink the concentrated wine, believing that to add water only adulterated this nectar and to do so was, therefore, a sin. The concentrated wine was called 'Brandewijn', which literally means burnt wine. With the passage of time, it has been shortened in English to brandy. Although the Dutch had provided much of the original capital, together with the stills and the technology necessary for distillation, local producers did not lag behind in the production of brandy and they were forced to seek new markets for their products. Elsewhere throughout France, but particularly in Armagnac, other wine producers had also begun to distil their wines, especially those of poor quality.

38.4 TECHNOLOGY FOR PRODUCTION OF BRANDY

Brandy is made from the distillation of wine and involves various unit operations (Figure 38.1).

38.4.1 Raw Materials for Brandy Production

The raw materials used in brandy production are liquids that contain any form of sugar. French brandies are made from the wine of the Colombard (or Folle Blanche) grapes. However, anything that ferments can be distilled and turned into a brandy. Grapes, apples, blackberries, plums, sugarcane, honey, milk, rice, wheat, corn, potatoes and rye are all commonly fermented and distilled. These raw materials are known to contain more acid, tannins and sugars and have a stronger aroma, making them a valuable raw material for the production of brandy. During World War II, Londoners made wine from cabbage leaves and carrot peel, which they subsequently distilled to produce what must have been a truly vile form of brandy. Sound fruits/vegetable (raw material) should be used in the production of distilling material; the use of

Raw material (fruits)
↓
Washing of fruits
↓
Juice extraction
↓
Sulphitation of juice
↓
Inoculation of starter culture
↓
Fermentation of juice
↓
Distillation
↓
Ageing
↓
Blending
↓
Bottling
↓
Labeling

FIGURE 38.1 Unit operations for the preparation of brandy

mouldy, partially fermented and otherwise rotten fruit should be avoided. The variety of fruits used as the raw material and the time of harvest have been found to influence the quality of the brandy produced. The wines of French Colombard produce brandies with a more distinctive aroma, but those of Thompson Seedless are significantly better. Late-harvested French Colombard produce wine distillates of lesser quality than earlier harvesting, but harvest time has little effect on the distillate quality of Thompson Seedless.

Generally, it is believed that the wines of high acidity grapes produce the best brandy. Out of grapes harvested at four stages of maturity, the best brandy was made from mature grapes. For brandy production, white varieties, clarified juice (to reduce fusel oil formation) and no sulphur dioxide (unless grapes are of poor quality) are recommended. Wines with high volatile acidity produce poor quality brandy. Pomace mash wines should only be used in the production of fortifying wine spirits.

38.4.2 Production of Wine

Winemaking involves three main categories of operation, viz. pre-fermentation, fermentation and post-fermentation. Pre-fermentation involves crushing the fruits and releasing the juice. In the case of white wine, the juice is separated from the skin, whereas in red wine, the skin, pulp and seeds are not separated from the juice. After crushing, the juice is clarified by sedimentation or centrifugation for white wine, a process not carried out for red wine. Yeast is added to the juice (or must as it is called for red wine because juice also contains pulp, skin and seeds). Fermentation involves a reaction that converts the sugars in the juice into alcohol and carbon dioxide:

$$C_6H_{12}O_6 \rightarrow 2C_2H_5OH + 2CO_2$$

Yeast utilises sugar during the fermentation period. Fermentation proceeds under anaerobic conditions and may be boosted with di-ammonium phosphate (DAP) to supplement the nitrogen required for yeast growth. Post-fermentation practices begin after fermentation has reached the desired stage or when fermentation is complete. Here, wine is racked off the yeast lees, usually in stainless steel vessels or in oak barrels. Various fining agents, such as enzymes, bentonite, diatomaceous earth and egg albumin, may be commercially added to aid in the clarification of wine. Separation occurs with minimal agitation to avoid re-suspending the particulate matter. The residue from racking may be filtered to recover wine otherwise lost with the lees or it may be used ‘as is’ for brandy production.

38.4.3 Distillation

Using heat, distillation is the process of separating and selecting specific volatile components from a liquid mixture such as wine to make brandy. Therefore, distillation purifies wine and removes diluting components such as water to increase the proportion of alcohol content. As distilled beverages contain more alcohol, they are considered ‘harder’. In North America, the term hard liquor is used to distinguish distilled from undistilled beverages.

38.4.3.1 Distillation Theory

The kinetic energy of molecules increases as the temperature is raised. The number and velocity of the molecules escaping from the surface of the liquid also increase, raising the vapour pressure. When the vapour pressure on the surface of a liquid equals the external atmospheric pressure, the liquid starts to ‘boil’. Substances require different degrees of heat to raise their temperature by one degree, *i.e.* their specific heat varies (that of alcohol is only about 0.6 compared to water at 1.0). The heat required to change a substance from a liquid to a vapour is known as the heat of vaporisation, which is approximately the same for substances with a similar molecular weight. The liquid–vapour system of ethanol and water follows Dalton’s and Henry’s laws. The vapours leaving a water–alcohol mixture contain a higher percentage of ethanol than the original liquid. This forms the basis for separating ethanol and water by fractional distillation. The degree of separation is also affected by the nature of the mixture and the method of distillation. Furthermore, binary mixtures may form maximum or minimum boiling point (BP) mixtures, thereby preventing separation of the two components at lower or higher temperatures. The boiling points of water and alcohol are 212°F (100°C) and 173.3°F (78.5°C), respectively, at atmospheric pressure (760 mm). Since only one liquid phase is present, this mixture is homogeneous. In addition, this is a binary azeotrope with a minimum boiling

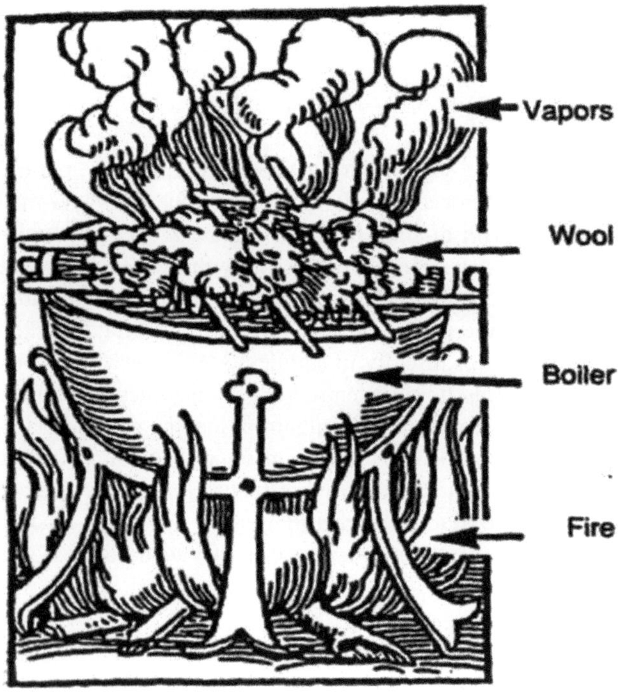

FIGURE 38.2 Early still. *Source:* Leaute (1990)

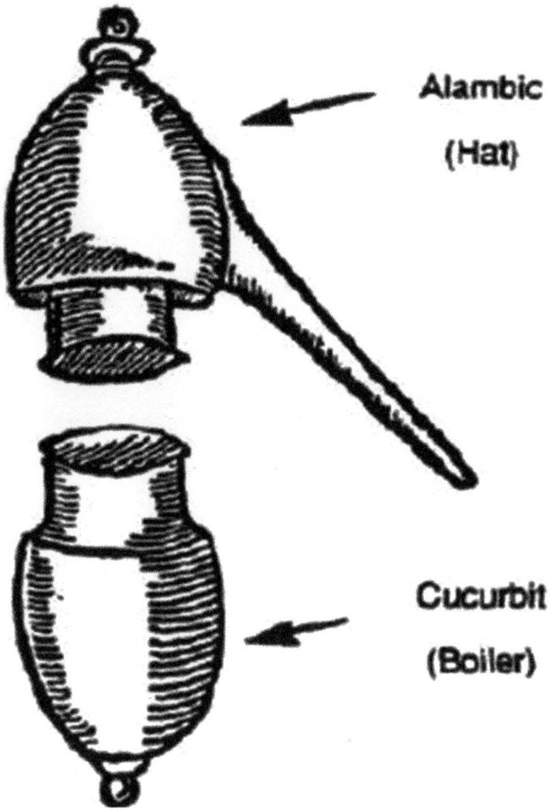

FIGURE 38.3 Later distillation equipment. *Source:* Leaute (1990)

point. The alcohol content of the distillate remains below 96% alcohol by volume (max. 80% alcohol/vol.) during Cognac distillation. The ethanol–water system has a minimum boiling point at 97.4% ethanol, so the separation of the two components above a boiling point of 78.1°C (126.6°F) is impossible at atmospheric pressure. The ethanol content of a constant boiling point mixture is higher at reduced pressures. The ratio between the percentage of ethanol in the vapour and that in the liquid is called the Sorel or k value, which varies with the ethanolic strength of the liquid. $CaCl_2$, $CoCl_2$, $CuCl_2$ and NaCl could also break the ethanol–water azeotrope, achieving hyper-azeotrope in alcohol obtained by extracting distillation. NaCl could break the azeotrope but it increases the energetic efficiency of the alcohol separation. The catalytic effect of $CuCl_2$ increases the concentration of ethyl acetate and, correspondingly, acetaldehyde concentration is decreased.

38.4.3.2 Distillation Apparatus

Stills were originally used to distil a liquid (alcohol) for medicinal purposes and to make perfumes. In early stills (Figure 38.2), parts of vapours were condensed in wool which was sometimes changed to obtain the distillate. Alchemists and monks made progressive improvements in both the technique of distillation (Figure 38.3) and the apparatus. In order to create a distillation apparatus, four basic items are necessary: (1) a heat source to heat the product; (2) a pot to hold the liquid; (3) a condenser to cool the vapours; and (4) a receiver to collect the distillate. Rudimentary stills were very simple, composed of bamboo, wood or clay which was very dangerous. The sketch in Figure 38.4 is an example of an ancient still used in India. In another traditional method in Himachal Pradesh (India), an empty ghee tin (*canister*) with about 15 kg capacity, was used for boiling the fermented liquid by heating it over a fire.

The modern still, incorporating safety features and a more efficient condensing system, was developed by Justin von Liebig and called the Liebig single surface condenser still. A thermometer was added and a bulb was placed just below the side arm where the vapour leaves the still. In this case, the desired temperature is 172–174°F. Normally, at or above 170°F, a water-cooled condenser is not required, as an elongated or helical (spiral) coil provides sufficient heat loss to condense the vapours. However, a water condenser accelerates the process. The still could be made of stainless steel, glass, copper or some other material except iron to avoid corrosion. These simple stills were relatively inefficient compared to fractional distillation stills. Different types of stills are discussed in Table 38.1.

At the industrial scale, of the various distillation systems used, copper pot stills and column stills are depicted in Plates 38.1 and 38.2.

Whatever distillation system is employed, the ethanol-containing vapour must be condensed back to a liquid. Condensers perform two main functions: conversion of vapour to a liquid and cooling the resultant liquid to room temperature. The second function removes only a small amount of the heat, thereby requiring a greater condenser area. The heat transfer between a liquid and a vapour is performed via the film of stationary liquid on the water side and the film of stationary vapour on the vapour side. As the vapour films are thinner, heat transfer is more rapid for an apparatus condensing a vapour than for the same apparatus cooling a liquid. The coefficient of heat transfer can be increased by increasing the velocity of the cooling water passing through the tubes, or by using specially designed condensers. Recent stills have started to use air-cooled condensers. Vapour from the top of the column still passes through a dephlegmator and a condenser. Initially, a portion of the vapour condenses and is returned to the column as the reflux. The condenser may also return a reflux to the column and the remainder is taken off as head. In this case, the actual product is removed as a side stream from the upper plates in the column. To reclaim the ethanol from the heads fraction, a modified still has been designed. To ensure the hydrolysis of acetals, the charge is diluted to 55/60% ethanol (v/v) and boiled under reflux at pH ≤2 with spraying by inert gas (CO_2 or N_2). The discharged vapours are vented under water and then to the exterior of the still house. When the acetaldehyde concentration in the charges reaches <200 mg/L, it is adjusted to pH >0 with a sodium hydroxide solution. The reflux continues for 20 minutes and is distilled at over 95%. A recovery of 99% has been reported. This reclaimed spirit is mixed in a ratio of 2:1 with wine spirits and used for fortification. A sensor for controlling the withdrawal of the ester-aldehyde cut from continuous brandy stills is also used.

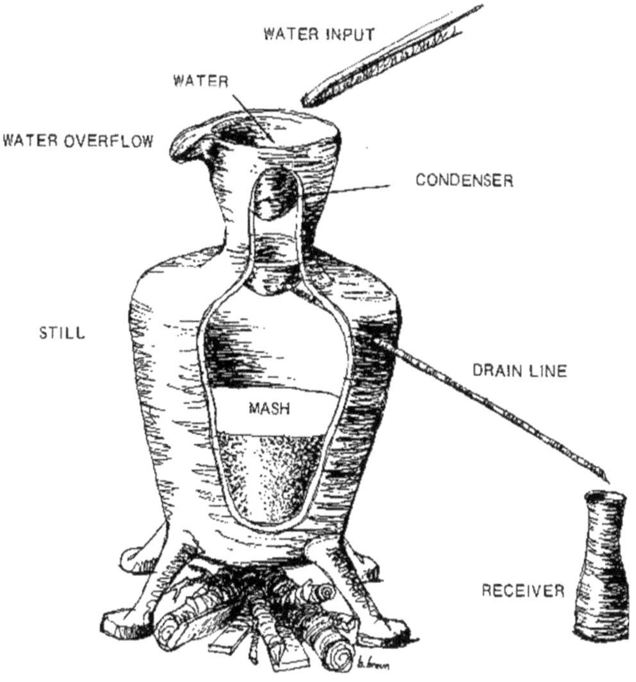

FIGURE 38.4 Very ancient still used in India. *Courtesy:* http://www.expats.org.uk/features_whitelightning.htm

38.4.3.3 Double Distillation Technique

In this case, the first mixture A is distilled to produce a second mixture B. When mixture B is sufficient in volume, it is redistilled to produce mixture C. This is called the double

TABLE 38.1
Different Types of Distillation Stills

Sr. no.	Distillation stills	Description
1.	**Reflux still**	The length and width of the reflux column determine the efficiency of the mechanism. Generally, the column (6 inches in height and 3–4 inches in diameter) is packed with stainless steel, wool, glass beads, marbles, broken chinaware, etc., to give the equivalent amount of alcohol. A column of 18 inches packed in the same fashion will help save time and effort. A reflux is still more efficient than other simple stills as it allows greater control of temperature, thereby negating the undesirable by-products of ordinary distillation caused by lower or higher temperatures.
2.	**Laboratory type of still**	The wine is loaded into a boiler which is heated by a gas burner for partial vaporisation. The vapours are condensed and collected into one or a series of fractions called a distillate.
3.	**Pot stills**	The original and least complex distillation is carried out in a pot still (Figure 38.5), a form of batch distillation. The pot still (Plate 38.1) consists of a receptacle into which the liquid is placed and heated. A long, tapered neck is attached at the top to collect the vapours that are formed when the liquid is heated. A spiral copper tube is attached to this neck and will pass through a cooling medium, usually water. The decrease in temperature condenses the vapours to a liquid again. The product is a distilled spirit but it is not necessarily a finished product. The operation of a pot still is considerably more complex. During distillation, wine is exposed to higher temperature resulting in many Maillard and Strecker degradation reactions. These produce heterocyclics, such as furans, pyridines and pyrazines, as well as hydrolysing non-volatile terpene glycosides and polyols to free volatile terpenes. The construction of pot stills from copper provides protection from long exposure to heat and the corrosive action of wine acids. In addition, copper reacts with sulphur and fatty acids, forming insoluble constituents, which can be removed from the distillate by filtration. However, organic copper compounds of butyric, caproic, caprylic, capric and lauric acid are formed during heating and are not distilled. Although these acids have a very disagreeable odour, the copper fixes these as their salts are insoluble. Cleaning pot stills every 8 days is recommended. Sodium phytate was found to be a useful means of removing the copper from brandy. Little analytical differences have been reported between pot- and column-distilled brandies involving the same wine (depending on what portions of the distillate are used). The difference between continuous- and pot-distilled brandies is through the thermal load during distillation leading to newly formed constituents in the pot still products. In California, it is customary to redistil the pot still product in a continuous still.

(*Continued*)

TABLE 38.1 (CONTINUED)
Different Types of Distillation Stills

Sr. no.	Distillation stills	Description
4.	**Cognac type of still – alembic**	The pot still used in the Cognac area is known as an alembic and is made of copper and bronze. Some parts affect the quality of the Cognac or the brandy and should be made of stainless steel, *e.g.* valves, fittings and the condenser tank. However, copper is considered the most efficient metal to build alembics. The advantages of using copper are that it is malleable in nature and has a good conduction of heat; resists corrosion from fire and wine; reacts with wine components such as sulphur components and fatty acids (desirable for Cognac or brandy quality); and acts as a catalyst for favourable reactions between wine components. A typical alembic still is composed of the following parts: The boiler is the main part of the alembic pot and consists of a pipe to fill the boiler; a vent and side glass; a sprinkler to clean the boiler; and a valve to empty the boiler. Depending on the specifications required by the distiller, the hat is about 10–12% the capacity of the boiler, and is located directly above the boiler. The shape and volume of the hat determine the concentration, selection and separation of the different volatile components. In an alembic still, the Swan's neck (*col de eygne*) part is curved and directs the vapour into the coil. Its height and curve are extremely important to the reflux process. A preheater is a cost-effective part of the alembic still and the Swan's neck runs through it and around its back. The preheater is refilled for the next batch of distillation during the first hours of distillation. The coil (serpentin) performs two functions: to condense vapours and to cool the distillate to the correct temperature for filtration. At the origin, the coil has a larger diameter to facilitate condensation. Progressively, the diameter of the coil reduces, until it reaches the hydrometer port. The coil is also made of copper. Copper reacts with the components of the distillates (*e.g.* sulphur and fatty acids) to give an insoluble combination during the condensation process. When these materials reach the hydrometer port, they are removed from the distillate by filtration. The condenser (*condenseur*) is a cylindrical tank that surrounds the condensing coil. It is made of copper or stainless steel and contains the copper coil pipe; it has a capacity of around 1300 gallons. During distillation, it is filled with water. Cold water enters the condenser from the bottom, while the hot water (heated during condensation) exits at the top of the condenser. The hydrometer port (*porte-alcoometer*) is made of copper with the purpose to filter the distillate, to monitor the temperature and the alcohol content of the future Cognac or brandy, to offer an access point for the distiller and to check the progress of the distillation. The heads tank is a small stainless steel tank of about 15 gal capacity, and is used to collect the first part of the distillate, called head. The gas burner is equipped with a pilot light along with a reliable security system. The fuels used are propane, butane and natural gas. The gas panel is located at the front of the alembic to monitor the burner. The temperature under the boiler reaches 1400–1600°F. This high temperature is necessary to heat the wine to create aromas in the distillation process.
5.	**Column stills**	A column still may consist of a series of interconnected, much modified, pot stills. It is much more efficient than pot stills. The first column still was invented by Scotsman Robert Stein in 1826, and the design was perfected and patented by Aeneas Coffey, Inspector General of Excise in Ireland, in 1830. The main advantages of a column still are its speed and increased productivity and that it does not need to be emptied and cleaned between batches as is done in pot stills. The modern column still is a cylindrical shell, divided into sections by a series of plates. The plates are perforated or may have openings covered by bubble caps (about 10% of their area is open) to allow the passage of vapour. Bubble cap plates with down pipes are used in the upper section of the column, though recent installations use valve or sieve trays. In the column, the plates are spaced 18–30 inches apart. If the plates are close together, the vapour capacity must be kept low. The alcohol-containing liquid is pumped into the top of a tall column or a tower called a rectifier. The liquid flows down twisting pipes to the bottom of the column. As it descends, the hot vapours rising through the column warm the alcohol liquid. The warm liquid that collects at the bottom is pumped to the top of a second column, the analyser. When the hot liquid enters at the top, it drains through the top plate onto the one below and so on until it reaches the bottom. The steam entering from the bottom causes vapourisation of the volatile elements of the liquid, as it travels down through the column. Only the water and least volatile compounds reach the bottom and are drained away. The vapours are drawn-off at the top and sent back to the first column. These vapours enter the rectifier at the bottom and rise. They are used to warm the incoming liquid. The hot vapours while warming the incoming liquid will cool the vapours resulting in their condensation back into a liquid. Distillates of varying strength can be drawn off, depending on the height at which this occurs. At relatively lower heights, less volatile elements will condense, while at higher levels, the more volatile fractions will turn back into a liquid. In this way, the distiller can exercise great control over the process and the distilling proof. There are different types of column stills (Plate 38.2). Single and double (or split) column stills may be operated either with an overhead product or a side draw. A still with an aldehyde section is particularly advantageous in producing neutral fortifying spirits. A continuous, double distillation system for brandy production produces 26% ethanol from the first column and 66.4% from the second. A copper batch column still similar to those found throughout Europe can be used in the production of fruit brandies. Use of whole fruit may yield a more distinct product. Plate 38.2 shows a column still with its different parts.

PLATE 38.1 Copper Pot Stills made in Germany by Adolf Adrian & Company specifically for Koening distillery. *Courtesy:* http://www.koenigdistillery.com/koenig_distillery.htm

PLATE 38.2 Column still with different parts. *Courtesy:* http://www.drinktec.com/jcdistill_x.html

distillation technique (Figure 38.6). Variations related to cuttings and mixtures (wine, heads, tails, brouillis, seconds and running time) of distillation tend to change the characteristics of Cognacs.

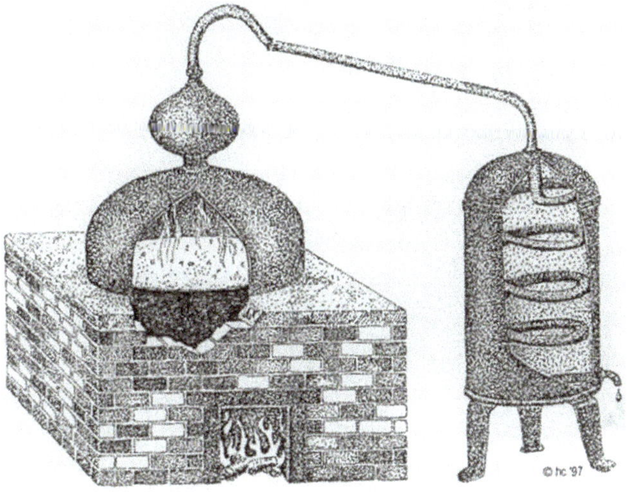

FIGURE 38.5 Diagram of a pot (alambic) brandy still. *Source:* Jackson (1999). With permission of the author

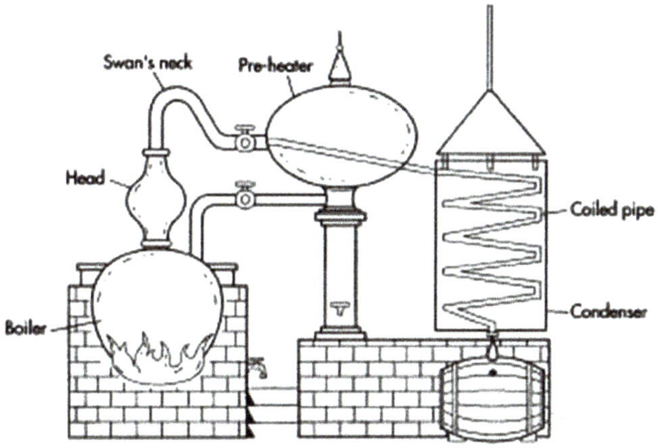

FIGURE 38.6 Double distillation unit of brandy. *Source:* http://www.madehow.com/Volume-7/Brandy.html

Suppose the wine is at 10% alcohol/volume. In the first distillation, the distillate is cut into three fractions, *i.e.* heads, heart or brouillis and tails. The alcohol content of the distillate is around 60% alcohol/volume in the first fraction, reaching 0% alcohol/volume at the end of the first distillation. In the second distillation, the distillate is cut into four fractions, *i.e.* heads; heart 1, or Cognac or brandy; heart 2 or seconds; and tails. In the first fraction, the alcohol content of the distillate is around 80% alcohol/volume, reaching 0% alcohol/volume at the end of the second distillation.

38.4.3.4 Distillation of Volatile Compounds of Wines

There are about 300 volatile compounds in wines. Other than for water and alcohol, it is very complex to calculate the volatility coefficient for each of them. Each of the volatile components will distil by following three criteria: (a) boiling point; (b) relationship with alcohol or water; and (c) the variation of alcohol content in the vapour during distillation. There are several possibilities with respect to the relationships with alcohol or water:

- The component is completely or partly soluble in alcohol and will distil when the vapour is rich in alcohol.
- The component is soluble in water and will distil when the vapour is low in alcohol.
- The component is soluble in both alcohol and water and will distil throughout the distillation.
- The component is not soluble in water, but the water vapour will carry over this component, *i.e.* hydro-distillation (Table 38.2).

During the second distillation, the alcohol content of the brouillis is increased. The heating programme established for the distillation of wine and brouillis certainly influences the concentration of the components in the distillates (Table 38.3). A higher heat is favourable for less volatile components, as increased heat will allow them to distil earlier and be present in the first fraction of the distillation in higher concentrations. The mixture remaining in the boiler after distillation is called stillage. This dealcoholised mixture must be treated to avoid pollution problems.

38.4.3.5 Fractional Distillation

Fractional distillation is more efficient than simple distillation wherein it is possible to separate volatile liquids that have boiling points close to one another. It is premised on the fact that no two liquid of different chemical composition have the same vapour pressure at all temperatures nor very often the same boiling point; however, every liquid has a definite vapour pressure at any given temperature. The whole aim of fractional distillation is to achieve the closest possible contact between the rising vapours and the descending liquid. The purification of the more volatile compounds by contact between such countercurrent streams of vapour and liquid is referred to as 'enrichment' or 'rectification'. The descending liquid is known as a reflux, thus the still is called a 'reflux still'.

38.4.3.6 Vacuum Distillation

Vacuum distillation produces an especially desirable brandy but with a non-alcohol coefficient of at least 280 as required under French law for beverage brandies. This eliminates such brandies from the beverage brandy market, which is

TABLE 38.2
Classification of Volatile Compounds

Type of volatile compounds	Description
Type 1	These are components of the first distillation. They have a low boiling point and are soluble in alcohol, *e.g.* acetaldehyde with a boiling point of 21°C (69.9°F), ethyl alcohol with a boiling point of 77°C (170.6°F). The majority of such components distil at the beginning of each distillation. Their concentration in the heads and at the beginning of the heart is very high.
Type 2	These are components that distil at the beginning of distillation. They have a relatively high boiling point and complete and partial solubility in alcohol and include fatty acids and fatty esters, *e.g.* ethyl caprylate with a BP of 208°C (406.4°F), ethyl caprate with a BP of 244°C (471.1°F), ethyl laurate with a BP of 269°C (516.2°F), ethyl caproate with a BP of 166.5°C (331.7°F) and isoamyl acetate with a BP of 137.5°C (279°F). Some of these components finish distilling in the middle of the heart. (*Source:* Laute, 1990)
Type 3	These are components in the heads and in the heart of the distillate and have a low boiling point (not above 200°C), are soluble in alcohol and are completely or partially soluble in water, *e.g.* methanol (BP 65.5°C [150°F]) and higher alcohols such as 1-propanol, isobutanol, methyl-2-butanol and methyl-3-butanol.
Type 4	These are components that start distilling during the middle of the heart, have a boiling point above that of water and are soluble or partially soluble in water, *e.g.* acetic acid (BP 110°C 230°F]), 2-phenyl ethanol, ethyl lactate and diethyl succinate are in the same class.
Type 5	The components that appear during the distillation have a high boiling point and are very soluble in water. They start distilling during the middle of the heart, *e.g.* furfural (BP 167°C [332.6°F]). The concentration of furfural increases in the middle of the heart to the tails.

TABLE 38.3
Volumes and Proof for Low-Wines Fractions and Brandy Fractions

		Redistillation of low wines			
Variety	**First distillation/low wines***	**Heads**	**Brandy**	**Tails**	**Residue**
Thompson Seedless volume (mL)	14,000	150	54.80	2550	5820
°proof	71.2	157.4	150.0	49.6	–
French Colombard volume (mL)	14,000	150	56.50	2500	5700
°proof	7	157.0	140.8	45.4	–

Source: Onishi *et al.* (1978)
* From distillation of 45 L (12 gal).

TABLE 38.4
Some of the Congeners and Their Content in Cognac Brandy

Component	Amount (mg/L)
Fusel oil	1544.00
Total acids (as acetic acid)	288
Esters (as ethyl acetate)	328
Aldehydes (as acetaldehyde)	60.800
Furfural	5.400
Total solids	5584.000
Tannins	200.000
Total congeners (w/v %)	0.239

Source: Joshi *et al.* (1999)

considered unfair. Additionally, the large vacuum stills required are expensive. Successful experiments with vacuum distillation have been carried out in Hungary; however, this process produces too-neutral distillates for Cognac.

38.4.3.7 Units of Measurements

The concentration of ethanol (ethyl alcohol) in distilled spirit is expressed differently in different countries. In Germany, the percentage of alcohol is expressed by weight (g/100 g), whereas in France it is expressed by volume (mL/100 mL). The French also use a system called Gay-Lussac (G.L.) which describes the alcohol percentage by degrees Gay-Lussac. A spirit with 40% alcohol by volume would be 40°G.L. In England and the United States, the commonly used unit is proof; however, the definition of proof differs in both the American and British system. In the United States, the official definition of 'proof' is the ethyl alcohol content of a liquid at 60°F, stated as twice the percentage of the ethyl alcohol volume. So, the liquid of pure alcohol would have to be 200° proof in the American system. In addition, '"proof spirits" shall mean that alcoholic liquor which contains 50 percent of ethyl alcohol by volume at 60 degrees Fahrenheit and which has a specific gravity of 0.93418 in air at 60 degrees Fahrenheit referred to water at 60 degrees Fahrenheit as unity'. From this, 'proof gallon' is defined as 'the alcohol equivalent of a United States gallon at 60°F, containing 50 percent of ethyl alcohol by volume'. In England, proof is measured using a Sike's hydrometer which indicates that the 'proof spirit' contains 49.28% of alcohol by weight or 57.10% by volume at 15.56°C (60°F). Alcohol percentages below the 'proof spirit' are termed 'underproofed' and above 57.10% as 'overproofed'.

38.4.3.8 Fusel Oil Removal

The separation of ethanol and water is not the only objective of distillation; various other desirable and undesirable congeners (Table 38.4) must be retained or removed (even partially). The term 'fusel oil' is used to describe higher alcohols and includes propanol, butanol and amyl alcohol which are congeners or by-products of ethyl alcohol fermentation, especially iso-amyl alcohol. The presence of fusel oil in alcohol beverages is known to cause headaches and hangovers. Distillates containing substantial quantities of fusel oil, aldehydes and other extraneous substances may be removed from the distillery system prior to fermenting or distilling the material. A decanter is a vessel used to separate the two-phase liquid. In a fusel oil decanter, an upper fusel oil phase is separated from a lower aqueous ethanol phase. Most continuous column stills have a fusel oil decanter. In their simple form, they consist of a try box into which the fusel oil is introduced through a pipe at the top. A smaller perforated water pipe is inside this pipe. Therefore, the incoming fusel oil is intimately mixed with water. The mixture is then passed through perforated screens and falls to the liquid surface in the try box. The boiling point of fusel oil is higher than ethanol and is generally removed by the distillation process to avoid accumulating in the rectifier. Fusel oils are only slightly soluble in water but have high volatility with steam and are thus distilled upward in the column. The moment they reach plates of higher ethanol content, they become soluble. Then, their volatility is normal and they tend to return down the column. As a result, they concentrate in the region of the still where the increase in the alcohol percentage is highest (about 135° proof or a boiling point of about 84°C [183°F]).

A minimum of four contiguous plates are connected to draw off the fusel oil. Temperature indicators should be installed on one or more of these plates and there should be provision for flow meters for the dilution water lines and 'low oils' cut. A ratio of 2 to 2.7 would improve fusel oil recovery by decantation. When a column is used to produce commercial brandy at 170° or less proof, the highest alcohol content on a plate is about 0.5%, indicating that higher alcohols can occur when a 'low oils' cut is made on one or two plates below those used for product removal. If a fusel oil cut (7–10% of the brandy flow rate) is taken, the fusel oil content will be reduced by about 5% which would be useful for brandies that are aged for a minimum period of 2 years. A proof of 160–170° distillation is recommended for beverage brandy. In the distillation of fortifying brandy, the fusel oil concentration in the column is sufficient to exceed its solubility at about 130° proof.

38.4.3.9 Aldehyde Removal

The removal of aldehydes is also desirable and this can be accomplished by using an aldehyde column. This high aldehyde product is most frequently used in rapidly fermenting musts where it is utilised for ethanol production. Brandies for ageing contain no more than 3 mg/100 mL of aldehydes or 50 mg/L of esters.

38.4.3.10 Wine Spirits and Neutral Spirits

Neutral spirit is purified odourless, tasteless and colourless ethanol which has been produced by distillation (at or above 190° proof) and rectification techniques that remove any significant amount of congeners. It is used in the production of various beverages. An alcohol spirit purified in a still to a minimum of 95% is of absolute alcohol purity. At that degree of proof (190°), the spirit is considered to have no important taste and little body or flavour under US laws. Generally, fortifying brandies with high fusel oil (over 600 mg/L), high aldehyde (over 200 mg/L) and poor sensory quality produce poor dessert wines. Neutral wine spirits are recommended for the production of early maturing dessert wines, but wine spirits for slower maturing wines should be produced from a lower proof brandy.

38.4.3.11 Rectification and Purification

If a brandy is purified or treated after distillation in any but the manner specified in the regulations, a special rectification tax must be paid and the treatment carried out in a separate building, and only after approval has been secured from the Bureau of Alcohol, Tobacco and Firearms. It is also permissible to continuously treat a brandy during distillation by passing it through oak shavings or treating it with an oxidising agent to improve its flavour or to remove sulphur dioxide or other impurities. Rectification systems also remove heavy material from light products.

38.4.4 Ageing

Following distillation, spirits are matured from several months to many years. They are aged in wooden barrels to develop their characteristic taste, colour and aroma. Wood is known to add value to the quality of the spirits during ageing. The harsh burning taste and unpleasant odour of newly distilled spirits are ameliorated by this ageing. The changes in spirits during ageing are caused by three types of reactions: (i) complex wood constituents are extracted by the liquid; (ii) oxidation of components originally present in the liquid as well as of material extracted from the wood; (iii) reaction between organic substances present in the liquid leading to the formation of more or new congeners.

38.4.4.1 Ageing Preparation

As brandy comes from a still at 170° proof or less, it is diluted to about 110° or more. Caramel syrup is usually added before water which is used for dilution. Water is the most natural additive and is normally used to cut excessive alcohol or heat. Sugar syrup is added if Armagnac is too tannic or to remove any rough edges it might contain. It is viscous and can be either dark or light. Legally, 2% of a Cognac's content can be sugar syrup. Caramel is a burnt sugar, dark in colour and slightly bitter in taste. It is not used to sweeten the brandy but to adjust its colour and establish consistency or to give the spirit the appearance of being older and smoother.

38.4.4.2 Oak Wood Chips Barrels

Oak is the most widely used wood in the manufacture of maturation barrels, although many other woods are used. Oak wood has been used for well over 2000 years to support the ageing of alcoholic beverages. The qualities of oak wood favouring its use in the ageing of spirits include its mechanical and working properties (durability, hardness, pliability and permeability) and the extractable compounds that it contains (tannins and aromatic compounds). The change in flavour of the maturing spirits is due to changes in the composition and concentration of the compounds influencing the taste and aroma, and may be caused by (1) direct extraction of wood compounds; (2) decomposition of wood macromolecules and subsequent extraction; (3) reactions between wood components and the constituents of the raw distillate; (4) reactions involving only the wood extractables; (5) reactions involving only the distillate components; and (6) the evaporation of volatile compounds. Two distinct types of oak are used in the manufacture of handmade barrels, either from the Limousin or Troncais region, but all are white oaks, members of the *Leucobalanus* or *Lepidobalanu* subgroup and, in Europe, consist almost exclusively of the two species *Quercus rubur* (pedunculate acorns) and *Quercus petraea* (sessile acorns). In North America, the predominant cooperage species is *Quercus alba*. Cognacs spend a year or more in small barrels of various ages and longer times in larger casks of European oak. The maximum solids extraction from oak is achieved at 55% ethanol and close to that is the traditional barrelling proof for spirit maturation. In France, it is legal to add wood chips or a liquid derived from a process involving wood chips to the barrels to enhance the characteristics of age. This process is called 'boise' which is created by boiling wood chips in water, then removing the chips and slowly reducing the remaining liquid which is dark brown and replete with wood flavour and tannin. It is used to give the impression of oak ageing to a final spirit. An aged boise is also available which is less bitter than straight boise and provides secondary wood aromas such as vanilla and grilled nuts as given by ageing to a spirit. It is estimated that around 80% of the flavour comes from the oak barrels used to store the spirits. Such flavours can be replicated by soaking spirits with oak chips or shavings. For more details, readers are referred to Chapter 28.

Depending on their respective concentrations, different wood-derived compounds confer different aromas/flavours that influence the quality of brandy (Table 38.5).

38.4.4.3 Effect of Toasting Intensity

The charring and toasting of the barrel wood produce large effects on the extraction and flavours contributing to the beverage being aged. European cooperages heat the raised barrel over a fire of wood shaving with various techniques of spraying or swabbing with water to enable the bending of the staves to the concave shape of a barrel without breaking–the bending phase. There is some degree of darkening, *i.e.* 'toasting', but it can be difficult to make it uniform depending on the winemaker's specifications. Charred wood becomes mostly carbon, with pyrolysis products, which are partly distilled away giving smoky flavours that are considered undesirable at recognisable levels in wines. On the other hand, the reuse of charred barrels for brandy and other spirits gives much less of the charred flavour, while toasting makes less drastic changes depending on the degree and depth.

38.4.4.4 Solera System of Ageing Brandy

The solera system is one of the methods used for maturing brandy and consists of a series of large casks called 'butts'. Each solera holds a slightly older brandy than the previous one. When the brandy is racked from the last butt, no more

TABLE 38.5
Aroma of Different Wood-Derived Compounds

Compounds	Aroma
Cis-/trans oak lactone(s)	Fresh oak wood (woody), sweet, coconut, celery, spicy
Eugenol	Cloves
Vanillin is known to be highly beneficial to brandy, whisky, wine, spirits and beverage quality	Vanilla
Guaiacol	Smoky
4-methylguaiacol	Smoky
4-ethylguaiacol	Smoky, spicy, floral, cured bacon like
Furfural	Almond like, grainy, contributes to 'hotness' of spirit
5-hydroxymethyl furfural	Odourless
Methyl furfural	Toasted almond
Phenolic acids	Contributes to overall bitterness and astringency
Phenolics: α-cresol, *p*-cresol	Reduce the harshness of brandy, improve the taste of brandy products, interact with flavour constituents and, as such, have a significant effect on the perception of food and beverage flavour.

Source: Jaarsveld and Hattingh (2012)

than a third of the volume is removed which is replenished with brandy drawn from the next butt in line all the way down the solera line to the first butt, where newly distilled brandy is added. This system of racking through a series of casks blends together a variety of vintages and results in the speeding up of the maturation process. Some solera have over 30 stages. The minimum ageing period is 3 years, but most are matured for more than 10 years. For further details, refer to Chapter 28.

38.4.4.5 Rapid Ageing

To age brandies more rapidly, various mechanical, physical and chemical procedures such as the use of mechanical vibration, variable temperatures, ultrasonics, adsorption, ion exchangers and ultraviolet and infrared radiation have been tried. Additionally, ozone, peroxide, permanganate, electrolysis and metallic and biological catalysers have also been employed. In the ageing of brandy, the use of oak chips (treated with alkali or untreated) is recommended. Optimum results have been obtained by treating oak chips with 0.063–0.075 N alkali at 10–15.56°C (50–60°F) for 2 days. Keeping the brandy with the treated oak chips at 20–25°C (68–77°F) for 6–8 months with the periodic introduction of oxygen (15–20 mg/L) is considered equivalent to 3–5 years of ageing in wood. Favourable effects have been observed by storing 140° proof brandy in 5 L oak barrels for 15 days at temperatures from 39 to 75°C (102.2–167°F). Heat treatment of young brandies for 20 days at 38–40°C (100.5–104.0°F) with or without oak chips (30 days) improved their sensory quality. Compounds such as syringic and vanillin as well as gallic acid are extracted from oak chips.

38.4.4.6 Changes During Ageing

The evolution of distillates during 3–5 years of storage in wooden barrels is due to aromatic compounds such as phenolics and aldehydes extracted from the alcoholysis of lignin. Variations in the major constituents of grappa were not found significant during ageing though it was established that water and ethanol escape from intact barrels during maturation. Brandies resulting from higher humidity storage of comparable barrels almost invariably had about a 15% higher content of nine small phenolic substances including vanillin and gallic acid. The irradiation of wooden barrels with UV or γ-rays tends to increase the oxidation reaction, enhance maturation and give higher free-radical products. Similar effects were observed when the barrels were heated with oxygen for 12 days. Heating oak in an autoclave at 120°C (248°F) for 100 hours at 15 atmospheric oxygen pressure reduced the cellulose and increased the lignin and aromatic aldehydes. Ethanol extract of treated wood was very high in aromatic aldehydes comparable to 20- to 50-year-old brandy. Very long–matured brandy has a higher methoxyl content (*i.e.* more lignin) than fresh oak extract, indicating slow reactions. The maximum content of lignin in aged spirits may be as high as 800 mg/L. However, with moderate analysis, 220 mg/L was found in very old Cognac. Acetate esters of isoamyl, n-hexyl and β-phenethyl alcohols decreased during ageing in oak barrels, while ethyl caproate, caprylate and caprate increased. The compounds derived partially or totally from oak were more abundant in brandies aged in American oak than in French oak. Brandies stored in reused barrels produced much smaller changes. More nitrogenous compounds were found in brandy stored in new barrels than in used barrel. The Limousin and American chip extracts were made and aged brandies were compared. The hydrolysis of brandies increased the concentration of all sugars. The American extracts had relatively high concentrations of diethyl succinate, 5-methyl furfural and β-methyl-γ-octalactone compared to French oak. Fatty acids and esters of distillates have been found to rise and fall during oak ageing. Volatile esters including ethyl acetate have pleasant fruity odours and reasonably low thresholds, but the development of a 'Cognac' flavour was best at a lower pH.

A rapid decrease in volatile esters took place. High surface/volume ratios accelerated ageing but resulted in unbalanced brandies. The 12 years of brandy ageing retained about 94% of the original lignin, but only 45% of the original tannins in the exposed barrel wood.

38.4.4.7 Removal from Storage

Brandy can be removed from storage at any time but may not be bottled in bond if it is less than 4 years of age or if it is not at 100° proof with a revenue stamp giving the dates of distillation and bottling. For sale, some brandies are reduced to 80° proof with deionised or distilled water while others are flavoured to give them their distinctive character though it is the trade secret of each producer. This usually consists of some sweetening agent with a variety of flavouring material added. Some flavoured brandies are too sweet and it would be preferable to limit treatment to a caramel colour and not over 1% sweetening. When young brandies are cut, they contain the sediment of caramel or oak extractives. To prevent this sediment getting into the bottle, chilling to −4 to −6.7°C (20–25°F), settling at a low temperature for a day or two and close filtration have been found useful. Brandies of more than 4 years of age are often cut to 100° proof, chilled, filtered and bottled. During filtration, the clarity may be controlled by a nephelometer operating at a 90° angle.

38.4.5 Blending, Bottling and Labelling

Blending is useful in the production of more uniform and higher-quality brandy. It is permitted under specific conditions while in other cases only rectification is allowed. Using a computer, a linear mathematical method has been devised to make blends to the desired ethanol, total extract, tannin, lignin, vanillin, aldehydes acetal, higher alcohols, esters, furfural, pH and sensory quality. Virtually all bottled Cognacs are blends of many different spirits, thus a particular very special old pale (VSOP) brand may be the result of blending up to 50 Cognacs. Hennessy was one of the first brands to introduce a system that would help consumers to differentiate between blends of brandies that have aged a minimum of 2 years, but many have aged considerably longer. Hennessy's very special (VS) includes brandies that are up to 10 years old, while VSOP refers to blends that are not less than 4 year old and the descriptor XO (extra old) denotes a blend of considerable age. In an attempt to emulate the qualities of Cognac, some German brandies are blended solely from grape spirits produced in pot stills. However, the best German brandy is blended from spirits derived from pot and continuous stills.

38.5 PRODUCTION OF TYPICAL BRANDIES

38.5.1 Cognac

The grapes are crushed and immediately pressed in a vertical or horizontal process, because fermentation on the skin produces less desirable wines for distillation. The must is fermented in cement tanks at a relatively high temperature (25°C) without prior sulphiting or clarification. The use of SO_2 is avoided in these fermentations because (i) the passage of SO_2 into the brandy leads to the formation of sulphuric acid which can greatly lower the pH; (ii) the combination of acetaldehyde and SO_2 has the properties of an acid sulphonate (strong acid) which corrodes the copper of the distilling columns; (iii) the reaction of ethanol and acetaldehyde to acetal, catalysed by H^+ ions, is favoured by SO_2 or bisulphite; and (iv) SO_2 favours the formation of acetaldehyde during the alcoholic fermentation and lowers the aromatic quality of Cognac. The fermentation of the must is spontaneous and has low concentrations of sugars (less than 180 g/L), tannins and high concentrations of malic acid. The wine produced has high acidity but with a relatively low ethanol concentration. These characteristics also help to protect the wine during storage prior to distillation and are suitable for the ultimate production of brandy. The yeast species present in the must and during the fermentation are *Saccharomyces uvarum*, *S. rasi*, *S. capensis*, *S. chevalieri*, *S. globosus* and quite often *Saccharomycodes ludwigii* besides the dominant *S. cerevisiae*. Sometimes, a surface film may develop resulting in the formation of ethyl acetate which detracts from the bouquet of the brandy. The surface film may include *Candida valida*, *Hansenula anomala* and *Pichia kluyveri*. Another problem which may develop is oxidative casse due to the oxidases of the grapes. Malo-lactic fermentation (MLF) usually occurs after the alcoholic fermentation, wherein the bacterial population of the musts consists of 60% *Leuconostoc mesenteroides*, 20% *L. oenos* and 20% *Lactobacillus plantarum*. MLF is facilitated by a higher K level, the relatively low alcohol concentration, the absence of antiseptics and the presence of lees consisting of yeast and grape particles. MLF increases the ethyl lactate concentration, but the quantity of esters does not directly affect the aroma of the brandy and the fermentation is considered desirable. The distillation of wines while they are still undergoing MLF yields brandies with a less desirable aroma. Until the wines are distilled, they are kept on the lees.

Distillation starts immediately after the vintage and continues until all the newly fermented wine is distilled. Wine that has been stored for as short a time as possible is preferred. Normally, only direct-fired pot stills of small capacity, not exceeding 30 hl (792 gal), are employed. The new wine with its lees (not more than 8% of added lees) is placed in the still and brought to boiling point. Distillation is allowed to continue until the vapours contain negligible amounts of alcohol. Distillation takes 8 hours or more and the main distillate (brouillis or low wines) contains about 24–32% alcohol. A tail fraction may be separated. The still is then emptied and refilled with fresh wine and often with the tails of the previous distillation. A second distillation is carried out followed by a third distillation. These three distillates are finally combined and then redistilled. This last distillation takes longer than the original distillation, *i.e.* about 14 hours or more. Approximately 1–2% of heads are separated. The main distillate (*coeur*) contains 58–60% ethanol and cannot exceed 72%.

A tails fraction (*seconde*) is also separated. To control the amount of each fraction, continuous testing using an alcoholic hydrometer and recycling of the heads and tails are practised. Distillation is carried out in such a way as to avoid mechanical entrainment, and no more than three distillations are recommended in 24 hours.

Brandy with 70% by volume alcohol is first stored in new oak barrels from the Limousin or the Forest of Troncais. The staves are prepared from 40- to 50-year-old trees. The trees are split, not sawed, and pieces of wood are used after prolonged exposure to the air. Later, these brandies are transferred to old casks to avoid any excessive enrichment in tannins. Very old brandies are kept in casks made from dense wood fibre. The dissolution of the soluble oak components, especially tannins, polyphenols, lignin, protein, pectins and minerals, is reflected in the colour of the brandy. The ageing process is schematically divided into two phases. During the first year, the acidity increases, acetal is formed and the extracted tannins are oxidised and intensify the colour of the brandy. The hemi-celluloses are hydrolysed, the lignins and esters are alcoholised and a vanilla-like or flowery odour appears. The aroma is greatly strengthened by the evaporation of water and ethanol between 10 and 30 years of storage. The taste of the brandy becomes sweeter due to the lower alcohol content and due to the formation of sugars by hydrolysis of the hemi-cellulose. All Cognac houses maintain inventories of old vintage Cognac to use in blending with live brands. The oldest Cognacs are removed from their casks in time and stored in glass demizohus (large jugs) to prevent further loss from evaporation and to limit excessively woody and astringent flavours. Luxury Cognacs are the very finest Cognacs of each individual Cognac house. For more current production of brandies, they are diluted with distilled water. However, this dilution can lead to a lack of clarity due to a reduction in the solubility of esters or higher alcohols or by the precipitation of calcium and copper salts. The Cognac is filtered at 5°C before being bottled; once bottled, it does not 'age' any more.

38.5.2 Armagnac

Armagnac is produced using a traditional process of fermentation. In its production, the wine is not clarified and the use of SO_2 and any other enological materials is avoided. It has an ethanol content of 8–9% by volume. The wines are stored on yeast lees which are removed before distillation. The new preheated wines are distilled in specially designed semi-continuous stills using the Verdier system. The process of distillation is carried out by (i) continuous distillation and (ii) discontinuous distillation or redistillation. The Armagnac-type still is made of copper having a continuous feed and with two or three heating vessels in series (vertically). The Armagnac produced by double distillation has a greater concentration of higher esters, particularly ethyl caprylate, caprate and laurate. The concentration of butanedial and ethyl laurate can be used to determine the method by which a young Armagnac has been distilled. Following distillation, the alcoholic concentration must not be higher than 72% by volume. The proof of distillation is very low, not exceeding 126°. The resulting brandy has a rustic, assertive character and aroma which require additional cask ageing to mellow out. Armagnac is aged for 20 years or more, but much is sold after only 5–8 years of ageing. The best Armagnac is aged in casks made from peduncular oak that is rich in tannins compared to the oak used in the ageing of Cognac. In recent years, Limousin and Troncais oak casks have been added to the mix of casks as suitable Monlezum oak becomes harder to find. Changes during ageing are similar as discussed in Section 38.4.4.6. Armagnac is sold with an alcohol concentration of 40% by volume; it is drier in taste than Cognac and its odour is less distinctive.

38.5.3 Pisco

As with wine, the process of Pisco, obtained from various varieties of grapes such as Quebranta, Moscatel and Albilla, starts in the vineyard. Pisco is generally obtained by the distillation of wine made from Muscat grapes. It is a traditional process and the different unit operations involved are shown in Plate 38.3. The grapes are crushed either mechanically or by the traditional method of stepping on the grapes. The must is kept in big deposits during fermentation (about 15 days) or until ready for distillation which is done either in batch distillers or in falcas (very old distillers). The first alcohol is separated and the process continues until it reaches 46° Gay-Lussac of alcohol, when the process is interrupted. This product is aged for 3–6 months, preserving all the primary and secondary flavours before bottling. Muscat grapes have been analysed for their aromatic potential; 54 components were separated and 18 of these components including esters, alcohols, terpenes, acids, aldehydes and miscellaneous compounds were detected for the first time in Pisco. These variations have hindered the introduction of this beverage to the international market. It is widely accepted that the peculiar flavour of wines made from Muscat varieties is mainly due to terpene alcohols and their derivatives. A distinctly Muscat fruit aroma distinguishes Pisco from other young distillates. All production vinification and distillation processes and technologies are oriented toward heightening the aroma of the final product. The vinification of Muscat grapes has progressively evolved from red to white vinification techniques to retain as much aroma as possible.

Four commercial categories of Pisco can also be described as relatively typical:

- Artisanal Pisco: Small producers using traditional technologies such as skin contact during winemaking, initiating complex products requiring some maturation or ageing.
- White Pisco: Young distillates with less than 6 months ageing, in general no contact with wood, frequently coming from single-varietal wines, only

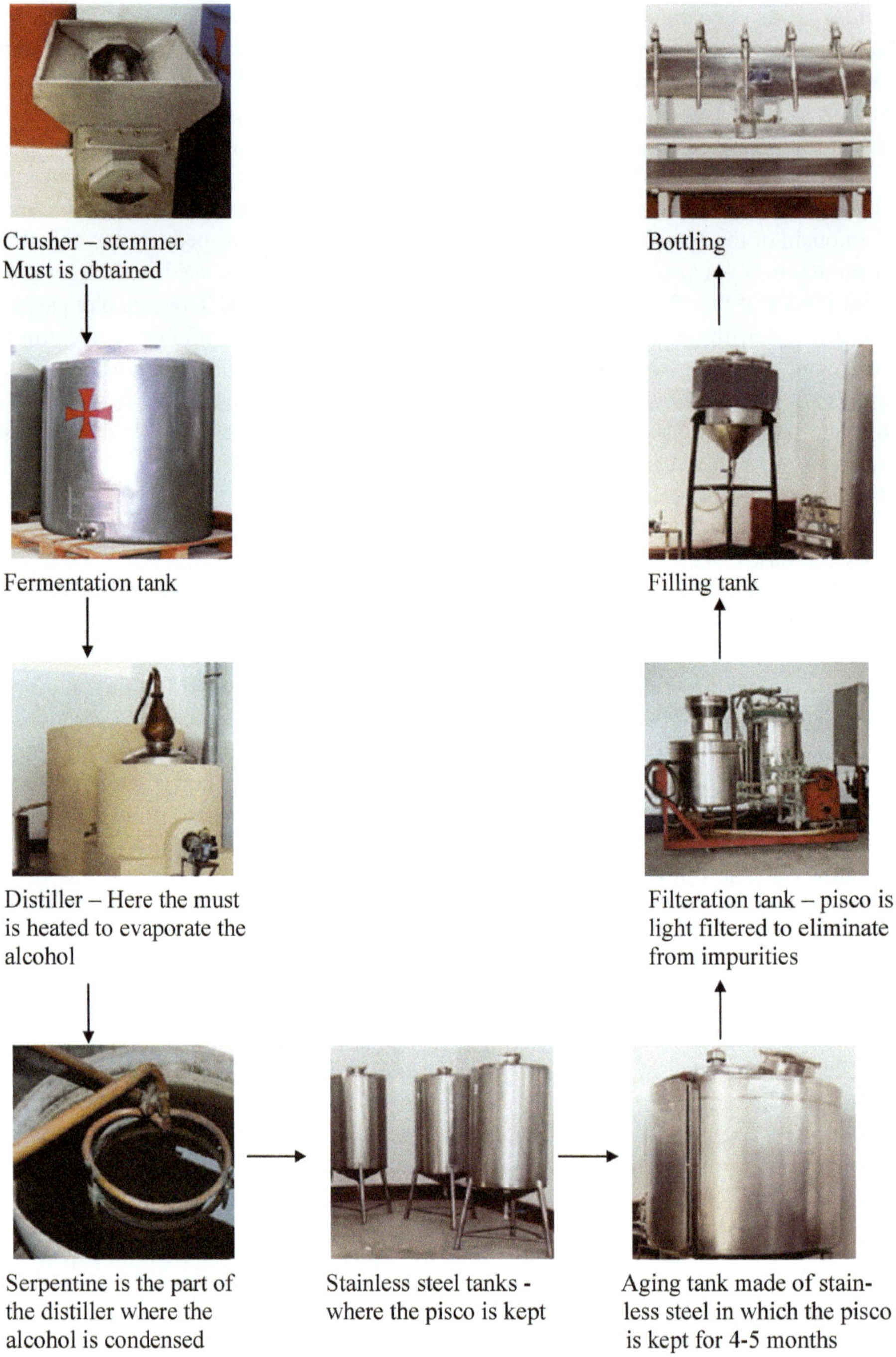

PLATE 38.3 Process of pisco production. *Courtesy:* http://www.piscopuro.com/i_elabo.htm; http://www.piscopuro.com/i_elabo2.html

'heart distillates' low in non-alcohol fractions. Can be triple distilled.

- Mature Pisco: Ageing between 6 and 12 months in used cooperage with a combination of fruity aromas and some wooden notes and light colour. Usually coming from a mixture of grape varieties.
- Aged Pisco: Over 12 months ageing in young American oak barrels. Amber in colour with wood-derived aromas combined with fruity aromas coming from grapes. Multi-varietal also.

38.5.4 Fruit Brandies

Fruit brandies are made from fermented fruit mash such as apple, plum, peach and pear. Some fruits, such as raspberries and blackberries, have a low sugar content and are not capable of fermentation. They can be soaked in neutral spirits and then redistilled. Most fruit brandies are aged in steel, glass or clay containers, which keeps them colourless and clear. Within the general category of fruit brandies, a distinction is made among distillates from pomaceous fruits, stone fruits, berries and other fruits (Table 38.6).

TABLE 38.6
Type of Fruit Brandies

Sr. no.	Fruit brandy	Description
1.	Apple brandy	The method of apple brandy (applejack or Calvados) production was described as early as 1939 and the distillation technology is similar to the grape brandy. Mostly cull fruits are used for apple brandy production. To make quality apple brandy, the use of fresh, mould-free and ripe apples for fermentation and distillation after clarification is recommended. The first distillation gives a product of 60° proof which is redistilled to 110–130° proof after cutting a suitable head and tail. Currently, continuous columns are used to produce apple brandy. All varieties of Calvados are aged in oak casks for a minimum period of 2 years. Apple brandies produced in the United States are aged in oak chips for a very short time.
2.	Pear brandy	For preparing pear brandy, tree-ripened Williams (Bartlett in America) from local north-west orchards are hand-picked, carefully fermented and distilled. Distillation is carried out in copper pot stills using centuries-old traditional methods. The resulting distillates have a rich aroma and fruit flavours that are typical of Williams pear, considered one of the most precious throughout Europe. Pear brandies (Poire) are now being produced in the United States, in California and Oregon.
3.	Peach brandy	The brandy produced from peach fruit is called peach brandy. The production of brandy can soak up excess production of peaches, especially the cullage, but the use of poor-quality waste fruit usually has little chance of making quality brandy. Peach brandy can be prepared in a similar way to grape brandy except for the method of wine preparation. Peach fruits are very pulpy so for fermentation, these have to be diluted. Consequently, the sugar content is reduced to a very low level, unlike grape and apple. However, dried peach has not been found suitable for winemaking nor would it make a brandy. A blend of tree-ripened, sunny slope peaches is carefully distilled to preserve the delicate aroma. Peach brandy is distilled in pot stills using centuries-old traditional methods and has a rich aroma and fruit flavour. Commercial-quality brandy can be made from cv. July Elberta. It is produced using the peach pulp of the July Elberta variety ameliorated with sugar, jaggary and molasses as the fermentable sugar sources in peach must and sugar-based peach have the highest ethanol content. It should be matured with *Quercus* wood chips to improve the quality characteristics.
4.	Plum brandy	Plum brandy is made in Serbia and Bosnia from blue plums and is very highly regarded. Other varieties include Mirabelle which is made from yellow plums, Ouetsch from Alsatian plums and light-green prunelic. Yugoslavia slivovitz is often aged in oak, mulberry or acacia cooperage. It is mostly light yellow in colour without distinct evidence of a wood-aged character. Plum brandies produced in the Alsace region of France are named for the variety of fruit used, *e.g.* Quetsch and Mirabelle. Although for brandy production, plums must have a high sugar content, the pattern of sugar is also important. Some plums contain sorbitol (0–36% of total sugar) which is not a fermentable sugar. If the sorbitol content is high, the sucrose content is low and vice versa. Currently, 50% of all plums grown in Yugoslavia belong to cv. Pozegaca and another 49.5% are local cultivars used for brandy production (Slivoviz). A comparison of five local varieties with Grase romanesti showed that *Gogosele de munte* gave the highest yields, was the hardiest and was found suitable for plum brandy production.
5.	Apricot brandy	Apricot fruits are used for making renowned apricot brandy. Brands include the French Abricotine and the Hungarian Barack Palinka. In 10 authentic laboratory brandies of stone fruit (sour and sweet cherry, damson plum and apricot), 0–6.4 mg of benzaldehyde per 100 mL of absolute alcohol and traces of 8.6 mg of hydrocyanic acid have been revealed. In 12 commercial samples, benzaldehyde ranged from traces to 13.4 and hydrocyanic acid from traces to about 10. λ-Undecalactone has also been used in certain apricot liqueurs. To make brandy from wild apricot, the pulp is diluted in a 1:1 ratio with water, fortified with 0.1% DAHP having TSS of 25°B and is fermented to completion. The wine is then distilled to make brandy and matured with slightly roasted oak wood chips. A comparison of the methanol content of brandy made experimentally with that prepared locally showed a considerable reduction in the quantity of this alcohol.
6.	Persimmon brandy	The distillate obtained from Kurokuma from different Diospyros kaki cultivars viz., *Thiene*, *lycopersicon*, *Kurokuma*, *Tipo*, *Fugi* and *Hachya*, fermented at 20–25°C with a selected starter yeast was found to contain the lowest amount of acids and aldehydes, and was adjudged the best.
7.	Banana brandy	The fruit should neither be over-ripe nor under-ripe for the preparation of banana brandy. Over-ripe fruit yields strongly flavoured brandy, whereas astringent brandy is produced from under-ripe fruit. The pulp of a ripe banana is clarified and converted into juice. The juice is allowed to ferment and is then distilled into a brandy with an alcohol content of 35–40%.
8.	Cashew brandy	In the Goa region of India, a type of brandy called 'Fenni' is made from ripe cashew apple on a cottage scale by fermenting the juice and then distilling it using an old and crude method. Mostly the juice is fermented to prevent its deterioration and made into a liquor with its characteristic flavour. The word 'feni' is derived from *feun* which means Proth in Konkani. The juice is extracted mechanically using a cashew apple expeller. The apples are fed into a small hopper and crushed between a wooden roller and a concave wooden board. The yield of apple juice extracted is 50–60%. Similar juice extractors are employed on a large scale. The extracted juice is kept for 2–3 days for fermentation with yeast which forms a film on the surface of the juice. The feni is obtained by distilling *Uraq* mixed with fermented juice (1:2). The alcohol is recovered from the fermentation brew by distillation in a pot and made into brandy with 60% alcohol. Distillation is carried out using column stills on a large scale. One ton of fruit is reported to yield an average of about 580 L of wine or about 74 L of brandy. The brandy is aged in oak wood barrels. The temperature of a cellar is kept at 15°C. After ageing, the alcohol content is diluted with the addition of water to 43%, followed by filtration, bottling and labelling. A product called 'konioi' similar to gin is made from cashew apple in Tanzania. In Brazil, a fascinating product is cashew apple-in-sugarcane brandy. The method includes removal of the nut when the peduncle is still small which is then introduced into a bottle and allowed to grow. When fully matured, the apple is separated from the main branch and sugarcane brandy is poured into the bottle.
9.	Guava brandy	The wine prepared from Guava fruit gives a characteristic flavour and can be distilled into brandy.

38.6 COMPONENTS OF BRANDY

Many compounds are present in grapes and several are formed during the alcoholic fermentation of grape must. During distillation, some compounds appear in the distillates, others are formed during ageing or are extracted from wood, while others are added to the beverage during processing. The bouquet of a brandy is a mixture of higher alcohols, esters, fatty acids, aldehydes, acetal and volatile acids produced during fermentation. In a sample of genuine Cognac, approximately 81 components were identified and evidence was found for 16 additional compounds which included 12 alcohols, 20 carbonyl compounds, 22 acids, 31 esters and 12 of a miscellaneous nature. Fourier transform infrared (FT-IR) spectroscopy was used in the characterisation and classification of brandy. Spanish, French and South African brandies as well as Cognacs and Armagnacs were characterised and a complete differentiation of the latter two types from the rest of the samples of distilled drinks was obtained. A concentration of 3-methyl-1-butanol can be used to distinguish Armagnac from whiskies. Characteristic differences are also found for higher alcohols depending on the fruit used in production. The flavour of brandy is due to carboxyls, higher alcohols, esters, phenolic compounds, lactones, nitrogen-containing compounds and some micro-nutrients.

38.6.1 Ethanol

Alcohol, more specifically ethanol, *i.e.* ethyl alcohol (C_2H_5OH), is the intoxicating compound in wine and other alcoholic beverages. It is a clear, colourless, flammable liquid with a density of 0.7939 at 100% alcohol or at 200° proof. Distilled beverages contain 37–42% v/v or more alcohol content. During the ageing of Cognac, the ethanol content decreases.

38.6.2 Other Alcohols

In addition to ethyl alcohol, brandy also contains other alcohols. Grapes with a high pectin content if mouldy may yield methanol, *i.e.* methyl alcohol. A small amount of methanol may be present in brandy (Table 38.7). It has physical properties similar to those of ethanol. A methanol content of 0.039–2.86% was reported in 37 uncut Piedment pomace brandies.

TABLE 38.7
Methanol Content of Some Brandies

Name of brandy	Quantity
Brandy	Traces–0.188%
Cognac	59 g/hac L
Plum	1200 g/hac L
Apple	100 g/hac L
Pear	700 g/hac L
Pomace	0.039–0.86%

Source: Joshi *et al.* (1999)

Fruit brandies contain a higher quantity of methanol than grape brandies. The methanol is toxic at high concentrations (4%) and may cause blindness or even death. When present in greater quantities, higher alcohols have an unfavourable effect on the bouquet of a brandy. The principal higher alcohols are 3-methyl-1-butanol, 2-methyl-1-propanol, 2-methyl-1-butanol and n-propanol.

In brandies at 50% by volume ethanol, the higher alcohols are generally within the range of 650–1000 mg/L, depending on the composition of the must, the microorganism used for fermentation and the condition employed during vinification. In California, commercial brandy distillates have 7.4–30.0 g/L at 100° proof of n-propanol, 6.8–25.0 of isobutanol and 20.0–87.5 of combined amyls sec-butanol. Employing statistical methods and using the data on the concentrations of fractions of higher alcohols and methanol, the origin of brandies has been determined. For example, Cognac and Armagnac contain little methanol; 2-buytanol, n-butanol and isopropyl alcohol are absent in quality brandies; but concentrations of 3-methyl-2-butanol are high. Components such as ethyl formate, acetaldehyde, acetal, ethyl acetate methanol, 2-butanol, 1-propanol and 3-methyl-1-butanol in Spanish brandies have been identified. Generally, Cognacs have more higher alcohols than other European brandies. If evaporation is taken into account, the higher alcohols do not increase during the ageing of Cognac. Better quality distillates have more higher alcohols and always contain 2-propanol (isopropanol), moderate amounts of 1-propanol (n-propanol) and 2-methyl-1-butanol (isoamyl) of about 100 mg/L. Pressing the juice from skins and lees, aerating musts, using the clearest must, lower fermentation temperatures (>20°C, 68°F) and using white grapes to reduce higher alcohol formation have been recommended.

The presence of 2-butanol is used as an indication of the growth of lactic acid bacteria in either the must or the fruits, though it has little effect on the quality of brandies. The yeast and lactic acid bacteria produce a high concentration of 1-propanol and can be considered an indication of bacterial spoilage of the must or the wines. Components present in low-quality wine were recovered in excess of 90% in the brandy product. But only 8% of β-phenethyl alcohol and 70–75% of its acetate esters were recovered in the brandy fraction. Compounds derived mostly or entirely from oak, furfural, 5-methyl furfural, diethyl succinate and the *cis* and *trans* isomers of β-methyl-y-octa-lactone (oak lactone -a and -b) were more abundant in brandies aged in US oak than in French oak and less amounts of oak-derived compounds were obtained in brandies from reused barrels.

A fraction (15%) with a boiling range higher than 3-methyl-1-butanol was separated from the fusel oil obtained from the distillation of wine made from Muscat of Alexandria raisins and was found to contain various components, including small amounts of ethyl penta decanoate, n-propyl caprylate, isobutyl caprylate, isoamyl caproate, active amyl caprylate, active amyl laurate and traces of active amyl caproate, acetic, caproic, caprylic, capric and isovaleric acids and 5% of 1-hexanol. Similar results from the esters and other compounds in Cognac were obtained. Important volatile components in

TABLE 38.8
Average Composition of Volatile Substances (mg/L) in Brandy

Component (mg/L)	Cognac	Armagnac
Acetaldehyde	32.4	50–70
1-Buanol	Trace	0–20
Phenyl ethanol	36.4	9–32
Ethyl acetate	268	500–600
Ethyl caprylate	13.6	8–100
Ethyl caprate	35.2	6–140
Ethyl laurate	36.8	5–70
Ethyl myristate	11.8	4–20
Ethyl lactate	268	100–500
Methanol	413	500–600
Isobutylalcohol	813	700–1000

Source: Lafon (1983)

Cognacs and Armagnacs are listed in Table 38.8. Fusel oils are more toxic than ethanol but are present in brandy in such low concentrations that they cause no danger to the average brandy consumer's health. Since they have higher boiling points than ethanol, they can be separated from it to a high degree during distillation and are collected on certain plates of the distillation column. Although their total content is usually less than 0.3%, they still constitute an important part of the flavour of brandies.

38.6.3 Aldehydes

In brandies, aldehydes, chiefly acetaldehyde, are present in small amounts while the presence of propanol, butyraldehyde and heptanal has also been reported. During fermentation, furfural (pyromucic aldehyde) is formed in the presence of lees containing lignin compounds. Furfural is constantly higher in brandies aged in American oak; in contrast, French oak barrels appear to constitute more extract and colour. Acetaldehyde is formed by the oxidation of ethanol while a part is also extracted from the wood of the barrels during ageing. Diethyl acetal present in all brandies is formed by the condensation of ethanol and acetaldehyde. Some lignin is alcoholysed during the ageing of brandies in barrels. The oxidation of the compounds formed results in the formation of aldehydes with a vanilla-type flavour that play an important role in the aroma of brandies. The free aldehyde content of Cognac is 38–112 mg/L, while the range for Italian brandies is 30–116 mg/L. A range of 0.3–3.8 g/100 L at 100° proof of aldehydes (as acetaldehyde) was shown in California brandies. In Armagnac, the concentration of aromatic aldehydes ranges about 1–4 mg/L and that of acids is 2.6–12.2 mg/L. Ethanol and acetaldehyde react slowly and form acetal, a compound with a pronounced odour, increasing the acetal content to about 20 mg/L, though higher has been found in Cognacs. The acetal in Cognacs varies from 26.5–77.5 mg/L; in young Italian brandies from 27 to 165 mg/L; and in old Italian brandies from 18–112 mg/L. Among others, 5-methyl furfural and B-methyl-y-octalactone were identified as substances derived from oak during ageing. These components were relatively high in concentration in brandies aged in American (US) oak but were quite low or apparently absent in French oak.

38.6.4 Esters

During distillation and ageing, ethanol reacts with acids to form small amounts of esters such as ethyl acetate. Esters of propyl and butyl alcohols are also formed, but in small concentrations they are not disagreeable. The volatile compounds in brandies are derived from the grapes and the fermentation; they are concentrated by distillation and modified during ageing. In the case of certain varieties, they reach their greatest intensity if they are grown in relatively cool districts, *e.g.* Ugni blanc grown in the Charentes. The brandy of Folle Blanche is richer in a wider range of esters than that of Saint Emilion or of Bacco 22A. Esters such as ethyl, hexyl and isopentyl myristate; monoethyl sciccinate; capryl enanthate, ethyl oleonoate and ethyl phenylcaproate are the major contributors to the aroma. The presence of the last-named compounds can be considered a sign of brandy quality. In 31 commercial California distillates, 2.6–12.8 g/100 L at 100° proof of total esters (as ethyl acetate) were obtained. The total amount of ethyl esters present in unaged brandy decreases during ageing in wood, especially ethyl laurate and caprate. A high concentration of ethyl n-butyrate and ethyl n-valerate which give brandy an undesirable 'lolly-like' odour was found. Important differences were also revealed in the concentration of ethyl acetate, n-propanol and n-hexanol. In Cognac, 124 esters were detected. The highly volatile esters are associated with the odour of Cognac. The better brandies were lower in esters and aldehydes and inexplicably higher in total acidifier. The amount of ethyl esters of fatty acids present in unaged brandy distillates diminished during ageing, particularly laurate and ethyl caprate. The 1-octanol acetate esters of isoamyl, n-hexyl and b-phenetyl alcohols decreased during ageing in oak barrels. The ethyl esters of fatty acids such as capric, caprylic and caproic increased during ageing while ethyl laurate changed little or slightly decreased. Important differences were revealed in the concentration of ethyl acetate, n-propanol and n-hexanol.

38.6.5 Other Constituents

Wine contains many other constituents such as acetic acid and lactic acid that are distilled in brandy in small amounts, usually less than 100 mg/L of 100° proof brandy. The free fatty acids of fermented solutions easily distil together with steam and alcohol and thus, may appear in appreciable amounts in distilled beverages. The concentration of fatty acids in brandies is directly correlated with the yeast biomass in wine at the time of distillation. Fatty acids of lower molecular weight are excreted by the yeasts into the wine, whereas those of higher molecular weight pass into the liquid at the time of distillation. The concentration of fatty acids in brandies depends on the production technique used. Caproic, caprylic, lauric, myristic and stearic acids are found in Cognac and play an important role in its aroma. The presence of propionic and butyric acids is considered undesirable. In contrast, only small concentrations of fatty acids higher than capric acid are found in Armagnac and they are largely esterified. When volatile fatty acids of Cognac were determined as free fatty acids excluding acetic acid, caprylic and capric acid are the main components, while caproic and lauric acid appear in a proportion about one–third that of caprylic and capric acid. Acetic acid regularly appears as the largest component, constituting 80–90% of the total fatty acids. Although propionic, isobutyric, butyric, 20 methylbutyric and isovaleric acids are relatively minor components in fatty acid composition, their individual odours are many times as strong as the odours of acetic acid, so they make a significant contribution to the aroma. The small quantities of lactone and phenolic compounds extracted from wood during ageing also contribute to the aroma. Animal experiments have shown that aroma compounds in alcoholic beverages are not harmful to health. Vanillic, ferullic and syringic acids have been isolated from aged brandies. In seven California commercial brandy distillates, 0.70, 0.59 and 0.24 g/100 L at 100° proof of caprylic (C_8), capric (C_{10}) and lauric (C_{12}) free fatty acids, respectively, were found. Terpene profiles could contribute much to the aroma of Pisco. Terpenes impart flowery characteristics to certain brandies. Propionic and butyric acids may be found in distillates, if they are distilled from spoiled wines. When young wines are distilled, they usually foam, resulting in the rapid loss of carbon dioxide, though foam-producing compounds have not been identified. If SO_2 is present in wine, it distils with alcohol and is present in brandy. High-proof brandy that is high in SO_2 should not be stored in metal tanks because it dissolves iron and other metals from pumps, pipelines and storage tanks. The SO_2 oxidises to sulphuric acid which is an undesirable constituent of beverage brandy. In the case of Cognac's media, the addition of caramel decreases the concentration of volatile sulphur compounds, and hydrogen sulphide and thiols are especially bounded. Sulphur may come from dusts applied to the vines and grapes, which may be reduced to hydrogen sulphide during fermentation. Sulphur also reacts with alcohols to form mercaptans, compounds of very disagreeable odours (garlic and skunk). Such wines should either be treated before distilling or be used for the production of wine spirits for fortification. Small amounts of acrolein (highly toxic, irritating to eyes and nose and has a horseradish odour) appear in Swiss fruit brandies but seldom in brandy distilled from wine or pomace. During storage, the amount decreases but can only be removed completely by redistillation in a multicolumn still. Sometimes, brandies have an excessive copper or iron content. Australian brandies contain 0.7–20 mg/L of copper which can be removed from unaged brandy by treating with cation-exchange. Brandies also contain 0.01–0.06 mg/L of lead besides ammonia and various nitrogenous degradation products.

38.7 LABELLING OF BRANDY

Brandy has a traditional age-grading system and these indicators are usually found on the label near the brand name.

- **VS** (very special) or *** (three stars) designates a blend in which the youngest brandy has been stored for at least 2 years in a cask.
- **VSOP** (very superior old pale), **Reserve** or ***** (five stars) designates a blend in which the youngest brandy has been stored for at least 4 years in a cask.
- **XO** (extra old) or **Napoleon** designates a blend in which the youngest brandy has been stored for at least 6 years.
- **Hors d'age** (beyond age) is a designation which is formally equal to XO for Cognac, but for Armagnac it designates brandy that is at least 10 years old. In practice, the term is used by producers to market a high-quality product beyond the official age scale.

Russian brandies, as well as brandies from many other post-Soviet states (except Armenia) use the traditional Russian grading system that is similar to the French one, but extends it significantly.

- 'Three stars' or *** designates a brandy with the youngest component cask-aged for at least 2 years, analogous to the French VS.
- 'Four stars' or **** is for blends where the youngest brandy has been aged for at least 3 years.
- 'Five star' or ***** means that the youngest brandy in the blend has been aged for 4 years, similar to the French VSOP.
- **KB/KV** (aged Cognac) is a designation corresponding to XO or Napoleon, meaning that the youngest spirit in the blend is at least 6 years old.
- **KBBK/KVVK** (aged Cognac, superior quality) designates 8-year-old blends and tends to be used only for the highest-quality vintages.
- **KC/KS** (old Cognac) indicates at least 10 years of ageing for the youngest spirit in the blend (similar to Armagnac's Hors d'age).
- **OC/OS** (very old), which is not in the French system, designates blends older than 20 years.

38.8 FIRE AND EXPLOSIVE HAZARDS

Recently emptied brandy barrels contain a highly explosive mixture of ethanol and air. Even a spark from a cigarette can cause these barrels to explode. Hence, distillery premises should be made of fireproof materials. Smoking or open fires should not be permitted near the still. Although the rich alcohol vapours have a tendency to settle, the ethanol flammable liquid can release vapours which form an explosive mixture with air at or above 13°C. The alcohol/air mixtures in the flammable range have a specific gravity very slightly greater than air.

Therefore, air currents will distribute such mixtures that can travel considerable distances to a source of ignition and flash back to leaking or open containers. They can also accumulate in confined spaces resulting in toxicity and flammable hazards. Closed containers may rupture violently when exposed to fire or excessive heat for a sufficient period of time. From a fire standpoint, it is understood that alcohol is almost as hazardous as gasoline. One can be inspired to distil one's own wine, and home distillation can either be very hazardous or reasonably safe depending on the degree of care taken. A little mishandling may result in burns or death and a great loss of property.

BIBLIOGRAPHY

Agosin, E, Belancic, A, Ibacahe, A, Baumes, R, Bordeu, E and Crawford, A (2000). Aromatic potential of certain Muscat grape varieties important for pisco production in Chile. *American Journal of Enology and Viticulture* 51(4):404–408.

Amerine, MA, Berg, HW, Kunkee, RE, Ough, CS, Singlaton, VL and Webb, AD. (1980). *The Technology of Wine Making*, 4th edn., AVI Publ. Co., Westport, CT.

Benda, I (1982). Wine and brandy. In: *Prescott and Dunn's Industrial Microbiology*, 4th edn., G. Reed (ed.), AVI Publ., Westport, CT, pp. 293–402.

Bordue, E, Agosin, E and Casaubon, G. (2012). Pisco: Production, flavor chemistry, sensory analysis and product development. In: *Alcoholic Beverages*, J. Piggot (ed.), Woodhead Publishing Series in Food Science, Technology and Nutrition, Cambridge, pp. 331–347.

Bougas, NV. (2014). *Factors Influencing the Style of Brandy*. PhD Thesis, Department of Viticulture and Oenology, Faculty of Agriscience, Stellenbosch University.

Cacho, J, Moncayo, L, Palma, J, Ferreira, V and Cullere, L (2012). Characterization of the aromatic profile of the Italia variety of Peruvian pisco by gas chromatography–olfactometry and gas chromatography coupled with flame ionization and mass spectrometry detection systems. *Food Research International* 49(1):117–125.

Caldeira, I, Belchior, AP and Canas, S (2013). Effect of alternative aging system on the wine brandy sensory profile. *Ciência e Técnica Vitivinícola* 28(1):9–18.

Canas, S, Caldeira, I, Belchior, AP, Spranger, MI, Clímaco, MC and Bruno-de-Sousa, R (2011). Chestnut wood: A sustainable alternative for the aging of wine brandies. In: *Food Quality: Control, Analysis and Consumer Concerns*, D.A. Medina and A.M. Laine (eds.), Science Publishers Inc, New York, pp. 181–228.

Claus, M and Berglund, K (2002). *Fruit Brandy Production Research*. Agricultural Engineering Department, Michigan State University, Newsletter, Nov/Dec, 2002.

De Lima, EQ, de Oliveira, E, Alves, HO and de Silva Leite, CF (2015). Obtaining physical-chemical analysis of the alcoholic distillate of cajarana (*Spondias* sp.) in semiarid Paraiba. *African Journal of Agricultural Research* 10(33):3271–3280.

Dhawan, SS, Kainsa, RL and Gupta, OP (1983). Screening of guava cultivars for wine and brandy making. *Haryana Agricultural University Journal of Research* 13(3):420–423.

Guly, H (2011). Medicinal brandy. *Resuscitation Pubmed Central* 82(7):951–954.

Jaarsveld, FP and Hattingh, S (2012). Rapid induction of aging character in brandy products: Ageing and general overview. *South African Journal for Enology and Viticulture* 33(2):225–251.

Jackson, RS (1999). Grape-based fermentation products. In: *Biotechnology: Food Fermentation*, V.K. Joshi and A. Pandey (eds.), Vol. II. Educational Publisher and Distributors, New Delhi, p. 583.

Jolly, NP and Hattingh, S (2001). A Brandy Aroma Wheel for South American Brandy. *South African Journal for Enology and Viticulture* 22(1):16–21.

Joshi, VK and Sandhu, DK (2000). Quality evaluation of naturally fermented alcoholic beverages, microbiological examination of source of fermentation and ethanolic productivity of the isolates. *Acta Alimentaria* 29(4):323–334.

Joshi, VK, Sandhu, DK and Thakur, NS (1999). Fruit-based alcoholic beverages. In: *Biotechnology: Food Fermentation*, Vol. II, V.K. Joshi and A. Pandey (eds.), Educational Publisher and Distributors, New Delhi, p. 647.

Kotecha, PM and Desai, BB (1995). Banana. In: *Handbook of Fruit Science and Technology*, D.K. Salunkhe and S.S. Kadam (eds.), Marcel Dekker, New York, p. 67.

Lafon, S (1983). Wine and brandy. In: *Biotechnology: Food and Feed Production with Microorganisms*, Vol. 5, H.J. Rehm and G. Reeds (eds.), Wiley-Blackwell, Hoboken, NJ, p. 148.

Leaute, R (1990). Distillation in alambic. *American Journal of Enology and Viticulture* 41:90–103.

Matos, ME, Medeiros, ABP, Pereira, GVM, Soccol, VT and Soccol, CR (2017). Production and characterization of a distilled alcoholic beverage obtained by fermentation of banana waste (*Musa cavendishii*) from selected yeast. *Fermentation* 62(3):3390–3397.

Mustapha, N (1997). Influence of complex media composition, Cognac's Brandy, or Cognac, on the gas chromatography analysis of volatile sulfur compounds – Preliminary results of matrix effect. *American Journal of Enology and Viticulture* 48(3):333–338.

Onishi, M, Crawell, EA and Guyman, JF (1978). Comparative composition of brandies from Thompson Seedless and three white-wine grape varieties. *American Journal of Enology and Viticulture* 29:54–59.

Palma, M and Barroso, CG (2002). Application of spectroscopy to the characterization and classification of wine, brandies and other distilled drinks. *Talanta* 58(2):265–271.

Patel, JD, Venkataramu, K and SubaRao, MS (1977). Studies on the preparation of cider and brandy from some varieties of India apples. *Indian Food and Packer* 31(6): 5–8.

Puech, JL and Moutounet, M (1992). Phenolic compounds in an ethanol–water extract of oak wood and in Brandy. *Lebensmittel-Wissenschaft und -Technologie* 251:350–352.

Rodriguez-Solana, R, Rodriguez, N, Dominguez, JM and Sandra, C (2012). Characterization by chemical and sensory analysis of commercial grape marc distillate (*Orujo*) aged in oak wood. *Journal of the Institute of Brewing* 118(2):205–212.

Rose, AH (1977). *Economic Microbiology*, Vol. VI. Alcoholic Beverages, Academic Press, London.

Swami, SB, Thakor, NJ and Divate, AD (2014). Fruit wine production: A review. *Journal of Food Research and Technology* 2(3):93–100.

Tsakiris, A, Kallithraka, S and Kourkoutas, T (2013). Grape brandy production, composition and sensory evaluation. *Journal of the Science of Food and Agriculture* 94(3):404–414.

Turk, J and Rozman, A (2002). A feasibility study of fruit brandy production. *Agricultura* 1:28–33.

Unwin, T (1991). *Wine and the Vine*, Routledge, London.

Urosevic, I, Nikicevic, N, Stankovic, L, Boban, A, Urosevic, T, Krstic, G and Vele, T (2014). Influence of yeast and nutrients on the quality of apricot brandy. *Journal of the Serbian Chemical Society* 79(10):1223–1234.

Vazque-Rowe, I, Caceres, AL, Torres-Garcia, JR, Quispe, I and Kahhat, R (2017). Life cycle assessment of the production of *pisco* in Peru. *Journal of Cleaner Production* 142(4):4369–4383.

Viekers, AK, Kuhn, ER and Ebale, SE (2002). Complete separation of fusel oils by Capllary GC. *Abstract No. 104 Poster, Gulf Coast Conference in 2002.*

Zavrathik, M and Hribar, J (1996). Quality parameters of fruit brandy made from various persimmon cultivars. *Sodobno-Kmetijstvo* 29(4):147–152.

Unit 7

Methods of Quality Evaluation

INTRODUCTION

Quality Evaluation Methods

The winemaking process involves a series of steps that must be carefully carried out in order to obtain a quality product. The process starts by selecting the grape variety or clone to be used, and then, selecting the optimal harvest time, thereby obtaining the best grape metabolic expression. Grapes are transported to the winery where fruit selection is carried out. Must is extracted and fermented to produce wine. Finally, wine is stabilised and subsequently, aged to obtain a quality wine.

Scientific knowledge has aided in the development of the wine industry. Important stages can now be monitored to avoid deviations and contaminations and to obtain the final desired product.

For any analyst, both grape and wine analysis are a challenge since they represent very complex matrices, in which compounds determining quality are present in very low concentrations (ng-µg/L). Additionally, potential contaminants, responsible for harmful effects, can also be present in low concentrations, complicating the quality control of the entire wine production process.

This unit presents a detailed description of the main analytical methods currently in use by the wine industry and/or research laboratories.

Chapter 39 presents the main analytical techniques useful for the evaluation of both volatile metabolites (solid-phase microextraction, gas chromatography, gas chromatography–mass spectrometry) and non-volatile components (high performance liquid chromatography, ion chromatography, matrix-assisted laser desorption ionisation mass spectrometry). Pesticide residue and contamination detection tools (GC, GC-MS, HPLC, MALDI), and metals content (atomic absorption and inductively coupled plasma emission) are also discussed.

The use of non-specific analytical techniques such as near-infrared spectrometry and nuclear magnetic resonance is discussed as analytical options, offering winemakers fast results for routine analysis. Applications involving flow injection analysis are also discussed.

In the search to provide diversity and innovation to consumers, and from a microbiological point of view, the selection of *Saccharomyces* and non-*Saccharomyces* yeasts and their flavour modifications contribute to the wine's quality.

Yeast selection requires the use of different analytical options ranging from the simplest, such as microscopy and enzymatic methods, to the increasingly used molecular methods, such as polymerase chain reaction and all its variables, bioluminescence and flow cytometry. These techniques also play a fundamental role in the detection and identification of contaminating micro-organisms during the fermentation process. Chapters 39 and 41 detail their applications in detecting contaminating micro-organisms and their metabolites.

Wine and brandy are products that are mainly consumed for their flavour and aroma, so the techniques associated with the sensory evaluation of these products have great importance at the level of the production and the development of new products as well as research.

Chapter 40 describes sensory evaluation and covers various aspects related to the anatomy and physiology of the senses involved including olfactory and gustatory as well as the requirements necessary to carry out a standard evaluation. The most common sensory evaluation methods are described using trained panels or affective testing by consumers. Efforts have been made in this chapter to illustrate the application of various techniques for evaluation of wine and brandy.

Dr Laura Fariña
Associate Professor
Área de Enología y Biotecnología de Fermentaciones
Departamento de Ciencia y Tecnología de los Alimentos
Facultad de Química de la Universidad de la República (UdelaR)
Uruguay

39 Techniques of Quality Analysis in Wine and Brandy

B.W. Zoecklein and B.H. Gump

CONTENTS

39.1 INTRODUCTION

Successful analyses of grapes and wines satisfy several winemaking and regulatory issues such as quality control: fruit ripening, processing and ageing; spoilage reduction and process improvement; blending: precise analyses leading to more precise blends; export certification: European Economic Community, Pacific Rim, Canada, global and regulatory requirements. Part of the winemaker's challenge is to adjust processing techniques to fit future consumer demands, including processing adaptations and quality assurance. This chapter discusses various analytical techniques used in wine and brandy production and their significance.

39.2 HAZARD ANALYSIS AND CRITICAL CONTROL POINTS (HACCP)

HACCP is a system for assuring product quality control from beginning to end, through the identification and monitoring of critical control points (CCPs) during processing. 'HACCP-like' plans are being developed to fit the wine industry's need to integrate chemical, physical, microbiological and sensorial analyses into quality- and style-control programmes. HACCP-like plans are designed to identify where CCPs occur and establish control and monitoring measures. Wine industry HACCP-like plans begin with establishing a processing chart or flow diagram, which starts in the vineyard and ends with the movement of wine through the distribution network. At each step, CCPs are identified and ranked in terms of their importance, and a corresponding list of analytical control measures is established. Properly prepared, the HACCP plan helps to answer several basic questions such as: When are specific analyses needed during the process? Why are they important, and where does each fit into the winemaker's processing protocol? How are analytical results interpreted, and what are the expected range of values for each? and What corrective measures are needed if results do not fall within specifications? Each CCP must be examined and evaluated using chemical, physical, microbiological, and sensorial methods. Results must be regularly assessed to determine if additional or corrective steps are required.

39.3 JUICE AND WINE ANALYTES

Juice and wine components can be divided into classes that are measured during the production process and after the wine is prepared for bottling (Table 39.1). The former includes soluble solids (sugar, extract, glucose, fructose and other carbohydrates);while the latter includes juice and wine aroma and aroma intensity, volatile compounds including H_2S, mercaptans; acidity (titratable and volatile acidity, pH, individual acids); alcohols (ethanol, methanol, fusel oils, glycerol); carbonyl compounds (acetaldehyde, hydroxymethyfurfural [HMF], diacetyl); esters (ethyl acetate, methyl anthranilate [*Vitis labruscana*]); nitrogen compounds (NH_4^+, amino acids, amines, proteins, methoxypyrazines, ethyl carbamate); phenolic compounds (total, phenolic fractions including anthocyanins, 4-ethylphenol, skin and seed tannin maturity); chemical additions (SO_2, ascorbic, sorbic and benzoic acids, dimethyldicarbonate); other (common and trace metals, oxygen, CO_2, fluoride, copper, iron, calcium, potassium) and microorganisms (yeast and bacteria).

TABLE 39.1
Analytical Techniques and Current Applications

Sensory evaluation, Grape maturity, Wine and distillate evaluation

- 'Wet' chemical methods, titratable acidity (TA), reducing sugar (RS), phenols and phenol fractions including anthocyanins, volatile acidity (VA)and SO_2
- GC: Alcohols, esters, a wide variety of volatile compounds
- GC/MS: EC, TCA, pesticide residues, various contaminants, etc.
- SPME techniques used with GC, GC/MS, HPLC and HPLC/MS analyses
- HPLC: Sugars, acids, phenolics, sulphur dioxide, etc.
- CE: Ethanol and various acids
- Enzymatic methods with UV spectroscopic detection: Various components
- AA and ICP spectroscopic techniques: Cu, Fe, Pb, Cd, Ca, K, etc.
- IR/NIR: A variety of infrared wavelengths for ethanol, sugars, etc.
- FIA/SIA: Flow injection analysis/sequential injection analysis with various reaction manifolds and detector systems (electrochemical, spectrometric, enzymatic, AA, etc.) for a variety of analytes
- Electrochemical techniques for measuring redox potential and as a detector system for a number of analytes
- Spectroscopic (visible and UV) techniques for colour/phenolics, etc.
- Microbiological (yeast, bacteria): Identification and enumeration

39.4 THE CHARACTERISATION OF JUICE, WINES AND DISTILLATES

The Association of Official Analytical Chemists (AOAC) and the *Organisation Internationale de la Vigne et du Vin* (OIV) define the analytical reference methods that are often used in the world wine industry. In addition to these approved methods, different wineries have adopted other generally accepted methods.

39.5 SENSORY EVALUATION

Sensory evaluation is often underutilised, but can provide many benefits to the winemaker who employs the techniques correctly. Properly utilised and interpreted, sensory evaluation can provide a basis on which viticultural and winemaking decisions can be made. Commonly used affective methods include paired preference, preference ranking and hedonic test methods. Descriptive evaluations provide a quantitative measure of wine characteristics that allows a comparison of the intensity among products, as well as a means of interpreting the results. Examples of descriptive test methods include quantitative descriptive analysis (QDA), flavour profile analysis, time-intensity descriptive analysis and free-choice profiling. Principal component analysis can be used to identify the smallest number of latent variables or principal components that explain the greatest amount of observed variability. A technique of free-choice profiling has been developed, which eliminates much of the time and training aspects. Free-choice profiling relies on the application of a generalised Procrustes analysis (GPA). For detailed information on the sensory evaluation of wine and brandy, see Chapter 40.

39.6 MATURITY EVALUATION AND GRAPE SAMPLING

39.6.1 Maturity Evaluation

Various fruit maturity gauges that should be considered include the following: assessment of varietal aroma/flavour and intensity of aroma/flavour (for example, changes from green herbaceous to fruit jam); colour; grape tannins (texture of skin tannins, suppleness); seed ripeness; sugar per berry; Brix, acidity and pH; general fruit condition, including skin extractability, berry softness, shrivelling; berry size/weight and ability to ripen further. It is important to view maturity evaluation in the context of stylistic goals. If the fruit does not contain varietal aroma/flavour character or is deficient in tannin, these characteristics will not be evident in the wine. See Chapter 9 for more information on grape maturity and wine quality.

39.6.2 Grape Sampling and Processing

39.6.2.1 Grape Sampling

Another important critical control point is accurate vineyard sampling. Grapes of the same variety and vineyard plot can, at a given moment, exhibit significantly different maturity profiles. The variation between the grower's estimation of maturity and the true value may have a significant and negative impact on the wine quality. Table 39.2 identifies sampling methods and the minimum quantity of fruit required for specific accuracies. Cluster sampling has the additional consideration of whether samples should be collected randomly from vines throughout the vineyard, or from one or more targeted vines.

The major factors influencing the rate of fruit maturity are heat, light, soil moisture and crop load. Of the variation in berry sampling, 90% is believed to come from the position of the cluster on the vine and the degree of sun exposure. The problem of collecting a representative sample is compounded by asynchronous berry, cluster or vine development. A crop with asynchronous maturity, for example, has a mixture of developmental stages, resulting in the portion of berries with optimal qualities being diluted by those that are inferior. The importance of asynchronous development is often overlooked, and can be a major factor in limiting the quality potential. The extent of uneven ripening can be determined by comparing the variations among sample replications.

39.6.2.2 Grape Sample Processing

The grape berry is frequently viewed as homogeneous when, in fact, it is not. Distinct fruit zones, which have different components, can be identified. Because of concentration gradients within the fruit, it is essential that growers and vintners standardise fruit sample processing. Without such standardisation, it is impossible to compare results.

39.6.2.3 Sensory Evaluation of Aroma/Flavour and Phenol Maturity

As fruit maturity increases, there is an increase in varietal character and a change in the tactile response arising from phenols. Aroma/flavour and tannin maturity may be evaluated using sensory methods (see Chapter 40 for details).

39.7 ANALYTICAL TECHNIQUES

39.7.1 Atomic Absorption and Inductively Coupled Plasma Emission (AA/ICP)

Flame atomic absorption (AA) and emission, as well as inductively coupled plasma (ICP) emission, is routinely used in wine analysis to detect the presence of metals, such as potassium, sodium, calcium, copper and iron. Electrothermal atomisation (graphite furnace) techniques allow the analyst to measure trace metals such as lead, cadmium and arsenic. In recent years, improvements, including more reliable platform surfaces in electrothermal vaporisation chambers and better auto-sampling techniques, have been made in AA techniques for the analysis of trace metals in wine. There are also ongoing attempts to improve the detection limits currently available with inductively coupled plasma methods. The primary attribute of a sensitive ICP method is its ability to perform simultaneous determinations of several metals in wine. Currently, detection limits are not adequate for the determination of certain important metals in the wine matrix. For example, aluminium levels in wine are generally below the

TABLE 39.2
Grape Sampling Considerations

Type	Number of sample	Method of collection
• Berry sampling		• Collect samples from both sides of the vine
±1.0°Brix	2 × 100 berries	• For each row, estimate the proportion of shaded bunches and sample accordingly
±0.5°Brix	5 × 100 berries	
• Cluster sampling		• Collect berries from top, middle and bottom of the cluster
±1.0°Brix	10 clusters	• Randomise the side of the cluster sampled
		• Maximum sample area should be less than 2 ha
		• Avoid edge rows and the first two vines in a row

normal working range of flame atomic absorption spectrometry, and above the linear working range for graphite furnace atomic absorption spectrometry. The analytical procedure requires dilution of the sample to reduce the aluminium concentration to within the linear working range of the graphite furnace technique. This also overcomes matrix effects in the analysis. Another AA method evaluates the concentrations of copper and iron. Two common techniques for the destruction of the organic matter present (dry and wet ashing) were tried; it was found that, while the precision and recovery data for both metals were not statistically different for the copper analysis, wet ashing produced low recovery values for iron. This is a subtle reminder that our analytical procedures are no better than the sample preparation schemes used to obtain prepared samples for analysis.

39.7.2 Gas Chromatography (GC)

Gas chromatography is routinely used for the determination of ethanol, fusel alcohols and methanol in wines and wine by-products, such as brandy. Various esters and other compounds, such as diethylene glycol, are analysed by GC for regulatory purposes. An analytical method for measuring the presence of aldehydes in wine samples has been developed, based on a derivatisation process involving the reactions of aldehydes with cysteamine (2-aminoethanethiol) at room temperature. A stable thiazolidine derivative is formed from each of the reacted aldehydes. The thiazolidine derivatives formed are extracted with chloroform, then quantified by GC, using a fused silica capillary column and a nitrogen-phosphorous detector (NPD). A series of short-chain, volatile and saturated aldehydes (C-1 to C-9) in wines, brandy and sherry can be quantitated. Compared to the results from the standard AOAC distillation/titration method for aldehydes in wine, the cysteamine derivatisation method is more accurate and gives higher recoveries.

With derivatisation chemistries, it is also possible to measure some generally non-volatile compounds such as malic and lactic acids. In the gas chromatographic method, Ce(IV)-oxidation of lactic acid to acetaldehyde, with subsequent headspace GC determination of acetaldehyde, is used. Wine carbohydrates have also been analysed by capillary GC of their acetates and aldononitrile acetates. A wide range of aldoses, polyols and disaccharides (29 compounds) can be analysed in a reasonably short time (<1 hour) using a single injection. While ribose and rhamnose co-elute, all of the other derivatives are well separated.

39.7.3 Gas Chromatography–Mass Spectrometry (GC/MS)

Gas chromatography–mass spectrometry is used in the determination of an increasing number of analytes of sensory and regulatory importance. These include ethyl carbamate, 4-ethylphenol, 2,4,6-trichloroanisole (TCA) and pesticide residues such as procymidone. GC/MS techniques have also been developed for the analysis of organic sulphides and other sensory compounds in wines. This is done to validate sensory examinations of wines with more specific, compound-based analytical information. TCA is known to contribute to a sensory characteristic known as cork taint or corkiness. The analytical detection threshold of this compound in wine is around two parts per trillion(ppt), thus presenting a significant analytical challenge.

GC/MS methods are also used for the analysis of 4-ethylphenol, which is associated with the presence of the spoilage yeasts *Brettanomyces/Dekkera* in wine. Volatile compounds in the hydroalcoholic extracts of French and American oak wood can also be analysed by means of GC/MS. Thirty-nine substances (including aldehydes, phenolic compounds, oak lactones, acids and furan derivatives) have been identified. Researchers have considered the possibility of using the relative concentrations of some of these compounds in wood in order to distinguish wood types. GC/MS has also been validated for the simultaneous analysis of *cis*- and *trans*-resveratrol in wine and juice. This method uses solid-phase extraction (SPE) of the resveratrol isomers on a C-18 column, followed by derivatisation with bis-[trimethylsilyl]-trifluoroacetamide. Selective ion monitoring is used (ion mass 444) for quantification, with the ions at mass 445 and 446 being qualifiers. This method requires only 1 mL of sample, has an instrument analysis time of 16 minutes and a detection limit of about 10 mg/L.

39.7.4 Near Infrared (NIR) Methods

Near-infrared spectroscopy techniques can determine ethanol in wine with equal or better accuracy than that attained using GC. NIR and, more recently, Fourier transform infrared (FTIR) methods have been developed for an increasing number of compounds present in grapes, must and wine. One of the advantages of Fourier transform instruments is that the entire infrared spectrum is collected simultaneously and transformed mathematically into a typical scanned spectrum. This allows for very rapid data collection and a quick throughput of samples. NIR and FTIR methods have been, and are being, developed for fermentable or reducing sugars, titratable acids (TAs), acids, sulphites, phenols, etc., in juices and wines. Since NIR/FTIR methods do not utilise reagents, they generate no waste. This development is particularly important given the increasing hazardous waste disposal requirements and costs for laboratories.

39.7.5 Solid-Phase Microextraction (SPME)

Solid-phase microextraction is a rapid, simple, solvent-free method for analysing volatile compounds. The technique combines extraction, concentration and chromatographic injection in one step, dramatically reducing labour, materials and waste disposal costs. Since its introduction a decade ago, the technique has rapidly grown in popularity for various applications. Because of the importance of volatile compounds to the aroma and flavour of wines and distillates, SPME combined with GC or GC/MS can be a significant tool for wine research and quality control analysis.

SPME was first developed as a technique to extract volatile compounds from liquid samples, by immersing a fused silica fibre coated with sorbent material into the sample, followed by desorption of the analytes in the injection port of a GC. Further development included placing the fibre into the headspace above the sample, in order to enhance mass transfer of the analytes into the sorbent, and the recognition that the sample matrix greatly affects the extraction. It has been found that salt and acid concentrations affected the response of volatiles, increasing the sample temperature caused faster equilibration and a lower response, and that the sample matrix affected the response. SPME–GC has been used for analysis of wine volatiles. The effects of temperature, time, salt concentration, ethanol concentration, pH and fibre type on the response of several terpenes using immersion sampling were examined. Immersion and headspace (HE-SPME) were compared with solvent extraction, and it was found that HE-SPME was preferable for the characterisation of a wide range of wine aroma compounds. HE-SPME conditions have also been evaluated and used to analyze wine for cork taint, sulphur compounds, methyl isothiocyanate, diacetyl, ethyl carbamate, methoxypyrazines and oak lactones.

39.7.6 High Performance Liquid Chromatography (HPLC)

High performance (or high pressure) liquid chromatography is a versatile analytical technique, due to the many separation mechanisms that can be utilised. In addition to the commonly used absorption mechanism (reversed phase C-18 columns), analysts can utilise ion exchange (ion chromatography [IC]), size exclusion, partition and ion exclusion techniques to tailor-make their required separations. The method has been successfully used in the separation and characterisation of various pigments, tannin groups and wine proteins. A number of other analytes, including fixed acids, alcohols and sulphur dioxide, have also been successfully measured using this technique. Many wine constituents are not volatile and are thus amenable to HPLC techniques. HPLC methods are also available for the determination of gluconic acid and for amino acids in musts and wines after derivatisation with phenylisothiocyanate or *ortho*-phthaldialdehyde.

Using alkaline phosphatase for the analysis of tannins in grapes and red wines, a HPLC method has been developed, wherein a microtiter plate assay (used for persimmon tannin) to detect grape and wine tannins was adapted. A normal phase HPLC separation of seed tannins provides the fractions subjected to analysis. The assay is easy to perform and requires only a spectrophotometer. Another separation technique used in HPLC is size exclusion, where molecules are separated on the basis of their size. The technique was used to analyse the soluble glycoproteins in red table wines. The proteins were isolated from the wine by precipitation with ammonium sulphate, fractionated by chromatography on Sephadex G-100 and characterised by preparative polyacrylamide gel electrophoresis (PAGE). Site exclusion was also used when the protein fraction of musts and wines was analysed by SDS-PAGE, isoelectric focusing (IEF) and HPLC. The protein fractions were recovered by HPLC and subjected to amino acid analysis.

39.7.7 Ion Chromatography

Another form of liquid chromatography, ion chromatography is a technique which uses ion exchange columns and ionic mobile phases. One method uses gradient ion chromatography, with conductivity detection and chemical suppression, to separate and quantify organic acids without prior sample clean-up. Tartaric, malic and citric acids can be separated on a Dionex OmniPac PAX-500 column, with an analysis time of 35 minutes. Results were comparable to other published colorimetric and enzymatic methods. In another application, the major anions in must were determined by IC with a conductivity detector. Inorganic anions were simultaneously analysed, with recoveries of the different anions ranging from 96.5 to 102.7%. A variation of a HPLC/MS method, using rapid qualitative and semi-quantitative determination of grape anthocyanins by extracting and concentrating from grape skins using solid-phase extraction has been developed. This is followed by analysis of extracts by a direct infusion electrospray injection into the mass spectrometer.

39.7.8 Matrix-Assisted Laser Desorption Ionisation Mass Spectrometry (MALDI)

An alternative approach was developed, employing matrix-assisted laser desorption ionisation mass spectrometry as a rapid and efficient screening technique for the presence and identification of anthocyanins in grape skins. The authors looked at the anthocyanin patterns of five red grape varieties and found that each exhibited a distinct mass spectrum. MALDI, coupled with time-of-flight mass spectrometry (MALDI-TOF), was used to obtain fast and accurate wine protein and peptide fingerprints. The wines were analysed by MALDI-TOF and surface-enhanced laser desorption/ionisation time-of-flight mass spectrometry (SELDI-TOF). It was found that most of the major proteins detected were common to all the wines analysed. In the SELDI-TOF technique, iminodiacetate-chelated copper ions were used to selectively bind proteins and peptides from the wine samples, effectively concentrating the minor components and permitting detection.

39.7.9 Flow Injection Analysis (FIA)

Flow injection analysis is an automatic, continuous flow technique for the determination of a large variety of analytes. Methods can be classified into two broad groups depending on whether or not they involve online separation or pre-concentration modes. Single, as well as multi-determinations, have been successfully developed. Flow injection analytical methods have demonstrated excellent sensitivity, selectivity, reproducibility and sampling frequency. With appropriate manifolds and working conditions, these methods can also be automated. The typical analytes detected include reducing sugars, tartaric, malic and lactic acids, sulphur dioxide and ethanol in juices and/or wines. Sampling frequencies of 50–70 per hour are routine with errors in precision (relative standard deviation) of less than 1%. Furfural has also been determined in a number of distillates, using the reaction between aniline and acetic acid with similar sample run times and precisions.

Metals have also been determined in wines by sequential injection analysis (SIA) using flame atomic absorption spectrometry as the detector system. Metals, including Zn, Mn, Fe and Cu in wines, is one such application of sequential injection analysis with flame atomic absorption spectrophotometric (SIA/FAAS) detection. Typical analyte concentrations of 0.2 mg/L and higher are routine. Trace levels of Cu have been pre-concentrated using commercial silica C-18 cartridges coupled to the system.

39.7.10 Capillary Electrophoresis (CE)

Capillary electrophoresis uses liquid mobile phases similar to those in HPLC, but obtains column flow from the osmotic forces developed in the system. Because capillary columns are utilised in this technique, very high efficiencies are obtained, which allow for very rapid and sensitive analyses. Capillary zone electrophoresis has also been used as a method for the analysis of wine proteins. In this method, both proteinaceous and non-proteinaceous ultraviolet (UV)-absorbing materials are detected when wine is analysed directly. If one obtains the protein fraction by ultrafiltration, it can be analysed separately. Using the capillary electrophoresis technique of micellar electrokinetic capillary chromatography (MECC), a rapid method has been developed for determining ethanol in wines. Sample times of less than 5 minutes (including capillary flushing, sample injection and analysis time), with a linear response over a range of 5–30% (v/v) ethanol, minimal interferences from glycerol and sugar and an average relative standard deviation of 1.38% can be obtained. This technique offers the speed of GC analysis and the versatility of CE, and has also been used for a number of other wine analytes.

Stable isotope analyses, especially site-specific natural isotope fractionation studied by nuclear magnetic resonance (SNIF-NMR), are physico-chemical methods for the characterisation of agricultural produce. The objective of these techniques is to provide impartial information on the composition and geographical origin of a variety of finished and raw materials in the food industry. Use of this technique in the wine industry has allowed the general area of production to be determined. To improve the accuracy of the SNIF-NMR method, trace element concentrations for five regions of France during the 1989 vintage were analysed. The trace element data for Zn, Ca, Sr and Mg, used in conjunction with stable isotope ratios, increased the overall classification from 78% (with isotope data only) to 89%.

Electrochemical methods have also been used for the analysis of juice, wines and distillates. Platinum *vs.* reference electrode systems have been used to follow the redox potential of a wine through processing and storage. Various selective ion electrodes have also been used in conjunction with FIA/SIA/HPLC analytical methods. Once an analyte (*e.g.* SO_2) is isolated, electrochemical methods (colorimetric and potentiometric) can be used for quantification. A method for the determination of lead, cadmium and zinc in table, sparkling and fortified wines by stripping potentiometry has been developed. Lead and zinc can be analysed in 1 minute or less. Due to its low concentration in wine, cadmium requires analysis times of up to 5 minutes per sample. The results compared favourably with those obtained by graphite furnace atomic absorption spectrophotometry (for lead and cadmium) and flame atomic absorption spectrophotometry (for zinc).

39.7.11 Spectrophotometry

Spectrophotometry has been a widely used technique in wine analysis. Classic colorimetric (visible region of spectrum) methods have been used to measure colour, phenolics, tartaric acid, nitrogen, laccase and other proteins. Newer analytical methods associated with enzymatic analyses, including nitrogen/arginine by *ortho*-phthaldialdehyde, have extended the range of wavelengths used in the UV. These wavelength regions are also used in detector systems for the HPLC, AA and ICP methods. Taking advantage of the selectivity of absorbance in the UV–visible spectroscopy (UV-vis) spectral regions has allowed researchers to better characterise the phenolic compounds in wines.

Many analytical methods utilising enzymes have been adapted for use with juices, wines and distillates. For most of these, an enzymatic reaction or a series of reactions is used to couple a particular analyte to a corresponding quantitative decrease in reduced nicotinamide-adenine dinucleotide (NADH). The initial and remaining amounts of NADH are then determined spectroscopically at 334, 340 or 365 nm. Although this third wavelength is on the edge of the visible, it is close to the 340 nm absorption band. But this wavelength does not work well with a simple visible colorimeter (*e.g.* Spectronic 20™); these methods work better using the other UV wavelengths given. Kits of reagents can be purchased from several suppliers. An interesting feature of these methods is that the reagents plus analyte are directly added to the sample cuvette, making the accuracy of the result dependent upon the ability to properly and accurately pipette volumes of sample and reagents as small as 0.01 mL. The exactness of these volumes is assumed when using the equation given (measured difference in absorbance, times a factor that includes the absorptivity, the volumes/concentrations used and the cell path length = concentration of analyte). On the positive side, these methods are quite selective and rapid to run for a large number of wine components. If one is concerned about accuracy, it is straightforward to run a standard. Typical kits measure close to 20 analytes (fixed and volatile acids, sugars, alcohols, urea/ammonia, acetaldehyde and sulphite) found in juices and wines. Reflectance spectrophotometers use emission photometry to measure the concentration of analytes. Systems are available that measure light intensity from enzyme-impregnated test strips. Light intensity is converted to analyte concentration. A variety of compounds measured by enzyme assay can be quantified by reflectance spectrophotometry.

39.8 GRAPE/WINE AROMA AND FLAVOUR COMPONENTS

More than 200 volatile aroma compounds have been identified in wines. Twenty to thirty common volatiles have been determined and studied, representing a broad diversity of organic compounds such as alcohols, organic acids, esters, terpenes, phenols, lactones, sulphur compounds, nitrogen compounds and heterocyclic compounds. Many of these volatiles are characteristic of specific varietal wines, and are desirable at high concentrations. Other organic volatiles, such as oak lactone, are introduced by subsequent processing to further improve the sensory quality of wines. Not all organic volatiles in wine contribute to its aesthetic characteristics. Some wine volatiles are detrimental to quality, including the mouldy scent of 2,4,6-trichloroanisole, associated with cork taint. *Brettanomyces* spp. contamination in the wine produces 4-ethylphenol and 4-ethylguaiacol, which at varying concentrations may be responsible for off-odours in wines. Other important volatiles, which do not necessarily contribute to the odour of wine, have also been measured. For example, analysis of ethyl carbamate, a natural fermentation product which may be carcinogenic, could easily become a priority when it becomes regulated in the future. Methyl isocyanate, a toxic fumigant, has also been successfully detected in wine using SPME and GC.

Grape-derived secondary metabolites are principal sources of wine aroma, flavour and colour. Aroma and flavour compounds are present as free volatiles and, in part, as sugar-bound non-volatile precursors, including glycosides. Glycosidically-bound compounds can undergo enzyme and/or acid-catalysed hydrolysis during processing and wine ageing to yield free volatiles. The majority of these are odourless, or possess only weak aromas or flavours. However, the sensory significance of some products of glycoside hydrolysis, relative to varietal wine aroma and flavour, has been established, providing justification and rationalisation for their quantification and study. A positive correlation between the concentration of grape glycoconjugates and wine quality has been suggested. As such, these conjugates have attracted much interest as precursors of aroma and flavour compounds in wine.

A variety of methods have been used to analyse the glycoside content of grapes and wines. Glycosides are isolated on the C-18 sorbent and, following hydrolysis, the resulting material is analysed by GC and GC/MS. Glycosides have also been extracted with XAD®-2 sorbent prior to hydrolysis and GC. Individual glycosides have also been isolated by counter current chromatography, and analysed by hydrolysis and GC/MS, NMR and MS, and HPLC followed by FAB-MS. A simple method for the determination of the total glycoside pool in a sample by quantitation of glycosyl-glucose (GG) was developed. The glycosyl-glucose method is based on the one-to-one molar ratio of glucose to glycosides found in grapes. The method involves the isolation of glycosides by C-18 solid-phase extraction, followed by hydrolysis and enzymatic determination of the released glucose. The method was further optimised. Because the glycoside fraction in red grapes is dominated by glycosylated anthocyanins, a spectrophotometric measurement of anthocyanins was also added, which was subtracted from the GG value to obtain an estimate of 'red-free' GG. The GG method was also modified by performing the glycoside extraction at pH 13, so that the phenolic glycosides would be ionised and not retained by the sorbent. This 'phenol-free' GG measurement (PFGG) is intended to give a more focused estimate of the pool of aroma- and flavour-active precursors. The GG methods have been employed to examine the effects of viticultural practices on grape quality, and to determine the effects of processing techniques on wine glycoside levels.

39.9 GRAPE AND WINE PHENOLICS

Variations in wine types and styles are largely due to the concentration and composition of wine phenols. From the vineyard to production and ageing, fine wines can be viewed in terms of the management of phenolic compounds. Phenols are responsible for red wine colour, astringency and bitterness; they may contribute to the olfactory profile, and serve as important oxygen reservoirs and as substrates for browning reactions. Grapes and wine contain a large array of phenolic compounds derived from the basic structure of phenol (hydroxybenzene). There is evidence that the extraction and retention of anthocyanin pigments are limited by the solubility of free anthocyanins and their co-pigmented forms. In varieties that typically have a light colour, there are usually no more than 150 mg/L of anthocyanins in the wine. To obtain the equivalent colour of 500–600 mg/L of anthocyanins, they must be co-joined to other non–pigmented compounds to form vertically stacked molecular aggregates. Once in stack form, the anthocyanins are more soluble, and are two to ten times more coloured. As the wine ages, the red-coloured anthocyanins combine with each other and other phenolic material to form polymeric pigments. These may be colourless or coloured red to yellow/brown. Polymeric pigments are not as sensitive to pH and SO_2 adjustment as are the free anthocyanins.

At any stage of a wine's development, only a small portion of the total pigment will be in the red-coloured form, reflecting the balance of the equilibrium in that particular wine. The absorbance measured at 520 nm gives an estimate of the concentration of red-coloured anthocyanins and red-coloured tannins present at natural wine conditions (*i.e.* at their actual wine pH and SO_2 levels). The addition of excess acetaldehyde, followed by the measurement of absorbance at 520 nm, provides an estimate of the concentration of red-coloured pigments (anthocyanins and tannins) present in the wine at that pH, but devoid of any bleaching effects of SO_2. A measure of the total red pigment colour is obtained by diluting a portion of the wine with 1 M HCl and reading the absorbance of the diluted sample at 520 nm. Under these low pH conditions, all the appropriate pigments are in the red-coloured form and all co-pigmentation is destroyed. The absorbance measured at 420 nm gives an estimate of the concentration of yellow/brown pigments under natural wine conditions. The

TABLE 39.3
Summary of Spectral Measurements

Wine colour density = $A_{520} + A_{420}$	Modified wine
	Colour density = $\left(A_{520}^{CH_3CHO} + A_{420}^{CH_3CHO}\right)$ pH 3.5
Wine colour hue = A_{420}/A_{520}	
Total red pigment colour = A_{528}	Modified wine colour hue = $\left(A_{420}^{CH_3CHO} + A_{520}^{CH_3CHO}\right)$ pH 3.5
Degree red	
Pigment colouration = $A_{520}^{HCl}/A_{520} \times 100\%$	
Total phenolics = A_{280} –4	Modified % red
	Pigment colouration = $\left(A_{520}^{CH_3CHO} + A_{520}^{CH_3CHO}\right)$ pH 3.5×100

TABLE 39.4
Summary of Co-pigmentation Calculations

Total phenols: A_{280} A_{20} – an estimation of momomeric and polymeric pigments	Fraction of colour due to co-pigmented anthocyanins: $(A_{acet} - A_{20})/A_{acet}$
Fraction of colour due to free anthocyanins: $A_{20} - A^{SO_2} A / A_{acet}$	Fraction of colour due to polymeric pigment: $A^{SO_2} A / A_{acet}$
Co-pigmented anthocyanin = $A_{acet} - A_{20}$	Total anthocyanin = $A_{20} - A^{SO_2}$
Polymeric pigment = A^{SO_2}	Flavone cofactor content = A_{365}

expression 'total phenols' takes into account all forms of phenolic compounds present in the wine.

Absorbance at 280 nm minus 4 gives an estimate of the concentration of all the phenolic material present in the wine. The subtraction of the value of 4 allows for the absorbance of non-phenolic material; however, not all procedures use this correction. The absorbance value (AV) for total phenols can be compared to the total phenolics measured by Folin–Ciocalteu reagent using the following relationship: $y = 29.5x + 210$, where y = total phenols expressed as gallic acid equivalents (GAE) mg/L, and x = total phenols expressed as 280–4 AV. All the absorbance values relate to a pathlength of 10 mm and include any dilution factors (*e.g.* with HCl measurements). The spectral measurements described are useful for comparing wines in their natural condition. However, when comparing wines with different pH and SO_2 values, interpretation can often be confusing. Since pH and SO_2 levels are adjustable in practice, it then appears logical that a comparison of the spectral measurements of the different wines should be made, not only in their natural conditions, but also under conditions of uniform pH and SO_2, thereby eliminating the effects of these variables on the spectral measures. Tables 39.3 and 39.4 summarise the spectral measurements of importance in defining wine colour.

39.10 MICROBIOLOGICAL IDENTIFICATION AND CHARACTERISATION IN WINE AND JUICE

Estimating the microbiological population density and diversity plays an important and often pivotal role at several junctures in the winemaking process. In recent years, microbiological characterisation and enumeration have evolved significantly, from classic and laborious bench-top methods to those capable of providing rapid and accurate responses. The development of newer methods has been driven by medical microbiologists but, increasingly, modified industrial applications are being developed. Microbiological critical control points in the winemaking process are summarised in Table 39.5. Estimating population density can be approached by either direct or indirect methods. Direct procedures have the advantage of providing an immediate response. A classic example of a direct technique is microscopic examination. More recently, flow cytometry (see below) has been used. Indirect procedures, by comparison, may require that a sample be collected, isolated and grown on solidified media. The number of colonies is then related to the number of individual viable cells in suspension. The presence of viable cells may also be detected and quantified by measurement of some metabolite indicative of active cell growth. One example is the lucerifase test for the contamination of sanitised surfaces.

39.10.1 Direct Estimations of Population Density and Diversity

39.10.1.1 Microscopy

Cell counting using bright-field, phase-contrast and fluorescence microscopy is used to estimate population density, preliminary and, in some cases, complete identification based upon cell morphology and the identification of unique cell surface antigens. Here, a volume of accurately diluted wine or

TABLE 39.5
Microbiological Critical Control Points in the Winemaking Process

CCP	Analytical quantification/characterisation
Harvest and transportation	
• Microbial populations	• Visual, HPLC, GC, FTIR, ELISA
Pre-fermentation processing	
• Crush pad, equipment, hose, cooperage sanitation	• Visual, olfactory, test kits, microbiological sampling
• Native *vs.* active dry yeast	• Microscopic, sensory
Yeast viability	
• Starter viability	• Microscopic
• Progress of fermentation	• Physical/chemical, microscopic methods, sensory
• Potential for stuck fermentation	• Yeast assimilable nitrogen concentration measurement, microscopic, sensory
Post-fermentation and malolactic fermentation	
• Bacterial starter preparation and viability	• Microscopic
• Completion of malolactic fermentation	• Chromatography (paper, enzymatic HPLC)
• Spoilage yeasts	• Plating, GC and GC/MS, ELISA
• Spoilage bacteria	• Visual, olfactory, wet chemical methods, GC and HPLC
Pre- and post-bottling	
• Packaging supplies QC	• Sensory (soak tests), GC and GC/MS, SPME
• Yeast and bacterial contamination	• Plating, epi- and immunofluorescent methods
• Case goods storage and distribution	• Recording thermographs, RH

juice is placed on the defined (gridded) area of a cell-counting slide. When the cover slip (provided in the kit) is applied, the volume of liquid is also defined. After several minutes to allow the cells to settle, cell counting is initiated. For counting and statistical purposes, the number of cells should typically lie within the range of 30–300 cells/field. Knowing the dilutions used in the original suspension and the volume on the slide, one can calculate cells/mL.

Although relatively rapid and inexpensive, the method has several significant pitfalls. Wine yeasts (*Saccharomyces* spp.) secrete a sticky capsule towards the end of fermentation that causes them to clump together in aggregations of a few to 20 or more cells. Unless separated, these present difficulties in both counting and statistical interpretation. Substantial effort is required to break apart cell aggregations and, thus, preparation may present a challenge. Within the laboratory, the consistency of the technique is crucial to successful microbial tracking. These concerns were addressed earlier. As a part of any total cell count, it may be necessary to separate viable from non-viable cells. Depending on the stage in the growth cycle, as well as the history of the sample, the ratio of viable to non-viable cells may vary considerably. Stains/dyes have traditionally been used to make this distinction. In wine/juice industries, the most widely used dye is methylene blue. Even when properly prepared, methylene blue is eventually toxic to the cell. Therefore, preparations must be examined within a few minutes after preparation or the percentage of cells scored as 'dead' significantly increases.

Epifluorescence microscopy is a cell-counting technique that relies on the fact that organic material, when stained with acridine dyes, absorbs light of one wavelength and re-emits light of another, a phenomenon known as fluorescence. Cells are collected on a polycarbonate membrane filter and stained with a fluorochrome dye, and the presence of fluorescing cells on a non-fluorescent membrane facilitates counting. Early methods required significant time to microscopically screen membranes. Modern methods utilise computer-assisted monitoring. Despite its laborious nature, reports indicate successful quantification, comparing favourably with plate count techniques. A major drawback of the method, however, is that acridine dyes react indiscriminately with organic materials, making the separation of viable from non-viable cells difficult. Interference has been observed from the wine preservative sorbic acid which, upon adherence to cells, fluoresces intermediate shades.

Immunofluorescent methods overcome many of the interpretational difficulties associated with the classic epifluorescent technique. Here, the dye is linked to species/strain-specific antibodies. Upon binding with the appropriate antigen at the cell surface, the fluorochrome tag can be visualised. Spectrophotometric methods for rapid population enumeration also have a long history. These are based on the principle that light passing through a suspension results in a portion being absorbed by the suspended material, and a portion scattered. Light scattering is proportional to the number of particles in suspension. This is also the basis of nephelometric analysis. Successful photometric applications in the microbiology laboratory require calibrating spectrophotometric results (in absorbency), with parallel samples that are plated, and colony-forming units (CFU) subsequently counted. Plotting one versus the other provides a basis for relating subsequent measurements of turbidity to cell density. For bacterial suspensions, a plot of absorbance versus cell mass is linear over much of the growth curve. However, two caveats must be considered. The linearity of the response is best at relatively low cell density and at shorter wavelengths within the visible spectrum. Most laboratories select 540 nm for such measurements. Since yeasts tend to aggregate, particularly towards the end of fermentation, the linearity of the response may be compromised, relative to that of bacteria. However, this is not the case for recently rehydrated yeast or young starter cultures. It is important to note that a clump of cells is equivalent to one colony-forming unit, when plating cultures.

39.10.1.2 Potentiometric Applications

For decades, attempts have been made to quantify microbial populations by measuring growth-associated changes in the electrical potential of a medium. Early versions utilised a set of electrodes to monitor changes in impedance when an alternating current was applied to the system. Here, monitoring relies on the fact that microbial utilisation of higher molecular weight uncharged species, such as carbohydrates, produces lower molecular weight products that may carry a charge, and that the electrical properties associated with utilisation are proportional to the cell density in a defined volume of medium.

Developed in the 1950s, the Coulter counter monitors changes in the electrical conductivity across a particulate suspension passing through a narrow-diameter (30 μm) glass orifice. The conductivity drops each time a microbe passes through the orifice. Since conductivity is proportional to cell number, quantification is based upon the difference between suspended particulates relative to that of the cell-free medium. An improved version (the Coulter Model B) measures the amplitude of the signal, which allows the microbiologist to both measure population density and sort cells by size distribution. The latter feature is useful in estimating population diversity and, in some cases, the stage in the life cycle. Classic methods utilising the electronic properties of cell suspensions are limited by the ability to measure only total cell density with, in some cases, estimation of size diversity within the population. Further, the signal improves with higher population densities and is thus of limited value for post-fermentation monitoring. Recent versions that measure the capacitance of the suspension have been developed that require neither sample dilution nor the use of dyes. Here, platinum electrodes create an electrical field in the sample cell. The capacitance of the intact cell membrane of a viable cell is different from non-viable cells or debris and thus can be detected. The signal is linear at cell densities ranging from 0 to more than 10^6 CFU/mL.

39.10.1.3 Flow Cytometry

Flow cytometry represents a significant advancement over classic photometric or potentiometric methods, in that the system can measure individual cells, one at a time, rather than populations. Additionally, viable and non-viable cells can be measured based upon differences in light-scattering properties. Immunofluorescent flow cytometry methods have also been developed that utilise highly specific reactions of monoclonal antibodies with target reactive groups (antigens) at the cell surface. Each antibody is coupled to a fluorescent tag which signals (fluoresces) upon reaction at the cell surface. This adaptation is useful for determining species or strain diversity within a cell population. An example of immunofluorescent techniques coupled with flow cytometry can be seen in Plate 39.1.

39.10.1.4 Microbial Identification

Classically, microbiologists characterise isolates by a combination of cell morphology, Gram stain reaction and a panel of tests to identify oxidative (assimilative) and fermentative utilisation of various carbon, nitrogen and other substrates. The results are scored with regard to growth and gas production. However, such methods are costly in terms of media and preparation time, as well as other expendables, and it may take from days to weeks to obtain results. One solution has been to utilise prepared panels, with panels of selected substrates that are pivotal in the separation of one species/strain from another. These may be as simple as 5–10 crucial substrates, to 90 or more. To facilitate handling, carbon and nitrogen substrates are pre-coated in wells on microtiter plates. A volume of suspended yeast or bacteria is then transferred to each of the wells and incubated. The results are read after 1, 2 and 3 days. Aside from increased turbidity indicative of growth, substrates can be prepared with a redox indicator, such as tetrazolium violet, that changes colour as the substrate is utilised. Although the results may be scored visually, most laboratories use computer-controlled microplate readers that scan wells photometrically (590 nm). Fully automated systems capable of carrying out the inoculation, incubation and scoring are also available. The results can be compared with large databases. Relative similarities versus differences are graphically compared using a dendrogram array (Figure 39.1).

39.10.1.5 Other Enzymatic Methods

Immunochemical techniques, such as enzyme-linked immunosorbent assay (ELISA), represent another rapid means of identifying microbes and other chemical substances. With detection levels of>10^4 CFU/mL, these semi-quantitative immunoassays utilise a 'sandwich' of the highly specific interactions between immobilised antibody and antigen (microbe) present in the sample, coupled with the addition of a secondary antibody–chromophore complex, which produces colour when the initial coupling is complete (Figure 39.2). The resultant colour development from the modified chromophore is proportional to the concentration (titre) of the primary antigen. The results (colour formation) can be read visually or via a colorimeter. Automated readers are also available. The ELISA methods that have been evaluated for use in the wine industry include the identification/enumeration of the spoilage yeast *Brettanomyces* and the detection of *Botrytis* on incoming grapes. Detection levels and the degree of cross-reactivity are within acceptable limits for most applications. More recently, the feasibility of commercialising the process has been investigated; however, monitoring kits are currently not available. Other applications of enzymatic methods

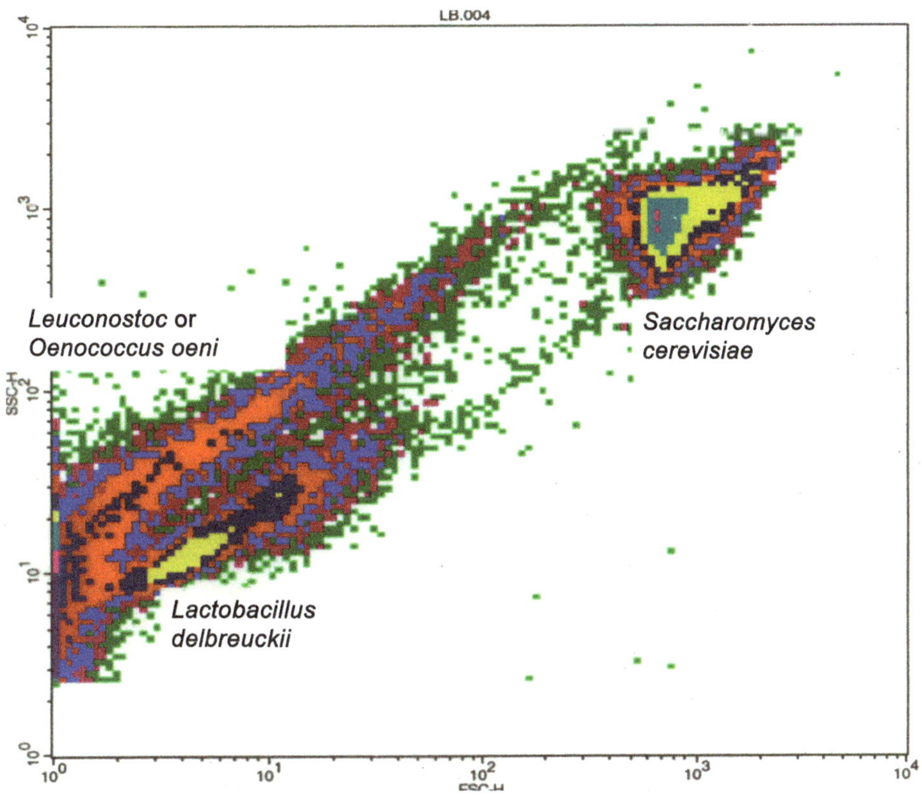

PLATE 39.1 Example of fluorescent-antibody technique for identifying species diversity (yeast and malolactic bacteria) in a fermentation sample. *Courtesy:* R. Thornton and S. Rodriguez

available to the wine microbiologist include *Botrytis* detection on grapes by laccase analysis. Unlike the ELISA methods, laccase analysis is available in kits for field assay.

39.10.1.6 Bioluminescent Techniques

Adenosine triphosphate (ATP), produced by living cells, rapidly degrades upon cell death. Therefore, measurement of the ATP concentration in a sample, theoretically, provides an estimate of the viable cell density. ATP, released upon boiling the sample in Tris buffer, is measured using the luciferin-luciferase assay. As seen in the following reaction, luciferin is oxidised in the presence of oxygen, the 'firefly' enzyme lucerifase, ATP and Mg^{2+} to the products noted below plus light:

$$\text{(a) Luciferin} + \text{enzyme} + \text{ATP} + Mg^{2+} \rightarrow \text{Luciferin} - \text{enzyme} - \text{AMP} + \text{ppi} \quad (39.1)$$

$$\text{(b) Luciferin} - \text{enzyme} - \text{AMP} + O_2 \rightarrow \text{oxyluciferin} + \text{enzyme} + \text{AMP} + CO_2 + \text{light} \quad (39.2)$$

The light produced is proportional to the amount of ATP present in the original sample. The results are measured using a luminometer. The method is reportedly very reproducible, and capable of detecting viable yeast cells at levels of 100 CFU/mL. For greater sensitivity (<5 CFU/mL), the sample must be incubated for several hours at optimal growth temperature. Concerns associated with bioluminescence methods include the reliability of the results. Since ATP production is related to cell physiological status, the results depend on the stage of growth. The results from populations in the early or decline phases may be distinctly different from samples taken during the growth phase. Further, in any stage of growth, ATP production will also be related to population diversity. During the growth phase, yeasts produce much more ATP than relatively slow-growing facultative anaerobes, such as lactic acid bacteria. Lastly, ATP may not degrade as rapidly as originally believed. The presence of measurable ATP after cell death may also pose interpretational problems. Despite the deficiencies noted, the application of ATP-bioluminescence techniques for sanitation monitoring, where results are required rapidly, continues to be a valuable tool.

39.10.1.7 Molecular Methods

Over the last 30 years, numerous approaches have evolved in attempts to characterise microbes, based on fundamental similarities or differences (polymorphism) in their genomes. These involve direct comparison at the gene level, or secondarily characterising proteins encoded by those gene(s). In either case, nucleic acids (DNA and RNA), as well as the proteins that they encode, are extracted, amplified and subsequently separated. The results are then compared to those from reference species, or to available databases.

Three strategies have evolved to directly compare similarities/differences between isolates by examining their respective genomes: (1) DNA harvested from isolates is digested using restriction enzymes to yield variously sized DNA

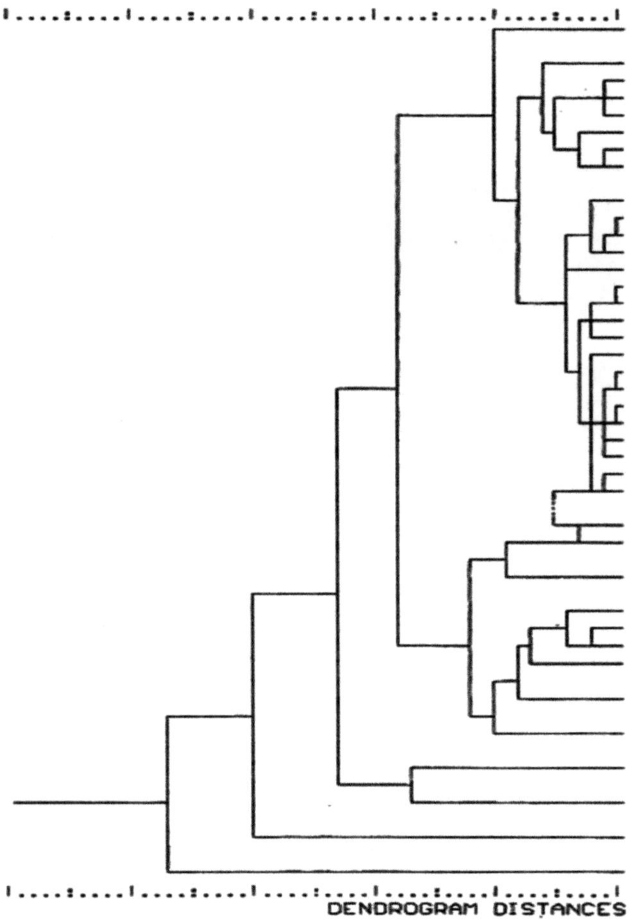

FIGURE 39.1 Dendrogram array based on the utilisation of carbon and nitrogen substrates. Each division is equivalent to the utilisation of five substrates. *Source:* Courtesy R. Thornton

fragments; (2) amplification of specific or randomly selected regions by polymerase chain reaction (PCR); and (3) nucleotide sequencing of selected amplified areas. In the first two, fragments are then separated electrophoretically and patterns compared against those of other isolates or databases.

39.10.1.7.1 Polymerase Chain Reaction®

Selected segments of a target nucleic acid may be amplified by several orders of magnitude and characterised in a matter of hours. Generally, the procedure involves thermal denaturation of native target DNA, binding two DNA primers (or 'probes') and replicating each strand using temperature-resistant DNA polymerase. Since each newly synthesised portion can, itself, be replicated, target DNA doubles at each cycle (Figure 39.3). PCR and electrophoretic separation have become an integral part of genomic characterisation.

39.10.1.7.2 Nucleic Acid Hybridisation Probes

DNA consists of millions of nucleotide sequences, each composed of four bases – guanine, cytosine, adenine and thymine – in an array unique to the species. Fundamental differences or similarities between organisms can be directly derived by characterisation of specific nucleotide sequences (<100 to several thousand bases) within the genome. Once a compatible probe has been developed, the fidelity of the probe–target interaction under controlled test conditions creates a highly specific diagnostic test system for comparison/identification of related strains or specific gene sequences within a single microbe.

39.10.1.7.3 Protein Characterisation

Proteins are the unique products of gene expression. They are either produced on a continuing basis (constitutive proteins) reflecting the ongoing or 'housekeeping' needs of all biological systems, or they are synthesised in response to specific environmental stimuli (inducible proteins) that may be unique to specific or closely related strains. Hence, they are 'fingerprints' of the nucleic acid segment (gene) that encoded them and, by extrapolation, the nucleic acids. Once extracted from the cell, protein separation is effected using 1- or 2-dimensional polyacrylamide gel electrophoresis (PAGE). A comparison of separation patterns can be used to identify similarities or differences between the isolates. Separation patterns may be compared visually or by computer-assisted technology. To facilitate detection, methods have been developed whereby antibody probes are hybridised to the protein(s) of interest. A comparison of bands with those of reference organisms provides useful information as to their relatedness. Recently, rapid analysis of protein and peptide fragments has been made possible by coupling separation techniques with mass spectrometry (see earlier discussions). Protein isolated from 1- and 2-D gels can be further analysed by MALDI-mass spectrometry. This greatly enhances sensitivity (femto- to picomole range) compared to traditional methods. The results can be compared with genomic databases. In that the expression of specific proteins may be closely tied to the physiological status of the cell, the conditions of cultivation must be carefully controlled.

Pulsed field gel electrophoresis (PFGE) utilises restriction enzymes to digest microbial DNA, which is then subject to electrophoretic separation. The method separates 10–15 high molecular weight (10–800 kb) fragments which may not be sufficient to resolve differences between the strains.

39.10.1.8 Random Amplified Polymorphic DNA-Polymeric Chain Reaction(RAPD-PCR)

RAPD-PCR utilises smaller oligonucleotide probes (9–12 bp) to amplify several regions of the genome as well as ribosomal and mitochondrial nucleic acids. The amplification products are then electrophoretically separated. The resolution depends upon the primer sequence and the reaction conditions. RAPD-PCR can be made more specific by utilising highly specific, instead of oligonucleotide, probes. Such probes may include poly GT nucleotides, which are of value in identifying genomic loci or 'hot spots' appropriate for recombinant work. In yeasts, ribosomal genes are organised as three tandemly repeated (50–200×) operons separated by two internal transcribed spacers (ITS). ITS regions exhibit relative variability within a relatively highly conserved system. Hence, they are potentially useful as phylogenetic markers for speciation purposes within the *Brettanomyces/Dekkera*

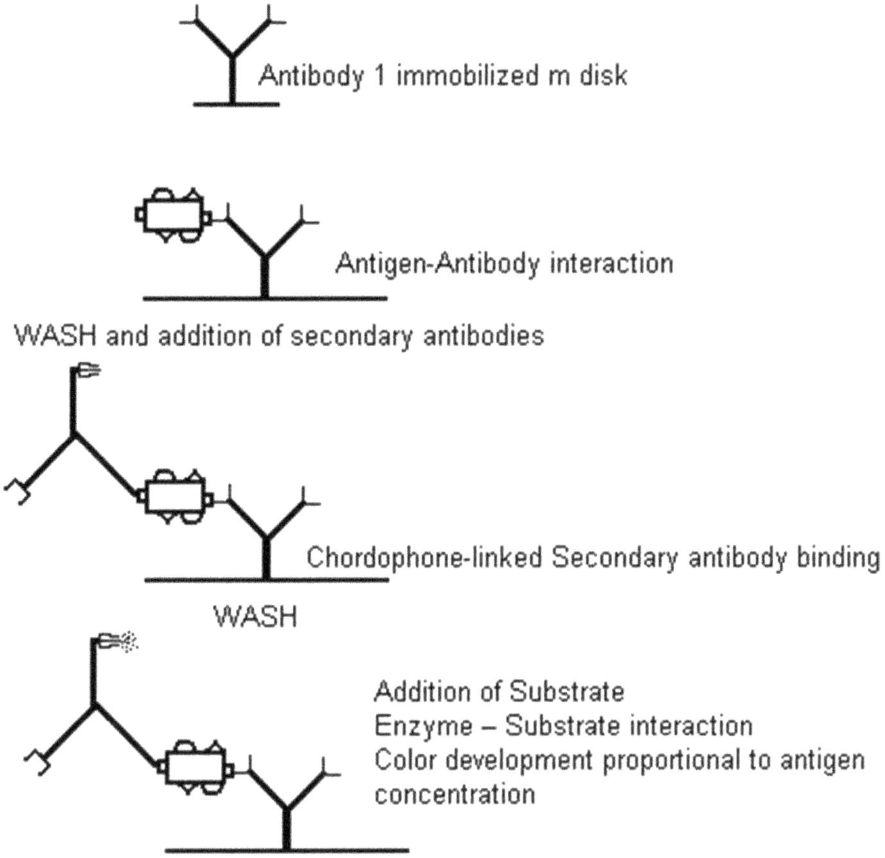

FIGURE 39.2 Enzyme-linked immunosorbent assay (ELISA)

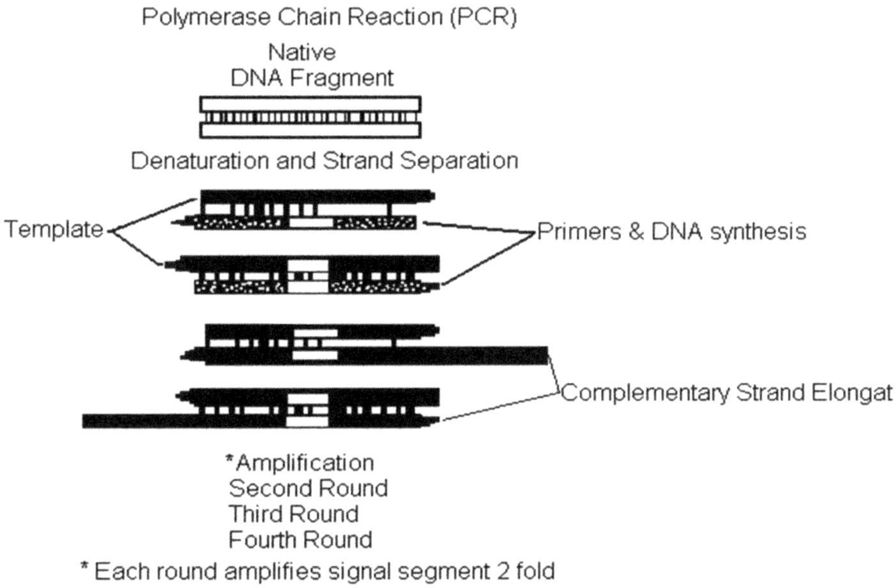

FIGURE 39.3 Polymerase chain reaction® (PCR)

group of yeasts. Species-specific peptide nucleic acid (PNA) probes complementary to 26S r-RNA have been developed for *Dekkera bruxellensis*. PNA probes can be coupled with innovative filter-based chemiluminescent hybridisation techniques (PNA-CISH) for rapid *in situ* detection and identification of *Brettanomyces* from winery air samples.

39.10.1.9 Repetitive Sequence-Based PCR (REP-PCR)

REP-PCR can be utilised to detect microbial genomes that have repetitive DNA sequences. Its value is limited only to regions of compatibility (average 10–15), and other polymorphic sites are missed.

39.10.1.10 Multi-Locus Sequencing Typing (MLST)

Several bacterial 'housekeeping' genes are compared on the basis of 450 bp internal fragments, resulting in each isolate being assigned a unique seven-digit allele combination. Interpretation is based upon the probability that identical allelic profiles will be detected by chance alone.

39.10.1.11 Extrachromosomal Elements (Satellites)

A wide range of living cells, including wine bacteria and yeast, are infected with extra-chromosomal nucleic acid elements called satellites or variable number of tandem repeats (VNTRs). Their importance to the host cell varies. Among mammalian cells, some encode for elevated expression of medically important proteins. Among yeast and bacteria, their role is less clear. One proposed role is in the formation of a killer protein among strains of *Saccharomyces cerevisiae* (see Chapter 17 for details). As such, these intracellular subviral elements are present as nucleic acid fragments, either single-stranded DNA or single-stranded or double-stranded RNA. Among animal cells, they may recur as tandem repeats (TR) of 15–30+ base pairs, which may be copied many times along the nucleic acid strand. As suggested, satellites do not contain sufficient nucleic acid to code for the proteins necessary to carry out their replicative function. Rather, they rely upon another 'helper' virus to carry out this function. Where the second virus doesn't co-exist, the satellite is, in essence, trapped in its host cell. In yeast, satellite nucleic acid is relatively miniscule, hence, these fragments are referred to as 'micro-satellites'. They exist as simple sequence tandem repeats (SSTR) of chromosome-associated DNA, ranging in size from 2–5 nucleotides, and capped on both sides by flanking regions composed of unordered DNA 30–50 base pairs in length. Since they are chromosome associated, replication is tied to the host. As the nucleic acid sequence is distinct from that of the host chromosome, micro-satellites are being evaluated as genetic markers and, thus, identification tools. However, micro-satellite loci exhibit significant variability, resulting from a high rate of mutation. Further, not all tandem repeats are polymorphic. As such, care must be taken in using these 'markers' as the basis of relatedness. At present, the separation and identification of satellite proteins require a suitably equipped and staffed molecular biology laboratory. However, methods may eventually be reduced to 'kit form', where most labs will have the capability to rapidly identify wine microbes using their satellite information.

39.11 THREE MICROBIOLOGICAL ISSUES OF IMPORTANCE TO THE WINEMAKER

39.11.1 *BRETTANOMYCES*

Worldwide, *Brettanomyces* species have historically been responsible for wine spoilage, and are responsible for costing the wine industry millions of dollars annually. The yeast can cause wines to develop unpleasant odours that have been variously described as band-aid, ammonia, mouse droppings, burnt beans and barnyard like. Despite this, some winemakers suggest that the presence of *Brettanomyces* spp. may contribute to a wine's complexity or accelerate wine ageing. Although *Brettanomyces bruxellensis* (formerly *B. intermedius*) is the species most frequently identified, most wines suspected of being contaminated with *Brettanomyces* spp. have not been adequately characterised in terms of the strain(s) involved. A significant variation in both the growth rate and stationary-phase population densities among eight *Brettanomyces* strains has been noted. Significant increases in the concentration of the 'marker' metabolite, 4-ethylphenol, occurred only after accumulated cell populations reached 2.5×10^5 CFU/mL. *Brettanomyces*-inoculated wines were found to have detectable concentrations of ethyl-2-methylbutyrate, isoamyl alcohol, ethyldecanoate, isovaleric acid, guaiacol, 2-phenylethanol, 4-ethylguaiacol and 4-ethylphenol, with significant differences in their concentrations among strains.

39.11.2 FERMENTABLE NITROGEN

Nitrogen is a crucial macro-nutrient found in grape juice that can impact the growth of *S. cerevisiae*, the conversion rate of sugar to alcohol and the formation of odour and flavour-active metabolites. Nitrogen deficiency slows yeast growth and may result in protracted or stuck fermentations (see Chapter 18 for more details), possibly due, in part, to the inhibition of sugar transporter synthesis. Juice nitrogen status may also play a role in the formation of fermentation aroma compounds, such as hydrogen sulphide, higher alcohols, esters and organic acids, and can impact the formation of ethyl carbamate. Fruit nitrogen can be influenced by grape-growing and winemaking parameters, including rootstock, climate and soil, fertilisation, irrigation practices, fruit maturity and grape variety. Of the various nitrogenous compounds found in grapes (amino acids, ammonium ions, peptides, proteins, nitrates and trace concentrations of vitamins), the only forms metabolically available to yeasts are ammonium salts and primary or free alpha-amino acids (FAN). The FAN fraction includes arginine, serine, threonine, lysine, gamma-amino butyric, aspartic and glutamic acids. Collectively, the concentrations of FAN and ammonium ions present in juice are referred to as yeast-assimilable nitrogen compounds (YANC). Levels of approximately 140 mg/L N have been suggested as minimal for satisfactory fermentation. This number is somewhat a function of the method of analysis. Rapid, accurate and precise analytical methods for assimilable nitrogen, involving simple sample preparation, minimal waste products and minimal instrumentation, would be valuable tools for winemakers. Currently, several analytical methods are in common use. The nitrogen by *ortho*-phthaldialdehyde (NOPA) procedure has been used to determine FAN nitrogen by derivatisation of primary amino groups with *o*-phthaldialdehyde. The resulting isoindole derivatives can be measured spectrophotometrically at 335 nm. As a result of the specificity of this reaction to primary amino acids, the amino acid proline cannot form a derivative and, therefore, does not contribute to the results. Ammonium ion, the second main source of assimilable nitrogen, is not measured by the NOPA procedure. A modification of NOPA, ARGOPA allows for the selective determination of arginine, quantitatively the most significant contributor to yeast nutrition and potential ethyl carbamate formation. This

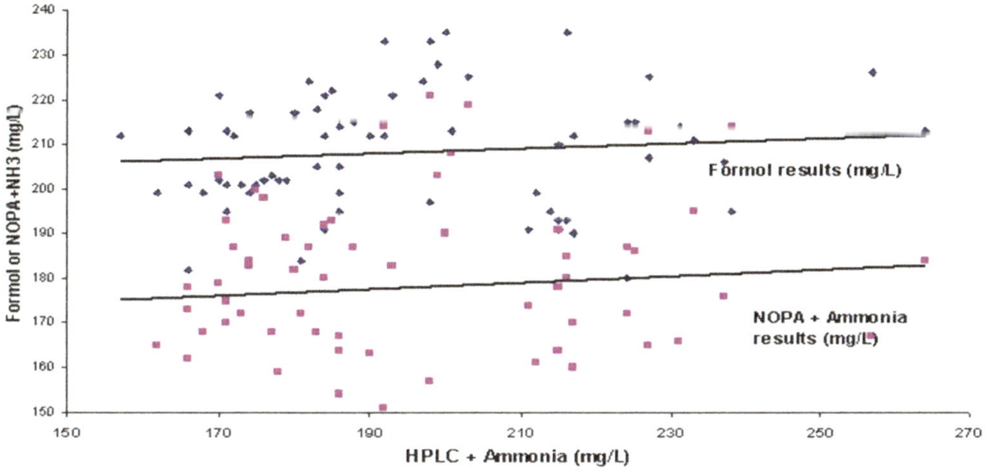

PLATE 39.2 Correlation between formol and HPLC plus NH_4^+, and NOPA plus NH_4^+, and HPLC plus NH_4^+

procedure uses cation exchange to isolate arginine from other amino acids, prior to analysis by NOPA. Formol titration is a simple and rapid determination of assimilable nitrogen. This method involves the addition of neutralised formaldehyde to liberate protons that are titrated directly with NaOH to a pH 8.2 end point. The analysis provides an approximate, but useful, index of the nutritional status of juice.

High performance liquid chromatographic methods have also been developed for quantifying individual amino acids. With automated, online derivatisation, these methods can result in rapid, accurate, sensitive and reproducible analyses of amino acids. It has been demonstrated that formol titration is positively correlated with NOPA plus ammonia (Plate 39.2), by enzymatic analysis, with a correlation coefficient of 0.82. The formol analysis method results in a 28.5% higher nitrogen value than the NOPA procedure, and an 18.4% higher value than NOPA plus ammonia. Both the NOPA and formol methods measure/titrate only one of arginine's four nitrogens. Since arginine hydrolyses to ornithine and urea, providing two assimilable nitrogens to the yeast, both methods understate the arginine content of a juice sample. The formol method compensates for this by partially titrating proline. While the numbers obtained may differ somewhat, both the NOPA and the formol methods provide rapid and useful estimations of the nutritional status of juice. The ARGOPA and HPLC methods provide measurements for arginine and other specific amino acids, respectively. As a result of its precursor role, arginine is of importance in ethyl carbamate formation. The ability of the HPLC method to provide measures of the entire suite of amino acids allows a better understanding of yeast fermentation dynamics and the impacts of vineyard and processing practices on grape amino acid composition (Plate 39.2).

39.11.3 Ethyl Carbamate

Ethyl carbamate is a natural by-product of fermentation. In 1985, a significant concentration of this human carcinogen was reported in Canadian-produced dessert wines. The US wine industry established a voluntary target for ethyl carbamate of 15 μg/L or less in table wines, and 60 μg/L or less in fortified wines. The US Food and Drug Administration is expected to establish recommendations for minimising ethyl carbamate in wine. The formation of ethyl carbamate in wine is related to urea and ethanol concentrations, and to temperature. The rate of formation increases exponentially with increased wine storage temperature. Urea is the principal precursor, and attempts to control the concentration of ethyl carbamate by controlling urea and the concentration of its precursor, arginine, have received significant attention. Winemaking factors which influence the levels of urea, and hence ethyl carbamate, include yeast selection, nitrogen supplementation, fermentation and wine storage temperature and, in the case of dessert wines, timing of fortification. Gas chromatography is the most commonly used method for determining ethyl carbamate in wine. The extraction of ethyl carbamate by methylene chloride, using solid-phase extraction columns, followed by GC/MS with selected ion monitoring (SIM), has been adopted by the AOAC for the analysis of alcoholic beverages. Alternative extraction procedures that have been used include steam distillation, followed by solvent extraction of the distillate, liquid–liquid extraction with ethyl ether and continuous extraction with ethyl ether. The common characteristics of all of these methods are that they are time-consuming and use large amounts of solvents. There is a rapid headspace solid-phase microextraction (HS-SPME) GC/MS method, using a carbowax/divinylbenzene fibre.

DEDICATION

This effort is dedicated to the memory of Professor Ken Fugelsang, friend, colleague and wine educator extraordinaire.

BIBLIOGRAPHY

Abbott, N.A., Williams, P.J. and Coombe, B.G. (1993). Measure of potential wine quality by analysis of glycosides. In: *Proceedings of the Eighth Australian Wine Industry Technical Conference*, C.S. Stockley, R.S. Johnstone, P.A. Leske, and T.H. Lee (Eds.), Winetitles, Adelaide, South Australia, p. 72.

Adams, D.O. and Harbertson, J.F. (1999). Use of alkaline phosphatase for the analysis of tannins in grapes and red wine. *Am. J. Enol. Vitic.*, 50: 247.

AOAC. (1995). *Official Methods of Analysis of AOAC International.* 16th edn. Vol. II. Ch. 28. Method 994.07. AOAC, Gaithersburg, MD.

Arthur, C.L. and Pawliszyn, J. (1990). Solid phase microextraction with thermal desorption using fused silica optical fibers. *Anal. Chem.*, 62(19): 2145.

Atlas, R.M. and Bartha, R. (1981). *Microbial Ecology: Fundamentals and Applications.* Addison-Wesley, Reading, MA.

Baumes, R., Cordonnier, R., Nitz, S. and Drawert, F. (1986). Identification and determination of volatile constituents in wine from different vine cultivars. *J. Sci. Food Agric.*, 37(9): 927.

Bely, M., Sablayrolles, J.M. and Barre, P. (1990). Automatic detection of assimilable nitrogen deficiencies during alcoholic fermentation in oenological conditions. *J. Ferment. Bioeng.*, 70(4): 246.

Buchholz, K.D. and Pawliszyn, J. (1994). Optimization of solid-phase microextraction conditions for determination of phenols. *Anal. Chem.*, 66(1): 160.

Buglass, A.J. and Garnham, S.C. (1991). A novel method for determination of lactic acid. Comparison lactic acid content of English and northern European wines. *Am. J. Enol. Vitic.*, 42: 63.

Butzke, C.E. and Austin, K.T. (1999). Concentration of arginine in grape juice: ARGOPA procedure. University of California (Davis) Cooperative Extension. http://wineserver.ucdavis.edu/oldsite/argopa99.pdf.

Butzke, C.E. and Bisson, L.F. (1997). *Ethyl Carbamate Preventive Action Manual.* US Food and Drug Administration, Center for Food Safety and Applied Nutrition. http://vm.cfsan.fda.gov/~frf/ecaction.html.

Butzke, C.E., Evans, T.H. and Ebeler, S.E. (1998). Detection of cork taint in wine using automated solid phase microextraction in combination with GC/MS-SIM. In: *Chemistry of Wine Flavor, ACS Symposium Series* 714, A.L.Waterhouse and S.E.Ebeler (Eds.), American Chemical Society, Washington, DC, p. 208.

Carevell, J.P. and Oddi, L. (2002). Using capacitance for rapid and reproducible estimation of live yeast cell mass during fermentation. *Austral. N. Z. Grapegrower Winemaker*, 466: 56.

Chatonnet, P., Dubourdieu, D., Boidron, J.N. and Pons, M. (1992). The origin of ethylphenols in wines. *J. Sci. Food Agric.*, 60(2): 165.

Chen, G.N., Scollary, G.R. and Vicente-Beckett, V.A. (1994). Potentiometric stripping determination of lead, cadmium, and zinc in wine. *Am. J. Enol. Vitic.*, 45: 305.

Cocolin, L., Bission, L.F. and Mills, D.A. (2000). Direct profiling of the yeast dynamics in wine fermentations. *FEMS Microbiol. Lett.*, 189(1): 81.

Collins, T.S., Miller, C.A., Altria, K.D. and Waterhouse, A.L. (1997). Development of a rapid method for the analysis of ethanol in wines using capillary electrophoresis. *Am. J. Enol. Vitic.*, 48: 280.

Connell, L., Stender, H. and Edwards, C.G. (2002). Rapid detection and identification of *Brettanomyces* from winery air samples based on peptide nucleic acid analysis. *Am. J. Enol. Vitic.*, 53: 322.

Costa, R.C.C., Cardoso, M.I. and Araújo, A.N. (2000). Metals determination in wines by sequential injection analysis with flame atomic absorption spectrometry. *Am. J. Enol. Vitic.*, 51: 131.

Day, M.P., Zhang, B.-L. and Martin, G.-R.J. (1994). The use of trace element data to complement stable isotope methods in the characterization of grape musts. *Am. J. Enol. Vitic.*, 45: 79.

De la Calle Garcia, D., Magnaghi, M., Reichenbacher, M. and Danzer, K. (1996). Systematic optimization of the analysis of wine bouquet components by solid-phase microextraction. *J. High Resol. Chromatogr.*, 19(5): 257.

De la Calle Garcia, D., Reichenbacher, M., Danzer, K., Hurlbeck, C., Bartzsch, C. and Feller, K.H. (1997). Investigations on wine bouquet components by solid-phase microextraction-capillary gas chromatography using different fibers. *J. High Resol. Chromatogr.*, 20(12): 665.

De la Calle Garcia, D., Reichenbacher, M., Danzer, K., Hurlbeck, C., Bartzsch, C. and Feller, K.H. (1998). Analysis of wine bouquet components using headspace solid-phase microextraction: Capillary gas chromatography. *J. High Resol. Chromatogr.*, 21(7): 373.

Dennis, M.J., Howarth, N., Massey, R.C., Parker, I., Scotter, M. and Startin, J.R. (1986). Method for the analysis of ethyl carbamate in alcoholic beverages by capillary gas chromatography. *J. Chromatogr.*, 369: 193.

Dizy, M. and Bisson, L.F. (1999). White wine protein analysis by capillary zone electrophoresis. *Am. J. Enol. Vitic.*, 50: 120.

Dubourdieu, D., Grassin, C., Deruche, C. and Ribereau-Gayon, P. (1984). Mise Au point d'une measure rapide de l'activite laccase, dans les mouts et dan les vins par methode a la syringaldizine. Application a l'appreciation de l'estat sanitaire envendages. *Conn. Vigne Vin*, 18(4): 237.

Dufour, C. and Bayonnove, C. (1999). Influence of wine structurally different polysaccharides on the volatility of aroma substance in model systems. *J. Agric. Food Chem.*, 47(2): 671.

Dukes, B.C. and Butzke, C.E. (1998). Rapid determination of primary amino acids in grape juice using an *o*-phthaldialdehyde/N-acetyl-L-cysteine spectrophotometric assay. *Am. J. Enol. Vitic.*, 49: 125.

Eddy, T. (2001). The production/marketing interface. In: *Successful Wine Marketing*, M. Lewis, and T. Young (Eds.), Aspen Publishers, Gaithersburg, MD, p. 57.

Egli, C.M. and Henick-Kling, T. (2001). Identification of *Brettanomyces/Dekkera* species based on polymorphism in the rRNA internal transcribed spacer region. *Am. J. Enol. Vitic.*, 53: 241.

Etievant, P., Schlich, P., Bouvier, J.-C., Symonds, P. and Bertrand, A. (1988). Varietal and geographical classification of French red wines in terms of elements, amino acids and aromatic alcohols. *J. Sci. Food Agric.*, 45(1): 25.

Fauhl, C. and Wittkowski, R. (1992). Determination of ethyl carbamate in wine by GC-SIM-MS after continuous extraction with diethyl ether. *J. High Resolut. Chromatogr.*, 15(3): 203.

Favretto, D. and Flamini, R. (2000). Application of electrospray ionization mass spectrometry to the study of grape anthocyanins. *Am. J. Enol. Vitic.*, 51: 55.

Ferreria, M.A. and Fernandes, J.O. (1992). The application of an improved GC-MS procedure to investigate ethyl carbamate behavior during the production of Madeira wines. *Am. J. Enol. Vitic.*, 43: 339.

Fleet, G.H. (Ed.). (1993). *Wine Microbiology and Biotechnology.* Harwood Academic Publishers, Chur, Switzerland, p. 395.

Fugelsang, K.C. (1997). *Wine Microbiology.* Chapman and Hall, New York. p. 245.

Fugelsang, K.C. and Zoecklein, B.W. (2003). Population dynamics and effects of *Brettanomyces bruxellensis* strain on Pinot noir (*Vitis vinifera* L.) wines. *Am. J. Enol. Vitic.*, 54(4): 294.

Funch, F. and Lisbjerg, S. (1988). Analysis of ethyl carbamate in alcoholic beverages. *Z. Lebensm.-Unters. Forsch.*, 186: 29.

Gandini, N. and Riguzzi, R. (1997). Headspace solid-phase microextraction of methyl isocyanate in wine. *J. Agric. Food Chem.*, 45: 3092.

Gower, J.C. (1973). Generalized Procrustes analysis. *Psychometrika*, 40(1): 33.

Gump, B.H. (Ed.). *Beer and Wine Production: Analysis, Characterization, and Technological Advances, ACS Symposium Series* 536. American Chemical Society, Washington, DC, p. 110.

Gump, B.H. and Kupina, S.A. (1979). Analysis of gluconic acid in Botrytised wines. In: *Liquid Chromatographic Analysis of Food and Beverages*, Vol. 2, G. Charalambous (Ed.), Academic Press, New York.

Gump, B.H., Saguaryteekul, S.M., Murray, G. and Villar, J.T. (1995). Determination of malic acid in wines by gas chromatography. *Am. J. Enol. Vitic.*, 36: 248.

Gump, B.H., Zoecklein, B.W. and Fugelsang, K.C. (2002). Prediction of prefermentation nutritional status of grape juice: The formol method. In: *Methods in Biotechnology: Food Microbiology Protocols*, J.F.T. Spencer, W.L. Whelan, J. Brown, and A. Ragout de Spencer (Eds.), Humana Press, Inc., Totowa, NJ, p. 283.

Gunata, Y.Z., Bayonove, C.L., Baumes, R.L. and Cordonnier, R.E. (1985). The aroma of grapes. I. Extraction and determination of free and glycosidically bound fractions of some grape aroma components. *J. Chromatogr.*, 331: 83.

Hartmann, P.J., McNair, H.M. and Zoecklein, B.W. (2002). Measurement of 3-alkyl-2-methoxypyrazine by headspace solid-phase microextraction in spiked model wines. *Am. J. Enol. Vitic.*, 53: 285.

Hayasaka, Y. and Bartowsky, E.J. (1999). Analysis of diacetyl in wine using solid-phase microextraction combined with gas chromatography–mass spectrometry. *J. Agric. Food Chem.*, 47(2): 612.

Henderson, J.W., Ricker, R.D., Bidlingmeyer, A. and Woodward, C. (2000). *Rapid, Accurate, Sensitive, and Reproducible HPLC Analysis of Amino Acids: Amino Acid Analysis Using Zorbax Eclipsee-AAA Columns and the Agilent 1100 HPLC*. Agilent Technologies, Wilmington, DC.

Hermandez, P., Orte, N., Guitart, A. and Cacho, J. (1997). HPLC after derivatization with phenylisothiocyanate. *Am. J. Enol. Vitic.*, 48: 229.

Hock, S. (1990). Coping with *Brettanomyces*. *Pract. Winery Vineyard*, 10(5): 26.

Iland, P.G., Cynkar, W., Francis, I.L., Williams, P.J. and Coombe, B.G. (1996). Optimisation of methods for the determination of total and red-free glycosyl glucose in black grape berries of *Vitis vinifera*. *Austral. J. Grape Wine Res.*, 2(3): 171.

Iland, P.G., E'Jart, A., Sitter, J., Markides, A. and Bruer, N. (2000). *Technique for Chemical Analysis and Quality Monitoring during Winemaking*. Tony Kitchener Printing, Adelaide, p. 111.

Ingledew, W.M. (1996). Nutrients, yeast hulls and proline in wine fermentation. *Wein Wiss.*, 51: 141.

Jordan, A.D. and Croser, B.J. (1983). Determination of grape maturity by aroma/flavor assessment. In: *Proceedings of the Fifth Australian Wine Industry Technical Conference* (Adelaide), T.H. Lee and T.C. Somers (Eds.), Australian Wine Research Institute, Urrbrae, South Australia, p. 261.

Kasimatis, A. and Vilas, E.P. (1985). Sampling for degrees Brix in vineyard plots. *Am. J. Enol. Vitic.*, 36: 207.

Koch, A.L. (1984). Turbidity measurement in microbiology. *ASM News*50: 473.

Koch, A.L. (1986). Estimating of size of bacteria low-angle light scattering measurements. *Theor. J. Micro. Methods*, 5(5–6): 221.

Kuniyuki, A.H., Rous, C. and Sanderson, J.C. (1984). Enzyme linked immunosorbent assay (ELISA) detection of *Brettanomyces* contamination in wine production. *Am. J. Enol. Vitic.*, 35: 143.

Kupina, S.A., Pohl, C.A. and Gannotti, J.L. (1991). Determination of tartaric, malic, and citric acids in grape juice and wine using gradient ion chromatography. *Am. J. Enol. Vitic.*, 42: 1.

Kyriakides, A.L. and Thurston, P.A. (1989). In: *Rapid Microbiological Methods for Food, Beverages and Pharmaceuticals*, C.J. Stannard, S.B. Petit, and F.A. Skinner (Eds.), Blackwell Scientific Publications, Oxford, UK, p. 101.

Lau, M.N., Ebeler, J.D. and Ebeler, S.E. (1999). Gas chromatographic analysis of aldehydes in alcoholic beverages using a cysteamine derivitization procedure. *Am. J. Enol. Vitic.*, 50: 324.

Licker, J.L., Acree, T.E. and Henick-Kling, T. (1999). What is 'Brett' (*Brettanomyces*) flavor?: A preliminary investigation. In: *Chemistry of Wine Flavor, ACS Symposium Series* 714, A.L.Waterhouse and S.E.Ebeler (Eds.), American Chemical Society, Washington, DC, p. 96.

Liu, S.-Q. and Davis, C.R. (1994). Analysis of wine carbohydrates using capillary gas liquid chromatography. *Am. J. Enol. Vitic.*, 45: 229.

Luque de Castro, M.D. and García-Mesa, J.A. (1992). Potential of flow injection methodology for beverage analysis. *Am. J. Enol. Vitic.*, 43: 93.

Marinos, V.A., Tate, M.E. and Williams, P.J. (1994). Protocol for FAB-MS/MS characterization of terpene disaccharides of wine. *J. Agric. Food Chem.*, 42(11): 2486.

Marois, J.J., Bledsoe, A.M., Ricker, R.W. and Bostock, R.M. (1993). Sampling for *Botrytis cinerea* in harvested grape berries. *Am. J. Enol. Vitic.*, 44: 261.

McKinnon, A.J., Cattrall, R.W. and Scollary, G.R. (1992). Aluminum in wine – Its measurement and identification of major sources. *Am. J. Enol. Vitic.*, 43: 166.

Meidell, J. (1987). Unsuitability of fluorescence microscopy for rapid detection of small numbers of yeast cells on a membrane filter. *Am. J. Enol. Vitic.*, 38: 159.

Mestres, M., Marti, M.P., Busto, O. and Guasch, J. (1999). Simultaneous analysis of thiols, sulphides and disulfides in wine aroma by headspace solid-phase microextraction-gas chromatography. *J. Chromatogr. A*, 849(1): 293.

Pérez-Cerrada, M., Casp, A. and Maquieira, A. (1993). Chromatographic determination of the anion content in Spanish rectified concentrated musts. *Am. J. Enol. Vitic.*, 44: 292.

Perez-Coello, M.S., Sanz, J. and Cabezudo, M.D. (1999). Determination of volatile compounds in hydroalcoholic extracts of French and American oak wood. *Am. J. Enol. Vitic.*, 50: 162.

Pollnitz, A.P., Jones, G.P. and Sefton, M.A. (1999). Determination of oak lactones in barrel-aged wines and in oak extracts by stable isotope dilution analysis. *J. Chromatogr. A*, 857(1–2): 239.

Puig-Deu, M., Lamuela-Raventos, R.M., Buxaderas, S. and Torre-Boronat, C. (1994). Determination of copper and iron in must: Comparison of wet and dry ashing. *Am. J. Enol. Vitic.*, 45: 25.

Santoro, M. (1995). Fractionation and characterization of must and wine proteins. *Am. J. Enol. Vitic.*, 46: 250.

Soleas, G.J., Goldberg, D.M., Diamandis, E.P., Karumanchiri, A., Yan, J. and Ng, E. (1995). A derivatized gas chromatographic-mass spectrometer method for the analysis of resveratrol in juice and wine. *Am. J. Enol. Vitic.*, 46: 346.

Stanier, R.Y., Adelberg, E.A. and Ingraham, J.L. (1976). *Introduction to the Microbial World*. Prentice-Hall, Englewood Cliffs, NJ, p. 870.

Steffen, A. and Pawliszyn, J. (1996). Analysis of flavor volatiles using headspace solid-phase microextraction. *J. Agric. Food Chem.*, 44(8): 2187.

Stender, H., Kurtzman, C., Hyldig-Nielsen, J.J., Sorensen, D., Broomer, A.J., Oliveira, H., Perry-O'Keefe, H., Sage, F., Young, B. and Coull, J. (2001). Identification of *Brettanomyces* (*Dekkera bruxellensis*) from wine by fluorescence in situ hybridization using peptide nucleic acid probes. *Appl. Environ. Microbiol.*, 67: 938.

Sugui, J.A., Wood, K.V., Yang, Z., Bonham, C.C. and Nicholson, R.L. (1999). Matrix-assisted laser desorption ionization mass spectrometry analysis of grape anthocyanins. *Am. J. Enol. Vitic.*, 50: 199.

Vas, G., Koteleky, K., Farkas, M., Dobo, A. and Vekey, K. (1998). Fast screening method for wine headspace compounds using solid-phase micro-extraction and capillary GC technique. *Am. J. Enol. Vitic.*, 49: 100.

Vernin, G., Boniface, C., Metzger, J., Fraisse, D., Doan, D. and Alamercery, S. (1987). Aromas of Syrah wines: Identification of volatile compounds by GC-MS-spectra data bank and classification by statistical methods. In: *Frontiers of Flavor. Proceedings of the 5th International Flavor Conference*, Porto Karras, Chalkidiki, Greece, p. 655.

Weiss, K.C., Yip, T., Hutchens, T.W. and Bisson, L.F. (1998). Rapid and sensitive fingerprinting of wine proteins by matrix-assisted laser desorption/ionization time-of-flight (MALDI-TOF) mass spectrometry. *Am. J. Enol. Vitic.*, 49: 231.

Wetmur, J.G. (1991). Applications of the principles of nucleic acid hybridization. *Crit. Rev. Biochem. Molec. Biol.*, 26(3–4): 227.

Whiton, R.S. and Zoecklein, B.W. (2000). Optimization of headspace solid-phase microextraction for analysis of wine aroma compounds. *Am. J. Enol. Vitic.*, 51: 379.

Whiton, R.S. and Zoecklein, B.W. (2002). Determination of ethyl carbonate in wine by solid-phase microextraction gas chromatography/mass spectrometry. *Am. J. Enol. Vitic.*, 53: 60.

Williams, D.J., Scudamore-Smith, P.D., Nottingham, S.M. and Petroff, M. (1992). A comparison of three methods for determining sulfur dioxide in white wine. *Am. J. Enol. Vitic.*, 43: 227.

Williams, D.J., Strauss, C.R., Wilson, B. and Massy-Westropp, R.A. (1982a). Novel monoterpene disaccharide glycosides of *Vitis vinifera* grapes and wines. *Phytochemistry*, 21(8): 2013.

Williams, D.J., Strauss, C.R., Wilson, B. and Massy-Westropp, R.A. (1982b). Use of C_{18} reversed-phase liquid chromatography for the isolation of monoterpene glycosides and norisoprenoid precursors from grape juice and wines. *J. Chromatogr.*, 235(2): 471.

Williams, P.J., Cynkar, W., Francis, I.L., Gray, J.D., Iland, P.G. and Coombe, B.G. (1995). Quantification of glycosides in grapes, juices, and wines through a determination of glycosyl glucose. *J. Agric. Food Chem.*, 43(1): 121.

Winterhalter, P. and Schreier, P.C. (1994). C_{13}-norisoprenoid glycosides in plant tissues: An overview on their occurrence, composition, and role as flavor precursors. *Flavor Fragr. J.*, 9(6): 281.

Winterhalter, P., Baderschneider, B. and Bonnländer, B. (1998). Analysis, structure, and reactivity of labile terpenoid aroma precursors in Riesling wine. In: *Chemistry of Wine Flavor, ACS Symposium Series* 714, A.L.Waterhouse and S.E.Ebeler (Eds.), American Chemical Society, Washington, DC, p. 1.

Winterhalter, P., Sefton, M.A. and Williams, P.J. (1990). Two-dimensional GC-DCCC analysis of the glycoconjugates of monoterpenes, norisoprenoids, and shikimate-derived metabolites from Riesling wine. *J. Agric. Food Chem.*, 38(4): 1041.

Yokotsuka, K., Nozaki, K. and Takayanagi, T. (1994). Characterization of soluble glycoproteins in red wine. *Am. J. Enol. Vitic.*, 45: 410.

Zhang, Z. and Pawliszm, J. (1993). Headspace solid-phase microextraction. *Anal. Chem.*, 65(14): 1843.

Zhang, Z., Yang, M.J. and Pawliszyn, J. (1994). Solid phase microextraction. *Anal. Chem.*, 66(17): 844A.

Zoecklein, B.W. (2013). Viognier wine balance. *Wines Vines*, Feb., 66–70.

Zoecklein, B.W., Devarajan, Y.S., Mallikarjunan, K. and Gardner, D.M. (2011). Monitoring the effect of ethanol spray on Cabernet Franc and Merlot grapes and wine volatiles using electronic nose systems. *Am. J. Enol. Vitic.*, 62(3): 351–358.

Zoecklein, B.W., Douglas, L.S. and Jasinski, Y.W. (2000). Evaluation of the phenol-free glycosyl-glucose determination. *Am. J. Enol. Vitic.*, 51: 420.

Zoecklein, B.W., Fugelsang, K.C. and Gump, B.H. (1999). *Wine Analysis and Production*. Chapman & Hall, New York, p. 621.

Zoecklein, B.W., Fugelsang, K.C. and Gump, B.H. (2011). Analytical techniques in wine and distillates. *Handbook of Enology*, Vol. 111. Technology of Production and Quality Control, V.K. Joshi (Ed.), Asia-Tech Publishers, New Delhi, p. 1287–1321.

Zoecklein, B.W., Gump, B.H. and Fugelsang, K.C. (2001). Nutritional status of grapes. In: *Methods in Biochemistry*, Vol. 14. F.T. Spencers and A. Ragout (Eds.), Humana Press, Totowa, NJ.

Zoecklein, B.W., Fugelsang, K.C., Gump, B.H. and Nury, F.S. (1990). *Production Wine Analysis*. Van Nostrand and Reinhold, New York, p. 475.

Zoecklein, B.W., Wolf, T.K., Duncan, S.E., Marcy, J.E. and Yasinski, Y.W. (1998b). Effect of fruit zone leaf removal on total glycoconjugates and conjugate fraction concentration of Riesling and Chardonnay (*Vitis vinifera* L.) grapes. *Am. J. Enol. Vitic.*, 49: 259–265.

Zoecklein, B.W., Wolf, T.K., Marcy, J.E. and Yasinski, Y.W. (1998a). Effect of fruit zone leaf thinning on glycosides and selected aglycone concentration of Riesling (*Vitis vinifera* L.) grapes. *Am. J. Enol. Vitic.*, 49: 35–43.

Zoecklein, B.W., Wolf, T.K., Pelanne, L., Miller, M.K. and Birkenmaier, S. (2008). Effect of vertical shoot positioned, Smart-Dyson and Geneva double curtain training systems on Viognier grape and wine composition. *Am. J. Enol. Vitic.*, 59: 11–21.

40 Sensory Evaluation of Wines and Brandies

General Concepts and Practices

V.K. Joshi, António Manuel Jordão and Fernanda Cosme

CONTENTS

40.1 INTRODUCTION AND IMPORTANCE

The quality of a food product including wine determines its ultimate acceptance by the consumer. Quality is directly related to consumer acceptance, which is difficult to quantify. Thus, it is important to understand what quality means. Quality is the composite of those characteristics that differentiate individual units of a product, which has significance in determining the degree of acceptability of that unit by the consumer. Sensory analysis is one of the most important methods of food quality evaluation. Determining the quality of a product is incomplete without a sensory evaluation as other methods such as physical, chemical, and microbiological methods fail in product development if a sensory evaluation is not carried out. Thus, sensory evaluation has an edge over other methods. However, a collective approach to the quality evaluation of a product is more acceptable today.

In the past, human sensory perception was the only means of assessing the quality of a product. With the development of instrumentation, the sensory analysis was overlooked in preference to instrumental methods as they were thought to provide absolute measures of quality. A product's quality was determined by physical, chemical, and microbiological criteria with the assumption that a product that met the prescribed specifications in the aforementioned parameters would automatically meet the desired sensory quality. However, it was quickly realised that without sensory evaluation, consumers' acceptability of a product could not be determined.

Consistency in production is essential for consumers' quality perception and is the basis of brand loyalty and identification. The integration of sensory analysis into a quality control programme makes it a powerful tool for product assessment, integrity, and consistency. Although many food processing plants are aware of this, few have attempted to make it a routine procedure to obtain objective and actionable data. This is perhaps due to the belief that sensory analysis is a 'subjective' method of evaluation, *i.e.* it is not repeatable or reproducible and is biased. This was so initially for want of objective sensory methods. Today, however, sensory analysis has developed into a full-fledged discipline. It has been defined as 'A scientific discipline used to evoke, measure, analyse and interpret reactions to those characteristics of foods and materials as they are perceived by the senses of sight, smell, taste, touch, and hearing'.

The main advantage of sensory analysis over traditional wine tasting is its ability to collect wine assessments in the least biased way. This is accomplished by performing the sensory analysis under standardised and controlled conditions that reduce the physiological and psychological factors known to affect sensory responses and using panellists who are highly sensitive to sensory stimuli and are able to evaluate their perception analytically and objectively. An analogy is usually made between a sensory panel and analytical instruments, such as gas chromatography–mass spectroscopy. In this way, it is expected that the collected data from the sensory panels are accurate, repeatable, sensitive, and reproducible. Sensory analysis is routinely used in the quality evaluation of foods and their products to determine the difference between the quality of different cultivars, their harvesting maturity level and the effects of processing methods.

With the advances in technology, especially in food production and processing, testing consumers' acceptance/preference for a product has become essential. However, it is necessary to differentiate between consumer acceptance and sensory analysis. Consumer acceptance is the reaction of the user to a product determined by personal factors such as age, like or dislike, familiarity, and economics.

Sensory analysis plays a very significant role in our lives (Table 40.1) as well as in oenology, the science of wine. It is the most important criterion leading to sound decisions in the winemaking process, and in the context of sensory responses, wine has several chemical compounds contributing to its aroma, taste, and other oral sensations. Some of the applications of sensory analysis in oenology include the screening of different varieties of various fruits such as grape varieties and their suitability to make wine; evaluating different methods and techniques to make wine; determining the quality of products, *e.g.* after maturation, the effect of treatments, *e.g.* the use of different yeasts strains, lactic acid bacteria (LAB), enzymes and maturation containers; finding the sensory properties of wine for consumer acceptance; developing various techniques and methods to evaluate wine including graphical, statistical and computer-based techniques. Thus, this chapter attempts to describe some of the issues.

40.2 SENSES AND SENSE ORGANS

Extensive research has been carried out on the stimuli, the sensation and the mode of transduction of the sensation.

TABLE 40.1
Sensory Evaluation: A Profile

- Man carries out a sensory evaluation every day.
- Everybody has a specific preference for a particular food.
- Ancient man could distinguish a good food from a bad food based on sensory testing.
- The sensory qualities of foods are affected by microbial contamination, improper harvesting, inadequate storage and packaging, improper processing, etc.
- A sensory evaluation uses the sense organs to evaluate food, sometimes called the subjective method.
- With the advancement of instrumental analysis, sensory evaluation was overlooked.
- It can be objective if properly conducted.

Source: Joshi (2006)

Transduction is the process whereby events of the physical world are converted into electrical activity in the sensory nerves through neurons. A neuron is a single cell composed of three basic parts: a sensitive tip called a dendrite; a cell body containing the cell nucleus and other related structures; and an *axon*, which serves as the output portion of the cell.

40.2.1 Eyesight

The eye is a visual receptor that works similarly to a camera (Figure 40.1). Light passes through the pupil of variable size depending on the light intensity. The iris around the pupil controls the size or contraction of the pupil and the amount of light that enters the eye. Behind the pupil and the iris lies a lens, which aided by the outer cornea, focuses the light rays in such a way that the focal length equals the distance between the lens and the light receptive areas at the back of the eye. The retina, with its layer of sensory cells covering the interior of the eyeball, is the focal point of light. The deepest layer of the retina (composed of a conglomerate of cells) is directly responsive to light. There are two types of specialised neurons: (i) elongated rods that have no colour-sensing capacity and are mostly found in the peripheral regions of the retina; and (ii) cones that are shorter and colour sensitive and are found in the centre of the retina, directly behind the lens. There are four types of visual pigments and all contain a derivative of vitamin A. Out of these four pigments, one is found in the rods which are very sensitive to low levels of light and have no colour-detecting capacity. Cones have three types of visual pigments, each responding to the red, blue, and green colouring pigment, respectively. The stimulation of more than one type of visual pigment at different intensities allows an individual to distinguish a whole series of intermediate colours.

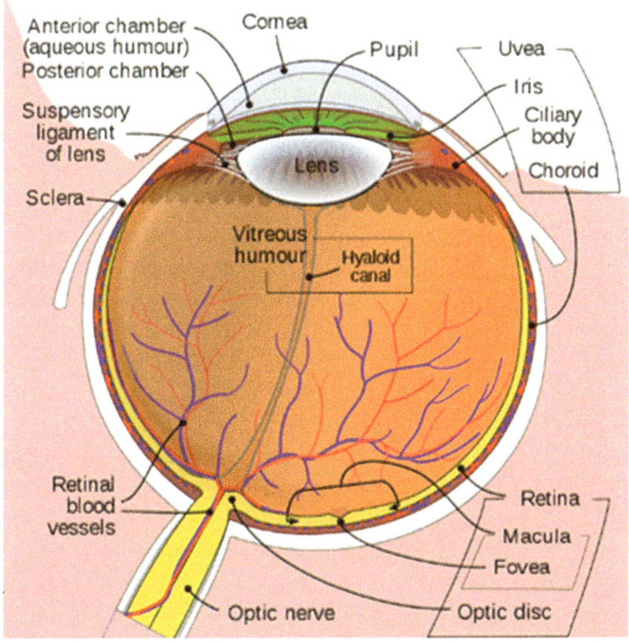

FIGURE 40.1 Human eye system. *Source:* https://en.wikipedia.org/wiki/File:Schematic_diagram_of_the_human_eye_en.svg

Vision begins when light causes a physical straightening of some of the visual pigment molecules, generating an action potential to two successive types of neurons in the retina, which is transmitted through the optic nerve to the thalamus of the brain and then, to the visual cortex at the back of the head. However, colour is a function of the light source, the optical properties of a food product, and the pigments in the retina which acts as photoreceptors and absorbs energy, give the neurological responses to the stimulus, and, finally, the interpretation based on the signals transmitted to the brain. The pathway for motion vision is depicted in Figure 40.2.

40.2.2 Hearing

It is known that any vibrating object causes bands of compression and rarefaction, and if the vibration is of sufficient intensity, it is detected as sound. Sound waves are intercepted by the eardrum at the base of the ear canal, causing it to vibrate. A set of three middle ear bones are attached to the eardrum, transmitting sound to the inner ear and serving to amplify the intensity of the vibration about 18 times above that received by the eardrum. The point of contact between the middle ear bones and the inner ear (the oval window) is coiled and filled with a fluid that receives the vibration (at an original frequency) with the membrane that divides the inner ear down the middle, vibrating at the same frequency. However, because of its structure, the sound of different frequencies causes maximum displacement at different points along the membrane. The membrane consists of two layers (not fused); during displacement, one layer slides over the other layer, deflecting tiny hairs (hair cells) that are connected to the auditory nerve. The deflection of the hairs generates an action potential, activating different fibres in the auditory nerve, which ultimately allows the brain to discriminate according to the sound frequency. The detection of complex sounds occurs by the simultaneous deflection of the hair cells responsible for the particular frequencies in the sound.

40.2.3 Touch

The sense of touch is perceived by tiny organs (specialised pressure receptors) surrounding an individual neuron, known as Pacinian corpuscles. As the rate of the action potential produced by the deflecting neuron/neurons increases, the magnitude of the stimulus is conveyed to the brain. The touch receptors, however, quickly fatigue.

40.2.4 Taste

Taste recognition is controlled by specialised epithelial taste receptor cells, which are arranged in groups of 50–100 in so-called taste buds, which have an onion-like shape. The taste buds are located in different areas on the tongue (Figure 40.3) and are specific to various types of basic tastes. Taste receptors are mainly situated on the surface of the tongue and, to a lesser extent, are spread over the oral cavity and down the oesophagus. Human taste buds are contained in three types of papilla: the foliate, fungiform, and circumvallate (Figure 40.4). On the

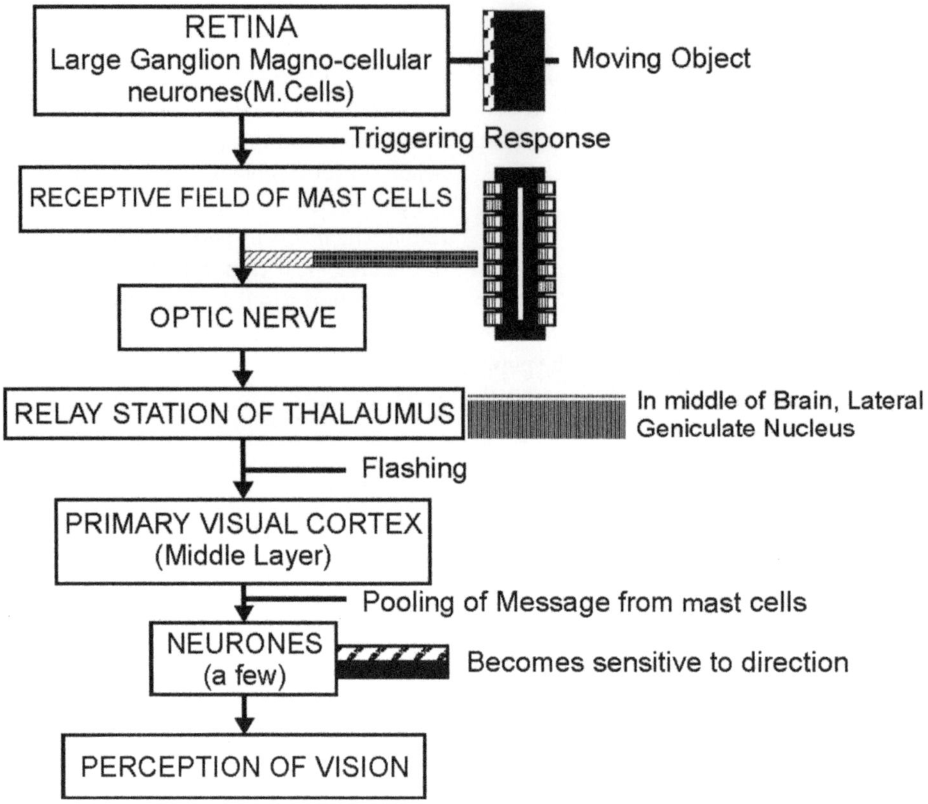

FIGURE 40.2 Steps in a motion vision pathway depicted diagrammatically. *Source:* Joshi (2006)

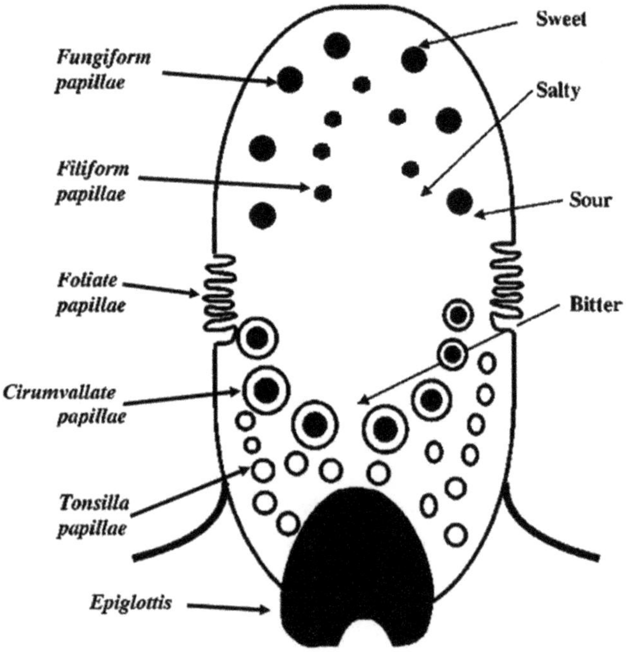

FIGURE 40.3 The tongue and its different areas for different taste perceptions

tongue, the fungiform papilla is mostly located at the front, the foliate papilla is at either side and the circumvallate papilla is at the back. Adults have about 2000 taste buds. Although all taste buds possess a common flask-like shape, the neuroepithelial receptors they contain fall into three morphologically and functionally different types. One class appears to support the activity of the other two (termed Type I). The other two are electrically excitable and possess distinct chemoreceptors on their cell membranes (Types II and III). Basal cells differentiate into the three, elongated, neuroepithelial categories.

The taste perception pathway is simpler than the olfactory pathway, but the mechanism by which it decodes in the brain is still unknown. The taste buds are a type of chemical receptors. Human beings have four kinds of taste buds for different tastes, viz. sweet, salty, sour, and bitter. The response of each taste bud may be equal to any of the four types of tastes but some respond more quickly to a particular taste than do the others.

40.2.5 Smell

The mechanism of odour detection is still not completely understood. Large numbers of cells in the upper reaches of the nasal passages in human beings are responsive to different types of odours. Such cells possess different types of receptive sites that respond to different types of odours based on the chemical structure of the molecules to be identified. All the olfactory cells are identical when viewed microscopically, and fatigue of these cells is rapid, thereby responding quickly to a strange odour, making the mechanism complex.

When volatile compounds bind to one or more of around 1000 protein receptors on the surface of the cells in the nasal passage, the sense of smell is activated. Both aroma and odour are the sensations triggered by the olfactory and trigeminal nerves. Additionally, a third nerve, the terminal nerve, in the nasal cavity is also involved in this perception (Figure 40.5).

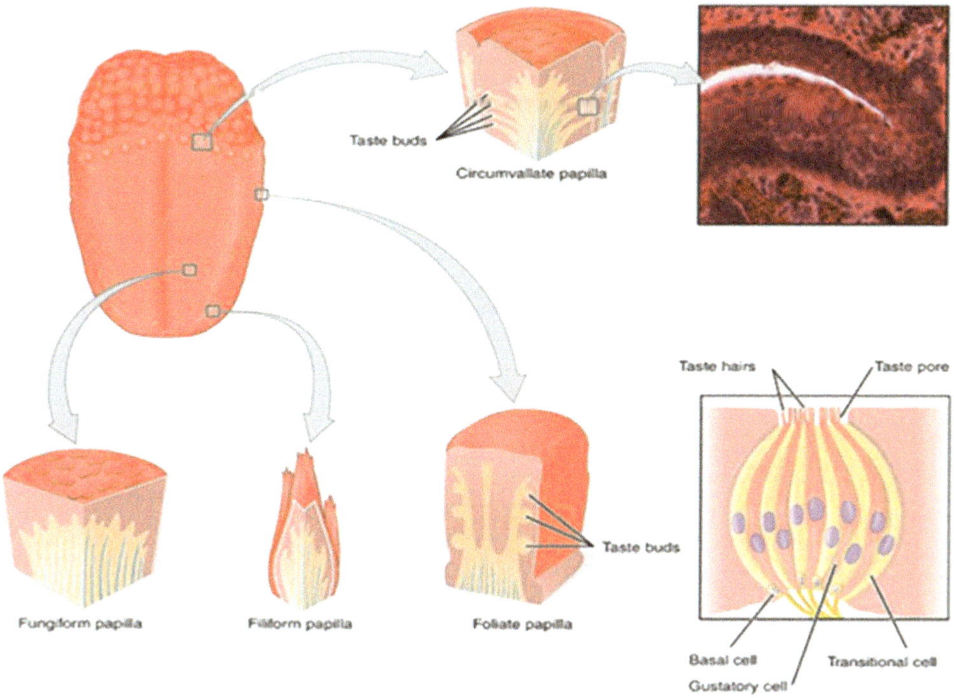

FIGURE 40.4 Human taste papilla system. *Source:* https://cnx.org/contents/FPtK1zmh@8.25:fEI3C8Ot@10/Preface

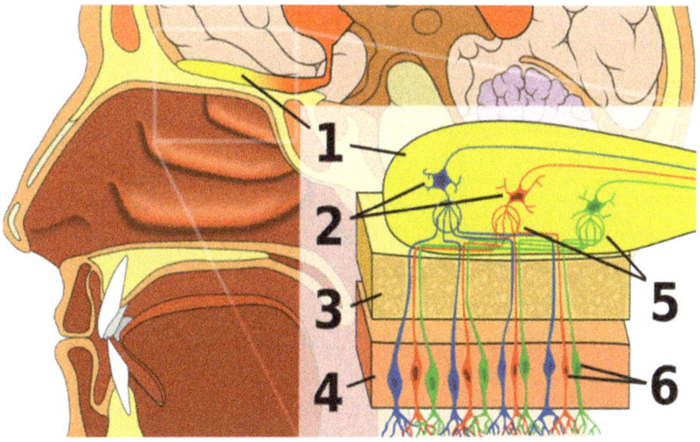

FIGURE 40.5 Human olfactory system: 1: olfactory bulb; 2: mitral cells; 3: bone; 4: nasal epithelium; 5: glomerulus; 6: olfactory receptor neurons. *Source:* https://commons.wikimedia.org/wiki/File:Olfactory_system.svg

During normal breathing, only approximately 5% of the respired air reaches the receptors; however, sniffing increases this to about 20%. Although several theories, such as hydrogen bonding and stereo-chemical oxidation, have been put forward to explain the mechanism by which the brain decodes a perceived sensation, none can fully explain the phenomenon.

Anatomical studies show that signals received from the human olfactory cells reach the olfactory area of the cortex (in the brain) after only a single relay in the olfactory bulb. It is said that the average human being can recognise around 10,000 separate odours. Neurons that sense these odour molecules lie high up in the nose, *i.e.* the olfactory epithelium consisting of around five million olfactory neurons. The olfactory epithelium is topped by at least 10 hair-like cilia that protrude into a thin bath of mucus at the cell surface. Cilia consist of a receptor protein that recognises and binds odorant molecules, thereby stimulating the cells to send signals to the brain. The pathway of signal transmission from the source to the brain is shown diagrammatically in Figure 40.6.

40.3 WINE SENSORY EVALUATION: PANEL SCREENING, SELECTION AND TRAINING

The components of a sensory evaluation programme are shown in Figure 40.7.

40.3.1 Sensory Evaluation Laboratory

Sensory evaluation requires laboratory equipment and staff. A sensory evaluation laboratory should be equipped with tables within booths/chambers and the booths should be configured so that one judge cannot see another judge. The booth should have a proper seat, a light arrangement, a sink, and a serving window through which the product is served to the judge. Each booth should also have a bell, to ask for anything required by the judge. Sensory analysis should be performed in a laboratory that precludes any visual, audio, or olfactory distractions. Therefore, the environment in which it is performed should be temperature controlled, odour-free, and serene.

40.3.2 Sensory Testing Programme

The objectives of sensory testing programmes are: quality control and assurance – for the maintenance of quality, process/equipment change, ingredient substitutions, and alternate supplier evaluation; product development or improvement – to

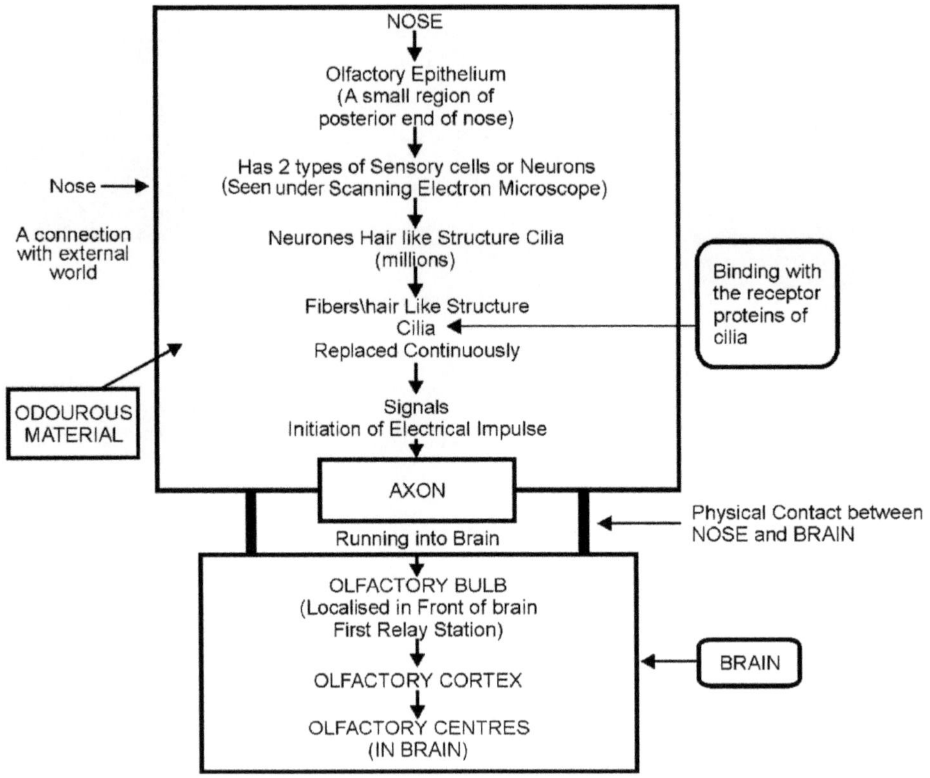

FIGURE 40.6 Schematic drawing of the functional parts of the sense of smell. *Source:* Joshi (2006)

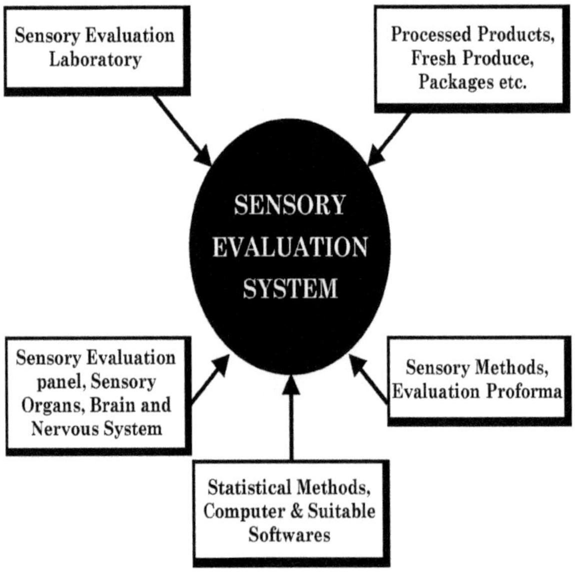

FIGURE 40.7 Components of a sensory evaluation system

match a competitive product and obtain or retain a share of the market; new product development; define packaging and storage profiles; and provide product information for marketing/management groups and quality grading/profiling.

40.3.3 Types of Sensory Panellists

Sensory panels are classified into three categories: trained, semi-trained, and consumer panels. A trained panel should be carefully selected and trained, and need not be expert panellists. The trained panel should be used to establish the intensity of a sensory character or the overall quality of a food product. A trained panel should comprise a small number of panellists, from 6 to 10, and may be used in all product developmental, processing, and storage studies. A semi-trained panel should be comprised of persons familiar with the food product. This panel is capable of discriminating differences and communicating their reactions, though it may not be formally trained. In a semi-trained panel, individual variations can be balanced out by including a greater number of panellists. The panel should normally consist of about 25–30 panellists, and should be used in the preliminary screening of a select few products for large-scale consumer trials. The members of a consumer or untrained panel should be selected at random and ensure due representation of different age, sex and income groups in the potential consumer population in the market area. More than 100 members are required to constitute a consumer panel.

40.3.4 Sensory Evaluation Panel

Sensory evaluation, involving panel members as instruments to measure product differences, characteristics, or preference levels, requires considerable care in screening, selecting and training panellists. The whole reliability of the analysis depends on these aspects, but no single general rule can be applied to calibrate a panel's sensitivity. Some of the more common steps are to select, screen, and train panellists for discriminative and/or descriptive tests.

TABLE 40.2
Taste Solutions Concentration Used to Examine the Acuity of Panellists

Material	Taste quality	Concentration in water at room temperature (g/L)
Caffeine	Bitter	0.27
Citric acid	Sour	0.60
Sodium chloride	Salty	2
Sucrose	Sweet	12

40.3.4.1 Panel Screening and Selection

General information regarding sex, age, food habits, food likes and dislikes, willingness to participate, knowledge about the product and, above all, their availability is the first few criteria for panel screening and selection.

The motivation of the panel is also an important aspect of panel screening and training. A panel member needs a basic interest or involvement in the product rather than a short-lived interest out of curiosity, or a change from a routine job or even for financial remuneration. Using a properly designed experiment, the number, and type of tests can be suitably selected.

Panellists are screened and selected using several tests. The most commonly used tests are to determine impairment of the primary senses (colour, vision), matching test for taste and odour substances, the ability to detect basic taste and odour acuity and performance in comparison with other candidates. Several examples of taste and odour solutions used for panellist accuracy during selection are shown in Tables 40.2 and 40.3. Tests for sensitivity to basic tastes and odours can help to group candidates into broad classes. Therefore, candidates are screened according to their sensory acuity, availability, and motivation. The assumption is that if the decision-making on quality is based on an in-house panel recommendation with its more analytical purposes, then it is safe for consumers. The panel performance, however, is very important as it is affected by their selection and screening.

40.3.4.2 Panel Training

The selected candidates require training to carry out consistent and objective sensory evaluations. Therefore, the purpose of panel training is to increase the panellists' sensory acuity and provide them with a rudimentary knowledge of the procedures used in sensory evaluation. Training ensures the systematic utilisation of the sensory responses to a product. Training prepares the panellist to provide an opinion about the test product and to perform the test without bias and with a certain degree of homogeneity. The ability to communicate a perceived sensation needs constant recalling from memory. The training is an individual rigorous exercise on each class and type of food and is specific to the attributes being examined. All panellists should receive subsequent training to maintain their sensory evaluation abilities.

TABLE 40.3
Odorous Material Used in Odour Recognition Tests

Material	Name most commonly associated with the odour
Benzaldehyde	Bitter almonds, cherry
Phenyl-2-ethyl acetate	Floral
Menthol	Peppermint
Butyric acid	Rancid butter
Acetic acid	Vinegar
Isoamyl acetate	Fruit, banana
Vanillin	Vanilla

Note: The concentration should be above the recommended threshold level.

The selection of a panel and their performance evaluations are constantly monitored. Sensitivity can be expressed in terms of repeatability and reproducibility and, occasionally, with respect to known compounds of taste or odour.

40.3.5 Wine Tasting Temperature

The temperature at which a wine is tasted has an immense impact on its taste. Serving cool wine will mask some imperfections, which is good for young or cheap wine, while a warmer wine temperature allows the expression of a wine's characteristics, which is best for older or more expensive wine. Room temperature will also affect the rate at which the wine warms up.

40.3.6 Wine Glass

As important as the wine tasting temperature is, the appropriate wine glass in which the wines will be tasted is equally important. The wine glass needs to be sufficient in size to enable it to be tilted sideways to inspect the colour at the rim of the glass and to swirl the wine to allow better judgement of the wine aroma. The wine glass should also have a stem that allows the taster to hold the glass, so that the wine is not unintentionally warmed up, or the clarity of the glass obscured. The shape of a wine glass can affect the wine taste, thus different wine types are tasted in different glasses. The three main types of wine glasses are: white wine glasses that are tulip shaped; red wine glasses that are more rounded and have a larger bowl; and sparkling wine flutes that are tall and thin. A suitable all-purpose wine glass should hold 10 oz of wine, be transparent to allow the taster to examine the wine colour and its body, and have a slight curve at the top to hold in the bouquet. Tasting should be carried out with a standardised (ISO) tulip-shaped glass, which is internationally recognised. The glass should only be partially filled to give a generous quantity of air-space above the glass from which the panellist can sniff the aroma escaping from the wine that is present just above the wine in the glass.

40.3.7 Wine Serving

The wine to be tasted in the glass needs to be at an adequate temperature. It is generally considered that lighter wines are served at lower temperatures and full red wines are served at room temperature. Wine decanting, particularly in red wines, is only needed when the wine presents with a deposit/sediment in the bottom of the bottle or a reductive aroma. Limpid wines do not need to be decanted and it is thought that aged wines or very fresh young wines, such as white wines, can easily lose their aroma as a result of reactions with oxygen. Oxygen should also be avoided in older wines and white wines. The room where the wine will be tested should be comfortable, clean and without smells, with adequate natural northern daylight to appreciate the colour. At all times, wine samples are presented uniformly in wine tasting glasses coded with random numbers to prevent trends or influences.

40.4 SENSORY EVALUATION METHODS

Sensory analysis methods (Figure 40.8) can be classified into two different groups: analytical (these tests treat the panellists as discriminative or descriptive measuring instruments and include sensitivity assessment; the interest is in the differences among the treatments or products or samples, the panels being the means of measurements; the quality attributes of the products are assessed, keeping the panel constant) and affective (based on hedonic considerations, the importance shifts from the products/samples to the target population; it should follow the sensory analysis and can take one or two optimised products and may involve the study of one or more of the following: the pattern of acceptance/preference of the product by various segments of a population; the effect of package, price, tradition and other such parameters on the acceptance; the effect of ethnic groups, geographical locations, family size and composition, food habits, on the acceptance; the prediction of market potential). Laboratory analysis carried out with trained or semi-trained panellists are analytical, and affective studies are carried out with untrained consumer panels.

Sensory testing mainly refers to discrimination, descriptive and affective tests. Discrimination and descriptive analyses are recognised as analytical tests to detect product differences or characteristics. In comparison, affective analysis is defined as a hedonic test, which explores consumer likings or preference levels for products. Compared to analytical tests, affective analysis requires a much larger panel size in order to have greater confidence in the interpretation of the results. The different methods used and the ideal number of samples required for the specific tests are described in Table 40.4 while different tests are used for different purposes are shown in Table 40.5.

The most commonly used analytical tests for sensory evaluation in the wine industry are discrimination (or difference) tests and descriptive tests. Discrimination tests can be used to determine if wines are different, if a given wine characteristic is different among samples or if one wine has more of a selected characteristic than another. Descriptive tests are used to provide more comprehensive profiles of wine by asking panellists to identify the different wine sensory characteristics in the product and quantify the characteristics. The different steps used to conduct the descriptive test are depicted in Figure 40.9.

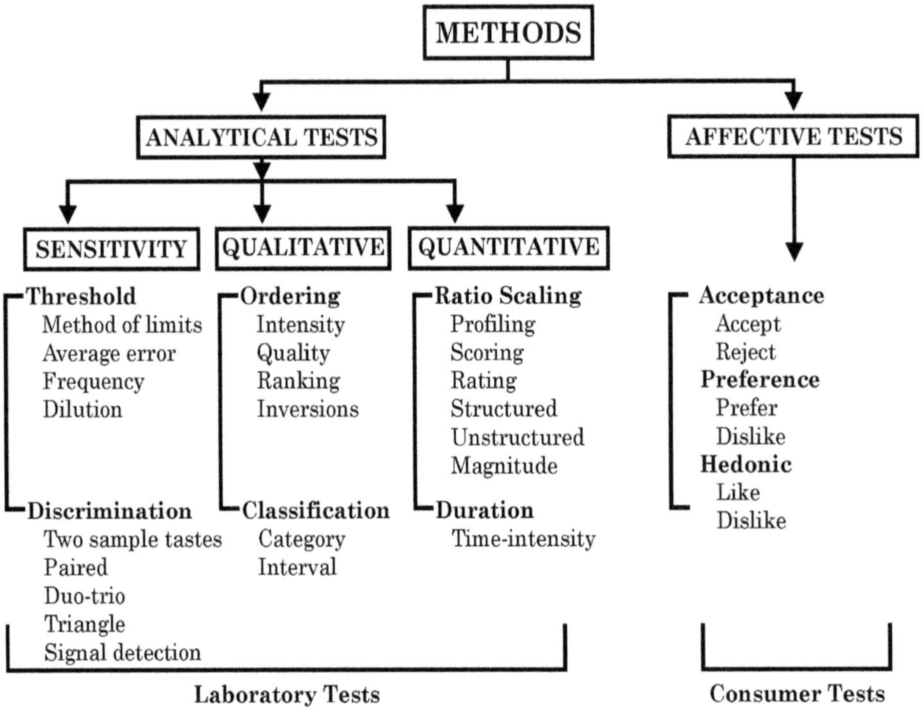

FIGURE 40.8 Classification of methods for the sensory analysis of food. *Source:* Narasimhan and Rajalakshmi (1999)

TABLE 40.4
Methods for the Sensory Evaluation and Acceptance Testing of Foods

Method and number of samples per test

1. Paired comparison (or two-paired preference) 2
2. Duo-trio 3 (2 identical, 1 different)
3. Triangle 3 (2 identical, 1 different)
4. Ranking 2–7
5. Rating difference/scalar 1–18 (the larger number only if different from control mild-flavoured or rated for texture only)
6. Threshold 5–15
7. Dilution 5–15
8. Attribute rating category 1–18 (the larger number only if scaling and ratio scaling mild-flavoured or rated for or magnitude estimation of texture only)
9. Flavour profile analysis 1
10. Texture profile analysis 1–5
11. Quantitative descriptive analysis 1–5
12. Hedonic (verbal or facial) 1–18 (the larger number only if scale rating mild-flavoured or rated for texture only)
13. Food action scale rating 1–18 (the larger number only if mild-flavoured or rated for texture only)

Source: IFT (1981)

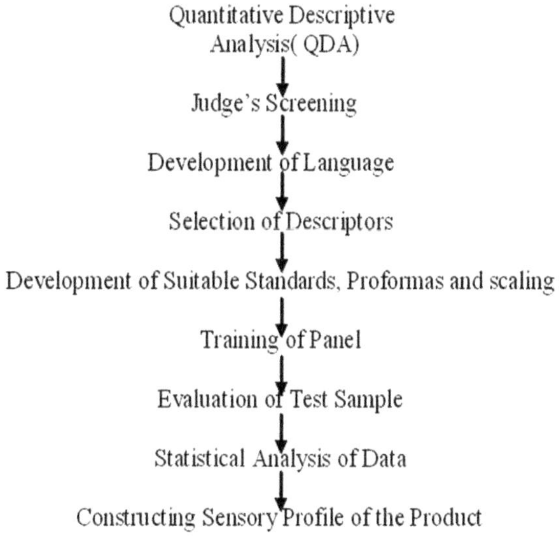

FIGURE 40.9 Flow sheet of the process to conduct a descriptive analysis. *Source:* Joshi (2006)

40.4.1 Sensitivity Tests

Sensitivity tests are a group of tests used to identify a stimulus. The two main groups under this classification are threshold and discrimination tests.

40.4.1.1 Threshold Tests

In the literature, various threshold values are reported for basic tastes and these are due to variations in the methodology and the increments of the dilutions tested. Threshold tests are always used with a qualifying term, defined here as the stimulus/recognition threshold or the detection threshold measures the minimum value of a stimulus needed to give rise to a sensation. The threshold value is the concentration of a compound at which a detectable difference in aroma or taste is found (detection threshold) or at which the characteristic odour or taste can be recognised (recognition threshold). The sensation need not be identified. The recognition threshold gives the minimum value of a sensory stimulus required for the identification of the sensation perceived, while the difference threshold gives the value of the smallest perceptible difference in the physical intensity of a stimulus. 'DL' (difference-limen) or 'JND' (just noticeable difference) sometimes designate the term-difference threshold. In the threshold method, each panellist evaluates a series of gradually increasing concentrations (either in geometric or arithmetic progression) of a material or an ingredient. The sample in which the panellist first detects a difference or a definite sensory characteristic is recorded.

TABLE 40.5
Application of Sensory Evaluation and Acceptance Testing Methods

Type of application	Appropriate test methods listed in Table 40.4
New product development	1, 2, 3, 4, 5, 8, 9, 10, 11, 12, 13
Product matching	1, 2, 3, 5, 8, 9, 10, 11, 12, 13
Product improvement	1, 2, 3, 5, 8, 9, 10, 11, 12, 13
Process change	1, 2, 3, 5, 8, 9, 10, 11, 12, 13
Cost reduction and/or selection of a new source of supply	1, 2, 3, 5, 8, 9, 10, 11, 12, 13
Quality control	1, 2, 3, 5, 8, 9, 10, 11
Storage stability	1, 2, 3, 4, 5, 8, 9, 10, 11, 12, 13
Product grading or rating	8
Consumer acceptance and/or opinions	1, 4, 12
Panellist selection and training	1, 2, 3, 4, 5, 6, 7, 8
Correlation of sensory with chemical and physical measurements	5, 8, 9, 10, 11

Source: IFT (1981)

40.4.1.2 Discrimination Tests

Discrimination tests, also called difference tests or similarity tests, are another category of sensitivity tests commonly used in sensory analysis. These are used to measure specific effects by discrimination but are not used to determine preference or acceptance. The difference tests are useful either as one-tailed or two-tailed tests and yield statistically analysable data. They are used to determine whether two different products are perceived as different by the human senses. These methods are used in quality control and as benchtop tests in research and development work to determine the possible effects of ingredient, processing, environmental change, etc. In oenology, they are frequently used to determine if different winemaking techniques have a sensory impact. Within this general class of discrimination tests are a variety of specific methods, such as paired comparison, two-sample differences, duo-trio, triangle, and multiple sample tests.

40.4.1.2.1 Paired Comparison Test

A pair of the sample representing the test variable is served and the panellist is asked to indicate if the samples are the same or different. This test is relatively easy to organise and implement. The number of pairs that can be tested at one session depends on fatigue interfering with judgement. To avoid order and duplication bias, all possible combinations of samples must be given. Panel performance can be monitored with built-in control of the same sample duplicated as one of the test pair. The panellist must answer, one way or the other. There is a chance probability of one half of the panellists placing the samples in a certain order. The binomial distribution method of analysis can be used to test the significance of difference identification. A paired comparison test is typically used in comparing new and old processing techniques, change of ingredients in a product, etc.

40.4.1.2.2 Two-Sample Difference Method

The two-sample difference method is a variation of the paired comparison method. Two coded samples in order of AA, BB, AB, BA (four pairs) are served simultaneously, and the panellist has to decide if there is any difference or not. Each pair consists of a known standard and an unknown test sample. In two pairs, the standards are duplicated and in the other two pairs, the test sample is paired with the standard. The panellist is asked to judge each pair independently. Because this method involves several sittings, it is not useful for precisely identifying 'how' and 'why' differences.

40.4.1.2.3 Duo-Trio Method

This test is a modified paired comparison test. The panellist is first presented with one sample identified as the reference for evaluation. Subsequently, two coded samples, one of which is identical to the reference, are presented. The panellist is asked to indicate which of the two samples served is identical to the sample served first (reference sample). The panellist must give an answer, even if it is a guess. The chance probability of correctly matching the samples is one half. The results can be statistically analysed as in the triangle test. The test is used for quality control.

40.4.1.2.4 Triangle Tests

Triangle tests are the most well known and most frequently used test. In this test, three samples, two alike and one odd are presented simultaneously. The panellist is asked to determine the odd sample using the sensory modality (vision, smell or taste), and an answer must be given even if it is a guess. The chance probability of selecting the correct odd sample is one third. The triangle test is a more difficult test because the panellist must recall the sensory characteristics of two products before evaluating the third and then, make a decision. Statistical analysis can determine if a significant difference exists.

40.4.2 Qualitative Tests

The next class of laboratory sensory tests is qualitative tests, which in addition to difference identification progress to the stage of answering the nature of the difference. They use the technique of ordering, in terms of intensity, quality or checking for inversions in ordering. Tests for classification based on a set of categories also fall under this class of qualitative tests.

40.4.2.1 Ranking Method

This is a non-parametric test that is very useful for quickly screening products wherein a comparison of samples and a test of inversions are possible.

40.4.2.1.1 Inversions (Individual's Sensitivity and Consistency)

A series of concentration differences of a known stimulus for an attribute, viz. colour, appearance, odour, aroma or taste, are prepared and the panellists are asked to rank the samples from lowest to highest intensity based on sensory perception. The number of inversions from the expected rank order at each concentration level for each panellist is then, calculated. The sum of inversions for each panellist and each level of concentration provides the data for determining the panellist's sensitivity and also provides information on responses at different concentrations.

40.4.2.1.2 Comparison of More Than Two Products

When the intensity or quality differences between several samples are to be compared, the ranking test is useful but the magnitude of the difference cannot be obtained. First, the ranked data are analysed to find out whether there is a significant difference among the set of samples. When a significant difference among the set of samples is established, the data will be analysed to segregate the samples into homogeneous subgroups. Commonly used data analysis techniques are Friedman's method followed by the Wilcoxon–Mann–Whitney U-test. Non-parametric methods, which are longitudinal in nature, apply only to the whole set at a given time. These scales can be made continuous by anchoring the categories/points with the description of the quality and a panel can be trained to use such scales.

40.4.3 Quantitative Tests

A parametric analysis of samples using a scoring procedure is one of the tests intended to determine the magnitude of the differences between samples. Numerical or category rating scales are interval scales in which the distance between the points is assumed to be equal and the scale has an arbitrary zero point. In this situation, it is not possible to claim that the absolute magnitude of the attributes or quality is measured. Sometimes, all the samples/treatments may not be available together for ranking and they need to be scored individually or in smaller sets. For example, in storage tests when the study involves a product, the analysis will be over a pre-fixed time frame and the testing has to be latitudinal. Scoring methods can provide absolute values as near as possible and are useful for comparisons over a period of time. The data generated are tested using a t-test or an analysis of variance followed by Duncan's multiple range test for testing the significance of differences, if any.

40.4.3.1 Magnitude Estimation Method

This method is one of the ratio scaling techniques. It assigns numerical values to stimuli that are proportional to match a reference called a modulus. The modulus either has a fixed score or is free, where the panellist can devise his/her score for the modules. The outcome is a set of scores concerning the reference modulus in single or more attributes. For example, a beverage could be twice as sweet, half as sour, etc., compared to the reference. Ratio scaling, in general, and magnitude estimation, in particular, are expected to eliminate most of the bias. It is expected to follow a log normal distribution and thus, is amenable to all types of parametric analysis.

40.4.3.2 Descriptive Profiling

Descriptive quality is a more advanced sensory test method and, with computing capability along with user-friendly software for data analysis, such tests are more comprehensive. Descriptive analysis is quantified in terms of a structured scale anchored at two end points and is classified as a quantitative descriptive analysis (QDA). In the literature, this is often referred to as quantitative sensory profiling (QSP). Examples of a descriptive test including a quantitative descriptive analysis are flavour profile analysis, time-intensity descriptive analysis, and free-choice profiling. For a profile analysis to be correct without ambiguity, a uniform understanding of the descriptors is essential. It is very effective if the description of a product is in the user's vernacular or popular language. The selection of descriptors is an important step. Initially, the panel leader will choose a few descriptors with reference to the food/beverage under study. Subsequently, the terminology generated by a panel in training is introduced and, during the procedure, a description of the perceived sensation is also alternately expressed using chemical names, for example, diacetyl for 'buttery'.

Frequently, a host of descriptors may be generated in describing foods or beverages, and these need to be concise and specific. Mathematical or statistical techniques help to reduce the dimensionality of the number of descriptors used. Currently, multivariate data analysis tools are frequently used in solving sensory analysis problems. An assessment of the aroma and mouth flavour of wines is shown in Table 40.6.

TABLE 40.6
Assessment of Aroma and Mouth Flavour of Wines (Descriptive Profiling of Wines)

Trained Panel Ballot Name :_________ Date ________
Sample Code ____________
Using the 9-point intensity goals shown below. Rate each sample for all attributes listed below:
FOR AROMA AND MOUTH FEEL
1 = None, 2 = Threshold, 3 = Slight, 4 = Slight to moderate, 5 = Moderate, 6 = Moderate to large, 7 = Large, 8 = Large to extreme, 9 = Extreme to large.
OVERALL INTENSITY ________
General group-specific terms
(Score on intensity) (Scored on appropriateness)
1. ALDEHYDIC—— Sulphur dioxide——/Acetaldehyde----
2. SHARP —— Acid ____/Vinegary/acetic Acid ____ Sharp ____ Acrid _____ Lactic_____
3. METALLIC ______Metallic _______
4. ALCOHOLIC ——Alcoholic___/Warming ___/FusellyAmyl alcohol like ______
5. WOODY —— Woody/Corky _____/Musty______ Dusty/Earthy_________
6. FRUITY ——— Processed black currant like ______ Raspberry like ____Plum like___Elderberry___Grape-berry like—
7. GREEN/UNRIPE FRUITY Green/Unripe ________
8. SCENTED ——— Scented _____________
9. SPICY ——— Spicy ______________
10. PEPPERY ———— Peppery
11. CREAMY ——— Vanilla ______/Creamy _______
12. CARAMEL ——Syrup ____/Sugary ___/Sweet ___Toffee_____/Sweet Sherry____/Burnt ___/Diacetyle __/raisin___
13. YEASTY ——— Yeast/Thiamine --------------------.
14. SULPHURY ___Hydrogen sulphide/- Cabbage__/ Rubbery____

In QDA and QSP, a line scale with anchoring points at either end representing the 'low' and 'high' intensity of the perceived sensations, has been developed. The marking made on this line indicates the intensity of the sensory note and the numerical value can be found by calculating from a real 'zero' at the beginning of the line. This scale facilitates a parametric analysis of the data and the significance of difference can be calculated with more precision. The QDA technique uses well-trained panellists to identify and quantify the sensory properties of a product in order of perception, *e.g.* appearance, odour and taste. The basic steps of this method include the development of a language, the selection and training of sensory subjects, individual data collection of repeated measurements, use of a line scale and the analysis of the results by parametric statistics. QDA makes replication and statistical analysis of the data mandatory. Another specification is that a line scale is used instead of category scaling. The line scale is generally 15 cm in length with descriptors anchored at 1.25 cm from either end of the line. From the viewpoint of product development, descriptive information is essential in finding out those product variables that are different and from which one can establish cause and effect relationships.

40.4.4 Time-Intensity Measurement

To characterise persistent sensations, such as bitterness or astringency, temporal procedures are used to monitor the perceived intensity of a flavour over time. To record the evolution of panellists' perception of a dominant sensation, a method called temporal dominance of sensations (TDS) has been developed. The TDS score is determined from the intensity and time of each attribute.

40.4.5 Affective or Consumer Tests

The second major category of sensory analysis is affective tests in which consumers participate. Often, these are confused with laboratory analysis. Consumer acceptance is defined as the reaction of the user to a product which is influenced by factors such as age, like-dislike, familiarity, location, and cost. The three classical methods are hedonic tests (Table 40.7), acceptance tests (Table 40.8), and preference tests. The combined bipolar scales consisting of hedonic root words such as acceptance/rejection, preference/no preference, and like/dislike are used. The source of the root words belongs to the same adverb. In some situations, adjectives such as extremely, moderately, or slightly are also used. Care should be taken to collect as many responses as possible, at least in the hundreds. A variety of approaches are employed in consumer product testing, and for each technique, there may be several major or minor variations. Expediency is often the watchword where procedures can be adjusted to a particular test and situation. Some commonly used approaches are: (i) home use testing, (ii) central location testing, (iii) mail panels, (iv) telephone interview, (v) focus groups, and (vi) market testing.

40.5 STATISTICAL ANALYSIS USED IN SENSORY DATA ANALYSIS

40.5.1 Numerical Data

Numerical data collected by trained tasters or groups of consumers need to be subjected to statistical analysis. For example, when consumer tests are carried out to find out which wine the consumer likes best, the main interest is the differences between the consumers and especially what their differences may be due to; for example, if the consumers prefer sweet wines to dry ones or prefer the bouquet of one wine to another one. Each panellist tastes the same sample in duplicate or triplicate.

40.5.2 Measurement Scales

Measurement scales used to quantify sensory information can be classified according to their type as nominal, ordinal,

TABLE 40.7
Performa for the Hedonic Rating of a Product

Name of the judge __________ Name of product____________ Date ______

You are provided with sample(s). Please check these samples for the extent of liking or disliking using the scale given below. Your judgement is extremely important in determining its acceptability. An honest expression is very valuable in the evaluation of the product.

Code	Code	Code	
9. Like extremely	--------	--------	--------
8. Like very much	--------	--------	--------
7. Like moderately	--------	--------	--------
6. Like slightly	--------	--------	--------
5. Neither like nor dislike	--------	--------	--------
4. Dislike slightly	--------	--------	--------
3. Dislike moderately	--------	--------	--------
2. Disliked very much	--------	--------	--------
1. Disliked extremely	--------	--------	--------
Reasons			

Source: IFT (1981)

TABLE 40.8
Performa for the Food Acceptance Rating of a Product

Proforma for the fact ratıng Score Fact Rating/Sample Codes/Value
7. I would drink this at every opportunity
6. I would drink this very often
5. I would like this and would drink it now and then
4. I would drink this if available but I would not go out of my way
3. I would drink this only occasionally
2. I would drink this only if there were no other choice
1. I would not drink this

Source: IFT (1981)

interval or ratio scales. Nominal scales are the simplest of all scales. In these types of scales, numbers represent labels or category names and have no real numerical value. In ordinal scales, the numbers represent ranks. Samples are ranked in order of magnitude. Ranks do not indicate the size of the difference between samples. Interval scales allow samples to be ordered according to the magnitude of a single product characteristic or according to acceptability or preference. The degree of difference between samples is indicated when interval scales are used. Ratio scales are similar to interval scales, except that a true zero exists. On an interval scale, the zero-end point is chosen arbitrarily and does not necessarily indicate the absence of the characteristic being measured. On a ratio scale, the zero points indicate the complete absence of the characteristic. The type of scale chosen in a sensory analysis will affect the type of statistical analysis done; therefore, the measurement scale should be chosen only after consideration of the objectives of the study. Data from nominal and ordinal scales are analysed using non-parametric statistical tests, while data from interval and ratio scales are analysed using parametric statistical tests. Nominal sensory data are usually analysed by binomial or chi-square tests.

40.5.3 Ranking Data

Ordinal or ranked sensory data are most frequently analysed using the Friedman Test. The most common parametric test for interval or ratio scale sensory data is the analysis of variance (ANOVA). The analysis of variance is often the first step in determining whether there are differences between the samples and whether the individual panellist's assessments are reproducible. The means of the various terms used to describe wine are calculated. Comparisons of wines for their mean scores can be done on a spider plot. Figure 40.10 shows an example of a sensory profile of different wines using a spider plot diagram.

40.5.4 Multiple comparisons

Once the presence of statistical differences has been confirmed using ANOVA, multiple comparisons of means tests, such as Duncan's new multiple range test, Tukey's Test, the least significant difference (LSD) test and Scheffe's Test, are used to identify samples that differ from each other.

The development of statistical techniques, including the so-called multidimensional statistical analysis has enabled the handling of sensory data and has greatly helped to progress sensory testing.

40.5.5 Use of Multivariate Data Analysis Techniques

Multivariate data analysis techniques can be used when relationships among a number of different measurements or tests are being investigated. Correlation and regression analysis,

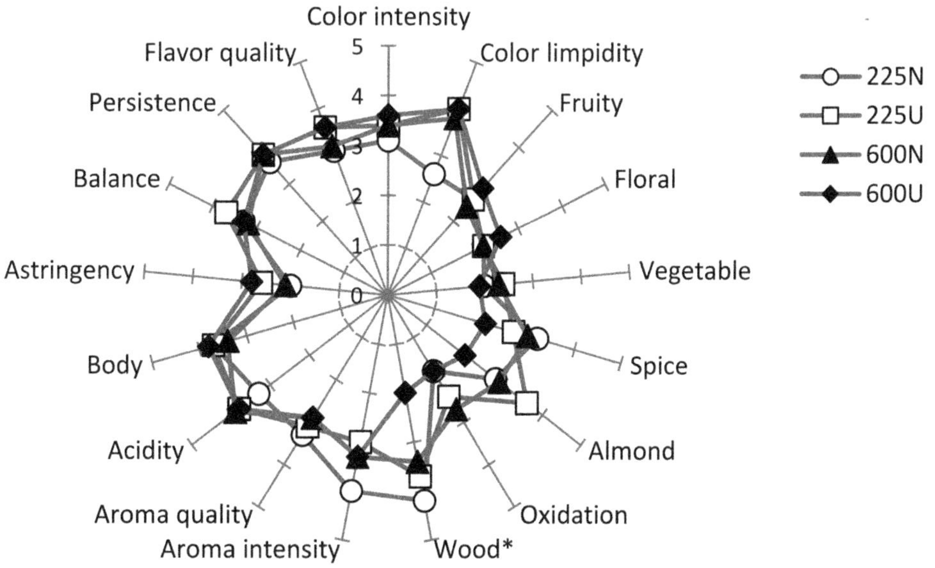

FIGURE 40.10 Sensorial profile of white wines aged in different oak wood barrels after 180 days of ageing. *Source:* Nunes *et al.* (2017)

TABLE 40.9
New Approaches in Sensory Analysis

Multiple Regression

- Is a popular and extensively used technique.
- Develops a working relationship between the dependent variable and several independent variables, ultimately into the model that is based on the correlation between the independent variable and the coefficient of variability appearing in the final model.

Factor Analysis/Principal Component Analysis (PCA)

- Is a powerful set of techniques to eliminate unwanted data from a set of samples of large size.
- The assumption is the population belonging to the same population whose large variable is studied.
- Identifies the subsets of all the variables.
- The method creates linear combinations that are orthogonal to the rest of the combinations.
- Characterise the group, find out association or differences.

Multivariate Analysis of Variance

- Is a logical extension of the univariate analysis of variance.

Canonical Discriminate Analysis (CDA)

- It works on multiple populations.
- Variables not useful in discriminating are removed.
- Works like PCA in orthogonal combinations.
- Data on same commodity found in different regions with same set of attributes.

Multiple Correspondence Analysis

- Is more generalised, working on a rectangular matrix through single value decomposition (SVD) or Echert–Young decomposition.
- Offers mapping of samples in the space of attributes and can be linked to PCA or CDA.

Generalised Procrustes Analysis

- Is a consensus-making technique.
- Samples measured by different agencies are considered and evaluated for consensus.
- It makes use of translation, rotation and scaling.
- Applicable to the data of free-choice profile.

Source: Anon (1987)

discriminant analysis, factor analysis and principal component analysis (PCA) are types of multivariate analysis frequently used in sensory studies (Table 40.9).

In several studies, where a comparison is required between many samples, conclusions cannot be drawn where a vast set of data are generated. To simplify the interpretation of the large data set, PCA is used to show the relationship between the descriptors. The first principal component (PC) is derived to explain the maximum amount of variance in the data. The second PC is extracted in the same way from the remaining variance. Each PC is a linear combination of the variables (descriptors), and plotting the loading for each descriptor as a vector facilitates the differences in the sensory properties of wines. Partial least square (PLS) is another statistical tool, which can relate data from two data sets, *e.g.* one set from soil conditions and another from a descriptive analysis was generated by a trained panel wherein an association of wines with a higher intensity of 'vegetable' aroma from the soil with a higher water-holding capacity was found. Conversely, fruitier wines from older and gravellier soil with limited water-holding capacity have also been identified.

Sensory analysis data can be easily analysed and interpreted using multivariate analytical techniques such as principal components analysis, cluster analysis and canonical variate analysis. Figure 40.10 shows an example of a sensory results analysis using different sensory descriptors of different wines aged in different oak wood barrels employing a principal component analysis. Another example is apple wines fermented with different standard *Saccharomyces cerevisiae* strains and with a natural yeast source evaluated by a trained panel using a descriptive analysis. PCA could differentiate between the natural source of fermentation and the standard yeast strains in the wine flavour profile.

40.6 WINE SENSORY EVALUATION

Wine sensory evaluation is carried out in the following sequence: (i) appearance – colour and wine limpidity when the wine glass is tilted and its contents viewed against a light source; (ii) aroma – the olfactory response; (iii) taste – first in the 'mouth' followed by the 'finish'. Based on these attributes, the tasters communicate their perception of the wine under review by descriptive words or phrases.

40.6.1 Appearance/Visual

The wine colour, as well as the wine limpidity in a clear glass, is examined by the eye, usually at the beginning of a tasting. The wine, which is held in front of a white background, should be limpid without a deposit. Wine colours change, even within the same type of wine. For example, the colour of white wines ranges from green to yellow to brown. In a white wine, more colours usually indicate more flavour and age, while a brown wine may indicate that it has gone bad. Time improves many red wines but ruins most white wines. Red wines are not just red; they range from a pale red to a deep brown-red, usually becoming lighter in colour as they age. Terms used to describe the colour of red wines include light, medium or dark red, purple, ruby, brick red and brown. The age of red wine is evaluated by observing its 'rim' in the glass. The glass is slightly tilted and the edge of the wine is observed. A purple tint may indicate a young wine while brick red, orange to brown or even yellow indicates a mature red wine. Some of the terms used to describe the colour of white wines include various depths (pale, mild or deep) of yellow, gold, amber or straw.

Colour and appearance give the first impression and are associated with wine quality. Acceptance or rejection depends almost entirely on what the eye picks up quickly. If the colour deviates from the expected, the other attributes are rarely judged. Sensory evaluation can play a vital role in defining, optimising and correlating the colour with instrumental

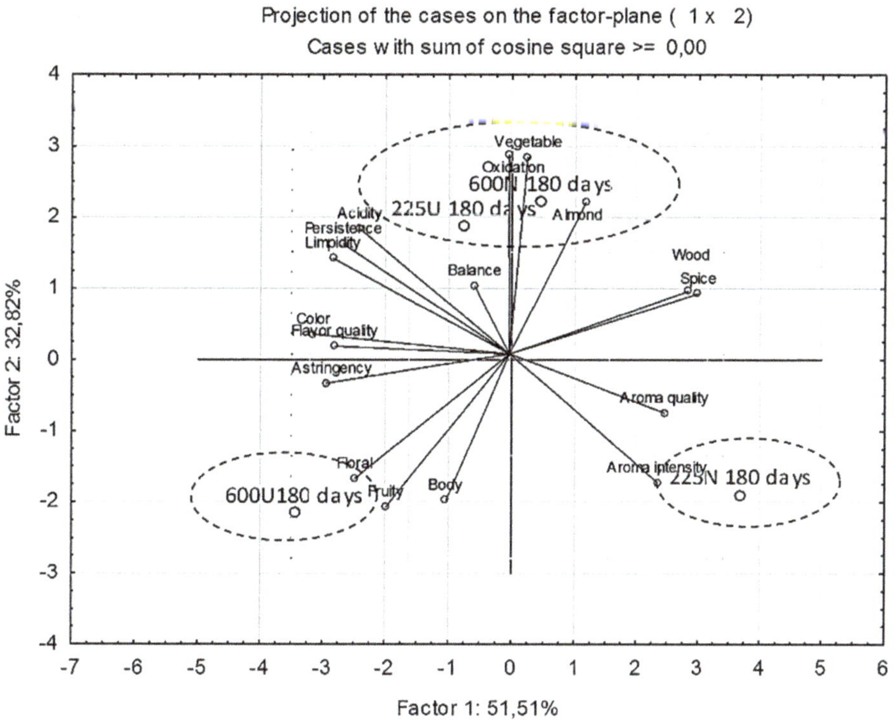

FIGURE 40.11 Principal component analysis score plot (PC1 and PC2) for the sensorial parameters of white wines aged in different oak wood barrels. *Source:* From Nunes *et al.* (2017)

analysis. To measure colour, the difference is another problem and for valid results, a 'just noticeable difference' should be established through sensory analysis. Several methods can be used to evaluate the colour difference, viz. matching with a standard colour/sample, interval scaling, descriptive methodology and ratio scaling. Figure 40.12 shows an example of a visual wine aspect analysis of red wines by a wine expert in an individual booth.

FIGURE 40.12 Modern individual booth where a wine expert is analysing the visual aspect of several samples of red wines samples. *Source:* Photo courtesy of António M. Jordão

40.6.2 Aroma/Smell

There are two routes, referred to as nasal and retronasal, by which wine volatile compounds reach the olfactory organ, which is located just behind the top of the nose where smells are somehow recognised, followed by signals to communicate this information to the brain to identify the smell. The first route means that the wine volatile compounds will reach the olfactory organ through the nostrils of the nose during the period of nosing the wine from the glass before any sample is placed in the mouth. The second route means that the smells released from the wine in the mouth go *via* the back of the mouth to the olfactory organ (retronasal perception). There is a temperature change, with the wine warming to body temperature. Swirling the wine releases the volatile compounds present in the wine, allowing the panellist to smell the volatile compounds, also called a bouquet. The two main techniques that wine tasters use are: (i) taking a quick whiff of the air-space above the wine glass, an initial impression is formulated, then taking a second deeper whiff; and (ii) taking only one deep whiff of the air-space above the wine glass. Either way, after smelling the wine, the panellist should sit back and contemplate the aroma. They are not supposed to 'taste' the wine yet, but concentrate only on what is smelt. Initially, this may be difficult to describe in words, but after trying many wines the judge will notice similarities and differences. Sometimes, a certain smell will be very strong with underlying hints of other smells. A panellist should take time to record an aroma. Wherever possible, a tasting glass with a narrow mouth should be used as it helps to retain the volatile aromas rising from the wine. It is essential to remember that we smell much of what we would describe as 'tastes'. Characteristics such as 'berries', 'pepper' or 'wood' are not tasted in the mouth, but registered through the nose. The wine's nose should be free of any obvious off-odours, rather it should have several pleasant smells, often identified with other memories – such as lemon, melon, vanilla or mint. This particular stage of the evaluation is crucial and many wines can be assessed by nose alone. Therefore, wine aroma is also an important evaluation attribute in sensory analysis, playing an important role in the acceptability of the product. Aroma is distinguishable from odour. The term 'aroma' is frequently referred to as the perception of smelling and is described as 'aromatic'. While the term 'odour' refers to perception by the nose, the term aroma refers to perception by mouth through the nasopharyngeal region. The term 'flavour' is a combination of aroma and taste and not simply a sense of smell. Quality assurance for aroma has two possible approaches: (i) instrumental analysis and (ii) sensory analysis. Instrumental analysis has the limitations that the total impact of an aroma is rarely from a single compound and no instrument has so far been devised to analyse all the possible components of an aroma. Aroma or a flavour profiling test can give actionable data on the quantitative, qualitative and hedonic aspects of food.

40.6.3 Taste/Flavour

The perceptions of taste and mouthfeel are derived from two distinct sets of chemoreceptors. Taste is associated with specialised receptors primarily located in the taste buds on the tongue and generate at least five distinct, receptor-mediated, gustatory sensations – sweet, umami, bitter, sour and salty. Mouthfeel is activated by free nerve endings, and gives rise to the sensations of astringency, dryness, viscosity, heat, coolness, prickling and pain. The only textural aspect associated with wine is generated by the bursting of a sparkling wine's bubbles. The four basic tastes that are well recognised are sweet, salty, bitter and sour sensations. The combined stimulation of the taste buds and the olfactory receptors is called the 'flavour', where non-volatile compounds contribute to taste and volatile compounds contribute to aroma.

To get the full taste of a wine, the three steps involved are: (i) initial taste (or first impression): this is where the wine awakens the senses and the taste buds respond to sensations; (ii) kinaesthetic sense: slosh the wine around in the mouth and draw in some air; then examine the body of the wine – is it light or rich? smooth or harsh?; and (iii) aftertaste: the taste that remains in the mouth after the wine is swallowed is noted here. The wine flavour is perceived when the wine is in the mouth and is a combination of three of the taste sensations from non-volatile substances perceived on the tongue and the aroma sensation from volatile substances perceived by the olfactory organs behind the nose. When the wine is in the mouth, it is warmed up and moved around, and there is the option of noisily sucking air through the mouth. All these actions help to release the volatile compounds from the wine which retronasally go to the olfactory organ via the back of the mouth. The increase in temperature in the mouth will enhance the diffusion coefficients, affecting the partition coefficients, of the volatile compounds. After tasting the wine, whether its overall flavour and balance are appropriate for that type of wine is decided. While the method of assigning a point score to wine can be useful, it has limitations in determining the quality of the wine. To get the most from the taste buds during the wine tasting, the wine should be swished around the mouth, allowing all the taste buds to participate in the detection of the wine flavours. The word to remember here is 'balance'. The amount of fruit must balance with the acids or the tannins.

As always with learning, it is necessary to go with the first impression. When tasting wines, some of the common terms used are: acidity (indicates sharpness or tartness to the taste, high acidity can have a drying effect on the mouth); astringent (the puckering sensation is usually caused by tannins derived from the grape skin and/or oak; high-quality red wine shows velvety astringency, while those aggressively astringent are termed harsh or too tannic); balance (harmonious taste balance of wine constituents, having the right proportions of sweetness and acidity); complexity (a wine is described as complex when it provides a diversity of desirable sensory sensations in harmony with each other which is derived from the various combinations of primary fruit characters, developed fruit characters and winemaking contributed characters); finish (the lingering taste, the flavour sensation after the wine has been swallowed).

Some words used to describe the finish of wine include: rich, alcoholic, hot, bitter, watery, sharp, crisp, acidic, dull, flat, clean, tannic and oaky; flavours (it is the combined taste and smell of a wine when held in the mouth); freshness (the odour of young wine is similar to fruit and a wine that portrays freshness will not only be crisp, but will also have associated stimulating and fresh fruit aromas and flavours); roundness (denotes the character of a wine which is soft with a pleasing finish, well balanced, showing body and fruit, not hard or aggressive); sweetness (sweet taste sensation is due mainly to the sugar content of the wine); and synergy (the combination of separate aromas/flavours to produce a new aroma and flavour sensations, quite different from the individual components present in all wines, the effects of synergy are most obvious when winemakers blend wines).

A wine's non-volatile compounds contribute to the overall taste quality of the wine. For example, wine acids contribute to the so-called 'wine acidity taste', and the residual sugars to 'dryness' or 'sweetness.' Other basic tastes, such as bitterness, are determined using a variety of compounds of different chemical compositions, while saltiness is determined using the mineral salt content, mainly sodium chloride, which is not important in wines, nor is the taste of umami. Astringency within the mouth as a whole is often an important characteristic; some compounds have a bitter taste as well as an astringent feel.

Sourness and saltiness are commonly called electrolytic tastes, as these are triggered by small soluble inorganic cations (positively charged ions). Salts induce membrane depolarisation which then activates the release of neurotransmitters from axonal endings that induce the firing of associated nerve fibres. Sourness is induced primarily by H^+ ions, as well as other cations to varying degrees, whereas saltiness is activated by metal and metalloid cations.

Tartaric, malic, lactic, and citric acids, account for some 90% of the total acidity, are generally recognised as the most important contributors to the quality and taste of a wine, though the minor role of lactic acid should be considered. Acetic acid is, however, a potential spoilage agent (*i.e.* 'vinegary' and 'sour'). Tartaric acid contributes to a wine's tart taste. Malic acid gives a green taste to the wine. Succinic, lactic, citric, and acetic acids are present in small very amounts in wines. In the particular case of succinic acid, originating during fermentation, its characteristics are attributed to the typical fermentation taste. Citric acid gives a fresh and slightly acid taste to the wine. Lactic acid formed in wine during the fermentation process gives a slightly sour taste.

Sweetness or dryness in wine is given by the residual amount of sugars; in dry wine, glucose is around 0.2–0.8 g/L and fructose is around 1.2 g/L, and for sweet wine, it may be up to 30 and 60 g/L, respectively.

40.6.4 Texture

The texture is one of the most important food characteristics, thus, it is a sensory attribute and is measurable directly by sensory means. The three senses of touch, sight, and hearing may be involved in the sensory assessment of texture. Texture characteristics are categorised into: (i) mechanical characteristics; and (ii) geometrical and other miscellaneous characteristics, related to the reaction of food to stress, the size, shape, and orientation of particles or those relating to the moisture and fat contents.

Normally, the sense organs involved in texture evaluation may be grouped into: (i) those in the superficial structure of the mouth; (ii) those around the roots of the teeth; and (iii) those in the muscles and tendons. Methods for measuring texture by sensory testing are finger feel, hedonic rating, descriptive tests and profiling tests. The best way is to measure a part of the components of the complex texture using suitable instruments and then, these instrumental measures can be correlated to the sensory perceptions of the texture.

40.6.5 Mouthfeel Sensations

The in-mouth sensory properties of wine are a complex mixture of taste (*e.g.* bitterness, acidity, sweetness, and saltiness) and mouthfeel sensations, mostly astringency, and flavour. Bitterness and astringency play an important role in the quality of red wine. Bitterness is a taste correlated with the presence of various structured receptors that are activated by a wide range of molecules, while astringency is a sensation of drying and puckering that is considered to be a mouth tactile response. Phenolic compounds in wine are the main class of compounds that take part in in-mouth sensory properties, in particular monomeric flavanols and their polymerised forms, usually referred to as proanthocyanidins.

The mouthfeel sensations perceived during the tasting of red wines are astringency, acidity, body, and heat. Bitterness and astringency are distinct, affecting different areas in the mouth and tongue. Bitterness is one of our basic tastes, while astringency is more a sensation in the mouth. Mouthfeel generally corresponds to 'body' and is a sensation perceived in the mouth as a whole, rather than just the tongue. The body is often correlated with viscosity or 'thickness' in the mouth.

Astringency refers to a complex of puckery, rough, dry, dust-in-the-mouth, occasionally velvety sensations, whose precise molecular origins are still in dispute. Nonetheless, in red wines, they are primarily activated by flavonoid (condensed) tannins. Anthocyanins can enhance the astringency induced by procyanidins, but do not directly contribute to astringency or bitterness. White wines show less astringency due to their lower phenolic concentrations. When astringency is detected in white wines, it probably arises due to high acidity. Bitterness and astringency are related to the flavan-3-ols present in wine. Bitterness is mainly produced by small amounts of monomeric flavan-3-ols such as catechins. Oligomeric and polymeric proanthocyanidins, with 2–5 monomeric units will tend to be more astringent than bitter in solution, these are mainly dimers such as procyanidins B1, B2, B3, and B4.

In wine, astringency is normally ascribed to the binding and precipitation of salivary proteins namely glycoproteins with phenolic compounds. Although flavonoid monomers and dimers do not effectively precipitate proteins, they may provoke astringency by structural protein deformation. Currently, it is accepted that astringent molecules form complexes with salivary proteins due to hydrophobic interactions, and hydrogen bonding precipitates the saliva protein, resulting in a lack of lubrication in the mouth. It may be added that other tannin–protein interactions are known, but are apparently of little significance in wine. As a consequence of polymerisation, proanthocyanidins mass, shape, and electrical properties change, potentially leading to protein denaturation, aggregation, and/or precipitation.

The interaction of tannins with salivary proteins is involved in astringency. Additionally, a breakdown of the mouth saliva film is detected by the increasing activation of mechano-receptors, and the precipitation of dead cells and other mouth debris that enhances the feeling of particles in the mouth. The focus is on saliva lining oral mucosa, the mucosal pellicle. Observation using microscopic techniques in a cell-based model reveals the impact of two tannins (epigallocatechin – EgC and epigallocatechingallate – EgCG) on the mucosal pellicle structure and properties at low (0.05 mM) tannin concentration, below the sensory detection threshold, in which the distribution of the salivary mucins MUC5B on cells remained unaffected, but at 0.5 and 1 mM, MUC5B–tannin aggregates could be seen and their size increased with tannin concentration and with galloylation. In addition, 3 mM EgCG resulted in higher friction forces measured by atomic force microscopy (AFM). In the presence of basic proline-rich proteins (bPRPs), the size distribution of the aggregates were greatly modified and tended to resemble that of the 'no tannin' condition, highlighting that bPRPs have a protective effect against the structural alteration induced by dietary tannins.

The sensory properties of different grape anthocyanin fractions in terms of their reactivity towards bovine serum albumin (BSA) and salivary proteins, and in tasting sessions to assess anthocyanins best estimated thresholds (BET) in a wine-like solution, were analysed. Anthocyanins react with both BSA and salivary proteins, but to different extents, as higher interaction between salivary proteins and anthocyanins could be found. Cinnamoylated anthocyanins (*p*-coumaroylated and caffeoylated) are the most reactive to salivary proteins. Tasting sessions suggested the involvement of anthocyanins as in-mouth contributors in wine, and the descriptors used for this study were astringency and bitterness.

40.7 FACTORS WITH IMPACT ON WINES SENSORY PROFILE

The sensory quality of wine is influenced by a number of factors and some of these has been discussed here.

40.7.1 Effect of Yeast Strains/Varieties

Using some basic analytical approaches, the role of yeast, grape or other fruit variety, the altitude where the fruit is grown and the content of the oxygen in the finished product have been determined. The selection of yeast is based on the aroma profile, GC analysis, and the importance of the ultimate aroma quality. In several varieties of grapes, the physicochemical characteristics, colour (intensity hue and limpidity), and aroma profile by a trained panel specifically in the production of the base wines for sparkling wine have been evaluated. The proportion of mixtures of different grapes grown in Southern Italy to obtain the end product was calculated. The effect of altitude on the quality of grapes used in winemaking, using Italian and French clones grown between altitudes of 250 and 700 m and applying the factor analysis (varimax rotation) method for data analysis, showed that a sensory panel was helpful in identifying that the grapes grown at high altitude gave a better body, sour-fruity, sweet-pungent, and varietal fruit character to the wine.

Sensory analysis data from a descriptive profile of wine were coupled with GC-olfactometry and checked for a mass spectral pattern in Vidal Blanca wines. Young wines were found to have a characteristic fruity odour as observed both by the sensory panel and GC-olfactometry/MS, which indicated that it was from terpenic compounds. The aged wines, however, showed an asparagus-like aroma by sensory test only, but not in GC-olfactometry and hence, none by mass spectra. A sensory test was able to give the vital information that Vidal wine was at its best only up to 2 years of ageing, while GC olfactometry/MS could not.

40.7.2 Effect of Malolactic Fermentation

Malolactic fermentation is the second fermentation that takes place in wines, particularly in red wines, to ensure its stability and harmony in the mouth and is carried out in a lactic acid bacteria–controlled manner to avoid possible sensory alterations in the wine. Red wines with a malolactic fermentation are much more stable aromatically and microbiologically over time. When red wine has a high pH, low alcohol, and low SO_2, malolactic fermentation is highly likely to be triggered spontaneously in uncontrolled conditions. The wine then acquires aromas of putrefaction (associated with molecules, *i.e.* putrescine), sweat aromas, or aromas of stagnant water (indicative of cadaverine). Both putrescine and cadaverine are derived from the enzymatic decarboxylation of some amino acids where lactic acid bacteria (LAB) utilise all of the acetaldehyde produced during the micro-oxygenation of those red wines that have been produced under this process. From initial aromas of freshly cut apple and a certain amount of bitterness in the mouth to a marked ripe fruit character is developed by the wine, with softer tannins and a less green impression.

Completion of malolactic fermentation is indicated by the presence of diacetyl, which in low amounts reduces the fruit character of the wine and in higher amounts releases notes of vanilla, milk and caramel. Some bacteria strains can also transform cinnamic acids present in the grapes into vinyl phenols, which are the substrate for the synthesis of ethyl phenols, which release odours of barnyard, leather and horse sweat in the case of ethyl phenol, and aromas of burnt gum and spices in the case of ethyl guaiacol.

40.8 DESCRIPTIVE SENSORY EVALUATION OF BRANDY

Brandy is an alcoholic beverage produced from the distillation of wine. After distillation, the freshly distilled brandy undergoes a period of maturation or ageing in oak or other wood such as chestnut barrels. A descriptive sensory evaluation using different descriptive terms can be used to describe the various aroma nuances in brandy. However, different people use different terms to describe the same aroma nuance. Normally, experts working in small groups within a brandy-producing company tend to develop their own company-specific terminology which can be difficult for outsiders to understand. Such tendency makes the interpretation of the results of research and product development of brandy evaluation data more difficult. A comparison of the sensory evaluation of brandies from Changyu XO (extra old) and Hennessy XO was performed by a panel of tasters. It revealed that floral, alcohol and rancid aroma descriptors achieved higher scores in Changyu XO and Hennessy XO, while a lime aroma seemed specific to Hennessy XO and herb, and almond aromas were specific to Changyu XO.

The brandy aroma descriptors are words most commonly used to describe brandy aroma and most of these descriptors have specific meanings within a brandy context that may be subtly different to their use in other industries. The terminology was developed to be applicable to all types of brandy. In evaluating brandy, the term 'herbaceous' aroma notes, *i.e.* 'minty', 'eucalyptus' and 'buchu', relates to natural herbs and plants that are sometimes added to medicinal brandies and are not natural brandy–derived aromas. The 'fruity' aroma notes are considered the most important in brandy and make it uniquely different and distinguishable from other distilled products such as different kinds of alcoholic beverages. The 'muscat' and 'floral' notes are found especially in brandies produced from aromatic Muscat grape types, *e.g.* Hanepoot grapes (Muscat d'Alexandrie). 'Woody' and 'toasted' notes are those derived during maturation from oak wood and the prior treatment of the barrels, respectively. The 'nutty', 'sweet associated' and 'spices' aroma notes are often associated with brandies that have undergone maturation for 15–20 years.

A 'Brandy Aroma Wheel' has been developed that includes brandy terminology currently in use, wine aroma terminology as well as new terminology specially selected for this purpose. The Brandy Aroma Wheel is a two-tier wheel with 18 first-tier and 75 second-tier descriptors. The first-tier descriptors provide a broader description of an aroma, while the second-tier descriptors provide a more precise definition. The descriptors are divided into positive and negative brandy associated aromas. There are 10 positive brandy aromas associated with the first-tier descriptors, sub-divided into 44 second-tier descriptors. The 8 negative descriptors are in turn sub-divided into 31 second-tier descriptors. The positive descriptors are arranged in a progression from aromas that occur most commonly in young distillates ('smooth associated' and 'herbaceous') to more mature aromas ('sweet associated', 'nutty' and 'spicy'), while negative descriptors are used to describe faults that may occur during the production process.

During brandy evaluation, the Brandy Aroma Wheel serves as an aid in two ways as the user may choose a principal term at the centre of the wheel and, working outwards, a more precise description of the brandy aroma can be found. The expert may find a descriptive term that comes spontaneously to mind when evaluating a brandy. This descriptor may then be 'keyed-in' on the second tier. By working back to the centre of the wheel, the principal aroma category may be found. By linking each descriptive term to an intensity scale, a pictorial aroma profile of a brandy can be made to illustrate how brandies differ from each other. While using the wheel analytically, sensory panel members can be trained and familiarised with the terminology of brandy aromas.

The results of sensory descriptive analysis of the addition of wood chips to accelerate the ageing process of Lourinhã brandy, in which wooden barrels and stainless steel tanks are filled with wood tablets or wood staves from Limousin oak wood (*Quercus robur* L.) and chestnut wood (*Castanea sativa* mill.), show a significant effect of the ageing system on a few sensory attributes. However, the most discriminant factors are the wood botanical species and the ageing time, which significantly affected several sensory attributes. Further, brandies aged in the presence of wood tablets present a lower intensity of the golden attribute and higher intensities of the topaz, greenish, toasted, flavour, complexity and persistence attributes in comparison with brandies aged in wooden barrels. Brandies aged in the presence of wood staves present an intermediate sensory profile. Nevertheless, the overall quality of the brandies is not affected by the ageing system. In another study, the sensory profile of commercial aged distillates from grape marc, Orujo, shows that the product aged over 60 months in wooden barrels from *Quercus robur* reached the highest overall quality, next to the distillate aged over 144 months in *Quercus petraea* oak wood. The tasters evaluated more positively the species of oak used in the ageing process than the time that the distillate remained in contact with the barrel. The principal components analysis clearly showed a good separation of the aged Orujo samples in terms of sensory descriptors according to the species and origin of the oak wood employed in the ageing process. Significant sensory differences can be attributed mainly to the origin and the species of the wood used in the ageing process and less to the contact time with the barrel.

40.9 WINE EVALUATION IN A WINERY

The attributes of a wine rely on the sensory acuity of the winemaker or the winemaker's team. Based on the winery operations or the style of wine made, the winemaker can be considered the expert crafting an artisan wine or producing a commercial alcoholic beverage designed to be accepted by many consumers. In contrast, an expert assessor is 'someone with a high degree of sensory acuity and that has the ability to make consistent and repeatable sensory assessments'. Wine experts have superior abilities to discriminate between different wines.

The wine industry values sensory data, but very few wineries use sensory techniques in their winery operations. The winemaker has extensive experience in a product category, and is able to perform evaluations to draw conclusions on the effects of raw materials, processing, storage, ageing, etc. To make use of the innovations and developments in the sensory evaluation of wine, it is suggested that the wine industry have a few 'should': wines should always be tasted blind; the tasting should be organised by a third party; tasters should not be informed of the purpose of the test; comments (as well as meaningful gestures or noises) should happen after individual data have been collected; emerging decisions should be based on the data and not on the opinion of the leader or a 'respected' taster; tasters should be required to conduct regular training sessions to keep their skills sharp. To perform these tasks, ideally, there is a need for a fully dedicated position with the endorsement of the winery management. Sensory techniques such as (1) focus groups, in which consumers discuss with a moderator why they like or dislike a product, and (2) hedonic tests that are employed in market research, in which consumers express their degree of liking using a hedonic scale – or product ranking – and then, answer 'diagnostic questions'. However, using such techniques raises the problem of language. To overcome this language discrepancy, new sensory/marketing techniques that would bypass the need for consumers to 'verbalise' their sensations have been developed. Two such techniques include (1) preference mapping and (2) reverse engineering.

40.10 SENSORY ANALYSIS AND CONSUMER REACTION

Consumer reaction to any food product forms the basis for the success or failure of any such product when launched in the market. Using surveys, consumer sensory analysis is generally carried out by the alcoholic beverages industry before commercial production and marketing of the food product. Generally, based on the research, one or two products are tested in a consumer survey. After a product has been optimised, it is always desirable to study the target population preferences using one or two formulations. In consumer analysis, food habits play a very significant role. Sensory evaluation indicates the various attributes of a food contributing to its acceptance. Either a written questionnaire is supplied or an interview is conducted to determine the acceptability of a food. The term 'survey' is used to refer to either of the methods. Of the two methods, the interview is labour intensive and time-consuming in data collection and data reduction.

Consumer analysis requires trained people as it provides only one opportunity without follow-up. The second method in which a written questionnaire is provided requires more knowledge about the subject matter than an interview. A questionnaire needs substantial time for its development; can be sent by mail; can be given to individuals or small groups; is at greater risk of losing data due to lack of survey returns; needs less trained persons to carry out the survey; and it can be used as a tool to collect a large amount of information at comparatively low costs. The tests should be very simple with any added questions on food habits and the demographic and economic details depending on the requirement of the experiment. Prior to statistically analysing the data, it is equally important to examine the data for linearity or non-linearity. Cost-related data could also be included to gain a better insight into optimising products. A consumer-oriented optimisation survey is a powerful tool that successful companies routinely use. In a well-known Australian wine region, differences in the behaviour dynamics and sensory preferences of consumer groups with the product preferences of consumers reveal that specific differences exist in the wine consumption behaviour and sensory preferences of males and females and between generational cohorts, specifically millennial and older consumers. The survey shows that the females drink less wine than males and drink higher amounts of white wine than their male counterparts; they also show a preference for a sweeter wine at a young age. Most important are fruit tastes and aromas, especially among females, as are the vegetative, wood/oak and mouthfeel characters. On the other hand, more males prefer the aged characters of the wine.

BIBLIOGRAPHY

Anon. (1987). *Basic Techniques in Sensory Analysis of Food: A Course Material*. Dept. of Sensory Analysis, Central Food Technological Research Institute, Mysore.

Arnold, R.A. and Noble, A.C. (1978). Bitterness and astringency of grape seed phenolics in a model wine solution. *Am. J. Enol. Vitic.* 29, 150–152.

Asselin, C., Morlat, R., Leon, H., Robichet, J., Remous, M. and Salette, J. (1983). Methods of sensory analysis suitable for characterization of red wines from vineyards with different types of soil. *Sci. Aliments* 3(2), 245.

ASTM 424. (1982). Annual ASTM standards American Society for-testing and materials, part 46.

ASTM Committee, E.18. (1981). *Guidelines for the Selection and Training of Sensory Panel Members, STP 758*. American Society for Testing and Materials, Philadelphia, PA.

Aubry, V., Etievant, P., Sauvageot, F. and Issanchou, S. (2007). Sensory analysis of Burgundy Pinot noir wines: A comparison of orthanasal and retronasal profiling. *J. Sens. Stud.* 14(1), 97–117.

Barrio-Galán, R., Cáceres-Mella, A., Medel-Maraboli, M. and Peña-Neira, Á. (2015). Effect of selected *Saccharomyces cerevisiae* yeast strains and different aging techniques on the polysaccharide and polyphenolic composition and sensorial characteristics of Cabernet Sauvignon red wines (wileyonlinelibrary.com). *J. Sci. Food Agric.* 95(10), 2132–2144.

Bell, G.A. (1999). Future technologies envisaged from molecular mechanisms of olfactory perceptions. In: *Tastes and Aromas*, A.B. Graham and A.J. Watson (Eds.), UNSW Press, Blackwell Science, Edinburgh, UK, p. 149.

Brossaud, F., Cheynier, V. and Noble, A.C. (2001). Bitterness and astringency of grape and wine polyphenols. *J. Grape Wine Res.* 7(1), 33–39.

Bruwer, J. and Saliba, A. (2011). Consumer behaviour and sensory preference differences: Implications for wine product marketing. *J. Con. Mark.* 28/1(1), 5–18.

Caldeira, I., Pedro Belchior, A. and Canas, S. (2013). Effect of alternative ageing systems on the wine brandy sensory profile. *Ciência Téc. Vitic.* 28(1), 9–18.

Chang, R.B., Waters, H. and Liman, E.R. (2010). A proton current drives action potentials in genetically identified sour taste cells. *PNAS* 107(51), 22320–22325.

Chaudhari, N. and Roper, S.D. (2010). The cell biology of taste. *J. Cell Biol.* 190(3), 285–296.

Chisholm, M.G., Guiher, L.A. and Zaczkiewicz, S.M. (1995). Aroma-characteristics of aged Vidal Blanca wine. *Am. J. Enol. Vitic.* 46(1), 56.

Duncan, S.E. (1995). Application of sensory evaluation in wine making. In: *Wine Analysis and Production*, B. Zoecklein, K.C. Fuigelsang and B. Gump (Eds.), Chapman & Hall, New York, p. 31.

Forss, D.A. (1981). Sensory characterization. In: *Flavour Research Recent Advances*, R. Teranishi, R.A. Flath and H. Sugisava (Eds.), Marcel Dekker, Inc., New York, p. 125.

Fretz, B.C., Luisier, J.L., Tominage, T. and Amado, R. (2005). 3-Mercaptohexanol: An aroma impact compound of Petite Arvine wine. *Am. J. Enol. Vitic.* 56(4), 407.

Gawel, R., Oberholster, A. and Francis, I.L.A. (2000). 'Mouth-feel wheel': Terminology for communicating the mouth-feel characteristics of red wine. *Aust. J. Grape Wine Res.* 6(3), 203.

Graham, A.B. and Watson, A.J. (Eds.) (1999). *Tastes and Aromas*. UNSW Press, Blackwell Science, Edinburgh, UK.

Guarreara, N., Campisi, S. and Asmundo, C.N. (2005). Identification of the odorants of two Passito wines by gas chromatography olfactometry and sensory analysis. *Am. J. Enol. Vitic.* 56(4), 349.

Guinard, J., Pangborn, R.M. and Lewis, M.J. (1986). The time-course of astringency in wine upon repeated ingestion. *Am. J. Enol. Vitic.* 37, 184–189.

Guinard, J., Pangborn, R.M. and Lewis, M.J. (1986). Preliminary studies on acidity–astringency interactions in model solutions and wines. *J. Sci. Food Agric.* 37(8), 811–817.

Haslam, E. and Lilley, T.H. (1988). Natural astringency in foods. A molecular interpretation. *Crit. Rev. Food Sci. Nutr.* 27(1), 1–40.

IFT. (1981). Sensory evaluation guide for testing food and beverage products. Sensory Evaluation Division. *Food Technol.* 35(11), 50.

Lesschaeve, I. (2007). Sensory evaluation of wine and commercial realities: Review of current practices and perspectives. *Am. J. Enol. Vitic.* 58(2), 252–257.

Jackson, R.S. (Ed.) (2004). Sensory perception and wine assessment. In: *Wine Science*, 4th ed. Elsevier Inc., New York, London, pp. 851–888.

Jackson, R.S. (2017). *Wine Tasting: A Professional Handbook*, 3rd ed. Academic Press, an imprint of Elsevier, London, New York.

Jordão, A.M., Vilela, A. and Cosme, F. (2015). From sugar of grape to alcohol of wine: Sensorial impact of alcohol in wine. *Beverages* 1(4), 292–310.

Joshi, V.K. (2006). *Sensory Science. Principles and Applications in Food Evaluation.* Agrotech Pub. Academy, Udaypur, India, p. 596.

Joshi, V.K. and Bhushan, S. (2000). Sensory evaluation of fruits, vegetables and their products. In: *Postharvest Technology of Fruits and Vegetables*, L.R. Verma and V.K. Joshi (Eds.), The Indus Publ., New Delhi, pp. 286–236.

Joshi, V.K. and Sandhu, D.K. (1997). Effect of different concentrations of initial soluble solids on physico-chemical and sensory qualities of apple wine. *Ind. J. Hort.* 54(2), 116–123.

Joshi, V.K., Sandhu, D.K., Thakur, N.S. and Walia, R.K. (2002). Effect of different sources of fermentation on flavor profile of apple wines by descriptive analysis. *Acta Aliment.* 31(3), 211.

Joshi, V.K. and Thakur, N.S. (2013). Sensory evolution of food. In: *Food Processing and Preservation*, A.K. Bakshi, V.K. Joshi, D. Vaidya and S. Sharma (Eds.), Jagmander Book Agency, New Delhi, pp. 851–866.

Joshi, V.K., Thakur, N.S. and Bhushan, S. (2012). Methods in sensory evaluation. In: *Food Biotechnology: Principles and Practices*, V.K. Joshi and R.S. Singh (Eds.), IK International Publishing House, New Delhi, pp. 861–880.

Key, B. (1999). Anatomy of the peripheral chemosensory systems, how they grow and age in humans. In: *Tastes and Aromas*, G.A. Bell and A.J. Watson (Eds.), UNSW Press, Blackwell Science Ltd., Edinburgh, UK, p. 138.

Lee, C.B. and Lawless, H.T. (1991). Time-course of astringent sensations. *Chem. Senses* 16(3), 225–238.

Mohony, O. (1985). *Sensory Evaluation of Food. Statistical Methods and Procedure.* Marcel Dekker, Inc., New York.

Narasimhan, S. and Rajalakshmi, D. (1999). Sensory evaluation of fermented foods. In: *Biotechnology: Food Fermentation*, Vol. I., V.K. Joshi and A. Pandey (Eds.), Educational Publishers & Distributors, Ernakulam, New Delhi, India, pp. 345–382.

Narasimhan, S. and Stephen, S.N. (2011). Sensory evaluation of wine and brandy. In: *Handbook of Enology*, Vol. 3 set., V.K. Joshi (Ed.), Asia Tech Publ and Distributor, New Delhi, pp. 1288–1331.

Nikićević, N. (2005). Terminology used in sensory evaluation of plum brandy sljivovica quality. *J. Agric. Sci.* 50(1), 89–99.

Noble, A.C. (1999). Using analytical sensory techniques to understand wine preference. In: *Tastes and Aromas*, G.A. Bell and A.J. Watson (Eds.), UNSW Press, Blackwell Science, Edinburgh, UK, p. 98.

Noble, A.C., Arnold, R.A., Buechsenstein, J., Leach, E.J., Schmidt, J.O. and Stern, P.M. (1987). Modification of a standardized system of wine aroma terminology. *Am. J. Enol. Vitic.* 38, 143–146.

Nunes, P., Muxagata, S., Correia, A.C., Nunes, F., Cosme, F. and Jordão, A.M. (2017). Effect of oak wood barrel capacity and utilization time on phenolic and sensorial profile evolution of an Encruzado white wine. *J. Sci. Food Agric.* 97(14), 4847–4856.

O'Mahony, M. (1986). *Sensory Evaluation of Food: Statistical Methods and Procedures.* Food Science and Technology, A Series of Monographs and Text Books. S.R. Tannebaum and P. Walstrs (Eds.), Marcel Dekker, Inc., New York.

Ossorio, P. and Torres, P.B. (2019). Sensory analysis of red wines for winemaking purposes. In: *Red Wine Technology*, Elsevier Inc., London, pp. 253–256.

Paissoni, M.A., Waffo-Teguo, P., Ma, W., Jourdes, M., Rolle, L.P. and Teissedre, L. (2018). Chemical and sensorial investigation of in-mouth sensory properties of grape anthocyanins. *Sci. Rep.* 8(1), 17098.

Piggott, J.R. (1988). *Sensory Analysis of Foods*, 2nd ed. Elsevier Applied Science, London and New York, p. 105.

Pomar, M. and Gonzalez-Mendoza, L.A. (2001). Changes in composition and sensory quality of red wine aged in American and French oak barrels. *J. Int. Sci. Vigne Vin* 35(1), 41–48.

Ranganna, S. (1986). *Handbook of Analysis of Fruits Vegetables, and Their Products.* Tata McGraw Hill Co., New Delhi.

Rodríguez-Solana, R., Rodríguez, N., Dominguez, J.M. and Cortés, S. (2012). Characterization by chemical and sensory analysis of commercial grape marc distillate (Orujo) aged in oak wood. *J. Inst. Brew.* 118(2), 205–212.

Ployon, S., Morzel, M., Belloir, C., Bonnotte, A., Bourillot, E., Briand, L., Lesniewska, E., Lherminier, J., Aybeke, E. and Canon, F. (2018). Mechanisms of astringency: Structural alteration of the oral mucosal pellicle by dietary tannins and protective effect of bPRPs. *Food Chem.*, 253, 79–87.

Sereni, A., Osborne, J. and Tomasino, E. (2016). Exploring retro-nasal aroma's influence on mouthfeel perception of Chardonnay wines. *Beverages* 2(1), 7–11.

Shah, P.K. and Joshi, V.K. (1999). Effect of different sugar sources and wood chips on the quality of Peach Brandy. *J. Sci. Indust. Res.* 58, 955–1004.

Stephen, S.N. (1999). *An Integrated Study of Chemical and Sensory Quality Parameters of Alcoholic Beverages*, PhD Thesis. University of Mysore.

Stone, H. (1992). Quantitative descriptive analysis (QDA). In: *Manual on Descriptive Analysis Testing for Sensory Evaluation*, R.C. Hootman (Ed.), American Society of Testing Material, Philadelphia, PA.

Stone, H. and Sidel, J. L. (2004). Sensory evaluation practices. Introduction to sensory evaluation. In: *Sensory Evaluation Practices*, 1–19.

Zhao, Y.P., Wang, L., Li, J.M., Pel, G.R. and Liu, Q.S. (2011). Comparison of volatile compounds in two brandies using HS-SPME Coupled with GC-O, GC-MS and sensory evaluation. *S. Afr. J. Enol. Vitic.* 32(1), 9–20.

41 Microbial Spoilage of Wine

Aline Lonvaud-Funel

CONTENTS

41.1 INTRODUCTION

Grape berries host a great variety of fungi, yeasts and bacteria. As soon as grapes are crushed and transferred to fermentation tanks, the initial complex microflora is submitted to new environmental conditions, which determine the natural selection of those micro-organisms responsible for alcoholic and malolactic fermentations, and sometimes spoilage. Only filamentous fungi do not tolerate the conditions. Of the diverse microflora on the surface of grapes, yeasts including *Saccharomyces* and non-*Saccharomyces*, are the best adapted to grow in must, but acetic acid and lactic acid bacteria are not eliminated and thus can transform the same grape must either into a great wine or into a common or a spoiled wine. Usually, fast selection occurs and *Saccharomyces cerevisiae* predominates. Alcoholic fermentation starts spontaneously with the indigenous microflora, but it can be induced by selected starters. Both lactic and acetic acid bacteria are strongly inhibited. Yeast activity results in increasing toxicity towards micro-organisms due to ethanol, fatty acids and other metabolites. Thus, the wine is microbiologically more stable than grape must. But even then, potential spoilage micro-organisms such as acetic acid bacteria or lactic acid bacteria and even yeasts of the *Saccharomyces* and non-*Saccharomyces* genera such as *Brettanomyces bruxellensis*, *Saccharomycodes ludwigii* and *Zygosaccharomyces bailii* species can survive. However, while acetic acid bacteria always alter wine, lactic acid bacteria also participate in winemaking by conducting malolactic fermentation, but some are prejudicial. Eventually, microbial spoilage of wine is induced either when desirable strains develop at inappropriate times, such as yeasts after alcoholic fermentation and bacteria before the end of alcoholic fermentation, or when undesirable strains grow. Even if the wine quality does not show any kind of alteration, it is obvious that all possible microbial spoilages do not have the same impact. Some alter the sensory quality by producing off-flavours, volatile acidity and ropiness, and others produce bioactive components with possible effects on consumer's health. Some are due to badly controlled fermentation processes and others occur during wine ageing before or after bottling. Today, increasing knowledge of genomes and the physiology and metabolism of wine micro-organisms provides efficient tools to prevent such spoilages. The possible risk of spoilage during fermentation or storage can be detected by microbiological analysis. It is important to know not only the level of contaminating micro-organisms but also their nature which could predict their possible growth according to the prevailing physico-chemical composition, and to choose the best preservation methods. With this aim, DNA tools are helping the winemaker to make the right decisions. This chapter reviews the present knowledge on the most important spoilages due to micro-organisms in wine.

41.2 YEASTS AS SPOILAGE MICRO-ORGANISMS AFTER ALCOHOLIC FERMENTATION

41.2.1 Refermentation of Sweet Wines

The alcoholic fermentation of dry wines ends spontaneously when less than 1 or 2 g/L of glucose and fructose remains unfermented. Several types of wines are semi-sweet or sweet wines because fermentation has been stopped, most often by the addition of sulphur dioxide (SO_2). In those wines, a concentration of up to 50–60 mg/L of free SO_2 is necessary to guarantee their stability. Some are produced from botrytised grapes and, depending on the vintage, high levels of SO_2 combine due to the composition of the must and too low free SO_2 remains. In these cases, refermentation occurs during storage in tanks, in barrels and even in bottled wines. It is induced either by *S. cerevisiae* strains or by other yeast species which did not develop before. *Zygosaccharomyces bailii* and *Saccharomycodes ludwigii* are well known for their high tolerance to SO_2, ethanol and high concentration of sugar. This is also the case with some *S. cerevisiae* strains resulting from the natural domestication process in the winemaking environment of sweet wines under SO_2 pressure. As they ferment, they produce acetaldehyde and, consequently, free SO_2 continuously decreases. At the same time, acetic acid is also produced, often in high levels depending on the yeast.

Sometimes, sulphiting, within the regulatory limits of the legislation, is not sufficient. Sorbic acid is the other chemical additive generally used that can be added at the maximum concentration of 200 mg/L. It reinforces the action of SO_2 which in any case must be maintained at around 40 mg/L to avoid oxidation. However, it is not permitted in organic wines. The other suitable treatment is heating wine at 50–55°C for 2–3 min. The heat resistance of several species has been determined in wines. *Z. bailii* is generally the most resistant of the spoiling yeasts. Thermal treatment is the most efficient method to stabilise such wines.

41.2.2 The 'Flower Disease'

Contrary to *Saccharomyces* sp. and other fermenting yeasts, some yeasts have an oxidative metabolism towards ethanol and develop on the surface of wine on contact with air. They form a biofilm layer and generate acetaldehyde as their main product. The yeast belong to *Candida*, *Pichia* and *Metschnikowia* genera and are normally present on grapes and in must, but cannot multiply during fermentation. They can develop after, when wine is not protected from oxygen. This yeast contamination is completely different from the biofilm development of the 'flor yeast strains' involved in the biological ageing of sherry-like wines, due to special *S. cerevisiae* 'flor' strains. By completely filling the tanks or barrels, it is easy to prevent the spoilage and, at the same time, inhibit acetic acid bacteria. Another interesting method is to fill the tank with inert gas such as argon, CO_2 and N_2 or a mixture of both, which replaces the air above the wine. All the micro-organisms that need oxygen are inhibited but are probably not killed, so that it is necessary to maintain the inert atmosphere. In practice, a low pressure of inert gas is maintained. Interestingly, storage under N_2 or CO_2 not only protects the wine from oxidative spoilage but it can also be used to optimise the dissolved CO_2 in wine.

41.2.3 *Brettanomyces bruxellensis* and the Volatile Phenols

The alteration of red wines by *B. bruxellensis* is today the most widespread problem in most of the world's wine-producing regions. Strains of this species produce volatile phenols which, in a concentration higher than the threshold of 500–700 µg/L according to the specific wine, give the wine an unpleasant taste and smell. Volatile phenols in wine comprise 4-vinylphenol, 4-vinylguaiacol, 4-ethylphenol (4-EP) and 4-ethylguaiacol (4-EG). In red wines, the most important concern is the accumulation of 4-EP and 4-EG produced after vinification, during ageing in tanks, in barrels and even in bottles. They result from the biochemical transformation of hydroxycinnamic acids, ferulic and p-coumaric acid, normal components of wine that are free or esterified with tartaric acid. Two successive reactions are involved: a decarboxylation and a reduction (Figure 41.1).

Of the micro-organisms that are able to develop in wine, *B. bruxellensis* is responsible for the production of volatile phenols. *B. bruxellensis* has been isolated from wines throughout the world. Its presence is not rare and it must be considered a normal micro-organism in the ecological niche considered. During alcoholic fermentation, it is normally outnumbered by *S. cerevisiae*. However, with the decline in yeast activity, it can develop as a real spoilage agent. The most sensitive period is when wine is stored in barrels, especially when the temperature of the cellar increases, since these conditions are most favourable for *B. bruxellensis* growth. However, spoilage can also occur at the end of alcoholic fermentation if its initial population in the grape must was high, and also in wine during storage in tanks or even in bottles if the population was sufficiently high at bottling time. In finished wines, sugars such as unfermented hexoses, minor oses (diosides, pentoses, etc.) and other wine components provide enough energy, carbon, nitrogen and growth factors to support the growth of a significant population.

The level of population necessary to produce excess ethyl phenol is usually above 10^3 CFU/mL. In fact, the presence of *B. bruxellensis* strains in any concentration is correlated to a risk of spoilage. As soon as they are detected, they must be inhibited. During ageing, it is difficult to totally eliminate them, but racking, filtration and even flash pasteurisation can serve the purpose. But the sulphiting of wine which stabilises it against other microbiological or chemical alterations is essential. In practice, it is necessary to maintain 0.5 mg/L of molecular SO_2 in addition to lowering the temperature. Sulphur dioxide is very effective against most *B. bruxellensis* strains, but it must be kept in mind that 0.5 mg/L is about 45 mg/L of free SO_2 at pH 3.75, which is a high concentration. As soon as the barrels are contaminated by wines that carry

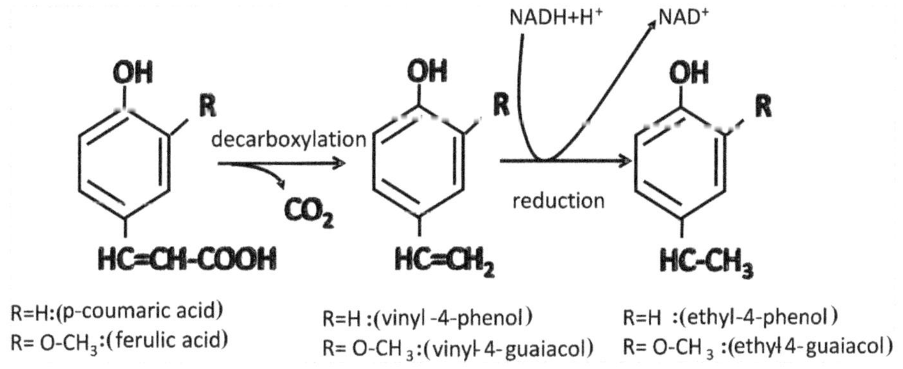

FIGURE 41.1 Pathway of the production of ethyl phenols in wine by *Brettanomyces bruxellensis*

Brettanomyces, the cellar is at risk because wood is difficult to clean and sanitise. The sanitisation of barrels at each racking, by burning sulphur and adjusting the free SO_2, is usually sufficient to contain the problem. However, it is now known that certain strains within the species are particularly tolerant to SO_2, which explains that spoilage may occur even when wines are correctly treated. Phylogenetic studies have identified the genome peculiarities of these strains and provided the molecular basis for their detection.

In all cases, but particularly in sensitive cellars, the detection and enumeration of *B. bruxellensis* should be performed. It is easier to prevent degradation if detection is done early, but it is not possible to eliminate the odour after it has been produced. The enumeration of *B. bruxellensis* is usually carried out by plate counting on a nutrient medium (yeast extract 10 g/L, peptone 20 g/L, glucose 20 g/L) which is made 'selective' by the addition of cycloheximide (500 mg/L). In reality, this medium is suitable for other non-*Saccharomyces* yeasts; however, normally in wine during ageing, only *B. bruxellensis* should be able to grow. Early detection is possible by polymerase chain reaction (PCR). The PCR protocol applied to a wine sample without any previous treatment can detect contamination at the level of 10^4 CFU/mL and no other yeast species is detected. The result on the possible contamination of wine *by B. bruxellensis* is known within a day instead of the culture method. As soon as the PCR test is positive (10^4 CFU/mL), the risk of the wine deteriorating is certain. PCR has been developed and improved to provide more precision to the winemaker by quantifying this population. Quantitative PCR is very well adapted to control the presence of *B. bruxellensis* in wines. Nowadays, it is widely used because it makes it possible to count very specifically and very quickly the *B. bruxellensis* in samples taken at any moment during the process. Phylogenetic studies show that the *B. bruxellensis* species comprises three groups. Interestingly, phenotyping, focusing on the response to sulphur dioxide, shows that the three genotypes can be assigned three levels of sensitivity: very sensitive, tolerant and resistant. Based on this result, PCR primers have been designed as new tools to evaluate the presence/absence of resistant strains. A simple duplex PCR followed by gel electrophoresis indicates if the wine contains *B. bruxellensis* and if it is (or not) resistant to sulphur dioxide. However, this recently patented method is not very widespread (TYPBrett). Other kits are available that are very simple to use. The principle is to put a sample of the wine in a culture medium (liquid or agar) conducive to the development of *B. bruxellensis* to which p-coumaric acid is added. If the wine contains yeast after an incubation period of several days, the culture has the characteristic odour of "Brett" spoiled wines due to synthesis of volatile phenols.

41.3 LACTIC ACID BACTERIA AS POSSIBLE SPOILAGE MICRO-ORGANISMS

For nearly all red wines and most white wines, winemaking is not finished if lactic acid bacteria does not complete malolactic fermentation. They multiply, metabolising malic acid and many other wine compounds. The quality of the wine is significantly improved. However, as with yeasts, bad bacterial strains produce undesirable molecules, mainly after winemaking. Even if growth is limited, the remaining micro-organisms continue to use substrates in order to gain energy and survive. Although all biochemical reactions are not yet known, some have been characterised and are implied in a wine's deterioration which can lower the wine's quality and even make it unmarketable. Additionally, under certain conditions, *Oenococcus oeni*, the most favoured bacteria for malolactic fermentation, can become an agent of alteration.

41.3.1 Excessive Volatile Acidity by *Oenococcus oeni*

O. oeni is a heterofermentative lactic acid bacterial species that is responsible for the malolactic fermentation that occurs after alcoholic fermentation. Using the pentose phosphate pathway, it ferments glucose which leads to the production of D-lactic acid, acetic acid, ethanol and CO_2 as the main products. Fructose is fermented in the same way but can also be reduced to mannitol. Glucose, fructose and pentose that have been left unfermented by yeast after alcoholic fermentation are available for any lactic acid bacteria. *O. oeni* can become a spoilage agent when it develops in a wine that still contains several grams per litre of glucose and fructose. In these conditions, volatile acidity may become excessive, it is one of the most redoubtable problems during winemaking.

Usually, lactic acid bacteria, which are already present in grape must, are inhibited by antagonism with the yeast during alcoholic fermentation. Yeast and bacteria compete for some essential nutrients, such as amino acids, but yeast are dominant. In addition, yeasts produce ethanol, fatty acids and some unknown inhibitory compounds against bacteria. In these conditions, the bacterial population is low and from the diverse microflora of the grape must, the species *O. oeni* is selected. In some cases, the antagonism during alcoholic fermentation is not efficient and relatively high concentrations of bacteria can remain, especially if the pH is high. In these circumstances, if alcoholic fermentation slows down for any reason, bacteria will actively grow. Competition then favours the bacteria. Unfermented sugars with concentrations of several grams per litre up to 10 g/L or more are fermented not by yeast to ethanol but by lactic acid bacteria to lactic acid and acetic acid. This can occur either because of a problem in the yeast survival at the end of alcoholic fermentation or because of an uncontrolled temperature.

However, *O. oeni* is not the only species implied in increased volatile acidity from sugar fermentation. Several heterofermentative lactobacilli can produce the same effect. This problem has been discussed and new species of heterofermentative lactobacilli, *Lactobacillus kunkee* and *L. nagelii*, associated with stuck or sluggish fermentation have been identified. Sulphiting the grapes as soon as they are harvested or crushed is the most efficient way to reduce the bacterial population during alcoholic fermentation. A very special case is the alteration of dessert wines, or fortified wines, where despite a very high ethanol concentration, bacteria can grow. As fermentable sugars are also present in high amounts, the concentration of lactic acid and acetic acid becomes so high that the wine is no longer marketable. The bacteria involved are predominantly heterofermentative lactobacilli (*L. fructivorans*, *L. hilgardii*) and *O. oeni*. The problem frequently occurs after bottling and the wine is lost.

41.3.2 Mousiness

A mousy off-flavour in wine is caused by a group of N-heterocyclic bases comprising 2-ethyltetrahydropyridine, 2-acetyltetrahydropyridine and 2-acetyl-1-pyrroline (Figure 41.2).

Brettanomyces sp. and heterofermentative lactic acid bacteria are responsible for this deviation in wines. Although *Brettanomyces* strains can synthetise such compounds, the yield depends on the strain and on the composition of the medium. Out of lactic acid bacteria, the ability is restricted to heterofermentative lactobacilli, *Leuconostoc* sp and *O. oeni*. However, many wines contain high levels of heterofermentative lactic acid bacteria after winemaking, without the typical mousy taint. Therefore, it is obvious that very particular conditions and/or particular strains must be met to trigger the alteration which still needs to be elucidated. The mechanism of biosynthesis was studied for a strain of *L. hilgardii*. N-heterocycle amines were produced from several substrates including ethanol, fructose and amino acids (L-ornithine and L-lysine) in the presence of Fe^{3+}. However, the practical conditions which led to the development of the off-flavour were not stated.

41.3.3 Ropiness

Among the 'diseases' of wine described by Pasteur in 1856, ropiness, which had disappeared for decades, is now increasingly frequent. Ropy wines are characterised by abnormally high viscosity and an oily or slimy appearance due to the presence of a particular exo-polysaccharide synthesised by lactic acid bacteria (Figure 41.3). The alteration is observed in tanks and barrels, but most often in bottles, long after bottling. Its development is very slow and may take several months, or even years, to make the wine ropy.

In practice, ropy wine can be quite easily recovered. If it is in a tank, it is only necessary to transfer it with mechanical

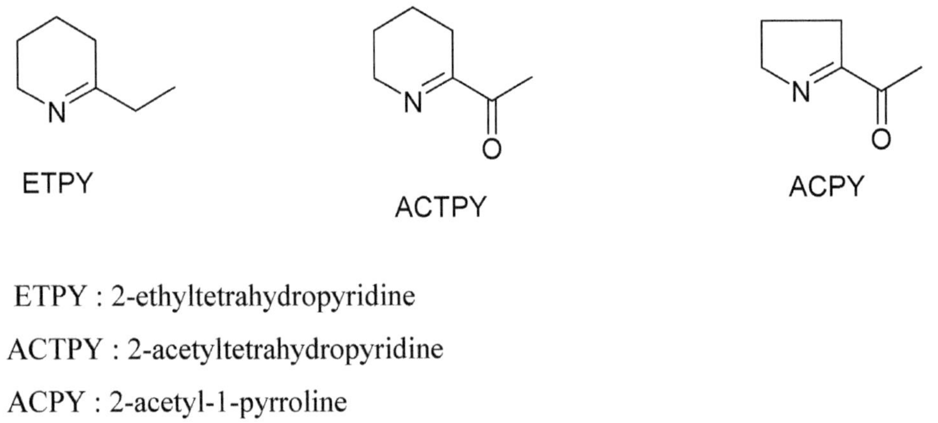

FIGURE 41.2 The N-heterocyclic bases responsible for the mousy taints of wines

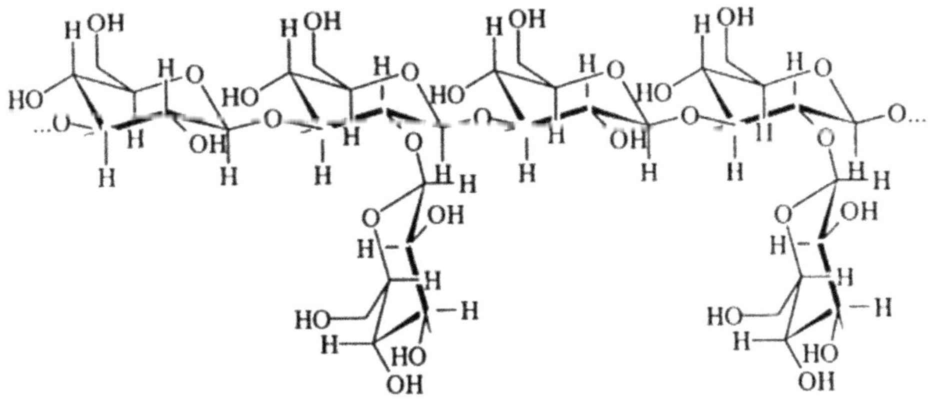

FIGURE 41.3 Structure of the exopolysaccharide, β-glucan, produced by ropy *Pediococcus damnosus* strains

action, to decrease the viscosity, and then to treat it in order to eliminate the bacteria responsible for the ropiness. However, the problem is frequently encountered in bottles, so that they must be drained into a tank, the wine agitated and then sterile filtered or treated with heat, in addition to correct sulphiting. Strains isolated from spoiled wines have been identified as *Pediococcus damnosus*. They grow very easily in wine and appear to be more tolerant to the adverse conditions of pH, SO_2 and ethanol than other lactic acid bacteria. When plated on a nutrient agar, they produce ropy colonies easily characterised by producing a slimy filament when picked up with a small wooden stick. Not all strains of *P. damnosus* harbour the phenotype. Many wines contain high levels of *P. damnosus* and are not spoiled. Only some strains can produce glucan and differ from non-ropy strains by the presence of an additional plasmid of 5.5 kb. The cultivation of such ropy strains by successive transfer in a growth medium without ethanol leads to the loss of the plasmid and the phenotype. On the contrary, ethanol and especially wine should as act a selection pressure which stabilises the plasmid in the strain.

Ropy *P. damnosus* strains produce the glucan from glucose not fermented either by yeast during alcoholic fermentation or by lactic acid bacteria during malolactic fermentation. Very small amounts of glucose are sufficient to produce glucan which, even at ≅100 mg/L, greatly increases viscosity by interacting with the wine matrix. The population of spoilage bacteria generally reaches 10^5–10^7 CFU/mL when ropiness is visible. The sequencing of the plasmid has revealed the presence of three putative genes encoding a replicating protein, a mobilisation protein and a glucosyltransferase. Molecular tools have been designed from knowledge of the plasmid. First, using DNA/DNA hybridisation, a fragment of the plasmid was used as a DNA probe to detect only the specific ropy *P. damnosus* strains in colony hybridisation. Then, PCR primers in the gene were chosen, which encode the glycosyltransferase; a population of 10^2–10^3 CFU/mL can be detected by visualisation of the amplicons after gel electrophoresis. Afterwards, a quantitative qPCR protocol was developed to rapidly and precisely determine the level of contamination. The detection and/or enumeration of ropy strains should be carried out in cellars where ropy wines have already been observed. Indeed, these strains are very well adapted to this ecological niche and the spoilage is often repeated in the same cellars. As for other microbial alterations, the higher the pH, the higher the risk of ropy wines. If detected just before bottling, the best method to stabilise the wine is heat treatment which kills the bacteria. Filtration may, however, leave a residual population.

41.3.4 Bitterness: Production of Acrolein

Glycerol is the second most important component in determining the quality of wine produced by yeast during alcoholic fermentation. Some wines are contaminated by lactic acid bacteria which can degrade glycerol by the glycerol dehydratase enzyme. The glycerol dehydratase pathway comprises glycerol dehydratase and 1,3-propanediol dehydrogenase (Figure 41.4). This pathway is not present in all lactic acid bacteria strains. It leads to the production of 1,3-propanediol with an intermediary product (3-hydroxypropionaldehyde [3-HPA]) which is easily dehydrated into acrolein. This metabolism should be classified undesirable both for the production of acrolein as an irritating and carcinogenic component, and as responsible for the deterioration of sensorial quality since acrolein combines with polyphenols and gives a strong bitter taste. So far, it has only been found in several *Lactobacillus* species such as *L. collinoides* isolated from ciders and *L. hilgardii* and *L. diolivorans* isolated from wines. The detection of such bacteria is now easily done using the PCR test based on the nucleotidic sequence of the gene encoding a subunit of the glycerol dehydratase.

41.4 ACETIC ACID BACTERIA: A CONTINUOUS RISK FROM TANK FILLING UNTIL BOTTLING

Acetic acid bacteria are constantly present, from the grape berries to the finished wine. The less they grow, the lower the final volatile acidity of wines. Acetic acid bacteria belong to the Acetobacteraceae family of bacteria which need oxygen for re-oxidation of the co-enzymes involved in substrate oxidation to gain the energy necessary for their growth. Following

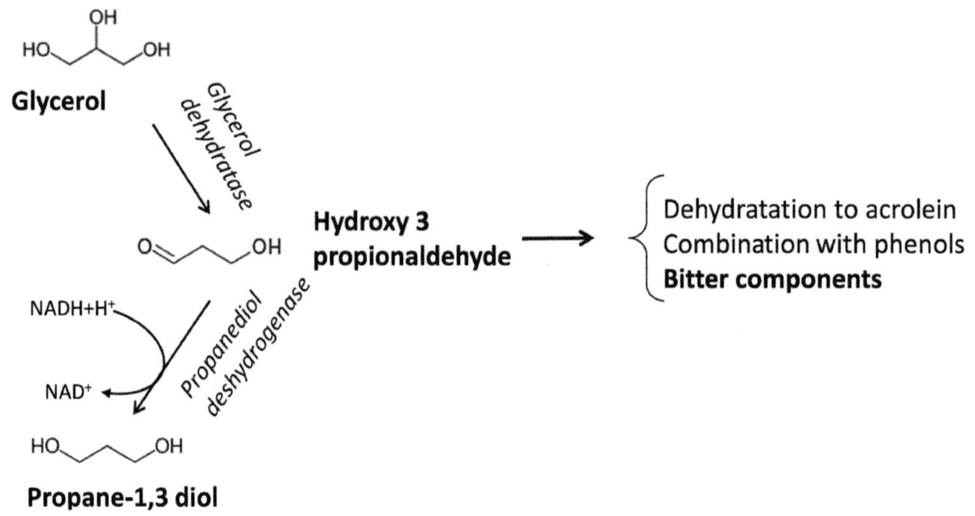

FIGURE 41.4 The glycerol dehydratase pathway of *Lactobacillus* species isolated from wine (under rarer conditions 3-HPA can be oxidized to 3-hydroxy propionic acid)

phylogenetic studies of the family structure, they have been reclassified into 32 genera of which only three are usually detected in the grape and wine environment: *Acetobacter*, *Gluconobacter* and *Gluconacetobacter*. On grapes and in wines, the most predominant species are *Gluconobacter oxydans*, *Acetobacter aceti* and *A. pasteurianus*.

Only *Acetobacter* can oxidise ethanol to acetic acid and overoxidise it to CO_2 and H_2O by the tricarboxylic acid cycle. However, overoxidation is not possible in wine conditions. Glucose is the preferred substrate for *Gluconobacter* and ethanol for *Acetobacter*. The consequences of these metabolisms on wine quality are of two orders: first, the damage due to *Gluconobacter* at the beginning of winemaking, or even on the grape berries; second, the wine spoilage by *Acetobacter* sp. (*A. aceti* or *A. pasteurianus*) at any time during winemaking if fermentation stops, and during storage and ageing, only on condition that oxygen is available.

41.4.1 Presence of Acetic Acid Bacteria on Grapes and Its Consequences

The population of acetic acid bacteria within the microbial biofilm developed on grapes greatly depends on the climatic conditions during ripening, and on the surface integrity of the berry. If insects or mechanical impacts have damaged the grape skin, moulds, yeasts and bacteria readily invade the berry. On the contrary, a high population with intact skin is very rare. The level of the total population, however, is also linked to the humidity of the atmosphere. On healthy and unspoiled berries, the population is around $10–10^3$ CFU/mL and *G. oxydans* is nearly exclusive. On damaged berries, and in the case of rain or fog, moulds and yeasts should be the first to attack, taking advantage of skin injuries. They grow and penetrate inside the berry which increases its availability of sugars and other nutrients to them and other micro-organisms of the microbiota. On rotten grapes, the acetic acid bacteria population is much higher than on healthy grapes. *G. oxydans* is not more predominant except in the case of 'noble rot'. On spoiled berries, the species *A. pasteurianus* and *A. aceti* can represent the total population.

The higher presence of *Gluconobacter* on healthy berries and of *Acetobacter* on damaged berries is explained by their metabolism. Sugars are available on the surface of healthy grapes by exudation and are oxidised by *Gluconobacter*. For *Acetobacter*, the preferred substrate is ethanol produced in microquantities by the yeasts which can ferment sugars inside the biofilm. Thus, the damaged grapes, where moulds have injured the skin and where yeasts ferment the grape juice, are a perfect habitat for *Acetobacter* sp.

G. oxydans mainly produces dihydroxyacetone, 5-oxofructose and gluconic acid and the latter can also be oxidised to 2-oxogluconic and then 2.5-dioxogluconic acid. Up to 85% of glucose is easily oxidised to gluconic acid. From glycerol, which originates from the metabolism of yeasts and fungi, *G. oxydans* and *A. aceti* produce dihydroxyacetone, which accumulates. Only if glycerol is completely exhausted is dihydroxyacetone consumed. Finally, the other important substrate is D-fructose, which is oxidised by all strains of *G. oxydans* isolated from grapes to 5-oxofructose. When all the hexose has been transformed, the 5-oxofructose disappears. The production of gluconic acid (in the form of its derivatives δ and γ gluconolactone), dihydroxyacetone and 5-oxofructose by *G. oxydans* explains the total ability of rotten grape must to combine SO_2. In low SO_2-combining musts, these compounds only represent 28% of the SO_2-combining compounds, which shows the crucial role of *G. oxydans* in the quality of grape musts. The concern is that if *G. oxydans* has developed before the harvest, the sulphite added to stop the alcoholic fermentation in the production of sweet wines is quickly inactivated by combination with ketonic compounds. In these wines, they may be in such a high concentration that it is difficult to maintain enough free SO_2 to stabilise the wine within the limits of the regulations (see end of Section 41.2.1).

41.4.2 Presence of Acetic Acid Bacteria in Wine

Acetobacter spp. are more adapted to wine conditions than *Gluconobacter* and can be detected from the start of winemaking until bottling. Their level depends on the aeration of wine, more precisely on the redox potential. As soon as alcoholic fermentation starts, the medium becomes hostile to aerobic micro-organisms in general, and acetic acid bacteria in particular. From an initial population that varies from 10^3 to 10^5 CFU/mL depending on the quality of the grapes, the level goes down and stabilises at around $10–10^2$ CFU/mL in fermenting must and wine.

In red winemaking, wine is run off and transferred to another tank or to barrels for malolactic fermentation. During this operation generally wine is aerated, oxygen dissolves and is available for acetic acid bacteria, which dramatically rises. According to the conditions prevailing, mainly the pH and temperature, the population can reach 10^4 or 10^5 CFU/mL during the following day, but again decreases markedly in the subsequent days and then remains steady at approximately $10^2–10^3$ CFU/mL or lower. This scheme is repeated along the ageing period, each time the wine is racked and transferred from one tank/barrel to another. Thus, each time the latent acetic acid bacteria population reaches a peak just after racking, it stabilises until the next one. The peak level mostly depends on the temperature, pH and the level of O_2, which could dissolve during transfer. This evolution cannot be avoided; the amplitude of variation can only be moderated by lowering contact with air and temperature. Finally, at the end of ageing, the wine is prepared for bottling. At this stage, acetic acid bacteria are still present. Except in the case of heating or sterile filtration, they remain alive in wine. Several months after bottling, it is still possible to enumerate a low population of acetic acid bacteria. However, they can no longer multiply since the wine is no longer aerated and the bacteria slowly die and disappear. Bottled wine is completely free of acetic acid bacteria, except in the case of corking failure.

The growth of an acetic acid bacterial population means that cells find energy and components to synthetise their biomass. Wine compounds are suitable substrates and oxygen is the other indispensable substrate. The most important reaction is the two-step oxidation of ethanol to acetaldehyde and then to acetic acid. The amount of acetic acid produced at each transfer of wine varies according to the increase in population. During ageing in barrels, since run-off is repeated every 3 months, the increase in volatile acidity is unavoidable though usually limited. When all the operations are properly controlled, 10–50 mg/L of acetic acid is produced in the total time of ageing. If O_2 is dissolved in large amounts, for example if the surface of the wine stays in contact with air, the substrate for the oxidation of ethanol is not restrictive. Finally, acetic acid accumulates and can completely spoil the wine, which becomes pricked. Acetic acid bacteria also produce ethyl acetate, whose perception threshold is around 160–180 mg/L by their esterase activity. During ageing in vats, but mainly in barrels, the increase in volatile acidity is obligatory, but it must be limited. Sulphur dioxide has a very weak effect on acetic acid bacteria especially in red wines, as described earlier. Even at a concentration of 25 mg/L of SO_2, acetic acid bacteria survive and develop each time the wine is aerated. The only means to limit their growth and their metabolism is to control the only omitted parameter, the temperature. In air-conditioned cellars at temperatures of less than 15°C, the total increase in volatile acidity can be limited.

41.5 SPECIFIC METABOLISMS ALTER THE HYGIENIC QUALITY OF WINE

In addition to the main transformations of sugar by yeast during alcoholic fermentation and malic acid by lactic acid bacteria during malolactic fermentation, and also the production of aroma and flavour, some other transformations lead to undesirable compounds such as ethylcarbamate (urethane) and biogenic amines. These and related aspects are described next.

41.5.1 Ethyl Carbamate (Urethane)

Ethylcarbamate is present in wine at concentrations ranging from 1.5 to 70 μg/L. Toxicological studies have tried to define limits in relation to the carcinogenic effect of ethylcarbamate. A virtually safe dose of about 20–80 ng/kg per day has been established for humans, while other doses have been found at about 300 ng/kg per day. For more details, see Chapter 6. In wine, ethylcarbamate is produced by chemical reactions between ethanol and the products of yeast and lactic acid bacteria which are urea, citrulline and carbamylphosphate. The reaction with ethanol is accelerated by heat. The main precursor is urea produced from arginine by yeast during alcoholic fermentation (Figure 41.5). The amount of ethylcarbamate formed by heating is strictly related to the concentration of urea. By heating, it is possible to evaluate the maximum concentration of ethylcarbamate that can be produced in wine with time during storage. There is a linear relationship between urea and ethyl carbamate, whatever the time and temperature of heating. During alcoholic fermentation, the urea concentration increases, reaching an optimum level and then decreasing. The final concentration of urea in wine depends not only on the grape must, but also on the yeast strain in relation to the activities of arginase and urea amidolyase. Fertilisation in the case of excessive nitrogen levels in grapes may increase the potential ethyl carbamate, *i.e.* the maximum amount that can be produced during storage. The only possibility to decrease the urea concentration of wine is to degrade it by the urease enzyme such as from *L. fermentum* that works even at the acidic pH of wine. A preparation of dead *L. fermentum* cells is now allowed for wine treatment that take effect 8–12 weeks after treatment, depending upon the wine. Another strategy is to find the yeast strains that are able to produce less urea during alcoholic fermentation by genetic engineering of yeast strains. The CAR1 gene (encoding arginase) has been disrupted so that the transformed strain is unable to degrade arginine and produce urea (see Chapter 17 for more details on this aspect). Citrulline, the other

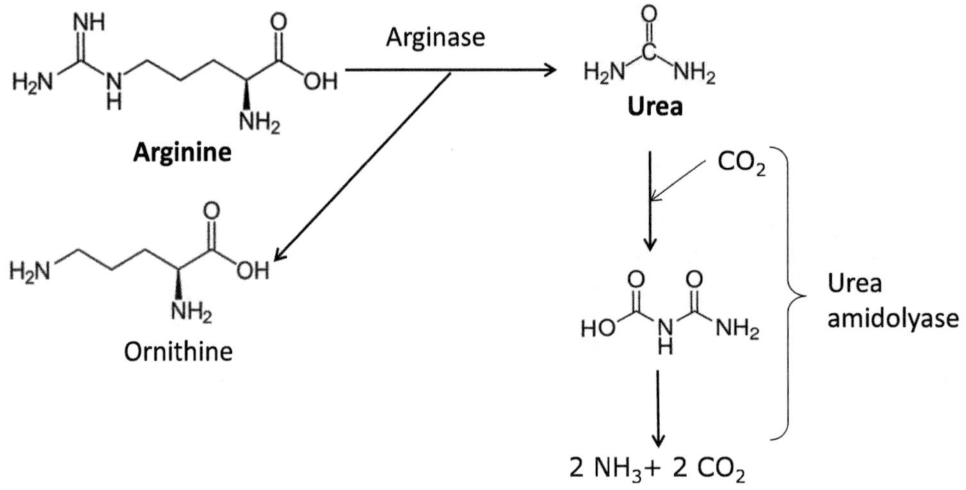

FIGURE 41.5 Schematic pathway of urea production by yeast

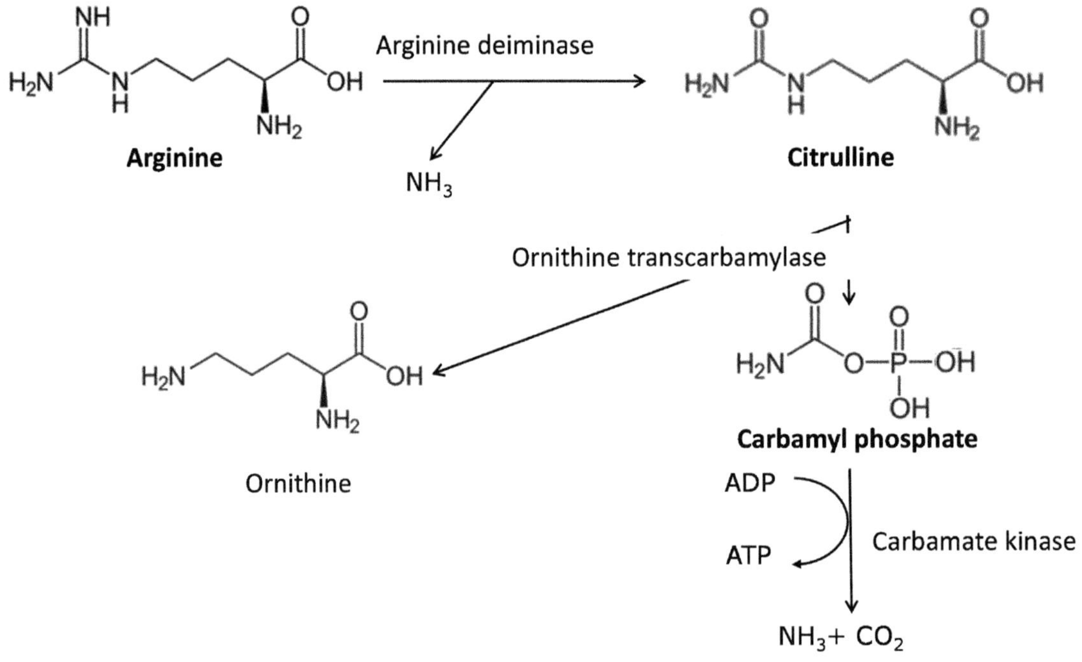

FIGURE 41.6 The arginine deiminase pathway by heterofermentative lactic acid bacteria, including *Oenococcus oeni*

precursor, is produced during the metabolism of arginine by lactic acid bacteria such as carbamyl phosphate. All the heterofermentative *Lactobacillus* sp. and some *O. oeni* strains degrade arginine by the arginine deiminase (ADI) pathway and can increase ethylcarbamate in wine, where they grow during or after winemaking. Three enzymes are involved (Figure 41.6). The genes coding them were studied in the species *O. oeni*. They belong to an operon only present in strains able to degrade arginine by this pathway.

41.5.2 The Biogenic Amines

Lactic acid bacteria decarboxylate amino acids to biogenic amines, which are considered undesirable because of their possible effect on human health, especially allergies and other variable diseases, depending on the person. In wines, histamine, tyramine and putrescine (diaminobutane), respectively, produced by the decarboxylation of histidine, tyrosine and ornithine decarboxylase, as well as from agmatine deiminase to putrescine, are the most important (Figure 41.7). Methylamine, phenylethylamine and diaminopentane (cadaverine) are also present (see Chapter 6). The concentration is very variable; some wines contain no biogenic amines while others have all of them. In general, the problem is more frequent in red wines than in white wines. In reality, only wines in which malolactic fermentation occurs are propitious to biogenic amines. Actually, only some strains of lactic acid bacteria can decarboxylate

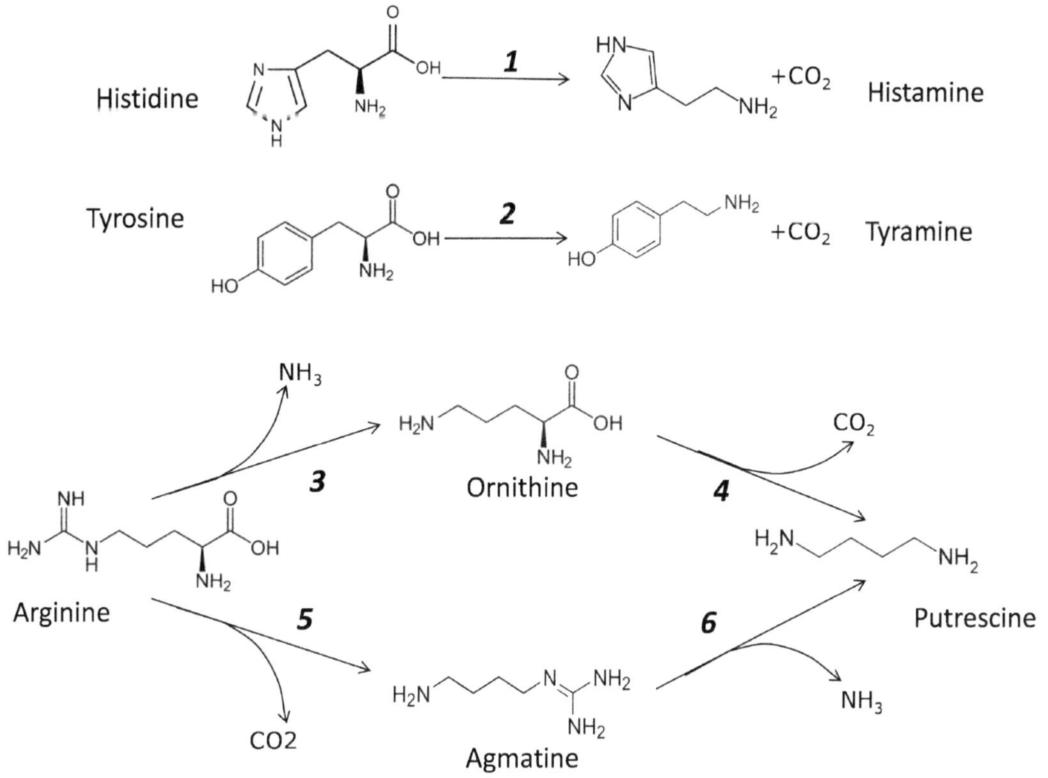

FIGURE 41.7 Six Schematic pathways of biogenic amines production by lactic acid bacteria. 1: histidine decarboxylase; 2: tyrosine decarboxylase; 3: arginine deiminase; 4: ornithine decarboxylase; 5: arginine decarboxylase; 6: agmatine deiminase

amino acids and desaminate agmatine, irrespective of their species. The necessary conditions for the optimal production of amines are first the presence of suitable strains and then favourable conditions for their growth (relatively high pH, temperature above 15°C and low SO_2 amounts). Moreover, if wine stays in contact with lees after winemaking, the amino acid concentration increases due to their release from dead cells and to the proteolytic activities. In consequence, if the concerned strains are present and still active, then amines are produced. Indeed, an increase in their concentration is frequent during the weeks or months following winemaking. The difference between a histamine-producing strain and a non-producing strain is the presence in the genome of the first of a gene cluster encoding a complete system comprising the histidine decarboxylase (HDC), an antiporter histidine/histamine and a putative regulator. A strict relationship exists between the presence of the gene and the ability to produce histamine. Based on the *hdc* gene sequence, primers have been designed and, using PCR amplification, it is possible to detect the presence of histidine decarboxylating strains. It is not species-dependent and is therefore relevant in detecting any histamine-producing bacteria in wine. Tyramine, the other most important biogenic amine in wines, is produced by decarboxylation of tyrosine by lactic acid bacteria that carry the tyrosine decarboxylase. As for histamine, a gene cluster encoding the tyrosine decarboxylase (TDC), the tyrosine/tyramine antiporter and putative regulators has been sequenced. Specific primers for the detection of tyramine-producing strains have been designed. Tyramine-producing strains belong to several different species that are present in wines. The interest in detection, targeting the *tdc* gene, is that whatever the species may be, the presence of undesirable strains can be established. Finally, it is completely analogous for putrescine-producing bacteria to decarboxylate ornithine and to deaminate agmatine for which the targets are *odc* and *agdi* genes.

Initial observations have already shown that some winemaking environments always host biogenic amines–producing strains but not others. Ecological studies are indispensable to knowing their origin and why they are selected in some wines and not in others. Currently, the winemaker has no specific tool against such strains. Obviously, SO_2 is the most efficient at eliminating lactic acid bacteria after winemaking. However, sometimes, due to the high pH of wine and to the special tolerance of bacteria, its effect may be limited. Interestingly, wines inoculated by selected strains of *O. oeni* to start the malolactic fermentation usually contain no, or little, biogenic amines than an uninoculated control. Thus, the massive inoculum of *O. oeni* could decrease the possibility of the growth of indigenous lactic acid bacteria, including the biogenic amines–producing strains. However, it does not kill indigenous bacteria which remain latent in wine and must be discarded or kept at a low population by any microbial stabilisation operation, *i.e.* sulphiting, filtration or heating.

41.6 BASIC EVIDENCE TO AVOID MOST FREQUENT MICROBIAL SPOILAGE

The oenological microbiota are composed of dozens of strains of yeasts and bacteria that come from the grape itself, or are added in the form of starters. They work in grape juice first, then in wine, through innumerable interactions resulting in the transient dominance of strains of yeasts and bacteria that carry out alcoholic fermentation and then, in many cases, malolactic fermentation. The result is an extraordinary complex of biochemical transformations. In most cases, under the supervision of the oenologist and the winemaker, the result is good quality wine. But the spoiling micro-organisms are part of the microbiota, and the evolution of the system does not always lead to their elimination. Some alterations in quality are difficult to predict and/or avoid. But, today, microbiological analysis, which now includes molecular methods based on genome knowledge of spoiling species and strains, is a first-rate aid to help make the right decisions in preventing or treating microbial alterations. It is indispensable and must be adapted to the different stages of the winemaking process.

The basic technological operations that have been known for decades must not be forgotten. During periods of storage and ageing, when the risks are the greatest, sulphiting remains the first protection tool, associated with the maintenance of a cool temperature, with racking to eliminate settled biomasses and topping up to avoid high redox potential. Filtration and, in some cases, heat treatment are also well-known oenological practices, which cannot be avoided. For white wines, the most frequent problem is the refermentation of sweet wines. For red wines, it is the off-flavours due to *B. bruxellensis*. Spoilage by lactic acid bacteria is more rare, but nevertheless probable when the rudimentary oenological rules are not respected. Early, specific and rapid detection of undesirable strains is possible and must point to appropriate solutions.

Curiously, in many traditionally wine-producing regions, incidents are increasingly frequent. It is not unlikely that strains of alteration evolve and are selected spontaneously in cellars over time. However, microbiological analysis can detect them very early, and the means to prevent their multiplication or elimination exist. But the main cause is undoubtedly the lack of vigilance intervention by some producers under the pretext of 'back to nature', perfectly highlighted by microbiological controls. Used well, the physical treatments and sulphiting to discard or decrease microbial populations have proven their suitability for a long time in the treatment of quality wines. There is, therefore, no reason to take risks that most often lead to the prevention of commercialisation.

BIBLIOGRAPHY

Cibrario, A., Avramova, M., Dimopoulou, M., Magani, M., Miot-Sertier, C., Mas, A., Portillo, M.C., Ballestra, P., Albertin, W., Masneuf-Pomarede, I. and Dols-Lafargue, M. (2019) *Brettanomyces bruxellensis* wine isolates show high geographical dispersal and long persistence in cellars. *Plos One*, 14: 12 e022274.

Claisse, O. and Lonvaud-Funel, A. (2001) Primers and a specific DNA probe for detecting lactic acid bacteria producing 3-hydroxypropionaldehyde from glycerol in spoiled ciders. *J. Food Prot.*, 64(6): 833.

Costello, P. J. and Henschke, P.A. (2002). Mousy off-flavor of wine: Precursors and biosynthesis of the causative N-Heterocycles 2-ethyltetrahydropyridine, 2-acetyltetrahydropyridine, and 2-acetyl-1-pyrroline by Lactobacillus hilgardii DSM 20176. *J Agric Food Chem* 50: 7079–7087.

Curtin, C., Varela, C. and Borneman, A. (2015) Harnessing improved understanding of *Brettanomyces bruxellensis* biology to mitigate the risk of wine spoilage. *Aust. J. Grape Wine Res.*, 21: 680–692.

Delaherche, A., Claisse, O. and Lonvaud-Funel, A. (2004) Detection and quantification of *Brettanomyces bruxellensis* and 'ropy' *Pediococcus damnosus* strains in wine by real-time polymerase chain reaction. *J. Appl. Microbiol.*, 97(5): 910–915.

International Organisation of Vine and Wine (2018) Compendium of international methods of analysis. Enumerating yeasts of the species *Brettanomyces bruxellensis* using qPCR. http://www.oiv.int/public/medias/1333/oiv-oeno-414-2011-en.pdf.

International Organisation of Vine and Wine (2018) Compendium of international methods of analysis. Microbiological analysis of wines and musts. http://www.oiv.int/public/medias/2590/oiv-ma-as4-01.pdf.

Lonvaud-Funel, A. (2001). Biogenic amines in wines. Role of lactic acid bacteria. *FEMS Microbiol. Lett.*, 199(1): 9–13.

Malfeito-Ferreira, M. (2018) Two decades of "horse sweat" taint and *Brettanomyces* yeasts in wine: Where do we stand now? *Beverages*, 4(2): 32; doi10.3390.

Nannelli, F., Claisse, O., Gindreau, E., de Revel, G., Lonvaud-Funel, A. and Lucas, P. (2008) Determination of lactic acid bacteria producing biogenic amines in wines by quantitative PCR methods. *Lett. Appl. Microbiol.*, 47(6): 594–599.

Pinto, L., Baruzzi, F., Cocolin, L. and Malfeito-Ferreira, M. (2020) Emerging technologies to control Brettanomyces spp. in wine: Recent advances and future trends. *Trends Food Sc Tech.* 99: 88–100.

Ribéreau-Gayon, P., Dubourdieu, D., Donèche, B. and Lonvaud, A. (2006) *Handbook of Enology: The Microbiology of Wine and Vinifications*. 2nd edition, Volume 1, pp. 115–192. John Wiley & Sons, Chichester, England.

Snowdon, E.M., Bowyer, M.C., Grbin, P.R. and Bowyer, P.K. (2006) Mousy off-flavor: A review. *J. Agric. Food Chem.*, 54(18): 6465–6474.

Spano, G., Russo, P., Lonvaud-Funel, A., Lucas, P., Alexandre, H., Grandvalet, C., Coton, E., Coton, M., Barnavon, L., Bach, B., Rattray, F., Bunte, A., Magni, C., Ladero, V., Alvarez, M., Fernández, M., Lopez, P., de Palencia, P.F., Corbi, A., Trip and H., Lolkema, (2010) Biogenic amines in fermented foods. *J. S. Eur. J. Clin. Nutr.*, 2010(64): S95–S100.

Stadler, E. and Fischer, U. (2020) Sanitization of Oak Barrels for Wine - A Review. *J Agric Food Chem* 68: 5283–5295.

Unit 8

Wine Industry

INTRODUCTION

Globally, the organic wine market is growing. Using a scientific approach, the consumers' motivations have been identified and analysed. Sensorial quality is of course the priority concern of consumers. But today the characteristics related to the impact on health occupy more and more place in the decision of purchase. For some consumers any treatment of vine, grapes and wine, must be eliminated. They want "natural wines" which have entered the market. But it must be recognized that most are tastefully altered. By contrast, organic wines are much more appreciated and rightly so. Relevant agricultural practices and appropriate winemaking result in good wines. The difference between organic wine and non-organic wine is low; therefore, it is difficult to justify its high price despite the cost of organic winemaking. The chapter describes the groups of consumers who were studied. For some, the interest in organic wine cannot justify the higher price, as opposed to others who are willing to pay more. Personal elements can influence behaviour in these different groups; individualism or peer orientation with nuances in between are related. The results are helpful in developing a marketing strategy for organic wines.

Wine tourism is an ongoing development in many wine-producing countries that have realised its immense value for their industry. It is playing a major role owing to its economic spin-offs in various sectors and the development of rural areas. It integrates and promotes several resources and activities. Wine tourism marketing is not solely about wine; it is also linked to the image of the region or the country. Wine tasting at a winery in the company of the winemaker is obviously a highlight, and promoting the benefits of wine clubs educates more people in wine tasting. With a list of best practices and strategic directions for the future, the author convinces us that wine tourism is growing or emerging according to the country, and is a key factor in the global tourism industry.

The cultivation of vines and winemaking have benefited from advances in knowledge and techniques. The aim of precision viticulture and improvements in grape varieties is to obtain better matured, healthier and aromatic grapes, while minimising the use of pesticides. The micro-organisms of fermentations and alterations are known in great detail. Owing to the management of physicochemical parameters, fermentation and ageing are better controlled. Cellar equipment throughout the production chain, from harvest to wine conditioning, facilitates the adjustment of key parameters, while decreasing sulphiting. Sensory analysis, combining the finest chemical methods and the neurosciences, is becoming increasingly important. The big challenge of innovation is to satisfy consumers who are increasingly demanding, more educated and sensitive to the protection of the environment.

Marketing keys are in line with the wishes of different classes of consumers. Young millennials have higher regard for sustainability, quality of life and ethical principles. One of the consequences is the growing success of organic wines. Innovation in packaging continues and certain forms are successful, mainly among millennials. The design of a brand's label is important to achieve the image it seeks. Computers and applications, artificial intelligence, augmented reality, social networks and 'apps' for smartphones are becoming increasingly sophisticated and 'chatbots' dedicated to wine are multiplying. The consumer has access to all details, from the vineyard to expert evaluations, in a simple and fast way.

Statistics on wine-producing and consuming countries and an overview of the wine industry at the beginning of the 21st century has been lucidly described in the last chapter of this unit. It then, addresses the changes expected due to global climate change and innovations of all kinds.

Dr Aline Lonvaud-Funel
Professor Emeritus of the University of Bordeaux
Institute of Vine and Wine Sciences (ISVV), France

42 International Market of Organic Wine

Daniela Callegaro-de-Menezes, Antonio D. Padula and Carlos A.M. Callegaro

CONTENTS

42.1 INTRODUCTION

Among the different transformations taking place in production systems and consumer behaviour, organic production is of particular to note. This tendency appears to be worldwide and has been shown to have considerable potential for growth. In Brazil, this phenomenon is still in the discovery and development phase, and the production and commercialisation are limited to local consumption only. The condition of local production and commercialisation has led to the enhanced value of elements related to the producer and the point of sale instead of other mechanisms that communicate elements of confidence, such as certification, for example. Thus, there is no obligation on the part of the producer to obtain any certification of organic origin. The situation is very different in the international market, where certification is often valued due to the distance between producers and consumers.

Recently, both local and international markets have shown growth in specific segments that seek organic products and perceive differentials in them that warrant the payment of differentiated prices, generally from 30% to 70% above those of the products originating from conventional agriculture. Within the various sectors, the production of organic wine, though still in its early stages, has been shown to be an interesting field of study due to the fact that wine has a high level of added and perceived value. Among the wine consumers, there is a strong tendency to value characteristics linked directly to the production process. This being the case, it can be expected that elements of personal values and motivations may be behind the purchasing behaviour of the wine consumer. It is, therefore, of strategic importance to determine the nature of the homogeneity within the consumer segments, in terms of purchase motivation and personal characteristics in order to identify the profile of the consumer market and offer an appropriate product mix. Moreover, considering the possibility of exporting, it is important that cultural differences are recognised and that similar values and behaviour are sought in order to furnish the productive chain with information relevant for the development of strategies suitable for local and international markets (Figure 42.1).

It is known that the national cultures are influencing factors in the determination of individuals' personal values. However, considering that the organic production under study is a proposal that is based on a philosophy that seeks a socially just, ecologically correct and economically sustainable agriculture, we believe that the option for organic wine is also the fruit of specific personal values. Therefore, it is opportune to measure the motivations and consumption habits of organic wine consumers based on their personal values thus, seeking to generate subsidies for an orientation of the organic wine production chain in its internationalisation process.

With this in mind, this chapter aims to identify and analyse the motivational factors and the personal values of consumers of organic wines in different countries. It was decided to interview consumers in Brazil (Porto Alegre), Spain (San Sebastian) and France (Montpellier), as these regions have a local production of organic wine and, because of this, there is a tendency for consumers to have more knowledge regarding the commercialisation and consumption of the product. At the end of the study, a comparison is made of the cultural aspects of the three consumer groups studied and it is assessed whether they have had any distinct influence on the manner of perception and behaviour of consumers of organic wine. The information gathered here offers potentially useful insights for the development of strategic actions directed at the organic wine chains interested in both local and international markets.

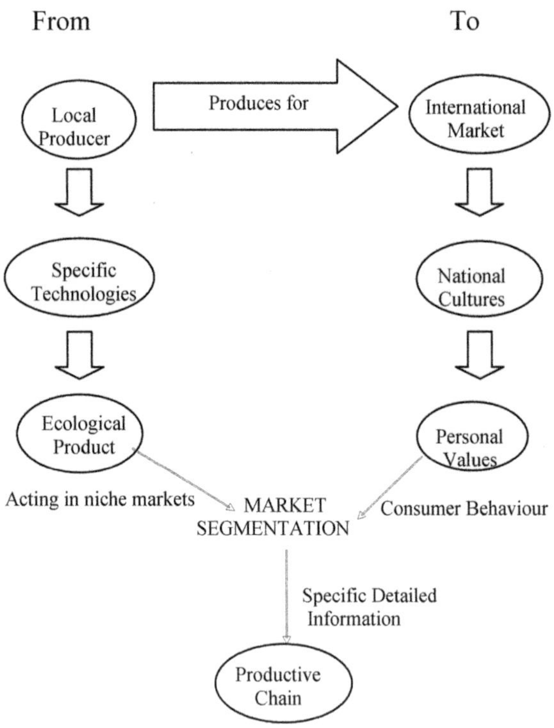

FIGURE 42.1 Internationalisation of the organic wine productive chain

42.2 CULTURE AS AN INFLUENTIAL ELEMENT OF HUMAN BEHAVIOUR

The concept of marketing proposes that the activities of organisations should be directed towards satisfying the desires and goals of people and organisations. This definition requires that such organisations know their consumers and, principally, their purchasing behaviour, identifying their motivations and perceptions.

In determining its market logic composition, the organisation comes across environmental elements capable of stimulating consumer behaviour. These elements form part of the macro-environment, in which the enterprise is inserted. Furthermore, they affect the consumers in the same way as they affect the organisation. That is, changes in this environment determine discreet or radical changes in the elements comprising the micro-environment.

These stimuli are directed towards the consumer and are perceived and absorbed in an individualised manner. Solomon points out that what determines this perception are groups of factors that compose the human being and that influence the purchasing process and behaviour, guiding the decision of the purchaser. The first group, identified by Kotler, is composed of culture, subculture and social class, which are the factors that determine the behaviour of large groups of people in their social environments.

Culture can be defined as a set of values, beliefs and behaviours that are passed from generation to generation and are shared by a large group of people. It is the transmission, over time, of lifestyles created by a group of individuals. It includes values, ideas, attitudes and conscious and unconscious symbols that mould human behaviour. According to Solomon, each culture is characterised by the endorsement, on the part of its members, of a value system, which is normally unanimous. Therefore, it is possible to identify a set of values that define the uniqueness of a culture.

Authors such as Schiffman and Kanuk point out that in order to know the degree of influence exerted by a culture on an individual, or group, it is necessary that other societies are known, with different cultures that permit a comparison of the behaviour in relation to a specific situation. This is because culture exists in order to satisfy the needs of the people within a society, offering order, direction and guidance in all stages of the solution of the human problem, by supplying 'tried and tested' methods of "satisfying personal and social psychological needs".

In attempting to reach a better understanding of the influence of culture on the individual and in making a comparison between national cultures, Hofstede developed a cultural classification model. Despite having been elaborated from research within the organisational field, his concept can be extrapolated to individuals within a society, and is of use in the identification of the cultural profile of consumers and their behaviour. The construct developed by Hofstede proposes four dimensions: (1) power distance; (2) uncertainty avoidance; (3) individualism; and (4) masculinity.

The dimension power distance (PDI) refers to the level of inequality that persons within a country consider acceptable. Where great distance from power is found, it characterises a society that accepts the differences in social, income and power distribution. The study shows that, of the three cultures studied, Brazil is the most tolerant of such inequality, France is in an intermediate position, while Spain is the culture that has the least tolerance of the three. The second dimension, denominated uncertainty avoidance (UAI), is defined as the extent to which the individuals of a certain culture prefer structured situations, with clear rules. The greater the need to avoid uncertainty, the greater the sensation of threat in relation to ambiguous situations, thus becoming more rigid and averse to risk.

The three cultures analysed present a high degree of uncertainty avoidance, the second dimension, having positions around 80 points on Hofstede's scale, demonstrating that the three countries need planned and organised situations.

The third dimension concerns the degree of integration of the individuals of a society in groups; it deals with the dichotomy between individualism and collectivism (IDV). This dimension describes the degree to which people prefer to act as individuals or as members of a group. In societies with a high degree of individualism, the focus is on the individual interest and his/her immediate family. While in societies with a high degree of collectivism, the focus is on the interests of the group. Lengler highlights that this behaviour is reflected in the decisions made regarding the consumption of the individual: 'in countries where there is greater orientation towards individual values, purchasing decisions are made on an individual basis. Countries with a more collectivist orientation have a purchase decision process with greater interference of individuals, other than the user and purchaser'. In relation to individualism, it was found that Brazil presents collectivist characteristics, obtaining the lowest rating of the three countries. Spain is located between Brazil and France and has a rating close to 50, while France appears as the most individualist country.

The fourth dimension deals with the importance of masculine or feminine values in a society (MAS). In countries with a masculine culture, the man is presented as the provider and strong, while the woman plays a domestic role and cares for the well-being of the children. Whereas feminine societies present a more balanced view with overlapping social roles of men and women. Kotabe and Helsen illustrate this situation, characterising more feminine cultures as more concerned with solidarity and environmental conservation. Hofstede's research identified that the country that attributed greater importance to masculine values was Brazil. The French and Spanish cultures were not very distant from that of Brazil, though more emphasis tends to be given to feminine values.

42.3 THE MOTIVATIONAL PROCESS

As seen in the previous section, culture as a determining factor of human behaviour, moulds the reactions of the individual to environmental stimuli. Of particular interest, among these reactions, is the motivation to action. As consumers, individuals seek motives to buy and consume. Given this, understanding this process is of strategic importance for organisations.

Consumer behaviour occurs as a result of the combination of the constituent elements of the psychological factors: motivation, learning, attitude and perception. Factors exterior to the human being influence the way in which these factors are going to interact, so determining individual behaviour.

Motivation is an internal process that occurs when a person perceives a need sufficiently strong to make him/her act in such a way as to solve the problem. The aim is to reduce an installed state of tension and re-establish the equilibrium that produces the sense of comfort in the individual.

Motivation is understood, then, as a state of tension induced by the necessity that exercises force over the individual so that he/she acts in a determined way, which he/she expects will satisfy the need. The actions taken by individuals are guided by objectives and are selected in accordance with their thought processes and previous learning.

42.4 PERSONAL VALUES

Personal values, in the study of marketing, are inserted within concepts of market segmentation and consumer behaviour. Psychography, as this area is known, is concerned with the characteristics by which individuals may be described in terms of their psychological and behavioural make-up.

Values are means-ends, as if they were the objectives by which persons live. They are concerned with modes of conduct and means-ends of existence. Schiffman and Kanuk define values as mental images that affect specific attitudes, and that, as a consequence, affect the manner of reaction to a specific attitude. Solomon points out, however, that as values are universal, what differentiates the value systems between different cultures is the relative importance that each culture attributes to these values. This subject, coming from the field of social psychology, has, as one of its main theoreticians, the psychologist Milton Rokeach, who, in 1973, after several years of study in the field, presented a classification of personal values in which ways of measuring such values are developed.

In the marketing literature dedicated to this issue, three methodological approaches of particular note are found: the VALS model – Values and Lifestyles System; the RVS – Rokeach's Value Survey; and the LOV – List of Values.

The List of Values, developed by Kahle and a group of researchers from the University of Chicago, emerged as a good alternative to Rokeach's model. As it is easier to use, can be completed in less time, is easier to translate and is applicable in transcultural situations, the LOV scale came to be considered more advantageous, in place of the RVS. The LOV scale aids in the identification of new consumer segments based on the values that they endorse and relates each value to differences in consumption behaviour. In a comparative study, Beatty *et al.* conclude that the set of 9 values proposed by Kahle more closely reflects daily activities than the 36 values proposed by Rokeach (see literature cited).

The LOV, then, has come to be widely accepted and used as a tool in the area of marketing as a way of understanding consumer behaviour and in the study of market segmentation. In this way, with several studies derived from the initial proposal, this scale has guaranteed its reliability and validity, being replicated in various studies applied in single cultures or in comparisons between cultures.

In order to measure personal values, LOV requires that the respondents evaluate a list of nine values based on a nine-point Likert scale, in which only one of the values can be given a mark of 9. This list is composed of nine terminal and instrumental values that are related to the main roles played by individuals in their day-to-day lives.

42.5 INTERNATIONAL MARKET FOR ORGANIC WINE: A SURVEY

The study consisted of two stages. The aim of the first stage was to supply subsidies for the elaboration of the applied questionnaire. In Brazil (Porto Alegre), France (Montpellier) and Spain (San Sebastian), 25 interviews were carried out with specialists in the production (3) and marketing (4) of organic products and consumers of organic wine (12), as well as with traders of organic wine (6), who were able to contribute with their experience of contact with both producers and consumers. An interview was also carried out with a certified producer of organic wine in Brazil. The data obtained from the interviews were compiled and analysed, based on their contents. Still within the first stage, a survey was made of secondary data and bibliographies based on existing research reports and literature on the subject.

The second stage of the study consisted of the collection of data obtained from a questionnaire elaborated based on the results from the first stage. The questionnaire used was divided into three blocks: the first block identified the personal values of the consumers and the motivations for choosing the organic wine; the second block sought to evaluate the purchasing and consumption habits of these consumers; the third block allowed some demographic characteristics of the interviewees to be identified. The instrument was prepared in Portuguese and then, translated into French and Spanish by specialised professionals. Assurance that the translations obtained were faithful to the original Portuguese was obtained by reversing the translation process, that is, the questionnaires were translated back into Portuguese by those people who were fluent in French and Spanish and a comparison was made with the original.

The questionnaires were supplied to consumers of organic products in ecological markets, specialised shops and production and marketing cooperatives of ecological products. In Porto Alegre, 200 completed questionnaires were obtained, in Montpellier 100 and in San Sebastian 51.

Considering the aims of this study, it was found that the most suitable techniques for analysing the results were descriptive statistics and the multivariate cluster analysis of statistical method. In order to identify the clusters, it was decided to perform the tests in combination, using the hierarchical process to identify the number of clusters and the central points of the clusters. In this stage, three distinct clusters were identified. Based on the identified central points, the non-hierarchical process (K-means) was used for a redefinition of the objects that formed each one of the three clusters previously identified.

Descriptive statistics were used to present the data in a concise and organised manner, allowing the data referring to the survey to be characterised as personal values, purchase motivations and purchasing and consumption habits, as well as demographic characteristics. In order to check the associations between the clusters, Fisher's exact test was used instead of the Chi-squared test. This choice was made because a necessary condition for use of the Chi-squared test, which determines that at least 5% of the cells in the table have fewer than five respondents, was not met. In Fisher's exact test, the frequency of the responses was not considered in the statistics, but the probability with which the results obtained occurred was considered (see the literature cited).

The variability between the variables on the personal values was what determined the identification of the clusters, being that the other variables served to describe them. The characterisation of the clusters became apparent as a result of the individual characteristics of the respondents, independently of the country from which the cases were collected. Thus, it was decided not to identify the clusters within the countries, as it was believed that national cultures would be strong determinants of personal values. Moreover, few responses were collected in Spain, which might produce bias in the responses.

42.5.1 Habits and Motivations for the Consumption of Organic Wine

The motivations for the consumption of organic wine can be identified and classified according to macro issues (society and the environment) and micro issues (personal and collective concerns) as well as with the purchasing process. These may be differentiated between individuals depending on their education and the personal values they prioritise.

During the interviews with consumers it became clear that there was some difficulty in distinguishing the motivations for purchasing organic wine from the motivations for purchasing organic products in general.

In evaluating the concept of organic production, it was found that the producers, specialists and consumers had common concerns with regard to the environment. The use of agrotoxins in treating agricultural crops leaves residues that have direct long-lasting effects on farmland, the surrounding environment, the animals living in the area and, as a consequence, the food chain. Still within the environmental focus, another issue related to the organic production approach is the assurance that is sought that the products are not genetically modified, representing, in this case, a counter-argument to the advent of transgenic products.

It is also worth mentioning the concern expressed regarding the rural producer and sustainability in the production chain. The socio-economic issue in relation to organic farming gains importance because it is seen to offer the possibility of adding value to the product offered and generating new alternatives for placing the product on the market directed towards niches, permitting the small farmer to take part in this process.

Consumers were also found to express concern with respect to social issues. This occurred most frequently with consumers who displayed more acute social awareness, discussing subjects such as social inequality, the economic problems of the small producer and the issue of the migration of rural producers to large urban centres due to lack of opportunity in the countryside. According to the consumers interviewed, 'organic production is more accessible to small farmers because it is a return to traditional techniques of agricultural production, as it used to be, and does not depend on chemical products that are often prohibitively expensive'.

The consumers were also found to value organic wine, together with other organically produced items. Those who consume the wine had probably tried and are the consumers of other organic products such as vegetables and cereals. With a few exceptions among the interviewees, of those who had come to consume organic wine while trying different types of wine, it was most common to find that the consumer of organic products in general had come to adopt organic wine.

When questioned on the main motivations for the consumption of organic wine, the points raised involved individual concerns related to health, and collective concerns related to the consumption habits of the family and reference groups. Certain differential characteristics attributed to organic wine give a perception of higher quality in relation to other wines. In particular, the colour, taste, aroma and texture were highlighted as marked differential elements that attract the interest of those who appreciate a stronger wine. 'We noticed it is as natural, it is very different', one consumer pointed out. Another consumer added 'it is not like table wine, it has a marked flavour'. This reference to table wine was made because a great majority of organic wine producers are not certified, thus they sell their wines in organic product markets and fairs, seeking to legitimate their products by placing them in the street markets. Consequently, the wine is sold directly by the producer, giving the idea that it is a table wine.

Regarding consumption habits, many of the interviewees reported that they consume the organic wine with meals and that they offer it on social occasions. As one of the interviewers pointed out, 'Whenever I go to a party I take my bottle of organic wine. There's not likely to be any at the place'. Yet another consumer stated that 'the wine is not that much more expensive than the conventional wines from the same region', while explaining why he also served organic wine to his friends at home. Others, moreover, believe that because they consume and offer organic wine, the people with whom they associate are going to respect their concern for health and the environment, so they accrue a certain status.

Finally, an element that showed up in the survey and that appears to be relevant to the producers and intermediaries in the organic wine chain. In addition to the difficulty in differentiating the motivations for purchasing organic wine from those for purchasing other organic products, wine in, general, is a valued product, enjoying a certain distinction among its users. With this, using the organic characteristic of the product to differentiate it from other types of wine and add value to it is a strategy with which much care should be taken. A case of such a strategy is illustrated by a French specialist, for whom 'the French wines are recognised throughout the world, they require no additional characteristic. However, in our region (Montpellier), as there is no tradition of wine, we chose to adopt organic cultivation in order to obtain distinctive feature'.

42.5.2 The Consumer of Organic Wine

The results obtained in the collection of the quantitative phase of the study are presented here. The clusters were determined from the significant difference between the means of the personal values of organic wine consumers, identified by the K-means method, calculating their distance from the centroid. Euclidean distance was used for this calculation. In order to better organise the presentation of the results obtained, it was decided to describe them by cluster. A significant difference between the wine consumption habits was found with the application of Fisher's exact test, in which all the results were $p < 0.01$. Table 42.1 shows the composition of the respondents by cluster and by country.

TABLE 42.1
Distribution of Respondents by Cluster and by Country

	Country			
Cluster	Brazil	France	Spain	Total
1	192	4	37	233
2	2	26	4	32
3	6	69	10	85
Total	200	99	51	350

42.5.2.1 Cluster 1

Cluster 1 is mainly composed of the Brazilian respondents, though a large part of the Spanish sample was also found in this group. Also, though small, there is some participation from the French respondents.

This cluster is initially characterised by their personal values. From Table 42.2, it can be seen that the members of Cluster 1 prioritise values linked to social relations. With the main focus on self-fulfilment and the sense of belonging, or participating, members of Cluster 1 are concerned with living in groups, be they social or familial. Highlighting the values of confidence and self-respect, it can be perceived that the behaviour of the members of this cluster is guided by the search to be a part of a group based on actions acceptable to their peers and for their own well-being.

With regard to food purchasing and consumption habits (Table 42.3), it can be seen that there is no concern with choosing the most healthy food. Consequently, members of this cluster are not accustomed to seek out information regarding the origin and nutritional characteristics of the consumed products. While the influence of third parties in the choice of consumer products is not a determinant for members of

TABLE 42.2
Personal Values Identified in Cluster 1

	Cluster 1				
	% Low	% Medium	% High	Average	% Grade 9
Sense of belonging	4.73	16.31	78.97	7.31	20.60
Excitement	5.15	32.19	62.66	6.61	0.43
Warm relations with others	3.01	13.73	83.26	7.37	9.87
Self-fulfilment	0	11.59	88.41	7.74	21.03
Being well respected	1.29	16.31	83.69	7.49	10.30
Fun and enjoyment in life	0.43	12.02	87.55	7.52	6.44
Security	2.15	6.87	80.99	7.76	16.31
Self-respect	0	3.00	97.00	7.84	4.72
Sense of accomplishment	1.29	13.73	84.98	7.38	2.57

TABLE 42.3
Purchasing and Consumption Habits of Cluster 1

	Cluster 1				
	1–3	4–6	7–9	Average	p
I am concerned with a healthy diet	60.94	12.45	26.61	3.66	<0.01*
I always try to find out the origin and nutritional characteristics of the products I consume	45.92	29.61	24.47	4.20	<0.01*
I am willing to pay more for products that are produced in an ecologically correct way/without agrotoxins	53.22	22.75	24.03	4.04	<0.01*
I always try to consume organic products	48.93	27.04	24.03	4.14	<0.01*
I always choose the healthiest option among the products I consume	53.65	20.60	25.75	4.06	<0.01*
I prefer products that contribute to the valorisation of the social classes most in need	39.91	32.62	27.47	4.7	<0.01*
I choose products recommended and consumed by friends, family and colleagues	30.9	34.76	34.33	5.15	<0.01*

* indicates the significance at the specified level of significance.

TABLE 42.4
Motivational Factors of Cluster 1

Variables	*F*
Health concerns	100
Attributes (taste, colour, texture)	45
Reduction in environmental impact	26
Generation of income for the rural producer	17
Maintenance of rural producer in the countryside	11
Assurance that the product has not been genetically modified	11
Different characteristics from conventional wine	9
Price	6
My friends and family consume this type of wine	4
Certification	2
The consumption of organic wine determines distinction among people	1
Ease of purchase	1

Cluster 1, there is a tendency towards evaluating the opinions of peers at the moment of choice. Furthermore, it was found that the concept of the organic product generally, in terms of conservation of the environment and valorisation of the social classes in need, is valued, but does not warrant the payment of a high price.

The motivations for the consumption of organic wine in this cluster are responses to individual stimuli, such as health concerns and the attributes of the product, as shown in Table 42.4. In their purchasing process, organic wine is valued because of the reduction in the impact on the environment associated with its production process.

42.5.2.2 Cluster 2

Cluster 2 is composed almost entirely of French respondents. This leads to the establishment of a bias towards the French culture in this cluster. This is confirmed by the identification of personal values that are in accordance with the orientation contained in Hofstede's classification of national cultures. (See the literature for further reading.)

Cluster 2 prioritises cultural values orientated towards the individual. Table 42.5 shows that the members seek self-fulfilment, a sense of success, to be admired and respected, and

TABLE 42.5
Personal Values Identified in Cluster 2

	Cluster 2				
	% Low	% Medium	% High	Average	% Grade 9
Sense of belonging	9.39	78.13	12.5	5.5	0
Excitement	9.38	65.63	25	5.28	0
Warm relations with others	3.13	56.26	40.63	6.16	6.25
Self-fulfilment	0	37.51	62.51	6.50	3.12
Being well respected	12.51	37.51	50.01	6.00	9.37
Fun and enjoyment in life	6.25	40.63	53.13	6.88	18.75
Security	18.75	43.75	37.50	5.56	0
Self-respect	3.13	53.13	43.75	6.13	0
Sense of accomplishment	9.37	31.25	59.37	6.31	6.25

TABLE 42.6
Purchasing and Consumption Habits of Cluster 2

	Cluster 2				
	1–3	**4–6**	**7–9**	**Average**	**p**
I am concerned with a healthy diet	21.88	31.25	46.88	5.47	<0.01*
I always try to find out the origin and nutritional characteristics of the products I consume	18.75	65.63	15.62	4.94	<0.01*
I am willing to pay more for products that are produced in an ecologically correct way/without agrotoxins	28.13	53.13	18.75	4.75	<0.01*
I always try to consume organic products	65.63	25	9.38	3.63	<0.01*
I always choose the healthiest option among the products I consume	21.88	28.13	50	5.63	<0.01*
I give a preference to products that contribute to the valorisation of the social classes most in need	28.13	62.5	9.38	4.69	<0.01*
I choose products recommended and consumed by friends, family and colleagues	18.75	40.63	40.63	5.59	<0.01*

* indicates the significance at the specified level of significance.

TABLE 42.7
Motivational Factors of Cluster 2

Variables	***F***
Reduction in environmental impact	11
Health concerns	8
Attributes (taste, colour, texture)	4
Price	4
Generation of income for the rural producer	1
Assurance that the product has not been genetically modified	1
Different characteristics from conventional wine	1
Certification	1
Maintenance of rural producer in the countryside	0
Ease of purchase	0
The consumption of organic wine determines distinction among people	0
My friends and family consume this type of wine	0

fun and enjoyment in life. That is, there is a total focus on concerns with oneself and one's own well-being, as well as with the image that one presents to one's peers.

With regard to food purchasing and consumption habits, a certain valorisation in relation to healthy food is noticeable. Among the products that they consume, they choose the healthiest option. Given this, there is a tendency to seek out information regarding the origin and characteristics of the consumed products, as shown in Table 42.6. In accordance with their individualist orientation, the tendency of members of this cluster is not to value products that contribute to the social classes most in need.

Table 42.7 shows that among the motivations for the consumption of organic wine, of particular note are the reduction in the environmental impact, health concerns and the attributes of organic wine, demonstrating that the organic wine for individual benefit approach may be of benefit in this market.

42.5.2.3 Cluster 3

French and Spanish respondents comprise the greater part of Cluster 3, with the French representing the majority.

Cluster 3 stands out from the others due to the high average given to the value 'warm relationships with others'. Moreover, it presents personal values orientated by individual and

TABLE 42.8
Personal Values Identified in Cluster 3

	Cluster 3				
	% Low	**% Medium**	**% High**	**Average**	**% Grade 9**
Sense of belonging	2.35	10.58	87.06	7.16	5.88
Excitement	12.95	23.53	63.53	6.32	3.53
Warm relations with others	1.18	8.24	90.59	7.47	4.70
Self-fulfilment	9.41	7.07	83.51	7.14	9.41
Being well respected	8.24	10.59	81.18	6.94	0
Fun and enjoyment in life	5.88	18.82	75.29	7.05	12.94
Security	10.59	5.88	83.53	7.06	8.23
Self-respect	3.53	7.06	89.41	7.56	11.76
Sense of accomplishment	9.41	10.59	80.00	7.34	34.12

TABLE 42.9
Purchasing and Consumption Habits of Cluster 3

	Cluster 3				
	1–3	4–6	7–9	Average	p
I am concerned with a healthy diet	9.41	37.65	52.94	6.02	<0.01*
I always try to find out the origin and nutritional characteristics of the products I consume	9.41	52.94	37.65	5.62	<0.01*
I am willing to pay more for products that are produced in an ecologically correct way/without agrotoxins	24.71	51.76	23.53	4.96	<0.01*
I always try to consume organic products	24.71	52.94	22.35	4.89	<0.01*
I always choose the healthiest option among the products I consume	8.24	31.76	60	6.48	<0.01*
I give a preference to products that contribute to the valorisation of the social classes most in need	11.76	63.53	24.71	5.24	<0.01*
I choose products recommended and consumed by friends, family and colleagues	11.76	9.41	78.82	6.79	<0.01*

* indicates the significance at the specified level of significance.

collective concerns, particularly the values like self-respect, fun and enjoyment in life, and a sense of success, as presented in Table 42.8.

The members of this group have a strong tendency to be concerned with a healthy diet and they obtain information regarding the origin and characteristics of the products, they consume. Cluster 3 attributes great importance to the issue of traceability and certification, though they do not represent purchase motivation factors.

In Cluster 3, there is a perceptibly higher valorisation of organic products and of products that value the social classes most in need, confirming the existence of a tendency towards paying more for those products produced in an ecologically correct manner. Moreover, members of this cluster seek to consume organic products, as shown in Table 42.9.

Members of this cluster admit that they are influenced by third parties in the process of choosing products and, despite consuming organic products, are generally not in the habit of consuming organic wine.

As shown in Table 42.10, the motivations for the choice of organic wine in Cluster 3 are health concerns and the reduction in the environmental impact. Price was of particular note in this cluster. Considering that they are willing to pay more for the organic product, it is understood that the price legitimates the offer. That is, given this fact, low price cannot be considered a motivational factor, but a price suited to the perception of value can be a motivational factor.

The attributes of organic wine and the influence of friends and family also represent determinant stimuli of purchasing motivation among members of this cluster.

TABLE 42.10
Motivational Factors of Cluster 3

Variables	*F*
Health concerns	23
Reduction in environmental impact	20
Price	18
Attributes (taste, colour, texture)	6
Assurance that the product has not been genetically modified	4
Generation of income for the rural producer	3
Certification	3
Maintenance of rural producer in the countryside	2
Ease of purchase	2
Different characteristics from conventional wine	2
The consumption of organic wine determines distinction among people	1
My friends and family consume this type of wine	1

42.5.2.4 Highlight of the Consumer Survey Results

Based on the analysis carried out, it is possible to highlight some of the results. Independent of nationality and thus, of the culture of the consumers, *health concerns* is a rather strong motivational factor for the consumption of organic wine. Another point that stood out was the *attributes* of organic wine in relation to taste, colour and texture, as well as its natural properties, which are considered one of the top three motivation factors for the consumption choice.

Initially, it was found that the division of the respondents into the identified clusters largely corresponded to the culture of the countries. This is due to the fact that they had been determined as a function of personal values, which, according to Solomon, are determined by the culture into which the individuals are inserted. However, there is no exact relationship between cluster and culture, as there is participation of interviewees from the three countries in each of the clusters. With this, it can be noted that there are elements of identification between cultures that lead the interviewees to be grouped despite their respective national cultures.

In attempting to identify the most important personal values (identified with grade or degree nine) for the listed clusters, it was found that in Cluster 1 there was a tendency towards peer orientation, in the sense that members were influenced by and concerned with the individuals with whom they socialise (Table 42.2). There was a high concentration of responses to the values 'feeling of belonging or participating' (20.60%), 'self-fulfilment' (21.03%) and 'confidence' (16.31%).

Cluster 2 has a concentration of Grade 9 in the value 'fun and enjoyment in life' (18.75%), indicating an individualist orientation. From Table 42.4, it can be seen that there was not a 100% response from members of this cluster regarding which personal value should be attributed Grade 9, which may, to a certain extent, generate bias in the interpretation of this result.

Cluster 3 also shows an individualist orientation, with a concentration of responses in the value 'sense of success' (34.12%). This demonstrates that the members of this cluster value the capacity of achieving their own projected goals. Table 42.7 shows that the values 'fun and enjoyment in life' (12.94%) and 'self-respect' (11.76%) with considerably lower percentages confirm the individualist tendency in this cluster.

42.6 SUMMARY OF INTERNATIONAL MARKET OF ORGANIC WINES

An analysis of the compiled data revealed some particularly interesting results. During the interviews, it became clear that there was some difficulty in differentiating the motivations for purchasing organic wine from those for purchasing organic products in general. Independent of nationality, and consequently of the culture of the consumers, *health concerns* is a strong motivational factor for the consumption of organic wine. Another point of note was that of the *attributes* of organic wine in terms of flavour, colour and texture, as well as its natural properties, which was placed among the three main motivational factors in choosing it for consumption. Differentiated orientations were found among the nationalities in relation to the nine personal values proposed in the LOV scale.

Using cluster analysis, this study permitted the identification of consumer groups and of differences in personal values and motivations in relation to the consumption of organic wine among the consumers interviewed in three countries. Among the identified clusters, it was seen that Cluster 1 had mass market characteristics, as insufficient value was perceived in organic wine to warrant the payment of a higher price. Cluster 3 is the cluster that most lends itself to the arguments of organic production, presenting relatively individualist orientations, but mainly valuing 'warm relationships with others'. Its members perceive value in organic wine and are willing to pay a higher price for it. This cluster is the most suitable to influence in a marketing sense for the purpose of maintaining the differential status of organic wine.

The results of this research brought some useful insights into the development of strategic actions directed at organic wine chains interested in both local and international markets. Some managerial implications for the marketing of organic wine can be summarized in 4Ps which stands for product, price, promotion, and place. The 4Ps is the traditional theoretical framework to implement the operational phase of the marketing-management process and can be applied to the marketing of organic wine. Regarding the product, consumers' low perception concerning the level of differentiation when compared to the wine produced in a conventional way was observed. This limits the use of the 'organic product' argument to determine a high degree of added value. Thus, the practice of high prices is difficult because there is not a high degree of differentiation, despite the costs and risks involved in the production and processing of organic wine.

To make the product accessible to international markets, it is understood that the use of an importer and/or distributor with a network of connections with local reputed retailers is the most advantageous for a wider market reach. In this sense, the participation of the organic producer in international fairs is essential in order to develop relationships with these intermediaries. In addition, as regards communication with the end consumer, the use of 'certification' as a credibility tool allows better positioning with the final consumer.

BIBLIOGRAPHY

Beaty, S. E.; Kahle, L. R.; Homer, P. M. and Misra, S. (1985). Alternative measurement approaches to consumer values: The List of Values and Rokeach Value Survey. *Psychology & Marketing* 3, 181.

Boone, L. E. and Kurtz, D. L. (1998). *Marketing contemporâneo.* 8th ed. Rio de Janeiro: LTC.

Butler, G.; Newton, H.; Bourlakis, M. and Leifert, C. (2004). Factors influencing supply and demand for organic food. In: M. Bourlakis; Weightman, P., (Eds.), *Food Supply Chain Management.* Oxford: Blackwell Publishing Ltd.

Churchill, G. A. and Peter, J. P. (2000). *Marketing: Criando Valor Para Os Clientes.* São Paulo: Saraiva.

Hofstede, G. (1980). Motivations, leadership and organization: Do American theories apply abroad? *Organizational Dynamics* 9(1), 42.

Kahle, L. (1983). *Social Values and Social Changes: Adaptation to Life in America.* New York: Praeger.

Kahle, L.; Beatty, S. and Homer, P. (1986). Alternative measurement approaches to consumer values: The List of Values (LOV) and Values and Life Style (VALS). *Journal of Consumer Research* 13(3), 405.

Kahle, L.; Rose, G. and Shoham, A. (2000). *Findings of LOV Throughout the World and Other Evidence of Cross-National Consumer Psycographics: Introduction.* Philadelphia, PA: The Haworth Press.

Keating, M. (1993). *Agenda 21 for Change.* Geneva: Our Common Future.

Keegan, W. J. and Green, M. C. (2000). *Princípios de marketing global.* São Paulo: Saraiva.

Kotabe, M. and Helsen, K. (2000). *Administração de marketing global.* São Paulo: Atlas.

Kotler, P. (2000). *Administração de marketing.* 10ª ed. São Paulo: Prentice Hall.

Kotler, P. and Armstrong, G. (2004). *Princípios de marketing.* São Paulo: Pearson.

Lengler, J. F. B. (2003). *Verificação da influência da nacionalidade sobre a hierarquia de valores, atitudes e comportamentos dos consumidores em shopping centers regionais dos Estados Unidos, Uruguai e Brasil: Um estudo comparativo.* Tese de Doutorado. Porto Alegre: PPGA/UFRGS.

Mitchel, A. (1983). *The Nine American Lifestyles*. New York: Warren Books.
Rockeach, M. (1968). *Crenças, atitudes e valores: Uma teoria de organização e mudança*. Rio de Janeiro: Ed. Interciência.
Rockeach, M. (1973). *The Nature of Human Values*. New York: Free Press.
Schiffman, L. G. and Kanuk, L. L. (1997). *Comportamento do consumidor*. 6th ed. Rio de Janeiro: LTC.
Schwartz, S. H. and Bilsky, W. (1987). Toward a universal psychological structure of human values. *Journal of Personality & Social Psychology* 53, 3.
Sheth, J. N.; Mittal, B. and Newman, B. I. (2001). *Comportamento do Cliente: Indo além do comportamento do consumidor*. São Paulo: Atlas.
Shim, S. and Eastlick, M. (1998). The hierarchical influence of personal values on mall shopping attitude and behavior. *Journal of Retailing* 74(1), 139.
Solomon, M. R. (2002). *O comportamento do consumidor: Comprando possuindo e sendo*. 5th ed. Porto Alegre: Bookman.
Stevenson, W. J. (2001). *Estatística aplicada à administração*. São Paulo: Harbra.

43 Global Wine Tourism

Current Trends and Future Strategies

Zoltán Szakál

CONTENTS

43.1 INTRODUCTION

Tourism is one of the engines in the economic development of a country, which in addition to its basic results, has numerous positive external effects. A type of multiplier mechanism starts that progresses to affect more sectors. The first step is the formation of a product and a service package, which is based on a country's capabilities but can also be bound to traditions. The "La Cité du Vin" is an example of such a project: 10 levels between the core and the tower, a permanent tour, three tasting areas including one multi-sensory immersive space, a reading room, a boutique concept store, two restaurants and a panoramic restaurant. The spectacular building is in Bordeaux. The next step is the provision of goods and services to tourists. The conditions necessary for a tourist industry are the creation of transport, accommodation, dining, safety, and health and hygiene services, which, depending on the requirements of the target groups, may have different qualities, and be different in space and time. Regional and local marketing and tourism development are carried out in parallel. Investment in the tourism sector is generally high, thus the speed of development depends on the economic situation in a given country, the government in power and the support of the economic union. This is the reason why the world's tourism is at different levels. Some countries are at the beginning of the road and others have developed their services to a professional level and are able to maintain constant awareness of changing dynamics.

The universal values of viticulture and wine culture mean something to people. Only certain countries have the ability to grow grapes. People all over the world are interested in wines, and are curious about how this special beverage is made in different places. The different values and capabilities of some cultures mean that different types of wine are produced, and thus a variety of product and service packages could be assembled. Wine can be showcased in restaurants, in rural tourism and as a gift product, but it also functions as a simple and safe drink. On the demand side of global tourism, there is a desire for wine and knowledge of its related activities, legends and specialities. This chapter outlines the general picture of the world's wine tourism, highlighting some good practices and formulating wine tourism trends and future directions. Some of the wine tourism strategies of those

PLATE 43.1 Wine tourism and festivals in Hungary. *Source:* Zoltán Szakál (2014)

countries that have the most developed wine tourism are discussed. Globally, wine festivals are very common. Plate 43.1 shows a typical wine festival.

The tourism literature ranks wine tourism as an industry in itself, noting, however, that it is inextricably linked to a number of other forms and parts of tourism. Getz, for instance, associates wine tourism with cultural tourism, rural tourism, festivals, events and more. Nevertheless, it seems that the majority of scholars adopt the definition given by Hall and Macionis: 'visitation to vineyards, wineries, wine festivals and wine shows for which grape wine tasting and/or experiencing the attributes of a grape wine region are the prime motivating factors for visitors'.

The wine tourism sector is developing around the world. In some countries, especially in certain wine countries, it is in its introductory phase; however, the phase of growth in other countries is increasing and is indicated by growing guest numbers, guest nights and spending. Outstanding professional and complex wine tourism services can be found in every continent of the world. This development is propelled by the changing requirements of the wine tourism consumer, the creativity of the actors in this complex system and the variety of products and services on the wine tourism's supply side.

43.2 THE PLACE AND ROLE OF WINE TOURISM

The importance of wine tourism could be explained by the supply capable of renewing and developing. In this chapter, the concept of wine tourism is investigated, supported by important literature on the topic.

Wine tourism is currently a key issue for the European wine industry. While substantial research has been carried out into wine tourism in the New World, much less has been done in Europe. Currently, however, there is growing interest in wine tourism in Europe, notably in France.

As global tourism is on the rise and competition between destinations grows, unique local and regional intangible cultural heritage has gradually become the discerning factor for the allure and enjoyment of tourists. Gastronomy tourism has emerged as particularly important in this regard, not only because food and drink are central to any tourist experience, but also because the concept of gastronomy tourism has evolved to encompass its cultural facets and links to local culture. Incorporated into the discourse of the World Tourism Organisation (UNWTO) are the ethical and sustainable values of a territory, its landscape, sea, local history, values and culture heritage. Wine tourism, as a crucial component of gastronomy tourism, has evolved into a key element for both emerging and mature tourism destinations in which tourists can experience the culture and lifestyle of destinations while fostering sustainable tourism development. Representatives of the UNWTO members states, UNWTO affiliate members, tourism administrations, international and regional organisations, the private sector, academia and civil society gathered for the First UNWTO Global Conference on Wine Tourism, organised by the UNWTO in collaboration with the Georgian National Tourism Administration in the Kakheti wine region, Georgiaon, in 2016 (Table 43.1).

According to Mintel's 'Global Food Tourism' publication, wine tourism is the collection of vineyards, wineries, festivals and wine tastings. The concept is usually part of the tourist activities carried out in the traditional grape-producing countries of Europe, and in the New World, Chile, Argentina, Canada, Australia and New Zealand.

The special characteristic of the continuously growing wine tourism is that the participating visitors use a mix of natural and man-made attractions which offer different tourist activities from viticulture to shopping for bottles of wines. The attractiveness of wine tourism – which, although part of gastronomic tourism, is also a single product – lies in the

TABLE 43.1
Contents of UNWTO Wine Tourism Declarations

- Wine tourism is a fundamental part of gastronomy tourism.
- Wine tourism can contribute to fostering sustainable tourism by promoting both the tangible and intangible heritage of the destination.
- Wine tourism is capable of generating substantial economic and social benefits for key players in each destination, in addition to playing an important role in terms of cultural and natural resource preservation.
- Wine tourism facilitates the linking of destinations around the common goal of providing unique and innovative tourism products, thereby maximising synergies in tourism development, surpassing traditional tourism subsectors.
- Wine tourism provides an opportunity for underdeveloped tourism destinations, in most cases rural areas, to mature alongside established destinations and enhance the economic and social impact of tourism on a local community.
- Wine tourism provides an innovative way to experience a destination's culture and lifestyle, responding to consumers' evolving needs and expectations.
- Wine tourism's potential will be heightened if implemented appropriately through a public–private collaboration strategy, promoted through effective communication across different sectors and involving the local community.

Source: UNWTO (2016)

PLATE 43.2 European wine exhibition. *Source:* Courtesy Zoltán Szakál (2011)

consumption of wine in authentic surroundings. In addition to enhancing the reputation of the wine and creating an image of the wine country, a wine route is part of the marketing policy that contributes to the successful realisation of targeted results in the field of tourism in a given area. Additionally, wine tourism is one possible method in the development of rural areas and the tourism development of villages, which greatly contributes to improving the quality of life in wine countries. From a business point of view, the objective of wine tourism is to encourage an awareness of local products, thereby enhancing their sale. In summary, the frame of wine tourism is to provide a theme based on the grape and wine culture, including the values of the production traditions inherited from generation to generation, the local gastronomy and the rural lifestyle. Plate 43.2 shows a European wine exhibition where modern marketing tools are used. Many people are attracted to such an event.

Tourism theory defines wine and gastronomic tourism as part of cultural tourism. Gastronomy – which is part of wine culture – is a man-made attraction. It involves trying out local and national cuisine and visiting famous catering facilities, festivals and tournaments related to food and drink.

The main motivation of wine tourism is to allow the visitor to become familiar with viticulture and the wine culture of the destination visited; to get experience of viticulture and wine consumption, for example through wine tastings, wine cellar visits, and wine festivals; and to participate in vintage or even to actively participate in the viticulture or wine-producing processes. Accordingly, the supply side of wine tourism includes the wine countries and wine routes, wineries, wine producers and wine cellars. Wine tourism is a tourism product specific to rural areas, but it does occur in an urban environment too. A wine route is the thematic linking of tourist services where the basis of the product and the theme of the route are the grape and the wine. The fundamental objective of wine route development is to introduce the local viticulture and wine culture and to popularise the related services and programmes. Gastronomic and wine tourism can be perfectly matched to other tourism products, such as cultural tourism, health tourism, active tourism and meetings, incentives, conventions and exhibitions (MICE) tourism.

According to the tourism product planning and development thesis of the Kempelen Farkas Student Information Centre, University of Pécs, the major objective in the development of gastronomy and wine tourism is to utilise the values of wine culture as tourism products and to ensure that restaurants meet the quality requirements of both foreign and domestic tourists, thereby improving the service level in many restaurants.

The nature and significance of event tourism experiences have four factors. Figure 43.1 shows Getz's

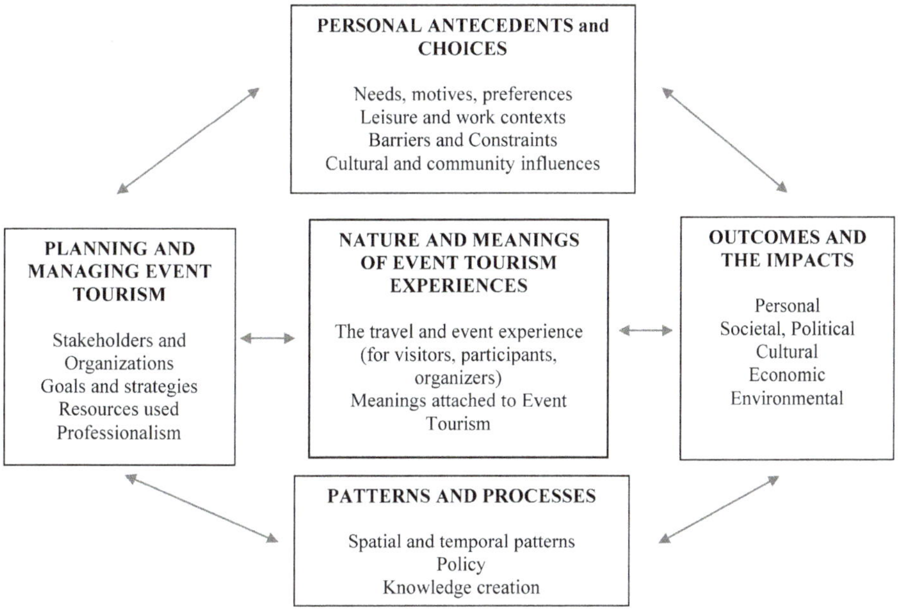

FIGURE 43.1 A framework for understanding and creating knowledge about event tourism. *Source:* Getz (2007) in Getz (2008)

interpretation. According to Getz, personal antecedents and tourism planning events, outcomes and patterns constitute the system.

As components of the tourism system, tourists are the driving force behind tourism. Their various motives and expectations induce tourism activity. Furthermore, tourist destinations, together with the features that promote the destination, are responsible for providing tourists with all the necessary facilities to accommodate their stay. However, natural resources, infrastructure and superstructure must be accompanied by a spirit of hospitality to enhance the destination's appeal. Together, all these elements constitute a tourism system that is perceived to be a significant contributor to national economies.

The research of Carlsen has highlighted the duality of the approach that has emerged over the last decade. It is hoped that future wine tourism research will integrate both the production orientation of the wine industry and the service orientation of the tourism industry in order to develop our understanding of the potentials and pitfalls of wine tourism. Wine tourism is not a new phenomenon, but research into the many factors that motivate wine tourists and indeed, wineries and wine regions, is yet to be fully developed.

Reviewing the international literature on wine tourism, it can be determined that the complex and both vertically and horizontally interpretable system has different definitions. All the resources agree that this special tourism field is important and can determine the overall tourism of some nations. The UNWTO accepted a unified declaration and declared statements and standpoints. Visits to vineyards, wineries, wine festivals, wine exhibitions and wine museums play an important role in the cultural landscape. A wine tourist's primary motivation for moving from one place to another is to visit wineries. Additionally, they visit and use all the services that are not necessarily associated with wine, but might be interesting to tourists.

43.3 TREND OF WINE TOURISM IN THE WORLD

The world's wine tourism market is composed of different trends and strategies because different countries have different capabilities and different value systems. All actors involved in tourism would like to achieve and maintain their effectiveness and awareness. There is huge competition to persuade consumers to spend their savings on their particular product. The common target is to provide experiences to all segments, but mainly to the primary target groups which are a sure basis and could be maintained in the long term. The wine tourism strategy of countries that enjoy significant wine tourism and other types of tourist attractions are discussed (Figure 43.2).

43.3.1 Wine Tourism's Development in Italy

Italy is among the leading tourism destinations. Its popularity is strongly linked to its vast pool of cultural attractions and its historical heritage. However, a significant amount of the country's rural regions are excluded from tourist flows. Hence, wine tourism is considered to be a good fit with the Italian rural regions. In particular, the whole length of the Italian Peninsula is covered by wine regions providing promising potential for rural development through wine tourism. Today in 2020, according to figures supplied by the Movimento Turismo del Vino, wine tourism in Italy yields about €2.5 million annually. Wine is ranked third in foreign tourists' motivation to visit Italy. Almost 3.5 million tourists visit Italy's wineries per year.

It is widely acknowledged that the features of a landscape can play a major role in determining tourism demand. Tuscany is a

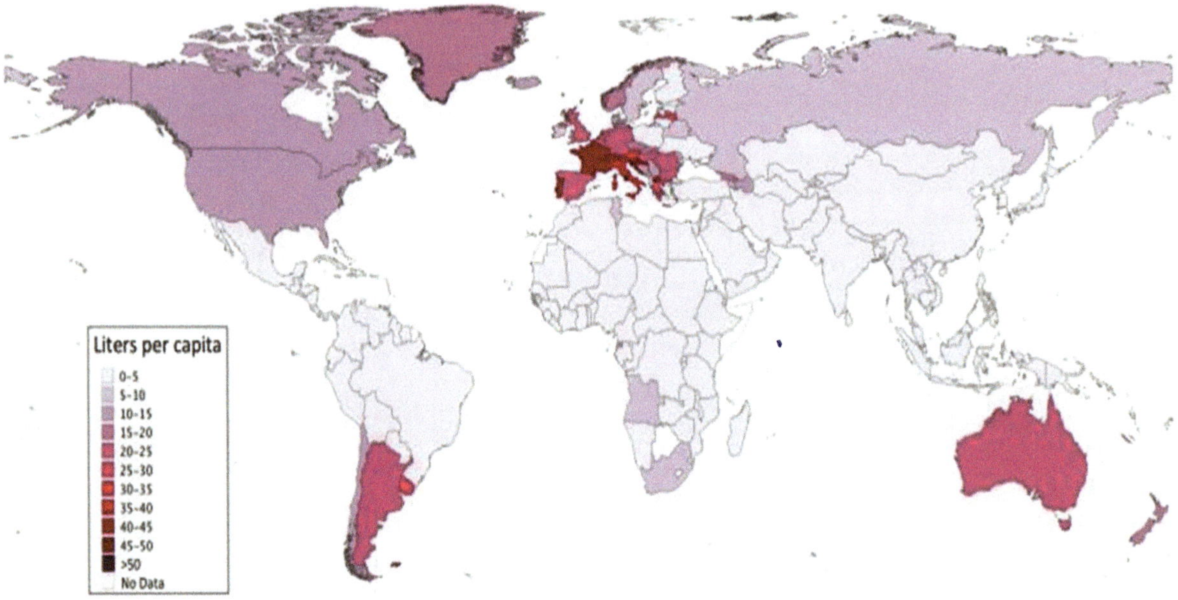

FIGURE 43.2 Wine consumption and wine tourism areas in the world. *Source:* Alonso (2014)

major tourist region in Italy, renowned for its pleasant climate and enchanting countryside. Its agricultural landscape and its production of quality wines represent a positive externality for tourism. It can have important policy implications for tourism promotion choices and allows a profile to be drawn of the average tourist visiting different parts of Tuscany. During the last decade, there has been a growing awareness in Italy of the importance of valorising and promoting the territory through the creation of thematic itineraries that can be considered as 'localising tourist packages'. Clearly, a good case in point of such a tourist package is represented by the Wine Routes whose aim is to promote rural areas as tourist destinations. This model derives from two of the most important productive sectors: vine cultivation and tourists. The Wine Routes were instituted by National law N. 268 of 1999 which defines them as itineraries created in geographical areas where quality wines are produced. The law aims to exploit the wine-growing areas and wineries, including cultural and natural resources, as well as allowing tourists to benefit from these. The identification of a territory devoted to wine has helped to create the various Wine Routes, recently including typical foods of the district. Consequently, many are now called Wine and Food Routes (WFRs).

43.3.2 Concept of Wine Tourism in France

France enjoys its leading position as the world's favourite tourism destination with the greatest number of international visitors. Additionally, French wines constitute a significant feature of the country's culture and are globally acknowledged for their quality and authenticity. In fact, viticulture in France yields around €20 billion per annum and it employs 500,000 personnel. Moreover, French wine production accounts for 40% of the value of the food processing industry in the country. However, the country's wine industry is more oriented to its wines rather than to attracting tourists and offering them a unique wine tourism experience. Moreover, despite the strong brand name of French wines, the country is facing competition from the emerging wine markets of the New World. In addition, there is a common perception that wine tourism in France is less developed and organised than that offered by the New World countries.

Perhaps it is the case, as Waller claims, that the burden of the wine industry's tradition hinders the transformation of wine into a leisure product. Nevertheless, wine tourism is not a new phenomenon in France. For example, the first wine route in Alsace was established in the 1950s. During the 1980s, more systematic attention was given to wine tourism as the 'crise viticole' forced French wine producers to deal with unfavourable conditions in the production and sale of wine. Thus, they turned to wine tourism to increase their sales at the cellar door. Ever since, France has led other countries in the provision of wine tourism experiences with the regions of Beaujolais, Bordeaux, Burgundy, Champagne, Côtes-du-Rhône and Provence being most successful. In 2004, for instance, France attracted about 7.5 million wine tourists of which 2.5 million were foreign wine tourists. Yet, the perception that wine tourism in France is relatively less developed remains. According to Hall *et al.*, wine tourism is composed of resources coming from both industries. Thus, under the supply side of wine tourism, one finds wineries, winegrowers, vineyards, oenological specialists, festivals, shows, wine tours, restaurants and accommodation. The rural landscape is also part of the wine tourism experience and should be included in the supply side. Moreover, governments may be involved in this practice but the extent of their involvement differs across countries.

PLATE 43.3 Exclusive French champagne available at the Eiffel Tower – wine tourism at tourist attractions. *Source:* Zoltán Szakál (2011)

Plate 43.3 shows an exclusive French champagne available at the Eiffel Tower, an example of combining creative wine tourism with a tourist attraction.

This is increasingly important in debates about the future of wine businesses and is often considered a good way to attract new investors. However, this development is less significant in the more rural regions of France, such as Champagne. On the one hand, some consider that the image provided by tourism cannot reflect the prestige and reputation of such great wines. On the other hand, when the demand exceeds supply, the producers of champagne think of wine tourism in terms of increased sales and thus, consider that it has little purpose for them, due to their great success since 2001. However, both these approaches ignore the possible enhancement of the product's image, an aspect of wine tourism that was demonstrated in the New World several years ago.

43.3.3 Wine Tourism in Spain

A tourism product club is a product development partnership established and led by tourism industry stakeholders including small and/or medium-sized companies. The group pools its resources to develop new market-ready products or to increase the value of existing products. Barriers to the productivity and efficiency of small and/or medium-sized companies around a tourist sector or activity can be solved through a tourism product club. The partnership can launch new or improved ready-for-sale products at the end of their development programme. Lodging companies, tour operators, administrators of tourism facilities, tourist associations, governments, other companies of the sector and even non-tourist companies can participate as members of a tourism product club. One of the advantages of a tourism product club is that it gives 'voice' to companies that develop their activities around a market segment or a tourist-specific activity regardless of their location. Following this model, wine tourism can contribute to creating a tourism product club around wine. In Spain, the international promotion of tourist products is carried out through Turespaña. 'Las Rutas del Vino en España' is a tourism product club developed by the Spanish government. It is an ambitious project that started in 2001. Now, on its fourth phase, it is a consolidated reality that structures wine tourism offerings in the most important Spanish wine regions. Basically, 'Las Rutas del Vino en España' works to integrate the resources and services of a wine area.

43.3.4 Wine Tourism Strategy of Australia

Australia is among the most popular tourist destinations in the world. However, the majority of tourists do not move from the boundaries of big cities such as Sydney and Perth. Hence, one of the difficulties faced by the country is to attract international visitors to its rural regions. The paramount performance of Australian wine exports has induced an international interest in the wine regions of the country. Flourishing wine exports reflect Australia's tradition in wine production and the fact that it constitutes one of the constantly growing industries in the country. Australia has long recognised wine as a strong tourism asset and a means of enhancing regional growth. Therefore, since the end of the 1990s, a national priority for the country has been the development of wine tourism by providing a grant of AUD$70,000 to the Winemakers' Federation of Australia to develop a national wine tourism strategy.

Today, the country lists more than 500 wineries and has more than 60 wine regions; the most important are located in New South Wales, South Australia, Victoria, Western Australia, Queensland and Tasmania. Furthermore, remarkable progress has been made in the wine tourism sector in the last decade. The total number of visitors to the 1,647 Australian wineries amounts to almost 5 million, while the total expenditure of wine tourists in the country is about AUD$7.1 million. There has also been rapid tourism growth in the Margaret River and Swan Valley wine regions. In addition, 11% of the 5 million international tourists to Australia visit the country's wineries. However, despite its growth, the wine tourism sector in Australia still lacks much in organisation, attitude and planning and infrastructure facilities. Carlsen and Dowling opine: Although wine tourism in Australia and internationally is well established, marketing research efforts to date have been negligible.

The supply side of Australian wine tourism consists of actors from both the wine and the tourism industries. Moreover, since wine tourism is seen as a tool for rural development, Australia's federal and state governments are also involved in this practice. Having recognised the benefits of promoting wine regions, regional communities, regional tourism associations and wine tourism bodies are also part of the supply side. However, due to a series of challenges identified by Beames, not all of these players are actively involved in wine tourism. Australia has a good supply factor, flexible winemaking practices, strong identifiable brands, a strong infrastructure of viticulturists and winemakers and good consumer research. The ranking of Australian food and wine improves after visiting the country. The top three interested nations are the UK, India and China.

43.3.5 Argentina Mendoza as Wine Tourism Good Practice

In recent times, wine heritage preservation and the development of wine tourism have grown in importance in Argentina. Wineries, technological artefacts and vineyard landscapes are becoming the main tourist attractions in Malbec country, and their enhancement is also an element favouring rural development. Mendoza has always been the main destination in Argentina for wine tourists. In 2013, Mendoza was visited by more than 1 million wine tourists, representing 70% of total visitors to wineries in Argentina. Malbec has become synonymous with Argentine wine and, for most wine lovers, the Mendoza province means Malbec. The three wine areas of Mendoza province are north of the province (sub-divided into Zona Alta or the Northern sub-region and Zona Este or the Eastern sub-region), the Uco Valley and the south of the province, three geographical areas where the Malbec grape variety is the main protagonist. Heritage is understood in a territorial context in which personal property, buildings and landscapes coexist and are shaped by economic development; this is a well-defined environment with multiple historical references, which together offer enormous possibilities for tourism activities as a development mechanism. The wine sector is presented as an integrating element of the historical, industrial, cultural and landscape heritage, with enormous possibilities for the development of wine tourism. The areas of wine tourism are the cultural and gastronomic activities offered around the wineries and their vineyard landscapes and the activities that wineries themselves offer, such as guided tours of their facilities, wine tasting and tasting courses, vineyard tours, and catering and hospitality. Mendoza has integrated wine tourism products, such as the 'Wine Routes', 'Tango within the Wine Routes' and the 'Grape Harvest Festival', echoing the efforts of the Argentine Wine Corporation (COVIAR) in their implementation of the 2014–2020 marketing plan for the promotion of wine tourism in Argentina and around the world. The focus is the development of wine tourism in the Malbec landscape areas and the improvement in the conditions of the industrial, cultural and natural heritage around wineries and vineyards.

43.3.6 Wine Tourism in India

Wine tourism is gaining significant momentum in India. Many vineyards in India have their own tasting rooms so that wine lovers can enjoy travelling to and exploring India's wine regions. India is a country with an ancient winemaking tradition but a very new and emerging wine-producing industry. The first vines were planted long before the 20th century and, as incredible as it sounds, winemaking has existed throughout most of India's history. In India, 80% of wine consumption (Figure 43.3) is confined to major cities such as Mumbai (39%) and Delhi (23%).

Maharashtra has always been viewed as a commercial state as it includes the commercial capital Mumbai, semi-urban areas of importance and historical monuments. Wine tourism, with its gaining popularity globally, is ideal for this location, especially with regard to meeting international demand. On the world wine map, Maharashtra is positioned as a New World wine tourism hub, using Napa Valley and the like as models. Karnataka is India's second-largest producer of grapes. This terroir is set to promote wine tourism, making Karnataka more popular as it is close to tourist hotspots such as Goa and heritage centres such as Bijapur and Belgaum.

43.3.7 China's Wine Tourism

Started in 2007, and aided by the high-speed development of the domestic wine industry, China's wine tourism industry has been 'warmed with wine'. Wine tourism was rethought in parallel with viticulture and oenological developments from 2000 onwards. Spectacular results have only appeared in the last decade. The wineries of the wine-producing areas have created a wine tourism brand. The 2019 measures against the Covid virus created a fundamentally new situation. The system of preferences for wine tourists is completely changing due to the shock to world tourism, so new strategies are needed. Recalling the ups and downs of 2008, wine tourism continues to maintain rapid growth.

Coastal Shandong Province has 'the world's seventh largest grape coast' resources, with the province's wine tourism resources currently located in Yantai, Weihai and Qingdao, as well as three regions, in particular, the Yantai region. Yantai is the birthplace of China's modern wine industry; it is also China's largest wine production base and its urban area is known for establishing the earliest of China's wine business – the former site of the winery Changyu. In recent years, the Changyu Pioneer Wine Company is committed to the development of wine cultural tourism: it built the Changyu Wine Culture Museum on their Changyu winery site, and the company's new plant. In June 2005, the Changyu Wine Culture Museum and the tourism sector in Yantai jointly launched a special tourist route, the first in Shandong Province, with 'wine' as the theme of the tourist routes that travel to Shandong Province. Wine tourists coming to China can meet viticulture and winemaking traditions. The modernization of winemaking has begun in recent decades, with tradition and innovation represented at the same time by development.

In addition, visitors to vineyards during the mature grape season can pick the grapes, and then experience the passion of brewing wine. Today, Penglai is listed as one of the world's seventh major grape coast. On both sides of the highway between Yantai and Penglai, a grape-planting base has been established, forming a spectacular 'grape promenade'. In developing wine tourism, the Shandong Provincial Tourism Administration is currently planning a Shandong wine culture and tourism product that will include the three cities of Yantai, Qingdao and Weihai, showing the coast of Yantai in Shandong and its wine resources to their best advantage. The focus will be on Yantai as 'Asia's only international grape wine city', to create a world-class wine and cultural tourism product.

Common to all the national and destination strategies discussed in this chapter is the focus on the consumer. In all the documents the target group analysis is tried to achieve by the

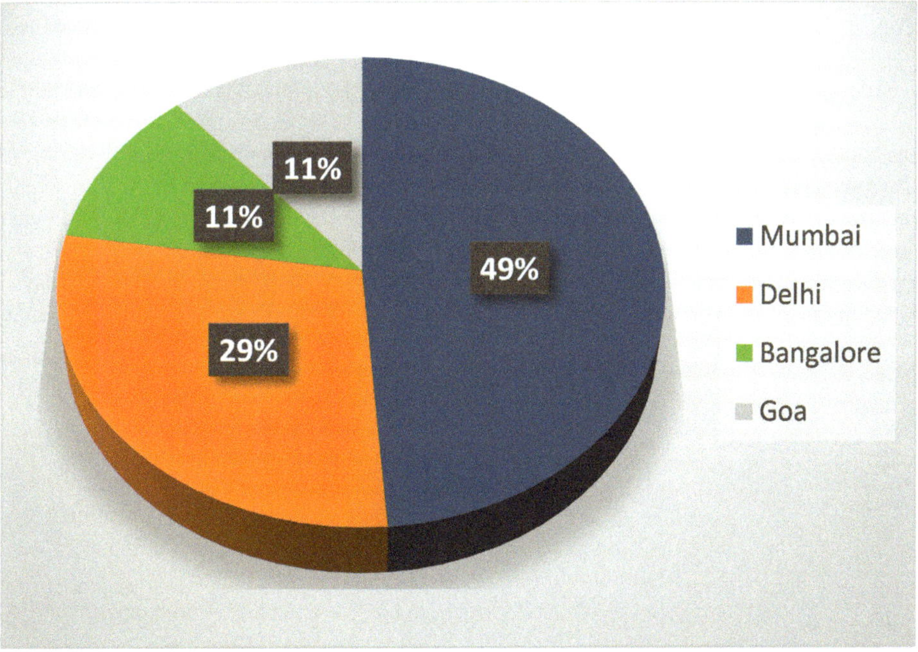

FIGURE 43.3 Consumption of wine in India. *Source:* Goel (2016)

help of professional wine tourism researches. For this task, representative questionnaires, focus group research and cluster analysis methods were implemented. In addition to the traditional above the line (ATL) communication elements, the mass media, e-mail, blog, movie sharing and event marketing solutions were emphasised. Wine tours and other non-governmental organisations play a huge role in the operation of the system. The support of this sector is also a part of the concept. In each country, wine marketing and wine tourism are centrally managed by the country's tourism authority and supplemented by the activity of the local destination's tourism management and other tourism organisations.

Differences in wine tourism strategies are derived from different capabilities and different value systems. A value system is formed through traditions, habits, the available financial framework, previous results and national creativity. Capability means the terroir, viticulture, winery, natural and cultural landscape systems. In line with the developments, and taking the requirements of the demand side into consideration, this creative positioning could mean competitive advantage. This will definitely be part of a future wine tourism strategy. The principle of 'cooperating in competition' is an element of a regional marketing strategy that must be implemented by the wine tourism actors in the future. In the strategy of the winemaking companies, the unique comparative advantages and the unique strength, *e.g.* the legends of the land, the vineyard and the area, must be focused on.

43.3.8 Wine Tourism in Hungary

Wine tourism has an important role in the communal wine marketing strategy of Hungary. Besides the popularization of wine culture, stimulation of consumption and image development, the popularization of wine tourism and the stimulation of the local-type HORECA sector got place in the strategy.

The main objectives of the strategy are to

- elaborate on issues and campaigns popularising wine tourism;
- modernise wine route websites;
- modernise wine route sign systems;
- upgrade the existing organisation of events, festivals and wine tastings;
- spread participation in wine expert courses and sommelier courses among local gastronomic actors.

In Hungary, wine tourism is in its infancy; however, significant development has taken place in recent years. It is argued that wine tourism offers considerable benefits both to the wine industry and the tourism industry for a number of reasons. Firstly, wine tourism can contribute to the dispersal of tourist flows from the established tourist centres. Secondly, it can enhance the image of a destination, and thirdly, it can create an awareness of the importance of the quality issues in wine tourism. Thus, it can be concluded that the next step in promoting and developing wine tourism must go beyond the initial marketing efforts and that appropriate market research and development policies are needed for the long-term development of a successful wine tourism industry in Hungary.

The prominent role of wine regions in Hungary's culture and tourism is shown by the fact that the Tokaj wine region, which includes the Tokaj-Hegyalja historical wine region cultural landscape, won the UNESCO World Heritage title in 2002. The Sopron wine region belongs to the Fertő-Neusiedlersee area which won the World Heritage title in 2001. Such awards reflect the attractiveness of wine routes. Currently, there are 30 registered wine routes operating in Hungary, of which 19 are members of the Hungarian Wine Route Association. Wine and gastronomic tourism have no primary motivation, but they represent an integral part of the travel experience. There are several World Heritage wine-growing regions in Europe, and many of them have had an impact on tourism, as summarised in Figure 43.4.

The marketing of gastronomy and wine is managed by the Hungarian Tourism Agency (HTA). In the field of wine marketing, the task is to create a unique Hungarian wine brand. The national wine marketing strategy and the sectoral strategy of the National Council of the Mountain Villages are closely related to each other. Both of them target improving the prestige of Hungarian wine and producers' market opportunities. Due to cooperation between sectors, the HTA has started to integrate the Hungarian wine marketing concept into the country's brand.

Wine festivals, local selling and related tourist services form the basis of Hungary's wine tourism. The latter include accommodation providing dining options and vineyard tours, wine museum visits and wine-tasting events. In the Hungarian capital, Budapest, more wine-related events are being organised, including the Budapest Wine Festival at Buda Castle. Events such as wine days, vintage days and wine tastings are organised in other parts of the country and attract many

Effects:
- ☐ more tourists
- ☐ more revenue
- ☐ more recognition
- ☐ more development
- ☐ more investors
- ☐ better quality of life
- ☐ development

FIGURE 43.4 European World Heritage vineyards and effects. *Source:* Lavaux (2013)

tourists. The 22 Hungarian wine areas cover more than 50% of the country. Wine is an integral part of Hungarian culture and an important pillar of the tourism supply.

43.4 THE IMPORTANCE OF WINE CONSUMER BEHAVIOUR AND MARKETING MIX

Tourism marketing consists of all marketing activities that influence the sale of tourism products and services.

The impact of wine tourism on a country's economy, gross domestic product, number of visitors, etc., is depicted in Table 43.2.

A field within tourism marketing is wine tourism, which focuses not only on selling wine but also on the sale of products and services in a winery, and on the different organisations in a particular wine area related to the wine, grape, wine culture and wine shops. At a higher level, taking advantage of a country's wine tourism could benefit a country's image. The characteristics of wine consuming behaviour greatly contribute to the demand for wine tourism products. Marketing provides an opportunity to introduce other values in addition to the wine, from which the local tourism organisation and the destination could benefit. Kotler's marketing theory can be adapted to wine tourism. The challenge of this marketing strategy is the complexity of the system and the harmonisation of the heterogeneous elements. The global wine culture and nations' relationship to the grape and wine originate from a country's religion, identity and entertainment values. Every wine country and winery proudly shows the wine produced and its manufacturing processes. The global mission of the wine tourism industry is to show variability and the complex harmony that a wine, a terroir, a dish and even a brand can represent. The philosophy is that a culture should be proud of the wine and the tourism that is built on it, and that the winemaker, the grape, the country, the nature and the people should be respected. More complex is the philosophy of wine tourism in which emotions, attitudes, ideas, beliefs and traditions are present.

43.4.1 Market Segmentation, Types of Wine Tourists and Their Consumer Behaviour

From Figure 43.4, three large groups can be observed and characterised as wine lovers, those who are interested in wine and those who are curious tourists.

There is no unanimous definition of the wine tourist. However, wine tourists are classified according to their motivation. Hall has identified three market segments of wine tourism: the wine lover, those interested in wine and the curious tourist. A description of each segment is illustrated in Figure 43.5. In Pratt's opinion there is a fourth profile segment: the 'disinterested wine tourist' who visits wineries as part of a group, and sees it as an alternative to a bar. Generally,

TABLE 43.2
The Impacts of Wine Tourism

Australia	AUD $ 7.1 bn (2009)
South Africa	Wine tourism contribute 2.2% of the national GDP.
Italy	4 million visitors (2004)
California	20 million visitors, $ 2.1 bn accounting for over 10,000 jobs in NAPA (2009)
USA	7.5 million visitors (2004)
France	$ 700 mn (2008)

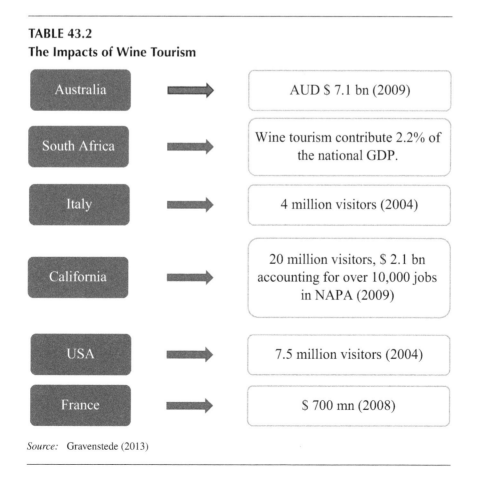

Source: Gravenstede (2013)

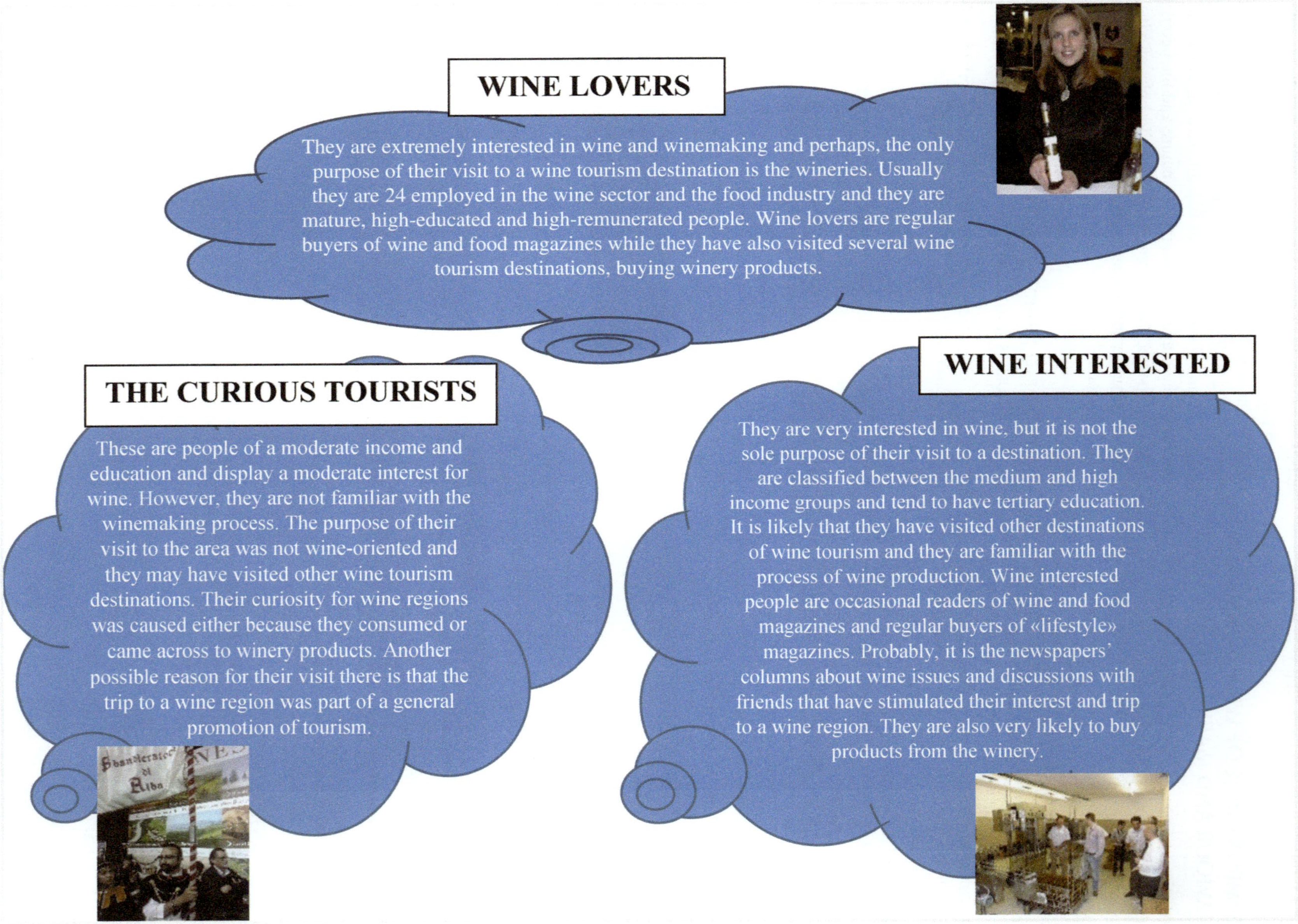

FIGURE 43.5 Segmentation of wine tourism. *Source:* Hall (1996)

PLATE 43.4 World Travel Market London exhibition – wine tasting for persons interested in wine. *Source:* Szakál (2010)

this segment is only concerned with drinking wine, and has no interest in learning about wine.

Plate 43.4, depicting the World Travel Market London exhibition and conference, shows that there is a great interest in wine events.

Market research identifies marketing problems. It is important to understand wine consumers and their needs, how much they are willing to pay and their shopping and communication habits. Figure 43.6 summarises the motivations and intentions of wine tourists.

In carrying out research, decisions need to be taken regarding the information sought, the research methods and the cost and timeline of the research. The available human resources and the tools required should also be taken into consideration. The required information should be gathered through observation, experiment or survey. Tasks to be completed include recording sampling procedures; carrying out sampling; preparing tests; editing the questionnaire; and selecting, training and calibrating the examiners. Market segmentation is the process of ranking potential consumers. In the marketing of a given product, groups of similar persons or organisations are created who react similarly to a given wine marketing strategy. By dividing into parts, segments can be created on the consumer side of the wine market.

43.4.2 Marketing Mix in the Wine Tourism Market

The wine marketing mix is the combination of marketing tools used in different market situations, corresponding to a given wine market situation. The data on market segmentation and the market research are stored and processed in the marketing information system. The previously listed things and the marketing perspective together determine the ratio of the marketing tools. In the case of production orientation the product policy, moreover the technology plays an important role, while in the case of strong market competition and price-sensitive consumer market the pricing policy has greater importance.

Although many food events, particularly community events, are free to participants and visitors in terms of entry costs, issues around price are significant at all levels of event management and marketing. Charges may cover costs, maximise profits, subsidise all participants or certain market segments, encourage competition or result in unanticipated consumer reactions. Prices should be set in accordance with a wide range of factors including past history, general economic conditions, ability to pay, revenue potential, costs, level of sponsorship and competition. Pricing may even be used as a mechanism to appeal to particular markets. For example, low-income consumers may be more price sensitive or have issues such as access to food because of lack of transportation or no grocery store in their neighbourhood.

Attracting and retaining non-resident visitors to wine-producing regions require the coordinated efforts of both the wine tourism and the tourism industry stakeholders. More targeted consumer research, exploring the wine tourist's expectations regarding the destination and their product preferences

FIGURE 43.6 Roudiere's opinion on the motivations of wine tourists including age, interest and nationality. *Source:* Roudiere (2012)

would provide a useful information base on which to explore this issue.

Developing a good cellar door experience is key to selling more wine! Cellar door sales are on the rise; 70% of wineries sell more than 20–50% of their wine at the cellar door; and 85% of wineries are involved in shipping wine directly to consumers. It's important to have a good cellar door wine tasting experience (with the winemaker). This is because people drink stories and wine; therefore, if they like you, they will buy your wine. Wine clubs offer a good incentive for people to become brand ambassadors, tell others about the wine and visit the winery. It is a popular tool in the New World wine countries. Revenue from wine club membership is higher in dollar/euro terms than from the average visitor. Set-up costs are low but investment may be needed at the point of sale. However, when it comes to technology and online bookings, wineries are not so advanced: 62% of suppliers don't have a booking calendar; 17% use online tools; 34% use e-mail, and 39% use the phone for bookings.

According to Karlsson, there are two types of wine tourism: general wine tourism and specialised wine tourism. BKWine is an organisation that deals with wine tourism, wine journalism and video making, and organises wine meetings. Today, modern communication is an interactive, user-friendly website, blog writing, e-mail marketing, YouTube videos and sharing on community platforms and social media. The success of modern communication is its continuity, fun and very powerful communication tools. Successful wine tourism requires substantial resources. Livesey explored wineries around the world in the context of wine tourism. Globally, 500 wineries had nearly equal representation in the survey. It was found that 35% of customers use social media for recommendations on travel and wine. Social media is an effective and relatively inexpensive way to promote a business online using the power of storytelling and images. Images are more important than text and video to attract a customer's attention and are easy to upload to Facebook, Instagram and Twitter. The favourite platforms are Facebook (91%), Twitter (52%) and Instagram (40%) (Figure 43.7). Currently, Facebook is in the lead.

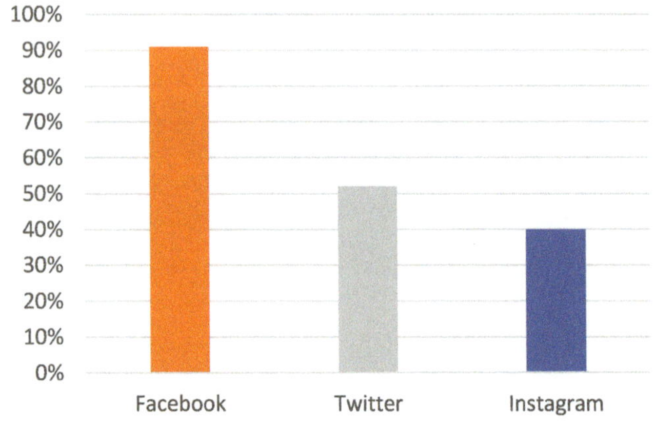

FIGURE 43.7 Favourite social media platforms. *Source:* Livesey (2016)

Of wineries, 43% only post two to three times a week on social channels and 23% post just twice a month. A blog is a great way to convey to customers and fans what is happening in the vineyard, how a vintage is progressing, the harvest and the tasting notes of the new vintage. But 80% of wineries don't have a blog. Online and events have proved to be the most popular and cost-effective channels of advertising. After trade, tourism and online marketing strategies are key priorities. In community terms, collaboration is a very important part of food and wine event marketing management, with a wide range of institutions and stakeholders.

The marketing aspect of wine tourism relates to general wine marketing. In summary, it can be stated that the internet and application-type services available through cell phone platforms absolutely determine current marketing activity. The development of artificial intelligence might change our lives, including wine tourism as well. It is highly likely that the production, communication, storage and analysis of certain information will no longer be human tasks. Such changes may take decades, and it can be taken for granted that in the field of wine tourism, winemakers, sommeliers, hosts and other actors will endure to the end to maintain the personal experience.

43.5 GOOD PRACTICE OF WINE TOURISM IN HUNGARY

In the following, good practice is introduced, and the best practices of the global wine tourism industry are summarised.

43.5.1 The 12 Best Practices of Global Wine Tourism

The most successful wine regions have adopted best practices that enable them to provide tourists with memorable experiences that inspire them to return time after time –bringing their relatives and friends. In Liz Thach's opinion, the following good practices can be formulated:

1. *Wine Roads*
   Any wine region that wants to be taken seriously must take the time to develop maps listing their wineries and provide information on their hours of operation, website, phone numbers and directions. In addition, wine maps should include local restaurants, hotels and other tourist sites. These maps are provided free on the web and in brochure format, and are very helpful for tourists planning a trip.
2. *Wine Community Partnerships*
   Successful wine regions work in partnership with local hotels, restaurants, airports and transportation companies to make sure that tourists have a way to find them. There is often an executive director of wine tourism and marketing for the region who is responsible for developing these community partnerships and tours. A good example is in the Hunter Valley of Australia where visitors are picked up from

Sydney airport and transported to the valley where they spend 4 days visiting wineries, accommodated in a hotel with meals. The wineries of Hunter Valley work together with local tour operators to create this beneficial partnership.

3. *Special Wine Events and Festivals*
   Many wine regions host special events and festivals, but the most innovative regions think 'out of the box' in developing unique events. For example, Lodi, California, has an annual 'Wine and Crane Festival', and Melton Wine Estate in New Zealand hosts a 'Cabaret and Wine Show' with comedians and singers.
4. *Experiential Wine Programmes*
   Related to special events is the recent practice of offering wine tourists unique experiential programmes. For example, in Napa and Sonoma Valleys of California, it is common for visitors to participate in wine blending seminars where they mix together different types of wine to create their own customised bottle – such as a Bordeaux blend with Merlot, Cabernet Sauvignon and Malbec. They design their own wine label and get to take the wine home with them. Another example of an experiential programme is 'Dog Walks in the Vineyard', similar to that offered by Martha Clare Winery in New York.
5. *Link Wine to Regional Tourism*
   Smart wine regions make sure to link to other local tourism sites. This is a win-win strategy for everyone involved because the more activities that can be advertised, the more likely the region will attract greater numbers of tourists. For example, tourists visiting Beijing for the first time always want to see the Great Wall and the Forbidden Palace, but now many also want to taste the local wine and visit famous wineries such as Chateau Changyu and Jinshanling.
6. *Unique Partnerships*
   Linking up with different types of partners, rather than just the usual marriages of food, wine, music and art, is another best practice of a successful wine region. For example, the Sonoma Mission Inn Spa in California has teamed up with local wineries to offer afternoon wine tastings for visitors who have spent the day at the spa enjoying such wine-related treatment as a Chardonnay Scrub and massage.
7. *Wine Villages*
   Some wine regions have committed their time and resources to creating a 'wine village'. This is a town in a wine region that is designed specifically around the theme of wine. Generally, there are multiple wine-tasting rooms within walking distance that tourists can visit – an example is the mountaintop wine village of Montalcino in the Brunello region of Italy.
8. *Focus on Art and Architecture*
   Some wineries attract visitors by adding art galleries, sculpture gardens or other unique art-related items. For example, both Bodegas O Fournier Winery outside of Mendoza, Argentina and the Hess Collection Winery in Napa Valley, have famous art collections that visitors can see while tasting wine.
9. *Food and Wine Matching*
   Another best practice is targeting tourists who enjoy the culinary aspects of wine tourism. Generally, this is implemented by a wine region organising special food and wine tours or events. A good example is the wine and paella event held every spring in Baja, Mexico, where the local wineries match their wines to many different types of paella rice dishes.
10. *'Green' or Ecotourism Focus*
    For wine tourists who seek organic and biodynamic wines, or those who enjoy the outdoors and being around nature, a newer best practice is an emphasis on 'green' or ecotourism aspects of wine. For example, some wineries offer special tours and educational programmes on how they craft organic and biodynamic wines.
11. *Unique Wine Tours*
    Another cutting-edge practice is offering very unique tours to winery visitors. These are usually targeted at the more adventurous wine consumer or those who have already visited a specific wine region and are looking for something different. An example is 'wine and kayaking' as offered by Chatham Winery in Virginia, or 'river-rafting and wine tasting' as offered by Southern Oregon Wineries working in partnership with a local tour company.
12. *Social Media for Wine Tourism*
    Finally, many wineries and regions are catching on to the benefits of using social media to attract wine tourists. This includes making sure that those tourists who use their mobile phones and the internet to seek information on which winery to visit can easily locate the winery. They do this by ensuring that the GPS directions are correct, and that they are easily found in search engines. Finally, savvy wineries have set up Facebook fan pages and work with other sites, such as Tripadvisor, to make sure they can interact with wine tourists.

43.6 THE WINE TOURISM OF THE HUNGARIAN TOKAJ WINE REGION, AS A WORLD HERITAGE SITE

Wine is a noble beverage that certain nations proudly prepare and consume. Viniculture and vinification have traditions. In addition to the traditional wine-growing countries, 'conquering' wine nations have also appeared on the international wine market. The wines of the 'New World' have acquired a market in countries such as France. Globalisation is inevitable in this sector, too. In addition to new market characters, a third 'wave' appears to be emerging, which although it is as yet not significant, in 10 years time it may become the largest wine exporter. This country is China, where current imports of wine are higher than exports; however, a process has started that may result in quality wine production, not only for internal consumption but also for export, too.

Hungary's most important wine region is Tokaj-Hegyalja where the unique Tokaj Wine Specialities are prepared. Tokaj Wine Specialities have several competitors, which are similar 'noble sweet' wines. The countries competing to produce such products are Austria, France, Germany, Canada, the United States, Australia, South Africa and Slovakia. Tokaj Wine Specialities belong to the dessert wine category, thus many other dessert wines are considered competitors. Owing to their sweet taste, they are not suitable for consumption in large quantities. Their value is their speciality. Hungary cannot produce them in large quantities, but, in addition to satisfying the domestic market, they are also produced for export.

During the decades preceding the change of regime in 1989–1990, the brand 'Tokaj' lost its credibility. However, over the past 27 years, thanks to the excellent marketing activity of some wineries, Tokaj Wine Specialities have started to improve. The cooperation between the characters of the wine districts has not been realised entirely yet. During the research work, the author, however, assessed consumers' habits regarding Tokaj Wine Specialities, and studied the wine producers' marketing strategy. The research was carried out in Hungary and Austria. A factor analysis and a cluster analysis were performed. During the research work, a questionnaire was administered and online marketing research was conducted, and information was obtained through in-depth investigations. In the Hungarian sample (n = 1179), five distinct clusters were defined (Figure 43.8).

Typically, all five clusters considered price intervals that were proportionate to the family's monthly net income. People who are interested in wine know more than the average wine consumer and take into consideration the price–value ratio – they will even buy a cheap wine. The first two clusters were price sensitive. Most Tokaj Wine Specialities are purchased at hypermarkets and mainly by women. Tokaj Wine Specialities are generally bought for birthdays, name-days and Christmas. Changes in consumers' incomes modify the minimum and maximum value of price intervals. Price increase reduces the risk of purchasing Tokaj Wine Specialities. The findings of the cluster analysis were supported with focus group surveys.

Some 60% of Viennese wine consumers interviewed in Austria had no knowledge of Tokaj Wine Specialities, which was the main reason for not purchasing them. Additionally, they did not like the taste, preferring the Austrian 'Auslese'-type wines (which are similar to Tokaj Wine Specialities). Respondents who knew Tokaj Wine Specialities purchased a small amount once or twice a year. The Austrians more frequently go to the HORECA sector, usually their income and life standard are higher.

According to the Austrian sample (n = 107), three main segments could be distinguished:

- Respondents with a low income.
- 'Wine adepts'.
- Visual consumers (with a low and a high income).

Austrian wineries could adapt to consumer and market demands to a great extent. The proportion of locally sold wines is higher, which is in addition to their other tourism services that are of a far higher standard, too. This development can also be observed at Tokaj-Hegyalja, although at a slower pace for the time being. The community wine market that operates well in Austria is already making its effects felt in wineries. In Hungary, this organisation has already been set up.

Conclusions can be summarised as follows:

- Consumers know their way about wine supply with difficulty.
- The current position of Tokaj Wine Specialities in gastronomy and culture makes their frequent consumption difficult.
- Wine is a confidential product. The attitude towards a wine already tasted, the positive feeling of satisfaction and/or memory, as well as the (perhaps opinion-shaping) recommendation of an influential person (group) are determinants when purchasing wine. The price in the form of a price interval acts as a

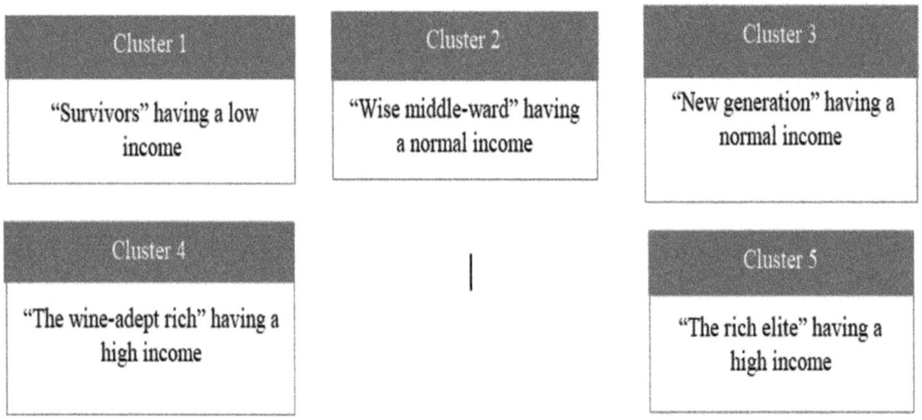

FIGURE 43.8 Five clusters of the Hungarian sample. *Source:* Szakál (2009)

kind of a filter, which correlates with the family's total monthly net income.

- In the survey, wine consumers of Tokaj Wine Specialities can be segmented into well-separated target groups in both the Hungarian and the Austrian markets. Therefore, the segment-specific marketing mixes can be elaborated, which largely improves efficiency, too.
- Hungary has a low-level wine culture, but it is on a developmental path. Consequently, associations that are improving the wine culture should be supported.
- The tourism developments of a given region would considerably promote the popularity of wine consumption onsite. Investment would considerably enhance the sale of Tokaj Wine Specialities.
- The Austrian example shows that knowledge of wine largely affects the sale of Tokaj Wine Specialities and similar products. A higher standard of living also favourably influences the turnover of Wine Specialities.
- Both the wine district and the national community of wine marketing considerably promote wine sales in Austria. However, the same cannot be said for Tokaj-Hegyalja yet, so more efficient marketing work is needed both at the national and the wine district levels.
- A local tourism strategy promotes sales on site. This is already operating well on the Burgenland side, while Tokaj-Hegyalja is striving to efficiently implement this strategy.
- Marketing information on the Tokaj-Hegyalja wineries is sparse, so research by the wine district's organisation is necessary. A marketing strategy and elements of the marketing mix can only be elaborated on with relevant market information. This is true at the corporate, wine district and national levels, too.
- At Tokaj-Hegyalja, the composition of export markets is very heterogeneous and adjusts itself to the system of relations of each winery. Presence is not always accompanied with Community marketing support. On the Austrian side the opposite is true, so it can operate in a far more efficient system. The diverse export market results in the consumption and crumbling of the resources, at the expense of efficiency.
- A reconsideration and reform of the sales channel policy would significantly help wineries gain higher profitability. A fairer distribution of the profits from sales should be a basic criterion among the participants.
- To date, no marketing information system at the wine district level has been established. Market research should be elaborated and operated along a more conscious strategy, otherwise no wine district strategy can be developed.

To summarise, it can be stated that Tokaj Wine Specialities are now on a developmental path, but there is much still to be done, and the development and implementation of a targeted marketing strategy largely determine the future of this specific market. Cooperation and consensus are essential for Tokaj-Hegyalja. At the same time, the strategy of the wine district should adapt to the basic principles and activities of the national wine marketing community.

To conclude the topic of good practice, it should be noted that unique features have unlimited possibilities, and their utilisation only depends on the actors. The role of leadership is to decide how to position wine tourism at the different levels. The responsibility of leadership is to provide the conditions for the development of tourism. The good practice shows the complexity and supra-structure of the system which assumes the multiplier effect.

43.7 NEW TRENDS IN THE GLOBAL WINE TOURISM

Some of the new trends in global tourism are summarised as follows:

- Experiential programmes (participation in the grape growing or winemaking process).
- Wine villages/education centre for museums.
- Innovative wine events (wine and wildlife, wine and wool-gathering festival, wine and culinary tourism events).
- Murder mystery tours; bargain bash (a spring clean, garage sale of wine and related items).
- Regional interactive website.
- Unique tour options (a jeep ride, organic wine tour, horseback riding, hot-air balloon ride); wine cruise with gourmet lunches and dinners with superb wines.
- Wine and opera escorted tours.
- New types of partnership (with golf, resorts, spas).
- Innovative collaboration (discount coupons).

In line with developments and taking the requirements of the demand side into consideration, creative positioning could mean a competitive advantage. It will definitely be part of a future wine tourism strategy. Strategies can also be devised at the organisational and community level. In both cases, such strategies could affect more sectors that will have to cooperate with each other. Wine tourism is a multiplayer with which stakeholders of others sectors can connect. The principle of 'cooperating in competition' is an element of the regional marketing strategy that must be implemented by the wine tourism actors in the future.

In the strategy of winemaking companies, the advantages, the unique strength – *e.g.* the terroir and stories. At the company level, the quality of human resources is an advantage, but retaining these human resources will be a challenge. Style has to be represented not only in the wines, but also in a company's philosophy and services. This harmony will be key to success. Community strategies are very important as they form the basis for future directions at the settlement, wine region, province, country and Economic Union level. The EU and the

United States have different interests; they also have different strengths and so draw up different strategies. Within this, for example Hungary with its 22 smaller wine regions also has to focus on different things. In all cases, the grape and wine and their related services determine the image of the different levels. In future, strategies on climate change and its consequences (affecting cultivation) and issues including terrorism (affecting tourism) will represent challenges. Additionally, an understanding of consumers' behaviour and how it effects consumption patterns, and the implementation of marketing strategies in the most cost-effective way would be appreciated. The already unlimited potential of social media and technical developments provide unlimited creativity and various solutions to the actors. Artificial intelligence, the spread and later the regulation of the virtual space in 30 years, at about 2050 maybe achieve the wine tourism sector. Due to the coronavirus, the process of digitization is accelerated. The coronavirus has a huge impact on international tourism. Consumer attitudes are rearranging. Wine tourism may be in trouble due to restrictive state measures, but these affect all sectors. At the same time, health precautions can be followed in wine tourism, so it can be considered safe. It will be another era which will globally raise questions. Wine will probably be present along with mankind, and the wine tourism itself will also be part of our lives – independently of the environmental changes.

43.8 DEVELOPMENTS OF EXPECTED GLOBAL WINE TOURISM

A summary of some of the developments expected in global wine tourism are as follows:

- Adapting to new aspects of tourism transformed by the coronavirus. Examining the transformation of new, innovative solutions and consumer needs. Creating health security, collaborations are inevitable.
- Infrastructural developments (airport, motorway, highway, railway and water transport, hygiene and parking, etc.).
- Telecommunication and internet network developments.
- Organisational developments (tourist, communication, marketing agencies, destination management organisations, tourist governmental organisations, tourist non-governmental sector, etc.).
- Legislation in the field of tourism, vineyards and wineries (the whitening of the sector, tax discipline, market regulation, market protection measures, protection of origin, brand protection regulations, cultivation, processing and mellowing descriptions, etc.).
- Accommodation developments (specifically regarding the needs of the target groups determined in the strategy, based on the expected tourist numbers).
- Restaurant developments (specifically regarding the needs of the target groups determined in the strategy, based on the expected tourist numbers).
- Wine-related tourist attraction development (interactive wine museums, visitor centres suitable for demonstrating highly important universal values, etc.).
- Other tourist attraction development (cultural, sport and recreation centre developments, active and passive tourism products, spa and wellness services, landscape improving investments, etc.).
- Technological development of viticulture and winery (support systems for grapevines, crop safety improvement methods, processing technologies, mellowing technique, bottling, etc.).
- Wine tourism developments related to viticulture and winery (establishing wine hotels and wine restaurants, installations for the presentation of winery, wine boutiques, etc.).
- Implementation of special, target group–oriented, modern tools of wine tourism marketing (community marketing, social media, online and offline marketing blended with the classic marketing mix elements, modern information sign systems, etc.).

Roudiere's opinion is that the government or regional strategy to promote or increase wine tourism is based on the economic returns it provides. There is a growing interest in environmental issues and in dining and cooking. Some regions have seen an increase in the number of wineries.

43.9 STRATEGIES FOR THE GLOBAL WINE TOURISM

Based on the professional literature and experiences, it can be stated that global wine tourism has huge potential that should be regenerated from time to time and that there is a constant demand for it. The marketing tool kit of wine tourism is increasing in parallel with the developments in marketing. In the future, the opinion formers of community sites will play a role in determining important wine regions. Consumers are looking for special experiences that cannot be found in every wine region, and attention should be drawn to these experiences. The number of wine regions on the tourist map will increase where professional tourism development is found. There is no unsalable product, only the right target group needs to be found. From the regional marketing point of view, one should think specifically of the target group. All possible supports should be utilised and it is important to fit into a country's strategy.

Taking Porter's competition model into consideration, the global wine tourist market shows a different picture. In certain countries and provinces in the Economic Community, the conditions for entering the market, the distance between competitors and the bargaining position of both suppliers and consumers are different. If the world's wine tourism were demonstrated on an aggregate product life-cycle function, it would be placed in the growth phase. Of course, there is a large spread, because both ends of wine tourism development are present in the markets.

The UNWTO Declaration on Wine Tourism sets out several future development points (Table 43.3). Such developments

TABLE 43.3
The UNWTO Declaration on Wine Tourism

Recognition of the importance of wine tourism as an integral part of cultural tourism and provision of the opportunity for sustainable development of wine tourism in many destinations.
Development of policies that facilitate both the promotion and preservation of wine tourism destinations as well as respect for the social and cultural values of a local community.
Fostering of public–private partnerships with an emphasis on local entrepreneurship within the tourism value chain, while taking into account the authenticity of the destination.
Engaging in pertinent sub-national, national, regional, international and multilateral dialogues in order to advance the implementation of the above-mentioned objectives.
Advancing research in wine tourism to boost the competitiveness of destinations through innovative product development within and beyond wine tourism activities.

Source: UNWTO (2016)

clearly point out towards improving the quality of goods and services. Investigation of the consumption behaviour and communication habits of the wine tourist is inevitable in order to understand their requirements. In optimal case, a given development in the wine tourist market will return in about 5–10 years, and it is hardly able to completely renew within this timeframe. However, consumer habits might change over the course of 5–10 years, for reasons such as a change in the segment's characteristics or rapidly changing global or local market surroundings. The time during which the segment is static depends on its basic characteristics and critical changes in the segments. In 2019, 2020, and 2021, the Covid virus fundamentally changed the world's life and tourism. In addition to large losses, preferential systems are also being transformed. Hygienic safety is expected to be essential, safekeeping away is also expected. Wine tourists will appear in smaller groups, avoiding the crowds. These demands are reflected in wineries, restaurants, accommodations, and other tourist attractions. It is expected that there will be less travel and wine shopping, but tourists will pay more for them. The actors will travel with the slogan "less, better quality".

BIBLIOGRAPHY

Alonso, J. F. 2014. El país donde se consume más vino per cápita es… el Vaticano. http://abcblogs.abc.es/proxima-estacion/public/post/viajar-vino-consumo-mundial-16365.asp/.

Alpár, L. *et al.* 2008. Közösségi Bormarketing stratégia 2009–2013. Community wine marketing strategy Magyar Bormarketing Kht., Budapest. p. 144. http://www.bormarketing.hu.

Asero, V. and Patti, S. 2009. From wine production to wine tourism experience: The case of Italy. http://www.wine-economics.org/aawe/wp-content/uploads/2012/10/AAWE_WP52.pdf.

Beames, G. 2003. The rock, the reef and the grape: The challenges of developing wine tourism in regional Australia. *Journal of Vacation Marketing*, 9(3), 205–212.

Brown, G. and Getz, D. 2005. Linking wine preferences to the choice of wine tourism destinations. *Journal of Travel Research*, 43(3), 266–276.

Bulletin. 2013. A bor és gasztronómia mint turisztikai termék. Wine and gastronomy as a tourism product *Turizmus Bulletin XV.* p. 45, Magyar Turizmus Zrt.

Carlsen, J. and Charters, S. 2006a. Conclusion: The future of wine tourism research, management and marketing. In: J. Carlsen and S. Charters (Eds.), *Global Wine Tourism: Research, Management and Marketing* (pp. 263–275). CAB International, Wallingford.

Carlsen, J. and Charters, S. (Eds.) 2006b. *Global Wine Tourism: Research, Marketing and Management.* CAB International, Wallingford.

Carlsen, J. and Dowling, R. 1998. Wine tourism marketing issues in Australia. MCB UP Limited. http://www.emeraldinsight.com/journals.htm?articleid=1658874&show=abstract.

Carlsen, P. J. 2007. A review of global wine tourism research. *Journal of Wine Research*, 15(1), 5–13.

Carmichael, B. 2005. Understanding the wine tourism experience for winery visitors in the Niagara region, Ontario, Canada. *Tourism Geographies*, 7(2), 185–204.

Cey-Bert, R. 2002. *A Bor Vallása.* Paginarum, Budapest.

Charters, S. and Ali-Knight, J. 2002. Who is the wine tourist? *Tourism Management*, 23, 311–319.

Charters, S. and Menival, D. 2011. Wine tourism in Champagne. *Journal of Hospitality and Tourism Research*, 35(1), 102–118.

Elsevier. 2008. Adapted from Getz, 2007; in Getz, 2008; Event tourism: definition, evolution, and research, Elsevier Ltd. www.sciencedirect.com, Tourism management, 29.

ETC & WTO. 2005. *City Tourism & Culture, The European Experience.* World Tourism Organisation, Madrid.

French Wines Bulletin. 2013. http://www.frenchwinesbulletin.com/uk/2013/edition 2/focus-on-the-french-wine-industry/4wine-tourism-a-major-growth-avenue-towards-a sustainable-future.aspx.

Gatti, S. *et al.* 2003. *Wine in the Old World. New Risks and Opportunities.* Franco Angeli s.r.l., Milano, Italy.

Getz, D. 1998. Wine tourism: Global overview and perspectives on its development. In: R. Dowling and J. Carlsen (Eds.), *Wine Tourism–Perfect Partners. Proceedings of the First Australian Wine Tourism Conference* (pp. 13–33). Bureau of Tourism Research, Canberra.

Getz, D. 1999. Wine tourism: Global overview and perspectives on its development. In: R. Dowling and J. Carlsen (Eds.), *Wine Tourism, Perfect Partners. Proceedings of the First Australian Wine Tourism Conference*, May 1998, Bureau of Tourism Research, Margaret River, Western Australia.

Getz, D. 2008. Event tourism: Definition, evolution, and research. *Tourism Management* 29(3), 403–428.

Getz, D. and Brown, G. 2006. Critical success factors for wine tourism regions: A demand analysis. *Tourism Management*, 27(3), 403–428.

Goel, S. 2016. Exploring wine tourism in India – A potential to be uncorked, Dissertation, New Delhi, p. 51. https://www.slideshare.net/SheetuGoel/wine-tourism-62745533.

Gomis, F. J. D. C. (Ed.) 2009. More authors. Wine tourism product clubs as a way to increase wine added value. The case of Spain. Universidad Miguel Hernández, Spain. http://www.wine-economics.org/aawe/wp-content/uploads/2013/07/67-Reims2009-DelCampo-Brugarolas.pdf.

Gravenstede, V. 2013. MA events marketing management. https://www.slideshare.net/VictoriaGravenstede/wine-tourism-slides?next_slideshow=1.

Hall, C. M. 1996. Wine tourism. In: *New Zealand*, J. Higham (Ed.), Proceedings of the Tourism Down Under II: A research conference. University of Otago, New Zealand.

Hall, C. M. and Macionis, N. 1998. Wine tourism in Australia and New Zealand. In: R. W. Butler, C. M. Hall and J. M. Jenkins (Eds.) *Tourism and Recreation in Rural Areas* (pp. 197–224), Wiley, Chichester.

Hall, C. M. and Sharples, L. 2008. *Food and Wine Festivals and Events Around the World*. Butterworth-Heinemann is an Imprint of Elsevier, Oxford, UK, Burlington, NJ, p. 369, ISBN: 9780750683807.

Hall, C. M. *et al.* 2000. *Wine Tourism Around the World: Development, Management and Markets*. Butterworth Heinemann, Oxford.

Holderness, D. 2012. Wine tourism: Exploring the potential, tourism Australia. https://www.slideshare.net/dholderness/wine-tourism-exploring-the-potential-15023784.

HTA 2017. http://borespiac.hu/2017/06/15/kozos-szakmai-rendezvenysorozatot-indit-az-mtu-es-a-hnt/.

Karlsson, B. 2011. Modern communication and wine tourism, an example of how to sell the wine tour destination with modern communication tools, BKWine AB. https://www.slideshare.net/bkwineper/bkwine-modern-communication-and-wine-tourism.

Kotler, P., Bowen, J. T. and Makens, J. C. 2010. *Marketing for Hospitality and Tourism*, 5th Edition. Pearson, Boston, MA. ISBN-13: 9780132453134.

Lavaux 2013. http://www.lavaux-unesco.ch/en/N5710/vitour-landscape-the-project.html.

Lewis, P. and Bibo, B. 2003. Hani-English, English-Hani dictionary; Wine, food, and tourism marketing. *Journal of Travel and Tourism Marketing*, 14(3/4).

Lirong, H. 2011. The prospect and forecast of China's wine tourism in 2011. *Energy Procedia* 5, 1616–1620, Elsevier Ltd., IACEED2010. www. sciencedirect.com.

Livesey, T. 2016. Best Practices in Wine and Food Tourism, London Wine Fair Trade. https://www.slideshare.net/TatianaLivesey/best-practices-in-wine-tourism-winerist-at-london-wine-fair-4-may2016.

Menival, D. and Charters, S. 2013. Wine tourism in Champagne: A solution to increase the value of a standard quality product. http://www.wine-economics.org/aawe/wp-content/uploads/2013/05/portland-046-Menival-Charters.pdf.

Michalkó, G. 2012. *Turizmológia*. Akadémiai Kiadó, Budapest.

Michalkó, G. and Vizi, I. 2006. http://itthon.hu/site/upload/mtrt/Turizmus_Bulletin/bulletin_2006_balaton/html/balaton_jborturizmus.html.

Mintel 2014. MINTEL 'Global Food Tourism' document. http://itthon.hu/documents/28123/11848364 /Mintel_Global_Food_Tourism.pdf/33848ec7-0865-4d21-8797-2ec10eef4d14.

Mitchell, R., Charters, S. and Albrecht, J. N. 2012. Cultural systems and the wine tourism product. *Annals of Tourism Research*, 39(1), 311–335.

Movimento 2012. Movimento turismo del vino. http://www.movimentoturismovino.it.En/Home/.

Nunes, P. A. and Loureiro, M. L. 2012. Agricultural landscape, vineyards and tourism flows in Tuscany, Italy. http://www.wine-economics.org/dt_catalog/working-paper-no-103/.

O'Neil, M. and Palmer, A. 2004. Wine production and tourism: Adding service to a perfect partnership. *Cornell Hotel and Restaurant Administration Quarterly*, 45(3), 269–284.

Official, C. C. I. V. 2005. Official Chamber of Commerce and industry of Valladolid. http://www.chamber-commerce.net/dir/3897/Camara-Oficial-de-Comercio-e-Industria-de-Valladolid-in-Valladolid.

Pratt, M. 2011. Profiling wine tourists, more than just demographics. *6th AWBR International Conference*, Bordeaux Management School, Bordeaux, France.

Pratt, M. 2014. Four wine tourist profiles, Griffith Business School, Griffith University, Australia. *8th International Conference Geisenheim, Germany, AWBR an HGU organization.*

Presensa, A., Minguzzi, A. and Petrillo, C. 2010. Managing wine tourism in Italy. *Journal of Tourism Consumption and Practice*, 2(1), 46–61.

PTE 2011. http://www.tankonyvtar.hu/hu/tartalom /tamop425/0051_Turisztikai_termektervezes_es_fejlesztes/ch08s05.html.

Romano, M. F. and Natilli, M. 2009. Wine tourism in Italy: New profiles, styles of consumption, ways of touring. *Tourism: An International Interdisciplinary Journal*, 57(4), 463–476.

Roudiere, K. 2012. Wine tourism. https://www.slideshare.net/caramany/wine-tourism.

Szakál, Z. 2009. A wine market and marketing analysis of wine specialities from the Tokaj-Hegyalja wine district. *Studies in Agricultural Economics*, 109, 85–101.

Szakál, Z. 2017. The opportunities of the Tokaj Wine Region as Word Heritage Destination on Global Wine tourism Market, 'Balance and Challenges' X. *International Scientific Conference*, Miskolc-Lillafüred, Hungary, 278 p. ISBN 978-963-358-140-7.

Szakál, Z. 2019. Wine tourism analysis of the Tokaj Wine Region. MAG Scholar Conference in Business, Marketing and Tourism. 2019 (Europe). "Reconnecting Asia with Eastern Europe": Conference Proceedings, Hungary, ISBN: 9789633581902, 156–162. Miskolc, Hungary.

Szakál, Z. *et al.* 2019. Borút, borturizmus integrált fejlesztési gyakorlat. Turisztikai és Vidékfejlesztési Tanulmányok, Wine route, wine tourism integrated development practice. *Tourism and Rural Development Studies*. 2677-0431 2498-6984, 21–35, Pécs, Hungary.

Szivas, E. 1999. The development of wine tourism in Hungary. *International Journal of Wine Marketing*, 11(2), 7–17. doi:10.1108/eb008692.

Thach, L. 2012. 12 best practices in global wine tourism; this article was originally published in Fine Wine & Liquor Magazine, Dec. 2012 and Jan. 2013 in both English and Chinese. https://lizthachmw.com/2013/12/06/12-best-practices-in-global-wine-tourism/.

UNWTO. 2016. UNWTO Declaration Wine Tourism – A growing tourism segment, Tbilisi, Georgia. http://cf.cdn.unwto.org/sites/all/files/pdf/georgia_declaration.pdf.

Valdani, E. and Ancarani, F. (Eds.) 2000. *Strategie di marketing del territorio*. Egea, Milano.

Villanueva, Girini 2015. Wine and vine Heritage Marketing in the Malbec landscape, American Association of Wine Economists (AAWE). *9th Annual Conference*. http://www.wine-economics.org/aawe/wp-content/uploads/2015/05/Mendoza-program-Villanueva-Girini.pdf.

Waller, D. 2006. Wine tourism: The case of Alsace. Master Thesis in European Tourism Management. Bournemouth University, UK.

Williams, P. W. and Dossa, K. B. 1995. *Journal of Travel and Tourism Marketing*, 14(3/4), C. Michael Hall (Eds.), The Haworth Hospitality Press, Philadelphia, PA. ISBN 0789000822. https://books.google.hu/books?id=IdFEAQAAQBAJ&pg=PA1&lpg=PA1&dq=Williams+-+Karim+B.+Dossa&source=bl&ots=iWAqFbONUZ&sig=bbxRDgbe5zhwCwAZ0BvLPZEDyK0&hl=hu&sa=X&ved=0ahUKEwjrhaTI0LTVAhWFbRQKHZNECL4Q6AEIJTAA#v=onepage&q=Williams%20-%20Karim%20B.%20Dossa&f=false.

Zisou, K. D. 2013. *Wine Tourism and Economic Development of Rural Areas, Erasmus University Rotterdam*. Erasmus School of Economics, Rotterdam.

44 Innovations in Wine Production

Ronald S. Jackson

CONTENTS

44.1 INTRODUCTION

Innovations in winemaking have ancient beginnings. Wine itself was an innovation, *albeit* likely accidentally discovered. A container full of grapes, left for more than a few days under warm conditions, will begin to self-ferment. The resultant weakening of the skins leads to rupture, juice release, and its fermentation by epiphytic yeasts. When the origins of this joyous accident were sufficiently realized to induce the intentional production of wine, is unknown. Nonetheless, evidence suggests it occurred at least 7000 years ago. Present evidence suggests wine's inception began in the northern parts of the Fertile Crescent and adjacent southern Caucasus, corresponding to regions where the spread of agriculture first overlapped the southernmost distribution of wild grapevines (*Vitis vinifera* f. *sylvestris*).

A settled agricultural lifestyle is a prerequisite for converting a seasonally available, mildly alcoholic mash, produced from fruit collected in the wild, into a semi-stable beverage, befitting the name wine. For wine production to become a cultural adornment, consistently available, also requires intentional cultivation. A nomadic existence is incompatible with planting vines that begin to produce a crop only several years after planting. The proximity of an oak forest was also serendipitous. The indigenous habitat of *Saccharomyces paradoxus*, the presumptive progenitor of the wine yeast (*S. cerevisiae*) occurs in the sap of oak trees. *S. cerevisiae* is not a typical member of the grape flora. Although the yeast flora of grapes can ferment grape juice, they usually do not completely metabolize all fermentable sugars. As a result, the product can be readily populated by spoilage yeasts or acetic acid bacteria—the latter converting nascent wine into vinegar. Although vinegar is useful to primitive cultures, vinegar is not a seraphic beverage. Thus, the presence of at least some *S. cerevisiae* in the vicinity was crucial to inoculate the juice to produce a semi-stable drinkable beverage. Microbial stability was aided by the natural acidity of grapes (inhibitory to most bacteria) and that the principal acid (tartaric acid) was atypical for fruit—tartaric acid being resistant to degradation by most microbes. Also favoring stability was the presence of a ready supply of antimicrobial phenolics, extractable during fermentation from grape seeds and skin.

Although grape cultivation probably originated along the northern extremities, contingent with the beginnings of farming, it eventually moved further south, into the Levant, Mesopotamia and Egypt. Although supplying wine for the ruling and religious elite, the suitability of grape cultivation was marginal, making its availability to the general populace unlikely. Archaeological and written evidence suggest that the common alcoholic beverage was beer—a product more easily produced from a product (grains) that could be dried and be available year round.

Once nascent viticulture had begun, the roguing of non-productive (male) vines would have unintentionally favored the selection of vines that were self-fertile (producing functionally monoecious flowers), and eventually the elimination of purely female vines. This would have been facilitated by the process of serendipitous clonal selection involving vegetative propagation (burying trailing canes in the soil). Buried canes readily root, sending up shoots that can eventually be transplanted. Although vegetative propagation retains the properties of the original plant, subsequent mutation would have generated visible or otherwise distinctive and desirable variants—their selection and propagation leading to the diversity found today (Figure 44.1).

The storage of wine for any length of time was a major problem. The issue was partially solved with lining the inner surfaces of the principal storage vessel (amphorae) with resin. Resin prevented wine from seeping through the porous vessel, but it also contributed a marked flavor unacceptable to most modern consumers. This situation changed only when technical advances in amphora production gave them a vitreous inner lining, about the beginning of the Imperial Roman period. The earlier adoption of cork as an amphora closure provided wine with a relatively oxygen-impermeable container in which improvement during aging was possible. In contrast, most wine had a short "shelf-life." To make it more

FIGURE 44.1 Grapevine heterogeneity as illustrated by the diversity in morphology, structure and coloration in grapes and grape clusters. *Source:* Photo courtesy of Julius Kühn-Institute, Institute for Grapevine Breeding Geilweilerhof, Siebeldingen, Germany

palatable, wine was often "doctored" with a host of ingredients, presumably to mask the wine's poor quality. The habit of adding flavorants was especially favored by unscrupulous proprietors, occasionally to the point of lethality. The practice continued unabated during medieval times, and periodically up to modern times. A classic example was the addition of lead salts to mask excessive acidity.

When the Roman Empire went into decline, wine production did likewise. Increasingly, wine was both fermented and stored in oak cooperage, a vessel less efficient than vitreous-lined amphorae at excluding oxygen and was readily contaminated by spoilage microbes. Thus, wine often became undrinkable within a year, especially when transported over the crude road systems of the time.

Other than some improvements in press design, the first significant late medieval innovation in improving wine stability was the introduction of the use of sulfur dioxide in the late 1400s. This involved burning sulfur-soaked wicks in barrels before adding wine. Nonetheless, the practice spread slowly due to unfounded fears that it might poison the wine. Another important advance in wine storage came with improvements in glass production in the mid-1600s. Bottles sufficiently strong to be closed with a tightly fitting cork became available, providing conditions under which wine could be aged for extended periods in the absence of oxygen. Due to developing industrialization, glass production costs fell, and there was a shift from both transporting and aging wine in barrels to glass bottles. This was also critical to the production and spreading popularity of sparkling wines.

Improvements in distillation also permitted the fortification (and stabilization) of wines exported from Southern Europe. Without refrigeration, overheating during fermentation often led to incomplete fermentation. The resultant residual sugar content, combined with low acidity (due to malic acid degradation before harvest), and unfamiliarity with the benefits or use of sulfur dioxide, generated conditions that often favored microbial spoilage during transport. The addition of wine spirits prevented such microbial spoilage and ultimately led to the evolution of today's fortified wines.

The improving economic conditions in Central Europe following the Renaissance led to the emergence of a burgeoning middle class, willing and able to pay for the changes required to improve wine production. This, in turn, encouraged the expansion of grape and wine production. Regrettably, equivalent conditions did not occur simultaneously in the warmer regions of Southern Europe. Thus, their local cultivars and wines languished largely unknown, except locally—a situation that partially still exists.

Economic developments fostered renewed interests in scientific advancements. Among these were developments in chemistry that led to understanding the role of sugar in fermentation—a feature crucial in determining the amount of sugar needed for the second, in-bottle fermentation of sparkling wines. Other advancements in chemistry provided the means of measuring grape maturity, and thus, when best to harvest grapes for optimum quality. Discoveries in microbiology demonstrated the role played by yeasts in fermentation, and the action of other microbes in wine spoilage. With this knowledge came an understanding of the beneficial role played by sulfur dioxide in limiting wine spoilage. In addition, as the microbial origins of most grapevine diseases were realized, the stage was set for a rational means of disease control. Furthermore, developments in plant and yeast physiology led to improved vine and yeast nutrition, crucial to enhanced grape and wine quality.

The spread of these innovations worldwide has democratized wine production. Thus, quality wines can and are being produced globally—no longer the preserve of a few regions blessed with conditions favoring grape and wine production, and positioned close to markets willing and able to afford their production. As markets have expanded, the demand for "artisanal" wines has encouraged experimentation, designed to enhance wine distinctiveness and character. Equally, major producers are tailoring their wines to specific consumer groups. Thus, all segments of wine production are being driven by innovations.

In the past few decades, an industry formerly steeped in tradition is increasingly embracing innovation. Although certain aspects of individuality may be lost, the ideal of a "fine wine for all on the supermarket shelf" is increasingly within reach. The winemaker is no longer just the equivalent of a midwife at the birth of a wine, they are its designer, almost as a weaver is of wool.

Control begins, as it must, in the vineyard. The state of the fruit arriving at the cellar door sets limits on the attributes ultimately possessed by the wine. Nonetheless, the conditions of fermentation and maturation direct how the wine develops and subsequently ages. Among recent innovations are experiments with different yeast and bacterial strains, how these are combined, or the use of local strains. Procedures that were once regional in use, notably *sur lies* maturation, *appassimento*, *saignée*, and cold maceration are spreading worldwide. This has been greatly aided by research on these techniques appearing not only in modern texts and scholarly journals, but also in trade and popular wine publications. Long gone are the days when knowledge was the prerogative of a few, or passed from father to son as proprietary information. Sensory procedures have also allowed a means of more precisely measuring the sensory influence of modifications conducted in the vineyard and/or winery. Marketers are also beginning to tap into this database to direct their advertising.

44.2 VINEYARD INNOVATIONS

Initially, most advances were based on astute observation. This began to change about 150 years ago with the increasing application of scientific knowledge to viticultural problems.

That wine is "made in the vineyard" is an oft quoted adage. However, Mother Nature is often fickle and the vagaries of the climate are becoming more extreme. Thus, the ingenuity of the vineyardist is often sorely tested, attempting to diminish the negative effects of climatic variability on fruit quality. Although everyone genuflects to the term quality, its definition is anything but simple. Like beauty, it is relative, often based on paradigms developed with experience, historic and personal. For grapes, this usually correlates with reaching full maturity, typically declining thereafter—maturity itself usually couched in terms of desirable sugar, acid, and phenolic contents. Thankfully, flavor development during ripening is usually closely correlated with these long-established harvest parameters (sugar and acid contents, their ratio, and grape coloration). However, because the correlation is imperfect, and can vary considerably from year to year, and region to region, direct assessments of flavorants known to be of varietal significance would be preferable. Regrettably, the crucial details of aromatic maturity are insufficiently known, variously reaching a peak in concentration and declining thereafter. In addition, determining their presence is complex due to their diversity, their presence in trace amounts, and that they occur in non-volatile complexes. These may be released only during fermentation, maturation, or aging. Because many of these non-volatile complexes are bound to sugars, there was hope that determining the grape glycosyl-glucose (G-G) content would yield a better indicator of aromatic maturity. Regrettably, the G-G content has often been no more an effective indicator of grape maturity than traditional and simpler methods. Other important flavorants may also accumulate as oxides (terpenes) or exist weakly bound to other constituents (*e.g.*, glutathione or cysteine). More modern approaches, such as measuring grape coloration with handheld optical sensors (Figure 44.2), do not necessarily correlate with extraction during fermentation and subsequent retention during maturation and aging. The same also applies to measurements of the grape tannin content. Additional indicators of maturity being investigated include features such as skin hardness. This property affects the grape cell-wall fragility, and thus the ease with which important skin phenolics may be released. Nonetheless, it is always important to realize that any theoretical "optimum" maturity state is style dependent—for example, late harvesting generates attributes that favor the production of sweet wines, whereas moderate immaturity provides the characteristics preferred for most sparkling wines.

FIGURE 44.2 A handheld spectrophotometer designed for phenolic maturity monitoring, detection of nitrogen deficiency, and early pathogen detection. The Multiplex® 3 illustrated measures flavonol, anthocyanin, and chlorophyll contents from leaves and fruit epidermis. Other versions can also assess stilbene contents. Its internal GPS permits geolocalization of blocks. *Source:* Photo courtesy of Force-A

To further direct viticultural efforts to enhance wine flavor, there is a major effort to understand the chemical origins of a wine's desirable fragrance. This also demands assessing the relative sensory significance of the various volatile compounds isolated. Initially, this involves determining their individual olfactory thresholds, and comparing them to their respective concentrations. The ratio, the odor activity value (OAV) provides an indicator of the potential sensory impact of each compound. Those with a ratio of >1 are potentially sensorially significant. While a useful first step, aromatic compounds often interact with one another, acting either synergistically or antagonistically. In addition, human sensory perception is not directly correlated with OAV values, generating different sensory qualities based on concentration, combination with other aromatics, the manner of presentation (*via* a glass or in-mouth), and the context of the tasting. Thus, the goal of correlating the concentration of a compound in the grapes with its sensory significance in a wine still remains a distant hope.

Another aspect of vineyard innovation involves measuring variability within grape clusters, location within the canopy, and throughout the vineyard. Limiting yield, up to a point, favors flavor development, but this property is nonlinear, as well as cultivar and clone dependent (Figure 44.3). Because limiting yield has most of any beneficial effect when conducted early in the season, an effective early predictor of grape yield is required. Previously, obtaining an estimate of yield by measuring flower number and fruit set in the vineyard was complex, expensive, and time-consuming. However,

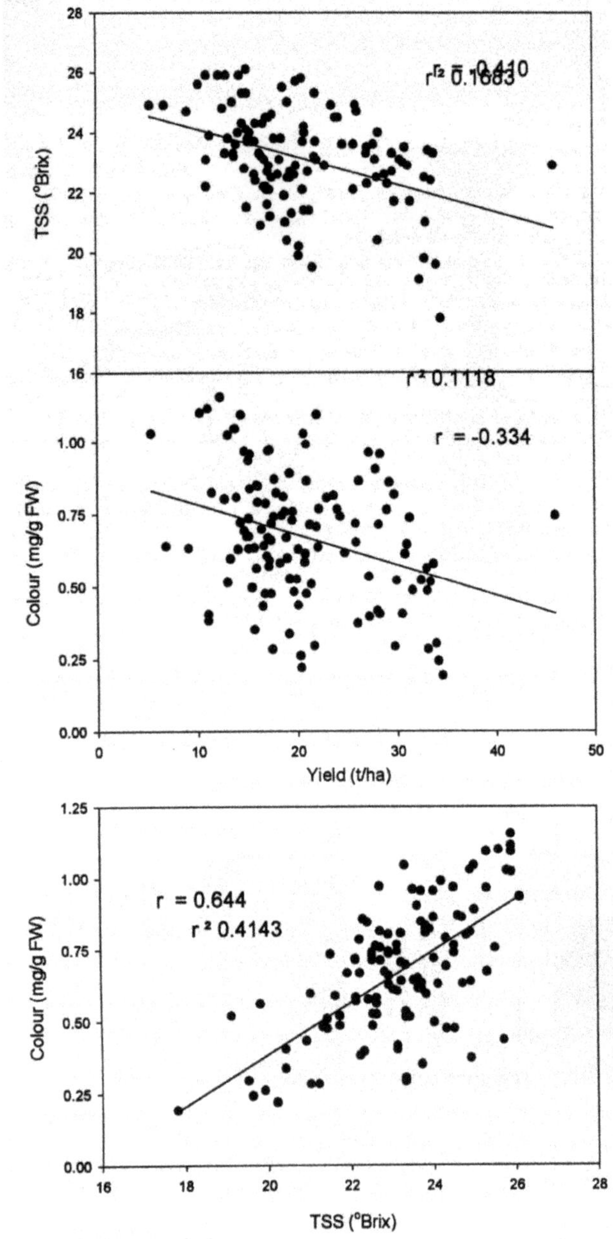

FIGURE 44.3 Variability between total soluble solids (TTS), yield, and color in "Shiraz" relative to yield and Brix values. *Source:* Reproduced from Holzapfel, B.P., Rogiers, S., Degaris, K., Small, G., 1999. Ripening grapes to specification: Effect of yield on colour development of Shiraz grapes in the Riverina. *Aust. Grapegrower Winemaker* 428, 24, 26–28, by permission

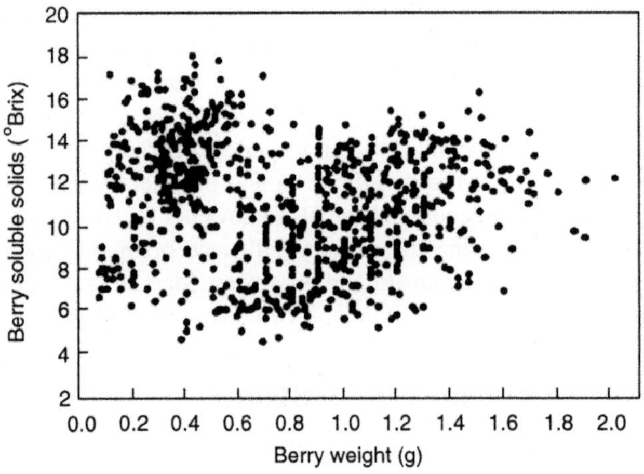

FIGURE 44.4 Variation in berry size in grape clusters and soluble solids of Chardonnay grapes. *Source:* Reproduced from Trought, M.C., 1996. The New Zealand terroir: Sources of variation in fruit composition in New Zealand vineyards. In: Henick-Kling, Geneva, NY. pp. I-23–27, with permission

a computer-based analysis of vineyard photographs appears to provide a simpler means of estimating potential grape yield by which fruit load can be adjusted to vine capacity. Whether a lack of synchrony in flower development, fertilization, and fruit ripening throughout the vineyard (Figure 44.4) will reduce the practical value of this technique has yet to be established. Much of the asynchrony in fruit development appears to be due to vineyard microclimatic non-uniformity. Thus, one of the most significant trends in viticulture has been the measurement of within vine and vineyard variability, as a guide to its reduction. Crop uniformity is often viewed, correctly or incorrectly, as crucial to wine quality.

Because crop uniformity is the result of multiple influences that occur throughout the growing season (and past seasons), the grape grower potentially has a multitude of options by which to compensate for or limit grape variability. Selective harvesting of parts of grape clusters or distinct sections of a vineyard can be used to enhance grape uniformity, where the financial returns justify the expense. In most instances, other techniques are more cost-effective. One option is planting several different clones or cultivars to cushion the vagaries of climate. More generally efficacious, though, is the application of precision viticulture (PV).

Precision viticulture often incorporates traditional onsite analyses (soil, disease, yield, grape composition) with ground or remote sensing. The purpose is to precisely locate the sources of vineyard variability that may cause non-uniform grape ripening and quality (Figure 44.5). As such data become more readily available, effectively limiting fruit heterogeneity may become one of the principal tasks of the vineyard manager.

Evidence indicates that considerable improvements in yield and maturity prediction are possible with data provided by PV. Remote sensing, often involving GPS-directed drones, is making the mapping of vineyard variability down to individual vines both feasible and increasingly affordable. It is also permitting the precise application of pesticides only where and when needed. Where considerable vineyard variability exists, and has not as yet been adequately corrected, harvesting can often be adjusted to minimize the impact of fruit non-uniformity. Ideally, improved fruit uniformity at the winery door should yield wines with a more predictable character. Nonetheless, wine quality is multifactorial, not easily correlated with objective measurements (Figure 44.6). In addition, fruit variability may not always be a negative feature. It could

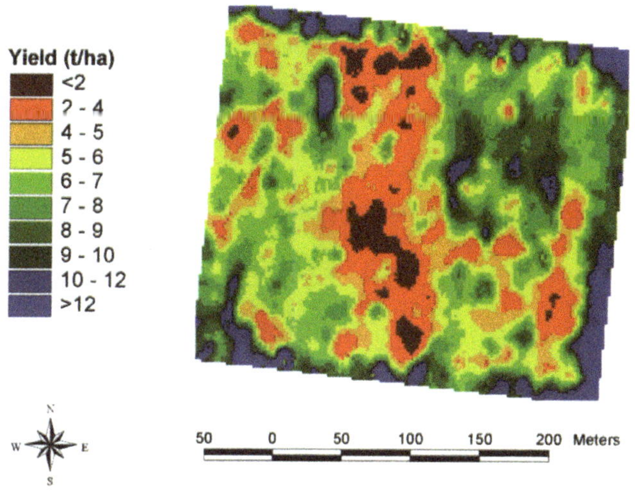

FIGURE 44.5 Variation in grape yield (1999), clearly showing the marked influence of microclimatic, terrestrial, and atmospheric conditions on vine yield. *Source:* Reproduced from Bramley, R., Proffitt, T., 1999. Managing variability in viticultural production. *Aust. Grapegrower Winemaker* 427, 11–12, 15–16, by permission

accentuate flavor complexity, the *sine qua non* of quality. In addition, aspects of vineyard variability may partially compensate for climatic vicissitudes—not all vines being equally affected.

Remote sensing with near-infrared spectroscopy and chlorophyll fluorescence can assess water-related photosynthetic efficiency, permitting the vineyardist to adjust irrigation relative to vine need. Water use efficiency can be enhanced using modern techniques, such as regulated deficit irrigation (RDI) and partial rootzone drying (PRD). Irrigation can also be combined with simultaneous fertilization and used as a conduit for some systemic disease–control agents. Remote sensing is also being tested as a means of early detection of mineral deficiencies and foliage diseases.

Mechanical harvesting was initially developed as an innovative labor-saving device. Subsequent improvements now permit its use with most cultivars, if the vines are appropriately trained. Combined with special screens, or optical sorters, essentially only healthy grapes and clusters pass on to the stemmer-crusher, or directly on to the press.

Innovations in pest control have helped reduce prophylactic pesticide use, for example with the use of microclimate-based disease models. Pest resistance may be delayed, if not prevented, by selectively alternating agents which have different modes of action. Some newer agents are not themselves toxic but activate inherent disease-regulating factors in the host. Generating microclimatic modifications with the canopy, such as basal-leaf removal and more open training systems, can often reduce the incidence of disease. These open the canopy, permitting better access of pest-control agents to fruit and foliage (thereby reducing the amount needed to be applied). Combined with developments in nozzle design, pesticide distribution has improved as well as minimizing run off or drift. Biological control is another option in some instances, often combined with specific ground covers to foster the survival

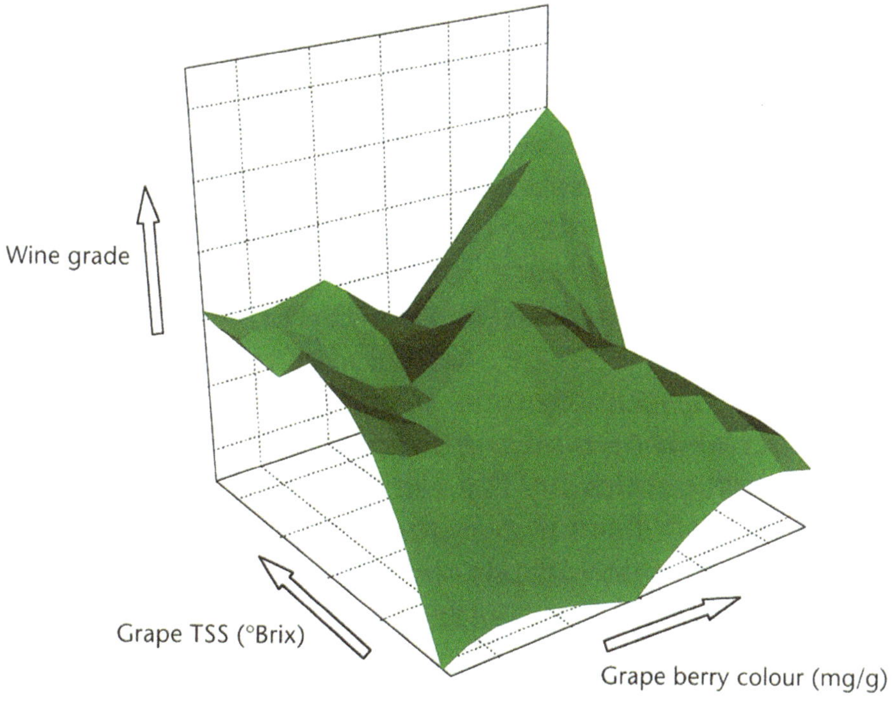

FIGURE 44.6 Schematic representation of the relationship between grape total soluble solids and berry color with the quality grading of the resultant wine. *Source:* Reproduced from Gishen, M., Iland, P.G., Dambergs, R.G., Esler, M.B., Francis, I.L., Kambouris, A., *et al.*, 2002. Objective measures of grape and wine quality. In: Blair, R.J., Williams, P.J., Høj, P.B. (Eds.), *11th Aust. Wine Ind. Tech. Conf. Oct. 7–11, 2001*, Adelaide, South Australia, Winetitles, Adelaide, Australia, pp. 188–194, by permission

of natural pest-control agents. Although incorporating genetic resistance has always been an option, traditional techniques were shown to have disrupted a cultivar's varietal characteristics. This may change with the introduction of clustered regularly interspaced short palindromic repeats (CRISPR) and related genetic techniques. They have the advantage of being much more specific, faster and less expensive to use, do not incorporate foreign genetic material, and, more significantly, do not change crucial varietal characteristics.

44.3 WINERY INNOVATIONS

For millennia, crushing grapes under foot was standard. Although gentle, it was labor-intensive, slow, and incomplete. Unbroken grapes underwent auto-fermentation. At a time when few wines aged well, it was a potential benefit—producing wines drinkable sooner. With the development of mechanical crushers, the workload was reduced and all grapes were crushed. Combining stem removal with crushing avoided the uptake of undesirable stem tannins during maceration. Only when grapes are pressed whole, such as in producing sparkling wines, are stems useful during pressing.

Occasionally, innovation can be a backward step. When oxidation was discovered to be the cause of premature browning of white wines, the solution seemed obvious, avoid all contact with oxygen during and after production. However, blanketing the crushed juice with carbon dioxide or nitrogen only aggravated the situation. Unknowingly, the practice retained readily oxidized phenolics that otherwise would have oxidized and precipitated during fermentation and clarification. Thus, a return to older white winemaking techniques has come into vogue. Another innovation is, in reality, a broader application of an old technique—cold pre-fermentative maceration. It is the modern version of a traditional Burgundian technique where red grapes are left on the seeds and skins (macerated) for several days at cool temperatures. The procedure favored the extraction and stability of pigments from Pinot Noir grapes. The technique is now being used experimentally with several other red cultivars, for example Syrah and Cabernet Sauvignon. Also reflecting on the return to old procedures is the renewed interest in spontaneous fermentation. Initially, all wine fermentations occurred spontaneously, starting with the action of the epiphytic grape flora. In most instances, fermentation was subsequently dominated by one or more strains of *Saccharomyces cerevisiae*, typically derived from cells remaining on winery equipment from previous vintages.

Once techniques were developed to select and store yeasts in a viable state, winemakers could purchase and inoculate their musts with strains with known characteristics. In some instances, this involved species closely related to *S. cerevisiae*, for example *S. bayanus* var. *bayanus* for fino sherries and sparkling wines, and *S. bayanus* var. *uvarum* for cool fermentations (*e.g.*, many white wines). The innovation of precise inoculation avoided several problems associated with uncontrolled spontaneous fermentations, and provided greater assurance that the fermentation would yield wines with attributes desired by the producer. However, this has been viewed by some as eliminating the unique characteristics of regional wines. Thus, to regain (or retain) some of this typicity, some producers have returned to spontaneous fermentation, or inoculation with regionally isolated strains. For others, spontaneous fermentation is viewed as a means of distinguishing their wines from local competitors. To obtain some control over the process, inoculation with several yeast strains, with or without other yeast species, may be used. Inoculation with non-*Saccharomyces* yeasts, such as *Metschnikowia pulcherrima*, is also a possible solution to the increasing alcohol content of table wines associated with delayed harvest and global warming. Another alternative is the use of interspecific yeast strains with some of the sensory complexity claimed for spontaneous fermentations, but without the associated risks, such as sensitivity to the action of killer-yeast factors or the accumulation of sulfur off-odors. Other forms of strain improvement may even involve over-expression of flavor genes or their insertion.

Another comparatively recent innovation is inoculation for malolactic fermentation. Because its occurrence, when desired, was always unpredictable, inoculation with known strains of *Oenococcus oeni* both solved this problem and gave greater control over the sensory consequences (besides deacidification). Inoculation also helped limit the action of undesirable lactic acid bacteria.

Innovations in the design of analytic instruments have generated sensors that can monitor fermentation parameters such as the temperature and pH, and the sugar, nitrogen, and alcohol contents. This has been particularly useful in regulating fermentation in gargantuan cooperages. Thus, winemakers have unprecedented opportunities to follow the progress of fermentation and institute quick corrective measures as needed or desired. Although nowhere near "autopilot" winemaking, such control that was impossible even a decade ago is now possible.

In the past, excessive temperature buildup during fermentation often caused stuck fermentation. Above 30°C, fermentation progressively slows and can eventually terminate. This problem came to an end with the deployment of tank refrigeration. In addition, cool fermentation became an option, not a feature dictated by the external climate. More effective means of automatically submerging the cap of seeds and skins that accumulates during the fermentation of red wines have reduced, if not eliminated, temperature differentials between the cap and the fermenting juice. In the past, this differential could easily be up to 10°C. Such procedures have also facilitated the earlier extraction of anthocyanins and flavorants from the skins, as well as providing another aspect of control over the wine's flavor profile (by influencing yeast metabolism).

Innovations have also involved fermentor volume as well as construction. Stainless steel tanks are now the industry standard, replacing large wooden or cement tanks and vats. Stainless steel tanks can be produced in almost any size or shape desired, they transfer heat rapidly (facilitating temperature control), are easier to clean and store empty, have the potential of being supplemental storage vessels, and are both inert and impervious (protecting the wine's attributes unmodified). In addition, rotary fermentors combine minimizing temperature stratification within the must with gentle mixing of the seeds and skins with the fermenting juice (Figure 44.7). The horizontal position of the fermentor also increases the contact between the juice and the pomace. Nonetheless, oak fermentors have certain advantages in producing particular wines. Typically, this has involved barrels, but new large oak fermentors are much easier to empty and clean than previously. Inserts also permit effective temperature control as well as facilitating pumping over. Replacing one of the oak staves with clear fiberglass permits direct visualization of the progress of fermentation (Figure 44.8).

Blending usually occurs at the end of maturation, involving samples derived from fermenting different press fractions, from different fermentors or storage vessels (notably barrels), from grapes harvested separately, or from distinct vineyard sites or locations. However, blending the must (or juice) of different cultivars before fermentation—termed co-fermentation—is another return to an old technique. It is viewed as improving the wine's flavor more than blending the finished wines. A variant of the technique is used with some red cultivars deficient in copigments. The juice from select white grapes is typically added to the red must to enhance color stability. Another unique form of co-fermentation, also intended to enhance the color of a red wine, is a by-product of rosé production. Once sufficient pigmentation has been extracted from the fermenting must for producing the rosé, the pomace is removed and added to the fermenting must of the red requiring a richer color. The technique is termed *saignée*.

In an attempt to achieve more flavorful wines, producers may resort to grape over-maturation, either delaying harvest or cool storage after harvest—termed *appassimento*. That partial dehydration raises the grape Brix (and the alcohol content of the resultant wine) is clear, but whether by itself *appassimento* sensorially enhances the perceived flavor is still contentious. Originally, the process in Veneto was associated

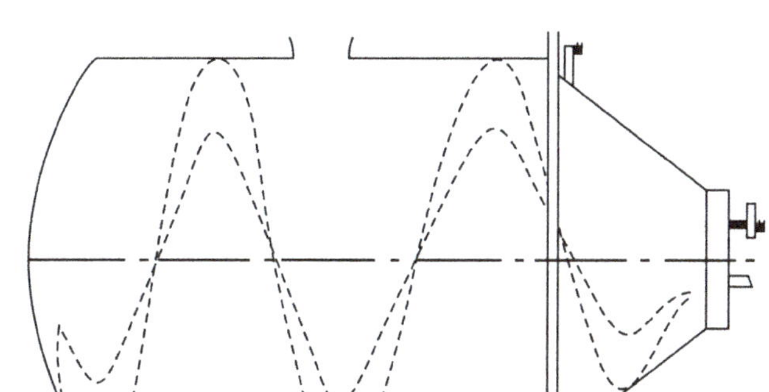

FIGURE 44.7 Diagram of a rotary fermentor. *Source:* Courtesy of Bucher Vaslin, France

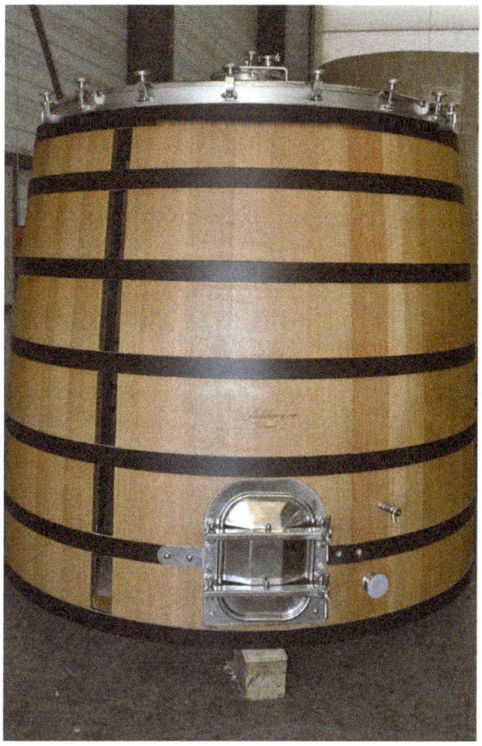

FIGURE 44.8 Modern version of an oak tank (Seguin Moreau, France) facilitating temperature control and observation of fermentation, cleaning. *Source:* Photo courtesy Ronald Jackson

with chemical changes induced by a slow infection of a portion of the grapes by *Botrytis cinerea*. These changes seem crucial to producing the features that characterize most traditional Amarone wines. In addition, some research suggests that higher alcohol contents reduce the volatilization of some wine flavorants, or distort wine flavor profiles, only occasionally enhancing fruit flavors. For other cultivars, such as Gewürztraminer, varietal aromatics degrade during over-maturation.

Most techniques capable of reducing the higher alcohol contents now typical of most wines have simultaneously removed wine aromatics. A novel approach that circumvents this issue involves picking a portion of the crop unripe. Once fermented, the wine is treated with bentonite and charcoal to produce a flavorless, low-alcohol, acidic wine. It is subsequently added to wine made from fully or over-mature grapes, thereby decreasing the blended wine's alcohol content. The resultant wine also possesses a fuller, more traditional flavor profile. The acidity supplied by the low-alcohol wine enhances the color intensity and freshness.

Other innovations designed to improve wine flavor promote the liberation of flavorants from non-volatile complexes. These complexes tend to break spontaneously during aging, but can be accelerated by heating or adding hydrolytic enzymes. Of these alternatives, enzymic hydrolysis is preferred. It is less associated with flavor loss or distortion. More efficient (but expensive) regulation can be achieved with enzyme immobilization on a column, through which the wine is slowly passed. For wines designed for early drinking (the majority), the process is probably valuable. However, for premium wines, designed for long in-bottle aging, it is a moot point as to whether the early release of aromatics in volatile form is judicious. Gases and volatiles do slowly escape from even the best sealed wines.

After fermentation, all wines require time for clarification and stabilization. Although effective, these natural processes often occur more slowly than currently desired. Correspondingly, a wide range of techniques have been developed to speed up such processes. Older clarification techniques include the addition of fining agents, such as tannins, egg whites, isling glass, or bentonite. Regrettably, these may also remove some flavorants or add allergens. Alternatives now include newer plant-based fining agents, centrifugation, or clarifying filters (notably cross-flow filters that are less prone to plugging). Other related innovations include microfiltration (for sterile filtration) or ultrafiltration (to remove undesirable colloidal and other macromolecules).

Despite the modern attention to winery hygiene, there is still the risk of microbial spoilage. A few microbes are capable of multiplying in the acidic, alcoholic, anaerobic conditions of wine. For this reason, wines are often given a small dose of sulfur dioxide at bottling. Not only is it antimicrobial, but it also acts as an efficient and useful antioxidant. While alternatives to sulfur dioxide are under investigation, no equivalently effective substitute has been found. Sterile filtration and hot bottling are often used in combination with sulfur dioxide when the likelihood of microbial spoilage is high, *e.g.*, in low alcohol or sweet wines.

When an oak character is desired, barrels are still the preferred source, partially due to tradition or appellation dictates. Nonetheless, lower-cost options are available. Modern options, without the expense and upkeep of barrels, include thin oak slats in a stainless steel frame, oak chips, oak "flour," and oak extract. Based on the desire of the vintner, the oak can come from American or European oaks, and be given a desired degree of toasting, equivalent to that supplied to barrels during their construction. In addition to direct cost savings, the duration of the contact with the oak can be reduced. What is missing, however, is the slow infusion of trace amounts of oxygen *via* the barrel, favoring color stability in red wines. This feature can be contributed by using diffusers (micro-oxidation). Uptake can also be regulated with precision using fiber-optic inserts connected to oxygen-measuring fluorescence spectroscopy. The rates of oxygen uptake in the range of 5 mL O_2/liter per month (at 15–20°C) are roughly equivalent to barrel uptake, and lower than the maximal rate of oxygen consumption by red wine. Another feature that may be missing with these alternative oak sources may come from the shorter duration of contact. While increased surface contact between the wine and oak promotes earlier flavorant uptake, it may not provide sufficient time for the inherently slow degradation and formation of oak flavorants (notably those derived from lignins).

For white wines, generally benefitting less from oak exposure, this may be supplied during in-barrel fermentation. While efficacious, any oak character given to a white wine is more frequently supplied by the use of oak slats, oak chips, or other less expensive and simpler options.

Where oak cooperage is preferred, but a limited uptake of oak flavorants is desired, one option is to use previously used barrels. Each barrel use progressively reduces and changes the nature of the flavorants extracted (Rous and Anderson, 1983). Alternatively, where oak cooperage is preferred, but oxygen uptake is to be restricted, large-volume oak cooperage may be used.

As noted above, a number of innovations in winemaking are actually the spread of formerly regional procedures. The current interest in *sur lies* maturation is an example. The process, once widespread, was retained in Burgundy to enhance the flavors of prestigious appellations. It involved allowing the wine to remain in the barrel in which it was fermented—in contact with the dead and dying yeast cells (lees). The procedure involves *batonnage* (periodic stirring of the wine). The oxygen uptake so permitted limited the development of a highly reductive lees layer (conducive to the generation of sulfur off-odors).

One of the innovations of the last several centuries (the use of glass bottles) is itself undergoing a transformation. A recent innovation, associated primarily with marketing, uses the abilities of various metallic oxides to increase the range of distinctive colors possible in glass. A less-expensive alternative is the application of a colored and/or textured coating. They can also provide improved abrasion and impact resistance, can protect against ultraviolet-induced wine oxidation, and facilitate recycling (the coating melts off, leaving colorless glass for processing).

While the advantages of glass still apply, its disadvantages (production costs, breakability, weight, and disposal issues) have encouraged the introduction of alternatives. Particularly successful has been the bag-in-box container. It can be stored more easily, provides a large surface for marketing information, and as the wine is removed, the collapse of the "bag" keeps the wine under anaerobic conditions. The major impediment to greater use relates to problems with oxygen ingress around the tap, limiting its shelf-life to usually less than one year. For smaller volumes, 1 L cartons save both on production and shipping costs. Another alternative is the aluminum can. One clear drawback to all these innovations is their lack of elegance. Nonetheless, for the market sector for which it is aimed, this could be an advantage.

Unlike the glass alternatives noted above, lack of a traditional shape is not an impediment to the adoption of polyethylene terephthalate (PET) bottles. It possesses most of the qualities of glass—moldability, transparency, impermeability—but is lighter, impact resistant, more readily recycled, and its production costs are less. PET plastic bottles can also be provided with a degree of oxygen impermeability similar to glass, with the incorporation of a gas barrier and/or the addition of an oxygen scavenger. Another glass alternative is poly-lactic acid (PLA). It has the additional advantage of being biodegradable.

The dominance of cork as the preferred closure is essentially over. Its demise coincides with the occurrence of an off-odor traced to the cork and inappropriately designated as a corky odor. One of the advantages of cork is its lack of an odor. Regrettably, use of a pesticide against insects on cork oak trees unwittingly became the source of cork's most nefarious fault. The metabolism of the pesticide by epiphytic fungi on the bark converted it to a highly aromatic, off-odor, 2,4,6-trichloroanisole (TCA). Although use of the pesticide seems to have stopped, the reputation of cork has been severely damaged. Consequently, studies into cork substitutes became inevitable. The first polyethylene corks were soon discovered to be excessively gas permeable, giving the wine a short shelf-life. Polyvinyl, or ethylene vinyl acetate, stoppers followed, having better gas impermeability, but the roll-on, aluminum screw cap has been the winner. The cap liner, based either on polyvinylidene chloride (Saran) or polytetrafluoroethylene (Teflon), can provide a range of oxygen permeabilities. This provides the oxygen permeability some producers consider essential to optimal wine aging (similar to natural cork). Screw caps also have consumer appeal, being easily opened and resealed. Another new entry into the closure market is a T-shaped ground-glass stopper, similar to those used to stopper bottles in chemistry labs. As with screw caps, the neck of the bottle must be specially molded to accept the stopper. To counter issues concerning cork's variability in oxygen permeability, cork producers began to promote the use of agglomerate cork. A variant, possessing the ease of opening and closing of screw caps, contains curved grooves to match treads in the bottle neck. It also requires a bottle designed specifically for its use.

Because premium wines require considerable bottle aging to reach their peak sensory quality, stock is often tied up for years before being released and demands additional aging by the purchaser. Although early investigations into accelerated aging were unsuccessful, more recent attempts with relatively short exposures to 45°C were reported to produce changes resembling those engendered by several years of in-bottle aging. Alternating current (600 V/cm for 3 min) has also been reported to improve the balance and mouth-feel, reduce aldehyde and higher alcohol contents, and slightly increase the ester content. Other techniques investigated have included exposure to ultrasonic and gamma rays, nanogold photocatalysis, and high pressure. Nonetheless, none of these treatments has achieved any industrial acceptance.

44.4 SPARKLING WINES

The development of sparkling wines owes much to innovations, starting in the late 1600s. Such innovations included the development of bottles capable of sustaining pressure up to 6 atm, the use of cork closures, prediction of the amount of sugar appropriate for the second fermentation, and effective means of removing the yeast lees from the bottle (riddling and disgorging). Modern innovations cover most aspects of production, from press to bottle closure.

Innovations in press design have involved the use of pneumatic pressure rather than the former tradition of a wide, shallow basket press. Pneumatic presses reduce the pressure needed by increasing the surface area over which pressure can be applied, thereby diminishing the extraction of pigments, tannins, and suspended solids from the grapes. Recent modifications make them easier to both load and empty, facilitate crumbling between pressings, simplify cleaning and maintenance, as well as taking up less floor space (*e.g.*, http://www.coquardpresses.com/uk-pai.php).

To develop the effervescence features that characterize sparkling wine, several years on lees have usually been required to extract mannoproteins. This period is also considered important to the flavor of champagnes. In the hopes of accelerating the process, breeders have developed yeast strains that undergo more rapid autolysis, at least shortening the maturation period required for mannoprotein liberation. Another innovation has been the chemical extraction and purification of mannoproteins from yeasts, allowing their addition at some stage prior to or just after disgorging. How rapid autolysis may influence the release of potentially important flavorants from lees appears to be unknown.

Traditionally, riddling, the movement of sedimented yeasts to the bottle neck, took 3–8 weeks, and was done manually. However, the design of automated riddling machines has obviated this arduous, monotonous, and potentially dangerous task. The result has been a reduction in the cost, economies in space utilization, minimized handling, and a foreshortening of riddling to about 7–10 days. Other potential innovations, designed to reduce the expense and complications of disgorging, such as yeast immobilization in a stable gel matrix (immobilization) or a membrane cartouche, have not been adopted by the industry. Adding magnetized nanoparticles to yeasts, permitting their rapid collection to the neck prior to disgorging is another option, whose acceptability is as yet unknown.

Innovations have also involved bottle closure. The development of the crown cap resulted in its replacing natural cork stoppers, held in place with a metal clamp (*agrafe*). Crown caps are not used on the finished wine primarily due to the image being inappropriate (resemblance to old-style soft-drink bottles). Nonetheless, the cork that seals the disgorged bottle incorporates the combination of a modern agglomerate cork exterior fused to two disks (*rondelles*) of natural cork, positioned next to the wine. A thoroughly modern development in closure design employs a plastic, resealable cap (Figure 44.9). It fits any standard sparkling wine bottle and retains the outward appearance of a traditionally sealed champagne. The cap liner provides excellent gas impermeability. Another alternative closure is a modified aluminum screw cap.

Older innovations, designed to reduce production costs, have included the transfer and Charmat (bulk) processes. However, riddling machines eliminated the economic

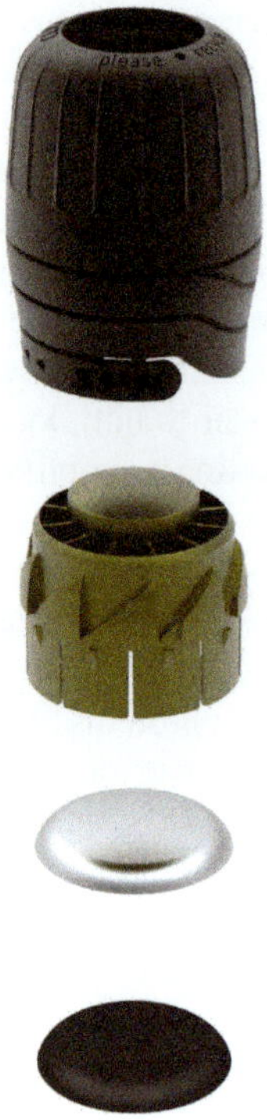

FIGURE 44.9 The SPK Zork cork for sparkling wines that offers the option to seal the bottle effectively once it has been opened. *Source:* Photo courtesy of Scholle Packaging

advantage of the transfer process. In contrast, the Charmat method still retains its appeal in the production of several sparkling wines. It substitutes the second, in-bottle fermentation with fermentation in a sealed tank, capable of sustaining the high pressures that develop. Riddling and disgorging are replaced by isobaric centrifugation or filtration.

44.5 FORTIFIED WINES

Many innovations were involved in the evolution of fortified wines, but the most central was the perfection of distillation. Crucial to the process was the ability to isolate fractions as they came out of the still, permitting the selection of the portions with the properties desired. Adding wine spirits (unmatured brandy) is usually a requirement for reaching the requisite alcohol contents. Raising the alcohol level of the base wine to above 18% effectively inhibits essentially all microbial growth and spoilage. Subsequent developments in different regions led to the three major styles—sherry, port, and madeira.

For sherries, fractional (solera) blending was the crucial next innovation. It involves periodic and sequential blending of portions of one set of barrels (a *criadera*) to another, older criadera. The process averages out vintage differences, and gives rise to a uniform product—ideal for producing brand-named products.

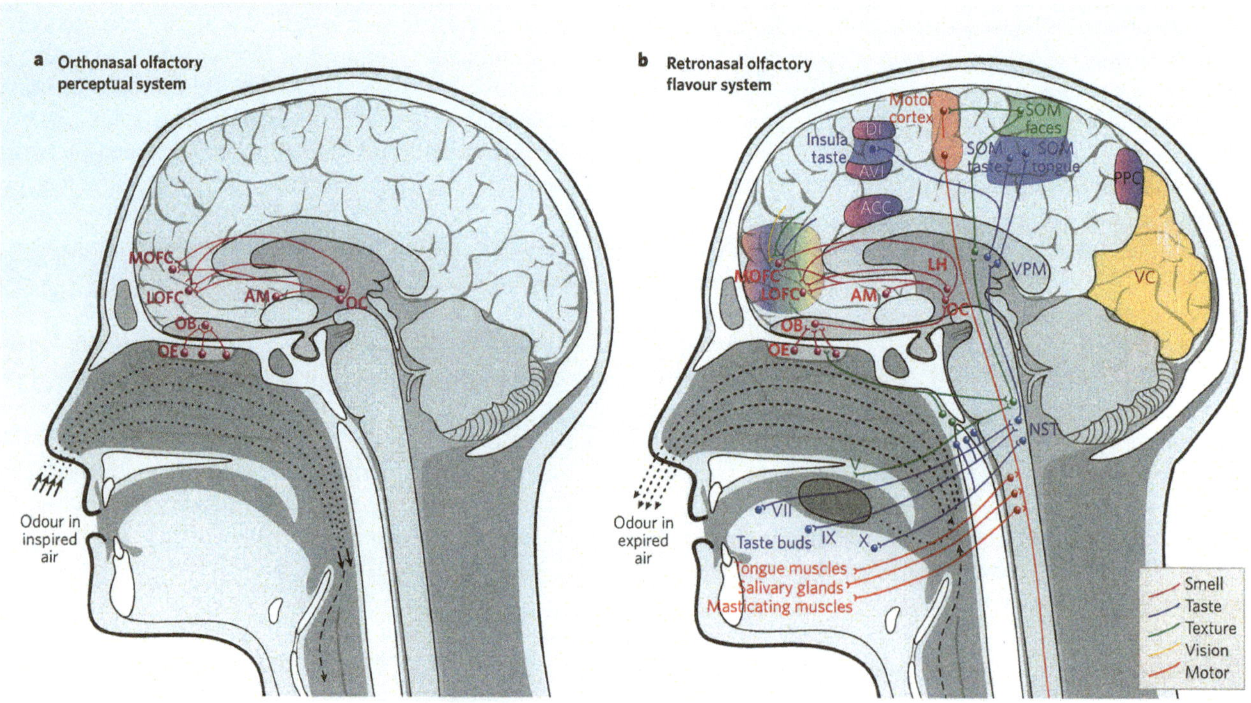

FIGURE 44.10 The dual olfactory system. (a) Brain systems involved in smell perception during orthonasal olfaction (sniffing in). (b) Brain systems involved in smell perception during retronasal olfaction (breathing out), with food in the oral cavity. Air flows indicated by dashed and dotted lines; dotted lines indicate air carrying odor molecules. ACC, nucleus accumbens; AM, amygdala; AVI, anterior ventral insular cortex; DI, dorsal insular cortex; LH, lateral hypothalamus; LOFC, lateral orbitofrontal cortex; MOFC, medial orbitofrontal cortex; NST, nucleus of the solitary tract; OB, olfactory bulb; OC, olfactory cortex; OE, olfactory epithelium; PPC, posterior parietal cortex; SOM, somatosensory cortex; V, VII, IX, X, cranial nerves; VC, primary visual cortex; VPM, ventral posteromedial thalamic nucleus. *Source:* Reprinted from Shephard, G.M., 2006. Smell images and the flavour system in the human brain. *Nature* 444, 316–321, by permission of Nature Publishing Group; Copyright Clearance Center #4093271244613, Apr 20, 2017

Recent innovations have primarily concentrated on increasing the precision with which a base wine's evolution can be directed. Other innovations have been aimed at shortening and improving the conditions traditionally used in partially dehydrating grapes prior to fermentation, and shortening the solera aging required.

In contrast to the addition of wine spirits after fermentation of the base wines, as with sherry, red port has the spirits added about halfway through fermentation. The final alcohol content is raised to between 18 and 20%, retaining about half of the original grape-sugar content. Subsequent maturation may involve a combination of storage in large and/or smaller oak cooperage, its duration, and the types and degree of blending before bottling. These specifics define the various types of ports. Except for the new production of rosé ports, port production has retained its traditional production methods almost intact over the past 100 years.

Madeira has also retained its traditional production methods, largely unaltered from their inception. The principal innovations have involved modernization of the methods and controls for the heating (*estufagem*) process. However, for special versions, heating may still occur in oak *butts* on the top floors of warehouses. Storage in this location can last for upwards of eight years, and expose the wine to alternating cycles of heat and cold. One "innovation" has been a return to using the grape variety to produce the category that bears the cultivar's name.

44.6 SENSORY EVALUATION

In recent years, the investigation of a wine's sensory attributes has possibly undergone the greatest transformation than any other aspect of wine production. Initially, wine tasting was hedonic—the prerogative of the aficionado and wine critic. Negociants, wine agents, brokers, and wholesalers used tasting to rank and select wines for sale to retailers. Marketers used comparative tastings to reach new clients and gain media attention. More recently, sensory scientists have begun to seriously investigate the origin and nature of a wine's flavor characteristics. Subsequently, attention has turned to measuring a wine's sensory attributes in terms of descriptors, and most recently, graphically representing the temporal dynamics of these attributes over the course of a tasting. With these developments, information relevant to the consumer is becoming available that can guide innovations not only in winemaking but also in grape production. Changes in production procedures aimed at producing wines that will appeal not only to a wider audience but also to the aficionado is the ultimate goal.

Most of the current stress is on the wine's flavor, primarily its fragrance in the glass as well as in the mouth. This is the source of the vast majority of a wine's pleasure-giving qualities. Admittedly, a wine's taste attributes are crucial, not only to consumers unacquainted with assessing the wine's flavor, but also to connoisseurs searching for the nebulous attributes termed balance and body. What sensory science or any other enological discipline cannot adequately define are a wine's extrinsic qualities, those based on an estate's or producer's prestige or provenance—what might be called connoisseur appeal. Such questions are the prerogative of the social scientist and psychologist.

One of the most important advancements in sensory science has been the realization that perception is not directly linked to sensation. The visual, gustatory, olfactory, and other contextual inputs from a wine are initially processed in a specific sensory part of the brain, then funneled through the emotional, and finally integrated with memory models in the orbitofrontal cortex (Figure 44.10). Memory traces that form early and throughout a tasting career create patterns, seemingly analogous to algorithms in computers. If these patterns are progressively reinforced, they speed up interpretation and tentative identification of sensory inputs. While useful in day-to-day functioning, such as the almost instantaneous identification of familiar faces, speeding up interpretation, there is a hierarchal dominance by which these inputs are recorded in memory. For example, vision is dominant over olfactory, and olfactory over taste sensations. The orbitofrontal cortex seems to compare new sensory inputs with the memory traces of equivalent past experiences. The result can be a distortion of the perception of sensory input to match preconceived patterns. For example, the color of a solution often results in

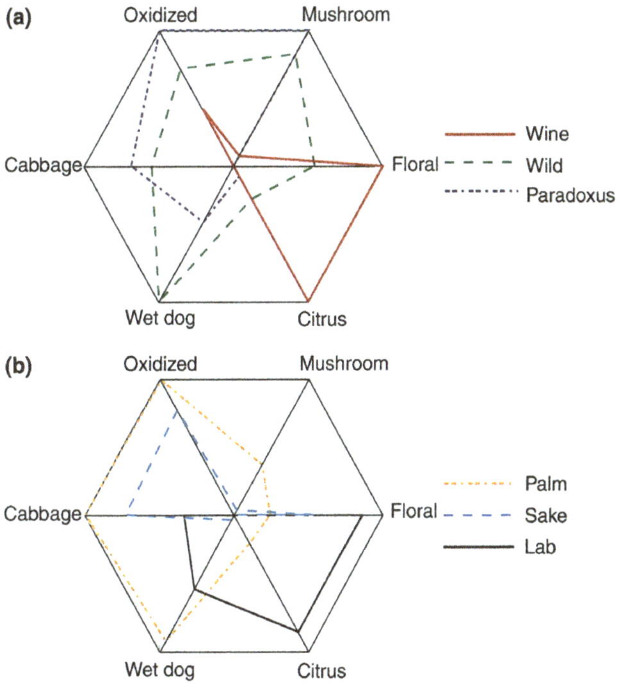

FIGURE 44.11 Illustration of the sensory differences attributable to wine and non-wine strains of *Saccharomyces cerevisiae*. (a) Class means for wine strains, wild strains, and *Saccharomyces paradoxus* strains. (b) Means for the palm, sake, and laboratory strains relative to six wine attributes that distinguish wine strains from non-wine strains. Means were scaled from 0 (center) to 1 (spokes), where 0 represents the lowest mean score, and 1 represents the highest mean score for any class. *Source:* Reprinted from Hyma, K. E., Saerens, S. M., Verstrepen, K. J., and Fay, J. C. (2011) Divergence in wine characteristics produced by wild and domesticated strains of *Saccharomyces cerevisiae*. *FEMS Yeast Res.* 11, 540–551., by permission of Oxford University Press; Copyright Clearance Center # 4093611151322 April 21, 2017

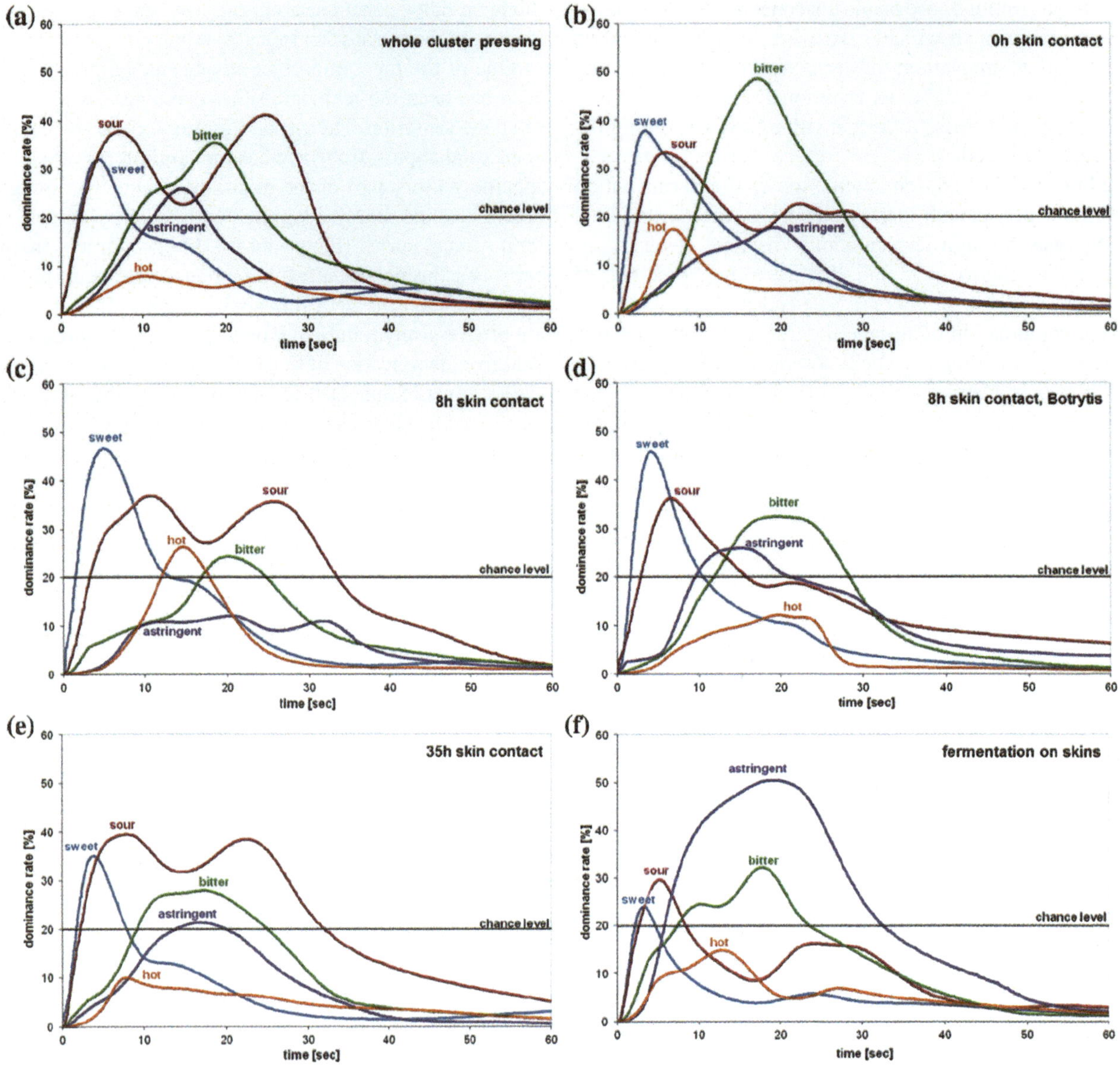

FIGURE 44.12 The effect of different skin-contact procedures as shown by temporal dominance of sensations (TDS) for 2010 Gewürztraminer wines from juice following (A) whole-cluster pressing, (B) 0 h skin contact, (C) 8 h skin contact, (D) 8 h skin contact of berries 30% infected by *Botrytis cinerea*, (E) 35 h skin contact, and (F) fermentation on skins. (Reprinted from Sokolowsky, M., Rosenberger, A., Fischer, U., 2015. Sensory impact of skin contact on white wines characterized by descriptive analysis, time–intensity analysis and temporal dominance of sensations analysis. *Food Qual. Pref.* 39, 285–297, with permission from Elsevier) Copyright Clearance Center #4093270901972, Apr 20, 2017

flagrant flavor misidentification, or fragrance can generate perceived tastes that were non-existent. Where a wine's color is likely to distort taster perception, wines are presented either by falsifying coloration (*e.g.*, pail red lighting) or in black glasses. The message is clear, one must doubt any interpretation, attempting to assure oneself that one's perception actually reflects the reality as detected by one's sensory neurons.

In the past, recording fleeting sensations during a tasting was difficult to impossible. This situation changed dramatically with the invention of the computer. Impressions can be recorded on a monitor as a bar moves across the screen at a constant speed. This has permitted studying the temporal dynamics of perception. Keyboard entry of data also avoids issues related to poor handwriting.

Where the sensory attributes of a series of wines need to be assessed, panel members are trained in the use of a particular set of terms—the members being used as substitute analytic instruments. Panels are used to minimize the natural variability that exists among tasters. Panel member selection reduces this variability but does not eliminate it. Individuals vary in their sensory acuity from day to day and throughout the day.

Presenting sensory data in a clear and easily comprehensible manner has always been difficult. Numerical scores, so popular in the retail trade, are valueless in wine assessment.

Too many, potentially opposed, sensory responses are subjugated into a single number. Polar (spider) plots were one of the early innovations presenting sensory data in a clear manner. They visualize the intensities of the different sensory attributes recorded, as well as their statistical significance (Figure 44.11). Because such plots are chosen to selectively highlight the differences between wines, based on production and regional or varietal origins, they are not intended to visually represent the totality of the sensory attributes of the wines. This is the equivalent of botanical floral diagrams that visualize the essential floral characteristics of a group of flowers, but give little impression of what the flowers would look like if viewed. Polar plots also fail to adequately represent the dynamically changing character of wines, one of the features of greatest interest to the aficionado.

The first development in representing temporal changes involved sapid sensations—time intensity (TI) analysis. However, it separately assesses single facets of a wine's sensory complexity, and in so doing can exaggerate their significance. To provide a more complete picture of a wine's sensory dynamism, the procedure termed temporal dominance of sensations (TDS) was developed. It provides information on how a range of attributes express themselves, in relative dominance, over the course of tasting. Data so provided can highlight the most significant attributes of a wine, and how these are affected by changes in production technique (*e.g.*, Figure 44.12). These and future developments in sensory techniques are likely to provide researchers with better means of correlating wine chemistry with wine flavor, as well as give winemakers information on how potential changes in the vineyard and winery will affect attributes of relevance to their customers.

Another innovation in sensory analysis has been the development of electronic (e-)noses and tongues. Although still comparatively limited in their detection range, they do have their advantages. They are more consistent (not subject to fatigue, adaptation, or psychological or contextual biases), there is no need for a panel to offset member variability, and they are always available, whenever needed. Theoretically, advances in design could produce instruments that are able to detect the typicity of various varietal, regional, and stylistic wines, as well as incorporate the sensory preferences of different consumer groups. While useful from a research standpoint, such assessments are likely of marginal value to the consumer. Part of the intrigue of wine is its variation, and the pleasure derived both from that variation and the option to make interesting and educational mistakes.

Improvements have also been made in how consumer testing is carried out. Trained panels, due to the changes induced by training, and the selection of only those with the pertinent skills, are unlikely to provide evaluations of consumer relevance. To better understand consumer preferences, selecting those likely to purchase the product is needed. In addition, tasting needs to be carried out under conditions that approximate real-life consumption conditions (not a store or a laboratory setting). It is also critical that questioning be designed not to be "leading," and is relevant to a specific situation (*e.g.*, home, restaurant, a special occasion).

BIBLIOGRAPHY

Allen, D. (2007) Prefermentative cryomaceration. *Aust. N.Z. Grapegrower Winemaker* 523, 59–64.

Anonymous (1986) The history of wine: Sulfurous acid – Used in wineries for 500 years. *German. Wine Rev.* 2, 16–18.

Belhaj, K., Chaparro-Garcia, A., Kamoun, S., Patron, N.J., and Nekrasov, V. (2015) Editing plant genomes with CRISPR/Cas 9. *Curr. Opin. Biotechnol.* 32, 76–84.

Bellon, J.R., Eglinton, J.M., Siebert, T.E., Pollnitz, A.P., Rose, L., de Barros Lopes, M., and Chambers, P.J. (2011) Newly generated interspecific wine yeast hybrids introduce flavour and aroma diversity to wines. *Appl. Microbiol. Biotechnol.* 91(3), 603–612.

Berovic, M., Berlot, M., Kralj, S., and Makovec, D. (2014) A new method for the rapid separation of magnetized yeast in sparkling wine. *Biochem. Engin. J.* 88, 77–84.

Best, M.R. (1976) The mystery of vintners. *Agric. Hist.* 50, 362–376.

Bony, M., Bidart, F., Camarasa, C., Ansanay, V., Dulau, L., Barre, P., and Dequin, S. (1997) Metabolic analysis of *Saccharomyces cerevisiae* strains engineered for malolactic fermentation. *FEBS Lett.* 410(2–3), 452–456.

Bortesi, L., and Fischer, R. (2015) The CRISPR/Cas9 system for plant genome editing and beyond. *Bioctechnol. Adv.* 33(1), 41–52.

Bramley, R.G.V. (2005) Understanding variability in winegrape production systems. 2. Within vineyard variation in quality over several vintages. *Aust. J. Grape Wine Res.* 11(1), 33–42.

Buzrul, S. (2012) High hydrostatic pressure treatment of beer and wine. A review. *Innov. Food Sci. Emerg. Technol.* 13, 1–12.

Caruso, G., Tozzini, L., Rallo, G., Primicerio, J., Morionco, M., Palai, G., and Gucci, R. (2017) Estimating biophysical and geometrical parameters of grapevine canopies ('Sangiovese') by an unmanned aerial vehicle (UAV) and VIS-NIR cameras. *Vitis* 56, 63–70.

Castellari, M., Simonato, B., Tornielli, G.B., Spinelli, P., and Ferrarini, R. (2004) Effects of different enological treatments on dissolved oxygen in wines. *Ital. J. Food Sci.* 16, 387–397.

Charpentier, C. (2010) Ageing on lees (*sur lies*) and the use of speciality inactive yeasts during wine fermentation. In: *Managing Wine Quality. Vol. 2. Oenology and Wine Quality.* (A.G. Reynolds, Ed.), pp. 164–187. Woodhead Publishing Ltd., Cambridge, UK.

Chinnici, F., Natali, N., and Riponi, C. (2014) Efficacy of chitosan in inhibiting the oxidation of (+)-catechin in white wine model solutions. *J. Agric. Food Chem.* 62(40), 9868–9875.

Coelho, E., Rocha, S.M., Barros, A.S., Delgadillo, I., and Coimbra, M.A. (2007) Screening of variety- and pre-fermentation-related volatile compounds during ripening of white grapes to define their evolution profile. *Anal. Chim. Acta* 597(2), 257–264.

Conner, J.M., Birkmyre, L., Paterson, A., and Piggott, J.R. (1998) Headspace concentrations of ethyl esters at different alcoholic strengths. *J. Sci. Food Agric.* 77(1), 121–126.

Contreras, A., Hidalgo, C., Henschke, P.A., Chambers, P.J., Curtin, C., and Varela, C. (2014) Evaluation of non-*Saccharomyces* yeasts for the reduction of alcohol content in wine. *Appl. Environ. Microbiol.* 80(5), 1670–1678.

Cozzolino, D. (2015) Sample presentation, source of error and future perspectives on the application of vibrational spectroscopy in the wine industry. *J. Sci. Food Agric.* 95(5), 861–868.

Cozzolino, D., Cynkar, W., Shah, N., and Smith, P. (2011) Technical solutions for analysis of grape juice, must, and wine: The role of infrared spectroscopy and chemometrics. *Anal. Bioanal. Chem.* 40(5), 1475–1484.

de Bei, R., Cozzolino, D., Sullivan, W., Cynkar, W., Fuentes, S., Dambergs, R., Pech, J., and Tyerman, S. (2011) Non-destructive measurement of grapevine water potential using near infrared spectroscopy. *Aust. J. Grape Wine Res.* 17(1), 1078–1086.

de Souza, C.R., Maroco, J.P., Chaves, M.M., dos Santos, T., Rodriguez, A.S., Lopes, C., *et al.* (2004) Effects of partial root drying on the physiology and production of grapevines composition. In: Vallone, R.C. (Ed.), *International Symposium on Irrigation and Water Relations in Grapevine and Fruit Trees. Acta Hortic.* 646, 121–126.

Diago, M.P., Sanz-Garcia, A., Millan, B., Blasco, J., and Tardaguila, J. (2014) Assessment of flower number per inflorescence in grapevine by image analysis under field conditions. *J. Sci. Food Agric.* 94(10), 1981–1987.

Dry, P.R., and Loveys, B.R. (1998) Factors influencing grapevine vigour and the potential for control with partial rootzone drying. *Aust. J. Grape Wine Res.* 4(3), 140–148.

Dubernet, M. (2010) Automatic analysers in Oenology. In: *Wine Chemistry and Biochemistry.* (M. Moreno-Arribas and C. Polo, Eds.), pp. 649–676. Springer Verlag, New York/Heidelberg.

Dubourdieu, D., Moine-Ledoux, V., Lavigne-Cruège, V., Blanchard, L., and Tominaga, T. (2000) Recent advances in white wine aging: The key role of lees. *Proc. ASEV 50th Anniv. Ann. Meeting, Seattle, WA*, June 19–23, 2000. American Society for Enology and Viticulture, Davis, CA, pp. 345–352.

Escalona, H., Piggott, J.R., Conner, J.M., and Paterson, A. (1999) Effects of ethanol strength on the volatility of higher alcohols and aldehydes. *Ital. J. Food Sci.* 11, 241–248.

Francis, I.L., Leino, M., Sefton, M.A., and Williams, P.J. (1993) Thermal processing of Chardonnay and Semillon juice and wine: Sensory and chemical changes. *Proc. 8th Aust. Wine Ind. Tech. Conf. Oct. 25–29, 1992*, Melbourne, Australia. (C.S. Stockley, R.S. Johnstone, P.A. Leske and T.H. Lee, eds.), pp. 158–160. Winetitles, Adelaide, Australia.

Garde-Cerdán, T., López, R., Garijo, P., González-Arenzana, L., Gutiérrez, A.R., López-Alfara, I., and Santamaría, P. (2014) Application of colloidal silver versus sulfur diocide during vinification and storage of Tempranillo red wines. *Aust. J. Grape Wine Res.* 20(1), 51–61.

García-Carpintero, E.G., Sánchez-Palomo, E., and González Viñas, M.A. (2010) Influence of co-winemaking technique in sensory characteristics of new Spanish red wines. *Food Qual. Pref.* 21(7), 705–710.

Ghozlen, B.N., Cerovic, Z.G., Germain, C., Toutain, S., and Latouche, G. (2010) Non-destructive optical monitoring of grape maturation by proximal sensing. *Sensors* 10(11), 10040–10068.

Giaramida, P., Ponticello, G., Di Maio, S., Squadrito, M., Genna, G., Barone, E., Scacco, A., Corona, O., Amore, G., di Stefano, R., and Oliva, D. (2013) *Candida zemplinina* for production of wines with less alcohol and more glycerol. *S. Afr. J. Enol. Vitic.* 34(2), 204–211.

Guymon, J.F., and Crowell, E.A. (1977) The nature and cause of cap-liquid temperature differences during wine fermentation. *Am. J. Enol. Vitic.* 28, 74–78.

Hallgarten, F. (1986) *Wine Scandal.* Wine Appreciation Guild, San Francisco, CA.

Harbertson, J.F., Mireles, M.S., Harwood, E.D., Weller, K.M., and Ross, C.F. (2009) Chemical and sensory effects of saignée, water addition, and extended maceration on high Brix must. *Am. J. Enol. Vitic.* 60, 450–460.

Heymann, H., LiCalzi, M., Conversano, M.R., Bauer, A., Skogerson, K., and Matthews, M. (2013) Effects of extended grape ripening with and without must and wine alcohol manipulations on Cabernet Sauvignon wine sensory characteristics. *S. Afr. J. Enol. Vitic.* 34(1), 86–99.

Holt, S., Cordente, A.G., Williams, S.J., Capone, D.L., Jitjaroen, W., Menz, I.R., Curtin, C., and Anderson, P.A. (2011) Engineering *Saccharomyces cerevisiae* to release 3-mercaptohexan-1-ol during fermentation through overexpression of an *S. cerevisiae* gene, *STR3*, for improvement of wine aroma. *Appl. Environ. Microbiol.* 77(11), 3626–3632.

Hyma, K.E., Saerens, S.M., Verstrepen, K.J., and Fay, J.C. (2011) Divergence in wine characteristics produced by wild and domesticated strains of *Saccharomyces cerevisiae. FEMS Yeast Res.* 11(7), 540–551.

Jallerat, E. (1990) Les nouvelles techniques de tirage. III. Le mini-fermenteur. Un nouveau dispositif pour la prise de mousse enbouteilles. *Vigneron Champenois* 10, 9–24.

Jin, X., Wu, X., Liu, X., and Liao, M. (2017) Varietal heterogeneity of textural characteristics and their relationship with phenolic ripeness of wine grapes. *Sci. Hortic.* 216, 205–214.

Karbowiak, T., Gougeon, R.D., Alinc, J.-B., Brachais, L., Debeaufort, F., Voilley, A., and Chassagne, D. (2010) Wine oxidation and the role of cork. *Crit. Rev. Food Sci. Nutrit.* 50(1), 20–52.

Keller, M. (2005) Deficit irrigation and vine mineral nutrition. *Am. J. Enol. Vitic.* 56, 267–283.

Koehler, C.G. (1986) Handling of Greek transport amphoras. In: *Recherches sur les Amphores Greques* (J.-Y. Empereur and Y. Garlan, Eds.), pp. 49–67. Bull. Correspondance Hellénique, supp. 13. École françaised'Anthènes, Paris, France.

Kontoudakis, N., Esteruelas, M., Fort, F., Canals, J.M., and Zamora, F. (2011) Use of unripe grapes harvested during cluster thinning as a method for reducing alcohol content and pH of wine. *Aust. J. Grape Wine Res.* 17(2), 230–238.

Lange, J., and Wyser, Y. (2003) Recent innovation in barrier technologies for plastic packaging – A review. *Packag. Technol. Sci.* 16(4), 149–158.

López-Rituerto, E., Cabredo, S., López, M., Avenoza, A., Busto, J.H., and Peregrina, J.M. (2009) A thorough study on the use of quantitative ^{1}H NMR in Rioja red wine fermentation processes. *J. Agric. Food Chem.* 57(6), 2112–2118.

Lorenzo, C., Pardo, F., Zalacain, A., Alonzo, G.L., and Salinas, M.R. (2005) Effect of red grapes co-winemaking in polyphenols and color of wines. *J. Agric. Food Chem.* 53(19), 7609–7616.

McCarthy, M.G., Loveys, B.R., Dry, P.R., and Stoll, M. (2002) Regulated deficity irrigation and partial rootzone drying as irrigation management techniques for grapevines. In: *Deficit Irrigation Practices*, pp. 79–88. FAO (Land and Water Division), Rome, Italy.

McGovern, P.E., Glusker, D.L., Exner, L.J., and Voigt, M.M. (1996) Neolithic resinated wine. *Nature* 381(6582), 480–481.

Messenger, S. (2007) New MOG removal system makes the grade. *Aust. N.Z. Grapegrower Winemaker* 516, 49–50.

Moss, R., Daniels, K., and Shasky, J. (2013) Effect of cold soak on the phenolic extraction of Syrah. *Aust. N.Z. Grapegrower Winemaker* 592, 59–62.

Moutounet, M., Mazauric, J.P., Saint-Pierre, B., and Hanocq, J.F. (1998) Gaseous exchange in wines stored in barrels. *J. Sci. Tech. Tonnellerie* 4, 131–145.

Moya, I., Camenen, L., Evain, S., Goulas, Y., Cerovic, Z.G., and Latouche, G. (2004) A new instrument for passive remote sensing. 1. Measurements of sunlight induced chlorophyll fluorescence. *Remote Sens. Environ.* 91(2), 186–197.

O'Kennedy, K., and Canal-Llaubères, R.-M. (2013a) The A–Z of wine enzymes: Part 1. *Aust. N.Z. Grapegrower Winemaker* 589, 57–61.

O'Kennedy, K., and Canal-Llaubères, R.-M. (2013b) The A–Z of wine enzymes: Part 2. *Aust. N.Z. Grapegrower Winemaker* 590, 42–46.

Österbauer, R.A., Matthews, P.M., Jenkinson, M., Beckmann, C.F., Hansen, P.C., and Calvert, G.A. (2005) Color of scents: Chromatic stimuli modulate odor responses in the human brain. *J. Neurophysiol.* 93(6), 3434–3441.

Paronetto, L., and Dellaglio, F. (2011) Amarone: A modern wine coming from an ancient production technology. *Adv. Food Nutr. Res.* 63, 285–306.

Pati, S., Mentana, A., La Notte, E., and Del Nobile, M.A. (2010) Biodegradable poly-lactic acid package for the storage of carbonic maceration wine. *LWT Food Sci. Technol.* 43(10), 1573–1579.

Pérez-Magariño, S., Martínez-Lapuente, L., Bueno-Herrera, M., Ortego-Heras, M., Guadalupe, Z., and Ayestarán, B. (2015) Use of commercial dry yeast products rich in mannoproteins for white and rosé sparkling wine elaboration. *J. Agric. Food Chem.* 63(23), 5670–5681.

Phaff, H.J. (1986) Ecology of yeasts with actual and potential value in biotechnology. *Microb. Ecol.* 12(1), 31–42.

Pineau, N., Schlich, P., Cordelle, S., Mathonnière, C., Issanchou, S., Imbert, A., Rogeaux, M., Eteévant, P., and Koster, E. (2009) Temporal dominance of sensations: Construction of the TDS curves and comparison with time-intensity. *Food Qual. Pref.* 20(6), 450–455.

Pozo-Bayón, M.A., and Moreno-Arribas, M.V. (2011) Sherry wines. *Adv. Food Nutrit. Res.* 63, 17–40.

Proffitt, T., and Malcolm, A. (2005) Zonal vineyard management through airborne remote sensing. *Aust. N.Z. Grapegrower Winemaker* 502, 25–26.

Proffitt, T., Bramley, R., Lamb, D., and Winter, E. (2006) *Precision Viticulture.* Winetitles Pty Ltd., Adelaide, Australia.

Raco, B., Dotsika, E., Poutoukis, D., Battaglini, R., and Chantzi, P. (2015) O-H-C isotope ratio determination in wine in order to be used as a fingerprint of its regional origin. *Food Chem.* 168, 588–594.

Röck, F., Barsan, N., and Weimar, U. (2008) Electronic nose: Current status and future trends. *Chem. Rev.* 108(2), 705–725.

Röcker, J., Strub, S., Ebert, K., and Grossmann, M. (2016) Usage of different aerobic non-*Saccharomyces* yeasts and experimental conditions as a tool for reducing the potential ethanol content in wines. *Eur. Food Res. Technol.* 242(12), 2051–2070.

Rolle, L., Segade, S.R., Torchio, F., Giacosa, S., Cagnasso, E., Marengo, F., and Gerbi, V. (2011) Influence of grape density and harvest date on changes in phenolic composition, phenol extractability indices, and instrumental texture properties during ripening. *J. Agric. Food Chem.* 59(16), 8796–8805.

Rolls, E.T., Critchley, H.D., Verhagen, J.V., and Kadohisa, M. (2010) The representation of information about taste and odor in the orbitofrontal cortex. *Chem. Percept.* 3(1), 16–33.

Romero, P., Fernández-Fernández, J.I., and Martinez-Cutillas, A. (2010) Physiological thresholds for efficient regulated deficit-irrigation management in winegrapes grown under semiarid conditions. *Am. J. Enol. Vitic.* 61, 300–312.

Rous, C., and Alderson, B. (1983) Phenolic extraction curves for white wine aged in French and American oak barrels. *Am. J. Enol. Vitic.* 34, 211–215.

Roussis, I.G., Patrianakou, M., and Drossiadis, A. (2013) Protection of aroma volatiles in a red wine with low sulphur dioxide by a mixture of glutathione, caffeic acid and gallic acid. *S. Afr. J. Enol. Vitic.* 34(2), 262–265.

Rustioni, L., Rossoni, M., Calatrioni, M., and Failla, O. (2011) Influence of bunch exposure on anthocyanins extractability from grapes skins (*Vitis vinifera* L.). *Vitis* 50, 137–143.

Schmid, F., Schadt, J., Jiranek, V., and Block, D.E. (2009) Formation of temperature gradients in large- and small-scale red wine fermentations during cap management. *Aust. J. Grape Wine Res.* 15(3), 249–255.

Singleton, V.L. (1962) Aging of wines and other spiritous products, acceleration by physical treatments. *Hilgardia* 32(7), 319–373.

Sokolowshy, M., Rosenberger, A., and Fischer, U. (2015) Sensory impact of skin contact on white wines characterized by descriptive analysis, time–intensity analysis and temporal dominance of sensations analysis. *Food Qual. Pref.* 39, 285–297.

Tao, Y., García, J.F., and Sun, D.-W. (2014) Advances in wine aging technologies for enhancing wine quality and accelerating wine aging process. *Crit. Rev. Food Sci. Nutr.* 54(6), 817–835.

Tempere, S., Cuzange, E., Malik, J., Cougeant, J.C., de Revel, G., and Sicard, G. (2011) The training level of experts influences their detection thresholds for key wine compounds. *Chem. Percept.* 4(3), 99–115.

Torresi, A., Frangipane, M.T., and Anelli, G. (2011) Biotechnologies in sparkling wine production. Interesting approaches for quality improvement: A review. *Food Chem.* 129(3), 1232–1241.

Usseglio-Tomasset, L. (1986) Riattivazione della fermentazione e prevenzione degli arresti fermentativi mediante l'impiego di pareti cellulari di lievito. *Enotecnico* 1, 53–57.

Williams, P.J., and Francis, I.L. (1996) *Sensory Analysis and Quantitative Determination of Grape Glycosides – The Contribution of these Data to Winemaking and Viticulture.* Biotechnol. Improved Foods Flavors. ACS Symposium Series 637. pp. 124–133. American Chemical Society, Washington, DC.

Wollan, D. (2010) Reducing wine alcohol: Some myths busted. *Aust. N.Z. Grapegrower Winemaker* 562, 54–59.

Zeng, X.A., Yu, S.J., Zhang, L., and Chen, X.D. (2008) The effects of AC electric field on wine maturation. *Innov. Food Sci. Emerg. Technol.* 9(4), 463–468.

45 The Wine Industry
An Overview of Threats, Opportunities, Innovations and Trends

Julie Kellershohn and Inge Russell

CONTENTS

45.1 INTRODUCTION

Wine, a beverage with a long and stable heritage, is now undergoing significant changes in where and how it is produced, how it is packaged, how it is marketed to different consumer segments (in particular the millennial consumers) and how consumers evaluate and select their wines. Compared to previous generations, millennial consumers, with their significant purchasing power, are more health-conscious, more concerned about the environment and sustainability and expect transparency and convenience, not only in how a product is packaged but also the ease with which it can be ordered and rapidly delivered. Also in flux are where the wine is produced and who the major purchasers are, as climate change means that traditional growing areas are being impacted and new markets are forming. This chapter examines the present state of the global wine industry in terms of threats and opportunities.

45.2 THE GLOBAL WINE INDUSTRY: A STATISTICAL OVERVIEW

45.2.1 Area Under Grape Cultivation and Wine Production

In 2016, 7.5 million hectares of world area were under cultivation for vines destined for wine grapes, table grapes and dried grapes, with five countries representing 50% of the world's

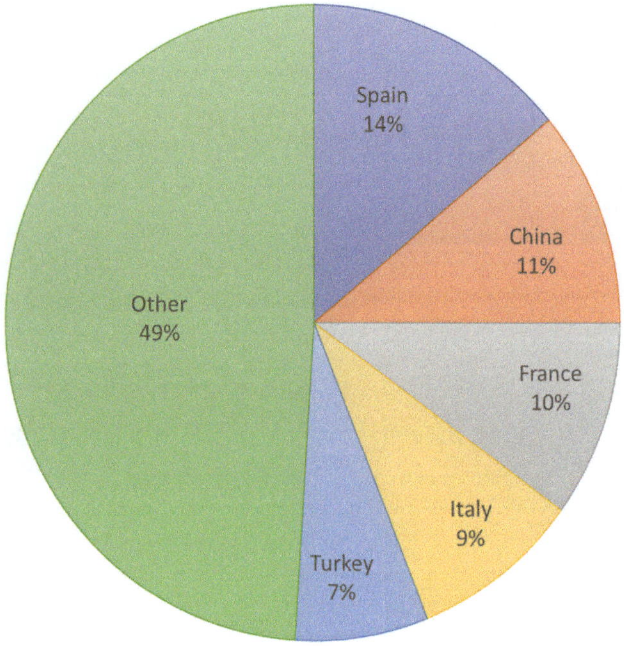

FIGURE 45.1 World area under cultivation for vines destined for wine grapes, table grapes and dried grapes. *Source:* OIV (2017)

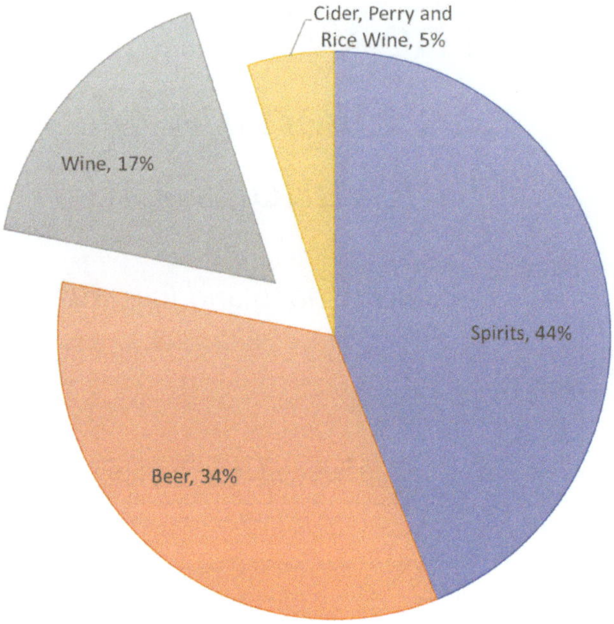

FIGURE 45.2 Worldwide alcoholic drink revenue. *Source:* Statista (2018b)

TABLE 45.1
Rank in Global Wine Consumption, Export and Import Volume

Rank	Global wine consumption	Global wine export volume	Global wine import volume
1	United States (13.2%)	Spain	Germany
2	France (11.2%)	Italy	United Kingdom
3	Italy (9.4%)	France	United States
4	Germany (8.1%)	Chile	France
5	China (7.2%)	Australia	China

vineyards (Figure 45.1). The top 12 wine producers by volume are Italy, France, Spain, the United States, Australia, China, South Africa, Chile, Argentina, Germany, Portugal and Russia.

Since 2012, China has had the largest increase in acreage, while Portugal and Australia have seen the largest decline in terms of volume of wine produced.

45.2.2 Wine Consumption, Export and Import

Based on the export and import volume of global wine consumption (Table 45.1), the country responsible for the most change in the industry is China. In 2016, China was the fifth-leading consumer of wine in the world. With a population of 1.4 billion, there is tremendous potential for growth in wine sales within China. China has the second-largest grape area under cultivation at 847 kha (thousands of hectares); however, most of these grapes are currently used for other purposes (edible grapes, grape juice, raisins). In 2016, China was the world's sixth-largest wine producer and fifth-largest wine importer. In China, the largest consumer market for imported wine (50%) is dominated by the younger female population.

45.2.3 Sale and Total Revenue

The total revenue of the alcoholic drinks market in 2017 was US$837 billion, representing a growth of 3.5% compared to 2016. Wine held 17% of the alcoholic drinks market by revenue, significantly behind spirits and beer (Figure 45.2).

Mergers and acquisitions are playing a central role in changes in the industry with the ongoing consolidation of major production and vineyard assets by large wine companies. In addition, distributors are merging with distribution companies, increasing in size and reach as a result.

Direct-to-consumer (DTC) sales is the key growth area for many smaller wineries. In 2017, DTC sales were 10% of the US market (US$2.69 billion in sales), with 90% of these sales originating from tourism and small domestic winery tasting rooms offering wines at under $39 per bottle. New sales avenues will continue to arise with emerging technologies, expanding access to the internet, and market deregulation.

45.3 A CHANGING ENVIRONMENT

45.3.1 Climate Change and Wildfires

A major concern for wine-producing countries is climate change and its effect on growth potential. In 2015–2016, wine production dropped in both Brazil and South Africa due to climate issues. The world wine map is changing. While there is disagreement as to exactly how climate change will impinge on the different areas, there is little doubt that there will be significant effects.

FIGURE 45.3 Vineyards in many parts of the world will have to deal with climate change. *Source:* Photograph courtesy of Pexels.co m, credit: Pixabay

Some areas of the world will become more suitable for grape cultivation, with areas in England, Michigan (US) and Tasmania already showing increased numbers of wineries. Wines from Central and Eastern Europe (*e.g.* Romania, Slovenia, Moldova) are also projected to grow in popularity in 2018 due to increasing harvest volumes and lower prices for production, making these wines very competitive in the global market.

New markets are also emerging, driven by local production in countries such as Ethiopia, which is well-suited to grape cultivation. For example, the Castel winery in Ethiopia takes advantage of the country's diverse landscape, with six climatic zones, ideal for grape growing. With an annual rainfall of about 650 mm, an average yearly temperature of 25°C, sandy soil and Ethiopia's proximity to the equator, two harvests per year are possible. As grape cultivation expands into new areas, this will create change in what was once a staid industry.

Changing weather conditions are forecast to have a large impact on the future of vineyards (Figure 45.3), with a major concern being a growing shortage of water. The development of new forms of irrigation, as well as advances in analytical technology tools will be needed. Advances in technology will also drive the role that automation will play in the future of viticulture.

Due to climate change and the lack of rain, there are increasing numbers of wildfires, resulting in smoke-tainted wines. Smoke taint occurs when grapes are exposed to smoke, resulting in wines with undesirable sensory characteristics (smoky, burnt, ashy, medicinal) due to latent phenolic compounds tainting the wine. The degree of taint varies and depends on a number of factors (*e.g.* most susceptible are grapes at the onset of ripening). More research is needed to better investigate ways of preventing smoke taint. Currently, there is no solution that prevents 100% of smoke taint and an increasing number of wineries are struggling with the problem.

45.3.2 Market Segmentation

Currently, the baby boomer population (ages 51–68) comprises 41% of wine sales followed by the Gen Xers (ages 39–50) at 33% of sales. The generation named 'matures' (consumers aged 69+) hold only a 10% share of wine sales. However, sales to boomers and matures are now in decline. The two cohorts for growth in wine purchases are the millennials (ages 22–38), currently only at 17% of sales, along with the Gen Xers (Figure 45.4).

A market segmentation study of wine purchases in California found that overall, the biggest spenders on wine tended to be male baby boomers. In contrast, millennials were looking for more moderately priced wines that represent good value, offer innovative packaging and are suitable for socialising outside of the home. Millennials seek a unique product and a positive experience; however, for on-premises

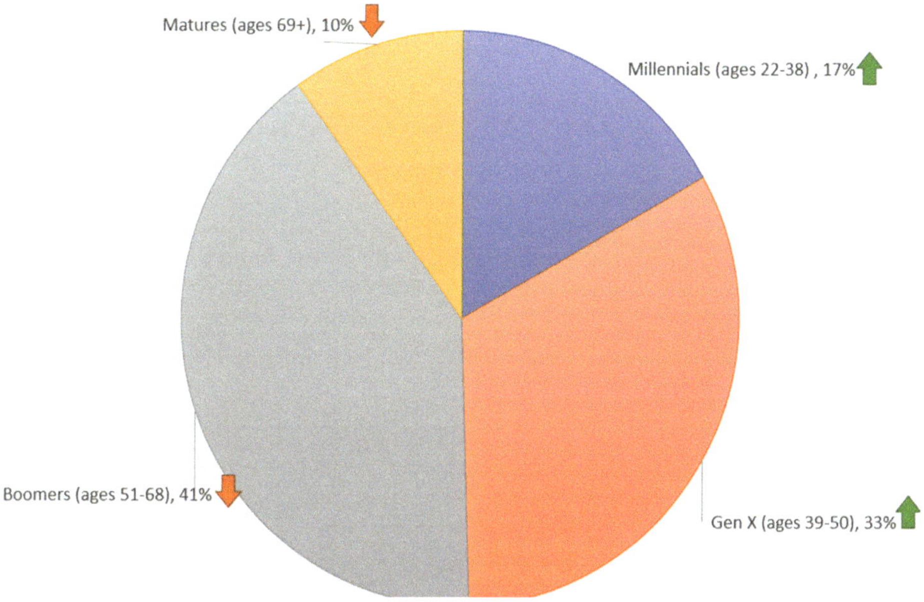

FIGURE 45.4 Wine sale shares in the United States in 2016, by demographic. *Source:* Silicon Valley Bank (2016)

occasions, they currently often substitute craft beer and spirits for wine. Nevertheless, as millennials age, they are expected to spend more on wine purchases, while retiring baby boomers will shift into more moderately priced products and lower volumes. Current US demographics are as follows: Gen Xers (~66 million people), baby boomers (~75 million) and millennials (~71 million). By 2026, millennials are expected to surpass Gen Xers in terms of fine wine consumption.

Younger consumers appear to be more willing to try a broader selection of brands. This is likely driven by the fact that their 'favourite' product has not yet been established and as such, they have a mindset that is more focused on flavour exploration than on habitual buying patterns. In a 2015 Nielsen study, almost one-quarter (23%) of wine drinkers aged 21–34 were found to have purchased 10 or more wine brands during the previous year.

While in past years, wine retailers have focused on what millennials look for in wine purchases, more recent consumer studies suggest that retailers should expand their focus to include Gen X consumers, as they move from being a cocktail consumer to a wine consumer. The Gen X generation, now in their 40s and 50s, are projected to surpass the baby boomers by 2021, not only in wine volume purchased, but in dollars spent as they are open to purchasing higher-priced products. In the United States, Gen X consumers are focusing on premium and ultra-premium priced wines and have an annual average income of $95,100 compared with the younger millennials with average incomes of $65,300 and retiring baby boomers with average incomes of $79,700.

45.3.2.1 Non-Grape-Based Wines

Another market segment that is gaining in importance is fruit wines. The exact definition of wine varies by the region of the world, and some countries insist that products labelled 'wine' must come from grapes, while other countries have no such restriction. Most non-grape wines are labelled with their main ingredient (*i.e.* elderberry wine). Fruit wines may include additional flavours from flowers and herbs (Figure 45.5). Although, technically, wine from apples is a fruit wine, for historical reasons it is usually excluded from the normal definition of fruit wine, as is mead (main fermented ingredient honey) and perry (main fermented ingredient pears). Some fruit wines in the past have suffered from an image problem as, depending on the fruit, the addition of water, sugar and yeast nutrients is often needed for a successful fermentation to produce a product with the right taste balance. Most fruit wines are best consumed fresh (within a year of bottling). An increase in sales of fruit wines and in wines with unusual flavour profiles is expected.

FIGURE 45.5 Today fruit wines are exploring a wide range of ingredients. *Source:* Photograph courtesy of Pexels.com, credit: Adonyi Gábor

45.3.3 Changing Venues

According to a Nielsen research study, when entering a store only 29% of consumers already know which brand they intend to buy, while the remaining 71% of consumers make their decisions as they look over products on the store shelf; 64% of consumers try a new product simply because the package has caught their eye. Typical wine store purchases are being replaced by consumer purchases in big box stores such as Costco. Costco is now the largest wine retailer in the United States, with close to $2 billion in wine sales annually and growing.

Retailers in France have seen success with the online 'click and drive' model, where shoppers use drive-through lanes to pick up their online orders. As driverless cars become more commonplace, it is unclear how this advance in technology will affect the industry once consumer 'drinking and driving' is no longer a concern, especially in terms of potential benefits to wine tourism.

Online wine sales continue to grow. The long-term role that large retailers, such as Amazon marketplace, will play in the future remains to be determined, but in some countries such as Germany, where retailers were slow to move to online sales, Amazon already attracts half of online wine sales. While online wine retailers may offer delivery in as little as one day, the introduction of wine apps on mobile phones that promise rapid delivery, some as fast as within an hour of ordering, may again change how consumers purchase wine.

Traditionally, value is the dominant factor when consumers are making purchase decisions, where value is defined as quality divided by price. However, with a fine wine purchase, value is defined as quality *plus experience* divided by price. The experience is not limited to the way that owning or consuming a particular product makes the consumer feel, but also includes the consumer's shopping experience (*i.e.* a leisurely visit to a boutique winery that includes a relaxed tasting versus paying a premium to have wine delivered to their home within an hour of placing the order).

The direct-to-consumer category is now at ~$3 billion in direct sales, with purchases growing. Using the winery location and wine tourism visits as the sole point of experience will no longer suffice to drive sales in the future. Smaller wineries will need to explore additional means of delivering a memorable experience, in order to stay competitive and continue to deliver value in ways that are relevant to their customer in an increasingly crowded market.

45.3.4 Transparency

Total transparency from vineyard to consumer is no longer an aspirational concept but can now be made readily available to consumers through the use of blockchain technology. Although the use of blockchain is still in its infancy, especially in the wine arena, it facilitates trust in the origin and in the processing steps of the wine, and increases transparency in the wine's journey to the final customer. Consumers with a smartphone can learn every detail about the wine from grape to final sale. The blockchain system creates an unalterable and transparent tracking system, where each record in the wine's production journey is marked with a timestamp, and the record cannot be altered once it is created. It is the format used by bitcoin and other cyber currencies. The chain consists of a continuously growing list of records, called blocks, which are linked and secured using cryptography. One of the first companies to use it for fine wines was Everledger with their Chai Wine Vault system. In 2016, a bottle of 2001 Château Margaux was authenticated as a block and now any time that this specific bottle changes hands, the transfer is permanently and automatically updated on the blockchain.

A number of high-profile fraud cases have illustrated how difficult it can be to detect deliberately mislabelled wines. Wine is one of the most counterfeited products and the counterfeiting of wines poses harm both to the wine's brand and to the end consumer, especially when poisonous counterfeit ingredients are used. It is estimated that 20% of wine globally is mislabelled or counterfeit and as much as 50% of the wine sold in China is mislabelled or counterfeit.

Tracking technologies are being developed that allow for the tracking of bottles from the vineyard to the point of sale, using microprinted labels, high-tech etching on bottles and embedded microchips to digitally authenticate the wine's provenance. The TagItWine project secures authenticity by utilising the Internet of Things (IoT), smart labels and blockchain-based protocol (http://www.tagitwine.me/). Amcor and Selinko use chip technology to help prevent bottles from being reused. The chip technology goes on the capsule of the bottle, and when the wine is opened, the system is automatically deactivated, so the bottle cannot be refilled and resold as the original product.

Technology innovations will continue to be developed in the coming years to ensure that the customer can trust that they are not purchasing a counterfeit product and, as the technology improves in response to consumer demand for product transparency, new innovations will offer inexpensive options that enable the monitoring of all wine products, not just high-end wines.

45.4 SUSTAINABILITY IN THE WINE INDUSTRY

Today, sustainability is a concept that must be integrated into grape growing, winemaking, distribution and retail. Consumers, particularly millennial consumers, are paying attention to whether a product is produced in a way that protects the environment and takes into consideration both societal well-being and ethical principles. The younger generation of customers is more eco-conscious, tend to be digital natives and do not demonstrate the same brand loyalty as previous generations. They are focused more on the product than on the brand that accompanies the product. Environmental certifications in various forms are becoming increasingly popular and wines designated with prestigious appellations command higher prices. According to the Global Wine SOLA report, organic wine was identified as a top growth opportunity in 2018. Organic is a term that consumers understand, and it is a term that appeals to many consumer groups. Terms such as sustainably produced, fair trade and environmentally friendly also hold potential as descriptors that appeal to prospective customers.

45.4.1 By-Products and Waste

Grape growing and wine production yield a number of by-products and waste including prunings from the vines, grape stalks, grape marc (pomace) and seeds. One litre of wine requires an average of 1.3 kg of grapes. For every 100 litres of wine produced, approximately 30 kg of waste is produced.

Finding ways to utilise some of the waste products is not new, for example using grape seeds to produce grapeseed oils for cooking. But other approaches, such as their inclusion in skincare products due to the high concentration of antioxidants, the production of nutritional supplements and food preservatives, and gluten-free flour substitutes are newer examples of grape waste valorisation. More advanced technologies are still needed to valorise even more of the waste into new products to take advantage of the valuable compounds in the grape waste.

45.4.2 Wine Pricing, Branding and the Consumer

Price can influence your taste buds! Although studies have been conducted on price influencing taste perception, researchers at Bonn University in 2015 took an approach to studying the effect of wine price and taste that used magnetic resonance imaging (MRI) brain scans and *identical* test wines. They found that it was possible to trick the brain into thinking that wine labelled as expensive tasted better. Seeing the higher prices activated areas of the brain associated with reward and motivation, which *actually increased the taste experience* for the drinker.

How consumers make purchase decisions is complex to decipher, as are the influencing effects of price, branding and recommendations. As Pitt discussed in his 2017 *Journal of Wine Research* editorial title, 'Ten Reasons Why Wine is a Magical Marketing Product', branded wine offerings in the average supermarket exhibit the greatest price range of any food product. His examples (from the American grocery store chain Trader Joe's) showed that comparing a box of the cheapest cereal to the most expensive cereal had a price factor of ~13 to 1, while in comparison, wine had a price factor of 197 to 1. A true scientific understanding of the price–quality relationship for the consumer that results in such a wide price range is a significant gap in current marketing knowledge.

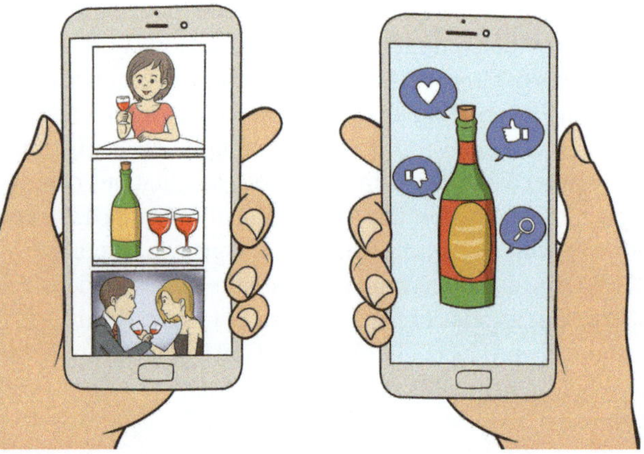

FIGURE 45.6 Millennials rely on social media to share their wine drinking experiences with others as well as to seek advice through social media on which wine to purchase. *Source:* Image courtesy of J. Kellershohn

In 2018, Dobele and associates examined the effect of conspicuous value indicators such as advertising, vineyard, region or brand, and inconspicuous value indicators such as friend referrals in terms of original purchase intent. Their research suggested that when the purchase was not for personal consumption, such as a gift, or if the purchase was for consumption in public, the conspicuous value indicators were key motivators. However, when the purchase was for personal consumption (at home), the less conspicuous indicators of personal recommendation played a much larger role, suggesting that the influence of social media and expert recommendations on purchase choice cannot be underestimated.

45.4.3 Social Media

Digital marketing allows for the targeting of specific consumer groups in a way that traditional marketing is unable to do. Consumers share personal experiences on sites such as Facebook, YouTube, Instagram and Twitter. The number of active social media users worldwide in January 2018 was 3.2 billion, up 13% year-on-year, with a 42% penetration rate. In 2017, millennial internet users spent an average of 223 minutes per day on mobile devices, a large increase from the 188 daily minutes in 2016. Facebook was the first social network to surpass 1 billion registered accounts. In 2018, there were 2.2 billion monthly active Facebook users. Over half of millennials discuss their wine-drinking experiences on Facebook and a third share their observations on YouTube, Twitter or Instagram (Figure 45.6).

45.4.4 Next Generation Marketing

Traditional marketing approaches, such as television advertisements, billboards and print, require a large budget and target a wide audience. For example, in the United States in 2018, a 30 second commercial during the Super Bowl cost ~$5 million, and the Super Bowl was watched by over 100 million American consumers. Australian wine brand Yellow Tail, a product fifth in global wine volume, was the first wine company to buy advertisement time during a Super Bowl. They circumvented a lack of access at the national level by buying 80 local Super Bowl spots (a more expensive option). Their light-hearted kangaroo-themed advertisements drew extensive attention, both positive and controversial, including on social media, thereby raising the wine brand's profile and increasing consumer discussion of their product.

Chilean wine giant Concha y Toro, the second-largest global wine brand, had great success with the company's Wine Legend advertising campaign. Their premium wine, Casillero del Diablo, had an elaborate ad campaign titled 'The Wine Legend', set up to look like a film trailer for a glamorous action-packed crime caper movie. They told a story about their wine that was rumoured to have been repeated many times in the past century, but they told it in a totally innovative way and in less than two minutes. They were able to capture the brand's message with suspense and emotion. Their advertising tied in well with the wine being positioned as a premium product and as 'the wine' to drink while watching a great movie.

Leveraging what makes your wine special is the approach taken by the New Zealand winery Stoneleigh. They launched an experiential campaign for their Stoneleigh Wild Valley range, which is distinguished by their use of 100% wild natural fermentation. One facet of their campaign was the production of an art film inspired by the wine: a 60-second film titled 'Original Art by Stoneleigh'. A visual artist (Susie Sie) and an experimental sound designer (Nikolai Von Sallwitz) created a sensory experience film, with every frame and sound included in the film inspired and sourced from the winery. The art film takes you both visually and audibly inside their wild yeast fermentation liquid for a very unique experiential glimpse (https://www.youtube.com/watch?v=5_GQhm7nACg).

Barefoot wines took a very different and again non-traditional approach to their wine advertising with 'worthy cause marketing'. During their first 20 years of growing the Barefoot wines brand, they never paid for commercial advertising. Rather, they chose community fundraisers and non-profits that resonated with their brand and which gave their members a social reason to buy their wine. They selected groups to help based on the consumers who were the most likely to purchase their products. Rather than becoming sponsors, they became partners in the various causes. Barefoot is now the world's largest wine brand.

45.4.5 Stunt Marketing

Stunt marketing, with a budget of less than $7,000, was accomplished by the Napa Valley Mira winery. They claimed that they were testing a new method of ageing wine, by submerging four cases of a 2009 Cabernet Sauvignon in the ocean after seeing reports of wine found in sunken ships having a unique taste. They submerged their wine in February 2013, just off the coast of Charleston (SC). A videographer filmed the process and made a two-minute video of the wine's retrieval in May. A sommelier compared the wines. All of this was broadcast to their wine club members and to the media *via* a conference call. The campaign prompted 640 media placements, traffic to the winery's website tripled and the wine club's enrolment rate doubled.

A one-day stunt marketing event takes place every November at Yunessan, a natural hot spring facility near Mt Fuji, with the launch of that year's Beaujolais Nouveau. Participants get to soak and play in a red wine–soaked natural hot spring.

Both examples illustrate how stunt marketing can catch the imagination of the consumer through short-term unique events designed to attract publicity and drive sales.

45.5 PACKAGING THE WINE: TRENDS AND INNOVATION

The desire for lighter and environmentally friendly packaging has led to innovation in an industry where the traditional bottle has historically seen little change. Rather than focusing exclusively on the label, wineries are now focusing on how to make the overall packaging more unique. Even when using traditional wine bottles, the type of closure (screw cap, synthetic or natural cork) has a significant effect on the perception of the product and can serve as a marketing tool, as it conveys visual, audible and tactile messages to the consumer. Artificial corks and screw caps had once been sorted into a 'lower-quality' perception bracket by consumers. However, in a recent UK survey, cork and screw caps were equally favoured, with younger drinkers significantly favouring screw caps and this group was more likely to reject natural cork closures than other wine drinkers in the UK, or compared to consumers surveyed in 2013.

Along with non-traditional wine flavours, artistic expression can be seen in new bottle graphic designs. For example, Friends Fun Wine worked with world-famous graffiti artist Miguel Paredes to create an amalgamation of graffiti, landscape and pop art for their 5.5% alcohol wine bottles. Designed for the New York consumer market, these wine bottles featured not just a label, but the entire wine bottle was covered with art. It was introduced by being featured on the Times Square vision billboard.

This is quite different from the more traditional way of thinking about wine label design where a large body of research usually focused on the colour in wine labels. In the past, more colours were deemed to be less sophisticated (*i.e.* a brighter label giving the perception of a cheaper wine). More white space was thought to give the perception of a higher-priced wine. How colours connect with the target consumer of a brand is therefore an area that requires careful research and the importance of quantitative feedback from consumers on label design cannot be overvalued. According to Nielsen, most brands evaluate only three or less design directions, and only 15% of brands explore more than four options. A very successful new market introduction by the E. & J. Gallo Winery was Dark Horse, where they describe their process as relentless, testing over 100 design directions to create the perfect image for their specific market target.

Distinctive approaches to labelling are not limited to visuals but may also include scent. Recently in the United States, Winc.com released a Californian rosé wine, Cocomero (translated as watermelon in Italian), with a label featuring watermelon art. The label also had a scratch and sniff aspect. When scratched, it released the scent of watermelon, preparing the consumer for the aroma of the wine with its notes of watermelon, honeysuckle and rhubarb.

45.5.1 Wine-on-Tap (Kegged Wine)

Although wine-on-tap is not a new concept, its consumer acceptance has been growing rapidly in the past few years. No longer considered a fad, it has become part of the sustainability effort in the evolving consumer mindset. Wine-on-tap requires an initial equipment investment for the restaurant and a training programme to ensure quality when managing such a system. However, once a keg is tapped, inert gas pressurises the wine to ensure that it does not come into contact with oxygen, thereby reducing the wastage problem of oxidised bottles.

Although wine bottles, labels, corks and boxes can all be recycled, the footprint of a 20-litre keg is much smaller as it replaces 26.7 bottles. Also, the keg can be reused many times. There are no oxidation or corkage issues. Tapped wine stays fresh for 3 months, untapped for 12 months, and this format easily allows for many different size options for serving. There has been a premiumisation of the wines put into kegs as the concept has become more mainstream.

A more difficult aspect to deal with for a restaurant is the 'snob factor', as seeing the actual bottle and cork on the table is missing from the single glass on the table experience. In addition, not all wines are candidates for kegging, as kegs do not allow wines to evolve. The offerings must be wines that are ready to be consumed now.

45.5.2 Self-Serve Wine Bars

Self-serve wine bars have great appeal for those who like to explore wine on their own in a restaurant or a hotel. Usually, a credit card is given in exchange for a card to use with the system and there is a choice of pour sizes from 1 oz to 4 oz, with the cost varying with the price of the bottle. At the end of the visit, the customer hands back the wine card and settles the tab with their credit card. The Italian-made Enomatic dispenser (developed in 2002) can hold up to 16 bottles, with different refrigeration zones and the system keeps bottles fresh for up to three weeks by replacing oxygen with argon/nitrogen gas. The Winestation (developed by Napa Technologies) advertises a 60-day preservation window. Wine on demand stations, holding two or more bottles, from various manufacturers for home use, are the latest 'must-haves' in the luxury home market, illustrating how the attitude to these systems has moved from an interesting concept in a boutique hotel, to an integral part of a high-end designer kitchen (Figure 45.7).

Innovations in how wines are offered are being developed both for commercial settings and for the home. South Korean

FIGURE 45.7 Example of a wine dispenser system (Shutterstock image)

electronics giant LG has branched out into wine fridges for the home, with a state-of-the-art voice-activated wine cellar fridge that holds 65 bottles, and in which the wines can be stored at different temperatures, meaning reds, whites and sparkling wines can all share the same cellar. With voice recognition technology, users can issue voice commands to turn on the cellar light and browse their collections without ever opening the fridge door.

45.5.3 Innovations in Bottle Types

Consumers interested in at-home wine delivery were often faced with the dilemma of how to accept the wine delivery when they are not at home, and a reluctance to go to the post office to pick up deliveries. This consumer purchase hurdle resulted in the development of a flat wine bottle that can be posted through a home's letterbox. The flat 750 mL bottle, made of 100% recycled polyethylene terephthalate (PET), which is recyclable after use, was developed by Garçon Wines (www.garconwines.com). Slightly taller than the average wine bottle but only half as thick, it comes packaged in a cardboard box (Figure 45.8). The product won the 2017 World Beverage Innovation Award for best new beverage concept. The bottles are some of the first produced from 100% recycled PET and are 87% lighter and 40% spatially smaller than traditional wine bottles. While the wine bottled in PET should be consumed within a 12-month window, it offers an additional benefit to the consumer, since plastic is more conducive for outdoor usage.

45.5.4 Fractional Bottles, Pouches and Cans

In the United States, there has been huge growth in wine packaged in cans, a seven-fold increase since 2012. Wine offered in cans is not a new concept. Wine was first canned in 1936, and then in the 1950s, but at that time the quality was a problem with acidity forming holes in the tin cans. It wasn't until the 1980s, that there was renewed packaging interest and success with wine in aluminium cans sold to airlines.

Today, cans of wine are especially appealing to the millennial consumer who is not tied to traditional packaging. This format appeals to millennials who are interested in wines that allow them to experience unique varieties from around the world in a convenient single-serve format. Wine in cans, rather than a fad, appears to represent a new wine category that is reaching new audiences. It allows both for portion control (without having to deal with leftover wine), while offering the environmental aspect of aluminium cans being 100% recyclable, less expensive, lighter and easier to ship than bottles.

Alternatives to cans include options such as the 187 mL foil-lined, bottle-shaped pouches (74% paper construction, topped with nitrogen gas to eliminate oxidation) developed by producer Oneglass™ in Northern Italy. Nuvino offers a single-serve portable premium wine in a pouch (with no box). These bags have an easy snap-and-lock feature, so you do not have to drink the entire pouch in one sitting. Stacked Wines

FIGURE 45.8 Garcon wine flat 750 mL bottle, made of 100% recycled PET. *Source:* Photographs courtesy of Garçon Wines

FIGURE 45.9 Example of typical bag-in-box wine packages sold in Canada. (A) Bag filled with red wine and sealed spout. *Source:* Photograph courtesy of I. Russell; (B) Bag inserted into box with pouring spout (Shutterstock image)

LLC offers portable single-serve PET wine packages consisting of four interlocking stackable stemless containers, with a 14-month shelf life, bundled in a shrink-sleeve label. These examples are a few of the many new offerings for consumer convenience, as well as embodying a supply chain–friendly footprint.

It is interesting to note that in Scandinavia, half of all the wine sold is bag-in-box. Part of this is ascribed to price competitiveness, stylish box design, availability of premium wines packaged this way and, most notably, the appeal of an environmentally friendly package in a region where this is considered a key purchase decision factor for consumers (Figure 45.9).

45.6 DIGITAL MARKETING TOOLS

Wine apps for smartphones have become very popular. One of the early and most successful wine apps is Vivino, launched in 2010. It is the most widely used wine app with an online wine community, database and mobile application that allow users to buy, rate and review wines. By taking a photo of any wine label with a smartphone, consumers can instantly see the rating of the wine, reviews and the average price – taken from the database of Vivino's 30 million users (Figure 45.10).

FIGURE 45.10 A scan of a wine's label is all that is needed to take you to the Vivino database. *Source:* Photograph courtesy of I. Russell

Crowdsourcing provides reviews on wines not rated by experts and allows consumers to add their own opinions. In early 2018, the Vivino database contained 96 million ratings, 32 million reviews and 609 million scanned labels (https://www.vivino.com). Within the first few months of launch in Hong Kong, the Vivino wine app had already gained 300,000 users and there are plans to add a Chinese language app in the near future.

Innovative digital marketing drives wine sales. An excellent example of this is Australian Treasury Wine Estates, with their Living Wine Labels app (https://www.livingwinelabels.com). Consumers can view brand-related entertaining digital content after scanning the bottle label with their smartphone. For example, the '19 Crimes app' was launched in 2017. Its augmented reality (AR) content was tailored to each brand's specific story (convicts turned Australian colonists) while the 'Walking Dead app', inspired by a TV series, had ties to the show's zombie theme. The Living Wine Label app has been downloaded over 1.3 million times and the company can use it to access consumer-use behaviour data, such as length of use and how they are sharing the experience.

'Bottles and Bubbles' is Moët Hennessy's initiative for US consumers to interact with Amazon's Alexa for an innovative and more immersive voice experience using voice artificial intelligence (AI). Once consumers enable the app, and say 'Alexa, open Bottles and Bubbles' to their Alexa-enabled device, they can engage in offerings such as recommended food and champagne pairings, hosting tips, champagne-inspired playlists and ambient sounds for a perfect champagne moment, bringing recommendations and environmental influences, such as music, directly into the home consumption environment.

Brock University, located in the wine-growing region of Niagara Falls, Canada, is coupling consumer behaviour with the technical tools of augmented and virtual reality. With a new $1 million VR laboratory dedicated to wine research, they will be using physical, augmented and virtual reality technology. This consumer laboratory will allow researchers to create a variety of environments in which people purchase and consume wines, allowing for the study of how a range of factors such as sights, sounds and smells impact choice and impressions of wine. Environments that can be simulated include wineries, liquor stores and dining rooms, and scenes can be interwoven with music, smells and other sensory information.

45.7 HEALTH FOCUS, LOWER-ALCOHOL PRODUCTS AND CANNABIS

As millennials and baby boomers age, there is an increased focus on healthy foods and healthy beverages as part of their lifestyle. Surveys indicate that 68% of millennials are willing to pay more for organic foods and 66% are willing to pay more for foods that are considered sustainably produced. When millennials were surveyed in an Ontario Canada study, 43% stated they were choosing wine as their alcoholic beverage more often as they believed it was better for their health.

In the wine category there has been an increase in lower-alcohol styles such as Moscato wines. Moscato wines, with alcohol levels between 5 and 10%, are white, semi-sweet, fun and fizzy wines with a target audience of the young and the hip.

Different types of alcohol and foods are being blended together to continue the development of new cross-category drinks such as Moscato grape–infused vodka and fanciful products such as Coconut Chardonnay™ and Cabernet Coffee Espresso™ with alcohol levels of only 5.5%. Fortnum & Mason have launched a new range of alcoholic and non-alcoholic sparkling beverages infused with tea, with white wine as a base and grape juice for the non-alcoholic bottles. Up to 13 types of teas are infused in several flushes to create what they describe as a 'unique tipple with a depth akin to a sparkling wine'. As the beer, wine and spirits categories become ever more blurred, businesses are increasingly combining one, two or all three together to create new products.

A previously unexpected challenge to the wine industry is the rapid growth of the recreational cannabis market in parts of the world. Its legalisation in Canada and parts of the United States now gives the wine consumer another product to choose from and creates increased competition for a share of the consumer's wallet. A probable target for cannabis sales is women and older wealthier individuals, a target group that also accounts for a high percentage of wine sales. Cannabis is a consumer product that may take on a health halo due to the high use of medical cannabis and may appeal as a zero-calorie product choice. Cannabis tours are already starting to be available, some even advertised as 'weed and wine' tours. Some wine-growing areas are starting to shift the usage of their acreage to cannabis cultivation, as cannabis harvesting requires less labour and offers potentially higher profits to the grower.

45.8 SOCIAL MARKETING AND CHATBOTS

Communicating with consumers via social media will be a key area for marketers as social marketing has become one of the most important and influential communication means in marketing. In 2018, the three social media channels with the greatest reach were Facebook, Instagram and Twitter, accounting for 96% of total online conversations. Facebook has over 2 billion active users, Instagram over 800 million and Twitter over 328 million.

Millennial wine drinkers are adventuresome and keen to try an extensive range of wines. Over 85% of millennial wine drinkers report that two to three times a month, they buy wine that they have not tried previously.

Wine is an 'experienced good'. Consumers need to smell and taste the wine in order to determine if they like it. Millennial women are more involved than men in wine purchase decisions. With so many wine purchase options, consumers are turning with increasing frequency to social media to seek advice from friends rather than from experts on which wines to purchase. Social media allows for conversations on wine instead of unidirectional communications on the topic.

Chatbots, which use artificial intelligence to mimic human conversations, are now being used to help consumers select wines that go with a specific meal. Chatbots allow for a two-way personalised interaction between the consumer and a brand and they typically attract a younger, more tech-savvy consumer. A survey from DigitasLBi found that more than one in three Americans would be willing to make a purchase through a chatbot, and nearly half of millennials have or would be willing to receive recommendations from a chatbot. An example specific to wine is the German grocer Lidl. They use a winebot called 'Margot' that picks up keywords to match a customer's food choices with recommended wine pairings.

In restaurants, wine pairing apps empower customers who prefer to make decisions without human interaction but still want advice. Chatbots can put customers at ease and narrow the restaurant wine list for them to consider with their chosen foods. The use of chatbots in many venues is expected to become widespread. The Gartner research company estimates that by 2020, 85% of customer relationships will be managed without human interaction.

Artificial intelligence algorithms, which send consumers wine recommendations, are expected to have a high impact on the evolving needs of wine consumers, as will the use of geo-location, which uses the consumer's location to target them for sales.

FIGURE 45.11 Supercomputer at Palmaz Vineyard in the Napa Valley monitoring the fermentation process. *Source:* Photograph courtesy of Christian Gastón Palmaz

45.9 THE WINERY OF THE FUTURE IS ALREADY HERE

An excellent example of where technology is used to ensure consistency and quality in a winery is the Palmaz Vineyard in the Napa Valley (www.palmazvineyards.com). Built to a depth of 240 feet, the height of the wine cave is equivalent to an 18-storey building underground, which allows for the use of gravity rather than pumps to move the wine. It holds the largest wine cave in the Napa Valley. The fermentation is monitored by a supercomputer to ensure conditions within the fermentation tank are ideal (Figure 45.11). The vineyards use infrared growth optical recognition (VOGOR). Data is obtained from a device attached to an aircraft. Infrared images from 10,000 feet taken of each plot are mapped to assess plant growth, via data on the amount of chlorophyll in the leaves (which translates into how much water is needed in the area at a given time) (Figure 45.12). This winery uses state-of-the-art technology for monitoring, from growing the grapes to bottling. Quality parameters are measured on a full-time basis but with a profound respect for tradition, combining the best of cutting-edge technology with the art of winemaking – a truly impressive example of what has already become possible in one winery and a glimpse into the future for other wineries.

45.10 FUTURE OUTLOOK

The world of wine is undergoing many changes. Will the yeast used in the future to produce wine be very different from what is used today? For some consumers, genetically modified (GMO) yeast is not something that they agree with in terms of wine fermentation, but the technology is advancing faster than the accompanying legislation. Using CRISPR/Cas9 technology, it is now possible to manipulate yeast so that no foreign DNA is present, leaving it unclear as to whether or not this should be identified as a GMO yeast, when CRISPR only edits the DNA already present rather than inserting foreign DNA. What different product profiles could such a yeast produce? This technology is not only being used to manipulate yeast cells but is also being investigated in areas such as the development of new strains of grape vines resistant to downy mildew.

Artificial intelligence and image processing are increasingly being used in vineyards. For example, in Australia, Professor Fuentes' team has worked on using image processing to make it easier to differentiate and identify grape varieties that are very similar in appearance. Grape vine cultivar classification uses photographs of the vine leaves and runs them through machine learning algorithms, which can both identify the cultivar and provide information on water and fertiliser stress. In addition, their work on judging the ripeness of grapes through images rather than by tasting or chemical analysis of sugar levels and acidity, is based on the measurement of the level of cell death of the grapes.

The percentage of dead fruit produces different aromas and flavours. Using a handheld device (near-infrared

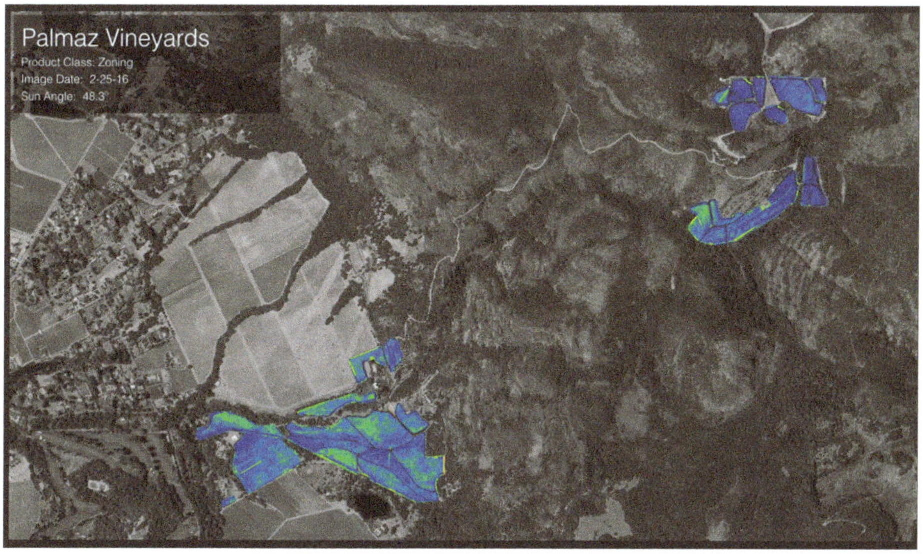

FIGURE 45.12 Aerial image of Palmaz Vineyard in the Napa Valley. *Source:* Photograph courtesy of Christian Gastón Palmaz

wavelengths) to measure grape death levels and machine learning algorithms, they can forecast the readiness of the crop for harvesting.

Innovation will be the important driver of future success including: the introduction of new and/or unfamiliar grape varieties; the technology to produce the wine; how wine is packaged; how, who and where the future customer will purchase wine; and, finally, how consumers will consume the product. The quality of the end product will always be one of the key parameters for repeat sales; however, today the transparency of the process and sustainability and ethical concerns play a much larger part in the consumer's decision-making process. Today's consumer, with easy and instant digital access to information in terms of price, reviews, source and speed of delivery, will be a much more difficult consumer from which to obtain repeat sales. With the wide array of choices available to them, pleasing this new generation of digital-savvy consumers will be a challenging task.

BIBLIOGRAPHY

Barber, N., Taylor, D.C. and Dodd, T. (2009). The importance of wine bottle closures in retail purchase decisions of consumers. *J. Hosp. Mark. Manag.*, 18(6): 597–614.

Bisson, L.F., Waterhouse, A.L., Ebeler, S.E., Walker, M.A. and Lapsley, J.T. (2002). The present and future of the international wine industry. *Nature*, 418(6898): 696.

Bloomberg News. (2017). Why it's so hard to tell if a $100,000 bottle of wine is fake. Sept. 6. Available from: https://www.bloomberg.com/news/articles/2017-09-06/why-it-s-so-hard-to-tell-if-a-100-000-bottle-of-wine-is-fake.

DigitasLBi. (2016). New DigitasLBi research shows more than 1 in 3 Americans are willing to make purchases via Chatbots. Available from: http://www.digitasdose.com/2016/12/new-digitaslbi-research-shows-1-3-americans-willing-make-purchases-via-chatbots/.

Dobele, A.R., Greenacre, L. and Fry, J. (2018). The impact of purchase goal on wine purchase decisions. *Int. J. Wine Bus. Res.*, 30(1): 19–41.

Forbes. (2018). Direct-to-consumer retail wine sales were impressively up in 2017. 28 January. Available from: https://www.forbes.com/sites/thomaspellechia/2018/01/28/direct-to-consumer-2017-retail-wine-sales-were-impressively-up/#7c6f1e4d6bcd.

Hannah, L., Roehrdanz, P.R., Ikegami, M., Shepard, A.V., Shaw, M.R., Tabor, G., Zhi, L., Marquet, P.A. and Hijmans, R.J. (2013). Climate change, wine, and conservation. *Proc. Natl. Acad. Sci. U.S.A.*, 110(17): 6907–6912.

Kosseva, M., Joshi, V.K. and Panesar, P.S. eds. (2016). *Science and Technology of Fruit Wine Production*. Academic Press, New York, NY.

LCBO (Liquor Control Board of Ontario). (2016). Wine trends and sales. Available from: https://wgao.ca/wp-content/uploads/2016/03/Shari_MogkEdwards_Presentation.pdf.

Li, Y. and Bardají, I. (2017a). Adapting the wine industry in China to climate change: Challenges and opportunities. *OENO One*, 51(2): 71–89.

Li, Y. and Bardají, I. (2017b). A new wine superpower? An analysis of the Chinese wine industry. *Cah. Agric.*, 26(6). Article 65002.

Maru/Matchbox. (2017). The future of food. Are you ready for the Millennials? Available from: http://marumatchbox.com/resources/lp-the-future-of-food-are-you-ready-for-the-millennials/.

McMillan, R. (2016). *State of the Wine Industry 2016*. Silicon Valley Bank, Wine Division, St. Helena, CA.

McMillan, R. (2017). State of the wine industry 2017. Available from: https://industry.oregonwine.org/wp-content/uploads/Rob-McMillan-State-of-the-Industry-1.pdf.

Mintel. (2014). *Mintel—Wine Overview 2014*. London, UK.

Moscovici, D. and Reed, A. (2018). Comparing wine sustainability certifications around the world: History, status and opportunity. *J. Wine Res.*, 29(1): 1–25.

Nielsen. (2015). Tried and true or adventurous: Do consumers stick to their favorite alcohol brands? Available from: http://www.nielsen.com/us/en/insights/news/2015/tried-and-true-or-adventurous-most-consumers-stick-to-their-favorite-alcohol-brands.html.

Nielsen. (2016). Design audit series. Available from: http://innovation.nielsen.com/wine-usa-2016.

Nielsen. (2017). How package design attracts today's wine consumer. Available from: http://www.nielsen.com/us/en/insights/news/2017/how-package-design-attracts-todays-wine-consumer.html.

OIV. (2017). The OIV 2017 statistical report on world vitiviniculture. Available from: http://www.oiv.int/public/medias/5479/oiv-en-bilan-2017.pdf.

Pew Research. (2018). Millennials 2018. Available from: http://www.pewresearch.org/topics/millennials/2018/.

Pitt, L. (2017). Ten reasons why wine is a magical marketing product. *J. Wine Res.*, 28(4): 255–258.

Plassmann, H., O'Doherty, J., Shiv, B. and Rangel, A. (2008). Marketing actions can modulate neural representations of experienced pleasantness. *Proc. Natl. Acad. Sci. U.S.A.*, 105(3): 1050–1054.

PR Daily. (2013). Winery looks outside the box with aging stunt. Available from: https://www.prdaily.com/Awards/SpecialEdition/232.aspx.

Reynolds, D., Rahman, I., Bernard, S. and Holbrook, A. (2018). What effect does wine bottle closure type have on perceptions of wine attributes? *Int. J. Hosp. Manag.* doi:10.1016/j.ijhm.2018.05.023

Schmidt, L., Skvortsova, V., Kullen, C., Weber, B. and Plassmann, H. (2017). How context alters value: The brain's valuation and affective regulation system link price cues to experienced taste pleasantness. *Sci. Rep.* 7(1). Article number 8098.

SOLA. (2018). Global wine SOLA report: Sustainable, organic and lower-alcohol wine opportunities 2018. Available from: http://www.wineintelligence.com/global-wine-sola-report-2018/.

Statista. (2017). Daily social media usage worldwide 2012–2017. Sourced from GlobalWebIndex. Available from: https://www.statista.com/statistics/433871/daily-social-media-usage-worldwide/.

Statista. (2018). Global social networks ranked by number of users 2018. Sourced from: We are Social; Kepios; SimilarWeb; TechCrunch; Apptopia; Fortune ID 272014. Available from: https://www.statista.com/statistics/272014/global-social-networks-ranked-by-number-of-users/.

Statista. (2018). Worldwide alcoholic drink revenue: Consumer market outlook, alcoholic drinks report Jan. 2018. Available from: https://www.statista.com/study/48819/alcoholic-drinks-report-cider-perry-and-rice-wine/.

SVB (Silicon Valley Bank). (2018). State of the wine industry 2018. Available from: https://www.svb.com.

Szolnoki, G., Dolan, R., Forbes, S., Thach, L. and Goodman, S. (2018). Using social media for consumer interaction: An international comparison of winery adoption and activity. *Wine Econ. Policy.*, 7(2): 109–119.

The Black Label. (2017). Scandinavia's boxed wine obsession. Available from: https://www.thebacklabel.com/scandinavias-boxed-wine/#.WyEpC6ZvQS8.

The Buyer. (2018). How FMCG consumer insights are giving Concha y Toro the edge. Available from: http://www.the-buyer.net/insight/how-fmcg-style-consumer-insights-are-driving-concha-y-toro-forward/.

The Drinks Business. (2018). Lidl has a new Facebook tool which helps you pair wines with food — here's how it works. Available from: https://www.thedrinksbusiness.com/2018/01/lidl-has-a-new-facebook-tool-which-helps-you-pair-wines-with-food-heres-how-it-works/.

US Dept. of Labor. (2017). Consumer expenditure survey 2016. Available from: https://www.bls.gov/cex/.

Van Leeuwen, C., Schultz, H.R., de Cortazar-Atauri, I.G., Duchêne, E., Ollat, N., Pieri, P., Bois, B., Goutouly, J.P., Quénol, H., Touzard, J.M. and Malheiro, A.C. (2013). Why climate change will not dramatically decrease viticultural suitability in main wine-producing areas by 2050. *Proc. Natl. Acad. Sci. U.S.A.*, 110(33): E3051–E3052.

Vinex. (2018). Why 2018 will be the year emerging central and Eastern Europe wine countries become key global players. Available from: https://en.vinex.market/articles/2018/01/10/why_2018_will_be_the_year_emerging_central_and_eastern_europe_wine_countries_become_key_global_players.

Wine Intelligence. (2018). Wine packaging formats and closures in the UK market 2018. Available from: http://www.wineintelligence.com/wine-packaging-formats-and-closures-in-the-uk-market-2018/.

Wine Spectator. (2016). The youngest millennial just turned 21. What does that mean for wine? Available from: http://www.winespectator.com/webfeature/show/id/52689.

Wine Spectator. (2018). Massive Rhône Valley wine fraud reported by French authorities. 19 March. Available from: http://www.winespectator.com/webfeature/show/id/Massive-Rhone-Valley-Wine-Fraud-Reported.

Wolf, M.M., Higgins, L.M., Wolf, M.J. and Qenani, E. (2018). Do generations matter for wine segmentation? *J. Wine Res.*, 29(3): 177–189.

Index

C

D

E

F

G

H

I

J

N

Q

R

S

T

U

V

W

X

Y

Yellow haze 410

Z